(LXXXIX) LAGRANGIAN

THE ADVA CALCULUS PROBLEM SOLVER®

REGISTERED TRADEMARK

A Complete Solution Guide to Any Textbook

**Staff of Research and Education Association
Dr. M. Fogiel, Director**

Research and Education Association
61 Ethel Road West
Piscataway, New Jersey 08854

THE ADVANCED CALCULUS
PROBLEM SOLVER®

Copyright © 1999, 1981 by Research & Education Association. All rights reserved. No part of this book may be reproduced in any form without permission of the publisher.

Printed in the United States of America

Library of Congress Catalog Card Number 99-70014

International Standard Book Number 0-87891-533-8

PROBLEM SOLVER is a registered trademark of Research & Education Association, Piscataway, New Jersey 08854

WHAT THIS BOOK IS FOR

Students have generally found advanced calculus a difficult subject to understand and learn. Despite the publication of hundreds of textbooks in this field, each one intented to provide an improvement over previous textbooks, students continue to remain perplexed as a result of the numerous conditions that must often be remembered and correlated in solving a problem. Various possible interpretations of terms used in advanced calculus have also contributed to much of the difficulties experienced by students.

In a study of the problem, REA found the following basic reasons underlying students' difficulties with advanced calculus taught in schools:

(a) No systematic rules of analysis have been developed which students may follow in a step-by-step manner to solve the usual problems encountered. This results from the fact that the numerous different conditions and principles which may be involved in a problem, lead to many possible different methods of solution. To prescribe a set of rules to be followed for each of the possible variations, would involve an enormous number of rules and steps to be searched through by students, and this task would perhaps be more burdensome than solving the problem directly with some accompanying trial and error to find the correct solution route.

(b) Textbooks currently available will usually explain a given principle in a few pages written by a professional who has an insight in the subject matter that is not shared by students. The explanations are often written in an abstract manner which leaves the students confused as to the application of the principle. The explanations given are not sufficiently detailed and extensive to make the student aware of the wide range of applications and different aspects of the principle being studied. The numerous possible variations of principles and their applications are usually not discussed, and it is left for the students to discover these for themselves while doing the exercises. Accordingly, the average student is expected to rediscover that which has long been known and practiced, but not published or explained extensively.

(c) The examples usually following the explanation of a topic are too few in number and too simple to enable the student to obtain a thorough grasp of the principles involved. The explanations do not provide sufficient basis to enable a student to solve problems that may be subsequently assigned for homework or given on examinations.

The examples are presented in abbreviated form which leaves out much material between steps, and requires that students derive the omitted material themselves. As a result, students find the examples difficult to understand--contrary to the purpose of the examples.

Examples are, furthermore, often worded in a confusing manner. They do not state the problem and then present the solution. Instead, they pass through a general discussion, never revealing what is to be solved for.

Examples, also, do not always include diagrams/graphs, wherever appropriate, and students do not obtain the training to draw diagrams or graphs to simplify and organize their thinking.

(d) Students can learn the subject only by doing the exercises themselves and reviewing them in class, to obtain experience in applying the principles with their different ramifications.

In doing the exercises by themselves, students find that they are required to devote considerably more time to advanced calculus than to other subject of comparable credits, because they are uncertain with regard to the selections and application of the theorems and principles involved. It is also often necessary for students to discover those "tricks" not revealed in their texts (or review books), that make it possible to solve problems easily. Students must usually resort to methods of trial-and-error to discover these "tricks," and as result they find that they may sometimes spend several hours to solve a single problem.

(e) When reviewing the exercises in classrooms, instructors usually request students to take turns in writing solutions on the boards and explaining them to the class. Students often find it difficult to explain in a manner that holds the interest of the class, and enables the remaining students to follow the material written on the boards. The remaining students seated in the class are, furthermore, too occupied with copying the material from the boards, to listen to the oral explanations and concentrate on the methods of solution.

This book is intended to aid students in advanced calculus to overcome the difficulties described, by supplying detailed illustrations of the solution methods which are usually not apparent to students. The solution methods are illustrated by problems selected from those that are most often assigned for class work and given on examinations. The problems are arranged in order of complexity to enable students to learn and understand a particular topic by reviewing the problems in sequence. The problems are illustrated with detailed step-by-step explanations, to save the student the large amount of time that is often needed to fill in the gaps that are usually found between steps of illustrations in textbooks or review/outline books.

The staff of REA considers advanced calculus a subject that is best learned by allowing students to view the methods of analysis and solution techniques themselves. This approach to learning the subject matter is similar to that practiced in various scientific laboratories, particularly in the medical fields.

In using this book, students may review and study the illustrated problems at their own pace; they are not limited to the time allowed for explaining problems on the board in class.

When students want to look up a particular type of problem and solution, they can readily locate it in the book by referring to the index which has been extensively prepared. It is also possible to locate a particular type of problem by glancing at just the material within the boxed portions. To facilitate rapid scanning of the problems, each problem has a heavy border around it. Furthermore, each problem is identified with a number immediately above the problem at the right-hand margin.

To obtain maximum benefit from the book, students should familiarize themselves with the section, "How To Use This Book," located in the front pages.

To meet the objectives of this book, staff members of REA have selected problems usually encountered in assignments and examinations, and have solved each problem meticulously to illustrate the steps which are usually difficult for students to comprehend. Special gratitude, for their effort and competence in this area, is due to Michael Freed, David Frenkel, Gerard Granata, Daniel Craig Jay, R. Kannan, Leonard Lubarsky, Tara Nanda, and Ralph Oberste-Vorth. Thanks are, furthermore, due to several contributors who devoted brief periods of time to this work.

Gratitude is also expressed to the many persons involved in the difficult task of typing the manuscript with its endless changes, and to the REA art staff who prepared the numerous detailed illustrations together with the layout and physical features of the book.

The difficult task of coordinating the efforts of all persons was carried out by Carl Fuchs. His conscientious work deserves much appreciation. He also trained and supervised art and production personnel in the preparation of the book for printing.

Finally, special thanks are due to Helen Kaufmann for her unique talents to render those difficult border-line decisions and constructive suggestions related to the design and organization of the book.

                                                  Max Fogiel, Ph. D.
                                                  Program Director

HOW TO USE THIS BOOK

This book can be an invaluable aid to students in advanced calculus as a supplement to their textbooks. The book is subdivided into 19 chapters, each dealing with a separate major topic. The subject matter is developed beginning with point set theory and extending through vector spaces, continuity, theorems of differentiation and intergration, sequences, and series. Sections on complex variables, Laplace transforms, Fourier transforms, and differential geometry have also been included. A very extensive number of applications have been included, since these appear to be most troublesome to students.

TO LEARN AND UNDERSTAND A TOPIC THOROUGHLY

1. Refer to your class text and read the section pertaining to the topic. You should become acquainted with the principles discussed there. These principles, however, may not be clear to you at this time.

2. Then locate the topic you are looking for by referring to the "Table of Contents" in the front of this book, "The Advanced Calculus Problem Solver."

3. Turn to the page where the topic begins and review the problems under each topic, in the order given. For each topic, the problems are arranged in order of complexity, from the simplest to the more difficult. Some problems may appear similar to others, but each problem has been selected to illustrate a different point or solution method.

To learn and understand a topic thoroughly and retain its contents, it will be generally necessary for students to review the problems several times. Repeated review is essential in order to gain experience in recognizing the principles that should be applied, and to select the best solution technique.

TO FIND A PARTICULAR PROBLEM

To locate one or more problems related to a particular subject matter, refer to the index. In using the index, be certain to note that the numbers given there refer to problem numbers, not to page numbers. This arrangement of the index is intended to facilitate finding a problem more rapidly, since two or more problems may appear on a page.

If a particular type of problem cannot be found readily, it is recommended that the student refer to the "Table of Contents" in the front pages, and then turn to the chapter which is applicable to the problem being sought. By scanning or glancing at the material that is boxed, it will generally be possible to find problems related to the one being sought, without consuming considerable time. After the problems have been located, the solutions can be reviewed and studied in detail. For this purpose of locating problems rapidly, students should acquaint themselves with the organization of the book as found in the "Table of Contents."

In preparing for an exam, it is useful to find the topics to be covered in the exam from the "Table of Contents," and then review the problems under those topics several times. This should equip the student with what might be needed for the exam.

CONTENTS

Chapter No.   **Page No.**

1. **POINT SET THEORY**   1
    - Sets and Sequences   1
    - Closed and Open Sets and Norms   13
    - Metric Spaces   25

2. **VECTOR SPACES**   38
    - Definitions   38
    - Properties   54
    - Invertibility   68
    - Diagonalization   74
    - Orthogonality   87

3. **CONTINUITY**   97
    - Showing that a Function is Continuous   98
    - Discontinuous Functions   123
    - Uniform Continuity and Related Topics   138
    - Paradoxes of Continuity   151

4. **ELEMENTS OF PARTIAL DIFFERENTIATION**   162
    - Partial Derivatives   163
    - Differentials and the Jacobian   167
    - The Chain Rule   175
    - Gradients and Tangent Planes   184
    - Directional Derivatives   192
    - Potential Functions   195

5. **THEOREMS OF DIFFERENTIATION**   201
    - The Mean Value Theorems   201
    - Taylor's Theorem   216
    - The Implicit Function Theorem   233

6. **MAXIMA AND MINIMA**   239
    - Relative Maximum and Relative Minimum   239
    - Extremes Subject to a Constraint   251
    - Extremes in a Region   256
    - Method of Lagrange Multipliers   261
    - Functions of Three Variables   272
    - Extreme Value in R^n 276

7 THEORY OF INTEGRATION   289
 Riemann Integrals   289
 Stieltjes Integrals   314

8 LINE INTEGRALS   340
 Method of Parametrization   340
 Method of Finding Potential Function (Exact Differential)   353
 Independence of Path   373
 Green's Theorem   392

9 SURFACE INTEGRALS   417
 Change of Variables Formula   417
 Area   428
 Integral Function over a Surface   435
 Integral Vector Field over a Surface   439
 Divergence Theorem   445
 Stokes's Theorem   451
 Differential Form   459

10 IMPROPER INTEGRALS   461
 Improper Integrals of the 1st, 2nd, and 3rd Kind   461
 Absolute and Uniform Convergence   478
 Evaluation of Improper Integrals   491
 Gamma and Beta Functions   497

11 INFINITE SEQUENCES   507
 Convergence of Sequences   508
 Limit Superior and Limit Inferior   529
 Sequence of Functions   538

12 INFINITE SERIES   553
 Tests for Convergence and Divergence   554
 Series of Functions   580
 Operations on Series   592
 Differentiation and Integration of Series   599
 Estimates of Error and Sums   612
 Cesaro Summability   621
 Infinite Products   626

13 POWER SERIES   629
 Interval of Convergence   630
 Operations on Power Series   642

14 FOURIER SERIES   666
 Definitions and Examples   667
 Convergence Questions   687
 Further Representations   700
 Applications   718

15 COMPLEX VARIABLES 728
Complex Numbers 728
Complex Functions and Differentiation 732
Series 742
Integration 750

16 LAPLACE TRANSFORMS 788
Definitions and Simple Examples 790
Basic Properties of Laplace Transforms 803
Step Functions and Periodic Functions 822
The Inversion Problem 837
Applications 862

17 FOURIER TRANSFORMS 880
Definition of Fourier Transforms 881
Properties of Fourier Transforms 893
Applications of Fourier Transforms 907

18 DIFFERENTIAL GEOMETRY 915
Curves 915
Surfaces 933

19 MISCELLANEOUS PROBLEMS AND APPLICATIONS 970
Miscellaneous Applications 970
Elliptic Integrals 982
Physical Applications 993

INDEX 1029

CHAPTER 1

POINT SET THEORY

This chapter develops the set theoretic and topological preliminaries necessary for the study of advanced calculus. The basic notions of set theory and point set (or general) topology are introduced and developed through the problems. The topics covered vary from countability through the Principle of Contraction Mappings.

The problems in this chapter are usually either well-known classical problems or propositions which are to be proved. Among the classical problems are the countability of the rationals, the uncountability of the reals, the Cantor set, the equivalence of metrics (or distance functions), and a connected set which is not path-connected.

Among the propositions are those indicating well-known properties such as the least upper bound property, the Archemedian axiom, the nature of limit points, the Bolzano-Weierstrass property, the nested interval property, the nature of open and closed sets, the Euclidean triangle inequality, the nature of compact sets, and the nature of compact sets (including an application to contraction mappings).

In addition, there are a few problems which simply develop topics needed to solve later problems. In this manner, sequences are introduced, but will be covered in more detail in a later chapter. Similarly, the continuity of functions is introduced and will be further developed in a later chapter.

SETS AND SEQUENCES

● **PROBLEM 1-1**

Show that the set Q of rational numbers x such that $0 < x < 1$ is countably infinite.

<u>Solution</u>: A set A is countably infinite if it is in a one to one correspondence with the natural numbers (i.e., $\{1,2,3, \ldots\}$). Construct a table of Q in the following manner:

$$\frac{1}{2} \quad \frac{1}{3} \quad \frac{1}{4} \quad \frac{1}{5} \quad \frac{1}{6} \quad \ldots$$

$$\frac{2}{3} \quad \frac{2}{5} \quad \frac{2}{7} \quad \frac{2}{9} \quad \frac{2}{11} \quad \ldots$$

$$\frac{3}{4} \quad \frac{3}{5} \quad \frac{3}{7} \quad \frac{3}{8} \quad \frac{3}{10} \quad \cdots$$

...

The numerators in successive rows of this table are 1,2,3, ... The denominators in each row are increasing but so that each fraction is proper (i.e., in Q) and in lowest terms (i.e., appears only once in the table). Now match the above table of Q with this table of the natural numbers:

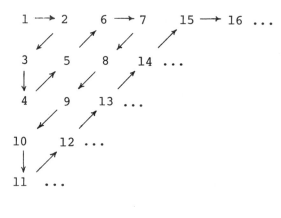

...

Consequently the set Q is in a one to one correspondence with the natural numbers.

Remark Let A_n = {p/n : p an integer} for $n \geq 1$, i.e., A_n is the set of all rational numbers with denominator n. Then

$$Q = \bigcup_{n=1}^{\infty} A_n$$

and by construction each A_n is countable. A countable union of countable sets is countable. Hence Q is countable.

This shows that Q is countably infinite.

● **PROBLEM 1-2**

Show that the set A = {x ∈ R | 0 < x < 1} is uncountable. Conclude that R is uncountable.

Solution: A set S is countable if it is in a one to one correspondence with a subset of the natural numbers

$$N = \{1,2,3,4, \ldots\} .$$

S is uncountable if it is not countable. Since

$$\{ \tfrac{1}{n+1} \mid n \in N \} \subset A,$$

A is at least countably infinite. Suppose A is countably infinite. Then we could list the members of A (represented as infinite decimals) as follows:

$$a_1 = .a_{11}\, a_{12}\, a_{13}\, a_{14} \ldots$$

$$a_2 = .a_{21}\, a_{22}\, a_{23}\, a_{24} \ldots$$

$$a_3 = .a_{31}\, a_{32}\, a_{33}\, a_{34} \ldots$$

$$\ldots$$

where

$$a_n = .a_{n1}\, a_{n2}\, a_{n3}\, a_{n4} \ldots = \sum_{i=1}^{\infty} a_{ni} 10^{-i}.$$

Let

$$b = .b_1\, b_2\, b_3\, b_4 \ldots$$

where

$$b_i = 5 \quad \text{if} \quad a_{ii} \neq 5$$

and

$$b_i = 6 \quad \text{if} \quad a_{ii} = 5.$$

Hence b, which differs from each a_i in the ith decimal place, is not in the list. Since $b \in A$, A is not countably infinite, therefore it is uncountable. If

$$f : A \to R$$

is defined as the one-to-one map onto the real numbers by

$$f(x) = \begin{cases} \dfrac{2x-1}{x}, & 0 < x < 1/2 \\[1em] \dfrac{2x-1}{1-x}, & 1/2 \leq x < 1, \end{cases}$$

then it is seen that R is of the same uncountable order as A. In order to check that f is onto, let c be any real number. If $c < 0$, let x be chosen so that

$$\frac{2x-1}{x} = c,$$

i.e., $2x-1 = cx$ or $x(2-c) = 1$ or $x = \dfrac{1}{2-c}$.

3

By construction $f(x) = c$. If $c \geq 0$, let x be such that

$$\frac{2x-1}{1-x} = c$$

or

$$2x + cx = 1 + c$$

or

$$x = \frac{1+c}{2+c} \; .$$

Again $f(x) = c$.

This shows that f is onto. To show f is one-one note that for

$$0 < x < \frac{1}{2} \leq y$$

$$f(x) < 0 \leq f(y) \; .$$

On the other hand if

$$0 < x < y < \frac{1}{2}$$

(or $\frac{1}{2} \leq x < y < 1$),

then

$$f(y) - f(x) = \frac{2y-1}{y} - \frac{2x-1}{x} = \frac{y-x}{xy} \neq 0$$

(respectively

$$\frac{2y-1}{1-y} - \frac{2x-1}{1-x} = \frac{2y - 2x}{(1-y)(1-x)} \neq 0 \;)$$

and this is positive, i.e., $f(y) > f(x)$.

This means that if

$$x \neq y, \quad f(x) \neq f(y)$$

showing f is one-one.

● **PROBLEM 1-3**

Define boundedness and state the property of real numbers concerning least upper bounds (or greatest lower bounds).

Solution: A non-empty set S is bounded from above if there exists an upper bound (i.e., a number M such that

$x \leq M$ for all $x \in S$). The least upper bound of S (i.e., if α is an upper bound for S, and if $x < \alpha$ then x is not an upper bound for S, then α is the least upper bound of S, written lub $S = \alpha$ or sup $S = \alpha$) is the smallest of the upper bounds. This is not necessarily the maximal member of S (written max S), but max $S = \sup S$ when max S exists and $x < \sup S$ for all $x \in S$ when max S does not exist. For example, if $S = (-\infty, 1)$, then sup $S = 1$ but max S does not exist, whereas if $S = (-\infty, 1]$, then max $S = \sup S = 1$. Similarly, S is bounded from below if there exists a lower bound (i.e., a number m such that $x \geq m$ for all $x \in S$). The greatest lower bound of S (i.e., if β is a lower bound for S, and if $x > \beta$ then x is not a lower bound for S, then β is the greatest lower bound of S, written glb $S = \beta$ or inf $S = \beta$) is the largest of the lower bounds. Again, inf $S = \min S$ (minimal member of S) when min S exists and inf $S < x$ for all $x \in S$ when min S does not exist. S is bounded if and only if S is both bounded from above and bounded from below. Note that the existence of sup S and inf S were assumed. This is one of the properties of the real numbers, i.e., every non-empty set S that is bounded from above (respectively below) has a least upper (respectively greatest lower) bound. Note also that if S is a bounded and non-empty subset of R and

$$(-S) = \{-x \mid x \in S\},$$

then

$$\sup(-S) = -\inf S$$

and

$$\max(-S) = -\min S$$

(when these values exist).

Remark   The least upper bound and the greatest lower bound of a set are unique whenever they exist. For example, if α and α' are two least upper bounds then by definition of upper bound $x \leq \alpha$ for every x in S and $x \leq \alpha'$ for every x in S. However α is the least of the upper bounds so $\alpha \leq \alpha'$. Similarly,

$$\alpha' \leq \alpha, \text{ and so } \alpha = \alpha'.$$

● **PROBLEM 1-4**

If $a > 0$ and $b > 0$, show that there exists an integer n such that $na > b$.

Solution: Suppose $S = \{na \mid n \in Z\}$ is bounded from above by b. Then there exists sup $S = b_0$ such that $na \leq b_0$ for all n ∈ Z. Then $b_0 - a < b_0$ is not an upper bound. So $n_0 a > b_0 - a$ for some n_0 in Z. Hence $(n_0 + 1) a > b_0$ which contradicts the fact that b_0 is an upper bound because $(n_0 + 1)a$ belongs to S. Therefore, S is unbounded and there exists an integer n such that $na > b$. This is called the Archemedian Axiom.

● **PROBLEM 1-5**

> Give an example of a subset C of R which contains no open interval and is not dense in any interval containing itself, but yet is uncountable.

Solution: It is known that a subset of R containing an open interval is uncountable, so such an example is somewhat unusual. To construct such an example, let

$$C_0 = [0,1].$$

Obtain C_1 by deleting the middle third of this interval. Obtain each successive C_n by deleting the middle thirds of all intervals in C_{n-1}. Thus,

$$C_1 = [0, \tfrac{1}{3}] \cup [\tfrac{2}{3}, 1]$$

$$C_2 = [0, \tfrac{1}{9}] \cup [\tfrac{2}{9}, \tfrac{1}{3}] \cup [\tfrac{2}{3}, \tfrac{7}{9}] \cup [\tfrac{8}{9}, 1]$$

... .
Let the Cantor set

$$C = \bigcap_{n=1}^{\infty} C_n.$$

The Cantor set can be described by using ternary (base 3) notation. Let

$$x = \sum_{k=1}^{\infty} \frac{x_k}{3^k} = .x_1 x_2 \ldots \quad \text{(base 3)},$$

where

$$x_k = 0, 1, \text{ or } 2.$$

Then for all x in C_1, $x_1 = 0$ or 2. [Note, for example, that

$$\tfrac{1}{3} = .1000 \ldots = .0222 \ldots \quad \text{(base 3).}]$$

6

For all x in C_2, x_1, $x_2 = 0$ or 2. In general, for all x in C_n, $x_m = 0$ or 2 for $m = 1, 2, \ldots n$. Therefore, C consists of all $x \in [0,1]$ whose ternary expansion contains only zeroes and twos. Define

$f : C \to C_0 = [0,1]$ by

$$f(x) \equiv f(.x_1 x_2 x_3 \ldots) = .y_1 y_2 y_3 \ldots$$

$$= \sum_{k=1}^{\infty} \frac{y_k}{2^k}, \quad \text{(base 2)}$$

where

$$y_k = \frac{1}{2} x_k \quad (k = 1, 2, 3, \ldots).$$

This function is surjective (i.e., this function is onto) and therefore, the Cantor set is uncountable.

Since C is the complement of an open set (the union of the deleted intervals), it is closed. From the ternary representation of C it is clear that if

$$x = \sum_{1}^{\infty} \frac{x_k}{3^k} \quad (x_k = 0 \text{ or } 2)$$

is in C and if $\varepsilon > 0$ is given, then the point

$$y = \sum_{k=1}^{n} \frac{x_k}{3^k} + \frac{2}{3^m}$$

(where m is such that

$$\frac{2}{3^m} < \frac{\varepsilon}{2}$$

and n is such that

$$\sum_{k=n+1}^{\infty} \frac{x_k}{3^k} \leq 2 \sum_{k=n+1}^{\infty} \frac{1}{3^k} < \frac{\varepsilon}{2} \Bigg)$$

is in C and

$$|y - x| < \varepsilon.$$

This means that every point of C is an accumulation point. Such a set is called a perfect set: namely a closed set for which every point is an accumulation point. Note that

7

C cannot contain any open interval because any open interval in [0,1] will contain a deleted interval, which is not in C. The same reasoning shows that if I is any interval contained in [0,1] and which contains C, then C cannot be dense in I. For density would imply that a deleted point is a limit point of C and must be therefore in C which is false, since by definition a set E is dense in X, (X is a metric space) if every point of X is a limit point of E, or a point of E (or both). Hence. since a neighborhood around 1/2 which is contained in [0,1] contains no point of C, it is not a limit point of C.

● **PROBLEM 1-6**

Prove the following given that the sequences $\{s_n\}$ and $\{t_n\}$ converge to s and t respectively:

a) $$\lim_{n \to \infty} (s_n + t_n) = s + t ;$$

b) $$\lim_{n \to \infty} cs_n = cs , \text{ for constant c} ;$$

c) $$\lim_{n \to \infty} (c + s_n) = c + s , \text{ for constant c} ;$$

d) $$\lim_{n \to \infty} s_n t_n = st ;$$

e) $$\lim_{n \to \infty} \frac{1}{s_n} = \frac{1}{s} , \text{ provided}$$

$s_n \neq 0$ (n = 1,2, ...), $s \neq 0$.

Solution: A sequence $\{p_n\}$ converges to a point p if for a given $\varepsilon > 0$ there is an integer N such that $n \geq N$ implies $|p_n - p| < \varepsilon$. Suppose then that $\varepsilon > 0$ is given. There exist integers N_1, N_2 such that

$$|s_n - s| < \varepsilon \quad \text{for } n \geq N_1$$

and

$$|t_n - t| < \varepsilon \quad \text{for } n \geq N_2 .$$

a) Let $N = \max(N_1, N_2)$, then $n \geq N$ implies

$$|(s_n + t_n) - (s + t)| \leq |s_n - s| + |t_n - t| < 2\varepsilon .$$

This proves (a).

b)
$$|(cs_n) - cs| = |c||s_n - s| < |c|\varepsilon \quad \text{for } n \geq N_1 .$$

Hence (b) follows.

c) Similarly (c) follows from

$$|(c + s_n) - (c + s)| = |s_n - s| < \varepsilon \quad \text{for } n \geq N_1 .$$

d) Since

$$s_n t_n - st = (s_n - s)(t_n - t) + s(t_n - t) + t(s_n - s) ,$$

if
$$N = \max(N_1, N_2) ,$$

then

$n \geq N$ implies

$$|s_n t_n - st| \leq |s_n - s||t_n - t| + |s||t_n - t| + |t||s_n - s|$$

$$< \varepsilon^2 + |s|\varepsilon + |t|\varepsilon = \varepsilon(\varepsilon + |s| + |t|)$$

and so (d) follows.

e) For $n \geq N_1$, since

$$|s| - |s_n| \leq |s_n - s| < \varepsilon$$

implies

$$|s_n| > |s| - \varepsilon ,$$

$$\left|\frac{1}{s_n} - \frac{1}{s}\right| = \left|\frac{s_n - s}{s_n s}\right| < \frac{\varepsilon}{|s|(|s| - \varepsilon)} .$$

Therefore, (e) holds.

Note that the final result is not bounded by ε in particular cases, but rather by something which is also arbitrarily small for arbitrarily small $\varepsilon > 0$.

● **PROBLEM 1-7**

Show that:

a) a bounded, monotonic real sequence converges;

b) the geometric series

$$\sum_{i=0}^{\infty} \alpha^i$$

converges if $|\alpha| < 1$.

Solution: a) Let $\{x_n\}$ be a bounded, monotonic real sequence. Assume without loss of generality that $\{x_n\}$ is non-decreasing (for if $\{x_n\}$ were non-increasing then look at $\{-x_n\}$). Since $\{x_n\}$ is bounded, sup $\{x_n\}$ exists. Hence,

$$x_n \leq x_{n+1} \leq \sup_{n \geq 1} \{x_n\} = x \ .$$

Then $x_n \leq x$ for $n = 1, 2, 3, \ldots$.

For every $\varepsilon > 0$, there exists an integer N such that

$$x - \varepsilon \leq x_N \leq x ,$$

for if this was not true then $x - \varepsilon$ would be an upper bound that is less than x, the least upper bound. Since $\{x_n\}$ increases, we have for $n \geq N$

$$x - \varepsilon < x_n \leq x$$

or

$$0 \leq x - x_n < \varepsilon \ .$$

Therefore, $\{x_n\}$ converges to x. (If $\{x_n\}$ were non-increasing, then $\{x_n\}$ would converge to

$$-\sup \{-x_n\} = \inf \{x_n\}.)$$

b) A series

$$\sum_{i=0}^{\infty} a_i$$

converges if its sequence of partial sums $\{s_n\}$, where

$$s_n = \sum_{i=0}^{n} a_i \quad \text{for } n \geq 0,$$

converges. For the series

$$\sum_{i=0}^{\infty} \alpha^i, \quad |\alpha| < 1,$$

the n-th partial sum is

$$s_n = \sum_{i=0}^{n} \alpha^i.$$

Note that

$$\alpha s_n = \sum_{i=1}^{n+1} \alpha^i$$

and

$$s_n - \alpha s_n = \sum_{i=0}^{\infty} \alpha^i - \sum_{i=1}^{n+1} \alpha^i = \alpha^0 - \alpha^{n+1}.$$

Therefore,

$$s_n = \frac{1 - \alpha^{n+1}}{1 - \alpha}.$$

Note that

$$\lim_{n \to \infty} s_n = \frac{1}{1-\alpha} - \lim_{n \to \infty} \frac{\alpha^{n+1}}{1-\alpha} = \frac{1}{1-\alpha}$$

since $|\alpha| < 1$.

Thus

$$\sum_{i=0}^{\infty} \alpha^i = \frac{1}{1-\alpha}.$$

• **PROBLEM 1-8**

Show that a sequence of real-valued functions $\{f_n\}$, defined on complete metric space X, is uniformly convergent if and only if for every $\varepsilon > 0$ there exists an integer N such that

$$m, n \geq N, \ t \in X$$

implies

$$|f_n(t) - f_m(t)| \leq \varepsilon.$$

(This is known as the Cauchy condition.)

Solution: $\{f_n\}$ converges uniformly on X to f if for every $\varepsilon > 0$ there exists an integer M such that $n \geq M$ implies

$$|f_n(t) - f(t)| \leq \varepsilon$$

for all $t \in X$. Suppose first that $\{f_n\}$ converges uniformly to f. Then, for $n, m \geq M$ and $t \in X$,

$$|f_n(t) - f_m(t)| \leq |f_n(t) - f(t)| +$$

$$|f(t) - f_m(t)| \leq 2\varepsilon.$$

This ends the first part of the proof. Conversely, suppose the Cauchy condition holds. Now for a sequence $\{p_n\}$ that is a Cauchy sequence in X, we know that $\{p_n\}$ converges to some point of X. Hence the sequence

$$\{f_n(t)\}$$

converges, for every t, to a limit which can be called $f(t)$. Thus $\{f_n\}$ converges to f on X. We now have to prove that the convergence is uniform. Now for $\varepsilon > 0$ choose N such that $n, m \geq N, \ t \in X$ implies

$$|f_n(t) - f_m(t)| \leq \varepsilon. \qquad (1)$$

Fix n and let $m \to \infty$ in (1). Since $f_m(t) \to f(t)$ as $m \to \infty$ this gives

$$|f_n(t) - f(t)| \leq \varepsilon$$

for every $n \geq N$ and every $t \in X$. Thus, the sequence $\{f_n\}$ defined on X is uniformly convergent.

CLOSED AND OPEN SETS AND NORMS

● **PROBLEM 1-9**

Show that every neighborhood of an accumulation point of a set S contains infinitely many points of S.

Solution: A point y is an accumulation point (or a limit point) of a set S if every neighborhood of y contains at least one point $x \neq y$ where x is in S (i.e., $x \in S$). The set

$$B_\varepsilon(y) = \{x \mid d(x,y) < \varepsilon\}$$

where $d(x,y)$ is the distance between x and y, is called an ε-neighborhood of y. In the case of any real number y, for example every neighborhood of y contains a rational number. Let y be an accumulation point of S. Choose any neighborhood B_1 of y. Then there is some $x_1 \in B_1$ such that $x_1 \in S$ and $x_1 \neq y$. Now choose any neighborhood B_2 of y such that x_1 is not in B_2. Such a neighborhood surely exists; take, for example,

where
$$B_\varepsilon(y) \cap B_1,$$
$$\varepsilon = d(x_1,y).$$

Now there is an $x_2 \in B_2$ such that $x_2 \in S$ and $x_2 \neq y$. Continuing in the same manner gives us neighborhoods

$$B_3, B_4, \ldots$$

and points

$$x_3, x_4, \ldots$$

all in S such that, for each k, $x_k \in B_k$, but $x_k \in B_{k+1}$ and $x_k \neq y$. All of the x_i (i = 1, 2, ...) are distinct and in S. Hence, the arbitrary neighborhood B_1 of y contains infinitely many points of S.

● **PROBLEM 1-10**

Let X be a topological space and let C and U be subsets of X. Define C to be closed if C contains all its limit points and define U to be open if every point $p \in U$ has a neighborhood which is contained in U. Assuming these definitions show that the following statements are equivalent for a subset S of X.

i) S is closed in X;

ii) X - S is open in X;

iii) $S = \bar{S}$.

Solution: (i) $\Rightarrow$ (ii): From i) follows that S contains all its limit points. Hence, any neighborhood of such a limit point contains a point of S other than itself. Now suppose X - S is not open. Then there exists some

$$x \in X - S$$

such that every neighborhood of x contains a $y \neq x$ such that

$$y \in X - (X - S) = S.$$

But then x is a limit point of S and therefore $x \in S$. So, X - S is open since x cannot belong to both S and X - S.

(ii) $\Rightarrow$ (iii): By definition $\bar{S} = S \cup S'$ where S´ is the set of all limit points of S. Evidently,

$$S \subseteq \bar{S}$$

for any set S. Let x be any limit point of S (i.e., $x \in S'$). Since any neighborhood of x contains a different point of S and X - S is open, $x \in S$. Hence, $S' \leq S$ and therefore $\bar{S} \subseteq S$. Thus $S = \bar{S}$.

(iii) $\Rightarrow$ (i): This is obvious, since by (iii) S contains all of its limit points so S is closed.

Note that $S' \subseteq S$ or even $\partial S \subseteq S$, where ∂S is the boundary of S, is equivalent to the conditions above. [$S' \subseteq \partial S$ since ∂S is the set of points whose neighborhoods have non-empty intersections with both S and X - S.] Therefore, $\bar{S}$ can be thought of as the smallest closed set containing S.

• **PROBLEM 1-11**

Show that: (a)

$$\bigcup_{i=1}^{n} A_i$$

is closed if all the A_i are;

b)

$$\bigcap_{i \in \Lambda} A_i$$

is closed if all the A_i are;

c)
$$\bigcap_{i=1}^{\infty} K_i$$

is not empty if the K_i are non empty closed intervals in R and $K_i \supseteq K_{i+1}$ for all $i \geq 1$. ($A_i \subseteq R^m$ for all i).

Solution: a) By De Morgan's law we have

$$\bigcup_{i=1}^{n} A_i = R^m - \bigcap_{i=1}^{n} (R^m - A_i).$$

Hence to show that

$$\bigcup_{i=1}^{n} A_i$$

is closed is equivalent to showing that

$$\bigcap_{i=1}^{n} (R^m - A_i)$$

is open. Let x be any point in

$$\bigcap_{i=1}^{n} (R^m - A_i).$$

Then
$$x \in R^m - A_i \quad (1 \leq i \leq n),$$

which is open. Therefore, for each i, there exists $\varepsilon_i > 0$ such that

$$B_{\varepsilon_i}(x) = \{y \in R^m \mid \|y-x\| < \varepsilon_i\} \subseteq R^m - A_i.$$

Let
$$\varepsilon = \min_{1 \leq i \leq n} \varepsilon_i > 0.$$

Since there are only a finitely many of the $\varepsilon_i > 0$, ε is positive. Then

$$B_\varepsilon(x) \subseteq R^m - A_i \qquad (1 \leq i \leq n).$$

Hence

$$B_\varepsilon(x) \subseteq \bigcap_{i=1}^{n} (R^m - A_i)$$

and

$$\bigcap_{i=1}^{n} (R^m - A_i)$$

is open.

b) We have

$$\bigcap_{i \in \Lambda} A_i = R^m - \bigcup_{i \in \Lambda} (R^m - A_i)$$

by De Morgan's law. Hence, it is enough to show that

$$\bigcup_{i \in \Lambda} (R^m - A_i)$$

is open. Let x be any point in

$$\bigcup_{i \in \Lambda} (R^m - A_i).$$

Then $\quad x \in R^m - A_n \quad$ for some n. So,

$$x \in B_\varepsilon(x) \subseteq R^m - A_n \subseteq \bigcup_{i \in \Lambda} (R^m - A_i)$$

for some > 0 since $R^m - A_n$ is open. Therefore,

$$\bigcup_{i \in \Lambda} (R^m - A_i) \text{ is open.}$$

c) Let $K_i = [a_i, b_i]$ $(i \geq 1)$. Note that the sequence $\{a_i\}$ (respectively $\{b_i\}$) is bounded from above (respectively below) by b_1 (respectively a_1). Also, since these intervals are nested (i.e., $K_i \supseteq K_{i+1}$ for all $i \geq 1$), the sequence $\{a_i\}$ (respectively $\{b_i\}$) is non-decreasing (respectively non-increasing).

If for some i

$$K_i = [a,a] = \{a\},$$

then
$$K_n = [a,a] = \{a\}$$

for all $n \geq i$. In this case

$$\bigcap_{n=1}^{\infty} K_n = \{a\}.$$

Assume then that

$$b_i - a_i > 0$$

for each i. The sequences $\{a_n\}$ and $\{b_n\}$ being monotonic and bounded, so

$$\lim_{n \to \infty} a_n = \sup_{n \geq 1} a_n = A$$

exists and so does

$$\lim_{n \to \infty} b_n = \inf_{n \geq 1} b_n = B.$$

Next note that $a_n < b_n$ for every n and hence for fixed m

$$a_n < b_n \leq b_m$$

for all $n \geq m$.

Hence

$$\sup_{n \geq 1} a_n \leq b_m,$$

i.e., $A \leq b_m$ for all m. Therefore

$$A \leq \inf_{m \geq 1} b_m = B.$$

We assert that the interval $[A, B] \subset \bigcap_{i=1}^{\infty} K_n$.

In order to see this note that $a_n \leq A \leq B \leq b_n$ so $[a_n, b_n]$ contains $[A,B]$ for every n. This completes the proof of (c).

<u>Remark</u>: If $b_n - a_n$ goes to zero, the intervals K_n are said to be nested. In this case given any $\varepsilon > 0$ there exists n_0 such that $b_n - a_n < \varepsilon$ if $n \geq n_0$,

i.e., $b_n < a_n + \varepsilon$ for every $n \geq n_0$. Hence

$$\inf_{n \geq 1} b_n \leq b_n < a_n + \varepsilon$$

for each $n \geq n_0$ or $B \leq a_n + \varepsilon$ for every $n \geq n_0$. Now $a_n \leq A$ for all n so $B \leq A + \varepsilon$. Since ε is arbitrary $B \leq A$ and $B = A$. In other words

$$\bigcap_{n=1}^{\infty} K_n$$

consists of only one point $\{A\}$. For if

$$C \in \bigcap_{n=1}^{\infty} K_n,$$

$a_n \leq C \leq b_n$ for all n and hence

$$A \leq C \leq B = A,$$

i.e., $C = A$.

This result is known as the Cantor's Theorem on nested intervals.

• **PROBLEM 1-12**

Prove: Every bounded infinite subset S of R^n has a limit point.

Solution: Let

$$S \subseteq [a,b]^n = \underset{i=1}{\overset{n}{X}} [a,b] = K_0$$

for some interval $[a,b]$. By subdividing $[a,b]$ into two halves in each dimension and choosing J_i as either half interval in the i-th dimension, construct

$$\underset{i=1}{\overset{n}{X}} J_i .$$

There are 2^n of these types of products. At least one of these contains an infinite subset of S. Call one of these K_1. Now assume that for $i = 1, 2, \ldots$, K_i has been chosen such that $K_i \cap S$ is infinite and $K_i \supseteq K_{i+1}$.

the length of each side of K_i is

$$\frac{1}{2^i} \|b-a\| \quad.$$

By the Nested Intervals Theorem, there is a single point

$$p \in \bigcap_{i=0}^{\infty} K_i \quad.$$

Let $\varepsilon > 0$ be given. Choose i so that

$$\frac{\|b-a\|}{2^i} < \frac{\varepsilon}{n^{\frac{1}{2}}} \quad.$$

Then

$$K_i \subseteq B_\varepsilon(p) = \{X \in R^n \mid \|X-p\| < \varepsilon\}$$

for if

$$X = (x_1, \ldots, x_n) \in K_i ,$$

then for all

$$1 \leq j \leq n, \quad |p_j - x_j| \leq \frac{\|b-a\|}{2^i}$$

and thus

$$\|X-p\| \leq \left[n \left(\frac{\|b-a\|}{2^i}\right)^2 \right]^{\frac{1}{2}} < \varepsilon \quad.$$

But $K_i \cap S$ is infinite so $B_\varepsilon(p) \cap S$ contains infinitely many points distinct from p. Hence, p is a limit point of S. This is known as the Bolzano-Wierestrass Theorem.

● **PROBLEM 1-13**

Prove the triangle inequality in R^n.

Solution: The triangle inequality states that for any

$$x, y \in R^n, \quad \|x+y\| \leq \|x\| + \|y\|,$$

where

$$\|x\| = \left(\sum_{i=1}^{n} x_i^2\right)^{\frac{1}{2}}$$

for

$$x = (x_1, x_2, \ldots, x_n),$$

is the norm of x. Since

$$\|x\| \geq 0$$

for all x, it suffices to show that

$$\|x + y\|^2 \leq (\|x\| + \|y\|)^2$$
$$= \|x\|^2 + 2\|x\| \|y\| + \|y\|^2 .$$

By definition of $\|x\|$,

$$\|x + y\|^2 = \sum_{i=1}^{n} (x_i + y_i)^2$$

$$= \sum_{i=1}^{n} x_i^2 + 2 \sum_{i=1}^{n} x_i y_i + \sum_{i=1}^{n} y_i^2$$

$$= \|x\|^2 + 2\langle x,y \rangle + \|y\|^2 ,$$

where

$$x = (x_1, \ldots, x_n), \quad y = (y_1, \ldots, y_n), \text{ and}$$

$$\langle x,y \rangle = \sum_{i=1}^{n} x_i y_i$$

is the usual inner product (or dot product).

Note that $\|x\| = \langle x,x \rangle^{1/2}$. For (1) to yield the desired result, the Cauchy-Schwartz inequality is needed. That is, if

$$\langle x,y \rangle \leq \|x\| \|y\| ,$$

then (1) yields

$$\|x + y\|^2 \leq \|x\|^2 + 2\|x\| \|y\| + \|y\|^2 .$$

To prove the Cauchy-Schwartz inequality note that for all real λ,

$$0 \leq \|x + \lambda y\|^2 = \sum_{i=1}^{n} (x_i + \lambda y_i)^2$$

$$= \sum_{i=1}^{n} x_i^2 + 2\lambda \sum_{i=1}^{n} x_i y_i + \lambda^2 \sum_{i=1}^{n} y_i^2$$

$$= \langle x,x \rangle + 2\lambda \langle x,y \rangle + \lambda^2 \langle y,y \rangle . \qquad (2)$$

Note that

$$\|x\| = \langle x,x \rangle^{1/2} .$$

Consider first the case $y = 0$. In this case it is clear that

$$\langle x,y \rangle = 0 = \|x\| \|y\| .$$

Assume then that $y \neq 0$ and let

$$\lambda = - \frac{\langle x,y \rangle}{\langle y,y \rangle} .$$

Then (2) gives

$$0 \leq \langle x,x \rangle - \frac{2|\langle x,y \rangle|^2}{\langle y,y \rangle} + \frac{|\langle x,y \rangle|^2}{\langle y,y \rangle} = \langle x,x \rangle - \frac{|\langle x,y \rangle|^2}{\langle y,y \rangle}$$

or

$$|\langle x,y \rangle|^2 \leq \langle x,x \rangle \langle y,y \rangle . \qquad (3)$$

Remark: Equality holds in (3) if and only if x and y are linearly dependent. To see this first suppose $y = \alpha x$ for some α. Then

$$\langle x,y \rangle = \alpha \|x\|^2$$
$$\langle y,y \rangle = |\alpha|^2 \|x\|^2$$

and

$$|\langle x,y \rangle|^2 = |\alpha|^2 \|x\|^2 \|x\|^2 = \langle x,x \rangle \langle y,y \rangle .$$

Conversely assume equality holds in (3) and assume $y \neq 0$. Put

$$\lambda = - \frac{\langle x,y \rangle}{\langle y,y \rangle} .$$

Then by (2)

$$\|x + \lambda y\|^2 = \langle x,x \rangle - \frac{|\langle x,y \rangle|^2}{\langle y,y \rangle} = 0 ,$$

i.e., $x + \lambda y = 0$ or $x = -\lambda y$. If $y = 0$, then $y = 0 \cdot x$ and x and y are linearly dependent.

As a consequence of the above remark it follows that equality holds in the triangle inequality if and only if

$$\langle x,y \rangle = \|x\| \, \|y\|$$

and this holds if and only if

$$y = \lambda x \text{ with } \lambda \geq 0 \,.$$

(This follows because

$$\langle x,y \rangle = \lambda \langle x,x \rangle \geq 0 \quad \text{so} \quad \lambda \geq 0) \,.$$

● **PROBLEM 1-14**

Show that any open spherical neighborhood (or ball) in R^n is an open convex set.

Solution: For any ε, x_0 let

$$B_\varepsilon(x_0) = \{x \mid d(x,x_0) < \varepsilon\}.$$

In R^n,

$$d(a,b) = \|a-b\| = \left(\sum_{i=1}^{n} (a_i - b_i)^2\right)^{\frac{1}{2}}$$

where

$$a = (a_1, a_2, \ldots, a_n)$$

and

$$b = (b_1, b_2, \ldots, b_n).$$

To show $B_\varepsilon(x_0)$ is open it must be shown that for any $x_1 \in B_\varepsilon(x_0)$ there exists a neighborhood of x_1, $B_{\varepsilon'}(x_1)$, contained in $B_\varepsilon(x_0)$. Let

$$\varepsilon' = \frac{1}{2}(\varepsilon - \|x_1 - x_0\|).$$

Using the triangle inequality, it is seen that for any $x_2 \in B_{\varepsilon'}(x_1)$

$$\|x_2 - x_0\| = \|x_2 - x_1 + x_1 - x_0\|$$

$$\leq \|x_2 - x_1\| + \|x_1 - x_0\|$$

$$< \frac{1}{2}(\varepsilon - \|x_1 - x_0\|) + \|x_1 - x_0\|$$

$$= \frac{1}{2}(\varepsilon + \|x_1 - x_0\|)$$

$$< \varepsilon$$

since

$$\|x_1 - x_0\| < \varepsilon \quad .$$

So,

$$x_2 \in B_\varepsilon(x_0)$$

and therefore

$$B_{\varepsilon'}(x_1) \subseteq B_\varepsilon(x_0) \quad .$$

This shows that $B_\varepsilon(x_0)$ is open. To show that $B_\varepsilon(x_0)$ is convex, it must be shown that for any $x_1, x_2 \in B_\varepsilon(x_0)$, $(tx_1 + (1-t)x_2) \in B_\varepsilon(x_0)$ for all $0 \le t \le 1$. Again, using the triangle inequality

$$\|tx_1 + (1-t)x_2 - x_0\| = \|(tx_1 - tx_0)$$

$$+ ((1-t)x_2 - x_0 + tx_0)\|$$

$$= \|t(x_1 - x_0) + (1-t)(x_2 - x_0)\|$$

$$\le t\|x_1 - x_0\| + (1-t)\|x_2 - x_0\|$$

$$< t\varepsilon + (1-t)\varepsilon = \varepsilon. \quad (0 \le t \le 1)$$

So,

$$(tx_1 + (1-t)x_2) \in B_\varepsilon(x_0)$$

for $0 \le t \le 1$ and therefore $B_\varepsilon(x_0)$ is convex.

• **PROBLEM 1-15**

Let T be a linear transformation of R^m to R^n. Show that there exists a number λ such that

$$\|TX\| \le \lambda \|X\|$$

for all $X \in R^m$.

<u>Solution</u>: T can be represented by the matrix

$$A = (a_{ij})$$

with the normal basis $\{e_i \mid i=1,2, \ldots, k\}$,
$e_i = (0, \ldots, 0,1,0, \ldots, 0)$ with the 1 in the i-th position, for R^k ($k = m,n$). Let $M = \max |a_{ij}|$. Then

$$\|TX\| = \|A(x_1, \ldots, x_m)^T\|$$

$$= \|(\sum_{j=1}^{m} a_{1j}x_j, \ldots, \sum_{j=1}^{m} a_{nj}x_j)\|$$

$$= \left[\sum_{i=1}^{n} (\sum_{j=1}^{m} a_{ij}x_j)^2\right]^{\frac{1}{2}}$$

$$\leq \left[\sum_{i=1}^{n} (M \sum_{j=1}^{m} |x_j|)^2\right]^{\frac{1}{2}} \tag{1}$$

$$\leq \left[\sum_{i=1}^{n} M^2 (m \|x\|^2)\right]^{\frac{1}{2}} \tag{2}$$

$$\leq \sqrt{nm}\, M \|x\|.$$

To see that (2) follows from (1) note that

$$(\sum_{j=1}^{m} |x_j|)^2 \leq (\sum_{j=1}^{m} 1 \cdot |x_j|)^2 \tag{3}$$

$$\leq \left[(\sum_{j=1}^{m} |x_j|^2)^{\frac{1}{2}} (\sum_{j=1}^{m} 1^2)^{\frac{1}{2}}\right]^2 \tag{4}$$

$$= m \sum_{i=1}^{m} x_j^2$$

$$= m \|x\|^2$$

where (4) follows from (3) by the Cauchy-Schwartz inequality. Therefore,

$$\lambda = \sqrt{nm}\, \max |a_{ij}|.$$

METRIC SPACES

• **PROBLEM 1-16**

Show that the following are equivalent metrics for R^n:

(i) $d_1(x,y) = \|x-y\|_1 = (\sum_{i=1}^{n} (x_i - y_i)^2)^{1/2}$;

(ii) $d_2(x,y) = \|x-y\|_2 = \max_{1 \leq i \leq n} |x_i - y_i|$;

(iii) $d_3(x,y) = \|x-y\|_3 = \sum_{i=1}^{n} |x_i - y_i|$.

Solution: For any of the above to define metrics they must obey the following for $x, y, z \in R^n$:

(a) $d(x,y) \geq 0$, and $d(x,y) = 0$ if and only if $x = y$;

(b) $d(x,y) = d(y,x)$ for every x and y

(c) $d(x,y) \leq d(x,z) + d(z,y)$ for every x, y and z.

A set together with a metric on the set is called a metric space. Hence, R^n with any metric is a metric space. For (i), (a) and (b) clearly hold and (c) is the usual triangle inequality. ((i) is called the usual or Euclidean metric and R^n with this metric is called Euclidean space). For (ii) and (iii), (a) and (b) clearly hold and (c) holds since it holds for absolute value in R. Two metrics are equivalent if for all $x \in R^n$ and $\varepsilon > 0$, there exist $\delta_1 > 0$ and $\delta_2 > 0$ such that

$$B^1_{\delta_1}(x) \subseteq B^2_{\varepsilon}(x) \quad \text{and} \quad B^2_{\delta_2}(x) \subseteq B^1_{\varepsilon}(x)$$

where

$$B^i_{\delta}(x) = \{y \in R^n \mid \|y-x\|_i < \delta\} .$$

In order to show that d_1 and d_2 are equivalent let $\varepsilon > 0$ be given and let

25

$$\delta_1 = \varepsilon, \quad \delta_2 = \frac{\varepsilon}{\sqrt{n}} \ .$$

Note that

$$B_\varepsilon^2(x) = \{y \in R^n \mid \max_{1 \le i \le n} |x_i - y_i| < \varepsilon\} \ .$$

Then

$$\|x-y\|_1 = \left(\sum_{i=1}^n (x_i - y_i)^2\right)^{\frac{1}{2}} < \delta_1 = \varepsilon$$

implies that

$$\|x-y\|_2 = \max_{1 \le i \le n} |x_i - y_i| < \varepsilon \ ,$$

and

$$\|x-y\|_2 = \max_{1 \le i \le n} |x_i - y_i| < \delta_2$$

implies that

$$\|x-y\|_1 = \left(\sum_{i=1}^n (x_i - y_i)^2\right)^{\frac{1}{2}}$$

$$\le \left(\sum_{i=1}^n (\max|x_i - y_i|)^2\right)^{\frac{1}{2}}$$

$$< (n\delta_2^2)^{\frac{1}{2}}$$

$$= \sqrt{n}\, \delta_2 = \varepsilon \ .$$

To show d_2 and d_3 are equivalent for a given ε let

$$\delta_1 = \varepsilon, \quad \text{and} \quad \delta_2 = \frac{\varepsilon}{n} \ .$$

Then,

$$\|x-y\|_3 = \sum_{i=1}^n |x_i - y_i| < \delta_1 = \varepsilon$$

implies

$$\|x-y\|_2 = \max_{1 \le i \le n} |x_i - y_i| < \varepsilon \ ,$$

and

26

$$\|x-y\|_2 = \max_{1\le i\le n} |x_i - y_i| < \delta_2 = \varepsilon/n$$

implies

$$\|x-y\|_3 = \sum_{i=1}^{n} |x_i - y_i|$$

$$\le \sum_{i=1}^{n} \max_{1\le i\le n} |x_i - y_i|$$

$$< n\delta_2 = \varepsilon.$$

Finally, the equivalence of d_1 and d_3 follows from above. Let $\varepsilon > 0$ be given. Since d_2 and d_3 are equivalent there exists δ_1 such that

$$B^2_{\delta_1}(x) \subseteq B^3_{\varepsilon}(x).$$

Again since d_1 and d_2 are equivalent there exists a δ such that

$$B^1_{\delta}(x) \subseteq B^3_{\varepsilon}(x).$$

Similarly, one can show that for some δ,

$$B^3_{\delta}(x) \subseteq B^1_{\varepsilon}(x)$$

showing that d_1 and d_3 are equivalent.

• **PROBLEM 1-17**

State and prove the Heine-Borel Theorem in R and show, by examples, that the conditions of the theorem are necessary.

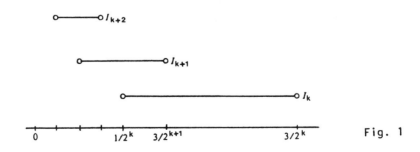

Fig. 1

Solution: A subset S of R is said to be compact if and only if for each collection

$$\{Q_\alpha \mid \alpha \in J\}$$

of open sets with union

$$\bigcup_{\alpha \in J} Q_\alpha$$

containing S, some finite subset $\{\alpha_1, \ldots, \alpha_n\}$

of J exists such that

$$S \subseteq \bigcup_{i=1}^{n} Q_{\alpha_i} .$$

a) Every compact subset S of R is bounded. Indeed, the open intervals $(-n, n)$ where $n = 1, 2, 3, \ldots$ cover the whole real line.

$$R = \bigcup_{n=1}^{\infty} (-n, n) .$$

In particular,

$$S \subset \bigcup_{n=1}^{\infty} (-n, n) ,$$

so for some n

$$S \subset (-n, n) ,$$

i.e., $|x| \leq n$ for every $x \in S$, i.e., S is bounded.

b) Every compact subset S of R is closed. Let $p \notin S$ and let

$$O_n = \{x \in R \mid |x-p| > \tfrac{1}{n} \} .$$

O_n is the complement of $[p - \tfrac{1}{n}, p + \tfrac{1}{n}]$ and is open. The union of all the O_n as n sums from $1, 2, 3 \ldots$ to infinity contains the whole real line except the point p. Hence,

$$S \subset \bigcup_{n=1}^{\infty} O_n .$$

Since S is compact

$$S \subset \bigcup_{i=1}^{k} O_{n_i}$$

for some k. Let $n = \max(n_1, \ldots, n_k)$ so that $O_{n_i} \subset O_n$

for $i = 1, 2, \ldots, k$ and

$$\bigcup_{i=1}^{k} O_{n_i} = O_n$$

and so some O_n contains S. Hence the interval

$$[p - \tfrac{1}{n},\ p + \tfrac{1}{n}] = O_n^c$$

is contained in S^c and p is an interior point of S^c showing that S^c is open and hence S is closed.

Properties a) and b) above characterize compact subsets of R and this is the content of the Heine-Borel Theorem which states that:

If S is a closed and bounded subset of R, then it is compact.

Proof: Let an infinite collection of open sets O_α contain S and assume (to get a contradiction) that no finite collection of the O_α's cover S. This necessarily means that S is infinite. Since S is bounded there exists an interval

$$I = [a, b]$$

so that

$$S \subset [a, b].$$

Divide the interval I into two equal parts I_1 and I_2. Then

$$S = (S \cap I_1) \cup (S \cap I_2) \equiv S_1 \cup S_2$$

so one of the S_1, S_2 cannot be covered by finite number of the O_α's. Call the interval $[a_1, b_1]$ for which $S \cap [a_1, b_1]$ cannot be covered by a finite number of the Q_α's. Continuing in this way one can define intervals $[a_n, b_n]$ such that $[a_n, b_n] \subset [a_{n-1}, b_{n-1}]$ and $S \cap [a_n, b_n]$ cannot be covered by a finite number of O_α. Since S is infinite and

$$b_n - a_n = \frac{b-a}{2^n}$$

goes to zero, by Cantor's theorem on nested intervals

$$\bigcap_{n=1}^{\infty} [a_n, b_n]$$

contains exactly one point p which must necessarily be an accumulation point of S. Since S is closed p ∈ S. Now any open interval around p contains $[a_n, b_n]$ for all but a finite number of n. But p ∈ $\bigcup_\alpha Q_\alpha$ so p ∈ O_α for some α and hence some open interval around p is contained in O_α. This means that O_α contains S∩$[a_n, b_n]$ for all but a finite number of n. This is contrary to the definition of $[a_n, b_n]$ and the theorem is proved.

Example 1. S = (0,1) is not compact because if

$$O_n = (0, 1 - \tfrac{1}{n}), \quad n = 2, 3 \ldots.$$

then

$$\bigcup_{n=1}^{\infty} O_n = (0,1)$$

and no finite collection of the O_n's can cover S. The reason that S is not compact is that it is not closed.

Example 2. S = (0,∞) is not compact. For the intervals (0,n) cover S but for no n is S ⊂ (0,n). Again, non-compactness of S is due to unboundedness.

• PROBLEM 1-18

Show that a subset of R^2 is connected if it is path connected, but the converse is not necessarily true.

Solution: Let A be path connected. Then for any x, y ∈ A there exists a continuous function f: [0,1] → A such that f(0) = x and f(1) = y. This is called a path and may be written

f: I, 0, 1 → A, x, y

indicating

F: I → A, f(0) = x, and f(1) = y.

f is continuous if for any open set

$U \subseteq A$, $f^{-1}(U) = \{x \in A | f(x) \in U\}$
is an open subset of $[0,1]$. (Note, for example, that $[0,\frac{1}{2})$ is open in $[0,1]$ because

$$[0,\frac{1}{2}) = [0,1] \cap (-\frac{1}{2}, \frac{1}{2}) \text{ and } (-\frac{1}{2}, \frac{1}{2})$$

is open in R). Suppose A is not connected. Then there exist non empty open subsets B, C of A such that $B \cup C = A$ and $B \cap C = \phi$. Choose $b \in B$ and $c \in C$ and a path

$$g: I, 0, 1 \to A, b, c.$$

Since

$$A = B \cup C, \quad g^{-1}(A) = [0,1] = g^{-1}(B) \cup g^{-1}(C)$$

where $g^{-1}(B)$, $g^{-1}(C)$ are non-empty and open. Also, since

$$B \cap C = \phi, \quad g^{-1}(B) \cap g^{-1}(c) = \phi.$$

However, $[0,1]$ is connected. Therefore, A is connected. To see that the converse is not necessarily true, construct the following counterexample in R^2. Let

$$S = \bigcup_{k=1}^{\infty} \overline{A_k A_{k+1}}$$

where $A_k = (\frac{1}{k}, 1)$ if k is odd and $A_k = (\frac{1}{k}, -1)$ if k is even. (See figure 1.) Since $A_{2k} \to (0,-1)$ and $A_{2k+1} \to (0,1)$ as $k \to \infty$ the closure

$$[S] = S \cup \{(0,y) | -1 \le y \le 1\}.$$

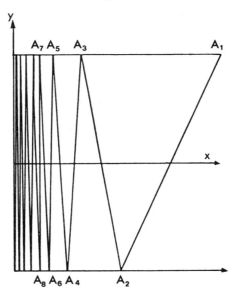

Fig. 1

Now S is path connected and hence connected. It is known that if X is connected and

$$X \subseteq Y \subseteq [X],$$

then Y and hence [X] is connected. Therefore, [S] is connected. However, [S] is not path connected since $A_k A_{k+1}$ does not intersect the y axis for any k.

● **PROBLEM 1-19**

Show that R is a complete metric space.

Solution: A sequence in R is a function from the positive integers Z^+ to R and is usually denoted by $\{x_n\}$ where $n \to x_n \in R$ for all $n \in Z^+$. A Cauchy sequence in R is a sequence $\{x_n\}$ such that for any $\varepsilon > 0$ there exists an N_ε such that $|x_n - x_m| < \varepsilon$ for all $n, m > N_\varepsilon$. A sequence converges in R if there is an $x \in R$ such that for every $\varepsilon > 0$ there is an integer N_ε such that $n \geq N$ implies that $|x_n - x| < \varepsilon$. Clearly, every convergent sequence is Cauchy. If the converse is true—i.e., if every Cauchy sequence in R converges to a point in R—then R is complete. In the above definitions, R and its distance (e.g., $|x-y|$) could be replaced by any metric space. Let $\{x_n\}$ be a Cauchy sequence. For some $N \in Z^+$ and $n, m \geq N$, $0 < |x_n - x_m| < \varepsilon$. For

$$n \geq N, \quad |x_n| \leq |x_n - x_N| + |x_N| = b$$

(triangle inequality). Let

$$c = \max_{n < N} (|x_n|, b).$$

So, $|x_n| \leq c$ for all n. Therefore, $\{x_n\}$ is bounded. So all $x_n \in I_1$ for some interval I_1. Take half of I_1 and let I_2 be the half with infinitely many of the x_n. Continue in this manner to get nested intervals

$$\cdots \supseteq I_m \supseteq I_{m+1} \supseteq \cdots$$

where each I_m contains infinitely many of the x_n. Hence,

$$\bigcap_{m=1}^{\infty} I_m \neq \phi,$$

so letting

$$x \in \bigcap_{m=1}^{\infty} I_m,$$

we have that there exists a subsequence $\{x_{n_j}\}$ which converges to x, by picking x_{n_j} from I_j. That is, for any $\varepsilon > 0$ there exists an integer M such that

$$|x_{n_j} - x| < \varepsilon$$

whenever $n_j \geq M$.

Now $|x_n - x| \leq |x_n - x_{n_j}| + |x_{n_j} - x| < 2\varepsilon$

for $n, n_j \geq \max(N, M)$. So, $\{x_n\}$ also converges to x.

● **PROBLEM 1-20**

Show that $C(X)$, with X compact and metric

$$\rho(f,g) = \sup_{t \in X} |f(t) - g(t)|$$

for all $f, g \in C(X)$, is a complete metric space.

Solution: $C(X)$ denotes the set of all continuous functions from X into R. Since

$$\rho(f,g) = \sup_{t \in X} |f(t) - g(t)|,$$

$\rho(f,g) \geq 0$ and $\rho(f,g) = 0$ if and only if $f = g$. Also,

$$\rho(f,g) = \sup_{t \in X} |f(t) - g(t)| = \sup_{t \in X} |g(t) - f(t)|$$

$$= \rho(g,f).$$

To prove the triangle inequality note that for all $t \in X$,

$$|f(t) - h(t)| \leq \sup_{t \in X} |f(t) - h(t)|.$$

Hence, for all $t \in X$,

$$|f(t) - g(t)| \leq |f(t) - h(t)| + |h(t) - g(t)|$$

$$\leq \sup_{t \in X} |f(t) - h(t)| + \sup_{t \in X} |h(t) - g(t)|$$

since the triangle inequality holds in R. Therefore,

$$\sup_{t \in X} |f(t) - g(t)| \leq \sup_{t \in X} |f(t) - h(t)| + \sup_{t \in X} |h(t) - g(t)|$$

and the triangle inequality holds in $C(X)$. To show that this metric space is complete is to show that any Cauchy sequence $\{f_n(t)\}$ converges to a function in $C(X)$. For every $\varepsilon > 0$ there exists N such that

$$|f_n(t) - f_m(t)| < \varepsilon$$

for $n, m > N$ and all $t \in X$. Thus the sequence $\{f_n(t)\}$ is uniformly convergent. Let $f(t)$ be the limit of this sequence. Clearly,

$$|f(t) - f(t_0)| \leq |f(t) - f_n(t)| + |f_n(t) - f_n(t_0)|$$

$$+ |f_n(t_0) - f(t_0)|,$$

where $t, t_0 \in X$. Then, by uniform convergence, for any $\varepsilon > 0$,

$$|f(t) - f_n(t)| < \varepsilon$$

and

$$|f_n(t_0) - f(t_0)| < \varepsilon$$

for all sufficiently large n. Since f_n is continuous on X for all n,

$$|f_n(t) - f_n(t_0)| < \varepsilon$$

for t_0 sufficiently close to t. Thus, $f(t)$ is continuous on X and $C(X)$ is complete.

● **PROBLEM 1-21**

Let R be a complete metric space with metric ρ. Prove that every contraction mapping $A: R \to R$ has a unique fixed point.

Solution: R is complete if every Cauchy sequence in R converges in R. A Cauchy sequence is a sequence $\{x_n\}$ such that for any $\varepsilon > 0$ there exists $N > 0$ such that

$$\rho(x_{n_1}, x_{n_2}) < \varepsilon$$

whenever $n_1, n_2 \geq N$. For A to be a contraction mapping means that there exists $\alpha < 1$ such that for any $x, y \in R$

$$\rho(Ax, Ay) \leq \alpha \rho(x, y);$$

To prove the theorem let x_0 be any point in R and

$$x_1 = Ax_0, \quad x_2 = Ax_1 = A^2 x_1,$$

and, in general,

$$x_n = Ax_{n-1} = A^n x_0.$$

Since

$$\rho(x_n, x_m) = \rho(A^n x_0, A^m x_0) \quad (n < m)$$

$$\leq \alpha^n \rho(x_0, x_{m-n})$$

$$\leq \alpha^n \left(\sum_{i=1}^{m-n} \rho(x_{i-1}, x_i) \right) \quad \text{(triangle inequality)}$$

$$\leq \alpha^n \left(\sum_{i=1}^{m-n} \alpha^{i-1} \rho(x_0, x_1) \right)$$

$$\leq \frac{\alpha^n \rho(x_0, x_1)}{1 - \alpha} \quad \text{(since } \sum_{i=0}^{\infty} \alpha^i = \frac{1}{1-\alpha} \text{)}$$

since $\alpha < 1$, $\rho(x_n, x_m)$ is arbitrarily small for sufficiently large n. Thus $\{x_n\}$ is a Cauchy sequence which converges to x since R is complete. Since A is continuous,

$$Ax = A \lim_{n \to \infty} x_n = \lim_{n \to \infty} Ax_n = \lim_{n \to \infty} x_{n+1} = x.$$

Thus, x is a fixed point. Suppose y is also a fixed point. Then

$$\rho(Ax, Ay) = \rho(x, y) \leq \alpha \rho(x, y)$$

where $\alpha < 1$. So $x = y$, therefore x is a unique fixed point. This theorem is called the Principle of Contraction Mappings or the Banach Fixed Point Theorem. It is easily generalized: If A is a continuous mapping of a complete metric space into itself and A^n is a contraction mapping for some integer $n \geq 1$, then A has a unique fixed point.

• **PROBLEM 1-22**

Determine the roots of the equation $F(x) = 0$ where F is a real-valued continuously differentiable function on [a,b] such that

$$F(a) < 0 < F(b)$$

and

$$0 < k_1 \leq F'(x) \leq k_2 \quad \text{on } [a,b].$$

Solution: Let $f(x) = x - \lambda F(x)$. Then a solution to $f(x) = x$, which is equivalent to $F(x) = 0$, is desired. Since $f'(x) = 1 - \lambda F'(x)$,

$$1 - \lambda k_2 \leq f'(x) \leq 1 - \lambda k_1.$$

Take $0 < \lambda \leq \frac{1}{k_2}$, then $0 \leq f'(x) < 1$

and hence $f(x)$ is a contraction of [a,b] into itself; To see this observe that

$$f(y) - f(x) = \int_x^y f'(t) dt$$

so

$$|f(y) - f(x)| \leq \sup_{a \leq t \leq b} |f'(t)| \; |y-x|$$

$$= \alpha |y-x|$$

where

$$\alpha = \sup_{a \leq t \leq b} |f'(t)| < 1.$$

Using the Principle of Contraction Mappings choose any $x_0 \in [a,b]$. Then the sequence

$$x_0, \; x_1 = f(x_0), \; x_2 = f(x_1) \ldots$$

converges to the unique root of $f(x) = x$, which is also

the unique root of F(x) = 0 on [a,b].

As an illustration, Figures 1 and 2 indicate the course of the successive approximations in the cases

$$0 < f'(x) < 1$$

and

$$-1 < f'(x) < 0 .$$

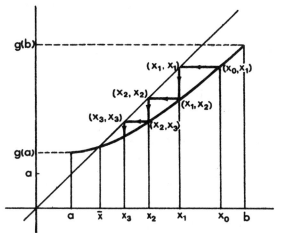

Fig. 1

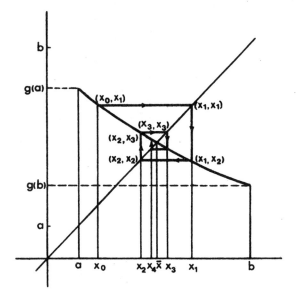

Fig. 2

CHAPTER 2

VECTOR SPACES

This chapter introduces some of the concepts of linear algebra which are of importance in the theory of advanced calculus.

Linear algebra is concerned with linear transformations on vector spaces where a vector space is a set with certain properties and a linear transformation is a function (or mapping) with certain properties (of linearity) between vector spaces.

These notions, although quite abstract for their own sake, are of importance in advanced calculus because R^n and C^n are vector spaces. Furthermore, recall that the derivative of a function at a point is the linear transformation satisfying some equation. Hence, to study differentiation and, similarly, to study the theory of integration, requires an underlying knowledge of the theory of linear transformations on vector spaces.

However, vector spaces do not include any notions about distance such as in Euclidean spaces. Again, the theory of linear analysis is quite important to vector spaces in which the notion of distance is included. Here the linear algebra of inner product spaces, where an inner product space is a vector space together with an inner product (some kind of bilinear function), is the underlying theory. Accordingly, derived from this inner product is the notion of distance.

Another use of the inner product is in the notion of orthogonality (or perpendicularity). A well-known result called the Gram-Schmidt orthogonalization process guarantees the existence of orthogonal coordinate systems (e.g., the usual coordinate axes).

The following problems should develop the notions of linear algebra necessary for the study of advanced calculus.

DEFINITIONS

● **PROBLEM 2-1**

Show that the space R^n (comprised of n-tuples of real numbers $(x_1, \ldots, x_n)$) is a vector space over the field R of real numbers. The operations are addition of n-tuples, i.e., $(x_1, \ldots, x_n) + (y_1, y_2, \ldots, y_n) = (x_1 + y_1, x_2 + y_2, \ldots, x_n + y_n)$, and scalar multiplication, $\alpha(x_1, x_2, \ldots, x_n) = (\alpha x_1, \alpha x_2, \ldots, \alpha x_n)$ where $\alpha \in R$.

Solution: Any set that satisfies the axioms for a vector space over a field is known as a vector space. We must show that R^n satisfies the vector space axioms. The axioms fall into two distinct categories:

A) the axioms of addition for elements of a set

B) the axioms involving multiplication of vectors by elements from the field.

1) Closure under addition

By definition, $(x_1, x_2, \ldots, x_n) + (y_1, y_2, \ldots, y_n)$
$= (x_1 + y_1, x_2 + y_2, \ldots, x_n + y_n)$.

Now, since $x_1, y_1, x_2, y_2, \ldots, x_n, y_n$ are real numbers, the sums of $x_1 + y_1, x_2 + y_2, \ldots, x_n + y_n$ are also real numbers. Therefore, $(x_1 + y_1, x_2 + y_2, \ldots, x_n + y_n)$ is also an n-tuple of real numbers; hence, it belongs to R^n.

2) Addition is commutative

The numbers $x_1, x_2, \ldots, x_n$ are the coordinates of the vector $(x_1, x_2, \ldots, x_n)$, and $y_1, y_2, \ldots, y_n$ are the coordinates of the vector $(y_1, y_2, \ldots, y_n)$.

Show $(x_1, x_2, \ldots, x_n) + (y_1, \ldots, y_n) = (y_1, \ldots, y_n) + (x_1, \ldots, x_n)$. Now, the coordinates $x_1 + y_1, x_2 + y_2, \ldots, x_n + y_n$ are sums of real numbers. Since real numbers satisfy the commutativity axiom, $x_1 + y_1 = y_1 + x_1$, $x_2 + y_2 = y_2 + x_2, \ldots, x_n + y_n = y_n + x_n$.
Thus,

$(x_1, x_2, \ldots, x_n) + (y_1, y_2, \ldots, y_n)$

$= (x_1 + y_1, x_2 + y_2, \ldots, x_n + y_n)$

(by definition)

$= (y_1 + x_1, y_2 + x_2, \ldots, y_n + x_n)$

(by commutativity of real numbers)

$= (y_1, y_2, \ldots, y_n) + (x_1, x_2, \ldots, x_n)$

(by definition).

We have shown that n-tuples of real numbers satisfy the commutativity axiom for a vector space.

3) Addition is associative:

$$(a+b) + c = a + (b+c).$$

Let
$$(x_1, x_2, \ldots, x_n), (y_1, y_2, \ldots, y_n)$$
and
$$(z_1, z_2, \ldots, z_n)$$
be three points in R^n.

Now,
$$\Big[(x_1, x_2, \ldots, x_n) + (y_1, y_2, \ldots, y_n)\Big]$$
$$+ (z_1, z_2, \ldots, z_n)$$
$$= (x_1 + y_1, x_2 + y_2, \ldots, x_n + y_n) + (z_1, z_2, \ldots, z_n)$$
$$= \Big[(x_1 + y_1) + z_1, (x_2 + y_2) + z_2, \ldots, (x_n + y_n) + z_n\Big]. \quad (1)$$

The coordinates $(x_i + y_i) + z_i$ $(i = 1, \ldots, n)$ are real numbers. Since real numbers satisfy the associativity axiom,
$$(x_i + y_i) + z_i = x_i + (y_i + z_i).$$

Hence, (1) may be rewritten as
$$\Big[x_1 + (y_1 + z_1), x_2 + (y_2 + z_2), \ldots, x_n + (y_n + z_n)\Big]$$
$$= (x_1, x_2, \ldots, x_n) + (y_1 + z_1, y_2 + z_2, \ldots, y_n + z_n)$$
$$= (x_1, x_2, \ldots, x_n) + \Big[(y_1, y_2, \ldots, y_n)$$
$$+ (z_1, z_2, \ldots, z_n)\Big].$$

4) Existence and uniqueness of a zero element.

The set R^n should have a member $(a_1, a_2, \ldots, a_n)$ such that for any point $(x_1, \ldots, x_n)$ in R^n, $(x_1, x_2, \ldots, x_n) + (a_1, a_2, \ldots, a_n) = (x_1, x_2, \ldots, x_n)$.

The point
$$\underline{(0, 0, 0, \ldots, 0)}$$
$$\text{n zeros}$$

where 0 is the unique zero of the real number system, satisfies this requirement.

5) Existence and uniqueness of an additive inverse.

Let
$$(x_1, x_2, \ldots, x_n) \in R^n.$$

An additive inverse of $(x_1, x_2, \ldots, x_n)$ is an n-tuple $(a_1, a_2, \ldots, a_n)$ such that

$$(x_1, x_2, \ldots, x_n) + (a_1, a_2, \ldots, a_n) = (0, 0, \ldots, 0).$$

Since $x_1, x_2, \ldots, x_n$ belong to the real number system, they have unique additive inverses

$$(-x_1), (-x_2), \ldots, (-x_n).$$

Consider $(-x_1, -x_2, \ldots, -x_n) \in R^n$.

$$(x_1, x_2, \ldots, x_n) + (-x_1, -x_2, \ldots, -x_n)$$
$$= \left[x_1 + (-x_1), x_2 + (-x_2), \ldots, x_n + (-x_n)\right]$$
$$= (0, 0, \ldots, 0).$$

We now turn to the axioms involving scalar multiplication.

6) Closure under scalar multiplication.

By definition,
$$\alpha(x_1, x_2, \ldots, x_n) = (\alpha x_1, \alpha x_2, \ldots, \alpha x_n)$$

where the coordinates αx_i are real numbers. Hence,
$$(\alpha x_1, \alpha x_2, \ldots, \alpha x_n) \in R^n.$$

7) Associativity of scalar multiplication.

Let α, β be elements of R. We must show that

$$(\alpha \beta)(x_1, x_2, \ldots, x_n) = \alpha(\beta x_1, \beta x_2, \ldots, \beta x_n).$$

But, since α, β and $x_1, x_2, \ldots, x_n$ are real numbers,

$$(\alpha \beta)(x_1, x_2, \ldots, x_n) = (\alpha \beta x_1, \ldots, \alpha \beta x_n)$$
$$= \alpha(\beta x_1, \beta x_2, \ldots, \beta x_n).$$

8) The first distributive law.

We must show that $\alpha(x+y) = \alpha x + \alpha y$ where x and y are vectors in R^n and $\alpha \in R$.

$$\alpha[(x_1, x_2, \ldots, x_n) + (y_1, y_2, \ldots, y_n)]$$
$$= \alpha\left[(x_1 + y_1), (x_2 + y_2), \ldots, (x_n + y_n)\right]$$
$$= \left[\alpha(x_1 + y_1), \alpha(x_2 + y_2), \ldots, \alpha(x_n + y_n)\right] \quad (2)$$

(by definition of scalar multiplication).

Since each coordinate is a product of a real number and the sum of two real numbers,

$$\alpha(x_i + y_i) = \alpha x_i + \alpha y_i.$$

Hence, (2) becomes

$$\left((\alpha x_1 + \alpha y_1), (\alpha x_2 + \alpha y_2), \ldots, (\alpha x_n + \alpha y_n)\right)$$
$$= (\alpha x_1, \alpha x_2, \ldots, \alpha x_n) + (\alpha y_1, \alpha y_2, \ldots, \alpha y_n)$$
$$= \alpha x + \alpha y.$$

9) The second distributive law.

We must show that $(\alpha + \beta) x = \alpha x + \beta x$ where $\alpha, \beta \in R$ and x is a vector in R^n. Since $\alpha + \beta$ is also a scalar,

$$(\alpha + \beta)(x_1, x_2, \ldots, x_n)$$
$$= \left[(\alpha + \beta) x_1, (\alpha + \beta) x_2, \ldots, (\alpha + \beta) x_n\right]. \quad (3)$$

Since α, β, x_i are all real numbers, then

$$(\alpha + \beta) x_i = \alpha x_i + \beta x_i.$$

Therefore, (3) becomes

$$\left[(\alpha x_1 + \beta x_1), (\alpha x_2 + \beta x_2), \ldots, (\alpha x_n + \beta x_n)\right]$$
$$= (\alpha x_1, \alpha x_2, \ldots, \alpha x_n) + (\beta x_1, \beta x_2, \ldots, \beta x_n)$$
$$= \alpha(x_1, x_2, \ldots, x_n) + \beta(x_1, x_2, \ldots, x_n).$$

10) The existence of a unit element from the field.

We require that there exist a scalar in the field R, call it "1", such that

$$1 (x_1, x_2, \ldots, x_n) = (x_1, x_2, \ldots, x_n).$$

Now the real number 1 satisfies this requirement.

Since a set defined over a field that satisfies (1) –

(10) is a vector space, the set R^n of n-tuples is a vector space when equipped with the given operations of addition and scalar multiplication.

Similarly, F^n is a vector space over F for any field F, where elements of F^n are n-tuples of elements of F.

● **PROBLEM 2-2**

Show that the zero mapping and the identity transformation are linear transformations.

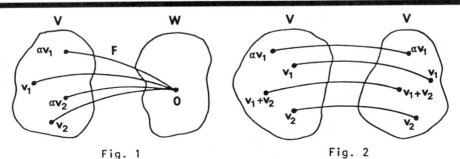

Fig. 1 Fig. 2

Solution: Let V and W be vector spaces over the field K. Then, the function

$$L: V \to W$$

is a linear transformation if:

i) $L(V_1 + V_2) = L(V_1) + L(V_2)$.

ii) $L(\alpha V_1) = \alpha L(V_1)$

for $V_1, V_2 \in V$ and $\alpha \in K$. An equivalent way of expressing i) and ii) is

$$L(\alpha v_1 + v_2) = \alpha L(v_1) + L(v_2).$$

Let $F: V \to W$ be the mapping that sends every element in V to the zero vector in W. F is the zero mapping. We see from Fig. 1 that

$$F(v_1 + v_2) = 0 = F(v_1) + F(v_2).$$

Furthermore,

$$F(\alpha v_1) = 0 = \alpha F(v_1).$$

Thus, the zero mapping is a linear transformation.

Next, let $I: V \to W$ be the function that assigns every element in V to itself. Thus, $W = V$. I is the identity mapping. (That is,

$v_1 \in V$ gives $I(v_1) = v_1$.)

We see from Fig. 2 that

$$F(v_1 + v_2) = (v_1 + v_2) = v_1 + v_2 = F(v_1) + F(v_2).$$

Also,

$$F(\alpha v_1) = \alpha v_1 = \alpha F(v_1).$$

Thus, I, the identity mapping, is a linear transformation.

● **PROBLEM 2-3**

Define affine transformation. Show that

$$T: R^n \to R^n$$

defined by

$$T(x) = x + a \qquad (1)$$

for a fixed $a \in R^n$ is an affine transformation according to this definition. Show that T has an inverse and find general conditions under which an affine map is invertible.

<u>Solution</u>: A linear transformation is a function $L: R^n \to R^m$ such that

$$L(x+y) = L(x) + L(y) \qquad (2)$$

and

$$L(cx) = cL(x) \qquad (3)$$

for $c \in R$; $x, y \in R^n$. A translation is a function $T: R^n \to R^n$ such that $T(x) = x + a$ for some fixed $a \in R^n$ and all $x \in R^n$. An affine map $A: R^n \to R^n$ is a composition of linear maps and translations. For example,

$$A = T \circ L: R^n \to R^n,$$

where

$$L: R^n \to R^n \quad \text{and} \quad T: R^n \to R^n$$

are a linear map and a translation respectively, is affine. Hence, any linear map, as well as any translation, is affine. For T defined by (1), T is a translation. Note that a linear transformation maps zero to zero, since from (3),

$$L(0) = L(0 \cdot x) = 0 \cdot L(x) = 0.$$

Hence, T is linear if and only if $a = 0$, showing that, other than the identity translation, translations are not linear. Let

$$S: R^n \to R^n$$

be another translation defined by
$$S(x) = x + b$$
for some $b \in R^n$. Then $T \circ S$ is defined by
$$(T \circ S)(x) = T[S(x)]$$
$$= T(x + b)$$
$$= x + b + a \qquad (4)$$
and $S \circ T$ is defined by
$$(S \circ T)(x) = S[T(x)]$$
$$= S(x + a)$$
$$= x + a + b . \qquad (5)$$
Hence, by commutativity of elements in R^n,
$$a + b = b + a$$
so that equating (4) and (5) gives
$$S \circ T = T \circ S ,$$
which means that translation multiplication (i.e., composition) is commutative. Now, if $b = -a$ in the definition of S, then
$$(S \circ T)(x) = (T \circ S)(x) = x + a - a = x$$
or
$$(S \circ T) = (T \circ S) = 1$$
where 1 is the identity map on R^n, thus showing that T has an inverse, namely,
$$S(x) = x - a .$$
Note further, that if $B = K \circ S$ and $A = L \circ T$, where K, L are linear maps, S, T are translations, and
$$A, B, K, L, S, T: R^n \to R^n ,$$
then the compositions $A \circ B$ and $B \circ A$ are also affine. That is,
$$A \circ B = (L \circ T) \circ (K \circ S)$$
$$= (L \circ K) \circ (T \circ S)$$
and
$$B \circ A = (K \circ S) \circ (L \circ T)$$
$$= (K \circ L) \circ (T \circ S) .$$

However, they are not necessarily equal unless

$$L \circ K = K \circ L.$$

Therefore, an affine map $A = L \circ T$ is invertible only when L is invertible (i.e., when $\det L \neq 0$).

● **PROBLEM 2-4**

Give examples of the following types of linear operators:

a) two commutative operators
b) two non-commutative operators.

Solution: Let V be a vector space over a field K. A linear operator T is a function $T: V \to V$ such that

i) $T(v_1 + v_2) = T(v_1) + T(v_2)$

ii) $T(\alpha v_1) = \alpha T(v_1)$

for $v_1, v_2 \in V$ and $\alpha \in K$.

Let $S: V \to V$ and $T: V \to V$ be functions from V into V. Then, for $V \in V$, $T(v)$, $S(v) \in V$, and we can operate on the results, i.e., $S(T(v))$ and $T(S(v))$ are defined.

a) Let $S: V \to V$ and $T: V \to V$ be two operators defined on V. S and T are commutative operators if, for $v \in V$,

$$S(T(v)) = T(S(v)). \qquad (1)$$

As an example of commutative operators, let $V = R^2$ and $K = R$. Let $S: R^2 \to R^2$ be a counterclockwise rotation in R^2 through the angle θ, and let $T: R^2 \to R^2$ be a counterclockwise rotation through the angle ϕ. Now, for a given vector $v \in R^2$, $S(T(v))$ is a counterclockwise rotation of v from its original position, first through an angle θ and then through an angle ϕ. Thus, $S(T(v))$ is a counterclockwise rotation through $\phi + \theta$.

On the other hand, $T(S(v))$ is a counterclockwise rotation of v, first through an angle of ϕ and then through an angle θ, i.e., the total rotation is $\theta + \phi$. Hence, $S(T(v)) = T(S(v))$, i.e., S and T commute.

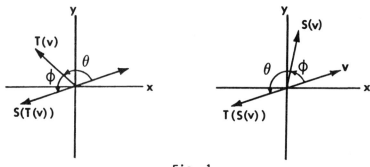

Fig. 1

b) Let $V = R^2$ and $K = R$. Let S be a counterclockwise rotation in R^2 through the angle $\pi/4$, and let T be a reflection in the y-axis. Thus, if $v = (x,y) \in V$, $T(x,y) = (-x,y)$.

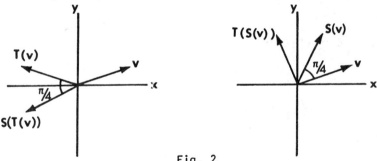

Fig. 2

We see from Fig. 2 above that $S(T(v)) \neq T(S(v))$, i.e., S and T are not commutative.

● **PROBLEM 2-5**

Let

$$(x_1, x_2, \ldots, x_n), (y_1, y_2, \ldots, y_n),$$

$$(z_1, z_2, \ldots, z_n)$$

be three vectors in R^n. Verify the following properties of the dot product using these vectors:

a) $X \cdot X \geq 0$; $X \cdot X = 0$ if and only if $X = 0$

b) $X \cdot Y = Y \cdot X$.

c) $(X+Y) \cdot Z = X \cdot Y + X \cdot Z$.

d) $(cX) \cdot Y = X \cdot (cY) = c(X \cdot Y)$.

<u>Solution</u>: First define the dot product. If

$$X = (x_1, x_2, \ldots, x_n)$$

and

$$Y = (y_1, y_2, \ldots, y_n)$$

are vectors in R^n, then their dot product is defined by:

$$X \cdot Y = x_1 y_1 + x_2 y_2 + \ldots + x_n y_n.$$

The dot product is also known as the inner product since it does satisfy properties a) - d), the properties required of an inner product.

a) $X \cdot X = (x_1, x_2, \ldots, x_n) \cdot (x_1, x_2, \ldots, x_n)$

$$= x_1^2 + x_2^2 + \ldots + x_n^2$$

$$= \sum_{i=1}^{n} x_i^2 \geq 0.$$

If $X = 0$ $\left[X = (0,0,0, \ldots, 0) \right]$

then $X \cdot X = 0$,

and if $X \neq 0$, $\exists\, x_i \neq 0$.

This implies

$$X \cdot X = \sum_{i=1}^{n} x_i^2 \neq 0.$$

b)

$$X \cdot Y = (x_1, x_2, \ldots, x_n) \cdot (y_1, y_2, \ldots, y_n)$$

$$= x_1 y_1 + x_2 y_2 + \ldots + x_n y_n$$

$$= y_1 x_1 + y_2 x_2 + \ldots + y_n x_n$$

$$= (y_1, y_2, \ldots, y_n) \cdot (x_1, x_2, \ldots, x_n) = Y \cdot X.$$

c)

$$(X+Y) \cdot Z = [(x_1, x_2, \ldots, x_n) + (y_1, y_2, \ldots, y_n)] \cdot$$

$$\cdot (z_1, z_2, \ldots, z_n)$$

$$= (x_1 + y_1, x_2 + y_2, \ldots, x_n + y_n) \cdot (z_1, z_2, \ldots, z_n)$$

$$= [(x_1 + y_1) z_1 + (x_2 + y_2) z_2 + \ldots + (x_n + y_n) z_n]$$

$$= [(x_1 z_1 + y_1 z_1) + (x_2 z_2 + y_2 z_2) + \ldots + (x_n z_n + y_n z_n)]$$

(by the associative property of real numbers)

$$= (x_1 z_1 + x_2 z_2 + \ldots + x_n z_n + y_1 z_1 + y_2 z_2 + \ldots + y_n z_n)$$

$$= (x_1 z_1 + x_2 z_2 + \ldots + x_n z_n) + (y_1 z_1 + y_2 z_2 + \ldots + y_n z_n)$$

$$= X \cdot Z + Y \cdot Z.$$

d) Here, c is a scalar from the field R over which R^n is defined.

$$(cX) \cdot Y = [c(x_1, x_2, \ldots, x_n)] \cdot (y_1, y_2, \ldots, y_n)$$

$$= (cx_1, cx_2, \ldots, cx_n) \cdot (y_1, y_2, \ldots, y_n)$$
$$= cx_1y_1 + cx_2y_2 + \ldots + cx_ny_n. \tag{1}$$

From (1), first obtain:

$$(cX) \cdot Y = x_1 cy_1 + x_2 cy_2 + \ldots + x_n cy_n = X \cdot (cY).$$

Returning to (1):

$$(cX) \cdot Y = cx_1y_1 + cx_2y_2 + \ldots + cx_ny_n = c(X \cdot Y).$$

The dot product is used to define the notions of length and distance in R^n. Thus,

$$\|X\| = (X \cdot X)^{\frac{1}{2}} \quad \text{and} \quad \|X-Y\| = [(X-Y) \cdot (X-Y)]^{\frac{1}{2}}.$$

The dot product is also useful for defining the angle between two vectors:

$$\cos \theta = \frac{X \cdot Y}{\|X\| \, \|Y\|} = \frac{(X \cdot Y)}{(X \cdot X)^{\frac{1}{2}} (Y \cdot Y)^{\frac{1}{2}}}.$$

Moreover, these can be generalized for any inner product space. In an arbitrary inner product space the inner product of two elements u and v is usually denoted by "$<u,v>$".

• PROBLEM 2-6

Let C^n denote the vector space of complex n-tuples. Define a suitable inner product for C^n and find $<u,v>$ and $\|u\|$ where $u = (2+3i, 4-i, 2i)$ and $v = (3-2i, 5, 4-6i)$.

Solution: In the vector space R^n, the inner or dot product is defined for

$$u = (u_1, u_2, \ldots, u_n)$$

and

$$v = (v_1, v_2, \ldots, v_n)$$

as

$$<u,v> = u_1v_1 + u_2v_2 + \ldots + u_nv_n = \sum_{i=1}^{n} u_i v_i.$$

This inner product function satisfies the following properties:

i) $<u,v> = <v,u>$

ii) $<au + bv, w> = a<u,w> + b<v,w>$ for $w \in R^n$ and a, b scalars from the field.

iii) $\langle u,u \rangle \geq 0$ and $\langle u,u \rangle = 0 \Leftrightarrow u = 0$.

Since the inner product is used for defining lengths and angles an inner product on C^n should, if it is to be useful, enjoy the same properties.

But, here we run into difficulties. For example, let $ix \in C$. Then

$$\|ix\|^2 = \langle ix, ix \rangle = [(ix)(ix)]$$
$$= (i^2 x^2) = (-x^2) = -\|x\|^2 \qquad (i^2 = -1).$$

Hence, for

$x \neq 0$, $\|ix\|^2 > 0$ implies $\|x\|^2 < 0$ and $\|ix\|^2 < 0$ implies $\|x\|^2 > 0$. But, by property iii) above, length is nonnegative. Hence, we redefine the dot product for complex vector spaces. Let $u, v \in C$. Then,

$$\langle u, v \rangle = u\bar{v}$$

where the bar denotes the complex conjugate, i.e., $\overline{a+ib} = a-ib$. With this definition of the dot product,

$$\|ix\| = \langle ix, ix \rangle^{1/2} = [(ix)(-ix)]^{1/2} = (-i^2 x^2)^{1/2}$$
$$= \|x\| \geq 0.$$

But now the property of symmetry, (i), is lost since

$$\langle u, v \rangle = \overline{\langle v, u \rangle} \neq \langle v, u \rangle.$$

This new inner product function satisfies the following conditions:

i) $\langle u, v \rangle = \overline{\langle v, u \rangle}$

ii) $\langle au + bv, w \rangle = a\langle u, w \rangle + b\langle v, w \rangle$

iii) $\langle u, u \rangle \geq 0$; $\langle u, u \rangle = 0 \Leftrightarrow u = 0$.

In C^n, i.e., the set of all n-tuples of complex numbers, if

$$u = (z_1, z_2, \ldots, z_n) \quad \text{and} \quad v = (w_1, w_2, \ldots, w_n),$$

$$\langle u, v \rangle = z_1 \bar{w}_1 + z_2 \bar{w}_2 + \ldots + z_n \bar{w}_n = \sum_{i=1}^n z_i \bar{w}_i,$$

and

$$\|u\| = \langle u, u \rangle^{1/2} = \sqrt{z_1 \bar{z}_1 + z_2 \bar{z}_2 + \ldots + z_n \bar{z}_n}$$
$$= \left(\sum_{i=1}^n |z_i|^2 \right)^{1/2}.$$

Turning to the given problem,

$$u = (2+3i,\ 4-i,\ 2i) \quad \text{and} \quad \bar{v} = (3+2i,\ 5,\ 4+6i).$$

$$\langle u,v \rangle = (2+3i)(3+2i) + (4-i)(5) + 2i(4+6i).$$

The product of two complex numbers $(a+ib)(c+id)$ is defined as

$$ac - bd + i(ad + bc).$$

Thus,

$$\langle u,v \rangle = (6-6) + i(13) + 20 - i5 + (i8 - 12).$$

Addition of complex numbers $(a+ib) + (c+id)$ is defined as $(a+c) + i(b+d)$. Thus,

$$\langle u,v \rangle = 8 + i16.$$

Next, find the norm of u.

$$\|u\| = \langle u,u \rangle^{\frac{1}{2}} = [(2+3i)(2-3i) + (4-i)(4+i) + 2i(-2i)]^{\frac{1}{2}}$$

$$= [13 + i(0) + 17 + 4]^{\frac{1}{2}} = \sqrt{34} \doteq 5.83.$$

● **PROBLEM 2-7**

Distinguish between n-dimensional Euclidean space and the vector space of n-tuples.

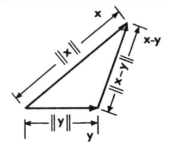

Solution: Let R^n denote the vector space of n-tuples of the form

$$x = (x_1,\ x_2,\ \ldots,\ x_n)$$

where the x_i are real numbers. The axioms for a vector space do not provide any definition of distance. If we think of the length of a vector x in R^n as a function of its distance from the zero vector 0, then, by the Pythagorean theorem,

$$d(x,0) = \sqrt{x_1^2 + x_2^2 + \ldots + x_n^2}. \tag{1}$$

Instead of the origin, or zero vector, consider the distance between any two vectors in R^n.

From the figure and using the Pythagorean theorem,

$$d(x,y) = \sqrt{(y_1-x_1)^2 + (y_2-x_2)^2 + \ldots + (y_n-x_n)^2}$$

where

$$x = (x_1, x_2, \ldots, x_n), \quad y = (y_1, y_2, \ldots, y_n).$$

The distance function in combination with the vector space R^n results in Euclidean n-dimensional space.

The distance function above is also known as the Euclidean norm. A norm is a nonnegative valued function of a vector $x \in V$, where V is a vector space, denoted by $\|x\|$ such that

a) $\|x\| > 0, \; x \neq 0$ and $\|x\| = 0 \Leftrightarrow x = 0$.

b) $\|kx\| = |k| \|x\|$ for any scalar k,

c) $\|x+y\| \leq \|x\| + \|y\|$ for $y \in V$.

Note that we can define other norms on R^n. The resulting structure will then no longer be Euclidean space. For example,

$$\|x\| = \max_i |x_i|,$$

where $|x_i|$ denotes the absolute value of x_i, is also a norm on R^n. However, different norms defined on the same set may be equivalent (i.e., may yield equivalent results).

The Complex analogue C^n can also be considered both as a vector space and as a normed space. In the latter case, it is not a Euclidean space but rather a unitary space.

• **PROBLEM 2-8**

What is a quadratic form?

Solution: A function of n variables $f(x_1, \ldots, x_n)$ is called homogeneous of the second degree if:

$$f(tx_1, tx_2, \ldots, tx_n) = t^2 f(x_1, x_2, \ldots, x_n). \tag{1}$$

A quadratic form is a homogeneous function of degree two that is of the form:

$$A = \sum_{i=1}^{n} \sum_{j=1}^{n} a_{ij} x_i x_j$$

where $a_{ij} = a_{ji}$ for $i = 1, \ldots, n$.

$$A = a_{11}x_1^2 + a_{12}x_1x_2 + a_{13}x_1x_3 + \ldots + a_{1n}x_1x_n$$

$$+ a_{21}x_2x_1 + a_{22}x_2^2 + a_{23}x_2x_3 + \ldots a_{2n}x_2x_n + \ldots$$

$$+ a_{n1}x_nx_1 + a_{n2}x_nx_2 + a_{n3}x_nx_3 + \ldots + a_{nn}x_n^2 . \qquad (2)$$

Equation (2) may be rewritten as:

$$A = a_{11}x_1^2 + a_{22}x_2^2 + \ldots + a_{nn}x_n^2$$

$$+ 2a_{12}x_1x_2 + 2a_{13}x_1x_3 + \ldots + 2a_{1n}x_1x_n$$

$$+ 2a_{23}x_2x_3 + 2a_{24}x_2x_4 + \ldots + 2a_{2n}x_2x_n + \ldots$$

$$+ 2a_{n-2,n-1}x_{n-2}x_{n-1} + 2a_{n-2,n}x_{n-2}x_n$$

$$+ 2a_{n-1,n}x_{n-1}x_n$$

$$= \sum_{i=1}^{n} a_{ii}x_i^2 + 2 \sum_{i=1}^{n-1} \sum_{j=i+1}^{n} a_{ij}x_ix_j ,$$

where we have used the symmetry of $a_{ij} = a_{ji}$. Note that A can be given a matrix form; i.e.,

$$A = [x_1 x_2 \ldots x_n] \begin{bmatrix} a_{11} & a_{12} & \cdots & a_{1n} \\ a_{21} & a_{22} & \cdots & a_{2n} \\ \vdots & \vdots & \vdots & \vdots \\ a_{n1} & a_{n2} & \cdots & a_{nn} \end{bmatrix} \begin{bmatrix} x_1 \\ x_2 \\ \vdots \\ x_n \end{bmatrix} = X^T \Phi X = <\Phi X, X>,$$

where X and Φ are defined in the obvious way and $<,>$ denotes the inner product. It is always possible to construct a quadratic form from a given real symmetric matrix. For example, let

$$A = \begin{bmatrix} 1 & 2 \\ 2 & 3 \end{bmatrix}, \quad B = \begin{bmatrix} 1 & 3 & 5 \\ 3 & 2 & -1 \\ 5 & -1 & 4 \end{bmatrix} .$$

We have

$$Q_A(x_1, x_2) = [x_1 x_2] \begin{bmatrix} 1 & 2 \\ 2 & 3 \end{bmatrix} \begin{bmatrix} x_1 \\ x_2 \end{bmatrix}.$$

Upon multiplying the above, the result is the quadratic form:

$$x_1^2 + 2x_1x_2 + 2x_1x_2 + 3x_2^2 = x_1^2 + 4x_1x_2 + 3x_2^2.$$

Similarly,

$$Q_B(x_1, x_2, x_3) = [x_1 x_2 x_3] \begin{bmatrix} 1 & 3 & 5 \\ 3 & 2 & -1 \\ 5 & -1 & 4 \end{bmatrix} \begin{bmatrix} x_1 \\ x_2 \\ x_3 \end{bmatrix}$$

$$= [x_1 + 3x_2 + 5x_3, \; 3x_1 + 2x_2 - x_3, \; 5x_1 - x_2 + 4x_3] \begin{bmatrix} x_1 \\ x_2 \\ x_3 \end{bmatrix}$$

$$= x_1^2 + 3x_1x_2 + 5x_1x_3 + 3x_1x_2 + 2x_2^2 - x_3x_2$$

$$+ 5x_1x_3 - x_2x_3 + 4x_3^2$$

$$= x_1^2 + 2x_2^2 + 4x_3^2 + 6x_1x_2 + 10x_1x_3 - 2x_2x_3.$$

When a quadratic form is represented by a diagonal matrix, the terms x_ix_j, $i \neq j$, vanish, and the form is said to be in canonical form. Every real quadratic form can be put into canonical form using a suitable basis for the matrix.

PROPERTIES

● **PROBLEM 2-9**

Show that the dimension of a finite-dimensional vector space X equals the order (or cardinality) of any of its bases.

Solution: A set of vectors $\{x_i\}$ is linearly independent if, for scalars $\{a_i\}$, $\Sigma a_i x_i = 0$ implies each $a_i = 0$; otherwise $\{x_i\}$ are said to be linearly dependent. The vector $\Sigma a_i x_i$ is called a linear combination for any set of scalars $\{a_i\}$. The set of all linear combinations of $S = \{x_i\}$ is called the span of S (i.e., S is said to span this set). A set $B = \{x_i\}$ is called a basis of X if B is linearly independent and B spans X. A vector space is called finite-dimensional if all of its bases are finite. The dimension of a finite-dimensional vector space X is n (i.e., dim X = n) if X contains a linearly independent set of n vectors but no linearly independent set of n+1 vectors. Hence, for any particular finite-dimensional vector space X,

$$\dim X \geq |B| = n$$

where $||$ indicates order (or cardinality) and

$$B = \{x_1, \ldots, x_n\}$$

is any basis of X. It is known that n exists since each basis is finite. Since B is a basis, B also spans X. In order to show that dim X = n, it must be shown that X contains no linearly independent set of n+1 elements. Let

$$S = \{y_1, \ldots, y_m\}$$

be any finite linearly independent set in X. Suppose dim S = m > n. Since B spans X, y_1 is a linear combination of B:

$$y_1 = a_1 x_1 + \ldots + a_n x_n. \tag{1}$$

Note that none of the elements of S are 0 for if 0 were in S then S would not be linearly independent. Hence, some $a_i \neq 0$ in (1). Now assume $a_n \neq 0$ for definiteness. Then

$$x_n = \frac{1}{a_n} y_1 - \frac{1}{a_n} (a_1 x_1 + \ldots + a_{n-1} x_{n-1}).$$

Thus

$$B_1 = \{x_1, \ldots, x_{n-1}, y_1\}$$

spans X and y_2 is a linear combination of elements of B_1

$$y_2 = b_1 x_1 + \ldots + b_{n-1} x_{n-1} + c y_1.$$

Since $y_2 \neq 0$, some $b_i \neq 0$ or $c \neq 0$. If $c \neq 0$ and $b_1 = \ldots = b_{n-1} = 0$, then $y_2 = c y_1$ which is impossible since S is linearly independent. Therefore, some $b_i \neq 0$. Assume that $b_{n-1} \neq 0$. Then

$$x_{n-1} = \frac{1}{b_{n-1}} y_2 - \frac{1}{b_{n-1}} (b_1 x_1 + \ldots + b_{n-2} x_{n-2} + c y_1)$$

and thus

$$B_2 = \{x_1, \ldots, x_{n-2}, y_1, y_2\}$$

spans X. Continuing in this manner, we have

$$B_n = \{y_1, \ldots, y_n\}$$

spans X. Since $m > n$,

$$y_{n+1} = d_1 y_1 + \ldots + d_n y_n,$$

which contradicts the linear independence of S. Hence, $n \geq m$. Therefore, $\dim X = n$. [Note that this implies that all bases are of the same order.]

• **PROBLEM 2-10**

Which of the following vectors form a basis for R^3?

i) $(1, 1, 1)$ and $(1, -1, 5)$

ii) $(1, 2, 3)$, $(1, 0, -1)$, $(3, -1, 0)$ and $(2, 1, -2)$

iii) $(1, 1, 1)$, $(1, 2, 3)$ and $(2, -1, 1)$

iv) $(1, 1, 2)$, $(1, 2, 5)$ and $(5, 3, 4)$.

Solution: A basis B for a vector space V has the following properties:

a) B is the smallest set of vectors that span V:

b) B is the largest set of linearly independent vectors in V;

c) the number of vectors in B is the dimension of V.

We may use properties a) - c) to test whether the given sets form a basis for R^3.

i) Since R^3 has dimension 3 (any point in R^3 can be expressed as a unique linear combination of the unit vectors $(1, 0, 0)$, $(0, 1, 0)$ and $(0, 0, 1)$), $(1, 1, 1)$ and $(1, -1, 5)$ cannot be a basis for R^3 by property c).

ii) Here we have 4 vectors in the alleged basis. Again, using property c), these vectors do not form a basis. That is, a basis for R^3 must contain only 3 vectors.

iii) Since there are 3 vectors in this set, it suffices to show that they span R^3. Let (x, y, z) be an arbitrary vector in R^3. Then,

$$(x, y, z) = c_1 (1, 1, 1) + c_2 (1, 2, 3) + c_3 (2, -1, 1)$$

$$(x, y, z) = (c_1, c_1, c_1) + (c_2, 2c_2, 3c_2) + (2c_3, -c_3, c_3)$$

$$(x, y, z) = (c_1 + c_2 + 2c_3, c_1 + 2c_2 - c_3, c_1 + 3c_2 + c_3)$$

$$x = c_1 + c_2 + 2c_3$$

$$y = c_1 + 2c_2 - c_3 \tag{1}$$

$$z = c_1 + 3c_2 + c_3.$$

The system (1) may be rewritten in matrix form as

$$\begin{bmatrix} x \\ y \\ z \end{bmatrix} = \begin{bmatrix} 1 & 1 & 2 \\ 1 & 2 & -1 \\ 1 & 3 & 1 \end{bmatrix} \begin{bmatrix} c_1 \\ c_2 \\ c_3 \end{bmatrix} \tag{2}$$

or

$X = AC$, where A is the matrix in (2). Since

$$\det A = (2 - 1 + 6) - (4 - 3 + 1) = 5,$$

A is invertible by the theorem which states that A^{-1} exists if and only if $\det A \neq 0$. Hence, (2) has a unique solution.

iv) If the given vectors are independent, they form a basis for R^3. Form the matrix whose rows are the given vectors, and row reduce to echelon form by applying the required row operations.

$$\begin{bmatrix} 1 & 1 & 2 \\ 1 & 2 & 5 \\ 5 & 3 & 4 \end{bmatrix} \text{ to } \begin{bmatrix} 1 & 1 & 2 \\ 0 & 1 & 3 \\ 0 & -2 & -6 \end{bmatrix} \text{ to } \begin{bmatrix} 1 & 1 & 2 \\ 0 & 1 & 3 \\ 0 & 0 & 0 \end{bmatrix} \tag{3}$$

From the row echelon matrix (3), we can see that there are only two linearly independent rows, i.e., the given set is not linearly independent. Thus, it cannot be a basis for R^3.

● **PROBLEM 2-11**

Let $L: V \to W$ be a linear transformation of the finite-dimensional vector space V into the vector space W. Show that dim V = dim ker L + dim Im L, where ker L is the kernel of L and Im L is the image (or range) of L.

Solution: First recall the definitions of the kernel and image of a transformation $T: V \to W$; ker T = the set of all vectors in V that are mapped by T to the zero vector of W; Im T = the set of all vectors in W that are images of vectors in V under T. Thus, if $L \equiv 0$, then ker $L = V$ and

Im L = 0 so that dim ker L = dim V, dim Im L = 0, and dim V = dim ker L + dim Im L. Suppose L $\not\equiv$ 0. Let n = dim V and k = dim ker L (which is also called the nullity of L while ker L is sometimes called the nullspace of L). Suppose $0 \leq k < n$. The desired result will follow if dim Im L = n - k. Choose a basis

$$\{x_1, \ldots, x_k\}$$

for ker L. Extend this to a basis

$$S = \{x_1, \ldots, x_n\}$$

for V. Let

$$T = \{Lx_{k+1}, \ldots, Lx_n\}$$

and y be any vector in Im L. Then y = Lx for some $x \in V$. Express x uniquely in terms of the basis S as the linear combination

$$a_1 x_1 + \ldots + a_n x_n .$$

Then

$$y = Lx = L(a_1 x_1 + \ldots + a_n x_n)$$

$$= a_1 Lx_1 + \ldots + a_n Lx_n$$

$$= a_{k+1} Lx_{k+1} + \ldots + a_n Lx_n$$

since

$$x_1, \ldots, x_k \in \ker L .$$

Hence, T spans Im L. Suppose

$$a_{k+1} Lx_{k+1} + \ldots + a_n Lx_n = 0 .$$

Then

$$L(a_{k+1} x_{k+1} + \ldots + a_n x_n) = 0$$

and hence

$$(a_{k+1} x_{k+1} + \ldots + a_n x_n) \in \ker L .$$

So, then

$$a_{k+1} x_{k+1} + \ldots + a_n x_n = b_1 x_1 + \ldots + b_k x_k$$

or

$$b_1 x_1 + \ldots + b_k x_k + a_{k+1} x_{k+1} + \ldots + a_n x_n = 0$$

for some scalars b_i. Since S is linearly independent,

$$b_1 = \ldots = b_k = a_{k+1} = \ldots = a_n = 0.$$

Hence, T is also linearly independent and therefore is a basis for Im L. Thus, dim Im $L = |T| = n-k$.

● **PROBLEM** 2-12

Let $T: V_1 \to V_2$; $S: V_2 \to V_3$; $R: V_3 \to V_4$ be linear transformations where V_1, V_2, V_3 and V_4 are vector spaces defined over a common field K. If we define multiplication of transformations by

$$S \circ T(v) = S(T(v)),$$

show that the multiplication is associative, i.e.,

$$(RS)T(v) = R(ST(v)), \quad \text{where } v \in V_1. \tag{1}$$

Solution: We first identify the domains and co-domains of the various transformations. T transforms V_1, the domain, into V_2. The domain of S is the co-domain of T. Similarly, the domain of R is the co-domain of S. Finally the co-domain of R is V_4.

On the left-hand side of (1), since S transforms V_2 into V_3 and R transforms V_3 into V_4, RS transforms V_2 into V_4. Then, (RS)T transforms V_1 into V_4. On the right-hand side, ST transforms V_1 into V_3 and R(ST) transforms V_1 into V_4. Thus, the domains of (RS)T and R(ST) are the same.

Now, let $v \in V_1$. Then, $((RS)T)v = (RS)(Tv) = R(S(Tv)) = R((ST)v) = (R(ST))v$, i.e., $R(ST) = (RS)T$. Thus, transformation multiplication is associative.

● **PROBLEM** 2-13

Show that in matrix arithmetic we can have the following:

a) $AB \neq BA$.

b) $A \neq 0$, $B \neq 0$, and yet, $AB = 0$.

c) $A \neq 0$ and $A^2 = 0$.

d) $A \neq 0$, $A^2 \neq 0$, and $A^3 = 0$.

e) $A^2 = A$ with $A \neq 0$ and $A \neq I$.

f) $A^2 = I$ with $A \neq -I$ and $A \neq I$.

59

Solution: a)

$$A = \begin{bmatrix} 2 & 1 \\ -1 & 0 \end{bmatrix} \qquad B = \begin{bmatrix} 1 & 0 \\ 3 & 1 \end{bmatrix}$$

$$AB = \begin{bmatrix} 2+3 & 0+1 \\ -1+0 & 0 \end{bmatrix} = \begin{bmatrix} 5 & 1 \\ -1 & 0 \end{bmatrix}$$

$$BA = \begin{bmatrix} 1 & 0 \\ 3 & 1 \end{bmatrix} \begin{bmatrix} 2 & 1 \\ -1 & 0 \end{bmatrix} = \begin{bmatrix} 2+0 & 1+0 \\ 6-1 & 3+0 \end{bmatrix} = \begin{bmatrix} 2 & 1 \\ 5 & 3 \end{bmatrix}.$$

Thus, $AB \neq BA$. Therefore, matrix multiplication and, hence, transformation multiplication are not commutative.

b) Let

$$A = \begin{bmatrix} 0 & 1 \\ 0 & 0 \end{bmatrix} \; ; \; B = \begin{bmatrix} 0 & 4 \\ 0 & 0 \end{bmatrix}.$$

$$AB = \begin{bmatrix} 0 & 1 \\ 0 & 0 \end{bmatrix} \begin{bmatrix} 0 & 4 \\ 0 & 0 \end{bmatrix} = \begin{bmatrix} 0+0 & 0+0 \\ 0+0 & 0+0 \end{bmatrix}$$

$$= \begin{bmatrix} 0 & 0 \\ 0 & 0 \end{bmatrix}$$

Thus, $AB = 0$, $A \neq 0$ and $B \neq 0$.

c)

Let $A = \begin{bmatrix} 0 & 1 \\ 0 & 0 \end{bmatrix}$,

$$A^2 = AA = \begin{bmatrix} 0 & 1 \\ 0 & 0 \end{bmatrix} \begin{bmatrix} 0 & 1 \\ 0 & 0 \end{bmatrix} = \begin{bmatrix} 0+0 & 0+0 \\ 0+0 & 0+0 \end{bmatrix} = \begin{bmatrix} 0 & 0 \\ 0 & 0 \end{bmatrix}.$$

Let $A = \begin{bmatrix} 1 & -1 \\ 1 & -1 \end{bmatrix}$,

then

$$A^2 = AA = \begin{bmatrix} 1 & -1 \\ 1 & -1 \end{bmatrix} \begin{bmatrix} 1 & -1 \\ 1 & -1 \end{bmatrix} = \begin{bmatrix} 1-1 & -1+1 \\ 1-1 & -1+1 \end{bmatrix} = \begin{bmatrix} 0 & 0 \\ 0 & 0 \end{bmatrix}.$$

Thus, $A \neq 0$ and $A^2 = 0$.

d) Let

$$A = \begin{bmatrix} 0 & 1 & 1 \\ 0 & 0 & 1 \\ 0 & 0 & 0 \end{bmatrix}$$

$$A^2 = AA = \begin{bmatrix} 0 & 1 & 1 \\ 0 & 0 & 1 \\ 0 & 0 & 0 \end{bmatrix} \begin{bmatrix} 0 & 1 & 1 \\ 0 & 0 & 1 \\ 0 & 0 & 0 \end{bmatrix}$$

$$= \begin{bmatrix} 0+0+0 & 0+0+0 & 0+1+0 \\ 0+0+0 & 0+0+0 & 0+0+0 \\ 0+0+0 & 0+0+0 & 0+0+0 \end{bmatrix} = \begin{bmatrix} 0 & 0 & 1 \\ 0 & 0 & 0 \\ 0 & 0 & 0 \end{bmatrix}.$$

$$A^3 = A^2 A = \begin{bmatrix} 0 & 0 & 1 \\ 0 & 0 & 0 \\ 0 & 0 & 0 \end{bmatrix} \begin{bmatrix} 0 & 1 & 1 \\ 0 & 0 & 1 \\ 0 & 0 & 0 \end{bmatrix}$$

$$= \begin{bmatrix} 0 & 0 & 0 \\ 0 & 0 & 0 \\ 0 & 0 & 0 \end{bmatrix}.$$

Thus, $A \neq 0$ and $A^2 \neq 0$ but $A^3 = 0$. (A matrix is called nilpotent if some power of itself is 0.)

e) Let

$$A = \begin{bmatrix} \tfrac{1}{2} & \tfrac{1}{2} \\ \tfrac{1}{2} & \tfrac{1}{2} \end{bmatrix}$$

$$A^2 = AA = \begin{bmatrix} \tfrac{1}{2} & \tfrac{1}{2} \\ \tfrac{1}{2} & \tfrac{1}{2} \end{bmatrix} \begin{bmatrix} \tfrac{1}{2} & \tfrac{1}{2} \\ \tfrac{1}{2} & \tfrac{1}{2} \end{bmatrix} = \begin{bmatrix} \tfrac{1}{4}+\tfrac{1}{4} & \tfrac{1}{4}+\tfrac{1}{4} \\ \tfrac{1}{4}+\tfrac{1}{4} & \tfrac{1}{4}+\tfrac{1}{4} \end{bmatrix} = \begin{bmatrix} \tfrac{1}{2} & \tfrac{1}{2} \\ \tfrac{1}{2} & \tfrac{1}{2} \end{bmatrix}$$

Thus, $A^2 = A$ with $A \neq 0$ and $A \neq I$. Any such matrix (including 0 and I) is called idempotent.

f) Let
$$A = \begin{bmatrix} -1 & 0 \\ 0 & 1 \end{bmatrix}$$

$$A^2 = AA = \begin{bmatrix} -1 & 0 \\ 0 & 1 \end{bmatrix} \begin{bmatrix} -1 & 0 \\ 0 & 1 \end{bmatrix}$$

$$A^2 = \begin{bmatrix} 1+0 & 0+0 \\ 0+0 & 0+1 \end{bmatrix} = \begin{bmatrix} 1 & 0 \\ 0 & 1 \end{bmatrix}.$$

Hence, $A^2 = I$ with $A \neq I$ and $A \neq -I$.

● **PROBLEM 2-14**

For n×n matrices A, B show that det AB = det A det B.

Solution: If A is not invertible (i.e., singular or det A = 0), then A is row equivalent to a matrix $D = E_k \ldots E_1 A$ with a row of zeros. Thus,

$$DB = E_k \ldots E_1 AB$$

and AB is row equivalent to DB which has a row of zeros. Hence, since det DB = 0, we have det AB = 0 since $\det(E_k C) = \det E_k \det C$ for any elementary matrix E_k. Thus,

$$\det AB = \det A \det B = 0$$

If A is invertible, A is row equivalent to I. Thus,

$$A = E_k \ldots E_1 I = E_k \ldots E_1$$

for some elementary matrices $E_1, \ldots, E_k$. Then,

$$\det A = \det E_k \ldots E_1$$
$$= \det E_k \ldots \det E_1.$$

Now
$$AB = E_k \ldots E_1 B$$

and therefore

$$\det AB = \det E_k \ldots E_1 B$$

$$= \det E_k (\det E_{k-1} \cdots E_1 B)$$

$$= \det E_k \cdots \det E_1 \det B$$

$$= \det A \det B,$$

since

$$\det EB = \det E \det B$$

for an elementary matrix E.

● **PROBLEM 2-15**

Find the volume of the parallelepiped determined by the vectors u = (2,3,5), v = (-4,2,6) and w = (1,0,3) in xyz-space.

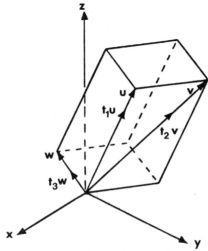

Solution: The parallelepiped can be represented as the set

$$\{t_1 u + t_2 v + t_3 w\}$$

where $0 \leq t \leq 1$ for i = 1,2,3, where the parallelepiped is outlined by the three vectors u, v, w.

In general, let μ(E), where E is an n-dimensional parallelepiped in the vector space R^n, denote the measure (or n-volume) of E. Let C be the n-cube (or n dimensional hyper cube)

$$I^n = [0,1]^n.$$

Now, let T be the linear operator which takes I^n onto E. It is known that

$$\mu(E) = \mu(T(I^n)) = |\det T| \, \mu(I^n) = |\det T|,$$

i.e., measure varies under a linear transformation by a factor of the absolute value of the determinant of the transformation. Hence, the formula for the volume of a parallelepiped is the absolute value of the 3x3 determinant of the matrix formed by using the three determining vectors as its rows. In this case

Volume = absolute value of

$$\begin{vmatrix} 2 & 3 & 5 \\ -4 & 2 & 6 \\ 1 & 0 & 3 \end{vmatrix} .$$ (1)

Recall that multiplying a row by a constant and adding the result to another row does not change the value of the determinant. Hence

$$\begin{vmatrix} 2 & 3 & 5 \\ -4 & 2 & 6 \\ 1 & 0 & 3 \end{vmatrix} = \begin{vmatrix} 0 & 3 & -1 \\ 0 & 2 & 18 \\ 1 & 0 & 3 \end{vmatrix} .$$

This equation is a result of the following two row operations:

1) Multiply the third row of the matrix in equation (1) by 4 and add it to the second row.

2) Multiply the third row by (-2) and add it to the first row.

Expanding along the first column yields:

$$\begin{vmatrix} 2 & 3 & 5 \\ -4 & 2 & 6 \\ 1 & 0 & 3 \end{vmatrix} = \begin{vmatrix} 0 & 3 & -1 \\ 0 & 2 & 18 \\ 1 & 0 & 3 \end{vmatrix}$$

$$= 0 + 0 + 1 \begin{vmatrix} 3 & -1 \\ 2 & 18 \end{vmatrix}$$

$$= 54 - (-2) = 56.$$

From equation (1) we obtain

Volume = 56.

• **PROBLEM 2-16**

Consider the vector space C[0,1] of all continuous functions defined on [0,1]. If

$f \in C[0,1]$,

show that

$$\left(\int_0^1 f^2(x)\,dx\right)^{1/2}$$

defines a norm on all elements of this vector space.

Solution: Since f is continuous, f^2 is continuous. If a function is continuous, it is integrable; thus,

$$\left(\int_0^1 f^2(x)\,dx\right)^{1/2} = \|f\|$$

is well-defined.

Let V be a vector space. A norm on V is a nonnegative number associated with every $v \in V$, denoted $\|v\|$, such that

a)
$$\|v\| \geq 0 \quad \text{and} \quad \|v\| = 0 \Longleftrightarrow v = 0,$$

b)
$$\|kv\| = |k|\,\|v\| \quad \text{for } k, \text{ a scalar},$$

c)
$$\|v+w\| \leq \|v\| + \|w\|, \quad \text{for } w \in V.$$

It is necessary to show that

$$\left(\int_0^1 f^2(x)\,dx\right)^{1/2}$$

satisfies conditions a) - c) above.

a) If $f(x) \in C[0,1]$, then $f^2(x) \geq 0$ on $[0,1]$. But this implies that

$$\left(\int_0^1 f^2(x)\,dx\right)^{1/2} \geq 0.$$

If

$$\|f\| = \left(\int_0^1 f^2(x)\,dx\right)^{1/2} = 0,$$

then

$$f^2(x) = 0$$

and, therefore, $f(x) = 0$ on $[0,1]$ since a nonnegative function continuous over the interval $[0,1]$ can have zero integral over $[0,1]$ only if that function is identically zero on $[0,1]$. So, $\|f\| = 0$ if and only if $f \equiv 0$.

b) Let k be a scalar from the field over which $C[0,1]$ is defined. Then,

$$\|kf\| = \left(\int_0^1 [kf(x)]^2 \, dx\right)^{\frac{1}{2}} = |k| \left(\int_0^1 f^2(x) \, dx\right)^{\frac{1}{2}} = |k| \, \|f\|.$$

c) First, show that

$$\|f+g\| \, , \quad f,g \in C[0,1] \text{ is well-defined.}$$

Now,

$$\|f+g\| = \left(\int_0^1 [f(x) + g(x)]^2 \, dx\right)^{\frac{1}{2}}$$

exists since sums and squares of continuous functions in $C[0,1]$ are also in $C[0,1]$ and, therefore, integrable over $[0,1]$. Note also that

$$fg = \tfrac{1}{2}[(f+g)^2 - f^2 - g^2]$$

is also in $C[0,1]$ and integrable over $[0,1]$. It is now possible to show

$$\|f+g\| \leq \|f\| + \|g\| \quad .$$

$$\|f+g\| \leq \|f\| + \|g\| \iff \|f+g\|^2 \leq (\|f\| + \|g\|)^2, \quad (1)$$

but

$$\|f+g\|^2 = \int_0^1 [f(x) + g(x)]^2 \, dx$$

$$= \int_0^1 f^2(x) \, dx + 2 \int_0^1 f(x)g(x) \, dx + \int_0^1 g^2(x) \, dx \quad (2)$$

and

$$(\|f\| + \|g\|)^2 = \|f\|^2 + 2\|f\|\|g\| + \|g\|^2$$

$$= \int_0^1 f^2(x) \, dx + 2\left(\int_0^1 f^2(x) \, dx\right)^{\frac{1}{2}} \left(\int_0^1 g^2(x) \, dx\right)^{\frac{1}{2}}$$

$$+ \int_0^1 g^2(x)\,dx \ . \tag{3}$$

Comparing (2) and (3), we see that (1) can hold if and only if

$$\int_0^1 f(x)g(x)\,dx \leq \left(\int_0^1 f^2(x)\,dx\right)^{\frac{1}{2}} \left(\int_0^1 g^2(x)\,dx\right)^{\frac{1}{2}}. \tag{4}$$

By the properties of the absolute value functions,

$$\int_0^1 f(x)g(x)\,dx \leq \int_0^1 |f(x)| \cdot |g(x)|\,dx \ .$$

It can be proven that

$$\int_0^1 |f(x)| \cdot |g(x)|\,dx \leq \|f\| \, \|g\| \ .$$

Let λ be a real variable and form

$$\int_0^1 \left(|f(x)| + \lambda|g(x)|\right)^2 dx \geq 0 \ .$$

But,

$$\int_0^1 \left(|f(x)| + \lambda|g(x)|\right)^2 dx =$$

$$\lambda^2 \int_0^1 g^2(x)\,dx + 2\lambda \int_0^1 |f(x)||g(x)|\,dx + \int_0^1 f^2(x)\,dx$$

is a nonnegative quadratic polynomial in λ. Hence, it has no real roots and its discriminant is nonpositive; i.e.,

$$4\left(\int_0^1 |f(x)| \, |g(x)|\,dx\right)^2 - 4 \int_0^1 g^2(x)\,dx \int_0^1 f^2(x)\,dx \leq 0$$

which implies

$$\int_0^1 |f(x)||g(x)|\,dx \leq \left(\int_0^1 f^2(x)\,dx\right)^{\frac{1}{2}} \left(\int_0^1 g^2(x)\,dx\right)^{\frac{1}{2}} ,$$

as was to be shown.

Thus,
$$\left(\int_0^1 f^2(x)\,dx\right)^{1/2}$$
defines a norm on C[0,1] known as the Euclidean norm.

● **PROBLEM 2-17**

Let A be an n×n Hermitian matrix (i.e., $A = A^* \equiv (A^T)^- = (A^-)^T$ where A^- is the conjugate of A). Show that the eigenvalues of A are real.

Solution: Let < , > denote the standard inner product. Note that since
$$\langle x,y \rangle = \overline{\langle y,x \rangle} \quad \text{and} \quad \langle Bu,v \rangle = \langle u, B^*v \rangle$$
for complex x,y and real u,v, and matrix B, then
$$\langle Ax,y \rangle = \langle x, A^*y \rangle.$$
Recall also the properties of inner products
$$\langle ax,y \rangle = a\langle x,y \rangle = \langle x, \bar{a}y \rangle$$
and
$$\langle x,x \rangle = 0 \Leftrightarrow x = 0.$$
Let λ be any eigenvalue of A. Hence there exists a corresponding eigenvector $v \neq 0$ such that $Av = \lambda v$. Therefore
$$\langle Av,v \rangle = \langle \lambda v,v \rangle = \lambda \langle v,v \rangle$$
and
$$\langle v,Av \rangle = \langle v,\lambda v \rangle = \bar{\lambda} \langle v,v \rangle.$$
Since
$$A = A^*, \quad \langle Av,v \rangle = \langle v, A^*v \rangle = \langle v,Av \rangle$$
and thus
$$\lambda = \bar{\lambda} \quad (\langle v,v \rangle \neq 0 \text{ since } v \neq 0).$$
This means that λ is real. Note also that a Hermitian quadratic form is real valued for
$$\langle Ax,x \rangle = \langle x,Ax \rangle = \overline{\langle Ax,x \rangle}.$$

INVERTIBILITY

● **PROBLEM 2-18**

Show that a linear operator A on a finite-dimensional vector space X is invertible if and only if it is one-to-one or onto.

Solution: If X were any vector space, then A being invertible (i.e., there exists a linear operator A^{-1} on X such that

$$AA^{-1}x = A^{-1}Ax = x$$

for all $x \in X$) is equivalent to A being one-to-one and onto. To see this suppose first that A is one-to-one and onto. Then each $x \in X$ has a unique $u \in X$ such that $Au = x$. Defining A^{-1} by $A^{-1}x = u$ for each x where $Au = x$, we have

$$AA^{-1}x = Au = x$$

and

$$A^{-1}Au = A^{-1}x = u .$$

Furthermore, if $x, y \in X$, then

$$A[A^{-1}(x+y)] = AA^{-1}(x+y) = x+y$$

and

$$A[A^{-1}x + A^{-1}y] = AA^{-1}x + AA^{-1}y = x+y .$$

Since A is one-to-one and onto, this means that

$$A^{-1}(x+y) = A^{-1}x + A^{-1}y .$$

Similarly,

$$A^{-1}(cx) = cA^{-1}x$$

for all $x \in X$ and scalar c. Hence A^{-1} is linear. Conversely, suppose A^{-1} is an inverse of A. Then $Ax = Ay$ for $x, y \in X$ implies that

$$A^{-1}Ax = x = y = A^{-1}Ay .$$

Also, $AA^{-1}x = x$ for all $x \in X$ implies that $Au = x$, where $u = A^{-1}x \in X$, for all $x \in X$. Hence, A is one-to-one and onto. If X is a finite-dimensional vector space, then A is still invertible if and only if it is one-to-one and onto. To show that A is invertible if and only if it is one-to-one or onto, it must be shown that (for a finite-dimensional X) A is one-to-one and onto if and only if it is one-to-one or onto—i.e., A is one-to-one if and only if it is onto. To show this, first choose a basis

$$\{x_1, \ldots, x_n\}$$

of X. Since A is linear its image, Im A, equals the span of

$$B = \{Ax_1, \ldots, Ax_n\} .$$

Therefore, Im A = X if and only if B spans X or, equivalently, if the elements of B are independent.

Moreover, A is onto if and only if B is independent. Suppose A is one-to-one and

$$\sum_{i=1}^{n} c_i A x_i = 0 \text{ for scalars } c_i.$$

Then

$$A\left(\sum_{i=1}^{n} c_i x_i\right) = 0.$$

But $A0 = 0$, so $\sum_{i=1}^{n} c_i x_i = 0$.

Hence

$$c_1 = \ldots = c_n = 0$$

and thus B is independent. Conversely, suppose B is independent and

$$A\left(\sum_{i=1}^{n} c_i x_i\right) = 0.$$

Then

$$\sum_{i=1}^{n} c_i A x_i = 0$$

and hence

$$c_1 = \ldots = c_n = 0.$$

Therefore, $Ax = 0$ if and only if $x = 0$. If $Ax = Ay$ then $Ax - Ay = A(x-y) = 0$. But then $x-y = 0$ or $x = y$. Thus, A is one-to-one.

• **PROBLEM 2-19**

Let T and S be linear operators on the vector space V. Show that the following are possible:

i)   T is one-to-one but not invertible;

ii) S is onto but not invertible.

Solution: If V is finite dimensional, i) and ii) cannot occur by the previous problem. However, a linear operator must be both one-to-one and onto (i.e., injective and surjective respectively or, equivalently, bijective) for it to be invertible. Thus, examples must be found on an infinite dimensional vector space where an operator T is one-to-one but not onto and an operator S is onto but not one-to-one.

Let

$$V = \{(x_1, x_2, \ldots) : x_i \in R, i = 1, 2, \ldots\}.$$

As a result V is the set of all sequences of real numbers. Defining addition of sequences and scalar multiplication as in R^n, V becomes an infinite dimensional vector space over the reals.

Define $T: V \to V$ and $S: V \to V$ by

$$T(x_1, x_2, x_3, \ldots) = (0, x_1, x_2, x_3, \ldots)$$

$$S(x_1, x_2, x_3, \ldots) = (x_2, x_3, \ldots)$$

for all $v \in V$. Both T and S are linear transformations since

$$T(v_1 + v_2) = T\bigl((x_1, x_2, \ldots) + (y_1, y_2, \ldots)\bigr)$$

$$= [(0, x_1, x_2, \ldots) + (0, y_1, y_2, \ldots)]$$

$$= (0, x_1 + y_1, x_2 + y_2, \ldots).$$

On the other hand, $T(v_1) = (0, x_1, x_2, \ldots)$,

$$T(v_2) = (0, y_1, y_2, \ldots)$$

and, hence,

$$T(v_1) + T(v_2) = (0, x_1 + y_1, x_2 + y_2, \ldots) = T(v_1 + v_2).$$

Also,

$$T(\alpha v_1) = T(\alpha x_1, \alpha x_2, \ldots) = (0, \alpha x_1, \alpha x_2, \ldots) = \alpha T(v_1).$$

Similarly, it can be shown that S is a linear transformation. Further, T one-to-one because $T(x) = T(y)$ implies

$$(0, x_1, x_2, \ldots) = (0, y_1, y_2, \ldots)$$

so
$x_i = y_i$ for every i   i.e., $x = y$.

But T is not onto since the vector $(1,0,0,\ldots)$, although in V, can never be the image of a vector in V under T. Thus, T is one-to-one but not onto.

S is onto since the range of $S(x_1, x_2, \ldots)$ is $(x_2, x_3, \ldots)$, again an infinite sequence of real numbers. But S is not one-to-one since $S(1,0,0,\ldots) = (0,0,\ldots) = 0$, i.e., the kernel of S contains other elements besides the zero vector. Thus, S is onto but not one-to-one.

● **PROBLEM 2-20**

Show that the n×n matrix A is invertible if and only if det $A \neq 0$.

Solution: The elementary row (or column) operations on a matrix A are (a) row (column) interchanging, (b) row (column) multiplication by a non-zero constant, and (c) row (column) addition [or non-zero multiple of row (column) addition]. An elementary matrix is then an n×n matrix obtained from the identity matrix by an elementary row or column operation. Two m×n matrices are equivalent if one can be obtained from the other by a finite sequence of elementary row and/or column operations. Since any matrix A is row equivalent to a matrix in reduced row echelon form, then, by elementary column operations (of type (a) above followed by some of type (c) above), A is equivalent to a matrix D of the form

$$\begin{pmatrix} I_k & 0 \\ 0 & 0 \end{pmatrix}$$

where I_k is the k×k identity and $k \leq m,n$. Thus, there exist elementary matrices $E_1, \ldots, E_p$ and $F_1, \ldots, F_r$ such that

$$A = E_p E_{p-1} \cdots E_1 D F_1, \ldots, F_r .$$

It is easily seen that det EA = Det AE = det E det A for an elementary matrix E. Hence

$$\det A = \det E_p \cdots \det E_1 \det D \det F_1 \cdots \det F_r .$$

Since $\det E_i \neq 0$ (i=1, ..., p) and $\det F_j \neq 0$ (j=1, ..., r), $\det A \neq 0$ if and only if $\det D \neq 0$. Moreover, $\det A \neq 0$ if and only if k = n. But A is invertible if and only if A is a product of elementary matrices and hence if and only if A is equivalent to I_n. Therefore, A is invertible if and only if $\det A \neq 0$.

• **PROBLEM 2-21**

Show that the system

$$a_{11}x_1 + \ldots + a_{1n}x_n = b_1$$
$$\vdots \quad\quad\quad (1)$$
$$a_{n1}x_1 + \ldots + a_{nn}x_n = b_n$$

has a unique solution if det A ≠ 0 where

$$A = \begin{pmatrix} a_{11} & \cdots & a_{1n} \\ \vdots & & \vdots \\ a_{n1} & \cdots & a_{nn} \end{pmatrix} \quad\quad (2)$$

Moreover, solve the system (1) for $x_1, \ldots, x_n$ (Cramer's Rule).

Solution: If $\det A \neq 0$, then A is invertible. Let (1) be expressed as

$$Ax = B$$

where A is as in (2), $x = (x_1, \ldots, x_n)^T$, and $B = (b_1, \ldots, b_n)^T$. Now

$$x = A^{-1}B = \frac{1}{\det A} \begin{pmatrix} A_{11} & \cdots & A_{n1} \\ \vdots & & \vdots \\ A_{1n} & \cdots & A_{nn} \end{pmatrix} \begin{pmatrix} b_1 \\ \vdots \\ b_n \end{pmatrix},$$

where A_{ij} is the cofactor of a_{ij} (i.e., if M_{ij} is A without the i-th row and j-th column, then

$$A_{ij} = (-1)^{i+j} \det M_{ij}) \quad .$$

Hence

$$x_i = \frac{1}{\det A} (A_{1i}b_1 + \ldots + A_{ni}b_n) \qquad (i = 1, \ldots, n).$$

Let

$$A_i = \begin{pmatrix} a_{11} & \cdots & a_{1i-1} & b_1 & a_{1i+1} & \cdots & a_{1n} \\ \vdots & & \vdots & \vdots & \vdots & & \vdots \\ a_{n1} & \cdots & a_{ni-1} & b_n & a_{ni+1} & \cdots & a_{nn} \end{pmatrix}$$

Evaluate $\det A_i$ by expanding about the i-th column to get:

$$\det A_i = A_{1i}b_1 + \ldots + A_{ni}b_n.$$

Hence

$$x_i = \frac{\det A_i}{\det A} \qquad (i = 1, \ldots, n). \qquad (3)$$

Note that $\det A \neq 0$ and hence, the solution (3) of (1) exists and is unique.

• **PROBLEM 2-22**

If A is a real invertible matrix, show that it is a product of a symmetric positive definite matrix S and an orthogonal matrix R.

Solution: Since, for two matrices B and C we have
$$(BC)^T = C^T B^T$$
and
$$<BCx, x> = <Cx, B^T x>,$$
then for $B = A$ and $C = A^T$
$$(AA^T)^T = (A^T)^T A^T = AA^T$$
and
$$<AA^T x, x> = <A^T x, A^T x> = ||A^T x||^2 > 0$$
for $x \neq 0$. Thus AA^T is symmetric and positive definite. By the Principal Axis Theorem, there exists an orthogonal matrix P such that
$$P^{-1} AA^T P$$
is diagonal and positive definite. Let T be the diagonal positive definite matrix such that
$$T^2 = P^{-1} AA^T P.$$

Hence,
$$S = PTP^{-1}$$
is symmetric and positive definite and
$$S^2 = PT^2 P^{-1} = AA^T.$$
So, let $R = S^{-1} A.$

Then
$$RR^T = S^{-1} AA^T (S^{-1})^T = S^{-1} S^2 (S^{-1})^T = S^{-1} S^2 S^{-1} = I,$$
since S is symmetric (i.e., $S = S^T$). Thus, R is orthogonal and $A = SR$.

DIAGONALIZATION

• **PROBLEM 2-23**

Construct an orthogonal matrix H from the eigenvectors associated with the symmetric matrix:

$$A = \begin{pmatrix} 5 & 1 & 1 \\ 1 & 5 & -1 \\ 1 & -1 & 5 \end{pmatrix}. \quad (1)$$

How does the transformation x = Hy affect the related quadric surface, where H is the orthogonal matrix?

Solution: An orthogonal matrix is a matrix H such that

$$H \cdot H^T = I,$$

where I is the identity matrix. If H is a matrix whose column vectors are mutually orthonormal, then H is an orthogonal matrix.

We know that any symmetric real n × n matrix is diagonalizable, and the eigenvectors of such a matrix form an orthogonal basis of R^n. Thus, from the given matrix A which is symmetric, we construct the associated orthogonal matrix H. This is done by finding the eigenvectors of A, normalizing them, and making them the column vectors of H.

Recall that these eigenvectors also form the basis of a coordinate system in which any quadric surface whose quadric form corresponds to A is in canonical form. That is, the transformation x = Hy will give us a coordinate transformation that diagonalizes the quadratic form.

The quadratic form associated with the symmetric matrix (1) is:

$$[x_1 x_2 x_3] \begin{pmatrix} 5 & 1 & 1 \\ 1 & 5 & -1 \\ 1 & -1 & 5 \end{pmatrix} \begin{pmatrix} x_1 \\ x_2 \\ x_3 \end{pmatrix}$$

$$= 5x_1^2 + 2x_1 x_2 + 5x_2^2 + 5x_3^2 + 2x_1 x_3 - 2x_2 x_3.$$

The eigenvalues associated with (1) are scalars λ that satisfy the matrix equation

$$Ax = \lambda x. \tag{2}$$

From (2), we obtain

$$[A - I\lambda]x = 0$$

which implies $\det[A - \lambda I] = 0$. Thus, the result is the characteristic equation:

$$\det \begin{pmatrix} 5-\lambda & 1 & 1 \\ 1 & 5-\lambda & -1 \\ 1 & -1 & 5-\lambda \end{pmatrix} = 0. \tag{3}$$

Expanding (3) along the first row,

$(5-\lambda)[(5-\lambda)^2 - 1] - 1[(5-\lambda) + 1] + 1[-1 - (5-\lambda)] = 0$

$$\lambda^3 - 15\lambda^2 + 72\lambda - 108 = 0 \qquad (4)$$

This can be factored as

$$(\lambda-x)(\lambda-y)(\lambda-z) = 0$$

Thus,

$$xyz = 108$$
$$x + y + z = 15.$$

By trial and error, we find x=3 and y = z = 6; hence (4) is

$$(\lambda-3)(\lambda-6)(\lambda-6) = 0.$$

Thus, the eigenvalues are

$$\lambda_1 = 3; \quad \lambda_2 = 6; \quad \lambda_3 = 6.$$

An eigenvector associated with $\lambda_1 = 3$ is found by solving the system:

$$\begin{pmatrix} 2 & 1 & 1 \\ 1 & 2 & -1 \\ 1 & -1 & 2 \end{pmatrix} \begin{pmatrix} x_1 \\ x_2 \\ x_3 \end{pmatrix} = \begin{pmatrix} 0 \\ 0 \\ 0 \end{pmatrix}.$$

One solution is [-1,1,1]. Next, find mutually orthogonal eigenvectors corresponding to the double root λ = 6. Form the system

$$\begin{pmatrix} -1 & 1 & 1 \\ 1 & -1 & -1 \\ 1 & -1 & -1 \end{pmatrix} \begin{pmatrix} x_1 \\ x_2 \\ x_3 \end{pmatrix} = \begin{pmatrix} 0 \\ 0 \\ 0 \end{pmatrix}.$$

Two solutions to this system are [1,1,0] and [1,-1,2]. Further, since [1,1,0] · [1,-1,2] = 0, they are orthogonal.

The next stage is to form an orthogonal matrix whose columns are the components of the normalized eigenvectors. The norms of [-1,1,1], [1,1,0] and [1,-1,2] are $\sqrt{3}$, $\sqrt{2}$ and $\sqrt{6}$, respectively. Hence, the required orthogonal matrix is

$$H = \begin{pmatrix} -1/\sqrt{3} & 1/\sqrt{2} & 1/\sqrt{6} \\ 1/\sqrt{3} & 1/\sqrt{2} & -1/\sqrt{6} \\ 1/\sqrt{3} & 0 & 2/\sqrt{6} \end{pmatrix}.$$

Since H is orthogonal, $H^{-1} = H^T$, and $H^{-1}AH = H^TAH$ is a 3 × 3 diagonal matrix with diagonal entries 3, 6, 6.

The transformation

$$\begin{pmatrix} x_1 \\ x_2 \\ x_3 \end{pmatrix} = H \begin{pmatrix} y_1 \\ y_2 \\ y_3 \end{pmatrix}$$

transforms the equation

$$5x_1^2 + 2x_1x_2 + 2x_1x_3 + 5x_2^2 - 2x_2x_3 + 5x_3^2 = k$$

in the (x_1, x_2, x_3) coordinate system to the equation

$$3y_1^2 + 6y_2^2 + 6y_3^2 = k$$

in the (y_1, y_2, y_3) system.

● **PROBLEM 2-24**

Find a matrix P that diagonalizes

$$A = \begin{pmatrix} 3 & -2 & 0 \\ -2 & 3 & 0 \\ 0 & 0 & 5 \end{pmatrix}$$

Solution: The following is the procedure for diagonalizing a diagonalizable n×n matrix A.

1. Find n linearly independent eigenvectors of

    A, $x_1, x_2, \ldots, x_n$.

2. Form the matrix P having

    $x_1, x_2, \ldots, x_n$

as its column vectors.

3. The matrix $P^{-1}AP$ will then be diagonal with

$$\lambda_1, \lambda_2, \ldots, \lambda_n$$

as its successive diagonal entries where λ_i is the eigenvalue corresponding to

$$x_i, \; i = 1, 2, \ldots, n .$$

The characteristic equation of A is

$$\det \begin{vmatrix} \lambda-3 & 2 & 0 \\ 2 & \lambda-3 & 0 \\ 0 & 0 & \lambda-5 \end{vmatrix} = 0 .$$

Expanding along the third row,

$$(\lambda - 5) \begin{vmatrix} \lambda-3 & 2 \\ 2 & \lambda-3 \end{vmatrix} = 0 ,$$

or $(\lambda-5)[(\lambda-3)^2 - 4] = 0$

or $(\lambda-5)(\lambda-1)(\lambda-5) = 0 .$

The eigenvalues of A are $\lambda_1 = 1$ and $\lambda_2 = 5$.

Solve the equation $(5I - A)y = 0$ to obtain the eigenvectors corresponding to $\lambda = 5$:

$$\begin{pmatrix} 2 & 2 & 0 \\ 2 & 2 & 0 \\ 0 & 0 & 0 \end{pmatrix} \begin{pmatrix} y_1 \\ y_2 \\ y_3 \end{pmatrix} = \begin{pmatrix} 0 \\ 0 \\ 0 \end{pmatrix}$$

Solving the above system yields:

$$y_1 = -s, \; y_2 = s, \; y_3 = t ,$$

or

$$y = \begin{pmatrix} -s \\ s \\ t \end{pmatrix} = \begin{pmatrix} -s \\ s \\ 0 \end{pmatrix} + \begin{pmatrix} 0 \\ 0 \\ t \end{pmatrix} = s \begin{pmatrix} -1 \\ 1 \\ 0 \end{pmatrix} + t \begin{pmatrix} 0 \\ 0 \\ 1 \end{pmatrix} .$$

Thus,

$$x_1 = \begin{pmatrix} -1 \\ 1 \\ 0 \end{pmatrix} \text{ and } x_2 = \begin{pmatrix} 0 \\ 0 \\ 1 \end{pmatrix}$$

are two linearly independent eigenvectors corresponding to $\lambda = 5$.

Similarly, we find that

$$x_3 = \begin{pmatrix} 1 \\ 1 \\ 0 \end{pmatrix}$$

is an eigenvector associated with $\lambda = 1$. Thus x_1, x_2 and x_3 are linearly independent vectors such that

$$P = \begin{pmatrix} -1 & 0 & 1 \\ 1 & 0 & 1 \\ 0 & 1 & 0 \end{pmatrix}.$$

To find P^{-1}, use the adjoint method:

$$P^{-1} = \frac{1}{\det P} \text{ adj. } P.$$

We find

$$P^{-1} = \begin{pmatrix} -\tfrac{1}{2} & \tfrac{1}{2} & 0 \\ 0 & 0 & 1 \\ \tfrac{1}{2} & \tfrac{1}{2} & 0 \end{pmatrix}.$$

It follows that $P^{-1}AP$ is the required diagonal matrix.

$$P^{-1}AP = \begin{pmatrix} -1/2 & 1/2 & 0 \\ 0 & 0 & 1 \\ 1/2 & 1/2 & 0 \end{pmatrix} \begin{pmatrix} 3 & -2 & 0 \\ -2 & 3 & 0 \\ 0 & 0 & 5 \end{pmatrix} \begin{pmatrix} -1 & 0 & 1 \\ 1 & 0 & 1 \\ 0 & 1 & 0 \end{pmatrix}$$

$$= \begin{pmatrix} 5 & 0 & 0 \\ 0 & 5 & 0 \\ 0 & 0 & 1 \end{pmatrix}.$$

There is no preferred order for the columns of P. Since the ith diagonal entry of $P^{-1}AP$ is an eigenvalue for the ith column vector of P, changing the order of the columns of P merely changes the order of the eigenvalues on the diagonal of $P^{-1}AP$.

Thus, had we written

$$P = \begin{pmatrix} -1 & 1 & 0 \\ 1 & 1 & 0 \\ 0 & 0 & 1 \end{pmatrix}$$

in the last example, we would have obtained

$$P^{-1}AP = \begin{pmatrix} 5 & 0 & 0 \\ 0 & 1 & 0 \\ 0 & 0 & 5 \end{pmatrix}.$$

Furthermore, if the columns of P (i.e., the eigenvectors of A) are normalized, then

$$P = \begin{pmatrix} -\frac{1}{\sqrt{2}} & 0 & \frac{1}{\sqrt{2}} \\ \frac{1}{\sqrt{2}} & 0 & \frac{1}{\sqrt{2}} \\ 0 & 1 & 0 \end{pmatrix}$$

for example. Now $P^T = P^{-1}$, so that P is orthogonal. Thus, in this case, and in the case of any symmetric n×n matrix A, there exists an orthogonal matrix P which diagonalizes A.

• **PROBLEM 2-25**

Diagonalize the transformation T on C^3, where T is represented by the matrix

$$T = \begin{pmatrix} 0 & 1 & -1 \\ 1/4 & 0 & 3/4 \\ 2 & 1 & 1 \end{pmatrix}$$

over the standard basis. That is, represent T as a diagonal matrix over some basis. Check the result.

Solution: To do this, find the eigenvalues of T (i.e., solve det (T - tI) = 0).

$$\det (T - tI) = \det \begin{pmatrix} -t & 1 & -1 \\ 1/4 & -t & 3/4 \\ 2 & 1 & 1-t \end{pmatrix}$$

$$= (t^2 - t^3 + \tfrac{3}{2} - \tfrac{1}{4}) - (2t - \tfrac{3}{4}t + \tfrac{1}{4} - \tfrac{1}{4}t)$$

$$= -t^3 + t^2 - t + 1$$

and hence the eigenvalues of T are the solutions of

$$-t^3 + t^2 - t + 1 = 0.$$

These are $\lambda_1 = 1$, $\lambda_2 = i$, and $\lambda_3 = -i$. To find the corresponding eigenvectors, solve

$$Tz = \lambda_i z \quad (i = 1, 2, 3).$$

First, for $z = (z_1, z_2, z_3) \in C^3$,

$$Tz = \begin{pmatrix} 0 & 1 & -1 \\ 1/4 & 0 & 3/4 \\ 2 & 1 & 1 \end{pmatrix} \begin{pmatrix} z_1 \\ z_2 \\ z_3 \end{pmatrix}$$

$$= \begin{pmatrix} z_2 - z_3 \\ (1/4)z_1 + (3/4)z_3 \\ 2z_1 + z_2 + z_3 \end{pmatrix} \qquad (1)$$

For $Tz = 1z$, (1) gives

$$z_1 = z_2 - z_3$$
$$z_2 = \tfrac{1}{4}z_1 + \tfrac{3}{4}z_3$$
$$z_3 = 2z_1 + z_2 + z_3 \quad .$$

This system yields the solution

$$z_1 = -\tfrac{1}{2}z_2 = -\tfrac{1}{3}z_3 \quad .$$

Thus, an eigenvector is $(1, -2, -3)$. Similarly, from $Tz = iz$, an eigenvector is $(-3 + 4i, 1 - 3i, 5)$ and from $Tz = -iz$, an eigenvector is $(-3 -4i, 1 + 3i, 5)$. Therefore, T can be represented by the diagonal matrix,

$$D = \begin{pmatrix} 1 & 0 & 0 \\ 0 & i & 0 \\ 0 & 0 & -i \end{pmatrix}$$

over the basis

$$B = \{(1, -2, -3), (-3 + 4i, 1 - 3i, 5),$$
$$(-3 -4i, 1 + 3i, 5)\}.$$

To check this let

$$u = \begin{pmatrix} 1 \\ -2 \\ -3 \end{pmatrix} x + \begin{pmatrix} -3 + 4i \\ 1 - 3i \\ 5 \end{pmatrix} y + \begin{pmatrix} -3-4i \\ 1 + 3i \\ 5 \end{pmatrix} z .$$

$$= \begin{pmatrix} x - (3-4i)y - (3 + 4i)z \\ -2x + (1-3i)y + (1+3i)z \\ -3x + 5y + 5z \end{pmatrix} .$$

Then,

$$Tu = \begin{pmatrix} 0 & 1 & -1 \\ 1/4 & 0 & 3/4 \\ 2 & 1 & 1 \end{pmatrix} \begin{pmatrix} x - (3-4i)y - (3+4i)z \\ -2x + (1-3i)y + (1+3i)z \\ -3x + 5y + 5z \end{pmatrix}$$

$$= \begin{pmatrix} -2x + (1-3i)y + (1+3i)z + 3x - 5y - 5z \\ \frac{1}{4}x - \left(\frac{3}{4} - i\right)y - \left(\frac{3}{4} + i\right)z - \frac{9}{4}x + \frac{15}{4}y + \frac{15}{4}z \\ 2x - (6-8i)y - (6+8i)z - 2x + (1-3i)y + (1+3i)z - 3x + 5y + 5z \end{pmatrix}$$

$$= \begin{pmatrix} x - (4+3i)y - (4-3i)z \\ -2x + (3+i)y + (3-i)z \\ -3x + 5iy - 5iz \end{pmatrix}$$

$$= \begin{pmatrix} 1 & -3+4i & -3-4i \\ -2 & 1-3i & 1+3i \\ -3 & 5 & 5 \end{pmatrix} \begin{pmatrix} x \\ iy \\ -iz \end{pmatrix}$$

$$= \begin{pmatrix} 1 & -3+4i & -3-4i \\ -2 & 1-3i & 1+3i \\ -3 & 5 & 5 \end{pmatrix} \begin{pmatrix} 1 & 0 & 0 \\ 0 & i & 0 \\ 0 & 0 & -i \end{pmatrix} \begin{pmatrix} x \\ y \\ z \end{pmatrix}$$

$$= ADu$$

where A is the change of basis from the basis B to the standard basis.

• **PROBLEM 2-26**

Reduce the quadratic form

$$Q(x_1, x_2, x_3) = x_1^2 + x_3^2 - 2x_1x_2 - 2x_1x_3 + 10x_2x_3$$

to the simplest form. What is the matrix of the transformation?

Solution: A quadratic form is in simplest form when all cross-product terms are eliminated. This is accomplished by completing the square and relabeling terms within parentheses by a change of variables. Geometrically, this relabeling corresponds to a coordinate transformation. We have,

$$Q(x_1, x_2, x_3) = x_1^2 + x_3^2 - 2x_1x_2 - 2x_1x_3 + 10x_2x_3 .$$

Now add and subtract

$$x_2^2 \text{ to } Q.$$

$$Q = (x_1^2 + x_2^2 + x_3^2 - 2x_1x_2 - 2x_1x_3 + 2x_2x_3) + 8x_2x_3 - x_2^2$$

Note that

$$(x_1 - x_2 - x_3)^2 = x_1^2 + x_2^2 + x_3^2 - 2x_1x_2 - 2x_1x_3 + 2x_2x_3 .$$

Therefore,

$$Q = (x_1 - x_2 - x_3)^2 + 8x_2x_3 - x_2^2 . \tag{1}$$

Next, introduce new variables

$$y_1 = (x_1 - x_2 - x_3); \quad y_2 = x_2; \quad y_3 = x_3 .$$

Then (1) becomes

$$Q' = y_1^2 - y_2^2 + 8y_2y_3 .$$

Subtract and add $16y_3^2$ to Q': So,

$$Q' = y_1^2 - y_2^2 + 8y_2y_3 - 16y_3^2 + 16y_3^2$$

$$= y_1^2 - (y_2^2 - 8y_2y_3 + 16y_3^2) + 16y_3^2$$

$$= y_1^2 - (y_2 - 4y_3)^2 + 16y_3^2 .$$

Introduce new variables once more:

$$z_1 = y_1; \quad z_2 = y_2 - 4y_3; \quad z_3 = 4y_3 .$$

Then Q' is transformed into

$$Q'' = z_1^2 - z_2^2 + z_3^2 .$$

The above transformations were all linear, i.e., they have matrices associated with them. Since

$$y_1 = x_1 - x_2 - x_3, \ y_2 = x_2, \text{ and } y_3 = x_3,$$

the transformation from

$$(x_1, x_2, x_3) \text{ to } (y_1, y_2, y_3)$$

is given by the matrix equation.

$$[x_1, x_2, x_3] \begin{bmatrix} 1 & 0 & 0 \\ -1 & 1 & 0 \\ -1 & 0 & 1 \end{bmatrix} = \begin{bmatrix} y_1 \\ y_2 \\ y_3 \end{bmatrix} .$$

Similarly, the transformation from

$$(y_1, y_2, y_3) \text{ to } (z_1, z_2, z_3)$$

is

$$T(y_1, y_2, y_3) = (y_1, y_2 - 4y_3, 4y_3) .$$

The matrix equation of the linear transformation is, therefore,

$$[y_1, y_2, y_3] \begin{bmatrix} 1 & 0 & 0 \\ 0 & 1 & 0 \\ 0 & -4 & 4 \end{bmatrix} = \begin{bmatrix} z_1 \\ z_2 \\ z_3 \end{bmatrix} .$$

The change of coordinates from (x_1, x_2, x_3) to (z_1, z_2, z_3) is, consequently, given by the product of the two matrices:

$$\begin{bmatrix} 1 & 0 & 0 \\ -1 & 1 & 0 \\ -1 & 0 & 1 \end{bmatrix} \begin{bmatrix} 1 & 0 & 0 \\ 0 & 1 & 0 \\ 0 & -4 & 4 \end{bmatrix} = \begin{bmatrix} 1 & 0 & 0 \\ -1 & 1 & 0 \\ -1 & -4 & 4 \end{bmatrix} .$$

● PROBLEM 2-27

Show that any $n \times n$ real symmetric matrix A can be diagonalized by an orthogonal matrix T such that the eigenvalues of A are on the diagonal of A. (i.e.,

$$T^{-1}DT = T^tDT = A$$

where D is diagonal.) Conclude that any real quadratic form is equivalent to a diagonal quadratic form (Principal Axis Theorem) and that the norm of corresponding vectors is preserved.

Solution: To prove this important result, the following lemma is needed:

Let $Q = ax^2 + 2bxy + cy^2$.

If the maximum value of Q evaluated on the unit circle S

(i.e., $x^2 + y^2 = 1$)

is at $x = 0$, $y = 1$, then $b = 0$. To prove this, note that $x^2 + y^2 = 1$ implies that $2x + 2yy' = 0$ and hence $y' = \frac{-x}{y}$. Also,

$$Q' = 2ax + 2by + 2bxy' + 2cyy' = 2ax + 2by - 2b\frac{x^2}{y} - 2cy\frac{x}{y}$$

on S. Therefore, at the maximum $x = 0$, $y = 1$, $Q' = 2b$. Since $Q' = 0$ at a maximum, $b = 0$.

Back to the problem, define the quadratic form

$$Q(x) = <Ax, x>,$$

where

$$x = (x_1, \ldots, x_n) \in R^n.$$

Since this is a polynomial, it is known to be continuous. Furthermore, Q is continuous and bounded by

$$\sum_{i,j=1}^{n} |a_{ij}|$$

on the compact sphere $S^m = \{x \in R^n \mid \|x\| = 1\}$.

Thus, Q has a maximum λ_1 on S^m.

Suppose that for $x_1 \in S^m$, $Q(x_1) = \lambda_1$. Letting $\alpha_1 = x_1$, choose a new orthonormal basis

$$B_\alpha = \{\alpha_1, \ldots, \alpha_n\}.$$

Now in terms of new coordinates $y_1, \ldots, y_n$ of x relative to B_α and a new symmetric matrix $B = (b_{ij})$, $Q(x) = <By, y>$

where $y = (y_1, \ldots, y_n)$. Note that $\alpha_1 = (1, 0, \ldots, 0)$ in terms of B_α, and hence the maximum value is

$$Q(\alpha_1) = \langle B\alpha_1, \alpha_1 \rangle = b_{11} = \lambda_1.$$

This is still the maximum if Q is restricted so that all but the two variables y_1 and y_i ($i \ne 1$) are zero. Thinking of this resticted Q as a function in the two variables y_1 and y_i, then Q attains its maximum on S at $y_1 = 1$, $y_i = 0$. So, by the above lemma, $b_{1i} = 0$. Applying this procedure for $i = 2, \ldots, n$, then $b_{1i} = 0$ for $i = 2, \ldots, n$. Hence,

$$Q(x) = \langle By, y \rangle = \lambda_1 y_1^2 + \sum_{i,j=2}^{n} b_{ij} y_i y_j.$$

Now set

$$Q_1(x) = Q(x) - \lambda_1 y_1^2 = \sum_{i,j=2}^{n} b_{ij} y_i y_j,$$

which is a quadratic form in n-1 variables. These variables are coordinates in

$$S^{m-1} \subset \text{Span } \{\alpha_2, \ldots, \alpha_n\} = \alpha_1^\perp$$

(Here, $\alpha_1^\perp$ is the orghogonal complement of $\alpha 1$).

Reapply the above method and continue until

$$Q(x) = \lambda_1 z_1^2 + \ldots + \lambda_n z_n^2$$

for

$$\lambda_1 \ge \ldots \ge \lambda_n.$$

Here $(z_1, \ldots, z_n)$ represents x relative to the orthonormal basis

$$B^* = \{\alpha_1^*, \ldots, \alpha_n^*\}$$

where $\alpha_1^* = \alpha_1$, $\alpha_2^* = \beta_2$ (from the second basis B_β),

$\alpha_3^* = \gamma_3$ (from the third basis B_γ), etc.

Thus,

$$Q(x) = \langle \Lambda z, z \rangle$$

where Λ is the diagonal matrix sought and z can be expressed by x over the orthonormal basis B^*. Since the basis for x can be thought of as the usual basis

$$E = \{e_1, \ldots, e_n\}$$

where e_i consists of all zeros except the number 1 in the ith place, let
$$T = (\alpha_1^* \ldots \alpha_n^*)$$
where the α_i^* are expressed in terms of E. This means that

$$\langle Ax, x \rangle = \langle \Lambda z, z \rangle$$
$$= \langle \Lambda Tx, Tx \rangle$$
$$= \langle T^t \Lambda Tx, x \rangle$$

and therefore $A = T^t \Lambda T$. But T is orthogonal since the α_i^* are orthogonal, so $A = T^{-1} \Lambda T$. Moreover

$$\langle Tx, Tx \rangle = \langle x, T^t Tx \rangle = \langle x, T^{-1} Tx \rangle = \langle x, x \rangle$$

and hence

$$\|Tx\| = \|x\| \quad .$$

Furthermore, since similar matrices have the same characteristic equation,

$$\Big[\det(A - \lambda I) = \det(T^{-1} \Lambda T - \lambda T^{-1} I T) =$$
$$\det(T^{-1}(\Lambda - \lambda I) T)$$
$$= \det T^{-1} \det(\Lambda - \lambda I) \det T = \det(\Lambda - \lambda I)\Big]$$

and hence same eigenvalues, the λ_i's are the eigenvalues of A and the α_i^*'s were chosen to be eigenvectors.

ORTHOGONALITY

• **PROBLEM 2-28**

Show geometrically that for each angle θ, the transformation

$$T_\theta : R^2 \to R^2,$$

defined by $T_\theta(x,y) = (x \cos \theta - y \sin \theta, x \sin \theta + y \cos \theta)$, is an orthogonal transformation.

<u>Solution</u>: Let V be an inner product space. A linear transformation $T: V \to V$ is said to be orthogonal if

$$\|Tv\| = \|v\| \quad ,$$

i.e., if it preserves the norms or magnitudes for each vector v in V. Thus, orthogonal transformations are those linear transformations which preserve distances.

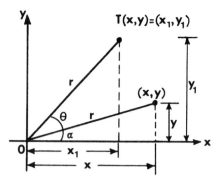

The matrix of the transformation with respect to the standard basis is

$$A_\theta = \begin{bmatrix} \cos\theta & -\sin\theta \\ \sin\theta & \cos\theta \end{bmatrix}.$$

From the figure, it is seen that $x = r\cos\alpha$, $y = r\sin\alpha$. Then,

$$(x_1, y_1) = T(x,y)$$

or

$$x_1 = r\cos(\alpha + \theta) = r\cos\alpha\cos\theta - r\sin\alpha\sin\theta$$

$$= x\cos\theta - y\sin\theta$$

$$y_1 = r\sin(\alpha + \theta) = r\cos\alpha\sin\theta + r\sin\alpha\cos\theta$$

$$= x\sin\theta + y\cos\theta.$$

Thus,

$$\begin{bmatrix} x_1 \\ y_1 \end{bmatrix} = \begin{bmatrix} \cos\theta & -\sin\theta \\ \sin\theta & \cos\theta \end{bmatrix} \begin{bmatrix} x \\ y \end{bmatrix} = T_\theta(x,y).$$

The figure describes a rotation through an angle θ, and from the figure we derive the given transformation. Hence, the given transformation is a rotation. A rotation preserves magnitudes; however, we can check explicitly that T_θ is orthogonal. Let

$$v = (x,y), \quad \|v\| = \sqrt{x^2 + y^2}.$$

Now,

$$T_\theta v = (x\cos\theta - y\sin\theta,\; x\sin\theta + y\cos\theta),$$

so

$$\|T_\theta v\| = \sqrt{(x\cos\theta - y\sin\theta)^2 + (x\sin\theta + y\cos\theta)^2}$$

$$= \sqrt{x^2 \cos^2 \theta + y^2 \sin^2 \theta - 2xy \cos \theta \sin \theta + x^2 \sin^2 \theta +}$$

$$\overline{+ y^2 \cos^2 \theta + 2xy \cos \theta \sin \theta}$$

$$= \sqrt{x^2 (\cos^2 \theta + \sin^2 \theta) + y^2 (\sin^2 \theta + \cos^2 \theta)}$$

$$= \sqrt{x^2 + y^2}$$

(since $\sin^2 \theta + \cos^2 \theta = 1$)

$$= \|v\| .$$

• **PROBLEM 2-29**

Let W be a subspace of the finite-dimensional inner product space V. Show that

$$V = W \oplus W^\perp \quad \text{and} \quad (W^\perp)^\perp = W.$$

(Projection Theorem).

Solution: $W^\perp$ denotes the set of all vectors in V which are orthogonal to every vector in W. That is, if $\langle \,,\, \rangle$ denotes the inner product, then

$$\langle w, v \rangle = 0$$

for all $w \in W$ and $v \in W^\perp$. [Note that $W^\perp$ is sometimes called "W perp"]. Also, if U and W are vector spaces (over the same field) then their direct sum $U \oplus W$ is the vector space whose underlying set is

$$U \times W = \{(x,y) \mid x \in U, y \in W\} \text{ with}$$

$$a(x_1, y_1) + b(x_2, y_2) = (ax_1 + bx_2, ay_1 + by_2).$$

For example, the vector space $R \oplus R$ is the vector space whose underlying set is

$$R \times R = R^2$$

and is therefore usually denoted by R^2. Note also that this is equivalent to saying that every vector z in $U \oplus W$ is determined uniquely by $z = x + y$ where $x \in U$ and $y \in W$. In order to prove the desired results, the following will be needed: If

$$S = \{x_1, \ldots, x_m\}$$

is any finite orthonormal set in an inner product space V, x is any vector in V, and

$$a_i = \langle x, x_i \rangle,$$

then

$$\sum_{i=1}^{m} |a_i|^2 \leq \|x\|^2$$

(Bessel's inequality).

Also, the vector

$$y = x - \sum_{i=1}^{m} a_i x_i$$

is orthogonal to each x_i and consequently to the span of S. To prove the first part of this statement:

$$0 \leq \|y\|^2 = \langle y, y \rangle = \langle x - \sum_{i=1}^{m} a_i x_i , x - \sum_{i=1}^{m} a_i x_i \rangle$$

$$= \langle x, x \rangle - \sum_{i=1}^{m} a_i \langle x_i, x \rangle - \sum_{i=1}^{m} \overline{a_i} \langle x, x_i \rangle + \sum_{i,j=1}^{m} a_i \overline{a_j} \langle x_i, x_j \rangle$$

$$= \|x\|^2 - \sum_{i=1}^{m} |a_i|^2 - \sum_{i=1}^{m} |a_i|^2 + \sum_{i=1}^{m} |a_i|^2$$

$$= \|x\|^2 - \sum_{i=1}^{m} |a_i|^2 .$$

Hence, Bessel's inequality is proven. To prove the second part we have:

$$\langle y, x_j \rangle = \langle x - \sum_{i=1}^{m} a_i x_i , x_j \rangle$$

$$= \langle x, x_j \rangle - \sum_{i=1}^{m} a_i \langle x_i, x_j \rangle$$

$$= a_j - a_j$$

$$= 0 .$$

Now, to prove the Projection Theorem, let

$S = \{x_1, \ldots, x_m\}$ be an orthonormal set that is complete in W (i.e., spans W) and let z be any vector in V. Set

$$x = \sum_{i=1}^{m} a_i x_i$$

where

$$a_i = \langle z, x_i \rangle.$$

The above result says that $y = z - x$ is orthogonal to W and hence in $W^\perp$. W and $W^\perp$ are disjoint (except 0), for if $x \in W$ and $x \in W^\perp$ then

$$\|x\|^2 = \langle x, x \rangle = 0.$$

Hence, $z = x + y$ where $x \in W$ and $y \in W^\perp$ and thus

$$V = W \oplus W^\perp.$$

Observe that

$$\langle z, x \rangle = \langle x+y, x \rangle = \|x\|^2 + \langle y, x \rangle = \|x\|^2$$

and

$$\langle z, y \rangle = \langle x+y, y \rangle = \|y\|^2.$$

If $z \in (W^\perp)^\perp$ so that $\langle z, y \rangle = 0$, then $\|y\|^2 = 0$ and $z = x \in W$. Thus $(W^\perp)^\perp \subseteq W$. But clearly $W \subseteq (W^\perp)^\perp$, so

$$W = (W^\perp)^\perp.$$

● **PROBLEM 2-30**

Let $S = \mathrm{Sp}\{(1,0,1),(0,2,1)\}$. Then S is a subspace of R^3. Find the orthogonal complement of S.

Solution: Let V be a vector space and W a subspace of V. A vector $v \in V$ is orthogonal to W if v is orthogonal to each vector in W. Then, the orthogonal complement of W in V is the set of all vectors of V that are orthogonal to W.

To find the orthogonal complement, first find a set that spans the subspace W. Then, a vector v is orthogonal to the subspace W if v is orthogonal to each vector in the spanning set.

Two vectors that span S are the given vectors, namely, $w_1 = (1,0,1)$ and $w_2 = (0,2,1)$. Let

$$v = (x_1, x_2, x_3) \in R^3.$$

Then, the condition that v be orthogonal to w_1 and w_2 is given by the solution set to the two homogeneous equations:

$$v \cdot w_1 = (x_1, x_2, x_3) \cdot (1,0,1) = 0,$$
$$v \cdot w_2 = (x_1, x_2, x_3) \cdot (0,2,1) = 0, \tag{1}$$

or

$$x_1 + x_3 = 0,$$
$$2x_2 + x_3 = 0,$$

or

$$x_1 = -x_3,$$
$$2x_2 = -x_3.$$

Let $x_3 = -2$. Then, $x_2 = 1$ and $x_1 = 2$. Thus, the orthogonal complement of S in R^3 is the set

$$S^\perp = \text{Sp}\{(2,-1,-2)\}$$ since the system (1) has only one linearly independent solution. Note: dim $S^\perp = 1$, and, since w_1 and w_2 are independent, dim S = 2. Consequently,

$$\dim(S^\perp \oplus S) = 3$$

and the set

$$\{(1,0,1), (0,2,1), (2,-1,-2)\}$$

is a basis for R^3.

● **PROBLEM 2-31**

Use the Gram-Schmidt process to transform

$$[(1,0,1), (1,2,-2), (2,-1,1)]$$

into an orthogonal basis for R^3. Assume the standard inner product.

Solution: The standard inner product for R^n is the dot product defined as follows:

Let

$$u = (u_1, u_2, \ldots, u_n), \quad v = (v_1, v_2, \ldots, v_n)$$

be two vectors in R^n. Then, the inner or dot product of u and v, denoted $<u,v>$, is given by

$$<u,v> = <(u_1, u_2, \ldots, u_n), (v_1, v_2, \ldots, v_n)>$$
$$= u_1v_1 + u_2v_2 + \ldots + u_nv_n$$
$$= \sum_{i=1}^{n} u_i v_i.$$

The dot product is useful for defining both orthogonalization and the length of a vector. The norm of u, denoted $\|u\|$, is given by

$$\|u\| = \langle u,u \rangle^{\frac{1}{2}}.$$

Two vectors u,v are orthogonal if $\langle u,v \rangle = 0$.

The Gram-Schmidt process constructs a set of orthogonal vectors from a given set of linearly independent vectors. Since a basis for R^3 must contain three linearly independent vectors, we check that the given vectors are linearly independent. Set

$$c_1(1,0,1) + c_2(1,2,-2) + c_3(2,-1,1) = 0.$$

Thus,

$$c_1 + c_2 + 2c_3 = 0$$

$$2c_2 - c_3 = 0$$

$$c_1 - 2c_2 + c_3 = 0.$$

The only solution to this system is

$$c_1 = c_2 = c_3 = 0,$$

i.e., the three vectors are linearly independent and form a basis for R^3.

Label the three vectors u_1, u_2, u_3 respectively. Choose u_1 and relabel it v_1. Thus,

$$v_1 = (1,0,1).$$

The set now becomes $\{v_1, u_2, u_3\}$. We know u_1 and u_2 are independent, and the vector space they span is also spanned by $\{v_1, u_2\}$. We wish to construct a vector v_2 in this two-dimensional subspace of R^3 which is perpendicular to v_1. Choose v_2 as the candidate for this transfiguration.

$$v_2 \in Sp\{u_1, u_2\}$$

implies $v_2 \in Sp\{v_1, u_2\}$ which, in turn, implies $v_2 = c_1 v_1 + c_2 u_2$ for some constants c_1 and c_2. The required orthogonality of v_1 and v_2 implies

$$\langle v_1, v_2 \rangle = 0.$$

Hence,

$$\langle v_1, c_1 v_1 + c_2 u_2 \rangle = \langle v_1, c_1 v_1 \rangle + \langle v_1, c_2 u_2 \rangle$$

$$= c_1 \langle v_1, v_1 \rangle + c_2 \langle v_1, u_2 \rangle = 0.$$

Assume that $c_2 = 1$. Then

$$c_1 = \frac{-\langle v_1, u_2 \rangle}{\langle v_1, v_1 \rangle} = \frac{-\langle v_1, u_2 \rangle}{\|v_1\|^2}$$

and hence,

$$v_2 = c_2 u_2 + c_1 u_1 = u_2 - \frac{\langle v_1, u_2 \rangle}{\|v_1\|^2} v_1.$$

Thus,

$$v_2 = (1,2,-2) - \frac{\langle (1,0,1), (1,2,-2) \rangle}{\|(1,0,1)\|^2} (1,0,1) = (1,2,-2) -$$

$$- (-\tfrac{1}{2})(1,0,1) = (3/2, 2, -3/2).$$

The set

$$\{v_1, v_2\}$$

is orthogonal and forms a basis for the subspace of R^3.

Now seek a third vector v_3 that is orthogonal to both v_1 and v_2. Again,

$$Sp\{u_1, u_2, u_3\} = Sp\{v_1, v_2, v_3\},$$

so

$$v_3 \in Sp\{u_1, u_2, u_3\}$$

implies v_3 can be written as

$$v_3 = b_1 v_1 + b_2 v_2 + b_3 u_3,$$

subject to

$$\langle v_3, v_1 \rangle = \langle v_3, v_2 \rangle = 0.$$

That is,

$$\langle b_1 v_1 + b_2 v_2 + b_3 u_3, v_1 \rangle = 0$$

$$\langle b_1 v_1 + b_2 v_2 + b_3 u_3, v_2 \rangle = 0.$$

By the rules of the dot product,

$$b_1 \langle v_1, v_1 \rangle + b_2 \langle v_2, v_1 \rangle + b_3 \langle u_3, u_1 \rangle = 0$$

$$b_1 \langle v_1, v_2 \rangle + b_2 \langle v_2, v_2 \rangle + b_3 \langle u_3, v_2 \rangle = 0.$$

Since
$$\langle v_2, v_1 \rangle = \langle v_1, v_2 \rangle = 0,$$
the above two equations reduce to
$$b_1 \langle v_1, v_1 \rangle + b_3 \langle u_3, v_1 \rangle = 0$$
$$b_2 \langle v_2, v_2 \rangle + b_3 \langle u_3, v_2 \rangle = 0.$$

Assume $b_3 = 1$:
$$b_1 = \frac{-\langle u_3, v_1 \rangle}{\|v_1\|^2} \quad \text{and} \quad b_2 = \frac{-\langle u_3, u_2 \rangle}{\|v_2\|^2}.$$

Thus,
$$v_3 = b_3 u_3 + b_2 v_2 + b_1 v_1 = u_3 - \frac{\langle u_3, v_2 \rangle v_2}{\|v_2\|^2} - \frac{\langle u_3, v_1 \rangle v_1}{\|v_1\|^2}$$

$$= (2,-1,1) - \frac{\langle (2,-1,1), (3/2,2,-3/2) \rangle}{34/4} (3/2,2,-3/2)$$

$$- \frac{\langle (2,-1,1), (1,0,1) \rangle}{2} (1,0,1)$$

$$= (2,-1,1) - \frac{(-1/2)}{34/4}(3/2,2,-3/2) - 3/2(1,0,1)$$

$$= (2,-1,1) + (3/34, 2/17, -3/34) + (-3/2, 0, -3/2)$$

$$= (10/17, -15/17, -10/17).$$

The required orthogonal basis is
$$\{(1,0,1), (3/2,2,-3/2), (10/17, -15/17, -10/17)\}.$$

Any scalar multiple of these basis vectors will have the same properties of orthogonality, and so, if we wish, we may take $\{(1,0,1), (3,4,-3), (2,-3,-2)\}$ as an orthogonal basis of R^3.

● **PROBLEM 2-32**

Orthonormalize the first four functions of the set
$$\{1, x, x^2, x^3, \ldots\}$$
over $[0,1]$. (Use the inner product $\langle f, g \rangle = \int_0^1 fg\,dx$)

Solution: This is an application of the Gram-Schmidt process. Let
$$a_1 = 1, \; a_2 = x, \; \ldots, \; a_n = x^{n-1}, \; \ldots$$

Let $\{b_1, b_2, \ldots\}$ be the orthogonal functions constructed from $\{a_1, a_2, \ldots\}$. Start by letting $b_1 = 1$.

Using the formula $b_n = a_n - \sum_{i=1}^{n-1} \frac{\langle a_n, b_i \rangle}{\langle b_i, b_i \rangle} b_i$, $n \geq 2$, we obtain

$$b_2 = a_2 - \frac{\langle a_2, b_1 \rangle}{\langle b_1, b_1 \rangle} b_1 = x - \frac{\langle x, 1 \rangle}{\langle 1, 1 \rangle} 1 = x - \left(\int_0^1 x\,dx\right) \bigg/ \left(\int_0^1 dx\right) = x - \tfrac{1}{2}$$

$$b_3 = a_3 - \left[\frac{\langle a_3, b_1 \rangle}{\langle b_1, b_1 \rangle} b_1 + \frac{\langle a_3, b_2 \rangle}{\langle b_2, b_2 \rangle} b_2\right] = x^2 - \left[\frac{\langle x^2, 1 \rangle}{\langle 1, 1 \rangle} 1 + \frac{\langle x^2, x - \tfrac{1}{2}\rangle}{\langle x - \tfrac{1}{2}, x - \tfrac{1}{2}\rangle}(x - \tfrac{1}{2})\right]$$

$$= x^2 - \left[\frac{\int_0^1 x^2 dx}{\int_0^1 dx} + \frac{\int_0^1 x^2(x-\tfrac{1}{2})dx}{\int_0^1 (x-\tfrac{1}{2})^2 dx}(x-\tfrac{1}{2})\right] = x^2 - \left[\tfrac{1}{3} + (x-\tfrac{1}{2})\right] = x^2 - x + \tfrac{1}{6}$$

For b_4, we have

$$b_4 = a_4 - \left[\frac{\langle a_4, b_1 \rangle}{\langle b_1, b_1 \rangle} b_1 + \frac{\langle a_4, b_2 \rangle}{\langle b_2, b_2 \rangle} b_2 + \frac{\langle a_4, b_3 \rangle}{\langle b_3, b_3 \rangle} b_3\right]$$

$$= x^3 - \left[\frac{\langle x^3, 1\rangle}{\langle 1, 1\rangle} 1 + \frac{\langle x^3, x-\tfrac{1}{2}\rangle}{\langle x-\tfrac{1}{2}, x-\tfrac{1}{2}\rangle}(x-\tfrac{1}{2}) + \frac{\langle x^3, x^2-x+\tfrac{1}{6}\rangle}{\langle x^2-x+\tfrac{1}{6}, x^2-x+\tfrac{1}{6}\rangle}(x^2-x+\tfrac{1}{6})\right]$$

$$= x^3 - \left[\frac{\int_0^1 x^3 dx}{\int_0^1 dx} + \frac{\int_0^1 x^3(x-\tfrac{1}{2})dx}{\int_0^1 (x-\tfrac{1}{2})^2 dx}(x-\tfrac{1}{2}) + \frac{\int_0^1 x^3(x^2-x+\tfrac{1}{6})dx}{\int_0^1 (x^2-x+\tfrac{1}{6})^2 dx}(x^2-x+\tfrac{1}{6})\right]$$

$$= x^3 - \left[\tfrac{1}{4} + \tfrac{\tfrac{1}{40}}{\tfrac{1}{12}}(x-\tfrac{1}{2}) + \tfrac{\tfrac{1}{120}}{(\tfrac{1}{5}-\tfrac{7}{36})}(x^2-x+\tfrac{1}{6})\right] = x^3 - \left[\tfrac{3}{2}x + (\tfrac{9}{10} - \tfrac{3}{2})x + (\tfrac{1}{2} - \tfrac{9}{20})\right]$$

$$\underset{\text{×}}{=} x^3 - \tfrac{3}{2}x^2 - \tfrac{3}{10}x + \tfrac{1}{20} \qquad -\tfrac{3}{5} \qquad +\tfrac{1}{20}$$
$$+ \tfrac{3}{6}$$

$$x^3 - \tfrac{3}{2}x^2 + \tfrac{3}{5}X - \tfrac{1}{20}$$

CHAPTER 3

CONTINUITY

The study of continuity is an integral part of mathematical analysis. Historically it arose from attempts to explain Zeno's paradoxes (Achilles and the tortoise, how does an arrow ever leave the bow, etc.). In calculus, continuity is defined in terms of limits; in advanced calculus alternative definitions are phrased using sequences and neighborhoods. The mose rarified approach is that of topology where a function is said to be continuous if the inverse image of every open set in the range is an open set in the domain. The underlying idea for all these definitions of continuity is that points which are near each other in a set are mapped by a continuous function to points that are near one another in some other set.

The problem of determining whether a function $f: R^n \to R^m$ is continuous at a point $x_0 \in R^n$ is solved by applying the following definition: Let $\varepsilon > 0$ be given. Then f is continuous at x_0 if there exists a $\delta > 0$ such that

$$||f(x) - f(x_0)|| < \varepsilon \quad \text{whenever} \quad ||x - x_0|| < \delta$$

Here $||\cdot||$ denotes the standard Euclidean norm on R^n or R^m, i.e.,

$$||x|| \equiv \sqrt{\langle x, x \rangle} \equiv (x_1^2 + \ldots + x_n^2)^{\frac{1}{2}}$$

if $x \in R^n$.

Remark   It must be observed that for a given $\varepsilon > 0$ and $x_0 \in R^n$ the δ will in general depend on both x_0 and ε. In order to stress this dependence it is better to write $\delta \equiv \delta(\varepsilon, x_0)$.

Functions may be discontinuous at isolated points in their domain or they may be discontinuous at more points than they are continuous.

The Dirichlet function

$$f(x) = \begin{cases} 0 \text{ if } x \text{ is rational} \\ 1 \text{ if } x \text{ is irrational} \end{cases}$$

is an example of the latter kind of function.

The notion of continuity, like that of convergence can be expanded and refined. Uniform continuity of a function occurs when the number δ depends only on ε and not on the point x_0; that is to say whenever

$$\| x_1 - x_2 \| < \delta \quad \text{implies} \quad \| f(x_1) - f(x_2) \| < \varepsilon .$$

Absolute continuity and semicontinuity are further variations on the theme. Finally, the concept of equicontinuity is used to prove the Arzéla-Ascoli theorem, an analogue of the Heine-Borel theorem, for spaces of continuous functions.

The 19th century pioneers of mathematical analysis uncovered some paradoxes of their own. The most famous is that of Weierstrass who discovered a function that, although continuous everywhere, fails to be differentiable anywhere. Peano's continuous space-filling curve which blurs the distinction between length and area is another study in mathematical pathology.

SHOWING THAT A FUNCTION IS CONTINUOUS

• **PROBLEM 3-1**

Show that

$$\lim_{n \to \infty} \left(1 + \frac{C}{n^2} \right)^n = 1 .$$

Solution: The idea of the limit of a sequence of numbers is useful in defining continuity. Let

$$\{a_n\}_{n=1}^{\infty} = (a_1, a_2, a_3, \ldots, a_n, \ldots)$$

be a sequence of real numbers. The sequence is said to have the limit L if, given $\varepsilon > 0$, there exists a positive integer n_0 such all the terms of the subsequence

$$\{a_m\}_{m=n_0+1}^{\infty}$$

remain at a distance less than ε from L, i.e.,

$$L - \varepsilon < a_m < L + \varepsilon$$

for all $m > n_0$.

To find

$$\lim_{n \to \infty} (1 + C/n^2)^n ,$$

use Newton's binomial theorem to find the expansion of $(1 + c/n^2)^n$ for finite n.

Now
$$\left(1 + \frac{c}{n^2}\right)^n = 1 + \sum_{r=1}^{n} \binom{n}{r}\left(\frac{c}{n^2}\right)^r \quad (1)$$

where
$$\binom{n}{r} = \frac{n!}{r!(n-r)!} .$$

But
$$\binom{n}{r} \leq \frac{n^r}{r!} .$$

To see this, observe that
$$\binom{n}{r} = \frac{n(n-1)(n-2) \ldots (n-(r-1))}{r!}$$

while
$$\frac{n^r}{r!} = \frac{\overbrace{n \cdot n \cdot n \ldots n}^{r\text{-times}}}{r!}$$

Hence
$$1 + \sum_{r=1}^{n} \binom{n}{r}\left(\frac{|c|}{n^2}\right)^r \leq 1 + \sum_{r=1}^{n} \frac{n^r}{r!}\left(\frac{|c|}{n^2}\right)^r$$

But
$$\frac{n^r}{n^{2r}} = \frac{1}{n^r}$$

which implies
$$1 + \sum \frac{n^r}{r!}\left(\frac{c}{n^2}\right)^r \leq 1 + \sum_{r=1}^{n} \left(\frac{|c|}{n}\right)^r$$

(because $r! \geq 1$)

Now compare this sum with the potential limit (i.e., 1). Thus
$$\left| 1 + \sum_{r=1}^{n} \left(\frac{c}{n}\right)^r - 1 \right| \leq \sum_{r=1}^{n} \left(\frac{|c|}{n}\right)^r .$$

Since $n > |C|$ as $n \to \infty$

$$\sum_{r=1}^{n} \left(\frac{|c|}{n}\right)^r \leq \frac{|c|/n}{1 - (|c|/n)}$$

(The sum of a geometric series is

$$\sum_{r=1}^{n} a^r = \frac{a - a^{n+1}}{1 - a} \leq \frac{a}{1 - a}$$

if $0 < a < 1$) .

Let $\varepsilon > 0$ be given. Chose n_0 so large that

$$|c|/n_0 < \min\left(\frac{1}{2}, \frac{\varepsilon}{2}\right) .$$

Then if

$$n > n_0 \qquad \frac{|c|}{n} < \frac{|c|}{n_0} < \min\left(\frac{1}{2}, \frac{\varepsilon}{2}\right)$$

hence

$$1 - \frac{|c|}{n} > 1 - \frac{1}{2} = \frac{1}{2}$$

and

$$\frac{|c|/n}{1 - |c|/n} < \varepsilon$$

Thus

$$\left| \left(1 + \frac{c}{n^2}\right)^n - 1 \right| \leq \frac{|c|/n}{1 - |c|n} < \varepsilon$$

if $n > n_0$ showing that

$$\lim_{n \to \infty} \left(1 + \frac{c}{n^2}\right)^n = 1 .$$

• **PROBLEM 3-2**

Find

a) $\lim_{x \to c} \dfrac{x^n - c^n}{x - c}$

b) $\lim_{x \to 0} \dfrac{x^2}{(x^2 + 1)^{\frac{1}{2}} - 1}$.

Solution: The function appearing in a) is of the form

$$\frac{P(z)}{Q(z)}$$

where P and Q are polynomials, i.e.,

$$P(z) = a_0 + a_1 z + a_2 z^2 + \ldots + a_n z^n,$$

$$Q(z) = b_0 + b_1 z + \ldots + b_k z^k.$$

The value of

$$\lim_{z \to a} \frac{P(z)}{Q(z)},$$

where a is a constant is different according to the values of P(a) and Q(a).

i) If

$$Q(a) \neq 0, \quad \lim_{z \to a} \frac{P(z)}{Q(z)} = \frac{P(a)}{Q(a)}$$

ii) If

$$Q(a) = 0, \, P(a) \neq 0, \quad \lim_{z \to a} \frac{P(z)}{Q(z)}$$

fails to exist.

iii) If

$$P(a) = Q(a) = 0,$$

$$\lim_{z \to a} \frac{P(z)}{Q(z)} = \lim_{z \to a} \frac{P(z)(z-a)^n}{Q(z)(z-a)^m}$$

where m,n are the first powers of $(z - a)$ such that

$$\frac{P(z)}{Q(z)}$$

is similar to cases i) or ii).

a)
$$\lim_{x \to C} \frac{x^n - C^n}{x - C}.$$

Here, at C, $P(C) = Q(C) = 0$ so case iii) applies.

$$P(x) = x^n - C^n = (x - C)(x^{n-1} + Cx^{n-2} + \ldots + C^{n-2}x + C^{n-1})$$

Then

$$\lim_{x \to C} \frac{(x-C)(x^{n-1} + Cx^{n-2} + \ldots + C^{n-2}x + C^{n-1})}{(x - C)}$$

$$= C^{n-1} + C^{n-1} + \ldots C^{n-1} = nC^{n-1}.$$

b) Limits of this type can be evaluated by rationalizing the denominator. Multiply numerator and denominator by

$$(x^2 + 1)^{1/2} + 1$$

to obtain:

$$\lim_{x \to 0} \frac{x^2}{(x+1)^{1/2} - 1} = \lim_{x \to 0} \frac{x^2 \{(x^2+1)^{1/2} + 1\}}{[(x^2+1)^{1/2} - 1][(x^2+1)^{1/2} + 1]}$$

$$= \lim_{x \to 0} \frac{x^2 \{(x^2+1)^{1/2} + 1\}}{(x^2+1) - 1} = \lim_{x \to 0} (x^2 + 1)^{1/2} + 1 = 2.$$

● **PROBLEM 3-3**

Let $f: R^2 \to R$ be given by

$$f(x,y) = x^2 + y^2.$$

Show that f is continuous at $(0,0)$.

Solution: A real-valued function

$$f: R^2 \to R$$

is continuous at a point (a_1, a_2) in its domain if

1)
$$\lim_{(x,y) \to (a_1, a_2)} f(x,y) \quad \text{exists}$$

2)
$$\lim_{(x,y) \to (a_1, a_2)} f(x,y) = f(a_1, a_2).$$

The definition of the limit of a function of two variables is as follows: A function $f(x,y)$ approaches a limit A as x approaches a and y approaches b,

$$\lim_{\substack{x \to a \\ y \to b}} f(x,y) = A$$

if, and only if, for each $\varepsilon > 0$, there is another $\delta > 0$ such that whenever $|x-a| < \delta$, $|y-b| < \delta$ and $0 < (x-a)^2 + (y-b)^2$, then $|f(x,y) - A| < \varepsilon$.

This means that when (x,y) is at any point inside a certain square with center at (a,b) and width 2δ (except at the center), $f(x,y)$ differs from A by less than ε.

The function whose limit is to be evaluated as (x,y)

approaches the origin is $f(x,y) = x^2 + y^2$. Let $\varepsilon > 0$ be given and choose $\delta = \sqrt{\varepsilon/2}$. Then the inequalities $|x| < \sqrt{\varepsilon/2}$, $|y| < \sqrt{\varepsilon/2}$ imply $(x^2 + y^2) < \varepsilon$. Hence,

$$\lim_{\substack{x \to 0 \\ y \to 0}} (x^2 + y^2) = 0$$

and condition 1) is satisfied.

For condition 2) note that

$$f(0,0) = 0^2 + 0^2 = 0$$

and thus f is continuous at $(0,0)$.

● **PROBLEM 3-4**

Show that

$$\lim_{(x,y) \to (0,0)} \frac{2x^3 - y^3}{x^2 + y^2} = 0$$

Solution: Let $f: R^2 \to R$ be a real-valued function from the plane to the real line. To show that f is continuous at a given point (a_1, a_2) in the plane it suffices to show:

1)
$$\lim_{(x,y) \to (a_1,a_2)} f(x,y) \text{ exists}$$

2)
$$\lim_{(x,y) \to (a_1,a_2)} f(x,y) = f(a_1,a_2)$$

If only the first condition is satisfied, f is said to have a removable discontinuity or a discontinuity of the first kind. If neither condition is satisfied f has an essential discontinuity or discontinuity of the second kind.

In the given problem it is required to show that f approaches 0 as $(x,y) \to (0,0)$. This means that if $\varepsilon > 0$ is given, there exists a $\delta > 0$ (depending on ε) such that

$$\left| \frac{2x^3 - y^3}{x^2 + y^2} \right| < \varepsilon \text{ if } 0 < (x^2 + y^2)^{\frac{1}{2}} < \delta \tag{1}$$

(Thus x,y are in an open disk of radius δ, with center at the origin; this point is deleted since what happens at $(0,0)$ is not of concern here.)

Now

$$|2x^3 - y^3| \le 2|x|^3 + |y|^3, \quad (|a-b| \le |a| + |b|)$$

$$2|x|^3 + |y|^3 = 2|x|x^2 + |y|y^2.$$

Now

$$|x| \le (x^2 + y^2)^{\frac{1}{2}}, \quad |y| \le (x^2 + y^2)^{\frac{1}{2}}$$

and thus

$$|2x^3 - y^3| \le 2(x^2 + y^2)^{\frac{1}{2}} x^2 + (x^2 + y^2)^{\frac{1}{2}} y^2$$

or,

$$|2x^3 - y^3| \le (x^2 + y^2)^{\frac{1}{2}} [2x^2 + y^2] \le 2(x^2 + y^2)^{3/2}$$

But

$$|2x^3 - y^3| \le 2(x^2 + y^2)^{3/2}$$

implies

$$\left| \frac{2x^3 - y^3}{x^2 + y^2} \right| \le 2(x^2 + y^2)^{\frac{1}{2}} \quad \text{if } 0 < x^2 + y^2. \tag{2}$$

Let $\varepsilon > 0$ given Choose $\delta = \varepsilon/2$. Then, comparing (1) and (2), observe that (1) will hold for all positive ε. Thus

$$\lim_{(x,y) \to (0,0)} \frac{2x^3 - y^3}{x^2 + y^2} = 0.$$

● **PROBLEM 3-5**

> Let $f(x,y) = x^2 + 3y$ and let $\varepsilon > 0$ be given. Prove that F is continuous in the whole plane by finding $\delta > 0$ such that for $|(x,y) - (x_0, y_0)| < \delta$
>
> $|F(x,y) - F(x_0,y_0)| < \varepsilon$ where x_0, y_0 is an arbitrary point in the plane.

Solution: The difference between $F(x,y)$ and $F(x_0, y_0)$ is:

$$F(x,y) - F(x_0, y_0) = (x^2 + 3y) - (x_0^2 + 3y_0)$$

$$= (x^2 - x_0^2) + (3y - 3y_0)$$

$$= (x + x_0)(x - x_0) + 3(y - y_0).$$

Given $\varepsilon > 0$, the problem is to find a neighborhood S of p_0 (x_0, y_0) such that

$$|F(p) - F(p_0)| < \varepsilon$$

(where $p = (x,y)$), for all points $p \in S$. The neighborhood can be in any shape but for convenience let S be a square of side 2δ centered at (x_0, y_0). Then

$$|x - x_0| < \delta \quad \text{and} \quad |y - y_0| < \delta$$

for any $p \in S$, where without any loss of generality it can be assumed that $0 < \delta < 1$ (δ is usually thought of as a small number). For $p \in U$ one has

$$|x + x_0| = |x - x_0 + 2x_0| \leq |x - x_0| + 2|x_0|$$
$$\leq 1 + 2|x_0|$$

and

$$|F(p) - F(p_0)| \leq |x - x_0||x + x_0| + 3|y - y_0|$$
$$\leq \delta(1 + 2|x_0|) + 3\delta$$
$$\leq (4 + 2|x_0|)\delta$$

The condition

$$|F(p) - F(p_0)| < \varepsilon \quad \text{holds when} \quad (4 + 2|x_0|)\delta < \varepsilon.$$

i.e.,

$$0 < \delta < \frac{\varepsilon}{(4 + 2|x_0|)}. \tag{1}$$

Thus given $\varepsilon > 0$, δ may be chosen according to (1) so that for any (x_0, y_0) in the plane

$$|(x,y) - (x_0, y_0)| < \delta$$

implies

$$|F(x,y) - F(x_0, y_0)| < \varepsilon.$$

This shows that F is continuous everywhere in the plane.

● PROBLEM 3-6

Let $F(x,y) = x^2 + 3y$ be defined on the unit square

$$S = \{(x,y) : 0 \leq x \leq 1, 0 \leq y \leq 1\}$$

Show that $F(x,y)$ is continuous on S.

Solution: Here $F:R^2 \to R$ is a real-valued function and the domain of definition is the unit square with the lower left vertex at the origin.

Since the problem is to show that F is everywhere continuous on S, let

$$p_0 = (x_0, y_0)$$

be an arbitrary point in S (p_0 may even be on the boundary of S). Let $F(x,y) = F(p)$. Then

$$F(p) - F(p_0) = F(x,y) - F(x_0, y_0)$$

$$= x^2 + 3y - (x_0^2 + 3y_0)$$

$$= (x^2 - x_0^2) + (3y - 3y_0)$$

$$= (x - x_0)(x + x_0) + 3(y - y_0) \quad (1)$$

Now (1) may be used to obtain an estimation of the variation of F around p_0. Since

$$0 \leq x \leq 1$$

and

$$0 \leq x_0 \leq 1, \quad 0 \leq x + x_0 \leq 2$$

and (1) may be rewritten as

$$|F(p) - F(p_0)| \leq 2|(x - x_0)| + 3|(y - y_0)| \quad (2)$$

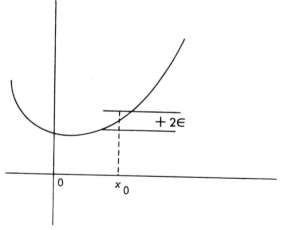

Fig. 1

Let $\varepsilon > 0$ be given. The inequality (2) may be used to determine a $\delta > 0$ such that

$$|F(p) - F(p_0)| < \varepsilon$$

whenever

$$|p - p_0| < \delta$$

Thus showing that $F(x,y)$ is continuous on S.

For example, suppose $\varepsilon = 0.03$. Then a neighborhood U about p_0 in which the variation of F is less than ε must be found. Since the problem is in rectangular coordinates the neighborhood can be chosen as a square of side η. Since the center of the square is (x_0, y_0),

$$|x - x_0| < \eta/2, \quad |y - y_0| < \eta/2.$$

Since

$$|F(p) - F(p_0)| \leq 2|x - x_0| + 3|y - y_0|$$

and

$$|F(p) - F(p_0)| < \varepsilon, \quad (= .03)$$

set

$$2|x - x_0| + 3|y - y_0| < .03$$

or

$$2(\eta/2) + 3(\eta/2) < .03$$

or

$$\eta < .0012.$$

In general, if the neighborhood U is a square of side $(.4)\varepsilon$, then

$$|x - x_0| < (.2)\varepsilon$$

and

$$|y - y_0| < (.2)\varepsilon$$

so that

$$|F(p) - F(p_0)| < 2(.2)\varepsilon + 3(.2)\varepsilon = \varepsilon.$$

Thus, given any $\varepsilon > 0$, $\delta > 0$ can be found so that

$$|(x,y) - (x_0,y_0)| < \delta$$

implies

$$|F(x,y) - F(x_0,y_0)| < \varepsilon$$

where $(x_0,y_0) \in S$.

This shows that $F(x,y)$ is continuous.

• **PROBLEM 3-7**

Let $f: R^2 \to R$ be a real valued function with domain R^2 which is of the form

$$f(x_1, x_2) = \frac{5x_1}{1 + x_2^2}.$$

Show that f is continuous at $(0,0)$.

Solution: Let R^n denote the set of all n-tuples,

$$R^n = \{(x_1, \ldots, x_n) : x_i \in R \; i=1, \ldots, n\}$$

Let $f : R^n \to R$ be a function from R^n to the set of real numbers. Then f is said to be continuous at $(x_1^{(0)}, \ldots, x_n^{(0)}) \in R^n$ if, given any $\varepsilon > 0$, there exists a $\delta > 0$ such that

$$|f(x_1, \ldots, x_n) - f(x_1^{(0)}, x_2^{(0)}, \ldots, x_n^{(0)})| < \varepsilon$$

for all $(x_1, \ldots, x_n) \in R^n$ for which

$$||(x_1, \ldots, x_n) - (x_1^{(0)}, x_2^{(0)}, \ldots, x_n^{(0)})|| < \delta$$

Here $||(x_1, \ldots, x_n)||$ denotes the Euclidean norm of $(x_1, \ldots, x_n)$ given by

$$||(x_1, \ldots, x_n)|| = \left[\sum_{i=1}^{n} x_i^2\right]^{\frac{1}{2}}.$$

The norm is used to measure the distance of a point from the origin or the distance between points. Since f is real-valued, $||f||$ can be written as $|f|$.

Applying this definition to the given function, note that $n = 2$ and $(x_1^{(0)}, \ldots, x_n^{(0)}) = (0,0)$. Thus $f(x_1, x_2)$ is continuous at $(0,0) \in R^2$ if, given any $\varepsilon > 0$, there exists a $\delta > 0$ such that

$$|f(x_1, x_2) - f(0,0)| < \varepsilon$$

whenever

$$||(x_1, x_2) - (0,0)|| < \delta .$$

Since

$$f(x_1, x_2) = \frac{5x_1}{1 + x_2^2}, \quad f(0,0) = \frac{5(0)}{1 + (0)^2} = 0.$$

$$|f(x_1, x_2) - f(0,0)| = |f(x_1, x_2)|.$$

Now

$$|f(x_1, x_2)| = \left|\frac{5x_1}{1 + x_2^2}\right| \leq 5|x_1|,$$

since $1 + x_2^2 \geq 1$ for all real x_2.

But

$$|x_1| \leq ||(x_1,x_2)|| \quad \text{for all } x_1, x_2 \quad \text{since}$$

$$||(x_1,x_2)|| = \sqrt{(x_1^2 + x_2^2)} \geq \sqrt{x_1^2} = |x|.$$

Thus,

$$\left|\frac{5x_1}{1 + x_2^2}\right| \leq 5||(x_1,x_2)||$$

Let $\varepsilon > 0$ be given and choose $\delta = \varepsilon/5$.

Then

$$||(x_1,x_2)|| < \delta$$

implies

$$\left|\frac{5x_1}{1 + x_2^2}\right| < 5\delta = \varepsilon$$

and $f(x_1,x_2)$ is indeed continuous at $(0,0) \in R^2$.

Note that δ is chosen according to how ε, the challenge number is chosen. In fact δ is a function of ε and the point

$$(x_1^0, x_2^0)$$

at which continuity is tested.

● **PROBLEM 3-8**

Let $f: R^3 \to R$ be given by $f(x,y,z) = y^3 z/(1 + x^2 + z^3)$. Show that f is continuous at $(0,0,0)$.

Solution: f is a real valued function from R^3 to R. If it is to be continuous at the origin then it must satisfy the definition of continuity at the origin.

$f: R^3 \to R$ is continuous at $(x_0, y_0, z_0) \in R^3$ if, given any $\varepsilon > 0$, there exists a $\delta > 0$ such that $||(x,y,z) - (x_0,y_0,z_0)|| < \delta$ implies $|f(x,y,z) - f(x_0,y_0,z_0)| < \varepsilon$, $(x,y,z) \in R^3$. In this problem, $(x_0, y_0, z_0) = (0,0,0)$ and $f(x,y,z) = y^3 z/1 + x^2 + z^2$. Suppose $\varepsilon > 0$ is given (say $\varepsilon = 1/10$). Then the problem is to find a $\delta > 0$ such that $||(x,y,z) - (0,0,0)|| = ||(x,y,z)|| < \delta$ implies $|f(x,y,z) - f(0,0,0)|$

$$= \left|\frac{y^3 z}{1 + x^2 + z^2} - \frac{0^3(0)}{1 + 0^2 + 0^2}\right| = \left|\frac{y^3 z}{1 + x^2 + z^2}\right|$$

$$< \varepsilon \ (=1/10).$$

109

The method used here to find δ is to show that $\left|\dfrac{y^3 z}{1 + x^2 + z^2}\right|$ is less than some function of the norm of (x,y,z). Once this is done, choosing (x,y,z) such that $||(x,y,z)|| < \delta$ will guarantee that $\left|\dfrac{y^3 z}{1 + x^2 + z^2}\right| < \varepsilon$.

Now

$$\left|\dfrac{y^3 z}{1 + x^2 + z^2}\right| \leq |y^3 z|$$

since

$$\dfrac{1}{1 + x^2 + z^2} \leq 1 \text{ for all } x, y.$$

But

$$|y^3 z| \leq ||(x,y,z)||^4$$

for all $(x,y,z) \in R^3$ since

$$||(x,y,z)|| = (x^2 + y^2 + z^2)^{1/2} \geq |z| \text{ and}$$

$$||(x,y,z)||^3 \geq |y|^3.$$

Thus, for each $\varepsilon > 0$, choose $\delta = \varepsilon^{1/4} > 0$. Then

$$||(x,y,z)|| < \delta \text{ implies } |f(x,y,z)| < \delta^4 = \varepsilon.$$

If

$$\varepsilon = 1/10, \quad \delta = \dfrac{1}{1.178} = .5623.$$

This means that if the coordinates (x,y,z) are chosen such that the distance from (x,y,z) to $(0,0,0)$ is less than $.5623$, $|f(x,y,z) - f(0,0,0)| < 1/10$.

Thus f is continuous at $(0,0,0)$.

• **PROBLEM 3-9**

Show that the function $f: R^n \to R$ given by $f(x_1, \ldots, x_n) = \max\{x_1, \ldots, x_n\}$ is continuous everywhere.

Solution: The given function f is a real-valued function on R^n. It is a special version of the projection function which sends an n-tuple to its ith coordinate. Since the projection function is continuous, the function sending an n-tuple to the coordinate with largest positive value (or least negative value) must also be continuous.

The coninuity of f may also be shown as follows: Let $\varepsilon > 0$ be given. Now,

$$|(x_1, x_2, \ldots, x_n) - (a_1, a_2, \ldots a_n)| < \delta$$

implies

$$|(x_1 - a_1, x_2 - a_2, \ldots, x_n - a_n)| < \delta .$$

In particular this means

$$|x_1 - a_1| < \delta, |x_2 - a_2| < \delta, \ldots,$$
$$|x_n - a_n| < \delta$$

and hence

$$\max(|x_1 - a_1|, \ldots, |x_n - a_n|) < \delta$$

Let $x \in R^n$ be arbitrary and let $x_k = \max(x_1, \ldots, x_n)$ and let $a_j = \max(a_1, \ldots, a_n)$

Then

$$\max(x_1, \ldots, x_n) - \max(a_1, \ldots, a_n) \leq x_k - a_k \leq$$
$$|x_k - a_k|$$
$$\max(a_1, \ldots, a_n) - \max(x_1, \ldots, x_n) \leq a_j - x_j \leq$$
$$|x_j - a_j|$$

so that

$$|\max(x_1, \ldots, x_n) - \max(a_1, \ldots, a_n)| \leq$$
$$\max(|x_1 - a_1|, \ldots |x_n - a_n|)$$

The above string of inequalities now imply that

$$\max(|x_1 - a_1|, \ldots, |x_n - a_n|) < \delta$$

if

$$||(x_1, \ldots, x_n) - (a_1, \ldots, a_n)|| < \delta$$

Letting $\delta = \varepsilon$ yields the continuity of the function f.

• **PROBLEM 3-10**

Let $f: R \to R$ be given by $f(x) = x^n$ where $n \in N$, the set of natural numbers. Prove that f is everywhere continuous.

Solution: A function $f: R \to R$ is continuous at a point $a \in R$ if i) $\lim_{x \to a} f(x)$ exists ii) $\lim_{x \to a} f(x) = f(a)$. An alternate definition would be that f is continuous at a if, given $\varepsilon > 0$, there exists a $\delta > 0$ such that if $|x - a| < \delta$ then $|f(x) - f(a)| < \varepsilon$. Both of the above definitions are local characterizations of a continuous function. However, if a is chosen arbitrarily in the domain then they become global definitions also.

For the given function, let x be close to a, say $|x - a| < 1$. Then

$$|f(x) - f(a)| = |x^n - a^n| = |(x - a)(x^{n-1} + ax^{n-2} + \ldots + a^{n-2}x + a^{n-1})|.$$

Now

$$|x - a| < 1 \text{ implies } |x| < 1 + |a|.$$

Substitution of

$$1 + |a| \text{ for } |x|$$

yields

$$|x^n - a^n| \leq |(x - a)((1 + |a|)^{n-1} + a(1 + |a|)^{n-2} + \ldots + a^{n-2}(1 + |a|) + a^{n-1})|.$$

But

$$|a|^r \leq (1 + |a|)^r$$

so

$$|a^r(1 + |a|)^{n-r-1}| \leq (1 + |a|)^{n-1}$$

Thus

$$|x^n - a^n| \leq |x - a| n (1 + |a|)^{n-1}$$

Set

$$\delta = \min\left(1, \frac{\varepsilon}{n(1 + |a|)^{n-1}}\right)$$

Then $|x - a| < \delta$ implies

$$|x^n - a^n| < \varepsilon$$

and f is continuous at a.

● **PROBLEM 3-11**

Show that the exponential function defined by

$$\exp x = \lim_{n \to \infty} \left(1 + \frac{x}{n}\right)^n$$

is everywhere continuous.

Solution: The problem falls naturally into three pieces:

(1) Show that exp x exists for all $x \in R$

(2) Prove the functional equation

$$\exp(x + y) = (\exp x)(\exp y)$$

(3) Use the functional equation in (2) to show that exp x is continuous everywhere.

(1) Exp $x = \lim_{n \to \infty} \left(1 + \frac{x}{n}\right)^n$.

$$\left(1 + \frac{x}{n}\right)^n = 1 + n\left(\frac{x}{n}\right) + \frac{n(n-1)}{2!}\left(\frac{x}{n}\right)^2 + \cdots$$

$$+ \frac{n(n-1)\cdots(n-(r-1))}{r!}\left(\frac{x}{n}\right)^r + \cdots$$

$$+ \left(\frac{x}{n}\right)^n \text{ (by Newton's binomial theorem)},$$

$$= 1 + \sum_{r=1}^{n} \binom{n}{r}\left(\frac{x}{n}\right)^r.$$

The general term $\dfrac{n(n-1)(n-2)\cdots(n-(r-1))}{r!}\left(\dfrac{x}{n}\right)^r$ may be rewritten as

$$\frac{n}{n}\left(\frac{n-1}{n}\right)\left(\frac{n-2}{n}\right)\cdots\frac{n-(r-1)}{n}\frac{x^r}{r!}$$

$$= \frac{1(1 - 1/n)(1 - 2/n) \cdots (1 - (r-1)/n)\, x^r}{r!}$$

Thus

$$\sum_{r=1}^{n} \binom{n}{r}\left(\frac{x}{n}\right)^r =$$

113

$$1 + \sum_{r=1}^{n} \frac{1(1-1/n)(1-2/n)(1-3/n)\ldots(1-(r-1)/n)}{r!} x^r \quad (2)$$

$$\leq 1 + \sum_{r=1}^{n} \frac{x^r}{r!}$$

since

$$1(1 - 1/n)(1 - 2/n) \ldots (1 - (r - 1)/n) \leq 1 \quad (3)$$

for

$$r \geq 1.$$

As n increases (2) increases monotonically in value for each positive x. For $0 \leq x < 1$, $\sum_{r=1}^{n} \frac{x^r}{r!} < \frac{1}{1-x}$ in (3). For $x \geq 1$, replace x by $\frac{x}{k}$ where $k > x$, k an integer. Then

$$\left(1 + \frac{x}{kn}\right)^n < \frac{1}{1 - (x/k)} \quad \text{(from (3))}$$

or,

$$\left(1 + \frac{x}{kn}\right)^{kn} < \left(1 - \frac{x}{k}\right)^{-k} \quad \text{(for all } n \in N\text{)}$$

or,

$$\left(1 + \frac{x}{n}\right)^n < \left(1 - \frac{x}{k}\right)^{-k} \quad \text{(since } \left(1 + \frac{x}{n}\right)^n \leq \left(1 + \frac{x}{kn}\right)^{kn}\text{).}$$

Thus, for $x \geq 0$, $\left(1 + \frac{x}{n}\right)^n$ is bounded and monotonically increasing. This shows that

$$\text{Exp}(x) = \sup_{n \in N} \left(1 + \frac{x}{n}\right)^n$$

exists for nonegative x. For $x < 0$, note that $-x > 0$ and

$$\lim_{n \to \infty} \left(1 + \frac{x}{n}\right)^n = \lim_{n \to \infty} \left(1 + \frac{x}{n}\right)^n \frac{\left(1 - \frac{x}{n}\right)^n}{\left(1 - \frac{x}{n}\right)^n}$$

$$\lim_{n \to \infty} \frac{\left(1 - \left(\frac{x}{n}\right)^2\right)^n}{\left(1 - \frac{x}{n}\right)^n} = \lim_{n \to \infty} \frac{1}{\left(1 - \frac{x}{n}\right)^n}$$

since

114

$$\lim_{n\to\infty} \left(1 - \frac{x^2}{n^2}\right)^n = 1.$$

Thus, for $x < 0$,

$$\exp(x) = \frac{1}{\exp(-x)}.$$

Exp x is therefore defined and exists for all $x \in R$.

(2)
$$(\text{Exp } x)(\text{Exp } y) = \lim_{n\to\infty}\left(1 + \frac{x}{n}\right)^n \lim_{n\to\infty}\left(1 + \frac{y}{n}\right)^n.$$

For convergent sequences the limit of the product is equal to the product of the limits. Thus

$$\lim_{n\to\infty}\left(1 + \frac{x}{n}\right)^n \lim_{n\to\infty}\left(1 + \frac{y}{n}\right)^n = \lim_{n\to\infty}\left(1 + \frac{x+y}{n} + \frac{xy}{n^2}\right)^n$$

$$= \lim_{n\to\infty}\left(1 + \frac{x+y}{n}\right)^n \left[1 + \frac{xy/n^2}{1 + \frac{(x+y)}{n}}\right]^n$$

But as $n\to\infty$ and for fixed x and y

$$\left(\frac{1 + xy/n^2}{1 + \frac{x+y}{n}}\right)^n \to 1.$$

Thus,

$$\lim_{n\to\infty}\left(1 + \frac{x}{n}\right)^n \lim_{n\to\infty}\left(1 + \frac{y}{n}\right)^n = \lim_{n\to\infty}\left(1 + \frac{x+y}{n}\right)^n$$

$$= \exp(x + y)$$

(3) To show that exp x is continuous everywhere, suppose $0 < x < 1$. Then

$$1 < \left(1 + \frac{x}{n}\right)^n < \frac{1}{1-x}$$

or $\quad 1 < \exp x \leq \frac{1}{1-x}$.

As $\quad x \to 0 \to \frac{1}{1-x} \to 1$ and thus

$$\lim_{\substack{x\to 0 \\ x>0}} \exp x = 1 = \exp 0$$

Also

$$\lim_{\substack{x \to 0 \\ x < 0}} \exp(x) = \lim_{\substack{x \to 0 \\ x > 0}} \exp(-x) = \lim_{\substack{x \to 0 \\ x > 0}} \frac{1}{\exp(x)} = 1$$

Therefore exp x is continuous at 0.

For any $x \in R$ and $h \in R$,

$$\lim_{h \to 0} \{\exp(x+h) - \exp(x)\} = \lim_{h \to 0} (\exp x)(\exp h - 1)$$

$$= \lim_{h \to 0} (\exp x)(1-1) = 0$$

and exp x is indeed continuous.

● **PROBLEM 3-12**

Let $f: R^n \to R^m$ be a linear transformation, i.e.,

$$f(ax + bz) = af(x) + bf(z),$$

where

$x = (x_1, \ldots, x_n)$, $z = (z_1, \ldots, z_n) \in R^n$,

$a, b \in R$

and

$f(x), f(z) \in R^m$.

Show that f is continuous at every point in its domain.

Solution: Every linear transformation has associated with it an m × n matrix with respect to the usual bases

$$\{e_1, \ldots, e_n\} \quad \{e_1, \ldots e_m\}$$

for R^n and R^m respectively.

Let $f(e_j) = z_j \qquad j = 1, 2, \ldots n$

Since $z_j \in R^m$ it can be expressed as

$$z_j = C_{1j} e_1 + C_{2j} e_2 + \ldots C_{mj} e_m = \sum_{\ell=1}^{m} C_{\ell j} e_\ell$$

If $x \in R^n$ and $x = x_1 e_1 + \ldots + x_n e_n$, it

116

follows by linearity of f that

$$f(x) = x_1 f(e_1) + \ldots + x_n f(e_n) = \sum_{k=1}^{n} x_k f(e_k)$$

$$= \sum_{k=1}^{n} x_k \sum_{\ell=1}^{m} c_{\ell k} e_\ell$$

$$= \sum_{\ell=1}^{m} \left(\sum_{k=1}^{n} c_{\ell k} x_k \right) e_\ell = \sum_{\ell=1}^{m} y_\ell e_\ell$$

where

$$y_\ell = \sum_{k=1}^{n} c_{\ell k} x_k \qquad (\ell = 1, 2, \ldots m) \qquad (1)$$

This is the general linear transformation from $R^n \to R^m$.

The matrix of the transformation is

$$\begin{bmatrix} c_{11} & c_{12} & \cdots & c_{1n} \\ c_{21} & c_{22} & \cdots & c_{2n} \\ \cdot & & & \\ \cdot & & & \\ \cdot & & & \\ c_{m1} & c_{m2} & & c_{mn} \end{bmatrix}$$

The Cauchy inequality may be used to prove that every linear transformation is everywhere continuous.

The Cauchy inequality

$$\sum |a_i b_i| \leq \left(\sum a_i^2 \right)^{\frac{1}{2}} \left(\sum b_i^2 \right)^{\frac{1}{2}}$$

may be restated as

$$\left| \sum a_i b_i \right|^2 \leq \sum a_i^2 \sum b_i^2$$

or,

$$|a_1 b_1 + a_2 b_2 + \ldots + a_n b_n|^2 \leq (a_1 + a_2 + \ldots + a_n)^2$$

$$(b_1 + b_2 + \ldots + b_n)^2$$

Applying this inequality to each equation in system (1) yields, for $1 \leq j \leq m$

$$|y_j|^2 \leq (|c_{j1}|^2 + |c_{j2}|^2 + \ldots + |c_{jn}|^2) \|x\|^2 = \sum_{i=1}^{n} |c_{ji}|^2 \|x\|^2$$

(since

$$|y_j|^2 = |c_{j1} x_1 + c_{j2} x_2 + \ldots + c_{jn} x_n|^2$$

$$\leq (|c_{j1}|^2 + |c_{j2}|^2 + \ldots + |c_{jn}|^2)(x_1^2 + x_1^2 + \ldots + x_n^2))$$

Summing the j equations $1 \leq j \leq m$

$$\|y\|^2 \leq \left\{ \sum_{j=1}^{m} \sum_{i=1}^{n} |c_{ji}|^2 \right\} \|x\|^2 \tag{2}$$

(where $\|x\|$ is the norm of x, i.e.,

$$\|x\| = \|(x_1, \ldots, x_n)\| = (x_1^2 + x_2^2 + \ldots + x_n^2)^{\frac{1}{2}}).$$

From (2) it follows that

$$\|y\| = \|f(x)\| \leq \left\{ \sum_{j=1}^{m} \sum_{i=1}^{n} |c_{ji}|^2 \right\}^{\frac{1}{2}} \|x\|$$

or, setting

$$\left\{ \sum_{j=1}^{m} \sum_{i=1}^{n} |c_{ji}|^2 \right\}^{\frac{1}{2}} = A,$$

$$\|f(x)\| \leq A \|x\| \tag{3}$$

Inequality (3) may be used to prove that every linear transformation is continuous, i.e., the following theorem:

Let $F: R^n \to R^m$ be a linear transformation. Then there exists a positive constant A such that for $u, v \in R^n$,

$$\|f(u) - f(v)\| \leq A \|u-v\|. \tag{4}$$

As a corollary note that if $\varepsilon > 0$ is given,

$$\|f(u) - f(v)\| < \varepsilon$$

whenever

$||u-v|| < \varepsilon/A$, i.e.,

f is continuous at any $V \in R^n$.

To prove the theorem, let $x = u-v$ in (3). Then

$$||f(x)|| = ||f(u-v)|| \le A||u-v||$$

Since f is a linear transformation,

$$||f(u-v)|| = ||f(u) - f(v)|| \le A||u-v||$$

which is (4). Thus the theorem is proved and according to the corollary every linear transformation from R^n to R^m is everywhere continuous.

● **PROBLEM 3-13**

Show that a continuous function of a continuous function is continuous by means of the following examples (i.e., show that g∘f is continuous):

(a) $f(x,y) = (1 + xy)^2$ $g(z) = \sin z$.

(b) $f: R^2 \to R^3$ is defined by

$$f(x,y) = (x, y, x(1 + x^2 + y^2)^{-3/2})$$

$g: R^3 \to R$ defined by

$$g(x,y,z) = x^2 + y^2 + z^2.$$

Solution: The theorem applied here is:

Let $D \subset R^n$ and let $f: D \to R^m$ be continuous. If the image of D under f is $S \subseteq R^m$, let $g: S \to R^p$ be continuous. Then the function $h: D \to R^p$ defined by

$$h(x) = (g \circ f)(x) = g(f(x)),$$

where $x = (x_1, \ldots, x_n) \in R^n$, is continuous in D.

a) The composition g∘f is

$$g(f(x,y)) = g([1 + xy]^2) = \sin(1 + xy)^2.$$

Now $f(x,y) = (1 + xy)^2$ is continuous everywhere in R, and $g(z) = \sin z$ is continuous everywhere in R. Thus by the theorem $(g \circ f)(x)$ is everywhere continuous in R.

b) Clearly f is continous everywhere. Similarly

$g(x,y,z)$

is everywhere continuous. Then
$$(g \circ f)(x,y) = g(f(x,y)) = g(x,y, x(1 + x^2 + y^2)^{-3/2})$$

$$= x^2 + y^2 + \frac{x^2}{(1 + x^2 + y^2)^3}$$

is an everywhere continuous function of $(x,y) \in R^2$.

● **PROBLEM 3-14**

Let $f: R \to R$ be a real-valued function. Then f is said to be continuous at a point $x_0 \in R$ (the domain) if, given $\varepsilon > 0$, there exists a $\delta > 0$ such that

$$|x - x_0| < \delta \quad \text{implies that} \quad |f(x) - f(x_0)| < \varepsilon.$$

This is a local definition of continuity. Suppose

$$f: A \to R, \quad A = [0,4] \text{ is given by } f(x) = 2x.$$

Show how a global definition of continuity can be applied to this function, and generalize to an n-dimensional domain.

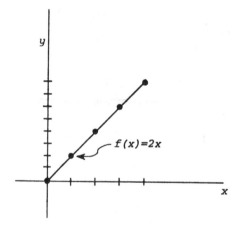

Fig. 1

Solution:
The domain of f is $[0,4]$. In order to apply the global definition, let G be any open set in R. It is to be shown that $f^{-1}(G)$ is an open set in D, i.e., $f^{-1}(G) = H \cap D$ where H is open in R. If $f^{-1}(G) = \phi$ (for example, if $G = (5,10)$ $f^{-1}(G) = \phi$) there is nothing to prove. Let $c \in f^{-1}(G)$ so that $f(c) = 2c \in G$.

There are two cases now

case (i) $0 < c < 4$.

In this case $f(c) = 2c$ so $0 < f(c) < 8$

Chose $\varepsilon > 0$ so that

$$0 < 2c - \varepsilon < 2c < 2c + \varepsilon < 8 \quad \text{(see fig.)}$$

```
|―――――――――+―――――+――+―――+――――|      Fig. 2
0        2c-ε   2c  2c+ε   8
```

since G is open and $f(c) = 2c \in G$, the ε may be chosen so that $(2c - \varepsilon, 2c + \varepsilon) \subset G$. With such a choice it follows that

$$\left(c - \frac{\varepsilon}{2}, c + \frac{\varepsilon}{2}\right) \subset f^{-1}(G)$$

and

$$\left(c - \frac{\varepsilon}{2}, c + \frac{\varepsilon}{2}\right) \subset [0,4].$$

In particular

$$\left(c - \frac{\varepsilon}{2}, c + \frac{\varepsilon}{2}\right) = \left(c - \frac{\varepsilon}{2}, c + \frac{\varepsilon}{2}\right) \cap [0,4]$$

is an open set in $[0,4]$, it contains C and is contained in $f^{-1}(G)$

case (ii) $c = 0$ or $c = 4$.

Suppose for the sake of argument $c = 4$. Then for some ε
$(8-\varepsilon, 8+\varepsilon) \subset G$

and

$$f^{-1}(8-\varepsilon, 8+\varepsilon) = \left(4 - \frac{\varepsilon}{2}, 4 + \frac{\varepsilon}{2}\right) \cap [0,4]$$

The set

$$\left(4 - \frac{\varepsilon}{2}, 4 + \frac{\varepsilon}{2}\right) \cap [0,4]$$

is open in $[0,4]$, contains 4 and is contained in $f^{-1}(G)$. The argument for $c = 0$ is similar showing that $f^{-1}(G)$ is open in $[0,4]$ whenever G is open in R.

The obvious generalization to n-dimensions is:

Let $f: D \to R$ be a function defined on a set $D \subset R^n$. Then f is said to be continuous if $f^{-1}(G)$ is open in D whenever G is open in R, i.e., $f^{-1}(G) = H \cap D$ where H is open in R^n.

● **PROBLEM 3-15**

Show the equivalence of the following three definitions of continuity.

Let (X,d_1) (Y,d_2) be two metric spaces. A function f from X into Y is said to be continuous at $a \in X$ if:

(1) For each $\varepsilon > 0$, there exists a $\delta > 0$ such that

$$f(\beta(a;\delta)) \subset \beta(f(a);\varepsilon)$$

(2) For each neighborhood N of f(a) in Y there exists a neighborhood M of a in X such that

$$f(M) \subset N$$

(3) For each neighborhood N of f(a) in Y, $f^{-1}(N)$ is a neighborhood of a in X.

Solution: A metric space is a set of points with a distance function defined on it. Thus, given any two points in a metric space it is possible to say how far apart they are. For example, R^n the set of all n-tuples of real numbers is a metric space if the distance between x and y is defined as

$$d(x,y) = \left[\sum_{i=1}^{n} (x_i - y_i)^2 \right]^{\frac{1}{2}}$$

The equivalence of the three given definitions of a continuous function from one metric space into another can be shown by proving the implications

(1) ⇒ (2) ⇒ (3) ⇒ (1).

(1) ⇒ (2):

The first definition states that if f is continuous then there exists a ball of radius δ around $a \in X$ which is mapped by f into a ball of radius ε around $f(a) \in Y$. Pictorially:

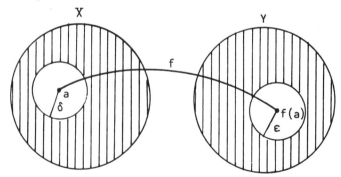

Fig. 1

In metric spaces, open balls of radius δ or ε are nothing but neighborhoods of the points which are their centers. Thus (1) states that for each neighborhood of radius ε around $f(a) \in Y$ there exists a neighborhood of radius δ around $a \in X$ such that the neighborhood of a is mapped by f into the neighborhood of f(a). But this is (2). Thus (1) ⟹ (2).

(2) ⟹ (3):

The second definition states that every neighborhood N of f(a) contains the image of some neighborhood M of a ∈ X. By the rules of compositions of functions and their inverses, since f(M) ⊂ N,

$$f^{-1}(f(M)) \subset f^{-1}(N)$$

or, since $f^{-1}(f(M)) \supset M$,

$$M \subset f^{-1}(N).$$

But M is a neighborhood of a in X. Thus $f^{-1}(N)$, since it contains a neighborhood of a is itself a neighborhood of a in X which proves (3).

(3) ⟹ (1):

Again, since $f(f^{-1}(N)) \subset N$, it follows that every neighborhood of f(a) in Y contains the image of a neighborhood of a in X. That is, every neighborhood of radius ε of f(a) contains the image of a neighborhood of radius δ of a in X. This is the first definition.

Thus the three definitions are equivalent.

DISCONTINUOUS FUNCTIONS

• PROBLEM 3-16

Show that

$$f(x,y) = \frac{x-y}{x+y}, \; x \neq -y, \; f(x,y) = 1 \; x = -y$$

is discontinuous at the origin by showing that

$$\lim_{(x,y) \to (0,0)} f(x,y)$$

does not exist.

Solution: Here f is a real-valued function from R^2. A real-valued function $f: R^n \to R$ is continuous at a point a in its domain if

i) $\lim_{x \to a} f(x)$ exists and ii) $\lim_{x \to a} f(x) = f(a)$.

In the given problem it is required to prove that i) does not hold. But there are two ways in which the limit of a function as it approaches a point may not exist:

a) The function may grow without bound as x approaches the point of interest.

b) A basic theorem regarding limits is that the limit of a function is unique. Thus if it can be shown that the value of f does not approach one and only one value as $x \to a$, the limit of f at a does not exist. This method is particularly useful when evaluating functions of two or more variables.

Let $f: R^2 \to R$ be given by $f(x,y) = 0$ and suppose it is of interest to find whether f is continuous at some $(a_1, a_2) \in R^2$, i.e., whether (firstly)

$$\lim_{(x,y) \to (a_1,a_2)} f(x,y)$$

exists.

A quick way of verifying that the limit does not exist is to let f approach (a_1, a_2) along two different paths (since in the plane (a_1, a_2) can be approached from an infinite number of directions). If the limits along the two routes are different, then f has no limit at (a_1, a_2).

Applying this technique to the given function, let y approach zero independently of x. Thus

$$\lim_{\substack{x \to 0 \\ y \to 0}} f(x,y) = \lim_{x \to 0}\left[\lim_{y \to 0} f(x,y)\right] = \lim_{x \to 0}\left[\lim_{y \to 0} \frac{x-y}{x+y}\right]$$

$$= \lim_{x \to 0} \frac{x}{x} = \lim_{x \to 0} 1 = 1 \qquad (1)$$

On the other hand, let x approach zero first independently of y. Then

$$\lim_{\substack{x \to 0 \\ y \to 0}} f(x,y) = \lim_{y \to 0}\left[\lim_{x \to 0} f(x,y)\right] = \lim_{y \to 0}\left[\lim_{x \to 0} \frac{x-y}{x+y}\right]$$

$$= \lim_{y \to 0} -\frac{y}{y} = -1$$

Thus

$$\lim_{\substack{x \to 0 \\ y \to 0}} f(x,y)$$

is not independent of the method of approach, i.e., the limit is not unique implying that the limit does not exist. Thus $f(x,y)$ is discontinuous at the origin.

● **PROBLEM 3-17**

Let $f(x,y)$ be given by

$$f(x,y) = \begin{cases} xy^2/(x^2 + y^4) & (x,y) \neq (0,0) \\ 0 & x = y = 0 \end{cases}$$

Is f continuous at the origin?

Solution: The problem is to determine whether

$$\lim_{\substack{x \to 0 \\ y \to 0}} f(x,y) = \lim_{\substack{x \to 0 \\ y \to 0}} \frac{xy^2}{x^2 + y^4} = 0 \quad ?$$

Continuity of a function may be defined in terms of neighborhoods or, equivalently, in terms of sequences. The sequential definition of continuity is as follows: Let $f: D \to R^m$ and let $\{p_n\}$ be any sequence of points in the domain D of f. Let p be the limit of the sequence, i.e., $\{p_n\}$ converges to $p \in D$. Then f is said to be continuous at p if and only if

$$\lim_{n \to \infty} f(p_n) = f(p) ,$$

for every sequence (p_n) which converges to p, i.e., the sequence $\{f(p_n)\}$ converges to $f(p)$.

It is interesting to see how this definition can actually be deduced as a theorem from the more elementary definition of continuity (f is continuous at x_0 in the domain if, given $\varepsilon > 0$, there exists a $\delta > 0$ such that $|x - x_0| < \delta$ implies $|f(x) - f(x_0)| < \varepsilon$ for x in the domain).

The theorem to be proved is: Let $f: D \to R^m$ where $D \subseteq R^n$. Let $\{p_n\}$ be a sequence of points in D converging to p. Then f is continuous at p if and only if the sequence $\{f(p_n)\}$ converges to $f(p)$.

To prove the necessity, assume f is continuous at p, i.e.,

$$|f(p_n) - f(p)| < \varepsilon$$

whenever $|p_n - p| < \delta$. Since

$$\lim_{n \to \infty} p_n = p ,$$

there exists an integer N such that $|p_n - p| < \delta$ for all $n \geq N$. Thus there exists an N such that $|f(p_n) - f(p)| < \varepsilon$ for $n \geq N$, i.e., $\{f(p_n)\}$ converges to $f(p)$.

For the reverse implication suppose that f is not continuous at p. Then there exists an $\varepsilon > 0$ and a sequence (p_n) in D such that $|p_n - p| < \frac{1}{n}$ but

$$|f(p_n) - f(p)| \geq \varepsilon .$$

The sequence p_n converges to p but $f(p_n)$ is always a distance at least ε from $f(p)$ and so cannot converge to $f(p)$. This is contrary to assumption and hence f must be continuous at p.

To check the given function for continuity at $(0,0)$, let $y = cx$ and consider all sequences of the form

$$p_n = \left(\frac{1}{n}, \frac{c}{n}\right).$$

As c takes on different values, the sequence p_n will approach $(0,0)$ along different lines. Since

$$f(p_n) = \left(\frac{1}{n}\right)\left(\frac{c^2}{n^2}\right) \bigg/ \left(\frac{1}{n^2} + \frac{c^4}{n^4}\right) = \frac{c^2}{n^3} \bigg/ \left(\frac{1}{n^2} + \frac{c^4}{n^4}\right)$$

$$= \frac{c^2}{n^3} \bigg/ \left[\frac{1}{n^2}\left(1 + \frac{c^4}{n^2}\right)\right] = \frac{c^2 n^2}{n^3\left(1 + \frac{c^4}{n^2}\right)} = \frac{c^2}{n\left(\frac{n^2 + c^4}{n^2}\right)}$$

$$= \frac{nc^2}{n^2 + c^4} \quad \text{and} \quad \lim_{n \to \infty} \frac{c^2 n}{n^2 + c^4} = 0 .$$

This shows that $f(x,y)$ is continuous when y is a linear function of x. However, if

$$x = y^2,$$

a new sequence

$$q_n = \left(\frac{1}{n^2}, \frac{1}{n}\right)$$

is formed and

$$f(q_n) = \left(\frac{1}{n^2}\right)\left(\frac{1}{n^2}\right)\bigg/\left(\frac{1}{n^4} + \frac{1}{n^4}\right) = \frac{1/n^4}{2/n^4} = 1/2 \neq 0 .$$

Note that

$$\lim_{n\to\infty} q_n = \lim_{n\to\infty} \left(\frac{1}{n^2}, \frac{1}{n}\right) = (0,0)$$

but

$$\lim_{n\to\infty} f(q_n) = 1/2 \neq f(0,0) .$$

Thus the sequences $\{q_n\}$ and $\{f(q_n)\}$ converge to values such that

$$\lim_{n\to\infty} q_n = q$$

but

$$\lim_{n\to\infty} f(q_n) \neq f(q) .$$

Thus f is not continuous at the origin.

● **PROBLEM 3-18**

Distinguish between removable and essential discontinuities. Decide, for the following functions whether the points of discontinuity are removable or discontinuous.

a) $f(x,y) = xy/x^2 + y^2$ $(x,y) \in R^2 - \{(0,0)\}$

b) $f(x,y) = x^2y^2/x^2 + y$ $(x,y) \in R^2 - \{(0,0)\}$

c) $f(x) = x^x$ $x \in R, x > 0$

d) $g(x) = \begin{cases} x & x \in R, x > 0 \\ 2 & x \in R, x = 0 \end{cases}$

e) $h(x) = \sin(1/x)$ $x \in R, x > 0$

<u>Solution</u>: Let $D \subseteq R^n$ and let $f: D \to R^m$. Let $p_0 \in D$. If $f(p_0)$ is defined and

$$L = \lim_{p \to p_0} f(p)$$

exists but $L \neq f(p_0)$, then f is said to possess a removable discontinuity at p_0. The discontinuity may be removed by redefining f at p_0 as L so that

$$\lim_{p \to p_0} f(p) = L$$

and f is continuous at p_0. A point of discontinuity is also removable if f is not defined at p_0 but

$$L = \lim_{p \to p_0} f(p)$$

exists. Here the domain of f can be extended so as to include p_0.

Now, suppose

$$\lim_{p \to p_0} f(p)$$

does not exist. Then p_0 is an essential discontinuity for f since no value can be assigned to $f(p_0)$ so that

$$\lim_{p \to p_0} f(p) = f(p_0) .$$

a)
$$f(x,y) = \frac{xy}{x^2 + y^2}$$

is discontinuous at the origin. To check the type of discontinuity let $y = mx$. Then

$$\lim_{(x,y) \to (0,0)} \frac{xy}{x^2 + y^2} = \frac{mx^2}{x^2(m^2+1)} = \frac{m}{m^2+1} .$$

Now let $x = y^2$ and consider the sequences

$$(1/n^2, 1/n) = (x,y)$$

for $n \in N$ the natural numbers. Then

$$\lim_{(x,y) \to (0,0)} \frac{xy}{x^2+y^2} = \lim_{n \to \infty} \frac{1/n^3}{\frac{1}{n^2}\left[\frac{1}{n^2}+1\right]}$$

$$= \lim_{n \to \infty} \frac{1}{n\left[\frac{1}{n^2}+1\right]} = \lim_{n \to \infty} \frac{1}{n + \frac{1}{n}} = 0$$

Thus

128

$$\lim_{(x,y) \to (0,0)} f(x,y)$$

is not unique and therefore does not exist. The origin is a point of essential discontinuity.

b)
The function

$$f(x,y) = \frac{x^2 y^2}{x^2 + y^2}$$

is undefined at the origin. To check whether

$$\lim_{p \to 0} f(p)$$

(where $p = (x,y)$) exists, note that

$$|x^2 y^2| \leq ||p||^2 \, ||p||^2 = ||p||^4$$

and

$$|x^2 + y^2| = ||p||^2$$

so that

$$|f(p)| \leq \frac{||p||^4}{||p||^2} = ||p||^2 \ .$$

But

$$\lim_{p \to 0} ||p||^2 = 0$$

since the norm function is continuous. Thus

$$\lim_{(x,y) \to (0,0)} f(x,y) = 0 \ .$$

Since the limit of f as (x,y) approaches (0,0) exists, f has a removable discontinuity at the origin. If f is

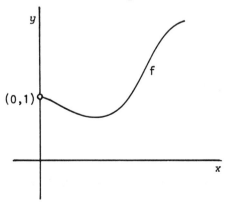

Fig. 1

defined as zero at the origin the domain f becomes all of R^2.

c)
The graph of x^x is

As x approaches zero

$$\lim_{x \to 0} x^x .$$

To show this, rewrite x^x as

$$e^{\log x^x} .$$

Then

$$\lim_{x \to 0} x^x = e^{\lim_{x \to 0} x \log x} .$$

But

$$\lim_{x \to 0} x \log x = 0$$

since x approaches zero faster than the decrease in log x, and

$$\lim_{x \to 0} x^x = e^0 = 1 .$$

The point $x = 0$ is therefore a removable discontinuity.

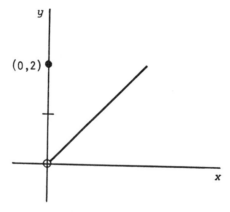

Fig. 2

d)
The graph of g is

The limit of g as x approaches 0 is

$$\lim_{x \to 0} g(x) = \lim_{x \to 0} x = 0 .$$

But $g(0) = 2 \neq 0$. Since the limit exists $x = 0$ is a

removable discontinuity. If g is redefined at 0 so that $g(0) = 0$ g is continuous for nonnegative reals.

e) At $x = 0$ the function

$$h(x) = \sin\left(\frac{1}{x}\right)$$

is not defined. Since $1/x$ gets larger and larger, $\sin 1/x$ oscillates more and more rapidly between zero and one. Thus

$$\lim_{x \to 0} \sin\left(\frac{1}{x}\right)$$

does not exist and the point $x = 0$ is an essential discontinuity.

● **PROBLEM 3-19**

What is meant by a discontinuity of the first kind? A discontinuity of the second kind? Show that monotonic functions have no discontinuities of the second kind.

Solution: Let $(a,b) \subset R$ and let $f: R \to R$. If x is a point in the domain of definition of f at which f is not continuous, we say that f is discontinuous or that f has a discontinuity at x. There are two kinds of discontinuity and they are defined using the concepts of right-hand and left-hand limits of f at a point y in its domain. The right-hand limit of f at y is denoted by $f(y+)$ while the left-hand limit is $f(y-)$.

The right-hand limit of f at y is defined as $f(y+) = q$ if $f(t_n) \to q$ as $n \to \infty$ for all sequences $\{t_n\}$ in (y,b) such that $t_n \to y$. Similarly, the left-hand limit of f at y is $f(y-) = p$ for $a < y \leq b$ if $f(t_n) \to p$ as $n \to \infty$ for all sequences $\{t_n\}$ in (a,y) such that $t_n \to y$. The function f is said to have a discontinuity of the first kind if $f(y+)$ and $f(y-)$ exist. In this case there are two possibilities

a) $f(y+) \neq f(y-)$

b) $f(y+) = f(y-) \neq f(y)$

(If $f(y+) = f(y-) = f(y)$ then f is not discontinuity at y).

The function f has a discontinuity of the second kind if $f(x+)$ or $f(x-)$ do not exist.

For example, the Dirichlet function

$$f(x) = \begin{cases} 1 & (x \text{ rational}) \\ 0 & (x \text{ irrational}) \end{cases}$$

has a discontinuity of the second kind at every point x since neither $f(x+)$ nor $f(x-)$ exist.

Now consider monotonic functions. Monotonic functions are defined on intervals (a,b). f is said to be monotonically increasing on (a,b) if $a < x < y < b$ implies $f(x) \leq f(y)$. f is said to be monotonically decreasing on (a,b) if $a < x < y < b$ implies $f(x) \geq f(y)$.

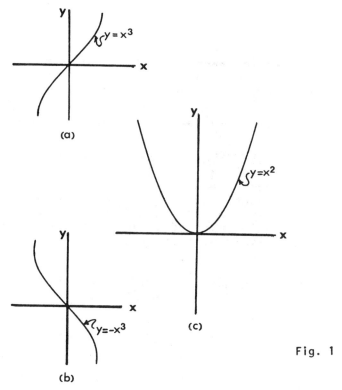

Fig. 1

In the given figure $y = x^3$ is monotonically in-increasing over R, $y = -x^3$ is monotonically decreasing and $y = x^2$ is neither over R.

A basic theorem on monotonic functions is the following: Let f be monotonically increasing on (a,b). Then $f(x+)$ and $f(x-)$ exist at every point x of (a,b). In fact,

$$\sup_{a<t<x} f(t) = f(x-) \leq f(x) \leq f(x+) = \inf_{x<t<b} f(t).$$

To prove the theorem note that, since f is monotonically increasing, the set

$$F = \{f(t): a < t < x\}$$

is bounded above by $f(x)$. By the properties of the real number system, F has a least upper bound, say A. Now $A \leq f(x)$ and it remains to show that $A = f(x-)$. To do this let $\varepsilon > 0$ be given. Since A is a least upper bound, there exists $\delta > 0$ such that $a < x - \delta < x$ and $A - \varepsilon < f(x - \delta) \leq A$. But

$$f(x - \delta) \leq f(t) \leq A \qquad (x - \delta < t < x)$$

(by monotonicity). Thus

$$A - \varepsilon < f(t) \leq A \quad (x - \delta < t < x)$$

or

$$|f(t) - A| < \varepsilon \qquad (x - \delta < t < x)$$

and thus

$$f(x-) = A .$$

The fact that $f(x+) = B$ is proved in the same way.

Since, by the theorem, $f(x+)$ and $f(x-)$ exist at every point in the domain of f it follows that f can have no discontinuities of the second kind.

If f is not monotonic, then $f(x+)$ and $f(x-)$ need not both exist. For example, let $f(x) = 1/x$. Then, at 0, $f(0+)$ and $f(0-)$ both do not exist.

● **PROBLEM 3-20**

Consider the function $f: [0,1] \rightarrow [0,1]$ given by

$f(x) = \quad 1/q, \; q > 0, \; x = p/q$, i.e., rational

$f(x) = \quad 0 \qquad$ x irrational, $x \in R - Q$

a) Show that f is discontinuous at any rational number.

b) Show that f is continuous at each irrational.

Solution: a) Let $x_0 = \dfrac{p}{q}$

be any rational number. Then

$$f\left(\frac{p}{q}\right) = \frac{1}{q} .$$

In order to show that f is discontinuous at x_0 let

$$\varepsilon = \frac{1}{2q} .$$

For any $\delta > 0$ $(x_0 - \delta, x_0 + \delta)$ contains an irrational number x. Hence
$$|f(x) - f(x_0)| = |0 - \frac{1}{q}| = \frac{1}{q}$$
can never be less than $\varepsilon = \frac{1}{2q}$ showing that f is discontinuous at x_0.

b) Let x_0 be irrational so that $f(x_0) = 0$. Let $\varepsilon > 0$ be given. Choose n so that $\frac{1}{n} < \varepsilon$. Let A_m be the set of those rational numbers which are of the form
$$\frac{p}{m} \quad p = 0, \pm 1, \pm 2, \ldots$$
and let
$$A = \bigcup_{m=1}^{n-1} A_m.$$

The set A_m has no limit points and hence A has no limit points. Therefore
$$\min_{a \in A} |x_0 - a| = \delta > 0.$$
Let
$$B = \bigcup_{m=n}^{\infty} A_m.$$

It follows that whenever x is rational and
$$|x - x_0| < \delta$$
then $x \notin A$ i.e., x must be in B and hence of the form
$$x = \frac{p}{j}$$
with $j \geq n$. Hence
$$|f(x) - f(x_0)| = |\frac{1}{j} - 0| \leq \frac{1}{n} < \varepsilon.$$

On the other hand, if x is irrational and $|x - x_0| < \delta$ then
$$|f(x) - f(x_0)| = 0 < \varepsilon$$
Thus f is continuous at x_0 if x_0 is irrational.

● **PROBLEM 3-21**

The Baire category theorem states that the real line is of the second category.

Show how the Baire category theorem can be used to prove the following theorem:

> There exists no $F: R \to R$ such that f is continuous at each rational but discontinuous at each irrational.

Solution: The term category (as used in analysis) is an application of the following definition:

Let $A \subseteq R$. Then A is nowhere dense in R if the closure of A, $\overline{A}$, contains no nonempty open interval of R. For example, the set of integers Z is nowhere dense as is the Cantor set.

Let $D \subseteq R$. Then D is said to be of the first category if $D = \bigcup_{n=1}^{\infty} E_n$ where E_n is nowhere dense as defined above. Thus D is of the first category if it is equal to the countable union of nowhere dense sets. If D is not of the first category it said to be of the second category.

Thus Baire's theorem states that R is not equal to the countable union of nowhere dense sets.

To prove the given theorem, note that R equals the disjoint union of the set of all rationals and the set of all irrationals. The set of all rationals, Q, is a set of the first category since $Q = \bigcup_{n=1}^{\infty} q_n$ where q_n is a rational number (i.e., a one-point set). Here $\{q_n\}$ is closed and nowhere dense for each n.

Now note that if A and B are sets of the first category then A U B is of the first category. To prove this let $A = \bigcup_{n=1}^{\infty} A_n$ and $B = \bigcup_{n=1}^{\infty} B_n$ where A_n, B_n are nowhere dense. Then A U B is the union of a countable number of nowhere dense sets, i.e., A U B is of the first category.

Since R is of the second category by Baire's theorem, it follows that the set of all irrationals is of the second category (otherwise R being the disjoint union of two sets of the first category would be of the first category).

It follows that if there indeed exists a function that is continuous at each rational and discontinuous at each irrational, the set of its discontinuous points must be of the second category. To prove that this is impossible requires a string of definitons and lemmas.

To begin with, let f be any real valued function. If J is any bounded open interval in R, the oscillation of f over J ($\omega[f;J]$) is defined as

$$\omega[f;J] = \underset{x \in J}{\text{l.u.b.}} f(x) - \underset{x \in J}{\text{g.l.b.}} f(x) .$$

Let a ∈ R. The oscillation of f at a is the infimium of all ω[f;J] where J contains a. Thus

$$\omega[f;a] = \text{g.l.b. } \omega[f;J] .$$

Since

$$\underset{x \in J}{\text{l.u.b}} \, f(x) \geq \underset{x \in J}{\text{g.l.b}} \, f(x),$$

$$\omega[f;J] \geq 0$$

and

$$\omega[f;a] \geq 0$$

for any a R.

Now suppose f is discontinuous at a. This implies that the oscillation of f on J is positive for some J containing a and thus $\omega[f;a] > 0$. This fact may be used to prove the following theorem:

Let $f: R \to R$. For $r > 0$, let G_r be the set of all $a \in R$ such that $\omega[f;a] \geq 1/r$. Then G_r is closed.

The theorem states that the set of all points at which f has positive oscillation (i.e., f is discontinuous) is a closed set. To prove the theorem, let x be a limit point of G_r, and let J be a bounded open interval containing x. Since J is an open set and x is a limit point of G_r every neighborhood of x in J must contain a point $y \in G_r$. But $\omega[f;J] \geq \omega[f;y]$ (since $\omega[f;y]$ is the g.l.b. over all $\omega[f;J]$ where J contains y). Since $y \in G_r$, $\omega[f;y] \geq 1/r$. Thus $\omega[f;J] \geq 1/r$. Since $\omega[f;x]$ is the g.l.b. of all $\omega[f;J]$ where J contains x, $\omega[f;x] \geq 1/r$ and $x \in G_r$. Since G_r contains its limit points, G_r is a closed set as was to be proved.

By definition $D \subseteq R$ is said to be of a type F_σ if $D = \bigcup_{n=1}^{\infty} F_n$ where each F_n is a closed set in R. (Note that the countable union of closed sets need not be closed e.g.,

$$A = \bigcup_{n=1}^{\infty} [1/n, n]).$$

Thus a set $D \subset R$ is of type F_σ if it is equal to a countable union of closed sets. Let f be a real-valued function defined on R. The set of points at which f is discontinuous is of type F_σ. To prove this, let $x \in D$, the set of discontinuities. Then

$$\omega[f;x] > 0,$$

i.e.,
$$\omega[f;x] \geq 1/n$$

and
$$D \subset \bigcup_{n=1}^{\infty} G_{1/n}$$

Conversely, if
$$x \in \bigcup_{n=1}^{\infty} G_{1/n}$$

then
$$\omega[f;x] > 0$$

and $x \in D$. Thus

$$D = \bigcup_{n=1}^{\infty} G_{1/n},$$

but each $G_{1/n}$ is closed by the previous theorem (the set of points at which f has positive oscillation is a closed set). Thus D is a countable union of closed sets, i.e., of type F_σ.

Note that the set of all irrationals R-Q is not of type F_σ for if it were then

$$R - Q = \bigcup_{n=1}^{\infty} F_n$$

where each F_n is closed. But each F_n contains only irrationals and thus no open interval (between any two irrationals there exists a rational). Thus each F_n is closed and nowhere dense, i.e., the set of irrationals is of the first category which is a contradiction. This shows that the set of all irrationals is not of type F_σ.

The conclusion is that for any $f: R \to R$ the set of discontinuities is of type F_σ while the set of all irrationals is not of type F_σ. This proves that there exists no $f: R \to R$ which is continuous at each rational and discontinuous at each irrational.

• **PROBLEM 3-22**

Prove that there is no continuous real-valued function f on R such that

f(x) is irrational if x is rational

f(x) is rational if x is irrational.

Solution: This problem requires the notion of connectedness. Let (X,ρ) be a metric space, i.e., a set of points with a distance function defined on it. Two sets

$$A, B \neq \phi \subset X$$

are said to be separated if $\bar{A} \cap B = \phi$, $A \cap \bar{B} = \phi$
(where $\bar{S}$ denotes the closure of a set). The metric space X is connected if there exist no separated nonempty sets A,B such that $X = A \cup B$.

If X is not connected, then there exist nonempty sets $A,B \subset X$ such that $X = A \cup B$, $\bar{A} \cap B = \phi$, $A \cap \bar{B} = \phi$
But this implies that $\bar{A} \subset B^c = A$ (since $A \cup B = X$, the whole space).

Since $A \subset \bar{A}$, A is closed. Similarly B is closed. Thus the definition of connectedness could also be phrased as follows: The space X is connected if there exists no pair of non-empty closed sets A and B in X such that $X = A \cup B$ and $A \cap B = \phi$.

The real line is a connected space.

The range of a continuous function defined on a connected set is connected.

Using these two facts the given problem may be solved. First, note that the domain of the given function is the set of real numbers, a connected set. Assume that there exists a continuous function from R into itself such that

$$f(x) = \begin{cases} \text{irrational}, & x = p/q \quad p,q \in Z \\ r/s, \ r,s \in Z & x \text{ irrational}. \end{cases} \quad (1)$$

(Note that there exist many such discontinuous functions. For example,

$$f(x) = \begin{cases} \pi & x \text{ rational} \\ 0 & x \text{ irrational}). \end{cases}$$

Now, for any non-degenerate interval in R (any interval in R is connected) the image of f should be connected. But the image of f consists of a set of rational numbers and a countable set of irrational numbers. In other words the range of f is a countable set which cannot be connected because the only connected subsets of R, besides ϕ and singletons, are intervals which are uncountable. Thus f cannot be continuous.

UNIFORM CONTINUITY AND RELATED TOPICS

• **PROBLEM 3-23**

Let $f: R \to R$ be given by

$$f(x) = \frac{1}{1 + x^2} \quad (-\infty < x < \infty).$$

Show that f attains a maximum value but does not attain a minimum value.

What is a sufficient condition for f to attain its maximum and minimum on an interval in R?

Solution: When $x = 0$, $f(x) = 1$. For any other value of x, $f(x) < 1$. Thus f attains its maximum at $x = 0$.

If $|x_2| > |x_1|$ $1 + x_2^2 > 1 + x_1^2$

hence

$$\frac{1}{1 + x_2^2} < \frac{1}{1 + x_1^2} \tag{1}$$

Therefore f cannot attain a minimum at any point x_0. For otherwise by (1) the value of f at any point x with $|x| > |x_0|$ will be less than $f(x_0)$ contradicting the definition of minimum.

A sufficient condition for f to attain its maximum and minimum on an interval I in R is that I must be closed and bounded.

● **PROBLEM 3-24**

To which class of continuous functions do the following functions belong?

a) Weierstrass's nowhere differentiable function

b) $f(x,y,z) = e^{x-y+z}$

c) $f(t) = t^3 e_1 + (\sin t) e_2 + t^{8/3} e_3$, $-\infty < t < \infty$.

Solution: Functions can be classified according to the number of times they can be differentiated with a continuous differential resulting.

Let $f: E \to R^m$, $E \subseteq R^n$ be a vector-valued function.

Then f is said to be of class C^k on E if the kth order derivative of f exists and is continuous on E, where $k \geq 0$.

a) Weierstrass's example of a continuous nowhere differentiable function was

$$f(x) = \sum_{n=0}^{\infty} a^n \cos(b^n \pi x).$$

This function, although continuous on the entire real line is nowhere differentiable there. Thus

$$f \in C^0 .$$

b) For any element in the set

$$R^3$$

e^{x-y+z} is continuous. Furthermore, it is infinitely differentiable, i.e., it is in C^∞.

c) $f(t) = t^3 e_1 + (\sin t) e_2 + t^{8/3} e_3 \quad -\infty < t < \infty$.
This is a vector function of the form $f: R \to R^3$. This belongs to C^m on the interval I if and only if its components $f_1(t)$, $f_2(t)$ and $f_3(t)$ belong to C^m on I. Now

$$f'(t) = 3t^2 e_1 + \cos t\, e_2 + \frac{8}{3} t^{5/3} e_3$$

$$f''(t) = 6t e_1 - \sin t\, e_2 + \frac{40}{9} t^{2/3} e_3$$

$$f'''(t) = 6 e_1 - \cos t\, e_2 + \frac{80}{27} t^{-1/3} e_3$$

The given function, its first derivative and its second derivative are all continuous on the interval $(-\infty < t < \infty)$. However, $f'''(t)$ does not exist at $t=0$, since

$$\frac{80}{27} t^{-1/3} e_3$$

is undefined there. Hence $f(t)$ is in C^2 on $-\infty < t < \infty$ but not in C^3. Note here that since a differentiable function is continuous, if a function belongs to C^m it belongs to C^j for $j \leq m$.

● **PROBLEM 3-25**

Prove the Intermediate Value Theorem for the derivative of a real differentiable function on [a,b].

Solution: Suppose f is such a function. Note that the derivative exists on [a,b] but is not necessarily continuous, Hence, it is not possible to appeal to the Intermediate Value Theorem for continuous functions. The theorem states that for any $\lambda \in (f'(a), f'(b))$ [or $\lambda \in (f'(b), f'(a))$ when appropriate], there exists an $x \in (a,b)$ such that $f'(x) = \lambda$. Let $g(t) = f(t) - \lambda t$. Then $g'(a) = f'(a) - \lambda < 0$ and $g'(b) = f'(b) - \lambda > 0$, so that for some $t_1, t_2 \in (a,b)$,

$g(t_1) < g(a)$ and $g(t_2) < g(b)$. Hence the continuous function g attains its minimum on [a,b] at some $x \in (a,b)$. Thus $g'(x) = f'(x) - \lambda = 0$ or $f'(x) = \lambda$. This means that f' has no simple (or removable or first kind) discontinuities but may have essential (or second kind) discontinuities. Thus this is a stronger version of the Intermediate Value Theorem (since the conditions are weaker). Note also that the intermediate value theorem for continuous functions is just a particular case of this version; for if (f(x) is continuous on [a,b], then it is just the derivative of

$$F(x) = \int_a^x f(t)\,dt$$

by the Fundamental Theorem of Integral Calculus.

● **PROBLEM 3-26**

Let $f: R \to R$ be given by

$$f(x) = [x] = \begin{cases} n: n \text{ is the greatest integer} \\ \text{less than or equal to } x. \end{cases}$$

(f is also known as the greatest integer function). Show that f is upper semi-continuous.

<u>Solution</u>: Upper and lower semicontinuity are defined as the inequalities that follow from the definition of continuity. Let f be a real-valued function defined on a subset D of R^n. Then f is continuous at $a \in D$ if, given $\varepsilon > 0$, there exist a $\delta > 0$ such that

$$|f(x) - f(a)| < \varepsilon \quad \text{for} \quad |x - a| < \delta \quad .$$

Removing the absolute-value signs yields:
$-\varepsilon < f(x) - f(a) < \varepsilon$, or $f(a) - \varepsilon < f(x) < f(a) + \varepsilon$
for $|x - a| < \delta$.

f is said to be upper semicontinuous if

$$f(x) < f(a) - \varepsilon, \quad |x - a| < \delta \tag{1}$$

Similarly, f is said to be lower semicontinuous if
$$f(x) > f(a) + \varepsilon, \quad |x - a| < \delta. \tag{2}$$
The graph of the greatest integer function is:

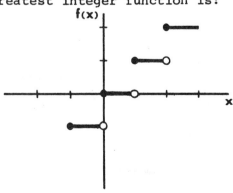

Fig. 1

To show that f is upper semicontinuous requires the notion of the nondeleted limit superior of a function at a point a.

The inequality (1) implies that

$$\sup\{f(x): |x - a| < \delta\} \leq f(a) + \varepsilon \qquad (3)$$

If the least upper bounds are taken in (3) for all $\delta > 0$, (3) may be rewritten as

$$\inf_{\delta>0} \sup\{f(x): |x - a| < \delta\} \leq f(a). \qquad (4)$$

But $|a - a| < \delta$ for all positive δ and hence

$$f(a) = \inf_{\delta>0} \sup\{f(x): |x - a| < \delta\}.$$

Since

$$\sup\{f(x): |x - a| < \delta\}$$

is nonincreasing as $\delta \to 0$ (the supremum of f over a larger neighborhood is always greater than or equal to the supremum over a smaller neighborhood),

$$\inf_{\delta>0} \sup\{f(x): |x - a| < \delta\} = \lim_{\delta \to 0^+} \sup\{f(x): |x - a| < \delta\}.$$

This is the limit superior of f at a and is denoted by $\overline{L}f(a)$.

A function is upper semicontinuous at a point a in its domain if and only if $f(a) = \overline{L}f(a)$

For the greatest integer function, let $a \in [n, n+1]$ where n, n + 1 are integers and a is a real number. Since

$$\lim_{\delta \to 0^+} \sup\{f(x): |x - a| < \delta\} = n = f(a)$$

for all a in the interval (n, n + 1), [x] is upper semicontinuous in this open interval. Since f is discontinuous at either end-point, these end-points must be considered separately. At n, every δ neighborhood of n yields n and n - 1 as values of f(x).

But $\sup\{n, n-1\} = n$ and thus $f(n) = \overline{L}f(n)$.

At n + 1, every δ neighborhood of n + 1 contains points at which $f(x) = n + 1$ and points at which $f(x) = n$. Thus $\sup\{n, n + 1\} = n + 1$ and since $f(n+1) = \overline{L}f(n+1)$, f is upper semi-continuous at n + 1. Thus f is upper semicontinuous on the real line.

• **PROBLEM 3-27**

Show that the function $f(x) = 1/x$ is not uniformly continuous on the half-open interval (0,1] but is uniformly continuous on [1,b] where $b \in R$. What is a sufficient condition for functions defined on subsets of R^n to be uniformly continuous?

Solution: Let $D \subseteq R^n$ and let $f: D \to R^m$. Then f is uniformly continuous on D if, given $\varepsilon > 0$, there exists a $\delta > 0$ such that $||x-y|| < \delta$ implies $||f(x) - f(y)|| < \varepsilon$ for all $x = (x_1, \ldots, x_n)$ $y = (y_1, \ldots, y_n) \in R^m$.
Here $|| z ||$ denotes the Euclidean norm of a point in R^n and equals

$$\left(\sum_{i=1}^{n} z_i\right)^{\frac{1}{2}}.$$

The given function $f: R \to R$ is defined on $(0,1]$.

To show that f is not uniformly continuous on this set, let $x_0 > 0$. Then, given $\varepsilon > 0$ choose δ such that

$$\left|\frac{1}{x} - \frac{1}{x_0}\right| < \varepsilon \text{ if } |x - x_0| < \delta. \tag{1}$$

As x_0 approaches zero $\frac{1}{x_0}$ becomes larger and larger. Thus in order to keep $\frac{1}{x}$ within ε units of $\frac{1}{x_0}$ the distance between x and x_0 must be made smaller and smaller. Since x_0 approaches 0 closer than any preassigned limit δ approaches zero in the same way and therefore there is no one value of δ independent of $x_0 \in (0,1]$ that will guarantee the validity of (1). Thus f is not uniformly continuous on $(0,1]$.
When $1 \leq x \leq b$, let $\delta = \varepsilon$. Then for $x, x_0 \in [1,b]$ and

$|x - x_0| < \varepsilon \ (=\delta)$

$$\left|\frac{1}{x} - \frac{1}{x_0}\right| = \frac{|x_0 - x|}{xx_0} \leq |x - x_0| < \varepsilon$$

Thus given any $\varepsilon > 0$, it is possible to find a $\delta > 0$ such that (1) is true regardless of the value of x_0.

From the above problem it appears that open sets can prevent continuous functions from being uniformly continuous. Therefore a natural sufficiency criterion for uniform continuity is:

Let S be a closed and bounded subset of R^n. Let $f: R^n \to R^m$ be defined and continuous at each point of R^n. Then f is uniformly continuous on S. Note that this is not a necessary condition for uniform continuity. For example, let $f(x) = x$ on the open interval $(0,1)$. Then, given any $\varepsilon > 0$, let $\delta = \varepsilon$, for

$|x - x_0| < \varepsilon$ implies $|f(x) - f(x_0)| = |x - x_0| < \varepsilon$

and f is uniformly continuous on $(0,1)$.

● **PROBLEM 3-28**

Let $f: R \to R$ be given by $f(x) = x^2$.

Show that although f is continuous at every point in the domain it is not uniformly continuous.

Solution: The first task is to verify that f is indeed continuous throughout its domain. Let $\varepsilon > 0$ be given. Then a $\delta > 0$ must be found such that $|x - x_0| < \delta$ implies that $|f(x) - f(x_0)| < \varepsilon$ for $x, x_0 \in R$. Now

$$|f(x) - f(x_0)| = |x^2 - x_0^2| = |(x + x_0)(x - x_0)|$$
$$= |x + x_0||x - x_0|.$$

Let I be an interval of length 2δ, $0 < \delta < 1$ on the real line centered at x_0. Then $|x - x_0| < \delta$ and $|x + x_0| = |x - x_0 + 2x_0| \le |x - x_0| + 2|x_0| < 1 + 2|x_0|$. Note that δ is chosen less than 1.

Thus

$$|x^2 - x_0^2| = |x + x_0||x - x_0|$$
$$\le (1 + 2|x_0|)\delta$$

Thus δ is to be chosen so that

$$(1 + 2|x_0|)\delta < \varepsilon$$

or,

$$\delta < \frac{\varepsilon}{1 + 2|x_0|} \qquad (1)$$

Since δ can be found both for all $\varepsilon > 0$ and for all $x_0 \in R$, it follows that f is continuous everywhere. From (1) observe that depends not only on ε but on $|x_0|$. For example, suppose $\varepsilon = 2$ and $|x_0| = 1$. Then

$$\delta < \frac{2}{1 + 2} = \frac{2}{3}.$$

Now suppose $\varepsilon = 2$ but $|x_0| = 100$. Then

$$\delta < \frac{2}{1 + 200}$$

or, $\delta < .00995$. This shows that for $f(x) = x^2$, δ is a function of both ε and x_0.

Contrast this with $f: R \to R$ given by $f(x) = 6x$. Let $\varepsilon > 0$ be given. Then $|f(x) - f(x_0)| = 6|x - x_0|$. Choose an interval around x_0 of length 2δ. Hence $|x - x_0| < \delta$ and δ is to be chosen so that

$$6|x - x_0| < 6\delta, \text{ i.e., } 6\delta < \varepsilon, \text{ or } \delta < \varepsilon/6.$$

Since δ depends only on ε and not on the point where continuity is being tested, $f(x) = 6x$ is said to be uniformly continuous on R.

The precise definition of uniform continuity is:

Let $D \subseteq R^n$ and let $f: D \to R^m$. Then f is uniformly continuous on D if, given $\varepsilon > 0$, there exists $\delta > 0$ such that $||x - a|| < \delta$ implies $||f(x) - f(a)|| < \varepsilon$ for all $a \in D$.

From the definition it follows that $f(x) = x^2$ is not uniformly continuous on the real line. Note, however that f is uniformly continuous on any bounded interval of the real line.

● **PROBLEM 3-29**

Which of the following functions are uniformly continuous on the specified domains?

a) $f(x) = x^3$ $(0 \leq x \leq 1)$

b) $f(x) = x^3$ $(0 \leq x < \infty)$

c) $f(x) = \sin x^2$ $(0 \leq x < \infty)$

d) $f(x) = 1/(1 + x^2)$ $(0 \leq x < \infty)$.

Solution: Let $A \subseteq R$ and let $f: A \to R$. The function f is continuous at a point x_0 in its domain if, given $\varepsilon > 0$ there exists a $\delta > 0$ such that $|f(x) - f(x_0)| < \varepsilon$ whenever $|x - x_0| < \delta$. In general, the number δ depends on both ε, the challenge number, and x_0, the point at which f is being tested for continuity. The function f is said to be continuous on A if it is continuous at every point of A.

Now suppose $\varepsilon > 0$ is given and a δ is found that satisfies the definition of continuity for all $X_0 \in A$, i.e., δ depends only on ε and not on x_0. Then f is uniformly continuous on A.

a) Let $\varepsilon > 0$ be given. Choose $\delta = \varepsilon^{1/3}$. Then for

$$|x - x_0| < \delta = \varepsilon^{1/3}, \quad |f(x) - f(x_0)|$$

$$= |x^3 - x_0^3| < |x^3 - x_0^3| + 3xx_0|x - x_0| = |x - x_0|^3$$

if

$$0 \leq x \leq 1 \quad \text{and} \quad 0 \leq x_0 \leq 1.$$

But

$$|x - x_0|^3 < (\varepsilon^{1/3})^3 = \varepsilon$$

and thus

$$|x^3 - x_0^3| < \varepsilon$$

Hence f is uniformly continuous on [0,1].

b) Since f is defined on an unbounded set it will not be uniformly continuous. What this means is that given any $\varepsilon > 0$, x_0, x_1 in the domain can be found such that although $|x_0 - x_1| < \delta$, $|f(x_0) - f(x_1)| \geq \varepsilon$.

To prove that $f(x) = x^3$ cannot be uniformly continuous, rewrite $|f(x_0) - f(x_1)| = |x_0^3 - x_1^3|$ as
$|x_0 - x_1||x_0^2 + x_0 x_1 + x_1^2| < \delta |x_0^2 + x_0 x_1 + x_1^2|$.

Now, let $\delta |x_0^2 + x_0 x_1 + x_1^2| < \varepsilon$

or

$$\delta < \frac{\varepsilon}{|x_0^2 + x_0 x_1 + x_1^2|}$$

But as x_0, $x_1 \to \infty$, $\delta \to 0$ and thus no δ can be found that will work for all x in the interval $[0, \infty]$.

c) Here $|\sin x^2| \leq 1$ regardless of the value of x. Furthermore, $f(x) = \sin x^2$ is uniformly continuous on the interval $[0, \sqrt{2\pi}]$. Since this is its period, it follows that $\sin x^2$ is uniformly continuous over any interval. The proof that $(f(x) = \sin x^2$ is uniformly continuous on the interval $[0, \sqrt{2\pi}]$ is an application of the following sufficiency criterion for uniform continuity: If the real valued function f is continuous on the closed, bounded interval [a,b] then f is uniformly continuous on [a,b].

d) If

$$f(x) = \frac{1}{1 + x^2} \quad (0 \leq x < \infty), \quad f(0) = 1$$

and

$$\lim_{x \to \infty} f(x) = 0.$$

Thus $|f(x)| \leq 1$. To see that

$$\lim_{x \to \infty} f(x) = 0,$$

let $\varepsilon > 0$ be given. Set $\frac{1}{1 + x^2} < \varepsilon$

Then when $x > \sqrt{1/\varepsilon - 1}$, $f(x) < \varepsilon$

Now the problem is, given ε, to show that there exists a $\delta > 0$ such that whenever $|x - a| < \delta$,

$$|f(x) - f(a)| < \varepsilon \quad \text{for all } a \in [0,\infty) .$$

$$\left| \frac{1}{1 + x^2} - \frac{1}{1 + a^2} \right| = \left| \frac{a^2 + 1 - x^2 + 1}{(1 + x^2)(1 + a^2)} \right|$$

$$= \left| \frac{(a + x)(a - x)}{(1 + x^2)(1 + a^2)} \right| < \delta \frac{|a + x|}{|(1 + x^2)(1 + a^2)|} .$$

Now if

$$\frac{\delta |a + x|}{|(1 + x^2)(1 + a^2)|} < \varepsilon ,$$

then $|f(x) - f(a)| < \varepsilon$. Thus δ must be chosen such that

$$\delta < \frac{\varepsilon}{|a + x|} |(1 + x^2)(1 + a^2)| .$$

As a and x get larger (always remaining less than δ apart),

$$\frac{\varepsilon}{|a + x|} |(1 + x^2)(1 + a^2)|$$

tends to infinity. Thus there is no difficulty in choosing δ such that $|x - a| < \delta$ implies $|f(x) - f(a)| < \varepsilon$. Hence

$$f(x) = \frac{1}{1 + x^2}$$

is uniformly continuous on the interval $[0,\infty]$.

• **PROBLEM 3-30**

Let f be the real-valued function defined on [0,1] given by $f(x) = x^2 \sin(1/x)$ for $x \in (0,1]$ and $f(0) = 0$. Show that f is absolutely continuous on [0,1].

Solution: The concept of absolute continuity is used in the theory of integration; specifically in the integration of a derivative. The definition of absolute continuity of a function is:

Let f be a real-valued function defined on an interval I. Then f is absolutely continuous if given $\varepsilon > 0$, there exists a $\delta > 0$ such that

$$\sum_{i=1}^{n} |f(b_i) - f(a_i)| < \varepsilon$$

for every finite pairwise disjoint collection $\{(a_i, b_i) : i = 1, \ldots, n\}$ of open intervals in I with

$$\sum_{i=1}^{n} (b_i - a_i) < \delta.$$

Note that an absolutely continuous function on I is uniformly continuous on I since the number of disjoint open intervals in I can be taken equal to one to yield $|f(x) - f(x_0)| < \varepsilon$ for every open interval in I such that $|x - x_0| < \delta, x, x_0 \in I$.

Also, if a function f satisfies the Lipschitz condition $|f(x) - f(y)| \leq c|x - y|$ $x, y \in I$ then f is absolutely continuous on I. To see this let $\varepsilon > 0$ be given. Then choose $\delta = \varepsilon/c$ so that if

$$\sum_{i=1}^{n} |x_i - y_i| < \delta = \varepsilon/c,$$

$$\sum |f(x_i) - f(y_i)| \leq c \sum |x_i - y_i| < c(\varepsilon/c) = \varepsilon$$

and f is absolutely continuous.

Thus there exist necessary (uniform continuity) and sufficient (f satisfies a Lipschitz condition) conditions for a function to be absolutely continuous.

To illustrate the concept of absolute continuity consider the given figure.

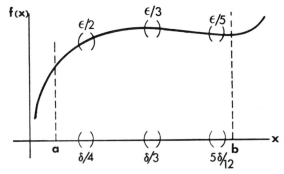

Fig. 1

Given $\varepsilon > 0$, three open disjoint intervals have been chosen in $I = [a,b]$ of length $\delta/4$, $\delta/3$ and $5\delta/12$ respectively. Corresponding to these intervals, f(x) varies in values of $\varepsilon/2$, $\varepsilon/3$ and $\varepsilon/5$ respectively. Thus

$$\sum_{i=1}^{3} (b_i - a_i) < \delta \text{ and } \sum |f(b_i) - f(a_i)| \leq \varepsilon.$$

Note that $f(x) = x^2 \sin 1/x$ is continuous throughout the half-open interval $(0,1]$. It is discontinuous at 0 but this is a removable discontinuity since

$$\lim_{x \to 0} x^2 \times \sin 1/x \leq \lim_{x \to 0} x^2 |1|$$

(since $|\sin 1/x| \leq 1$ and $\lim_{x \to 0} x^2 = 0$).

Thus f may be defined as zero at 0 making it continuous throughout the closed interval [0,1].

The sufficiency criterion (existence of a Lipschitz constant) may be used to show that f is absolutely continuous in [0,1]. By the mean-value theorem:

$$\frac{|f(x) - f(y)|}{|x - y|} = |f'(\xi)| \text{ for some } \xi \text{ between } x \text{ and } y$$

or,

$$|f(x) - f(y)| = |f'(\xi)||x - y| .$$

Replace $|f'(\xi)|$ by $\sup_{x \in [0,1]} |f'(x)|$ to obtain the inequality

$$|f(x) - f(y)| \leq A|x - y|$$

where

$$A = \sup_{x \in [0,1]} |f'(x)| .$$

It remains to show that $\sup_{x \in [0,1]} |f'(x)|$ exists, i.e. is equal to a positive constant. Now $f'(x) = 2x \sin 1/x - \cos 1/x$ and this function is bounded on the interval (0,1]. Furthermore it has value 0 at 0. Thus $f'(x)$ is bounded on [0,1] i.e. $\sup_{x \in [0,1]} |f'(x)| = A < \infty$. This shows that $f(x) = x^2 \sin 1/x$ satisfies a Lipschitz condition on [0,1] and thus it is absolutely continuous in that interval.

● **PROBLEM 3-31**

The Arzèla-Ascoli theorem states that a set E in $C(X,R)$ is compact if and only if it is closed, bounded and equicontinuous.

Let $C(X,R^m)$ be the metric space consisting of all continuous functions from the compact metric space X into R^n with distance function given by $d(f,g) = \sup \{||f(x) - g(x)|| : x \in X\}$. State the Arzèla-Ascoli theorem for this space.

Solution: Let (X,d_1) be a metric space, and let $E \subset X$. An open cover of E is a collection of open sets in X whose union contains E. Thus

$E \subset \bigcup_{\gamma \in \Lambda} E_\gamma$ where $\{E_\gamma\}_{\gamma \in \Lambda}$ is a collection of open sets indexed by Λ. E is said to be compact if every open cover of E contains a finite collection of sets whose union covers E (known as a finite subcover).

The set $C(X,R)$ consists of all continuous real-valued functions defined on the set X. Usually X is taken to be the unit interval $[0,1]$ or the closed interval $[a,b]$. If X is compact the subset E consists of functions that are not only continuous but uniformly continuous. Thus for each $f \in E$, and each $\varepsilon > 0$, there exists a $\delta > 0$ such that

$$d(x,y) < \delta \text{ implies } |f(x) - f(y)| < \varepsilon \qquad (1)$$

for $x, y \in [a,b]$.

Now suppose that a δ can be found such that (1) holds for all $f \in E$. Then E is known as a family of equicontinuous functions. The definition of equicontinuity is as follows: A set E in $C(X,R)$ is said to be equicontinuous if for each $\varepsilon > 0$ there exists a $\delta > 0$ such that

$$|f(x_1) - f(x_2)| < \varepsilon \text{ for } x_1, x_2 \in X, \; d(x_1, x_2) < \delta$$

and $f \in E$.

The Arzèla-Ascoli theorem gives necessary and sufficient conditions for a subset of $C(X,R)$ to be compact, where $C(X,R)$ is a metric space whose elements are continuous real-valued functions defined on the compact metric space X. The given metric space is a generalization of $C(X,R)$. (It may be thought of as $C(X,R^m)$). Since X is compact the uniform metric

$$d(f,g) = \sup_{x \in X} \{||f(x) - g(x)||\}$$

may be replaced by

$$d(f,g) = \max \{||f(x) - g(x)|| : x \in X\}.$$

The metric space $C(X, R^m)$ is complete with this metric defined on it. The Arzéla-Ascoli theorem for this space states:

A set E in $C(X, R^m)$ is compact if and only if it is closed, bounded and equicontinuous.

• **PROBLEM 3-32**

Show that any finite set of continuous functions is equicontinuous.

Solution: Let $C(X,R)$ denote the space of continuous real-valued functions on a compact metric space X. A set E in $C(X,R)$ is said to be equicontinuous if, given $\varepsilon > 0$, there exists a $\delta > 0$ such that

$$|f(x_1) - f(x_2)| < \varepsilon$$

whenever $d(x_1,x_2) < \delta$, $\forall\, x_1, x_2 \in X$ and $f \in E$. Here $d(x_1,x_2)$ denotes the distance between x_1 and x_2 according to whichever metric is imposed on X.

Let $E = \{f_1, \ldots, f_n\}$ be a finite family of functions in $C(X,R)$. Since each f_i, $1 \leq i \leq n$, is a continuous function defined on a compact space, each f_i is uniformly continuous. This implies that, given $\varepsilon > 0$, there exist $\delta_i > 0$, $1 \leq i \leq n$, such that

$$|f_i(x_1) - f_i(x_2)| < \varepsilon \qquad 1 \leq i \leq n$$

whenever

$$d(x_1,x_2) < \delta_i \qquad 1 \leq i \leq n$$

for all $x_1, x_2 \in X$.

Pick $\delta = \min\limits_{1 \leq i \leq n} \delta_i$

Then $|f_i(x_1) - f_i(x_2)| < \varepsilon \qquad 1 \leq i \leq n$

whenever

$$d(x_1,x_2) < \delta, \; \forall x_1, x_2 \in X.$$

This shows that $\{f_1, \ldots, f_n\}$ is equicontinuous. An alternative method of solving the problem is to use the Arzèla-Ascoli theorem, which states that a subset E of $C(X,R)$ is compact if and only if it is closed, bounded and equicontinuous. Since $\{f_1, \ldots, f_n\}$ is a finite set, it is compact. Hence, by the Arzèla-Ascoli theorem it is equicontinuous.

PARADOXES OF CONTINUITY

• **PROBLEM 3-33**

Let $I = [0,1]$, the closed unit interval on the real line. Then $I \times I = I^2$ is the unit square in R^2. Show that there exists a continuous function $f: I \to I$ which is surjective, i.e. the image of f is the square.

Solution: A curve is defined as follows: Let $[a,b] \subset R$. A curve defined on $[a,b]$ is a mapping $f: [a,b] \to R^n$. Thus the problem asks one to find a curve whose image is the unit square.

The map f is constructed as the limit of a sequence of continuous functions, $\{f_n\}$ with domain I and co-domain I^2.

The sequence of continuous functions is set up as follows:

i) f_0 is the map that sends $[0,1]$ to $[0,1] \times [0,1]$ as drawn

in Fig. 1(a)

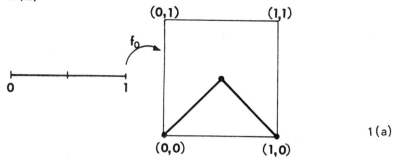

1(a)

ii) f_1 is made by breaking up [0,1] into 8 equal intervals and mapping to four quadrants of the square in Fig. 1(a) (see Fig. 1(b)):

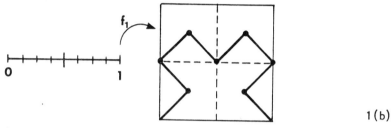

1(b)

iii) f_2 is constructed from f_1 in a similar manner:

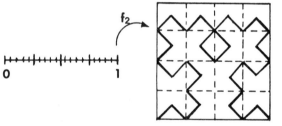

1(c)

iv) Note that at the nth step, f_n is a path made up of 4^n triangular paths ($f_0: 4^0 = 1$, $f_1: 4^1 = 4$, $f_2: 4^2 = 16$). Furthermore, each triangular path lies in a square of edge length $1/2^n$.

v) The function f_{n+1} is obtained by applying i) to each triangular path, replacing each one by four smaller triangular paths.

It remains to show that I) $\{f_n\}$ converges to a function f which is continuous. II) f is surjective, i.e. every point in I^2 is the image of some point in I. Let $d((x_1,x_2), (y_1,y_2))$ denote the square metric on R^2,

$$d((x_1,x_2)(y_1,y_2)) = \max\{|x_1 - y_1|, |x_2 - y_2|\}.$$

Using this metric it is possible to define a supremum metric on the space of all continuous functions from I to I^2 ($C(I,I^2)$). Let $f, g \in C(I,I^2)$.

Then

$$\rho(f,g) = \text{lub } \{d(f(t), g(t)): t \in I\} \quad (1)$$

Equipped with the metric (1) the space $C(I,I^2)$ is a complete metric space, i.e. every Cauchy sequence converges to an element in the space.

Consider the sequence of functions $\{f_n\}$ given by iv) and v) above. If it can be shown that $\{f_n\}$ is a Cauchy sequence, then the limit function $f \in C(I,I^2)$, since $C(I,I^2)$ is complete. To show this, note that each small triangular path in f_n lies in a square of edge length $1/2^n$. f_{n+1} is obtained from f_n by replacing one triangular path by four triangular paths that lie in the same square. Therefore the distance between $f_n(t)$ and $f_{n+1}(t)$ is at most $1/2^n$ using the square metric on I^2. This is shown for f_0 and f_1 in Fig. 2.

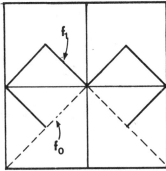

Fig. 2

Thus $\rho(f_n, f_{n+1}) \leq 1/2^n$ and

$$\rho(f_n, f_{n+m}) \leq \frac{1}{2^n} + \frac{1}{2^{n+1}} + \cdots + \frac{1}{2^{n+m-1}}$$

But

$$\frac{1}{2^n} + \frac{1}{2^{n+1}} + \cdots + \frac{1}{2^{n+m-1}} = \frac{1}{2^n}(1 + \frac{1}{2} + \cdots + \frac{1}{2^{m-1}})$$

$$< \frac{1}{2^n}(2) .$$

Thus, $\rho(f_n, f_{n+m}) < \frac{2}{2^n}$, for all m and n and hence $\{f_n\}$ is a Cauchy sequence which converges to a continuous function $f: I \to I^2$. Next, to prove that f is surjective, let $(x_1,x_2) \in I^2$. For a given n, the path f_n comes within a distance of $1/2^n$ of (x_1,x_2) as may be seen from Fig. 3 where n = 2.

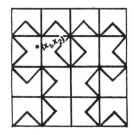

Fig. 3

153

Let $\varepsilon > 0$ be given and choose N large enough so that $\rho(f_N, f) < \varepsilon/2$ and $\frac{1}{2^N} < \varepsilon/2$. There is a point $t_0 \in I$ such that $d((x_1, x_2), f_N(t_0)) \leq \frac{1}{2^n}$, where $f_N(t_0)$ is the point on the path closest to (x_1, x_2). Then

$$d((x_1, x_2), f(t_0)) \leq d((x_1, x_2), f_N(t)) +$$

$$d((f(t), f_N(t)) \leq \frac{1}{2^n} + \varepsilon/2 < \varepsilon/2 + \varepsilon/2 = \varepsilon.$$

Thus, given $\varepsilon > 0$, there exists an ε neighborhood of (x_1, x_2) which intersects $f(I)$ and (x_1, x_2) belongs to $\overline{f(I)}$ (the closure of $f(I)$). Since I is compact $f(I)$, its continuous image is compact and therefore closed, i.e., $\overline{f(I)} = f(I)$ and $(x_1, x_2) \in f(I)$, which shows that f is surjective.

● PROBLEM 3-34

Show why the function,

$$f(x) = \sum_{n=0}^{\infty} a^n \cos(b^n \pi x) \quad 0 < a < 1 \quad b = 2k-1 \quad (1)$$

constructed by Weierstrass, is continuous everywhere but differentiable nowhere.

<u>Solution</u>: Comparing (1) with the series

$$\sum_{n=0}^{\infty} a^n \quad (0 \leq a < 1)$$

note that

$$|a^n \cos(b^n \pi x)| \leq a^n \quad (n = 0, 1, 2, \ldots).$$

By the Weierstrass M-test the series given by (1) converges uniformly to $f(x)$ and since convergence is uniform the limit function $f(x)$ is continuous everywhere.

Note that if

$$|ab| < 1, \quad \sum_{n=0}^{\infty} |ab|^n < \infty,$$

then the series representing $f(x)$ may be differentiated term by term and the resulting series converges uniformly to $f'(x)$; thus $f(x)$ is continuously differentiable if $|ab| < 1$.

Weierstrass showed that if $|ab|$ is large enough, then f is nowhere differentiable. To find an inequality involving $|ab|$, first construct a sequence x_n tending to x by setting $b^n x = u_n + c_n$ where u_n is an integer and $-\frac{1}{2} < c_n \leq \frac{1}{2}$, defining x_n by $b^n x_n = u_n + 1$. Thus $b^n(x-x_n) = c_n - 1$ and

$$|x - x_n| \leq b^{-n}|c_n - 1| \quad \text{or} \quad |x - x_n| \leq b^{-n}(3/2) . \tag{2}$$

This inequality holds for x_n sufficiently close to x but also depends on b. The derivative of f at x is the limit of the difference quotient

$$\frac{f(x) - f(x_n)}{x - x_n} . \tag{3}$$

But

$$f(x_n) = \sum_{m=0}^{\infty} a^m \cos(b^m \pi x_n) .$$

Thus (3) equals

$$\left(\sum_{m=0}^{n-1} + \sum_{m=n}^{\infty} \right) \frac{a^m \left(\cos b^m \pi x - \cos b^m \pi x_n \right)}{x - x_n} \tag{4}$$

For the first sum

$$\sum_{m=0}^{n-1} \frac{a^m \left(\cos b^m \pi x - \cos b^m \pi x_n \right)}{x - x_n}$$

note that according to the mean value theorem,

$$\frac{\cos b^m \pi x - \cos b^m \pi x_n}{x - x_n} = -b^m \pi \sin(c) .$$

Thus

$$|\cos b^m \pi x - \cos b^m \pi x_n| = |-b^m \pi \sin c| \, |x - x_n| .$$

But

$$|\sin c| \leq 1 \text{ and } |-b^m \pi \sin c| \leq b^m \pi .$$

Hence,

$$|\cos b^m \pi x - \cos b^m \pi x_n| \leq b^m \pi |x - x_n|$$

and

$$\left| \sum_{m=0}^{n-1} \frac{a^m \cos b^m \pi x - \cos b^m \pi x_n}{x - x_n} \right| \leq \sum_{m=0}^{n-1} a^m b^m \pi .$$

But
$$\sum_{m=0}^{n-1} a^m b^m \pi = \pi \left[1 + ab + (ab)^2 + \ldots + (ab)^{n-1} \right].$$

Now
$$1 + ab + (ab)^2 + \ldots + (ab)^{n-1} = \frac{(ab)^n - 1}{(ab) - 1}$$

(since $(ab)^n - 1 = (ab - 1)[(ab)^{n-1} + (ab)^{n-2} + \ldots + ab + 1]$)

Thus
$$\sum_{m=0}^{n-1} a^m b^m \pi = \pi \frac{(ab)^n - 1}{(ab - 1)} < \frac{\pi (ab)^n}{ab - 1}, \tag{5}$$

assuming that $ab > 1$.

The second summation must be treated differently,
$$b^m \pi x_n = b^{m-n} (b^n \pi x_n) = b^{m-n} \pi (u_n + 1)$$

(since $b^n x_n = u_n + 1$). Since b is odd, b^{m-n} is odd and $b^{m-n} \pi (u_n + 1)$ is an even multiple of π if and only if $u_n + 1$ is even. Thus
$$\cos (b^m \pi x_n) = (-1)^{u_n + 1}$$

($\cos 2k\pi = +1$, $\cos (2k - 1)\pi = -1$ k an integer).

Similarly,
$$\cos b^m \pi x = \cos \{b^{m-n} \pi (u_n + c_n)\},$$

since
$$u_n + c_n = b^n x.$$

But
$$\cos \{b^{m-n} \pi (u_n + c_n)\} = (-1)^{u_n} \cos b^{m-n} \pi c_n.$$

Thus,
$$\frac{\cos b^m \pi x - \cos b^m \pi x_n}{x - x_n} = \frac{(-1)^{u_n} \cos b^{m-n} \pi c_n - (-1)^{u_n + 1}}{x - x_n}$$

$$= (-1)^{u_n} \left\{ \frac{1 - \cos(b^{m-n} \pi c_n)}{x - x_n} \right\}.$$

This shows that

$$\sum_{m=n}^{\infty} a^m \frac{\cos b^m \pi x - \cos b^m \pi x_n}{x - x_n} =$$

$$\sum_{m=n}^{\infty} \frac{a^m \{1 + \cos (b^{m-n} \pi c_n)\}}{|x - x_n|} \quad . \tag{6}$$

But

$$\frac{1}{|x - x_n|} \geq \frac{2}{3} b^n$$

(from (2)). Thus (6) becomes

$$\sum_{m=n}^{\infty} a^m \frac{\{1 + \cos (b^{m-n} \pi c_n)\}}{|x - x_n|} \geq \sum_{m=n}^{\infty} \frac{2}{3} a^m b^n \{1 + \cos (b^{m-n} \pi c_n)\}$$

$$\frac{2}{3} a^n b^n \{1 + \cos \pi c_n\} + \sum_{m=n+1}^{\infty} \frac{2}{3} a^m b^n \{1 + \cos b^{m-n} \pi c_n\}$$

$$\geq \frac{2}{3} a^n b^n \{1 + \cos \pi c_n\} \geq \frac{2}{3} a^n b^n$$

(since $|c_n| \leq \frac{1}{2}$ implies that $|\pi c_n| \leq \frac{1}{2} \pi$

which implies that $\cos \pi c_n \geq 0$). $\tag{7}$

Substituting these two inequalities into (4):

$$\left| \frac{f(x) - f(x_n)}{x - x_n} \right| \geq - \left| \sum_{m=0}^{n-1} \frac{a^m \cos b^m \pi x - \cos b^m \pi x_n}{x - x_n} \right|$$

$$+ \left| \sum_{m=n}^{\infty} a^m \cos b^m \pi x - \cos b^m \pi x_n \right|$$

$$\geq \frac{-\pi (ab)^n}{ab - 1} + \frac{2}{3} a^n b^n \quad \text{(from (5) and (7))}$$

$$= (ab)^n \left[\frac{2}{3} - \frac{\pi}{ab - 1} \right] .$$

This difference quotient grows without bounds as n gets larger (i.e., as x_n approaches x) if $ab > 1$ and $\frac{2}{3} > \frac{\pi}{ab - 1}$. Thus if $ab > 1 + \frac{3\pi}{2}$, f is not differentiable at x, i.e., $f'(x)$ does not exist. Since x was arbitrarily chosen from the domain this shows that

$$f(x) = \sum_{n=0}^{\infty} a^n \cos(b^n \pi x)$$

is nowhere differentiable if $ab > 1 + \frac{3\pi}{2}$. In contrast, if $|ab| < 1$, $f(x)$ is continuously differentiable.

• **PROBLEM 3-35**

Construct an everywhere continuous function that is nowhere differentiable from the following function: $K(x)$ is the distance from x to the nearest integer, $K: R \to R$.

Solution: The simplest example of a function that is continuous everywhere but is not differentiable at a point in its domain is the absolute value function, $f: R \to R$ given by $f(x) = |x|$.

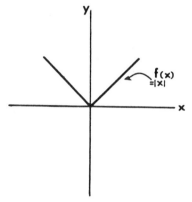

Fig. 1

Note that f is continuous at the origin, since given $\varepsilon > 0$, we can set $\delta = \varepsilon$ so that

$$|x - 0| < \delta \text{ implies } |f(x) - 0| < \varepsilon.$$

But f is not differentiable at $x = 0$. To see this, note that when $x < 0$, $f(x) = -x$ and $f'(x) = -1$ and that when $x > 0$, $f(x) = x$ and $f'(x) = 1$.

Thus

$$\lim_{x \to 0^-} \frac{|f(x+h) - f(x)|}{|h|} = -1$$

and

$$\lim_{x \to 0^+} \frac{|f(x+h) - f(x)|}{|h|} = 1$$

A function is differentiable at a point only if the derivative is unique at that point. Thus $f(x) = |x|$ is not differentiable at $x = 0$.

The same principle (a function being continuous at a point with different values of the derivative depending on the approach to the point) is used to construct everywhere continuous but nowhere differentiable functions.

Note that $K(x)$ has period 1, i.e., $K(x+1) = K(x)$.

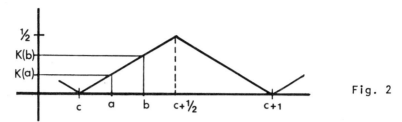

Fig. 2

It can be seen from Fig. 2 that $K(x)$ increases continuously from 0 to ½ and then decreases continuously from ½ to 0. Note that K is not differentiable at $C + ½$ (where C is an integer). Also note that if a,b both lie in either $[C, C + ½]$ or $[C + ½, C + 1]$ then

$$|K(b) - K(a)| = |b - a| .$$

Define $v_i(x) = 10^{-i} K(10^i x)$ for $i = 0,1,2,\ldots$.

Thus,

$$v_0(x) = K(x), \quad v_1(x) = \tfrac{1}{10} K(10x) \quad v_2(x) = \tfrac{1}{100} K(100x)$$

.... . Now $v_1(x)$ shrinks the height of $K(x)$ in Fig. 2 by a factor of 10, $v_2(x)$ by a factor of 100, etc.

Let $H(x) = \sum_{i=0}^{\infty} v_i(x)$, $x \in R$.

Since $0 \le v_i(x) \le 10^{-i}$ for all x, the series $H(x)$ is uniformly convergent. This can be deduced from the comparison test for uniform convergence (Weierstrass M-test). Every $v_i(x)$ is everywhere continuous and $v_i(x+1) = v_i(x)$. Thus, $H(x)$ is everywhere continuous and has period 1.

It remains to show that $H(x)$ is nowhere differentiable. Since H is periodic, let $[a,b] = [0,1]$ and let $d \in [0,1]$. Then

$$H'(d) = \lim_{x \to d} \frac{H(x) - H(d)}{x - d} \tag{1}$$

is the derivative of H at d.

To show that (1) exists nowhere in $[0,1]$, suppose d has the infinite decimal expansion

$$d = .d_1 d_2 d_3 \ldots = \sum_{i=1}^{\infty} \frac{d_i}{10^i}$$

where, as usual, $.c_1 \ldots 999 \ldots$ is written as $.c_1 \ldots 1000 \ldots$

Let x_n be defined near d by

$$x_n = \begin{cases} d + 10^{-n} & \text{if } d_n \neq 4 \text{ or } 9 \\ d - 10^n & \text{if } d_n = 4 \text{ or } 9 \end{cases}$$

If $d_n = \frac{3}{10^n}$, for example, then x_n has

$$\frac{3}{10^n} + \frac{1}{10^n} = \frac{4}{10^n}$$

at the n-th decimal position and hence $10^j x_n$ and $10^j d$ lie in the same half of an interval $[m, m+1]$ when

$$0 \leq j \leq n-1 .$$

If $d_n = \frac{4}{10^n}$, for example, then x_n has $\frac{3}{10^n}$ at the n-th decimal position, so once again $10^j x_n$ and $10^j d$ lie in the same half of an interval $[m, m+1]$ when $0 \leq j \leq n-1$.

This shows that

$$|K(10^j x_n) - K(10^j d)| = |10^j x_n - 10^j d| = 10^{j-n} \quad 0 \leq j \leq n-1$$

and hence

$$|v_j(x_n) - v_j(d)| = 10^{-j} |K(10^j x_n) - K(10^j d)|$$

$$= 10^{-n} = |x_n - d|, \quad 0 \leq j \leq n-1$$

Also for $j \geq n$, $10^j x_n$ and $10^j d$ differ by an integer. By the periodicity of v_j it follows then that

$$|v_j(x_n) - v_j(d)| = 0 \quad \text{for} \quad j \geq n$$

Adding over $j = 0, 1, \ldots, n-1$:

$$\frac{H(x_n) - H(d)}{x_n - d} = \pm \frac{10^{-n}}{10^{-n}} \pm \ldots \pm \frac{10^{-n}}{10^{-n}} \quad (n \text{ terms})$$

The kth term is positive or negative depending on the decimal expansion of d. However, regardless of the signs

$$\pm 1 \pm 1 \pm 1 \ldots \pm 1$$

is even when n is even and odd when n is odd.

Therefore

$$\lim_{x_n \to d} \frac{H(x_n) - H(d)}{x_n - d}$$

does not exist. i.e., $H(x)$ is not differentiable at d. Since d was arbitrarily chosen, $H(x)$ is not differentiable on $[0,1]$. Since $H(x)$ is periodic with period 1, $H(x)$ is nowhere differentiable.

CHAPTER 4

ELEMENTS OF PARTIAL DIFFERENTIATION

The concept of "rate of change" of a dependent variable with respect to an independent variable is the cornerstone of differential calculus. If a function, say the distance z of a projectile above the earth depends only on one variable, e.g., the position p of the projectile on the axis along which it was shot (say the x-axis) then the derivative of $z(x)$ with respect to x is simply

$$z'(x) = \frac{dz}{dx} = \lim_{\Delta x \to 0} \frac{\Delta z}{\Delta x} = \lim_{\Delta x \to 0} \frac{z(x + \Delta x) - z(x)}{\Delta x}.$$

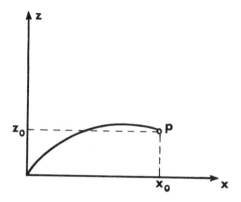

Fig. 1

In other words, $z'(x)$ is the ratio of the change in z to the change in x, where the change is infinitesimally small. Now if there is a breeze blowing perpendicular to the xz-plane, the projectile will not travel along the x-axis and z will depend on both x and y. (Figure 2). Now there are several rates of change which are of interest: 1) at what rate is p moving along the x-axis at a given point (x_0, y_0)?; 2) at what rate is p moving along the y-axis at (x_0, y_0)?; 3) at what rate does p move along its direct path, C (assuming the breeze is constant)?; and 4) if the breeze is not constant and depends on time, t, at what rate does p move along some axis, say the x-axis? Questions 1) and 2) lead to the concept of partial derivatives, $\partial z/\partial x$ and $\partial z/\partial y$. Question 3) leads to the concept of directional derivative, $D_{\vec{C}} z(x,y)$, and question 4) leads to the chain rule for functions of several variables for finding

$$\frac{dz(x(t), y(t))}{dt}.$$

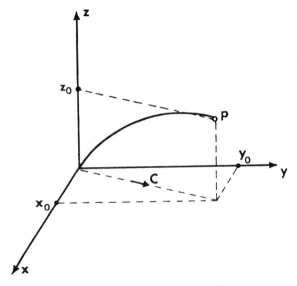

Fig. 2

Another quantity, called the differential of a function, $df(x,y)$, which may have seemed superfluous to the student when it was defined in single-variable calculus as $df(x) = f'(x)dx$ will be found to be of great importance in multi-variable calculus.

PARTIAL DERIVATIVES

● **PROBLEM 4-1**

> Let $u = u(x,y)$ be implicitly defined as a function of x and y by the equation $u + \ln u = xy$. Find
> $$\frac{\partial u}{\partial x}, \frac{\partial u}{\partial y}, \frac{\partial^2 u}{\partial x \partial y} \text{ and } \frac{\partial^2 u}{\partial y \partial x}.$$

Solution: The partial derivative of a function is defined as follows: Let $z = f(x,y)$ be defined in a domain D of the xy-plane and let (x_1, y_1) be a point of D. Then $f(x, y_1)$ is a function depending only on x and if its derivative at the point x_1 exists, it is called the partial derivative of f with respect to x at (x_1, y_1) and is denoted by

$$\frac{\partial f}{\partial x}(x_1, y_1) \text{ or } \left.\frac{\partial z}{\partial x}\right|_{(x_1, y_1)}.$$

If the point (x_1, y_1) is now allowed to vary, one obtains a new function of x and y (wherever the derivative exists) denoted by

$$\frac{\partial f}{\partial x}(x,y) = f_x(x,y) = \frac{\partial z(x,y)}{\partial x}.$$

Apparently, then, $\frac{\partial f}{\partial x}(x,y)$ may be obtained by simply treating y as a constant and differentiating f with respect to its only remaining variable, x. In the case at hand, u is not defined explicitly as a function of x and y but the partial derivatives may be obtained by differentiating both sides of the defining equation with respect to x or y recalling that u depends on both x and y. Thus

$$\frac{\partial u(x,y)}{\partial x} + \frac{\partial \ln[u(x,y)]}{\partial x} = \frac{\partial(xy)}{\partial x} . \qquad (1)$$

Using the chain rule of single variable calculus, (1) becomes

$$\frac{\partial u}{\partial x} + \frac{d[\ln u]}{du} \frac{\partial u}{\partial x} = y \qquad (2)$$

or $u_x + \frac{1}{u} u_x = y$. Therefore, $u_x\left(\frac{u+1}{u}\right) = y$, so that,

$$u_x = \frac{\partial u}{\partial x} = \frac{uy}{u+1} = \frac{u(x,y) \cdot y}{u(x,y) + 1} . \qquad (3)$$

Also, from the defining equation,

$$\frac{\partial u(x,y)}{\partial y} + \frac{\partial \ln[u(x,y)]}{\partial y} = \frac{\partial(xy)}{\partial y} ,$$

and the single variable chain rule can be used again to obtain

$$u_y + \frac{1}{u} u_y = x ,$$

so that

$$u_y = \frac{\partial u}{\partial y} = \frac{ux}{u+1} = \frac{u(x,y) \cdot x}{u(x,y) + 1} . \qquad (4)$$

The second partial derivative $\partial^2 u/\partial x \partial y$ is defined as $\partial(u_y)/\partial x$; that is, the partial derivative of the new function $\partial u/\partial y$ with respect to x. Thus, from (4), one obtains

$$\frac{\partial^2 u}{\partial x \partial y} = \frac{\partial}{\partial x}\left[\frac{ux}{u+1}\right] = \frac{u}{u+1} + x \frac{\partial}{\partial x}\left[\frac{u(x,y)}{u(x,y)+1}\right]$$

$$= \frac{u}{u+1} + x \frac{d}{du}\left[\frac{u}{u+1}\right] \cdot u_x$$

$$= \frac{u}{u+1} + \frac{x}{(u+1)^2} \cdot u_x = \frac{u}{u+1} + \frac{uxy}{(u+1)^3} . \qquad (5)$$

Similarly,

$$\frac{\partial^2 u}{\partial y \partial x} = \frac{\partial}{\partial y}\left[\frac{uy}{u+1}\right] = \frac{u}{u+1} + y \frac{\partial}{\partial y}\left[\frac{u(x,y)}{u(x,y)+1}\right]$$

$$= \frac{u}{u+1} + y \frac{d}{du}\left[\frac{u}{u+1}\right] \cdot u_y$$

$$= \frac{u}{u+1} + \frac{y}{(u+1)^2} \cdot u_y = \frac{u}{u+1} + \frac{uxy}{(u+1)^3} \cdot \qquad (6)$$

It can be seen from (5) and (6) that $\frac{\partial^2 u}{\partial x \partial y} = \frac{\partial^2 u}{\partial y \partial x}$. This is a relation that can be proved to be true for all continuous functions $u(x,y)$.

● **PROBLEM 4-2**

Let $u(x,y)$ and $v(x,y)$ be defined as functions of x and y by the equations

$$u \cos v - x = 0, \qquad (1)$$

$$u \sin v - y = 0. \qquad (2)$$

Find

$$\frac{\partial u}{\partial x} \quad \text{and} \quad \frac{\partial v}{\partial x}.$$

Solution: Two methods will be used.
Method 1: The derivatives can be calculated directly if u and v can be found explicitly in terms of x,y. Squaring (1) and (2) yields

$$u^2 \cos^2 v = x^2, \quad u^2 \sin^2 v = y^2.$$

Adding these two equations gives

$$u^2 \cos^2 v + u^2 \sin^2 v = x^2 + y^2 \quad \text{or} \quad u^2(\cos^2 v + \sin^2 v) = x^2 + y^2.$$

Thus

$$u^2 = x^2 + y^2$$

or

$$u = \pm \sqrt{x^2 + y^2}. \qquad (3)$$

Now substitute this value for u into (1) and (2) to obtain

$$\cos v = \frac{x}{\pm \sqrt{x^2 + y^2}}, \quad \sin v = \frac{y}{\pm \sqrt{x^2 + y^2}}$$

so that

$$\tan v = \frac{y}{x} \qquad (4)$$

if $x \neq 0$. Note that v has an infinite number of possible values differing by multiples of π and that u has two distinct values for each x,y (differing only in sign). Thus, from (3),

$$\frac{\partial u}{\partial x} = \frac{x}{\pm \sqrt{x^2 + y^2}} \qquad (5)$$

and from (4)

$$\frac{\partial v}{\partial x} = \frac{\partial}{\partial x} \tan^{-1}\left(\frac{y}{x}\right) = \frac{d \tan^{-1}(y/x)}{d(y/x)} \cdot \frac{\partial(y/x)}{\partial x}$$

$$= \frac{1}{1+(y/x)^2} \cdot \frac{-y}{x^2} = \frac{-y}{x^2+y^2}. \qquad (6)$$

This equation is only valid on a specified branch of $v(x,y) = \tan^{-1}(y/x)$.
Method 2: A second method is to differentiate each side of equations (1) and (2), keeping in mind that u and v are each functions of x and y. Also, we must at this point simply assume that these equations do implicitly define certain functions, and that these functions do indeed have partial derivatives. The theoretical con-

siderations raised by this method are discussed in connection with the implicit function theorem, which will be discussed in the next chapter. Now, assuming everything is valid, differentiate (1) and (2) to yield

$$\cos v \frac{\partial u}{\partial x} - u \sin v \frac{\partial v}{\partial x} - 1 = 0,$$

$$\sin v \frac{\partial u}{\partial x} + u \cos v \frac{\partial v}{\partial x} - 0 = 0.$$

These two linear equations in $\frac{\partial u}{\partial x}$ and $\frac{\partial v}{\partial x}$ may be solved by Cramer's rule, i.e.,

$$\frac{\partial v}{\partial x} = \frac{\begin{vmatrix} \cos v & 1 \\ \sin v & 0 \end{vmatrix}}{\begin{vmatrix} \cos v & -u \sin v \\ \sin v & u \cos v \end{vmatrix}} = \frac{-\sin v}{u \cos^2 v + u \sin^2 v}$$

or

$$\frac{\partial v}{\partial x} = \frac{-\sin v}{u}. \qquad (7)$$

Similarly,

$$\frac{\partial u}{\partial x} = \frac{\begin{vmatrix} 1 & -u \sin v \\ 0 & u \cos v \end{vmatrix}}{\begin{vmatrix} \cos v & -u \sin v \\ \sin v & u \cos v \end{vmatrix}} = \frac{u \cos v}{u \cos^2 v + u \sin^2 v}$$

or

$$\frac{\partial u}{\partial x} = \cos v. \qquad (8)$$

Finally, from (1) and (2) it is seen that

$$\cos v = \frac{x}{u} = \frac{x}{\pm\sqrt{x^2 + y^2}} = \frac{\partial u}{\partial x},$$

$$\frac{-\sin v}{u} = \frac{-y}{u^2} = \frac{-y}{x^2 + y^2} = \frac{\partial v}{\partial x};$$

which brings (7) and (8) into agreement with the answers obtained by Method 1.

● **PROBLEM 4-3**

Define the partial derivative of a real valued function, f. Then compute $\partial f/\partial x$ and $\partial f/\partial y$ where

a) $f(x,y) = xy$, b) $f(x,y) = \cos[x(1+y)]$, c) $f(x,y) = x^y$.

<u>Solution</u>: Let f be a real-valued function defined in a neighborhood of

$$\vec{x}_0 = \begin{pmatrix} x_{1_0} \\ x_{2_0} \\ \vdots \\ x_{n_0} \end{pmatrix}$$

in R^n, i.e., $f: R^n \to R$. Then the partial derivative of f with respect to x_i at the point $\vec{x}_0$ is defined as

$$\frac{\partial f}{\partial x_i}(\vec{x}_0) = \lim_{t \to 0} \frac{f(x_{1_0}, \ldots, x_{i_0} + t, \ldots, x_{n_0}) - f(x_{1_0}, \ldots, x_{n_0})}{t}. \quad (1)$$

Thus, the partial derivative is computed simply by considering all but the relevant variable as constant.

a) $\frac{\partial f}{\partial x}(x,y) = \frac{\partial (xy)}{\partial x} = y \cdot \frac{dx}{dx} = y$,

$\frac{\partial f}{\partial y}(x,y) = \frac{\partial (xy)}{\partial y} = x \cdot \frac{dy}{dy} = x$.

b) $\frac{\partial f}{\partial x}(x,y) = \frac{\partial \{\cos[x(1+y)]\}}{\partial x} = -\sin[x(1+y)] \frac{\partial [x(1+y)]}{\partial x}$

$= -(1+y) \sin[x(1+y)]$,

$\frac{\partial f}{\partial y}(x,y) = \frac{\partial \{\cos[x(1+y)]\}}{\partial y} = -\sin[x(1+y)] \frac{\partial [x(1+y)]}{\partial y}$

$= -x \sin[x(1+y)]$.

c) $\frac{\partial f}{\partial x}(x,y) = \frac{\partial (x^y)}{\partial x} = yx^{y-1}$.

Here, to find $\partial f/\partial y$, first take the natural logarithm of $f(x,y) = x^y$, that is,

$$\ln f = \ln(x^y) = y \ln x .$$

Now take the partial of both sides of this equation with respect to y to yield

$$\frac{1}{f} \frac{\partial f}{\partial y} = \ln x \quad \text{or} \quad \frac{\partial f}{\partial y} = f \ln x = x^y \ln x .$$

DIFFERENTIALS AND THE JACOBIAN

• **PROBLEM 4-4**

Define the terms differentiable and differential for a function $f: R^2 \to R$ of the variables x and y. Then find the differentials, df, of the functions
a) $f(x,y) = x^2 - y^2$
b) $f(x,y,z) = xy/z$.

Solution: A real-valued function f of two variables x,y is said to be differentiable at (x_0, y_0) if f is defined in some neighborhood of (x_0, y_0) and if there exist numbers A,B such that

$$\lim_{(dx,dy) \to (0,0)} \frac{|\Delta f(x_0, y_0) - (A \, dx + B \, dy)|}{\sqrt{dx^2 + dy^2}} = 0 \quad (1)$$

where $\Delta f(x_0, y_0) = f(x_0 + dx, y_0 + dy) - f(x_0, y_0)$. This definition is easily extended to vector-valued functions of n variables $\vec{f}: R^n \to R^m$ by changing the norms in (1), i.e.,

$$\lim_{(dx_1,dx_2,\ldots,dx_n) \to (0,0,\ldots,0)} \frac{\|\Delta \vec{f} - (\vec{A}_1 dx_1 + \ldots + \vec{A}_n dx_n)\|}{(dx_1^2 + dx_2^2 + \ldots + dx_n^2)^{\frac{1}{2}}} = 0,$$

where $\vec{A}_i$, $1 \leq i \leq n$, and $\Delta \vec{f}$ are now both vectors and $\|\cdot\|$ is the norm in R^m.

In any case, the quantity subtracted from Δf is called the differential of f at (x_0, y_0) if the quantities A, B or $\vec{A}_1, \ldots, \vec{A}_n$ exist. It is easily proven that if these numbers do exist, they are simply related to the partial derivatives of f at the point in question. In the case of two variables (see equation (1)) the relations are

$$A = \frac{\partial f}{\partial x}(x_0, y_0), \quad B = \frac{\partial f}{\partial y}(x_0, y_0), \tag{2}$$

so that for a general point (x, y), the differential is usually written

$$df = \frac{\partial f}{\partial x} dx + \frac{\partial f}{\partial y} dy. \tag{3}$$

a) $\frac{\partial f}{\partial x} = 2x$, $\frac{\partial f}{\partial y} = -2y$, hence

$$df = \frac{\partial f}{\partial x} dx + \frac{\partial f}{\partial y} dy = 2x\,dx - 2y\,dy.$$

b) There are three independent variables here, however, the following extension of (3) is obvious:

$$df = \frac{\partial f}{\partial x} dx + \frac{\partial f}{\partial y} dy + \frac{\partial f}{\partial z} dz \tag{4}$$

and since

$$\frac{\partial f}{\partial x} = \frac{y}{z}, \quad \frac{\partial f}{\partial y} = \frac{x}{z}, \quad \frac{\partial f}{\partial z} = \frac{-xy}{z^2}$$

equation (4) gives

$$df = \frac{y}{z} dx + \frac{x}{z} dy - \frac{xy}{z^2} dz.$$

• **PROBLEM 4-5**

Find the differential of the vector function $\vec{f}: R^3 \to R^3$ defined by

$$\vec{f}(x_1, x_2, x_3) = \begin{pmatrix} f_1(x_1, x_2, x_3) \\ f_2(x_1, x_2, x_3) \\ f_3(x_1, x_2, x_3) \end{pmatrix} = \begin{pmatrix} x_1^2 + x_2^2 - x_3^2 \\ x_1^2 - x_2^2 + x_3^2 \\ -x_1^2 + x_2^2 + x_3^2 \end{pmatrix}. \tag{1}$$

Evaluate $\vec{f}(2,1,1)$ and then find $\vec{f}(2.01, 1.03, 1.02)$ exactly and approximately using differentials.

Solution: The differential $d\vec{f}$ of an m-dimensional vector function $\vec{f}$ of n variables $(x_1, \ldots, x_n)$ is given by

$$d\vec{f} = \widetilde{J}_f \, d\vec{x} \tag{2}$$

168

where $d\vec{x} = \begin{pmatrix} dx_1 \\ dx_2 \\ \vdots \\ dx_n \end{pmatrix}$, $d\vec{f} = \begin{pmatrix} df_1 \\ df_2 \\ \vdots \\ df_m \end{pmatrix}$ and $\tilde{J}_f$ is the Jacobian

matrix of $\vec{f}$ given by

$$\tilde{J}_f = \begin{pmatrix} \frac{\partial f_1}{\partial x_1} & \frac{\partial f_1}{\partial x_2} & \cdots & \frac{\partial f_1}{\partial x_n} \\ \frac{\partial f_2}{\partial x_1} & & & \vdots \\ \vdots & & & \vdots \\ \frac{\partial f_m}{\partial x_1} & \cdots & \cdots & \frac{\partial f_m}{\partial x_n} \end{pmatrix} \quad (3)$$

In the case at hand, $m = n = 3$ and $\tilde{J}_f$ may be computed using equation (1). Thus, taking the derivatives indicated in (3),

$$\tilde{J}_f = \begin{pmatrix} 2x_1 & 2x_2 & -2x_3 \\ 2x_1 & -2x_2 & 2x_3 \\ -2x_1 & 2x_2 & 2x_3 \end{pmatrix}$$

and equation (2) gives

$$\begin{pmatrix} df_1 \\ df_2 \\ df_3 \end{pmatrix} = \begin{pmatrix} 2x_1 & 2x_2 & -2x_3 \\ 2x_1 & -2x_2 & 2x_3 \\ -2x_1 & 2x_2 & 2x_3 \end{pmatrix} \begin{pmatrix} dx_1 \\ dx_2 \\ dx_3 \end{pmatrix} \quad (4)$$

or

$$d\vec{f} = \begin{pmatrix} 2x_1 dx_1 + 2x_2 dx_2 - 2x_3 dx_3 \\ 2x_1 dx_1 - 2x_2 dx_2 + 2x_3 dx_3 \\ -2x_1 dx_1 + 2x_2 dx_2 + 2x_3 dx_3 \end{pmatrix} \quad (5)$$

From (1) it is found that

$$\vec{f}(2,1,1) = \begin{pmatrix} f_1(2,1,1) \\ f_2(2,1,1) \\ f_3(2,1,1) \end{pmatrix} = \begin{pmatrix} 2^2 + 1^2 - 1^2 \\ 2^2 - 1^2 + 1^2 \\ -2^2 + 1^2 + 1^2 \end{pmatrix} = \begin{pmatrix} 4 \\ 4 \\ -2 \end{pmatrix} \quad (6)$$

and similarly $\vec{f}(2.01, 1.03, 1.02) = (4.0606, 4.0196, -1.9388)$. Now recall that the differentials df_1, df_2, df_3 of the components of

$\vec{f}$ are linear approximations at each point $\vec{x} = (x_1, x_2, x_3)$ to

$$\Delta f_i = f_i(x_1+\Delta x_1, x_2+\Delta x_2, x_3+\Delta x_3) - f(x_1, x_2, x_3)$$

where $i = 1, 2, 3$, respectively. In fact, the important point about df_i is that it is the only linear function of (dx_1, dx_2, dx_3) which arbitrarily approximates Δf_i. By "arbitrarily" it is meant that the error $|\Delta f_i - df_i| = \epsilon \|(dx_1, dx_2, dx_3)\|$ can be made arbitrarily small by making $\|(dx_1, dx_2, dx_3)\|$ small enough. Here ϵ is a function of a real-variable t such that $\epsilon(t) \to 0$ as $t \to 0$ and $\|\cdot\|$ is the norm in R^3. Thus, the following approximation is valid

$$\vec{f}(2.01, 1.03, 1.02) \simeq \vec{f}(2,1,1) + d\vec{f}(\vec{x}_0, d\vec{x}_0) \ . \tag{7}$$

Here $\vec{x}_0 = (2,1,1)$ and $d\vec{x}_0 = (.01, .03, .02)$ so that

$$d\vec{f}(2,1,1,.01,.03,.02) = \tilde{J}_f(2,1,1) d\vec{x}_0$$

$$= \begin{pmatrix} 4 & 2 & -2 \\ 4 & -2 & 2 \\ -4 & 2 & 2 \end{pmatrix} \begin{pmatrix} .01 \\ .03 \\ .02 \end{pmatrix} = \begin{pmatrix} .06 \\ .02 \\ .06 \end{pmatrix}$$

Therefore, from (7) and (6),

$$\vec{f}(2.01, 1.03, 1.02) \simeq (4, 4, -2) + (.06, .02, .06)$$

which means that

$$\vec{f}(2.01, 1.03, 1.02) \simeq (4.06, 4.02, -1.94).$$

Note how close this is to the exact value computed above (just after equation (6)).

● **PROBLEM 4-6**

Find the Jacobian matrix and the differential of the vector function $\vec{f}: R^3 \to R^3$ defined by

$$\vec{f}(x, y, z) = \begin{pmatrix} u(x,y,z) \\ v(x,y,z) \\ w(x,y,z) \end{pmatrix} = \begin{pmatrix} x^2 + y - z \\ xyz^2 \\ 2xy - y^2 z \end{pmatrix} \ . \tag{1}$$

Solution: The differential $d\vec{f}$ of an m-dimensional vector function $\vec{f}$ of n variables $(x_1, x_2, \ldots, x_n)$ is given by

$$d\vec{f} = \tilde{J}_f \, d\vec{x} \tag{2}$$

where

$$d\vec{x} = \begin{pmatrix} dx_1 \\ dx_2 \\ \vdots \\ dx_n \end{pmatrix}, \quad d\vec{f} = \begin{pmatrix} df_1 \\ df_2 \\ \vdots \\ df_m \end{pmatrix} \quad \text{and} \quad \tilde{J}_f$$

is the Jacobian matrix of $\vec{f}$ and is given by

$$\tilde{J}_f = \begin{pmatrix} \frac{\partial f_1}{\partial x_1} & \frac{\partial f_1}{\partial x_2} & \cdots & \frac{\partial f_1}{\partial x_n} \\ \frac{\partial f_2}{\partial x_1} & & & \vdots \\ \vdots & & & \vdots \\ \frac{\partial f_m}{\partial x_1} & & \cdots & \frac{\partial f_m}{\partial x_n} \end{pmatrix} \quad . \tag{3}$$

In the case at hand $m = n = 3$, $f_1 = u$, $f_2 = v$, $f_3 = w$ and $\tilde{J}_f$ may be computed using equation (1). Thus, taking the derivatives indicated in (3),

$$\tilde{J}_f = \begin{pmatrix} 2x & 1 & -1 \\ yz^2 & xz^2 & 2xyz \\ 2y & 2x-2yz & -y^2 \end{pmatrix} \quad . \tag{4}$$

Finally, using equations (2) and (4) to obtain the differential of $\vec{f}$, we have,

$$d\vec{f} = \begin{pmatrix} du \\ dv \\ dw \end{pmatrix} = \begin{pmatrix} 2x & 1 & -1 \\ yz^2 & xz^2 & 2xyz \\ 2y & 2x-2yz & -y^2 \end{pmatrix} \begin{pmatrix} dx \\ dy \\ dz \end{pmatrix}$$

or

$$d\vec{f} = \begin{pmatrix} 2xdx + dy - dz \\ yz^2 dx + xz^2 dy + 2xyzdz \\ 2ydx + (2x - 2yz)dy - y^2 dz \end{pmatrix} \quad .$$

● **PROBLEM 4-7**

Find the Jacobian matrix and the differential of the vector function $\vec{f}: R^2 \to R^3$ defined by

$$\vec{f}(u,v) = \begin{pmatrix} x(u,v) \\ y(u,v) \\ z(u,v) \end{pmatrix} = \begin{pmatrix} \cos u \cos v \\ \cos u \sin v \\ \sin u \end{pmatrix} \quad . \tag{1}$$

Solution: The Jacobian matrix of $\vec{f}$ is given by

$$\tilde{J}_f = \begin{pmatrix} \frac{\partial x}{\partial u} & \frac{\partial x}{\partial v} \\ \frac{\partial y}{\partial u} & \frac{\partial y}{\partial v} \\ \frac{\partial z}{\partial u} & \frac{\partial z}{\partial v} \end{pmatrix} \quad . \tag{2}$$

If these partial derivatives are computed it is found that

$$\tilde{J}_f = \begin{pmatrix} -\sin u \cos v & -\cos u \sin v \\ -\sin u \sin v & \cos u \cos v \\ \cos u & 0 \end{pmatrix}. \quad (3)$$

The differential of $\vec{f}$ is defined to be

$$d\vec{f} = \tilde{J}_f \, d\vec{u} \quad (4)$$

so that in this case, with $\vec{u} = (u,v)$, we find

$$d\vec{f} = \begin{pmatrix} -\sin u \cos v & -\cos u \sin v \\ -\sin u \sin v & \cos u \cos v \\ \cos u & 0 \end{pmatrix} \begin{pmatrix} du \\ dv \end{pmatrix}$$

or

$$d\vec{f} = \begin{pmatrix} -\sin u \cos v \, du - \cos u \sin v \, dv \\ -\sin u \sin v \, du + \cos u \cos v \, dv \\ \cos u \, du \end{pmatrix}.$$

● **PROBLEM 4-8**

Compute the Jacobian determinant of the transformation from
(a) cylindrical coordinates to Cartesian coordinates,
(b) spherical coordinates to Cartesian coordinates.

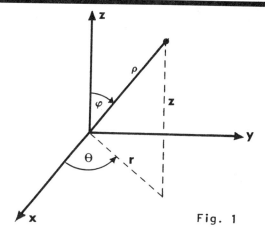

Fig. 1

Solution: (a) Examination of Figure 1 shows that the cylindrical coordinates (r,θ,z) are related to the Cartesian coordinates x,y,z by the equations $x = r \cos \theta$, $y = r \sin \theta$, $z = z$. Thus, the assignment of Cartesian coordinates to points represented by the cylindrical coordinates (r,θ,z) may be viewed as a transformation or vector function $\vec{f}: R^3 \to R^3$ where $\vec{f}$ is given by

$$\vec{f}(r,\theta,z) = \begin{pmatrix} x(r,\theta,z) \\ y(r,\theta,z) \\ z(r,\theta,z) \end{pmatrix} = \begin{pmatrix} r \cos \theta \\ r \sin \theta \\ z \end{pmatrix}. \quad (1)$$

The Jacobian matrix of such a transformation is given by

$$\tilde{J}_f = \begin{pmatrix} \frac{\partial x}{\partial r} & \frac{\partial x}{\partial \theta} & \frac{\partial x}{\partial z} \\ \frac{\partial y}{\partial r} & \frac{\partial y}{\partial \theta} & \frac{\partial y}{\partial z} \\ \frac{\partial z}{\partial r} & \frac{\partial z}{\partial \theta} & \frac{\partial z}{\partial z} \end{pmatrix}. \qquad (2)$$

Therefore, using (1) and taking the derivatives indicated in (2), we have

$$\tilde{J}_f = \begin{pmatrix} \cos\theta & -r\sin\theta & 0 \\ \sin\theta & r\cos\theta & 0 \\ 0 & 0 & 1 \end{pmatrix}. \qquad (3)$$

Finally, the Jacobian determinant or simply the Jacobian, which is denoted by $\frac{\partial(x,y,z)}{\partial(r,\theta,z)}$, is defined by $\det \tilde{J}_f = |\tilde{J}_f|$. Hence, from (3) we find that

$$\frac{\partial(x,y,z)}{\partial(r,\theta,z)} = |\tilde{J}_f| = \begin{vmatrix} \cos\theta & -r\sin\theta & 0 \\ \sin\theta & r\cos\theta & 0 \\ 0 & 0 & 1 \end{vmatrix}.$$

Expanding about the last row gives

$$\frac{\partial(x,y,z)}{\partial(r,\theta,z)} = 1 \cdot (r\cos^2\theta + r\sin^2\theta) = r. \qquad (4)$$

(b) In this case the transformation may be represented by the vector function $\vec{g}: R^3 \to R^3$ given by

$$\vec{g}(\rho,\theta,\varphi) = \begin{pmatrix} x(\rho,\theta,\varphi) \\ y(\rho,\theta,\varphi) \\ z(\rho,\theta,\varphi) \end{pmatrix} = \begin{pmatrix} \rho\sin\varphi\cos\theta \\ \rho\sin\varphi\sin\theta \\ \rho\cos\varphi \end{pmatrix}. \qquad (5)$$

Hence, the Jacobian determinant is given by

$$\frac{\partial(x,y,z)}{\partial(\rho,\theta,\varphi)} = |\tilde{J}_g| = \begin{vmatrix} \frac{\partial x}{\partial \rho} & \frac{\partial x}{\partial \theta} & \frac{\partial x}{\partial \varphi} \\ \frac{\partial y}{\partial \rho} & \frac{\partial y}{\partial \theta} & \frac{\partial y}{\partial \varphi} \\ \frac{\partial z}{\partial \rho} & \frac{\partial z}{\partial \theta} & \frac{\partial z}{\partial \varphi} \end{vmatrix}$$

$$= \begin{vmatrix} \sin\varphi\cos\theta & -\rho\sin\varphi\sin\theta & \rho\cos\varphi\cos\theta \\ \sin\varphi\sin\theta & \rho\sin\varphi\cos\theta & \rho\cos\varphi\sin\theta \\ \cos\varphi & 0 & -\rho\sin\varphi \end{vmatrix}$$

Expanding this determinant about the last row gives

$$\frac{\partial(x,y,z)}{\partial(\rho,\theta,\varphi)} = \cos\varphi(-\rho^2\sin^2\theta\sin\varphi\cos\varphi - \rho^2\sin\varphi\cos\varphi\cos^2\theta)$$
$$- \rho\sin\varphi(\rho\sin^2\varphi\cos^2\theta + \rho\sin^2\varphi\sin^2\theta)$$
$$= -\rho^2\sin^2\theta\cos^2\varphi\sin\varphi - \rho^2\cos^2\varphi\sin\varphi\cos^2\theta$$

$$-\rho^2 \sin^2\varphi \cos^2\theta \sin\varphi - \rho^2 \sin^2\varphi \sin^2\theta \sin\varphi$$
$$= -\rho^2 [\cos^2\varphi(\sin^2\theta + \cos^2\theta)\sin\varphi$$
$$\quad + \sin^2\varphi(\cos^2\theta + \sin^2\theta)\sin\varphi]$$
$$= -\rho^2 \sin\varphi(\cos^2\varphi + \sin^2\varphi)$$
$$= -\rho^2 \sin\varphi .$$

● **PROBLEM 4-9**

Let $\vec{u}(x,y) = \begin{pmatrix} u(x,y) \\ v(x,y) \end{pmatrix} = \begin{pmatrix} x+y \\ x-y \end{pmatrix}$ and $\vec{\xi}(u,v) = \begin{pmatrix} \xi(u,v) \\ \eta(u,v) \end{pmatrix} = \begin{pmatrix} uv \\ u+v \end{pmatrix}$ so that $\vec{\xi}(u(x,y),v(x,y)) = \begin{pmatrix} x^2 - y^2 \\ 2x \end{pmatrix}$. Verify the theorem which states that if T_1 and T_2 are successive transformations with the resultant transformation T_3, then the product of the Jacobians of T_1 and T_2 is equal to the Jacobian of T_3.

Solution: First compute the Jacobian of $\vec{u}(x,y)$ (that is, the determinant of the Jacobian matrix):

$$\frac{\partial(u,v)}{\partial(x,y)} = \begin{vmatrix} \frac{\partial u}{\partial x} & \frac{\partial u}{\partial y} \\ \frac{\partial v}{\partial x} & \frac{\partial v}{\partial y} \end{vmatrix} = \begin{vmatrix} 1 & 1 \\ 1 & -1 \end{vmatrix}$$

$$= -1 - 1 = -2 . \tag{1}$$

Now, the Jacobian of the transformation $\vec{\xi}(u,v)$ is

$$\frac{\partial(\xi,\eta)}{\partial(u,v)} = \begin{vmatrix} \frac{\partial \xi}{\partial u} & \frac{\partial \xi}{\partial v} \\ \frac{\partial \eta}{\partial u} & \frac{\partial \eta}{\partial v} \end{vmatrix} = \begin{vmatrix} v & u \\ 1 & 1 \end{vmatrix}$$

$$= v - u . \tag{2}$$

Finally, the Jacobian of the composite transformation $\vec{\xi}(x,y) = \vec{\xi}(u(x,y),v(x,y))$ is

$$\frac{\partial(\xi,\eta)}{\partial(x,y)} = \begin{vmatrix} \frac{\partial \xi}{\partial x} & \frac{\partial \xi}{\partial y} \\ \frac{\partial \eta}{\partial x} & \frac{\partial \eta}{\partial y} \end{vmatrix} = \begin{vmatrix} 2x & -2y \\ 2 & 0 \end{vmatrix}$$

$$= 4y . \tag{3}$$

Now observe that

$$v - u = v(x,y) - u(x,y) = (x-y) - (x+y) = -2y$$

so that from (2)

$$\frac{\partial(\xi,\eta)}{\partial(u,v)} = -2y . \tag{4}$$

Combining (1), (3), and (4) yields

$$\frac{\partial(\xi,\eta)}{\partial(u,v)} \cdot \frac{\partial(u,v)}{\partial(x,y)} = (-2y)(-2) = 4y = \frac{\partial(\xi,\eta)}{\partial(x,y)} . \tag{5}$$

Thus, the theorem is verified for these transformations.

THE CHAIN RULE

• **PROBLEM 4-10**

Let $F(x,y)$ be differentiable in x and y and introduce polar coordinates r, θ by writing $x = r \cos \theta$, $y = r \sin \theta$. Find $\partial F/\partial r$ and $\partial F/\partial \theta$ in terms of $\partial F/\partial x$ and $\partial F/\partial y$.

Solution: The chain rule for composite functions states that if $F = F(x,y)$ and $x = g(u,v)$, $y = h(u,v)$, then at the points where all of the derivatives exist,

$$\frac{\partial F}{\partial u} = \frac{\partial F}{\partial x}\frac{\partial x}{\partial u} + \frac{\partial F}{\partial y}\frac{\partial y}{\partial u}, \tag{1}$$

$$\frac{\partial F}{\partial v} = \frac{\partial F}{\partial x}\frac{\partial x}{\partial v} + \frac{\partial F}{\partial y}\frac{\partial y}{\partial v}. \tag{2}$$

In the case at hand, $x = x(r,\theta) = r \cos \theta$ and $y = y(r,\theta) = r \sin \theta$, so that

$$\frac{\partial x}{\partial r} = \cos \theta, \quad \frac{\partial x}{\partial \theta} = -r \sin \theta, \quad \frac{\partial y}{\partial r} = \sin \theta, \quad \frac{\partial y}{\partial \theta} = r \cos \theta. \tag{3}$$

Thus, replacing u and v with r and θ in equations (1) and (2) and using the results of (3) gives

$$\frac{\partial F}{\partial r} = \frac{\partial F}{\partial x} \cos \theta + \frac{\partial F}{\partial y} \sin \theta \tag{4}$$

$$\frac{\partial F}{\partial \theta} = -\frac{\partial F}{\partial x} r \sin \theta + \frac{\partial F}{\partial y} r \cos \theta. \tag{5}$$

• **PROBLEM 4-11**

Let $\vec{Y}(\vec{U}) = \begin{pmatrix} y_1(u_1,u_2,u_3) \\ y_2(u_1,u_2,u_3) \end{pmatrix} = \begin{pmatrix} u_1 u_2 - u_1 u_3 \\ u_1 u_3 + u_2^2 \end{pmatrix}$ and

$\vec{U}(\vec{X}) = \begin{pmatrix} u_1(x_1,x_2) \\ u_2(x_1,x_2) \\ u_3(x_1,x_2) \end{pmatrix} = \begin{pmatrix} x_1 \cos x_2 + (x_1 - x_2)^2 \\ x_1 \sin x_2 + x_1 x_2 \\ x_1^2 - x_1 x_2 + x_2^2 \end{pmatrix}$

Find

$\left.\dfrac{\partial y_1}{\partial x_1}\right|_{(1,0)}$, $\left.\dfrac{\partial y_1}{\partial x_2}\right|_{(1,0)}$, $\left.\dfrac{\partial y_2}{\partial x_1}\right|_{(1,0)}$, $\left.\dfrac{\partial y_2}{\partial x_2}\right|_{(1,0)}$

Solution: The general chain rule for vector functions of several variables states that

$$\tilde{J}_{\vec{Y}(\vec{X})} = \left(\tilde{J}_{\vec{Y}(\vec{U})}\right)\left(\tilde{J}_{\vec{U}(\vec{X})}\right) \tag{1}$$

where $\tilde{J}$ is the Jacobian matrix of the indicated transformation. Therefore, to find the desired quantities the Jacobian matrices of

(1) will be computed at the point $(x_1, x_2) = (1,0)$. However,

$$\mathcal{J}_{\vec{Y}(\vec{U})} = \begin{pmatrix} \frac{\partial y_1}{\partial u_1} & \frac{\partial y_1}{\partial u_2} & \frac{\partial y_1}{\partial u_3} \\ \frac{\partial y_2}{\partial u_1} & \frac{\partial y_2}{\partial u_2} & \frac{\partial y_2}{\partial u_3} \end{pmatrix} = \begin{pmatrix} u_2 - u_3 & u_1 & -u_1 \\ u_3 & 2u_2 & u_1 \end{pmatrix}, \quad (2)$$

$$\mathcal{J}_{\vec{U}(\vec{X})} = \begin{pmatrix} \frac{\partial u_1}{\partial x_1} & \frac{\partial u_1}{\partial x_2} \\ \frac{\partial u_2}{\partial x_1} & \frac{\partial u_2}{\partial x_2} \\ \frac{\partial u_3}{\partial x_1} & \frac{\partial u_3}{\partial x_2} \end{pmatrix}$$

$$= \begin{pmatrix} \cos x_2 + 2(x_1 - x_2) & -x_1 \sin x_2 - 2(x_1 - x_2) \\ \sin x_2 + x_2 & x_1 \cos x_2 + x_1 \\ 2x_1 - x_2 & 2x_2 - x_1 \end{pmatrix} \quad (3)$$

Now multiplying the matrix of (2) by that of (3) gives, by equation (1)

$$\mathcal{J}_{\vec{Y}(\vec{X})} = \begin{pmatrix} \frac{\partial y_1}{\partial x_1} & \frac{\partial y_1}{\partial x_2} \\ \frac{\partial y_2}{\partial x_1} & \frac{\partial y_2}{\partial x_2} \end{pmatrix}.$$

For example,

$$\frac{\partial y_1}{\partial x_1} = (u_2 - u_3)[\cos x_2 + 2(x_1 - x_2)] + u_1(\sin x_2 + x_2) - u_1(2x_1 - x_2).$$

To compute $\mathcal{J}_{\vec{Y}(\vec{X})}$ at a particular point, however, it is more convenient to first find $\mathcal{J}_{\vec{Y}(\vec{U})}$ and $\mathcal{J}_{\vec{U}(\vec{X})}$ at this point and then do the matrix multiplication. Thus, at $(x_1, x_2) = (1, 0)$, note that $(u_1, u_2, u_3) = (2, 0, 1)$ from the defining equation of $\vec{U}(\vec{X})$ so that

$$\mathcal{J}_{\vec{Y}(\vec{U})}\bigg|_{\substack{u_1=2 \\ u_2=0 \\ u_3=1}} = \begin{pmatrix} -1 & 2 & -2 \\ 1 & 0 & 2 \end{pmatrix},$$

$$\mathcal{J}_{\vec{U}(\vec{X})}\bigg|_{\substack{x_1=1\\x_2=0}} = \begin{pmatrix} 3 & -2 \\ 0 & 2 \\ 2 & -1 \end{pmatrix}$$

so that

$$\mathcal{J}_{\vec{Y}(\vec{X})}\bigg|_{\substack{x_1=1\\x_2=0}} = \mathcal{J}_{\vec{Y}(\vec{U})}\bigg|_{\substack{u_1=2\\u_2=0\\u_3=1}} \times \mathcal{J}_{\vec{U}(\vec{X})}\bigg|_{\substack{x_1=1\\x_2=0}}$$

$$= \begin{pmatrix} -1 & 2 & -2 \\ 1 & 0 & 2 \end{pmatrix} \begin{pmatrix} 3 & -2 \\ 0 & 2 \\ 2 & -1 \end{pmatrix}$$

$$= \begin{pmatrix} -7 & 8 \\ 7 & -4 \end{pmatrix} \qquad (4)$$

But
$$\mathcal{J}_{\vec{Y}(\vec{X})}\bigg|_{(1,0)} = \begin{pmatrix} \frac{\partial y_1}{\partial x_1}\bigg|_{(1,0)} & \frac{\partial y_1}{\partial x_2}\bigg|_{(1,0)} \\ \frac{\partial y_2}{\partial x_1}\bigg|_{(1,0)} & \frac{\partial y_2}{\partial x_2}\bigg|_{(1,0)} \end{pmatrix}$$

so that from (4)

$$\frac{\partial y_1}{\partial x_1}\bigg|_{(1,0)} = -7, \quad \frac{\partial y_1}{\partial x_2}\bigg|_{(1,0)} = 8$$

$$\frac{\partial y_2}{\partial x_1}\bigg|_{(1,0)} = 7 \text{ and } \frac{\partial y_2}{\partial x_2}\bigg|_{(1,0)} = -4 \ .$$

● **PROBLEM 4-12**

Let $F = F(x,y)$ have second partial derivatives in a region of the xy-plane and introduce polar coordinates in this region by writing $x = r \cos \theta$, $y = r \sin \theta$. Find $\partial^2 F/\partial r \partial \theta$ in terms of derivatives of F with respect to x and y.

Solution: Here we have a composite function $F = F(x(r,\theta),y(r,\theta))$ so that the chain rule must be used. First, the chain rule gives

$$\frac{\partial F}{\partial \theta} = \frac{\partial F}{\partial x}\frac{\partial x}{\partial \theta} + \frac{\partial F}{\partial y}\frac{\partial y}{\partial \theta}$$

$$= -\frac{\partial F}{\partial x} r \sin \theta + \frac{\partial F}{\partial y} r \cos \theta \ . \qquad (1)$$

Now differentiate both sides of (1) with respect to r, keeping in mind that $\partial F/\partial x$ and $\partial F/\partial y$ depend on x and y and are thus composite functions of r and θ. Therefore, the chain rule must be used when differentiating these functions with respect to r. Hence

$$\frac{\partial^2 F}{\partial r \partial \theta} = \frac{\partial}{\partial r}\left(\frac{\partial F}{\partial \theta}\right) = \frac{\partial}{\partial r}\left(-\frac{\partial F}{\partial x} r \sin \theta\right) + \frac{\partial}{\partial r}\left(\frac{\partial F}{\partial y} r \cos \theta\right)$$

$$= -\frac{\partial F}{\partial x}\frac{\partial}{\partial r}(r \sin \theta) - \frac{\partial}{\partial r}\left(\frac{\partial F}{\partial x}\right) r \sin \theta$$

$$+ \frac{\partial F}{\partial y}\frac{\partial}{\partial r}(r \cos \theta) + \frac{\partial}{\partial r}\left(\frac{\partial F}{\partial y}\right) r \cos \theta . \qquad (2)$$

Now, the chain rule must be used on the composite functions

$\frac{\partial F}{\partial x}[x(r,\theta),y(r,\theta)]$ and $\frac{\partial F}{\partial y}[x(r,\theta),y(r,\theta)]$. Hence, it is found that

and

$$\frac{\partial}{\partial r}\left(\frac{\partial F}{\partial x}\right) = \frac{\partial}{\partial x}\left(\frac{\partial F}{\partial x}\right) \cdot \frac{\partial x}{\partial r} + \frac{\partial}{\partial y}\left(\frac{\partial F}{\partial x}\right) \cdot \frac{\partial y}{\partial r} \qquad (3)$$

$$\frac{\partial}{\partial r}\left(\frac{\partial F}{\partial y}\right) = \frac{\partial}{\partial x}\left(\frac{\partial F}{\partial y}\right) \cdot \frac{\partial x}{\partial r} + \frac{\partial}{\partial y}\left(\frac{\partial F}{\partial y}\right) \cdot \frac{\partial y}{\partial r} . \qquad (4)$$

Thus, (3) and (4) become

$$\frac{\partial}{\partial r}\left(\frac{\partial F}{\partial x}\right) = \frac{\partial^2 F}{\partial x^2} \cos \theta + \frac{\partial^2 F}{\partial y \partial x} \sin \theta , \qquad (5)$$

$$\frac{\partial}{\partial r}\left(\frac{\partial F}{\partial y}\right) = \frac{\partial^2 F}{\partial x \partial y} \cos \theta + \frac{\partial^2 F}{\partial y^2} \sin \theta . \qquad (6)$$

Using (5) and (6) in (2) yields

$$\frac{\partial^2 F}{\partial r \partial \theta} = -\frac{\partial F}{\partial x} \sin \theta - \left(\frac{\partial^2 F}{\partial x^2} \cos \theta + \frac{\partial^2 F}{\partial y \partial x} \sin \theta\right) r \sin \theta$$

$$+ \frac{\partial F}{\partial y} \cos \theta + \left(\frac{\partial^2 F}{\partial x \partial y} \cos \theta + \frac{\partial^2 F}{\partial y^2} \sin \theta\right) r \cos \theta . \qquad (7)$$

Now, if the second partial derivatives of a function F are continuous, as shall be assumed in this problem, then a theorem which states that $\partial^2 F/\partial x \partial y = \partial^2 F/\partial y \partial x$ is valid. Using this relation, equation (7) may be simplified to yield

$$\frac{\partial^2 F}{\partial r \partial \theta} = r \sin \theta \cos \theta \left(\frac{\partial^2 F}{\partial y^2} - \frac{\partial^2 F}{\partial x^2}\right) - \sin \theta \frac{\partial F}{\partial x}$$

$$+ \cos \theta \frac{\partial F}{\partial y} + r(\cos^2 \theta - \sin^2 \theta) \frac{\partial^2 F}{\partial x \partial y} .$$

• **PROBLEM 4-13**

Use differentials to compute

(a) $\frac{\partial z}{\partial x}, \frac{\partial z}{\partial y}$ where $z = \frac{x^2 - 1}{y}$

(b) $\frac{\partial r}{\partial x}, \frac{\partial r}{\partial y}, \frac{\partial x}{\partial r}$ where $r = \sqrt{x^2 + y^2}$

(c) $\frac{\partial z}{\partial x}, \frac{\partial z}{\partial y}$ where $z = \arctan(y/x)$.

Solution: The important concept here is that if $z = f(x,y,t,...)$ then
$$dz = \frac{\partial z}{\partial x} dx + \frac{\partial z}{\partial y} dy + \frac{\partial z}{\partial t} dt + ... \quad (1)$$

holds whether $x, y, t, ...$ are independent variables or functions of other variables, in which case $dx, dy, ...$ are also differentials of dependent variables. Thus, any equation in differentials which is correct for one choice of independent and dependent variables remains true for any other choice. An application of this idea is that in order to compute partial derivatives we can first compute differentials as if all variables were functions of a single variable, t, so that all of the rules of ordinary differential calculus apply. From the resulting differential formula the partial derivatives may be obtained.

(a) $z(x,y) = \frac{x^2 - 1}{y}$. Now assume that x and y depend on t so that
$$z(x,y) = \frac{[x(t)]^2 - 1}{y(t)} = \frac{f(t)}{g(t)} = z(x(t), y(t)) = z(t). \quad (2)$$

By ordinary differential calculus
$$\frac{dz}{dt} = \frac{f'(t)g(t) - f(t)g'(t)}{[g(t)]^2} = \frac{2x\frac{dx}{dt} y - (x^2 - 1)\frac{dy}{dt}}{y^2}$$

so that
$$dz = \frac{2xy\, dx - (x^2 - 1)\, dy}{y^2}. \quad (3)$$

By (1) it can be seen that
$$\frac{\partial z}{\partial x} = \frac{2x}{y}, \quad \frac{\partial z}{\partial y} = \frac{1 - x^2}{y^2}. \quad (4)$$

This might seem to be a rather circuitous route, but it is convenient when the partial derivatives cannot be easily computed directly.

(b) Again, let $x = x(t)$, $y = y(t)$ so that $r = r(t)$. Then $[r(t)]^2 = [x(t)]^2 + [y(t)]^2$ and upon differentiating both sides of this expression with respect to t it is found that
$$2r\frac{dr}{dt} = 2x\frac{dx}{dt} + 2y\frac{dy}{dt}$$

so that
$$2r\, dr = 2x\, dx + 2y\, dy \quad (5)$$

or
$$dr = \frac{x}{r} dx + \frac{y}{r} dy. \quad (6)$$

From (1) and (6) it is seen that
$$\frac{\partial r}{\partial x} = \frac{x}{r} = \frac{x}{\sqrt{x^2 + y^2}}, \quad \frac{\partial r}{\partial y} = \frac{y}{r} = \frac{y}{\sqrt{x^2 + y^2}} \quad (7)$$

Equation (5) can also be arranged to give $dx = r/x\, dr - y/x\, dy$ so that
$$\frac{\partial x}{\partial r} = \frac{r}{x} = \frac{\sqrt{x^2 + y^2}}{x}.$$

(c) Introducing the variable t gives $z(t) = \arctan(y(t)/x(t))$ so that
$$\frac{dz}{dt} = \frac{1}{1 + (y/x)^2} \frac{d(y(t)/x(t))}{dt}$$

$$= \frac{1}{1+(y/x)^2} \frac{\frac{dy}{dt}x(t) - y(t)\frac{dx}{dt}}{[x(t)]^2}$$

$$= \frac{x(t)dy/dt - y(t)dx/dt}{x^2 + y^2}$$

and
$$dz = \frac{xdy - ydx}{x^2 + y^2} \ . \tag{8}$$

Using (1) and (8) it is seen that

$$\frac{dz}{dx} = \frac{-y}{x^2+y^2} \ , \quad \frac{dz}{dy} = \frac{x}{x^2+y^2} \ .$$

• **PROBLEM 4-14**

Let $u = F(x,y)$ have second partial derivatives in a region of the xy-plane and introduce the change of variables

$$x = s + t \ , \quad y = s - t \tag{1}$$

in that region so that $u = F(x(s,t), y(s,t))$. Find

$$\frac{\partial^2 u}{\partial x^2} - \frac{\partial^2 u}{\partial y^2}$$

in terms of derivatives with respect to s and t.

Solution: It is interesting to note that the equation

$$\frac{\partial^2 u}{\partial x^2} - \frac{\partial^2 u}{\partial y^2} = 0 \tag{2}$$

is called the wave equation and is important in applied mathematics. The change of variables (1) is frequently used to solve equations such as (2). First, solve (1) for s and t to find

$$s = \tfrac{1}{2}(x+y) \ , \quad t = \tfrac{1}{2}(x-y) \ . \tag{3}$$

Now the chain rule may be used to obtain

$$\frac{\partial u}{\partial x} = \frac{\partial u}{\partial s}\frac{\partial s}{\partial x} + \frac{\partial u}{\partial t}\frac{\partial t}{\partial x} = \tfrac{1}{2}\frac{\partial u}{\partial s} + \tfrac{1}{2}\frac{\partial u}{\partial t} \ . \tag{4}$$

Similarly,

$$\frac{\partial u}{\partial y} = \frac{\partial u}{\partial s}\frac{\partial s}{\partial y} + \frac{\partial u}{\partial t}\frac{\partial t}{\partial y} = \tfrac{1}{2}\frac{\partial u}{\partial s} - \tfrac{1}{2}\frac{\partial u}{\partial t} \ . \tag{5}$$

Differentiating (4) with respect to x gives

$$\frac{\partial^2 u}{\partial x^2} = \tfrac{1}{2}\frac{\partial}{\partial x}\left(\frac{\partial u}{\partial s}\right) + \tfrac{1}{2}\frac{\partial}{\partial x}\left(\frac{\partial u}{\partial t}\right) \ . \tag{6}$$

But use of the chain rule on the composite function $\partial u/\partial s$ gives

$$\frac{\partial}{\partial x}\left(\frac{\partial u}{\partial s}\right) = \frac{\partial}{\partial s}\left(\frac{\partial u}{\partial s}\right) \cdot \frac{\partial s}{\partial x} + \frac{\partial}{\partial t}\left(\frac{\partial u}{\partial s}\right) \cdot \frac{\partial t}{\partial x} \ , \tag{7}$$

and use of the chain rule on $\partial u/\partial t$ gives

$$\frac{\partial}{\partial x}\left(\frac{\partial u}{\partial t}\right) = \frac{\partial}{\partial s}\left(\frac{\partial u}{\partial t}\right) \cdot \frac{\partial s}{\partial x} + \frac{\partial}{\partial t}\left(\frac{\partial u}{\partial t}\right) \cdot \frac{\partial t}{\partial x} \ . \tag{8}$$

Using (3), we have (7) and (8) may be rewritten as

$$\frac{\partial}{\partial x}\left(\frac{\partial u}{\partial s}\right) = \frac{1}{2}\frac{\partial^2 u}{\partial s^2} + \frac{1}{2}\frac{\partial^2 u}{\partial t\, \partial s} \tag{9}$$

and

$$\frac{\partial}{\partial x}\left(\frac{\partial u}{\partial t}\right) = \frac{1}{2}\frac{\partial^2 u}{\partial s\, \partial t} + \frac{1}{2}\frac{\partial^2 u}{\partial t^2}. \tag{10}$$

Using (9) and (10) in (6) gives

$$\frac{\partial^2 u}{\partial x^2} = \frac{1}{4}\frac{\partial^2 u}{\partial s^2} + \frac{1}{4}\frac{\partial^2 u}{\partial t\, \partial s} + \frac{1}{4}\frac{\partial^2 u}{\partial s\, \partial t} + \frac{1}{4}\frac{\partial^2 u}{\partial t^2}. \tag{11}$$

Differentiating (5) with respect to y gives

$$\frac{\partial^2 u}{\partial y^2} = \frac{1}{2}\frac{\partial}{\partial y}\left(\frac{\partial u}{\partial s}\right) - \frac{1}{2}\frac{\partial}{\partial y}\left(\frac{\partial u}{\partial t}\right). \tag{12}$$

Again, the chain rule is used on the composite functions $\partial u/\partial s$ and $\partial u/\partial t$ to give

$$\frac{\partial}{\partial y}\left(\frac{\partial u}{\partial s}\right) = \frac{\partial}{\partial s}\left(\frac{\partial u}{\partial s}\right)\cdot \frac{\partial s}{\partial y} + \frac{\partial}{\partial t}\left(\frac{\partial u}{\partial s}\right)\cdot \frac{\partial t}{\partial y} = \frac{1}{2}\frac{\partial^2 u}{\partial s^2} - \frac{1}{2}\frac{\partial^2 u}{\partial t\, \partial s} \tag{13}$$

and

$$\frac{\partial}{\partial y}\left(\frac{\partial u}{\partial t}\right) = \frac{\partial}{\partial s}\left(\frac{\partial u}{\partial t}\right)\cdot \frac{\partial s}{\partial y} + \frac{\partial}{\partial t}\left(\frac{\partial u}{\partial t}\right)\cdot \frac{\partial t}{\partial y} = \frac{1}{2}\frac{\partial^2 u}{\partial s\, \partial t} - \frac{1}{2}\frac{\partial^2 u}{\partial t^2}. \tag{14}$$

Therefore, (12) may be written as

$$\frac{\partial^2 u}{\partial y^2} = \frac{1}{4}\frac{\partial^2 u}{\partial s^2} - \frac{1}{4}\frac{\partial^2 u}{\partial s\, \partial t} - \frac{1}{4}\frac{\partial^2 u}{\partial t\, \partial s} + \frac{1}{4}\frac{\partial^2 u}{\partial t^2}. \tag{15}$$

The assumption will be made that all of the second partial derivatives, involved are continuous so that

$$\frac{\partial^2 u}{\partial s\, \partial t} = \frac{\partial^2 u}{\partial t\, \partial s}.$$

Then equations (11) and (15) become, respectively,

$$\frac{\partial^2 u}{\partial x^2} = \frac{1}{4}\frac{\partial^2 u}{\partial s^2} + \frac{1}{2}\frac{\partial^2 u}{\partial t\, \partial s} + \frac{1}{4}\frac{\partial^2 u}{\partial t^2}, \tag{16}$$

$$\frac{\partial^2 u}{\partial y^2} = \frac{1}{4}\frac{\partial^2 u}{\partial s^2} - \frac{1}{2}\frac{\partial^2 u}{\partial t\, \partial s} + \frac{1}{4}\frac{\partial^2 u}{\partial t^2}. \tag{17}$$

Subtracting (17) from (16) gives the desired result,

$$\frac{\partial^2 u}{\partial x^2} - \frac{\partial^2 u}{\partial y^2} = \frac{\partial^2 u}{\partial t\, \partial s}.$$

● **PROBLEM** 4-15

If $G(s,t) = F(e^s \cos t, e^s \sin t)$, show that $G_{11} + G_{22} = e^{2s}(F_{11}+F_{22})$, where G_{11} and G_{22} are evaluated at (s,t) and F_{11} and F_{22} are evaluated at $(e^s \cos t, e^s \sin t)$. Note: G_{11}, G_{22} are abbreviations for

$$\frac{\partial^2 G}{\partial s^2}, \ \frac{\partial^2 G}{\partial t^2}$$

respectively and F_{11}, F_{22} are abbreviations for $\dfrac{\partial^2 F}{\partial s^2}$ and $\dfrac{\partial^2 F}{\partial t^2}$, respectively.

<u>Solution</u>: The problem is to find the second partial derivatives of a composite function. For such a function, the chain rule must be employed. Here,
$$G(s,t) = F(e^s \cos t, e^s \sin t). \qquad (1)$$
Letting $f(s,t) = e^s \cos t$ and $g(s,t) = e^s \sin t$, $\qquad (2)$

(1) can be rewritten as
$$G(s,t) = F(f(s,t), g(s,t)). \qquad (3)$$
Then, by the chain rule
$$G_1(s,t) = \frac{\partial G}{\partial s} = \frac{\partial F}{\partial f}\frac{\partial f}{\partial s} + \frac{\partial F}{\partial g}\frac{\partial g}{\partial s}$$
which is written as
$$G_1(s,t) = F_1(f(s,t),g(s,t))f_1(s,t) + F_2(f(s,t),g(s,t))g_1(s,t). \qquad (4)$$
Similarly from (3), by the chain rule,
$$G_2(s,t) = F_1(f(s,t),g(s,t))f_2(s,t) + F_2(f(s,t),g(s,t))g_2(s,t). \qquad (5)$$
Next, $\dfrac{\partial^2 G}{\partial s^2}$, $\dfrac{\partial^2 G}{\partial t^2}$ must be computed from (4) and (5) respectively. Again, employing the chain rule, this yields
$$G_{11}(s,t) = \frac{\partial^2 G}{\partial s^2} = \frac{\partial G_1(s,t)}{\partial s} =$$
$$= \frac{\partial}{\partial s}[F_1(f(s,t),g(s,t))f_1(s,t) + F_2(f(s,t),g(s,t))g_1(s,t)] \qquad (6)$$

Then, using the abbreviations
$$f_1 = f_1(s,t),\ f_2 = f_2(s,t),\ g_1 = g_1(s,t),\ g_2 = g_2(s,t),$$
$$F_1 = F_1(f(s,t),g(s,t)),\ F_2 = F_2(f(s,t),g(s,t)),$$
$$F_{11} = F_{11}(f(s,t),g(s,t))\ \text{and}\ F_{22} = F_{22}(f(s,t),g(s,t)),$$
(6) becomes
$$G_{11} = F_1 f_{11} + (F_{11}f_1 + F_{12}g_1)f_1 + F_2 g_{11} + (F_{21}f_1 + F_{22}g_1)g_1. \qquad (7)$$
Similarly,
$$G_{22} = F_1 f_{22} + (F_{11}f_2 + F_{12}g_2)f_2 + F_2 g_{22} + (F_{21}f_2 + F_{22}g_2)g_2. \qquad (8)$$
However by (2),
$$f(s,t) = e^s \cos t \quad \text{and} \quad g(s,t) = e^s \sin t.$$
Therefore
$$f_1(s,t) = e^s \cos t \qquad g_1(s,t) = e^s \sin t$$
$$f_{11}(s,t) = e^s \cos t \qquad g_{11}(s,t) = e^s \sin t$$
$$f_2(s,t) = -e^s \sin t \qquad g_2(s,t) = e^s \cos t$$

$$f_{22}(s,t) = -e^s \cos t \qquad g_{22}(s,t) = -e^s \sin t$$

Hence, by (7) and (8),

$$G_{11} = F_1(e^s \cos t) + (F_{11}(e^s \cos t) + F_{12}(e^s \sin t))e^s \cos t + F_2(e^s \sin t)$$

$$+ (F_{21}(e^s \cos t) + F_{22}(e^s \sin t))e^s \sin t,$$

$$G_{22} = F_1(-e^s \cos t) + (F_{11}(-e^s \sin t) + F_{12}(e^s \cos t))(-e^s \sin t) + F_2(-e^s \sin t)$$

$$+ (F_{21}(-e^s \sin t) + F_{22}(e^s \cos t))e^s \cos t.$$

Thus

$$G_{11} + G_{22} = F_1(e^s \cos t) + F_{11}(e^{2s} \cos^2 t) + F_{12}(e^{2s} \sin t \cos t) + F_2(e^s \sin t)$$

$$+ F_{21}(e^{2s} \sin t \cos t) + F_{22} e^{2s} \sin^2 t - F_1(e^s \cos t) + F_{11}(e^{2s} \sin^2 t)$$

$$- F_{12}(e^{2s} \cos t \sin t) - F_2(e^s \sin t) - F_{21}(e^{2s} \sin t \cos t)$$

$$+ F_{22}(e^{2s} \cos^2 t).$$

Which simplifies to

$$G_{11} + G_{22} = F_{11}(e^{2s} \cos^2 t + e^{2s} \sin^2 t) + F_{22}(e^{2s} \sin^2 t + e^{2s} \cos^2 t)$$

$$G_{11} + G_{22} = F_{11} e^{2s} + F_{22} e^{2s}$$

$$G_{11} + G_{22} = e^{2s}(F_{11} + F_{22}).$$

This is the desired answer.

● **PROBLEM 4-16**

Show that if a function $F: R^2 \to R$ satisfies $\partial F/\partial x = \partial F/\partial y$, then $F(x,y) = f(x+y)$ where f is some differentiable function of one variable, $s = x+y$.

Solution: First note that this result gives useful information to someone trying to solve the differential equation $\partial F/\partial x - \partial F/\partial y = 0$ for $F(x,y)$ (where F has some physical significance). To establish the result, make the substitutions

$$s = x + y, \quad t = x - y \qquad (1)$$

so that

$$x = \tfrac{1}{2}(s+t), \quad y = \tfrac{1}{2}(s-t). \qquad (2)$$

Then F becomes the composite function $F(x(s,t), y(s,t))$. The chain rule may then be used on F to yield

$$\frac{\partial F}{\partial t} = \frac{\partial F}{\partial x}\frac{\partial x}{\partial t} + \frac{\partial F}{\partial y}\frac{\partial y}{\partial t} = \frac{\partial F}{\partial x} \cdot \tfrac{1}{2} + \frac{\partial F}{\partial y} \cdot \left(\frac{-1}{2}\right). \qquad (3)$$

But the hypothesis is that $\partial F/\partial x - \partial F/\partial y = 0$ so that (3) becomes

$$\frac{\partial F}{\partial t} = 0. \qquad (4)$$

That is, F must be independent of t and hence be some function

of s alone. Thus
$$F(x,y) = f(s) \tag{5}$$
and to finish the problem it remains only to show that $f(s)$ is differentiable. But this is simple for when $y = 0$, (5) becomes
$$F(x,0) = f(s)_{y=0} = f(x+y)_{y=0} = f(x) \tag{6}$$
and since $\partial F/\partial x$ exists everywhere (in particular at $y = 0$), (6) implies that $f'(x)$ exists for all x. Of course, x is just a name for a real variable which may take on all values. Hence, this is equivalent to saying that $f'(s)$ exists for all s.

• **PROBLEM 4-17**

Find the rate of change of the temperature $T(x,y,z)$ at the point (x,y,z) experienced by a particle moving along the curve given by
$$\vec{r}(t) = \begin{pmatrix} x(t) \\ y(t) \\ z(t) \end{pmatrix} = \begin{pmatrix} t \\ 3t^2 \\ 2\cos t \end{pmatrix}. \tag{1}$$
Assume that the temperature at (x,y,z) is given by
$$T(x,y,z) = xy + yz + zx. \tag{2}$$

Solution: As far as the particle is concerned, the temperature at any time t is a composite function given by
$$T(x(t),y(t),z(t)) = x(t)y(t) + y(t)z(t) + z(t)x(t). \tag{3}$$
Therefore, the particle experiences a change in temperature as time goes by at the rate
$$\frac{dT}{dt} = T'(t). \tag{4}$$
The chain rule tells us that
$$\frac{dT}{dt} = \frac{\partial T}{\partial x}\frac{dx}{dt} + \frac{\partial T}{\partial y}\frac{dy}{dt} + \frac{\partial T}{\partial z}\frac{dz}{dt}, \tag{5}$$
so that using (1) and (2) in (5) yields
$$\frac{dT}{dt} = (y+z)\cdot 1 + (x+z)\cdot 6t + (y+x)(-2\sin t). \tag{6}$$
However $x = t$, $y = 3t^2$ and $z = 2\cos t$ and substituting this information into (6) gives
$$\frac{dT}{dt} = 3t^2 + 2\cos t + (t + 2\cos t)6t + (3t^2 + t)(-2\sin t)$$
$$= 9t^2 + (12t + 2)\cos t - 2(t + 3t^2)\sin t$$
as the rate of change of temperature observed by the particle.

GRADIENTS AND TANGENT PLANES

• **PROBLEM 4-18**

Show that the line normal to the surface given by $F(x,y,z) = 0$ at a point (x_0,y_0,z_0) has direction ratios

$$\left. \frac{\partial F}{\partial x} \right|_{(x_0,y_0,z_0)} : \left. \frac{\partial F}{\partial y} \right|_{(x_0,y_0,z_0)} : \left. \frac{\partial F}{\partial z} \right|_{(x_0,y_0,z_0)} .$$

What are these ratios when $z = f(x,y)$ is a solution to $F(x,y,z) = 0$?

Solution: The equation

$$F(x,y,z) = 0 \tag{1}$$

represents some surface S in 3-dimensional space consisting of all points (x,y,z) which satisfy it. That is, the points on the surface S are given by $S = \{(x,y,z) | F(x,y,z) = 0\}$. Let (x_0,y_0,z_0) be some point on this surface (i.e., $F(x_0,y_0,z_0) = 0$), and let

$$\vec{r}(t) = \begin{pmatrix} x(t) \\ y(t) \\ z(t) \end{pmatrix}$$

be some curve passing through

$$\vec{r}_0 = \vec{r}(t_0) = \begin{pmatrix} x_0 \\ y_0 \\ z_0 \end{pmatrix} = \begin{pmatrix} x(t_0) \\ y(t_0) \\ z(t_0) \end{pmatrix} . \quad \text{Then} \tag{2}$$

$F(x(t),y(t),z(t)) = F(t) = 0$ for all t in the domain of the vector function $\vec{r}$. By the chain rule, it is then seen that at t_0

$$\left. \frac{dF}{dt} \right|_{t_0} = \left. \frac{\partial F}{\partial x} \right|_{r_0} \left. \frac{dx}{dt} \right|_{t_0} + \left. \frac{\partial F}{\partial y} \right|_{r_0} \left. \frac{dy}{dt} \right|_{t_0} + \left. \frac{\partial F}{\partial z} \right|_{r_0} \left. \frac{dz}{dt} \right|_{t_0} = 0 . \tag{3}$$

This can be written in vector form as

$$\vec{\nabla} F(\vec{r}_0) \cdot \vec{r}'(t_0) = 0 , \tag{4}$$

where $\vec{\nabla}F(\vec{r}_0)$ is the gradient of F at $\vec{r}_0$ and is given by

$$\vec{\nabla}F(\vec{r}_0) = \begin{pmatrix} F_x(\vec{r}_0) \\ F_y(\vec{r}_0) \\ F_z(\vec{r}_0) \end{pmatrix} , \tag{5}$$

while

$$\vec{r}'(t_0) = \begin{pmatrix} x'(t_0) \\ y'(t_0) \\ z'(t_0) \end{pmatrix} .$$

Recall now that $\vec{r}'(t_0)$ is the "velocity vector" and is tangent to the curve. When the dot product of two vectors is zero, they must be perpendicular. Hence, we conclude from (4) that $\vec{\nabla}F(\vec{r}_0)$ must be perpendicular to the tangent at $\vec{r}_0$. That is, $\vec{\nabla}F(\vec{r}_0)$ is normal to the curve and therefore to the surface given by $F(\vec{r}) = 0$, at (x_0,y_0,z_0), so that the direction numbers for the normal are

$$(F_x(\vec{r}_0), F_y(\vec{r}_0), F_z(\vec{r}_0)) .$$

Obviously, since the chosen curve and the point $\vec{r}_0$ were arbitrary, the above conclusion holds for all points (x,y,z) on the surface. It is important to note the following point; in equation (3) we could have multiplied through by dt to obtain (dropping the subscripts)

$$\frac{dF}{dt} dt = \frac{\partial F}{\partial x}\left(\frac{dx}{dt} dt\right) + \frac{\partial F}{\partial y}\left(\frac{dy}{dt} dt\right) + \frac{\partial F}{\partial z}\left(\frac{dz}{dt} dt\right) = 0 . \quad (6)$$

Then since the differential of $x(t)$ is defined as $dx = x'(t)dt = dx/dt\, dt$, and similarly for the other functions of t, (6) may be written as

$$dF = \frac{\partial F}{\partial x} dx + \frac{\partial F}{\partial y} dy + \frac{\partial F}{\partial z} dz = 0 . \quad (7)$$

This reasoning is frequently used in texts and at first glance it may be confusing. It seems as though we have shown that if $F(x,y,z) = 0$ then the differential of F at (x,y,z) is zero, and this is certainly not a true statement. The confusion arises because whether we specify a curve $\vec{r}(t)$ or not, it has been assumed that (dx,dy,dz) can vary only along the surface $F(x,y,z) = 0$. That is, the reasoning used in going directly from (1) to (7) is that if $F(x,y,z) = 0$ and (dx,dy,dz) can only vary in directions such that $F(x+dx, y+dy, z+dz) = F(x,y,z) = 0$, then $dF(x,y,z) = 0$.

Finally, if the equation of the surface can be written as $z = f(x,y)$, then $F(x,y,z) = f(x,y,z)-z = 0$ so that the direction ratios $F_x : F_y : F_z$ at a point (x,y,z) become $f_x : f_y : -1$.

• **PROBLEM 4-19**

Find the equation of the tangent plane to the surface described by the equation $F(x,y,z) = 0$ at the point (x_0, y_0, z_0). What is this equation if $z = f(x,y)$ is a solution to $F(x,y,z) = 0$?

Solution: The key point in the argument to be given is that the gradient vector of a function $F(x,y,z)$ at a point is normal to the surface given by $F(x,y,z) = 0$ at that point. This result was proved in a previous problem but the proof may be sketched as follows: let (x_1, y_1, z_1) be a point on the surface S given by $F(x,y,z) = 0$ and let

$$\vec{r}(t) = \begin{pmatrix} x(t) \\ y(t) \\ z(t) \end{pmatrix}$$

be a position vector which traces out a curve C on S with

$$\vec{r}(t_0) = \vec{r}_0 = \begin{pmatrix} x_0 \\ y_0 \\ z_0 \end{pmatrix} .$$

Then $F(x(t),y(t),z(t)) = F(t) = 0$ for all t in the domain of C and by the chain rule

$$\left.\frac{dF}{dt}\right|_{\vec{r}_0} = \left.\frac{\partial F}{\partial x}\right|_{\vec{r}_0} x'(t)\bigg|_{t_0} + \left.\frac{\partial F}{\partial y}\right|_{\vec{r}_0} y'(t)\bigg|_{t_0} + \left.\frac{\partial F}{\partial z}\right|_{\vec{r}_0} z'(t)\bigg|_{t_0} . \quad (1)$$

This may be written in vector notation as

$$\vec{\nabla}F(\vec{r}_0) \cdot \vec{r}'(t_0) = 0, \qquad (2)$$

where $\vec{\nabla}F = (F_x, F_y, F_z)$ is the gradient of F and $\vec{r}'(t) = (x'(t), y'(t), z'(t))$ is the velocity vector of C. Since $\vec{r}'(t)$ is always tangent to C, (2) implies that $\vec{\nabla}F$ is normal or perpendicular to C at $\vec{r}_0$ since the dot product of two nonzero vectors is zero only if they are perpendicular. But C can be any curve on S through $\vec{r}_0$, so (2) implies that $\vec{\nabla}F$ is normal to S at $\vec{r}_0$.

Now, a point (x,y,z) will be on the plane tangent to S at $\vec{r}_0 = (x_0, y_0, z_0)$ if the vector directed from (x_0, y_0, z_0) to (x,y,z) is perpendicular to the normal $\vec{\nabla}F(x_0, y_0, z_0)$. That is, (x,y,z) is on the tangent plane if

$$\vec{\nabla}F(\vec{r}_0) \cdot (\vec{r} - \vec{r}_0) = 0 \qquad (3)$$

or, writing this out, the equation of the tangent plane to S at $\vec{r}_0$ is given by

$$\left.\frac{\partial F}{\partial x}\right|_{\vec{r}_0}(x - x_0) + \left.\frac{\partial F}{\partial y}\right|_{\vec{r}_0}(y - y_0) + \left.\frac{\partial F}{\partial z}\right|_{\vec{r}_0}(z - z_0) = 0. \qquad (4)$$

If the surface is described by $z = f(x,y)$, then $F(x,y,z) = f(x,y) - z = 0$ so that

$$\frac{\partial F}{\partial x} = \frac{\partial f}{\partial x}, \quad \frac{\partial F}{\partial y} = \frac{\partial f}{\partial y}, \quad \frac{\partial F}{\partial z} = -1 \qquad (5)$$

and (4) becomes

$$\left.\frac{\partial f}{\partial x}\right|_{(x_0,y_0)}(x - x_0) + \left.\frac{\partial f}{\partial y}\right|_{(x_0,y_0)}(y - y_0) - (z - z_0) = 0. \qquad (6)$$

● **PROBLEM 4-20**

Find the line orthogonal to the graph of $f(x,y) = xy$ at the point $(x_0, y_0, z_0) = (-2, 3, -6)$.

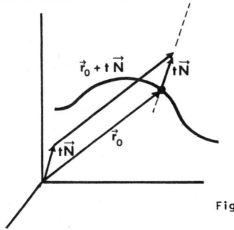

Fig. 1

Solution: It is possible to solve this problem if one of two facts is known. The first fact, which has been derived in a previous problem, is that the equation of the tangent plane to the graph of $f(x,y) = xy$ at a point $(x_0, y_0, z_0) = \vec{r}_0$ is given by

$$\left.\frac{\partial f}{\partial x}\right|_{(x_0,y_0)} (x - x_0) + \left.\frac{\partial f}{\partial y}\right|_{(x_0,y_0)} (y - y_0) - (z - z_0) = 0. \quad (1)$$

In the case at hand,

$$\left.\frac{\partial f}{\partial x}\right|_{(x_0,y_0)} = \left.\frac{\partial f}{\partial x}\right|_{(-2,3)} = y\bigg|_{(-2,3)} = 3 \quad (1a)$$

$$\left.\frac{\partial f}{\partial y}\right|_{(x_0,y_0)} = \left.\frac{\partial f}{\partial y}\right|_{(-2,3)} = x\bigg|_{(-2,3)} = -2, \quad (1b)$$

so that from (1)

$$3(x + 2) - 2(y - 3) - (z + 6) = 0$$

or

$$z = 3x + 6 - 2y \quad (2)$$

is the equation of the tangent plane to the graph of $f(x,y) = xy$ at the point $(-2, 3, -6)$. Equation (2) can also be written as

$$3x - 2y - z + 6 = 0. \quad (3)$$

Now recall that for a plane given by the equation $Ax + By + Cz + D = 0$ a vector normal to this plane is given by $\vec{n} = (A,B,C)$. Hence, it can be seen from (3) that the vector $\vec{N} = (3,-2,-1)$ is normal to this tangent plane. Thus, the line orthogonal to the graph of $f(x,y) = xy$ at $\vec{r}_0$, that is to say the line orthogonal to the tangent plane to this graph at $\vec{r}_0$, is given by

$$(x,y,z) = \vec{r}_0 + t\vec{N} = (-2, 3, -6) + t(3, -2, -1) \quad (4)$$

or

$$\begin{pmatrix} x \\ y \\ z \end{pmatrix} = \begin{pmatrix} -2 + 3t \\ 3 - 2t \\ -6 - t \end{pmatrix}. \quad (5)$$

Justification of (4) may be made if one remembers the parallelogram rule for adding vectors, as shown in Figure 1.
The second fact which gives an immediate solution to the problem and which was also derived in a previous problem is that the vector which is normal to a surface given by $f = f(x,y)$ at a point (x_0, y_0) has direction numbers $(f_x(x_0, y_0), f_y(x_0, y_0), -1)$. Thus, from (1a) and (1b), the normal to the surface $f(x,y) = xy$ at $(-2, 3, -6)$ has direction numbers

$$\vec{N} = (3, -2, -1).$$

This is the same result as was found before, so that using equation (4) will give the same result (i.e., equation (5)), for the line orthogonal to $f(x,y) = xy$ at $(-2, 3, -6)$ as we previously found.

• PROBLEM 4-21

Find the equation of the tangent plane to the graph of
$$z = f(x,y) = x^2 + 2y^2 - 1$$
at the points
(a) $(x_0, y_0) = (0,0)$
(b) $(x_0, y_0) = (1,1)$.

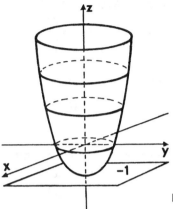

Fig. 1

Solution: As has been shown in previous problems, since points (x,y,z) on the tangent plane to the graph of $F(x,y,z) = 0$ at (x_0,y_0,z_0) must be such that the line segment or vector connecting (x,y,z) with (x_0,y_0,z_0) is perpendicular to the normal to this surface, the equation of this plane is

$$\vec{\nabla}F(x_0,y_0,z_0) \cdot \begin{pmatrix} x - x_0 \\ y - y_0 \\ z - z_0 \end{pmatrix} = 0 . \qquad (1)$$

This is so because the gradient of F at (x_0,y_0,z_0), i.e., $\vec{\nabla}F(x_0,y_0,z_0)$, is indeed normal to the surface described by $F(x,y,z) = 0$. In the particular case where the surface is described by $z = f(x,y)$, we have
$$F(x,y,z) = f(x,y) - z = 0$$
so that writing out (1) gives

$$\vec{\nabla}F(x_0,y_0,z_0) \cdot \begin{pmatrix} x - x_0 \\ y - y_0 \\ z - z_0 \end{pmatrix} = \left.\frac{\partial F}{\partial x}\right|_0 (x - x_0) + \left.\frac{\partial F}{\partial y}\right|_0 (y - y_0) + \left.\frac{\partial F}{\partial z}\right|_0 (z - z_0)$$

$$= \left.\frac{\partial f}{\partial x}\right|_0 (x - x_0) + \left.\frac{\partial f}{\partial y}\right|_0 (y - y_0) - (z - z_0)$$

$$= 0 \qquad (2)$$

where the symbol $|_0$ means that the partial derivative is to be evaluated at the point (x_0,y_0,z_0). In the case at hand, (2) yields

$$2x\Big|_0(x - x_0) + 4y\Big|_0(y - y_0) - (z - z_0) = 0 \qquad (3)$$

for the equation of the tangent plane at the point (x_0, y_0, z_0).

(a) Here $(x_0, y_0, z_0) = (0, 0, -1)$ so that (3) gives

$$z = -1 \qquad (4)$$

as the equation of the tangent plane at $(0, 0, -1)$. This is a horizontal plane as shown in the figure.

(b) Here $(x_0, y_0, z_0) = (1, 1, 2)$ so that (3) gives

$$2(x-1) + 4(y-1) - (z-2) = 0$$

or

$$z = 2x + 4y - 4 . \qquad (5)$$

Equation (5) then gives the equation of the tangent plane to the graph of
$$z = f(x,y) = x^2 + 2y^2 - 1 \quad \text{at the point } (1,1,2).$$

• **PROBLEM 4-22**

(a) Let $f(x,y) = r = \sqrt{x^2 + y^2}$. Find grad $f(\vec{X})$ where $\vec{X} = (x,y)$.
(b) Let $f(x,y) = \sin r = \sin\sqrt{x^2 + y^2}$. Find grad $f(\vec{X})$.
(c) State a general theorem which encompasses the results of (a) and (b).

Solution: (a) The gradient of a function $f(\vec{X}) = f(x,y)$ is given by

$$\vec{\nabla} f = \text{grad } f = \left(\frac{\partial f}{\partial x}, \frac{\partial f}{\partial y}\right) . \qquad (1)$$

In the case at hand, $f(\vec{X}) = \sqrt{x^2 + y^2}$ (which is the distance function in R^2). Hence, taking the partial derivatives of f with respect to x and y gives

$$\frac{\partial f}{\partial x} = 2x \cdot \frac{(x^2 + y^2)^{-\frac{1}{2}}}{2} = \frac{x}{\sqrt{x^2 + y^2}} = \frac{x}{r} \qquad (2)$$

and

$$\frac{\partial f}{\partial y} = 2y \cdot \frac{(x^2 + y^2)^{-\frac{1}{2}}}{2} = \frac{y}{\sqrt{x^2 + y^2}} = \frac{y}{r} . \qquad (3)$$

Therefore

$$\text{grad } f(x,y) = \text{grad } r(x,y) = \left(\frac{x}{r}, \frac{y}{r}\right) \qquad (4)$$

which may be conveniently written as

$$\text{grad } r = \frac{1}{r} \vec{X} \qquad (5)$$

where $\vec{X}$ is the vector (x,y).

(b) Again $f(x,y) = f(r(x,y))$ is a function of distance alone. From the chain rule

$$\frac{\partial f}{\partial x} = \frac{df}{dr} \frac{\partial r}{\partial x} = (\cos r) \cdot \tfrac{1}{2}(x^2 + y^2)^{-\frac{1}{2}} 2x$$

or
$$\frac{\partial f}{\partial x} = (\cos r)\left(\frac{x}{r}\right). \tag{6}$$

Also,
$$\frac{\partial f}{\partial y} = \frac{df}{dr}\frac{\partial r}{\partial y} = (\cos r)\tfrac{1}{2}(x^2 + y^2)^{-\tfrac{1}{2}} 2y$$

or
$$\frac{\partial f}{\partial y} = (\cos r)\left(\frac{y}{r}\right). \tag{7}$$

Substituting (6) and (7) into (1) yields

$$\vec{\nabla}f = \left((\cos r)\left(\frac{x}{r}\right), (\cos r)\left(\frac{y}{r}\right)\right)$$

$$= \frac{\cos r}{r}(x,y)$$

or using $\vec{X} = (x,y)$,

$$\vec{\nabla}f = \frac{\cos r}{r} \vec{X} . \tag{8}$$

(c) In both (a) and (b) the function f depended only on distance $r = \sqrt{x^2 + y^2}$; that is, $f = f(r(x,y))$. The gradient of any such function is given by

$$\vec{\nabla}f = \left(\frac{\partial f}{\partial x}, \frac{\partial f}{\partial y}\right) = \left(\frac{df}{dr}\frac{\partial r}{\partial x}, \frac{df}{dr}\frac{\partial r}{\partial y}\right)$$

$$= (f'(r) x/r, f'(r) y/r)$$
$$= \frac{f'(r)}{r} (x,y) .$$

Writing this in vector form with $\vec{X} = (x,y)$, gives

$$\vec{\nabla}f(r(x,y)) = \frac{f'(r)}{r} \vec{X} .$$

This is the desired general result and it is easily seen that (a) and (b) are examples of its use.

● **PROBLEM 4-23**

Let $u = r^3$ be a scalar field where r is the distance OP from the origin O to a variable point P, in R^3. Find the gradient of u at P without resorting to rectangular coordinates.

Solution: Recall two important facts about the gradient:
(i) The rate of change of a scalar field $u = f(P)$ per unit distance at P is the component of the gradient, $\vec{\nabla}u$, at that point. The direction of $\vec{\nabla}u$ is that in which the rate of increase of u is greatest.
(ii) The gradient of u at P is perpendicular to the level surface through P.

With these two pieces of information the problem can be reasoned through, without making too many calculations. The level surfaces of $u = r^3$ are those surfaces for which $u = r^3 = c$ where c can be any constant. Thus, these surfaces are given by $r = c^{\tfrac{1}{3}} = C$; that is, they are spheres of radius C (where C is any constant greater than zero) with centers at O. By (ii) we know that $\vec{\nabla}u$ is normal to the sphere through P and by (i) we see that its direction is outward from the sphere since $u = r^3$ increases as r grows larger.

To find an expression for $\vec{\nabla}u$, note that $du/dr = 3r^2$ is the rate of increase of u per unit distance in the radial direction. Therefore, by (ii) $\vec{\nabla}u$ has length $3r^2$. A unit vector in the direction of $\vec{OP}$ is $1/r\, \vec{OP}$ so that the final expression for $\vec{\nabla}u$ is

$$\vec{\nabla}u = |\vec{\nabla}u|\,\frac{\vec{OP}}{r} = \frac{3r^2}{r}\,\vec{OP}$$

or

$$\vec{\nabla}u = 3r\,\vec{OP}.$$

DIRECTIONAL DERIVATIVES

● **PROBLEM 4-24**

Define the term directional derivative and use the definition to compute the derivative of $f(x,y) = x^2 + 3xy$ in the direction $\vec{\beta} = \left(\frac{1}{\sqrt{2}}, -\frac{1}{\sqrt{2}}\right)$ at the point $\vec{P}_0 = (2,0)$.

Solution: Let f be a real valued function of n variables, $f: R^n \to R$, defined and continuous in a neighborhood of the point represented by the direction vectors $\vec{P}_0$, and let $\vec{\beta}$ be a unit vector in R^n; that is, $|\vec{\beta}| = 1$. Then the directional derivative of f at $\vec{P}_0$ in the direction $\vec{\beta}$ is defined to be

$$D_{\vec{\beta}}\, f(\vec{P}_0) = \lim_{t \to 0} \frac{f(\vec{P}_0 + t\vec{\beta}) - f(\vec{P}_0)}{t}. \tag{1}$$

Note that if $\vec{\beta}$ is a vector in the direction of one of the axes, e.g., $\vec{\beta} = (0,0,\ldots,1,0,\ldots,0)$ where the 1 is in the ith place, then (1) reduces to the partial derivative of f with respect to x_i. That is, with $\vec{P}_0 = (p_{1_0}, p_{2_0}, \ldots, p_{n_0})$, (1) becomes

$$D_{x_i}\, f(\vec{P}_0) = \lim_{t \to 0} \frac{f(p_{1_0}, \ldots, p_{i_0}+t, \ldots, p_{n_0}) - f(p_{1_0}, \ldots, p_{n_0})}{t}$$

or

$$D_{x_i}\, f(\vec{P}_0) = \left.\frac{\partial f}{\partial x_i}\right|_{P_0}.$$

Hence, the directional derivative, (1), gives the rate of change of f at $\vec{P}_0$ in the direction $\vec{\beta}$ and may be interpreted as the slope of a tangent line to the graph of f in the direction $\vec{\beta}$. Now, for the given function, point, and direction, (1) gives

$$D_{(1/\sqrt{2},\, -1/\sqrt{2})}(x^2+3xy)\Big|_{(2,0)} = \lim_{t \to 0} \frac{f(2+t/\sqrt{2},\, 0-t/\sqrt{2}) - f(2,0)}{t}. \tag{2}$$

Since $f(2 + t/\sqrt{2},\, -t/\sqrt{2}) = (2 + t/\sqrt{2})^2 + 3(2 + t/\sqrt{2})(-t/\sqrt{2})$

$$= 4 - 2/\sqrt{2}\, t - t^2 = 4 - \sqrt{2}\, t - t^2$$

and $f(2,0) = 4$. By (2),

$$D_{\vec{V}}(1/\sqrt{2},\ -1/\sqrt{2})(x^2+3xy)\Big|_{(2,0)} = \lim_{t\to 0} \frac{(4-\sqrt{2}\,t-t^2)-4}{t}$$

$$= \lim_{t\to 0} \frac{-\sqrt{2}\,t-t^2}{t}$$

$$= \lim_{t\to 0}(-\sqrt{2}-t)$$

so that

$$D_{\vec{V}}(1/\sqrt{2},\ -1/\sqrt{2})(x^2+3xy)\Big|_{(2,0)} = -\sqrt{2}\ .$$

• **PROBLEM 4-25**

Derive the formula

$$D_{\vec{V}} F(\vec{X}) = \vec{\nabla} F \cdot \vec{V} = \text{grad } F \cdot \vec{V} \tag{1}$$

where F is a real valued function of n variables, $F: R^n \to R$; $\vec{V}$ is a unit vector in R^n, i.e., $|\vec{V}| = 1$; $\vec{X} = (x_1, x_2, \ldots, x_n)$ is the directed vector of the point $(x_1, \ldots, x_n)$ and $D_{\vec{V}} F(\vec{X})$ is the directional derivative of F at $\vec{X}$ in the direction $\vec{V}$. Then use formula (1) to find the directional derivative of $F(\vec{X}) = F(x,y) = x^2 + y^3$ at $(-1,3)$ in the direction of $\vec{V} = (1,2)$.

Solution: The directional derivative described above is defined by

$$D_{\vec{V}} F(\vec{X}) = \lim_{t\to 0} \frac{F(\vec{X}+t\vec{V}) - F(\vec{X})}{t} \tag{2}$$

which may be written out as

$$D_{\vec{V}} F(\vec{X}) = \lim_{t\to 0} \frac{F(x_1+tV_1,\ldots,x_n+tV_{x_n}) - F(x_1,\ldots,X_n)}{t}. \tag{3}$$

The assumption is made that F is differentiable at $\vec{X}$; hence, it has a differential there. By the definition of a differential,

$$\lim_{(dx_1,\ldots,dx_n)\to 0} \frac{\Delta F(\vec{X}) - dF(\vec{X})}{\sqrt{dx_1^2+\ldots+dx_n^2}} = 0 \tag{4}$$

we see that

$$\lim_{(dx_1,\ldots,dx_n)\to 0} \frac{\Delta F(\vec{X})}{\sqrt{dx_1^2+\ldots+dx_n^2}} = \lim_{(dx_1,\ldots,dx_n)\to 0} \frac{dF(X)}{\sqrt{dx_1^2+\ldots+dx_n^2}} \tag{5}$$

Changing the variables in the limit to $dx_1 = tV_{x_1},\ldots,dx_n = tV_{x_n}$ the conclusion from (5) is that

$$\lim_{t\vec{V}\to 0} \frac{F(x_1+tV_{x_1},\ldots,x_n+tV_{x_n}) - F(x_1,\ldots,x_n)}{|t\vec{V}|} = \lim_{t\vec{V}\to 0} \frac{\Delta F(X)}{|t\vec{V}|}$$

$$= \lim_{t\vec{V}\to 0} \frac{dF(\vec{X})}{t|\vec{V}|} = \lim_{t\vec{V}\to 0} \frac{dF(\vec{X})}{t} \; . \tag{6}$$

Using (3) and recalling that

$$dF(\vec{X}) = \frac{\partial F}{\partial x_1} tV_{x_1} + \ldots + \frac{\partial F}{\partial x_n} tV_{x_n} = \text{grad } F \cdot (t\vec{V})$$

equation (6) gives

$$D_{\vec{V}} F(\vec{X}) = \lim_{t\vec{V}\to 0} \frac{\frac{\partial F}{\partial x_1} tV_1 + \ldots + \frac{\partial F}{\partial x_n} tV_{x_n}}{t}$$

or

$$D_{\vec{V}} F(\vec{X}) = \frac{\partial F}{\partial x_1} V_1 + \ldots + \frac{\partial F}{\partial x_n} V_{x_n} = \text{grad } F \cdot \vec{V} \tag{7}$$

which is the desired result.

For the specific problem with $F = x^2 + y^3$ at $(-1,3)$ note that the direction vector $\vec{V} = (1,2)$ is not a unit vector. Therefore, to use (7), it must be converted to the unit vector $\vec{V}_u = \frac{\vec{V}}{|\vec{V}|}$.

Since $|\vec{V}| = \sqrt{1^2 + 2^2} = \sqrt{5}$, $\vec{V}_u = \left(\frac{1}{\sqrt{5}}, \frac{2}{\sqrt{5}}\right)$

and

$$D_{\vec{V}} F(\vec{X}) = D_{\vec{V}_u} F(\vec{X}) = \frac{\partial F}{\partial x}\bigg|_{(-1,3)} \cdot \frac{1}{\sqrt{5}} + \frac{\partial F}{\partial y}\bigg|_{(-1,3)} \cdot \frac{2}{\sqrt{5}} \; . \tag{8}$$

Now $\frac{\partial F}{\partial x} = 2x$ and $\frac{\partial F}{\partial y} = 3y^2$ so that (8) gives

$$D_{\vec{V}} F(\vec{X}) = 2 \cdot (-1) \cdot \frac{1}{\sqrt{5}} + (3 \cdot 3^2) \cdot \frac{2}{\sqrt{5}} = \frac{52}{\sqrt{5}} \; .$$

● **PROBLEM 4-26**

Compute the directional derivative of $f(x,y) = x^2 + y^3$ at the point $(-1,3)$ in the direction of maximal increase of f. Explain your reasoning.

Solution: As was proven in a previous problem, the directional derivative of a real-valued function f of n variables at the point P in the direction $\vec{A}$ is given by

$$D_{\vec{A}} f(P) = \text{grad } f(P) \cdot \vec{A} \tag{1}$$

where $\vec{A}$ must be a unit vector. This dot product may be written as

$$\text{grad } f(P) \cdot \vec{A} = \|\text{grad } f(P)\| \|\vec{A}\| \cos \theta \tag{2}$$

where θ is the angle between the vectors $\text{grad } f(P)$ and $\vec{A}$. Since $\vec{A}$ is a unit vector, $\|\vec{A}\| = 1$ and (1) and (2) give

$$D_{\vec{A}} f(P) = \|\text{grad } f(P)\| \cos \theta \; . \tag{3}$$

The directional derivative is thus seen to be a maximum when $\theta = 0$; i.e., the direction of maximal increase of f is the direction of the gradient and the directional derivative in that direction is simply

$$D_{\vec{A}} f(P)\Big|_{max} = \|\text{grad } f(P)\| \ . \tag{4}$$

Now in the case at hand, $f = x^2 + y^3$ so $\partial f/\partial x = 2x$, $\partial f/\partial y = 3y^2$ and at the point $(-1,3)$,

$$\|\text{grad } f\| = \sqrt{\left(\frac{\partial f}{\partial x}\right)^2 + \left(\frac{\partial f}{\partial y}\right)^2} = \sqrt{(2x)^2 + (3y^2)^2}$$

$$= \sqrt{4x^2 + 9y^4} = \sqrt{4(-1)^2 + 9(3)^4}$$

$$= \sqrt{733} \ .$$

POTENTIAL FUNCTIONS

• **PROBLEM 4-27**

Give a definition of a potential function, φ, of a vector field $\vec{F}(\vec{X})$. Determine whether or not $\vec{F}(x,y) = (e^{xy}, e^{x+y})$ has a potential function.

Solution: Let $\vec{F}(\vec{X})$ be a vector field defined on an open set U in n-dimensional space. That is with each vector $\vec{X} = (x_1, \ldots, x_n)$ which represents a point $P = (x_1, \ldots, x_n)$ in n-space, we associate the vector $\vec{F} = \vec{F}(\vec{X})$, also of dimension n. Then if $\varphi(\vec{X})$ is a differentiable real valued function on U such that $\vec{F} = \text{grad } \varphi$, it is said that φ is a potential function for $\vec{F}$. In this case, it is said that the vector field $\vec{F}$ is conservative.

To determine whether $\vec{F}(x,y) = (e^{xy}, e^{x+y})$ is conservative (i.e., has a potential function), a simple theorem will be derived. Suppose

$$\vec{F}(x,y) = (f(x,y), g(x,y))$$

has a potential function, φ. Then by definition

$$\vec{\nabla}\varphi = \vec{F}$$

or

$$\frac{\partial \varphi}{\partial x} = f \ , \quad \frac{\partial \varphi}{\partial y} = g \tag{1}$$

so that upon differentiating (1), we have

$$\frac{\partial f}{\partial y} = \frac{\partial^2 \varphi}{\partial y \partial x} \ , \quad \frac{\partial g}{\partial x} = \frac{\partial^2 \varphi}{\partial x \partial y} \ . \tag{2}$$

Assuming that these second partials are continuous,

$$\frac{\partial^2 \varphi}{\partial y \partial x} = \frac{\partial^2 \varphi}{\partial x \partial y}$$

so that

$$\frac{\partial f}{\partial y} = \frac{\partial g}{\partial x} . \tag{3}$$

The conclusion is then reached that if $\vec{F}(x,y)$ has a potential function, then (3) holds. The negation of this statement, also holds: if

$$\frac{\partial f}{\partial y} \neq \frac{\partial g}{\partial x}$$

then $\vec{F}$ has no potential function. To apply this result to

$$\vec{F}(x,y) = (f(x,y), g(x,y)) = (e^{xy}, e^{x+y}) \tag{4}$$

simply note that

$$\frac{\partial (e^{xy})}{\partial y} = x\, e^{xy} , \quad \frac{\partial (e^{x+y})}{\partial x} = e^{x+y} .$$

Thus $\partial f/\partial y \neq \partial g/\partial x$ and therefore $\vec{F}$ has no potential function.

● **PROBLEM 4-28**

Determine whether the following vector fields have potential functions and if so find them:

a) $F(x,y) = (3xy, \sin xy^3)$

b) $F(x,y) = (2xy, x^2 + 3y^2)$

Solution: a) We use the theorem: Given that f, g are differentiable functions having continuous partial derivatives on an open set V in R^2 if $\partial f/\partial y \neq \partial g/\partial x$ then the vector field, $F(x,y) = (f(x,y), g(x,y))$, does not have a potential function.

Since $F(x,y) = (3xy, \sin xy^3)$, we have $\partial f/\partial y = 3x$ and $\partial g/\partial x = y^3 \cdot \cos xy^3$. Hence, since $\partial f/\partial y \neq \partial g/\partial x$ the vector field does not have a potential function.

b) We use the theorem: Given that f, g are differentiable functions having continuous first partial derivatives at all point of some open rectangle, if $\partial f/\partial y = \partial f/\partial x$ then the vector field, $F(x,y) = (f(x,y), g(x,y))$, has a potential function.

We are given $F(x,y) = (2xy, x^2 + 3y^2)$. Let

$$f(x,y) = 2xy \quad \text{and} \quad g(x,y) = x^2 + 3y^2 .$$

Then $\partial f/\partial y = 2x$; $\partial g/\partial x = 2x$ and so, $\partial f/\partial x = \partial g/\partial y$. Hence, a potential function does exist. We want to find $\varphi(x,y)$ such that

$$\partial \varphi/\partial x = 2xy \quad \text{and} \quad \partial \varphi/\partial y = x^2 + 3y^2 . \tag{1}$$

Solving the first equation with respect to x we have $\partial \varphi = 2xy\, \partial x$ and $\varphi(x,y) = \int 2xy\, dx + h(y) = x^2 y + h(y) \tag{2}$

where $h(y)$ is some function of y. Now taking the partial derivative of $\varphi(x,y)$ with respect to y, we have

$$\frac{\partial \varphi}{\partial y} = x^2 + \frac{dh}{dy} \tag{3}$$

and upon equating (3) with (1) we have

$$\frac{dh}{dy} = 3y^2$$

which gives $h = y^3 + C$, upon integrating with respect to g.

Then (2) gives $\varphi(x,y) = x^2y + y^3 + C$. This is a potential function for $\vec{F}(x,y)$.

● **PROBLEM 4-29**

Find a potential function for the vector field
$$\vec{F}(x,y,z) = (y\cos(xy), x\cos(xy) + 2yz^3, 3y^2z^2) \quad . \tag{1}$$

Solution: First the existence of a potential function for $\vec{F}$ must be verified. This is not difficult if one uses the following theorem: for $\vec{F}(x,y,z) = (f_1(x,y,z), f_2(x,y,z), f_3(x,y,z))$ defined on a rectangular box in $\mathbb{R}^3$, with f_1, f_2 and f_3 having continuous partial derivatives on the box, if

$$\frac{\partial f_2}{\partial x} = \frac{\partial f_1}{\partial y}, \quad \frac{\partial f_3}{\partial x} = \frac{\partial f_1}{\partial z}, \quad \frac{\partial f_3}{\partial y} = \frac{\partial f_2}{\partial z}, \tag{2}$$

then $\vec{F}$ has a potential function; i.e., $\vec{F}$ is conservative. For the given problem, we have the vector field $\vec{F}$ given by (1), is defined everywhere and is such that f_1, f_2, f_3 have continuous partial derivatives everywhere. In addition, we have

$$\frac{\partial f_2}{\partial x} = \cos xy - xy\sin xy = \frac{\partial f_1}{\partial y} \tag{3}$$

$$\frac{\partial f_3}{\partial x} = 0 = \frac{\partial f_1}{\partial z} \tag{4}$$

$$\frac{\partial f_3}{\partial y} = 6yz^2 = \frac{\partial f_2}{\partial z} \tag{5}$$

hold everywhere; therefore, we conclude from the theorem that $\vec{F}$ has a potential function which is valid on all of $\mathbb{R}^3$. ($\mathbb{R}^3$ is, of course, a limiting case of a rectangular box).

Now, the problem of finding the potential function of $\vec{F}$ is, by definition, that of finding the function $\varphi(x,y,z)$ such that

$$\vec{\nabla}\varphi = \vec{F} \tag{6}$$

That is, if φ is a potential function of $\vec{F}$, then

$$\frac{\partial \varphi}{\partial x} = f_1, \quad \frac{\partial \varphi}{\partial y} = f_2, \quad \frac{\partial \varphi}{\partial z} = f_3 \quad . \tag{7}$$

To proceed, first take the integral of the first of equations (7) with respect to x. This gives

$$\int \frac{\partial \varphi}{\partial x} dx = \int y \cos xy \, dx \tag{8}$$

or

$$\varphi(x,y,z) = \sin xy + \psi(y,z) \tag{9}$$

where $\psi(y,z)$ is a function of integration analagous to a constant of integration in ordinary integration. It is included since the antiderivatives of equation (8) may differ by an arbitrary function of y and z, since integration with respect to x treats y and z as constants. Differentiating (9) with respect to y gives

$$\frac{\partial \varphi}{\partial y} = x \cos xy + \frac{\partial \psi}{\partial y}$$

but the second condition in (7) states that

$$\frac{\partial \varphi}{\partial y} = x \cos xy + \frac{\partial \psi}{\partial y} = f_2 = x \cos(xy) + 2yz^3 ,$$

which implies that

$$\frac{\partial \psi}{\partial y} = 2yz^3 \tag{10}$$

Integrating (10) with respect to y yields

$$\psi(y,z) = \int 2y \, z^3 dy = y^2 z^3 + u(z) \tag{11}$$

where $u(z)$ is a function of integration with respect to y. Using (11) in (9) gives

$$\varphi(x,y,z) = \sin xy + y^2 z^3 + u(z) . \tag{12}$$

The final step is to use the third condition in (7). Thus, we integrate (12) with respect to z and use this condition to find

$$\frac{\partial \varphi}{\partial z} = 3y^2 z^2 + u'(z) = f_3 = 3y^2 z^2 . \tag{13}$$

Hence

$$u(z) = c \tag{14}$$

where c is any constant.

Therefore, a potential function for $\vec{F}$ on R^3 is

$$\varphi(x,y,z) = \sin xy + y^2 z^3 + c \tag{15}$$

where c is some constant. Recall that if g and h are potential functions of $\vec{F}$ then $g = h + c_0$ for some constant c_0. That is, any potential function of a vector field $\vec{F}$ can differ at most by a constant from any other potential function of that vector field. Thus, (15) gives all possible potential functions of $\vec{F}$.

● **PROBLEM 4-30**

Determine whether or not the vector field

$$F(x,y) = \left(\frac{e^r x}{r} , \frac{e^r y}{r} \right) , \quad r = \sqrt{x^2 + y^2} ,$$

has a potential function (i.e., if it is conservative). If a potential function exists, find it.

Solution: A sufficient condition for the potential function of $F(x,y) = (f_1(x,y), f_2(x,y))$ to exist is that if f_1 and f_2 have continuous partial derivatives in some rectangular box on which $\vec{F}$ is defined, then $\vec{F}$ has a potential function if

$$\frac{\partial f_1}{\partial y} = \frac{\partial f_2}{\partial x}. \qquad (1)$$

Clearly, $f_1 = \frac{e^r x}{r}$ and $f_2 = \frac{e^r y}{r}$ both have continuous partials everywhere except at the origin. In addition for $(x,y) \neq (0,0)$,

$$\frac{\partial f_1}{\partial y} = \frac{x[r \frac{\partial(e^r)}{\partial y} - e^r \frac{\partial(r)}{\partial y}]}{r^2}.$$

However,

$$\frac{\partial(r)}{\partial y} = \frac{\partial(x^2+y^2)^{\frac{1}{2}}}{\partial y} = \frac{1}{2}(x^2+y^2)^{-\frac{1}{2}}(2y) = \frac{y}{r}$$

and

$$\frac{\partial(e^r)}{\partial y} = e^r \frac{\partial r}{\partial y} = \frac{y}{r} e^r ,$$

so that

$$\frac{\partial f_1}{\partial y} = \frac{x[ye^r - \frac{y}{r}e^r]}{r^2} = \frac{xye^r(1 - 1/r)}{r^2}.$$

It can be shown analogously that

$$\frac{\partial f_2}{\partial x} = \frac{xye^r}{r^2}(1 - 1/r).$$

Therefore $\frac{\partial f_1}{\partial y} = \frac{\partial f_2}{\partial x}$, $(x,y) \neq (0,0)$, and we see that $\vec{F}$ has a potential function on any rectangular box not containing the origin. To find this function, $\varphi(x,y)$, recall that the definition of potential function requires that

$$\vec{\nabla}\varphi(x,y) = (f_1, f_2)$$

or, in this case

$$\frac{\partial \varphi}{\partial x} = \frac{e^r x}{r}, \quad \frac{\partial \varphi}{\partial y} = \frac{e^r y}{r}. \qquad (2)$$

Integrating the first of these equations with respect to x gives

$$\int \frac{\partial \varphi}{\partial x} dx = \int \frac{e^r x}{r} dx = \int e^r dr \qquad (3)$$

where the variable of integration has been changed by using

$$r = (x^2+y^2)^{\frac{1}{2}}, \quad y \text{ constant}$$

so that

$$dr = \frac{x}{(x^2+y^2)^{\frac{1}{2}}} dx = \frac{x}{r} dx.$$

Hence, from (3)

$$\varphi(x,y) = e^r + \psi(y) \qquad (4)$$

where ψ is an arbitrary (for now) function of y. Notice that ψ arises due to the fact that the antiderivatives of (3) may differ by an arbitrary function of y, since the integration treats y as constant. Differentiating (4) with respect to y yields

$$\frac{\partial \varphi}{\partial y} = \frac{\partial e^r}{\partial r} \frac{\partial r}{\partial y} + \psi'(y)$$

or

$$\frac{\partial \varphi}{\partial y} = \frac{e^r y}{r} + \psi'(y) . \tag{5}$$

From the second equation of (2) we see that

$$\frac{\partial \varphi}{\partial y} = \frac{e^r y}{r} \tag{6}$$

and equating (5) and (6) gives

$$\psi'(y) = 0$$

or

$$\psi(y) = C \tag{7}$$

where C is an arbitrary constant. Therefore, from (4) we conclude that

$$\varphi(x,y) = e^r + C = e^{(x^2+y^2)^{\frac{1}{2}}} + C \tag{8}$$

is a potential function of $F(x,y)$ for any choice of C. In fact, since any two potential functions of $\vec{F}$ can differ at most by a constant, (8) gives all possible potential functions of $\vec{F}$.

A simpler way to proceed in this problem would be to recall that if a function $\varphi(x,y)$ depends only on distance, $r = \sqrt{x^2+y^2}$, i.e., if $\varphi(x,y) = g(r)$, then

$$\text{grad } \varphi = \frac{g'(r)}{r} \vec{X} = \left(\frac{g'(r)x}{r}, \frac{g'(r)y}{r}\right) \tag{9}$$

Thus, finding φ for $F(x,y) = \left(\frac{e^r x}{r}, \frac{e^r y}{r}\right)$ amounts to solving

$$g'(r) = e^r .$$

Thus

$$g(r) = e^r + C$$

and

$$\varphi(x,y) = \varphi(r) = g(r) = e^r + C = e^{(x^2+y^2)^{\frac{1}{2}}} + C.$$

CHAPTER 5

THEOREMS OF DIFFERENTIATION

This chapter is devoted to the development of the theory of differentiation. The three major results are the Mean Value Theorem, Taylor's Theorem, and the Implicit Function Theorem. Each is an existence theorem (that is, each theorem guarantees the existence of something). In addition, both Taylor's Theorem and the Implicit Function Theorem depend upon the Mean Value Theorem while the Implicit Function Theorem also depends upon the Principle of Contraction Mappings. Hence, these results are all related.

In order to prove the Mean Value Theorem, a special case--Rolle's Theorem--is proved first. Then good approximations arise by applying the Mean Value Theorem, while L'Hospital's Rule also evolves from this theorem, as does the important result on the mixed higher order partial derivatives.

Taylor's Theorem is a generalized version of the Mean Value Theorem (just as the latter generalized Rolle's Theorem) and the notion of real analytic functions is directly related to Taylor's Theorem. The Chain Rule and Euler's Theorem for homogeneous functions are derived from Taylor's Theorem.

Furthermore, the Inverse Function Theorem arises from the Mean Value Theorem and the Principle of Contraction Mappings (another existence theorem), and in turn gives rise to the Implicit Function Theorem. Once again, the Inverse Function Theorem can be proved as a corollary of the Implicit Function Theorem, which can be proved directly. Consequently, a sufficient condition for functional dependence arises from this.

THE MEAN VALUE THEOREMS

● **PROBLEM 5-1**

State and prove Rolle's Theorem.

Solution: Let f be a continuous function on a closed interval $J = [a,b]$ and suppose that $f'(x)$ exists for all $x \in (a,b)$ where $f(a) = f(b) = 0$. Then Rolle's Theorem states that there exists a point c in (a,b) such that $f'(c) = 0$.

To prove the theorem, first note that if f is identically zero, then $f'(x) = 0$ for all $x \in (a,b)$ and the theorem is proved. If $f(x) \not\equiv 0$ in $[a,b]$ then since f is continuous in $[a,b]$, there must be points at which $f(x)$ attains its maximum or minimum values denoted by M and m respectively (Maximum Value Theorem). Since $f \not\equiv 0$, one of M and m is non zero. Suppose $M \neq 0$ and $f(c) = M$ (Figure 1). Then

$$f(c + h) \leq f(c) \text{ and if } h > 0,$$

$$\frac{f(c + h) - f(c)}{h} \leq 0$$

so that

$$f_r'(c) = \lim_{h \to 0^+} \frac{f(c + h) - f(c)}{h} \leq 0 . \qquad (1)$$

If $h < 0$ then $f(c + h) \leq f(c)$ and

$$f_\ell'(c) = \lim_{h \to 0^-} \frac{f(c + h) - f(c)}{h} \geq 0 \qquad (2)$$

But since $f'(x)$ exists for all $x \in (a,b)$, $f'(c)$ must exist so that the right hand derivative of (1) and the left hand derivative of (2) must be equal. This can only be true if they are both 0, i.e., $f'(c) = 0$. It is seen by Figure 1 that c is not the only point with $f'(x) = 0$.

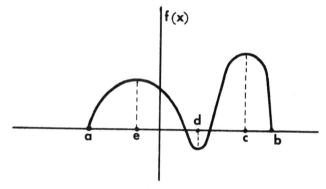

Fig. 1

The proof if $M = 0$ and $m \neq 0$ is similar.

● **PROBLEM 5-2**

(a) State and prove the Mean Value Theorem for the derivative of a real valued function of a single real variable.

(b) Give a geometrical interpretation to this result.

<u>Solution</u>: (a) Let f be a real valued function of a real variable, x, which is continuous on a closed interval [a,b] and has a derivative in the open interval (a,b). Then the Mean Value Theorem states that there exists a point c in (a,b) such that

$$f(b) - f(a) = f'(c)(b - a) . \qquad (1)$$

To prove this theorem, consider the function

$$\phi(x) = f(x) - \left[f(a) + \frac{f(b) - f(a)}{b - a} (x - a) \right] . \qquad (2)$$

As can be seen from the Figure, φ is the difference of f and the linear function whose graph consists of the line segment passing through the points (a,f(a)) and (b,f(b)). Since f is continuous on [a,b], so is φ and since f has a derivative at all points in (a,b), so does φ. Furthermore, φ(a) = φ(b) = 0 so that all of the conditions of Rolle's Theorem are satisfied for the function φ(x). Hence the conclusion of that theorem holds; i.e., there is a point c ∈ (a,b) such that

$$\phi'(c) = f'(c) - \frac{f(b) - f(a)}{b - a} = 0$$

or

$$f'(c) = \frac{f(b) - f(a)}{b - a}. \tag{3}$$

Thus, the theorem is proved.

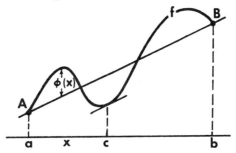

(b) The geometrical interpretation of (3) can be seen in the Figure. The equation states that there is a point c whose tangent line has the same slope as (i.e., is parallel to) the line connecting A and B.

• **PROBLEM 5-3**

Obtain an approximate value for $\sqrt{105}$ to within .01 by using the Mean Value Theorem.

Solution: Let $f(x) = \sqrt{x}$, a = 100, and b = 105. Then the Mean Value Theorem tells us that there exists a c between 100 and 105 such that

$$\frac{f(b) - f(a)}{b - a} = f'(c).$$

That is, since $f'(c) = \frac{1}{2\sqrt{c}}$,

we have

$$\frac{\sqrt{105} - \sqrt{100}}{105 - 100} = \frac{1}{2\sqrt{c}} \tag{1}$$

Now since $f(x) = \sqrt{x}$ is monotonically increasing,

$$\sqrt{a} < \sqrt{c} < \sqrt{b}$$

or

$$10 < \sqrt{c} < \sqrt{105} < \sqrt{121} = 11 \qquad (2)$$

But (1) implies that

$$\sqrt{105} - 10 = \frac{5}{2\sqrt{c}} < \frac{5}{2(10)} . \qquad (3)$$

On the other hand

$$\frac{5}{2(11)} < \frac{5}{2\sqrt{c}} = \sqrt{105} - 10 \qquad (4)$$

so that

$$\frac{5}{2(11)} < \sqrt{105} - 10 < \frac{5}{2(10)}$$

and so,

$$10.22 < \sqrt{105} < 10.25 . \qquad (5)$$

Now that we have obtained this rough estimate, equation (2) can be improved since it is now know that $\sqrt{105} < 10.25$ so that $\sqrt{c} < \sqrt{105} < 10.25$. Using this in (4) yields

$$\frac{5}{2(10.25)} < \frac{5}{2\sqrt{c}} = \sqrt{105} - 10 .$$

Hence

$$10.243 < \sqrt{105} ,$$

so that by (5) we have $10.243 < \sqrt{105} < 10.250$.

This is the required result to the desired accuracy. Increased accuracy can be obtained upon iteration of the above steps.

● **PROBLEM 5-4**

State and prove the Cauchy Mean Value Theorem.

Solution: Let f and g be real valued functions of a single variable and let each be continuous on [a,b] and have derivatives on (a,b). Then the Cauchy Mean Value Theorem states that there exists a point c in (a,b) such that

$$f'(c)[g(b) - g(a)] = g'(c)[f(b) - f(a)] \qquad (1)$$

The proof is analogous to that of the ordinary Mean Value Theorem. If $g(b) = g(a)$, then the result follows immediately if we choose c such that $g'(c) = 0$ (that this can be done is shown by Rolle's Theorem). If $g(b) \neq g(a)$, then define the function ϕ on [a,b] by

$$\phi(x) = f(x) - \left[f(a) + \frac{f(b) - f(a)}{g(b) - g(a)} [g(x) - g(a)] \right]$$

The continuity of f and g imply that ϕ is continuous on [a,b] and the differentiability of f and g imply that ϕ is differentiable on (a,b).

In addition, since $\phi(b) = \phi(a) = 0$, we have all of the conditions required for Rolle's Theorem to be satisfied. Thus there exists a c in (a,b) such that

$$\phi'(c) = f'(c) - \frac{f(b) - f(a)}{g(b) - g(a)} g'(c) = 0$$

This is the desired result.

● **PROBLEM** 5-5

State and prove L'Hospital's Rule for the indeterminant forms $\frac{0}{0}$; $\frac{\infty}{\infty}$.

Solution: If $f(x)$ and $g(x)$ are differentiable at all points in the interval (a,b) except possibly at a point x_0 in (a,b) and $g(x) \neq 0$ in some deleted neighborhood of x_0, if $f(x) \to 0$ and $g(x) \to 0$ as $x \to x_0$ or if $|f(x)| \to \infty$ and $|g(x)| \to \infty$ as $x \to x_0$, then

$$\lim_{x \to x_0} \frac{f(x)}{g(x)} = \lim_{x \to x_0} \frac{f'(x)}{g'(x)} \qquad (1)$$

Equation (1) is called L'Hospital's Rule and can easily be extended to cases where $x \to \infty$ or $-\infty$ or where $x_0 = a$ or b, or in which only one sided limits such as $x_0 \to a^+$ or $x_0 \to b^-$ are involved.

To prove the theorem for the case in which

$$\lim_{x \to x_0} f(x) = 0, \quad \lim_{x \to x_0} g(x) = 0,$$

assume f and g are continuous at x_0 and $f(x_0) = g(x_0) = 0$. Now for any $x \to x_0$ the Cauchy Mean Value Theorem states that there exists a $c \in (x_0, x)$ such that

$$\frac{f(x)}{g(x)} = \frac{f(x) - f(x_0)}{g(x) - g(x_0)} = \frac{f'(c)}{g'(c)}. \qquad (2)$$

Therefore, since $c \to x_0^+$ as $x \to x_0^+$

$$\lim_{x \to x_0^+} \frac{f(x)}{g(x)} = \lim_{x \to x_0^+} \frac{f'(c)}{g'(c)} = \lim_{x \to x_0^+} \frac{f'(x)}{g'(x)} \qquad (3)$$

If $x < x_0$ then we take the left hand limit of (2) which yields

$$\lim_{x \to x_0^-} \frac{f(x)}{g(x)} = \lim_{x \to x_0^-} \frac{f'(c)}{g'(c)} = \lim_{x \to x_0^-} \frac{f'(x)}{g'(x)} \qquad (4)$$

where $c \in (x, x_0)$. Since the right and left hand limits are equal,

$$\lim_{x \to x_0} \frac{f(x)}{g(x)} = \lim_{x \to x_0} \frac{f'(x)}{g'(x)}.$$

To prove (1) for the case in which $\lim_{x \to x_0} f(x) = \infty$ and $\lim_{x \to x_0} g(x) = \infty$ where $x_0 \in (a,b)$, let x_1 be such that $a < x_0 < x < x_1 < b$. By the Cauchy Mean Value Theorem, there exists a number c such that

$$\frac{f(x) - f(x_1)}{g(x) - g(x_1)} = \frac{f'(c)}{g'(c)} \qquad c \in (x, x_1).$$

Hence,

$$\frac{f(x) - f(x_1)}{g(x) - g(x_1)} = \frac{f(x)}{g(x)} \cdot \frac{1 - f(x_1)/f(x)}{1 - g(x_1)/g(x)} = \frac{f'(c)}{g'(c)}$$

• **PROBLEM 5-6**

Use L'Hospital's Rule to evaluate

(a) $\lim_{x \to 2} \frac{2x^2 - 4x}{x-2}$

(b) $\lim_{x \to 0} x \ln x$

(c) $\lim_{x \to 0} \frac{1}{\sin x} - \frac{1}{x}$

(d) $\lim_{x \to \infty} (1 + x)^{\frac{1}{x}}$

Solution: (a) The function takes the form $0/0$, and therefore we can apply L'Hospital's rule to obtain:

$$\lim_{x \to 2} \frac{4x - 4}{1} = 4$$

We can also solve the problem in a different way by noting that the numerator can be factored.

$$\lim_{x \to 2} \frac{2x^2 - 4x}{x - 2} = \lim_{x \to 2} \frac{2x(x - 2)}{x - 2}$$

$$= \lim_{x \to 2} 2x$$

$$= 4$$

(b) Since $\ln x \to -\infty$ when $x \to 0$, the product $x \ln x$ takes the indeterminate form $0 \cdot \infty$ when $x \to 0$. We must, therefore, write the expression in the form of a ratio, so that we can apply L'Hospital's rule. If we write

$$x \ln x = \frac{\ln x}{\frac{1}{x}},$$

this fraction takes the indeterminate form ∞/∞ when $x \to 0$. Applying L'Hospital's rule, therefore,

$$\lim_{x \to 0} (x \ln x) = \lim_{x \to 0} \left[\frac{\ln x}{\frac{1}{x}} \right] = \lim_{x \to 0} \left[-\frac{1}{x} \bigg/ \frac{1}{x^2} \right]$$

$$= \lim_{x \to 0} (-x) = 0.$$

(c) Since

$$\frac{1}{\sin x} \to \infty \quad \text{and} \quad \frac{1}{x} \to \infty \quad \text{when } x \to 0,$$

the given function takes the indeterminate form $\infty - \infty$. We should, consequently, write the function in the form of a ratio which takes an indeterminate form so that L'Hospital's rule can be applied. Therefore,

$$\frac{1}{\sin x} - \frac{1}{x} = \frac{x - \sin x}{x \sin x},$$

which becomes $0/0$ when $x \to 0$. Applying L'Hospital's rule twice, we obtain

$$\lim_{x \to 0} \left(\frac{1}{\sin x} - \frac{1}{x} \right) = \lim_{x \to 0} \frac{x - \sin x}{x \sin x}$$

$$= \lim_{x \to 0} \frac{1 - \cos x}{x \cos x + \sin x}$$

$$= \lim_{x \to 0} \frac{\sin x}{-x \sin x + \cos x + \cos x} = 0.$$

A simpler approach is to recognize that, for small angles, $\sin x = x$. The original expression therefore becomes

$$\lim_{x \to 0} \left(\frac{1}{x} - \frac{1}{x} \right) = 0 .$$

(d) The function takes the form ∞^0, which is indeterminate. We proceed to manipulate the function to arrive at a determinate form or one to which L'Hospital's rule may be applied. Since a power in $1/x$ is present, we attempt to use logarithms. Thus,

$$\ln y = \lim_{x \to \infty} \frac{\ln (1 + x)}{x}$$

The function now takes the form ∞/∞, so that we can apply L'Hospital's rule.

$$\ln y = \lim_{x \to \infty} \frac{\frac{1}{1+x}}{1} = 0$$

$$y = e^0 = 1 .$$

or

$$\frac{f(x)}{g(x)} = \frac{f'(c)}{g'(c)} \cdot \frac{1 - g(x_1)/g(x)}{1 - f(x_1)/f(x)} \quad (5)$$

Assume that

$$\lim_{x \to x_0} \frac{f'(x)}{g'(x)} = \lim_{x \to x_0^+} \frac{f'(x)}{g'(x)} = L.$$

Then (5) becomes

$$\frac{f(x)}{g(x)} = \left[\frac{f'(c)}{g'(c)} - L \right] \left[\frac{1 - g(x_1)/g(x)}{1 - f(x_1)/f(x)} \right] + L \left[\frac{1 - g(x_1)/g(x)}{1 - f(x_1)/f(x)} \right] \quad (6)$$

Now, since

$$\lim_{x \to x_0^+} f(x) = \infty = \lim_{x \to x_0^+} g(x) ,$$

it is seen that

$$\lim_{x \to x_0^+} L \left[\frac{1 - g(x_1)/g(x)}{1 - f(x_1)/f(x)} \right] = L \cdot 1 = L .$$

Also, since x_1 can be chosen arbitrarily close to x_0,

$$\lim_{x \to x_0^+} \left| \left(\frac{f'(c)}{g'(c)} - L \right) \right| = 0$$

so that (6) becomes

$$\lim_{x \to x_0^+} \frac{f(x)}{g(x)} = L = \lim_{x \to x_0^+} \frac{f'(x)}{g'(x)} .$$

A similar argument shows that the left hand limits are also equal, hence, the proof is complete.

● **PROBLEM 5-7**

State and prove the Mean Value Theorem for a real valued function, f, of n real variables $x_1, x_2, \ldots, x_n$.

Solution: Let f be continuous at distinct points

$$\bar{X}_1 = (x_{1_1}, x_{2_1}, \ldots, x_{n_1})$$

and $\bar{X}_2 = (x_{1_2}, x_{2_2}, \ldots, x_{n_2})$,

and differentiable on the line segment L joining them. Then the Mean Value Theorem states that there is a point $\bar{X}_0$ on L such that

$$f(\bar{X}_2) - f(\bar{X}_1) = \left(d_{\bar{X}_0} f\right)(\bar{X}_2 - \bar{X}_1) \tag{1}$$

where $d_{\bar{X}_0} f$ is the differential of f at $\bar{X}_0$.

To prove the theorem, define the unit vector

$$U = \frac{\bar{X}_2 - \bar{X}_1}{\|\bar{X}_2 - \bar{X}_1\|}$$

and let

$$g(t) = f(\bar{X}_1 + t U), \quad (0 \leq t \leq \|\bar{X}_2 - \bar{X}_n\|).$$

Since f is continuous on L, g is continuous on

$$[0, \|\bar{X}_2 - \bar{X}_1\|],$$

and it will now be shown that g has a derivative on $(0, \|\bar{X}_2 - \bar{X}_1\|)$.

Let $t_0 \in (0, \|\bar{X}_2 - \bar{X}_1\|)$. Then

$$\frac{g(t) - g(t_0)}{t - t_0} = \frac{f(\bar{X}_1 + t U) - f(\bar{X}_1 + t_0 U)}{t - t_0}$$

or with

$$t - t_0 = \tau$$

and

$$\hat{\bar{X}} = \bar{X}_1 + t_0 U,$$

$$\frac{g(t) - g(t_0)}{t - t_0} = \frac{f(\hat{\bar{X}} + \tau U) - f(\hat{\bar{X}})}{\tau}.$$

Since f is differentiable at $\hat{\underline{X}}$, the right hand side approaches the directional derivative

$$D_U f(\hat{\underline{X}})$$

as τ approaches zero. Therefore, $g'(t_0)$ exists and

$$g'(t_0) = \lim_{t \to t_0} \frac{g(t) - g(t_0)}{t - t_0}$$

$$= \lim_{t \to t_0} \frac{f(\hat{\underline{X}} + \tau U) - f(\hat{\underline{X}})}{\tau} \qquad (2)$$

$$= D_U f(\hat{\underline{X}}) = (d_{\hat{\underline{X}}} f)(U)$$

where $D_U(f(\hat{X}))$ is the directional derivative of f at $\hat{\underline{X}}$ in the direction U. The last equality was proved in the previous chapter.

Thus, g has a derivative in $(0, \|\underline{X}_2 - \underline{X}_1\|)$ and the ordinary Mean Value Theorem for single variable functions holds. Thus there exists an $S_0 \in (0, \|\underline{X}_2 - \underline{X}_1\|)$ such that

$$g(\|\underline{X}_2 - \underline{X}_1\|) - g(0) = \|\underline{X}_2 - \underline{X}_1\| g'(S_0) . \qquad (3)$$

But $g(\|\underline{X}_2 - \underline{X}_1\|) = f(\underline{X}_2)$ and $g(0) = f(\underline{X}_1)$. Thus, using (2) with $t_0 = S_0$ and $\underline{X}_0 = \underline{X}_1 + S_0 U$ in (3) gives

$$f(\underline{X}_2) - f(\underline{X}_1) = \|\underline{X}_2 - \underline{X}_1\|(d_{\underline{X}_0} f)(U)$$

$$= \left(d_{\underline{X}_0} f\right)(\underline{X}_2 - \underline{X}_1) ,$$

which completes the proof since $\underline{X}_0$ is on L.

• **PROBLEM 5-8**

Let $f(x,y) = x^2 + y^2 + x^3$. Find a suitable (u,v) on the line segment connecting (a,b) with (c,d) such that

$$f(c,d) - f(a,b) = \frac{\partial f}{\partial x}(u,v)(c-a) + \frac{\partial f}{\partial y}(u,v)(d-b) \qquad (1)$$

if $(a,b) = (1,2)$ and $(c,d) = (1+h, 2+k)$.

Solution: If ϕ is continuous at points α and β and differentiable on the line segment $\overline{\alpha\beta}$, then the Mean Value Theorem says that the folloiwng holds for some γ on $\overline{\alpha\beta}$:

$$\phi(\beta) - \phi(\alpha) = \phi'(\gamma)(\beta - \alpha) .$$

In the case of $f(x,y) = x^2 + y^2 + x^3$, $\beta = (c,d)$ and $\alpha = (a,b)$, the theorem says that for some $(u,v) \in \overline{\alpha\beta}$

$$f(c,d) - f(a,b) = f'(u,v)(c-a, d-b)$$

$$= \frac{\partial f}{\partial x}(u,v)(c-a) + \frac{\partial f}{\partial y}(u,v)(d-b) \tag{2}$$

Hence, there exists a (u,v) on the segment between (a,b) and (c,d) such that (1) is satisfied. To find (u,v) note that

$$\frac{\partial f}{\partial x}(x,y) = 2x + 3x^2$$

and

$$\frac{\partial f}{\partial y}(x,y) = 2y$$

and substitute for

$$\frac{\partial f}{\partial x}(u,v), \quad \frac{\partial f}{\partial y}(u,v), \quad f(c,d), \text{ and } f(a,b) \text{ into}$$

(2) to obtain

$$(c^2 + d^2 + c^3) - (a^2 + b^2 + a^3) =$$

$$(2u + 3u^2)(c-a) + (2v)(d-b) \tag{3}$$

Now, with $a = 1$, $b = 2$, $c = 1 + h$, and $d = 2 + k$, (3) becomes

$$(1+h)^2 + (2+k)^2 + (1+h)^3 - 1^2 - 2^2 - 1^3 =$$

$$(2u + 3u^2)(1+h - 1) + (2v)(2+k - 2)$$

or $(5h + 4h^2 + h^3) + (4k + k^2) = (2u + 3u^2)h + (2v)k$. \quad (4)

Since (u,v) is on the line segment from $(1,2)$ to $(1+h, 2+k)$, $(u,v) = (1 + \theta_1 h, 2 + \theta_2 k)$ for some $0 < \theta_1 < 1$, $0 < \theta_2 < 2$. Hence, (4) becomes

$$(5 + 4h + h^2)h + (4 + k)k = (2(1 + \theta_1 h) + 3(1 + \theta_1 h)^2)h$$
$$+ (2(2 + \theta_2 k))k$$

which can be separated into

$$(5 + 4h + h^2)h = \left(2(1 + \theta_1 h) + 3(1 + \theta_1 h)^2\right)h \tag{5}$$

and

$$(4 + k)k = 2(2 + \theta_2 k)k . \tag{6}$$

Solving (5) for θ_1:

$$5 + 4h + h^2 = 2 + 2\theta_1 h + 3(1 + 2\theta_1 h + \theta_1^2 h^2)$$

211

$$5 + 4h + h^2 = 2 + 2\theta_1 h + 3 + 6\theta_1 h + 3\theta_1^2 h^2$$

or

$$\theta_1^2 (3h^2) + \theta_1 (8h) - (h^2 + 4h) = 0.$$

Dividing both sides of this equation by h gives

$$\theta_1^2 (3h) + \theta_1 (8) - (h + 4) = 0.$$

Using the quadratic formula,

$$\theta_1 = \frac{-8 \pm \sqrt{8^2 - 4(3h)(-h-4)}}{2(3h)}$$

$$= \frac{-8 \pm \sqrt{64 + 12h^2 + 48h}}{6h}$$

$$= \frac{-4 \pm \sqrt{3h^2 + 12h + 16}}{3h}.$$

Since $\theta_1 > 0$, $\theta_1 = \frac{-4 + \sqrt{3h^2 + 12h + 16}}{3h}$ is the desired solution. Note that this exists for all $h \neq 0$. Thus

$$u = 1 + \theta_1 h = \frac{-1 + \sqrt{3h^2 + 12h + 16}}{3}.$$ Solving (6) for θ_2:

$$4 + k = 2(2 + \theta_2 k)$$

$$4 + k = 4 + 2\theta_2 k$$

$$\theta_2 = \frac{1}{2}.$$

Thus, $v = 2 + \theta_2 k = 2 + \frac{1}{2} k$. Hence,

$$(u, v) = \left(\frac{-1 + \sqrt{3h^2 + 12h + 16}}{3}, 2 + \frac{1}{2} k \right)$$

satisfies (3) and hence (1).

● **PROBLEM 5-9**

Show that if a function $f: V \to R$, $V \subseteq R^n$, is C^2 locally at a, then

$$\frac{\partial^2 f}{\partial x_i \partial x_j} (a) = \frac{\partial^2 f}{\partial x_j \partial x_i} (a)$$

for all i,j between 1 and n inclusive.

Solution: The fact that f belongs to C^2 means that in some neighborhood $N \cap V \subseteq R^n$, all of the second order partial derivatives of f (i.e., $\frac{\partial^2 f}{\partial x_i \partial x_j}$ for any $1 \leq i, j \leq n$) exist and are continuous. Without loss of generality n can be assumed to be 2. Letting f be a function of x and y, it must now be shown that

$$\frac{\partial^2 f}{\partial x \partial y}(a) = \frac{\partial^2 f}{\partial y \partial x}(a).$$

Take $a = (x_0, y_0)$. Define $A(t)$ [$|t| > 0$], $g(x)$, and $h(y)$ as follows:

$$A(t) = \frac{1}{t^2}[f(x_0+t, y_0+t) - f(x_0+t, y_0)$$
$$- f(x_0, y_0+t) + f(x_0, y_0)] \qquad (1)$$

$$g(x) = f(x, y_0+t) - f(x, y_0) \qquad (2)$$

$$h(y) = f(x_0+t, y) - f(x_0, y). \qquad (3)$$

From (1) and (2) note that

$$A(t) = \frac{1}{t^2}[g(x_0+t) - g(x_0)]. \qquad (4)$$

By the Mean Value Theorem (MVT), for $0 < |t_1| < |t|$, (4) yields

$$A(t) = \frac{1}{t^2}[(x_0+t) - x_0][g'(x_0+t_1)]$$

$$= \frac{1}{t}\left[\frac{\partial f}{\partial x}(x_0+t_1, y_0+t) - \frac{\partial f}{\partial x}(x_0+t_1, y_0)\right]. \qquad (5)$$

Again, by the MVT, for

$$0 < |t_2| < |t|,$$

(5) yields

$$A(t) = \frac{1}{t}\left[(y_0+t) - y_0\right]\left[\frac{\partial}{\partial y}\left(\frac{\partial f}{\partial x}\right)(x_0+t_1, y_0+t_2)\right]$$

$$= \frac{\partial^2 f}{\partial x \partial y}(x_0+t_1, y_0+t_2). \qquad (6)$$

Similarly, for appropriate $0 < |t_3|, |t_4| < |t|$, (1), (3), and the MVT give

$$A(t) = \frac{1}{t^2}[h(y_0+t) - h(y_0)]$$

$$= \frac{1}{t} h'(y_0+t_3)$$

$$= \frac{1}{t}\left[\frac{\partial f}{\partial y}(x_0+t, y_0+t_3) - \frac{\partial f}{\partial y}(x_0, y_0+t_3)\right]$$

$$= \frac{\partial^2 f}{\partial y \partial x} (x_0 + t_4, y_0 + t_3). \qquad (7)$$

Since the second order partial derivatives of f are continuous on N∩V, it is possible to choose $|t|$ small enough so that for any $\varepsilon > 0$

$$\left| \frac{\partial^2 f}{\partial x \partial y} (x_0, y_0) - \frac{\partial^2 f}{\partial x \partial y} (x,y) \right| < \frac{\varepsilon}{2}, \qquad (8)$$

$$\left| \frac{\partial^2 f}{\partial x \partial y} (x_0, y_0) - \frac{\partial^2 f}{\partial x \partial y} (x,y) \right| < \frac{\varepsilon}{2}, \qquad (9)$$

and $(x,y) \in N \cap V$ whenever

$$|x - x_0| < |t| \quad \text{and} \quad |y - y_0| < |t|. \qquad (10)$$

Since $(x_0 + t_1, y_0 + t_2)$ and $(x_0 + t_4, y_0 + t_3)$ are in the rectangle defined by (10), (8) and (9) yield

$$\left| \frac{\partial^2 f}{\partial x \partial y} (x_0, y_0) - \frac{\partial^2 f}{\partial x \partial y} (x_0, y_0) \right| < \varepsilon$$

by the triangle inequality. Hence

$$\frac{\partial^2 f}{\partial x \partial y} (a) = \frac{\partial^2 f}{\partial x \partial y} (a).$$

This can clearly by generalized so that if f is C^k, then

$$\frac{\partial^k f(a)}{\partial x_{i_1} \ldots \partial x_{i_k}} = \frac{\partial^k f(a)}{\partial x_{j_1} \ldots \partial x_{j_k}}$$

where $(i_1, \ldots, i_k)$ and $(j_1, \ldots, j_k)$ are permutations of each other. Hence, in this case, it is possible to rewrite this with no redundancies:

$$\frac{\partial^k f}{\partial x_{i_1}^{k_1} \ldots \partial x_{i_m}^{k_m}} (a), \quad \sum_{i=1}^m k_i = k.$$

● **PROBLEM 5-10**

(a) Let $f: R^2 \to R$ be defined by

$$f(x,y) = \begin{cases} 2xy \dfrac{x^2 - y^2}{x^2 + y^2}, & x^2 + y^2 \neq 0 \\ 0, & x = y = 0. \end{cases}$$

Show that

$$\frac{\partial^2 f}{\partial x \partial y} \neq \frac{\partial^2 f}{\partial x \partial y}$$

and explain why.

(b) Does there exist a function f with continuous second partial derivatives (i.e., an element of C^2) such that

$$\frac{\partial f}{\partial x} = x^2$$

and

$$\frac{\partial f}{\partial y} = xy \ ?$$

Solution: (a) Calculate the partial derivatives:

$$\frac{\partial f}{\partial x}(x,y) = 2y \frac{x^2 - y^2}{x^2 + y^2} + 2xy \frac{4xy^2}{(x^2 + y^2)^2}, \quad x^2 + y^2 \neq 0$$

$$\frac{\partial f}{\partial y}(x,y) = 2x \frac{x^2 - y^2}{x^2 + y^2} - 2xy \frac{4xy^2}{(x^2 + y^2)^2}, \quad x^2 + y^2 \neq 0$$

$$\frac{\partial f}{\partial x}(0,0) = \lim_{h \to 0} \frac{f(h,0) - f(0,0)}{h} = \lim_{h \to 0} \frac{0}{h} = 0$$

$$\frac{\partial f}{\partial x}(0,0) = \lim_{k \to 0} \frac{f(0,k) - f(0,0)}{k} = \lim_{k \to 0} \frac{0}{k} = 0$$

$$\frac{\partial^2 f}{\partial x \partial y}(0,0) = \lim_{h \to 0} \frac{\frac{\partial f}{\partial y}(h,0) - \frac{\partial f}{\partial x}(0,0)}{h} = \lim_{h \to 0} \frac{2h}{h} = 2$$

and

$$\frac{\partial^2 f}{\partial x \partial y}(0,0) = \lim_{k \to 0} \frac{\frac{\partial f}{\partial x}(0,k) - \frac{\partial f}{\partial x}(0,0)}{k} = \lim_{k \to 0} -\frac{2k}{k} = -2$$

Therefore,

$$\frac{\partial^2 f}{\partial x \partial y}(0,0) = 2 \neq -2 = \frac{\partial^2 f}{\partial x \partial y}(0,0),$$

or simply,

$$\frac{\partial^2 f}{\partial y \partial x} \neq \frac{\partial^2 f}{\partial x \partial y}.$$

This seemingly contradicts the theorem which states that the mixed partial derivatives are equal when they are continuous. However, neither $\frac{\partial^2 f}{\partial x \partial y}$ nor $\frac{\partial^2 f}{\partial y \partial x}$ is continuous at (0,0). (For example, it is easily seen that $\lim_{(x,y)\to(0,0)} \frac{\partial^2 f}{\partial x \partial y}(x,y) \neq -2$ and, in fact, does not exist.)

(b) Since $\frac{\partial f}{\partial x} = x^2$ and $\frac{\partial f}{\partial y} = xy$, then

$$\frac{\partial^2 f}{\partial x \partial y} = y \neq 0 = \frac{\partial^2 f}{\partial y \partial x}.$$

If f was an element of C^2, this would be impossible. Therefore, no such C^2 function exists.

TAYLOR'S THEOREM

● **PROBLEM 5-11**

Prove Taylor's Theorem for $f \in C^r(E)$ where $E \subseteq R^n$ is an open convex set.

Solution: Let $f: E \to R$ where $E \subseteq R^n$ is open and convex have continuous partial derivatives through order r (i.e., $f \in C^r(E)$). Let a be any point in E, then there exists an $\bar{x} \in E$ such that

$$0 < \|\bar{x} - a\| < \|x - a\|$$

and

$$f(x) = f(a) + J_f(a)(x-a) + \frac{1}{2}\langle H_f(a)(x-a), x-a \rangle + \ldots$$

$$= f(a) + \sum_{k=1}^{r-1} \frac{1}{k!} \sum_{\substack{i_j=1 \\ 1 \le j \le k}}^{n} \frac{\partial^k f}{\partial x_{i_1} \cdots \partial x_{i_k}}(a)(x_{i_1} - a_{i_1}) \cdots (x_{i_k} - a_{i_k}) +$$

$$\frac{1}{r!} \sum_{\substack{i_j=1 \\ 1 \le j \le r}}^{n} \frac{\partial^r f}{\partial x_{i_1} \cdots \partial x_{i_r}}(\bar{x})(x_{i_1} - a_{i_1}) \cdots (x_{i_r} - a_{i_r}),$$

where $J_f(a)$ and $H_f(a)$ are, respectively, the Jacobian and Hessian matrices evaluated at a, and $\langle,\rangle$ denotes the Euclidean inner product. This is the version of Taylor's Theorem which is to be proved. In order to prove this version, the one-dimensional version of the theorem is needed: Let

$$f \in C^r[a,b]$$

and let $f^{(r+1)}$ exist in (a,b). If $\alpha, \beta \in [a,b]$, then there exists $\gamma \in (\alpha, \beta)$ such that

$$f(\beta) = f(\alpha) + f'(\alpha)(\beta-\alpha) + \frac{1}{2}f''(\alpha)(\beta-\alpha)^2 + \ldots + \frac{1}{r!}f^{(r)}(\alpha)(\beta-\alpha)^r$$

$$+ \frac{1}{(r+1)!} f^{(r+1)}(\gamma)(\beta-\alpha)^{r+1}.$$

To prove the latter version, let $\zeta \in \mathbb{R}$ be defined by (1)

$$\frac{(\beta-\alpha)^{r+1}}{(r+1)!} \zeta = f(\beta) - [f(\alpha) + f'(\alpha)(\beta-\alpha) + \ldots + \frac{1}{r!} f^{(r)}(\alpha)(\beta-\alpha)^r]$$

and define ϕ on $[a,b]$ by

$$\phi(x) = f(\beta) - \left[f(x) + f'(x)(\beta-x) + \ldots + \frac{1}{r!} f^{(r)}(x)(\beta-x)^r \right.$$
$$\left. + \frac{\zeta}{(r+1)!} (\beta-x)^{r+1} \right]$$

which is continuous on $[a,b]$ and differentiable on (a,b).

Evidently $\phi(\beta) = 0$ and, by the definition of ζ, $\phi(\alpha) = 0$. Applying Rolle's Theorem, there exists $\gamma \in (\alpha, \beta)$ such that $\phi'(\gamma) = 0$. Differentiating ϕ gives

$$\phi'(x) = - \left[f'(x) - f'(x) + f''(x)(\beta-x) - \ldots - \frac{(\beta-x)^{r-1}}{(r-1)!} f^{(r)}(x) \right.$$
$$\left. + \frac{1}{r!} f^{(r+1)}(x)(\beta-x)^r - \frac{\zeta}{r!}(\beta-x)^r \right]$$

$$= \frac{\zeta - f^{(r+1)}(x)}{r!} (\beta - x)^r \qquad (2)$$

Since $\phi'(\gamma) = 0$, (2) implies that $\zeta = f^{(r+1)}(\gamma)$ and hence (1) is the desired result. Now to apply the n-dimensional version of the theorem,

Let $F \in C^r[0,1]$ be defined by

$$F(t) = f(a + t(x - a))$$

By the version which was just proved, there exists a
$$t_0 \in (0,1)$$
such that
$$F(1) = F(0) + F'(0) + \frac{1}{2} F''(0) + \ldots + \frac{1}{(r-1)!} F^{(r-1)}(0)$$
$$+ \frac{1}{r!} F^{(r)}(t_0).$$

Note from the above definition of $F(t)$ that $F(1) = f(x)$, $F(0) = f(a), \ldots,$

$$F^{(k)}(0) = \sum_{\substack{i_j=1 \\ 1 \le j \le k}}^{n} \frac{\partial^k f}{\partial x_{i_1} \ldots \partial x_{i_k}}(a)(x_{i_1} - a_{i_1}) \ldots (x_{i_k} - a_{i_k}),$$

and

$$F^{(r)}(t_0) = \sum_{\substack{i_j=1 \\ 1 \le j \le r}}^{n} \frac{\partial^r f}{\partial x_{i_1} \ldots \partial x_{i_r}}(a+t_0(x-a))(x_{i_1}-a_{i_1})\ldots(x_{i_r}-a_{i_r}).$$

Since E is convex,

$$\bar{x} = a + t_0(x-a) \in E \text{ and } 0 < \|\bar{x} - a\| < \|x - a\|$$

● **PROBLEM 5-12**

Define "real analytic" at a point. Also, suppose $f \in C^{\infty}(E)$ where $E \subset R^n$ is convex and bounded. Give criteria for f to be real analytic locally at $a \in E$ and globally on E.

Solution: A function $f \in C^{\infty}(E)$ (that is, all partial derivatives exist and are continuous on E) is real analytic at $a \in E$ if the Taylor series about a converges in some neighborhood of a. Furthermore, if the Taylor series about every $a \in E$ converges on E, then f is real analytic on E. The Taylor series expansion of f about $a \in E$ is

$$f(x) = f(a) + \sum_{i=1}^{n} \frac{\partial f}{\partial x_i}(a)(x_i - a_i) + \ldots +$$

$$\frac{1}{k!} \sum_{\substack{i_j = 1 \\ 1 \le j \le k}}^{n} \frac{\partial^k f}{\partial x_{i_1} \ldots \partial x_{i_k}}(a)(x_{i_1} - a_{i_1}) \ldots (x_{i_k} - a_{i_k}) + \ldots \quad (1)$$

$$= f(a) + J_f(a)(x-a) + \frac{1}{2} \langle H_f(a)(x-a), x-a \rangle + \ldots$$

where $J_f(a)$ and $H_f(a)$ are the Jacobian and Hessian matrices, respectively, and $\langle,\rangle$ denotes the Euclidean inner product. To show that (1) converges note that, by Taylor's Theorem,

$$f(x) = f(a) + J_f(a)(x-a) + \frac{1}{2}\langle H_f(a)(x-a), x-a\rangle + \ldots + R_k(x,a)$$

where

$$R_k(x,a) = \frac{1}{k!} \sum_{\substack{i_j=1 \\ 1 \le j \le k}}^{n} \frac{\partial^k f}{\partial x_{i_1} \ldots \partial x_{i_k}}(\bar{x})(x_{i_1} - a_{i_1}) \ldots (x_{i_k} - a_{i_k}), \quad (2)$$

for some $\bar{x}$ such that $0 < \|\bar{x} - a\| < \|x - a\|$, and that (1) converges if and only if $R_k(x,a) \to 0$ as $k \to \infty$. Suppose that

$$\sup_{x \in E} \frac{\partial^k f(x)}{\partial x_{i_1} \ldots \partial x_{i_k}} \le c \, \alpha^k \, k! \quad (3)$$

218

for non-negative real constants c and α. Then, from (2) and (3),

$$|R_k(x,a)| \leq \frac{1}{k!} \left| \sum_{\substack{i_j=1 \\ 1 \leq j \leq k}}^{n} c \alpha^k k! \, \|x-a\|^k \right|$$

$$= \frac{c \alpha^k k!}{k!} \|x-a\|^k n^k$$

$$= c(n\alpha \|x-a\|)^k$$

Therefore, when $\|x-a\| < \frac{1}{n\alpha}$, $R_k(x-a) \to 0$ as $k \to \infty$.

Hence, (3) is a sufficient criterion for f to be real analytic locally at any $a \in E$. Now suppose that

$$\sup_{x \in E} \left| \frac{\partial^k f(x)}{\partial x_{i_1} \cdots \partial x_{i_k}} \right| \leq c \alpha^k \qquad (4)$$

for non-negative real constants c and α. Then (2) and (4) give

$$|R_k(x,a)| \leq \frac{1}{k!} \sum_{\substack{i_j=1 \\ 1 \leq j \leq k}}^{n} c \alpha^k \|x-a\|^k$$

$$= \frac{c \alpha^k \|x-a\|^k n^k}{k!}$$

$$= c \frac{(n\alpha \|x-a\|)^k}{k!} . \qquad (5)$$

Since the power series whose k^{th} term is the right hand side of (5) converges by the ratio test, $R_k(x,a) \to 0$ as $k \to \infty$.

Hence, (4) is a sufficient criterion for f to be real analytic globally on E.

● **PROBLEM 5-13**

Give a Taylor expansion of $f(x,y) = e^x \cos y$ on some compact convex domain E containing $(0,0)$.

Solution: Let $f(x,y) = \phi(x) \psi(y)$ where $\phi(x) = e^x$ and $\psi(y) = \cos y$. Also let $E = E_x \times E_y$ where E_x and E_y are compact convex subsets of R, i.e., closed and bounded intervals. Note that

$$\sup_{x \in E_x} \left\| \frac{d^k \phi(x)}{dx^k} \right\| = e^{\max E_x}$$

and
$$\sup_{y \in E_y} \left\| \frac{d^k \psi(y)}{dy^k} \right\| \leq 1 .$$

Hence $\phi(x)$ and $\psi(y)$ are real analytic on E.

Furthermore, note that

$$e^x = \sum_{i=0}^{\infty} \frac{x^i}{i!} \frac{d^i e^x}{dx^i} \bigg|_{x=0} = \sum_{i=0}^{\infty} \frac{x^i}{i!} \quad (1)$$

and

$$\cos y = \sum_{i=0}^{\infty} \frac{x^i}{i!} \frac{d^i \cos y}{dy^i} \bigg|_{y=0} = \sum_{i=0}^{\infty} (-1)^i \frac{y^{2i}}{(2i)!} . \quad (2)$$

Since $f = \phi\psi$ and ϕ and ψ are real analytic on E, f is real analytic on E:

$$\sup_{x,y \in E} \left\| \frac{\partial^k f(x,y)}{\partial x^i \partial y^{k-i}} \right\| \leq e^{\max E_x} .$$

hence, from (1) and (2)

$$e^x \cos y = \left(\sum_{i=0}^{\infty} \frac{x^i}{i!} \right) \left(\sum_{j=0}^{\infty} (-1)^j \frac{y^{2j}}{(2j)!} \right)$$

$$= \sum_{k=0}^{\infty} \left(\sum_{\substack{i+j=k \\ i,j \geq 0}} (-1)^j \frac{x^i y^{2j}}{i!(2j)!} \right) . \quad (2)$$

The first three terms corresponding to k=0, k=1, and k=2 give the approximation:

$$e^x \cos y \approx 1 + x - \frac{1}{2} y^2 + \frac{1}{2} x^2 - \frac{1}{2} xy^2 + \frac{1}{24} y^4$$

which is known to be accurate near (0,0).

● **PROBLEM 5-14**

Rewrite the polynomial $\sum_{i=0}^{n} a_i t^i$ as a polynomial in $x = t - 1$. Verify this for the polynomial $1 + t + 3t^4$.

<u>Solution</u>: To rewrite the polynomial $f(t) = \sum_{i=0}^{n} a_i t^i$
as a polynomial in $x = t - 1$, i.e., as $g(x) = \sum_{i=1}^{m} b_i x^i$,
use Taylor's Theorem. Note that f is C^∞ and that

$$f'(t) = \sum_{i=0}^{n} i\alpha_i t^{i-1}, \quad f''(t) = \sum_{i=0}^{n} i(i-1)\alpha_i t^{i-2}, \ldots,$$

$$f^{(n)}(t) = \sum_{i=0}^{n} i(i-1) \cdots (i-n+1)\alpha_i t^{i-n}, \text{ and}$$

$$0 = f^{(n+1)}(t) = f^{(n+2)}(t) = \ldots$$

To get a polynomial in t-1, expand about 1. Hence,

$$f(1) = \sum_{i=0}^{n} \alpha_i, \quad f'(1) = \sum_{i=1}^{n} i\alpha_i, \quad f''(1) = \sum_{i=2}^{n} i(i-1)\alpha_i, \ldots,$$

$$f^{(k)}(1) = \sum_{i=k}^{n} i(i-1) \cdots (i-k+1)\alpha_i, \ldots,$$

$$f^{(n)}(1) = n!\alpha_n.$$

Therefore, the Taylor expansion

$$f(t) = \sum_{j=0}^{n} \frac{f^{(j)}(1)}{j!} (t-1)^j$$

$$= \sum_{j=0}^{n} \frac{1}{j!} \sum_{i=j}^{n} i(i-1) \cdots (i-j+1)\alpha_i (t-1)^j$$

$$= \sum_{j=0}^{n} \frac{1}{j!} \sum_{i=j}^{n} \frac{i!}{(i-j)!} \alpha_i (t-1)^j$$

$$= \sum_{j=0}^{n} \left(\sum_{i=j}^{n} \frac{i!}{(i-j)!j!} \alpha_i \right) (t-1)^j$$

$$= \sum_{j=0}^{n} \left(\sum_{i=j}^{n} \binom{i}{j} \alpha_i \right) (t-1)^j \tag{1}$$

$$= \sum_{j=0}^{n} b_j x^j = g(x).$$

Note that the degree of the new polynomial (1) is also n and the j-th coefficient is in terms of the last n-j+1 of the α_i. As an example, take

$$f(t) = 1 + t + 3t^4.$$

According to this method, $\alpha_0 = \alpha_1 = 1$, $\alpha_2 = \alpha_3 = 0$, and $\alpha_4 = 3$. Hence

$$f(t) = \sum_{j=0}^{4} \left[\sum_{i=j}^{4} \frac{i!}{(i-j)!j!} \alpha_i \right] (t-1)^j$$

$$= \left[\frac{0!}{0!0!}\alpha_0 + \frac{1!}{1!0!}\alpha_1 + \frac{2!}{2!0!}\alpha_2 + \frac{3!}{3!0!}\alpha_3 + \frac{4!}{4!0!}\alpha_4\right](t-1)^0 +$$

$$\left[\frac{1!}{0!1!}\alpha_1 + \frac{2!}{1!1!}\alpha_2 + \frac{3!}{2!1!}\alpha_3 + \frac{4!}{3!1!}\alpha_4\right](t-1)^1 +$$

$$\left[\frac{2!}{0!2!}\alpha_2 + \frac{3!}{1!2!}\alpha_3 + \frac{4!}{2!2!}\alpha_4\right](t-1)^2 + \left[\frac{3!}{0!3!}\alpha_3 + \right.$$

$$\left.\frac{4!}{1!3!}\alpha_4\right](t-1)^3 + \frac{4!}{0!4!}\alpha_4(t-1)^4$$

$$= (1+1+0+0+3) + (1+0+0+12)(t-1) + (0+0+18)(t-1)^2$$
$$+ (0+12)(t-1)^3 + 3(t-1)^4$$
$$= 5 + 13(t-1) + 18(t-1)^2 + 12(t-1)^3 + 3(t-1)^4 . \qquad (2)$$

To show that (2) is actually equal to f(t) expand:

$$f(t) = 5 + 13(t-1) + 18(t^2-2t+1) + 12(t^3-3t^3+3t-1) +$$
$$\qquad 3(t^4-4t^3+6t^2-4t+1)$$
$$= (5-13+18-12+3) + t(13-36+36-12) + t^2(18-36+18)$$
$$\qquad + t^3(12-12) + t^4(3)$$
$$= 1 + t + 3t^4 .$$

So, indeed,

$$1 + t + 3t^4 = 5 + 13x + 18x^2 + 12x^3 + 3x^4$$

where $x = t - 1$.

● **PROBLEM 5-15**

Using Taylor's Theorem, approximate $\sqrt{40}$ to three decimal places.

Solution: Let f and its first n derivatives be continuous on [a,b]. Also, let its $(n+1)^{st}$ derivative exist on (a,b). Then, for some $c \in (a,b)$,

$$f(b) = \sum_{i=0}^{n} \frac{f^{(i)}(a)}{i!}(b-a)^i + R_{n+1} \qquad (1)$$

where

$$R_{n+1}(x) = \frac{f^{(n+1)}(c)}{(n+1)!}(b-a)^{n+1}$$

is called the remainder. That is, (1) is the Taylor expansion of f about a. In this problem, let $f(x) = \sqrt{x}$, $a = 36$, $b = 40$. Note that

$$f(x) = \sqrt{x} \Longrightarrow f(36) = 6,$$

$$f'(x) = \frac{1}{2} x^{-\frac{1}{2}} \Longrightarrow f'(36) = \frac{1}{12},$$

$$f''(x) = -\frac{1}{4} x^{-3/2} \Longrightarrow f''(36) = \frac{-1}{4(6)^3},$$

$$f'''(x) = \frac{3}{8} x^{-5/2} \Longrightarrow f'''(36) = \frac{3}{8(6)^5}$$

and

$$f^{(4)}(x) = -\frac{15}{16} x^{-7/2} \Longrightarrow |f^{(4)}(x)| < |f^{(4)}(36)| = \frac{15}{16(6)^7}$$

on (a,b). Hence,

$$f(x) = \frac{6}{0!}(x-36)^0 + \frac{1}{12(1!)}(x-36)^1 + \frac{-1}{4 \cdot 6^3 \cdot 2!}(x-36)^2 +$$

$$\frac{3}{8 \cdot 6^5 \cdot 3!}(x-36)^3 + R_4(x)$$

$$= 6 + \frac{1}{12}(x-36) - \frac{1}{8 \cdot 6^3}(x-36)^2 + \frac{1}{16 \cdot 6^5}(x-36)^3 + R_4(x).$$

Also, since

$$|f^{(4)}(x)| < \frac{15}{16 \cdot 6^7} \quad \text{on } (a,b), \quad |R_4(x)| = \left|\frac{f^{(4)}(c)}{4!}(x-36)^4\right|$$

$$< \frac{15}{16 \cdot 6^7 \cdot 4!}(x-36)^4 \quad \text{on } (a,b).$$

This implies that

$$\left|f(40) - \left[6 + \frac{1}{12}(4) - \frac{1}{8 \cdot 6^3}(4)^2 + \frac{1}{16 \cdot 6^5}(4)^3\right]\right| = |R_4(x)|$$

$$< \left(\frac{15}{16 \cdot (6)^7 \cdot 4!}\right)(4)^4$$

$$= \frac{10}{6^7}$$

$$< .00004 .$$

Therefore, to three places,

$$f(40) = \sqrt{40} = 6 + \frac{1}{3} - \frac{2}{6^3} + \frac{4}{6^5} = 6.325$$

● **PROBLEM 5-16**

Find the quadratic form associated with $f(x,y) = e^{-(x^2+y^2)}$ at $(0,0)$.

Solution: From Taylor's Theorem

$$F(x) = F(a) + J_F(a)(x-a) + \frac{1}{2}\left\langle H_F(a)(x-a), x-a\right\rangle + R$$

where R is "small" when $\|x-a\|$ is "small" (enough). Hence, for $\|x-a\|$ "small"

$$F(x) \approx F(a) + J_F(a)(x-a) + \frac{1}{2}\left\langle H_F(a)(x-a), x-a\right\rangle \qquad (1)$$

is a "good" approximation. That is, (1) is a local approximation at a. For

$$f(x,y) = e^{-(x^2+y^2)},$$

we have $f(0,0) = 1$,

$$\frac{\partial f}{\partial x} = -2xe^{-(x^2+y^2)} \Longrightarrow \frac{\partial f}{\partial x}(0,0) = 0,$$

$$\frac{\partial f}{\partial y} = -2ye^{-(x^2+y^2)} \Longrightarrow \frac{\partial f}{\partial y}(0,0) = 0,$$

$$\frac{\partial^2 f}{\partial x^2} = (4x^2-2)e^{-(x^2+y^2)} \Longrightarrow \frac{\partial^2 f}{\partial x^2}(0,0) = -2,$$

$$\frac{\partial^2 f}{\partial x \partial y} = \frac{\partial^2 f}{\partial y \partial x} = 4xye^{-(x^2+y^2)} \Longrightarrow \frac{\partial^2 f}{\partial x \partial y}(0,0) = \frac{\partial^2 f}{\partial y \partial x}(0,0) = 0,$$

and

$$\frac{\partial^2 f}{\partial y^2} = (4y^2-2)e^{-(x^2+y^2)} \Longrightarrow \frac{\partial^2 f}{\partial y^2}(0,0) = -2$$

Now (1) implies that for small $\|(x,y)\|$,

$$e^{-(x^2+y^2)} \approx 1 + (0,0)\begin{pmatrix}x\\y\end{pmatrix} + \frac{1}{2}\left\langle \begin{pmatrix}-2 & 0\\ 0 & -2\end{pmatrix}\begin{pmatrix}x\\y\end{pmatrix}, \begin{pmatrix}x\\y\end{pmatrix}\right\rangle$$

$$= 1 + \left\langle \begin{pmatrix}-1 & 0\\ 0 & -1\end{pmatrix}\begin{pmatrix}x\\y\end{pmatrix}, \begin{pmatrix}x\\y\end{pmatrix}\right\rangle$$

$$= 1 + \left\langle \begin{pmatrix}-x\\-y\end{pmatrix}, \begin{pmatrix}x\\y\end{pmatrix}\right\rangle$$

$$= 1 - (x^2+y^2). \qquad (2)$$

The quadratic form associated with $F(x)$ at a is

$$\frac{1}{2}\left\langle H_F(a)(x-a), x-a\right\rangle.$$

For $f(x,y) = e^{-(x^2+y^2)}$ the associated quadratic form at $(0,0)$ is $-(x^2 + y^2)$. Note that (1), and hence (2), are only approximations which are only good close to a and $(0,0)$ respectively.

• **PROBLEM 5-17**

Calculate e^4 within an error of 10^{-3}.

Solution: Since

$$\frac{d}{dx}(e^x) = e^x, \quad \frac{d^k}{dx^k}(e^x) = e^x$$

for all $k \geq 0$.

By Taylor's Theorem

$$e^x = \sum_{i=0}^{n} \frac{1}{i!} \frac{d^{(i)}}{dx^i}(e^x)\bigg|_{x=0} \cdot (x-0)^i +$$

$$\frac{1}{(n+1)!} \frac{d^{(n+1)}}{dx^{n+1}}(e^x)\bigg|_{(\bar{x})} (x-0)^{n+1}$$

for some $\bar{x} \in (0,x)$. Hence,

$$e^x = \sum_{i=0}^{n} \frac{x^i}{i!} + e^{\bar{x}} \frac{x^{n+1}}{(n+1)!}.$$

In order to approximate e^4, note that since

$$0 < \bar{x} < 4, \quad e^{\bar{x}} < e^4 < 3^4.$$

To approximate e^4 to three decimal places, choose n such that

$$e^{\bar{x}} \frac{4^{n+1}}{(n+1)!} < \frac{3^4 \cdot 4^{n+1}}{(n+1)!} \leq 10^{-3}.$$

Note that

$$\frac{3^4 \cdot 4^{n+1}}{(n+1)!} < \frac{4^4 \cdot 4^{n+1}}{(n+1)!} \quad (1)$$

$$\leq \frac{4^{n+5}}{8^{n-7}} \quad (2)$$

$$= \frac{2^{2n+10}}{3^{3n-2}}$$

$$= \frac{1}{2^{n-31}} \qquad (3)$$

$$< \frac{1}{10^{(3/10)(n-31)}} \qquad (4)$$

Hence, choose n such that

$$\frac{3}{10}(n-31) \geq 3$$

or

$$n \geq 41.$$

We have, (2) follows from (1) since $8^{n-7} \leq (n+1)!$ for $n \geq 1$. [To see this note that this is certainly true for $1 \leq n \leq 7$. Then for $n \geq 8$, $8^{n-8} \leq n!$ implies $8^{n-7} \leq 8n! \leq (n+1)!$ Thus the result.]. Additionally, (4) follows from (3) since $10^{3/10} < 2$. Therefore, the following is a good approximation within 10^{-3}:

$$e^4 \approx \sum_{i=0}^{41} \frac{4^i}{i!} \approx 54.598.$$

● **PROBLEM 5-18**

State and prove the Chain Rule.

Solution: Let $V \subseteq R^p$ and $W \subseteq R^n$ be open sets. If $f: V \to R^n$ is differentiable at a and $g: W \to R^m$ is differentiable at $f(a)$, then the composition $g \circ f: V \to R^m$ is differentiable at a and

$$J_{g \circ f}(a) = J_g(f(a)) J_f(a)$$

where $J_\phi(\alpha)$ is the Jacobian matrix of ϕ evaluated at α. This is the Chain Rule. To prove it note that by Taylor's Theorem, for appropriate ε and δ,

$$f(x) = f(a) + J_f(a)(x-a) + \varepsilon(x) \| x-a \| \qquad (1)$$

and

$$g(f(x)) = g(f(a)) + J_g(f(a))(f(x) - f(a))$$
$$+ \delta(f(x)) \| f(x) - f(a) \| \qquad (2)$$

Substituting from (1) into (2) gives

$$g(f(x)) - g(f(a)) = J_g(f(a)) \Big(J_f(a)(x-a) + \varepsilon(x) \| x-a \| \Big) +$$
$$\delta(f(x)) \left(\left\| J_f(a)(x-a) + \varepsilon(x) \| x-a \| \right\| \right)$$

$$= J_g(f(a))J_f(a)(x-a) + J_g(f(a))\varepsilon(x)\|x-a\| +$$

$$\delta(f(x))\left\|J_f(a)(x-a) + \varepsilon(x)\|x-a\|\right\|.$$

Taking norms, this yields

$$\frac{\|g(f(x)) - g(f(a)) - J_g(f(a))J_f(a)(x-a)\|}{\|x-a\|} \leq \|J_g(f(a))\| \ \|\varepsilon(x)\|$$

$$+ \|\delta(f(x))\| \ \|J_f(a)\| + \|\varepsilon(x)\| \ \|\delta(f(x))\|.$$

However, from the definition of the derivative for a function $f: E \to R^m$ ($E \subseteq R^n$), we have that f is differentiable at $x \in E$ if there exists a linear transformation A of R^n into R^m such that

$$\lim_{h \to 0} \frac{\|f(x+h) - f(x) - Ah\|}{\|h\|} = 0.$$

In addition, $f'(x) = A$. Therefore, since $\varepsilon, \delta \to 0$ as $x \to a$, this shows that

$$J_{g \circ f}(a) = J_g(f(a))J_f(a)$$

which is the desired result.

● **PROBLEM 5-19**

Let the vector-valued function $f: R^n \to R^m$ be defined by $f(x_1, \ldots, x_n) = (f_1(x_1, \ldots, x_n), \ldots, f_m(x_1, \ldots, x_n))$.

(a) Show that f is differentiable at $a \in R^n$ if and only if each f_i ($1 \leq i \leq m$) is differentiable at a and $J_f(a) = (J_{f_1}(a), \ldots, J_{f_m}(a))$.

(b) Show that this derivative $J_f(a)(x-a)$ is unique. (Note: (b) does not depend on (a).)

Solution: (a) If each f_i is differentiable at a and

$$\lambda = \left(J_{f_1}(a), \ldots, J_{f_m}(a)\right), \text{ then}$$

$$f(x) - f(a) - \lambda(x-a) = \left(f_i(x) - f_i(a) - J_{f_i}(a)(x-a)\right),$$

$$(1 \leq i \leq m).$$

Since

$$\lim_{x \to a} \frac{\|f_i(x) - f_i(a) - J_{f_i}(a)(x-a)\|}{\|x-a\|} = 0$$

for all $1 \leq i \leq m$, then

$$\lim_{x \to a} \frac{\|f(x) - f(a) - \lambda(x-a)\|}{\|x - a\|} \leq$$

$$\lim_{x \to a} \sum_{i=1}^{m} \frac{\|f_i(x) - f_i(a) - J_{f_i}(a)(x-a)\|}{\|x - a\|} = 0$$

Hence, $\lambda = J_f(a)$. Conversely, suppose f is differentiable at a. Then $f_i = \pi_i \circ f$ $(1 \leq i \leq m)$, where $\pi_i : R^m \to R$ is defined by $\pi_i(x_1, \ldots, x_m) = x_i$. Since π_i is also differentiable, the Chain Rule says that each f_i is differentiable at a and

$$J_{f_i}(a) = J_{\pi_i}(f(a)) J_f(a) \tag{1}$$

$$= \pi_i(J_f(a)) \tag{2}$$

which is the same as $J_f(a) = \left(J_{f_1}(a), \ldots, J_{f_m}(a) \right)$.
Note that (2) follows from (1) since π_i is a linear transformation and for any linear transformation T, $J_T(a) = T$. [To see this let $\pi_i = (0 \ldots 010 \ldots 0)$ where the 1 is in the i-th position and notice:

$$\lim_{x \to a} \frac{\|T(x) - T(a) - T(x-a)\|}{\|x - a\|} =$$

$$\lim_{x \to a} \frac{\|T(x) - T(a) - T(x) + T(a)\|}{\|x - a\|}$$

$$= 0 \;].$$

(b) Suppose that μ is a linear transformation satisfying

$$\lim_{x \to a} \frac{\|f(x) - f(a) - \mu(x-a)\|}{\|x - a\|} = 0.$$

Then, letting $\delta(x) = f(x) - f(a)$,

$$\lim_{x \to a} \frac{\|J_f(a)(x-a) - \mu(x-a)\|}{\|x - a\|} =$$

$$= \lim_{x \to a} \frac{\|J_f(a)(x-a) - \delta(x) + \delta(x) - \mu(x-a)\|}{\|x - a\|}$$

$$\leq \lim_{x \to a} \frac{\|J_f(a)(x-a) - \delta(x)\|}{\|x - a\|} + \lim_{x \to a} \frac{\|\delta(x) - \mu(x-a)\|}{\|x - a\|}$$

$$= 0.$$

Since $tx \to 0$ as $t \to 0$, then, for $x \neq 0$,

$$0 = \lim_{t \to 0} \frac{\|J_f(a)(tx) - \mu(tx)\|}{\|tx\|} = \frac{\|J_f(a)(x) - \mu(x)\|}{\|x\|}.$$

Hence

$$J_f(a)(x) = \mu(x), \text{ i.e., } J_f(a) \text{ is unique.}$$

● **PROBLEM 5-20**

(a) Define "homogeneous of degree n" and "positively homogeneous of degree n" for a function of two variables $F(x,y)$ and give a geometric interpretation of homogeneity.

(b) Determine whether the following functions are homogeneous:

(i) $x^2 y \log(y/x)$;

(ii) $x^{1/3} + xy^{-2/3}$;

(iii) $\dfrac{x^2 - y^2}{x^2 + y^2}$;

(iv) Ar^n where A is any constant and $r = (x^2 + y^2)^{1/2}$

Solution: (a) $F(x,y)$ is said to be homogeneous of degree n if

$$F(tx, ty) = t^n F(x,y) \tag{1}$$

for all values of x,y and t for which $F(x,y)$ and $F(tx, ty)$ are defined. Here n is any real constant. If (1) is true only when t is nonnegative then it is said that F is positively homogeneous of degree n.

Suppose the value of $F(x,y)$ is known at some point (x_0, y_0). Then according to (1) the value of F at all points on the line consisting of all points (tx_0, ty_0) is easily computed from $F(x_0, y_0)$ (see Figure).

(b) (i) $F(x,y) = x^2 y \log(y/x)$ so,

$$F(tx,ty) = (tx)^2 (ty) \log(ty/tx)$$
$$= t^3 (x^2 y \log y/x)$$
$$= t^3 F(x,y) .$$

Therefore,

$x^2 y \log(y/x)$ is homogeneous of degree 3.

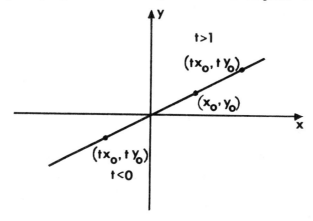

(ii) $F(x,y) = x^{1/3} + xy^{-2/3}$ so,

$$F(tx,ty) = (tx)^{1/3} + (tx)(ty)^{-2/3}$$
$$= t^{1/3} x^{1/3} + t^{1/3} xy^{-2/3}$$
$$= t^{1/3} (x^{1/3} + xy^{-2/3})$$
$$= t^{1/3} F(x,y) .$$

Therefore $x^{1/3} + xy^{-2/3}$ is homogeneous of degree 1/3.

(iii)

$$F(x,y) = \frac{x^2 - y^2}{x^2 + y^2} \text{ so,}$$

$$F(tx,ty) = \frac{(tx)^2 - (ty)^2}{(tx)^2 + (ty)^2} = \frac{t^2(x^2 - y^2)}{t^2(x^2 + y^2)} = F(x,y) .$$

Therefore

$$\frac{x^2 - y^2}{x^2 + y^2}$$

is homogeneous of degree 0.

(iv) $F(x,y) = Ar^n = A(x^2 + y^2)^{n/2}$ so,

$$F(tx,ty) = A((tx)^2 + (ty)^2)^{n/2} = A[t^2(x^2 + y^2)]^{n/2}$$

$$= t^n A(x^2 + y^2)^{n/2} = t^n Ar^n, \quad t \geq 0 .$$

Therefore, Ar^n is positively homogeneous of degree n since by definition the radical sign calls for the nonnegative square root of the radicand,

$$(t^2)^{n/2} = \left((t^2)^{1/2}\right)^n = t^n$$

where $t \geq 0$.

● **PROBLEM 5-21**

(a) State and prove Euler's Theorem on positively homogeneous functions of two variables.

(b) Let $F(x,y)$ be positively homogeneous of degree 2 and $u = r^m F(x,y)$ where $r = (x^2 + y^2)^{1/2}$. Show that

$$\frac{\partial^2 u}{\partial x^2} + \frac{\partial^2 u}{\partial y^2} = r^m \left(\frac{\partial^2 F}{\partial x^2} + \frac{\partial^2 F}{\partial y^2} \right) + m(m + 4) r^{m-2} F . \quad (1)$$

Solution: (a) Let $F(x,y)$ be positively homogeneous of degree n. Then at any point where F is differentiable

$$x \frac{\partial F}{\partial x} + y \frac{\partial F}{\partial y} = nF(x,y) . \quad (2)$$

This is the statement of Euler's Theorem. To prove it, write $u = tx$, $v = ty$ and use the chain rule on the composite function

$$F(u(t,x,y), v(t,x,y)) . \quad \text{Thus}$$

$$\frac{\partial F}{\partial t} = \frac{\partial F}{\partial u} \frac{\partial u}{\partial t} + \frac{\partial F}{\partial v} \frac{\partial v}{\partial t} = x \frac{\partial F}{\partial u} + y \frac{\partial F}{\partial v} \quad (3)$$

provided that F is differentiable at (tx,ty).

Now, since F is positively homogeneous of degree n, we have by definition that

$$F(u,v) = F(tx,ty) = t^n F(x,y), \quad t \geq 0 . \quad (4)$$

Differentiating (4) with respect to t gives

$$\frac{\partial F(u,v)}{\partial t} = \frac{\partial (t^n F(x,y))}{\partial t} = nt^{n-1} F(x,y) . \quad (5)$$

Combining (3) and (5) gives

$$x \frac{\partial F}{\partial u} + y \frac{\partial F}{\partial v} = nt^{n-1} F(x,y) . \quad (6)$$

and letting $t=1$ so that $u=x$, $v=y$, (6) becomes

$$x \frac{\partial F}{\partial x} + y \frac{\partial F}{\partial y} = nF(x,y)$$

231

and the theorem is proved.

(b) The expression $\frac{\partial^2 u}{\partial x^2} + \frac{\partial^2 u}{\partial y^2}$ is called the Laplacian of u and is important in the subject of partial differential equations. Thus, if u is of the form $r^m F(x,y)$, then the conversion of the Laplacian of u to an expression in terms of F as in (1) may be useful. To prove (1), note that

$$\frac{\partial r}{\partial x} = \frac{x}{(x^2+y^2)^{1/2}} = \frac{x}{r}$$

so that

$$\frac{\partial u}{\partial x} = r^m \frac{\partial F}{\partial x} + mr^{m-1} \frac{\partial r}{\partial x} F = r^m \frac{\partial F}{\partial x} + mr^{m-2} xF . \quad (7)$$

Differentiating (7) with respect to x yields

$$\frac{\partial^2 u}{\partial x^2} = r^m \frac{\partial^2 F}{\partial x^2} + mr^{m-1} \frac{\partial r}{\partial x} \frac{\partial F}{\partial x} + mr^{m-2} x \frac{\partial F}{\partial x}$$

$$+ mr^{m-2} F + m(m-2) r^{m-3} \frac{\partial r}{\partial x} x F$$

$$= r^m \frac{\partial^2 F}{\partial x^2} + 2mr^{m-2} x \frac{\partial F}{\partial x} + mr^{m-2} F$$

$$+ m(m-2) r^{m-4} x^2 F . \quad (8)$$

Since x and y occur symmetrically in $r(x,y)$, it is clear that analogously to (8) we have

$$\frac{\partial^2 u}{\partial y^2} = r^m \frac{\partial^2 F}{\partial y^2} + 2mr^{m-2} y \frac{\partial F}{\partial y} + mr^{m-2} F$$

$$+ m(m-2) r^{m-4} y^2 F . \quad (9)$$

Adding (8) and (9) yields

$$\frac{\partial^2 u}{\partial x^2} + \frac{\partial^2 u}{\partial y^2} = r^m \left(\frac{\partial^2 F}{\partial x^2} + \frac{\partial^2 F}{\partial y^2} \right) + 2mr^{m-2} \left(x \frac{\partial F}{\partial x} + y \frac{\partial F}{\partial y} \right)$$

$$+ 2mr^{m-2} F + m(m-2) r^{m-4} (x^2+y^2) F . \quad (10)$$

Now recall that $r^2 = x^2 + y^2$ and that since F is positively homogeneous of order 2, by Euler's Theorem we know that

$$x \frac{\partial F}{\partial x} + y \frac{\partial F}{\partial y} = 2F .$$

Using these results in (10) gives

$$\frac{\partial^2 u}{\partial x^2} + \frac{\partial^2 u}{\partial y^2} = r^m \left(\frac{\partial^2 F}{\partial x^2} + \frac{\partial^2 F}{\partial y^2} \right) + 4mr^{m-2} F + 2mr^{m-2} F$$

$$+ m(m-2) r^{m-2} F$$

$$= r^m \left(\frac{\partial^2 F}{\partial x^2} + \frac{\partial^2 F}{\partial y^2} \right) + m(m+4) r^{m-2} F.$$

which is the desired result.

THE IMPLICIT FUNCTION THEOREM

● **PROBLEM** 5-22

State and prove the Inverse Function Theorem.

Solution: Let $E \subseteq R^n$ be an open set. Suppose $f: E \to R^n$ is C^1 (has continuous first order partial derivatives), $f'(a)$ is invertible for some $a \in E$, and $b = f(a)$. Then there exist open sets U and V in R^n such that $a \in U$, $b \in V$, f is one-to-one (i.e., invertible) on U, $f(U) = V$, and $f^{-1}: V \to U$ is C^1. This is the Inverse Function Theorem. Let $A = f'(a)$ and choose λ such that

$$2\lambda \|A^{-1}\| = 1 \tag{1}$$

Since f' is continuous, there exists a neighborhood $U \subseteq E$ containing a, such that for $x \in U$,

$$\|f'(x) - A\| < \lambda. \tag{2}$$

For each $y \in R^n$ define ϕ on E by

$$\phi(x) = x + A^{-1}(y - f(x)). \tag{3}$$

Note that $f(x) = y$ if and only if x is a fixed point of ϕ (i.e., $\phi(x) = x$). Since $\phi'(x) = 1 - A^{-1} f'(x) = A^{-1}(A - f'(x))$, by (1) and (2),

$$\|\phi'(x)\| = \|A^{-1}\| \, \|A - f'(x)\| = \frac{1}{2\lambda} \lambda = \frac{1}{2}$$

on U. So, by the Mean Value Theorem

$$\|\phi(x_1) - \phi(x_2)\| \leq \frac{1}{2} \|x_1 - x_2\| \tag{4}$$

for $x_1, x_2 \in U$. Hence ϕ has at most one fixed point in U and $f(x) = y$ for at most one $x \in U$. Therefore, f is one-to-one in U. Now let $V = f(U)$ and $y_0 = f(x_0) \in V$ for $x_0 \in U$. Choose an open ball B about x_0 with radius ρ such that the closure $[B] \subseteq U$. To prove that V is open, it is enough to show that $y \in V$ whenever $\|y - y_0\| < \lambda \rho$. So, fix y such that $\|y - y_0\| < \lambda \rho$. With ϕ as in (3),

$$\|\phi(x_0) - x_0\| = \|A^{-1}(y - y_0)\| < \|A^{-1}\| \lambda \rho = \frac{\rho}{2}.$$

If $x \in [B] \subseteq U$, then by (4)

$$\|\phi(x) - x_0\| \leq \|\phi(x) - \phi(x_0)\| + \|\phi(x_0) - x_0\|$$

$$< \frac{1}{2}\|x - x_0\| + \rho/2$$

$$\leq \rho.$$

That is, $\phi(x) \in [B]$. Thus, for $x_1, x_2 \in [B]$, by (4), ϕ is a contraction of the complete space $[B]$ into itself [since a closed subspace of a complete space, such as R^n, is complete]. Hence, by the Principle of Contraction Mappings, ϕ has a unique fixed point $x \in [B]$. Thus,

$$y = f(x) \in f([B]) \subseteq f(U) = V. \text{ Therefore, } f^{-1}: V \to U \text{ exists.}$$

It remains to show that f^{-1} is C^1. Choose $y_1, y_2 \in V$. There exist $x_1, x_2 \in U$ such that $f(x_1) = y_1$, $f(x_2) = y_2$. With ϕ as in (3),

$$\phi(x_2) - \phi(x_1) = x_2 - x_1 + A^{-1}(f(x_1) - f(x_2))$$

$$= (x_2 - x_1) - A^{-1}(y_2 - y_1)$$

By (4), $\|(x_2 - x_1) - A^{-1}(y_2 - y_1)\| \leq \frac{1}{2}\|x_2 - x_1\|$. So,

$$\frac{1}{2}\|x_2 - x_1\| \leq \|A^{-1}(y_2 - y_1)\| \leq \frac{1}{2\lambda}\|y_2 - y_1\| \quad \text{or}$$

$$\|x_2 - x_1\| \leq \frac{1}{\lambda}\|y_2 - y_1\|. \tag{5}$$

From (1), (2) it follows that $(f')^{-1}$ exists locally about a. Since

$$f^{-1}(y_2) - f^{-1}(y_1) - (f')^{-1}(y_1)(y_2 - y_1) =$$

$$= (x_2 - x_1) - (f')^{-1}(y_1)(y_2 - y_1)$$

$$= -(f')^{-1}(y_1)\left[-f'(x_1)(x_2 - x_1) + f(x_2) - f(x_1)\right],$$

(5) implies

$$\frac{\|f^{-1}(y_2) - f^{-1}(y_1) - (f')^{-1}(y_1)(y_2 - y_1)\|}{\|y_2 - y_1\|}$$

$$\leq \frac{\|(f')^{-1}\|}{\lambda} \cdot \frac{\|f(x_2) - f(x_1) - f'(x_1)(x_2 - x_1)\|}{\|x_2 - x_1\|}$$

As $y_2 \to y_1$, $x_2 \to x_1$ by (5). Hence $(f^{-1})'(y) = [f'(f^{-1}(y))]^{-1}$ for $y \in V$. Since f^{-1} is differentiable, it is continuous. Also, f' is continuous and its inversion, where it exists, is continuous. Therefore $(f^{-1})'$ is continuous or f^{-1} is C^1 or V.

● **PROBLEM 5-23**

(a) Give an example of a C^1 function f which has a C^0 inverse about f(a) despite det f'(a) = 0. Also, show that for any C^1 function f, if det f'(a) = 0, then f^{-1} is not differentiable at f(a).

(b) Let $f \in C^1(E, R^n)$, E open in R^n. If f'(x) is invertible for all $x \in E$, show that f(W) is open in R^n for all open $W \subseteq E$.

Solution: (a) Define f: R → R by $f(x) = x^3$. Then f'(0) = 0. This f is C^∞, hence it is C^1. Despite not satisfying the conditions of the Inverse Function Theorem at zero, f has a continuous inverse: $f^{-1}(y) = y^{1/3}$. However, this inverse is not C^1 since $(f^{-1})'(0)$ does not exist. In fact, this is true in general. Let f be any C^1 function for which det f'(a) = 0 for some a. If f^{-1} were differentiable at f(a), applying the Chain Rule to $I = f \cdot f^{-1}$ would give

$$f'(a) \cdot (f^{-1})'(f(a)) = I$$

or

$$\det f'(a) \cdot \det (f^{-1})'(f(a)) = 1$$

which contradicts det f'(a) = 0. Therefore, f^{-1} is indeed not differentiable at a.

(b) To show that f(W) is open, it must be shown that for any $b \in f(W)$, there exists an open set V containing b which is contained in f(W). Since f is C^1 and f' is invertible on E and hence on W, f is one-to-one. Therefore, for any $b \in f(W)$ there exists one $a \in W$ such that f(a) = b. In addition there also exists open sets U and V such that $a \in U$, $b \in V$, where $U \subseteq W$, and f(U) = V. Hence, $b \in V = f(U) \subseteq f(W)$. Hence, f is an open map.

● **PROBLEM 5-24**

State and prove the Implicit Function Theorem.

Solution: Let $E \subseteq R^{n+m}$ be an open set containing (a,b) [$a \in R^n$, $b \in R^m$]. Suppose $f = (f_1, \ldots, f_m): E \to R^m$ is C^1 and f(a,b) = 0. Define the m × m matrix

$$M = \left(\frac{\partial f_i}{\partial x_{n+j}} \right),$$

for $1 \leq i, j \leq m$. If det M ≠ 0, there exists an open set $A \subseteq R^n$ containing a and an open set $B \subseteq R^m$ containing b such that for each $x \in A$ there exists a unique $g(x) \in B$ such that $f(x, g(x)) = 0$ and g is C^1. This is called the Implicit Function Theorem since g is defined implicitly by the above conditions. To prove this, start by defining $F: R^{n+m} \to R^{n+m}$ by $F(x,y) = (x, f(x,y))$.

Note that since

$$F'(a,b) = \begin{pmatrix} \left(\dfrac{\partial x_i}{\partial x_j}\right)_{1 \le i,j \le n} & \left(\dfrac{\partial x_i}{\partial x_{n+j}}\right)_{\substack{1 \le i \le n \\ 1 \le j \le m}} \\ \left(\dfrac{\partial f_i}{\partial x_j}\right)_{\substack{1 \le i \le m \\ 1 \le j \le n}} & \left(\dfrac{\partial f_i}{\partial x_{n+j}}\right)_{1 \le i,j \le m} \end{pmatrix}$$

$$= \begin{pmatrix} I & O \\ N & M \end{pmatrix},$$

where I and O are the $n \times n$ identity and $n \times m$ zero matrices respectively, then

$$\det F'(a,b) = \det M \ne 0.$$

By the Inverse Function Theorem there exists an open set $V \subseteq R^{n+m}$ containing $F(a,b) = (a,0)$ and an open set of the form $A \times B \subseteq E$ containing (a,b), such that $F: A \times B \to V$ has a C^1 inverse $F^{-1}: V \to A \times B$. F^{-1} is of the form $F^{-1}(x,y) = (x, \phi(x,y))$ for some C^1 function ϕ. Define the projection $\pi: R^{n+m} \to R^m$ by $\pi(x,y) = y$. Then $\pi \circ F(x,y) = f(x,y)$.

Therefore

$$\begin{aligned} f(x, \phi(x,y)) &= f \circ F^{-1}(x,y) \\ &= (\pi \circ F) \circ F^{-1}(x,y) \\ &= \pi \circ (F \circ F^{-1})(x,y) \\ &= \pi(x,y) \\ &= y \end{aligned}$$

and $f(x, \phi(x,0)) = 0$. So, define $g: A \to B$ by $g(x) = \phi(x,0)$.

● PROBLEM 5-25

Represent the contour line

$$y - xe^y = 1 \tag{1}$$

near $(-1,0)$ as a function $x = \psi(y)$. Compute $\phi'(x)$ and $\phi'(-1)$ for x near -1, where $y = \phi(x)$.

Solution: Let $f(x,y) = y - xe^y$. Then $\frac{\partial f}{\partial x}(x,y) = -e^y$ and $\frac{\partial f}{\partial y}(x,y) = 1 - xe^y$. So, $\frac{\partial f}{\partial x}(-1,0) = -1$ and $\frac{\partial f}{\partial y}(-1,0) = 2$.
Since both of the partial derivatives are non-zero, the contour (1) can be represented either as $y = \phi(x)$ or as $x = \psi(y)$. In fact, we have $x = \psi(y) = (y-1)e^{-y}$ from (1). The Implicit Function Theorem guarantees the existence of ϕ, which, however, can not be defined explicitly. The theorem also says that ϕ is differentiable. From (1),

$$\phi(x) - xe^{\phi(x)} = 1 \text{ and hence}$$

$$\phi'(x) - e^{\phi(x)} - x\phi'(x)e^{\phi(x)} = 0$$

or

$$\phi'(x) = \frac{e^{\phi(x)}}{1 - xe^{\phi(x)}},$$

for x near -1. At $x = -1$, since $(-1,0)$ is on the contour, $\phi(-1) = 0$ and therefore

$$\phi'(-1) = \frac{e^0}{1 - (-1)e^0} = \frac{1}{2}.$$

● PROBLEM 5-26

Show that the functions $f, g \in C^1(E)$, E open in R^2, are functionally dependent (i.e., there exists a function F such that $g = F \circ f$) if $\det J_\Phi(x,y) = 0$ for $\Phi = (f,g)$ and (x,y) in some neighborhood of (a,b), where $\frac{\partial f}{\partial x}(a,b) \neq 0$.

Solution: Let $c = f(a,b)$. By the Implicit Function Theorem, the equation

$$z = f(x,y)$$

can be solved for x. In particular, there exists a function $\phi(y,z)$ such that $x = \phi(y,z)$ in a neighborhood $D \leq E \times R$ containing (a,b,c). Moreover,

$$\frac{\partial \phi}{\partial y} = -\frac{\partial f/\partial y}{\partial f/\partial x}.$$

Consider $g(x,y)$ as a function ψ in y and z, i.e., $g = \psi(\phi(y,z),y)$. Then, in some neighborhood N containing (a,b),

$$\frac{\partial \psi}{\partial y} = \frac{\partial g}{\partial x}\frac{\partial \phi}{\partial y} + \frac{\partial g}{\partial y} = \frac{\det J_\phi}{\partial f/\partial x} = 0.$$

Thus, $\psi(\phi(y,z),y)$ is independent of y. Hence let $F(z) = \psi(\phi(y,z),y) = g(x,y)$. Since $x = \phi(y,z)$ is equivalent to $z = f(x,y)$ on D, $F(z) = \psi(\phi(y,z),y)$ is equivalent to

$$F(f(x,y)) = g(x,y)$$

when $N \subseteq \pi(D)$, where $\pi: R^3 \to R^2$ is the projection $\pi(x,y,z) = (x,y)$. This proves that f and g are functionally dependent in a neighborhood of (a,b). Note that the condition $\frac{\partial f}{\partial x}(a,b) \neq 0$ could be replaced by $\frac{\partial f}{\partial y}(a,b) \neq 0$. Also, the proposition could be generalized to n functions of n variables. For example, with $n = 3$, the conditions $\Phi = (f,g,h)$, $\det J_\Phi$ near (a,b,c), and at least one of the expansions by minors along the last row is not zero at (a,b,c) imply that

$$h(x,y,z) = F(f(x,y,z),g(x,y,z))$$

in a neighborhood of (a,b,c). As in illustration of functional dependence take $f(x,y) = x^2y^2$ and $g(x,y) = -xy$. Then about $(1,1)$, $F(z) = -\sqrt{z}$ (so that $-\sqrt{x^2y^2} = -xy<0$). But about $(-1,1)$, $F(z) = \sqrt{z}$ (so that $\sqrt{x^2y^2} = -xy>0$).

CHAPTER 6

MAXIMA AND MINIMA

This chapter is concerned with the determination of the maximum and minimum values of functions of several variables.

In one variable, maximum and minimum values of a function $y = f(x)$, were found by first finding points at which the derivative, dy/dx, equals 0. These points, called critical points, were then found, by inspection, to be maxima, minima, or saddle points (i.e., neither a maxima nor a minima) of the function.

For functions of several variables a similar procedure can be applied. However, the procedure now becomes more complicated with the vanishing of all partial derivatives replacing the condition that the derivative (dy/dx) be equal to zero. Also, if a point is found to be a maximum (or minimum), it must be distinguished between a relative maximum (minimum) or an absolute maximum (minimum).

Interesting problems arise when a function is restricted by some constraint (i.e., given a function $f(x,y,z)$, the variables x,y,z are not independent of each other, but are restricted by the relation $G(x,y,z) = k$). For this type of problem, the method of direct elimination which involves solving for certain variable(s) in terms of the others, the method of implicit functions, or the method of Lagrange multipliers can be applied to find possible maxima or minima. Other extremal problems deal with the extreme values of a function in a closed region R ; this is done by inspection of the function on the interior of R and on the boundary of R.

Finally, the theory can be applied to many physical situations and can be used to derive such important formulas as Cauchy's Inequality and Hölder's Inequality.

RELATIVE MAXIMUM AND RELATIVE MINIMUM

● **PROBLEM 6-1**

Find the point of the plane $2x - 3y - 4z = 25$ which is nearest the point $(3,2,1)$.

Solution: Let D_1 be the distance from the point (x,y,z) of the plane to the point $(3,2,1)$. The distance formula in R^3 is

$$D = \sqrt{(x - x_0)^2 + (y - y_0)^2 + (z - z_0)^2} \qquad (1)$$

where $P = (x,y,z)$; $P_0 = (x_0,y_0,z_0)$ and D is the distance between these two points. Therefore, using (1), this yields

$$D_1 = \sqrt{(x-3)^2 + (y-2)^2 + (z-1)^2}$$

so that
$$D_1^2 = (x-3)^2 + (y-2)^2 + (z-1)^2 . \qquad (2)$$
The plane given is $2x - 3y - 4z = 25$ which means $z = \frac{1}{4}(2x - 3y - 25)$. Substituting this into (2) eliminates z and gives
$$D_1^2 = (x-3)^2 + (y-2)^2 + (\tfrac{1}{4}(2x - 3y - 25)-1)^2 .$$
Simplifying gives
$$D_1^2 = (x-3)^2 + (y-2)^2 + (\tfrac{1}{2}x - \tfrac{3}{4}y - 29/4)^2 . \qquad (3)$$

The point of the plane nearest to the point (3,2,1) is the point where D_1^2, the square of the distance between the two points, has its minimum value as x, y range through all possible values. To find this point the following theorem is used:

Let f be a function defined on some region R, and let f have a relative extreme (maximum or minimum) at the point (a,b) of R. Assume that (a,b) is an interior point of R (not on the boundary), and that the first partial derivatives of f at (a,b) exist. Then these derivatives are zero at that point (i.e., $f_1(a,b) = 0$, $f_2(a,b) = 0$).

By definition a function f has a relative maximum at the point (a,b) if there is some neighborhood of (a,b) such that $f(x,y) \le f(a,b)$ for all points (x,y) of R which are in this neighborhood. Also, by definition, given f defined in a region R, and S any part (or all) of R, if there is in S a point (a,b) such that $f(x,y) \le f(a,b)$ for all points (x,y) in S, then on the set S, the function f has an absolute maximum at (a,b). It should be noted that the vanishing of the first partial derivatives is not a sufficient condition for a relative extremum. That is, there may exist a point satisfying the condition which is not a relative extremum but rather is a "saddle point."

In this problem all points are interior points of the region R (the whole xy-plane), and D_1^2 has partial derivatives at all points. Therefore, the points to look for are those that satisfy the condition
$$\frac{\partial(D_1^2)}{\partial x} = \frac{\partial(D_1^2)}{\partial y} = 0 .$$
From (3) this gives the equations
$$\frac{\partial(D_1^2)}{\partial x} = 2(x-3) + 2(\tfrac{1}{2}x - \tfrac{3}{4}y - 29/4)\cdot\tfrac{1}{2} = 0 \qquad (4)$$

$$\frac{\partial(D_1^2)}{\partial y} = 2(y-2) + 2(\tfrac{1}{2}x - \tfrac{3}{4}y - 29/4)\cdot(-\tfrac{3}{4}) = 0 . \qquad (5)$$

Simplifying (4), (5) gives the equations
$$10x - 3y - 53 = 0, \quad -6x + 25y + 55 = 0$$
which, when solved simultaneously, give the solution $x = 5$, $y = -1$. Substituting in the equation of the plane, $2x - 3y - 4z = 25$, yields $z = -3$. Hence the point $(5, -1, -3)$ is a relative minimum of $D_1^2(x,y)$. Since there is only one such point it is the only minimum of $D_1^2(x,y)$. The geometric nature of the problem (to find a point on a plane closest to another, given, point) indicates that it is the absolute minimum (for otherwise the problem would have no solution).

● PROBLEM 6-2

For the following quadratic forms, tell by inspection whether the origin is a maximum or minimum:
a) $q(x,y) = x^2 + y^2$
b) $q(x,y) = x^2 - y^2$
c) $q(x,y) = xy$.

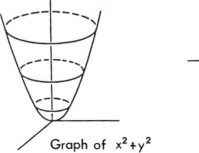

Graph of x^2+y^2 Level curves

Fig. 1

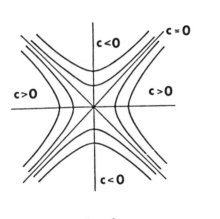

Fig. 2

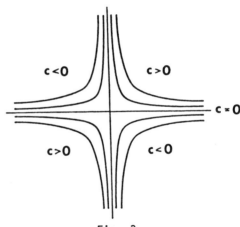

Fig. 3

Solution: a) The graph of the function $q(x,y) = x^2 + y^2$ and the level curves look like those in figure 1.

From this graph it is seen that the origin is a relative minimum point for the form. (A level curve of a function is a curve along which the function is constant in value).

b) For the function $q(x,y) = x^2 - y^2$, the level curves are hyperbolas, determined for each number c by the equation $x^2 - y^2 = c$. The graph of the level curves is drawn in figure 2. For $c = 0$ note that we get two straight lines through the origin. Here the origin is a critical point (i.e., the first partial derivatives vanish), but it is neither a relative maximum or relative minimum. Such a point is called a degenerate critical point or "saddle point".

c) For the function $q(x,y) = xy$ the level curves are again hyperbolas (similar to example b) but rotated). Here again, as in part b), the origin is neither a relative maximum nor a relative minimum.

● **PROBLEM 6-3**

Let the function $f(x,y)$ be continuous and have continuous first and second partial derivatives in a region R. Let (x_0, y_0) be an interior point of R for which $\frac{\partial f}{\partial x} = 0$, $\frac{\partial f}{\partial y} = 0$. Given the condition

$$[f_{12}(x_0,y_0)]^2 - f_{11}(x_0,y_0) f_{22}(x_0,y_0) < 0 \quad \text{and} \quad f_{11}(x_0,y_0) < 0$$

prove that $f(x,y)$ has a relative maximum at (x_0, y_0).

Solution: By Taylor's theorem, if the nth partial derivatives of $f(x,y)$ are continuous in a closed region and if the $(n+1)^{st}$ partial derivatives exist in an open region then

$$f(x_0 + h, y_0 + k) = f(x_0,y_0) + h f_1(x_0,y_0) + k f_2(x_0,y_0)$$
$$+ \frac{1}{2!}(h^2 f_{11}(x_0,y_0) + 2hk f_{12}(x_0,y_0) + k^2 f_{22}(x_0,y_0)) + \ldots$$
$$+ \frac{1}{n!}(h\frac{\partial}{\partial x} + k\frac{\partial}{\partial y})^n f(x_0,y_0) + R_n$$

where R_n, the remainder after n terms, is

$$R_n = \frac{1}{n+1!}[(h\frac{\partial}{\partial x} + k\frac{\partial}{\partial y})^{n+1} f(x,y)]_{x=x_0+\theta h, y=y_0+\theta k} \quad \text{and}$$

$0 < \theta < 1$.

For this problem, it is given $f_1(x_0,y_0) = f_2(x_0,y_0) = 0$, so that by Taylor's theorem (for $n = 2$)

$$f(x_0+h, y_0+k) - f(x_0,y_0) = \tfrac{1}{2}[h^2 f_{11}(x_0 + \theta h, y_0 + \theta k) + 2hk f_{12}(x_0+\theta h, y_0+\theta k)$$
$$+ k^2 f_{22}(x_0+\theta h, y_0+\theta k)] \quad (1)$$

where $0 < \theta < 1$. Next, we abbreviate (1) by the following

$$f(x_0+h, y_0+k) - f(x_0,y_0) = \tfrac{1}{2}[h^2 f_{11} + 2hk f_{12} + k^2 f_{22}] \quad (2)$$

which can be rewritten as

$$f(x_0+h, y_0+k) - f(x_0,y_0) = \tfrac{1}{2} f_{11}[h^2 + 2hk \frac{f_{12}}{f_{11}} + k^2 \frac{f_{22}}{f_{11}}]. \quad (3)$$

Then, by the method of completing the square, (3) becomes

$$f(x_0+h, y_0+k) - f(x_0,y_0) = \tfrac{1}{2} f_{11}\left[h^2 + 2hk\frac{f_{12}}{f_{11}} + \frac{k^2 f_{12}^2}{f_{11}^2} + \frac{k^2 f_{22}}{f_{11}} - \frac{k^2 f_{12}^2}{f_{11}^2}\right]$$

or

$$f(x_0+h, y_0+k) - f(x_0,y_0) = \tfrac{1}{2} f_{11}\left[\left(h + \frac{k f_{12}}{f_{11}}\right)^2 + \left(\frac{f_{11} f_{22} - f_{12}^2}{f_{11}^2}\right) k^2\right]. \quad (4)$$

Now the given condition can be rewritten as

$$f_{11}(x_0,y_0) f_{22}(x_0,y_0) - [f_{12}(x_0,y_0)]^2 > 0 \quad \text{and}$$
$$f_{11}(x_0,y_0) < 0.$$

Then, by hypothesis, $f_{11}(x,y) < 0$ for some neighborhood of (x_0,y_0). Also, the sum of the terms in the brackets in (4) is positive since the first term is squared and for the second term it is given

$f_{11}f_{22} - f_{12}^2 > 0$. Hence, since $f_{11}(x_0,y_0) < 0$ the right-hand side is negative, so that $f(x_0+h,y_0+k) \le f(x_0,y_0)$ for all sufficiently small h and k; thereby showing that $f(x_0,y_0)$ is a relative maximum.

Note: The same procedure can be applied for conditions of a relative minimum if $f_{11}(x_0,y_0) > 0$ (i.e., $f(x_0+h,y_0+k) \ge f(x_0,y_0)$ for all sufficiently small h and k).

● **PROBLEM 6-4**

Find the critical points of the function $f(x,y) = 4xy - 2x^2 - y^4$. Then determine whether each is a relative maximum, relative minimum or a saddle point of $f(x,y)$.

Solution: By definition, a point at which all the first partial derivatives of a function are equal to zero is called a critical point. To solve this problem, the following theorem is used: Let $f(x,y)$ be a differentiable function throughout a region R. Let P_0 be an interior point of R and suppose that the first partial derivatives of f vanish at this point ($f_1(P_0) = 0$, $f_2(P_0) = 0$). Also suppose $f_{11}(P_0), f_{12}(P_0), f_{22}(P_0)$ exist and let

$$\Delta = (f_{12}(P_0))^2 - f_{11}(P_0) f_{22}(P_0) . \tag{1}$$

Then:

(i) If $\Delta > 0$, P_0 is a saddle point for $f(x,y)$ (P_0 is neither a maximum point nor a minimum point of $f(x,y)$).

(ii) If $\Delta < 0$ and $f_{11}(P_0) > 0$, $f(x,y)$ has a relative minimum at P_0.

(iii) If $\Delta < 0$ and $f_{11}(P_0) < 0$, $f(x,y)$ has a relative maximum at P_0.

(iv) If $\Delta = 0$, no conclusion on the nature of P_0 is determined by this test. (Any of the above behaviors in parts (i)-(iii) may occur).

Since $f(x,y) = 4xy - 2x^2 - y^4$, we have that $\frac{\partial f}{\partial x} = 4y - 4x$ and $\frac{\partial f}{\partial y} = 4x - 4y^3$. Now the critical points of f occur where $\frac{\partial f}{\partial x} = 0$ and $\frac{\partial f}{\partial y} = 0$. This gives the equations $4y - 4x = 0$, $4x - 4y^3 = 0$, which upon solving give the solutions $(0,0)$, $(1,1)$, and $(-1,-1)$. Hence, $P_1 = (0,0)$; $P_2 = (1,1)$, and $P_3 = (-1,-1)$ are critical points of the function $f(x,y)$.

Now, to determine the nature of each of these critical points, $f_{12}(x,y)$, $f_{11}(x,y)$ and $f_{22}(x,y)$ must be computed:

$$f_{12} = \frac{\partial^2 f}{\partial x \partial y} = 4 ,$$

$$f_{11} = \frac{\partial^2 f}{\partial x^2} = -4 ,$$

$$f_{22} = \frac{\partial^2 f}{\partial y^2} = -12y^2 .$$

Then for P_1, $f_{12} = 4$, $f_{11} = -4$ and $f_{22} = 0$, so that upon substitution into (1) $\Delta = 16 - (-4)(0) = 16 > 0$. Therefore, by the theorem (part (i)), $P_1 = (0,0)$ is a saddle point for f.

For P_2, $f_{12} = 4$, $f_{11} = -4$ and $f_{22} = -12$, so by equation (1),
$\Delta = 16 - (-4)(-12) = 16 - 48 = -32 < 0$. Then since $f_{11}(P_2) = -4 < 0$, by part (iii) of the theorem, P_2 is a relative maximum of $f(x,y)$.

For P_3, $f_{12} = 4$, $f_{11} = -4$, and $f_{22} = -12$ and by equation (1)
$\Delta = 16 - (4)(-12) = -32 < 0$. Then since $f_{11}(P_3) = -4 < 0$, again by part (iii) of the theorem, P_3 is a relative maximum of $f(x,y)$.

• **PROBLEM 6-5**

Let f be the function given by $f(x,y) = x^2 + y^3 - 3xy$. Find the critical points of $f(x,y)$. Then determine whether each critical point is a relative maximum, relative minimum, or saddle point of $f(x,y)$.

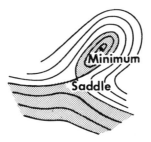

Fig. 1

Solution: By definition a point at which all the first partial derivatives of a function are equal to zero is called a critical point of the function. Since $f(x,y) = x^2 + y^3 - 3xy$, the critical points occur where $\frac{\partial f}{\partial x} = 0$, $\frac{\partial f}{\partial y} = 0$. Computing yields the equations

$$\frac{\partial f}{\partial x} = 2x - 3y = 0; \quad \frac{\partial f}{\partial y} = 3y^2 - 3x = 0. \tag{1}$$

To solve these equations, first notice that $x = 3y/2$, $y^2 = x$. (2)
Thus, $y^2 = 3y/2$, or $y^2 - 3y/2 = 0$. Factoring, $y(y - 3/2) = 0$ and $y = 0$, or $y = 3/2$, and by equation (2), $y = 0$ gives $x = 0$ and $y = 3/2$ gives $x = 9/4$. Hence, the critical points of $f(x,y)$ are $P_1 = (0,0)$, $P_2 = (9/4, 3/2)$. To determine the nature of each point compute

$$\Delta = [f_{12}(P_0)]^2 - f_{11}(P_0) f_{22}(P_0). \tag{3}$$

Then if $\Delta > 0$, P_0 is a saddle point for f. If $\Delta < 0$ and $f_{11}(P_0) < 0$, $f(x,y)$ has a relative maximum at P_0. If $\Delta < 0$ and $f_{11}(P_0) > 0$, $f(x,y)$ has a relative minimum at P_0. If $\Delta = 0$, no conclusion on the nature of P_0 is determined by this test.

Computing from (1), $f_{11}(x,y) = \frac{\partial^2 f}{\partial x^2} = 2$, $f_{12}(x,y) = \frac{\partial^2 f}{\partial x \partial y} = -3$ and
$f_{22}(x,y) = \frac{\partial^2 f}{\partial y^2} = 6y$.
Then for $P_1 = (0,0)$, $f_{11}(P_1) = 2$, $f_{12}(P_1) = -3$ and $f_{22}(P_1) = 0$
and by equation (3), $\Delta = (-3)^2 - (2)(0) = 9 > 0$. Therefore, since
$\Delta > 0$, $P_1 = (0,0)$ is a saddle point for $f(x,y)$.

For $P_2 = (9/4, 3/2)$, $f_{11}(P_2) = 2$, $f_{12}(P_2) = -3$ and $f_{22}(P_2) = 9$ and
by equation (3), $\Delta = (-3)^2 - (2)(9) = -9 < 0$. Therefore, since
$\Delta < 0$ and $f_{11}(P_2) > 0$, P_2 is a relative minimum for $f(x,y)$
(the level curves for $f(x,y)$ are shown in fig. 1).

● **PROBLEM 6-6**

a) Find whether the origin is a relative maximum or minimum, or neither for the function $f(x,y) = \log(1 + x^2 + y^2)$.

b) Find the critical points of the function $f(x,y) = x^2 - 12y^2 + 4y^3 + 3y^4$ and determine whether each is a relative maximum or minimum or saddle point of $f(x,y)$.

Solution: a) To determine whether the origin is a relative maximum or minimum or saddle point, it must first be determined if the origin is a critical point. Therefore since $f(x,y) = \log(1 + x^2 + y^2)$,

$$\frac{\partial f}{\partial x} = \frac{2x}{1 + x^2 + y^2} \; ; \; \frac{\partial f}{\partial y} = \frac{2y}{1 + x^2 + y^2}. \tag{1}$$

Therefore, at the origin $\frac{\partial f}{\partial x} = \frac{0}{1} = 0$, $\frac{\partial f}{\partial y} = \frac{0}{1} = 0$; showing the origin to be a critical point of $f(x,y)$. To determine the nature of the origin first compute

$$\Delta = [f_{12}(0,0)^2] - f_{11}(0,0) f_{22}(0,0) \tag{2}$$

Then if $\Delta > 0$, the origin is a saddle point for $f(x,y)$. If $\Delta < 0$, and $f_{11}(0,0) < 0$, the origin is a relative maximum for $f(x,y)$. If $\Delta < 0$ and $f_{11}(0,0) > 0$, the origin is a relative minimum for $f(x,y)$. Hence from (1),

$$f_{11}(x,y) = \frac{\partial^2 f}{\partial x^2} = \frac{2(1+x^2+y^2) - (2x)(2x)}{(1+x^2+y^2)}$$

$$= \frac{2 - 0}{1} = 2,$$

$$f_{12}(x,y) = \frac{\partial^2 f}{\partial x \partial y} = -\frac{(2x)(2y)}{(1+x^2+y^2)} = 0 \text{ at } (0,0),$$

and

$$f_{22}(x,y) = \frac{\partial^2 f}{\partial y^2} = \frac{2(1+x^2+y^2) - (2y)(2y)}{(1+x^2+y^2)} = 2 \text{ at } (0,0).$$

Then by equation (2), $\Delta = (0)^2 - (2)(2) = -4 < 0$.
Thus, since $\Delta < 0$ and $f_{11}(0,0) > 0$, the origin is a relative minimum for $f(x,y)$.

b) To determine the critical points of $f(x,y) = x^2 - 12y^2 + 4y^3 + 3y^4$, compute $\frac{\partial f}{\partial x}$, $\frac{\partial f}{\partial y}$ and solve the equations $\frac{\partial f}{\partial x} = 0$, $\frac{\partial f}{\partial y} = 0$. Doing this gives $\frac{\partial f}{\partial x} = 2x = 0$, which gives the solution $x = 0$, and

$\frac{\partial f}{\partial y} = -24y + 12y^2 + 12y^3 = 12y(y^2 + y - 2) = 0$, so that solutions for y are $y = 0$, $y = -2$, $y = 1$. From this it follows that the critical points of $f(x,y)$ are $P_1 = (0,0)$, $P_2 = (0,-2)$, $P_3 = (0,1)$. To determine the nature of each point let

$$\Delta = [f_{12}(P_0)]^2 - f_{11}(P_0) f_{22}(P_0). \qquad (3)$$

For this problem, $f_{11}(x,y) = 2$, $f_{12}(x,y) = 0$, $f_{22}(x,y) = -24 + 24y + 36y^2$. Then for $P_1 = (0,0)$, by equation (3),

$$\Delta = 0 - (2)(-24) = 48 > 0.$$

Hence, since $\Delta > 0$, $P_1 = (0,0)$ is a saddle point for $f(x,y)$.

For $P_2 = (0,-2)$, by equation (3),

$$\Delta = 0 - (2)[-24+24(-2) + 36(-2)^2] = (-2)(-72+144) = -144 < 0.$$

Therefore, since $\Delta < 0$ and $f_{11}(P_2) = 2 > 0$, the point $P_2 = (0,-2)$ is a relative minimum for $f(x,y)$.

For $P_3 = (0,1)$, $\Delta = (0) - (2)(-24+24+36) = -72 < 0$ so that, since $\Delta < 0$ and $f_{11}(P_3) = 2 > 0$, the point $P_3 = (0,1)$ is a relative minimum for $f(x,y)$.

• **PROBLEM 6-7**

Find the critical points and the nature of each critical point (i.e., relative maximum, relative minimum, or saddle point) for:
a) $f(x,y) = x^2 - 2xy + 2y^2 + x - 5$
b) $f(x,y) = (1 - x)(1 - y)(x + y - 1)$.

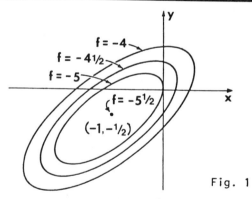

Fig. 1

Solution: a) For the critical points, $\frac{\partial f}{\partial x} = 0$, $\frac{\partial f}{\partial y} = 0$. So the critical points for $f(x,y) = x^2 - 2xy + 2y^2 + x - 5$ are the solutions of $\frac{\partial f}{\partial x} = 2x - 2y + 1 = 0$ and $\frac{\partial f}{\partial y} = -2x + 4y = 0$. To solve, add the equations $2x - 2y = -1$, $-2x + 4y = 0$ and obtain $y = -\frac{1}{2}$, and then by substitution into one of the two equations obtain $x = -1$, as solutions. Therefore, $f(x,y)$ has a critical point at $P_1 = (-1, -\frac{1}{2})$.

To determine the nature of this critical point find $f_{11}(P_1); f_{12}(P_1); f_{22}(P_1)$ then let $\Delta = [f_{12}(P_1)]^2 - f_{11}(P_1) f_{22}(P_1)$. (1)

Then if $\Delta > 0$, P_1 is a saddle point. If $\Delta < 0$ and $f_{11}(P_1) > 0$, P_1 is a relative minimum. If $\Delta < 0$ and $f_{11}(P_1) < 0$, P_1 is a relative maximum. If $\Delta = 0$, the nature of P_1 is not determined by this test. Computing gives $f_{11}(x,y) = 2$; $f_{12}(x,y) = -2$, $f_{22}(x,y) = 4$ then by equation (1), $\Delta = (-2)^2 - (2)(4) = 4 - 8 < 0$. Therefore, since $\Delta < 0$ and $f_{11}(x,y) > 0$, $P_1 = (-1,-\frac{1}{2})$ is a relative minimum. The level curves are shown in figure 1.

b) We have $f(x,y) = (1-x)(1-y)(x+y-1) = 2x + 2y - x^2 - y^2 - 3xy + x^2y + xy^2 - 1$. Therefore,

$$\frac{\partial f}{\partial x} = 2 - 2x - 3y + 2xy + y^2; \frac{\partial f}{\partial y} = 2 - 3x - 2y + 2xy + x^2.$$

Setting $\frac{\partial f}{\partial x} = 0$; $\frac{\partial f}{\partial y} = 0$ gives the solutions $P_1 = (1,0)$, $P_2 = (0,1)$, $P_3 = (1,1)$, $P_4 = (2/3, 2/3)$ as critical points. These solutions are found as follows. Factor:

$$2-2x-3y+2xy+y^2 = (y+2x-2)(y-1) = 0 \quad (2)$$

and

$$2-3x-2y+2xy+x^2 = (x+2y-2)(x-1) = 0. \quad (3)$$

From (2), $y = 1$ or $y = 2 - 2x$. By substituting into (3) $y = 1$ gives $x = 1$ or $x = 0$, and $y = 2 - 2x$ gives $x = 1$ and $x = 2/3$ so that $y = 0$ and $y = 2/3$ respectively. Hence the solutions above. Next,

$$f_{11}(x,y) = -2 + 2y; \; f_{12}(x,y) = -3 + 2x + 2y; \; f_{22}(x,y) = -2 + 2x.$$

For $P_1 = (1,0)$, $\Delta = (-3+2)^2 - (-2)(0) = 1 > 0$. Therefore, since $\Delta > 0$, P_1 is a saddle point for $f(x,y)$.

For $P_2 = (0,1)$, $\Delta = (-3+2)^2 - (0)(-2) = 1 > 0$ and for

$P_3 = (1,1)$, $\Delta = (-3+2+2)^2 - (0)(0) = 1 > 0$. Consequently both P_2 and P_3 are saddle points for $f(x,y)$.

For $P_4 = (2/3, 2/3)$, $\Delta = (-3+4/3+4/3)^2 - (-2+4/3)(-2+4/3)$

$$\Delta = 1/9 - (-2/3)(-2/3) = 1/9 - 4/9 = -3/9 < 0.$$

Therefore, since $\Delta < 0$ and $f_{11}(P_4) = -2/3 < 0$, $P_4 = (2/3, 2/3)$ is a relative maximum for $f(x,y)$.

● PROBLEM 6-8

Find the critical points of the function $f(x,y) = x^4 + y^4 - x^2 - y^2 + 1$. Then determine if each critical point is a relative maximum, relative minimum, or saddle point.

<u>Solution</u>: The critical points of $f(x,y) = x^4 + y^4 - x^2 - y^2 + 1$ occur where $\frac{\partial f}{\partial x} = 0$, $\frac{\partial f}{\partial y} = 0$. Computing gives the equations

$\frac{\partial f}{\partial x} = 4x^3 - 2x = 0$, $\frac{\partial f}{\partial y} = 4y^3 - 2y = 0$. Therefore, $x(4x^2 - 2) = 0$

and $x = 0$ or $x = \pm \frac{1}{\sqrt{2}}$; $y(4y^2 - 2) = 0$ and $y = 0$ or $y = \pm \frac{1}{\sqrt{2}}$.

Then the critical points of $f(x,y)$ are $P_1 = (0,0)$, $P_2 = \left(\frac{1}{\sqrt{2}}, -\frac{1}{\sqrt{2}}\right)$,

$P_3 = \left(-\frac{1}{\sqrt{2}}, \frac{1}{\sqrt{2}}\right)$, $P_4 = \left(\frac{1}{\sqrt{2}}, \frac{1}{\sqrt{2}}\right)$, $P_5 = \left(-\frac{1}{\sqrt{2}}, -\frac{1}{\sqrt{2}}\right)$, $P_6 = \left(0, \frac{1}{\sqrt{2}}\right)$,

$P_7 = \left(0, -\frac{1}{\sqrt{2}}\right)$, $P_8 = \left(\frac{1}{\sqrt{2}}, 0\right)$, $P_9 = \left(-\frac{1}{\sqrt{2}}, 0\right)$. Next

$f_{11}(x,y) = 12x^2 - 2$, $f_{12}(x,y) = 0$, $f_{22}(x,y) = 12y^2 - 2$. Then letting

$\Delta = [f_{12}(P_0)]^2 - f_{11}(P_0)f_{22}(P_0)$, it is known if $\Delta > 0$, P_0 is a saddle point. If $\Delta < 0$ and $f_{11}(P_0) > 0$, P_0 is a relative minimum. If $\Delta < 0$ and $f_{11}(P_0) < 0$, P_0 is a relative maximum. Finally, if $\Delta = 0$, the test is not applicable. For $P_1 = (0,0)$, $f_{11}(P_1) = -2$, $f_{22} = -2$, $f_{12} = 0$, then $\Delta = 0 - (4) = -4$. Hence, since $\Delta < 0$ and $f_{11}(P_1) = -2 < 0$, $P_1 = (0,0)$ is a relative maximum for $f(x,y)$. For P_2, P_3, P_4, P_5, $f_{11}(P_i) = 4$, $f_{22}(P_i) = 4$, $f_{12}(P_i) = 0$ and

$\Delta = 0 - (4)(4) = -16$. Therefore since $\Delta < 0$ and $f_{11}(P_i) > 0$,

P_2, P_3, P_4, P_5 are relative minima for $f(x,y)$. For P_6, P_7,

$f_{11}(P_j) = -2$, $f_{22}(P_j) = \frac{12}{2} - 2 = 4$, $f_{12}(P_j) = 0$. Then

$\Delta = 0 - (-2)(4) = -(-8) = 8$. Therefore, since $\Delta > 0$, P_6, P_7 are saddle points of $f(x,y)$. For P_8, P_9, $f_{11}(P_j) = 4$, $f_{22}(P_j) = -2$,

$f_{12}(P_j) = 0$. Then $\Delta = 0 - (4)(-2) = 8$. Hence, since $\Delta > 0$,

P_8, P_9 are saddle points of $f(x,y)$. Thus summarizing, we have

P_1 is a relative maximum of $f(x,y)$; P_2, P_3, P_4, P_5 are relative minima of $f(x,y)$; and P_6, P_7, P_8, P_9 are saddle points of $f(x,y)$.

• **PROBLEM 6-9**

Suppose $(a_1,b_1),\ldots,(a_n,b_n)$ are n points of $\mathbb{R}^2$. Which line $y = mx + b$ gives the best least square fit? (Assume not all of the a_i are equal.)

Solution: Let $y_i = ma_i + b$. Then the problem is to minimize the function

$$f(m,b) = \sum_{i=1}^{n} |y_i - b_i|^2$$

$$= \sum_{i=1}^{n} (ma_i + b - b_i)^2$$

$$= \sum_{i=1}^{n} [(ma_i + b)^2 - 2(ma_i + b)b_i + b_i^2]$$

$$= \sum_{i=1}^{n} [(ma_i)^2 + 2mba_i + b^2 - 2ma_ib_i - 2bb_i + b_i^2]$$

$$= m^2 \sum_{i=1}^{n} a_i^2 + 2mb \sum_{i=1}^{n} a_i + nb^2 - 2m \sum_{i=1}^{n} a_ib_i - 2b \sum_{i=1}^{n} b_i + \sum_{i=1}^{n} b_i^2.$$

Hence,

$$\frac{\partial f}{\partial m}(m,b) = 2m \sum_{i=1}^{n} a_i^2 + 2b \sum_{i=1}^{n} a_i - 2 \sum_{i=1}^{n} a_ib_i,$$

$$\frac{\partial f}{\partial b}(m,b) = 2m \sum_{i=1}^{n} a_i + 2nb - 2 \sum_{i=1}^{n} b_i.$$

Therefore, for the (m,b) which minimizes f, the system

$$\varphi(m,b) = m \sum_{i=1}^{n} a_i^2 + b \sum_{i=1}^{n} a_i - \sum_{i=1}^{n} a_ib_i = 0$$

$$\psi(m,b) = m \sum_{i=1}^{n} a_i + nb - \sum_{i=1}^{n} b_i = 0 \qquad (1)$$

must be satisfied. To see that this system does in fact have a solution, note that

$$\det \begin{pmatrix} \frac{\partial \varphi}{\partial m}(m,b) & \frac{\partial \varphi}{\partial b}(m,b) \\ \frac{\partial \psi}{\partial m}(m,b) & \frac{\partial \psi}{\partial b}(m,b) \end{pmatrix} = \det \begin{pmatrix} \sum_{i=1}^{n} a_i^2 & \sum_{i=1}^{n} a_i \\ \sum_{i=1}^{n} a_i & n \end{pmatrix}$$

$$= n \sum_{i=1}^{n} a_i^2 - \left(\sum_{i=1}^{n} a_i\right)^2$$

must be non-zero for (1) to have a solution. By the Cauchy-Schwartz inequality

$$\left(\sum_{i=1}^{n} a_i\right)^2 = \left(\sum_{i=1}^{n} a_i \cdot 1\right)^2 < \left(\sum_{i=1}^{n} a_i^2\right)\left(\sum_{i=1}^{n} 1^2\right) = n \sum_{i=1}^{n} a_i^2$$

(since the a_i's are not all equal). Hence, $n \sum_{i=1}^{n} a_i^2 - \left(\sum_{i=1}^{n} a_i\right)^2 > 0$

and consequently, (1) can be solved. To clarify the algebraic manipulations which follow let

$s = \sum_{i=1}^{n} a_i$, $t = \sum_{i=1}^{n} b_i$, $z = \sum_{i=1}^{n} a_i b_i$, $u = \sum_{i=1}^{n} a_i^2$, and $v = \sum_{i=1}^{n} b_i^2$. Then

(1) becomes

$$mu + bs = z, \qquad (2)$$
$$ms + bn = t, \qquad (3)$$

which is to be solved for m and b. Subtract u times (3) from s times (2) to get

$$bs^2 - bnu = sz - tu$$

and hence

$$b = \frac{sz - tu}{s^2 - nu} \qquad (4)$$

since $s^2 - nu < 0$ (from the preceding discussion). From (3) and (4) we have

$$m = \frac{t - bn}{s}$$

$$= \frac{t - \left(\frac{sz-tu}{s^2-nu}\right)n}{s}$$

$$= \frac{ts^2 - nut - nsz + nut}{s(s^2 - nu)}$$

$$= \frac{st - nz}{s^2 - nu} \qquad (5)$$

Substituting back into (4) and (5), the candidate for a minimum is

$$(m,b) = \left(\frac{\sum_{i=1}^{n} a_i \sum_{i=1}^{n} b_i - n \sum_{i=1}^{n} a_i b_i}{\left(\sum_{i=1}^{n} a_i\right)^2 - n \sum_{i=1}^{n} a_i^2}, \frac{\sum_{i=1}^{n} a_i \sum_{i=1}^{n} a_i b_i - \sum_{i=1}^{n} b_i \sum_{i=1}^{n} a_i^2}{\left(\sum_{i=1}^{n} a_i\right)^2 - n \sum_{i=1}^{n} a_i^2} \right)$$

To see that this is in fact a minimum, calculate the Hessian $H_f(m,b)$:

$$H_f(m,b) = \begin{pmatrix} \frac{\partial^2 f}{\partial m^2}(m,b) & \frac{\partial^2 f}{\partial m \partial b}(m,b) \\ \frac{\partial^2 f}{\partial b \partial m}(m,b) & \frac{\partial^2 f}{\partial b^2}(m,b) \end{pmatrix}$$

$$= \begin{pmatrix} 2\sum_{i=1}^{n} a_i^2 & 2\sum_{i=1}^{n} a_i \\ 2\sum_{i=1}^{n} a_i & 2n \end{pmatrix}$$

and note that it is constant and positive definite since

$$2 \sum_{i=1}^{n} a_i^2 > 0 \quad \text{and} \quad 4n \sum_{i=1}^{n} a_i^2 - 4\left(\sum_{i=1}^{n} a_i\right)^2 > 0$$

from above. Hence, the solution found is indeed a local (or relative) minimum and a global (or absolute) minimum since the Hessian is constant.

EXTREMES SUBJECT TO A CONSTRAINT

• **PROBLEM 6-10**

Find the point (x,y,z) obeying $g(x,y,z) = 2x + 3y + z - 12 = 0$ for which $f(x,y,z) = 4x^2 + y^2 + z^2$ is a minimum.

Solution: This problem is an extremal problem subject to a constraint. For the function $f(x,y,z)$, the variables x,y,z are not independent of each other, but are restricted by some relation existing between them, this relation being expressed by an equation $g(x,y,z) = 0$. There are several methods to deal with these extremal problems with constraints. For this problem we use the method of direct elimination. This method uses the equation or equations of constraint to express certain of the variables in terms of the remaining variables. Then these latter are regarded as the independent variables and the function $f(x,y,z)$ is then expressed in terms of these independent variables only. The solution for the extreme values of the function is then carried out by standard methods. For this problem the constraint is $g(x,y,z) = 2x + 3y + z - 12 = 0$. Solving for z, then expresses z as a function of x,y, the latter being regarded as the independent variables. This gives $z = 12 - 2x - 3y$, (1)
so that $f(x,y,z) = 4x^2 + y^2 + z^2$ is rewritten as

$$F(x,y) = 4x^2 + y^2 + (12 - 2x - 3y)^2 \tag{2}$$

Hence, the problem now is to find the values of x,y for which $F(x,y)$ is a minimum; then substitution of these values into (1) finds the value of z for the minimum of $f(x,y,z)$. Therefore, from equation (2), the critical points are solutions of

$$F_1(x,y) = 8x + 2(12 - 2x - 3y)(-2) = 0 \tag{3}$$

and

$$F_2(x,y) = 2y + 2(12 - 2x - 3y)(-3) = 0. \tag{4}$$

Simplifying gives the two equations $16x + 12y = 48$; $12x + 20y = 72$, which when solved simultaneously give one solution, namely $P_0 = (6/11, 36/11)$. Next let $\Delta = [F_{12}(P_0)]^2 - F_{11}(P_0)F_{22}(P_0)$. (5)

From (3), (4); $F_{11}(x,y) = 16$ so that $F_{11}(P_0) = 16$,

$F_{22}(x,y) = 20$ so that $F_{22}(P_0) = 20$,

and

$F_{12}(x,y) = 12$ so that $F_{12}(P_0) = 12$.

Then by equation (5); $\Delta = (12)^2 - (16)(20) = 144 - 320 < 0$. Hence, since $\Delta < 0$ and $F_{11}(P_0) = 16 > 0$, the point $P_0 = (6/11, 36/11)$ is a relative minimum for $F(x,y)$. To find z, substitution in (1) yields $z = 12 - 2(6/11) - 3(36/11)$; giving $z = 12/11$. Thus, the point (x,y,z) for which $f(x,y,z)$ is minimum subject to the constraint $g(x,y,z) = 0$ is $P_1 = (6/11, 36/11, 12/11)$.

● **PROBLEM 6-11**

Find the values of (x,y,z) that minimize $F(x,y,z) = xy + 2yz + 2xz$ given the condition $G(x,y,z) = xyz = 32$.

Solution: This is an extremal problem given the constraint $G(x,y,z) = k$ (k is a given constant). The method of direct elimination can be used to find the minimum of $f(x,y,z)$ given the constraint. However, an alternate procedure can be applied, namely the method of implicit functions. Assume that $F(x,y,z)$ and $G(x,y,z)$ are given functions with continuous first partial derivatives and that $\partial G/\partial z \neq 0$. If the equation

$$G(x,y,z) - k = 0 \tag{1}$$

has a solution $z = f(x,y)$, then it is desired to make the quantity $u = F(x,y,f(x,y))$ a maximum or minimum. So, it is needed that the equations

$$\frac{\partial u}{\partial x} = 0 \quad \text{and} \quad \frac{\partial u}{\partial y} = 0 \tag{2}$$

be solved. Now,

$$\frac{\partial u}{\partial x} = \frac{\partial F}{\partial x} + \frac{\partial F}{\partial z}\frac{\partial f}{\partial x} \; ; \; \frac{\partial u}{\partial y} = \frac{\partial F}{\partial y} + \frac{\partial F}{\partial z}\frac{\partial f}{\partial y} \tag{3}$$

where z is replaced by $f(x,y)$ after the differentiations are performed. But, since $G(x,y,f(x,y)) - k = 0$, it follows that

$$\frac{\partial G}{\partial x} + \frac{\partial G}{\partial z}\frac{\partial f}{\partial x} = 0 \; ; \; \frac{\partial G}{\partial y} + \frac{\partial G}{\partial z}\frac{\partial f}{\partial y} = 0. \tag{4}$$

Then solving (4) for $\partial f/\partial x$ and $\partial f/\partial y$ and substituting into (3) yields

$$\frac{\partial u}{\partial x} = \frac{\frac{\partial F}{\partial x}\frac{\partial G}{\partial z} - \frac{\partial F}{\partial z}\frac{\partial G}{\partial x}}{\frac{\partial G}{\partial z}}$$

(and a similar equation for $\frac{\partial u}{\partial y}$). Equations (2) now become

$$\frac{\partial F}{\partial x}\frac{\partial G}{\partial z} - \frac{\partial F}{\partial z}\frac{\partial G}{\partial x} = 0 \; ; \; \frac{\partial F}{\partial y}\frac{\partial G}{\partial z} - \frac{\partial F}{\partial z}\frac{\partial G}{\partial y} = 0 \tag{5}$$

in which after the differentiations are performed, $f(x,y)$ replaces z. Now, for the implicit function method, z is not solved for at the outset as in the method of direct elimination. Instead, equations (5) are arrived at as equations in all three variables. These two equations, along with the constraint give three equations which when solved simultaneously as equations in x,y,z give points among which the required points of extreme are present.

For this problem, think of y as a function of x and z (note: the preceding discussion utilized z as a function of x and y). The given function is $f(x,y,z) = xy + 2yz + 2xz$, $G(x,y,z) = xyz = 32$. (6) Differentiating both equations with respect to x gives

$$0 = \frac{\partial f}{\partial x} = y + x\frac{\partial y}{\partial x} + 2z\frac{\partial y}{\partial x} + 2z; \; 0 = \frac{\partial G}{\partial x} = yz + xz\frac{\partial y}{\partial x}.$$

252

Now eliminating $\partial y/\partial x$ between the two equations yields

$$\frac{\partial y}{\partial x} = \frac{-y}{x} \text{ (from the second equation), and}$$

so that
$$0 = 2z + x\left(\frac{-y}{x}\right) + y + 2z\left(\frac{-y}{x}\right)$$

$$0 = 2z - \frac{2zy}{x} \text{ from which } x = y. \tag{7}$$

Then differentiating both equations in (6) with respect to z gives

$$0 = \frac{\partial f}{\partial z} = x\frac{\partial y}{\partial z} + 2y + 2z\frac{\partial y}{\partial z} + 2x; \quad 0 = \frac{\partial G}{\partial z} = xy + xz\frac{\partial y}{\partial z}.$$

Now eliminating $\partial y/\partial z$ yields $\partial y/\partial z = -y/z$ so that

$$0 = x\left(\frac{-y}{z}\right) + 2y + 2z\left(\frac{-y}{z}\right) + 2x \text{ which simplifies to } 0 = \frac{-xy}{z} + 2x$$

from which $z = y/2$. \hfill (8)

Now to get values of x, y, z substitute into $G(x,y,z) = xyz = 32$ using (7), (8) and obtain $x \cdot x \cdot x/2 = 32 \Rightarrow x^3 = 64$. Therefore $x = 4$, $y = 4$, and $z = 2$. Hence, the value of (x,y,z) that minimizes $f(x,y,z)$ under the constraint $xyz = 32$ is $(4,4,2)$.

● **PROBLEM 6-12**

Find the maximum and minimum values of $F(x,y,z) = x^2 + y^2 + z^2$ on the surface of the ellipsoid

$$G(x,y,z) = \frac{x^2}{64} + \frac{y^2}{36} + \frac{z^2}{25} = 1.$$

Solution: The problem is to find the extreme values of $F(x,y,z)$ subject to the condition $G(x,y,z) = k$ (k a given constant; F and G are given functions with continuous first partial derivatives). The method of elimination could be used; however, an alternate procedure can be applied. Since $F(x,y,z)$ is the square of the distance from the point (x,y,z) to the origin, the problem is to find points on the ellipsoid at maximum and minimum distances from the center of the ellipsoid. The equation of the ellipsoid is given in the form

$$\frac{x^2}{a^2} + \frac{y^2}{b^2} + \frac{z^2}{c^2} = 1 \ ;$$

with center at the origin and end points for each principal axis $(\pm a, 0, 0)$, $(0, \pm b, 0)$, $(0, 0, \pm c)$. So,

$$G(x,y,z) = \frac{x^2}{8^2} + \frac{y^2}{6^2} + \frac{z^2}{5^2} = 1$$

has its center at the origin and end points of each principal axis to be $(\pm 8, 0, 0)$, $(0, \pm 6, 0)$, $(0, 0, \pm 5)$. Therefore, the longest principal axis has endpoints $(\pm 8, 0, 0)$, this being the maximum, and the shortest principal axis has endpoints $(0, 0, \pm 5)$; this being the minimum.

Note: We have the theorem — Given F and G have continuous first partial derivatives throughout a certain region of space. Let $G(x,y,z) = k$ define a surface S, every point of which is in the interior of the region, and suppose that the three partial derivatives

G_1, G_2, G_3 are never simultaneously zero at a point of S. Then a necessary condition for the values of $F(x,y,z)$ on S to attain an extreme value (either absolute or relative) at a point of S is that F_1, F_2, F_3 be proportional to G_1, G_2, G_3 at that point. If C is the value of F at that point, and if the constant of proportionality is not zero, the geometrical meaning of the proportionality is that the surface S and the surface $F(x,y,z) = C$ are tangent at the point in question. Applying this theorem to the maximum point $(8,0,0)$ gives the ratios $G_1 : G_2 : G_3$ and $F_1 : F_2 : F_3$ as $\frac{1}{4} : 0 : 0$ and $16 : 0 : 0$, respectively. The value of F at this point is 64, and the surface $F(x,y,z) = 64$ is a sphere. The sphere and the ellipsoid are tangent at $(8,0,0)$, as asserted in the theorem. However, this example shows that the tangency of the surfaces, or the proportionality of the two sets of ratios, is a necessary but not a sufficient condition for a minimum or maximum value of F. This is because the proportionality of the two sets of ratios exists at the points $(0,\pm 6,0)$, which are the endpoints of the principal axis of intermediate length, and F is neither a maximum nor a minimum at this point.

● **PROBLEM 6-13**

Find the maximum value of $f(x,y) = xy$; $(xy > 0)$ subject to the constraint $x^2 + y^2 = 8$ by drawing the level curves and by another method.

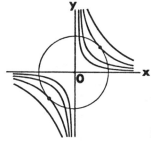

Fig. 1

Solution: For this problem, the function $f(x,y)$ is subject to the constraint $G(x,y) = k$. (The situation is shown in fig. 1). Here we have the fixed circle $x^2 + y^2 = 8$ and the function $f(x,y) = xy$. Then the level curves $xy = c$ (c a positive constant) are hyperbolas. From figure 1 it is seen that the hyperbola $xy = 4$ is tangent to the circle at the points $(-2,-2)$ and $(2,2)$ and these are the points at which the maximum value is attained. This shows that at a point of extremal value of f, the curve $f(x,y) = c$ through the point will be tangent to the curve defined by the constraint.

By the method of elimination, solving for y on the constraint gives

$y = \pm \sqrt{8 - x^2}$, then substituting this value into $f(x,y) = xy$ gives

$f(x) = x\sqrt{8 - x^2}$ $(f(x) > 0)$ which is a function of one variable. Then to find the maximum of the function, proceed as in elementary calculus. The critical points are $(\pm 2, \pm 2)$ since

$$f'(x) = (8 - x^2)^{\frac{1}{2}} - \frac{x^2}{(8 - x)^{\frac{1}{2}}} = 0$$

has the values $x = \pm 2$ as solutions; then substitution of this value into $y = \pm \sqrt{8 - x^2}$ gives $y = \pm 2$. But since $xy > 0$; the maximum value is at $(-2,-2)$ and $(2,2)$. The maximum value is $f(2,2) = f(-2,-2) = 4$.

• **PROBLEM 6-14**

For the following quadratic form
$$f(x,y) = ax^2 + 2bxy + cy^2,$$
state conditions for $f(x,y)$ to have a minimum and maximum value, using eigenvalues and the side condition $x^2 + y^2 = 1$.

Solution: The quadratic form $f(x,y) = ax^2 + 2bxy + cy^2$ can be written as

$$f(x,y) = [x \ y]\begin{bmatrix} a & b \\ b & c \end{bmatrix}\begin{bmatrix} x \\ y \end{bmatrix}$$

where [ ] are used as symbols to denote a matrix. It is given by definition that if A is an n×n matrix, then for some nonzero column vector $v = \text{col}(v_1, \ldots, v_n)$ it may happen that, for some scalar λ, $Av = \lambda v$. If this occurs, λ is called an eigenvalue of A and v is an eigenvector of A, associated with the eigenvalue λ. However, $Av = \lambda v$ can be rewritten as $Av = \lambda I(v)$ where I is the identity matrix. From this it follows $(A - \lambda I)v = 0$ (since v is a nonzero column matrix). This implies $\det|A - \lambda I| = 0$. Therefore, it follows

$a_n \lambda^n + a_{n-1} \lambda^{n-1} + \ldots + a_0 = 0$; which can be solved for λ. For this problem $A = \begin{bmatrix} a & b \\ b & c \end{bmatrix}$; and $A - \lambda I = 0$ is written

$$\begin{bmatrix} a & b \\ b & c \end{bmatrix} - \lambda \begin{bmatrix} 1 & 0 \\ 0 & 1 \end{bmatrix} = \begin{bmatrix} a & b \\ b & c \end{bmatrix} - \begin{bmatrix} \lambda & 0 \\ 0 & \lambda \end{bmatrix} = 0.$$

Hence, the eigenvalues of A are the solutions of $\begin{vmatrix} a - \lambda & b \\ b & c - \lambda \end{vmatrix} = 0$ or equivalently $\lambda^2 - (a+c)\lambda + ac - b^2 = 0$. Solving for λ, by the quadratic formula, gives the solution,

$$\lambda = \frac{(a+c) \pm \sqrt{(a+c)^2 - 4(ac - b^2)}}{2} = \frac{(a+c) \pm \sqrt{(a-c)^2 + 4b^2}}{2} \quad (1)$$

Therefore, by the last form, it is seen that the roots are always real. If $a = c$ and $b = 0$, the roots are equal and are equal to

$$\frac{a+c \pm \sqrt{0^2 + 4(0)^2}}{2} = \frac{(a+c)}{2}.$$

For any other condition the larger root is obtained from the plus sign and the smaller from the minus sign. Also, if $(ac-b^2) > 0$, both roots will have the same sign as $(a+c)$. Hence if $(ac-b^2) > 0$ and $(a+c) < 0$, then both roots are negative, if $(ac-b^2) > 0$ and $(a+c) > 0$, then both roots are positive.

Since $x^2 + y^2 = 1$ is the given side condition, the parametrization $x = \cos \theta$, $y = \sin \theta$ can be applied and $f(x,y)$ becomes

$$f(\cos \theta, \sin \theta) = a \cos^2 \theta + 2b \sin \theta \cos \theta + c \sin^2 \theta. \quad (2)$$

Therefore, when both roots are negative, $ac - b^2 > 0$ and $a + c < 0$.

255

Then $f(\cos\theta,\sin\theta)$ is negative for all θ. To prove this let $\theta = \pm \pi/2$, so that $f(\cos \pm \pi/2, \sin \pm \pi/2) = c < 0$, for if $c \geq 0$, then $a + c < 0$, implies $a < 0$, so that $ac \leq 0$; however this contradicts the condition $b^2 - ac < 0$. For $\theta \neq \pm \pi/2$, (2) can be rewritten

$f(\theta) = \cos^2\theta(a + 2b \tan\theta + c \tan^2\theta)$. Thus $f(\theta)$ is positive, negative, or 0, according to the quadratic expression $Q(u) = cu^2 + 2bu + a$, ($u = \tan\theta$) being positive, negative, or 0. Since $b^2 - ac < 0$, $Q(u)$ has no real roots; thus $Q(u)$ is always positive or always negative. For $u = 0$, $Q = a < 0$. Hence $Q(u)$ is always negative and the same holds for $f(\theta)$. Accordingly, the assertion is proved. Thus f has a negative maximum and remains negative for all θ when both roots are positive, $ac - b^2 > 0$ and $a + c > 0$. Then $f(\cos\theta, \sin\theta)$ is positive for all θ (this assertion can be proven in the same way as the first assertion) and f has a positive minimum.

EXTREMES IN A REGION

● **PROBLEM 6-15**

Find the maximum of the function $f(x,y) = x^2 y$ on the square with vertices $(0,0), (1,0), (1,1), (1,0)$.

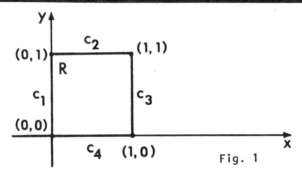

Fig. 1

Solution: The square is drawn in figure 1.
The problem here is to find the maximum point of a function on a closed and bounded set. We know a maximum point exists because f is a continuous function on this set (call the set S) and by the Weierstrass theorem, f has a maximum and minimum in this set. When given this situation of finding an extreme point for a function $f(x,y)$ in some region R, first determine the critical points of $f(x,y)$ in the interior of R and if an extremal point lies in the interior, it must be among these critical points.

Then the function should be investigated on the boundary of the region. Frequently, by parametrization of the boundary, the problem of finding an extremal value on the boundary becomes one of a lower-dimensional problem, to which the method of critical points can also be applied.

Finally, the possible extremes of $f(x,y)$ on the boundary and in the interior are compared to determine which points give the extremal values.

Hence, to follow this procedure first find the critical points of $f(x,y)$ on the interior of the square (denoted by R). By definition, the critical points of $f(x,y)$ on R are solutions of the equations:

$$\frac{\partial f}{\partial x} = 2xy = 0 \;;\; \frac{\partial f}{\partial y} = x^2 = 0 \tag{1}$$

Solving yields (1) to be true if and only if $(x,y) = (0,y)$ with an arbitrary value of y. Note the x-coordinate of a critical point must be 0, and when that happens $f(0,y) = 0$. Therefore, the critical points do not occur in the interior of the square, but do occur on some portion of the boundary. Hence, the maximum of $f(x,y)$ must occur on the boundary.

The boundary consists of the four segments labeled c_1 through c_4. The function $f(x,y)$ is now evaluated on these four segments to test where the maximum lies. On the segment c_1, $x = 0$ and $f(0,y) = 0$. On the segment c_2, $y = 1$ and $f(x,1) = x^2$. The maximum on this segment occurs when $x = 1$, in which case $f(1,1) = 1$. On the segment c_3, $x = 1$ and $f(0,y) = y$ and the maximum on c_3 occurs when $y = 1$, that is $f(1,1) = 1$ again. On the segment c_4, $y = 0$ and $f(x,0) = 0$. Comparing all these points then shows that the maximum value is at the point $(1,1)$ and the maximum value of $f(x,y)$ on the given square is $f(1,1) = 1$.

● **PROBLEM 6-16**

Find the maximum and minimum of $z = x^2 + 2y^2 - x$ on the set $x^2 + y^2 \leq 1$.

Solution: The function given is
$$z = f(x,y) = x^2 + 2y^2 - x \quad (1)$$
from which
$$\frac{\partial z}{\partial x} = 2x - 1 \; ; \; \frac{\partial z}{\partial y} = 4y \; .$$
Setting each partial derivative equal to zero and solving gives the critical point $p_0 = (\frac{1}{2}, 0)$, at which $z = (\frac{1}{2})^2 + 2 \times 0 - (\frac{1}{2}) = -\frac{1}{4}$.

However, for this problem, the function is subject to the constraint $x^2 + y^2 \leq 1$. So, on the boundary of the region E enclosed by this constraint we have $x^2 + y^2 = 1$. Here,

$$z = x^2 + 2(1-x^2) - x = 2 - x - x^2, \; -1 \leq x \leq 1.$$

For this function $dz/dx = -1 - 2x$, which when set equal to zero gives the critical point $x = -\frac{1}{2}$. So, at this point the function (on the boundary) has an absolute maximum value of $2\frac{1}{4}$; the absolute minimum is 0, at the end $x = 1$. Thus, the absolute maximum value is $2\frac{1}{4}$, occurring at $(-\frac{1}{2}, \pm \sqrt{3}/2)$, (from substitution into (1));

and the absolute minimum value is $-\frac{1}{4}$, occurring at $(\frac{1}{2}, 0)$.
Note: The function $z = f(x,y) = x^2 + 2y^2 - x$ on the set $x^2 + y^2 \leq 1$ has an absolute maximum and absolute minimum by the following theorem: Let D be a bounded domain in the xy-plane. Let $f(x,y)$ be defined and continuous in the closed region E formed by D plus its boundary. Then $f(x,y)$ has an absolute maximum and minimum in E. Also note that since the minimum occurs inside E, it is also a relative minimum of the function defined on the plane. This can be checked by letting $\Delta = [f_{12}(p_0)]^2 - f_{11}(p_0) f_{22}(p_0)$. Since

$$f_{11}(p_0) = 2; \; f_{12}(p_0) = 0; \; f_{22}(p_0) = 4;$$

$$\Delta = 0 - (2)(4) = -8.$$

Thus, since $\Delta < 0$ and $f_{11}(p_0) > 0$; $p_0 = (-\frac{1}{2}, 0)$ is a relative minimum.

• **PROBLEM 6-17**

Find the maximum and minimum of $f(x,y) = xy - y + x - 1$ in the closed disk $D = \{P: |P| \leq 2\}$ (P is a point in the xy-plane).

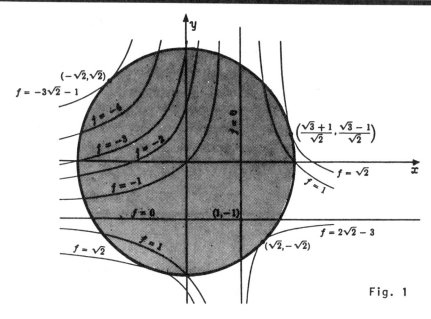

Fig. 1

Solution: To find the extreme values of the function given the constraint, first look for a relative extreme in the open disk $\{P: |P| < 2\}$. Since $\partial f/\partial x = y + 1$ and $\partial f/\partial y = x - 1$ the only critical point is at $P_1 = (1,-1)$. Then since $F_{11}(P_1) = 0$; $F_{22}(P_1) = 0$; $F_{12}(P_1) = 1$, $\Delta = F_{12}^2 - F_{11}F_{22}$ yields $\Delta = 1 > 0$ at P_1. Hence, P_1 is a saddle point, and there can be no maximum or minimum in the open disk; if they exist at all, the maximum and minimum must occur on the boundary. To check this, represent the boundary of D parametrically by $x = 2 \cos \theta$; $y = 2 \sin \theta$, then seek the maximum and minimum of $f(x,y) = f(2 \cos \theta, 2 \sin \theta) = 4 \cos \theta \sin \theta - 2 \sin \theta + 2 \cos \theta - 1$. This can be done by the elementary one-variable methods. Then among all points on the circle $\{P: |P| = 2\}$ ($x^2 + y^2 = 4$), f assumes a maximum at

$$\left(\frac{1 + \sqrt{3}}{\sqrt{2}}, \frac{\sqrt{3} - 1}{\sqrt{2}}\right) \text{ and at } \left(\frac{1 - \sqrt{3}}{\sqrt{2}}, \frac{-\sqrt{3} + 1}{\sqrt{2}}\right);$$

and f assumes a minimum at $(-\sqrt{2}, \sqrt{2})$. Now we have shown that there is no maximum or minimum in the open disk, but that considering only the points on the boundary circle, f assumes a maximum

$$f\left(\frac{1 + \sqrt{3}}{\sqrt{2}}, \frac{\sqrt{3} - 1}{\sqrt{2}}\right) = \sqrt{2}$$

and a minimum of $f(-\sqrt{2},\sqrt{2}) = -3\sqrt{2} - 1$. Thus it follows that $\max_D f = \sqrt{2}$; $\min_D f = -3\sqrt{2} - 1$, because every value assumed by f in D is actually assumed on the boundary. (As seen in fig. 1, every level curve having a point in D must intersect the boundary of D).

• **PROBLEM 6-18**

Find the maximum value of the function $f(x,y) = 4xy - 2x^2 - y^4$ in the square $D = \{(x,y) : |x| \leq 2, |y| \leq 2\}$.

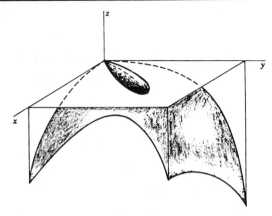

Graph of $z = 4xy - 2x^2 - y^4$ for $0 \leq x \leq 2, 0 \leq y \leq 2$.

<u>Solution</u>: Since $f_1(x,y) = 4y - 4x$ and $f_2(x,y) = 4x - 4y^3$, setting the partial derivatives equal to zero gives the equations $4y - 4x = 0$; $4x - 4y^3 = 0$ so that the critical points are $P_1 = (0,0)$, $P_2 = (1,1)$, $P_3 = (-1,-1)$. Then the maximum value of f in D must be attained at one of these points or on the boundary of D. However, it is not needed to compute the values of $f(x,y)$ on all the edges of this square; since we are looking for a maximum value, the parts of the boundary of D lying in the second and fourth quadrants where the term xy, which occurs in $f(x,y)$, is negative may be discarded. Moreover $f(x,y) = 4xy - 2x^2 - y^4$ is an even function ($f(-x,-y) = f(x,y)$), so that $f(x,y)$ takes equivalent values at symmetric points in the first and third quadrants. Therefore, the parts of D to be considered for a maximum value of $f(x,y)$ reduce to the values of $f(x,y)$ on the line $x = 2$, $0 \leq y \leq 2$, and the line $y = 2$; $0 \leq x \leq 2$. For the former, $f(2,y) = 8y - 8 - y^4$ for $0 \leq y \leq 2$. This is a function of one variable; to find it's maximum value set $f'(2,y) = 8 - 4y^3 = 0$ so that $y = (2)^{\frac{1}{3}}$. Then this function has its largest value at $y = 0$; $y = 2$; or $y = (2)^{\frac{1}{3}}$. Substitution into the function $f(2,y)$ then shows that $y = (2)^{\frac{1}{3}}$ gives the largest value which is

$$f(2,(2)^{\frac{1}{3}}) = 6(2)^{\frac{1}{3}} - 8 \cong -.44 \qquad (1)$$

On the second part of the boundary of D, $f(x,2) = 8x - 2x^2 - 16$, $0 \leq x \leq 2$. Then $f'(x,2) = 8 - 4x = 0$ so that $x = 2$ and this function has its largest value at $x = 0$ or $x = 2$. Substitution into the function $f(x,2)$ shows that $x = 2$ gives the largest value, which is $f(2,2) = -8$. This shows that the maximum value of $f(x,y)$ on the boundary of D is approximately $-.44$ (from equation (1)). Next, compare this value with the values $f(x,y)$ has at the three critical points P_1, P_2, P_3. At $P_1 = (0,0)$, $f(x,y) = f(0,0) = 0$. At $P_2 = (1,1)$ and $P_3 = (-1,-1)$, $f(x,y) = 1$. Therefore the absolute maximum value of $f(x,y)$ in D is 1 attained at the two points $(1,1)$ and $(-1,-1)$. Note: Similarly, for the minimum value of $f(x,y)$ in D, the parts of the boundary of D in the first and third quadrants are discarded since the term xy, which occurs in $f(x,y)$, is positive. Moreover, since $f(-x,y) = f(x,-y)$, $f(x,y)$ takes the same value at symmetric points in the second and fourth quadrants. So the parts of D considered for a minimum value of $f(x,y)$ are on the line $x = -2$, $0 \le y \le 2$ and the line $y = 2$; $-2 \le x \le 0$. Preceding in the same fashion as that of finding the maximum value of $f(x,y)$ in D, shows that the minimum value of $f(x,y)$ in D is -40, attained at the two boundary points $(2,-2)$ and $(-2,2)$ (since $f(x,-y) = f(-x,y)$).

• **PROBLEM 6-19**

Find the points which might furnish relative maxima and minima of the function
$$f(x,y) = 2xy - (1 - x^2 - y^2)^{3/2}$$
in the closed region $x^2 + y^2 \le 1$.

Solution: It is given that f is defined in R and (a,b) is an interior point of R. If (a,b) are relative extremes of R, then $f_1(a,b) = 0$, $f_2(a,b) = 0$. To apply this theorem first compute $\partial f/\partial x$; $\partial f/\partial y$, and set each equal to zero. This gives the equations:

$$\frac{\partial f}{\partial x} = 2y + 3x(1 - x^2 - y^2)^{\frac{1}{2}} = 0$$

$$\frac{\partial f}{\partial y} = 2x + 3y(1 - x^2 - y^2)^{\frac{1}{2}} = 0$$

The interior points of the region are those for which $x^2 + y^2 < 1$. Of these points, those that might furnish a relative maximum or minimum are among those found by solving the equations:

$$3x(1 - x^2 - y^2)^{1/2} = -2y \qquad (1)$$

$$3y(1 - x^2 - y^2)^{1/2} = -2x \qquad (2)$$

A possible solution of these equations is $x = 0$, $y = 0$. However, restricting x,y to be values $\ne 0$; equation (1) can be divided by equation (2) to yield

$$\frac{x}{y} = \frac{y}{x} \quad \text{or} \quad x^2 = y^2. \qquad (3)$$

Hence, substituting into (1), and squaring both sides, this gives $9x^2(1 - 2x^2) = 4x^2$, which simplifies to $9 - 18x^2 = 4$. Solving for x gives $x^2 = 5/18$, so $y^2 = 5/18$ by (3). Then $(1 - x^2 - y^2)^{\frac{1}{2}} = (1 - 5/18 - 5/18)^{\frac{1}{2}} = (4/9)^{\frac{1}{2}} = \frac{2}{3}$. Substituting this value into (1) gives $3x(\frac{2}{3}) = -2y$ or $x = -y$. Note that $x = y$ is ruled out. Therefore, there are three points in all which satisfy (1), (2). They are

$$P_0 = (0,0), \quad P_1 = \left(\frac{1}{3}\sqrt{\frac{5}{2}}, -\frac{1}{3}\sqrt{\frac{5}{2}}\right), \quad P_2 = \left(-\frac{1}{3}\sqrt{\frac{5}{2}}, \frac{1}{3}\sqrt{\frac{5}{2}}\right).$$

At P_0, $f(x,y) = -1$ at P_1 and P_2, $f(x,y) = -23/27$. \qquad (4)

Note: the theorem that was applied does not assert that the function has relative extrema at all three of these points, it only states that if any relative extrema occur at interior points, such extrema are to be found among these three points. However, before drawing any conclusions about absolute extrema, the behavior of the function on the boundary of the region must be investigated. The boundary is $x^2 + y^2 = 1$ so that here $f(x,y) = 2xy$. To look for extreme values of f on the boundary use the parametrization $x = \cos \theta$, $y = \sin \theta$ for the boundary circle ($0 \leq \theta \leq 2\pi$). Then $2xy = 2 \cos \theta \sin \theta = \sin 2\theta$. Therefore, the values range between -1 (at $\theta = 3\pi/4$ or $7\pi/4$) and $+1$ (at $\theta = \pi/4$ or $5\pi/4$). Now this adds four points to the other three to be considered when looking for the absolute minimum and maximum of the function in the closed region. When comparing the values ± 1 with the values in (4), it is seen that the function has the absolute maximum value $+1$, and the absolute minimum value -1. The maximum occurs at the two boundary points $(\sqrt{2}/2, \sqrt{2}/2)$, $(-\sqrt{2}/2, -\sqrt{2}/2)$. [This is because for the value $+1$, $\theta = \pi/4$ or $5\pi/4$ and for one point $x = \cos \pi/4$; $y = \sin \pi/4$ and for the other $x = \cos 5\pi/4$; $y = \sin 5\pi/4$]. The minimum occurs at the interior P_0 and at the two boundary points $(\sqrt{2}/2, -\sqrt{2}/2)$, $(-\sqrt{2}/2, \sqrt{2}/2)$. The interior points P_1, P_2 are saddle points.

METHOD OF LAGRANGE MULTIPLIERS

• **PROBLEM 6-20**

Find the extrema for the function $x^2 + y^2 + z^2$ subject to the constraint $x^2 + 2y^2 - z^2 - 1 = 0$.

Solution: The problem is to find the extreme values of a function $f(x,y,z)$ given a constraint $G(x,y,z) = k$ (k a constant). The methods of direct elimination or implicit functions can be tried to solve this problem. However, another method is applied, namely Lagrange's method. This method consists of the formation of the function $u = f(x,y,z) + \lambda G(x,y,z)$ subject to the conditions

$$\frac{\partial u}{\partial x} = 0, \frac{\partial u}{\partial y} = 0, \frac{\partial u}{\partial z} = 0 . \qquad (1)$$

Here x,y,z are treated as independent variables and λ is a constant, independent of x,y,z, called Lagrange's multiplier.

For this method solve the three equations in (1) along with the equation of constraint $G(x,y,z) = k$ to find the values of x,y,z,λ. More than one point (x,y,z) may be found in this way, but among the points so found will be the points of extremal values of F. For this problem
$f(x,y,z) = x^2 + y^2 + z^2$ and $G(x,y,z) = x^2 + 2y^2 - z^2 = 1$.

To apply the method of Lagrange let

$$u = x^2 + y^2 + z^2 + \lambda (x^2 + 2y^2 - z^2).$$

Then

$$\frac{\partial u}{\partial x} = 2x + \lambda 2x = 0 \qquad (2)$$

$$\frac{\partial u}{\partial y} = 2y + \lambda 4y = 0 \qquad (3)$$

$$\frac{\partial u}{\partial z} = 2z - \lambda 2z = 0 \qquad (4)$$

Next, let (x_0, y_0, z_0) be a solution. If $z_0 \neq 0$, then from equation (4), $\lambda = 1$. For this to be true in (2) and (3), it is needed $x = y = 0$.

In that case from (1), this gives $z_0^2 = -1$ which is impossible. Hence, any solution must have $z_0 = 0$.

If $x \neq 0$, then from (2), $\lambda = -1$ and from (3) and (4) for this to be true $y = z = 0$. Then, from (1), this yields $x_0^2 = 1$; $x_0 = \pm 1$. Therefore, two solutions satisfying the conditions have been obtained, namely $(1,0,0)$ and $(-1,0,0)$. Similarly if $y \neq 0$, then from (3), $\lambda = -\frac{1}{2}$ and from (2) and (4), this gives $x = z = 0$. Hence, from (1), $y_0^2 = \frac{1}{2}$; $y_0 = \pm\sqrt{\frac{1}{2}}$ and two more solutions, namely $(0, \sqrt{\frac{1}{2}}, 0)$ and $(0, -\sqrt{\frac{1}{2}}, 0)$, are found. These four points are therefore the extrema of the function $f(x,y,z)$ subject to the constraint g. If a minimum of f is desired, then direct computation shows that the two points $(0, \pm\sqrt{\frac{1}{2}}, 0)$ are the only possible solutions. This is because the function $f(x,y,z) = x^2 + y^2 + z^2$ is the square of the distance from the origin. Since the constraint defines a surface, a minimum for $f(x,y,z)$ given the constraint is a point on the surface which is at a minimum distance from the origin. Hence because $1 > \frac{1}{2}$ the points $(0, \pm\sqrt{\frac{1}{2}}, 0)$ are a shorter distance from the origin than the points $(\pm 1, 0, 0)$.

● **PROBLEM 6-21**

Let the number 12 equal the sum of three parts x, y, z. Find values of x, y, z so that xy^2z^2 shall be a maximum (given the first condition and that $x, y, z > 0$).

<u>Solution</u>: For this problem let $f(x,y,z) = xy^2z^2$ and $G(x,y,z) = x + y + z = 12$. Then, the problem is to maximize the function $f(x,y,z)$ given the constraint $G(x,y,z) = 12$. The method to be used is the method of Lagrange. For this method first construct the function:

$$u = xy^2z^2 + \lambda(x + y + z)$$

The variable λ is called a multiplier. By incorporating the constraint into the function to be optimized, the problem now becomes one of unconstrained optimization. Then

$$\frac{\partial u}{\partial x} = y^2z^2 + \lambda = 0 \qquad (1)$$

$$\frac{\partial u}{\partial y} = 2xyz^2 + \lambda = 0 \qquad (2)$$

$$\frac{\partial u}{\partial z} = 2xy^2z + \lambda = 0 \qquad (3)$$

The next step is to find a solution for λ. To do this multiply (from (1),(2),(3)) $\partial u/\partial x$ by x; $\partial u/\partial y$ by y; $\partial u/\partial z$ by z. This yields,

$$xy^2z^2 + \lambda x = 0;\quad 2xy^2z^2 + \lambda y = 0;\quad 2xy^2z^2 + \lambda z = 0 .$$

Adding these equations gives

$$5xy^2z^2 + \lambda x + \lambda y + \lambda z = 5xy^2z^2 + \lambda(x + y + z) = 0 \qquad (4)$$

However, it is given that $G(x,y,z) = x + y + z = 12$. Therefore, on the constraint (4) becomes $5xy^2z^2 + 12\lambda = 0$ or $\lambda = -5/12\, xy^2z^2$. Then equation (1), $y^2z^2 + \lambda = 0$ becomes $y^2z^2 - 5/12\, xy^2z^2 = 0$ which simplifies to $x = 12/5$.

Similarly equation (2), $2xyz^2 + \lambda = 0$ becomes $2xyz^2 - 5/12\, xy^2z^2 = 0$ (5)
which simplifies to
$$y = 24/5.$$
Finally, equation (3), $2xy^2z + \lambda = 0$ becomes $2xy^2z - 5/12\, xy^2z^2 = 0$ (6)
which gives the solution
$$z = 24/5.$$ (7)

Therefore, by (5),(6),(7) the values of x,y,z that maximize xy^2z^2 given the constraint $x + y + z = 12$ are

$$x = 12/5;\quad y = 24/5;\quad z = 24/5$$

(The values give a maximum since it was given that $x,y,z > 0$).

• **PROBLEM 6-22**

Find the maximum of $f(x,y) = xy$ on the curve $G(x,y) = (x+1)^2 + y^2 = 1$, assuming that such a maximum exists.

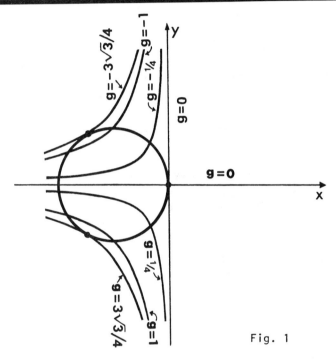

Fig. 1

Solution: Since the problem is to find the maximum of a function $f(x,y)$ given a constraint $G(x,y)$ the method of Lagrange can be applied. Since $f(x,y) = xy$ and $G(x,y) = (x+1)^2 + y^2 = 1$, let
$$u = xy + \lambda((x+1)^2 + y^2).$$ (1)
Then to use this method compute
$$\frac{\partial u}{\partial x}\,;\ \frac{\partial u}{\partial y}$$
and set each equal to zero. This gives
$$\frac{\partial u}{\partial x} = y + 2\lambda(x+1) = 0$$ (2)
and
$$\frac{\partial u}{\partial y} = x + 2\lambda y = 0.$$ (3)
Now the equations (2),(3) above are to be solved simultaneously with

the constraint curve to give solutions of x,y such that extremal values are among them. One solution is $x = 0$; $y = 0$ so that $\lambda = 0$; hence $P_1 = (0,0)$ is one point of possible extrema. For any other solution of (2),(3) $y \neq 0$; thus from (3), $\lambda = -x/2y$; then by (2)

$$y = (-2)\left(\frac{-x}{2y}\right)(x+1) = \frac{x^2 + x}{y} \quad \text{so that} \quad y^2 = x^2 + x.$$

Solving this simultaneously with the constraint

$(x+1)^2 + y^2 = 1$ gives $(x+1)^2 + x^2 + x = 1$ or $2x^2 + 3x = 0$, which equals $x(2x+3) = 0$ and has solutions $x = 0$; $x = -3/2$. The first value, $x = 0$, gives $y = 0$.
The other solution $x = -3/2$ gives the two points $(-3/2, \pm \sqrt{3}/2)$ (from substitution into $G(x,y,z)$). Therefore, the extreme values of the function are among the values given by the points $P_1 = (0,0)$; $P_2 = (-3/2, -\sqrt{3}/2)$ and $P_3 = (-3/2, \sqrt{3}/2)$. But direct substitution into $f(x,y) = xy$ shows the point $(-3/2, -\sqrt{3}/2)$ to give a maximum, the point $(-3/2, \sqrt{3}/2)$ to give a minimum, and $(0,0)$ to give neither. This is because $f(0,0) = 0$, $f(-3/2, -\sqrt{3}/2) = 3\sqrt{3}/4$ and $f(-3/2, \sqrt{3}/2) = -3\sqrt{3}/4$. Level curves of $f(x,y)$ are given in fig. 1.

● **PROBLEM 6-23**

a) Find the maxima and minima of $f(x,y) = x^2 + y^2$ on the ellipse $G(x,y) = 2x^2 + 3y^2 = 1$.

b) Find the maximum value of $f(x,y,z) = xyz$ on the plane $(x/a) + (y/b) + (z/c) = 1$ $(a,b,c > 0)$.

Solution: Both a) and b) are problems of finding extremal values of a function given a constraint. Hence, the method to be used is the method of Lagrange.

a) Let
$$u = x^2 + y^2 + \lambda(2x^2 + 3y^2) \tag{1}$$

where λ is a constant to be determined. Then, applying the Lagrange method $\partial u/\partial x$, $\partial u/\partial y$ are computed and set equal to zero. This yields from (1),

$$\frac{\partial u}{\partial x} = 2x + \lambda 4x = 0, \tag{2}$$

$$\frac{\partial u}{\partial y} = 2y + \lambda 6y = 0. \tag{3}$$

The next step is to solve equations (2),(3) simultaneously with the constraint $G(x,y)$, thus obtaining values for x,y,λ. To do this, first let $x \neq 0$, then from (1), $\lambda = -\frac{1}{2}$ and upon substitution into (2) this gives $y = 0$. Then, by $G(x,y) = 2x^2 + 3y^2 = 1$; at $y = 0$ this yields $x^2 = \frac{1}{2}$; $x = \pm 1/\sqrt{2}$. Thus two possible points of extremal value are at $P_1 = (1/\sqrt{2}, 0)$ and $P_2 = (-1/\sqrt{2}, 0)$. (4)
Similarly, if $y \neq 0$, from (2), $\lambda = -\frac{1}{3}$ and upon substitution into (3) this gives $x = 0$. Then by $G(x,y)$ at $x = 0$ this yields $3y^2 = 1$ or $y = \pm 1/\sqrt{3}$ so that two more possible extremes are at $P_2 = (0, 1/\sqrt{3})$; $P_3 = (0, -1/\sqrt{3})$. (5)
Therefore, if a maximum (or minimum) exists it is at one of the points in (4) or (5).

But since $x^2 + y^2$ is the square of the distance from (x,y) to the origin and $\frac{1}{2} > \frac{1}{3}$, it is concluded that the maximum occurs at $(\pm 1/\sqrt{2}, 0)$ and the minimum occurs at $(0, \pm 1/\sqrt{3})$.

b) Here $f(x,y,z) = xyz$ and $G(x,y,z) = (x/a) + (y/b) + (z/c) = 1$, $(a,b,c > 0)$. Thus, let $u = xyz + \lambda[x/a + y/b + z/c]$. Then

$$\frac{\partial u}{\partial x} = yz + \lambda/a = 0;\ \frac{\partial u}{\partial y} = xz + \lambda/b = 0;\ \frac{\partial u}{\partial z} = xy + \lambda/c = 0. \quad (6)$$

The next step is to find a solution for λ. To do this multiply (in equation (6))

$$\frac{\partial u}{\partial x}\ \text{by}\ x,\ \frac{\partial u}{\partial y}\ \text{by}\ y,\ \frac{\partial u}{\partial z}\ \text{by}\ z.$$

This yields $xyz + \lambda(x/a) = 0;\ xyz + \lambda(y/b) = 0;\ xyz + \lambda(z/c) = 0$. Adding these equations gives

$$3xyz + \lambda[(x/a) + (y/b) + (z/c)] = 0.$$

However, it is given that $G(x,y,z) = (x/a) + (y/b) + (z/c) = 1$. Therefore, on the constraint we have $3xyz + \lambda = 0$ or $\lambda = -3xyz$. However by (6), $yz + \lambda 1/a = 0;\ xz + \lambda 1/b = 0;\ xy + \lambda 1/c = 0$. Substituting $\lambda = -3xyz$, $yz - (3xyz)/a = 0$ which means $ayz - 3xyz = 0$ and simplifies to $x = a/3$. Similarly $xz - (3xyz)/b = 0$, $xy - (3xyz)/c = 0$ and $y = b/3$, $z = c/3$. So if a maximum of $f(x,y,z)$ exists, it exists at $(a/3, b/3, c/3)$ and has value

$$f(x,y,z) = a/3 \times b/3 \times c/3 = (abc)/27.$$

(The value is a maximum since $a,b,c > 0$ and thus $x,y,z > 0$).

• **PROBLEM 6-24**

Find the maximum value of xyz on the unit sphere $x^2 + y^2 + z^2 = 1$ by the Lagrange method.

Solution: For this problem, the function to be maximized is $f(x,y,z) = xyz$ and the constraint is $G(x,y,z) = x^2 + y^2 + z^2 = 1$. For the method of Lagrange first formulate a function u such that $u = f(x,y,z) + \lambda G(x,y,z)$ where λ is a constant to be determined. Here, this gives

$$u = xyz + \lambda(x^2 + y^2 + z^2).$$

The next step is to take the partial derivatives $\frac{\partial u}{\partial x}; \frac{\partial u}{\partial y}; \frac{\partial u}{\partial z}$ and set each equal to zero. This yields

$$\frac{\partial u}{\partial x} = yz + 2x\lambda = 0, \quad (1)$$

$$\frac{\partial u}{\partial y} = xz + 2y\lambda = 0, \quad (2)$$

$$\frac{\partial u}{\partial z} = xy + 2z\lambda = 0. \quad (3)$$

These equations are to be solved simultaneously with $G(x,y,z) = 1$. First solving for λ gives $\lambda = -yz/2x$ (from (1)); $\lambda = -xz/2y$ (from (2)); $\lambda = -xy/2z$ (from (3)). Equating each value for λ gives

$$-\frac{yz}{2x} = -\frac{xz}{2y} = -\frac{xy}{2z}$$

which is the same as $\frac{yz}{x} = \frac{xz}{y} = \frac{xy}{z}$. This can be written as

$z = 0$ or $x = 0$ or $y = 0$ or $y/x = x/y$; $z/y = y/z$. Thus, either one of the coordinates is zero or $x^2 = y^2 = z^2$. However, near any point where one of the coordinates is zero, $f(x,y,z)$ changes sign (since it is continuous) so these points are discarded from the points at which a maximum may be attained. Hence, upon solving $G(x,y,z) = x^2 + y^2 + z^2 = 1$ by the condition $x^2 = y^2 = z^2$ yields $3x^2 = 1$; $x = \pm 1/\sqrt{3}$ and $y = \pm 1/\sqrt{3}$; $z = \pm 1/\sqrt{3}$. Therefore the points of extremum (if they exist) are among the following 8 points $(\pm 1/\sqrt{3}, \pm 1/\sqrt{3}, \pm 1/\sqrt{3})$. Substituting any of these points into $f(x,y,z) = xyz$ gives the value $\pm 3^{-3/2}$ (the sign of the value depends upon the sign of the points substituted into $f(x,y,z)$). Thus, the value $3^{-3/2}$ is the maximum value attained at the points $(1/\sqrt{3}, 1/\sqrt{3}, 1/\sqrt{3})$, $(1/\sqrt{3}, -1/\sqrt{3}, -1/\sqrt{3})$, $(-1/\sqrt{3}, 1/\sqrt{3}, -1/\sqrt{3})$, and $(-1/\sqrt{3}, -1/\sqrt{3}, 1/\sqrt{3})$.

• **PROBLEM 6-25**

Find the maximum and minimum values of $f(x,y,z) = x^2 + y^2 + z^2$ subject to the constraints $x^2/4 + y^2/5 + z^2/25 = 1$ and $x + y - z = 0$.

Solution: The problem is to find the extremal values of a function $f(x,y,z) = x^2 + y^2 + z^2$ on two constraints, namely $G(x,y,z) = x^2/4 + y^2/5 + z^2/25 = 1$ and $H(x,y,z) = x + y - z = 0$. To do this, a generalization on the method of Lagrange on one constraint is applied. For one constraint, the method of Lagrange sets up the function

$$u = f(x,y,z) + \lambda G(x,y,z)$$

where $f(x,y,z)$ is the function to be maximized; $G(x,y,z)$ is the constraint; and λ is a constant to be determined known as Lagrange's multiplier. Next, the three first partials of u are computed and each equated to zero. Then these three equations, plus the equation of the constraint are solved simultaneously to give values for x,y,z,λ and among the values for x,y,z are those where the function attains extremal values.
The generalization of this method can be expressed as follows: Given the problem of finding the extremal values of a function $f(x_1, x_2, \ldots, x_n)$ subject to the constraints $G_1(x_1, \ldots, x_n) = 0$, $G_2(x_1, \ldots, x_n) = 0, \ldots G_k(x_1, \ldots, x_n) = 0$, form the function

$$u(x_1, x_2, \ldots, x_n) = F + \lambda_1 G_1 + \lambda_2 G_2 + \ldots + \lambda_k G_k$$

which is subject to the conditions $\frac{\partial u}{\partial x_1} = 0, \frac{\partial u}{\partial x_2} = 0, \ldots, \frac{\partial u}{\partial x_n} = 0$.

(Note $\lambda_1, \lambda_2, \ldots, \lambda_n$ are independent of $x_1, \ldots, x_n$; they are called Lagrangian multipliers). Then solve these equations simultaneously with the equations of constraints. The values where the function attains extremes are among these solutions.
For this problem there are two constraints, hence two Lagrangian multipliers λ_1, λ_2 are used. Therefore let

$$u(x,y,z) = f(x,y,z) + \lambda_1 G(x,y,z) + \lambda_2 H(x,y,z),$$

$$u(x,y,z) = x^2 + y^2 + z^2 + \lambda_1 (x^2/4 + y^2/5 + z^2/25 - 1)$$

Then
$$+ \lambda_2(x + y - z).$$

$$\frac{\partial u}{\partial x} = 2x + \frac{2x\lambda_1}{4} + \lambda_2 = 0, \quad (1)$$

$$\frac{\partial u}{\partial y} = 2y + \frac{2y\lambda_1}{5} + \lambda_2 = 0, \quad (2)$$

$$\frac{\partial u}{\partial z} = 2z + \frac{2z\lambda_1}{25} - \lambda_2 = 0. \quad (3)$$

Solving (1),(2),(3) for x,y,z respectively yields

$$x = \frac{-2\lambda_2}{\lambda_1+4}; \quad y = \frac{-5\lambda_2}{2\lambda_1+10}; \quad z = \frac{25\lambda_2}{2\lambda_1+50} \quad (4)$$

However, by the second condition $x + y - z = 0$, so (4) is substituted to give

$$-\frac{2\lambda_2}{\lambda_1+4} - \frac{5\lambda_2}{2\lambda_1+10} - \frac{25\lambda_2}{2\lambda_1+10} = 0$$

which equals

$$\frac{2}{\lambda_1+4} + \frac{5}{2\lambda_1+10} + \frac{25}{2\lambda_1+50} = 0$$

($\lambda_2 \neq 0$ because if so, we would have $x = 0, y = 0, z = 0$ and this does not satisfy the first constraint; therefore the equation can be divided by λ_2). Then combining fractions and simplifying gives the result

$17\lambda_1^2 + 245\lambda_1 + 750 = 0$ which can be factored into $(17\lambda_1 + 75)(\lambda_1+10) = 0$. Then $\lambda_1 = -10$ or $\lambda_1 = -75/17$.

For $\lambda_1 = -10$, substitution into (4) gives $x = \lambda_2/3$; $y = \lambda_2/2$; $z = 5\lambda_2/6$.

Substitution of these values into the first constraint,

$$\frac{x^2}{4} + \frac{y^2}{5} + \frac{z^2}{25} = 1,$$

gives

$$\frac{\lambda_2^2}{36} + \frac{\lambda_2^2}{20} + \frac{25\lambda_2^2}{900} = 1$$

which means

$$\lambda_2^2 = \frac{900}{95} \quad \text{or} \quad \lambda_2 = \pm 6\sqrt{5/19}.$$

This gives

$$x = \pm 2\sqrt{5/19}; \quad y = \pm 3\sqrt{5/19}; \quad z = \pm 5\sqrt{5/19}$$

as possible solutions. However, by the second condition it is given $x + y = z$, therefore this gives the two critical points

$(2\sqrt{5/19}, 3\sqrt{5/19}, 5\sqrt{5/19})$, $(-2\sqrt{5/19}, -3\sqrt{5/19}, -5\sqrt{5/19})$.

The value for $f(x,y,z) = x^2 + y^2 + z^2$ for these critical points is $(20 + 45 + 125/19) = 10$.

For $\lambda_1 = -75/17$, substitution into (4) gives $x = (34/7)\lambda_2$; $y = (-17/4)\lambda_2$; $z = (17/28)\lambda_2$, and substitution of these values into the first constraint gives

$$\frac{\left((34/7)\lambda_2\right)^2}{4} + \frac{\left((-17/4)\lambda_2\right)^2}{5} + \frac{\left((17/28)\lambda_2\right)^2}{25} = 1$$

which means $\lambda_2 = \pm 140/(17\sqrt{646})$. This gives

$$x = \pm \frac{34}{7}\left(\frac{140}{17\sqrt{646}}\right) ; \quad y = \pm \frac{17}{4}\left(\frac{140}{17\sqrt{646}}\right) ; \quad z = \pm \frac{17}{28}\left(\frac{140}{17\sqrt{646}}\right)$$

as possible solutions. However, by the second condition it is given $x + y = z$, therefore the two critical points are

$$\left(40/\sqrt{646}, -35/\sqrt{646}, 5/\sqrt{646}\right), \left(-40/\sqrt{646}, 35/\sqrt{646}, -5/\sqrt{646}\right).$$

The value for $f(x,y,z) = x^2 + y^2 + z^2$ is $(1600 + 1225 + 25)/646 = 75/17$.

Therefore, the maximum value is 10 and the minimum value is 75/17.

● **PROBLEM 6-26**

Minimize $f(x,y) = x^2 + y^2$ subject to $g(x,y) = (x-1)^3 - y^2 = 0$,
a) graphically
b) using the Lagrangian multiplier method.

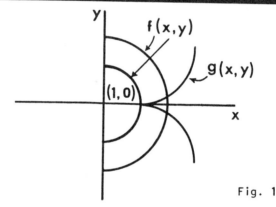

Fig. 1

<u>Solution</u>: a) $f(x,y)$ is the equation of a circle for various values of the radius. The constraint has its cusp at the point (1,0). Thus the function is minimized when $(x^*, y^*) = (1,0)$.

b) Let the Lagrangian be

$$L = x^2 + y^2 - \lambda[(x-1)^3 - y^2].$$

$$\frac{\partial L}{\partial x} = 2x - 3\lambda(x-1)^2 = 0, \quad (1)$$

$$\frac{\partial L}{\partial y} = 2y + 2\lambda y = 0, \quad (2)$$

$$\frac{\partial L}{\partial \lambda} = -(x-1)^3 + y^2 = 0. \quad (3)$$

From (2), $\lambda = -1$; from (1),

$$3x^2 - 4x + 3 = 0 \quad \text{or} \quad x = \frac{4 \pm \sqrt{16 - 4(3)(3)}}{6}.$$

Thus no real x can satisfy the first-order conditions and, according

to the Lagrangian procedure, the function and constraint system has no minimum.
Actually Lagrange's method can be generalized to handle this possibility. Let G be the m×n matrix

$$\left(\frac{\partial g_i}{\partial x_j}\right) = (\nabla g_i)$$

and

$$L(x,\lambda) = \lambda_0 f(x) - \sum_i \lambda_i [g_i(x) - b_i] \quad \text{where } x \in R^n.$$

The multiplier λ_0 is zero or one according to the following conditions:

i) If $\text{rank}\begin{bmatrix} G^* \\ \nabla f^* \end{bmatrix} = r(G^*) = m$, set $\lambda_0 = 1$. The other λ's are uniquely determined. Here $G^*, \nabla f^*$ mean that G and ∇f are evaluated at the extremum point.

ii) If $r\begin{bmatrix} G^* \\ \nabla f^* \end{bmatrix} = r(G^*) < m$, set $\lambda_0 = 1$. The λ's are not unique.

iii) If $r\begin{bmatrix} G^* \\ \nabla f^* \end{bmatrix} > r(G^*) = m - 1$, set $\lambda_0 = 0$. The λ's are not unique and not all zero.

iv) If $r\begin{bmatrix} G^* \\ \nabla f^* \end{bmatrix} > r(G^*)$ and $r\begin{bmatrix} G^* \\ \nabla f^* \end{bmatrix} < m$ set $\lambda_0 = 0$. The λ's, not all zero, are not unique.

Since letting $\lambda_0 = 1$ gave no real solutions let $\lambda_0 = 0$ in the modified Lagrangian $L(x,\lambda_0,\lambda) = \lambda_0(x^2 + y^2) - \lambda[(x-1)^3 - y^2]$

$$L(x,0,\lambda) = -\lambda[(x-1)^3 - y^2]$$

$$\frac{\partial L}{\partial x} = -3\lambda(x-1)^2 = 0$$

$$\frac{\partial L}{\partial y} = 2\lambda y = 0$$

$$\frac{\partial L}{\partial \lambda} = -(x-1)^3 + y^2 = 0.$$

Since $\lambda \neq 0$, $-3\lambda(x-1)^2 = 0$ implies that $x = 1$. $2\lambda y = 0$ implies that $y = 0$. Thus $(x^*, y^*) = (1,0)$.
To check whether iii) or iv) holds:

$$G = \begin{bmatrix} \frac{\partial g}{\partial x}, & \frac{\partial g}{\partial y} \end{bmatrix} = [3(x-1)^2, -2y], \quad G^* = (0,0)$$

$$\nabla f = [2x, 2y], \quad \nabla f^* = [2,0]$$

$$r\begin{bmatrix} G^* \\ \nabla f^* \end{bmatrix} = r\begin{bmatrix} 0 & 0 \\ 2 & 0 \end{bmatrix} = 1$$

$$r[G^*] = r[0 \quad 0] = 0$$

Hence iii) holds, $r\begin{bmatrix} G^* \\ \nabla f^* \end{bmatrix} > r(G^*) = 0$.

The second-order conditions for a strong local minimum are that the (n-m) principal minors of the bordered Hessian, (where m is the number of constraints) be of sign $(-1)^m$. Here $m = 1$, $n = 2$ and the bordered Hessian is:

$$= \begin{bmatrix} H_L & (\nabla g)^T \\ \nabla g & 0 \end{bmatrix}$$

$$= \begin{bmatrix} -6\lambda^*(x^*-1) & 0 & 3(x^*-1)^2 \\ 0 & 2\lambda^* & -2y^* \\ 3(x^*-1)^2 & -2y^* & 0 \end{bmatrix} \quad (4)$$

Since there is only one constraint, the determinant of (4) should be negative. But, at (1,0),

$$\det \begin{bmatrix} -6\lambda^*(x^*-1) & 0 & 3(x^*-1)^2 \\ 0 & 2\lambda^* & -2y^* \\ 3(x^*-1)^2 & -2y^* & 0 \end{bmatrix} = 0$$

Thus, the form is not positive definite so the second-order conditions are inconclusive. The fact that the function is minimized at (1,0) can only be shown graphically.

• **PROBLEM 6-27**

Maximize $u = \sqrt{q_1 q_2}$ subject to the constraint $p_1 q_1 + p_2 q_2 = M$ where $p_1 = 1$, $p_2 = 2$ and $M = 10$. Check the second-order conditions to verify that the solution is indeed a maximum.

Solution: The extrema of u subject to the constraint M can be found by the method of Lagrangian multipliers. Forming the objective function,

$$L = u(q_1, q_2) + \lambda(M - (p_1 q_1 + p_2 q_2))$$

$$L = \sqrt{q_1 q_2} + \lambda(10 - (q_1 + 2q_2))$$

$$\frac{\partial L}{\partial q_1} = \tfrac{1}{2} q_1^{-\frac{1}{2}} q_2^{\frac{1}{2}} - \lambda = 0$$

$$\frac{\partial L}{\partial q_2} = \tfrac{1}{2} q_1^{\frac{1}{2}} q_2^{-\frac{1}{2}} - 2\lambda = 0$$

The two equations may be solved for λ and then for q_1 in terms of q_2:

$$\tfrac{1}{2} q_1^{\frac{1}{2}} q_2^{-\frac{1}{2}} = q_1^{-\frac{1}{2}} q_2^{\frac{1}{2}}$$

or

$$q_1 = 2q_2.$$

Thus the extremum point occurs when $2q_2 + 2q_2 = 10$, or $q_2 = 5/2$ and $q_1 = 5$ and the value of u is $\sqrt{25/2}$.

To verify that this is indeed a maximum point requires the use of certain facts regarding the extrema of functions subject to constraints.

Suppose the problem is to maximize $f: R^n \to R$ subject to the m constraints $g_i(x_1,\ldots,x_n) = b_i$. Then x^* is a strong local maximum if $f(x^* + h) - f(x^*) < 0$ with $g_i(x^* + h) = b_i$, $i = 1,\ldots,m$ for sufficiently small $|h_i|$. A point x^* is a strong local maximum if $f(x^*) > f(x)$ for all x such that $|x-x^*| < |h|$. It is a weak local maximum if $f(x^*) \geq f(x)$. An equivalent statement is the following: Let

$$H_L = \left(\frac{\partial^2 L}{\partial x_i \partial x_j}\right)$$

be an n×n matrix and let

$$G = \left(\frac{\partial h_i}{\partial x_j}\right) = \begin{pmatrix} \nabla g_1 \\ \vdots \\ \nabla g_m \end{pmatrix}$$

be an m×n matrix. If $h^T H_L(x^*, \lambda^*) h$ is negative definite with $\nabla g_i(x^*) h = 0$, $i = 1,\ldots,n$, then f takes on a strong local maximum at x^*, subject to the constraints. Note that $h^T H h$ is a quadratic form. Furthermore, letting

$$h_{ij} = \frac{\partial^2 L}{\partial x_i \partial x_j}$$

and

$$g_{ij} = \frac{\partial g_i}{\partial x_j},$$

then it is negative definite if, and only if, the sequence of (n-m) principal minors of the matrix

$$\begin{bmatrix} H_L - I_n \lambda & G^T \\ G & 0_m \end{bmatrix},$$

i.e., the sequence

$$\begin{vmatrix} h_{11} & \cdots & h_{1\,m+1} & g_{11} & \cdots & g_{m1} \\ \vdots & & \vdots & \vdots & & \vdots \\ h_{m+1\,1} & \cdots & h_{m+1\,m+1} & g_{1\,m+1} & \cdots & g_{m\,m+1} \\ g_{11} & \cdots & g_{1\,m+1} & 0 & \cdots & 0 \\ \vdots & & \vdots & \vdots & & \vdots \\ g_{m1} & \cdots & g_{m\,m+1} & 0 & \cdots & 0 \end{vmatrix}, \begin{vmatrix} h_{11} & \cdots & h_{1\,m+2} & g_{11} & \cdots & g_{m1} \\ \vdots & & \vdots & \vdots & & \vdots \\ h_{m+2\,1} & \cdots & h_{m+2\,m+2} & g_{1\,m+2} & \cdots & g_{m\,m+2} \\ g_{11} & \cdots & g_{1\,m+2} & 0 & \cdots & 0 \\ \vdots & & \vdots & \vdots & & \vdots \\ g_{m1} & \cdots & g_{m\,m+2} & 0 & \cdots & 0 \end{vmatrix},$$

$$\cdots, \begin{vmatrix} h_{11} & \cdots & h_{1n} & g_{11} & \cdots & g_{m1} \\ \vdots & & \vdots & \vdots & & \vdots \\ h_{n1} & \cdots & h_{nn} & g_{1n} & \cdots & g_{mn} \\ g_{11} & \cdots & g_{1n} & 0 & \cdots & 0 \\ \vdots & & \vdots & \vdots & & \vdots \\ g_{m1} & \cdots & g_{mn} & 0 & \cdots & 0 \end{vmatrix},$$

alternates in sign, the first being of sign $(-1)^{m+1}$. The function takes on a strong local minimum at x^* ($f(x^*) < f(x)$ for all x in some deleted neighborhood of x^*) if the above bordered principal minors are all of the same sign $(-1)^m$. The quadratic form $h^T H_L h$ is then said to be positive definite.

In the given problem,

$$H_L(q^*, \lambda^*) = \begin{bmatrix} \dfrac{\partial^2 L^*}{\partial q_1^2} & \dfrac{\partial^2 L^*}{\partial q_1 \partial q_2} \\ \dfrac{\partial^2 L^*}{\partial q_1 \partial q_2} & \dfrac{\partial^2 L^*}{\partial q_2^2} \end{bmatrix}$$

$$= \begin{bmatrix} -\tfrac{1}{4} q_1^{-3/2} q_2^{1/2} & \tfrac{1}{4} q_1^{-1/2} q_2^{-1/2} \\ \tfrac{1}{4} q_1^{-1/2} q_2^{-1/2} & -\tfrac{1}{4} q_1^{1/2} q_2^{-3/2} \end{bmatrix}$$

$$= \begin{bmatrix} -\dfrac{1}{20\sqrt{2}} & \dfrac{\sqrt{2}}{20} \\ \dfrac{\sqrt{2}}{20} & -\dfrac{2\sqrt{2}}{20} \end{bmatrix} \qquad (1)$$

Since $g(q_1, q_2) = q_1 + 2q_2$, $\nabla g(q^*) = [1, 2]$. Adjoining this to (1) to form the bordered Hessian:

$$\begin{vmatrix} H_L & (\nabla g)^1 \\ \nabla g & 0 \end{vmatrix} = \begin{vmatrix} -\dfrac{1}{20\sqrt{2}} & \dfrac{\sqrt{2}}{20} & 1 \\ \dfrac{\sqrt{2}}{20} & -\dfrac{2\sqrt{2}}{20} & 2 \\ 1 & 2 & 0 \end{vmatrix} > 0.$$

Here $n = 2$ and $m = 1$. Thus, there is only one bordered principal minor to be evaluated. Its sign is $(-1)^{m+1}$ and thus the quadratic form is negative definite and q^* is indeed a maximum point.

FUNCTIONS OF THREE VARIABLES

● **PROBLEM 6-28**

Let f be the following quadratic form $f(x,y,z) = x^2 + y^2 + 3z^2 - xy + 2xz + yz$. Find a relative minimum of $f(x,y,z)$.

Solution: A quadratic form in three variables is defined as:

$$F(x_1, x_2, x_3) = \sum_{i=1}^{3} \sum_{j=1}^{3} a_{ij} x_i x_j$$

$$= a_{11} x_1^2 + a_{12} x_1 x_2 + a_{13} x_1 x_3$$

$$+ a_{21}x_2x_1 + a_{22}x_2^2 + a_{23}x_2x_3$$
$$+ a_{31}x_3x_1 + a_{32}x_3x_2 + a_{33}x_3^2$$

For this problem $f(x,y,z)$ is a quadratic form with certain of the coefficients equal to zero. To find a relative minimum for $f(x,y,z)$, the following theorem is applied:

If: i) All partial derivatives up to order 2 are continuous (i.e., f is a C^2 function).

ii) $f_1 = f_2 = f_3 = 0$ at some point $P_0 = (x_0, y_0, z_0)$.

iii) $f_{11} > 0$, $\begin{vmatrix} f_{11} & f_{12} \\ f_{21} & f_{22} \end{vmatrix} > 0$, and $\begin{vmatrix} f_{11} & f_{12} & f_{13} \\ f_{21} & f_{22} & f_{23} \\ f_{31} & f_{32} & f_{33} \end{vmatrix} > 0$

at P_0, then $f(x,y,z)$ has a relative minimum at P_0. Hence, the procedure is to first find a point P_0, that satisfies the condition $f_1(P_0) = f_2(P_0) = f_3(P_0) = 0$. From $f(x,y,z) = x^2 + y^2 + 3z^2 - xy + 2xz + yz$ this gives the equations:

$$f_1(x,y,z) = 2x - y + 2z = 0 \tag{1}$$
$$f_2(x,y,z) = -x + 2y + z = 0 \tag{2}$$
$$f_3(x,y,z) = 2x + y + 6z = 0 \tag{3}$$

Solving these equations simultaneously gives the solution $x = 0$, $y = 0$, $z = 0$. Hence, $P_0 = (0,0,0)$ and if the conditions in (iii) are met, then P_0 is a relative minimum for $f(x,y,z)$. From (1), (2), (3)

$$f_{11}(P_0) = 2; \ f_{12}(P_0) = -1; \ f_{13}(P_0) = 2$$
$$f_{21}(P_0) = -1; \ f_{22}(P_0) = 2; \ f_{23}(P_0) = 1$$
$$f_{31}(P_0) = 2; \ f_{32}(P_0) = 1; \ f_{33}(P_0) = 6.$$

Then by iii),

$$f_{11}(P_0) = 2 > 0, \quad \begin{vmatrix} 2 & -1 \\ -1 & 2 \end{vmatrix} = 4 - 1 = 3 > 0$$

and

$$\begin{vmatrix} 2 & -1 & 2 \\ -1 & 2 & 1 \\ 2 & 1 & 6 \end{vmatrix} = 22 - 8 - 10 = 4 > 0.$$

Therefore, since the conditions of the theorem are met; $P_0 = (0,0,0)$ is a relative minimum for $f(x,y,z)$, and $f(x,y,z) \geq f(0,0,0) = 0$.

Note: The theorem used can be generalized to R^n. It states:
a) If $f(\overline{X})$ $[\overline{X} \equiv (x_1, x_2, \ldots, x_n)]$ has a relative extreme at A, and if all $f_i(\overline{X})$ exist at A, then all $f_i(A) = 0$, that is, $\nabla f(\overline{X}) = 0$ at A.

b) Assume all $f_i(A) = 0$ and all $f_{i,j}$ are continuous at A. Denote by D_i, $1 \leq i \leq n$ the determinant

$$D_i = \begin{vmatrix} f_{1,1}(A) & \cdots & f_{1,i}(A) \\ f_{2,1}(A) & & f_{2,i}(A) \\ \vdots & & \vdots \\ f_{i,1}(A) & \cdots & f_{i,i}(A) \end{vmatrix}$$

If $D_i > 0$ for $1 \le i \le n$, then f has a relative minimum at A. If $(-1)^i D_i > 0$ for $1 \le i \le n$ (i.e., D_n is positive definite), then f has a relative maximum at A. If for some i, $1 \le 2i \le n$, we have $D_{2i} < 0$, then A is a saddle point for f (i.e., a degenerate critical point.)

● **PROBLEM 6-29**

Find the extremal values of the function $f(x,y,z) = xyz(1-x-y-z)$ on the set
$$D = \{(x,y,z) \mid 0 \le x, \; 0 \le y, \; 0 \le z, \; x+y+z \le 1\}.$$

Solution: The problem is to find the extremal values of the function $f(x,y,z) = xyz(1-x-y-z)$ on the given region D. Note that D is compact and $f(x,y,z)$ is continuous everywhere on D. On the boundary of D, $x = 0$, $y = 0$, $z = 0$, or $x+y+z = 1$, therefore

$$f(x,y,z) = xyz(1-(x+y+z)) = 0. \tag{1}$$

Hence, the value of $f(x,y,z)$ anywhere on the boundary of D is zero. On the interior of D; $x,y,z > 0$ and $x+y+z < 1$; therefore $f(x,y,z) = xyz(1-(x+y+z))$ is always positive on the interior of D. Then by (1), 0 is the minimum value for $f(x,y,z)$ on D.

The maximum value of $f(x,y,z)$ on D is attained at an interior point of D. At this point $f_1 = f_2 = f_3 = 0$, that is

$$f_1(x,y,z) = yz - 2xyz - y^2 z - yz^2 = 0 \tag{2}$$

$$f_2(x,y,z) = xz - 2xyz - x^2 z - xz^2 = 0 \tag{3}$$

$$f_3(x,y,z) = xy - 2xyz - x^2 y - xy^2 = 0. \tag{4}$$

The solutions of (2), (3), (4) yield points, among which is the point where the maximum value of $f(x,y,z)$ on D is attained. Rewriting (2), (3), (4) as $yz(1-2x-y-z) = 0$, $xz(1-x-2y-z) = 0$ $xy(1-x-y-2z) = 0$, yields, since $x,y,z \ne 0$, the equations $1-2x-y-z = 0$, $1-x-2y-z = 0$, $1-x-y-2z = 0$, which when solved simultaneously give the solution $x = y = z = \frac{1}{4}$. Thus, the maximum value of $f(x,y,z)$ on D is achieved at the point $(\frac{1}{4}, \frac{1}{4}, \frac{1}{4})$. Since $f(\frac{1}{4}, \frac{1}{4}, \frac{1}{4}) = (\frac{1}{4})^4$, the value $(\frac{1}{4})^4$ is the maximum value of $f(x,y,z)$ on D. Note that since the maximum is inside D, it is also a relative maximum. To check this, apply the test: If $f_1 = f_2 = f_3 = 0$ at P_0 and if $D_1 = f_{11} < 0$, $D_2 = \begin{vmatrix} f_{11} & f_{12} \\ f_{21} & f_{22} \end{vmatrix} > 0$,

$$D_3 = \begin{vmatrix} f_{11} & f_{12} & f_{13} \\ f_{21} & f_{22} & f_{23} \\ f_{31} & f_{32} & f_{33} \end{vmatrix} < 0, \text{ (i.e., } D_3 \text{ is negative definite)}$$

then $f(x,y,z)$ has a relative maximum at P_0. Computation gives at $P_0 = (\frac{1}{4},\frac{1}{4},\frac{1}{4})$, $f_{11} = -2/16$, $f_{12} = -1/16$, $f_{13} = -1/16$, $f_{21} = -1/16$, $f_{22} = -2/16$, $f_{23} = -1/16$, $f_{31} = -1/16$, $f_{32} = -1/16$, $f_{33} = -2/16$, then

$$-2/16 < 0 \qquad \begin{vmatrix} \frac{-2}{16} & \frac{-1}{16} \\ \frac{-1}{16} & \frac{-2}{16} \end{vmatrix} = 3/(16)^2 > 0, \qquad \begin{vmatrix} \frac{-2}{16} & \frac{-1}{16} & \frac{-1}{16} \\ \frac{-1}{16} & \frac{-2}{16} & \frac{-1}{16} \\ \frac{-1}{16} & \frac{-1}{16} & \frac{-2}{16} \end{vmatrix} = \frac{-4}{(16)^3} < 0$$

Therefore, since the conditions of the test are satisfied at $P_0 = (\frac{1}{4},\frac{1}{4},\frac{1}{4})$; P_0 gives a relative maximum for $f(x,y,z)$.

• **PROBLEM 6-30**

Find a relative minimum for

$$f(x,y,z) = 2x^2 + 2y^2 + 2z^2 - 2xz - 2yz - 6x + 2y + 8z + 14 \tag{1}$$

Solution: Given (1); computing the partial derivatives and setting each equal to zero gives the following equations:

$$f_1(x,y,z) = 4x - 2z - 6 = 0 \tag{1}$$

$$f_2(x,y,z) = 4y - 2z + 2 = 0 \tag{2}$$

$$f_3(x,y,z) = 4z - 2x - 2y + 8 = 0 . \tag{3}$$

Solving these equations simultaneously gives the solution $x = 0$, $y = -2$, $z = -3$. Hence $P_0 = (0,-2,-3)$ is a critical point for $f(x,y,z)$. For P_0 to be a relative minimum the following conditions must be met at P_0:

$$f_{11} > 0, \quad \begin{vmatrix} f_{11} & f_{12} \\ f_{21} & f_{22} \end{vmatrix} > 0 \quad \text{and} \quad \begin{vmatrix} f_{11} & f_{12} & f_{13} \\ f_{21} & f_{22} & f_{23} \\ f_{31} & f_{32} & f_{33} \end{vmatrix} > 0 \tag{4}$$

From (1), (2), (3), $f_{11}(P_0) = 4$, $f_{12}(P_0) = 0$, $f_{13}(P_0) = -2$

$$f_{21}(P_0) = 0, \quad f_{22}(P_0) = 4, \quad f_{23}(P_0) = -2$$
$$f_{31}(P_0) = -2, \quad f_{32}(P_0) = -2; \quad f_{33}(P_0) = 4 .$$

From which it follows

$$f_{11}(P_0) = 4 > 0, \quad \begin{vmatrix} 4 & 0 \\ 0 & 4 \end{vmatrix} = 16 > 0 \quad \text{and} \quad \begin{vmatrix} 4 & 0 & -2 \\ 0 & 4 & -2 \\ -2 & -2 & 4 \end{vmatrix} = 32 > 0.$$

Hence, since the conditions in (4) are satisfied, the point $P_0 = (0,-2,-3)$ is a relative minimum for $f(x,y,z)$. In fact, it is an absolute minimum, as can be shown by Taylor's theorem. Expanding $f(x,y,z)$ in powers of $X = x$, $Y = y+2$, $Z = z+3$ by

Taylor's theorem yields

$$f(x,y,z) = f(0,-2,-3) + 2X^2 + 2Y^2 + 2Z^2 - 2XZ - 2YZ = f(0,-2,-3)$$
$$+ [2X^2 - 2XZ] + 2Z^2 + 2Y^2 - 2YZ.$$

Then completing the square for the term in the bracket gives

$$f(x,y,z) = f(0,-2,-3) + 2(X^2 - XZ + Z^2/4) + 2Z^2 - Z^2/2 + 2Y^2 - 2YZ$$
$$= f(0,-2,-3) + 2(X - (1/2)Z)^2 + [2Y^2 - 2YZ] + 3/2\ Z^2.$$

Again completing the square for the term in the brackets finally yields $f(x,y,z) = f(0,-2,-3) + 2(X - (1/2)Z)^2 + 2(Y - (1/2)Z)^2 + Z^2$.

Now, the three terms to the right of $f(0,-2,-3)$ are always positive (squared terms). Therefore, for any (x,y,z); $f(x,y,z) \geq f(0,-2,-3)$, thus $f(0,-2,-3)$ is an absolute minimum.

EXTREME VALUE IN R^n

● **PROBLEM 6-31**

Let A denote a real symmetric matrix of order n. Find the maximum of $f(X) = X^tAX$ (X is a column matrix, and X^t is its transpose) on the sphere $X^tX = R^2$ ($\neq 0$).

Solution: $f(X) = X^tAX = [x_1,\ldots,x_n]\begin{pmatrix} a_{11} & a_{12} & \cdots & a_{1n} \\ \vdots & & & \\ a_{n1} & & \cdots & a_{nn} \end{pmatrix}\begin{pmatrix} x_1 \\ x_2 \\ \vdots \\ x_n \end{pmatrix}$

is the function to be maximized, on the constraint

$$G(X) = [x_1,\ldots,x_n]\begin{pmatrix} x_1 \\ \vdots \\ x_n \end{pmatrix} = R^2.$$

(Note $G(X)$ can also be written as

$$G(X) = x_1^2 + x_2^2 + \ldots + x_n^2 = \sum_{i=1}^{n} x_i^2 = R^2).$$

To apply Lagrange multipliers, note that $\nabla(X^tAX) = 2AX$ where ∇ represents the gradient. [If $X = (x_1, x_2, \ldots, x_n)^t$ and a_{ij} is the ij-th entry in A, then $f(X) = \sum_{i,j=1}^{n} a_{ij}x_ix_j = \sum_{i=1}^{n} a_{ii}x_i^2 + 2\sum_{\substack{j=1 \\ i<j}}^{n} a_{ij}x_ix_j$. So,

$$\frac{df}{dx_k}(X) = 2a_{kk}x_k + 2\sum_{i<k} a_{ik}x_i + 2\sum_{k<j} a_{kj}x_j = 2\sum_{i=1}^{n} a_{ik}x_i.$$ Therefore, $\nabla f(X)$

$= 2AX$]. Also $\nabla(X^tX) = 2X$. Therefore by the method of Lagrange there is a real λ such that $2AX = 2\lambda X$ or $AX = \lambda X$. \hfill (1)

This can be rewritten as $AX = \lambda I(X)$, which implies $(A - \lambda I)X = 0$ with $X \neq 0$. Then it follows that

$$\det(A - \lambda I) = 0 \hfill (2)$$

Therefore, this gives an equation of the form $a_n\lambda^n + a_{n-1}\lambda^{n-1} + \ldots + a_0 = 0$, which can be solved for λ. Hence to each real root λ of equation (2), there corresponds some X satisfying equation (1). It is a fact that for all real symmetric A, equation (2) has all roots real. (One proof begins with the existence of the root λ corresponding to the maximum of $X^t A X$ on $X^t X = R^2$). For this problem, since $AX = \lambda X$ (equation (1)) this yields $X^t A X = \lambda X^t X = \lambda R^2$ on the sphere. Therefore, the maximum value is λR^2 where λ is the largest eigenvalue of A (i.e., largest root of equation (2)).

● **PROBLEM 6-32**

Obtain the Cauchy-Schwartz inequality by extremalizing $a_1 x_1 + \ldots + a_n x_n$ on the compact surface $x_1^2 + \ldots + x_n^2 = c$. (Assuming $\vec{a} \neq 0$).

Solution: The Cauchy-Schwartz inequality is $|u \cdot v| \leq |u||v|$, where $u \cdot v$ is the dot product of u, v and $|\;|$ symbolize absolute value. To do this problem, Lagrange's theorem is used: Suppose the real-valued function $f(\vec{X})$ has a relative extremum on the set $\{\vec{X} | g(\vec{X}) = c\}$ at the point P_0. Assume that f and g are continuous and have continuous first partial derivatives around P_0.

Then the vectors $\vec{\nabla} f$ and $\vec{\nabla} g$ (∇ represents gradient) are parallel at P_0. Therefore, by the theorem, at the points where the maximum and minimum are attained, the gradients are parallel. Hence, letting $\vec{X} = (x_1, x_2, \ldots, x_n)$, $\vec{a} = (a_1, a_2, \ldots, a_n)$,

$$\vec{X} = k\vec{a} \quad (\text{i.e., } x_1 = ka_1, x_2 = ka_2, \ldots, x_n = ka_n) \quad (1)$$

Then
$$c = x_1^2 + x_2^2 + \ldots + x_n^2 = k^2 a_1^2 + k^2 a_2^2 + \ldots + k^2 a_n^2 = k^2 \vec{a} \cdot \vec{a},$$

or
$$\pm \sqrt{c} = \sqrt{k^2 \vec{a} \cdot \vec{a}} = k|\vec{a}|, \text{ so that } \pm \sqrt{c}|\vec{a}| = k|\vec{a}||\vec{a}|$$
$$= k\vec{a} \cdot \vec{a} \quad (2)$$

But from (1), $\vec{a} \cdot \vec{X} = k\vec{a} \cdot \vec{a}$, so that by (2)
$$\vec{a} \cdot \vec{X} = \pm \sqrt{c}|\vec{a}|. \quad (3)$$

However, since $c = x_1^2 + \ldots + x_n^2$, $\sqrt{c} = \sqrt{x_1^2 + \ldots + x_n^2} = |\vec{X}|$. Therefore, the extreme values of $\vec{a} \cdot \vec{X}$ are
$$\vec{a} \cdot \vec{X} = -|\vec{X}||\vec{a}| \quad \text{and} \quad \vec{a} \cdot \vec{X} = |\vec{X}||\vec{a}|.$$

Thus
$$-|\vec{a}||\vec{X}| \leq \vec{a} \cdot \vec{X} \leq |\vec{a}||\vec{X}|$$
with equality if and only if $\vec{X}$ and $\vec{a}$ are parallel (i.e., at the extreme points). Finally, this can be rewritten

$$|\vec{a} \cdot \vec{X}| \leq |\vec{a}||\vec{X}|$$

which is the Cauchy-Schwartz inequality.

• **PROBLEM 6-33**

Let p,q be positive real numbers such that $p^{-1} + q^{-1} = 1$. Consider the function $f = \vec{a} \cdot \vec{x} = \sum_{i=1}^{n} a_i x_i$ for any $\vec{a}$, with all $a_i > 0$, on the compact set $S = \{\vec{x} | \text{all } x_i \geq 0,\ x_1^p + \ldots + x_n^p = 1\}$. Show that the maximum value of $\vec{a} \cdot \vec{x}$ occurs at a point where all $x_i > 0$. Then, using this method, derive Hölder's inequality.

Solution: If $f(\vec{x})$ has a relative extremum on the set $\{\vec{x} | g(\vec{x}) = c\}$ at the point P_0, then the vectors $\nabla \vec{f}$ and $\nabla \vec{g}$ are parallel at P_0. For the case all $x_i > 0$, since $\nabla \vec{f} = (a_1, a_2, \ldots, a_n)$ and $\nabla \vec{g} = (px_1^{p-1}, \ldots, px_n^{p-1})$, at the extreme points,

$$x_i^{p-1} = \lambda a_i \quad (1 \leq i \leq n) \tag{1}$$

for some positive λ. Multiplying by x_i gives $x_i^p = \lambda a_i x_i$, $1 \leq i \leq n$. Then adding gives $x_1^p + x_2^p + \ldots + x_n^p = \lambda a_1 x_1 + \lambda a_2 x_2 + \ldots + \lambda a_n x_n$. So,

$$1 = \lambda \sum_{i=1}^{n} a_i x_i \quad \text{or} \quad \sum_{i=1}^{n} a_i x_i = 1/\lambda. \tag{2}$$

However, raising each side of equation (1) to the power $\frac{p}{(p-1)} = q$ (because $p^{-1} + q^{-1} = 1$) gives the equation $x_i^p = \lambda^q a_i^q$ $(1 \leq i \leq n)$. Then adding gives

$$x_1^p + x_2^p + \ldots + x_n^p = \lambda^q(a_1^q + a_2^q + \ldots + a_n^q) \quad \text{or} \quad 1 = \lambda^q \sum_{i=1}^{n} a_i^q.$$

Now $\sum_{i=1}^{n} a_i^q = 1/\lambda^q$ can be rewritten as $\left(\sum_{i=1}^{n} a_i^q\right)^{1/q} = 1/\lambda. \tag{3}$

From (2), $\vec{a} \cdot \vec{x}$ has a maximum of $1/\lambda$ or, $\left(\sum a_i^q\right)^{1/q} = 1/\lambda$.

The other possibilities for the maximum are extreme points where some $x_i = 0$. For this case (some $x_i = 0$), it is sufficient to take $x_1 \ldots x_m \neq 0$, $x_{m+1} = x_{m+2} = \ldots = x_n = 0$ (then permute the a_i). Therefore, the above analysis which was applied to R^n can now be applied to R^m. This gives the value $\left(\sum_{i=1}^{m} a_i^q\right)^{1/q}$, which is smaller in value than $\left(\sum a_i^q\right)^{1/q}$. Therefore the value where all $x_i > 0$ is a maximum.

The very important Hölder's inequality can be derived from the above problem. Equations (2) and (3) are:

$$\sum a_i x_i = 1/\lambda \quad \text{and} \quad 1 = \lambda^q \sum a_i^q.$$

Now $1 = \left(|\lambda|^q\right)^{1/q}\left(\sum |a_i|^q\right)^{1/q}$ or, $\frac{1}{|\lambda|} = \left(\sum |a_i|^q\right)^{1/q}$ or,

$$\Sigma |a_i x_i| \le \left(\Sigma |a_i|^q\right)^{1/q}. \tag{4}$$

Since $\left(\Sigma |a_i|^q\right)^{1/q}$ is the maximum of $\vec{a} \cdot \vec{x} = \Sigma |a_i x_i|$, the inequality (4) is true for all x. But $\left(\Sigma |x_i|^p\right)^{1/p} = 1$. Thus, it may be multiplied with (4) to yield

$$\Sigma |a_i x_i| \le \left(\Sigma |a_i|^q\right)^{1/q} \left(\Sigma |x_i|^p\right)^{1/p} \tag{5}$$

which is Hölder's inequality $(1/p + 1/q = 1)$.

The inequality is an equality if and only if $|x_i|^p = t|a_i|^q$ for then

$$\left(\Sigma |a_i|^q\right)^{1/q} \left(\Sigma |x_i|^p\right)^{1/p}$$

$$= \left(\Sigma |a_i|^q\right)^{1/q} \left(\Sigma t|a_i|^q\right)^{1/p} = t^{1/p} \left(\Sigma |a_i|^q\right)^{1/q} \left(\Sigma |a_i|^q\right)^{1/p}$$

while

$$\Sigma |a_i| |x_i| = \Sigma \left(|a_i|^q\right)^{1/q} \left(t|a_i|^q\right)^{1/p}$$

$$= t^{1/p} \Sigma \left(|a_i|^q\right)^{1/q} \left(|a_i|^q\right)^{1/p} = t^{1/p} \left(\Sigma |a_i|^q\right)^{1/q} \left(\Sigma |a_i|^q\right)^{1/p}$$

since $\dfrac{1}{p} + \dfrac{1}{q} = 1$.

● **PROBLEM 6-34**

Use Lagrange's method of finding the maximum of functions subject to constraints to develop
a) Cauchy's inequality
b) Hölder's inequality
c) Minkowski's inequality.

<u>Solution</u>: Let $F(x,y)$ be a real valued function with domain R^2. The unconstrained extrema of F occur at points where $\nabla F = (0,0)$. Now suppose $F(x,y)$ is constrained to move along some surface $G(x,y) = 0$. Then, according to Lagrange, the extrema of $F(x,y)$ subject to $G(x,y)$ is found by forming the objective function,

$$H(x,y) = F(x,y) + \lambda(G(x,y)),$$

and finding points at which $\nabla H = (0,0)$.

a) Suppose $F(x,y) = x^2 y^2$ and $G(x,y) = x^2 + y^2 - R^2$ where R is a constant. Let

$$H(x,y) = F(x,y) + \lambda(G(x,y)),$$

$$H(x,y) = x^2 y^2 + \lambda(x^2 + y^2 - R^2) \tag{1}$$

$$\frac{\partial H}{\partial x} = 2xy^2 + 2\lambda x = 0 \tag{2}$$

$$\frac{\partial H}{\partial y} = 2x^2 y + 2\lambda y = 0 \tag{3}$$

$$\frac{\partial H}{\partial \lambda} = x^2 + y^2 - R^2 = 0 \tag{4}$$

From (2) and (3): $\lambda = -y^2 = -x^2$ or $x^2 = y^2 = a$. Then, from (4),

$2a = R^2$ or $a = (R^2/2)$. Substituting into $F(x,y)$, the value of F at the maximum is

$$F(x,y) = x^2 y^2 = (R^2/2)(R^2/2) = (R^2/2)^2 .$$

This result may be generalized to n variables, i.e., the maximum value of $F(x_1,\ldots,x_n) = x_1^2 x_2^2 \ldots x_n^2$ subject to

$$G(x_1,\ldots,x_n) = x_1^2 + x_2^2 + \ldots + x_n^2 - R^2$$

occurs when $x_1^2 = x_2^2 \ldots = x_n^2$ and is equal to $(R^2/n)^n$. Since this is the maximum the inequality $(x_1^2 x_2^2 \ldots x_n^2) \leq (R^2/n)^n$ is always satisfied for all values of $(x_1, x_2, \ldots, x_n)$. But this may be rewritten as

$$(x_1^2 x_2^2 \ldots x_n^2) \leq \left[\frac{(x_1^2 + \ldots + x_n^2)}{n}\right]^n$$

or

$$(x_1^2 x_2^2 \ldots x_n^2)^{1/n} \leq \frac{(x_1^2 + \ldots + x_n^2)}{n}$$

or

$$(a_1 a_2 \ldots a_n)^{1/n} \leq \frac{a_1 + a_2 + \ldots + a_n}{n} \tag{5}$$

where $a_i = x_i^2$, $1 \leq i \leq n$, $a_i \geq 0$. This shows that the geometric mean of n positive numbers is always less than or equal to the arithmetic mean of the n numbers. Note that equality holds when $a_1 = a_2 = \ldots = a_n$ since

$$(a_k^n)^{1/n} = a_k = \frac{n a_k}{n} .$$

Suppose $n = 2$ in (5) and let $a_1 = u$, $a_2 = v$. Then

$$(uv)^{\frac{1}{2}} \leq \frac{u+v}{2}$$

or $\quad u^{\frac{1}{2}} v^{\frac{1}{2}} \leq \frac{1}{2}u + \frac{1}{2}v.$ \hfill (6)

Now suppose u_i and v_i are actually coordinates on n-tuples in R^n, specifically

$$u_i = \frac{a_i^2}{\Sigma a_i^2}, \quad v_i = \frac{b_i^2}{\Sigma b_i^2},$$

where $(a_1,\ldots,a_n)$, $(b_1,\ldots,b_n)$ are points in R^n. Thus Σa_i^2 denotes the square of the Euclidean norm in R^n. Summing over all i and noting that (6) holds for each i:

$$\Sigma \cdot \frac{|(a_i^2)^{\frac{1}{2}}|}{(\Sigma a_i^2)^{\frac{1}{2}}} \cdot \frac{|(b_i^2)^{\frac{1}{2}}|}{(\Sigma b_i^2)^{\frac{1}{2}}} \leq \frac{1}{2} \cdot \frac{\Sigma a_i^2}{\Sigma a_i^2} + \frac{1}{2} \cdot \frac{\Sigma b_i^2}{\Sigma b_i^2}$$

or

$$\Sigma \frac{|(a_i^2)^{\frac{1}{2}}|}{(\Sigma a_i^2)^{\frac{1}{2}}} \cdot \frac{|(b_i^2)^{\frac{1}{2}}|}{(\Sigma b_i^2)^{\frac{1}{2}}} \leq \frac{1}{2} + \frac{1}{2} = 1$$

or
$$\Sigma |(a_i^2)^{\frac{1}{2}}| \; |(b_i^2)^{\frac{1}{2}}| \leq (\Sigma \, a_i^2)^{\frac{1}{2}} (\Sigma \, b_i^2)^{\frac{1}{2}}$$

or
$$\Sigma |a_i| \, |b_i| \leq (\Sigma \, a_i^2)^{\frac{1}{2}} (\Sigma \, b_i^2)^{\frac{1}{2}} \, .$$

Thus, $\left|\Sigma \, a_i b_i\right| \leq (\Sigma \, a_i^2)^{\frac{1}{2}} (\Sigma \, b_i^2)^{\frac{1}{2}} \, .$ (7)

Inequality (7) is known as Cauchy's inequality.

b) To derive Hölder's inequality, define the weighted arithmetic and geometric means of $x_1, \ldots, x_n \geq 0$ as $C = \alpha_1 x_1 + \alpha_2 x_2 + \ldots + \alpha_n x_n$ and $x_1^{\alpha_1} x_2^{\alpha_2} \ldots x_n^{\alpha_n}$ respectively with $\sum_{i=1}^{n} \alpha_i = 1$, and C a positive constant.

The weighted geometric mean is always less than or equal to C for positive $x_1, \ldots, x_n$. To see this maximize

$$F = x_1^{\alpha_1} x_2^{\alpha_2} \ldots x_n^{\alpha_n}$$

subject to the constraint $\alpha_1 x_1 + \alpha_2 x_2 + \ldots + \alpha_n x_n = C$.

$$H = x_1^{\alpha_1} \ldots x_n^{\alpha_n} + \lambda \left(\alpha_1 x_1 + \alpha_2 x_2 + \ldots + \alpha_n x_n \right)$$

$$\frac{\partial H}{\partial x_i} = \alpha_i x_1^{\alpha_1} \ldots x_i^{\alpha_i - 1} \ldots x_n^{\alpha_n} + \lambda \alpha_i = 0, (i = 1, \ldots, n)$$

$$\frac{\partial H}{\partial \lambda} = \alpha_1 x_1 + \ldots + \alpha_n x_n \, .$$

Thus $x_1 = x_2 = \ldots = x_n = C$ at the maximum point. Then

$$F = C^{\alpha_1 + \ldots + \alpha_n} = C \quad (\text{since } \sum_{i=1}^{n} \alpha_i = 1).$$

Also note that $\alpha_1 C + \alpha_2 C + \ldots + \alpha_n C = (\alpha_1 + \ldots + \alpha_n) C = C$. Then $x_1^{\alpha_1} \ldots x_n^{\alpha_n} \leq C = \alpha_1 x_1 + \ldots + \alpha_n x_n$ for all positive values of $x_1, \ldots, x_n$.

Now, if u, v, p and q are positive and $1/p + 1/q = 1$ then

$$u^{1/p} v^{1/q} \leq \left(1/p\right) u + \left(1/q\right) v$$

from the above result. Let

$$u = \frac{|a_i|^p}{\sum_{i=1}^{n} |a_i|^p} \, , \quad v = \frac{|b_i|^q}{\sum_{i=1}^{n} |b_i|^q}$$

Then
$$u^{1/p} v^{1/q} = \frac{|a_i|}{\left(\Sigma |a_i|^p\right)^{1/p}} \frac{|b_i|}{\left(\Sigma |b_i|^q\right)^{1/q}}$$

and
$$\frac{1}{p} u + \frac{1}{q} v = \frac{1}{p} \frac{|a_i|^p}{\Sigma |a_i|^p} + \frac{1}{q} \frac{|b_i|^q}{\Sigma |b_i|^q} .$$

Summing over all i, $1 \le i \le n$:
$$\Sigma \frac{|a_i|}{\left(\Sigma |a_i|^p\right)^{1/p}} \frac{|b_i|}{\left(\Sigma |b_i|^q\right)^{1/q}} \le \frac{1}{p}\left(\frac{\Sigma |a_i|^p}{\Sigma |a_i|^p}\right) + \frac{1}{q}\left(\frac{\Sigma |b_i|^q}{\Sigma |b_i|^q}\right) = \frac{1}{p} + \frac{1}{q} = 1$$

or
$$\Sigma |a_i||b_i| \le \left(\Sigma |a_i|^p\right)^{1/p} \left(\Sigma |b_i|^q\right)^{1/q}$$

or
$$\Sigma |a_i b_i| \le \left(\Sigma |a_i|^p\right)^{1/p} \left(\Sigma |b_i|^q\right)^{1/q} . \tag{8}$$

Inequality (8) is an extension of the Cauchy inequality which states that, for two vectors in R^n, $\vec{u} = (a_1,\ldots,a_n)$, $\vec{v} = (b_1,\ldots,b_n)$,
$$|\vec{u}\cdot\vec{v}| \le |\vec{u}||\vec{v}| \quad \text{or,} \quad |\Sigma a_i b_i| \le (\Sigma |a_i|^2)^{\frac{1}{2}} (\Sigma |b_i|^2)^{\frac{1}{2}} . \tag{9}$$

Comparing (9) and (8) note that (8) equals (9) when $p = q = 2$. Inequality (8) is called Hölder's inequality and is used in the theory of integration to define norms for the L_p spaces.

c) Minkowski's inequality states that
$$\left(\sum_{i=1}^n |x_i + y_i|^p\right)^{1/p} \le \left(\sum_{i=1}^n |x_i|^p\right)^{1/p} + \left(\sum_{i=1}^n |y_i|^p\right)^{1/p}, \quad p > 1 .$$

It may be derived using Hölder's inequality. Consider the n-tuples $(x_1,\ldots,x_n)$, $(y_1,\ldots,y_n)$ and taking only the ith coordinate, note that
$$|x_i + y_i|^p = |x_i + y_i||x_i + y_i|^{p/q}, \quad (1/p + 1/q = 1).$$
But, since $(|a+b|)|c| \le |a||c| + |b||c|$
$$(|x_i+y_i|)|x_i+y_i|^{p/q} \le |x_i||x_i+y_i|^{p/q} + |y_i||x_i+y_i|^{p/q}$$

Thus
$$|x_i+y_i|^p \le |x_i||x_i+y_i|^{p/q} + |y_i||x_i+y_i|^{p/q} .$$

Summing over all i, $i = 1,\ldots,n$:
$$\Sigma |x_i+y_i|^p \le \Sigma |x_i||x_i+y_i|^{p/q} + \Sigma |y_i||x_i+y_i|^{p/q} . \tag{10}$$

Now, applying Hölder's inequality to the r.h.s. of (10):
$$\Sigma |x_i||x_i+y_i|^{p/q} \le \left(\Sigma |x_i|^p\right)^{1/p} \left(\Sigma \left(|x_i+y_i|^{p/q}\right)^q\right)^{1/q}$$
and
$$\Sigma |y_i||x_i+y_i|^{p/q} \le \left(\Sigma |y_i|^p\right)^{1/p} \left(\Sigma \left(|x_i+y_i|^{p/q}\right)^q\right)^{1/q} .$$

Thus

$$\Sigma |x_i||x_i+y_i|^{p/q} + \Sigma |y_i||x_i+y_i|^{p/q} \le \left[\left(\Sigma |x_i|^p\right)^{1/p}\right.$$
$$\left. + \left(\Sigma |y_i|^p\right)^{1/p}\right]\left[\left(\Sigma \left(|x_i+y_i|^{p/q}\right)^q\right)^{1/q}\right]$$

and from (10)

$$\Sigma |x_i+y_i|^p \le \left[\left(\Sigma |x_i|^p\right)^{1/p} + \left(\Sigma |y_i|^p\right)^{1/p}\right]\left(\Sigma |x_i+y_i|^p\right)^{1/q}$$

or

$$\left(\Sigma |x_i+y_i|^p\right)^{1-1/q} \le \left(\Sigma |x_i|^p\right)^{1/p} + \left(\Sigma |y_i|^p\right)^{1/p}.$$

But, since $1/p + 1/q = 1$, $1 - 1/q = 1/p$. Thus,

$$\left(\Sigma |x_i+y_i|^p\right)^{1/p} \le \left(\Sigma |x_i|^p\right)^{1/p} + \left(\Sigma |y_i|^p\right)^{1/p}$$

which is Minkowski's inequality.

• **PROBLEM 6-35**

Let $f \in C^1(E)$, the set of all continuously differentiable functions with domain E, E open in R^n, and let $D_v f(p)$ denote the directional derivative in the direction v at $p \in E$. What is $\max_{\|v\|=1} |D_v f(p)|$ and for which v's of norm 1 is the maximum achieved?

Solution: Recall that the gradient of f at p is defined by

$$\nabla f(p) = \left(\frac{\partial f}{\partial x_1}(p), \ldots, \frac{\partial f}{\partial x_n}(p)\right)$$

and the directional derivative is defined by

$$D_v f(p) = \lim_{t \to 0} \frac{f(p+tv) - f(p)}{t}. \qquad (1)$$

In order to understand (1), start by defining the curve $h: (a,b) \to E$ and $g = f \circ h: (a,b) \to R$ where each is C^1. By the Chain Rule

$$g'(t) = J_f(h(t)) J_h(t)$$
$$= \nabla f(h(t)) \begin{pmatrix} h'_1(t) \\ \vdots \\ h'_n(t) \end{pmatrix}$$
$$= \sum_{i=1}^{n} \frac{\partial f}{\partial x_i}(h(t)) h'_i(t)$$
$$= \langle \nabla f(h(t)), J_h(t) \rangle \qquad (2)$$

where $h_1, \ldots, h_n$ are the components of the curve h and $\langle \ \rangle$ denotes the Euclidean inner product. Let $h(t) = x + tv$ for any fixed $x \in E$, where v is any unit vector in R^n (i.e., $\|v\| = 1$), be defined on R. Hence, (2) shows that

$$g'(0) = \langle \nabla f(h(0)), J_h(0) \rangle$$
$$= \langle \nabla f(x), v \rangle. \qquad (3)$$

Since $g(t) - g(0) = f(x + tv) - f(x)$, (3) gives

$$\lim_{t \to 0} \frac{f(x+tv) - f(x)}{t} = \lim_{t \to 0} \frac{g(t) - g(0)}{t}$$

$$= \langle \nabla f(x), v \rangle . \qquad (4)$$

From (1) and (4),
$$D_v f(p) = \langle \nabla f(p), v \rangle \qquad (5)$$

Hence,
$$\max_{\|v\|=1} |D_v f(p)| = \max_{\|v\|=1} |\langle \nabla f(p), v \rangle|$$

$$= \max_{\|v\|=1} \langle \nabla f(p), v \rangle$$

for some fixed $p \in E$. That is, the function

$$F(v) = \langle \nabla f(p), v \rangle$$
$$= \sum_{i=1}^{n} \frac{\partial f}{\partial x_i}(p) v_i , \qquad (6)$$

where $v = (v_1, \ldots, v_n)$, is to be maximized under the constraint

$$\|v\| = \left(\sum_{i=1}^{n} v_i^2 \right)^{\frac{1}{2}} = 1$$

or, equivalently,

$$g(v) = \sum_{i=1}^{n} v_i^2 = 1 . \qquad (7)$$

By the method of Lagrange multipliers, there exists a real λ such that for some v

$$\nabla F(v) = \lambda \nabla g(v) .$$

This yields the $n+1$ equations

$$\frac{\partial F}{\partial v_i}(v) = \lambda \frac{\partial g}{\partial v_i}(v) \qquad (1 \le i \le n) \qquad (8)$$

and (7) in $n+1$ variables $(\lambda, v_1, \ldots, v_n)$. From (6), equations (7), (8) yield

$$\frac{\partial f}{\partial x_i}(p) = \lambda(2v_i) \qquad (1 \le i \le n).$$

Assuming a maximum exists, $\lambda \ne 0$ and hence

$$v_i = \frac{1}{2\lambda} \frac{\partial f}{\partial x_i}(p) \qquad (1 \le i \le n). \qquad (9)$$

Substituting (9) into the constraint (7) gives

$$\frac{1}{4\lambda^2} \sum_{i=1}^{n} \left[\frac{\partial f}{\partial x_i}(p) \right]^2 = 1$$

and therefore

$$\lambda = \pm \left[\frac{1}{4} \sum_{i=1}^{n} \frac{\partial f}{\partial x_i}^2 (p) \right]^{\frac{1}{2}}$$

$$= \pm \tfrac{1}{2} \|\nabla f(p)\| .$$

Thus the values

$$v_i = \pm \frac{1}{\|\nabla f(p)\|} \frac{\partial f}{\partial x_i}(p) \quad (1 \leq i \leq n)$$

or

$$v = \pm \frac{\nabla f(p)}{\|\nabla f(p)\|} \tag{10}$$

extremalize (6) subject to (7). Clearly, the positive choice in (10) maximizes while the negative one minimizes. Therefore,

$$\max_{\|v\|=1} |D_v f(p)| = \max_{\|v\|=1} \langle \nabla f(p), v \rangle$$

$$= \left\langle \nabla f(p), \frac{\nabla f(p)}{\|\nabla f(p)\|} \right\rangle$$

$$= \frac{1}{\|\nabla f(p)\|} \langle \nabla f(p), \nabla f(p) \rangle$$

$$= \|\nabla f(p)\|.$$

The maximum point $v = \dfrac{\nabla f(p)}{\|\nabla f(p)\|}$ can be interpreted as the unit vector in the direction of the gradient.

● **PROBLEM 6-36**

Show that any two hyperplanes in R^n either intersect or are parallel by using the method of Lagrange multipliers.

<u>Solution:</u> Let

$$H_x = \{x \in R^n \mid \langle x, \alpha \rangle + \beta = 0\}$$

and

$$H_y = \{y \in R^n \mid \langle y, \gamma \rangle + \delta = 0\},$$

where $\alpha = (\alpha_1, \ldots, \alpha_n) \neq 0$, $\gamma = (\gamma_1, \ldots, \gamma_n) \neq 0$, $\alpha, \gamma \in R^n$, $\beta, \delta \in R$, and $\langle \rangle$ denotes the Euclidean inner product, be any two hyperplanes. The distance between H_x and H_y at $p \in H_x$ (or $b \in H_y$) is the minimal distance between p and any $y \in H_y$ (or between b and any $x \in H_x$). H_x and H_y intersect if the minimum distance between them is zero and are parallel if the distance between them is constant. Moreover, H_x and H_y are parallel if $\alpha = \gamma$. Hence, to prove the desired result, use Lagrange multipliers to minimize the distance. Since distance is non-negative, this is the same as minimizing the square of the distance. Let

$$d(x,y) = \|x-y\|^2 = \sum_{i=1}^{n} (x_i - y_i)^2, \tag{1}$$

$$f(x,y) = \langle x, \alpha \rangle - \beta = \left(\sum_{i=1}^{n} x_i \alpha_i\right) - \beta = 0 \tag{2}$$

and

$$g(x,y) = \langle y, \gamma \rangle - \delta = \left(\sum_{i=1}^{n} y_i \gamma_i\right) - \delta = 0, \tag{3}$$

where $x = (x_1, \ldots, x_n)$, $y = (y_1, \ldots, y_n) \in R^n$. By minimizing (1) under the constraint equations (2) and (3), the square of the distance between H_x and H_y will be minimized. Applying the method of Lagrange multipliers, the following must hold for some real λ_1, λ_2

$$\nabla d(a,b) = \lambda_1 \nabla f(a,b) + \lambda_2 \nabla g(a,b)$$

or

$$\frac{\partial d}{\partial x_i}(a,b) = \lambda_1 \frac{\partial f}{\partial x_i}(a,b) + \lambda_2 \frac{\partial g}{\partial x_i}(a,b) \quad (1 \leq i \leq n)$$

and

$$\frac{\partial d}{\partial y_i}(a,b) = \lambda_1 \frac{\partial f}{\partial y_i}(a,b) + \lambda_2 \frac{\partial g}{\partial y_i}(a,b) \quad (1 \leq i \leq n)$$

for an extremum (a,b). From (1), (2), and (3) this becomes

$$2(a_i - b_i) = \lambda_1 \alpha_i \quad (1 \leq i \leq n)$$

and \hfill (4)

$$-2(a_i - b_i) = \lambda_2 \gamma_i \quad (1 \leq i \leq n),$$

where $a = (a_1, \ldots, a_n)$ and $b = (b_1, \ldots, b_n)$. A solution to the $2n$ simultaneous equations (4) is

$$a = b, \quad \lambda_1 = \lambda_2 = 0. \tag{5}$$

Since $a \in H_x$ and $b \in H_y$, (5) shows that H_x and H_y intersect for this solution. Assume now that they do not intersect. Then, without loss of generality, assume $\alpha_1 = \gamma_1 = 1$. [For suppose $\alpha_1 \neq 1$ and $\alpha_1 \neq 0$. Then the constraint $f(x,y) = 0$ is equivalent to $1/\alpha_i\, f(x,y) = 0$. If $\alpha_1 = 0$, relabel (i.e., let α_1 be the first non-zero α_i. At least one such $\alpha_i \neq 0$ exists since $\alpha \neq 0$). Likewise for γ_1]. Hence (4) is

$$2(a_1 - b_1) = \lambda_1$$
$$-2(a_1 - b_1) = \lambda_2 \tag{6}$$

$$2(a_i - b_i) = \lambda_1 \alpha_i \quad (2 \leq i \leq n) \tag{7}$$
$$-2(a_i - b_i) = \lambda_2 \gamma_i \quad (2 \leq i \leq n)$$

From (6), $\lambda_1 = -\lambda_2$, so that (7) becomes

$$2(a_i - b_i) = \lambda_1 \alpha_i \quad (2 \leq i \leq n)$$

$$-2(a_i - b_i) = -\lambda_1 \gamma_i \quad (2 \leq i \leq n).$$

Therefore $\alpha_i = \gamma_i$, $(2 \leq i \leq n)$ and hence $\alpha = \gamma$. So, H_x and H_y are parallel if they do not intersect.

● **PROBLEM 6-37**

State and prove a necessary and sufficient condition for

$$f \in C^3(E), \quad E \subseteq R^n$$

open, to have a
(a) local (or relative) minimum or
(b) strict local minimum.

Solution: Let $f: E \to R$ be C^3 and let f have a critical point at $p \in E$. Then i) p is a local minimum if and only if $H_f(p)$ is positive semi-definite, and ii) p is a strict local minimum if and only if $H_f(p)$ is positive definite.

a) To prove (i) note that by Taylor's theorem for $x \neq p$

$$f(x) - f(p) = \tfrac{1}{2} \langle H_f(p)(x-p), x-p \rangle + \epsilon(x)\|x-p\|^2 \geq 0$$

for all $x \in U$ where U is some neighborhood of p. Let $x-p = u \neq 0$. Then, with $\delta(u) = \epsilon(u+p) = \epsilon(x)$,

$$\langle H_f(p)u, u \rangle + 2\delta(u)\|u\|^2 \geq 0$$

for $\|u\|$ sufficiently small. For $0 < |t| \leq 1$, $t \in R$,

$$\langle H_f(p)(tu), tu \rangle + 2\delta(tu)\|tu\|^2 \geq 0$$

or

$$t^2 \langle H_f(p)u, u \rangle + 2\delta(tu)t^2\|u\|^2 \geq 0.$$

Thus

$$\langle H_f(p)u, u \rangle \geq -2\delta(tu)\|u\|^2.$$

Since the right hand side can be made arbitrarily small absolutely (i.e., less than any $\epsilon > 0$), then for u such that $x = u + p \in U$,

$$\langle H_f(p)u, u \rangle \geq 0.$$

Suppose that the quadratic form $H_f(p)$ is diagonalized to $D(p)$. Then

$$\langle H_f(p)u, u \rangle = \langle D(p)v, v \rangle \geq 0$$

for appropriate v such that $\|v\| > 0$ is sufficiently small as before. Let

$$D(p) = \begin{pmatrix} \lambda_1 & & & 0 \\ & \lambda_2 & & \\ & & \ddots & \\ 0 & & & \lambda_n \end{pmatrix}$$

where the λ_i's are the eigenvalues of $H_f(p)$. Then

$$0 \leq \langle H_f(p)u, u \rangle = \sum_{i=1}^{n} \lambda_i v_i^2 \leq \min_{1 \leq i \leq n} \lambda_i \|v\|$$

and hence

$$\min_{1 \leq i \leq n} \lambda_i \geq 0.$$

So all of the eigenvalues of $H_f(p)$ are non-negative. That is the same as saying that $H_f(p)$ is positive semi-definite. To prove the converse, again use Taylor's theorem and $u = x - p \neq 0$ to get

$$f(x) - f(p) = \tfrac{1}{2} <H_f(p)u,u> + \delta(u)\|u\|^2 .$$

Diagonalize $H_f(p)$ to get $D(p)$ where all the eigenvalues, λ_i, are positive. Then for appropriate small v

$$\begin{aligned} f(x) - f(p) &= \tfrac{1}{2} <D(p)v,v> + \delta(u)\|u\|^2 \\ &= \tfrac{1}{2} \sum_{i=1}^{n} \lambda_i v_i^2 + \delta(u)\|u\|^2 \\ &\geq \tfrac{1}{2} \min_{1 \leq i \leq n} \lambda_i \|v\|^2 + \delta(u)\|u\|^2 \\ &= \left(\tfrac{1}{2} \min_{1 \leq i \leq n} \lambda_i + \delta(u) \frac{\|u\|^2}{\|v\|^2} \right) \|v\|^2 . \end{aligned}$$

Hence, by choosing a neighborhood of p such that

$$\left| \delta(u) \frac{\|u\|^2}{\|v\|^2} \right| < \tfrac{1}{2} \min_{1 \leq i \leq n} \lambda_i$$

$\Big[$this exists since $\delta(u) \to 0$ as $u \to 0$ or $x \to p$ and $\|u\| = \|v\|\Big]$, or

$$\left| \delta(u) \frac{\|u\|^2}{\|v\|^2} \right| = 0$$

if the minimum eigenvalue equals zero. (i.e., $f(x) - f(p) \geq 0$). This proves (i), as was desired.

b) To prove (ii) follow the same procedure as a) but note that strict local minimum means that

$$f(x) - f(p) > 0$$

for $\|x - p\| \neq 0$ for some small neighborhood around p. Then notice that this implies that

$$\min_{1 \leq i \leq n} \lambda_i > 0$$

and also notice that the converse is true.

CHAPTER 7

THEORY OF INTEGRATION

This chapter is concerned with the definition and development of the Riemann integral and a brief nontheoretical introduction to the concept of Stieltjes integration. However, since the Riemann integral depends on the order structure of the real line we discuss only the integration of real-valued (both continuous and discontinuous) functions on intervals. Accordingly, the theory concerning complex and vector-valued functions on intervals and integration over sets other than intervals has not been included in this chapter.

In addition, since the theory of integration is a broad subject, we make note to the student of advanced calculus that this chapter covers only the very basic principles concerning this important and extensively developed topic.

RIEMANN INTEGRALS

● PROBLEM 7-1

Let $f: R \to R$ be bounded and defined on the closed interval $[a,b]$. Define the Riemann integral using the concept of partitions.

Solution: A logical but unsatisfactory definition of the definite integral is the following:

Let $F: R \to R$ be bounded and defined on the closed interval $[a,b]$. Subdivide $[a,b]$ into n subintervals (not necessarily of equal length). Evaluate f at the rightmost endpoint of each sub-interval and form the sum

$$\sum_{i=1}^{n} f(x_i)[x_i - x_{i-1}].$$

Let

$$\Delta x_i = x_i - x_{i-1}.$$

Then

$$\sum_{i=1}^{n} f(x_i) \Delta x_i$$

is an approximation to the area under f from a to b.

The approximation is improved as the number of subintervals is increased. Passing to the limit (i.e., as n goes to infinity), the integral of f over [a,b] is defined as:

$$\int_a^b f\,dx = \lim_{\substack{n \to \infty \\ \max|\Delta x_i| \to 0}} \sum_{i=1}^n f(x_i)\Delta x_i. \qquad (1)$$

The trouble with definition (1) is that no matter how small an interval is chosen $f(x_i)$ can take on infinitely many values. Thus the integral is not unique. Each infinite subdivision of [a,b] could produce a different value for the integral.

To ameliorate these difficulties the Riemann integral is introduced.

Let $f: R \to R$ be bounded and defined on the closed interval [a,b]. A partition P of [a,b] is a finite set of points

$$\{x_0, x_1, \ldots, x_n\}$$

such that

$$a = x_0 \leq x_1 \leq x_2 \leq \ldots \leq x_n = b.$$

Let $[x_{i-1}, x_i]$ be the ith subinterval of [a,b] where $x_{i-1}, x_i \in P$. Write

$$\Delta x_i = x_i - x_{i-1}.$$

Put

$$M_i = \sup_{x \in [x_{i-1}, x_i]} f(x)$$

and

$$m_i = \inf_{x \in [x_{i-1}, x_i]} f(x)$$

where sup f(x) and inf f(x) represent the least upper bound and the greatest lower bound, respectively, of f(x) and form the sums

$$U(P,f) = \sum_{i=1}^n M_i \Delta x_i \qquad L(P,f) = \sum_{i=1}^n m_i \Delta x_i. \qquad (2)$$

Now consider all partitions P of [a,b] and for each partition form the upper and lower sums as defined by (2). The set of real numbers given by U(P,f) for all possible P is bounded below (since f is bounded). Hence there exists a greatest lower bound. Let

$$\overline{\int_a^b} f\, dx = \inf_{P \in P[a,b]} U(P,f) \qquad (3)$$

where P[a,b] denotes all the possible partitions of [a,b]. Definition (3) is the upper integral of f over [a,b].

Similarly, the least upper bound of the set

{L(VP,f): P is a partition of f over [a,b]}

yields the lower integral of f over [a,b]:

$$\underline{\int_a^b} f\, dx = \sup_{P \in P[a,b]} L(P,f). \qquad (4)$$

If (3) and (4) are equal the Riemann integral is defined to be their common value, and denoted by

$$\int_a^b f(x)\, dx = \underline{\int_a^b} f(x)\, dx = \overline{\int_a^b} f(x)\, dx. \qquad (5)$$

When (5) is true, f is said to be Riemann integrable over [a,b].

● **PROBLEM 7-2**

Suppose f is defined on [0,2] as follows:

$$f(x) = \begin{cases} 1 & \text{for } 0 \leq x < 1 \\ 2 & \text{for } 1 \leq x \leq 2 \end{cases}$$

Show that f is Riemann integrable.

Solution: By definition a function which can be integrated according to Riemann's definition is called integrable. Note that a function does not have to be continuous to be integrable (i.e., on [a,b] all continuous functions and certain discontinuous functions are integrable).

For the given problem we need the following theorem: Suppose f is bounded on [a,b] and suppose corresponding to each positive ε there is a partition of [a,b] such that the corresponding upper and lower sums (represented as U(P,f) and L(P,f), respectively) satisfy the inequality

$$U(P,f) - L(P,f) < \varepsilon;$$

then f is integrable. To prove this theorem, we start with the condition

$$U(P,f) < \varepsilon + L(P,f).$$

Since, by definition,

$$\inf_{P \in P[a,b]} U(P,f) \leq U(P,f)$$

and

$$L(P,f) \leq \sup_{P \in P[a,b]} L(P,f),$$

combining inequalities we have

$$\inf_{P \in P[a,b]} U(P,f) \leq U(P,f) < L(P,f) + \varepsilon \leq$$

$$\sup_{P \in P[a,b]} L(P,f) + \varepsilon$$

or

$$\inf_{P \in P[a,b]} U(P,f) \leq \sup_{P \in P[a,b]} L(P,f) + \varepsilon \quad .$$

Since this conclusion is valid for every $\varepsilon > 0$, we infer that

$$\inf_{P \in P[a,b]} U(P,f) \leq \sup_{P \in P[a,b]} L(P,f) \quad .$$

But, by a lemma, we know that

$$\sup_{P \in P[a,b]} L(P,f) \leq \inf_{P \in P[a,b]} U(P,f)$$

is always true. Thus,

$$\sup_{P \in P[a,b]} L(P,f) = \inf_{P \in P[a,b]} U(P,f)$$

which means f is integrable. This completes the proof.

For the given problem suppose $\varepsilon > 0$. Let h be a positive number such that

$$h < 1 \quad \text{and} \quad h < \frac{\varepsilon}{2} \quad .$$

Consider the partition defined by

$$x_0 = 0, \ x_1 = 1-h, \ x_2 = 1+h, \ x_3 = 2 \ .$$

From fig. 1 it is seen that

$$m_1 = M_1 = 1, \ m_2 = 1, \ M_2 = 2, \ \text{and} \ m_3 = M_3 = 2,$$

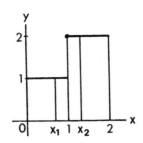

Fig. 1

where m_i and M_i represent the greatest lower bound and least upper bound, respectively, of $f(x)$, defined on the subinterval from x_{i-1} to x_i.

Therefore,

$$L(P,f) = m_1(x_1 - x_0) + m_2(x_2 - x_1) + m_3(x_3 - x_2)$$

or

$$L(P,f) = 1 \cdot (1-h) + 1 \cdot 2h + 2(1-h) = 3 - h.$$

Similarly,

$$U(P,f) = M_1(x_1 - x_0) + M_2(x_2 - x_1) + M_3(x_3 - x_2)$$

or

$$U(P,f) = 1 \cdot (1-h) + 2 \cdot 2h + 2(1-h) = 3 + h.$$

Accordingly

$$U(P,f) - L(P,f) = 2h < \varepsilon$$

since

$$h < \frac{\varepsilon}{2}.$$

Thus, by the theorem, f is integrable.

(Note that the function is discontinuous at $x = 1$.)

● **PROBLEM 7-3**

Let $f(x) = \begin{cases} 1, & x \text{ rational} \\ 0, & x \text{ irrational} \end{cases}$

be defined in the interval $[a,b]$.

Show that f is not Riemann integrable.

Solution: Let
$$[x_{i-1}, x_i]$$
be the ith subinterval of $[a,b]$ (i.e., $[a,b]$ is divided into n subintervals by the points
$$\{x_0, x_1, \ldots, x_n\}$$
such that
$$a = x_0 \leq x_1 \leq x_2 \leq \ldots \leq x_n = b).$$
In addition let $M_i = \sup f(x)$ where
$$x \in [x_{i-1}, x_i]$$
and let $m_i = \inf f(x)$ where
$$x \in [x_{i-1}, x_i].$$
Then we have $M_i = 1$, $m_i = 0$ since the rational and irrational numbers are dense in any subinterval.

Now let
$$\Delta x_i = x_i - x_{i-1}$$
and form the sums
$$U(P,f) = \sum_{i=1}^{n} M_i \Delta x_i$$
and
$$L(P,f) = \sum_{i=1}^{n} m_i \Delta x_i.$$
That is $U(P,f)$ and $L(P,f)$ are the upper and lower sums corresponding to any partition P.

Therefore for the given f we have
$$U(P,f) = \sum_{i=1}^{n} 1 \cdot \Delta x_i = (x_1 - x_0) + (x_2 - x_1) + \ldots + (x_n - x_{n-1}) = x_n - x_0 = b - a,$$
and
$$L(P,f) = \sum_{i=1}^{n} 0 \cdot \Delta x_i = 0.$$

Then let

   I = the greatest lower bound of the values of U(P,f) for all possible partitions and let J = the least upper bound of the values of L(P,f) for all possible partitions.

This means that

$$I = \overline{\int_a^b} f(x)\,dx \quad \text{and} \quad J = \underline{\int_a^b} f\,dx$$

so that if I = J then f is Riemann integrable on [a,b] and if I ≠ J then f is not Riemann integrable on [a,b].

For this problem we have that

$$I = b - a, \quad J = 0$$

and therefore since I ≠ J, f is not Riemann integrable.

● **PROBLEM 7-4**

Suppose f is defined on [0,2] by f(x) = x if 0 ≤ x < 1, f(x) = x - 1 if 1 ≤ x ≤ 2. Show that f is integrable.

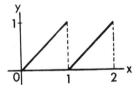

Fig. 1

Solution: The graph of f is shown in fig. 1. Note that this function is bounded on [0,2] and is continuous except at x=1. To show that this function is integrable we need the following theorems:

A) Given that f is bounded on [a,b] and that corresponding to each positive ε there is a partition of [a,b] such that the corresponding upper and lower sums satisfy the inequality U(P,f) - L(P,f) < ε, then f is integrable.

B) If f is integrable on [a,b], and if ε > 0, there is a partition with upper and lower sums such that

$$U(P,f) - L(P,f) < \varepsilon$$

(This is the converse of Theorem A.)

C) If a function f is continuous at each point of [a,b], it is integrable on that interval.

For the given function, suppose ε > 0. Form a partition of [0,2] by taking

$1 - \left(\frac{\varepsilon}{4}\right)$ and $1 + \left(\frac{\varepsilon}{4}\right)$

as two consecutive points in the partition. The remaining points are taken between

$[0, 1 - \left(\frac{\varepsilon}{4}\right)]$ and $[1 + \left(\frac{\varepsilon}{4}\right), 2]$

in the following manner. Note that f is continuous on the intervals

$[0, 1 - \left(\frac{\varepsilon}{4}\right)]$, $[1 + \left(\frac{\varepsilon}{4}\right), 2]$,

since the point of discontinuity has been enclosed in the subinterval

$[1 - \left(\frac{\varepsilon}{4}\right), 1 + \left(\frac{\varepsilon}{4}\right)]$.

This subinterval has length $\frac{\varepsilon}{2}$. Now consider the upper and lower sums, $U(P,f)$ and $L(P,f)$. Let $L_1(P,f)$ represent the part of $L(P,f)$ coming from the subintervals of

$[0, 1 - \left(\frac{\varepsilon}{4}\right)]$,

and let $L_3(P,f)$ represent the part from the subintervals of

$[1 + \left(\frac{\varepsilon}{4}\right), 2]$.

In addition, let $L_2(P,f)$ represent the part from the single subinterval

$[1 - \left(\frac{\varepsilon}{4}\right), 1 + \left(\frac{\varepsilon}{4}\right)]$

so that

$$L(P,f) = L_1(P,f) + L_2(P,f) + L_3(P,f).$$

Similarly for the upper sums,

$$U(P,f) = U_1(P,f) + U_2(P,f) + U_3(P,f).$$

Now when

$$1 - \frac{\varepsilon}{4} \leq x \leq 1 + \frac{\varepsilon}{4},$$

the least upper bound of $f(x)$ is 1, and the least value is 0. Hence,

$$L_2(P,f) = 0 \left(\frac{\varepsilon}{2}\right) = 0, \quad U_2(P,f) = 1 \left(\frac{\varepsilon}{2}\right) = \frac{\varepsilon}{2}.$$

Since f is continuous on

$[0, 1 - \left(\frac{\varepsilon}{4}\right)]$

and

$$[1 + \left(\frac{\varepsilon}{4}\right), 2],$$

by Theorems B and C, we can choose the part of the partition in these intervals so that

$$U_1(P,f) - L_1(P,f) < \frac{\varepsilon}{4}$$

and

$$U_3(P,f) - L_3(P,f) < \frac{\varepsilon}{4}.$$

Then

$$U(P,f) - L(P,f) = [U_1(P,f) - L_1(P,f)] +$$

$$[U_2(P,f) - L_2(P,f)] + [U_3(P,f) - L_3(P,f)]$$

$$< \frac{\varepsilon}{4} + \frac{\varepsilon}{2} + \frac{\varepsilon}{4} = \varepsilon.$$

Thus, by theorem A of this problem, the given function is integrable on $[0,2]$.

● **PROBLEM 7-5**

Given that a bounded function $f(x)$ is Riemann integrable in $[a,b]$ if and only if given any $\varepsilon > 0$ there exists a partition with upper and lower sums U and L such that $U - L < \varepsilon$, prove that a continuous function $f(x)$ in $[a,b]$ is Riemann integrable in $[a,b]$.

Solution: From the theorems on continuous functions we know that if a function f is defined and continuous at each point of a closed and bounded interval, then f is uniformly continuous on this interval. Therefore, for the given problem, $f(x)$ is uniformly continuous in $[a,b]$. This means that to each $\varepsilon > 0$ there corresponds a $\delta > 0$ such that

$$f(x) - f(x_0)$$

whenever x and x_0 are any points of $[a,b]$ such that

$$|x - x_0| < \delta.$$

Hence, let x^1 and x^2 be any two points of an interval (x_{i-1}, x_i). Then given $\varepsilon > 0$, there exists $\delta > 0$ such that

$$|f(x^1) - f(x^2)| < \frac{\varepsilon}{b-a}$$

whenever

$$|x^1 - x^2| < \delta .$$

Now we can choose points of subdivision so that

$$M_i - m_i < \frac{\varepsilon}{b-a} .$$

(M_i, m_i are defined as in the previous problems).

Then if the upper and lower sums corresponding to this partition are given by

$$U = \sum_{i=1}^{n} M_i \Delta x_i , \qquad L = \sum_{i=1}^{n} m_i \Delta x_i ,$$

we have

$$U - L = \sum_{i=1}^{n} (M_i - m_i) \Delta x_i < \sum_{i=1}^{n} \frac{\varepsilon}{b-a} \Delta x_i .$$

However,

$$\sum_{i=1}^{n} \frac{\varepsilon}{b-a} \Delta x_i = \frac{\varepsilon}{b-a} [(x_1 - x_0) + (x_2 - x_1) + \ldots + (x_n - x_{n-1})]$$

$$= \frac{\varepsilon}{b-a} [b-a] = \varepsilon .$$

Thus, $U - L < \varepsilon$ and therefore $f(x)$ is Riemann integrable in $[a,b]$.

• **PROBLEM 7-6**

Using the same given information as in the previous problem prove that a monotonic function $f(x)$ in $[a,b]$ is Riemann integrable in $[a,b]$.

Solution: Assume that $f(x)$ is monotonically increasing (the case where $f(x)$ is monotonically decreasing can be proved similarly or by considering $-f(x)$ in place of $f(x)$).

Hence, since $f(x)$ is monotonically increasing in $[a,b]$ we have for the given partition

$$a = x_0 < x_1 < \ldots < x_n = b$$

that

$$f(x) = f(x_0) \leq f(x_1) \leq f(x_2) \leq \ldots \leq f(x_n) = f(b).$$

Therefore $M_i = f(x_i)$ and $m_i = f(x_{i-1})$ so that

$$U = \sum_{i=1}^{n} f(x_i) \Delta x_i \qquad L = \sum_{i=1}^{n} f(x_{i-1}) \Delta x ;$$

where U and L denote the upper and lower sums corresponding to the given partition.

Hence,

$$U - L = \sum_{i=1}^{n} (f(x_i) - f(x_{i-1})) \Delta x_i . \qquad (1)$$

Now if a partition is chosen so that, assuming

$$f(b) \neq f(a)$$

(which must be true, since $f(x)$ is monotonically increasing in $[a,b]$),

$$\Delta x_i < \frac{\varepsilon}{f(b) - f(a)}, \quad \varepsilon > 0 ,$$

we then have from (1), since

$$f(x_i) \geq f(x_{i-1})$$

that

$$U - L < \frac{\varepsilon}{f(b) - f(a)} \sum_{i=1}^{n} [f(x_i) - f(x_{i-1})]$$

This means that

$$U - L < \frac{\varepsilon}{f(b) - f(a)} [f(b) - f(a)] = \varepsilon$$

Thus by the given, $f(x)$ is Riemann integrable on $[a,b]$.

● **PROBLEM 7-7**

Given that $f_1(x)$ and $f_2(x)$ are Riemann integrable on $[a,b]$ and that $I \geq J$ (where I and J represent the upper and lower integrals, respectively, of any Riemann integrable function), prove that $f_1(x) + f_2(x)$ is Riemann integrable on $[a,b]$ and that

$$\int_a^b [f_1(x) + f_2(x)] \, dx = \int_a^b f_1(x) \, dx + \int_a^b f_2(x) \, dx .$$

Solution: For the subinterval (x_{i-1}, x_i) let

$$M_i^{(1)}, \quad M_i^{(2)} \quad \text{and} \quad M_i$$

be the least upper bounds of $f_1(x)$, $f_2(x)$ and $f_1(x) + f_2(x)$ respectively. In addition let

$$M_i^{(1)}, \; M_i^{(2)} \text{ and } m_i$$

be the greatest lower bounds of

$$f_1(x), \; f_2(x) \text{ and } f_1(x) + f_2(x)$$

respectively, in this same subinterval. Then

$$M_i \leq M_i^{(1)} + M_i^{(2)} \quad \text{and} \quad m_i \geq m_i^{(1)} + m_i^{(2)}. \tag{1}$$

Now let

$U^{(1)}$, $U^{(2)}$ and U be the upper sums and let $L^{(1)}$, $L^{(2)}$ and L be the lower sums of

$$f_1(x), \; f_2(x) \text{ and } f_1(x) + f_2(x),$$

respectively. Then by (1) we have

$$U \leq U^{(1)} + U^{(2)} \quad \text{and} \quad L \geq L^{(1)} + L^{(2)}. \tag{2}$$

Therefore, letting $I^{(1)}$, $I^{(2)}$ and I be the corresponding upper integrals and $J^{(1)}$, $J^{(2)}$ and J be the corresponding lower integrals, we have by (2) that

$$I \leq I^{(1)} + I^{(2)} \quad \text{and} \quad J \geq J^{(1)} + J^{(2)}.$$

However, since $f_1(x)$ and $f_2(x)$ are Riemann integrable, we have

$$I^{(1)} = J^{(1)}, \; I^{(2)} = J^{(2)}.$$

Hence

$$I \leq I^{(1)} + I^{(2)} \quad \text{and} \quad J \geq I^{(1)} + I^{(2)}$$

so that $I \leq J$. But we are given that $I \geq J$ always. This means that

$$I = J = I^{(1)} + I^{(2)} = J^{(1)} + J^{(2)}.$$

Thus $f_1(x) + f_2(x)$ is Riemann integrable and

$$\int_a^b [f_1(x) + f_2(x)] dx = \int_a^b f_1(x) dx + \int_a^b f_2(x) dx$$

● **PROBLEM 7-8**

> Prove the following:
>
> a) If $f(x) \leq g(x)$ and $f(x)$ and $g(x)$ are Riemann integrable on $[a,b]$ then
>
> $$\int_a^b f(x)\,dx \leq \int_a^b g(x)\,dx .$$
>
> b) If $f(x)$ is bounded and Riemann integrable and if c is any point such that $c \in [a,b]$ then
>
> $$\int_a^b f(x)\,dx = \int_a^c f(x)\,dx + \int_c^b f(x)\,dx .$$

Solution: a) Let $M_i^{(1)}$, $M_i^{(2)}$ be the least upper bounds and $m_i^{(1)}, m_i^{(2)}$ be the greatest lower bounds in (x_{i-1}, x_i) corresponding to $f(x)$ and $g(x)$ respectively. In addition let $U^{(1)}, L^{(1)}$ and $U^{(2)}, L^{(2)}$ be the corresponding upper and lower sums. Then since

$$f(x) \leq g(x),$$

we have

$$M_i^{(1)} \leq M_i^{(2)} \quad \text{and} \quad m_i^{(1)} \leq m_i^{(2)} .$$

This means that

$$U^{(1)} \leq U^{(2)} \quad \text{and} \quad L^{(1)} \leq L^{(2)} .$$

Now, letting the upper and lower integrals be $I^{(1)}, J^{(1)}$ for $f(x)$ and $I^{(2)}, J^{(2)}$ for $g(x)$, we have

$$I^{(1)} \leq I^{(2)}, \quad J^{(1)} \leq J^{(2)} . \qquad (1)$$

However, we know that (since $f(x)$ and $g(x)$ are Riemann integrable in $[a,b]$),

$$I^{(1)} = J^{(1)} = \int_a^b f(x)\,dx ,$$

$$I^{(2)} = J^{(2)} = \int_a^b g(x)\,dx .$$

301

This means that (from (1))

$$\int_a^b f(x)\,dx \le \int_a^b g(x)\,dx$$

b) Let $U^{(1)}$ denote the upper sum for the interval (a,c) and let $U^{(2)}$ denote the upper sum for the interval (c,b). Before proceeding, note that unless C is a point of subdivision in defining the upper and lower sums, by a theorem, if we add c to the partition, the upper sum U is not increased and the lower sum L is not decreased. Therefore, we could now write

$$U \ge U^{(1)} + U^{(2)} \ge \int_a^c f(x)\,dx + \int_c^b f(x)\,dx .$$

Similarly for the lower sums,

$$L \le L^{(1)} + L^{(2)} \le \int_a^c f(x)\,dx + \int_c^b f(x)\,dx .$$

Now let I be the upper integral and let J be the lower integral corresponding to $f(x)$. This means that

$$U \ge I \ge \int_a^c f(x)\,dx + \int_c^b f(x)\,dx \ge J \ge L . \qquad (2)$$

However, since in $[a,b]$ the function $f(x)$ is Riemann integrable we have

$$I = J = \int_a^b f(x)\,dx .$$

Therefore, applying this to (2) we have

$$\int_a^b f(x)\,dx = \int_a^c f(x)\,dx + \int_c^b f(x)\,dx .$$

● **PROBLEM 7-9**

(a) Prove that if $f(x)$ is Riemann integrable on $[a,b]$ then $|f(x)|$ is Riemann integrable on the same interval.

b) Prove that if $f(x)$ is Riemann integrable on $[a,b]$ then

$$\left| \int_a^b f(x)\,dx \right| \leq \int_a^b |f(x)|\,dx .$$

Solution: a) First we define the two functions $F(x)$, $H(x)$ by

$$F(x) = \begin{cases} f(x), & f(x) \geq 0 \\ 0, & f(x) < 0 \end{cases}$$

and

$$H(x) = \begin{cases} -f(x), & f(x) \leq 0 \\ 0, & f(x) > 0 . \end{cases}$$

This means that

$$f(x) = F(x) - H(x) \tag{1}$$

and

$$|f(x)| = F(x) + H(x) . \tag{2}$$

Then we have, since $f(x)$ is Riemann integrable, that $F(x)$ and $H(x)$ are also Riemann integrable. However there is a theorem which states that if $f(x)$ and $g(x)$ are Riemann integrable on $[a,b]$ then $f(x) + g(x)$ is Riemann integrable on this same interval with

$$\int_a^b [f(x) + g(x)]\,dx = \int_a^b f(x)\,dx + \int_a^b g(x)\,dx$$

(See problem 7 of this chapter.)

Thus, $F(x) + H(x)$ is Riemann integrable on $[a,b]$ and by (2) this means that $|f(x)|$ is Riemann integrable on $[a,b]$.

b) We can prove this theorem by two methods. For the first method we refer to equations (1) and (2) of part (a) of this problem. Hence, we have

$$\int_a^b f(x)\,dx) = \int_a^b F(x)\,dx - \int_a^b H(x)\,dx$$

and

$$\int_a^b |f(x)|\,dx = \int_a^b F(x)\,dx + \int_a^b H(x)\,dx .$$

However this means that

$$\left| \int_a^b f(x)\,dx \right| = \left| \int_a^b F(x)\,dx - \int_a^b H(x)\,dx \right| \leq \left| \int_a^b F(x)\,dx \right|$$

$$+ \left| \int_a^b H(x)\,dx \right| = \int_a^b F(x)\,dx + \int_a^b H(x)\,dx =$$

$$\int_a^b |f(x)|\,dx .$$

For the second method we use the fact that if

$$f(x) \leq g(x) \quad \text{on } [a,b]$$

($f(x)$, $g(x)$ Riemann integrable on $[a,b]$), then

$$\int_a^b f(x)\,dx \leq \int_a^b g(x)\,dx .$$

Then since

$$f(x) \leq |f(x)| \quad \text{and} \quad -|f(x)| \leq f(x)$$

we simply have that

$$-\int_a^b |f(x)|\,dx \leq \int_a^b f(x)\,dx \leq \int_a^b |f(x)|\,dx ,$$

so that

$$\left| \int_a^b f(x)\,dx \right| \leq \int_a^b |f(x)|\,dx .$$

This is so since

$$-y \leq x \leq y \quad \text{means that} \quad |x| \leq y ,$$

where

$$x = \int_a^b f(x)\,dx$$

and

$$y = \int_a^b |f(x)|\,dx .$$

● PROBLEM 7-10

Find the derivatives of

a)
$$\int_1^x t^2 \, dt \quad \text{with respect to } x.$$

b)
$$\int_1^{t^2} \sin(x^2) \, dx \quad \text{with respect to } t.$$

Solution: a) Let $F(x)$ denote

$$\int_1^x t^2 \, dt.$$

The problem is to find $F'(x)$ where x is a parameter occurring in the limits of integration. To solve this problem, the chain-rule for composite functions of several variables and the theorem on the derivative of an integral which is a function of its upper limit of integration must be used.

Let

$$F(x_1, x_2) = F(x_1(t), x_2(t)),$$

i.e., F is a differentiable function of x_1 and x_2 where x_1 and x_2 are themselves differentiable functions of a parameter t. Then, according to the chain-rule:

$$\frac{dF}{dt} = \frac{\partial F}{\partial x_1} \frac{dx_1}{dt} + \frac{\partial F}{\partial x_2} \frac{dx_2}{dt}.$$

Next, suppose

$$G(y) = \int_a^y g(t) \, dt.$$

Then, if g is continuous on $[a, y]$, $G'(y) = g(y)$. To prove this theorem, let $y_0 \in [a, y]$ be a point at which g is continuous. Given $\varepsilon > 0$, choose δ such that

$$|g(y_1) - g(y_0)| < \varepsilon$$

whenever

$$|y_1 - y_0| < \delta .\tag{1}$$

Now, $G'(y) = g(y)$ implies that, by the definition of the derivative,

$$\lim_{h \to 0} \left| \frac{G(y_0 + h) - G(y_0)}{h} - g(y_0) \right| = 0 .\tag{2}$$

But

$$G(y_0 + h) - G(y_0) = \int_a^{y_0+h} g(t)\,dt - \int_a^{y_0} g(t)\,dt$$

$$= \int_{y_0}^{y_0+h} g(t)\,dt \quad \text{and} \quad g(y_0) = \frac{1}{h}\int_{y_0}^{y_0+h} g(y_0)\,dt .$$

Therefore,

$$\frac{G(y_0 + h) - G(y_0)}{h} - g(y_0) =$$

$$\frac{1}{h}\int_{y_0}^{y_0+h} [g(t) - g(y_0)]\,dt \tag{3}$$

Now let $0 < |h| < \delta$. Then the expression under the integral sign on the right side of (3) is in absolute value less than ε, by (1) since

$$|t - y_0| \leq |h| < \delta .$$

Accordingly, the entire right side of (3) is less than ε when $0 < |h| < \delta$. This proves (2) and thus proves the theorem. Hence for

$$F(x) = \int_1^x t^2\,dt, \quad F'(x) = x^2 .$$

b)

$$\int_1^{t^2} \sin(x^2)\,dx .$$

Here let $u = t^2$, so that $\frac{du}{dt} = 2t$. Then

$$\frac{dF}{dt} = \frac{dF}{du}\frac{du}{dt} = \sin u^2 \frac{du}{dt} = 2t \sin t^4 .$$

● **PROBLEM 7-11**

Let
$$F(y) = \int_0^\pi \sin(xy)\,dx.$$

Use Leibniz's rule to find $F'(y)$.

<u>Solution</u>: This is a problem in differentiating an integral which depends on a parameter; here y is the parameter.

Here $F(y)$ is of the form
$$\int_a^b f(x,y)\,dx.$$

Note that f is defined for the infinite strip
$$S = 0 \le x \le \pi,\ y \in R$$
and f is continuous on S. It follows that
$$F(y) = \int_a^b f(x,y)\,dx$$
is continuous for each $y \in R$. This is useful in proving Leibniz's rule: Let $f(x,y)$ be an integrable function of x for each value of y. Assume that
$$F(y) = \int_a^b f(x,y)\,dx$$
and that
$$D_2 f \left(= \frac{\partial f}{\partial y}\right)$$
exists and is a continuous function of x and y in the region $a \le x \le b,\ c \le y \le d$. Then
$$F'(y) = \int_a^b \frac{\partial f(x,y)}{\partial y}\,dx. \tag{1}$$

To prove Leibniz's rule, use the definition of the derivative of a function on (1) to obtain
$$\lim_{h \to 0} \left| \frac{F(y+h) - F(y)}{h} - \int_a^b D_2 f\,dx \right| = 0. \tag{2}$$

Since
$$F(y+h) = \int_a^b f(x,y+h)\,dx,$$

$$F(y+h) - F(y) = \int_a^b f(x,y+h)\,dx - \int_a^b f(x,y)\,dx$$

$$= \int_a^b [f(x,y+h) - f(x,y)]\,dx. \tag{3}$$

But
$$\frac{f(x,y+h) - f(x,y)}{h} = D_2 f(x,y+\xi h)$$

by the law of the mean, where $0 < \xi < 1$.
Applying this to (3):

$$\frac{F(y+h) - F(y)}{h} = \frac{1}{h}\int_a^b [f(x,y+h) - f(x,y)]\,dx$$

$$= \int_a^b D_2 f(x, y + \xi h)\,dx$$

and

$$\frac{F(y+h) - F(y)}{h} - \int_a^b D_2 f(x,y)\,dx$$

$$= \int_a^b [D_2 f(x,y+\xi h) - D_2 f(x,y)]\,dx \tag{4}$$

By hypothesis, $D_2 f$ exists and is continuous in the closed rectangle $P = a \leq x \leq b,\ c \leq y \leq d$. Hence, it is uniformly continuous on P. Let $\varepsilon > 0$ be given. Choose δ so that

$$|D_2 f(x_1,y_1) - D_2 f(x_0,y_0)| < \varepsilon$$

whenever
$$\|(x_1,y_1) - (x_0,y_0)\| < \delta.$$

This implies that
$$|D_2 f(x,y+\xi h) - D_2 f(x,y)| < \varepsilon$$

for $|h| < \delta$ (since $0 < \xi < 1$).

Substituting ε into (4) and evaluating,

$$\left| \frac{F(y+h) - F(y)}{h} - \int_a^b D_2 f(x,y)\, dx \right| < \varepsilon (b-a). \tag{5}$$

Since (5) is valid for any $\varepsilon > 0$, this proves (2) and hence (1), Leibniz's rule. In the given problem, $\sin(xy)$ satisfies the hypotheses of the theorem since it is integrable for each value of y and

$$D_2 \sin(xy) = x \cos(xy)$$

is a continuous function of x and y in the region

$$S = [0, \pi] \times R.$$

In fact, although S is not a closed rectangle, $\sin(xy)$ is uniformly continuous there and the proof of the theorem remains valid. It follows that

$$F'(y) = \int_0^\pi D_2(\sin xy)\, dx$$

$$= \int_0^\pi x \cos(xy)\, dx.$$

Applying integration by parts, let

$$u = x; \quad dv = \cos(xy)\, dx.$$

Then

$$du = dx \text{ and } v = \frac{1}{y} \sin(xy).$$

$$F'(y) = uv \Big|_0^\pi - \int_0^\pi v\, du$$

$$= x \frac{1}{y} \sin(xy) \Big|_0^\pi - \frac{1}{y} \int_0^\pi \sin(xy)\, dx$$

$$= \frac{\pi \sin(\pi y)}{y} + \frac{1}{y^2} \cos(xy) \Big|_0^\pi$$

$$= \frac{\pi \sin(\pi y)}{y} + \frac{1}{y^2} [\cos(\pi y) - 1].$$

• PROBLEM 7-12

Differentiate the integral

$$F(y) = \int_0^{y^2} x^5 (y-x)^7 \, dx$$

Solution: Integrals are differentiated according to Leibniz's rule which states:

Let $f(x,y)$ be an integrable function of x for each value of y. Suppose the partial derivative

$$\frac{\partial f(x,y)}{\partial y}$$

exists and is a continuous function of x and y in some rectangle R then $F(y)$ (i.e.,

$$F(y) = \int_a^b f(x,y) \, dx \, .$$

has a derivative given by

$$F'(y) = \int_a^b \frac{\partial f(x,y) \, dx}{\partial y} \, .$$

The given integral, however, has the parameter y appearing in the limits of integration as well as in the integrand. Since the integral is a function of its upper and lower limits, the following theorem needs to be used: Let

$$F(x) = \int_{x_0}^{x} f(t) \, dt \, .$$

Then the formula

$$F'(x) = f(x)$$

holds at each point where f is continuous.

Applying the two theorems to the given integral, define

$$G(u,v) = \int_0^u x^5 (v-x)^7 \, dx$$

and put

$$u = y^2, \quad v = y \, .$$

310

Then, treating $G(u,v)$ as an integral which is a function of its upper parameter u and applying the second theorem,

$$\frac{\partial G}{\partial u} = u^5 (v-u)^7.$$

Next, applying Leibniz's rule to $G(u,v)$ as a function of v:

$$\frac{\partial G}{\partial v} = \int_0^u \frac{\partial G}{\partial v} dx = \int_0^u 7x^5 (v-x)^6 dx$$

But $G(u,v)$ is a function of u and v where u and v are themselves functions of y. According to the chain-rule,

$$\frac{dF}{dy} = \frac{\partial G}{\partial u} \frac{du}{dy} + \frac{\partial G}{\partial v} \frac{dv}{dy}.$$

Thus

$$\frac{dF}{dy} = u^5 (v-u)^7 \frac{du}{dy} + \int_0^u 7x^5 (y-x)^6 dx \frac{dv}{dy}.$$

Since

$$u = y^2; \quad \frac{du}{dy} = 2y;$$

since

$$v = y, \quad \frac{dv}{dy} = 1.$$

Therefore,

$$F'(y) = y^{10}(y-y^2)^7 \cdot 2y + \int_0^{y^2} 7x^5 (y-x)^6 dx.$$

● **PROBLEM 7-13**

Find $F'(x)$ where

a)
$$F(x) = \int_0^x e^{-x^2 t^2} dt$$

b)
$$F(x) = \int_{x^2}^{\sin x} (x^2 - t^2)^n dt$$

Solution: The functions to be differentiated with respect to x are integrals whose integrands contain a parameter, x. In addition, the limits of integration are functions of the parameter. To solve the given integrals, we need the following theorem known as Leibnitz's rule which states:

Given that $f(x,y)$ is an integrable function of x for each value of y, and that the partial derivative

$$\frac{\partial f(x,y)}{\partial y}$$

exists and is a continuous function of x and y in the rectangle $R = [a,b] \times [c,d]$. Then $F(y) = \int_a^b f(x,y)dx$ has a derivative given by

$$F'(y) = \int_a^b \frac{\partial f(x,y)}{\partial y} dx \quad . \tag{1}$$

In addition, if

$$F(x) = \int_{x_0}^x f(t) dt,$$

then $F'(x) = f(x)$ at each point where f is continuous.

a)
$$F(x) = \int_0^x e^{-x^2 t^2} dt$$

For this integral define

$$G(u,v) = \int_0^u e^{-vt^2} dt$$

where $u = x$, $v = x^2$. Then, by the theorems,

$$\frac{\partial G}{\partial u} = e^{-vu^2}$$

and

$$\frac{\partial G}{\partial v} = \int_0^u -t^2 e^{-vt^2} dt$$

Additionally,

$$\frac{du}{dx} = 1 \quad \text{and} \quad \frac{dv}{dx} = 2x \quad .$$

By the chain rule,

$$\frac{dF}{dx} = \frac{\partial G}{\partial u} \frac{du}{dx} + \frac{\partial G}{\partial v} \frac{dv}{dx} \quad .$$

Therefore,
$$\frac{dF}{dx} = e^{-x^4} - 2x \int_0^x t^2 e^{-x^2 t^2} \, dt .$$

b)
$$F(x) = \int_{x^2}^{\sin x} (x^2 - t^2)^n \, dt .$$

Here define
$$G(u,v,w) = \int_u^v (w - t^2)^n \, dt$$

where
$$u = x^2 , \quad v = \sin x, \quad w = x^2$$

so that
$$\frac{du}{dx} = 2x , \quad \frac{dv}{dx} = \cos x , \quad \frac{dw}{dx} = 2x .$$

Now,
$$\frac{\partial G}{\partial u} = - (w - u^2)^n$$

(value is negative since u is lower limit).

$$\frac{\partial G}{\partial v} = (w - v^2)^n ,$$

and
$$\frac{\partial G}{\partial w} = \int_u^v n(w - t^2)^{n-1} \, dt .$$

By the chain rule
$$\frac{dF}{dx} = \frac{\partial G}{\partial u} \frac{du}{dx} + \frac{\partial G}{\partial v} \frac{dv}{dx} + \frac{\partial G}{\partial w} \frac{dw}{dx} .$$

Hence,
$$\frac{dF}{dx} = - 2x(x^2 - x^4)^n + \cos x (x^2 - \sin^2 x)^n$$
$$+ 2x \int_{x^2}^{\sin x} n(x^2 - t^2)^{n-1} \, dt .$$

STIELTJES INTEGRALS

• **PROBLEM 7-14**

Evaluate the following Stieltjes integrals:

a)
$$\int_a^b f(x)\,dc\ ,$$

where c is a constant function

b)
$$\int_a^b 1\,dx$$

c)
$$\int_a^b dx^2$$

d)
$$\int_a^b dg(x)$$

where g(x) is any bounded function.

Solution: The Stieltjes integral is an extension of the Riemann integral. It involves two functions f and g defined on a closed interval [a,b] and its general form is

$$\int_a^b f(x)\,dg(x)\ .$$

The function g is called the integrator.

The Stieltjes integral is defined as the limit of a sum in much the same way as the Riemann integral is defined.

Let

$$P = \{x_0,\ x_1,\ \ldots,\ x_n\}$$

314

be a partition of $[a,b]$, and let

$$\{z_1, \ldots, z_n\}$$

be respective points in the respective subintervals

$$(x_0,x_1), (x_1,x_2), \ldots, (x_{n-1},x_n).$$

The mesh of P, denoted by $|P|$ is

$$\max\{(x_1-x_0), (x_2-x_1), \ldots, (x_n-x_{n-1})\}$$

(i.e., the length of the largest sub-interval). Then

$$\int_a^b f(x)\,dg(x) = \int_a^b f\,dg = \lim_{|P|\to 0} \sum_{i=1}^n f(z_i)[g(x_i) - g(x_{i-1})]$$

if the sum converges to a unique limit for all partitions such that

$$|P| < \delta,$$

where $\delta > 0$ is a given challenge number.

a)
$$\int_a^b f(x)\,dc = \lim_{|P|\to 0} \sum_{i=1}^n f(z_i)[c(x_i) - c(x_{i-1})]$$

$$= \lim_{|P|\to 0} \sum_{i=1}^n f(z_i)[c-c] = 0.$$

Thus the Stieltjes integral of any function with respect to a constant function is zero.

b)
$$\int_a^b 1\cdot dx = \lim_{|P|\to 0} \sum_{i=1}^n 1[x_i - x_{i-1}]$$

$$= \lim_{|P|\to 0} \{(x_1-x_0) + (x_2-x_1) + \ldots + (x_n-x_{n-1})\}$$

$$= x_n - x_0 = b-a.$$

c)
$$\int_a^b dx^2 = \lim_{|P|\to 0} \sum_{i=1}^n 1[x_i^2 - x_{i-1}^2]$$

$$= \lim_{|P| \to 0} \{(x_1^2 - x_0^2) + (x_2^2 - x_1^2) + \ldots + (x_n^2 - x_{n-1}^2)$$

$$= x_n^2 - x_0^2 = b^2 - a^2.$$

d)

$$\int_a^b dg(x) = \lim_{|P| \to 0} \sum_{i=1}^n [g(x_i) - g(x_{i-1})]$$

$$= \left(g(x_1) - g(x_0)\right) + \left(g(x_2) - g(x_1)\right) + \ldots + \left(g(x_n) - g(x_{n-1})\right)$$

$$= g(x_n) - g(x_0) = g(b) - g(a).$$

The Stieltjes integral of any constant function c with respect to an integrator g(x) over [a,b] equals

$$c[g(b) - g(a)].$$

● **PROBLEM 7-15**

Proceed from the definition of the Stieltjes integral to show that the function f given by

$$f(x) = g(x) = \begin{cases} 0 & 0 \leq x \leq 1 \\ 1 & 1 < x \leq 2 \end{cases}$$

is not Stieltjes integrable with respect to g.

<u>Solution</u>: The Stieltjes integral is defined as an extension of the Riemann integral. Let f be defined on [a,b]. A partition P of [a,b] is a set of points

$$\{x_k\}_0^n$$

such that

$$a = x_0 < x_1 < x_2 < \ldots < x_n = b.$$

The norm of a partition P is

$$\|P\| = \max (x_1 - x_0, x_2 - x_1, \ldots, x_n - x_{n-1}).$$

Now let g be another real-valued monotonic function defined on [a,b]. Form the sum

$$S(f, g, P) = \sum_{k=1}^n f(z_k)(g(x_k) - g(x_{k-1}))$$

where

$$z_k \in (x_{k-1}, x_k),$$

the ith subinterval of the partition P.

The Stieltjes integral of $f(x)$ with respect to $g(x)$ from a to b is:

$$\int_a^b f(x)\,dg(x) = \lim_{||P|| \to 0} \sum_{k=1}^n f(z_k)[g(x_k) - g(x_{k-1})]. \quad (1)$$

Note that the integral is defined by considering all partitions P of [a,b] and selecting only those whose norm tends to zero. This means that, given any $\varepsilon > 0$, there exists a $\delta > 0$ such that for all partitions P with $||P|| < \delta$ and for all choices of $z_k \in (x_{k-1}, x_k)$

$$\left| \int_a^b f(x)\,dg(x) - \sum_{k=1}^n f(z_k)[g(x_k) - g(x_{k-1})] \right| < \varepsilon.$$

Applying definition (1) to the given problem, for any partition of [0,2],

$$\sum_{k=1}^n f(z_k)[g(x_k) - g(x_{k-1})]$$

has only one term that is different from zero, namely when f and g change values from 0 to one. For this term $g(x_k) - g(x_{k-1})$ has the value one but, since (x_{k-1}, x_k) is an open interval around the point one, $f(z_k)$ can take the value 0 or 1 depending on the way z_k is chosen. Thus

$$\sum_{k=1}^n f(z_k)[g(x_k) - g(x_{k-1})] = 1 \text{ or } 0$$

for every partition of [0,2] and (1) can not be satisfied if $\varepsilon < 1$. Thus, the Stieltjes integral of the given functions does not exist.

● **PROBLEM 7-16**

Evaluate the Stieltjes integral

$$\int_{-1}^1 x\,d|x|.$$

Solution: The integral is in the form

$$\int_a^b f\,dg$$

where f is monotonically increasing and g is continuous on $[a,b]$. Hence the following theorem applies:

1. $f(x)$ is monotonically increasing on $[a,b]$.
2. $g(x)$ is continuous on $[a,b]$.

Then

3.
$$\int_a^b f(x)\,dg(x) + \int_a^b g(x)\,df(x) = g(b)f(b) - g(a)f(a).$$

To prove this theorem let

$$P = \{x_0, \ldots, x_n\}$$

be a partition of $[a,b]$. Form the sum

$$S(f,g,P) = \sum_{i=1}^{n} f(z_i)[g(x_i) - g(x_0)] \tag{1}$$

This sum may be rearranged:

$$\sum_{i=1}^{n} f(z_i)[g(x_i) - g(x_{i-1})] = f(z_1)[g(x_1) - g(x_0)]$$

$$+ f(z_2)[g(x_2) - g(x_1)] + \ldots + f(z_n)[g(x_n) - g(x_{n-1})]$$

$$= g(x_1)[f(z_1) - f(z_2)] + g(x_2)[f(z_2) - f(z_3)]$$

$$+ \ldots + g(x_{n-1})[f(z_{n-1}) - f(z_n)] - f(z_1)g(x_0)$$

$$+ f(z_n)g(x_n)$$

$$= \sum_{i=1}^{n} g(x_i)[f(z_i) - f(z_{i+1})] - f(z_1)g(x_0) + f(z_n)g(x_n). \tag{2}$$

Comparing (1) and (2) observe that they are nearly the same except for the interchanging of f and g. One important difference is that the points

$$\{z_i\}_{i=0}^{n}$$

do not, in general, form a partition of $[a,b]$ unless

$z_0 = a$ and $z_n = b$.

To obviate this difficulty,

let $z_0 = a$ and $z_{n+1} = b$.

Adding and subtracting the terms

$g(x_0)f(z_0) = g(a)f(a)$

and

$g(x_n)f(z_{n+1}) = g(b)f(b)$

to the right hand side of (2) yields

$$S(f,g,P) = \sum_{i=0}^{n} g(x_i)[f(z_i) - f(z_{i+1})]$$

$$- f(z_1)g(x_0) + f(z_n)g(x_n)$$

$$- g(x_n)[f(z_n) - f(z_{n+1})]$$

$$- g(x_0)[f(z_0) - f(z_1)] =$$

$$\sum_{i=0}^{n} g(x_i)[f(z_i) - f(z_{i+1})]$$

$$- f(z_1)g(x_0) + f(z_n)g(x_n) - g(x_n)f(z_n)$$

$$+ g(x_n)f(z_{n+1}) - g(x_0)f(x_0) + g(x_0)f(z_1).$$

The $f(z_1)g(x_0)$ terms and the $f(z_n)g(x_n)$ terms cancel and so,

$$S(f,g,P) = \sum_{i=0}^{n} g(x_i)[f(z_i) - f(z_{i+1})]$$

$$+ g(x_n)f(z_{n+1}) - g(x_0)f(x_0)$$

or

$$S(f,g,P) = \sum_{i=0}^{n} g(x_i)[f(z_i) - f(z_{i+1})]$$

$$+ g(b)f(b) - g(a)f(a). \tag{3}$$

Let the mesh of P approach zero, i.e.,

$$\max \{(x_1 - x_0), (x_2 - x_1), \ldots, (x_n - x_{n-1})\} \to 0.$$

Then the subdivision formed by the points

$$\{z_i\}_{i=0}^{n+1} \quad \text{also} \to 0.$$

But by definition, the limit of (1) over all partitions with mesh tending to zero is the Stieltjes integral

$$\int_a^b f \, dg.$$

Since $f(x)$ is monotonically increasing on $[a,b]$,

$$f(z_i) - f(z_{i+1})$$

is negative and so, by the same reasoning, (3) becomes

$$-\int_a^b g \, df + g(b) \, f(b) - g(a) \, f(a).$$

Since (1) and (3) are rearrangements of each other (and finite sums), they are equal, i.e.,

$$\int_a^b f \, dg = -\int_a^b g \, df + g(b) \, f(b) - g(a) \, f(a).$$

or

$$\int_a^b f \, dg + \int_a^b g \, df = g(b) \, f(b) - g(a) \, f(a)$$

as was to be shown.

Applying the formula provided by the theorem to the given problem,

$$\int_{-1}^1 x \, d|x| + \int_{-1}^1 |x| \, dx = x|x| \Big|_{-1}^1$$

or

$$\int_{-1}^1 x \, d|x| = |x|x \Big|_{-1}^1 - \int_{-1}^1 |x| \, dx. \tag{4}$$

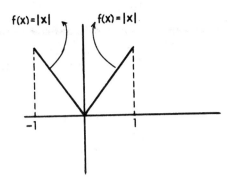

The value of the integral

$$\int_{-1}^{1} |x|\,dx$$

equals the value of

$$2\int_{0}^{1} x\,dx \qquad \text{(see Fig.)}.$$

Hence (4) is

$$\int_{-1}^{1} x\,d|x| = 1 - (-1) - \left.\frac{2x^2}{2}\right|_{0}^{1} = 2 - 1 = 1.$$

An alternative method of solution is to proceed from the definition of the Stieltjes integral to deduce the formula

$$\int_{a}^{b} f(x)\,dg(x) = \int_{a}^{c} f(x)\,dg(x) + \int_{c}^{b} f(x)\,dg(x),$$

where $a < c < b$.

Then,

$$\int_{-1}^{1} x\,d|x| = \int_{-1}^{0} x\,d(-x) + \int_{0}^{1} x\,dx$$

$$= -\int_{-1}^{0} x\,dx + \int_{0}^{1} x\,dx = \tfrac{1}{2} + \tfrac{1}{2} = 1.$$

• **PROBLEM 7-17**

Let f be a function from [a,b] into R which is continuous at $c \in [a,b]$ and let X_c be the characteristic function of c, i.e.,

$$X_c(x) = \begin{cases} 1 & x = c \\ 0 & x \neq c \end{cases}$$

Show, using the definition of the Stieltjes integral that

$$\int_a^b f \, dX_c = \begin{cases} 0 & c \in (a,b) \\ -f(a) & c = a \\ f(b) & c = b \end{cases}$$

Solution: The definition of the Stieltjes integral is as follows:

Let

$$P = \{(x_{k-1}, x_k) : k = 1, \ldots n\}$$

be a partition of [a,b] and let f,g be defined on [a,b]. Then

$$\int_a^b f \, dg = \lim_{\|P\| \to 0} \sum_{k=1}^n f(z_k)[g(x_k) - g(x_{k-1})]$$

where only partitions with sub-intervals less than any preassigned δ are considered.

Let $E \subseteq R$, i.e., E is a subset of the real line. The characteristic function of x with respect to E is

$$X(x) = \begin{cases} 1 & x \in E \\ 0 & x \notin E \end{cases}$$

In the given problem E is the singleton set $\{c\}$.

There are three separate cases to consider:

i) $c \in (a,b)$, ii) $c = a$, the left endpoint of the interval, and iii) $c = b$, the right endpoint of the interval.

i) Suppose $c \in (a,b)$ and that $x_k \neq c$ for

$k = 0, \ldots, n$.

Then

$$S(f, X_c, P) = \sum_{k=1}^{n} f(z_k)[X_c(x_k) - X_c(x_{k-1})]$$

$$= \sum_{k=1}^{n} f(z_k)[0-0] = 0.$$

On the other hand, if $x_k = c$, i.e., one of the endpoints of the sub-intervals is c, then, since $X_c(x_i) = 1$,

$$S(f, X_c, P) = f(z_i)(X_c(x_i) - X_c(x_{i-1}))$$

$$+ f(z_{i+1})(X_c(x_{i+1}) - X_c(x_i))$$

$$= f(z_i)(1-0) + f(z_{i+1})(0-1) = f(z_i) - f(z_{i+1}).$$

The function f being continuous at c implies that for each $\varepsilon > 0$, there exists $\delta > 0$ such that

$$|f(c) - f(x)| < \varepsilon/2$$

whenever $|c - x| < \delta$. Then for any partition P with $\|P\| < \delta$,

$$|S(f, X_c, P)| = |f(z_i) - f(z_{i+1})|$$

$$\leq |f(z_i) - f(c)| + |f(c) - f(z_{i+1})|$$

$$< \varepsilon/2 + \varepsilon/2 = \varepsilon$$

Hence

$$\int_a^b f \, dX_c = \lim_{\|P\| \to 0} S(f, X_c, P) = 0.$$

ii) When $c = a$, the left endpoint of the interval $[a,b]$,

$$S(f, X_c, P) = f(z_1)(X_c(x_1) - X_c(x_0))$$

$$= f(z_1)(0-1) = -f(z_1) = -f(a),$$

and

$$\int_a^b f \, dX_c = \lim_{\|P\| \to 0} S(f, X_c, P) = -f(a)$$

iii) When $c = b$, the right endpoint of the interval $[a,b]$,

$$S(f, X_c, P) = f(z_n)(X_c(x_n) - X_c(x_{n-1}))$$

$$= f(z_n)(1-0) = f(c) = f(b).$$

● PROBLEM 7-18

Let f, g be functions from $[a,b]$ into R. Suppose $[a,b]$ is divided into N parts by the partition

$$P = \{a_0, a_1, \ldots, a_N\}$$

where $a_0 = a$, $a_N = b$. Let g be a step-function on $[a,b]$ with $g(x) = c_i$, $1 \leq i \leq N$, on the sub-intervals

$$(a_0, a_1), (a_1, a_2) \ldots (a_{N-1}, a_N).$$

Furthermore, suppose f is continuous at each a_i. What is the value of

$$\int_a^b f \, dg \ ?$$

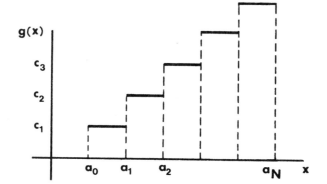

Fig. 1

Solution: The function g can be graphed as shown in Figure 1.

These are two ways of evaluating

$$\int_a^b f \, dg$$

when f and g satisfy the stated conditions.

The first method uses the fact that each sub-interval in the partition may be treated separately if the integrals over these partitions exist. Now, in the given problem f is continuous at each point where g is discontinuous (the endpoints of each sub-interval). Also g is nondecreasing on each sub-interval. Thus

$$\int_{a_{i-1}}^{a_i} f \, dg$$

is integrable and the integral exists. Hence since

$$\int_a^b f \, dg = \int_a^c f \, dg + \int_c^b f \, dg,$$

for $a < c < b$,

$$\int_a^b f \, dg = \int_a^{a_1} f \, dg + \int_{a_1}^{a_2} f \, dg + \ldots + \int_{a_{N-1}}^{a_N} f \, dg \, . \tag{1}$$

Consider the ith integral on the right hand side of (1), i.e.,

$$\int_{a_{i-1}}^{a_i} f \, dg \, .$$

By the definition of the Stieltjes integral,

$$\int_{a_{i-1}}^{a_i} f \, dg = \lim_{|P| \to 0} \sum_{j=1}^{n} f(z_j)[g(x_j) - g(x_{j-1})],$$

where $|P|$ denotes the mesh of the partition P.

Since $g(x) = c_i$ over the open interval (a_{i-1}, a_i) for any partition P of (a_{i-1}, a_i), $g(x_j) - g(x_{j-1}) = c_i - c_i = 0$ except when

$$x_0 = a_{i-1} \quad \text{and} \quad x_n = a_i \, .$$

Thus

$$\lim_{|P| \to 0} \sum_{j=1}^{n} f(z_j)[g(x_j) - g(x_{j-1})] =$$

$$\lim_{|P| \to 0} [f(z_1)[g(x_1) - g(x_0)]$$

$$+ f(z_2)[g(x_2) - g(x_1)] + f(z_3)[g(x_3) - g(x_2)]$$

$$+ \ldots + f(z_n)[g(x_n) - g(x_{n-1})] \, . \tag{2}$$

However, all the terms of the form

$g(x_i) - g(x_{i-1})$, $i = 1, 2, \ldots, n-1$,

are zero since the points x_i and x_{i-1} fall in the open interval from a_{i-1} to a_i, where

$$g(x_i) = g(x_{i-1}) = c_i .$$

In addition, as $|P| \to 0$, $z_1 \to a_{i-1}$ and $z_n \to a_i$. Hence, since f is continuous at each a_i,

$$f(z_1) \to f(a_{i-1}) \quad \text{and} \quad f(z_n) \to f(a_i), \text{ as } |P| \to 0.$$

Therefore,

$$\int_{a_{i-1}}^{a_i} f\, dg = f(a_{i-1})[c_i - g(a_{i-1})]$$
$$+ f(a_i)[g(a_i) - c_i] .$$

Summing the integrals

$$\int_{a_{i-1}}^{a_i} f\, dg \quad \text{from } i = 1 \text{ to } N,$$

$$\int_a^b f\, dg = \int_a^{a_1} f\, dg + \int_{a_1}^{a_2} f\, dg$$

$$+ \ldots + \int_{a_{N-1}}^b f\, dg$$

$$= f(a)[c_1 - g(a)] + f(a_1)[g(a_1) - c_1]$$

$$+ f(a_1)[c_2 - g(a_1)] + f(a_2)[g(a_2) - c_2]$$

$$+ \ldots + f(a_{N-1})[c_N - g(a_{N-1})]$$

$$+ f(b)[g(b) - c_N]$$

$$= f(a)[c_1 - g(a)] + f(a_1)[c_2 - c_1]$$

$$+ f(a_2)[c_3-c_2] + \ldots + f(b)[g(b) - c_N]$$

$$= f(a)[c_1-g(a)] + \sum_{i=1}^{N-1} f(a_i)[c_{i+1} - c_i]$$

$$+ f(b)[g(b) - c_N] \quad . \tag{3}$$

Another method for calculating

$$\int_a^b f \, dg$$

involves the notion of characteristic functions. Let E be a subset of R^n. Then the characteristic function of E is

$$X_E(x) = \begin{cases} 1 & x \in E \\ 0 & x \notin E \end{cases}.$$

The step-function g can be expressed as a sum of characteristic functions. Let

$$I_1 = (a_0, a_1), \quad I_2 = (a_1, a_2), \quad \ldots, \quad I_i = (a_{i-1}, a_i)$$

$$, \ldots, \quad I_N = (a_{N-1}, a_N) \quad .$$

Then,

$$g = \sum_{i=1}^{N} c_i X_{I_i} + \sum_{i=1}^{N} g(a_i) X_{a_i} \quad .$$

The second sum is necessary since the intervals I_i do not include a_i. Thus

$$\int_a^b f \, dg = \int_a^b f \, d\left(\sum_{i=1}^{N} c_i X_{I_i} + \sum_{i=1}^{N} g(a_i) X_{a_i}\right) \quad . \tag{4}$$

The set of Stieltjes integrals on an interval [a,b] forms a vector space. In particular, this means that

$$\int_a^b f d(\alpha g_1 + \beta g_2) = \alpha \int_a^b f \, dg_1 + \beta \int_a^b f \, dg_2 \quad . \tag{5}$$

Applying (5) to (4):

$$\int_a^b fd\left(\sum_{i=1}^N c_i X_{I_i} + \sum_{i=1}^N g(a_i) X_{a_i}\right)$$

$$= \sum_{i=1}^N c_i \int_a^b f\, d\, X_{I_i} + \sum_{i=1}^N g(a_i) \int_a^b f\, d\, X_{a_i} \quad \cdot\bigg) \qquad (6)$$

To evaluate (6), note that $I_i \subset [a,b]$ $(i = 1, \ldots, N)$.

Hence

$$\int_a^b fd\, X_{I_i} = f(a_{i-1}) - f(a_i).$$

Also, if a_i is a singleton set,

$$\int_a^b f\, d\, X_{a_i} = \begin{cases} 0 & a_i \in (a,b) \\ -f(a) & a_i = a \\ f(b) & a_i = b \end{cases}.$$

Substituting all of this into (6):

$$\int_a^b f\, dg = \sum_{i=1}^N c_i(f(a_{i-1}) - f(a_i)) + g(b)f(b) - g(a)f(a). \qquad (7)$$

But

$$\sum_{i=1}^N c_i(f(a_{i-1}) - f(a_i)) = c_1(f(a) - f(a_1))$$

$$+ c_2(f(a_1) - f(a_2)) + c_3(f(a_2) - f(a_3)) + \ldots +$$

$$c_N(f(a_{N-1}) - f(b))$$

$$= f(a_1)(c_2 - c_1) + f(a_2)(c_3 - c_2) + \ldots + f(a_{N-1})(c_N - c_{N-1})$$

$$+ c_1 f(a) - c_N f(b)$$

$$= \sum_{i=1}^{N-1} f(a_i)(c_{i+1} - c_i) + c_1 f(a) - c_N f(b). \qquad (8)$$

Substituting (8) into (7):

$$\int_a^b f\,dg = f(a)(c_1 - g(a)) + \sum_{i=1}^{N-1} f(a_i)(c_{i+1} - c_i) +$$

$$f(b)(g(b) - c_N) \qquad (9)$$

which is the same result as by the first method.

● **PROBLEM** 7-19

Suppose g is a continuously differentiable monotonically increasing function on [a,b] and f is bounded on [a,b]. Prove that

$$\int_a^b f\,dg = \int_a^b f(x)g'(x)\,dx .$$

Use this result to find the total mass of a linear distribution on [a,b] with a continuous density function $\rho(x)$.

Solution: The required proof may be divided into three parts: i) the definition of the upper and lower Stieltjes integrals, ii) subsidiary lemmas, and iii) the actual proof.

i) Let

$$P = \{a = x_0, x_1, \ldots, x_n = b\}$$

be a partition of [a,b]. Let

$$\Delta g_i = g(x_i) - g(x_{i-1})$$

$$M_i = \sup f(x) \quad x \in (x_{i-1}, x_i)$$
$$m_i = \inf f(x) \quad x \in (x_{i-1}, x_i).$$

Then

$$L(P,f,g) = \sum_{i=1}^n m_i \Delta g_i$$

$$U(P,f,g) = \sum_{i=1}^n M_i \Delta g_i .$$

The upper and lower Stieltjes integrals of f with respect

to g are defined by

$$\overline{\int_a^b} f\, dg = \inf U(P,f,g) \tag{1}$$

$$\underline{\int_a^b} f\, dg = \sup L(P,f,g) \tag{2}$$

respectively where the inf and sup are taken over all partitions P of [a,b]. If (1) and (2) are equal, f is Stieltjes integrable with respect to g on [a,b] and the Stieltjes integral is

$$\int_a^b f\, dg\ .$$

ii)

$$\int_a^b fdg$$

exists, if and only if, given $\varepsilon > 0$, there exists a partition P such that

$$U(P,f,g) - L(P,f,g) < \varepsilon\ . \tag{3}$$

To prove this, note that for every partition P,

$$L(P,f,g) \leq \underline{\int} fdg \leq \overline{\int} fdg \leq U(P,f,g)\ . \tag{4}$$

In particular, for a partition P such that

$$U(P,f,g) - L(P,f,g) < \varepsilon\ ,\ (4) \text{ implies that}$$

$$0 \leq \overline{\int} f\, dg - \underline{\int} f\, dg < \varepsilon\ .$$

Since this is true for every $\varepsilon > 0$,

$$\overline{\int} fdg = \underline{\int} fdg = \int fdg \text{ and f is}$$

Stieltjes integrable.

For the converse, assume $\int fdg$ exists and let $\varepsilon > 0$ be given. Then there exist partitions P_1 and P_2 such that

$$U(P_2,f,g) - \int fdg < \varepsilon/2 \tag{5}$$

and

$$\int fdg - L(P_1,f,g) < \varepsilon/2 . \tag{6}$$

Let P* be the common refinement of P_1 and P_2, i.e.,

$$P^* = P_1 \cup P_2 .$$

Thus, every point of P_1 and every point of P_2 is a point of P*. It follows that

$$L(P_1,f,g) \leq L(P^*,f,g)$$

and

$$L(P_2,f,g) \leq L(P^*,f,g).$$

Similarly,

$$U(P_1,f,g) \geq U(P^*,f,g)$$

and

$$U(P_2,f,g) \geq U(P^*,f,g).$$

Then using (5) and (6):

$$U(P^*,f,g) \leq U(P_2,f,g) < \int fdg + \varepsilon/2 < L(P_1,f,g) + \varepsilon$$

$$\leq L(P^*,f,g) + \varepsilon$$

or,

$$U(P^*,f,g) - L(P^*,f,g) < \varepsilon .$$

Thus P* is the required partition.

As a second lemma suppose that (3) holds for given ε and some partition

$$P = \{x_0, \ldots, x_n\} .$$

If s_i, t_i are arbitrary points in (x_{i-1}, x_i) then

$$\sum_{i=1}^{n} |f(s_i) - f(t_i)| \Delta g_i < \varepsilon .$$

This lemma states that if $\int fdg$ exists, the difference in f within any subinterval is negligible. To prove this result, note that since both $f(s_i)$ and $f(t_i)$ lie in $[m_i, M_i]$ (the inf and sup of f in (x_{i-1}, x_i) respectively,

$$|f(s_i) - f(t_i)| \leq M_i - m_i .$$

Thus

$$\sum_{i=1}^{n} |f(s_i) - f(t_i)| \Delta g_i \leq \sum_{i=1}^{n} (M_i - m_i) \Delta g_i$$

$$= U(P,f,g) - L(P,f,g) < \varepsilon$$

by assumption. This proves the lemma.

iii) To prove the main theorem, let $\varepsilon > 0$ be given. Applying the first lemma to g': There is a partition

$$P = \{x_0, \ldots, x_n\} \text{ of } [a,b] \text{ such that}$$

$$U(P,g') - L(P,g') < \varepsilon .$$

By the mean-value theorem, there exist $t_i \in [x_i, x_{i-1}]$ such that

$$\frac{g(x_i) - g(x_{i-1})}{x_i - x_{i-1}} = g'(t_i) \qquad (i = 1, \ldots, n),$$

or

$$\Delta g_i = g'(t_i) \Delta x_i .$$

If $s_i \in [x_{i-1}, x_i]$, then, by the second lemma:

$$\sum_{i=1}^{n} |g'(s_i) - g'(t_i)| \Delta x_i < \varepsilon . \tag{7}$$

Put $M = \sup |f(x)|$. This exists since f is bounded. Since

$$\sum_{i=1}^{n} f(s_i) \Delta g_i = \sum_{i=1}^{n} f(s_i) g'(t_i) \Delta x_i ,$$

$$\left| \sum_{i=i}^{n} f(s_i) \Delta g_i - \sum_{i=1}^{n} f(s_i) g'(s_i) \Delta x_i \right| \tag{8}$$

$$= \sum_{i=1}^{n} |f(s_i) g'(t_i) - f(s_i) g'(s_i)| \Delta x_i$$

$$= \sum_{i=1}^{n} |f(s_i) (g'(t_i) - g'(s_i))| \Delta x_i$$

332

$$\leq M \sum_{i=1}^{n} |g'(t_i) - g'(s_i)| \Delta x_i$$

$$\leq M\varepsilon.$$

This implies that

$$\sum_{i=1}^{n} f(s_i) \Delta g_i \leq U(P, fg') + M\varepsilon$$

for all choices of $s_i \in [x_{i-1}, x_i]$

so that

$$U(P, f, g) \leq U(P, fg') + M\varepsilon.$$

Similarly, from (8), since only the distance between

$$\sum_{i=1}^{n} f(s_i) \Delta g_i$$

and

$$\sum_{i=1}^{n} f(s_i) g'(s_i) \Delta x_i$$

is involved,

$$U(P, fg') \leq U(P, f, g) + M\varepsilon.$$

Thus,

$$|U(P, f, g) - U(P, fg')| \leq M\varepsilon$$

and

$$\left| \overline{\int_a^b} f\,dg - \overline{\int_a^b} f(x)g'(x)\,dx \right| \leq M\varepsilon.$$

But ε is arbitrary and therefore

$$\overline{\int_a^b} f\,dg = \overline{\int_a^b} f(x)g'(x)\,dx$$

for any function bounded on $[a,b]$.

In the same way, equality of the lower integrals follows from (8). Hence,

$$\int_a^b f\,dg = \int_a^b f(x)g'(x)\,dx.$$

Turning to the second part of the problem, the mass of the distribution on the interval [a,x] is given by

$$m(x) = \int_a^x \rho(t)\,dt \ .$$

By the Fundamental Theorem of Calculus,

$$m'(x) = \rho(x).$$

Hence, letting $m(x) = g(x)$,

$$\int_a^b f\,dm = \int_a^b f(x)m'(x)\,dx$$

and hence the total mass is

$$M = \int_a^b 1 \cdot dm(x) = \int_a^b \rho(x)\,dx \ .$$

($f(x)$ in this case is a constant function that equals 1).

● **PROBLEM 7-20**

Discuss the existence of

$$\int_0^1 x\,dx^2$$

and find its value.

Solution: The first task in the evaluation of an integral is to show that the integral exists. The given integral is of the form

$$\int_a^b f(x)\,dg(x)$$

where $g(x)$ is monotonically increasing on [a,b] and $f(x)$ is continuous on [a,b]. A fundamental criterion for integrability is the following: A function $f: [a,b] \to R$ is integrable with respect to a nondecreasing function

$$g: [a,b] \to R$$

if an only if for all $\varepsilon > 0$ there is a partition P_ε of

$[a,b]$ such that if P'_ε is a refinement of P_ε then

$$\sum_{i=1}^{n} (M_i - m_i)[g(x_i) - g(x_{i-1})] < \varepsilon$$

where

$$M_i = \sup \{f(x): x \in [x_{i-1}, x_i]\}$$

and

$$m_i = \inf \{f(x): x \in [x_{i-1}, x_i]\}$$

for $i = 1, \ldots, n$.

From this result a theorem on the integrability of a continuous function with respect to a monotonically increasing integrator may be deduced.

To prove this latter theorem for the given integral

$$\int_0^1 x \, dx^2,$$

let $f(x) = x$ and $g(x) = x^2$.

Then, since f is uniformly continuous on $[0,1]$, there exists a $\delta(\varepsilon) > 0$ such that

$$|x-y| < \delta(\varepsilon)$$

implies

$$|f(x) - f(y)| < \varepsilon \quad .$$

Let

$$P_\varepsilon = \{w_0, w_1, \ldots, w_k\} \text{ be a partition of } [a,b]$$

such that $\sup\{w_k - w_{k-1}\} < \delta(\varepsilon)$.

If P'_ε is a refinement of P_ε, then

$$\sup \{x_k - x_{k-1}\} < \delta(\varepsilon)$$

and thus $M_i - m_i < \varepsilon$, which implies that

$$\sum_{i=1}^{n} (M_i - m_i)\{g(x_i) - g(x_{i-1})\} \leq \varepsilon (g(b) - g(a)) \quad .$$

Since $\varepsilon > 0$ is arbitrary, the fundamental criterion of integrability, (1), may be applied. Thus

$$\int_0^1 x dx^2$$

exists.

The next task is to evaluate this integral. Using the definition of the integral as the limit of a sum,

$$\int_0^1 f(x) dg(x) = \lim_{n \to \infty} \sum_{j=1}^{n} f(x_j) [g(x_j) - g(x_{j-1})] ,$$

Let

$$x_j = \frac{j}{n} ,$$

then

$$\int_0^1 f(x) dg(x) = \lim_{n \to \infty} \sum_{j=1}^{n} f\left(\frac{j}{n}\right) \left[g(j/n) - g\left(\frac{j-1}{n}\right)\right]$$

or,

$$\int_0^1 x dx^2 = \lim_{n \to \infty} \sum_{j=1}^{n} \frac{j}{n} \left[\frac{j^2}{n^2} - (j-1)^2/n^2\right]$$

$$= \lim_{n \to \infty} \sum_{j=1}^{n} \frac{j}{n} \frac{2j-1}{n} \frac{1}{n} = \lim_{n \to \infty} \sum_{j=1}^{n} \frac{1}{n^3} [2j^2 - j]$$

$$= \lim_{n \to \infty} \sum_{j=1}^{n} \frac{1}{n^3} 2j^2 - \lim_{n \to \infty} \sum_{j=1}^{n} \frac{j}{n^3}$$

$$= \lim_{n \to \infty} \frac{1}{n^3} \left[\frac{n(n+1)(2n+1)}{3} - \frac{n(n+1)}{2}\right]$$

$$= \lim_{n \to \infty} \frac{1}{n^3} \left[\frac{4n^3}{6} + \frac{3n^2}{6} - \frac{n}{6}\right]$$

$$= 2/3 .$$

Thus

$$\int_0^1 x \, dx^2 = 2/3 .$$

● **PROBLEM 7-21**

Suppose $(c,d) \subset [a,b]$, f is a function from $[a,b]$ into R which is continuous at c and d and $X_{(c,d)}$ is the characteristic function of (c,d). Show that

$$\int_a^b f\,dX_{(c,d)} = f(c) - f(d).$$

Solution: Since $X_{(c,d)}$ is the characteristic function of (c,d),

$$X_{(c,d)} = \begin{cases} 1 & x \in (c,d) \\ 0 & x \notin (c,d) \end{cases}.$$

Let P be a partition of $[a,b]$, i.e.,

$$P = \{x_k\}_{k=0}^n, \quad a = x_0 < x_1 < \ldots < x_n = b.$$

The Stieltjes integral with respect to f and g over $[a,b]$ is

$$\int_a^b f\,dg = \lim_{\|P\| \to 0} \sum_{k=1}^n f(z_k)[g(x_k) - g(x_{k-1})] \qquad (1)$$

where

$$\|P\| = \max\{x_1 - x_0, x_2 - x_1, \ldots, x_n - x_{n-1}\}.$$

Definition (1) is interpreted as follows:

Let $\varepsilon > 0$ be given. Then f is Stieltjes integrable with respect to g, the integrator function, if for all partitions whose norm

$$\|\cdot\|$$

is less than δ and any $z_k \in (x_{k-1}, x_k)$ we have

$$\left| \int_a^b f\,dg - \sum_{k=1}^n f(z_k)[g(x_k) - g(x_{k-1})] \right| < \varepsilon.$$

To show that

$$\int_a^b f\,dX_{(c,d)} = f(c) - f(d),$$

consider partitions of $[a,b]$ with $\|P\| < d-c$. Then if

$$c \in [x_{i-1}, x_i] \quad \text{and} \quad d \in [x_{j-1}, x_j]$$

$$S(f, X_{(c,d)}, P) = \sum_{k=1}^{n} f(z_k)(X_{(c,d)}(x_k) - X_{(c,d)}(x_{k-1}))$$

$$= f(z_1)[X_{(c,d)}(x_1) - X_{(c,d)}(x_0)] +$$

$$f(z_2)[X_{(c,d)}(x_2) - X_{(c,d)}(x_1)]$$

$$+ \ldots + f(z_i)[X_{(c,d)}(x_i) - X_{(c,d)}(x_{i-1})] + \ldots +$$

$$f(z_j)[X_{(c,d)}(x_j) - X_{(c,d)}(x_{j-1})]$$

$$+ \ldots + f(z_n)[X_{(c,d)}(x_n) - X_{(c,d)}(x_{n-1})]$$

$$= f(z_1)[0-0] + f(z_2)[0-0] + \ldots + f(z_i)[1-0]$$

$$+ f(z_{i+1})[1-1] + \ldots + f(z_j)[0-1]$$

$$= f(z_i) - f(z_j).$$

Since f is continuous at $c \in [a,b]$, given any $\varepsilon > 0$ there exists a $\delta > 0$ such that

$$|x - c| < \delta$$

implies

$$|f(x) - f(c)| < \varepsilon.$$

Thus, for all partitions with norm less than δ,

$$|f(x) - f(c)| < \varepsilon,$$

i.e.,

$$f(z_i) = f(c).$$

Similarly,

$$f(z_j) = f(d)$$

and

$$\int_a^b f \, dX_{(c,d)} = f(c) - f(d).$$

CHAPTER 8

LINE INTEGRALS

A line integral is an integral of a function which is defined along a curve. A curve is defined as a continuous mapping $C:[a,b] \to R^n$ (in this chapter n=2 or n=3). This curve, oriented with a positive direction and a negative direction, has an initial point A and a final point B. Also this curve can form many shapes, crossing itself a number of times or not at all. A curve whose initial point A and terminal point B coincide is termed a closed curve (e.g. circle).

Line integrals have the form $\int_C p(x,y)dx + q(x,y)dy$ where $p(x,y)$, $q(x,y)$ are functions of x and y, and C denotes the curve we are integrating along.

Line integrals can be solved by the method of parameterization, which reduces the line integral to an ordinary integral.

Another method uses the fact that certain line integrals are independent of the curve C (these line integrals have the same value between two points in space along any curve). This method evaluates line integrals by finding an exact differential of $\vec{F} = (p(x,y), q(x,y))$ and by computing at the initial point and terminal point of C.

Finally, another method to solve line integrals is by Green's Theorem which transforms the line integral around some curve to a double integral on a region bounded by this curve.

METHOD OF PARAMETRIZATION

● PROBLEM 8-1

Find the values of:

(a) $\int_C (xy + y^2 - xyz)dx$

(b) $\int_C (x^2 - xy)dy$

if C is the arc of the parabola $y = x^2$, $z = 0$ from $(-1,1,0)$ to $(2,4,0)$.

<u>Solution</u>: Given a differentiable curve $C:[a,b] \to R^2$ and a

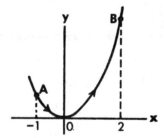

Fig. 1

vector field $\vec{F}$ which is continuous at every point on C, the definition of the integral of the vector field $\vec{F}$ along the curve C is

$$\int_C \vec{F} \cdot d\vec{c} = \int_a^b \vec{F}(\vec{c}(t)) \cdot \frac{d\vec{c}}{dt} dt \qquad (1)$$

where $\frac{d\vec{c}}{dt}$ is the tangent vector to the curve.

Given $\vec{c} = (x(t), y(t))$ then $\frac{d\vec{c}}{dt} = (x'(t), y'(t))$

Thus (1) can be rewritten:

$$\int_a^b \left[f(x(t), y(t)) \frac{dx}{dt} + g(x(t), y(t)) \frac{dy}{dt} \right] dt$$

$$= \int_{C(a)}^{C(b)} f(x,y) dx + g(x,y) dy \qquad (2)$$

$$= \int_C (f(x,y), g(x,y)) \cdot (dx, dy) = \int_C \vec{F} \cdot d\vec{c} \qquad \text{where}$$

$\vec{F}$ is the vector field $\vec{F}(x,y) = (f(x,y), g(x,y))$ and $d\vec{c} = (dx, dy)$. This may be extended to $C:[a,b] \to R^3$.

For these line integrals $\vec{F}$ is only one dimensional, i.e. a scalar field and x is used as the parameter.

(a) Here $f(x,y,z) = xy + y^2 - xyz$ and $g(x,y,z) = 0$ in (2). Let $x=x$; $y=x^2$, $z=0$, so that $dx=dx$, $dy=2xdx$; and $dz=0$. Then integral (a) becomes

$$\int_C (xy+y^2-xyz) dx = \int_{-1}^{2} (x^3+x^4) dx = \frac{x^4}{4}\Big|_{-1}^{2} + \frac{x^5}{5}\Big|_{-1}^{2}$$

$$= \frac{15}{4} + \frac{33}{5} = \frac{207}{20} .$$

(b) $$\int_C (x^2-xy) dy = \int_{-1}^{2} (x^2-x^3) 2x dx = \int_{-1}^{2} (2x^3-2x^4) dx$$

$$= \frac{2x^4}{4}\Big|_{-1}^{2} - \frac{2x^5}{5}\Big|_{-1}^{2} = \frac{30}{4} - \frac{66}{5} = \frac{-57}{10} .$$

• **PROBLEM 8-2**

Evaluate: a) the line integral $\int_C (x^3-y^3)dy$ where C is the semicircle $y = \sqrt{1-x^2}$ shown in Figure 1.
b) $\int_C \vec{F}\cdot d\vec{C}$ where $\vec{F}(x,y) = (x^2,xy)$ and C is $x=y^2$ between $(1,-1)$ and $(1,1)$.

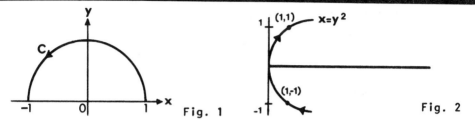

Fig. 1   Fig. 2

Solution: The curve C can be represented parametrically by $x = \cos\theta$ and $y = \sin\theta$, $0 \le \theta \le \pi$. Therefore the integral becomes $\int_0^\pi (\cos^3\theta - \sin^3\theta)\cos\theta d\theta$ $(dy = \cos\theta d\theta)$

$$= \int_0^\pi \cos^4\theta d\theta - \int_0^\pi \sin^3\theta\cos\theta d\theta. \quad (1)$$

Evaluate the first integral by parts letting $u = \cos^3\theta$, $dv = \cos\theta d\theta$ so that $du = -3\cos^2\theta\sin\theta$ and $v = \sin\theta$. Hence the first integral in (1) becomes:

$$\cos^3\theta\sin\theta\Big|_0^\pi + 3\int_0^\pi \sin^2\theta\cos^2\theta d\theta$$

$$= 0 + 3\int_0^\pi \sin^2\theta\cos^2\theta d\theta.$$

Using $2\sin\theta\cos\theta = \sin 2\theta$ and $\sin^2\theta = \frac{1-\cos 2\theta}{2}$ the above integral is computed to equal $3\int_0^\pi \left(\frac{1}{8} - \frac{\cos 4\theta}{8}\right)d\theta = \frac{3\pi}{8}$.

To evaluate the second integral, let $u = \sin\theta$, $du = \cos\theta$, then $-\int_0^\pi u^3 du = 0$. Therefore $\int_C (x^3-y^3)dy = \frac{3\pi}{8}$.

b) Since $\int \vec{F}\cdot d\vec{C}$ is being evaluated, first compute $\vec{F}\cdot d\vec{C} = (x^2,xy)\cdot(dx,dy) = x^2 dx + xy dy$. To parameterize let $y=y$; $x=y^2$ so that $dy=dy$; $dx=2ydy$ and

$$\int_C \vec{F}\cdot d\vec{C} = \int_C x^2 dx + xy dy = \int_{-1}^1 y^4 \cdot 2y dy + y^3 dy$$

$$= \frac{2y^6}{6} + \frac{y^4}{4}\Big|_{-1}^{1} = \left(\frac{1}{3} + \frac{1}{4}\right) - \left(\frac{1}{3} + \frac{1}{4}\right) = 0.$$

● **PROBLEM 8-3**

Compute the value of $\int_C xzdx + xdy - yzdz$ along the curve in Figure 1 consisting of a quarter circle in the xz-plane, and line segments in the xy-plane and yz-plane.

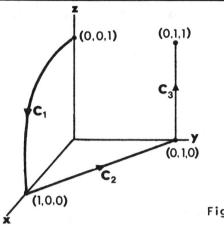

Fig. 1

Solution: By definition if a curve C is the union of a finite set of curves, $C = C_1 + C_2 + \ldots + C_n$, then

$$\int_C \vec{F} = \int_{C_1} \vec{F} + \int_{C_2} \vec{F} + \ldots + \int_{C_n} \vec{F}.$$ Therefore, label the three parts of the curve C as C_1, C_2, C_3 respectively. Then compute the value of the integral on each curve C_i and adding these values obtain the value of the integral on C. C_1 is the quarter circle from $(0,0,1)$ to $(1,0,0)$. Hence C_1 is in the xz-plane so choose x as the parameter.

Then $x = x$; $y = 0$; $z = \sqrt{1-x^2}$ so that

$$\int_{C_1} xzdx + xdy - yzdz = \int_{C_1} xzdx + x \cdot 0 - 0 \cdot dz$$

$$= \int_0^1 x\sqrt{1-x^2}\, dx \quad \text{which can be solved by}$$

letting $u = 1-x^2$; $du = -2x$ and changing the limits of integration to values of u which yields

$$-\frac{1}{2}\int_1^0 u^{\frac{1}{2}}\, du = \frac{1}{3}.$$

C_2 is the line segment in the xy-plane from $(1,0,0)$ to $(0,1,0)$. Here use y as the parameter so that $x = 1-y$;

$y=y$; $z=0$ and $dx=-dy$; $dy=dy$ and $dz=0$. Hence

$$\int_{C_2} xzdx + xdy - yzdz = \int_{C_2} 0\cdot dx + xdy - 0$$

$$= \int_0^1 xdy = \int_0^1 (1-y)dy = 1 - \frac{1}{2} = \frac{1}{2}.$$

C_3 is the line segment in the yz-plane from the point $(0,1,0)$ to $(0,1,1)$. Then use z as the parameter so that $x=0$, $y=1$; $z=z$ and $dx=0$, $dy=0$; and $dz=dz$. Hence

$$\int_{C_3} xzdx + xdy - yzdz = \int_{C_3} 0 + 0 - zdz$$

$$= -\int_0^1 zdz = -\frac{1}{2}.$$

Therefore since $\int_C = \int_{C_1} + \int_{C_2} + \int_{C_3}$ we have

$$\int_C xzdx + xdy - yzdz = \frac{1}{3} + \frac{1}{2} - \frac{1}{2} = \frac{1}{3}.$$

● **PROBLEM 8-4**

Evaluate the following line integrals:

a) $\oint_C y^2dx + x^2dy$ where C is the triangle with vertices $(1,0),(1,1),(0,0)$ (Figure 1). (1)

b) $\int_C x^2dx + xydy$ where C is the straight line segment from $(1,0)$ to $(2,3)$. (2)

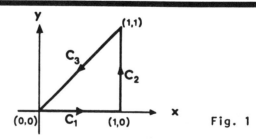

Fig. 1

Solution: a) To compute $\int_C y^2dx + x^2dy$ compute $\int_{C_i} y^2dx + x^2dy$ along 3 different C_i ($i = 1$ to 3). Then $\oint_C y^2dx + x^2dy$ equals the sum of the integrals along each C_i. Along C_1, $y=0$ and $dy=0$; therefore

$$\int_{C_1} y^2 dx + x^2 dy = \int_{C_1} 0 \cdot dx + x^2 \cdot 0 = 0. \qquad (3)$$

Along C_2 use y as the parameter and $x=1$ so that $dx=0$. Therefore

$$\oint_{C_2} y^2 dx + x^2 dy = \int_0^1 y^2 \cdot 0 + dy = \int_0^1 dy = 1. \qquad (4)$$

For C_3 use x as the parameter so that $x=x$ and $y=x$ and $dx=dx$; $dy=dx$.

$$\int_{C_3} y^2 dx + x^2 dy = \int_1^0 x^2 dx + x^2 dx = \int_1^0 2x^2 dx = -\frac{2}{3}. \qquad (5)$$

Hence (1) equals the sum of (3), (4), (5) so that

$$\oint_C y^2 dx + x^2 dy = 0 + 1 - \frac{2}{3} = \frac{1}{3}.$$

b) To compute (2), note that the line segment is of the form $P = P_1 + t(P_2-P_1)$, $(0 \leq t \leq 1)$, and can be parameterized accordingly. Since the given line segment goes from $(1,0)$ to $(2,3)$, by substitution:

$$x = 1 + t(2-1) = 1 + t;\ y = 0 + t(3-0) = 3t\ (0 \leq t \leq 1).$$

Hence substituting into (2)

$$\oint_C x^2 dx + xy dy = \int_0^1 (1+t)^2 dt + (1+t) 3t \cdot 3 dt$$

$$= \int_0^1 (1+2t+t^2+9t+9t^2) dt = \int_0^1 (1+11t+10t^2) dt$$

$$= 1 + \frac{11}{2} + \frac{10}{3} = \frac{59}{6}.$$

• **PROBLEM 8-5**

Evaluate the following line integrals:

a) $\oint_C \vec{F} \cdot d\vec{C}$ where $\vec{F}$ is the vector field $\vec{F}(x,y) = (x^2,xy)$ and C consists of the segment of the parabola $y=x^2$ between $(0,0)$ and $(1,1)$ and the line segment from $(1,1)$ to $(0,0)$. (Figure 1). (1)

b) $\int_C \vec{F} \cdot d\vec{C}$ where $\vec{F} = (x\left(\frac{1-y^2}{y^2+z^2}\right)^{\frac{1}{2}}, 0, 0)$ and C is the portion of the curve (in the first octant) of the intersection of the plane x=y and the cylinder $2y^2+z^2=1$ from $(0,0,1)$ to $(\frac{\sqrt{2}}{2}, \frac{\sqrt{2}}{2}, 0)$. (2)

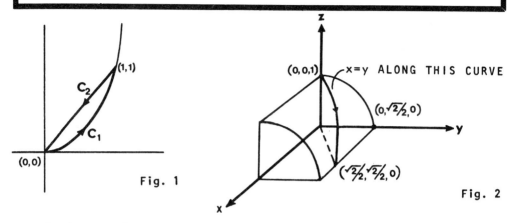

Fig. 1

Fig. 2

Solution: a) To compute (1), the line integral must be computed over two curves each with a different parameterization. First compute the integral over the parabola $y=x^2$. Using x as our parameter, we have $x=x$; $y=x^2$ and $dx=dx$; $dy=2xdx$. Hence,

$$\int_{C_1} \vec{F} \cdot d\vec{C} = \int_{C_1} (x^2, xy) \cdot (dx, dy) = \int_0^1 (x^2, x^3) \cdot (dx, 2xdx)$$

$$= \int_0^1 x^2 dx + x^3 \cdot 2xdx = \frac{x^3}{3}\Big|_0^1 + \frac{2x^5}{5}\Big|_0^1$$

$$= \frac{1}{3} + \frac{2}{5} = \frac{11}{15} \, .$$

To parameterize the line segment use the equation $P = P_1+t(P_2-P_1)$ so that $(0 \leq t \leq 1)$

$$x = 1 + t(0-1) = 1 - t \quad \text{and} \quad dx = -dt$$

$$y = 1 + t(0-1) = 1 - t \quad \text{and} \quad dy = -dt.$$

Therefore $\int_{C_2} \vec{F} \cdot d\vec{C} = \int_{C_2} x^2 dx + xy dy$

$$= \int_0^1 (1-t)^2(-1)dt + (1-t)(1-t)(-1)dt$$

$$= \int_0^1 (1-2t+t^2)(-1)dt + (1-2t+t^2)(-1)dt$$

$$= -2\int_0^1 (1-2t+t^2)dt = -2(t-t^2+\frac{t^3}{3})\Big|_0^1 = -\frac{2}{3} \, .$$

Since $\int_C \vec{F} = \int_{C_1} \vec{F} + \int_{C_2} \vec{F}$ (1) equals

$$\int_C \vec{F} \cdot d\vec{C} = -\frac{2}{3} + \frac{1}{3} + \frac{2}{5} = -\frac{1}{3} + \frac{2}{5} = \frac{1}{15} .$$

b) For (2) first compute $\vec{F} \cdot d\vec{C}$

$$= \left(x \left(\frac{1-y^2}{y^2+z^2} \right)^{1/2}, 0, 0 \right) \cdot (dx, dy, dz)$$

$$= x \left(\frac{1-y^2}{y^2+z^2} \right)^{1/2} dx + 0 \cdot dy + 0 \cdot dz .$$

Therefore (2) equals $\int_C x \left(\frac{1-y^2}{y^2+z^2} \right)^{1/2} dx.$ (3)

Since C is part of the curve of intersection of the plane (see Figure 2) x=y and the cylinder $2y^2+z^2=1$ use the parameterization x=x; y=x; $z=\sqrt{1-2y^2} = \sqrt{1-2x^2}$ so that (3) becomes

$$\int_C x \left(\frac{1-x^2}{x^2+1-2x^2} \right)^{\frac{1}{2}} dx$$

$$= \int_0^{\frac{\sqrt{2}}{2}} x \left(\frac{1-x^2}{1-x^2} \right)^{\frac{1}{2}} dx = \int_0^{\frac{\sqrt{2}}{2}} x\,dx = \left. \frac{x^2}{2} \right|_0^{\frac{\sqrt{2}}{2}} = \frac{2}{2 \cdot 4} = \frac{1}{4} .$$

● **PROBLEM 8-6**

a) Let $\vec{F}$ be a vector field on an open set V and C a curve in V defined on the interval [a,b]. Prove $\int_{C^-} \vec{F} = - \int_C \vec{F}$, where C^- is the reverse path of the curve C.

b) Then evaluate $\int_C \vec{F} \cdot d\vec{C}$ where $\vec{F}(x,y) = (x^2, xy)$ along the line segment from the point (1,1) to (0,0) using the reverse path.

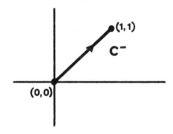

Solution: a) From the definition of the line integral

$$\int_C \vec{F} \cdot d\vec{C} = \int_a^b \vec{F}(\vec{C}(t)) \cdot \vec{C}'(t) \, dt \text{ where C is parameterized}$$

as $\vec{C}(t)$ and $d\vec{C} = \vec{C}'(t)\,dt$. Hence

$$\int_{C^-} \vec{F} \cdot d\vec{C} = \int_a^b \vec{F}(\vec{C}^-(t)) \cdot \frac{d\vec{C}^-}{dt}\,dt. \tag{1}$$

Now using the definition $\vec{C}^-(t) = \vec{C}(a+b-t)$, $\frac{d\vec{C}^-}{dt} = -\vec{C}'(a+b-t)$ and substituting these values into (1) yields

$$\int_{C^-} \vec{F} \cdot d\vec{C} = \int_a^b \vec{F}(\vec{C}(a+b-t))\vec{C}'(a+b-t)(-1)\,dt. \tag{2}$$

Now make a change of variables, letting $u = a+b-t$, $du = -dt$. Note that $t=a$ gives $u=b$ and $t=b$ gives $u=a$. Thus (2) becomes

$$\int_{C^-} \vec{F} \cdot d\vec{C} = \int_b^a \vec{F}(C(u)) \cdot \vec{C}'(u)\,du$$

$$= -\int_a^b \vec{F}(C(u)) \cdot \vec{C}'(u)\,du$$

$$= -\int_C \vec{F} \cdot d\vec{C} \qquad \text{which was to be proved.}$$

This result shows that if the vector field is integrated along the opposite direction of C, then the value of the line integral is the negative of the value obtained by integrating the vector field along C itself.

b) $\int_C (x^2, xy) \cdot (dx, dy)$ is to be evaluated using the reverse path of the line segment from the point $(1,1)$ to $(0,0)$. The line segment showing C^- is drawn in Figure 1. Now, using the result in part (a)

$$\int_C x^2\,dx + xy\,dy = -\int_{C^-} x^2\,dx + xy\,dy. \tag{3}$$

To evaluate (3) use x as the parameter and let $x=y$ since C^- is the line segment from $(0,0)$ to $(1,1)$. Hence (3) equals:

$$-\int_{C^-} x^2\,dx + x^2\,dx = -\int_0^1 2x^2\,dx = \left.\frac{-2x^3}{3}\right|_0^1 = \frac{-2}{3}.$$

Note that if integration had been along C the line segment

would have been parameterized as x=1-x and y=1-x, which would have made integration a more difficult process. Therefore integrating over the reverse path, particularly for line segments, can be more convenient.

● **PROBLEM 8-7**

Find the integral of the vector field $\vec{F}(x,y) = \left(\dfrac{-y}{x^2+y^2}, \dfrac{x}{x^2+y^2}\right)$ around the circle of radius 2 counterclockwise from the point $(2,0)$ to the point $(\sqrt{2}, \sqrt{2})$ (Figure 1). Repeat for the circle of radius 1 counterclockwise from $(1,0)$ to $\left(\dfrac{\sqrt{2}}{2}, \dfrac{\sqrt{2}}{2}\right)$ (Figure 2).

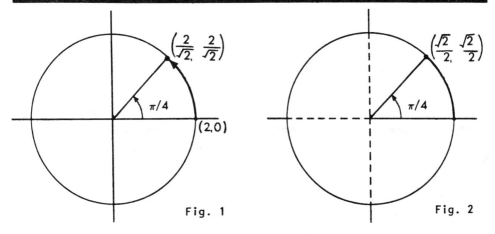

Fig. 1   Fig. 2

Solution: We want to compute

$$\int_C \vec{F} \cdot d\vec{C}. \qquad (1)$$

Therefore since $d\vec{C} = (dx, dy)$, (1) becomes

$$\int_C \left(\dfrac{-y}{x^2+y^2}\right) dx + \left(\dfrac{x}{x^2+y^2}\right) dy. \qquad (2)$$

To solve (2), first change to polar coordinates letting $x = 2\cos\theta$ and $y = 2\sin\theta$. Note that

$$\tan^{-1}\left(\dfrac{\sqrt{2}}{\sqrt{2}}\right) = \dfrac{\pi}{4};$$

therefore the parameter θ varies between 0 and $\dfrac{\pi}{4}$. Substituting these values into (2) and noticing $dx = -2\sin\theta\,d\theta$ and $dy = 2\cos\theta\,d\theta$ the desired integral in (2) is

$$\int_0^{\frac{\pi}{4}} \left[\dfrac{-2\sin\theta}{4}(-2\sin\theta) + \dfrac{2\cos\theta}{4}(2\cos\theta)\right] d\theta$$

$$= \int_0^{\frac{\pi}{4}} (\sin^2\theta + \cos^2\theta)\,d\theta = \int_0^{\frac{\pi}{4}} d\theta = \frac{\pi}{4}$$

For the circle of radius 1 use the parameterization given by $x = \cos\theta$; $y = \sin\theta$, $0 \le \theta \le \frac{\pi}{4}$ since

$$\tan^{-1}\left(\frac{\frac{\sqrt{2}}{2}}{\frac{\sqrt{2}}{2}}\right) = 1.$$

Hence the integral (2) becomes

$$\int_0^{\frac{\pi}{4}} \left[\frac{-\sin\theta}{1}(\sin\theta) + \frac{\cos\theta}{1}(\cos\theta)\right]d\theta$$

$$= \int_0^{\frac{\pi}{4}} (\sin^2\theta + \cos^2\theta)\,d\theta = \int_0^{\frac{\pi}{4}} d\theta = \frac{\pi}{4}.$$

• **PROBLEM 8-8**

Let C be the helix given parametrically as $\vec{C}(t) = \cos t\,\hat{\imath} + \sin t\,\hat{\jmath} + t\,\hat{k}$, $0 \le t \le 2\pi$.

a) Let $\vec{F}(x,y,z) = (x^2\hat{\imath} + y^2\hat{\jmath} + z^2\hat{k})$. Compute the line integral of $\vec{F}$ over the oriented curve C.

b) Let $f(x,y,z) = (x+y+z)$. Compute the line integral of f over the oriented curve C.

Solution: a) The problem is to find the line integral of a vector field over a curve. By definition this is

$$\int_C \vec{F}\cdot d\vec{C}$$ where $\vec{F}$ is the vector field and $d\vec{C} = \vec{C}'(t)dt$.

Since $\vec{F} = (x^2, y^2, z^2)$ and $\vec{C}(t) = (\cos t, \sin t, t)$, this gives on C $x = \cos t$; $y = \sin t$; $z = t$ so that

$$\int_C \vec{F}\cdot d\vec{C} = \int_C (\cos^2 t, \sin^2 t, t^2)\cdot(-\sin t, \cos t, 1)\,dt$$

$$= \int_C [-\sin t\,\cos^2 t + \sin^2 t\,\cos t + t^2]\,dt. \quad (1)$$

For the limits of integration it is given that $0 \le t \le 2\pi$ so that (2) becomes

$$\int_0^{2\pi} -\sin t\,\cos^2 t\,dt + \int_0^{2\pi} \sin^2 t\,\cos t\,dt$$

$$+ \int_0^{2\pi} t^2 dt. \qquad (2)$$

To solve the first integral let $u = \cos t$ $du = -\sin t\, dt$ and upon substitution this yields $\int_1^1 u^2 du = 0$. For the second integral let $u = \sin t$ $du = \cos t\, dt$ so that upon substitution this gives $\int_0^0 u^2 du = 0$. The third integral,

$$\int_0^{2\pi} t^2 dt = \left.\frac{t^3}{3}\right|_0^{2\pi} = \frac{8\pi^3}{3}.$$

Hence (2) equals $\frac{8\pi^3}{3}$, which is the value of the line integral.

b) In part (b) the problem is to find the line integral of a function over a curve. By definition this is $\int_C f\, d\vec{c}$ where f is the function and $d\vec{c} = \vec{C}'(t) dt$. Since $f(x,y,z) = (x+y+z)$ we have

$$\int_C f\, d\vec{c} = \int_0^{2\pi} (\cos t + \sin t + t)(-\sin t\, \vec{i} + \cos t\, \vec{j} + \vec{k}) dt$$

$$= \left[\int_0^{2\pi} -\sin t(\cos t + \sin t + t) dt\right]\vec{i}$$

$$+ \left[\int_0^{2\pi} \cos t(\cos t + \sin t + t) dt\right]\vec{j}$$

$$+ \left[\int_0^{2\pi} (\cos t + \sin t + t) dt\right]\vec{k} \qquad (3)$$

Solving the first integral

$$-\int_0^{2\pi} \sin t \cos t\, dt - \int_0^{2\pi} \sin^2 t\, dt - \int_0^{2\pi} t \sin t\, dt$$

$$= 0 - \left(\int_0^{2\pi} \frac{1-\cos 2t}{2} dt + \int_0^{2\pi} t \sin t\, dt\right) \text{ since } \sin^2 t = \frac{1-\cos 2t}{2}$$

$$= -\left(\pi + \int_0^{2\pi} t \sin t\, dt\right) \text{ which can be solved by}$$

parts letting $u = t$; $dv = \sin t\, dt$ and $du = dt$; $v = -\cos t$ so that this equals

$$-\left(\pi - t \cos t\Big|_0^{2\pi} + \int_0^{2\pi} \cos t\, dt\right)$$

$$= -[\pi - 2\pi + 0] = \pi. \tag{4}$$

Solving the second integral in (3),

$$\int_0^{2\pi} \cos^2 t \, dt + \int_0^{2\pi} \cos t \sin t \, dt + \int_0^{2\pi} t \cos t \, dt$$

$$= \int_0^{2\pi} \frac{1+\cos 2t}{2} dt + 0 + \int_0^{2\pi} t \cos t \, dt$$

$$= \pi + \int_0^{2\pi} t \cos t \, dt = \pi + \left[t \sin t \Big|_0^{2\pi} - \int_0^{2\pi} \sin t \, dt \right]$$

$$= \pi + 0 = \pi. \tag{5}$$

Solving the last integral in (3)

$$\int_0^{2\pi} \cos t \, dt + \int_0^{2\pi} \sin t \, dt + \int_0^{2\pi} t \, dt = 0 + 0 + \frac{t^2}{2} \Big|_0^{2\pi}$$

$$= 2\pi^2. \tag{6}$$

Hence upon substitution of (4), (5), and (6) in (3), we have (3) equals $\pi \hat{i} + \pi \hat{j} + 2\pi^2 \hat{k}$ which is the value of the line integral.

Notice that the value of the line integral of a vector field over a curve is a scalar and that the value of the line integral of a real-valued function over a curve is a vector.

• **PROBLEM 8-9**

Let $\vec{F}(x,y) = (cxy, x^6 y^2)$ where c is a positive constant. Let a,b be numbers >0. Find a value of a in terms of c such that the line integral of the vector field $\vec{F}$ along $y = ax^b$ from (0,0) to the line x = 1 is independent of b.

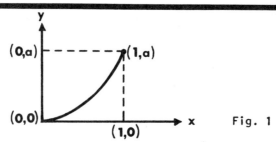

Fig. 1

Solution: The problem is to compute $\int_C \vec{F} \cdot d\vec{c}$ where C is the curve $y = ax^b$ from (0,0) to x = 1. The curve is graphed in Figure 1 and at $x = 1$, $y = a$.

Thus the integral of the vector field is on the curve from $(0,0)$ to $(1,a)$. Since $y = f(x)$ use x as the parameter, letting $x=x$, $y=ax^b$ so that $dx=dx$ and $dy=bax^{b-1}dx$ and

$$\int_C \vec{F} \cdot d\vec{c} = \int_C (cxy, x^6y^2) \cdot (dx, dy) = \int_C cxydx + x^6y^2 dy$$

$$= \int_0^1 cx(ax^b)dx + x^6(ax^b)^2 bax^{b-1}dx$$

$$= \int_0^1 cax^{b+1}dx + a^3bx^{2b}x^6x^{b-1}dx$$

$$= \int_0^1 cax^{b+1}dx + a^3bx^{2b+b-1+6}dx$$

$$= \int_0^1 (acx^{b+1} + a^3bx^{3b+5})dx$$

$$= \left[\frac{acx^{b+2}}{b+2} + \frac{a^3bx^{3b+6}}{3b+6}\right]_0^1 = \frac{ac}{b+2} + \frac{a^3b}{3b+6}$$

$$= \frac{ac}{b+2} + \frac{a^3b}{3(b+2)} = \frac{3ac + a^3b}{3(b+2)} . \qquad (1)$$

Now the problem is to find a value of a in terms of c such that the line integral is independent of b. The function $\frac{3ac + a^3b}{3(b+2)} \equiv I(b)$ converges to $\frac{a^3}{3}$ as b tends to $+\infty$ and converges to $\frac{ac}{2}$ as b tends to 0. We need $I(b) =$ constant for all $b>0$ hence $\lim_{b\to 0+} I(b) = \lim_{b\to\infty} I(b) =$ constant.

This gives $\frac{a^3}{3} = \frac{ac}{2}$ or $a = \sqrt{\frac{3c}{2}}$ (since $a>0$).

METHOD OF FINDING POTENTIAL FUNCTION (EXACT DIFFERENTIAL)

• **PROBLEM 8-10**

a) Let $\vec{F}$ be a vector field on some open set V. Assume that for some function ϕ on V, $\vec{F} = \overrightarrow{\text{grad}\phi}$. Let P and Q be two points in V and let C be a curve joining these two points. Prove

$$\int_C \vec{F} \cdot d\vec{c} = \phi(Q) - \phi(P).$$

b) Evaluate $\int_C \dfrac{x}{x^2-y^2} dx + \dfrac{y}{y^2-x^2} dy$ where C is a curve from (1,0) to (5,3) and lies between the lines y=x and y=-x, by the above method.

Solution: a) Let C be a curve on the interval [a,b], such that C(a) = P and C(b) = Q. By the definition of the line integral

$$\int_{C^P}^Q \vec{F} \cdot d\vec{c} = \int_a^b \vec{F}(\vec{C}(t)) \cdot \vec{C}'(t) dt \qquad (1)$$

because C is parametrized as $\vec{C}(t)$ and $d\vec{c} = \vec{C}'(t)dt$ since $\vec{C}'(t) = \dfrac{d\vec{c}}{dt}$. From the given information $\vec{F} = \overrightarrow{\text{grad}}\phi$ so that $\vec{F}(\vec{C}(t)) = \overrightarrow{\text{grad}}\phi(\vec{C}(t))$. Thus (1) equals

$$\int_a^b \overrightarrow{\text{grad}}\phi(\vec{C}(t)) \cdot \vec{C}'(t) dt. \qquad (2)$$

Next define a function g by $g(t) = \phi(\vec{C}(t))$, so that the derivative of g(t) with respect to t is: $g'(t) = \overrightarrow{\text{grad}}\phi(\vec{C}(t)) \cdot \vec{C}'(t)dt$ by the Chain Rule. Then by noticing that the expression inside the integral equals g'(t), (2) becomes

$$\int_a^b g'(t) dt = g(b) - g(a) = \phi(C(b)) - \phi(C(a))$$

$$= \phi(Q) - \phi(P) \qquad \text{which is the}$$

desired result and thus completes the proof. ϕ is known as the potential function of the vector field and the above method evaluates a line integral when the vector field has a potential function.

b) To evaluate this line integral by the above method first find a potential function for the vector field $\vec{F}(x,y) = \left(\dfrac{x}{x^2-y^2}, \dfrac{y}{y^2-x^2}\right)$. Since $\vec{F}(x,y) = \overrightarrow{\text{grad}}\phi(x,y)$ a function ϕ is required such that $\dfrac{\partial \phi}{\partial x} = \dfrac{x}{x^2-y^2}$ and $\dfrac{\partial \phi}{\partial y} = \dfrac{y}{y^2-x^2}$. $\qquad (3)$

Hence integrate with respect to x first and obtain $\phi(x,y)$
$$= \int \dfrac{x}{x^2-y^2} dx + H(y), \text{ where } H(y) \text{ is some function of y.}$$

Then this can be rewritten as $\phi(x,y) = \int x(x^2-y^2)^{-1} dx + H(y)$ which can be evaluated by letting $u = (x^2-y^2)$; $du = 2xdx$. This yields

$$\phi(x,y) = \frac{1}{2}\int u^{-1} du + H(y) = \frac{1}{2}\log(x^2-y^2) + H(y). \quad (4)$$

Now from (4) $\quad \frac{\partial \phi}{\partial y} = \frac{-y}{x^2-y^2} + H'(y) = \frac{y}{y^2-x^2} + H'(y)$.

But from (3) $\frac{\partial \phi}{\partial y} = \frac{y}{y^2-x^2}$. Hence take $H(y) = 0$ and therefore the potential function is $\phi(x,y) = \frac{1}{2}\log(x^2-y^2)$ (5)

(Note: for $\phi(x,y)$ we need $x^2 > y^2$). Hence by the method in part (a) this gives:

$$\int \frac{x}{x^2-y^2} dx + \frac{y}{y^2-x^2} dy = \phi(Q) - \phi(P) = \phi(5,3) - \phi(1,0)$$

which equals by use of equation (5):

$$\frac{1}{2}\log(25-9) - \frac{1}{2}\log(1-0)$$

$$= \frac{1}{2}\log 16 - 0 = \frac{1}{2}\log(4)^2 = \log 4.$$

Remark: Observe that we actually had $H'(y) = 0$ so $H =$ constant and the constant was chosen to be zero. However the value of the constant is irrelevant because the final answer involves only the difference in ϕ.

• **PROBLEM 8-11**

Does the vector field $\vec{F}(x,y) = \left(\frac{-y}{x^2+y^2}, \frac{x}{x^2+y^2}\right)$ have a potential function? Explain why or why not. Is this vector field exact?

Solution: A vector field $\vec{F}(x,y)$ is said to have a potential function if there exists a function $\phi(x,y)$ such that $\vec{\nabla}\phi = \vec{F}$ (where $\vec{\nabla}\phi = \left(\frac{\partial \phi}{\partial x}, \frac{\partial \phi}{\partial y}\right)$). Since $\vec{F}: R^2 \to R^2$, it may be rewritten as $\vec{F}(x,y) = (f(x,y), g(x,y))$. Then

$$\frac{\partial f}{\partial y} = \frac{-(x^2+y^2) + 2y^2}{(x^2+y^2)^2} = \frac{y^2-x^2}{(x^2+y^2)^2}$$

$$\frac{\partial g}{\partial x} = \frac{(x^2+y^2) - 2x^2}{(x^2+y^2)^2} = \frac{y^2-x^2}{(x^2+y^2)^2}.$$

Therefore since $\frac{\partial f}{\partial y} = \frac{\partial g}{\partial x}$ it would seem that this vector field has a potential function according to the following theorem: Let f,g be differentiable functions on an open set of the plane. If the open set is a rectangle or the entire plane, if the partial derivatives of f,g exist and are continuous and if $\frac{\partial f}{\partial y} = \frac{\partial g}{\partial x}$ then the vector field $\vec{F}(x,y) = (f(x,y),g(x,y))$ has a potential function; but this is not true since this vector field is not defined at the origin and does not have partial derivatives there. Hence the above theorem does not apply. Another theorem may be applied which states: Let $\vec{F}$ be a vector field on an open set V. If $\vec{F}$ has a potential function, then $\oint_C \vec{F} = 0$ for all closed paths in V. If there exists a closed path C in V such that $\oint_C \vec{F} \neq 0$ then $\vec{F}$ does not have a potential function.

This theorem can be proven by letting C be a closed path whose beginning and end point is the same point Q. If a potential function ϕ exists then

$$\int_C \vec{F} \cdot d\vec{c} = \phi(Q) - \phi(Q) = 0. \quad \text{Thus if } \int_C \vec{F} \cdot d\vec{c} \neq 0 \text{ then}$$

there cannot exist a potential function.

Therefore integrate $\vec{F}$ around a circle $\vec{C}(t) = $ (r cos t, r sin t) $0 \leq t \leq 2\pi$ to obtain:

$$\int_C \vec{F} \cdot d\vec{c} = \int_0^{2\pi} \left(\frac{-r \sin t}{r^2}, \frac{r \cos t}{r^2} \right) \cdot (-r \sin t, r \cos t) dt$$

$$= \int_0^{2\pi} \left(\frac{r^2 \sin^2 t}{r^2} + \frac{r^2 \cos^2 t}{r^2} \right) dt = \int_0^{2\pi} dt = 2\pi.$$

By the above theorem $\vec{F}(x,y) = \left(\frac{-y}{x^2+y^2}, \frac{x}{x^2+y^2} \right)$ does not have a potential function since there exists a closed path C such that $\oint_C \vec{F} \cdot d\vec{c} \neq 0$.

To determine if the vector field is exact apply the theorem: Let $\vec{F}$ be a continuous vector field defined in an open connected S. Then $\vec{F}$ is exact if and only if for every connected path C in S, $\int_C \vec{F}$ is determined by the endpoints of C. Also if C is a closed path and if $\vec{F}$ is exact then $\int_C \vec{F} = 0$.

Thus $\vec{F}$ is not exact since around a circle (closed path) the integral is not equal to 0.

● **PROBLEM 8-12**

Let $\vec{F}(x,y) = (x^2y, xy^2)$.

a) Determine whether this vector field has a potential function.

b) Evaluate the integral of $\vec{F}$ from O to the point P on Figure 1, along the line segment from (0,0) to $\left(\frac{\sqrt{2}}{2}, \frac{\sqrt{2}}{2}\right)$.

c) Evaluate the integral of $\vec{F}$ from O to P along the path that consists of the line segment from (0,0) to (1,0) and along the circular arc from (1,0) to P.

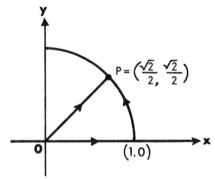

Fig. 1

Solution: a) Since $\vec{F}(x,y) = (x^2y, xy^2)$, $\frac{\partial f}{\partial y} = x^2$ and $\frac{\partial g}{\partial x} = y^2$. Therefore since $\frac{\partial f}{\partial y} \neq \frac{\partial g}{\partial x}$ there cannot be a potential function.

b) Here integration is over the line segment that is part of the line x=y, from (0,0) to $\left(\frac{\sqrt{2}}{2}, \frac{\sqrt{2}}{2}\right)$. So apply the parametrization x=x, y=x, dx=dx, dy=dx to obtain

$$\int_C x^2y\,dx + xy^2\,dy = \int_0^{\frac{\sqrt{2}}{2}} x^3\,dx + x^3\,dx = \left.\frac{2x^4}{4}\right|_0^{\frac{\sqrt{2}}{2}}$$

$$= \frac{\left(\frac{\sqrt{2}}{2}\right)^4}{2} = \frac{4}{32} = \frac{1}{8}.$$

c) Here C consists of two paths. The first path is the line segment y=0 from (0,0) to (1,0). Therefore use the parameterization x=x, y=0 so that dx=dx, dy=0 and

$$\int_{C_1} 0 \cdot dx + 0 = 0. \qquad (1)$$

357

The second segment is the arc of the circle $x^2+y^2=1$ from the point $(1,0)$ to $\left(\frac{\sqrt{2}}{2}, \frac{\sqrt{2}}{2}\right)$. Here apply the parameterization $x = \cos\theta$, $y = \sin\theta$. To find the range of θ,

$$\tan\theta = \frac{y}{x} \quad \text{so at } (1,0) \quad \tan\theta = \frac{0}{1} = 0; \quad \theta = 0$$

$$\text{at } \left(\frac{\sqrt{2}}{2}, \frac{\sqrt{2}}{2}\right) \quad \tan\theta = \frac{\frac{\sqrt{2}}{2}}{\frac{\sqrt{2}}{2}} = 1, \quad \theta = \frac{\pi}{4}.$$

Thus,

$$\int_{C_2} x^2 y\,dx + xy^2\,dy = \int_0^{\frac{\pi}{4}} \cos^2\theta \sin\theta(-\sin\theta)d\theta + \cos\theta \sin^2\theta \cos\theta\, d\theta$$

$$= \int_0^{\frac{\pi}{4}}(-\cos^2\theta \sin^2\theta + \cos^2\theta \sin^2\theta)d\theta = \int_0^{\frac{\pi}{4}} 0\, d\theta = 0. \qquad (2)$$

Hence $\int_C \vec{F}\cdot d\vec{c}$ equals $(1) + (2) = 0 + 0 = 0$.

Note that the value of the line integral over the path in part (c) is different from the value found in part (b). Therefore the line integral is not independent of the path and this is because $\vec{F}$ is not exact.

● **PROBLEM 8-13**

a) Find the value of the line integral of the vector field $\vec{F}(x,y) = (y,x)$ over the curve $\vec{C}(t) = (r\cos t, r\sin t)$, $0 \le t \le \frac{\pi}{4}$; both directly and by finding a potential function.

b) Repeat for $\vec{C}(t) = (3\cos t, 3\sin t)$, $0 \le t \le \frac{\pi}{6}$.

Solution: First find $\int_C \vec{F}\cdot d\vec{c}.$ (1)

Here $\vec{F}(x,y) = (y,x)$ and $\vec{C}(t) = (r\cos t, r\sin t)$ so that $\vec{C}'(t) = (-r\sin t, r\cos t)dt$. Hence upon substitution into (1)

$$\int_C (y,x)\cdot(dx,dy) = \int_C (r\sin t, r\cos t)\cdot(-r\sin t, r\cos t)dt$$

$$= \int_0^{\frac{\pi}{4}}(-r^2\sin^2 t + r^2\cos^2 t)dt$$

$$= -r^2 \int_0^{\frac{\pi}{4}} \sin^2 t \, dt + r^2 \int_0^{\frac{\pi}{4}} \cos^2 t \, dt. \tag{2}$$

To solve the first integral use the trigonometric substitution $\sin^2 t = \frac{1-\cos 2t}{2}$ to obtain

$$-r^2 \int_0^{\frac{\pi}{4}} \sin^2 t \, dt = -r^2 \left[\int_0^{\frac{\pi}{4}} \frac{1}{2} dt - \int_0^{\frac{\pi}{4}} \frac{\cos 2t}{2} dt \right]$$

$$= -r^2 \left[\frac{\pi}{8} - \frac{\sin 2t}{4} \right]_0^{\frac{\pi}{4}} = -r^2 \left[\frac{\pi}{8} - \frac{1}{4} \right] = \frac{-r^2 \pi}{8} + \frac{r^2}{4}.$$

For the second integral use the substitution $\cos^2 t = \frac{1+\cos 2t}{2}$ to obtain

$$r^2 \int_0^{\frac{\pi}{4}} \cos^2 t \, dt = r^2 \left[\int_0^{\frac{\pi}{4}} \frac{1}{2} dt + \int_0^{\frac{\pi}{4}} \frac{\cos 2t}{2} dt \right]$$

$$= r^2 \left[\frac{\pi}{8} + \frac{\sin 2t}{4} \right]_0^{\frac{\pi}{4}} = r^2 \left[\frac{\pi}{8} + \frac{1}{4} \right] = \frac{r^2 \pi}{8} + \frac{r^2}{4}.$$

Therefore (2) equals $\frac{-r^2 \pi}{8} + \frac{r^2}{4} + \frac{r^2 \pi}{8} + \frac{r^2}{4} = \frac{r^2}{2}$.

Since $\vec{F}(x,y) = (y,x)$ $\quad \frac{\partial f}{\partial y} = 1$ and $\frac{\partial g}{\partial x} = 1$;

Hence since $\frac{\partial f}{\partial y} = \frac{\partial g}{\partial x}$ a potential function exists. The problem is to find a function $\phi(x,y)$ such that $\vec{F} = \text{grad} \phi$
$\vec{\nabla} \phi = \left(\frac{\partial \phi}{\partial x}, \frac{\partial \phi}{\partial y} \right),$

$$\frac{\partial \phi}{\partial x} = y \quad \text{and} \quad \frac{\partial \phi}{\partial y} = x. \tag{3}$$

Integrating with respect to x first gives $\phi(x,y) = \int y \, dx + h(y)$ which equals

$$\phi(x,y) = yx + h(y). \tag{4}$$

Now from (3) $\frac{\partial \phi}{\partial y} = x$ and from (4) $\frac{\partial \phi}{\partial y} = x + h'(y)$.

Letting $h(y) = 0$ $\quad\quad \phi(x,y) = xy \tag{5}$

Therefore since $\vec{F} = \text{grad} \phi$:

$$\int_C \vec{F} \cdot d\vec{c} = \int_P^Q \vec{F} \cdot d\vec{c} = \phi(Q) - \phi(P). \qquad (6)$$

To find the points Q,P, $0 \le t \le \frac{\pi}{4}$ and $P = (x_0, y_0) = $ (r cos 0, r sin 0) = (r,0) and $Q = (x_1, y_1) = (r \cos \frac{\pi}{4},$ r sin $\frac{\pi}{4}) = \left(\frac{r}{\sqrt{2}}, \frac{r}{\sqrt{2}}\right)$. Substituting these values into (5) and (6)

$$\int_C \vec{F} \cdot d\vec{c} = \phi\left(\frac{r}{\sqrt{2}}, \frac{r}{\sqrt{2}}\right) - \phi(r,0) = \frac{r}{\sqrt{2}} \cdot \frac{r}{\sqrt{2}} - 0 \cdot r = \frac{r^2}{2},$$

which agrees with the value found directly.

b) Repeating for $\vec{C}(t) = $ (3cos t, 3sin t), $0 \le t \le \frac{\pi}{6}$

$$\int_C \vec{F} \cdot d\vec{c} = \int_C (y,x) \cdot (dx, dy)$$

$$= \int_C (3\sin t, 3\cos t) \cdot (-3\sin t, 3\cos t) dt$$

$$= \int_0^{\frac{\pi}{6}} (-9\sin^2 t + 9\cos^2 t) dt$$

$$= -9 \int_0^{\frac{\pi}{6}} \sin^2 t \, dt + 9 \int_0^{\frac{\pi}{6}} \cos^2 t \, dt$$

$$= -9 \left[\int_0^{\frac{\pi}{6}} \frac{dt}{2} - \int_0^{\frac{\pi}{6}} \frac{\cos 2t}{2} dt \right] + 9 \left[\int_0^{\frac{\pi}{6}} \frac{dt}{2} + \int_0^{\frac{\pi}{6}} \frac{\cos 2t}{2} dt \right]$$

$$= -9 \frac{\pi}{12} + 9 \frac{\sqrt{3}}{8} + 9 \frac{\pi}{12} + 9 \frac{\sqrt{3}}{8} = \frac{18\sqrt{3}}{8} = \frac{9\sqrt{3}}{4}. \qquad (7)$$

Since $\phi(x,y) = xy$ and $P = (x_0, y_0) = (3\cos 0, 3\sin 0) = (3,0)$ and $Q = (x_1, y_1) = \left(3\cos \frac{\pi}{6}, 3\sin \frac{\pi}{6}\right) = \left(\frac{3\sqrt{3}}{2}, \frac{3}{2}\right)$ from (5),(6)

$$\int_P^Q \vec{F} \cdot d\vec{c} = \phi(Q) - \phi(P) = \phi\left(\frac{3\sqrt{3}}{2}, \frac{3}{2}\right) - \phi(3,0)$$

$$= \frac{3\sqrt{3}}{2} \cdot \frac{3}{2} - 3 \cdot 0 = \frac{9\sqrt{3}}{4}$$ which agrees with (7), the

value found directly.

• PROBLEM 8-14

a) Find the integral of the vector field $\vec{F}(x,y) = \left(\dfrac{x}{r^3}, \dfrac{y}{r^3}\right)$, where $r = (x^2+y^2)^{\frac{1}{2}}$, along the curve $\vec{C}(t) = (e^t \cos t, e^t \sin t)$ from the point $(1,0)$ to the point $(e^{2\pi}, 0)$.

b) Does $\vec{F}$ admit a potential function? If so, compute the integral in a) by this method.

Solution: a) The integral $\int_C \vec{F} \cdot d\vec{c}$ is to be found. Since $\vec{F}(x,y) = \left(\dfrac{x}{r^3}, \dfrac{y}{r^3}\right)$ where $r = (x^2+y^2)^{\frac{1}{2}}$ the required integral is

$$\int_{C(t)} \dfrac{x}{(x^2+y^2)^{\frac{3}{2}}} dx + \dfrac{y}{(x^2+y^2)^{\frac{3}{2}}} dy. \qquad (1)$$

The curve is given parametrically as $x = e^t \cos t$, $y = e^t \sin t$. Hence $dx = (-e^t \sin t + e^t \cos t)dt$ and $dy = (e^t \cos t + e^t \sin t)dt$. The parameter t, varies from 0 to 2π for $e^t \cos t = 1$ when $t = 0$ and $e^t \cos t = e^{2\pi}$ when $t = 2\pi$. Substituting these values into (1) (with

$$(x^2+y^2)^{\frac{3}{2}} = [e^{2t}(\cos^2 t + \sin^2 t)]^{\frac{3}{2}}\bigg)$$

$$\int_0^{2\pi} \dfrac{e^t \cos t}{(e^{2t})^{\frac{3}{2}}} (-e^t \sin t + e^t \cos t)dt$$

$$+ \dfrac{e^t \sin t}{(e^{2t})^{\frac{3}{2}}} (e^t \cos t + e^t \sin t)dt$$

$$= \int_0^{2\pi} \dfrac{-e^{2t} \cos t \sin t + e^{2t} \cos^2 t + e^{2t} \sin t \cos t + e^{2t} \sin^2 t}{(e^{2t})^{\frac{3}{2}}} dt$$

$$= \int_0^{2\pi} \dfrac{e^{2t}(\cos^2 t + \sin^2 t)}{(e^{2t})^{\frac{3}{2}}} dt = \int_0^{2\pi} \dfrac{e^{2t}}{e^{3t}} dt$$

$$= \int_0^{2\pi} \dfrac{1}{e^t} dt = \int_0^{2\pi} e^{-t} dt = -e^{-t}\bigg|_0^{2\pi} = -e^{-2\pi} - (-1)$$

$$= 1 - e^{-2\pi}. \qquad (2)$$

b) Since the curve is continuous from $(1,0)$ to $(e^{2\pi},0)$ and there are no holes in the region, for the vector field to admit a potential function $\frac{\partial f}{\partial y}$ must equal $\frac{\partial g}{\partial x}$.

Since $F = \left(\dfrac{x}{(x^2+y^2)^{\frac{3}{2}}}, \dfrac{y}{(x^2+y^2)^{\frac{3}{2}}} \right) = (f, g)$

$$\frac{\partial f}{\partial y} = \frac{-\frac{3}{2}x\,(x^2+y^2)^{\frac{1}{2}}2y}{(x^2+y^2)^3} \quad \text{and} \quad \frac{\partial g}{\partial x} = \frac{-\frac{3}{2}y\,(x^2+y^2)^{\frac{1}{2}}2y}{(x^2+y^2)^3}$$

Thus the vector field does admit a potential function.

To find it let $\dfrac{\partial \phi}{\partial y} = \dfrac{y}{(x^2+y^2)^{\frac{3}{2}}}$ and $\dfrac{\partial \phi}{\partial x} = \dfrac{x}{(x^2+y^2)^{\frac{3}{2}}}$.

Integrating for x first $\phi(x,y) = \displaystyle\int \dfrac{x}{(x^2+y^2)^{\frac{3}{2}}} dx + h(y)$.

To solve the integral let $u = x^2+y^2$; $du = 2x\,dx$ and

$$\phi(x,y) = \frac{1}{2}\int u^{-\frac{3}{2}} du + h(y) = -u^{-\frac{1}{2}} + h(y).$$

$$\phi(x,y) = -(x^2+y^2)^{-\frac{1}{2}} + h(y). \tag{3}$$

But since $\dfrac{\partial \phi}{\partial y} = \dfrac{y}{(x^2+y^2)^{3/2}}$ and from (3) $\dfrac{\partial \phi(x,y)}{\partial y} =$

$\dfrac{y}{(x^2+y^2)^{3/2}} + h'(y)$, or $h(y) = 0$. Thus the potential

function is $\phi(x,y) = -(x^2+y^2)^{-\frac{1}{2}}$. Now using the theorem that states that if a potential function for the vector field $\vec{F}$ exists then $\displaystyle\int_C \vec{F}\cdot d\vec{c} = \phi(Q) - \phi(P)$,

$$\int_{(1,0)}^{(e^{2\pi},0)} \vec{F}\cdot d\vec{c} = \phi(e^{2\pi},0) - \phi(1,0)$$

$$ = -(e^{4\pi}+0)^{\frac{1}{2}} - \left[-(1+0)^{-\frac{1}{2}}\right] = -e^{-2\pi} + 1$$

$$ = 1 - e^{-2\pi}.$$

• PROBLEM 8-15

Let $\vec{F}(x,y) = \left(\dfrac{-y}{x^2+y^2}, \dfrac{x}{x^2+y^2}\right)$ on U, the plane from which the origin has been deleted. Find the integral of $\vec{F}$ along the path shown in Figure 1, between the points (1,0) and (0,1) by two different methods.

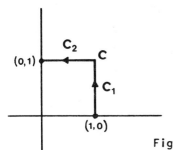

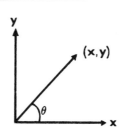

Fig. 2

Fig. 1

Solution: For the first method evaluate $\int_C \vec{F} \cdot d\vec{c}$ directly.
Since $C = \{C_1, C_2\}$

$$\int_C \vec{F} \cdot d\vec{c} = \int_{C_1} \vec{F} \cdot d\vec{c} + \int_{C_2} \vec{F} \cdot d\vec{c}$$

Along C_1 apply the parametrization x=1, y=y so that dx=0, dy=dy and

$$\int_{C_1} \vec{F} \cdot d\vec{c} = \int_{C_1} \dfrac{-y}{x^2+y^2} dx + \dfrac{x}{x^2+y^2} dy = \int_0^1 \dfrac{-y}{1+y^2} \cdot 0 + \dfrac{1}{1+y^2} dy$$

$$= \int_0^1 \dfrac{1}{1+y^2} dy.$$

Now since $\int \dfrac{1}{a^2+y^2} dy = \dfrac{1}{a} \tan^{-1}\left(\dfrac{y}{a}\right)$

$$\int_{C_1} \vec{F} \cdot d\vec{c} = \tan^{-1}\left(\dfrac{y}{1}\right)\Big|_0^1 = \tan^{-1}(1) - \tan^{-1}(0) = \dfrac{\pi}{4} - 0 = \dfrac{\pi}{4}. \quad (1)$$

Along C_2, use the parametrization x=x, y=1 and dx=dx; dy=0 so that

$$\int_{C_2} \vec{F} \cdot d\vec{c} = \int_{C_2} \dfrac{-y}{x^2+y^2} dx + \dfrac{x}{x^2+y^2} dy = \int_1^0 \dfrac{-1}{x^2+1} dx$$

$$= -\int_1^0 \dfrac{1}{x^2+1} dx = -\tan^{-1}\left(\dfrac{x}{1}\right)\Big|_1^0 = -\tan^{-1} 0 + \tan^{-1}(1)$$

$$= 0 + \dfrac{\pi}{4} = \dfrac{\pi}{4}. \quad (2)$$

Summing (2) and (3)

$$\int_C \vec{F}\cdot d\vec{c} = \frac{\pi}{4} + \frac{\pi}{4} = \frac{\pi}{2}. \tag{3}$$

For the second method we try to find a function ϕ such that $\vec{F} = \text{grad}\phi$ and

$$\int_P^Q pdx + qdy = \phi(Q) - \phi(P). \tag{4}$$

Since the origin is not included in the plane U, if $\frac{\partial f}{\partial y} = \frac{\partial g}{\partial x}$, then $\vec{F}$ has a potential function ϕ on U. Let

$$f(x,y) = \frac{-y}{x^2+y^2}, \quad g(x,y) = \frac{x}{x^2+y^2}, \quad \frac{\partial f}{\partial y} = \frac{-(x^2+y^2)+2y^2}{(x^2+y^2)^2}$$

$$= \frac{y^2-x^2}{(x^2+y^2)^2}$$

$$\frac{\partial g}{\partial x} = \frac{(x^2+y^2)-2x^2}{(x^2+y^2)^2} = \frac{y^2-x^2}{(x^2+y^2)^2} \quad \text{so} \quad \frac{\partial f}{\partial y} = \frac{\partial g}{\partial x} \quad \text{and a}$$

potential function ϕ exists on U. Note that

$$\frac{\partial \phi}{\partial x} = \frac{-y}{x^2+y^2}, \quad \frac{\partial \phi}{\partial y} = \frac{x}{x^2+y^2} \tag{5}$$

Integrating with respect to y,

$$\phi(x,y) = \int \frac{x}{x^2+y^2} dy + H(x) \quad (H(x) \text{ some function of } x)$$

Hence $\phi(x,y) = \tan^{-1}\left(\frac{y}{x}\right) + H(x)$.

Now $\quad \frac{\partial \phi(x,y)}{\partial x} = \frac{-y}{x^2+y^2} + H'(x) \quad$ by the chain rule,

letting $u = \frac{y}{x}$ and using $\frac{d(\tan^{-1}u)}{du} = \frac{1}{1+u^2}$.

Hence we can take $H(x) = 0$ so the above equals $\frac{\partial \phi}{\partial y}$ in (5). Therefore

$$\phi(x,y) = \tan^{-1}\left(\frac{y}{x}\right). \tag{6}$$

Now if θ is defined as in Figure 2

$$\tan\theta = \frac{y}{x} \quad 0 \le \theta \le \frac{\pi}{2}, \quad \theta = \tan^{-1}\left(\frac{y}{x}\right).$$

Thus (6) may be rewritten as $\phi(x,y) = \theta \quad 0 \le \theta \le \frac{\pi}{2}$.

Then by (4)

$$\int_{(1,0)}^{(0,1)} \frac{-y}{x^2+y^2} dx + \frac{x}{x^2+y^2} dy = \phi(0,1) - \phi(1,0) = \frac{\pi}{2} - 0 = \frac{\pi}{2}$$

which agrees with (3), the answer by the first method.

● **PROBLEM 8-16**

Evaluate the following line integrals:

a) $\int_{(1,-2)}^{(3,4)} \frac{ydx-xdy}{x^2}$ on the line $y = 3x-5$

b) $\int_{(0,2)}^{(1,3)} \frac{3x^2}{y} dx - \frac{x^3}{y^2} dy$ on the parabola $y = 2+x^2$

c) $\int_{(0,0)}^{(2,8)} \vec{\nabla}f \cdot d\vec{c}$ where $\vec{\nabla}f$ is grad f and f is the function $f(x,y) = x^2-y^2$. C is the curve $y = x^3$.

Solution: a) The function $\frac{ydx-xdy}{x^2}$ may be expressed as

$$\vec{F}(x,y) = \left(\frac{y}{x^2}, \frac{-x}{x^2}\right) = (f(x,y), g(x,y)).$$ If there exists a function $\phi(x,y)$ such that $\vec{\nabla}\phi = \left(\frac{\partial\phi}{\partial x}, \frac{\partial\phi}{\partial y}\right) = (f(x,y), g(x,y))$ then the given vector field has a potential function and

$$\int_{(1,-2)}^{(3,4)} \frac{ydx-xdy}{x^2} = \phi(3,4) - \phi(1,2).$$

Since $\frac{\partial^2\phi}{\partial y \partial x} = \frac{\partial^2\phi}{\partial x \partial y}$, an equivalent way of proving the existence of the potential function is to show that $\frac{\partial f}{\partial y} = \frac{\partial g}{\partial x}$. For the given vector field, $\vec{F}=(f,g)=\left(\frac{y}{x^2}, \frac{-x}{x^2}\right)$, $\frac{\partial f}{\partial y} = \frac{1}{x^2}$

and $\frac{\partial g}{\partial x} = \frac{1}{x^2}$. Therefore $\frac{\partial f}{\partial y} = \frac{\partial g}{\partial x}$. Thus the given vector field has a potential function $\phi(x,y)$ on the given segment of the line ($x \neq 0$, since $1 \leq x \leq 3$). To find this function $\phi(x,y)$ first note that

$$\frac{\partial\phi}{\partial x} = \frac{y}{x^2} ; \quad \frac{\partial\phi}{\partial y} = \frac{-x}{x^2} = \frac{-1}{x}. \tag{1}$$

Then integrate with respect to x to obtain

$$\phi(x,y) = \int \frac{y}{x^2} dx + H(y) \quad (H(y) \text{ some function of } y).$$

Then $\phi(x,y) = \frac{-y}{x} + H(y)$. Next find $\frac{\partial \phi}{\partial y}$: (2)

$\frac{\partial \phi}{\partial y} = \frac{-1}{x} + H'(y)$. But from (1) $\frac{\partial \phi}{\partial y} = \frac{-1}{x}$.

Setting $H'(y)$ equal to zero, $\phi(x,y) = \frac{-y}{x}$ (assuming $H(y) = 0$). (3)

Hence $\int_{(1,-2)}^{(3,4)} \frac{ydx - xdy}{x^2} = \phi(3,4) - \phi(1,-2)$

$= \frac{-4}{3} - 2 = \frac{-10}{3}$ (by substitution into (3)).

b) Since $\vec{F} = \left(\frac{3x^2}{y}, \frac{-x^3}{y^2}\right)$, $\frac{\partial f}{\partial y} = \frac{-3x^2}{y^2}$, $\frac{\partial g}{\partial x} = \frac{-3x^2}{y^2}$.

Therefore since $\frac{\partial f}{\partial y} = \frac{\partial g}{\partial x}$, $\vec{F}$ has a potential function

$\phi(x,y)$ such that $\frac{\partial \phi}{\partial x} = \frac{3x^2}{y}$, $\frac{\partial \phi}{\partial y} = \frac{-x^3}{y^2}$. (4)

(Note: $\vec{F}$ is not defined at $y=0$. However, the line integral is from the points $(0,2)$ to $(1,3)$ along $y = 2+x^2$, so $y \neq 0$ on this path). Integrate $\phi(x,y)$ with respect to x first to obtain $\phi(x,y) = \int \frac{3x^2}{y} dx + H(y) = \frac{x^3}{y} + H(y)$. (5)

Then $\frac{\partial \phi}{\partial y} = \frac{-x^3}{y^2} + H'(y)$. But from (4) $\frac{\partial \phi}{\partial y} = \frac{-x^3}{y^2}$.

Thus $H'(y) = 0$ in (5) so that $\phi(x,y) = \frac{x^3}{y}$. (6)

Then $\int_{(0,2)}^{(1,3)} \frac{3x^2}{y} dx - \frac{x^3}{y^2} dy = \phi(1,3) - \phi(0,2)$

$= \frac{1}{3} - 0 = \frac{1}{3}$ (by substitution into (6)).

c) $\int_{(0,0)}^{(2,8)} \vec{\nabla} f \cdot d\vec{c} = f(2,8) - f(0,0)$

$= 4 - 64 - 0 = -60$.

• **PROBLEM 8-17**

Evaluate the following line integrals

a) $\int_C \frac{1+y^2}{x^3} dx - \frac{y+x^2 y}{x^2} dy$ from $(1,0)$ to $(5,2)$

b) $\int_C \frac{-x\,dx}{y\sqrt{y^2-x^2}+y^2-x^2} + \frac{dy}{\sqrt{y^2-x^2}}$ from $(3,5)$ to $(5,13)$.

Solution: a) The approach is to test the integrand for a potential function i.e. to find a function ϕ such that $\vec{F} = \text{grad}\phi$. This is equivalent to calculating $\frac{\partial f}{\partial y}$, $\frac{\partial g}{\partial x}$ and if they are equal, this function ϕ exists. However, note that $\vec{F}(x,y)$ is not differentiable on the line $x=0$, therefore this part of the plane is deleted from the path C. Since

$$\vec{F}(x,y) = \left(\frac{1+y^2}{x^3}, \frac{-(y+x^2 y)}{x^2} \right),$$

$$\frac{\partial f}{\partial y} = \frac{2y}{x^3}, \quad \frac{\partial g}{\partial x} = \frac{-x^2(2xy) + 2x(y+x^2 y)}{x^4}.$$

Simplifying, $\frac{\partial g}{\partial x} = \frac{-2yx^3 + 2xy + 2yx^3}{x^4} = \frac{2xy}{x^4} = \frac{2y}{x^3}$.

Therefore $\frac{\partial f}{\partial y} = \frac{\partial g}{\partial x}$ and there exists a potential function $\phi(x,y)$, such that

$$\frac{\partial \phi}{\partial x} = \frac{1+y^2}{x^3}, \quad \frac{\partial \phi}{\partial y} = \frac{-(y+x^2 y)}{x^2}. \tag{1}$$

Integrating with respect to x first

$\phi(x,y) = \int \frac{1+y^2}{x^3} dx + H(y)$ (H(y) some function of y).

Then $\phi(x,y) = \frac{-(1+y^2)}{2x^2} + H(y)$ (2)

from which $\frac{\partial \phi}{\partial y} = \frac{-2y}{2x^2} + H'(y) = \frac{-y}{x^2} + H'(y)$. (3)

But from (1), $\frac{\partial \phi}{\partial y} = \frac{-(y+x^2 y)}{x^2} = \frac{-y}{x^2} - y$.

367

Therefore, take $H(y) = \frac{-y^2}{2}$ (assuming $C=0$ in $H(y) = \frac{-y^2}{2} + C$) then $H'(y) = -y$ and (3) will have the same value as (1). Thus (2) then becomes

$$\phi(x,y) = \frac{-(1+y^2)}{2x^2} - \frac{y^2}{2} . \tag{4}$$

Then, since $\int_P^Q \vec{F} \cdot d\vec{c} = \phi(Q) - \phi(P)$, the line integral equals

$$\int_{(1,0)}^{(5,2)} \frac{1+y^2}{x^3} dx - \frac{y+x^2y}{x^2} dy = \phi(5,2) - \phi(1,0)$$

which equals, using (4), $= \left(\frac{-(1+4)}{50} - 2\right) - \left(-\frac{1}{2} - 0\right)$

$$= \frac{-5}{50} - 2 + \frac{1}{2} = \frac{-16}{10} = \frac{-8}{5} .$$

b) For this integral we again try the method of finding a potential function, but first compute $\frac{\partial f}{\partial y}$; $\frac{\partial g}{\partial x}$ to see if this function exists. Again note that $\vec{F}(x,y)$ is not differentiable at the point $(0,0)$, therefore we restrict this point from our path C. Since

$$\vec{F}(x,y) = \left(\frac{-x}{y\sqrt{y^2-x^2}+y^2-x^2}, \frac{1}{\sqrt{y^2-x^2}}\right)$$

$\frac{\partial f}{\partial y} = \frac{x}{(y^2-x^2)^{\frac{3}{2}}} = \frac{\partial g}{\partial x}$, therefore there exists a potential function $\phi(x,y)$ such that

$$\frac{\partial \phi}{\partial x} = \frac{-x}{y\sqrt{y^2-x^2}+y^2-x^2} ; \frac{\partial \phi}{\partial y} = \frac{1}{\sqrt{y^2-x^2}} . \tag{5}$$

Integrating with respect to y first

$$\phi(x,y) = \int \frac{dy}{\sqrt{y^2-x^2}} + H(x) \quad (H(x) \text{ some function of } x).$$

Then $\phi(x,y) = \log[y+(y^2-x^2)^{\frac{1}{2}}] + H(x) \tag{6}$

from which $\frac{\partial \phi}{\partial x} = \frac{1}{y+(y^2-x^2)^{\frac{1}{2}}} \cdot \frac{-2x}{2(y^2-x^2)^{\frac{1}{2}}} + H'(x)$

$$= \frac{-x}{y(y^2-x^2)^{\frac{1}{2}}+y^2-x^2} + H'(x).$$

Hence, taking $H(x) = 0$; $H'(x) = 0$; the above equals $\frac{\partial \phi}{\partial y}$ from (5). Therefore,

$$\phi(x,y) = \log[y + (y^2-x^2)^{\frac{1}{2}}]. \tag{7}$$

Hence the line integral becomes

$$\int_{(3,5)}^{(5,13)} \frac{-xdx}{y\sqrt{y^2-x^2}+y^2-x^2} + \frac{dy}{\sqrt{y^2-x^2}} = \phi(5,13) - \phi(3,5).$$

Using (7), the above equals

$$\log[13 + (169-25)^{\frac{1}{2}}] - \log[5 + (25-9)^{\frac{1}{2}}]$$

$$= \log[13+12] - \log[5+4]$$

$$= \log 25 - \log 9 = \log\left(\frac{25}{9}\right) = 2\log\left(\frac{5}{3}\right).$$

• **PROBLEM 8-18**

Let $\vec{F}(\vec{X}) = \frac{k\vec{X}}{r^3}$ where $r = \|\vec{X}\|$; $\vec{X} = (x,y,z)$ and k is a constant. Find $\int_P^Q \vec{F} \cdot d\vec{c}$ where $P = (1,1,1)$ and $Q = (1,2,-1)$.

Solution: This is the vector field inversely proportional to the square of the distance from the origin. We see this because in 3-space the distance formula is given by

$$r = \|\vec{X}\| = \|(x,y,z)\| = \sqrt{x^2 + y^2 + z^2} \tag{1}$$

To solve this line integral try the method of finding a potential function. First, rewrite

$$\vec{F}(\vec{X}) = \frac{k(\vec{X})}{r^3} \quad \text{as} \quad \vec{F}(x,y,z) = \frac{k(x,y,z)}{(x^2+y^2+z^2)^{\frac{3}{2}}}$$

which also can be rewritten as

$$\vec{F}(x,y,z) = \left(\frac{kx}{(x^2+y^2+z^2)^{\frac{3}{2}}}, \frac{ky}{(x^2+y^2+z^2)^{\frac{3}{2}}}, \frac{kz}{(x^2+y^2+z^2)^{\frac{3}{2}}}\right).$$

To see if a potential function exists apply the following theorem: Let $\vec{F} = (f_1, f_2, f_3)$ be a vector field on a rectangular box in 3-space, such that each function f_1, f_2, f_3 has continuous partial derivatives. If $D_1 f_2 = D_2 f_1$; $D_1 f_3 = D_3 f_1$ and $D_2 f_3 = D_3 f_2$ then F has a potential function. Computing,

$$D_1 f_2 = \frac{-3kxy(x^2+y^2+z^2)^{\frac{1}{2}}}{(x^2+y^2+z^2)^3} \quad \text{and}$$

$$D_2 f_1 = \frac{-3kxy(x^2+y^2+z^2)^{\frac{1}{2}}}{(x^2+y^2+z^2)^3} \quad \text{so} \quad D_1 f_2 = D_2 f_1.$$

Also
$$D_1 f_3 = \frac{-3kzx(x^2+y^2+z^2)^{\frac{1}{2}}}{(x^2+y^2+z^2)^3} \quad \text{and}$$

$$D_3 f_1 = \frac{-3kzx(x^2+y^2+z^2)^{\frac{1}{2}}}{(x^2+y^2+z^2)^3}.$$

Thus $D_1 f_3 = D_3 f_1$ and in the same manner $D_2 f_3 = D_3 f_2$. Hence, $\vec{F}(x,y,z)$ does have a potential function and we can use this method to evaluate the line integral.

Now try to find a function $\phi(x,y,z)$ such that $\vec{F} = \text{grad}\phi$ and

$$\frac{\partial \phi}{\partial x} = \frac{kx}{(x^2+y^2+z^2)^{\frac{3}{2}}} \; ; \; \frac{\partial \phi}{\partial y} = \frac{ky}{(x^2+y^2+z^2)^{\frac{3}{2}}} \; ; \; \frac{\partial \phi}{\partial z} = \frac{kz}{(x^2+y^2+z^2)^{\frac{3}{2}}} \quad (2)$$

To do this first integrate with respect to x:

$$\phi(x,y,z) = \int \frac{kx}{(x^2+y^2+z^2)^{\frac{3}{2}}} dx + H(y,z)$$

$$= k \int x(x^2+y^2+z^2)^{-\frac{3}{2}} dx + H(y,z)$$

To solve this integral let $u = x^2+y^2+z^2$, $du = 2xdx$ and

$$\phi(x,y,z) = \frac{k}{2}(x^2+y^2+z^2)^{-\frac{1}{2}}(-2) + H(y,z). \quad (3)$$

However, $\dfrac{\partial(-k(x^2+y^2+z^2)^{-\frac{1}{2}})}{\partial y} = \dfrac{ky}{(x^2+y^2+z^2)^{\frac{3}{2}}}$ which equals

$\frac{\partial \phi}{\partial y}$ in (2). We find the same for $\dfrac{\partial(-k(x^2+y^2+z^2)^{-\frac{1}{2}})}{\partial z} =$

$\dfrac{kz}{(x^2+y^2+z)^{\frac{3}{2}}}$ which equals $\frac{\partial \phi}{\partial z}$ in (2). Hence in (3) take

$H(y,z) = 0$ and thus $\phi(x,y,z) = -k(x^2+y^2+z^2)^{-\frac{1}{2}} = \dfrac{-k}{r}$.

Therefore by the method of potential functions,

$$\int_P^Q \vec{F} \cdot d\vec{c} = \phi(Q) - \phi(P) = \frac{(-k)}{\|Q\|} - \frac{(-k)}{\|P\|}. \qquad (4)$$

Since we are given $P = (1,1,1)$ and $Q = (1,2,-1)$, upon substitution into (4):

$$\int_P^Q \vec{F} \cdot d\vec{c} = \phi(1,2,-1) - \phi(1,1,1)$$

$$= \frac{(-k)}{\sqrt{(1)^2 + (2)^2 + (-1)^2}} - \frac{(-k)}{\sqrt{(1)^2 + (1)^2 + (1)^2}}$$

$$= -k\left(\frac{1}{\sqrt{6}} - \frac{1}{\sqrt{3}}\right).$$

Note that if P and Q are two points at the same distance from the origin (i.e. lying on the same circle, centered at the origin), then letting r_1 be the distance from the origin to the point Q and letting r_2 be the distance from the origin to the point P,

$$\int_P^Q \vec{F} \cdot d\vec{c} = \phi(Q) - \phi(P) = \frac{(-k)}{r_1} - \frac{(-k)}{r_2} = 0$$

since $\phi(x,y,z) = \frac{-k}{r}$ where r is the distance from the origin of the point (x,y,z). Because the two points are the same distance from the origin then $r_1 = r_2$ and $\frac{-k}{r_1} + \frac{k}{r_2} = 0$.

● **PROBLEM 8-19**

Let $r = (x^2 + y^2 + z^2)^{\frac{1}{2}}$. Compute

$$\int_P^Q \frac{\cos r}{r}(xdx + ydy + zdz)$$ where P is the point $(\pi, -\pi, \frac{\pi}{2})$ and Q is the point $(\frac{2\pi}{3}, \frac{2\pi}{3}, \frac{-\pi}{3})$.

Solution: To solve this line integral, apply the method of finding a potential function $\phi(x,y,z)$, such that $\vec{F} = \text{grad}\,\phi(x,y,z)$. Then the line integral can be evaluated by

$$\int_P^Q \vec{F} \cdot d\vec{c} = \phi(Q) - \phi(P). \qquad (1)$$

However for this potential function ϕ to exist, the conditions $\frac{\partial f}{\partial y} = \frac{\partial g}{\partial x}$; $\frac{\partial f}{\partial z} = \frac{\partial h}{\partial x}$; and $\frac{\partial g}{\partial z} = \frac{\partial h}{\partial y}$ must be met.

Upon computing, since $f = \frac{x \cos r}{r}$, $g = \frac{y \cos r}{r}$, $h = \frac{z \cos r}{r}$,

$$\frac{\partial f}{\partial y} = \frac{r\left(-x \sin r \cdot \frac{y}{r}\right) - \left(x \cos r \cdot \frac{y}{r}\right)}{r^2} = \frac{-xy \sin r}{r^2} - \frac{xy \cos r}{r^3}$$

$$\frac{\partial g}{\partial x} = \frac{r\left(-y \sin r \cdot \frac{x}{r}\right) - \left(y \cos r \cdot \frac{x}{r}\right)}{r^2} = \frac{-xy \sin r}{r^2} - \frac{xy \cos r}{r^3}$$

Therefore $\frac{\partial f}{\partial y} = \frac{\partial g}{\partial x}$. Similarly it can be shown

$\frac{\partial f}{\partial z} = \frac{\partial h}{\partial x}$, $\frac{\partial g}{\partial z} = \frac{\partial h}{\partial y}$. Hence there is a function $\phi(x,y,z)$

such that

$$\frac{\partial \phi}{\partial x} = \frac{x \cos r}{r}, \quad \frac{\partial \phi}{\partial y} = \frac{y \cos r}{r}, \quad \frac{\partial \phi}{\partial z} = \frac{z \cos r}{r}. \quad (2)$$

To find this function, integrate with respect to x first:

$$\phi(x,y,z) = \int \frac{x \cos(x^2+y^2+z^2)^{\frac{1}{2}}}{(x^2+y^2+z^2)^{\frac{1}{2}}} dx + H(y,z). \quad \text{Then}$$

$$\phi(x,y,z) = \sin(x^2+y^2+z^2)^{\frac{1}{2}} + H(y,z) \quad \text{from which}$$

$$\frac{\partial \phi}{\partial y} = \frac{y \cos r}{r} + \frac{\partial H}{\partial y} ; \quad \frac{\partial \phi}{\partial z} = \frac{z \cos r}{r} + \frac{\partial H}{\partial z}. \quad (3)$$

If $H(y,z)$ is taken equal to zero, then (2) and (3) will be identical for $\frac{\partial \phi}{\partial y}$, $\frac{\partial \phi}{\partial z}$. Thus, it follows that

$$\phi(x,y,z) = \sin(x^2+y^2+z^2)^{\frac{1}{2}}. \quad (4)$$

Then by (1):

$$\int_P^Q \frac{\cos r}{r} (xdx + ydy + zdz) = \phi(Q) - \phi(P). \quad (5)$$

It is given $Q = \left(\frac{2\pi}{3}, \frac{2\pi}{3}, \frac{-\pi}{3}\right)$ and $P = \left(\pi, -\pi, \frac{\pi}{2}\right)$.

Then by (4), equation (5) becomes

$$\phi(Q) - \phi(P) = \sin\left(\frac{4\pi^2}{9} + \frac{4\pi^2}{9} + \frac{\pi^2}{9}\right)^{\frac{1}{2}} - \left(\sin \pi^2 + \pi^2 + \frac{\pi^2}{4}\right)^{\frac{1}{2}}$$

$$= \sin \pi - \sin \frac{3\pi}{2} = 0 - (-1) = 1.$$

INDEPENDENCE OF PATH

• **PROBLEM 8-20**

Prove, in U (an open connected set) that:

$$\oint_C \vec{F} \cdot d\vec{c} = 0 \quad \text{for every closed path if and only if}$$

$$\int_P^Q \vec{F} \cdot d\vec{c} \quad \text{depends only on P and Q, i.e.} \quad \int_P^Q \vec{F} \cdot d\vec{c} \quad \text{is independent of the path C connecting P and Q.}$$

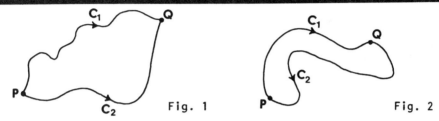

Fig. 1   Fig. 2

Solution: There are two implications to prove. For the first assume $\oint_C \vec{F} \cdot d\vec{c} = 0$ for every closed path and prove $\int_P^Q \vec{F} \cdot d\vec{c}$ is independent of the path. To do this let P,Q be points in U. Let C_1 and C_2 be paths from P to Q in U as in Figure 1. Then the closed curve C consists of C_1, C_2^{-}. Hence,

$$\oint_C \vec{F} \cdot d\vec{c} = \int_{C_1 \cup C_2^{-}} \vec{F} \cdot d\vec{c}$$

$$\oint_C \vec{F} \cdot d\vec{c} = \int_{P, C_1}^Q \vec{F} \cdot d\vec{c} + \int_{Q, C_2^{-}}^P \vec{F} \cdot d\vec{c} . \tag{1}$$

However by assumption, $\oint_C \vec{F} \cdot d\vec{c} = 0$, so (1) becomes

$$\int_{P, C_1}^Q \vec{F} \cdot d\vec{c} + \int_{Q, C_2^{-}}^P \vec{F} \cdot d\vec{c} = 0, \quad \text{from which it follows that}$$

$$\int_{P, C_1}^Q \vec{F} \cdot d\vec{c} = -\int_{Q, C_2^{-}}^P \vec{F} \cdot d\vec{c} = \int_{P, C_2}^Q \vec{F} \cdot d\vec{c}.$$

Therefore since $\int_{P, C_1}^Q \vec{F} \cdot d\vec{c} = \int_{P, C_2}^Q \vec{F} \cdot d\vec{c}$ the line integral is independent of the path and the value depends only on

P and Q, not C.

For the second condition assume the line integral is independent of the path and prove $\oint_C \vec{F} \cdot d\vec{c} = 0$ for every closed path. Letting P,Q be points in U, also let C_1 and C_2 be paths from P to Q in U as in Figure 2. Since we assume the line integral is independent of the path,

$$\int_{P,C_1}^{Q} \vec{F} \cdot d\vec{c} = \int_{P,C_2}^{Q} \vec{F} \cdot d\vec{c} \qquad \text{from which}$$

$$\int_{P,C_1}^{Q} \vec{F} \cdot d\vec{c} = -\int_{Q,C_2}^{P} \vec{F} \cdot d\vec{c}.$$

Thus it follows that $\int_{P,C_1}^{Q} \vec{F} \cdot d\vec{c} + \int_{Q,C_2}^{P} \vec{F} \cdot d\vec{c} = 0.$

Letting $C = C_1 \cup C_2$

$$\oint_C \vec{F} \cdot d\vec{c} = \int_{C_1 \cup C_2} \vec{F} \cdot d\vec{c} = 0.$$

Hence the integral around every closed path C is zero. Note that if the paths cross a finite number of times, a repetition of the same agrument gives the proof; this reasoning covers the case when C_1 and C_2 are broken lines. If the paths cross infinitely often, it is necessary to show that the integrals can be closely approximated by integrals on broken line paths, a limiting process then gives the same result.

• **PROBLEM 8-21**

Prove in an open connected set U that

$\int_{P,C}^{Q} \vec{F} \cdot d\vec{c}$ is independent of the path C if, and only

if, $\vec{F}$ has a potential function ($\vec{F} = \text{grad}\phi$ for some scalar function ϕ).

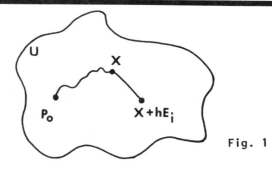

Fig. 1

Solution: There are two implications to prove. For the first assume $\vec{F}$ has a potential function (there exists ϕ such that $\vec{F} = \text{grad}\phi$) and prove $\int_{P,C}^{Q} \vec{F} \cdot d\vec{c}$ is independent of the path C. To prove this condition use the fact that if a potential function ϕ exists then $\int_{P,C}^{Q} \vec{F} \cdot d\vec{c} = \phi(Q) - \phi(P)$, from which we note that the value of the line integral depends only on P and Q not C (the path taken). Hence, the line integral is independent of the path.

For the second assume $\int_{P,C}^{Q} \vec{F} \cdot d\vec{c}$ is independent of the path and prove $\vec{F}$ has a potential function. To do this select some fixed point P_o in U. Then for an arbitrary point X in U, define

$$\phi(X) = \int_{P_o}^{X} \vec{F} \cdot d\vec{c}, \quad \text{along any path from } P_o \text{ to X. By}$$

assumption the line integral is independent of the path, so any path can be chosen. Also recall that U is assumed to be an open connected region. It must be shown that the partial derivatives $D_i\phi(X)$ exist for all P in U and if $\vec{F}$ has coordinate functions

$\vec{F} = (f_1, \ldots, f_n)$ then $D_i\phi(X) = f_i(X)$ $(i=1;n)$ (i.e. if $\vec{F} = (f,h)$ we wish to show $\frac{\partial \phi(x,y)}{\partial x} = f$ and $\frac{\partial \phi(x,y)}{\partial y} = h$).
Let $\vec{E}_1$ be the unit vector with 1 in the ith component and 0 in all the other components and use the relation

$$\vec{F}(X) \cdot \vec{E}_i = f_i(X). \tag{1}$$

Then by definition

$$D_i\phi(X) = \lim_{h \to 0} \frac{\phi(X + h\vec{E}_i) - \phi(X)}{h} \tag{2}$$

Looking at Figure 1, (2) equals

$$D_i\phi(X) = \lim_{h \to 0} \frac{1}{h} \left[\int_{P_o}^{X+h\vec{E}_i} \vec{F} - \int_{P_o}^{X} \vec{F} \right]$$

$$D_i\phi(X) = \lim_{h \to 0} \frac{1}{h} \int_{X}^{X+h\vec{E}_i} \vec{F}(c) \cdot d\vec{c} \tag{3}$$

Note that (3) is along any curve C between X and $X+h\vec{E}_i$. Taking C as the straight line segment as in Figure 1, apply the parametrization $\vec{C}(t) = X + th\vec{E}_i$ with $0 \leq t \leq 1$. The parametrization of a line segment from P to Q is given by $P + t(Q-P)$ and since $P = X$, $Q = X + h\vec{E}_i$,

$P + t(Q-P) = X + t(X+h\vec{E}_i - X)$, $\vec{C}(t) = X + th\vec{E}_i$;

at $P = X$ $X = X+th\vec{E}_i$ so $t = 0$; at $Q = X+h\vec{E}_i$ $X+h\vec{E}_i = X+th\vec{E}_i$ so $t = 1$. Then $\vec{C}'(t) = \dfrac{d\vec{c}}{dt} = h\vec{E}_i$ and by (1)

$$\vec{F}(\vec{C}(t)) \cdot \vec{C}'(t) = f_i(X + th\vec{E}_i)h \tag{4}$$

and since $\displaystyle\int_C \vec{F}(c) \cdot d\vec{c} = \int_C \vec{F}(\vec{C}(t)) \, \vec{C}'(t)\, dt$

(3) becomes (using (4))

$$D_i\phi(X) = \lim_{h \to 0} \frac{1}{h}\int_0^1 f_i(X+th\vec{E}_i)h\, dt$$

letting $U = th$ and $du = h\,dt$

$$D_i\phi(X) = \lim_{h \to 0} \frac{1}{h}\int_0^h f_i(X+u\vec{E}_i)\, du. \tag{5}$$

Now if we let $g(u) = f_i(X+u\vec{E}_i)$, we have an ordinary function of one variable u and (5) has the form

$$D_i\phi(X) = \lim_{h \to 0} \frac{1}{h}\int_0^h g(u)\, du. \tag{6}$$

Then by the fundamental theorem of calculus we know that if G is an indefinite integral for g ($G'=g$) then

$$\int_0^h g(t) = G(h) - G(0) \quad \text{and}$$

$$\lim_{h \to 0} \frac{1}{h}\int_0^h g(t)\, dt = \lim_{h \to 0} \frac{G(h)-G(0)}{h} = G'(0) = g(0).$$

Thus $\displaystyle\lim_{h \to 0} \frac{1}{h}\int_0^h g(u)\, du = g(0).$

Now, from (6): $D_i\phi(X) = g(0)$. Since $g(u) = f_i(X+u\vec{E}_i)$

$g(0) = f_i(X)$ and $D_i\phi(X) = f_i(X)$.

Therefore, $\vec{F} = \text{grad}\,\phi$ which was required to prove. Thus, since both implications are satisfied, the proof is complete.

● **PROBLEM 8-22**

Show that the following functions are independent of the path in the xy-plane and evaluate them:

a) $\displaystyle\int_{(1,1)}^{(x,y)} 2xy\,dx + (x^2-y^2)\,dy$

b) $\displaystyle\int_{(0,0)}^{(x,y)} \sin y \, dx + x \cos y \, dy.$

Solution: To determine if the line integral is independent of the path apply the following method. First, calculate the following partial derivatives of $F(x,y)$, $\frac{\partial f}{\partial y}$; $\frac{\partial g}{\partial x}$. Then check to see if the condition $\frac{\partial f}{\partial y} = \frac{\partial g}{\partial x}$ is satisfied. If it is, then we know that for this vector field (defined on a simply connected region) there exists some function $\phi(x,y)$ such that $\vec{F} = \text{grad}\phi$ (ϕ is called a potential function). In this case,
$$\int_P^Q \vec{F} \cdot d\vec{c} = \phi(Q) - \phi(P),$$
which means the value of the line integral depends only on the endpoints of the interval and not the path taken. Therefore the line integral is independent of the path.

a) $\vec{F}(x,y) = (2xy, x^2-y^2)$. Computing $\frac{\partial f}{\partial y} = 2x$; $\frac{\partial g}{\partial x} = 2x$ and $\frac{\partial f}{\partial y} = \frac{\partial g}{\partial x}$. Therefore there exists a potential function $\phi(x,y)$ and the line integral is thus independent of the path.

To evaluate this line integral, find $\phi(x,y)$ such that
$$\frac{\partial \phi}{\partial x} = 2xy \quad \text{and} \quad \frac{\partial \phi}{\partial y} = x^2 - y^2. \tag{1}$$

Integrating with respect to x first, $\phi(x,y) = \int 2xy \, dx + H(y)$

where $H(y)$ is some function of y. Then
$$\phi(x,y) = x^2 y + H(y). \tag{2}$$

From (2) $\frac{\partial \phi}{\partial y} = x^2 + H'(y)$. But from (1) $\frac{\partial \phi}{\partial y} = x^2 - y^2$. This implies that $H'(y) = -y^2$, so let $H(y) = \frac{-y^3}{3}$ and (2) becomes
$$\phi(x,y) = x^2 y - \frac{y^3}{3}. \tag{3}$$

Then since $\vec{F} = \text{grad}\phi$

$$\int_P^Q \vec{F} \cdot d\vec{c} = \int_{(1,1)}^{(x,y)} 2xy \, dx + (x^2-y^2) \, dy = \phi(x,y) - \phi(1,1) \tag{4}$$

Substituting (3) in (4)

$$\int_{(1,1)}^{(x,y)} 2xy \, dx + (x^2-y^2) \, dy = (x^2 y - \frac{y^3}{3}) - (1 - \frac{1}{3}) = x^2 y - \frac{y^3}{3} - \frac{2}{3}$$

$$= x^2 y - \frac{1}{3}(y^3 + 2).$$

b) $\vec{F}(x,y) = (\sin y, x \cos y)$. Computing, $\frac{\partial f}{\partial y} = \cos y$; $\frac{\partial g}{\partial x} = \cos y$. Therefore since $\frac{\partial f}{\partial y} = \frac{\partial g}{\partial x}$, there exists a potential function $\phi(x,y)$ and the line integral is independent of the path.

To evaluate this line integral, find a function $\phi(x,y)$ such that $\quad \dfrac{\partial \phi}{\partial x} = \sin y; \quad \dfrac{\partial \phi}{\partial y} = x \cos y.$ (5)

Integrating with respect to x first

$\phi(x,y) = \displaystyle\int \sin y \, dx + H(y) \quad$ where $H(y)$ is some function of y. Then

$$\phi(x,y) = x \sin y + H(y) \qquad (6)$$

From (6) $\dfrac{\partial \phi}{\partial y} = x \cos y + H'(y).$ But from (5) $\dfrac{\partial \phi}{\partial y} = x \cos y.$

Thus, letting $H(y) = 0$ in (6),

$$\phi(x,y) = x \sin y \qquad (7)$$

Hence $\displaystyle\int_P^Q \vec{F} \cdot d\vec{c} = \int_{(0,0)}^{(x,y)} \sin y \, dx + x \cos y \, dy$

$$= \phi(x,y) - \phi(0,0).$$

Substituting this into (7)

$$\int_{(0,0)}^{(x,y)} \sin y \, dx + x \cos y \, dy = x \sin y - 0 \cdot \sin 0 = x \sin y.$$

• **PROBLEM 8-23**

Let $\vec{F}$ be the vector field $\vec{F}(x,y) = (2xy^3, 3x^2y^2)$. Find the integral of $\vec{F}$ from (0,0) to (2,2) along each path in Figure 1. Is the line integral independent of the path? Evaluate the line integral by the method of potential functions.

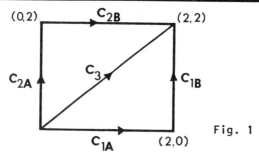

Fig. 1

Solution: We want to find $\displaystyle\int_C \vec{F} \cdot d\vec{c}$ along 3 different paths. First compute

$$\int_{C_1} (2xy^3, 3x^2y^2) \cdot (dx, dy) = \int_{C_1} 2xy^3 \, dx + 3x^2y^2 \, dy.$$

378

Here C_1 consists of two line segments.

Since $\int_{C_1} \vec{F} = \int_{C_{1A}} \vec{F} + \int_{C_{1B}} \vec{F}$:

Along C_{1A} $y = 0$; $x = x$ so that $dx = dx$; $dy = 0$ and

$$\int_{C_{1A}} (2xy^3 dx + 3x^2 y^2 dy) = \int_0^2 0 \cdot dx + 0 = 0.$$

Along C_{1B} $x = 2$; $y = y$ so that $dx = 0$; $dy = dy$ and

$$\int_{C_{1B}} 2xy^3 dx + 3x^2 y^2 dy = \int_0^2 4y^3 \cdot 0 + 12y^2 dy = 12 \cdot \frac{8}{3} = 32.$$

Therefore $\int_{C_1} 2xy^3 dx + 3x^2 y^2 dy = 32.$ (1)

For $\int_{C_2} 2xy^3 dx + 3x^2 y^2 dy$ evaluate in the same manner as C_1.

Along C_{2A} $x = 0$, $y = y$ so that $dx = 0$; $dy = dy$ and

$$\int_{C_{2A}} \vec{F} = \int_0^2 0 + 0 \cdot dy = 0$$

Along C_{2B} $x = x$, $y = 2$ so that $dx = dx$; $dy = 0$ and

$$\int_{C_{2B}} \vec{F} = \int_0^2 16x dx + 0 = 16 \cdot 2 = 32.$$

Hence $\int_{C_2} 2xy^3 dx + 3x^2 y^2 dy = 32.$ (2)

For $\int_{C_3} 2xy^3 dx + 3x^2 y^2 dy$ use x as the parameter to obtain
$x = x$, $y = x$; so that $dx = dx$; $dy = dx$ and

$$\int_{C_3} \vec{F} = \int_0^2 2x^4 dx + 3x^4 dx = 5 \int_0^2 x^4 dx$$

$$\int_{C_3} \vec{F} = 5 \left. \frac{x^5}{5} \right|_0^2 = 5 \cdot \frac{32}{5} = 32.$$ (3)

To determine if the line integral is independent of the path calculate $\frac{\partial f}{\partial y} = 6xy^2$ and $\frac{\partial g}{\partial x} = 6xy^2$; which are equal.

Hence, since $\frac{\partial f}{\partial y} = \frac{\partial g}{\partial x}$ the vector field has a potential function and the line integral is independent of the path. Notice we have the same value for the line integral computed over 3 different paths (equations (1), (2), (3)).

To evaluate the line integral by the method of potential functions let

$$\frac{\partial \phi}{\partial x} = 2xy^3 \quad \text{and} \quad \frac{\partial \phi}{\partial y} = 3x^2y^2. \tag{4}$$

Integrating for x $\phi(x,y) = x^2y^3 + h(y)$. (5)

Now taking $\frac{\partial \phi}{\partial y}$ of (5) $\frac{\partial \phi}{\partial y} = 3x^2y^2 + h'(y)$, but from (4),

$\frac{\partial \phi}{\partial y} = 3x^2y^2$. Thus, it follows that we can take $h(y) = 0$

and $\phi(x,y) = x^2y^3$. By the method of potential functions

$$\int_{P,C}^{Q} \vec{F} \cdot d\vec{c} = \phi(Q) - \phi(P),$$

$$\int_{(0,0)}^{(2,2)} 2xy^3 dx + 3x^2y^2 dy = \phi(2,2) - \phi(0,0)$$

$$= 2^2 2^3 - 0 = 4 \cdot 8 = 32 \quad \text{which agrees}$$

with the previous values.

• **PROBLEM 8-24**

Show that $\int_A^B \frac{xdx + ydy}{x^2 + y^2}$ is independent of the paths in the simply connected domain shown in Figure 1(a), and the doubly connected domain in 1(b). Then find the value of the line integral for each.

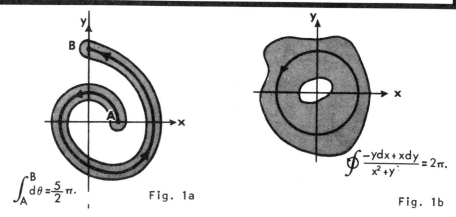

$\int_A^B d\theta = \frac{5}{2}\pi.$ Fig. 1a

$\oint \frac{-ydx + xdy}{x^2+y^2} = 2\pi.$ Fig. 1b

Solution: The path shown in Figure 1(a) is in a simply connected domain not containing the origin. In plain terms a domain is simply connected if it has no "holes." Since the domain is simply connected, the following theorem may be used. Let $P(x,y)$ and $Q(x,y)$ have continuous derivatives in D and let D be simply connected. If $\frac{\partial P}{\partial y} = \frac{\partial Q}{\partial x}$ in D, then $\int Pdx + Qdy$ is independent of the path in D. Computing $\frac{\partial P}{\partial y} = \frac{-2xy}{(x^2+y^2)^2} = \frac{\partial Q}{\partial x}$. Thus the line integral is independent of the path in the simply connected domain shown in Figure 1(a).

The path shown in Figure 1(b) is the circle $x^2+y^2 = 1$ and the domain is doubly connected. A doubly connected domain is a domain with just one hole. Since this is a doubly connected domain the above theorem does not guarantee independence of the path. However, for this example one has additional information: Defining a function $\phi(x,y)$ such that $\vec{F} = \text{grad}\,\phi$, yields the relation

$$\int_A^B \frac{xdx+ydy}{x^2+y^2} = \phi(B) - \phi(A). \tag{1}$$

Since we want $\frac{\partial \phi}{\partial x} = \frac{x}{x^2+y^2}$ and $\frac{\partial \phi}{\partial y} = \frac{y}{x^2+y^2}$, \hfill (2)

integrate with respect to x first to obtain

$\phi(x,y) = \int \frac{x}{x^2+y^2}\,dx + H(y)$ where $H(y)$ is some function of y. Then $\phi(x,y) = \frac{1}{2}\log(x^2+y^2) + H(y)$ \hfill (3)

so that $\frac{\partial \phi}{\partial y} = \frac{y}{x^2+y^2} + H'(y)$. But if $H(y) = 0$ is taken then we have the same value for $\frac{\partial \phi}{\partial y}$ as in (2). Thus (3) becomes $\phi(x,y) = \log\sqrt{x^2+y^2}$. \hfill (4)

Therefore, since $\log\sqrt{x^2+y^2}$ is well-defined (single-valued) with continuous derivatives except at the origin, the line integral is independent of the path in the doubly connected domain consisting of the xy-plane minus the origin (as in Figure 1(b)).
For the value of the line integral, applying (1), (4),

$$\int_A^B \frac{xdx+ydy}{x^2+y^2} = \log\sqrt{x^2+y^2}\Big|_A^B.$$

Letting $\sqrt{x^2+y^2} = r$, $\log r \Big|_A^B = \log r_B - \log r_A$ for Figure 1(a) and equals 0 for the circle excluding the origin as in Figure 1(b).

● **PROBLEM 8-25**

Let $\vec{F}$ be the following vector fields:

a) $\vec{F}_1(x,y,z) = (1-yz, 1-zx, -xy)$

b) $\vec{F}_2(x,y,z) = (yz, xz, xy)$

c) $\vec{F}_3(x,y,z) = (\log(xy), x, y)$

Determine, for each vector field, if $\int \vec{F}\cdot d\vec{c}$ is independent of the path and if so integrate the vector field over the curve from the point $P = (1,6,5)$ to the point $Q = (4,3,2)$.

<u>Solution:</u> To determine if the line integral is independent of the path compute $\frac{\partial f}{\partial y}$, $\frac{\partial g}{\partial x}$, $\frac{\partial f}{\partial z}$, $\frac{\partial h}{\partial x}$, $\frac{\partial g}{\partial z}$, and $\frac{\partial h}{\partial y}$.

If we have the conditions

$$\frac{\partial f}{\partial y} = \frac{\partial g}{\partial x}, \quad \frac{\partial f}{\partial z} = \frac{\partial h}{\partial x}, \quad \frac{\partial g}{\partial z} = \frac{\partial h}{\partial y} \tag{1}$$

then there exists a function ϕ such that $\vec{F} = \text{grad}\,\phi$. Then

$$\int_P^Q f\,dx + g\,dy + h\,dz = \phi(Q) - \phi(P),$$

thus showing that the value of the line integral depends only on the endpoints and not on the path chosen. Hence the line integral is independent of the path.

a) $\vec{F}_1(x,y,z) = (1-yz, 1-zx, -xy)$. Computing

$$\frac{\partial f}{\partial y} = -z = \frac{\partial g}{\partial x}, \quad \frac{\partial f}{\partial z} = -y = \frac{\partial h}{\partial x}, \text{ and } \frac{\partial g}{\partial z} = -x = \frac{\partial h}{\partial y}.$$

Therefore since the conditions in (1) are satisfied $\int_C \vec{F}_1 \cdot d\vec{c}$ is independent of the path. To evaluate this line integral look for a function ϕ such that $\vec{F} = \text{grad}\,\phi$,

i.e. $\frac{\partial \phi}{\partial x} = 1 - yz, \quad \frac{\partial \phi}{\partial y} = 1 - zx, \quad \frac{\partial \phi}{\partial z} = -xy.$ (2)

Integrating with respect to x first

$$\phi(x,y,z) = \int (1-yz)\,dx + H(y,z) \quad \text{so that}$$

$$\phi(x,y,z) = x - yzx + H(y,z) \tag{3}$$

Differentiating (3) with respect to y and z:

$$\frac{\partial \phi}{\partial y} = -xz + \frac{\partial H}{\partial y} \quad \text{and} \quad \frac{\partial \phi}{\partial z} = -xy + \frac{\partial H}{\partial z}.$$

But from (2) $\frac{\partial \phi}{\partial y} = 1 - xz$ and $\frac{\partial \phi}{\partial z} = -xy.$

Take $H(y,z) = y$ so that $\frac{\partial H}{\partial y} = 1$ and $\frac{\partial H}{\partial z} = 0.$

Therefore from (3), $\phi(x,y,z) = x + y - xyz$

$$\int_{(1,6,5)}^{(4,3,2)} (1-yz)\,dx + (1-zx)\,dy - xy\,dz = \phi(4,3,2) - \phi(1,6,5)$$

$$= -17 - (-23) = 23 - 17 = 6.$$

b) $\vec{F}_2(x,y,z) = (yz, xz, xy)$. Computing, $\frac{\partial f}{\partial y} = z = \frac{\partial g}{\partial x}$;

$\frac{\partial f}{\partial z} = y = \frac{\partial h}{\partial x}$, $\frac{\partial g}{\partial z} = x = \frac{\partial h}{\partial y}$. Therefore the conditions in (1) are satisfied and $\int_C \vec{F}_2 \cdot d\vec{c}$ is independent of the path. Thus ϕ is required such that

$$\frac{\partial \phi}{\partial x} = yz, \quad \frac{\partial \phi}{\partial y} = xz, \quad \frac{\partial \phi}{\partial z} = xy \tag{4}$$

Integrating with respect to x first

$\phi(x,y,z) = \int yz + H(y,z) = xyz + H(y,z)$. Differentiating with respect to y and z, $\frac{\partial \phi}{\partial y} = xz + \frac{\partial H}{\partial y}$ and $\frac{\partial \phi}{\partial z} = xy + \frac{\partial H}{\partial z}$.

But from (4) $\frac{\partial \phi}{\partial y} = xz$, $\frac{\partial \phi}{\partial z} = xy$. Therefore take $H(y,z) = 0$ and from (5) $\phi(x,y,z) = xyz$. Then

$$\int_{(1,6,5)}^{(4,3,2)} yzdx + xzdy + xydz = \phi(4,3,2) - \phi(1,6,5)$$
$$= 24 - 30 = -6.$$

c) $\vec{F}_3(x,y,z) = (\log(xy), x, y)$. Computing, $\frac{\partial f}{\partial y} = \frac{1}{y}$ and $\frac{\partial g}{\partial x} = 1$. Since $\frac{\partial f}{\partial y} \neq \frac{\partial g}{\partial x}$ there is a condition of (1) that is not satisfied so $\int_C \vec{F}_3 \cdot d\vec{c}$ is not independent of the path.

• **PROBLEM 8-26**

Evaluate $\int_C \vec{F} \cdot d\vec{c}$ where $\vec{F}$ is the vector field $\vec{F}(x,y,z) = (y^2+2z^2x-1, 2xy, 2x^2z)$ from the point $(0,0,0)$ to the point $(2,2,2)$ where C is the curve shown in Figure 1, by two different methods. What would happen to the value of the line integral if a different path C* were used?

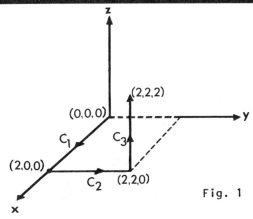

Fig. 1

Solution: For the first method seek some function ϕ such that $\vec{F} = \text{grad}\phi$. Then the line integral is independent of the path and

$$\int_P^Q \vec{F}\cdot d\vec{c} = \phi(Q) - \phi(P). \qquad (1)$$

First checking to see if this function exists compute $\frac{\partial f}{\partial y}$, $\frac{\partial g}{\partial x}$, $\frac{\partial f}{\partial z}$, $\frac{\partial h}{\partial x}$, $\frac{\partial g}{\partial z}$, and $\frac{\partial h}{\partial y}$. If the conditions $\frac{\partial f}{\partial y} = \frac{\partial g}{\partial x}$, $\frac{\partial f}{\partial z} = \frac{\partial h}{\partial x}$, $\frac{\partial g}{\partial z} = \frac{\partial f}{\partial y}$ hold then there exists a potential function ϕ such that $\vec{F} = \text{grad}\phi$. Computing $f = y^2 + 2z^2x - 1$, $g = 2xy$, $h = 2x^2z$ and

$$\frac{\partial f}{\partial y} = 2y = \frac{\partial g}{\partial x}, \quad \frac{\partial f}{\partial z} = 4xz = \frac{\partial h}{\partial x}, \quad \frac{\partial g}{\partial z} = 0 = \frac{\partial h}{\partial y}.$$

Hence since the conditions are satisfied a potential function ϕ does exist. To find this function apply the conditions

$$\frac{\partial \phi}{\partial x} = y^2 + 2z^2x - 1, \quad \frac{\partial \phi}{\partial y} = 2xy, \quad \frac{\partial \phi}{\partial z} = 2x^2z. \qquad (2)$$

Integrating with respect to x first

$$\phi(x,y,z) = \int (y^2 + 2z^2x - 1)\,dx + H(y,z) \quad \text{where } H(y,z) \text{ is}$$

some function of y,z. Then,

$$\phi(x,y,z) = y^2x + z^2x^2 - x + H(y,z). \qquad (3)$$

From (3) $\frac{\partial \phi}{\partial y} = 2xy + \frac{\partial H}{\partial y}$ and $\frac{\partial \phi}{\partial z} = 2zx^2 + \frac{\partial H}{\partial z}$.

But from (2) $\frac{\partial \phi}{\partial y} = 2xy$ and $\frac{\partial \phi}{\partial z} = 2x^2z$. Then if $H(y,z)$ is taken equal to zero, $\phi(x,y,z) = xy^2 + x^2z^2 - x$
(In other words $\vec{F}(x,y,z)$ is the differential of ϕ). Then by equation (1)

$$\int_C \vec{F}\cdot d\vec{c} = \int_{(0,0,0)}^{(2,2,2)} (y^2+2z^2x-1)dx + 2xy\,dy + 2x^2z\,dz$$

$$= \phi(2,2,2) - \phi(0,0,0)$$

$$= [2(2^2) + (2^2)(2^2) - 2] - 0 = 22. \qquad (4)$$

For the second method, compute the line integral directly along the curve C. Since C consists of the line segments C_1, C_2, C_3, $\int_C \vec{F} = \int_{C_1} \vec{F} + \int_{C_2} \vec{F} + \int_{C_3} \vec{F}$. Therefore evaluate $\int_{C_1} \vec{F}\cdot d\vec{c}$ first. The parametrization used is

$x=x$; $y=0$; $z=0$, $dx=dx$; $dy=0$; $dz=0$ so that

$$\int_{C_1} \vec{F}\cdot d\vec{c} = \int_{C_1} (y^2+2z^2x-1)dx + 2xydy + 2x^2zdz$$

$$= \int_{C_1} (0-1)dx + 0 + 0 = -\int_0^2 dx = -2. \tag{5}$$

Along C_2 use the parameterization $x=2$, $y=y$, $z=0$ so that $dx=0$, $dy=dy$, $dz=0$ and

$$\int_{C_2} \vec{F}\cdot d\vec{c} = \int_{C_2} (y^2-1)\cdot 0 + 4ydy + 0 = 4\int_0^2 ydy = 8. \tag{6}$$

Along C_3 $x=2$, $y=2$, $z=z$ so that $dx=0$, $dy=0$, $dz=dz$ and

$$\int_{C_3} \vec{F}\cdot d\vec{c} = \int_{C_3} 0 + 0 + 8zdz = 8\int_0^2 zdz = 16. \tag{7}$$

Hence adding (5), (6), and (7)

$$\int_C \vec{F}\cdot d\vec{c} = -2 + 8 + 16 = 22$$

which agrees with (4) (the answer by the first method).

If a different path is used, the value of the line integral would remain the same since the line integral is independent of the path. We know this because there exists a potential function $\phi(x,y,z)$ and from this method the value of the line integral depends only on the endpoints and not the path taken.

• **PROBLEM 8-27**

Let F be the vector field $\vec{F}(x,y,z) = (yz, xz, xy)$. Compute $\int_P^Q \vec{F}\cdot d\vec{c}$ where Q is the point $(1,2,3)$ and P is the point $(1,0,-1)$, first by choosing a broken line path with segments parallel to the axes. Then find a function $\phi(x,y,z)$ such that $\vec{F} = \text{grad}\phi$.

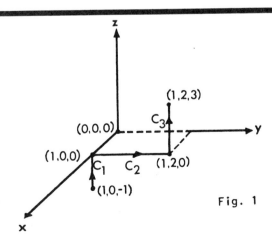

Fig. 1

Solution: $\int_P^Q \vec{F} \cdot d\vec{c}$ is to be evaluated along some path C.

But first compute the following partial derivatives of $\vec{F}(f,g,h)$, $\frac{\partial f}{\partial y}$, $\frac{\partial g}{\partial x}$, $\frac{\partial f}{\partial z}$, $\frac{\partial h}{\partial x}$, $\frac{\partial g}{\partial z}$, and $\frac{\partial h}{\partial y}$. Then if the conditions $\frac{\partial f}{\partial y} = \frac{\partial g}{\partial x}$, $\frac{\partial f}{\partial z} = \frac{\partial h}{\partial x}$, and $\frac{\partial g}{\partial z} = \frac{\partial h}{\partial y}$ hold, the line integral is independent of the path C. Computing, $\frac{\partial f}{\partial y} = z = \frac{\partial g}{\partial x}$, $\frac{\partial f}{\partial z} = y = \frac{\partial h}{\partial x}$, and $\frac{\partial g}{\partial z} = x = \frac{\partial h}{\partial y}$, hence the line integral is independent of the path chosen. Therefore any path C can be chosen, and the line integral will have the same value for each C.

For the given points P,Q choose the path C which consists of the line segments C_1, C_2, C_3 as shown in Figure 1. Then to evaluate $\int_C \vec{F} \cdot d\vec{c}$ use the relation

$$\int_C \vec{F} = \int_{C_1} \vec{F} + \int_{C_2} \vec{F} + \int_{C_3} \vec{F} \quad \text{where } C = \{C_1, C_2, C_3\}. \quad (1)$$

To compute $\int_{C_1} \vec{F} \cdot d\vec{c}$ use the parameterization x=1, y=0, z=z so that dx=0, dy=0, dz=dz and

$$\int_{C_1} \vec{F} \cdot d\vec{c} = \int_{C_1} (yz, xz, xy) \cdot (dx, dy, dz)$$

$$= \int_{C_1} yzdx + xzdy + xydz$$

$$= \int_{C_1} 0 + 0 + 0 \cdot dz = 0. \quad (2)$$

Along C_2 use the parameterization x=1; y=y; z=0 so that dx=0, dy=dy, dz=0 and

$$\int_{C_2} \vec{F} \cdot d\vec{c} = \int_{C_2} yzdx + xzdy + xydz$$

$$= \int_{C_2} 0 + 0dy + 0 = 0. \quad (3)$$

Along C_3 use x=1; y=2; z=z so that dx=0; dy=0; dz=dz and

$$\int_{C_3} yzdx + xzdy + xydz = \int_{C_3} 0 + 0 + 2dz$$

$$= 2\int_0^3 dz = 6. \quad (4)$$

Then by (1), $\int_C \vec{F} \cdot d\vec{c}$ equals the addition of (2),(3),(4).

Thus
$$\int_C \vec{F} \cdot d\vec{c} = 0 + 0 + 6 = 6. \tag{5}$$

Note that this value will be obtained for any path chosen since the line integral is independent of the path.

To find the value of this integral by finding a function ϕ such that $\vec{F} = \text{grad}\,\phi$, first verify if this function exists. This is easily done by using the following theorem: $\int_P^Q \vec{F} \cdot d\vec{c}$ is independent of the path C if and only if $\vec{F}$ has a potential function. Therefore, since this line integral is independent of the path C it follows that $\vec{F}$ has a potential function ϕ such that $\vec{F} = \text{grad}\,\phi$. For this function $\phi(x,y,z)$, it is required that

$$\frac{\partial \phi}{\partial x} = yz, \quad \frac{\partial \phi}{\partial y} = xz, \quad \frac{\partial \phi}{\partial z} = xy. \tag{6}$$

First integrating with respect to x

$$\phi(x,y,z) = \int yz\,dx + H(y,z) = xyz + H(y,z) \quad \text{where } H(y,z) \text{ is}$$

some function of y,z only. Then computing $\frac{\partial \phi}{\partial y}$, $\frac{\partial \phi}{\partial z}$ from this equation $\frac{\partial \phi}{\partial y} = xz + \frac{\partial H(y,z)}{\partial y}$ and $\frac{\partial \phi}{\partial z} = xy + \frac{\partial H(y,z)}{\partial z}$.

But from (6) $\frac{\partial \phi}{\partial y} = xz$ and $\frac{\partial \phi}{\partial z} = xy$; thus take $H(y,z) = 0$ to obtain $\phi(x,y,z) = xyz$. Therefore $\int_P^Q \vec{F} \cdot d\vec{c} = \phi(Q) - \phi(P)$ since $\vec{F} = \text{grad}\,\phi$, or

$$\int_{(1,0,-1)}^{(1,2,3)} yz\,dx + xz\,dy + xy\,dz = \phi(1,2,3) - \phi(1,0,-1)$$

$$= 6 - 0 = 6 \quad \text{which agrees with the}$$

value of the line integral over C (equation (5)).

• **PROBLEM 8-28**

Evaluate $\int_P^Q \dfrac{(z^3+zy^2)\,dx + xz^2\,dy + (xz^2+xy^2-xyz)\,dz}{xz^3 + xy^2 z}$ where Q is the point $(e,0,1)$ and P is the point $(1,1,1)$, by two different methods.

Solution: The first method is to try to find some function $\phi(x,y,z)$ such that $\vec{F} = \text{grad}\,\phi$ and use the relation $\int_P^Q \vec{F} \cdot d\vec{c} = \phi(Q) - \phi(P)$. But first check to see if this function ϕ exists. To do this compute the following

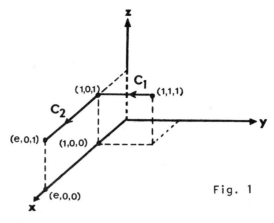

Fig. 1

derivatives of $\vec{F} = (f,g,h)$,

$\frac{\partial f}{\partial y}$, $\frac{\partial g}{\partial x}$, $\frac{\partial f}{\partial z}$, $\frac{\partial h}{\partial x}$, $\frac{\partial g}{\partial z}$, and $\frac{\partial h}{\partial y}$, and if $\frac{\partial f}{\partial y} = \frac{\partial g}{\partial x}$,

$\frac{\partial f}{\partial z} = \frac{\partial h}{\partial x}$, $\frac{\partial g}{\partial z} = \frac{\partial h}{\partial y}$, then there exists a function ϕ such that $\vec{F} = \text{grad}\phi$. That is, the differential is exact.

Computing, $f = \frac{(z^3+zy^2)}{(xz^3+xy^2z)} = \frac{1}{x}$; $g = \frac{xz^2}{xz^3+xy^2z} = \frac{z}{z^2+y^2}$;

$h = \frac{xz^2+xy^2-xyz}{xz^3+xy^2z} = \frac{-y}{y^2+z^2} + \frac{1}{z}$ and $\frac{\partial f}{\partial y} = 0 = \frac{\partial g}{\partial x}$;

$\frac{\partial f}{\partial z} = 0 = \frac{\partial h}{\partial x}$; $\frac{\partial g}{\partial z} = \frac{y^2-z^2}{(y^2+z^2)^2} = \frac{\partial h}{\partial y}$. Hence since the

conditions are satisfied a potential function ϕ does exist. To find it we want

$$\frac{\partial \phi}{\partial x} = \frac{1}{x}, \quad \frac{\partial \phi}{\partial y} = \frac{z}{z^2+y^2}, \quad \frac{\partial \phi}{\partial z} = \frac{-y}{y^2+z^2} + \frac{1}{z}. \tag{1}$$

Therefore integrating with respect to x first

$\phi(x,y,z) = \int \frac{dx}{x} + H(y,z)$ where $H(y,z)$ is some function of (y,z). Thus

$$\phi(x,y,z) = \log x + H(y,z). \tag{2}$$

Now $H(y,z)$ should be such that

$$\frac{\partial H(y,z)}{\partial y} = \frac{z}{z^2+y^2}; \quad \frac{\partial H}{\partial z} = \frac{-y}{y^2+z^2} + \frac{1}{z}. \tag{3}$$

Integrating with respect to y

$H(y,z) = \int \frac{z}{z^2+y^2} dy + G(z) = \tan^{-1}\left(\frac{y}{z}\right) + G(z)$ but from (3)

$\frac{\partial H}{\partial z} = \frac{1}{z} - \frac{y}{y^2+z^2}$ and since $\frac{d(\tan^{-1}u)}{du} = \frac{1}{1+u^2}$ by the chain rule $\dfrac{\partial \tan^{-1}\left(\frac{y}{z}\right)}{\partial z}$ may be found. Let $u = \dfrac{y}{z}$ then $\dfrac{du}{dz} = \dfrac{-y}{z^2}$

and $\dfrac{\partial \tan^{-1}\left(\frac{y}{z}\right)}{\partial z} = \dfrac{1}{1+\frac{y^2}{z^2}} \cdot \dfrac{-y}{z^2} = \dfrac{z^2}{z^2+y^2} \cdot \dfrac{-y}{z^2} = \dfrac{-y}{z^2+y^2}$.

So $\dfrac{\partial \tan^{-1}\left(\frac{y}{z}\right)}{\partial z} = \dfrac{-y}{y^2+z^2}$ and all we need is a $G(z)$ such that

$G'(z) = \dfrac{1}{z}$. Take $G(z) = \int \dfrac{dz}{z} = \log z$. Then the potential

function (2) equals $\phi(x,y,z) = \log x + \tan^{-1}\left(\dfrac{y}{z}\right) + \log z$

$$\phi(x,y,z) = \log(xz) + \tan^{-1}\left(\dfrac{y}{z}\right). \tag{4}$$

Hence $\displaystyle\int_{(1,1,1)}^{(e,0,1)} \dfrac{(z^3+zy^2)dx + xz^2 dy + (xz^2+xy^2-xyz)dz}{(xz^3+xy^2 z)}$

$= \phi(e,0,1) - \phi(1,1,1)$ which equals upon substitution into

(4): $\log e + \tan^{-1}\left(\dfrac{0}{1}\right) - \log 1 - \tan^{-1}\left(\dfrac{1}{1}\right)$

$$= 1 + 0 - 0 - \dfrac{\pi}{4} = 1 - \dfrac{\pi}{4}. \tag{5}$$

The second method used is to compute the integral along some curve C. Since a potential function exists the line integral is independent of the path C. Therefore choose any C and the line integral will still have the same value. For the given points P,Q choose the path C, which consists of the line segments C_1, C_2 as shown in Figure 1. Then

$$\int_C \vec{F} = \int_{C_1} \vec{F} + \int_{C_2} \vec{F} \quad \text{where } C = \{C_1, C_2\}.$$

To compute $\displaystyle\int_{C_1} \vec{F} \cdot d\vec{c}$ use the parameterization $x=1$, $y=y$, $z=1$, so that $dx=0$, $dy=dy$, $dz=0$ and

$$\int_{C_1} \vec{F} \cdot d\vec{c} = \int_{C_1} 0 + \dfrac{dy}{1+y^2} + 0 = \int_1^0 \dfrac{dy}{1+y^2}. \tag{6}$$

From a table of integrals $\displaystyle\int \dfrac{dx}{a^2+x^2} = \dfrac{1}{a}\tan^{-1}\left(\dfrac{x}{a}\right)$. Thus

(6) becomes $\displaystyle\int_1^0 \dfrac{dy}{1+y^2} = \tan^{-1}(y)\Big|_1^0 = \tan^{-1}(0) - \tan^{-1}(1)$

$$= 0 - \dfrac{\pi}{4}. \tag{7}$$

Along C_2 use $x=x$, $y=0$, $z=0$ and $dx=dx$, $dy=0$, $dz=0$ so that

$$\int_{C_2} \vec{F}\cdot d\vec{c} = \int_{C_2} \frac{(z^3+zy^2)}{x(z^3+zy^2)} dx + 0 + 0 = \int_1^e \frac{dx}{x}$$

$$= \log x \Big|_1^e = \log e - \log 1 = 1 - 0. \tag{8}$$

Therefore $\int_C \vec{F}\cdot d\vec{c}$ equals $(7) + (8) = \frac{-\pi}{4} + 1 = 1 - \frac{\pi}{4}$

which agrees with the answer using the first method (equation (5)).

• **PROBLEM 8-29**

A force F is called conservative if it is exact. Show that the force (vector field) $\vec{F}(x,y) = (y \cos xy, x \cos xy)$ is conservative. Then find the work done by this force in moving a particle from the origin to the point (3,8).

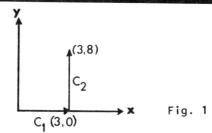

Fig. 1

Solution: To determine if $\vec{F}(x,y)$ is exact use the following theorem: Given the differential form $fdx + gdy$, if f and g have continuous first partial derivatives at all points of some open rectangle, the differential form is exact at each point of the rectangle if and only if the condition $\frac{\partial f}{\partial y} = \frac{\partial g}{\partial x}$ is satisfied throughout the rectangle. When this condition is satisfied, there is a function ϕ such that $d\phi = fdx + gdy$ and $\phi(x,y) = \int_C f(x,y)dx + g(x,y)dy$ where C goes from P_0 to $P = (x,y)$.

In this problem $\vec{F}(x,y) = (y \cos xy, x \cos xy)$. Letting $f = y \cos xy$ and $g = x \cos xy$

$$\frac{\partial f}{\partial y} = \frac{\partial(y \cos xy)}{\partial y} = xy(-\sin xy) + \cos xy$$

and $\frac{\partial f}{\partial x} = \frac{\partial(x \cos xy)}{\partial x} = xy(-\sin xy) + \cos xy$.

Thus $\frac{\partial f}{\partial y} = \frac{\partial g}{\partial x}$, and by the theorem $(y \cos xy)dx +$

$(x \cos xy) dy$ is exact. Hence the force is conservative.

To find the work done by this force $\vec{F}$ in moving the particle, compute the line integral from the origin to the point $(3,8)$. Since $\frac{\partial f}{\partial y} = \frac{\partial g}{\partial x}$ there exists some function ϕ such that $\vec{F} = \text{grad} \phi$ and

$$\int_P^Q f(x,y) dx + g(x,y) dy = \phi(Q) - \phi(P). \quad (1)$$

Since $\vec{F}(f,g) = \text{grad} \phi(x,y)$, $\frac{\partial \phi}{\partial x} = f$ and $\frac{\partial \phi}{\partial y} = g$; so

$$\frac{\partial \phi}{\partial x} = y \cos xy \quad \text{and} \quad \frac{\partial \phi}{\partial y} = x \cos xy. \quad (2)$$

Integrating with respect to x

$$\phi(x,y) = \int y \cos xy \, dx + H(y) \quad (H(y) \text{ is a function of } y)$$

$$\phi(x,y) = \sin(xy) + H(y). \quad (3)$$

Now $\frac{\partial \phi}{\partial y} = x \cos xy + H'(y)$. But from (1) $\frac{\partial \phi}{\partial y} = x \cos(xy)$. If $H(y)$ is taken equal to zero,

$$\phi(x,y) = \sin(xy). \quad (4)$$

Then upon substitution into (1) the work done by the force is

$$\int_{(0,0)}^{(3,8)} (y \cos xy) dx + (x \cos xy) dy = \phi(3,8) - \phi(0,0)$$

which equals using (4), $\sin(24) - \sin 0 = \sin 24$.

The work expended may also be calculated by evaluating the line integral over a path C from $(0,0)$ to $(3,8)$. Since there exists ϕ such that $\vec{F} = \text{grad} \phi$, the line integral is independent of the path. Thus any C may be chosen to evaluate the line integral. Letting C be the path in Figure 1, now C consists of C_1, C_2. Then along C_1

$$x=x, \ y=0 \text{ and } \int_{C_1} \vec{F} \cdot d\vec{c} = \int_{C_1} 0 dx + 0 = 0.$$

Along C_2 $x=3$, $y=y$ and

$$\int_{C_2} \vec{F} \cdot d\vec{c} = \int_{C_2} 0 + 3 \cos 3y \, dy = 3 \int_0^8 \cos 3y \, dy$$

$$= 3 \left. \frac{\sin 3y}{3} \right|_0^8 = \sin 24.$$

Thus,

$$\int_C \vec{F} \cdot d\vec{c} = 0 + \sin 24 = \sin 24.$$

GREEN'S THEOREM

• **PROBLEM** 8-30

Prove Green's Theorem in the plane if the region R is representable in both of the forms

$$a \leq x \leq b, \quad f_1(x) \leq y \leq f_2(x),$$
$$c \leq y \leq D, \quad g_1(y) \leq x \leq g_2(y) \quad \text{as in Fig. 1.}$$

Let R be the region in R^2 and let C be the curve bounding R as given in the figure. C is oriented in such a way that R is always to the left of C. Let $P(x,y)$, $Q(x,y)$ be real-valued differentiable functions defined on R. Then Green's Theorem in the plane states that

$$\oint_C Pdx + Qdy = \iint_R \left(\frac{\partial Q}{\partial x} - \frac{\partial P}{\partial y}\right) dxdy.$$

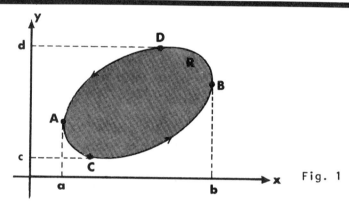

Fig. 1

Solution: In Figure 1, R is the region bounded by C. Let the equation of the curves ACB and BDA be $y = f_1(x)$ and $y = f_2(x)$ respectively. The double integral
$\iint_R \frac{\partial P}{\partial y} dxdy$ can be written as:

$$\iint_R \frac{\partial P}{\partial y} dxdy = \int_a^b \left[\int_{f_1(x)}^{f_2(x)} \frac{\partial P}{\partial y} dy \right] dx$$

$$= \int_a^b P(x,y) \Big|_{y=f_1(x)}^{f_2(x)} dx$$

$$= \int_a^b \Big(P(x, f_2(x)) - P(x, f_1(x)) \Big) dx$$

$$= -\int_a^b P(x, f_1(x)) dx - \int_b^a P(x, f_2(x)) dx$$

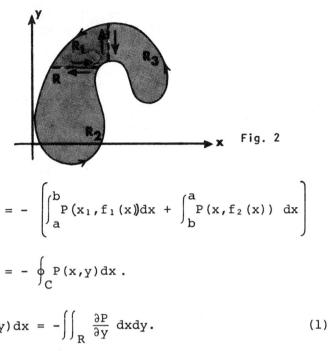

Fig. 2

$$= -\left[\int_a^b P(x_1, f_1(x))dx + \int_b^a P(x, f_2(x))\,dx\right]$$

$$= -\oint_C P(x,y)\,dx.$$

Thus
$$\oint_C P(x,y)\,dx = -\iint_R \frac{\partial P}{\partial y}\,dxdy. \qquad (1)$$

For the second half of the proof let the equations of the curves CAD and CBD be $x = g_1(y)$ and $x = g_2(y)$ respectively. Then

$$\iint \frac{\partial Q}{\partial x}\,dxdy = \int_c^d \left[\int_{g_1(y)}^{g_2(y)} \frac{\partial Q}{\partial x}\,dx\right] dy$$

$$= \int_c^d Q(x,y)\Big|_{x=g_1(y)}^{x=g_2(y)}\,dy$$

$$= \int_c^d \Big(Q(g_2(y),y) - Q(g_1(y),y)\Big)\,dy$$

$$= \int_c^d Q(g_2(y),y)\,dy - \int_c^d Q(g_1(y),y)\,dy$$

$$= \int_c^d Q(g_2(y),y)\,dy + \int_d^c Q(g_1(y),y)\,dy$$

$$= \oint_C Q(x,y)\,dy.$$

Thus
$$\oint_C Q(x,y)\,dy = \iint_R \frac{\partial Q}{\partial x}\,dxdy \qquad (2)$$

Then adding (1) and (2):

$$\oint_C P(x,y)dx + Q(x,y)dy = \iint_R \left(\frac{\partial Q}{\partial x} - \frac{\partial P}{\partial y}\right)dxdy$$

which is Green's Theorem in the plane.

Note: If a region is given in which a line parallel to the coordinate axes may cut C in more than two points, then R must be decomposed into separate regions (as in Fig. 2) having the properties of the region first considered in this problem.

If there are R_1, R_2, ..., R_n regions bounded by C_1, C_2, ..., C_n respectively, then

$$\oint_{C_1}(Pdx+Qdy) + \oint_{C_2}(\quad) + \ldots + \oint_{C_n}(\quad)$$

$$= \iint_{R_1}\left(\frac{\partial Q}{\partial x} - \frac{\partial P}{\partial y}\right)dxdy + \ldots + \iint_{R_n}(\quad)dxdy$$

However the sum on the left is just $\oint_C Pdx + Qdy$ where $C = \{C_1, C_2, \ldots, C_n\}$. For the integrals along the added arcs are taken once in each direction, thereby cancelling each other out; the remaining integrals then add up to precisely the integral around C in the positive direction. On the right side, the integrals add up to $\iint_R \left(\frac{\partial Q}{\partial x} - \frac{\partial P}{\partial y}\right)dxdy$. Thus

$$\oint_C Pdx + Qdy = \iint_R \left(\frac{\partial Q}{\partial x} - \frac{\partial P}{\partial y}\right)dxdy.$$

● **PROBLEM 8-31**

Let C be the ellipse $x^2 + 4y^2 = 4$. Compute $\oint_C (2x-y)dx + (x+3y)dy$ by Green's Theorem.

Solution: Green's Theorem states: Let p,q be functions on a region R, which is the interior of a closed path C (parametrized counterclockwise). Then

$$\int_C pdx + qdy = \iint_R \left(\frac{\partial q}{\partial x} - \frac{\partial p}{\partial y}\right)dxdy.$$

In the given problem,

$$\oint_C (2x-y)dx + (x+3y)dy \qquad (1)$$

where C is the ellipse $x^2 + 4y^2 = 4$, is the integral. To use Green's Theorem let $p = 2x-y$ and $q = (x+3y)$. Then $\frac{\partial p}{\partial y} = -1$ and $\frac{\partial q}{\partial x} = 1$ so that (1) equals

$$\iint_R (1+1)\,dxdy = 2\iint_R dxdy = 2 \times \text{(area of the ellipse)}. \qquad (2)$$

Hence rewriting the ellipse as $\frac{x^2}{4} + y^2 = 1$, and using the formula Area $= \pi ab$ where $\frac{x^2}{a^2} + \frac{y^2}{b^2} = 1$, the area of the given ellipse is 2π. Thus, (2) becomes $2 \times 2\pi = 4\pi$. Thus the value of the line integral is 4π.

• **PROBLEM 8-32**

Verify Green's Theorem where $\vec{F}$ is the vector field $\vec{F}(x,y) = (xy, x^2)$ and $R = \{(x,y): a \leq x \leq b, c \leq y \leq d\}$. (Figure 1.)

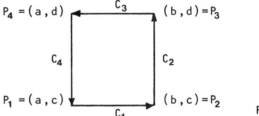

Fig. 1

Solution: Green's Theorem states:

$$\oint_C p\,dx + q\,dy = \iint_R \left(\frac{\partial q}{\partial x} - \frac{\partial p}{\partial y}\right) dxdy \qquad \text{where } p,q \text{ are functions}$$

on a region R which is the interior of a closed curve C (orientated counterclockwise).

To verify Green's Theorem first compute the integral $\oint_C \vec{F} \cdot d\vec{c} = \oint_C (p,q) \cdot (dx,dy)$ where $\vec{F} = (p,q)$. Hence we must compute

$$\oint_C xy\,dx + x^2\,dy. \qquad (1)$$

Here C consists of the four straight line segments in Figure 1. To compute (1) apply the definition

$$\oint_C \vec{F} = \int_{C_1} \vec{F} + \int_{C_2} \vec{F} + \int_{C_3} \vec{F} \cdot \int_{C_4} \vec{F} \qquad \text{where } C = \{C_1, C_2, C_3, C_4\}.$$

First compute

$$\int_{C_1} xy\,dx + x^2\,dy. \qquad (2)$$

Here use x as the parameter letting $x=x$, $y=c$ so that $dx=dx$ and $dy=0$. Then (2) becomes

$$\int_a^b cx\,dx + x^2 \cdot 0 = c\int_a^b x\,dx = \tfrac{1}{2}c(b^2-a^2).$$

For $\int_{C_2} xy\,dx + x^2\,dy$ \qquad (3)

use y as the parameter letting $y=y$, $x=b$ so that $dy=dy$ and $dx=0$. Then (3) becomes

$$\int_c^d by\cdot 0 + b^2\,dy = b^2(d-c).$$

For $\int_{C_3} xy\,dx + x^2\,dy$ \qquad (4)

let $x=x$, $y=d$ so that $dx=dx$, $dy=0$ and (4) becomes

$$d\int_b^a x\,dx = \tfrac{1}{2}d(a^2-b^2).$$

Finally for $\int_{C_4} xy\,dx + x^2\,dy$ \qquad (5)

let $x=a$, $y=y$ so that $dx=0$, $dy=dy$ and (5) becomes

$$a^2\int_d^c dy = a^2(c-d).$$

Hence substituting the values of (2), (3), (4), and (5), (1) equals

$$\tfrac{1}{2}c(b^2-a^2) + b^2(d-c) + \tfrac{1}{2}d(a^2-b^2) + a^2(c-d)$$

$$= \tfrac{1}{2}c(b^2-a^2) - \tfrac{1}{2}d(b^2-a^2) + b^2(d-c) - a^2(d-c)$$

$$= \tfrac{1}{2}(c-d)(b^2-a^2) + (b^2-a^2)(d-c)$$

$$= (b^2-a^2)(d-c) - \tfrac{1}{2}(d-c)(b^2-a^2)$$

$$= \tfrac{1}{2}(d-c)(b^2-a^2). \qquad (6)$$

By Green's Theorem $p=xy$ and $q=x^2$ so that $\frac{\partial p}{\partial y} = x$; $\frac{\partial q}{\partial x} = 2x$ and (1) equals

$$\iint_R (2x-x)\,dxdy = \int_c^d \int_a^b x\,dxdy$$

$$= \frac{1}{2}(b^2-a^2)\int_c^d dy = \frac{1}{2}(b^2-a^2)(d-c) \quad \text{which}$$

agrees with (6) and thus verifies Green's Theorem for this problem.

• **PROBLEM 8-33**

a) Let $\vec{F}(x,y) = (3xy, x^2)$. Verify Green's Theorem where R is the rectangle in Figure 1 (counterclockwise).

b) Let C be the closed curve consisting of the graphs of $y = \sin x$ and $y = 2\sin x$ for $0 \le x \le \pi$, oriented counterclockwise. Find

$$\int_C (1+y^2)\,dx + y\,dy \qquad (1)$$

directly and by Green's Theorem.

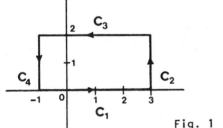
Fig. 1

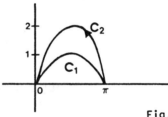
Fig. 2

Solution: Green's Theorem states:

$$\oint_C p\,dx + q\,dy = \iint_R \left(\frac{\partial q}{\partial x} - \frac{\partial p}{\partial y}\right) dxdy \quad \text{where C is a closed curve}$$

and the boundary of the region R.

a) To verify Green's Theorem first compute

$$\oint_C (p,q) \cdot (dx,dy) = \oint_C p\,dx + q\,dy \qquad (1)$$

Substituting $\vec{F}(x,y) = (3xy, x^2)$ into (1)

$$\oint_C 3xy\,dx + x^2\,dy. \qquad (2)$$

To compute this integral, observe that C consists of 4 segments. The integral is calculated over each segment and the integral over C is given by the relation

$$\oint_C \vec{F} = \int_{C_1} \vec{F} + \int_{C_2} \vec{F} + \int_{C_3} \vec{F} + \int_{C_4} \vec{F} \quad \text{where } C = \{C_1, C_2, C_3, C_4\}.$$

Along C_1 x=x, y=0 and dx=dx, dy=0 so that

$$\int_{C_1} 3xy\,dx + x^2\,dy = \int_{-1}^{3} 0\,dx + x^2 0 = 0. \tag{1}$$

Along C_2 x=3, y=y and dx=0, dy=dy so that

$$\int_{C_2} 3xy\,dx + x^2\,dy = \int_0^2 9y0 + 9\,dy = 9\int_0^2 dy = 18. \tag{2}$$

Along C_3 x=x, y=2 so that

$$\int_{C_3} \vec{F} \cdot d\vec{c} = \int_3^{-1} 6x\,dx + x^2 0 = 6\int_3^{-1} x\,dx$$

$$= 6\left(\frac{(-1)^2}{2} - \frac{(3)^2}{2}\right) = 6\left(\frac{1}{2} - \frac{9}{2}\right) = -24. \tag{3}$$

Along C_4 x=-1, y=y so that

$$\int_{C_4} \vec{F} \cdot d\vec{c} = \int_2^0 0 + (-1)^2\,dy = \int_2^0 dy = -2. \tag{4}$$

Now $\oint_C \vec{F}$ equals the sum of (1), (2), (3), (4). Hence

$$\oint_C 3xy\,dx + x^2\,dy = 0 + 18 - 24 - 2 = -8. \tag{5}$$

By Green's Theorem p=3xy, q=x^2 so that $\frac{\partial p}{\partial y} = 3x$; $\frac{\partial q}{\partial x} = 2x$ and

$$\oint_C 3xy\,dx + x^2\,dy = \iint_R (2x-3x)\,dx\,dy$$

$$= \int_{-1}^{3}\int_0^2 (-x)\,dy\,dx = \int_{-1}^{3} (-x)2\,dx = -(9-1) = -8$$

which agrees with equation (5) and thus verifies Green's Theorem for this problem.

b) First find $\int_C (1+y^2)\,dx + y\,dy$ directly. Drawing the curves in Figure 2, C_1 ($y = \sin x$) ranges from 0 to π and C_2 ($y = 2\sin x$) ranges from π to 0. Hence parametrize C_1 by letting $x=x$, $y=\sin x$ and for C_2 $x=x$ and $y=2\sin x$.

Since $\oint_C \vec{F} = \int_{C_1} \vec{F} + \int_{C_2} \vec{F}$ where $C = \{C_1, C_2\}$

$$\oint_C (1+y^2)dx + ydy = \int_{C_1} \left((1+\sin^2 x) + \sin x \cos x\right)dx$$

$$+ \int_{C_2} \left((1+4\sin^2 x) + 4\sin x \cos x\right)dx$$

$$= \int_0^\pi (1+\sin^2 x + \sin x \cos x)dx$$

$$+ \int_\pi^0 (1+4\sin^2 x + 4\sin x \cos x)dx$$

$$= \int_0^\pi (1+\sin^2 x + \sin x \cos x)dx$$

$$- \int_0^\pi (1+4\sin^2 x + 4\sin x \cos x)dx$$

$$= \int_0^\pi [-3\sin^2 x - 3\sin x \cos x]dx$$

$$= -3\int_0^\pi \sin^2 x \, dx - 3\int_0^\pi \sin x \cos x \, dx$$

$$= -3\int_0^\pi \frac{1-\cos 2x}{2} dx - 0 \quad \left(\text{since } \sin^2 x = \frac{1-\cos 2x}{2}\right)$$

$$= -3\int_0^\pi \frac{dx}{2} + 3\int_0^\pi \frac{\cos 2x}{2} dx = \frac{-3\pi}{2} + 0 = \frac{-3\pi}{2}.$$

Hence $\oint_C (1+y^2)dx + ydy = \frac{-3\pi}{2}.$ \hfill (6)

By Green's Theorem $p = 1+y^2$, $q=y$ and $\frac{\partial p}{\partial y} = 2y$, $\frac{\partial q}{\partial x} = 0$ so that

$$\oint_C (1+y^2)dx + ydy = \iint_R -2y \, dydx \hfill (7)$$

Where R is the region $\sin x \leq y \leq 2\sin x$ and $0 \leq x \leq \pi$.

399

Then (7) becomes

$$\int_0^\pi \int_{\sin x}^{2\sin x} -2y \, dy dx = \int_0^\pi \left(\frac{-y^2}{2}\bigg|_{\sin x}^{2\sin x}\right) dx$$

$$= \int_0^\pi -3\sin^2 x \, dx = -3 \int_0^\pi \frac{1-\cos 2x}{2} dx$$

$$= -3 \int_0^\pi \frac{dx}{2} + 3 \int_0^\pi \frac{\cos 2x}{2} dx$$

$$= \frac{-3\pi}{2} + 0 = \frac{-3\pi}{2}$$

which agrees with (6), thereby verifying Green's Theorem.

● **PROBLEM 8-34**

a) Verify Green's Theorem for the vector field $\vec{F}(x,y) = (-y,x)$ where $\vec{C}(t) = (r\cos t, r\sin t)$ $-\pi \le t \le \pi$ and $R = \{P: |P| \le r\}$.

b) Evaluate $\oint_C (x^2 + 2y^2) dx$ where C is the square with vertices $(1,1)$, $(1,-1)$, $(-1,-1)$, $(-1,1)$.

Solution: Green's Theorem states:

$$\oint_C p\, dx + q\, dy = \iint_R \left(\frac{\partial q}{\partial x} - \frac{\partial p}{\partial y}\right) dx dy$$ where C is a closed curve

and the boundary of R. Letting $p = -y$ and $q = x$, $\frac{\partial p}{\partial y} = -1$, $\frac{\partial q}{\partial x} = 1$ and by Green's Theorem:

$$\int_C -y\, dx + x\, dy = \iint_R (1 - (-1)) dx dy = 2 \iint_R dx dy \quad (1)$$

$= 2[\text{area of } R]$ where R is the circle of radius r since C, the boundary of R, is given parametrically as $\vec{C}(t) = (r\cos t, r\sin t)$. Hence (1) equals

$$2\pi r^2. \quad (2)$$

To verify Green's Theorem compute $\oint_C p\, dx + q\, dy$ and check the result with (2). Since $x = r\cos t$; $y = r\sin t$ on C, $dx = -r\sin t$; $dy = r\cos t$ and

$$\oint_C (-y)dx + xdy = \int_{-\pi}^{\pi} (-r\sin t)(-r\sin t)dt$$

$$+ \int_{-\pi}^{\pi} (r\cos t)(r\cos t)dt$$

$$= \int_{-\pi}^{\pi} (r^2\sin^2 t + r^2\cos^2 t)dt = \int_{-\pi}^{\pi} r^2 dt$$

$$= r^2(\pi+\pi) = 2\pi r^2 \quad \text{which agrees with (2) and thus}$$

verifies Green's Theorem.

b) This line integral could be evaluated along four different paths C_i where $C = \{C_1, C_2, C_3, C_4\}$, and each C_i is one side of the square. But since this is a closed curve, and C can be taken to be oriented counterclockwise, Green's Theorem may be applied.

Let $p = (x^2 + 2y^2)$ and $q = 0$; then $\frac{\partial p}{\partial y} = 4y$, $\frac{\partial q}{\partial x} = 0$

and $\oint_C (x^2+2y^2)dx + 0dy = \iint_R (0-4y)dxdy$

$$- 4\iint_R y \, dxdy \qquad (3)$$

which can be solved in two ways:

I) Compute the integral directly over R to obtain

$$-4 \int_{-1}^{1}\int_{-1}^{1} y \, dxdy = -8 \int_{-1}^{1} y \, dy = -8 \left. \frac{y^2}{2} \right|_{-1}^{1}$$

$$= -4(1-1) = 0.$$

II) Note that $\iint_R y \, dxdy = A\bar{y}$ where A is the area of the square and $(\bar{x}, \bar{y})$ is the centroid of the square (the point of the center of mass). Thus (3) equals

$$-4\iint_R y \, dxdy = -4A\bar{y} = 0 \qquad \text{since the centroid}$$

$(\bar{x}, \bar{y})$ is at $(0,0)$.

• **PROBLEM 8-35**

Use Green's Theorem to find:

a) $\int_C y^2 dx - xdy$ clockwise around the triangle whose vertices are at $(0,0)$, $(0,1)$, $(1,0)$.

b) The integral of the vector field $\vec{F}(x,y) = (y+3x, 2y-x)$ counterclockwise around the ellipse $4x^2 + y^2 = 4$.

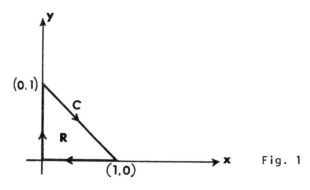

Fig. 1

Solution: Green's Theorem states:

$$\oint_C p\,dx + q\,dy = \iint_R \left(\frac{\partial q}{\partial x} - \frac{\partial p}{\partial y}\right) dx\,dy \tag{1}$$

where C is the boundary of the region R.

a) From Figure 1 the region lies to the right of the path. Therefore use C^- instead of C so that the region lies to the left. Then use the fact that

$$\int_C \vec{F} = -\int_{C^-} \vec{F}.$$ Hence

$$\int_C y^2\,dx - x\,dy = -\int_{C^-} y^2\,dx - x\,dy. \tag{2}$$

Now apply Green's Theorem to the right hand side of equation (2). Let $p = y^2$ and $q = -x$ so that $\frac{\partial q}{\partial x} = -1$ and $\frac{\partial p}{\partial y} = 2y$ to obtain

$$\int_C y^2\,dx - x\,dy = -\iint_R (-1-2y)\,dx\,dy.$$

From Figure 1 R is the region bounded by the x-axis, the y-axis, and the line x+y=1. Thus (2) becomes

$$\int_C y^2\,dx - x\,dy = -\int_0^1 \int_0^{1-y} -(1+2y)\,dx\,dy$$

$$= \int_0^1 \left(x+2yx \Big|_0^{1-y}\right) dy = \int_0^1 [1-y + 2y(1-y)]\,dy$$

$$= \int_0^1 (1+y-2y^2)\,dy = y + \frac{y^2}{2} - \frac{2y^3}{3}\Big|_0^1 = 1 + \frac{1}{2} - \frac{2}{3}$$

$$= \frac{5}{6}.$$

b) Since $\vec{F}(x,y)$ is being integrated counterclockwise over the ellipse $4x^2 + y^2 = 4$, we have the region R to the left of C with C being a closed curve. Given

$\vec{F}(x,y) = (y+3x, 2y-x)$; if we let $p = y+3x$ and $q = 2y-x$ then Green's Theorem may be applied. First calculate: $\frac{\partial p}{\partial y} = 1$ and $\frac{\partial q}{\partial x} = -1$, then substitute into (1) to obtain

$$\int_C (y+3x)dx + (2y-x)dy = \iint_R (-1-1)dxdy \qquad (3)$$

$$= -2 \iint_R dxdy = -2 \text{ (area of the ellipse)}.$$

If the ellipse is rewritten in the form $\frac{x^2}{a^2} + \frac{y^2}{b^2} = 1$ we could use the equation for area of the ellipse, $A = \pi ab$. Since the ellipse is given as $4x^2 + y^2 = 4$ rewrite it as $\frac{x^2}{1^2} + \frac{y^2}{2^2} = 1$ to obtain $A = \pi(1)(2) = 2\pi$. Thus

$$\int_{\text{ellipse}} \vec{F} = -2 \times 2\pi = -4\pi.$$

● **PROBLEM 8-36**

Verify Green's Theorem for $\int_C (x^2-y^2)dx + 2xy\,dy$, where C is the clockwise boundary of the square formed by the lines $x=0$, $x=2$, $y=0$, and $y=2$.

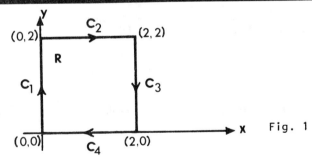

Fig. 1

Solution: Draw the square in Figure 1. To verify Green's Theorem for this line integral first compute $\int_C \vec{F}\cdot d\vec{c}$. To do this note that C consists of four paths C_1, C_2, C_3 and C_4. Then evaluate the line integral over each path using the equality:

$$\int_C \vec{F} = \int_{C_1} \vec{F} + \int_{C_2} \vec{F} + \int_{C_3} \vec{F} + \int_{C_4} \vec{F} \qquad \text{where}$$

$$C = \{C_1, C_2, C_3, C_4\}. \qquad (1)$$

403

Computing the line integral over C_1 first, use the parametrization $x=0$, $y=y$ so that $dx=0$, $dy=dy$ and

$$\int_{C_1} \vec{F} \cdot d\vec{c} = \int_{C_1} (0-y^2)0 + 0\,dy = 0. \tag{2}$$

Along C_2 use the parametrization $x=x$, $y=2$ so that $dx=dx$, $dy=0$ and

$$\int_{C_2} \vec{F} \cdot d\vec{c} = \int_{C_2} (x^2-4)\,dx + 4x \cdot 0 = \int_0^2 (x^2-4)\,dx$$

$$= \frac{8}{3} - 8 = -\frac{16}{3}. \tag{3}$$

Along C_3 use $x=2$, $y=y$, and $dx=0$, $dy=dy$ so that

$$\int_{C_3} \vec{F} \cdot d\vec{c} = \int_{C_3} (4-y^2) \cdot 0 + 4y\,dy = \int_2^0 4y\,dy$$

$$= 2y^2 \Big|_2^0 = -8. \tag{4}$$

Along C_4 use $x=x$, $y=0$ and $dx=dx$, $dy=0$ so that

$$\int_{C_4} \vec{F} \cdot d\vec{c} = \int_{C_4} (x^2-0)\,dx + 0 = \int_2^0 x^2\,dx = \frac{-8}{3}. \tag{5}$$

Now using equation (1) the integral of $\vec{F}$ over C equals the addition of (2), (3), (4), and (5). Thus

$$\int_C \vec{F} \cdot d\vec{c} = 0 - \frac{16}{3} - 8 - \frac{8}{3} = -16. \tag{6}$$

To verify Green's Theorem look at Figure 1 and notice that the curve C is oriented so that the region R is to the right as we advance along C. However, Green's Theorem states

$$\int_C p\,dx + q\,dy = \iint_R \left(\frac{\partial q}{\partial x} - \frac{\partial p}{\partial y} \right) dx\,dy \quad \text{where C is oriented}$$

so that the region R is to the left as we advance along C. So to apply Green's Theorem use C^- and the relation

$$\int_C \vec{F} \cdot d\vec{c} = -\int_{C^-} \vec{F} \cdot d\vec{c}. \quad \text{Therefore, let } p = x^2 - y^2,\ q = 2xy,$$

so that $\frac{\partial p}{\partial y} = -2y$ and $\frac{\partial q}{\partial x} = 2y$ and by Green's Theorem:

$$\int_C p\,dx + q\,dy = -\int_{C^-} p\,dx + q\,dy = -\iint_R (2y - (-2y))dx\,dy$$

$$= -4\iint_R y\,dx\,dy \qquad (7)$$

Equation (7) may be evaluated in two ways.

I) First compute it directly over R with $0 \le x \le 2$ and $0 \le y \le 2$ so that

$$-4\int_0^2\int_0^2 y\,dx\,dy = -8\int_0^2 y\,dy = -16 \qquad \text{which agrees with}$$

(6) and thus verifies Green's Theorem.

II) Second, by using the equation $\iint_R y\,dx\,dy = A\bar{y}$ where A is the area of R and $(\bar{x},\bar{y})$ is the centroid of R. Hence, since in this case, Area = 4 and $(\bar{x},\bar{y}) = (1,1)$,

$$-4\iint_R y\,dx\,dy = -4A\bar{y} = -4(4)(1) = -16$$

which agrees with (6).

● **PROBLEM 8-37**

Let $F(x,y) = \left(\dfrac{-y}{x^2+y^2},\ \dfrac{x}{x^2+y^2}\right)$. Find the integral of $\vec{F}$ over the path C shown in Figure 1.

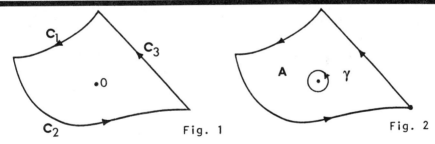

Fig. 1   Fig. 2

Solution: The path C consists of the three curves C_1, C_2, C_3 around the origin. However $\vec{F}(x,y)$ is not defined at the origin. Also, if the origin is not enclosed by the path C then the integral equals 0 since

$$\frac{\partial q}{\partial x} - \frac{\partial p}{\partial y} = \frac{(x^2+y^2) - 2x^2}{(x^2+y^2)^2} - \frac{-(x^2+y^2) + 2y^2}{(x^2+y^2)^2}$$

$$= \frac{2(x^2+y^2) - 2x^2 - 2y^2}{(x^2+y^2)^2} = 0.$$

Therefore by Green's Theorem,

$$\oint_C pdx + qdy = \iint_R \left(\frac{\partial q}{\partial x} - \frac{\partial p}{\partial y}\right) dxdy, \text{ or } \iint_R 0 dxdy = 0.$$

But the path C does enclose the origin and when computed has a value $\neq 0$. This integral can be computed separately over each curve C_i and using the fact that

$$\int_C \vec{F} = \int_{C_1} \vec{F} + \int_{C_2} \vec{F} + \int_{C_3} \vec{F} \text{ the desired value may be found.}$$

However this can be difficult due to the shape of each curve. But we can use Green's Theorem to simplify this problem. Draw a small circle γ around the origin, oriented counterclockwise. Therefore A is the region between the circle and the path. Letting γ^- denote the reverse path of γ^-, the boundary of A consists of the curves $\{C_1, C_2, C_3, \gamma^-\}$. Use γ^- to give the correct direction so that the region A lies to the left of each one of the curves. Now apply Green's Theorem to obtain

$$\int_C \vec{F} + \int_{\gamma^-} \vec{F} = \iint_A \left(\frac{\partial q}{\partial x} - \frac{\partial p}{\partial y}\right) dxdy = 0 \text{ by the computation}$$

earlier in this problem. Hence $\int_C \vec{F} = -\int_{\gamma^-} \vec{F} = \int_\gamma \vec{F}$.

Thus $\int_C \vec{F}$ equals the same value k for all curves C enclosing the hole at the origin. Thus whatever C is chosen, we will arrive at the same value. Therefore, let C be the circle $x^2 + y^2 = 1$ and using the parametrization $x = \cos\theta$, $y = \sin\theta$, $0 \leq \theta \leq 2\pi$

$$\int_C \vec{F} \cdot dc = \int_0^{2\pi} \frac{-\sin\theta}{1}(-\sin\theta)d\theta + \int_0^{2\pi}\frac{\cos\theta}{1}(\cos\theta)d\theta$$

$$= \int_0^{2\pi}(\sin^2\theta + \cos^2\theta)d\theta = \int_0^{2\pi} d\theta = 2\pi.$$

Thus for all curves C enclosing the origin, the integral of $\vec{F}(x,y) = \left(\frac{-y}{x^2+y^2}, \frac{x}{x^2+y^2}\right)$ over C equals 2π.

● **PROBLEM 8-38**

Evaluate $\int_C \frac{y^3 dx - xy^2 dy}{(x^2+y^2)^2}$ around the ellipse $x^2 + 3y^2 = 1$.

Solution: To evaluate this line integral first notice that the vector field is not defined at the origin. Therefore, the integral of this vector field over a closed path including the origin does not equal zero. This integral could be computed directly over the given

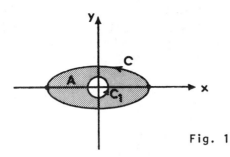

Fig. 1

ellipse, but the integration would be awkward. Thus a different method that involves Green's Theorem is used. Draw a small circle C_1 around the origin, oriented counterclockwise, inside the ellipse (see Figure 1). Let A be the region between the circle and the ellipse. Then the boundary of A consists of the curves $\{C, C_1^-\}$ where C_1^- is the reverse path of C_1.

Now, since $\frac{\partial p}{\partial y}$, $\frac{\partial q}{\partial x}$ are defined at all points in A and using C_1^- the region A lies to the left of each of the curves and Green's Theorem may be applied. Then

$$\int_C \vec{F} + \int_{C_1^-} \vec{F} = \iint_A \left(\frac{\partial q}{\partial x} - \frac{\partial p}{\partial y}\right) dxdy. \tag{1}$$

Since $\frac{\partial q}{\partial x} = \frac{-y^2(x^2+y^2)^2 + 4x^2y^2(x^2+y^2)}{(x^2+y^2)^4}$

$$= \frac{-y^2(x^4+y^4+2x^2y^2) + 4x^4y^2 + 4x^2y^4}{(x^2+y^2)^4}$$

$$\frac{\partial q}{\partial x} = \frac{3x^4y^2 + 2x^2y^4 - y^6}{(x^2+y^2)^4}$$

and $\frac{\partial p}{\partial y} = \frac{3y^2(x^2+y^2)^2 - 4y^4(x^2+y^2)}{(x^2+y^2)^4}$

$$= \frac{3y^2x^4 + 3y^6 + 6x^2y^4 - 4x^2y^4 - 4y^6}{(x^2+y^2)^4}$$

$\frac{\partial p}{\partial y} = \frac{3y^2x^4 + 2x^2y^4 - y^6}{(x^2+y^2)^4}$. Thus $\frac{\partial q}{\partial x} = \frac{\partial p}{\partial y}$ and $\frac{\partial q}{\partial x} - \frac{\partial p}{\partial y} = 0$. Hence (1) becomes

$$\int_C \vec{F} + \int_{C_1^-} \vec{F} = 0 \quad \text{so that} \quad \int_C \vec{F} = -\int_{C_1^-} \vec{F} = \int_{C_1} \vec{F} \tag{2}$$

Observe that the integral around the ellipse has the same value as the integral around the circle (or for any other closed curve). Therefore replace the ellipse by the circle $x^2+y^2=1$, which also enclosed the point of discontinuity (the origin). With the parametrization $x = \cos\theta$, $y = \sin\theta$, $0 \le \theta \le 2\pi$, the integral on the circle is

$$\int_0^{2\pi} \frac{(-\sin^3\theta\sin\theta - \cos\theta\sin^2\theta\cos\theta)}{(\cos^2\theta + \sin^2\theta)^2} d\theta$$

$$= \int_0^{2\pi} (-\sin^4\theta - \cos^2\theta\sin^2\theta) d\theta$$

$$= \int_0^{2\pi} -\sin^2\theta (\sin^2\theta + \cos^2\theta) d\theta$$

$$= \int_0^{2\pi} -\sin^2\theta \, d\theta.$$

Using $\sin^2\theta = \frac{1-\cos 2\theta}{2}$

$$\int_0^{2\pi} \left(-\frac{1}{2} + \frac{\cos 2\theta}{2}\right) d\theta = -\int_0^{2\pi} \frac{d\theta}{2} + \int_0^{2\pi} \frac{\cos 2\theta}{2} d\theta$$

$$= -\pi + 0 = -\pi.$$

Thus $\int_C \frac{y^3 dx - xy^2 dy}{(x^2+y^2)^2}$ where C is the ellipse $x^2 + 3y^2 = 1$

equals $-\pi$.

• **PROBLEM 8-39**

Use Green's Theorem to deduce the integral formula:

$$\iint_R \left(\frac{\partial^2 u}{\partial x^2} + \frac{\partial^2 u}{\partial y^2}\right) dxdy = \int_C \frac{\partial u}{\partial n} ds \qquad (1)$$

where s is the arc length along C and n is the outer normal to C.

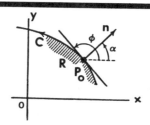

Solution: To derive (1) start by applying Green's Theorem which states

$$\oint_C p\,dx + q\,dy = \iint_R \left(\frac{\partial q}{\partial x} - \frac{\partial p}{\partial y}\right) dx\,dy$$

where C is a closed curve. Therefore letting $q = \frac{\partial u}{\partial x}$ and $p = -\frac{\partial u}{\partial y}$

$$\iint_R \left(\frac{\partial^2 u}{\partial x^2} + \frac{\partial^2 u}{\partial y^2}\right) dx\,dy = \int_C -\frac{\partial u}{\partial y} dx + \frac{\partial u}{\partial x} dy \qquad (2)$$

Hence we have to show that on C

$$-\frac{\partial u}{\partial y} dx + \frac{\partial u}{\partial x} dy = \frac{\partial u}{\partial n} ds \quad \text{which can be rewritten as}$$

$$-\frac{\partial u}{\partial y}\frac{dx}{ds} + \frac{\partial u}{\partial x}\frac{dy}{ds} = \frac{\partial u}{\partial n}. \qquad (3)$$

To do this look at Figure 1. Let P_0 be a point on the boundary of R and let n be the outward unit normal at P_0. Let α be the angle counterclockwise from the positive x-direction to the direction of n. Given that u is a function defined in R with first partials that are continuous in R, then the outer normal derivative of u at P_0 is defined to be

$$\frac{\partial u}{\partial n} = \frac{\partial u}{\partial x} \cos\alpha + \frac{\partial u}{\partial y} \sin\alpha \qquad (4)$$

where $\frac{\partial u}{\partial x}$ and $\frac{\partial u}{\partial y}$ are evaluated at P_0. Also $dx = ds \cos\phi$ and $dy = ds \sin\phi$ where ϕ is the angle which the positive tangent to C makes with the positive x-axis. Also from the Figure, the curve is directed so that the interior is always to the left and the outward normal lags behind the tangent by $\frac{\pi}{2}$ radians. Therefore $\alpha = \phi - (\frac{\pi}{2})$ and since $\cos\phi = \frac{dx}{ds}$ and $\sin\phi = \frac{dy}{ds}$ we have $\sin\alpha = -\frac{dx}{ds}$ and $\cos\alpha = \frac{dy}{ds}$. Hence substituting into (4),

$$\frac{\partial u}{\partial n} = \frac{\partial u}{\partial x}\frac{dy}{ds} - \frac{\partial u}{\partial y}\frac{dx}{ds} \quad \text{which is precisely (3)}.$$

Thus equation (2) now reads

$$\iint_R \left(\frac{\partial^2 u}{\partial x^2} + \frac{\partial^2 u}{\partial y^2}\right) dx\,dy = \int_C \frac{\partial u}{\partial n} ds \quad \text{which is}$$

equation (1).

● **PROBLEM 8-40**

Show that Green's Theorem can be used to prove the following theorem of Pappus: Let R be a regular region lying entirely on one side of the x-axis. If R is revolved about the x-axis, the volume of the solid that is generated equals $2\pi A\bar{y}$ where A is the area of R, and $(\bar{x},\bar{y})$ is the centroid of R.

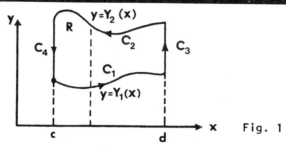

Fig. 1

Solution: First a regular region must be defined. Let R be a bounded closed region in the xy-plane. Let a finite number of simple closed curves which do not intersect each other, and each of which is sectionally smooth, comprise the boundary of R. Let C represent the aggregate of these curves, each oriented so that the region is to the left of the direction of each curve. Then R is a regular region. For this problem consider the case of the regular region pictured in Figure 1 and assume the reasoning can be extended to the most general regular region. When the region R is revolved about the x-axis, the volume of R equals the volume generated by revolving the upper curve ($y = Y_2(x)$) minus the volume generated by revolving the lower curve ($y = Y_1(x)$). From elementary calculus if a length Δx is taken and revolved about the x-axis, the volume generated equals $\int_a^b \pi y^2 dx$ where a and b are the endpoints of the length Δx. Therefore,

$$V = \int_c^d \pi (Y_2(x))^2 dx - \int_c^d \pi (Y_1(x))^2 dx$$

$$= \pi \left[\int_c^d (Y_2(x))^2 dx - \int_c^d (Y_1(x))^2 dx \right]$$

$$= -\pi \left[\int_c^d (Y_1(x))^2 dx - \int_c^d (Y_2(x))^2 dx \right]$$

$$V = -\pi \left[\int_c^d (Y_1(x))^2 dx + \int_d^c (Y_2(x))^2 dx \right] \tag{1}$$

Letting $y = Y_1(x)$ be curve C_1 and $y = Y_2(x)$ be curve C_2 and using the relation $\int_C F = \int_{C_1} F + \int_{C_2} F + \int_{C_3} F + \int_{C_4} F$

where $C = \{C_1, C_2, C_3, C_4\}$ (1) becomes

$$V = -\pi \left[\int_{C_1} y^2 dx + \int_{C_2} y^2 dx \right] = -\pi \int_C y^2 dx \quad \text{with } C$$

oriented counterclockwise. Observe that along C_3 and C_4 x is constant so $dx = 0$ and hence

$$\int_{C_3} F dx = \int_{C_4} F dx = 0.$$

Therefore, Green's Theorem may now be applied. Since

$$V = -\pi \int_C y^2 dx \quad \text{let} \quad P = -\pi y^2, \quad Q = 0. \quad \text{Thus by Green's}$$

Theorem:

$$V = -\pi \int_C y^2 dx = \iint_R (0 - (-2\pi y)) dx dy$$

$$V = 2\pi \iint_R y \, dx dy. \tag{2}$$

Now if a lamina (thin sheet) is specified to shape and position by a region R in the xy-plane, and if σ represents the constant area density, the coordinates of its center of gravity $(\bar{x}, \bar{y})$ are:

$$m\bar{x} = \iint_R x \sigma dA, \quad m\bar{y} = \iint_R y \sigma dA \tag{3}$$

where m is the mass of the lamina. Letting A represent the area of R, $m = \sigma A$. Therefore substituting into (3),

$$\sigma A \bar{x} = \iint_R x \sigma dA; \quad \sigma A \bar{y} = \iint_R y \sigma dA \quad \text{which become}$$

$$A\bar{x} = \iint_R x \, dA \quad \text{and} \quad A\bar{y} = \iint_R y \, dA.$$

Hence from equation (2)

$$V = 2\pi A \bar{y}, \quad \text{which is the desired result.}$$

● **PROBLEM 8-41**

Find the area of a triangle by use of line integrals.

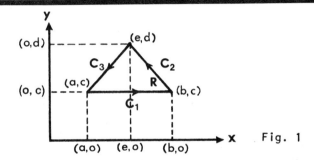

Fig. 1

Solution: From elementary calculus it is known that the area of a region R_1 is given by the following double integral,

$$A = \iint_{R_1} dx\, dy. \tag{1}$$

From Green's Theorem it is known that

$$\oint_C p\,dx + q\,dy = \iint_R \left(\frac{\partial q}{\partial x} - \frac{\partial p}{\partial y}\right) dx\,dy \tag{2}$$

where C is a closed curve boundary of R, oriented counterclockwise. Assume R_1 is a region to which Green's Theorem is applicable and let C^* be the boundary of R_1, oriented counterclockwise. Let $q = x$ and $p = 0$ in (1); then by (2)

$$\oint_{C^*} 0\cdot dx + x\,dy = \iint_{R_1} (1-0)\,dx\,dy \quad \text{since } \frac{\partial x}{\partial x} = 1.$$

Then

$$\oint_{C^*} x\,dy = \iint_{R_1} dx\,dy = \text{Area of } R_1. \tag{3}$$

For this problem let R_1 be the triangle in Figure 1. By (3) the area of this triangle is $\oint_{C^*} x\,dy$. However C^* consists of the paths C_1, C_2, C_3. Hence

$$\oint_{C^*} x\,dy = \int_{C_1} x\,dy + \int_{C_2} x\,dy + \int_{C_3} x\,dy \tag{4}$$

To compute $\int_{C_1} x\,dy$, use the parametrization given for a line segment, $P_0 = P + t(Q-P)$.

Along C_1 $\quad x = a + t(b-a)$, $y = c + t(c-c) = c$ and

$$dx = (b-a)\,dt, \quad dy = 0\cdot dt \quad 0 \leq t \leq 1.$$

Thus

$$\int_{C_1} x\,dy = \int_0^1 (a + t(b-a))\cdot 0 = 0. \tag{5}$$

For $\int_{C_2} x\,dy$ again use the parametrization $P_0 = P + t(Q-P)$.

Here $x = b + t(e-b)$, $y = c + t(d-c)$ so that $dy = (d-c)\,dt$, $0 \leq t \leq 1$. Then

$$\int_{C_2} x\,dy = \int_0^1 [b + t(e-b)][d - c]\,dt$$

$$= \int_0^1 [b(d-c) + t(e-b)(d-c)]\,dt$$

$$= \int_0^1 b(d-c)\,dt + \int_0^1 (e-b)(d-c)t\,dt$$

$$= b(d-c)t\Big|_0^1 + (e-b)(d-c)\frac{t^2}{2}\Big|_0^1$$

$$= b(d-c) + \frac{(e-b)(d-c)}{2}. \tag{6}$$

For $\int_{C_3} x\,dy$ use the same parametrization as C_1, C_2. Then $x = e + t(a-e)$, $y = d + t(c-d)$ so that $dy = (c-d)\,dt$, $0 \le t \le 1$. Thus it follows that

$$\int_{C_3} x\,dy = \int_0^1 [e + t(a-e)][c-d]\,dt$$

$$= \int_0^1 e(c-d)\,dt + \int_0^1 (a-e)(c-d)t\,dt$$

$$= e(c-d)t\Big|_0^1 + (a-e)(c-d)\frac{t^2}{2}\Big|_0^1$$

$$= e(c-d) + \frac{(a-e)(c-d)}{2}. \tag{7}$$

Then by (4), (5), (6), and (7),

$$\oint_{C^*} x\,dy = 0 + b(d-c) + \frac{(e-b)(d-c)}{2} + e(c-d) + \frac{(a-e)(c-d)}{2}$$

$$= \frac{2b(d-c) + e(d-c) - b(d-c) + 2e(c-d) + a(c-d) - e(c-d)}{2}$$

$$= \frac{b(d-c) + e(d-c) - 2e(d-c) - a(d-c) + e(d-c)}{2}$$

$$= \frac{b(d-c) - a(d-c)}{2} = \frac{(b-a)(d-c)}{2}. \tag{8}$$

However, from Fig. 1, $(b-a)$ is the base of the triangle;

(d-c) is the height. Thus letting B = (b-a); H = (d-c), (8) can be rewritten as

$$A = \frac{1}{2} BH ; \quad \text{which is precisely the area of}$$

a triangle.

● **PROBLEM 8-42**

Applying Green's Theorem to a plane flow, show the rate of flow out of S equals the integral over S of the divergence of the flow.

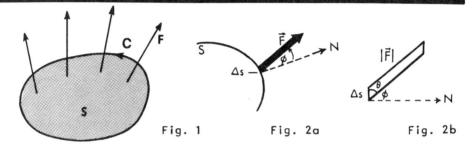

Fig. 1   Fig. 2a   Fig. 2b

Solution: Given a vector field $\vec{F}(x,y) = (f(x,y), g(x,y))$ the divergence of $\vec{F}$ (written $\nabla \cdot \vec{F}$) equals $\left(\frac{\partial f(x,y)}{\partial x} + \frac{\partial g(x,y)}{\partial y} \right)$. Therefore, we want to prove that the rate of flow out of S equals

$$\iint_S \nabla \cdot \vec{F} \, dA = \iint_S \left(\frac{\partial f(x,y)}{\partial x} + \frac{\partial g(x,y)}{\partial x} \right) dxdy. \tag{1}$$

In Figure 1 a curve C is oriented counterclockwise around a region S. To determine the rate of flow out of S (the amount of substance that flows out in a given time) let $\vec{F}(x,y)$ be the velocity of the substance at (x,y). Momentarily concentrating on one point of C (in Figure 2), the area of the arrow $\vec{F}$ is how much substance flows out of S. Using the equation for the area of a parallelogram A = ab sinθ, take Δs to be small enough so that $\vec{F}$ can be approximated by a parallelogram to obtain A = $|\vec{F}|$Δs sinθ from Figure 2(b). However since θ is the complement of φ

$$A = |\vec{F}| \Delta s \cos\phi = \vec{F} \cdot \vec{N} \Delta s \quad \text{where } \vec{N} \text{ represents the}$$

unit normal vector. Now taking the limit as Δs goes to zero, and summing over all points of S

$$\lim_{\Delta s \to 0} \Sigma \vec{F} \cdot \vec{N} \Delta s = \int_{S_0}^{S_1} \vec{F} \cdot \vec{N} \, ds. \tag{2}$$

Given a curve $c = c(x,y)$ $\left(\frac{dx}{ds}, \frac{dy}{ds}\right)$ is a unit tangent vector. Using the fact that a normal vector is perpendicular to a tangent vector and that the dot product of two perpendicular vectors equals zero, the unit normal is $\left(\frac{dy}{ds}, -\frac{dx}{ds}\right)$ since $\left(\frac{dx}{ds}, \frac{dy}{ds}\right) \cdot \left(\frac{dy}{ds}, -\frac{dx}{ds}\right) = 0$. Therefore (2) becomes

$$\int_{S_0}^{S_1} (f,g) \cdot \left(\frac{dy}{ds}, -\frac{dx}{ds}\right) ds$$

$$= \int_{S_0}^{S_1} \left(f(x,y)\frac{dy}{ds} - g(x,y)\frac{dx}{ds}\right) ds$$

$$= \int_C -g(x,y)\,dx + f(x,y)\,dy \qquad (3)$$

which is the rate of flow out of S. Letting $p = -g(x,y)$ and $q = f(x,y)$, by Green's Theorem (3) equals

$$\iint_S \left[\frac{\partial f(x,y)}{\partial x} - \left(\frac{-\partial g(x,y)}{\partial x}\right)\right] dxdy. \quad \text{Thus,}$$

Rate of flow $= \iint_S \left(\frac{\partial f(x,y)}{\partial x} + \frac{\partial g(x,y)}{\partial x}\right) dxdy$ which is precisely equation (1).

• **PROBLEM 8-43**

Show that if the flow $\vec{F}$ is defined in a rectangle R, and has zero divergence at every point of R, then the rate of flow across every closed path in R is zero.

Solution: Given a vector field $\vec{F}(x,y) = (f(x,y), g(x,y))$
$\text{div}\vec{F} = \frac{\partial f(x,y)}{\partial x} + \frac{\partial g(x,y)}{\partial y}$. The rate of flow across a closed curve C is

$$\oint_C -g(x,y)\,dx + f(x,y)\,dy. \qquad (1)$$

By hypothesis, the vector $\vec{F}(x,y)$ (defined in the rectangle R) is the flow and has zero divergence at every point of R. Therefore

$$\text{div}\vec{F} = \frac{\partial f}{\partial x} + \frac{\partial g}{\partial y} = 0 \qquad (2)$$

so that $\quad \dfrac{\partial f}{\partial x} = -\dfrac{\partial g}{\partial y}\qquad$ (3)

Using the above it can be shown that the rate of flow across every closed path in R is zero in two ways.

I) For the first let C be any closed curve in R and let S be the region bounded by C. Take C to be oriented counterclockwise. Then by Green's Theorem and equation (1), the rate of flow across C equals

$$\oint_C -g(x,y)\,dx + f(x,y)\,dy = \iint_S \left[\dfrac{\partial f(x,y)}{\partial x} - \left(-\dfrac{\partial g(x,y)}{\partial y}\right)\right] dxdy$$

$$= \iint_S \left(\dfrac{\partial f(x,y)}{\partial x} + \dfrac{\partial g(x,y)}{\partial y}\right) dxdy$$

$$= \iint_S 0\, dxdy = 0 \quad \text{by equation (2). Hence the}$$

rate of flow across any closed path C in R = 0.

II) For the second method, since (3) is true $\dfrac{\partial f}{\partial x} = -\dfrac{\partial g}{\partial y}$, it follows that the vector field $\vec{F}(x,y) = (-g,f)$ is exact in R. This may be seen from the theorem: Given $\vec{F}(f,g) = fdx + gdy$ (f,g functions). If f and g have continuous first partial derivatives at all points of some open rectangle, then $\vec{F}(f,g)$ is exact at each point of the rectangle if and only if the condition $\dfrac{\partial f}{\partial y} = \dfrac{\partial g}{\partial x}$ is satisfied throughout the rectangle. Hence since $\vec{F}=(-g,f)$ is exact in R, $\vec{F}=\vec{\nabla}\phi$ where ϕ is some single-valued differentiable function such that $\dfrac{\partial \phi}{\partial x} = -g$ and $\dfrac{\partial \phi}{\partial y} = f$. Thus

$$\oint_C -gdx + fdy = \int_{P_1}^{P_2} \left(\dfrac{\partial \phi}{\partial x}, \dfrac{\partial \phi}{\partial y}\right)\cdot (dx,dy)$$

$$= \phi(P_2) - \phi(P_1) \qquad (4)$$

where P_2 and P_1 are the endpoints of the path. But $P_2 = P_1$ since C is a closed path and (4) equals 0. Hence $\oint_C -gdx + fdy = 0$ for every closed path C in R.

CHAPTER 9

SURFACE INTEGRALS

A surface integral is an integral of a function which is defined on a surface. In elementary calculus we performed a double integral of a function which was defined over a region R in the xy-plane. This plane region R, however, was a flat surface, thereby showing double integrals to be a special case of surface integrals.

Surface integrals have two forms. One is the integral of a function over a surface denoted by $\iint_S f\, dA$, where f is a function on S. The second is the integral of a vector field over a surface denoted by $\iint_S \vec{F}\cdot \hat{n}\, dA$, where $\vec{F}$ is a vector field on S and $\hat{n}$ is the outward unit normal to the surface.

The evaluation of a surface integral is frequently accomplished by reducing it to a double integral. This is done by various methods depending upon the representation of the surface. Surface integrals can also be evaluated by the Divergence Theorem, Stokes's Theorem, and the Change of Variables Formula.

Surface Integrals are useful in the formulation of physical concepts such as finding area, centers of gravity, moments of inertia of curved laminas, and other important physical quantities.

CHANGE OF VARIABLES FORMULA • PROBLEM 9-1

Let U be open in R^2, and let R^* be a region which is the interior of a closed path C^* contained in U. Let $G: U \to R^2$ such that $\Delta_G > 0$ (Δ_G = Jacobian of G).
Let C be the image of the path C^* under the mapping G, and let R denote the interior of the region enclosed by C. Let $\varphi = \varphi(x,y)$ be a continuous function on R such that $\varphi(x,y) = \partial q/\partial x$ for some continuously differentiable function $q(x,y)$. Prove:

$$\iint_R \varphi(x,y)\, dy\, dx = \iint_{R^*} \varphi(G(u,v))\, \Delta_G(u,v)\, du\, dv\ .$$

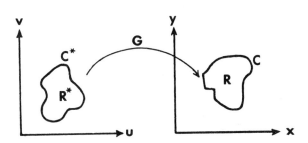

Fig. 1

Solution: Recall Green's Theorem which states $\oint_C p(x,y)dx + q(x,y)dy = \iint_R \left(\frac{\partial q}{\partial x} - \frac{\partial p}{\partial y}\right) dx\, dy$ (C is a closed path). Letting $p(x,y) = 0$ and $q = q(x,y)$, it follows that

$$\int_C q(x,y)\, dy = \int_R \frac{\partial q}{\partial x}(x,y)\, dx\, dy. \tag{1}$$

Since the Jacobian of G is non-zero, by the inverse function theorem G is locally one-to-one and $G(U)$ is an open set.

Write $G(u,v) = (G_1(u,v), G_2(u,v))$ and let C^* be parametrized by $(u(s), v(s))$, $a \le s \le b$, with $(u(a), v(a)) = (u(b), v(b))$. Then by definition C is parametrized as $(x(s), y(s)) \equiv (G_1(u(s), v(s)), G_2(u(s), v(s)))$, $a \le s \le b$. Let $Q(u,v) = q(x,y) \equiv q(G_1(u,v), G_2(u,v))$. Then

$$\int_C q(x,y)\, dy$$

$$= \int_a^b q(x(s), y(s))\frac{dy}{ds}\, ds$$

$$= \int_a^b Q(u(s), v(s))\left[\frac{\partial G_2}{\partial u}(u(s),v(s))\frac{du}{ds} + \frac{\partial G_2}{\partial v}(u(s),v(s))\frac{dv}{ds}\right] ds$$

$$= \int_{C^*} F_1(u,v)\, du + F_2(u,v)\, dv \tag{2}$$

where $F_1(u,v) = Q(u,v)\frac{\partial G_2}{\partial u}(u,v)$

$F_2(u,v) = Q(u,v)\frac{\partial G_2}{\partial v}(u,v)$.

Again applying Green's theorem to the right side of (2) gives

$$\int_C q(x,y)\, dy = \int_{R^*}\left[\frac{\partial F_2}{\partial u}(u,v) - \frac{\partial F_1}{\partial v}(u,v)\right] du\, dv. \tag{1'}$$

Now $\frac{\partial F_2}{\partial u}(u,v) = \frac{\partial Q}{\partial u}(u,v)\frac{\partial G_2}{\partial v}(u,v) + Q(u,v)\frac{\partial^2 G_2}{\partial v \partial u}(u,v)$. \quad (3)

By the inverse function theorem G is locally invertible, hence

$$\frac{\partial Q}{\partial u}(u,v) = \frac{\partial}{\partial u} q(x,y) = \frac{\partial q}{\partial x}\frac{\partial x}{\partial u} + \frac{\partial q}{\partial y}\frac{\partial y}{\partial u}$$

(one should bear in mind here that $(x,y) = (G_1(u,v), G_2(u,v))$)

or

$$\frac{\partial Q}{\partial u}(u,v) = \frac{\partial q}{\partial x}\frac{\partial G_1}{\partial u}(u,v) + \frac{\partial q}{\partial y}\frac{\partial G_2}{\partial u}(u,v), \tag{4}$$

similarly, $\frac{\partial F_1}{\partial v} = \frac{\partial Q}{\partial v}\frac{\partial G_2}{\partial u} + Q(u,v)\frac{\partial^2 G_2}{\partial u \partial v}(u,v)$ \quad (5)

and

$$\frac{\partial Q}{\partial v}(u,v) = \frac{\partial q}{\partial x}\frac{\partial G_1}{\partial v} + \frac{\partial q}{\partial y}\frac{\partial G_2}{\partial v}. \tag{6}$$

Substituting (4) in (3) and (6) in (5) yields

$$\frac{\partial F_2}{\partial u}(u,v) - \frac{\partial F_1}{\partial v}(u,v) = \frac{\partial Q}{\partial u}(u,v)\frac{\partial G_2}{\partial v}(u,v) - \frac{\partial Q}{\partial v}\frac{\partial G_2}{\partial u}$$

$$= \frac{\partial q}{\partial x}\frac{\partial G_1}{\partial u}\frac{\partial G_2}{\partial v} + \frac{\partial q}{\partial y}\frac{\partial G_2}{\partial u}\frac{\partial G_2}{\partial v}$$

$$- \frac{\partial q}{\partial x}\frac{\partial G_1}{\partial v}\frac{\partial G_2}{\partial u} - \frac{\partial q}{\partial y}\frac{\partial G_2}{\partial v}\frac{\partial G_2}{\partial u}$$

$$= \frac{\partial q}{\partial x}\begin{vmatrix} \dfrac{\partial G_1}{\partial u} & \dfrac{\partial G_1}{\partial v} \\ \dfrac{\partial G_2}{\partial u} & \dfrac{\partial G_2}{\partial v} \end{vmatrix}$$

$$= \varphi(x,y)\, \Delta_G(u,v)$$

$$= \varphi(G_1(u,v), G_2(u,v))\, \Delta_G(u,v)$$

or

$$\frac{\partial F_2}{\partial u} - \frac{\partial F_1}{\partial v} = \varphi(G(u,v))\, \Delta_G(u,v). \qquad (7)$$

Finally, using (7) in (1') and making use of equation (1) we get the desired result

$$\int_C q(x,y)dy = \int_R \varphi(x,y)dx\, dy = \int_{R^*} \varphi(G(u,v))\Delta_G(u,v)du\, dv.$$

• **PROBLEM 9-2**

Evaluate $\iint_{R_{xy}} (x+y)^3\, dx\, dy$ where R_{xy} is the parallelogram shown in Figure 1.

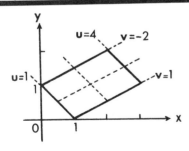
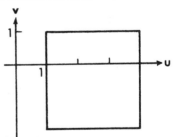

Fig. 1

Solution: From the diagram observe that the sides of R_{xy} are straight lines having equations of form $x+y = k_1$ and $x-2y = k_2$ for appropriate choices of k_1, k_2. Therefore, introduce as new coordinates $u = x+y$; $v = x - 2y$. The region R_{xy} then corresponds to the rectangle $1 \le u \le 4$, $-2 \le v \le 1$ in the uv-plane (a one-to-one correspondence).
Next, use the change of variables formula which states

$$\iint_{G(R)} f(x,y)\,dxdy = \iint_R f(G(u,v))|\Delta_G(u,v)|\,du\,dv . \qquad (1)$$

$$|\Delta_G(u,v)| = \left|\frac{\partial(x,y)}{\partial(u,v)}\right|$$

can be computed in two ways. One way is by solving for x and y in terms of u and v and finding $y = (u-v)/3$ and $x = (2u+v)/3$.

Hence $\left|\dfrac{\partial(x,y)}{\partial(u,v)}\right| = \left|\begin{matrix}\frac{2}{3} & \frac{1}{3}\\ \frac{1}{3} & -\frac{1}{3}\end{matrix}\right| = \left|-\frac{3}{9}\right| = \frac{1}{3}$. An alternative is to use the reciprocal:

$$\left|\frac{\partial(x,y)}{\partial(u,v)}\right| = \frac{1}{\left|\frac{\partial(u,v)}{\partial(x,y)}\right|} = \frac{1}{\left\|\begin{matrix}1 & 1\\ 1 & -2\end{matrix}\right\|} = \left|-\frac{1}{3}\right| = \frac{1}{3} .$$

Therefore, substitution into (1) yields:

$$\iint_{R_{xy}} (x+y)^3\,dxdy = \iint_{R_{uv}} u^3 \cdot \tfrac{1}{3}\,du\,dv$$

$$= \int_{-2}^{1}\int_{1}^{4} \frac{u^3}{3}\,du\,dv = \int_{1}^{4}\frac{u^3\,du}{3}\int_{-2}^{1}dv$$

$$= \frac{256-1}{12}\cdot(3) = \frac{255}{4} = 63\tfrac{3}{4} .$$

● **PROBLEM 9-3**

a) Let $f(x,y)$ be a continuous function in the xy-plane. Find an integral expression for $\iint_{G(R)} f(x,y)\,dxdy$ in polar coordinates using the change of variables formula.

b) Repeat for $f(x,y,z)$ and $\iiint f(x,y,z)\,dxdydz$ represented by spherical coordinates.

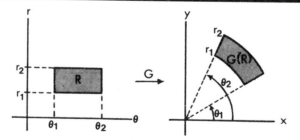

Fig. 1

Solution: a) The change of variables formula in two dimensions is:

$$\iint_{G(R)} f(x,y)\,dxdy = \iint_R f(G(u,v))|\Delta_G(u,v)|\,du\,dv .$$

Letting $x = r\cos\theta$ and $y = r\sin\theta$, $(x,y) = G(r,\theta) = (r\cos\theta, r\sin\theta)$. Therefore

$$|\Delta_G(r,\theta)| = \left|\frac{\partial(x,y)}{\partial(r,\theta)}\right| = \left\|\begin{matrix}\cos\theta & -r\sin\theta\\ \sin\theta & r\cos\theta\end{matrix}\right\| = r\cos^2\theta + r\sin^2\theta = r.$$

Hence $\iint_{G(R)} f(x,y)\,dx\,dy = \iint_R f(r\cos\theta, r\sin\theta)\,r\,dr\,d\theta .$

The mapping $G(r,\theta)$ for the rectangle $0 \le r_1 \le r \le r_2$ and $0 \le \theta_1 \le \theta \le \theta_2 \le 2\pi$ is shown in Figure 1.

b) The change of variables formula in 3 dimensions is:

$$\iiint_{G(R)} f(x,y,z)\,dx\,dy\,dz = \iiint_R f(G(u,v,w))\,|\Delta_G(u,v,w)|\,du\,dv\,dw .$$

Letting $x = r \sin\varphi \cos\theta$, $y = r \sin\varphi \sin\theta$, and $z = r \cos\varphi$

$$G(r,\varphi,\theta) = (r \sin\varphi \cos\theta,\ r \sin\varphi \sin\theta,\ r \cos\varphi) .$$

Therefore,

$$|\Delta_G(r,\varphi,\theta)| = \left|\frac{\partial(x,y,z)}{\partial(r,\varphi,\theta)}\right| = \begin{Vmatrix} \sin\varphi \cos\theta & r\cos\varphi \cos\theta & -r\sin\varphi \sin\theta \\ \sin\varphi \sin\theta & r\cos\varphi \sin\theta & r\sin\varphi \cos\theta \\ \cos\varphi & -r\sin\varphi & 0 \end{Vmatrix}$$

$= |\sin\varphi \cos\theta\,(r^2 \sin^2\varphi \cos\theta) + r\cos\varphi \cos\theta\,(r\sin\varphi \cos\varphi \cos\theta)|$
$\quad - r\sin\varphi \sin\theta\,(-r\sin^2\varphi \sin\theta - r\cos^2\varphi \sin\theta)|$

$= |r^2\sin\varphi(\sin^2\varphi + \cos^2\theta) + r^2\sin\varphi(\cos^2\varphi\cos^2\theta) - r\sin\varphi \sin\theta(-r\sin\theta)|$

$= |r^2\sin\varphi(\sin^2\varphi \cos^2\theta + \cos^2\varphi\cos^2\theta + \sin^2\theta)|$

$= |r^2\sin\varphi((\cos^2\varphi + \sin^2\varphi)\cos^2\theta + \sin^2\theta)| = |r^2\sin\varphi| = r^2\sin\varphi$ since $0 \le \varphi \le \pi$.

Hence $\iiint_{G(R)} f(x,y,z)\,dx\,dy\,dz = \iiint_R f(r\sin\varphi\cos\theta, r\sin\varphi\sin\theta,$

$r\cos\varphi)\,r^2\sin\varphi\,dr\,d\varphi\,d\theta .$

The mapping $G(r,\varphi,\theta)$ for $0 \le r,\ 0 \le \varphi \le \pi,\ 0 \le \theta \le 2\pi$ is shown in Figure 2.

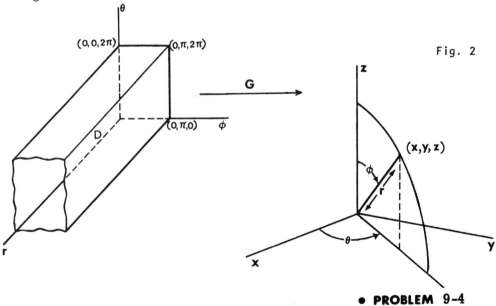

Fig. 2

• PROBLEM 9-4

Let S be the volume defined by $S = \{x^2+y^2 \le 1,\ 0 \le z \le 2\}$. Find the integral $\iiint_S (x^2+y^2+z^2)\,dx\,dy\,dz$.

Solution: Since $0 \le z \le 2$, the integral

$$\iiint_S (x^2+y^2+z^2)\,dx\,dy\,dz = \int_0^2 \left[\iint_{x^2+y^2\le 1} (x^2+y^2+z^2)\,dx\,dy\right] dz. \tag{1}$$

First compute the inner double integral using the change of variables formula in 2 dimensions:

$$\iint_{G(B)} f(x,y)\,dx\,dy = \iint_B f(G(u,v))\,|\Delta_G(u,v)|\,du\,dv$$

where f is a function on $G(B)$ and $|\Delta_G(u,v)|$ is the absolute value of the Jacobian of the transformation. Hence, letting $x = r\sin\theta$, $y = r\cos\theta$,

$$|\Delta_G(r,\theta)| = \left\|\begin{matrix} \sin\theta & \cos\theta \\ r\cos\theta & -r\sin\theta \end{matrix}\right\| = |-r\sin^2\theta - r\cos^2\theta| = |-r| = r.$$

Therefore,

$$\iint_{x^2+y^2\le 1} (x^2+y^2+z^2)\,dx\,dy = \int_0^{2\pi}\int_0^1 (r^2\sin^2\theta + r^2\cos^2\theta + z^2) r\,dr\,d\theta$$

$$= \int_0^{2\pi}\int_0^1 (r^2+z^2) r\,dr\,d\theta$$

Thus (1) equals

$$\int_0^2 \int_0^{2\pi}\int_0^1 (r^3+z^2 r)\,dr\,d\theta\,dz$$

$$\int_0^2 \int_0^{2\pi} \left[\frac{r^4}{4} + z^2\frac{r^2}{2}\right]_0^1 d\theta\,dz = \int_0^2\int_0^{2\pi} \left(\frac{1}{4}+\frac{z^2}{2}\right) d\theta\,dz$$

$$= 2\pi \int_0^2 \left(\frac{1}{4}+\frac{z^2}{2}\right)dz = 2\pi\left(\frac{z}{4}+\frac{z^3}{6}\right)\Big|_0^2 = 2\pi\left(\frac{2}{4}+\frac{8}{6}\right) = 11\frac{\pi}{3}.$$

● **PROBLEM 9-5**

Find the integral $\iiint_S x^2 y^4\,dV$ where S is the unit ball.

Solution: The volume $S = x^2+y^2+z^2 \le 1$ is the interior of the sphere of radius 1. To compute this integral change to spherical coordinates by use of the change of variables formula in 3 dimensions which states:

$$\iiint_{G(A)} f(x,y,z)\,dz\,dy\,dx = \iiint_A f(G(u,v,w))\,|\Delta_G(u,v,w)|\,du\,dv\,dw. \tag{1}$$

Therefore, letting $x = r\sin\varphi\cos\theta$; $y = r\sin\varphi\sin\theta$; $z = r\cos\varphi$,

$$|\Delta_G(r,\theta,\varphi)| = \left|\frac{\partial(x,y,z)}{\partial(r,\theta,\varphi)}\right| = \left\|\begin{matrix} \sin\varphi\cos\theta & r\cos\varphi\cos\theta & -r\sin\varphi\sin\theta \\ \sin\varphi\sin\theta & r\cos\varphi\sin\theta & r\sin\varphi\cos\theta \\ \cos\varphi & -\sin\varphi & 0 \end{matrix}\right\|$$

which is equal to $r^2\sin\varphi$. Hence by (1):

$$\iiint_S x^2 y^4\,dV = \int_0^{2\pi}\int_0^{\pi}\int_0^1 r^2\sin^2\varphi\cos^2\theta \cdot r^4\sin^4\varphi\sin^4\theta \cdot r^2\sin\varphi\,dr\,d\varphi\,d\theta$$

$$= \int_0^{2\pi} \int_0^{\pi} \int_0^1 r^8 \sin^7\varphi \sin^4\theta \cos^2\theta \, dr \, d\varphi \, d\theta$$

$$= \int_0^1 r^8 \, dr \int_0^{2\pi} \sin^4\theta \cos^2\theta \, d\theta \int_0^{\pi} \sin^7\varphi \, d\varphi$$

$$= \frac{1}{9} \int_0^{2\pi} \sin^4\theta \cos^2\theta \, d\theta \int_0^{\pi} \sin^7\varphi \, d\varphi \,.$$

Referring to a table of integrals the above can be reduced to

$$= \frac{1}{9} \cdot \frac{\pi}{8} \cdot \frac{32}{35} = \frac{4\pi}{315} \,.$$

● **PROBLEM 9-6**

Let $x = e^u \cos v$
$y = e^u \sin v$.

Find $\iint_{R^*} x^2 \, dx \, dy$ where R^* is the image of the rectangle R in uv-space consisting of $\{(u,v) \mid 0 \le u \le 1 \text{ and } 0 \le v \le \pi\}$.

Solution: The two equations define a local invertible mapping between (x,y) and (u,v). The region R is given but integration must be performed over R^* in the xy-plane. There are two ways of attacking the problem: a) Find the functions $u = f_1(x,y)$, $v = f_2(x,y)$ corresponding to $x = g_1(u,v)$ and $y = g_2(u,v)$ and then find R^* from R; b) Change the variables of integration to u and v and integrate over R. The change of variables formula in 2-dimensions is

$$\iint_{G(D)} f(x,y) \, dx \, dy = \iint_D f(G(u,v)) |\Delta_G(u,v)| \, du \, dv$$

where $|\Delta_G(u,v)|$ is the Jacobian of the transformation G. The latter approach is preferable in this problem.
The first step is to find the Jacobian of this transformation.

$$|J(x,y)| = \left|\frac{\partial(x,y)}{\partial(u,v)}\right| = \begin{Vmatrix} e^u \cos v & -e^u \sin v \\ e^u \sin v & e^u \cos v \end{Vmatrix}$$

$$= |e^{2u}(\cos^2 v + \sin^2 v)| = e^{2u} \,.$$

By the change of variables theorem

$$\iint_{R^*} x^2 \, dx \, dy = \iint_R (e^u \cos v)^2 \left|\frac{\partial(x,y)}{\partial(u,v)}\right| du \, dv$$

$$= \int_0^{\pi} \int_0^1 e^{4u} \cos^2 v \, du \, dv$$

$$= \int_0^{\pi} \cos^2 v \, dv \int_0^1 e^{4u} \, du$$

$$= \int_0^\pi \cos^2 v \, dv \, (\tfrac{1}{4} e^{4u} \Big|_0^1)$$

$$= \tfrac{1}{4}(e^4 - 1) \int_0^\pi \cos^2 v \, dv \ .$$

$\int_0^\pi \cos^2 v \, dv$ can be calculated as follows:

$$\int_0^\pi \cos^2 v \, dv = \tfrac{1}{2} \int_0^\pi (1 + \cos 2v) \, dv$$

$$= \tfrac{1}{2} \int_0^\pi dv + \tfrac{1}{2} \int_0^\pi \cos 2v \, dv$$

$$= \tfrac{\pi}{2} + \tfrac{1}{4} \sin 2v \Big|_0^\pi = \tfrac{\pi}{2} + 0 = \tfrac{\pi}{2} \ .$$

The final value for $\iint_{R^*} x^2 \, dx \, dy$ is therefore $\tfrac{\pi}{8}(e^4 - 1)$.

● **PROBLEM 9-7**

Find the volume of the ellipsoid

$$\tfrac{x^2}{a^2} + \tfrac{y^2}{b^2} + \tfrac{z^2}{c^2} \leq 1 \quad \text{by the change of variables formula.}$$

Solution: The change of variables formula in 3 dimensions states:

$$\iiint_A f(G(u,v,w)) \, |\Delta_G(u,v,w)| \, du\,dv\,dw = \iiint_{G(A)} f(x,y,z) \, dz \, dy \, dx \qquad (1)$$

where A is a bounded region in R^3 contained in some open set U and $G: U \to R^3$. Also f is a function on G(A) and $|\Delta_G(u,v,w)|$ is the absolute value of the Jacobian of the transformation. It is given that $S = \left\{ f(x,y,z) = \tfrac{x^2}{a^2} + \tfrac{y^2}{b^2} + \tfrac{z^2}{c^2} \leq 1 \right\}$. To find the volume of this ellipsoid compute

$$V_S = \iiint_S dx \, dy \, dz \ . \qquad (2)$$

To do this make a change of variables, letting $u = x/a$, $v = y/b$ and $w = z/c$. Hence we have $f(x,y,z) = f(G(u,v,w)) =$ $u^2 + v^2 + w^2 \leq 1$. Now to use the change of variables formula (equation (1))

$$|\Delta_G(u,v,w)| = \left| \tfrac{\partial(x,y,z)}{\partial(u,v,w)} \right|$$

must be computed. Since $x = au$; $y = vb$; and $z = wc$,

$$|\Delta_G(u,v,w)| = \begin{Vmatrix} a & 0 & 0 \\ 0 & b & 0 \\ 0 & 0 & c \end{Vmatrix} = abc.$$

Thus, substituting into (1) and (2)

$$V_S = \iiint_S dx \, dy \, dz = \iiint_{S^{-1}} abc \, du \, dv \, dw \quad \text{where } S^{-1} \text{ is the}$$

ball $u^2 + v^2 + w^2 \leq 1$. Therefore,

$$V_S = abc \iiint_{S^{-1}} du\, dv\, dw = abc \times [\text{volume of the ball } S^{-1}].$$

Thus $V_S = abc \cdot \frac{4}{3}\pi = \frac{4\pi}{3} abc$. Another method to compute $\iiint_{S^{-1}} du\, dv\, dw$ is to use the change of variables formula again by changing to spherical coordinates. Let $u = r \sin \varphi \cos \theta$; $v = r \sin \varphi \sin \theta$; and $z = r \cos \varphi$. Compute $|\Delta_G(r,\varphi,\theta)| = r^2 \sin \varphi$.

Evaluating,

$$abc \int_0^{2\pi}\!\!\int_0^{\pi}\!\!\int_0^1 r^2 \sin \varphi \, dr\, d\varphi\, d\theta = 2\pi \frac{abc}{3} \int_0^\pi \sin\varphi\, d\varphi = \frac{4\pi}{3} abc.$$

● **PROBLEM 9-8**

Find $\iint_{D^*} \exp\left(\frac{x-y}{x+y}\right) dx\, dy$ where D^* is the region bounded by the lines $x = 0$, $y = 0$, and $x+y = 1$.

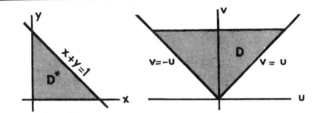

Fig. 1

Solution: The required integral is computed using the change of variables formula which states: Let T be a continuously differentiable one-to-one mapping of an open set R in the plane onto $T(R)$. Let f be a function defined on $T(R)$ which is integrable. Then

$$\iint_R (f \circ T) |\Delta_T| = \iint_{T(R)} f,$$

i.e., $\iint_R f(T(x,y)) |\Delta_T(x,y)| dx\, dy = \iint_{T(R)} f(x,y) dx\, dy.$

In the problem given let T be defined by $T(x,y) = (u,v)$ where $u = x-y$, $v = x+y$.
T is linear in the following sense, $T(ax,ay) = aT(x,y)$ and $T(x_1+x_2, y_1+y_2) = T(x_1,y_1) + T(x_2,y_2)$. It is clear that T is one-to-one. In order to find the image D of D^* under T one can exploit the linearity of T.
To this end observe that $T(0,1) = (-1,1)$, $T(1,0) = (1,1)$, and $T(0,0) = (0,0)$.
Any point P on the line $y = 0$ can be expressed as $P = \alpha(0,0) + (1-\alpha)(0,1)$. Thus $T(P) = \alpha T(0,0) + (1-\alpha) T(0,1)$ by linearity, i.e., P gets mapped onto a point which lies on the line segment joining $(0,0)$ and $(-1,1)$ in the uv-plane. This way it is clear that the line $x = 0$ is mapped onto the line $v = u$, the line $x+y = 1$ gets mapped onto $v = 1$ and $y = 0$ gets mapped onto $v = -u$

(see figure). The resulting region D is shown in the figure and integration on D can be performed by fixing v and letting u vary from $-v$ to v. After carrying out the u integration, the v integration is carried out by letting v vary from 0 to 1. One has

$$\Delta_T = \begin{vmatrix} \dfrac{\partial u}{\partial x} & \dfrac{\partial u}{\partial y} \\ \dfrac{\partial v}{\partial x} & \dfrac{\partial v}{\partial y} \end{vmatrix} = \begin{vmatrix} 1 & -1 \\ 1 & 1 \end{vmatrix} = 2 \; .$$

Let $f(x,y) = e^{x/y}$. Then by change of variables

$$\int_{D^*} \exp\left(\frac{x-y}{x+y}\right) dx\, dy = \int_{D^*} f(T(x,y))\, dx\, dy$$

$$= \int_{D^*} f(T(x,y)) \left| \frac{\Delta_T(x,y)}{\Delta_T(x,y)} \right| dx\, dy = \tfrac{1}{2} \int_{D^*} f(T(x,y)) |\Delta_T(x,y)|\, dx\, dy$$

$$= \tfrac{1}{2} \int_D f(x,y)\, dx\, dy$$

$$= \tfrac{1}{2} \int_0^1 dy \int_{-y}^{y} e^{x/y}\, dx$$

$$= \tfrac{1}{2} \int_0^1 \left([e^{x/y}\, y]\Big|_{-y}^{y} \right) dy$$

$$= \tfrac{1}{2} \int_0^1 (e - 1/e) y\, dy$$

$$= \frac{e - 1/e}{4} \; .$$

● **PROBLEM 9-9**

Consider the transformation

$$x = u + v \qquad y = v - u^2 \; .$$

Let D be the set in the $u - v$ plane bounded by the lines $u = 0$, $v = 0$, and $u + v = 2$.
Find the area of D^*, the image of D, directly and by a change of variables.

Solution: First compute the area of D^* directly. In order to do this, the region D^* must be found. Note that the straight line $u + v = 2$ gets mapped into $x = 2$, the lines $u = 0$ and $v = 0$ are mapped into $y = x$ and $y = -x^2$ respectively. The map $T: (u,v) \to (x,y)$, $x = u + v$, $y = v - u^2$ is one-to-one. For if

$$u_1 + v_1 = u_2 + v_2$$
$$v_1 - u_1^2 = v_2 - u_2^2$$

426

then $\quad v_1 - v_2 = u_2 - u_1 = u_1^2 - u_2^2$

so $\quad u_1 = u_2$ or $u_1 + u_2 = -1$;

but u_1, u_2 are ≥ 0, so $u_1 = u_2$ and hence $v_1 = v_2$. Thus D^* is the region bounded by the lines $y = x$, $x = 2$ and the parabola $y = -x^2$. The area of D^* is therefore

$$\iint_{D^*} dx\, dy = \int_0^2 dx \int_{-x^2}^x dy = \int_0^2 (x + x^2) dx = 14/3 .$$

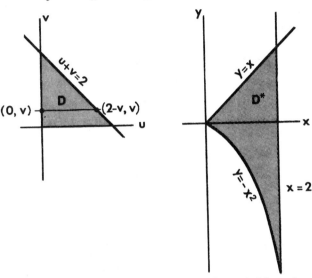

Fig. 1

Next compute the area by using the change of variables formula according to which: Let T be a one to one continuously differentiable mapping of an open set D in the plane onto $D^* = T(D)$. Let f be a function defined on D^* which is integrable. Then

$$\iint_D (f \circ T)(u,v) |\Delta_T| du\, dv = \iint_{D^*} f(x,y) dx\, dy .$$

Here Δ_T is the Jacobian of the map T. Let $f(x,y) = 1$. Since $T: (u,v) \to T(u,v) = (x,y) = (u+v, v-u^2)$,

$$\Delta_T(u,v) = \begin{vmatrix} \dfrac{\partial x}{\partial u} & \dfrac{\partial x}{\partial v} \\ \dfrac{\partial y}{\partial u} & \dfrac{\partial y}{\partial v} \end{vmatrix} = \begin{vmatrix} 1 & 1 \\ -2u & 1 \end{vmatrix} = 1 + 2u.$$

$$\iint_{D^*} f(x,y) dx\, dy = \text{area of } D^*$$

$$\iint_D (f \circ T)(u,v) |\Delta_T| du\, dv = \iint_D (1 + 2u) du\, dv \quad \text{so the area of } D^* \text{ is}$$

$$\int_0^2 (1 + 2u) du \int_0^{2-u} dv$$

$$= \int_0^2 (1 + 2u)(2 - u) du$$

$$= \int_0^2 (2 + 3u - 2u^2)\, du$$

$$= 4 + 6 - 16/3$$

$$= 14/3,$$

and this is the same as the value calculated directly.

AREA

● **PROBLEM 9-10**

Compute the area of the paraboloid given by the equation $z = x^2 + y^2$, with $0 \le z \le 2$.

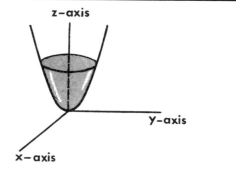

Fig. 1

<u>Solution</u>: Let a surface be defined by the relation $z = f(x,y)$ where (x,y) are in some region R of the plane. Let $(x_0, y_0) \in R$ and consider the curve on this surface defined by $x = x_0 + s$, $y = y_0$, $z = f(x_0 + s, y_0)$ for sufficiently small s. The tangent vector is

$$\left(\frac{dx}{ds}, \frac{dy}{ds}, \frac{dz}{ds}\right) = \left(1, 0, \frac{\partial f}{\partial x}(x_0, y_0)\right)$$

at the point (x_0, y_0, z_0). Similarly by considering the curve $x = x_0$, $y = y_0 + s$, $z = f(x_0, y_0 + s)$ one gets another tangent vector $(0, 1, \frac{\partial f}{\partial y}(x_0, y_0))$. These vectors are linearly independent and if one considers an infinitesimal length of arcs along these two directions and writing these vectors as

$$(dx, 0, \frac{\partial f}{\partial x}\, dx), \quad (0, dy, \frac{\partial f}{\partial y}\, dy)$$

then the area of the "parallelogram" spanned by these vectors is

$$\left\| (1, 0, \frac{\partial f}{\partial x}) \times (0, 1, \frac{\partial f}{\partial y}) \right\| dx\, dy = \left\| \left(-\frac{\partial f}{\partial x}, -\frac{\partial f}{\partial y}, 1\right) \right\| dx\, dy.$$

This small patch of area lies on the surface. Summing over all such patches one sees that the total area (in magnitude) of the surface is

$$A = \iint_R \sqrt{1 + \left(\frac{\partial f}{\partial x}\right)^2 + \left(\frac{\partial f}{\partial y}\right)^2}\, dx\, dy \tag{1}$$

Having motivated the discussion, one takes (1) as the definition of the area of a surface defined by $z = f(x,y)$.

In this problem the region R is the disc of radius $\sqrt{2}$. Therefore the area of the surface $= \iint_R \sqrt{1 + (2x)^2 + (2y)^2}\, dx\, dy$.

To solve this integral, use polar coordinates letting $x = r \cos \theta$, $y = r \sin \theta$ and $dx\, dy = r\, dr\, d\theta$. Then by the change of variables formula

$$\text{area} = \int_0^{2\pi} \int_0^{\sqrt{2}} \sqrt{1 + 4(r^2 \cos^2 \theta + r^2 \sin^2 \theta)}\; r\, dr\, d\theta$$

$$= \int_0^{2\pi} \int_0^{\sqrt{2}} \sqrt{1 + 4r^2}\; r\, dr\, d\theta$$

Substitute, letting $w = 1 + 4r^2$ and $dw = 8r\, dr$, with w ranging from 1 to 9.

$$\text{area} = \frac{1}{8} \int_0^{2\pi} \int_0^9 w^{\frac{1}{2}}\, dw\, d\theta$$

$$= \frac{1}{8} \left[\int_0^{2\pi} d\theta \int_0^9 w^{\frac{1}{2}}\, dw \right]$$

$$= \frac{2\pi}{8} \int_0^9 w^{\frac{1}{2}}\, dw$$

$$= \frac{\pi}{4} \left[\frac{2}{3} w^{3/2} \Big|_1^9 \right] = \frac{\pi}{4} \left[\frac{54}{3} - \frac{2}{3} \right] = \frac{13\pi}{3}$$

● **PROBLEM 9-11**

Find the area of the upper hemisphere of the sphere given by the equation $x^2 + y^2 + z^2 = 3^2$.

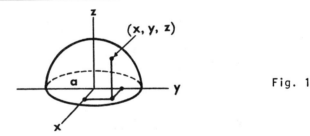

Fig. 1

Solution: The surface $x^2 + y^2 + z^2 = 3^2$ can be defined as an equation of the form $g(x,y,z) = 0$. Using the implicit function theorem, this can be rewritten as $g(x,y,f(x,y)) = 0$ letting $z = f(x,y)$. Then f has first partial derivatives given by

$$\frac{\partial f}{\partial x} = - \frac{\partial g / \partial x}{\partial g / \partial z} \qquad (1)$$

and

$$\frac{\partial f}{\partial y} = - \frac{\partial g / \partial y}{\partial g / \partial z}. \qquad (2)$$

To see this, note that $g_1 = \frac{\partial g}{\partial x} + \frac{\partial g}{\partial z} \frac{\partial f}{\partial x}$. Since $g(x,y,f(x,y)) = 0$, $g_1 = 0$ and therefore $\frac{\partial g}{\partial x} + \frac{\partial g}{\partial z} \frac{\partial f}{\partial x} = 0$. Solving for $\frac{\partial f}{\partial x}$, (1) is obtained. Similarly (2) is obtained for $\frac{\partial f}{\partial y}$. Next substitute (1) and (2) into the equation for the area of a surface given by

$$A = \iint_R \sqrt{1 + \left(\frac{\partial f}{\partial x}\right)^2 + \left(\frac{\partial f}{\partial y}\right)^2}\, dx\, dy\ .$$

Thus

$$A = \iint_R \sqrt{1 + \left(\frac{-\partial g/\partial x}{\partial g/\partial z}\right)^2 + \left(\frac{-\partial g/\partial y}{\partial g/\partial z}\right)^2}\, dx\, dy$$

$$A = \iint_R \sqrt{1 + \frac{(\partial g/\partial x)^2 + (\partial g/\partial y)^2}{(\partial g/\partial z)^2}}\, dx\, dy$$

$$A = \iint_R \frac{\sqrt{(\partial g/\partial x)^2 + (\partial g/\partial y)^2 + (\partial g/\partial z)^2}}{|\partial g/\partial z|}\, dx\, dy \qquad (3)$$

In this problem $\partial g/\partial x = 2x$, $\partial g/\partial y = 2y$, $\partial g/\partial z = 2z$. (4)

Solving for z in terms of x and y, $z = \sqrt{3^2 - x^2 - y^2}$. Observe that the surface is the upper hemisphere of the disc of radius 3 in the xy-plane. Substituting (4) into (3)

$$\text{Area Hemisphere} = \iint_R \frac{\sqrt{(2x)^2 + (2y)^2 + (2z)^2}}{|2z|}\, dx\, dy$$

$$= \iint_R \frac{\sqrt{4(x^2 + y^2 + z^2)}}{2z}\, dx\, dy =$$

$$= \iint_R \frac{\sqrt{x^2 + y^2 + z^2}}{z}\, dx\, dy\ .$$

Substituting $x^2 + y^2 + z^2 = 9$ this integral equals

$$\iint_R \frac{3}{z}\, dx\, dy = \iint_R \frac{3}{\sqrt{9 - (x^2 + y^2)}}\, dx\, dy\ .$$

Changing to polar coordinates, let $x = r\cos\theta$ and $y = r\sin\theta$. Then the area of the hemisphere of radius 3 is equal to

$$3\int_0^{2\pi}\int_0^3 \frac{1}{\sqrt{9 - (r^2\cos^2\theta + r^2\sin^2\theta)}}\, r\, dr\, d\theta = 3\int_0^{2\pi}\int_0^3 \frac{1}{\sqrt{9 - r^2}}\, r\, dr\, d\theta$$

$$= 3\cdot 2\pi \int_0^3 \frac{1}{\sqrt{9 - r^2}}\, r\, dr\ .$$

Let $u = 9 - r^2$, then $du = -2r\, dr$ and changing the limits of integration

$$A = \frac{-6\pi}{2}\int_9^0 u^{-\frac{1}{2}}\, du = -3\pi\left(2u^{\frac{1}{2}}\Big|_9^0\right) = -3\pi(0 - 6) = 18\pi\ .$$

● **PROBLEM 9-12**

Compute the total area of the torus given parametrically by
$x = (a + b\cos\varphi)\cos\theta$; $y = (a + b\cos\varphi)\sin\theta$; $z = b\sin\varphi$, $0 < b < a$.

430

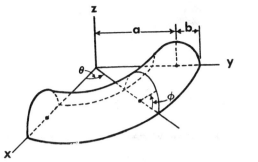

Fig. 1

<u>Solution</u>: The formula for the area of a parametrized surface S is

$$\text{Area} = \iint_S dA = \iint_R \left\| \frac{\partial X}{\partial \varphi} \times \frac{\partial X}{\partial \theta} \right\| d\varphi \, d\theta. \tag{1}$$

Since $\frac{\partial X}{\partial \varphi} = (-b \cos \theta \sin \varphi, -b \sin \theta \sin \varphi, b \cos \varphi)$ and $\frac{\partial X}{\partial \theta} = (-(a+b \cos \varphi) \sin \theta, (a+b \cos \varphi) \cos \theta, 0)$

Then $\left(\frac{\partial X}{\partial \varphi} \times \frac{\partial X}{\partial \theta} \right) = \begin{vmatrix} \hat{i} & \hat{j} & \hat{k} \\ -b \cos \theta \sin \varphi & -b \sin \theta \sin \varphi & b \cos \varphi \\ -(a+b \cos \varphi) \sin \theta & (a+b \cos \varphi) \cos \theta & 0 \end{vmatrix}$

$= [-b(a+b \cos \varphi) \cos \varphi \cos \theta, +b(a+b \cos \varphi) \cos \varphi \sin \theta,$

$\quad -b(a+b \cos \varphi) \cos^2 \theta \sin \varphi - b(a+b \cos \varphi) \sin^2 \theta \sin \varphi]$

$= [-b(a+b \cos \varphi) \cos \varphi \cos \theta, +b(a+b \cos \varphi) \cos \varphi \sin \theta,$

$\quad -b(a+b \cos \varphi) \sin \varphi].$

Therefore $\left(\frac{\partial X}{\partial \varphi} \times \frac{\partial X}{\partial \theta} \right) = -b(a+b \cos \varphi)(\cos \varphi \cos \theta, -\cos \varphi \sin \theta, \sin \varphi).$

Computing the norm

$\left\| \frac{\partial X}{\partial \varphi} \times \frac{\partial X}{\partial \theta} \right\| = \sqrt{(-b)^2 (a+b \cos \varphi)^2 (\cos^2 \varphi \cos^2 \theta + \cos^2 \varphi \sin^2 \theta + \sin^2 \varphi)}$

$= \sqrt{(-b)^2 (a+b \cos \varphi)^2 (\cos^2 \varphi + \sin^2 \varphi)}$

$= b(a+b \cos \varphi).$

Substituting this value into (1) and noticing that for the parametrization given $0 \leq \varphi \leq 2\pi$, $0 \leq \theta \leq 2\pi$, (the figure shows the part of the torus in the first octant only. The entire surface is a closed surface with both θ and φ ranging from 0 to 2π),

$\text{Area} = \int_0^{2\pi} \int_0^{2\pi} b(a+b \cos \varphi) \, d\varphi \, d\theta$

$= ab \int_0^{2\pi} \int_0^{2\pi} d\varphi \, d\theta + b^2 \int_0^{2\pi} \int_0^{2\pi} \cos \varphi \, d\varphi \, d\theta$

$= ab \, 2\pi \, 2\pi + b^2 \int_0^{2\pi} d\theta \int_0^{2\pi} \cos \varphi \, d\varphi$

$$= 4\pi^2 ab + 2\pi b^2 \sin \varphi \Big|_0^{2\pi}$$

$$= 4\pi^2 ab + 2\pi b^2 (0-0)$$

$$= 4\pi^2 ab .$$

• **PROBLEM 9-13**

Compute the area of the part of the sphere $x^2 + y^2 + z^2 = 1$ inside the cone $x^2 + y^2 = z^2$.

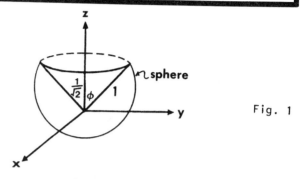

Fig. 1

Solution: The equation for area is:

$$\text{Area} = \iint_S dA = \iint_R \left\| \frac{\partial X}{\partial u} \times \frac{\partial X}{\partial v} \right\| du\, dv . \quad (1)$$

Since the given surface is a unit sphere, it may be parametrized by $x = \sin \varphi \cos \theta$; $y = \sin \varphi \sin \theta$; $z = \cos \varphi$. Hence $\partial X/\partial \varphi \times \partial X/\partial \theta = \sin^2 \varphi \cos \theta\, \hat{i} + \sin^2 \varphi \sin \theta\, \hat{j} + \cos \varphi \sin \varphi\, \hat{k}$ and $\|\partial X/\partial \theta \times \partial X/\partial \varphi\| = \sin \varphi$. Substituting into (1)

$$A = \iint_S dA = \iint_R \sin \varphi\, d\varphi\, d\theta \quad (2)$$

where R is the part of the unit sphere inside the cone $x^2+y^2 = z^2$. To find R first find the curve of intersection of the two surfaces. Since $x^2+y^2 = z^2$ for the cone, substitute z^2 for x^2+y^2 in the equation of the sphere and obtain $z^2+z^2 = 1$. Solving for z, $z = 1/\sqrt{2}$. Next, since φ is the angle with the z-axis (from the given parametrization), Fig. 1 shows that $\cos \varphi = 1/\sqrt{2}$. Therefore $\varphi = 45° = \pi/4$. So R is the part of the sphere where $0 < \varphi < \pi/4$ and $0 < \theta < 2\pi$. Substituting these values into (2),

$$A = 2 \int_0^{2\pi} \int_0^{\pi/4} \sin \varphi\, d\varphi\, d\theta.$$

(Note the factor 2 appears because the cone encloses 2 caps of the sphere).

$$= 4\pi \int_0^{\pi/4} \sin \varphi\, d\varphi = 4\pi(-\cos \varphi)\Big|_0^{\pi/4}$$

$$= 4\pi(-1/\sqrt{2} + 1) = 4\pi(1 - 1/\sqrt{2}) .$$

● **PROBLEM 9-14**

Find the area of the torus whose parametrization is given by

$x = (R - \cos v)\cos u$ $-\pi \leq u \leq \pi$

$y = (R - \cos v)\sin u$ $-\pi \leq v \leq \pi$

$z = \sin v$ where $R > 1$.

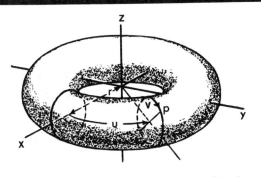

Fig. 1

Solution: Let S be a surface defined parametrically by

$$x = x(u,v) \quad y = y(u,v) \quad z = z(u,v)$$

where u and v belong to a domain D in the plane. The area of such a surface is by definition equal to

$$A = \iint_D \left\| \left(\frac{\partial x}{\partial u}, \frac{\partial y}{\partial u}, \frac{\partial z}{\partial u}\right) \times \left(\frac{\partial x}{\partial v}, \frac{\partial y}{\partial v}, \frac{\partial z}{\partial v}\right) \right\| du\, dv .$$

In the given problem

$\frac{\partial x}{\partial u} = -(R - \cos v)\sin u$ $\frac{\partial x}{\partial v} = \cos u \sin v$

$\frac{\partial y}{\partial u} = (R - \cos v)\cos u$ $\frac{\partial y}{\partial v} = \sin v \sin u$

$\frac{\partial z}{\partial u} = 0$ $\frac{\partial z}{\partial v} = \cos v$

$\left(\frac{\partial x}{\partial u}, \frac{\partial y}{\partial u}, \frac{\partial z}{\partial u}\right) \times \left(\frac{\partial x}{\partial v}, \frac{\partial y}{\partial v}, \frac{\partial z}{\partial v}\right)$

$= \Big((R - \cos v)\cos u \cos v, (R - \cos v)\sin u \cos v,$
$\quad - (R - \cos v)\sin^2 u \sin v - (R - \cos v)\cos^2 u \sin v\Big)$

$= (R - \cos v)(\cos u \cos v, \sin u \cos v, -\sin v)$

and the norm of this vector is $R - \cos v$.

The region of integration is $\{(u,v) | -\pi \leq u \leq \pi$ and $-\pi \leq v \leq \pi\}$. Therefore, the area is

$$\text{Area} = \int_{-\pi}^{\pi} \int_{-\pi}^{\pi} (R - \cos v)\, du\, dv$$

$$= \int_{-\pi}^{\pi} du \left(\int_{-\pi}^{\pi} R - \cos v \, dv \right)$$

$$= \int_{-\pi}^{\pi} du \cdot (Rv - \sin v) \Big|_{-\pi}^{\pi}$$

$$= \int_{-\pi}^{\pi} du (R\pi - \sin \pi - R(-\pi) + \sin(-\pi))$$

$$= \int_{-\pi}^{\pi} 2R\pi \, du = (2\pi)^2 R \, .$$

● **PROBLEM 9-15**

Let $f(x)$ be positive and continuously differentiable on $[a,b]$. Find a parametrization for the surface generated by revolving $z = f(x)$ about the x-axis, and compute the surface area.

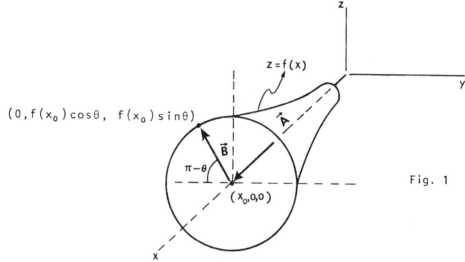

Fig. 1

Solution: Figure 1 illustrates how the intersection of the surface with a plane x = constant is a circle. Every point on the surface can be expressed as a vector sum $\vec{A} + \vec{B}$ where $\vec{A} = (x,0,0)$ and $\vec{B} = (0, f(x)\cos \theta, f(x)\sin \theta)$ for suitable values of x and θ. A parametrization of this surface is:

$x = x$ $\quad\quad\quad\quad\quad\quad\quad\quad\quad\quad\quad a \le x \le b$

$y = f(x)\cos \theta$ $\quad\quad\quad\quad\quad\quad\quad 0 \le \theta \le 2\pi$

$z = f(x)\sin \theta \, .$

The tangents, with respect to u and v, to the surface have components

$$\left(\frac{\partial x}{\partial x}, \frac{\partial y}{\partial x}, \frac{\partial z}{\partial x} \right) \quad \text{and} \quad \left(\frac{\partial x}{\partial \theta}, \frac{\partial y}{\partial \theta}, \frac{\partial z}{\partial \theta} \right) .$$

The area of a small segment of the surface is given by

$$\left\| \left(\frac{\partial x}{\partial x}, \frac{\partial y}{\partial x}, \frac{\partial z}{\partial x} \right) \times \left(\frac{\partial x}{\partial \theta}, \frac{\partial y}{\partial \theta}, \frac{\partial z}{\partial \theta} \right) \right\| .$$

The area of the total surface is found by integrating the norm of the cross-products of the tangent vectors to the surface at a point. Thus,

$$A = \iint_S \left\|\left(\frac{\partial x}{\partial x}, \frac{\partial y}{\partial x}, \frac{\partial z}{\partial x}\right) \times \left(\frac{\partial x}{\partial \theta}, \frac{\partial y}{\partial \theta}, \frac{\partial z}{\partial \theta}\right)\right\| d\theta \, dx .$$

$$\left\|\left(\frac{\partial x}{\partial x}, \frac{\partial y}{\partial x}, \frac{\partial z}{\partial x}\right) \times \left(\frac{\partial x}{\partial \theta}, \frac{\partial y}{\partial \theta}, \frac{\partial z}{\partial \theta}\right)\right\| d$$

$$= \|(1, f'(x)\cos\theta, f'(x)\sin\theta) \times (0, -f(x)\sin\theta, f(x)\cos\theta)\|$$

$$= \|(f(x)f'(x)\cos^2\theta + f(x)f'(x)\sin^2\theta, -f(x)\cos\theta, -f(x)\sin\theta)\|.$$

Factoring out $f(x)$ (and taking account of the fact that $f(x) \geq 0$) yields

$$= f(x)\|(f'(x), \cos\theta, -\sin\theta)\|$$

$$= f(x)((f'(x))^2 + \cos^2\theta + \sin^2\theta)^{\frac{1}{2}}$$

$$= f(x)(1 + (f'(x))^2)^{\frac{1}{2}} .$$

To find the area integrate over the region $\{(x,\theta) \,|\, a \leq x \leq b$ and $0 \leq \theta \leq 2\pi\}$ to obtain:

$$\int_a^b \int_0^{2\pi} f(x)(1 + (f'(x))^2)^{\frac{1}{2}} \, d\theta \, dx$$

$$= 2\pi \int_a^b f(x)\left(1 + (f'(x))^2\right)^{\frac{1}{2}} dx .$$

This integral can be converted into a line integral if we notice that $(1 + (f'(x))^2)^{\frac{1}{2}} dx = ds$, the line element. When $z = f(x)$, the usual line element of $z = f(x,y)$,

$$ds = \sqrt{1 + \left(\frac{\partial f}{\partial x}\right)^2 + \left(\frac{\partial f}{\partial y}\right)^2} \, dx \, dy,$$

degenerates to just

$$ds = \sqrt{1 + (f')^2} \, dx .$$

Therefore the area is $2\pi \int_a^b f(x) \, ds .$

INTEGRAL FUNCTION OVER A SURFACE

● **PROBLEM 9-16**

Let S be the surface defined by $x^2 + y^2 + z^2 = 1$ and let the unit normal vector function have representations directed away from the origin. Compute the integral of the function $f(x,y,z) = a$ over S.

Solution: Consider the following definition: Let f be a function on a surface S and let R be a region in the plane. Let $X(t,u)$ be the parametrization of S by a smooth mapping X. Then, provided f is sufficiently smooth, the integral of f over S is defined by the formula

$$\iint_S f \, dA = \iint_R f(X(t,u)) \left\|\frac{\partial X}{\partial t} \times \frac{\partial X}{\partial u}\right\| dt \, du . \tag{1}$$

In this problem the surface is parametrized by θ and φ.
$x = \sin\varphi \cos\theta$; $y = \sin\varphi \sin\theta$; $z = \cos\varphi$ with $0 \le \varphi \le \pi$ and $0 \le \theta \le 2\pi$.
First find the cross-product $\dfrac{\partial X}{\partial \varphi} \times \dfrac{\partial X}{\partial \theta}$

$$\dfrac{\partial X}{\partial \varphi} \times \dfrac{\partial X}{\partial \theta} = \begin{vmatrix} \hat{i} & \hat{j} & \hat{k} \\ \cos\theta \cos\varphi & \sin\theta \cos\varphi & -\sin\varphi \\ -\sin\theta \sin\varphi & \cos\theta \sin\varphi & 0 \end{vmatrix}$$

$$= \sin^2\varphi \cos\theta\, \hat{i} + \sin^2\varphi \sin\theta\, \hat{j}$$

$$+ (\cos\varphi \sin\varphi \cos^2\theta + \sin^2\theta \sin\varphi \cos\varphi)\hat{k}$$

$$= (\sin^2\varphi \cos\theta)\hat{i} + (\sin^2\varphi \sin\theta)\hat{j} + (\cos\varphi \sin\varphi)\hat{k}.$$

Next take the norm $\left\| \dfrac{\partial X}{\partial \varphi} \times \dfrac{\partial X}{\partial \theta} \right\| = \sin\varphi$. Substituting into (1),

$$\iint_S f\, dA = \int_0^{2\pi}\int_0^{\pi} f(\sin\varphi \cos\theta, \sin\varphi \sin\theta, \cos\varphi) \sin\varphi\, d\varphi\, d\theta.$$

Noticing that f is a constant function,

$$\iint_S f\, dA = \int_0^{2\pi}\int_0^{\pi} a \sin\varphi\, d\varphi\, d\theta$$

$$= a \int_0^{2\pi} d\theta \int_0^{\pi} \sin\varphi\, d\varphi$$

$$= 2\pi a \left(-\cos\varphi \Big|_0^{\pi}\right) = (2\pi a)(-(-1-1)) = -2\pi a[-2]$$

$$= 4\pi a.$$

Remark: If $a = 1$, the above integral gives the surface area of the sphere of radius 1.

● **PROBLEM 9-17**

Find the integral of the function $f(x,y,z) = x$ over the surface $z = x^2 + y$ with x,y satisfying the inequalities $0 \le x \le 1$ and $-1 \le y \le 1$.

Solution: By definition, the integral of a function over a surface is given by the formula

$$\iint_S f\, dA = \iint_R f(X(t,u)) \left\| \dfrac{\partial X}{\partial t} \times \dfrac{\partial X}{\partial u} \right\| dt\, du. \tag{1}$$

In this problem notice that the surface is of the form $z = f(x,y)$. Parametrize this surface by $X(x,y) = (x,y,f(x,y))$ and find

$\dfrac{\partial X}{\partial x} = (1,0,\dfrac{\partial f}{\partial x})$ and $\dfrac{\partial X}{\partial y} = (0,1,\dfrac{\partial f}{\partial y})$. Thus $\dfrac{\partial X}{\partial x} \times \dfrac{\partial X}{\partial y} = (-\dfrac{\partial f}{\partial x}, -\dfrac{\partial f}{\partial y}, 1)$

and

$$\left\| \dfrac{\partial X}{\partial x} \times \dfrac{\partial X}{\partial y} \right\| = \sqrt{1 + \left(\dfrac{\partial f}{\partial x}\right)^2 + \left(\dfrac{\partial f}{\partial y}\right)^2}. \tag{2}$$

Since $z = x^2 + y$, $\frac{\partial f}{\partial x} = 2x$ and $\frac{\partial f}{\partial y} = 1$. Hence, substituting (2) into (1),

$$\iint_S f \, dA = \iint_R x \sqrt{1 + (2x)^2 + 1} \, dx \, dy$$

$$= \int_{-1}^{1} \int_0^1 x \sqrt{2 + 4x^2} \, dx \, dy$$

(since $0 \le x \le 1$ and $-1 \le y \le 1$)

$$= \int_{-1}^{1} dy \int_0^1 x \sqrt{2 + 4x^2} \, dx$$

$$= 2 \int_0^1 x (4x^2 + 2)^{\frac{1}{2}} \, dx .$$

To evaluate this integral let $u = 2 + 4x^2$ and $du = 8x \, dx$. Therefore,

$$2 \int_0^1 x(4x^2 + 2)^{\frac{1}{2}} \, dx = \frac{2}{8} \int_0^1 (8x)(4x^2 + 2)^{\frac{1}{2}} \, dx$$

which upon the above substitution gives

$$\iint_S x \, dA = \frac{1}{4} \int_2^6 u^{\frac{1}{2}} \, du = \frac{1}{4}\left(\frac{2}{3} u^{3/2} \Big|_2^6 \right)$$

$$= \frac{1}{6}\left(6^{3/2} - 2^{3/2}\right) .$$

• **PROBLEM 9-18**

Let S be the hemisphere given by $S : \{(x,y,z) \mid x^2+y^2+z^2 = 1, z > 0\}$. Let f be the function $f(x,y,z) = x^2 y^2 z$. Compute the integral

$$\iint_S f \, dA .$$

Solution: We again use the definition:

$$\iint_S f \, dA = \iint_R f(t,u) \left\| \frac{\partial X}{\partial t} \times \frac{\partial X}{\partial u} \right\| dt \, du . \tag{1}$$

Parametrize the surface by spherical coordinates letting $x = \sin\varphi \cos\theta$; $y = \sin\varphi \sin\theta$; $z = \cos\varphi$. First compute

$$\frac{\partial X}{\partial \varphi} \times \frac{\partial X}{\partial \theta} = \begin{vmatrix} \hat{i} & \hat{j} & \hat{k} \\ \cos\theta \cos\varphi & \sin\theta \cos\varphi & -\sin\varphi \\ -\sin\varphi \sin\theta & \sin\varphi \cos\theta & 0 \end{vmatrix}$$

$$= \sin^2\varphi \cos\theta \, \hat{i} + \sin^2\varphi \sin\theta \, \hat{j} + \sin\varphi \cos\varphi \, \hat{k} .$$

Hence

$$\left\| \frac{\partial X}{\partial \varphi} \times \frac{\partial X}{\partial \theta} \right\| = \sqrt{\sin^4\varphi \cos^2\theta + \sin^4\varphi \sin^2\theta + \sin^2\varphi \cos^2\varphi}$$

$$= \sqrt{\sin^4\varphi (\cos^2\theta + \sin^2\theta) + \sin^2\varphi \cos^2\varphi}$$

$$= \sqrt{\sin^2 \varphi \sin^2 \varphi + \sin^2 \varphi \cos^2 \varphi}$$

$$= \sqrt{\sin^2 \varphi} = \sin\varphi .$$

Substituting into (1), $(f(x,y,z) = x^2 y^2 z)$

$$\iint_S f \, dA = \iint_R \sin^4 \varphi \cos^2 \theta \sin^2 \theta \cos\varphi \sin\varphi \, d\varphi \, d\theta . \qquad (2)$$

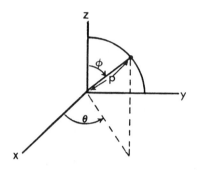

Fig. 1

Since we are integrating over the upper hemisphere of the sphere, $0 < \theta < 2\pi$; $0 < \varphi < \pi/2$ (see figure) and (2) becomes

$$\int_0^{2\pi} \int_0^{\pi/2} \sin^5 \varphi \cos^2 \theta \sin^2 \theta \cos\varphi \, d\varphi \, d\theta$$

$$= \int_0^{\pi/2} \sin^5 \varphi \cos\varphi \, d\varphi \int_0^{2\pi} \cos^2 \theta \sin^2 \theta \, d\theta . \qquad (3)$$

To evaluate the first integral, let $u = \sin\varphi$, $du = \cos\varphi \, d\varphi$ and change the limits of integration to values of u to obtain

$$\int_0^1 u^5 \, du = \frac{u^6}{6} \Big|_0^1 = \frac{1}{6} .$$ To evaluate the second integral use the trigonometric identity $2 \cos\theta \sin\theta = \sin 2\theta$. Thus $\cos^2 \theta \sin^2 \theta = \frac{\sin^2 2\theta}{4}$ (by squaring both sides of the equality). Next use the trigonometric identity $\frac{\sin^2 2\theta}{4} = \frac{1 - \cos 4\theta}{8}$ (by using the equality $\sin^2 \theta = \frac{1 - \cos 2\theta}{2}$). Hence

$$\int_0^{2\pi} \cos^2 \theta \sin^2 \theta \, d\theta = \int_0^{2\pi} \left(\frac{1}{8} - \frac{\cos 4\theta}{8}\right) d\theta$$

$$= \frac{1}{8} \int_0^{2\pi} d\theta - \frac{1}{8} \int_0^{2\pi} \cos 4\theta \, d\theta = \frac{2\pi}{8} - \frac{\sin 4\theta}{32} \Big|_0^{2\pi} = \frac{\pi}{4} - 0 = \frac{\pi}{4} .$$

Thus the integral expression (3) is equal to $\frac{1}{6}\left(\frac{\pi}{4}\right) = \frac{\pi}{24} .$

• **PROBLEM 9-19**

Integrate the function z over the surface $z = x^2 + y^2$ with $x^2 + y^2 \leq 1$.

Solution: By definition, (φ is a continuous function):

$$\iint_S \varphi \, dA = \iint_R \varphi(X(t,u)) \left\| \frac{\partial X}{\partial t} \times \frac{\partial X}{\partial u} \right\| dt \, du \qquad (1)$$

In this problem the surface is given in the form $z = f(x,y)$; hence

$$X(x,y) = (x,y,f(x,y)) = (x,y,x^2+y^2) \text{ so that } \frac{\partial X}{\partial x} = \left(1,0,\frac{\partial f}{\partial x}\right) \text{ and}$$

$\frac{\partial X}{\partial y} = \left(0,1,\frac{\partial f}{\partial y}\right)$. Hence $\frac{\partial X}{\partial x} \times \frac{\partial X}{\partial y} = \left(-\frac{\partial f}{\partial x}, -\frac{\partial f}{\partial y}, 1\right)$ and

$$\left\| \frac{\partial X}{\partial x} \times \frac{\partial X}{\partial y} \right\| = \sqrt{1 + \left(\frac{\partial f}{\partial x}\right)^2 + \left(\frac{\partial f}{\partial y}\right)^2},$$

so that (1) may be rewritten as:

$$\iint_S \varphi \, dA = \iint_R \varphi(X(x,y)) \sqrt{1 + \left(\frac{\partial f}{\partial x}\right)^2 + \left(\frac{\partial f}{\partial y}\right)^2} \, dx \, dy. \qquad (2)$$

Since $z = x^2+y^2$, $\frac{\partial f}{\partial x} = 2x$ and $\frac{\partial f}{\partial y} = 2y$. Substituting these values into (2),

$$\iint_S z \, dA = \iint_R (x^2+y^2) \sqrt{1 + 4x^2 + 4y^2} \, dx \, dy \qquad (3)$$

Next, use the change of variables formula, letting $x = r \cos \theta$, $y = r \sin \theta$, and replacing $dx \, dy$ with $r \, dr \, d\theta$ in equation (3) to obtain

$$\iint_S z \, dA = \iint_R (r^2 \cos^2 \theta + r^2 \sin^2 \theta) \sqrt{1+4r^2\cos^2\theta+4r^2\sin^2\theta} \, r \, dr \, d\theta. \qquad (4)$$

Since $x^2+y^2 \le 1$ is given, R is the region where $0 \le \theta \le 2\pi$ and $0 \le r \le 1$. Therefore (4) becomes

$$\iint_S z \, dA = \int_0^{2\pi} \int_0^1 r^2 \sqrt{1+4r^2} \, r \, dr \, d\theta = 2\pi \int_0^1 r^2 \sqrt{1+4r^2} \, r \, dr.$$

Letting $u = 1+4r^2$, $du = 8r \, dr$ and changing the limits of integrations to values of u,

$$\iint_S z \, dA = \frac{2\pi}{8} \int_1^5 \frac{u-1}{4} \cdot u^{1/2} du = \frac{\pi}{4} \int_1^5 \left(\frac{u^{3/2}}{4} - \frac{u^{1/2}}{4}\right) du$$

$$= \frac{\pi}{4} \left(\frac{u^{5/2}}{10} - \frac{u^{3/2}}{6}\right)\bigg|_1^5 = \frac{\pi}{4}\left[\left(\frac{5^{5/2}}{10} - \frac{5^{3/2}}{6}\right) - \left(\frac{1}{10} - \frac{1}{6}\right)\right]$$

$$= \frac{\pi}{4}\left(\frac{25\sqrt{5}}{10} - \frac{5\sqrt{5}}{6} + \frac{2}{30}\right) = \frac{\pi}{8}\left(\frac{10\sqrt{5}}{3} + \frac{2}{15}\right).$$

INTEGRAL VECTOR FIELD OVER A SURFACE

• **PROBLEM 9-20**

Compute the integral of the vector field $\vec{F}(x,y,z) = (y,-x,z^2)$ over the paraboloid $z = x^2+y^2$ with $0 \le z \le 1$.

<u>Solution</u>: Given that $X(t,u)$ parametrizes a surface, and letting $\vec{n}$ represent the outward normal unit vector to the surface, the integral of a vector field over a surface is defined by

$$\iint_S \vec{F} \cdot \vec{n} \, dA = \iint_R \vec{F} \cdot \vec{n} \left\| \frac{\partial X}{\partial t} \times \frac{\partial X}{\partial u} \right\| dt \, du$$

where $\vec{F}$ is a vector field in some open set in R^3. Furthermore, using the fact that $\dfrac{\vec{N}}{\|\vec{N}\|} = \vec{n}$, where $\vec{N}$ is any normal in the same direction as $\vec{n}$ (unit normal) and $\|\vec{N}\|$ is its norm,

$$\vec{n} \left\| \frac{\partial X}{\partial t} \times \frac{\partial X}{\partial u} \right\| = \pm \frac{\partial X}{\partial t} \times \frac{\partial X}{\partial u} \, .$$

Thus

$$\iint_S \vec{F} \cdot \vec{n} \, dA = \iint_R \vec{F}(X(t,u)) \cdot \left[\pm \left(\frac{\partial X}{\partial t} \times \frac{\partial X}{\partial u} \right) \right] dt \, du \, . \tag{1}$$

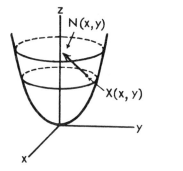

Fig. 1

In this problem the surface $z = x^2+y^2$ with $0 \le z \le 1$ is given. Parametrize the surface by $X(x,y) = (x,y,x^2+y^2)$. Computing

$$\vec{N}(x,y) = \left(\frac{\partial x}{\partial x}, \frac{\partial y}{\partial x}, \frac{\partial z}{\partial x} \right) \times \left(\frac{\partial x}{\partial y}, \frac{\partial y}{\partial y}, \frac{\partial z}{\partial y} \right)$$

we find $\vec{N}(x,y) = (1,0,2x) \times (0,1,2y)$

$$= \begin{vmatrix} \hat{i} & \hat{j} & \hat{k} \\ 1 & 0 & 2x \\ 0 & 1 & 2y \end{vmatrix} = -2x\hat{i} - 2y\hat{j} + \hat{k}$$

$$= (-2x,-2y,1).$$

Note that $\vec{N}(x,y)$ points inward and hence the outward normal direction is that of $-\vec{N}$. Next compute

$$\vec{F}(X(x,y)) \cdot \vec{N}(x,y) = (y,-x,z^2) \cdot (-2x,-2y,1)$$
$$= -2xy + 2xy + z^2 = z^2$$
$$= (x^2+y^2)^2 \, .$$

Substituting this value into (1) along with the negative sign due to the inward orientation

$$\iint_S \vec{F} \cdot \vec{n} \, dA = -\iint_R (x^2+y^2)^2 \, dx \, dy$$

where R is the disc $x^2+y^2 \le 1$ in the xy-plane. To evaluate this integral change to polar coordinates, letting $x = r \cos \theta$, $y = r \sin \theta$ and $dx \, dy = r \, dr \, d\theta$ to obtain

$$-\int_0^{2\pi} \int_0^1 (r^2 \cos^2\theta + r^2 \sin^2\theta)^2 \, r \, dr \, d\theta$$

$$= -2\pi \int_0^1 r^4 r \, dr = -2\pi \int_0^1 r^5 \, dr$$

$$= -2\pi \frac{r^6}{6}\Big|_0^1 = -\frac{2\pi}{6} = -\frac{\pi}{3}.$$

● **PROBLEM 9-21**

Compute the integral of the vector field $\vec{F}(x,y,z) = (x,y,0)$ over the sphere $x^2+y^2+z^2 = a^2$ ($a > 0$).

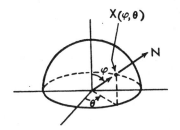

Fig. 1

Solution: Use the definition for the integral of a vector field over a surface:

$$\iint_S \vec{F} \cdot \vec{n} \, dA = \iint_R \vec{F} \cdot \vec{n} \left\|\frac{\partial X}{\partial t} \times \frac{\partial X}{\partial u}\right\| dt \, du$$

$$= \iint_R \vec{F}(X(t,u)) \cdot \left(\frac{\partial X}{\partial t} \times \frac{\partial X}{\partial u}\right) dt \, du \quad (1)$$

where $\vec{n}$ represents the outward unit normal and $X(t,u)$ parametrizes the surface. First, let $x = a \sin \varphi \cos \theta$; $y = a \sin \varphi \sin \theta$; $z = a \cos \varphi$ and letting $\vec{N}(\varphi,\theta)$ represent $(\partial X/\partial \varphi \times \partial X/\partial \theta)$, compute

$$\vec{N}(\varphi,\theta) = \begin{vmatrix} \hat{i} & \hat{j} & \hat{k} \\ a\cos\varphi\cos\theta & a\cos\varphi\sin\theta & -a\sin\varphi \\ -a\sin\varphi\sin\theta & a\sin\varphi\cos\theta & 0 \end{vmatrix}$$

$$= (a^2\sin^2\varphi\cos\theta)\hat{i} + a^2\sin^2\varphi\sin\theta \, \hat{j} + (a^2\cos^2\theta\sin\varphi\cos\varphi + a^2\sin^2\theta\cos\varphi\sin\varphi)\hat{k}$$

$$= (a^2\sin^2\varphi\cos\theta, a^2\sin^2\varphi\sin\theta, a^2\sin\varphi\cos\varphi)$$

$\vec{N}(\varphi,\theta) = a\sin\varphi(a\sin\varphi\cos\theta, a\sin\varphi\sin\theta, a\cos\varphi)$

$= a\sin\varphi \, X(\varphi,\theta)$.

Thus $\vec{N}(\varphi,\theta)$ is a positive multiple of the position vector $\vec{X}(\varphi,\theta)$ and thus points outward. Next compute

$$\vec{F}(X(\varphi,\theta)) \cdot \left(\frac{\partial X}{\partial \theta} \times \frac{\partial X}{\partial \varphi}\right).$$

Since $\vec{F}(x,y) = (x,y,0)$, $\vec{F}(X(\varphi,\theta)) = (a\sin\varphi\cos\theta, a\sin\varphi\sin\theta, 0)$ and thus

$$\vec{F}(X(\varphi,\theta)) \cdot \left(\frac{\partial X}{\partial \theta} \times \frac{\partial X}{\partial \varphi}\right) = (a^2\sin^2\varphi\cos\theta, a^2\sin^2\varphi\sin\theta, a^2\sin\varphi\cos\varphi) \cdot (a\sin\varphi\cos\theta,$$

$$a\sin\varphi\sin\theta, 0)$$

$$= a^3 \sin^3\varphi \cos^2\theta + a^3 \sin^3\varphi \sin^2\theta + 0$$

$$= a^3 \sin^3\varphi (\cos^2\theta + \sin^2\theta) = a^3 \sin^3\varphi . \qquad (2)$$

Thus substituting (2) into (1)

$$\iint_S \vec{F} \cdot \vec{n} \, dA = \iint_R a^3 \sin^3\varphi \, d\varphi \, d\theta$$

$$= a^3 \int_0^{2\pi} \int_0^{\pi} \sin^3\varphi \, d\varphi \, d\theta \quad \text{since} \quad 0 \le \varphi \le \pi; \, 0 \le \theta \le 2\pi$$

$$= a^3 \int_0^{2\pi} d\theta \int_0^{\pi} \sin^3\varphi \, d\varphi$$

$$= 2\pi a^3 \int_0^{\pi} \sin\varphi (1 - \cos^2\varphi) \, d\varphi$$

$$= 2\pi a^3 \left[\int_0^{\pi} \sin\varphi \, d\varphi - \int_0^{\pi} \sin\varphi \cos^2\varphi \, d\varphi \right] . \qquad (3)$$

Evaluating the first integral: $\int_0^{\pi} \sin\varphi \, d\varphi = 2$. To evaluate the second integral let $u = \cos\varphi$ and $du = -\sin\varphi \, d\varphi$. Changing the limits of integration to values for u makes the integral equal to

$$-\int_1^{-1} u^2 \, du = \int_{-1}^1 u^2 \, du = u^3/3 \Big|_{-1}^1 = 2/3 .$$

Substituting the evaluations of the two integrals into (3),

$$\iint_S \vec{F} \cdot \vec{n} \, dA = 2\pi a^3 [2 - \tfrac{2}{3}] = 2\pi a^3 \cdot \tfrac{4}{3} = \tfrac{8\pi a^3}{3} .$$

● **PROBLEM 9-22**

> Compute the integral of the vector field $\vec{F}(x,y,z) = (2x, y^2, z^2)$ over the unit sphere.

<u>Solution</u>: Solution of this problem requires the divergence theorem in 3-space which states:

$$\iint_S \vec{F} \cdot \vec{n} \, dA = \iiint_u \text{div} \, \vec{F} \, dV \quad \text{where} \quad \vec{F} \text{ is a vector field and } \vec{n}$$

is the outward unit normal to the surface S.

Since $\vec{F}(x,y,z) = (2x, y^2, z^2)$,

$$\text{div} \, \vec{F} = \frac{\partial F_1}{\partial x} + \frac{\partial F_2}{\partial y} + \frac{\partial F_3}{\partial z} = 2 + 2y + 2z$$

so that

$$\iint_S \vec{F} \cdot \vec{n} \, dA = \iiint_u (2 + 2y + 2z) \, dx \, dy \, dz . \qquad (1)$$

Since integration is over the volume $x^2 + y^2 + z^2 \le 1$ use the change of variables formula and change to spherical coordinates. Letting $x = r \sin\varphi \cos\theta$; $y = r \sin\varphi \sin\theta$; $z = r \cos\varphi$, replace $dx \, dy \, dz$ by $r^2 \sin\varphi \, dr \, d\theta \, d\varphi$ ($0 \le \theta \le 2\pi; \, 0 \le \varphi \le \pi$). Substituting into (2)

$$\iint_S \vec{F}\cdot\vec{n}\, dA = \int_0^\pi \int_0^{2\pi}\int_0^1 (2+2r\sin\varphi\sin\theta + 2r\cos\varphi)r^2\sin\varphi\, dr\, d\theta\, d\varphi$$

$$= \int_0^\pi \int_0^{2\pi}\int_0^1 (2r^2\sin\varphi + 2r^3\sin^2\varphi\sin\theta + 2r^3\cos\varphi\sin\varphi)dr\, d\theta\, d\varphi$$

$$= \int_0^\pi \int_0^{2\pi} (\tfrac{2}{3}\sin\varphi + \tfrac{1}{2}\sin^2\varphi\sin\theta + \tfrac{1}{2}\cos\varphi\sin\varphi)d\theta\, d\varphi$$

$$= \int_0^\pi \left(\tfrac{2}{3}\sin\varphi\, \theta\Big|_0^{2\pi} - \tfrac{1}{2}\sin^2\varphi\cos\theta\Big|_0^{2\pi} + \tfrac{1}{2}\cos\varphi\sin\varphi\, \theta\Big|_0^{2\pi}\right)d\varphi$$

$$= \int_0^\pi \left(\tfrac{4\pi}{3}\sin\varphi - \tfrac{1}{2}\sin^2\varphi(1-1) + \tfrac{2\pi}{2}\cos\varphi\sin\varphi\right)d\varphi$$

$$= \tfrac{4\pi}{3}\int_0^\pi \sin\varphi\, d\varphi + \pi \int_0^\pi \cos\varphi\sin\varphi\, d\varphi .$$

The first integral equals $\tfrac{4\pi}{3}(-\cos\varphi)\Big|_0^\pi = \tfrac{4\pi}{3}\cdot 2 = \tfrac{8\pi}{3}$.

For the second integral let $u = \sin\varphi$ and $du = \cos\varphi\, d\varphi$ and it equals $\pi\, \tfrac{\sin^2\varphi}{2}\Big|_0^\pi = 0$. Thus $\iint_S \vec{F}\cdot\vec{n}\, dA = \tfrac{8\pi}{3} + 0 = \tfrac{8\pi}{3}$.

• **PROBLEM 9-23**

Let S be the unit sphere and let
$\vec{F}(x,y,z) = x\hat{i} + y\hat{j} + z\hat{k}$. Compute $\iint_S \vec{F}$.

Solution: The following theorem may be used: Let S be the trace of the continuously differentiable surface $\vec{r}$, where $\vec{r}$ is a function on the connected set D, into R^3 for which there exists a continuous unit normal vector function; then for $\vec{F}$ a function on S into R^3,

$$\iint_S \vec{F} = \iint_D \vec{F}(r)\|\vec{r}_1 \times \vec{r}_2\| \tag{1}$$

Since S is the surface $x^2+y^2+z^2 = 1$, parametrize it by $x = \sin\varphi\cos\theta$; $y = \sin\varphi\sin\theta$, $z = \cos\varphi$ and $\vec{r}(\varphi,\theta) = (\sin\varphi\cos\theta)\hat{i} + (\sin\varphi\sin\theta)\hat{j} + (\cos\varphi)\hat{k}$, $0 \le \varphi \le \pi$ and $0 \le \theta \le 2\pi$.

$\vec{r}_1(\varphi,\theta) = (\cos\theta\cos\varphi)\hat{i} + (\sin\theta\cos\varphi)\hat{j} - (\sin\varphi)\hat{k}$ and

$\vec{r}_2(\varphi,\theta) = (-\sin\theta\sin\varphi)\hat{i} + (\cos\theta\sin\varphi)\hat{j} + 0\hat{k}$. Hence

$\vec{r}_1(\varphi,\theta) \times \vec{r}_2(\varphi,\theta) = (\sin^2\varphi\cos\theta)\hat{i} + (\sin^2\varphi\sin\theta)\hat{j} + (\cos\varphi\sin\varphi)\hat{k}$ and

$\|\vec{r}_1(\varphi,\theta) \times \vec{r}_2(\varphi,\theta)\| = \sin\varphi$.

Therefore upon substitution into (1)

$$\iint_S \vec{F} = \iint_D \vec{F}(r)\sin\varphi\, d\varphi\, d\theta .$$

Since $\vec{F}(x,y,z) = x\hat{i} + y\hat{j} + z\hat{k}$
$\vec{F}(\vec{r}) = (\sin\varphi\cos\theta)\hat{i} + (\sin\varphi\sin\theta)\hat{j} + (\cos\varphi)\hat{k}$.

So that $\iint_S \vec{F} = \int_0^{2\pi}\int_0^{\pi} [(\sin\varphi\cos\theta)\hat{i} + (\sin\varphi\sin\theta)\hat{j} + (\cos\varphi)\hat{k}]\sin\varphi\, d\varphi\, d\theta$

$$= \left[\int_0^{2\pi}\int_0^{\pi} \sin^2\varphi\cos\theta\, d\varphi\, d\theta\right]\hat{i} + \left[\int_0^{2\pi}\int_0^{\pi} \sin^2\varphi\sin\theta\, d\varphi\, d\theta\right]\hat{j}$$

$$+ \left[\int_0^{2\pi}\int_0^{\pi} \sin\varphi\cos\varphi\, d\varphi\, d\theta\right]\hat{k}.$$

The first integral equals $\int_0^{2\pi} \cos\theta\, d\theta \int_0^{\pi} \sin^2\varphi\, d\varphi\, d\theta$

$$= \sin\theta\Big|_0^{2\pi} \int_0^{\pi} \sin^2\varphi\, d\varphi = 0 \int_0^{\pi} \sin^2\varphi\, d\varphi = 0$$

Evaluating the second integral similarly

$$\int_0^{2\pi} \sin\theta\, d\theta \int_0^{\pi} \sin^2\varphi\, d\varphi$$

$$= -\cos\theta\Big|_0^{2\pi} \int_0^{\pi} \sin^2\varphi\, d\varphi = 0 \int_0^{\pi} \sin^2\varphi\, d\varphi = 0.$$

Finally the third integral equals

$$\int_0^{2\pi} d\theta \int_0^{\pi} \sin\varphi\cos\varphi\, d\varphi = 2\pi \int_0^{\pi} \sin\varphi\cos\varphi\, d\varphi.$$

Which equals, upon substituting $u = \sin\varphi$ and $du = \cos\varphi\, d\varphi =$

$$= 2\pi \int u\, du = 2\pi \frac{u^2}{2} = 2\pi \frac{\sin^2\varphi}{2}\Big|_0^{\pi} = 2\pi(0-0) = 0.$$

Thus $\iint_S \vec{F} = 0\hat{i} + 0\hat{j} + 0\hat{k}$,
$= 0$.

• **PROBLEM 9-24**

Let the unit hemisphere be parametrized by
$x = \cos u \sin v$ $0 < u < \pi$
$y = \sin u \sin v$ $0 < v < \pi$
$z = \cos v$
Find the normal $\vec{N}$, the area, and the surface integral $\iint \vec{F}\cdot\vec{n}$ where $\vec{F}(x,y,z) = (0,2,0)$.

Solution: Given a parametrized surface, compute using the equation

$\vec{N} = \frac{\partial X}{\partial u} \times \frac{\partial X}{\partial v}$ where $\vec{F}(X) = \vec{F}(x,y,z)$.

$$\vec{N} = \left(\frac{\partial x}{\partial u}, \frac{\partial y}{\partial u}, \frac{\partial z}{\partial u}\right) \times \left(\frac{\partial x}{\partial v}, \frac{\partial y}{\partial v}, \frac{\partial z}{\partial v}\right)$$

$$= (-\sin u \sin v, \cos u \sin v, 0)$$

$$\times (\cos u \cos v, \sin u \cos v, -\sin v)$$
$$= (-\cos u \sin^2 v, -\sin u \sin^2 v, -\sin v \cos v)$$
$$= -\sin v(\cos u \sin v, \sin u \sin v, \cos v).$$

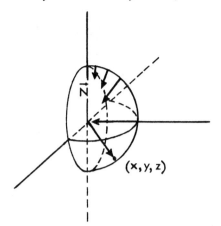

Fig. 1

Figure 1 illustrates how the normal is the negative of the position vector (x,y,z) multiplied by sin v.

To find the area use the formula
$$\text{Area} = \iint_S |\vec{N}| = \iint_R \left\| \frac{\partial X}{\partial u} \times \frac{\partial X}{\partial v} \right\| du\, dv.$$

Therefore, integrate
$|\vec{N}|$ over $\{(u,v) \mid 0 < u < \pi,\ 0 < v < \pi\}$

$|\vec{N}| = \sin v(\cos^2 u \sin^2 v + \sin^2 u \sin^2 v + \cos^2 v) = \sin v$

so the area is $\int_0^\pi \left(\int_0^\pi \sin v\, dv \right) du$

$$= \int_0^\pi \left(-\cos v \Big|_0^\pi \right) dv = \int_0^\pi 2\, dv = 2\pi.$$

Finally compute the surface integral $\iint \vec{F}\cdot\vec{n}$ where $\vec{F}(x,y,z) = (0,2,0)$.

$\iint \vec{F}\cdot\vec{n} = \iint (0,2,0)\cdot(-\cos u \sin^2 v, -\sin u \sin^2 v, -\sin v \cos v)\, dA$

$$= -\int_0^\pi \int_0^\pi 2 \sin u \sin^2 v\, du\, dv$$

$$= -2 \int_0^\pi \sin^2 v (-\cos u \Big|_0^\pi)\, dv$$

$$= -4 \int_0^\pi \sin^2 v\, dv$$

$$= -4 \cdot \frac{\pi}{2} = -2\pi.$$

DIVERGENCE THEOREM

• **PROBLEM 9-25**

Using the divergence theorem compute the integral $\iint_S \vec{F}\cdot\vec{n}\, dA$ where $\vec{F}$ is the vector field $\vec{F}(x,y,z) = (xy, yz, x)$ and S is the boundary of the domain $x^2 + y^2 < z$, $0 < z < 1$.

Solution: The Divergence Theorem states: Let S be a closed surface bounding an open set U and let $\vec{n}$ be the outward unit normal vector to S. Then $\iint_S \vec{F} \cdot \vec{n} \, dA = \iiint_U \text{div} \, \vec{F} \, dV$ where $\vec{F}$ is a vector field on an open set containing U and S. Next, to use the divergence theorem compute the divergence of $\vec{F}$. Since

$$\text{div} \, \vec{F} = \frac{\partial F_1}{\partial x} + \frac{\partial F_2}{\partial y} + \frac{\partial F_3}{\partial z}, \quad \text{div} \, \vec{F} = (y+z)$$ for the given vector field.

Therefore,

$$\iint_S \vec{F} \cdot \vec{n} \, dA = \int_0^1 \iint_{x^2+y^2 \le z} (y+z) \, dx \, dy \, dz \qquad (1)$$

To compute this triple integral first compute the inner double integral

$$\iint_{x^2+y^2 \le z} (y+z) \, dx \, dy . \qquad (2)$$

To do this change to polar coordinates letting $x = r\cos\theta$; $y = r\sin\theta$; and $z = z$. Thus (2) equals $\int_0^{2\pi} \int_0^{\sqrt{z}} (r\sin\theta + z) r \, dr \, d\theta$

$$= \int_0^{2\pi} \int_0^{\sqrt{z}} r^2 \sin\theta \, dr \, d\theta + z \int_0^{2\pi} \int_0^{\sqrt{z}} r \, dr \, d\theta$$

$$= \frac{(\sqrt{z})^3}{3} \int_0^{2\pi} \sin\theta \, d\theta + z \frac{(\sqrt{z})^2}{2} \int_0^{2\pi} d\theta$$

$$= 0 + \frac{z^2}{2} \cdot 2\pi = \pi z^2 .$$

Hence (1) equals $\pi \int_0^1 z^2 \, dz = \pi \frac{z^3}{3} \Big|_0^1 = \frac{\pi}{3}$.

● **PROBLEM 9-26**

Let $\vec{F}(x,y,z) = (0,0,z)$ and let $V = \{(x,y,z); a \le x \le b; c \le y \le d; e \le z \le f\}$. Note the oriented boundary of V consists of the six rectangles $\sigma^1, \sigma^2, \ldots, \sigma^6$ in Figure 1. Check the divergence theorem in this case.

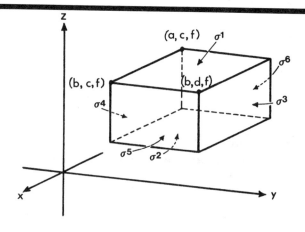

Fig. 1

<u>Solution</u>: The divergence theorem states $\iiint_V (\text{div } \vec{F}) dV = \iint_S \vec{F} \cdot \vec{n} \, ds$

where $\vec{F}$ is a vector field and $\vec{n}$ is the unit outward normal. In this problem

$$\iint_S \vec{F} \cdot \vec{n} \, ds = \iint_{\sigma^1} \vec{F} \cdot \vec{n} \, ds + \iint_{\sigma^2} \vec{F} \cdot \vec{n} \, ds + \iint_{\sigma^3} \vec{F} \cdot \vec{n} \, ds + \ldots + \iint_{\sigma^6} \vec{F} \cdot \vec{n} \, ds.$$

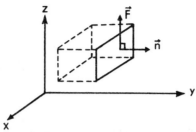

Fig. 2

Since $\vec{F} \cdot \vec{n} = |\vec{F}||\vec{n}| \cos\theta$; and $\vec{F}(x,y,z) = (0,0,z)$; $\vec{F} \cdot \vec{n}_\sigma j = 0$, for each surface except σ^1 and σ^2. Using σ^3 as an example

$\vec{F} \cdot \vec{n} = |\vec{F}||\vec{n}| \cos \frac{\pi}{2} = 0$ (see figure 2). This also applies to surfaces σ^4, σ^5, and σ^6. Hence the integral over the boundary of V,

$$\iint_S \vec{F} \cdot \vec{n} \, ds = \iint_{\sigma^1} \vec{F} \cdot \vec{n} \, ds + \iint_{\sigma^2} \vec{F} \cdot \vec{n} \, ds.$$ Along σ^2, $z = e$; so $\vec{F}$ has

the constant value $(0,0,e)$ on σ^2. Further the outward unit normal on σ^2 is directed in the $-z$ axis direction; therefore $\vec{n}_2 = (0,0,-1)$ and

$$\iint_{\sigma^1} \vec{F} \cdot \vec{n} \, ds = \iint_{\sigma^2} (0,0,e) \cdot (0,0,-1) ds = -e \iint_{\sigma^2} ds$$

$$= -e(\text{area of } \sigma^2).$$

Along σ^1, $z = f$; so $\vec{F}$ has the constant value $(0,0,f)$ on σ^1 and the outward unit normal on σ^1 is directed in the $+z$ axis direction giving $\vec{n}_1 = (0,0,1)$. Hence

$$\iint_{\sigma^1} \vec{F} \cdot \vec{n} \, ds = \iint_{\sigma^1} (0,0,f) \cdot (0,0,1) ds = f \iint_{\sigma^1} ds = f(\text{area of } \sigma^1).$$

Therefore

$$\iint_S \vec{F} \cdot \vec{n} \, ds = f(\text{area of } \sigma^2) - e(\text{area of } \sigma^2). \tag{1}$$

But the area of σ^1 is $(b-a)(d-c)$ and the area of σ^2 is also $(b-a)(d-c)$. Hence substituting these values into (1)

$$\iint_S \vec{F} \cdot \vec{n} \, ds = f(b-a)(d-c) - e(b-a)(d-c) = (b-a)(d-c)(f-e).$$

Thus,

$$\iint_S \vec{F} \cdot \vec{n} \, ds = \text{Volume of } V.$$

Now to verify the divergence theorem we need to show $\iiint_V (\text{div } \vec{F}) ds =$ Volume of V. By definition $\text{div } \vec{F} = \frac{\partial F_1}{\partial x} + \frac{\partial F_2}{\partial y} + \frac{\partial F_3}{\partial z}$. Since $\vec{F}(x,y,z) = (0,0,z)$, $\text{div } \vec{F} = 1$ and $\iiint_V (\text{div } \vec{F}) ds = \iiint_V ds =$ Volume of V. Thus, the divergence theorem checks in this case.

• **PROBLEM 9-27**

Let S be the surface $S = \{x^2+y^2+z^2 = 1\}$ and let f be the function $f(x,y,z) = x^2+y^2+z^2$. Let ∇f represent the gradient of $f(x,y,z)$. Compute the integral $\iint_S \nabla f \cdot \vec{n} \, dA$.

Solution: The divergence theorem in 3-space states that

$$\iint_S \vec{F} \cdot \vec{n} \, dA = \iiint_U \text{div } \vec{F} \, dV \tag{1}$$

where $\vec{F}$ is a vector field and $\vec{n}$ is the outward unit normal of S. Since the gradient makes a vector out of a scalar, ∇f is a vector field, and (1) may be applied (with $\vec{F} = \nabla f$).

$\nabla f = \left(\frac{\partial f}{\partial x}, \frac{\partial f}{\partial y}, \frac{\partial f}{\partial z}\right) = (2x, 2y, 2z)$ and since by definition

$\text{div } \vec{F} = \frac{\partial F_1}{\partial x} + \frac{\partial F_2}{\partial y} + \frac{\partial F_3}{\partial z} = 2 + 2 + 2 = 6$, substitution into (1) yields:

$$\iint_S \nabla f \cdot \vec{n} \, dA = \iiint_U 6 \, dV = 6 \iiint_U dV .$$

which can be evaluated in two ways. One way is by changing to spherical coordinates letting $x = r \sin\varphi \cos\theta$, $y = r \sin\varphi \sin\theta$, and $z = r \cos\varphi$; $0 \leq \varphi \leq \pi$; $0 \leq \theta \leq 2\pi$. By the change of variables formula $dx \, dy \, dz = r^2 \sin\varphi \, dr \, d\varphi \, d\theta$. Hence

$$6 \iiint_U dx \, dy \, dz = 6 \int_0^{2\pi} \int_0^\pi \int_0^1 r^2 \sin\varphi \, dr \, d\varphi \, d\theta$$

$$= 6 \cdot \frac{1}{3} \int_0^{2\pi} \int_0^\pi \sin\varphi \, d\varphi \, d\theta = 2 \cdot 2\pi \int_0^\pi \sin\varphi \, d\varphi = 4\pi \cdot 2 = 8\pi .$$

A second way is by noticing that $\iiint_U dV$ is just the volume of a sphere which equals $\frac{4}{3} \pi a^3$; where a is the radius of the sphere. In this problem the sphere is of radius 1 therefore,

$$6 \iiint_U dV = 6 \cdot \frac{4}{3} \pi = 8\pi .$$

● **PROBLEM 9-28**

Given that S is the surface of a region U for which the divergence theorem is applicable, let $P(x,y,z)$ be any point of S and let O be any fixed point in space. Show that the volume of U is given by $V = \frac{1}{3} \iint_S r \cos\varphi \, dA$, where φ is the angle between the directed line OP and the outer normal S at P, and r is the distance OP.

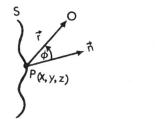

Fig. 1

Solution: The situation is shown in figure 1.
The divergence theorem can be used since it is given that it is applicable on S. Stating the theorem,

$$\iint_S \vec{F} \cdot \vec{n} \, dA = \iiint_U \text{div} \, \vec{F} \, dV \tag{1}$$

Now by definition of the dot product $\vec{F} \cdot \vec{n}$ is the product of the norm of $\vec{n}$ and the projection of $\vec{F}$ on the directed line of the vector $\vec{n}$. (See figure 2).

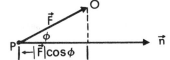

Fig. 2

So $\vec{F} \cdot \vec{n} = |\vec{F}| |\vec{n}| \cos\varphi$. But $|\vec{F}| = r$ and $|\vec{n}| = 1$ (unit vector).
Therefore $\vec{F} \cdot \vec{n} = r \cos\varphi$. (2)

Next, by definition $\text{div} \, \vec{F} = \frac{\partial P}{\partial x} + \frac{\partial Q}{\partial y} + \frac{\partial R}{\partial z}$ and taking $P = x$, $Q = y$, and $R = z$ we derive

$$\text{div} \, \vec{F} = \frac{\partial x}{\partial x} + \frac{\partial y}{\partial y} + \frac{\partial z}{\partial z} = 1 + 1 + 1 = 3 \tag{3}$$

Hence substituting (2) and (3) into equation (1)

$$\iint_S r \cos\varphi \, dA = \iiint_U 3 \, dV$$

and

$$\iiint_U dV = \frac{1}{3} \iint_S r \cos\varphi \, dA .$$

Noticing that $\iiint_U dV$ is the volume of U,

$$V = \frac{1}{3} \iint_S r \cos\varphi \, dA .$$

● **PROBLEM 9-29**

Using the divergence theorem find the integral of the vector field $\vec{F} = \vec{F}(x,y,z) = (x,y,z)$ taken over the sphere of radius a.

Solution: The divergence theorem states: Let U be a region in 3-space, forming the inside of a smooth surface S. Let $\vec{F}$ be a vector field defined on an open set containing U and S. Let $\vec{n}$ be the unit outward normal vector to S. Then

$$\iint_S \vec{F} \cdot \vec{n} \, d\sigma = \iiint_U \text{div} \, \vec{F} \, dV .$$

First compute the divergence of $\vec{F}$ and then use the divergence theorem.

Writing $\vec{F}$ as (F_1, F_2, F_3)

$$\text{div} \, \vec{F} = \frac{\partial F_1}{\partial x} + \frac{\partial F_2}{\partial y} + \frac{\partial F_3}{\partial z} .$$

In this case, $\text{div} \, \vec{F} = \frac{\partial x}{\partial x} + \frac{\partial y}{\partial y} + \frac{\partial z}{\partial z} = 3$. Therefore,

$$\iint_S \vec{F} \cdot \vec{n} \, da = \iiint_B 3 \, dV = 3 \iiint_B dV .$$

But $\iiint_B dV$ is just the volume of the sphere of radius a, which is $4/3 \, \pi \, a^3$. So $\iint \vec{F} \cdot \vec{n} \, da = 3 \cdot 4/3 \, \pi \, a^3 = 4\pi a^3$.

• **PROBLEM 9-30**

Let U be the interior of a closed surface S. Let f,g be functions. Let ∇f be the gradient of f and $\nabla^2 f$ be the divergence of the gradient of f. Prove:

a) $\iint_S f(\nabla g) \cdot \vec{n} \, da = \iiint_U (f \nabla^2 g + \nabla f \cdot \nabla g) \, dV .$

b) $\iint_S (f \nabla g - g \nabla f) \cdot \vec{n} \, da = \iiint_U (f \nabla^2 g - g \nabla^2 f) \, dV .$

Solution: a) This problem requires the use of the divergence theorem which states $\iint_S \vec{F} \cdot \vec{n} \, da = \iiint_U \text{div} \, \vec{F} \, dV$.

Now grad g is the vector field $\left(\frac{\partial g}{\partial x}, \frac{\partial g}{\partial y}, \frac{\partial g}{\partial z}\right)$ and $\vec{F} = f \, \text{grad} \, g$ is the vector field $\left(f \frac{\partial g}{\partial x}, f \frac{\partial g}{\partial y}, f \frac{\partial g}{\partial z}\right)$ so

$$\text{div} \, \vec{F} = \frac{\partial}{\partial x}\left(f \frac{\partial g}{\partial x}\right) + \frac{\partial}{\partial y}\left(f \frac{\partial g}{\partial y}\right) + \frac{\partial}{\partial z}\left(f \frac{\partial g}{\partial z}\right)$$

$$= \frac{\partial f}{\partial x}\frac{\partial g}{\partial x} + \frac{\partial f}{\partial y}\frac{\partial g}{\partial y} + \frac{\partial f}{\partial z}\frac{\partial g}{\partial z} + f\left(\frac{\partial^2 g}{\partial x^2} + \frac{\partial^2 g}{\partial y^2} + \frac{\partial^2 g}{\partial z^2}\right)$$

$$= \nabla f \cdot \nabla g + f \text{ div grad } g$$

i.e., $\text{div } \vec{F} = \nabla f \cdot \nabla g + f \nabla^2 g$.

Applying the divergence theorem to $\vec{F}$ gives

$$\iint_S \vec{F} \cdot \vec{n} \, da = \iint_S (f \text{ grad } g) \cdot \vec{n} \, da = \iiint_U \text{div } \vec{F} \, dV = \iiint_U (f \nabla^2 g + \nabla f \cdot \nabla g) dV \quad (1)$$

This establishes (a). To prove (b) interchange f and g in (1) to get

$$\iint_S (g \nabla f) \cdot \vec{n} \, da = \iiint_U (g \nabla^2 f + \nabla f \cdot \nabla g) dV \quad (2)$$

Subtract (2) from (1) to get

$$\iint_S (f \nabla g - g \nabla f) \cdot n \, da = \iiint_U (f \nabla^2 g - g \nabla^2 f) \, dV$$

which proves (b).

STOKES'S THEOREM • PROBLEM 9-31

Verify Stokes's Theorem for the vector field $\vec{F}(x,y,z) = (z,x,y)$ where S is defined by $z = 4 - x^2 - y^2$, $z \geq 0$.

Solution: Stokes's Theorem states: Let S be a smooth surface bounded by a closed non-intersecting curve C. Let S be orientable and assume that the boundary curve is oriented so that S lies to the left of C. Let $\vec{F}$ be a vector field in a region of space including S and its boundary. Then,

$$\iint_S (\text{curl } \vec{F}) \cdot \vec{n} \, dA = \oint_C \vec{F} \cdot d\vec{s}, \quad (1)$$

($\vec{n}$ is the outward unit normal of S). To verify Stokes's Theorem first compute $\oint_C \vec{F} \cdot d\vec{s}$ where C is the circle $x^2 + y^2 = 4$ ($z = 0$).

Parametrize the circle with $x = 2 \cos\theta$; $y = 2 \sin\theta$; $z = 0$. Thus, $dx = -2 \sin\theta$, $dy = 2 \cos\theta$; $dz = 0$. Since $\vec{F}(x,y,z) = (z,x,y) = (0, 2 \cos\theta, 2 \sin\theta)$ and

$$d\vec{s} = (dx, dy, dz) = (-2 \sin\theta, 2 \cos\theta, 0) \, d\theta,$$

$\vec{F} \cdot d\vec{s} = (0, 2 \cos\theta, 2 \sin\theta) \cdot (-2 \sin\theta, 2 \cos\theta, 0) \, d\theta = (0, 4\cos^2\theta, d\theta)$.

Therefore $\oint_C \vec{F} \cdot d\vec{s} = \int_0^{2\pi} 4 \cos^2\theta \, d\theta = 4 \int_0^{2\pi} (\tfrac{1}{2} + \tfrac{\cos 2\theta}{2}) \, d\theta$

$$= 4 \left[\int_0^{2\pi} \frac{d\theta}{2} + \int_0^{2\pi} \frac{\cos 2\theta}{2} \, d\theta \right] = 4[\pi - 0] = 4\pi.$$

Hence $\oint_C \vec{F} \cdot d\vec{s} = 4\pi$. \quad (2)

Now compute $\iint_S (\text{curl } \vec{F}) \cdot \vec{n} \, dA$ and check to see if it agrees with (2). By definition,

$$\text{Curl } \vec{F} = \begin{vmatrix} \hat{i} & \hat{j} & \hat{k} \\ \frac{\partial}{\partial x} & \frac{\partial}{\partial y} & \frac{\partial}{\partial z} \\ f_1 & f_2 & f_3 \end{vmatrix} \quad \text{where } \vec{F} = (f_1, f_2, f_3).$$

Hence

$$\text{Curl } \vec{F} = \begin{vmatrix} \hat{i} & \hat{j} & \hat{k} \\ \frac{\partial}{\partial x} & \frac{\partial}{\partial y} & \frac{\partial}{\partial z} \\ z & x & y \end{vmatrix} = \left(\frac{\partial y}{\partial y} - \frac{\partial x}{\partial z}, \frac{\partial z}{\partial z} - \frac{\partial y}{\partial x}, \frac{\partial x}{\partial x} - \frac{\partial z}{\partial y} \right)$$

$$= (1,1,1).$$

Since S is the portion of the paraboloid $z = 4 - x^2 - y^2$ with $z \geq 0$, this is a surface in the form $g(x,y,z) = z - f(x,y) = 0$. Now, $\vec{n}$ (the unit normal to the surface) can be expressed as

$\frac{\nabla g}{\|\nabla g\|}$ where ∇g is the gradient of $g(x,y,z)$. This becomes for

$f = -x^2 - y^2 + 4$: $\vec{n} = \frac{(2x, 2y, 1)}{\|(2x, 2y, 1)\|} = \frac{1}{(4x^2 + 4y^2 + 1)^{\frac{1}{2}}} (2x, 2y, 1)$. But

since $\iint_S \vec{F} \cdot \vec{n} \, dA = \iint_R \vec{F} \cdot \vec{N} \, dx, dy$ and $\vec{n} = \frac{\vec{N}}{\|\vec{N}\|}$ so that $\vec{N} = \vec{n} \|\vec{N}\|$,

$$\iint_S (\text{curl } \vec{F}) \cdot \vec{n} \, dA = \iint_R (1,1,1) \cdot (2x, 2y, 1) \, dx \, dy \quad (3)$$

$$= \iint_R (2x + 2y + 1) \, dx \, dy$$

where R is the disc of radius two. To evaluate this integral change to polar coordinates letting $x = r \cos\theta$; $y = r \sin\theta$ and $dx \, dy = r \, dr \, d\theta$ to obtain for (3):

$$\int_0^{2\pi} \int_0^2 (2r \cos\theta + 2r \sin\theta + 1) r \, dr \, d\theta$$

$$= \int_0^{2\pi} \int_0^2 2r^2 \cos\theta \, dr \, d\theta + \int_0^{2\pi} \int_0^2 2r^2 \sin\theta \, dr \, d\theta + \int_0^{2\pi} \int_0^2 r \, dr \, d\theta$$

$$= 2 \int_0^{2\pi} \cos\theta \, d\theta \int_0^2 r^2 \, dr + 2 \int_0^{2\pi} \sin\theta \, d\theta \int_0^2 r^2 \, dr + \int_0^{2\pi} d\theta \int_0^2 r \, dr. \quad (4)$$

Noticing that $\int_0^{2\pi} \cos\theta \, d\theta = 0$ and $\int_0^{2\pi} \sin\theta \, d\theta = 0$, (4) equals

$0 + 0 + 2\pi \int_0^2 r \, dr = 2\pi \cdot 2 = 4\pi$. Thus $\iint_S (\text{curl } \vec{F}) \cdot \vec{n} \, dA = 4\pi$ which agrees with (2) and verifies Stokes's Theorem (equation (1)).

• **PROBLEM 9-32**

Verify Stokes's Theorem for the vector field
$$\vec{F}(x,y,z) = (z-y, x+z, -(x+y))$$
and the surface bounded by the paraboloid $z = 4 - x^2 - y^2$ and the plane $z = 0$.

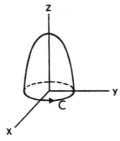

Fig. 1

Solution: Stokes's Theorem states that under routine assumptions of continuity and differentiability,

$$\iint_S (\text{curl } \vec{F}) \cdot \vec{n} \, dA = \oint_C \vec{F} \cdot d\vec{s}.$$

First compute $\oint_C \vec{F} \cdot d\vec{s}$ where C is the circle: $x^2 + y^2 = 4$.

Parametrizing the circle as $x = 2\cos\theta$, $y = 2\sin\theta$, $z = 0$.
$$\vec{F} = (z-y, x+z, -(x+y)) = (-2\sin\theta, 2\cos\theta, -(2\cos\theta + 2\sin\theta))$$
and
$$d\vec{s} = (dx, dy, dz) = (-2\sin\theta \, d\theta, 2\cos\theta \, d\theta, 0).$$
$$\vec{F} \cdot d\vec{s} = (4\sin^2\theta + 4\cos^2\theta) \, d\theta = 4 \, d\theta$$
so
$$\oint_C \vec{F} \cdot d\vec{s} = \int_0^{2\pi} 4 \, d\theta = 8\pi.$$

Now compute $\int_S \nabla \times \vec{F} \cdot \vec{n} \, dA$ where $(\nabla \times \vec{F})$ is the curl of f, and S is the portion of the paraboloid above the plane $z = 0$. $\nabla \times \vec{F}$ can be found by expanding the following determinant along the top row

$$\begin{vmatrix} \hat{i} & \hat{j} & \hat{k} \\ \frac{\partial}{\partial x} & \frac{\partial}{\partial y} & \frac{\partial}{\partial z} \\ z-y & x+z & -x-y \end{vmatrix} = \left(\frac{\partial}{\partial y}(-x-y) - \frac{\partial}{\partial z}(x+z), \frac{\partial}{\partial z}(z-y) - \frac{\partial}{\partial x}(-x-y), \right.$$
$$\left. \frac{\partial}{\partial x}(x+z) - \frac{\partial}{\partial y}(z-y)\right)$$
$$= (-2, 2, 2).$$

When the surface is given by $g(x,y,z) = z - f(x,y) = 0$ an expression for the normal is $\frac{\nabla g}{|\nabla g|}$. This becomes for $f = -x^2 - y^2 + 4$

$$\frac{(2x, 2y, 1)}{\|(2x, 2y, 1)\|} = \frac{1}{(4x^2 + 4y^2 + 1)^{\frac{1}{2}}} (2x, 2y, 1). \text{ Therefore } \int \nabla \times \vec{F} \cdot \vec{n} \, dA =$$

$\iint_R (-4x + 4y + 2)\,dx\,dy$ where R is the disc of radius two. To evaluate this notice that

$$\iint_R -4x\,dx\,dy = -4 \int_0^2 \int_0^{2\pi} r\cos\theta\, r\, dr\, d\theta$$

$$= -4 \int_0^2 r^2\, dr \int_0^{2\pi} \cos\theta\, d\theta = 0$$

since $\int_0^{2\pi} \cos\theta\, d\theta = 0$. Similarly $\iint_R 4y\,dx\,dy = 0$ and all that remains is $\iint_R 2\,dx\,dy = 2 \iint dx\,dy = 2 \cdot$ area of disc
$= 2 \cdot \pi(2)^2 = 8\pi$

which agrees with $\oint_C \vec{F} \cdot d\vec{s}$.

• **PROBLEM 9-33**

Let $\vec{F} = (F_1, F_2, F_3)$ be a vector field that satisfies the following conditions

$$\frac{\partial F_2}{\partial z} = \frac{\partial F_3}{\partial y}\,;\quad \frac{\partial F_3}{\partial x} = \frac{\partial F_1}{\partial z}\,;\quad \frac{\partial F_2}{\partial x} = \frac{\partial F_1}{\partial y}\,,$$

on a region bounded by a curve c. (See Fig. 1). Prove, using Stokes's Theorem that $\oint_C \vec{F} \cdot d\vec{s} = 0$.

Fig. 1

Solution: By Stokes's Theorem $\oint_C \vec{F} \cdot d\vec{s} = \int_S \nabla \times \vec{F} \cdot \vec{n}\, da$. Let us compute $\nabla \times \vec{F}$ (the curl of F).

$$= \begin{vmatrix} \hat{i} & \hat{j} & \hat{k} \\ \frac{\partial}{\partial x} & \frac{\partial}{\partial y} & \frac{\partial}{\partial z} \\ F_1 & F_2 & F_3 \end{vmatrix}$$

$$= \left(\frac{\partial F_3}{\partial y} - \frac{\partial F_2}{\partial z},\ \frac{\partial F_1}{\partial z} - \frac{\partial F_3}{\partial x},\ \frac{\partial F_2}{\partial x} - \frac{\partial F_1}{\partial y} \right).$$

This is the zero vector since each component is zero by our assumptions. Therefore $\oint_C \vec{F} \cdot d\vec{s} = \int_S \nabla \times \vec{F} \cdot \vec{n}\, dA = \int_S 0\, dA = 0$.

• **PROBLEM 9-34**

Verify Stokes's Theorem for $\vec{F}(x,y,z) = (3y, -xz, yz^2)$ where S is the surface $2z = x^2 + y^2$ bounded by $z = 2$ and C is its boundary.

Solution: Stokes's Theorem states:

$$\oint_C \vec{F} \cdot d\vec{r} = \iint_S (\text{curl } \vec{F}) \cdot \vec{n} \, dA .$$

To verify Stokes's Theorem first compute

$$\oint_C \vec{F} \cdot d\vec{r} \qquad (1)$$

Since $d\vec{r} = (dx, dy, dz)$ and $\vec{F}(x,y,z) = (3y, -xz, yz^2)$, $\vec{F} \cdot d\vec{r} = 3y \, dx - xz \, dy + yz^2 \, dz$. Hence (1) becomes

$$\oint_C 3y \, dx - xz \, dy + yz^2 \, dz . \qquad (2)$$

The boundary C of S is $x^2 + y^2 = 4$, $z = 2$. Let $x = 2 \cos\theta$; $y = 2 \sin\theta$; $z = 2$. Substituting into (2)

$$\int_{2\pi}^{0} [3(2 \sin\theta)(-2 \sin\theta) - (2 \cos\theta)(2)(2 \cos\theta) + (2 \sin\theta)(2) \, 0] \, d\theta$$

(integral goes from 2π to 0 since the boundary curve is oriented so that the surface lies to the left of the curve. See figure 1).

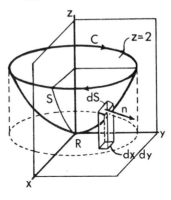

Fig. 1

$$= \int_{2\pi}^{0} (-12 \sin^2\theta - 8 \cos^2\theta) \, d\theta$$

$$= \int_{0}^{2\pi} 12 \sin^2\theta \, d\theta + \int_{0}^{2\pi} 8 \cos^2\theta \, d\theta \qquad (3)$$

The first integral can be evaluated by using the equality $\sin^2\theta = \frac{1 - \cos 2\theta}{2}$. Therefore

$$12 \int_{0}^{2\pi} \sin^2\theta \, d\theta = 12 \int_{0}^{2\pi} \tfrac{1}{2} \, d\theta - \int_{0}^{2\pi} \cos 2\theta \, d\theta = 12\pi + 0 = 12\pi .$$

To evaluate the second integral use $\cos^2\theta = \frac{1 + \cos 2\theta}{2}$; $8 \int_{0}^{2\pi} \cos^2\theta \, d\theta =$

$8\int_0^{2\pi} \frac{d\theta}{2} + \int_0^{2\pi} \cos 2\theta \, d\theta = 8\pi + 0 = 8\pi$. Hence (3) equals 20π. So

$$\oint_C \vec{F} \cdot d\vec{r} = 20\pi \tag{4}$$

To verify Stokes's Theorem we now need to show

$$\iint_S (\text{curl } \vec{F} \cdot \vec{n}) \, dA = 20\pi \ .$$

Using the definition

$$\nabla \times \vec{F} = \begin{vmatrix} \hat{i} & \hat{j} & \hat{k} \\ \frac{\partial}{\partial x} & \frac{\partial}{\partial y} & \frac{\partial}{\partial z} \\ f_1 & f_2 & f_3 \end{vmatrix}$$

where $\nabla \times \vec{F}$ represents the curl $\vec{F}$ and $\vec{F} = (f_1, f_2, f_3)$,

$$\nabla \times \vec{F} = \begin{vmatrix} \hat{i} & \hat{j} & \hat{k} \\ \frac{\partial}{\partial x} & \frac{\partial}{\partial y} & \frac{\partial}{\partial z} \\ 3y & -xz & yz^2 \end{vmatrix}$$

$$= \left(\frac{\partial(yz^2)}{\partial y} - \frac{\partial(-xz)}{\partial z}\right)\hat{i} - \left(\frac{\partial(yz^2)}{\partial x} - \frac{\partial(3y)}{\partial z}\right)\hat{j} + \left(\frac{\partial(-xz)}{\partial x} - \frac{\partial(3y)}{\partial y}\right)\hat{k}$$

$$= (z^2+x)\hat{i} - (z+3)\hat{k} = (z^2+x, 0, -z-3).$$

Since S is the paraboloid $2z = x^2+y^2$ bounded by $z = 2$ this is a surface in the form $g(x,y,z) = z - f(x,y) = 0$. The unit normal can be expressed as

$$\frac{\nabla(x^2+y^2-2z)}{\|\nabla(x^2+y^2-2z)\|}$$

where ∇ is the gradient. Therefore

$$\vec{n} = \frac{(2x, 2y, -2)}{\sqrt{4x^2+4y^2+4}} = \frac{(x,y,-1)}{\sqrt{x^2+y^2+1}} \ .$$

But since $\iint_S (\text{curl } \vec{F}) \cdot \vec{n} \, dA = \iint_R (\text{curl } \vec{F}) \cdot \vec{N}$ and $\vec{n} = \frac{\vec{N}}{\|\vec{N}\|}$

$$\iint_S (\text{curl } \vec{F}) \cdot \vec{n} \, dA = \iint_R (z^2+x, 0, -z-3) \cdot (x, y, -1) \, dx \, dy \tag{5}$$

$$= \iint_R (xz^2 + x^2 + z + 3) \, dx \, dy \ ,$$

which, when changed to polar coordinates, becomes

$$\int_0^{2\pi} \int_0^2 \left(r \cos\theta \cdot \frac{r^4}{4} + r^2 \cos^2\theta + \frac{r^2}{2} + 3\right) r \, dr \, d\theta$$

$$= \int_0^{2\pi} \int_0^2 \frac{r^6}{4} \cos\theta \, dr \, d\theta + \int_0^{2\pi} \int_0^2 r^3 \cos^2\theta \, dr \, d\theta + \int_0^{2\pi} \int_0^2 \left(\frac{r^3}{2} + 3r \, dr\right) d\theta \ .$$

The first integral equals 0 since $\int_0^{2\pi} \cos\theta \, d\theta = 0$. The second integral can be computed by the equality
$\cos^2\theta = \frac{1+\cos 2\theta}{2}$ so that $\int_0^{2\pi}\int_0^2 r^3 \cos^2\theta \, dr \, d\theta$

$$= 4\int_0^{2\pi} \left(\tfrac{1}{2} + \frac{\cos 2\theta}{2}\right) d\theta = 4\pi.$$

The last integral $\int_0^{2\pi}\int_0^2 \left(\frac{r^3}{2} + 3r\right) dr \, d\theta$

$$= 2\pi \int_0^2 \left(\frac{r^3}{2} + 3r\right) dr = \left[\frac{16}{8} + \frac{12}{2}\right] 2\pi = 16\pi.$$

So (5) equals $4\pi + 16\pi = 20\pi$ which checks with (4) and verifies Stokes's Theorem.

• **PROBLEM 9-35**

Let $\vec{F}(x,y,z) = \left(\frac{-y}{x^2+y^2}, \frac{x}{x^2+y^2}, 0\right)$. Evaluate $\oint_C \vec{F} \cdot d\vec{r}$ where C is the circle $x^2+y^2 = 1$. Also evaluate $\iint_S (\text{curl } \vec{F}) \cdot \vec{n} \, dA$ and explain the results.

Solution: First compute $\oint_C \vec{F} \cdot d\vec{r}$ where C is the circle $x^2+y^2 = 1$.

Since $d\vec{r} = (dx, dy, dz)$

$$\oint_C \vec{F} \cdot d\vec{r} = \oint_C \left(\frac{-x}{x^2+y^2}, \frac{x}{x^2+y^2}, 0\right) \cdot (dx, dy, dz) \quad (1)$$

$$= \oint_C \frac{-y}{x^2+y^2} dx + \frac{x}{x^2+y^2} dy.$$

Changing to polar coordinates and letting $x = \cos\theta$, $y = \sin\theta$, the above integral equals

$$\int_0^{2\pi} (-\sin\theta)(-\sin\theta) d\theta + (\cos\theta)(\cos\theta) d\theta$$

$$= \int_0^{2\pi} (\sin^2\theta + \cos^2\theta) d\theta = \int_0^{2\pi} d\theta = 2\pi.$$

To evaluate $\iint_S (\text{curl } \vec{F}) \cdot \vec{n} \, dA$ first compute curl $\vec{F}$.

$$\text{curl } \vec{F} = \begin{vmatrix} \hat{i} & \hat{j} & \hat{k} \\ \frac{\partial}{\partial x} & \frac{\partial}{\partial y} & \frac{\partial}{\partial z} \\ \frac{-y}{x^2+y^2} & \frac{x}{x^2+y^2} & 0 \end{vmatrix} = 0\hat{i} + 0\hat{j} + \left(\frac{\partial (x/(x^2+y^2))}{\partial x} + \frac{\partial (y/(x^2+y^2))}{\partial y}\right)\hat{k}$$

$$= 0\hat{i} + 0\hat{j} + \left(\frac{(x^2+y^2)-2x^2}{(x^2+y^2)^2} + \frac{(x^2+y^2)-2y^2}{(x^2+y^2)^2}\right)\hat{k}$$

$$= 0\hat{i} + 0\hat{j} + 0\hat{k} = 0 \quad \text{(in any region excluding (0,0))}.$$

Therefore $\iint_S (\text{curl } \vec{F}) \cdot \vec{n} \, dA = 0$ since curl $\vec{F} = 0$. This seems to contradict Stokes's Theorem because $\oint_C \vec{F} \cdot d\vec{r} \neq \iint_S (\text{curl } \vec{F}) \cdot \vec{n} \, dA$.

Also since curl $\vec{F} = 0$, $\partial P/\partial y = \partial Q/\partial x$ and using Green's Theorem we have $\oint_C \vec{F} \cdot d\vec{r} = \iint_R \left(\frac{\partial Q}{\partial x} - \frac{\partial P}{\partial y} \right) dx \, dy = 0$ which also contradicts our evaluation of (1). But notice that the curve C includes the origin and $F = \left(\frac{-y}{x^2+y^2}, \frac{x}{x^2+y^2}, 0 \right)$ does not have continuous derivatives throughout any region including (0,0). Hence, since both Green's Theorem and Stoke's Theorem assume continuous partial derivatives exist in a region of space including S, no contradiction exists.

• **PROBLEM 9-36**

Prove: $\oint_C \vec{F} \cdot d\vec{r} = 0$ for every closed curve C if and only if curl $\vec{F} = 0$.

Solution: Necessary and sufficient conditions must be proved. For the necessity assume curl $\vec{F} = 0$ and prove $\oint_C \vec{F} \cdot d\vec{r} = 0$. To do this use Stokes's theorem

$$\int_C \vec{F} \cdot d\vec{r} = \iint_S (\text{curl } \vec{F}) \cdot \vec{n} \, dA = 0 \quad \text{since curl } \vec{F} = 0.$$

Therefore $\oint_C \vec{F} \cdot d\vec{r} = 0$. For sufficiency, assume $\oint_C \vec{F} \cdot d\vec{r} = 0$ for every closed path C and prove curl $\vec{F} = 0$. For this proof assume curl $\vec{F} \neq 0$ for some point P and show a contradiction to the hypothesis. Additional assumption is that curl $\vec{F}$ is continuous and then there is a region containing P where curl $\vec{F} \neq 0$. Define S to be a surface in this region whose unit normal $\vec{n}$ at each point has the same direction as curl $\vec{F}$. Then curl $\vec{F} = k\vec{n}$ (where k is a positive constant). Then letting C be the boundary of S, apply Stoke's Theorem and obtain

$$\oint_C \vec{F} \cdot d\vec{r} = \iint_S (\text{curl } \vec{F}) \cdot \vec{n} \, dA = k \iint_S \vec{n} \cdot \vec{n} \, dA > 0.$$

Therefore $\oint_C \vec{F} \cdot d\vec{r} > 0$. This contradiction shows that curl $\vec{F} = 0$.

Thus, since both conditions are satisfied, $\oint_C \vec{F} \cdot d\vec{r} = 0$ for every closed curve C if and only if curl $\vec{F} = 0$.

DIFFERENTIAL FORM

• **PROBLEM 9-37**

Show that the 2-form

$$\sigma = \frac{x\, dy\, dz + y\, dz\, dx + z\, dx\, dy}{(x^2 + y^2 + z^2)^{3/2}}$$

satisfies $d\sigma = 0$ but that σ is not exact. Do this by proving that $\iint_S \sigma$, where S is the unit sphere, is not zero.

Solution: For an arbitrary 2-form

$$\omega = f(x,y,z)\,dy\,dz + g(x,y,z)\,dz\,dx + h(x,y,z)\,dx\,dy,$$

by definition $d\omega = \left(\dfrac{\partial f}{\partial x} + \dfrac{\partial g}{\partial y} + \dfrac{\partial h}{\partial z}\right) dx\, dy\, dz$. In this case

$$f = x(x^2 + y^2 + z^2)^{-3/2}$$

$$g = y(x^2 + y^2 + z^2)^{-3/2}$$

$$h = z(x^2 + y^2 + z^2)^{-3/2}$$

$$\frac{\partial f}{\partial x} = (x^2+y^2+z^2)^{-3/2} + x(-3/2)(x^2+y^2+z^2)^{-5/2}(2x)$$

$$\frac{\partial g}{\partial y} = (x^2+y^2+z^2)^{-3/2} + y(-3/2)(x^2+y^2+z^2)^{-5/2}(2y)$$

$$\frac{\partial h}{\partial z} = (x^2+y^2+z^2)^{-3/2} + z(-3/2)(x^2+y^2+z^2)^{-5/2}(2z).$$

Adding these three equations

$$\frac{\partial f}{\partial x} + \frac{\partial g}{\partial y} + \frac{\partial h}{\partial z} = 3(x^2+y^2+z^2)^{-3/2} - 3x^2(x^2+y^2+z^2)^{-5/2}$$

$$- 3y^2(x^2+y^2+z^2)^{-5/2} - 3z^2(x^2+y^2+z^2)^{-5/2}$$

$$= 3(x^2+y^2+z^2)^{-3/2} - 3(x^2+y^2+z^2)(x^2+y^2+z^2)^{-5/2}$$

$$= 0.$$

To compute $\iint_S \sigma$ parametrize:

$$x = \sin\varphi \cos\theta \qquad 0 \le \varphi \le \pi$$
$$y = \sin\varphi \sin\theta \qquad 0 \le \theta \le 2\pi$$
$$z = \cos\varphi$$

then by definition

$$dx\, dy = \frac{\partial(x,y)}{\partial(\varphi,\theta)} d\varphi\, d\theta$$

$$dz\, dx = \frac{\partial(z,x)}{\partial(\varphi,\theta)} d\varphi\, d\theta$$

and

$$dy\, dz = \frac{\partial(y,z)}{\partial(\varphi,\theta)} d\varphi\, d\theta.$$

Thus

$$dx\, dy = \begin{vmatrix} \dfrac{\partial x}{\partial \varphi} & \dfrac{\partial x}{\partial \theta} \\ \dfrac{\partial y}{\partial \varphi} & \dfrac{\partial y}{\partial \theta} \end{vmatrix} d\varphi\, d\theta$$

$$= \begin{vmatrix} \cos\varphi\, \cos\theta & -\sin\varphi\, \sin\theta \\ \cos\varphi\, \sin\theta & \sin\varphi\, \cos\theta \end{vmatrix} d\varphi\, d\theta$$

$$= (\cos^2\theta\, \cos\varphi\, \sin\varphi + \sin^2\theta\, \sin\varphi\, \cos\varphi)\, d\varphi\, d\theta$$

$$= \cos\varphi\, \sin\varphi\, d\varphi\, d\theta .$$

Similarly $dz\, dx = \sin^2\varphi\, \sin\theta\, d\varphi\, d\theta$ and $dy\, dz = \sin^2\varphi\, \cos\theta\, d\varphi\, d\theta$.

Now

$$\iint \frac{x\, dy\, dz + y\, dz\, dx + z\, dx\, dy}{(x^2 + y^2 + z^2)^{3/2}}$$

becomes, with the above substitutions,

$$\int_0^{2\pi}\!\!\int_0^{\pi}\!\Big[(\sin\varphi\, \cos\theta)(\sin^2\varphi\, \cos\theta) + (\sin\varphi\, \sin\theta)(\sin^2\varphi\, \sin\theta)$$
$$+ (\cos\varphi)(\cos\varphi\, \sin\varphi)\Big] d\varphi\, d\theta$$

$$= \int_0^{2\pi}\!\!\int_0^{\pi}\!\Big[\sin^3\varphi\, \cos^2\theta + \sin^3\varphi\, \sin^2\theta + \cos^2\varphi\, \sin\varphi\Big] d\varphi\, d\theta$$

$$= \int_0^{2\pi}\!\!\int_0^{\pi} \sin\varphi\, d\varphi\, d\theta$$

$$= 2\pi \int_0^{\pi} \sin\varphi\, d\varphi$$

$$= 2\pi \left(-\cos\varphi \Big|_0^{\pi}\right) = 2\pi(2) = 4\pi \neq 0 .$$

Therefore σ is not exact.

CHAPTER 10

IMPROPER INTEGRALS

By definition an integral is improper if at least one of the limits is infinite, or if the integrand has one or more points of discontinuity in the interval of integration, or if both of the above conditions exist. Respectively, the above defines improper integrals of the first, second, and third kind.

An important idea to note is that there are many analogies between the theory of improper integrals and that of infinite series. That is, improper integrals can be termed convergent or divergent in the same manner as infinite series (i.e., the comparison test, quotient test, and certain limit tests exist in the theory of improper integrals). Furthermore, the notions of absolute convergence and conditional convergence exist for improper integrals and in addition, if an improper integral defines a function, it can be determined whether or not this integral is uniformly convergent on a given interval.

Finally, from the theory of improper integrals, properties for certain functions such as the gamma function and beta function, both of which have many applications, can be developed.

IMPROPER INTEGRALS OF THE 1st, 2nd, AND 3rd KIND

● **PROBLEM 10-1**

Determine if the following integrals are proper or improper. If an integral is improper, determine if it is of the first, second or third kind.

a) $\int_{-\infty}^{-1} \frac{dx}{x(x-1)}$ b) $\int_{0}^{\infty} \frac{e^{-at} - e^{-bt}}{t} dt$ c) $\int_{0}^{\frac{1}{2}} \frac{dx}{x(x-1)}$

d) $\int_{4}^{8} \frac{x\,dx}{(x-3)^2}$ e) $\int_{0}^{\infty} \frac{e^{-x}}{\sqrt{x}} dx$.

<u>Solution</u>: Given that a function is bounded, and the intervals or regions of integration are bounded, if this function is integrable, the integral is then said to be a proper integral.
An integral is improper if it is an integral over an unbounded interval or region or if it is the integral of an unbounded function or both. Furthermore, the integral $\int_{a}^{b} f(x)\,dx$ is:

(i) an improper integral of the first kind if $a = -\infty$ or $b = \infty$ or both.
(ii) an improper integral of the second kind if $f(x)$ is unbounded at one or more points of $a \leq x \leq b$. These points are termed singularities of $f(x)$.
(iii) an improper integral of the third kind if both conditions (i) and (ii) are satisfied.

a) $\int_{-\infty}^{-1} \frac{dx}{x(x-1)}$ This is an improper integral of the first kind, since one of the integration limits is infinite and the integrand is bounded.

b) $\int_0^\infty \frac{e^{-at} - e^{-bt}}{t} dt$. This is an improper integral of the first kind,

for $\lim_{t \to 0} \frac{e^{-at} - e^{-bt}}{t} = b - a$. That is, as $t \to 0$, the integrand approaches a finite limit. Hence, by condition (i), the improper integral is of the first kind.

c) $\int_0^{1/2} \frac{dx}{x(x-1)}$. This is an improper integral of the second kind since the integral is unbounded at $x = 0$.

d) $\int_4^8 \frac{x \, dx}{(x-3)^2}$. This is a proper integral since, even though the integrand becomes unbounded at $x = 3$, this value for x is outside the interval of integration $4 \le x \le 8$.

e) $\int_0^\infty \frac{e^{-x}}{\sqrt{x}} dx$. This is an improper integral of the third kind by condition (iii) since one of the limits of integration is infinite and the integrand is unbounded at $x = 0$.

● **PROBLEM 10-2**

Determine if the following improper integrals of the first kind converge or diverge:

a) $\int_1^\infty \frac{1}{x^p} dx$ b) $\int_0^\infty e^{-rx} dx$ c) $\int_0^\infty \sin x \, dx$.

Solution: Let $f(x)$ be continuous in the interval $a \le x < \infty$. Then $\int_a^\infty f(x) dx$ converges if and only if $\lim_{R \to \infty} \int_a^R f(x) dx = A$ exists and is finite; in such a case one defines $\int_a^\infty f(x) dx = A$. If $\lim_{R \to \infty} \int_a^R f(x) dx$ does not exist, then $\int_a^\infty f(x) dx$ is said to diverge. Applying these definitions we have:

a) For $\int_1^\infty \frac{1}{x^p} dx$, $\lim_{R \to \infty} \int_1^R \frac{1}{x^p} dx = \lim_{R \to \infty} \frac{x^{1-p}}{1-p} \Big|_1^R$ $(p \ne 1)$

$$= \lim_{R \to \infty} \left[\frac{R^{1-p}}{1-p} - \frac{1}{1-p} \right],$$

which exists and equals $1/(1-p)$ for $p > 1$ and does not exist for $p < 1$. At $p = 1$ we have

$$\lim_{R \to \infty} \int_1^R \frac{dx}{x} = \lim_{R \to \infty} [\ln R - \ln 1]$$

which does not exist. Hence, $\int_1^\infty \frac{1}{x^p} dx$ converges if and only if

$p > 1$ and diverges if $p \leq 1$.

b) For $\int_0^\infty e^{-rx} dx$,

$$\lim_{R\to\infty} \int_0^R e^{-rx} dx = \lim_{R\to\infty} \left. \frac{-e^{-rx}}{r} \right|_0^R \quad (r \neq 0)$$

$$= \lim_{R\to\infty} \left[\frac{-e^{-Rr}}{r} + \frac{1}{r} \right]$$

which exists and equals $1/r$ if $r > 0$ and does not exist if $r < 0$. At $r = 0$

$$\lim_{R\to\infty} \int_0^R dx = \lim_{R\to\infty} R \text{ which does not exist. Therefore}$$

$\int_0^\infty e^{-rx} dx$ converges if $r > 0$ and diverges if $r \leq 0$.

c) For $\int_0^\infty \sin x \, dx$,

$$\lim_{R\to\infty} \int_0^R \sin x \, dx = \lim_{R\to\infty} (1 - \cos R)$$

which does not exist. Thus $\int_0^\infty \sin x \, dx$ diverges.

● **PROBLEM 10-3**

Test for convergence:

a) $\int_2^\infty \frac{x^2 dx}{\sqrt{x^7+1}}$ b) $\int_2^\infty \frac{x^3 dx}{\sqrt{x^7+1}}$ c) $\int_2^\infty \frac{x^2+4x+4}{(\sqrt{x}-1)^3 \sqrt[3]{x^3-1}} dx$.

Solution: To test the given integrals for convergence, apply the following theorem (Comparison test):

Let $\int_a^\infty f(x) dx$ and $\int_b^\infty g(x) dx$ be two integrals of the first kind with nonnegative integrands. Suppose that $f(x) \leq g(x)$ for all values of x beyond a certain point $x = c$. Then if $\int_b^\infty g(x) dx$ is convergent, so is $\int_a^\infty f(x) dx$, and if $\int_a^\infty f(x) dx$ is divergent, so is $\int_b^\infty g(x) dx$.

a) $\int_2^\infty \frac{x^2 dx}{\sqrt{x^7+1}}$. By the theorem, to prove that this integral is divergent we must find an integrand which is less than the one given whose integral we know to diverge; to prove it is convergent we must find an integrand greater than the one given whose integral we know to converge. Proceeding, for $2 \leq x < \infty$, we know that $x^4 x^3 < x^7 + 1$ so that $x^4/(x^7+1) < 1/x^3$ and, taking square roots,

$$0 < \frac{x^2}{\sqrt{x^7+1}} < \frac{1}{x^{3/2}} \,. \tag{1}$$

Furthermore, $\int_2^\infty \frac{1}{x^{3/2}}$ converges since it has previously been shown that $\int_2^\infty \frac{1}{x^p}$ converges for $p > 1$. Hence, by the comparison test and by (1), since $\int_2^\infty \frac{1}{x^{3/2}}$ converges, $\int_2^\infty \frac{x^2 dx}{\sqrt{x^7+1}}$ converges.

b) $\int_2^\infty \frac{x^3}{\sqrt{x^7+1}} dx$. Since $\frac{x^3}{\sqrt{x^7+1}} = \frac{1}{\sqrt{x^7+1}/\sqrt{x^6}} = \frac{1}{\sqrt{x+x^{-6}}} = \frac{1}{\sqrt{x}\sqrt{1+x^{-7}}} \geq \frac{1}{\sqrt{x}\sqrt{1+2^{-7}}}$ for $2 \leq x < \infty$, and $\frac{1}{\sqrt{1+2^{-7}}} \int_2^\infty \frac{1}{\sqrt{x}} dx$ diverges (since $\int_0^\infty \frac{1}{x^p} dx$ diverges for $p \leq 1$), by the comparison test, $\int_2^\infty \frac{x^3}{\sqrt{x^7+1}} dx$ diverges.

c) $\int_2^\infty \frac{x^2+4x+4}{(\sqrt{x}-1)^3 \sqrt[3]{x^3-1}} dx$. Since $\frac{x^2+4x+4}{(\sqrt{x}-1)^3 \sqrt[3]{x^3-1}} \geq \frac{x^2+4x+4}{(\sqrt{x})^3 \sqrt[3]{x^3}} = \frac{x(x+4+4/x)}{(\sqrt{x})^3 x\sqrt{x}}$

$= \frac{x+4+4/x}{x^2} = \frac{1+4/x+4/x^2}{x} \geq \frac{1}{x}$ for $2 \leq x < \infty$ and $\int_2^\infty \frac{1}{x} dx$ diverges, by the comparison test, the given integral diverges.

● **PROBLEM 10-4**

Test for convergence:

a) $\int_1^\infty \frac{x^2 dx}{2x^4-x+1}$ b) $\int_1^\infty \frac{x \, dx}{3x^4+6x^2+1}$.

Solution: a) To test this integral for convergence, use the following test, known as the Quotient test: Suppose $\int_a^\infty f(x) dx$ and $\int_b^\infty g(x) dx$ are integrals of the first kind with nonnegative integrands:

i) If the limit $\lim_{x \to \infty} \frac{f(x)}{g(x)} = L$ exists (finite) and is not zero, then either both integrals are convergent or both are divergent.

ii) If $L = 0$ in (1) and $\int_b^\infty g(x) dx$ converges, then $\int_a^\infty f(x) dx$ converges.

iii) If $L = \infty$ in (i) and $\int_b^\infty g(x) dx$ diverges, then $\int_a^\infty f(x) dx$ diverges.

To apply the test, first observe that for large values of x the integrand is approximately $1/2x^2$, since $2x^4 - x + 1$ will be approximately equal to $2x^4$. Specifically, taking

$$f(x) = \frac{x^2}{2x^4-x+1}, \quad g(x) = \frac{1}{x^2}$$

we find $\lim_{x\to\infty} \frac{f(x)}{g(x)} = \lim_{x\to\infty} \frac{x^4}{2x^4-x+1} = 1/2$.

Now the integral $\int_1^\infty \frac{1}{x^2} dx$ is convergent since $\int_1^\infty \frac{1}{x^p} dx$ is convergent for $p > 1$. Therefore, since both integrals are either convergent or divergent by (i), and because it has been shown that $\int_1^\infty \frac{1}{x^2} dx$ is convergent, then both are convergent. Thus,

$$\int_1^\infty \frac{x^2 \, dx}{2x^4-x+1} \text{ is convergent.}$$

b) $\int_1^\infty \frac{x}{3x^4+6x^2+1} dx$. This integral will be tested for convergence by two methods.

Method 1: The Comparison Test.

For large x, the integrand is approximately equal to
$$x/3x^4 = 1/3x^3.$$

Since $\frac{x}{3x^4+6x^2+1} \le \frac{1}{3x^3}$ and $\frac{1}{3}\int_1^\infty \frac{1}{x^3} dx$ converges (because $\int_1^\infty \frac{1}{x^p} dx$ converges for $p > 1$), by the comparison test $\int_1^\infty \frac{x}{3x^4+6x^2+1} dx$ converges.

Method 2: The Quotient Test.

Let $f(x) = \frac{x}{3x^4+6x^2+1}$, $g(x) = \frac{1}{x^3}$. Then $\lim_{x\to\infty} \frac{f(x)}{g(x)} = \lim_{x\to\infty} \frac{x^4}{3x^4+6x^2+1} = \frac{1}{3}$.

Now, the integral $\int_1^\infty \frac{1}{x^3} dx$ is convergent (p integral with $p = 3$).

Therefore, since $\lim_{x\to\infty} \frac{f(x)}{g(x)} = \frac{1}{3}$ and $\int_1^\infty g(x) dx$ converges,

$\int_1^\infty f(x) dx = \int_1^\infty \frac{x}{3x^4+6x^2+1} dx$ converges by the quotient test.

• **PROBLEM 10-5**

Test the following improper integrals of the first kind for convergence:

a) $\int_1^\infty \frac{\ln x}{x+a} dx$ b) $\int_0^\infty \frac{1-\cos x}{x^2} dx$ c) $\int_{-\infty}^\infty \frac{x^3+x^2}{x^6+1} dx$.

Solution: To test the given improper integrals for convergence, the following theorem is used:

Given $\int_a^\infty f(x)dx$, let $\lim_{x\to\infty} x^p f(x) = A$. Then:

(i) $\int_a^\infty f(x)dx$ converges if $p > 1$ and A is finite.

(ii) $\int_a^\infty f(x)dx$ diverges if $p \le 1$ and $A \ne 0$, (A may be infinite).

a) $\int_1^\infty \frac{\ln x}{x+a} dx$. To apply the theorem to this integral take $x^p = x$ (i.e., $p = 1$) and $f(x) = \frac{\ln x}{x+a}$. Then,

$$\lim_{x \to \infty} x^p f(x) = \lim_{x \to \infty} \frac{x \ln x}{x+a} = \lim_{x \to \infty} \frac{1+ \ln x}{1} = \infty$$

by L'Hospital's rule. Therefore, by (ii) in the theorem, with $p = 1$ and $A = \infty$, the given integral diverges.

b) $\int_0^\infty \frac{1-\cos x}{x^2} dx$. Here, to apply the theorem, let

$$\int_0^\infty \frac{1-\cos x}{x^2} dx = \int_0^\pi \frac{1-\cos x}{x^2} dx + \int_\pi^\infty \frac{1-\cos x}{x^2} dx \quad (1)$$

Now, the first integral on the right side of equation (1) is a proper integral. This is so since

$$\lim_{x \to 0^+} \frac{1-\cos x}{x^2} = \lim_{x \to 0^+} \frac{\sin x}{2x} = \frac{1}{2}$$

by L'Hospital's rule (therefore, there is no singularity at $x = 0$). For the second integral on the right side of equation (1) take $x^p = x^{3/2}$ ($p = 3/2$) and $f(x) = (1-\cos x)/x^2$. Then

$$\lim_{x \to \infty} x^p f(x) = \lim_{x \to \infty} x^{3/2} \left(\frac{1-\cos x}{x^2} \right) = \lim_{x \to \infty} \frac{1-\cos x}{\sqrt{x}} = 0.$$

Hence, since $p = 3/2$ and $A = 0$, by (i) in the theorem the integral converges. Thus, by (1) the given integral converges.

c) $\int_{-\infty}^\infty \frac{x^3+x^2}{x^6+1} dx$. To apply the theorem to this integral, let

$$\int_{-\infty}^\infty \frac{x^3+x^2}{x^6+1} dx = \int_{-\infty}^0 \frac{x^3+x^2}{x^6+1} dx + \int_0^\infty \frac{x^3+x^2}{x^6+1} dx \quad (2)$$

Then let $x = -t$ for the first integral on the right side of (2); this yields

$$-\int_0^\infty \frac{t^3-t^2}{t^6+1} dt. \quad (3)$$

Now apply the theorem to this integral, letting $t^p = t^3$ (i.e., $p = 3$) and $f(t) = \frac{t^3-t^2}{t^6+1}$. So that

$$\lim_{t \to \infty} t^p f(t) = \lim_{t \to \infty} t^3 \frac{(t^3-t^2)}{t^6+1} = \lim_{t \to \infty} \frac{1-1/t}{1+1/t^6} = 1.$$

Hence since $p = 3$ and $A = 1$, the integral (3) converges by (i) in the theorem.

Similarly, for the second integral on the right side of (2), take $p = 3$ and $f(x) = (x^3+x^2)/(x^6+1)$. Then,

$$\lim_{x \to \infty} x^p f(x) = \lim_{x \to \infty} x^3 \left(\frac{x^3+x^2}{x^6+1}\right) = 1.$$

Therefore, by (i), this integral also converges. Thus by (2), the given integral converges.

● **PROBLEM 10-6**

Test the following integrals of the first kind for convergence:

a) $\int_0^\infty (1/t) \sin t \, dt$ (1)

b) $\int_0^\infty \sin u^2 \, du$ (2)

Solution: a) $\int_0^\infty (1/t) \sin t \, dt$. At first glance the integrand shows a singularity at $t = 0$ so that it is an improper integral of the third kind. However, closer inspection of the integral shows this to be untrue since by L'Hospital's rule,

$$\lim_{x \to 0} \frac{\sin x}{x} = 1.$$

Therefore, there is no singularity at $t = 0$ and the integral is of the first kind. To test this integral for convergence the following theorem is necessary: let

$$\int_a^\infty \varphi(t) f(t) dt$$

be an improper integral of the first kind. Given that the functions φ and f satisfy the following conditions:

(i) $\varphi'(t)$ is continuous, $\varphi'(t) \leq 0$ and $\lim_{t \to \infty} \varphi(t) = 0$,

(ii) $f(t)$ is continuous, and bounded if the integral $F(x) = \int_a^x f(t) dt$ is bounded for all $x \geq a$,

then the integral $\int_a^\infty \varphi(t) f(t) dt$ is convergent.

To apply this theorem, take $\varphi(t) = 1/t$, $f(t) = \sin t$ (from (1)). Then $\varphi'(t) = -1/t^2 \leq 0$, $\lim_{t \to \infty} 1/t = 0$ and $F(x) = \int_0^x \sin t \, dt = 1-\cos x$,

so that $0 \leq F(x) \leq 2$. ($F(x)$ is bounded for all $x \geq a$). Therefore, since the conditions of the theorem are satisfied, the given integral (1) is convergent.

b) $\int_0^\infty \sin u^2 \, du$. To test this integral for convergence, apply the theorem mentioned in part a) of this problem. However, to do this, first make the following change of variable:

$u = \sqrt{t}$ so that $du = \dfrac{dt}{2\sqrt{t}}$.

Then

$$\int_0^x \sin u^2 \, du = \frac{1}{2} \int_0^{x^2} \frac{\sin t}{\sqrt{t}} \, dt. \qquad (3)$$

Therefore, letting $x \to \infty$ in (2), we have

$$\int_0^\infty \sin u^2 \, du = \frac{1}{2} \int_0^\infty \frac{\sin t}{\sqrt{t}} \, dt. \qquad (4)$$

(Note there is no singularity at $t = 0$ since $\lim_{t \to 0} \dfrac{\sin t}{\sqrt{t}} = 0$, by L'Hospital's rule).

Now to apply the theorem to the integral on the right side of equation (4) take

$$\varphi(t) = t^{-1/2}, \quad f(t) = \sin t.$$

Then $\varphi'(t) = -\dfrac{1}{2t^{3/2}} \le 0$ for $0 \le t < \infty$, $\lim_{t \to \infty} 1/\sqrt{t} = 0$ and

$$F(x) = \int_0^x \sin t \, dt = 1 - \cos x$$

so that $0 \le F(x) \le 2$. Thus, by the theorem, $\dfrac{1}{2} \int_0^\infty \dfrac{\sin t}{\sqrt{t}} \, dt$ is convergent, so that by (4), the integral (2) is convergent.

• **PROBLEM 10-7**

Determine if the following improper integrals of the second kind are convergent:

a) $\displaystyle\int_{-1}^4 \dfrac{dx}{(x+1)^{1/3}}$, b) $\displaystyle\int_{-1}^1 \dfrac{dx}{x}$, c) $\displaystyle\int_0^1 \ln\left(\dfrac{1}{1-x}\right) dx$.

Solution: An improper integral of the second kind is an integral of the form $\int_a^b f(x)dx$ where $f(x)$ has one or more singularities in the interval $a \le x \le b$.

a) $\displaystyle\int_{-1}^4 \dfrac{dx}{(x+1)^{1/3}}$. To determine if this integral is convergent, the following information is needed:

(i) if $f(x)$ has a singularity at the end point $x = a$ of the interval $a \le x \le b$, then by definition

$$\int_a^b f(x)dx = \lim_{\epsilon \to 0^+} \int_{a+\epsilon}^b f(x)dx \qquad (1)$$

Furthermore, if the limit on the right of (1) exists, then the integral on the left is convergent (divergent otherwise). Similarly, if $f(x)$ has a singularity at the end point $x = b$ in the same interval, then

$$\int_a^b f(x)dx = \lim_{\epsilon \to 0^+} \int_a^{b-\epsilon} f(x)dx. \qquad (2)$$

Again, if the limit exists in (2), the integral on the left is convergent (divergent otherwise).

(ii) If $f(x)$ has a singularity at an interior point $x = x_0$ of the interval $a \leq x \leq b$, then

$$\int_a^b f(x)dx = \lim_{\epsilon_1 \to 0^+} \int_a^{x_0-\epsilon_1} f(x)dx + \lim_{\epsilon_2 \to 0^+} \int_{x_0+\epsilon_2}^b f(x)dx . \qquad (3)$$

As before, if the limits exist in (3), the integral on the left is convergent (otherwise it is divergent).
Note, however, that if the limits on the right of (3) do not exist, it is possible that if we choose $\epsilon_1 = \epsilon_2 = \epsilon$ in (3), the limits do exist. If this is the case, this value is the Cauchy principal value of the integral on the left side of (3).

For the given integral, the intergrand has a singularity at $x = -1$, which is one of the end points of the interval. Hence, applying (i),

$$\int_{-1}^4 \frac{dx}{(x+1)^{1/3}} = \lim_{\epsilon \to 0^+} \int_{-1+\epsilon}^4 \frac{dx}{(x+1)^{1/3}} = \lim_{\epsilon \to 0^+} \frac{(x+1)^{2/3}}{2/3} \Big|_{-1+\epsilon}^4$$

$$= \lim_{\epsilon \to 0^+} \frac{5^{2/3}}{2/3} - \frac{3}{2}\epsilon^{2/3} = \frac{3}{2} 5^{2/3} .$$

Therefore, since the limit exists, the integral converges and equals $\left(\frac{3}{2}\right)5^{2/3}$.

b) $\int_{-1}^1 \frac{dx}{x}$. The integrand has a singularity at $x = 0$. Hence

$$\int_{-1}^1 \frac{dx}{x} = \lim_{\epsilon_1 \to 0^+} \int_{-1}^{-\epsilon_1} \frac{dx}{x} + \lim_{\epsilon_2 \to 0^+} \int_{\epsilon_2}^1 \frac{dx}{x}$$

$$= \lim_{\epsilon_1 \to 0}(\ln|-\epsilon_1| - \ln|-1|) + \lim_{\epsilon_2 \to 0}(\ln(1) - \ln(\epsilon_2)) .$$

However, these limits do not exist, so the integral does not converge in the usual sense. For the Cauchy value we have

$$\lim_{\epsilon \to 0^+}\left(\int_{-1}^{-\epsilon} \frac{dx}{x} + \int_\epsilon^1 \frac{dx}{x}\right) = \lim_{\epsilon \to 0^+}(\ln(-\epsilon) - \ln(-1) + \ln(1) - \ln(\epsilon))$$

$$= \lim_{\epsilon \to 0^+}(\ln(-\epsilon/\epsilon) - \ln(-1) + \ln(1))$$

$$= \lim_{\epsilon \to 0^+}(\ln(-1) - \ln(-1) + \ln(1)) = 0$$

Thus the Cauchy value of the integral is zero.

c) $\int_0^1 \ln\left(\frac{1}{1-x}\right) dx$. The integrand has a singularity at $x = 1$, therefore this improper integral is of the second kind. To determine if this integral converges, first set $y = (1-x)^{-1}$ so that $dx = y^{-2}dy$ and the integral becomes

$$\int_1^\infty \frac{\ln y}{y^2} dy , \qquad (1)$$

which is an improper integral of the first kind (i.e., the improper integral of the second kind was transformed into an improper integral of the first kind).

Now, by theorem, given two integrals of the first kind with positive integrands such that

$$\lim_{x \to \infty} \frac{f(x)}{g(x)} = L = 0,$$

then if

$$\int_b^\infty g(x)dx$$

is convergent, so is

$$\int_a^\infty f(x)dx.$$

For the given integral, let $g(y) = y^{-3/2}$. Then $\int_b^\infty g(y)dy = \int_1^\infty y^{-3/2} dy$,

which is convergent (p integral, with $p > 1$). In addition,

$$\lim_{y \to \infty} \frac{f(y)}{g(y)} = \lim_{y \to \infty} \frac{\ln y}{y^2} \cdot y^{3/2} = \lim_{y \to \infty} \frac{\ln y}{y^{1/2}}$$

$$= \lim_{y \to \infty} \frac{2}{y^{1/2}} = 0 \text{ by L'Hospital's rule.}$$

Thus (1) is convergent by the theorem and consequently, so is $\int_0^1 \ln\left(\frac{1}{1-x}\right)dx$, since this integral equals (1).

● **PROBLEM 10-8**

Test the following improper integrals of the second kind for convergence:

a) $\int_0^1 \frac{dx}{x^{\frac{1}{2}}(x+2x^2)^{\frac{1}{3}}}$

b) $\int_1^2 \frac{8-x^3}{(2x-x^2)^2} dx$.

Solution: For improper integrals of the second kind, there is a comparison test analogous to that for integrals of the first kind. Here we give the test for the case where $f(x)$ has a singularity at $x = a$ (the test is similar if $f(x)$ has a singularity at $x = b$ or $x = x_0$ for $a < x_0 < b$).
If $g(x) \geq 0$ for $a < x \leq b$ and if $\int_a^b g(x)dx$ converges, then if $0 \leq f(x) \leq g(x)$ for $a < x \leq b$, $\int_a^b f(x)dx$ converges. However, if $\int_a^b g(x)dx$ diverges, then if $f(x) \geq g(x)$ for $a < x \leq b$, $\int_a^b f(x)dx$ diverges.
Before applying this test, however, we make note that the basic, standard, reference integrals for improper integrals of the second kind are

$$\int_a^b \frac{dx}{(b-x)^p}, \quad \int_a^b \frac{dx}{(x-a)^p}. \tag{1}$$

These integrals are convergent for $p < 1$ divergent if $p \geq 1$, and if $p \leq 0$ they are proper integrals, since there would be no singularities in the integrand.

a) $\int_0^1 \frac{dx}{x^{\frac{1}{2}}(x+2x^2)^{\frac{1}{3}}}$. For this integral, $0 < \frac{1}{x^{\frac{1}{2}}(x+2x^2)^{\frac{1}{3}}} \leq \frac{1}{x^{5/6}}$ for

$x > 0$. Since the singularity is at $x = 0$, use the second integral in (1). Therefore,

$$\int_0^1 \frac{dx}{x^{5/6}}$$

is an integral of the second form in (1) with $a = 0$, $b = 1$ and $p = 5/6$; hence, it is convergent. Thus, by the comparison test,

$$\int_0^1 \frac{dx}{x^{\frac{1}{2}}(x+2x^2)^{\frac{1}{3}}}$$

is convergent.

b) $\int_1^2 \frac{8-x^3}{(2x-x^2)^2}\, dx$. Here, the integrand is nonnegative for $1 \le x < 2$ (the singularity is at $x = 2$). Now, using the first integral in (1) we have

$$\int_1^2 \frac{dx}{(2-x)}$$

is divergent (i.e., $b = 2$, $a = 1$, $p = 1$). In addition

$$\frac{8-x^3}{(2x-x^2)^2} = \frac{1}{x^2} \cdot \frac{(2-x)(4+2x+x^2)}{(2-x)^2} = \frac{(4+2x+x^2)}{x^2(2-x)} = \frac{\left(\frac{4}{x^2}\right)+\left(\frac{2}{x}\right)+1}{(2-x)}$$

$$\ge \frac{1}{(2-x)} \quad \text{for } 1 \le x < 2.$$

Hence, by the comparison test, since $f(x) \ge g(x)$ and $\int_a^b g(x)dx$ diverges, the integral

$$\int_1^2 \frac{8-x^3}{(2x-x^2)^3}\, dx$$

diverges.

• **PROBLEM 10-9**

Determine if the following improper integrals of the second kind are convergent:

a) $\int_0^1 \frac{dx}{(1-x^3)^{\frac{1}{3}}}$ b) $\int_0^\pi \frac{\sin x}{x^3}\, dx$ c) $\int_0^1 \ln\left(\frac{1}{1-x}\right) dx$.

Solution: a) $\int_0^1 \frac{dx}{(1-x^3)^{\frac{1}{3}}}$. For this integral, we use a quotient test for integrals with positive integrands analogous to that for improper integrals of the first kind. Here, assuming a singularity at $x = a$ (for singularities at $x = b$ or $x = x_0$, $a < x_0 < b$ the test is similar), if $f(x) \ge 0$ and $g(x) \ge 0$ for $a < x \le b$ and if $\lim_{x \to a} \frac{f(x)}{g(x)} = L \ne 0$ or ∞, then the two integrals $\int_a^b f(x)dx$ and $\int_a^b g(x)dx$ either both converge or both diverge. Furthermore, if $L = 0$ and $\int_a^b g(x)dx$ converges, then $\int_a^b f(x)dx$ converges, and if

$L = \infty$ and $\int_a^b g(x)dx$ diverges then $\int_a^b f(x)dx$ diverges.

To apply this test to the given integral let $f(x) = \dfrac{1}{(1-x^3)^{\frac{1}{3}}}$ and since the integrand has a singularity at the upper limit, let $g(x)$ be an integrand of the form $\dfrac{1}{(b-x)^p}$ because it is known that $\int_a^b \dfrac{dx}{(b-x)^p}$ converges for $p < 1$ and diverges for $p \geq 1$. Let $g(x) = \dfrac{1}{(1-x)^{\frac{1}{3}}}$. Now,

$$\lim_{x \to b^-} \frac{f(x)}{g(x)} = \lim_{x \to 1^-} \frac{(1-x)^{\frac{1}{3}}}{(1-x^3)^{\frac{1}{3}}}. \tag{1}$$

However, since

$$\frac{1}{(1-x^3)^{\frac{1}{3}}} = \frac{1}{(1-x)^{\frac{1}{3}}(1+x+x^2)^{\frac{1}{3}}},$$

(1) becomes

$$\lim_{x \to 1^-} \frac{1}{(1+x+x^2)^{\frac{1}{3}}} = (\tfrac{1}{3})^{\frac{1}{3}}.$$

In addition,

$$\int_0^1 \frac{dx}{(1-x)^{\frac{1}{3}}}$$

is convergent since $p = \tfrac{1}{3} < 1$. Thus, by the quotient test, since $L \neq 0$ or ∞, both integrals are either convergent or divergent. However, since $\int_a^b g(x)dx$ is convergent, $\int_a^b f(x)dx$ (the given integral) is also convergent.

b) $\int_0^\pi \dfrac{\sin x}{x^3} dx$. The test for convergence of this integral involves making use of the following test: If $f(x)$ has a singularity at the lower limit, let $\lim_{x \to a^+} (x-a)^p f(x) = L$ (for the case of a singularity at the upper limit b, let $\lim_{x \to b^-} (b-x)^p f(x) = L$); then

(i) $\int_a^b f(x) dx$ converges if $p < 1$ and L is finite.

(ii) $\int_a^b f(x)dx$ diverges if $p \geq 1$ and $L \neq 0$ (however, L may be infinite).

To apply this test, note that the given integral has a singularity at the lower limit, $x = 0$. Hence, choosing $p = 2$, $\lim_{x \to 0^+} x^2 \dfrac{\sin x}{x^3} = \lim_{x \to 0^+} \dfrac{\sin x}{x} = 1$ by L'Hospital's rule. Now, since $p > 1$, by (ii) the given integral diverges.

c) $\int_0^1 \ln\left(\frac{1}{1-x}\right)dx$. This integral was proven convergent in a previous problem by transforming it to an improper integral of the first kind. However, by the test given in part (b) of this problem, we can determine the convergence or divergence of this integral directly. To do this let $f(x) = \ln\left(\frac{1}{1-x}\right)$ and since the integral has a singularity at the upper limit, $x = 1$, let

$$\lim_{x \to b^-} (b-x)^P f(x) = \lim_{x \to 1^-} (1-x)^{\frac{1}{2}} \ln\left(\frac{1}{1-x}\right) = \lim_{x \to 1^-} \frac{\ln\left(\frac{1}{1-x}\right)}{(1-x)^{-\frac{1}{2}}} = \lim_{x \to 1^-} \frac{-\ln(1-x)}{(1-x)^{-\frac{1}{2}}} .$$

Then by L'Hospital's rule this equals

$$\lim_{x \to 1^-} \frac{1/(1-x)}{-\frac{1}{2}(1-x)^{-3/2}} = \lim_{x \to 1^-} -2(1-x)^{\frac{1}{2}} = 0 .$$

Therefore, by (i) since $p = \frac{1}{2} < 1$ and since L is finite, the given integral converges.

● **PROBLEM 10-10**

Determine for what values of x the integral

$$\int_0^1 e^{-t} t^{x-1} dt \qquad (1)$$

is a) proper; b) improper, but convergent.

Solution: a) The integral (1) is proper for $x \geq 1$. This results from the fact that for these values, the exponent $x-1$ is not negative, and thus t^{x-1} has no singularity at $t = 0$. However, if $x < 1$, the integral is improper, for the exponent $x-1$ would now be negative and t^{x-1} would have a singularity at $t = 0$.

b) For $x < 1$ the integral has a singularity at its lower limit, $t = 0$. To determine for what values of x this improper integral is convergent, use the following limit test:

Let $\lim_{x \to a^+} (x-a)^P f(x) = L$. Then

(i) $\int_a^b f(x) dx$ converges if $p < 1$ and L is finite.

(ii) $\int_a^b f(x) dx$ diverges if $p \geq 1$ and $L \neq 0$ (L may be infinite).

For the given problem, to apply this test consider

$$\lim_{t \to 0^+} t^P e^{-t} t^{x-1} . \qquad (2)$$

Then, at $x = 0$, (2) becomes $\lim_{t \to 0^+} \frac{t^P e^{-t}}{t} . \qquad (3)$

Then letting $p = 1$ in (3) yields

$$\lim_{t \to 0^+} \frac{t e^{-t}}{t} = \lim_{t \to 0^+} e^{-t} = 1 .$$

Hence, by (ii) the integral diverges at $x = 0$. At $0 < x < 1$, let $p = 1-x$ so that in this interval, $p < 1$ always. (2) now becomes

$$\lim_{t \to 0^+} t^{1-x} e^{-t} t^{x-1} = \lim_{t \to 0^+} e^{-t} = 1.$$ Thus, by (i) the integral converges for $0 < x < 1$. Finally, for $x < 0$, let $p = 1-x$ again so that in this interval $p > 1$. As before, $\lim_{t \to 0^+} t^{1-x} e^{-t} t^{x-1} = 1$. However, here $p > 1$, so that the integral diverges for $x < 0$ by (ii). Thus, summarizing the results we have, the integral

$$\int_0^1 e^{-t} t^{x-1} \, dt$$

is 1) proper for $x \geq 1$.
   2) improper but convergent for $0 < x < 1$.
   3) improper but divergent for $x \leq 0$.

● **PROBLEM 10-11**

Test the following integrals for convergence:

a) $\int_0^\infty \dfrac{dx}{(1+x)\sqrt{x}}$ b) $\int_{-\infty}^\infty \dfrac{dx}{x(x-1)}$.

Solution: The above integrals are integrals of the third kind (i.e., a combination of integrals of the first and second kind). To test an integral of this type for convergence, given that the integrand has at most a finite number of singularities, separate the integral into a finite number of integrals, each of the first or second kind. Then this integral of the third kind is convergent if each of these "pure" integrals of the first and second kind are convergent. However, if one or more of these "pure" integrals of the first and second kind are divergent, then the integral of the third kind is divergent.

a) $\int_0^\infty \dfrac{dx}{(1+x)\sqrt{x}}$. This integral has an infinite upper limit and a singularity of the integrand at the lower limit, $x = 0$. Since there are no other singularities, write the integral as:

$$\int_0^\infty \frac{dx}{(1+x)\sqrt{x}} = \int_0^1 \frac{dx}{(1+x)\sqrt{x}} + \int_1^\infty \frac{dx}{(1+x)\sqrt{x}} \quad . \tag{1}$$

Now, the first integral in (1) is an integral of the second kind. To test this integral for convergence let $f(x) = 1/((1+x)\sqrt{x})$ and $g(x) = 1/\sqrt{x}$. Now, if

$$\lim_{x \to a} \frac{f(x)}{g(x)} = L$$

exists and is not zero, then either both $\int_a^\infty f(x)\,dx$, $\int_a^\infty g(x)\,dx$ are convergent or both are divergent. Proceeding, we find

$$\lim_{x \to 0} \frac{\sqrt{x}}{(1+x)\sqrt{x}} = 1 \ .$$

Also, $\int_0^1 \frac{dx}{\sqrt{x}}$ is an integral of the form $\int_a^b \frac{dx}{(x-a)^p}$ with $a = 0$ and $p = \frac{1}{2} < 1$, hence this integral converges and consequently so does the first integral in (1). The second integral in (1) is an integral of the first kind and this integral is convergent, as can be proved by the comparison test since in the interval $0 < x < \infty$,

$$\frac{1}{(1+x)\sqrt{x}} < \frac{1}{x^{3/2}} \quad \text{and} \quad \int_1^\infty \frac{dx}{x^{3/2}}$$

is convergent. Thus, the integral of the third kind in (1) is convergent.

b) $\int_{-\infty}^\infty \frac{dx}{x(x-1)}$. This integral can be written as the sum of six other integrals corresponding to the intervals $(-\infty,-1), (-1,0), (0,\frac{1}{2}), (\frac{1}{2},1), (1,2), (2,\infty)$. The types are of the first, second, second, second, second and first kind, respectively. To test for convergence, each integral must converge for the given integral to be convergent. However, if at least one of the integrals over the corresponding interval diverges, then the given integral diverges. Hence, since

$\int_0^{\frac{1}{2}} \frac{dx}{x(x-1)}$ diverges (if we take $g(x) = x^{-1}$, this gives

$$\lim_{x \to 0} \frac{f(x)}{g(x)} = \lim_{x \to 0} \frac{x}{x(x-1)} = -1 \quad \text{and} \quad \int_0^{\frac{1}{2}} \frac{dx}{x}$$

diverges since it is an integral of the form $\int_a^b \frac{dx}{(x-a)^p}$ with $a = 0$ and $p = 1$. Hence, the integral $\int_0^{\frac{1}{2}} f(x)dx$ diverges since $\int_0^{\frac{1}{2}} g(x)dx$ diverges), we know that the given integral of the third kind diverges without having to inspect the integral over each of the other five intervals.

● **PROBLEM 10-12**

Given n is a real number, show that

$$\int_0^\infty x^n e^{-x} dx \qquad (1)$$

converges when $n > -1$ and diverges when $n \leq -1$.

Solution: First, rewrite (1) as

$$\int_0^1 x^n e^{-x} dx + \int_1^\infty x^n e^{-x} dx . \qquad (2)$$

Now if $n \geq 0$, the first integral in (2) converges since the integrand has no singularities. Next, if $-1 < n < 0$, the first integral in (2) is an improper integral of the second kind at the lower limit $x = 0$. However, by theorem, given that $\lim_{x \to a^+}(x-a)^p f(x) = L$, the integral $\int_a^b f(x)dx$ converges if $p < 1$ and L is finite. To apply this theorem, let $p = -n$, $a = 0$ to obtain $\lim_{x \to 0^+} x^{-n} x^n e^{-x} = 1$. Hence, the integral converges by the theorem. Thus, the first integral converges for $n > -1$.

For $n > -1$, the second integral is an improper integral of the first kind. By theorem, for an integral of the first kind, given $\lim_{x \to \infty} x^p f(x) =$

L, the integral
$$\int_a^\infty f(x)\,dx$$
converges if $p > 1$ and L is finite. To apply this theorem let $p = 2$, to obtain
$$\lim_{x \to \infty} x^2 x^n e^{-x} = \lim_{x \to \infty} \frac{x^{n+2}}{e^x} = 0$$
(by L'Hospital's rule). Therefore, by the theorem, the second integral in (2) also converges for $n > -1$. Consequently, the given integral converges for $n > -1$.

For $n < -1$, the first integral in (2) is an improper integral of the second kind. By theorem, given
$$\lim_{x \to a^+} (x-a)^p f(x) = L,$$
the integral, $\int_a^b f(x)\,dx$ diverges if $p \geq 1$ and $L \neq 0$. To apply this theorem take $p = 1$, $a = 0$ to obtain
$$\lim_{x \to 0^+} x \, x^n e^{-x} = \lim_{x \to 0^+} \frac{x^{n+2}}{e^x} = \infty$$
(since $n < -1$). Therefore, by the theorem, the first integral in (2) diverges for $n < -1$. In addition, at $n = -1$ the first integral in (2) becomes
$$\int_0^1 \frac{e^{-x}}{x}\,dx$$
which diverges since, in the interval $0 \leq x \leq 1$,
$$\frac{e^{-x}}{x} \geq \frac{1}{ex} \quad \text{and} \quad \int_0^1 \frac{1}{x}\,dx = \lim_{x \to 0}(-\log x) = \infty.$$
Hence, the above shows the first integral in (1) diverges for $n \leq -1$.

Thus, the given integral diverges for $n \leq -1$, since at least one of the integrals in (2) is divergent (therefore, there is no need to examine the second integral in (1)).

Summarizing the results, (1) is convergent for $n > -1$ and divergent for $n \leq -1$.

● **PROBLEM** 10-13

a) Given that $\int_1^\infty [f(t)]^2 \, t^{-2}\,dt$ converges, prove $\int_1^\infty |f(t)| t^{-2}\,dt$ converges.

b) Given that $f(t)$ has a singularity at $t = 0$ and that $\int_{0^+}^1 |f(t)|^b\,dt$ converges, prove that $\int_{0^+}^1 |f(t)|^a\,dt$ converges, where $0 < a < b$.

Solution: a) In the chapter on Infinite Series the inequality of Cauchy was applied to prove that a series was convergent. For this problem, the integral analogue of Cauchy's inequality, called the Schwarz inequality, will be utilized. This inequality states:

If $F(t), g(t)$ are continuous, for $a \leq x \leq b$ then
$$\left(\int_a^b |F(t)g(t)|\,dt\right)^2 \leq \int_a^b [F(t)]^2\,dt \int_a^b [g(t)]^2\,dt. \tag{1}$$

Note that if a function $F(t)$ is treated as the continuous analogue of a vector, we define the L_p-norm of $F(t)$ as

$$\|F(t)\|_p = \left[\int_a^b [F(t)]^p dt\right]^{1/p}$$

and the inner product of $F(t)$ and $g(t)$ as $<F(t), g(t)> = \int_a^b F(t)g(t)dt$. Using this notation (1) can be rewritten as

$$<|F(t)|, |g(t)|> \le \|F(t)\|_2 \|g(t)\|_2 .$$

To apply this inequality to the integral

$$\int_1^\infty |f(t)|t^{-2} dt \qquad (2)$$

take $a = 1$, $b = R$, $g(t) = 1/t$, and $F(t) = \dfrac{|f(t)|}{t}$, in (1). This yields:

$$\left(\int_1^R \frac{|f(t)|}{t} \cdot \frac{1}{t} dt\right)^2 \le \int_1^R \frac{[f(t)]^2}{t^2} dt \int_1^R \frac{dt}{t^2} .$$

Now take the limit as $R \to \infty$, to give

$$\left(\int_1^\infty \frac{|f(t)|}{t} \cdot \frac{1}{t} dt\right)^2 \le \int_1^\infty \frac{[f(t)]^2 dt}{t^2} \left(\lim_{R \to \infty} -(R^{-1} - 1)\right)$$

which implies

$$\left(\int_1^\infty \frac{|f(t)|}{t} \cdot \frac{1}{t} dt\right)^2 \le \int_1^\infty \frac{[f(t)]^2}{t^2} dt$$

or

$$\int_1^\infty \frac{|f(t)|}{t^2} dt \le \left[\int_1^\infty \frac{[f(t)]^2}{t^2} dt\right]^{\frac{1}{2}} ,$$

which converges. Thus, by the comparison test, the integral (2) converges.

b) As the Cauchy inequality for infinite series has an integral analogue, so does the Hölder inequality. This inequality states:
If $f(t), g(t) \in C$ (the set of continuous functions) for $a \le x \le b$ and if $1/p + 1/q = 1$ with $p > 1$, $q > 1$, then

$$\int_a^b |f(t)g(t)| dt \le \left(\int_a^b |f(t)|^p dt\right)^{1/p} \left(\int_a^b |g(t)|^q dt\right)^{1/q} . \qquad (3)$$

To apply this inequality to the integral

$$\int_{0^+}^1 |f(t)|^a dt \qquad (4)$$

take $g(t) = 1$ for $0 < \epsilon < 1$ in (3) to yield

$$\int_\epsilon^1 |f(t)|^a dt \le \left(\int_\epsilon^1 |f(t)|^{ap} dt\right)^{1/p} \left(\int_\epsilon^1 1^q dt\right)^{1/q} .$$

Now take the limit as $\epsilon \to 0^+$ to give

$$\int_{0^+}^1 |f(t)|^a \le \left(\int_{0^+}^1 |f(t)|^{ap} dt\right)^{1/p} \left(\lim_{\epsilon \to 0^+}(1-\epsilon)\right)^{1/q}$$

which is

$$\int_{0^+}^1 |f(t)|^a dt \le \left(\int_{0^+}^1 |f(t)|^{ap} dt\right)^{1/p} .$$

Finally, choose p so that $ap = b$. Then, since $0 < a < b$, we have p > 1 as required by the inequality. This gives

$$\int_{0^+}^1 |f(t)|^a \, dt \le \left(\int_{0^+}^1 |f(t)|^b \, dt\right)^{1/p} \le \left(\int_{0^+}^1 |f(t)|^b dt\right)^{a/b}$$

Thus, by the given inequality and the comparison test, integral (4) converges.

ABSOLUTE AND UNIFORM CONVERGENCE

• **PROBLEM 10-14**

Determine if the following improper integrals converge absolutely:

a) $\int_1^\infty \frac{\sin x}{x^2} \, dx$ b) $\int_1^\infty \frac{\cos x}{x} \, dx$ c) $\int_0^\infty \frac{\sin x}{x} \, dx$.

Solution: By definition, the integral $\int_a^\infty f(x) dx$ is absolutely convergent if $\int_a^\infty |f(x)| \, dx$ converges. If $\int_a^\infty f(x) dx$ converges and $\int_a^\infty |f(x)| \, dx$ diverges, then $\int_a^\infty f(x) dx$ is said to be conditionally convergent.

a) $\int_1^\infty \frac{\sin x}{x^2} \, dx$. By definition this integral is absolutely convergent

if $\int_1^\infty \left|\frac{\sin x}{x^2}\right| dx$ (1)

is convergent. Since $\left|\frac{\sin x}{x^2}\right| \le \frac{1}{x^2}$, $\int_1^\infty \left|\frac{\sin x}{x^2}\right| dx \le \int_1^\infty \frac{dx}{x^2}$, which

converges (p integral with p = 2). Hence, by the comparison test the integral (1) converges and thus the given integral is absolutely convergent.

b) $\int_1^\infty \frac{\cos x}{x} dx$. For this integral one method would be to use the

series test which states: for integrals with nonnegative integrands, the integral

$$\int_a^\infty f(x) dx$$

converges or diverges according to whether or not Σa_n, where $a_n = f(n)$, converges or diverges, respectively. However, we use another method. First, note that

$$\int_1^\infty \left|\frac{\cos x}{x}\right| dx \ge \sum_{n=1}^\infty \int_{2\pi n - \pi/3}^{2\pi n + \pi/3} \frac{\cos x}{x} \, dx .$$

Then, between $2\pi n - \pi/3$ and $2\pi n + \pi/3$,

$$\frac{\cos x}{x} \ge (2\pi n + \pi/3)^{-1} \tfrac{1}{2} ,$$

(since $\cos \pi/3 = \tfrac{1}{2}$ and in the given interval $\cos x \ge \tfrac{1}{2}$ and $x \le 2\pi n + \pi/3$). Thus,

478

$$\int_1^\infty \left|\frac{\cos x}{x}\right| dx \geq \sum_{n=1}^\infty \frac{1}{2} \cdot \frac{1}{(2\pi n + \pi/3)} \cdot \frac{2\pi}{3} = \infty$$

i.e., the series diverges. Therefore, the integral diverges and the given integral is not absolutely convergent.

c) $\int_0^\infty \frac{\sin x}{x} dx = \int_0^\pi \frac{\sin x}{x} dx + \int_\pi^{2\pi} \frac{\sin x}{x} dx + \ldots + \int_{n\pi}^{(n+1)\pi} \frac{\sin x}{x} dx + \ldots$

$$= \sum_{n=0}^\infty \int_{n\pi}^{(n+1)\pi} \frac{\sin x}{x} dx \ .$$

Therefore,

$$\int_0^\infty \left|\frac{\sin x}{x}\right| dx = \sum_{n=0}^\infty \int_{n\pi}^{(n+1)\pi} \left|\frac{\sin x}{x}\right| dx \ . \tag{2}$$

Then, letting $x = z + n\pi$, (2) becomes

$$\int_0^\infty \left|\frac{\sin x}{x}\right| dx = \sum_{n=0}^\infty \left|(-1)^n\right| \int_0^\pi \frac{\sin z}{z+n\pi} dz = \sum_{n=0}^\infty \int_0^\pi \frac{\sin z}{z+n\pi} dz \ . \tag{3}$$

Since $\frac{1}{z+n\pi} \geq \frac{1}{(n+1)\pi}$ for $0 \leq z \leq \pi$, the integral

$$\int_0^\pi \frac{\sin z}{z+n\pi} dz \geq \frac{1}{(n+1)\pi} \int_0^\pi \sin z \, dz = \frac{1-(-1)}{(n+1)\pi}$$

(the absolute value sign can be removed since $\frac{\sin z}{z+n\pi}$ is positive in the interval from $z = 0$ to $z = \pi$). Hence,

$$\sum_{n=0}^\infty \int_0^\pi \frac{\sin z}{z+n\pi} dz \geq \sum_{n=0}^\infty \frac{2}{(n+1)\pi}$$

which diverges. Thus, the series $\sum_{n=0}^\infty \int_0^\pi \frac{\sin z}{z+n\pi} dz$ diverges, so that by (3) and the comparison test, the integral $\int_0^\infty \left|\frac{\sin x}{x}\right| dx$ diverges.

This shows that the given integral is not absolutely convergent.
Note: the integral is convergent; therefore given this, we have shown the integral to be conditionally convergent.

• **PROBLEM 10-15**

Prove the following theorems (assume $f(x) \in C$):

a) For $a \leq x < \infty$, if $\int_a^\infty |f(x)| dx$ converges then $\int_a^\infty f(x) dx$ converges.

b) For $a \leq x < \infty$, if $\lim_{x \to \infty} x^p f(x)$ exists for $p > 1$, then $\int_a^\infty |f(x)| dx$ converges. It is assumed that f is bounded on $[a,c]$ for every $c > a$.

c) For $a < x \leq b$, if $\lim_{x \to a^+} (x-a)^p f(x)$ exists for $0 < p < 1$, and f is bounded on $(a,b]$ then $\int_{a+}^b |f(x)| dx$ converges.

Solution: a) Since $-|f(x)| \leq f(x) \leq |f(x)|$ for $a \leq x < \infty$ we have

$0 \leq f(x) + |f(x)| \leq 2|f(x)|$ so that
$$\int_a^\infty [f(x) + |f(x)|] dx \leq 2\int_a^\infty |f(x)| dx .$$

Since $\int_a^\infty |f(x)| dx$ converges, then $\int_a^\infty [f(x) + |f(x)|] dx$ converges. Then subtracting $\int_a^\infty |f(x)| dx$, which converges, this proves that $\int_a^\infty f(x) dx$ converges. This result shows that an absolutely convergent integral converges.

b) Given $\lim_{x \to \infty} x^p f(x) = L$ (exists), this implies that $\lim_{x \to \infty} x^p |f(x)| = |L|$. Hence, there exists a positive number b such that
$$x^p |f(x)| \leq |L| + 1 \quad \text{for} \quad b \leq x < \infty$$
or
$$|f(x)| \leq \frac{|L| + 1}{x^p} . \tag{1}$$

However, $\int_b^\infty \frac{|L| + 1}{x^p} dx$ is an integral that converges for $p > 1$.

Thus, by the comparison test and (1), $\int_b^\infty |f(x)| dx$ converges, whence the result $\int_a^\infty |f(x)| dx$ converges follows.

c) Given $\lim_{x \to a+} (x-a)^p f(x) = L$ (exists), this implies that $\lim_{x \to a+} (x-a)^p |f(x)| = |L|$. There exists a number c such that
$$(x-a)^p |f(x)| \leq |L| + 1 \quad \text{for} \quad a < x \leq c < b$$
or
$$|f(x)| \leq \frac{|L| + 1}{(x-a)^p} . \tag{2}$$

However, $\int_{a+}^c \frac{|L| + 1}{(x-a)^p} dx$ is a standard test integral for improper integrals of the second kind that converges for $0 < p < 1$ (it is proper for $p \leq 0$). Thus by the comparison test and (2),
$$\int_{a+}^c |f(x)| dx$$
converges, whence it follows that $\int_{a+}^b |f(x)| dx$ converges.

• **PROBLEM 10-16**

Given that:
(i) $g(x) \in C$ (the set of continuous functions) for $a \leq x < \infty$

(ii) $g(x)$ is a nonincreasing function for $a \leq x < \infty$

(iii) $\lim_{x \to \infty} g(x) = 0$.

Prove that:
a) $\int_a^\infty g(x) \sin x \, dx$ converges.

b) $\int_a^\infty g(x) |\sin x| dx$ converges if $\int_a^\infty g(x) dx$ converges and it diverges if $\int_a^\infty g(x) dx$ diverges.

Solution: a) First notice that since $g(x)$ is a nonincreasing function (by (ii)) and approaches zero (by (iii)), we have $g(x) \geq 0$. Let $a < m\pi < n\pi < R \leq (n+1)\pi$, where m and n are integers. Then,

$$\int_a^R g(x)\sin x \, dx = \int_a^{m\pi} g(x)\sin x \, dx + \sum_{k=m}^{n-1} \int_{k\pi}^{(k+1)\pi} g(x)\sin x \, dx$$

$$+ \int_{n\pi}^R g(x)\sin x \, dx . \qquad (1)$$

Now keeping m fixed, let R, and consequently n, become infinite. Since

$$\left|\int_{n\pi}^R g(x)\sin x \, dx\right| \leq g(n\pi) \int_{n\pi}^{(n+1)\pi} |\sin x| \, dx = 2g(n\pi)$$

we have (using (iii)) that the last term on the right side of (1) approaches 0 as R becomes ∞. Hence, to prove the desired result it is therefore sufficient to prove that the second term on the right side of (1) approaches a limit; that is, that the series

$$\sum_{k=m}^\infty \int_{k\pi}^{(k+1)\pi} g(x) \sin x \, dx \quad \text{converges.} \qquad (2)$$

To prove this, make use of the following theorem: given that V_k is nonincreasing and nonnegative for $k = 1, 2, \ldots$, if $\lim_{k\to\infty} V_k = 0$ then the infinite series $\sum_{k=1}^\infty (-1)^k V_k$ converges.

To apply this result, first notice that $\sin x$ does not change sign for $k\pi \leq x \leq (k+1)\pi$, so

$$V_k = \left|\int_{k\pi}^{(k+1)\pi} g(x)\sin x \, dx\right| = \int_{k\pi}^{(k+1)\pi} g(x) |\sin x| \, dx .$$

However, since $g(x)$ is a nonincreasing function, we have

$$g(k\pi+\pi) \int_{k\pi}^{(k+1)\pi} |\sin x| \, dx \leq V_k \leq g(k\pi) \int_{k\pi}^{(k+1)\pi} |\sin x| \, dx$$

and $\quad 2g(k\pi) \leq V_{k-1} \leq 2g(k\pi - \pi) .\qquad (3)$

Combining these inequalities yields

$$0 \leq V_k \leq V_{k-1} \leq 2g(k\pi - \pi),$$

which shows that V_k is nonincreasing and $\lim_{k\to\infty} V_k = 0$. Hence, the series (2) is convergent, and therefore by the discussion, so is $\int_a^\infty g(x)\sin x \, dx$. This completes the proof.

b) <u>Case 1</u>: Given $\int_a^\infty g(x) \, dx$ diverges. First replace $\sin x$ by $|\sin x|$ in (1) to yield

$$\int_a^R g(x) |\sin x| \, dx = \int_a^{m\pi} g(x) |\sin x| \, dx + \sum_{k=m}^{n-1} \int_{k\pi}^{(k+1)\pi} g(x) |\sin x| \, dx$$

$$+ \int_{n\pi}^R g(x) |\sin x| \, dx . \qquad (4)$$

Now, as before, the last term on the right still approaches 0 as $R \to \infty$. Since $g(x)$ is nonincreasing, using (3) yields

$$V_{k-1} \geq 2g(k\pi) \geq 2\int_{k\pi}^{(k+1)\pi} g(x)dx$$

(where as before

$$V_k = \int_{k\pi}^{(k+1)\pi} g(x)|\sin x|dx \;),$$

so that

$$\sum_{k=m+1}^{n} V_{k-1} \geq 2\int_{(m+1)\pi}^{(n+1)\pi} g(x)dx. \tag{5}$$

Now, as R (and consequently n) becomes infinite, this integral diverges. Hence, the series (5) diverges. Since

$$\sum_{k=m+1}^{n} V_{k-1} = \sum_{k=m}^{n-1} V_k ,$$

we have by (4), that

$$\int_a^\infty g(x)|\sin x|dx$$

diverges.

Case 2: Given $\int_a^\infty g(x)dx$ converges. Since $|\sin x| \leq 1$, we have

$$g(x)|\sin x| \leq g(x)$$

so that

$$\int_a^\infty g(x)|\sin x|dx \leq \int_a^\infty g(x) \, dx,$$

which converges. Hence, $\int_a^\infty g(x)|\sin x|dx$ converges.

This completes the proof.

● **PROBLEM 10-17**

Determine if the following integrals converge absolutely:

a) $\int_0^\infty \dfrac{\cos x}{\sqrt{1+x^3}} dx$ b) $\int_0^\infty \sin x^2 \, dx$.

Solution: a) To test this integral for absolute convergence, apply the following theorem:

If $f(x)$ is continuous for $a \leq x < \infty$ and if $\lim_{x \to \infty} x^p f(x) = A$ for $p > 1$, then the integral $\int_a^\infty |f(x)|dx$ converges, from which it follows that $\int_a^\infty f(x)dx$ is absolutely convergent.

To apply the theorem to the given problem take p such that $1 < p < 3/2$. Therefore, taking $p = 5/4$ yields

$$\lim_{x \to \infty} x^p f(x) = \lim_{x \to \infty} x^{5/4} \frac{\cos x}{\sqrt{1+x^3}}$$

$$= \lim_{x \to \infty} \frac{x^{5/4} \cos x}{x^{3/2}(1+x^{-3})^{\frac{1}{2}}} = \lim_{x \to \infty} \frac{\cos x}{x^{\frac{1}{4}}(1+x^{-3})^{\frac{1}{2}}} = 0 .$$

Hence since $p = 5/4 > 1$ and $A = 0$, by the theorem, $\int_0^\infty \left|\dfrac{\cos x}{\sqrt{1+x^3}}\right| dx$ converges. Thus, it follows that $\int_0^\infty \dfrac{\cos x}{\sqrt{1+x^3}} dx$ converges absolutely.

b) $\int_0^\infty \sin x^2 \, dx$. This integral has previously been shown to be convergent. To test it for absolute convergence, again as before, begin by setting $x = \sqrt{t}$, so that $dx = 1/(2\sqrt{t}) \, dt$. Then

$$\int_0^\infty |\sin x^2| \, dx = \frac{1}{2} \int_0^\infty \frac{|\sin t|}{\sqrt{t}} \, dt, \qquad (1)$$

which is an improper integral of the first kind.

To proceed, apply the following theorem: If $g(x)$ is continuous and is a nonincreasing function, and if $\lim_{x \to \infty} g(x) = 0$, then the integral $\int_a^\infty g(x) |\sin x| \, dx$ converges if $\int_a^\infty g(x) \, dx$ converges. Similarly, $\int_a^\infty g(x) |\sin x| \, dx$ diverges if $\int_a^\infty g(x) \, dx$ diverges. To apply the theorem, let $a = 1$, $g(t) = t^{-\frac{1}{2}}$. Then $g(t)$ is a nonincreasing function and

$$\lim_{t \to \infty} g(t) = \lim_{t \to \infty} \frac{1}{t^{\frac{1}{2}}} = 0.$$

Furthermore

$$\int_1^\infty \frac{1}{t^{\frac{1}{2}}} \, dt = \lim_{b \to \infty} \int_1^b t^{-\frac{1}{2}} \, dt = \lim_{b \to \infty} 2b^{\frac{1}{2}} - 2 = \infty.$$

Hence,

$$\int_1^\infty \frac{|\sin t|}{\sqrt{t}} \, dt$$

diverges by the theorem, since $\int_a^\infty g(x) \, dx$ diverges. Thus by (1), $\int_0^\infty |\sin x^2| \, dx$ is divergent and because the given integral was proven to be convergent, this shows it to be conditionally convergent.

• **PROBLEM 10-18**

Determine if the following integrals converge absolutely, conditionally or not at all.

a) $\int_{0+}^{\frac{1}{2}} (\log 1/x)^n \, dx$ for $-\infty < n < \infty$.

b) $\int_{0+}^{1} \frac{\sin 1/x}{x^{3/2}} \, dx$.

Solution: a) $\int_{0+}^{\frac{1}{2}} (\log 1/x)^n \, dx$. This is an improper integral of the second kind where the integrand has a singularity at the lower limit, $x = 0^+$. To determine if this integral converges absolutely, we need the following theorem: Given $f(x)$ is continuous for $a < x \leq b$, if $\lim_{x \to a^+} (x-a)^p f(x) = L$ (L finite) for $0 < p < 1$, then $\int_{a+}^b |f(x)| \, dx$ converges. Before applying this theorem note, however, that for $n \leq 0$, the integrand approaches a limit when $x \to 0^+$; that is,

$$\lim_{x \to 0^+} (\log 1/x)^n$$

exists for $n \leq 0$, therefore for these values of n the integral is proper. Hence, to apply the theorem, let $p = \frac{1}{2}$, $a = 0$ for $n > 0$

to yield $\lim_{x \to 0^+} x^{\frac{1}{2}} \log(1/x)^n = 0$. Thus, since $p = \frac{1}{2} < 1$ and $L = 0$, the integral

$$\int_{0^+}^{\frac{1}{2}} |(\log 1/x)^n| \, dx \text{ converges.}$$

Hence, it follows that the integral $\int_{0^+}^{\frac{1}{2}} (\log 1/x)^n \, dx$ converges absolutely.

b) $\int_{0^+}^{1} \frac{\sin 1/x}{x^{3/2}} \, dx$. This is an improper integral of the second kind where the integrand has a singularity at $x = 0^+$. To test this integral, we first determine if it converges. To do this we use the following theorem: Given that $g(x)$ is continuous for $a < x \leq b$, if $g(x)(x-a)^2$ is a nondecreasing function for $a < x \leq b$ and $\lim_{x \to a^+} g(x)(x-a)^2 = 0$, then $\int_{a^+}^{b} g(x) \sin \frac{1}{x-a} \, dx$ converges.

To apply this theorem to the given integral, take $g(x) = x^{-3/2}$, $a = 0$. Then

$$g(x)(x-a)^2 = x^{-3/2} x^2 = \sqrt{x},$$

which is a nondecreasing function for $0 < x \leq 1$. Then

$$\lim_{x \to 0^+} x^{-3/2} x^2 = \lim_{x \to 0^+} x^{\frac{1}{2}} = 0.$$

Thus, the integral $\int_{0^+}^{1} x^{-3/2} \sin \frac{1}{x} \, dx$ converges. Now, to test this integral for absolute convergence, apply the following theorem: Given that $g(x)$ is continuous and $g(x)$ is a nonincreasing function for $a \leq x < \infty$, if $\lim_{x \to \infty} g(x) = 0$ and $\int_{a}^{\infty} g(x) \, dx$ diverges, then $\int_{a}^{\infty} g(x) |\sin x| \, dx$ diverges.

To apply this theorem, let $x = t^{-1}$ in the given improper integral. This yields

$$\int_{0^+}^{1} \frac{\sin 1/x}{x^{3/2}} \, dx = \int_{1}^{\infty} \frac{\sin t}{\sqrt{t}} \, dt, \qquad (1)$$

which is an improper integral of the first kind. Here, let $g(t) = t^{-\frac{1}{2}}$ which is a nonincreasing function for $1 \leq t < \infty$. Furthermore, $\lim_{t \to \infty} t^{-\frac{1}{2}} = 0$ and $\int_{1}^{\infty} \frac{dt}{\sqrt{t}}$ is divergent (p integral with $p = \frac{1}{2} < 1$). Hence by the theorem, $\int_{1}^{\infty} \frac{|\sin t|}{\sqrt{t}} \, dt$ is divergent, so that by (1),

$\int_{0^+}^{1} \frac{|\sin 1/x|}{x^{3/2}} \, dx$ is divergent. Thus, the integral $\int_{0^+}^{1} \frac{\sin 1/x}{x^{3/2}} \, dx$ is not absolutely convergent, but is conditionally convergent.

● PROBLEM 10-19

a) Define uniform convergence for improper integrals.

b) Show that $\int_0^\infty e^{-xt} dt$ converges uniformly to $1/x$ in the interval $1 \le x \le 2$.

c) Show that $\int_0^\infty xe^{-xt} dt$ does not converge uniformly in the interval $0 \le x \le 1$.

Solution: a) Suppose that the integral

$$\int_a^\infty f(t,x) dt \qquad (1)$$

converges for each fixed x in the interval $A \le x \le B$ and has the value $F(x)$. In addition let

$$S_R(x) = \int_a^R f(t,x) dt.$$

Then, by definition, the integral (1) converges uniformly to $F(x)$ in the interval $A \le x \le B$ if for each $\epsilon > 0$ we can find a number N depending on ϵ but independent of x such that $|F(x) - S_R(x)| < \epsilon$ for all $R > N$ and for all x in $A \le x \le B$.
Note that this definition of uniform convergence is written in terms of improper integrals of the first kind. However, an analogous definition can be given for improper integrals of the second and third kind. (i.e., for improper integrals of the second kind let

$$F(x) = \int_{a^+}^b f(x,t) dt, \quad S_R(x) = \int_R^b f(t,x) dt,$$

and replace the condition $R > N$ in the definition by the condition $R - a < \delta$ for some δ, and $R < b$.

b) $\int_0^\infty e^{-xt} dt$. Here, for $x > 0$, the integral converges to $1/x$.
Hence,

$$F(x) = \lim_{b \to \infty} \int_0^b e^{-xt} dt = \lim_{b \to \infty} \left(\frac{-e^{-bx}}{x} + \frac{1}{x} \right) = \frac{1}{x} \quad (x > 0).$$

Let $S_R(x) = \int_0^R e^{-xt} dt$. Then

$$|F(x) - S_R(x)| = \left| \frac{1}{x} - S_R(x) \right| = \left| \frac{1}{x} + \frac{e^{-Rx}}{x} - \frac{1}{x} \right|$$

$$= \frac{e^{-xR}}{x} \le e^{-R} < \epsilon \quad \text{in } 1 \le x \le 2.$$

Thus to show uniform convergence it is necessary to show that the above condition exists for some N independent of x, such that $R > N$. Hence, since $1/e^R < \epsilon$ we have $e^R > 1/\epsilon$ or $R > \ln 1/\epsilon$, so choose $N = \ln 1/\epsilon$ to show that the integral is uniformly convergent in the given interval.

c) $\int_0^\infty xe^{-xt} dt$. Here

$$F(x) = \lim_{b \to \infty} \int_0^b xe^{-xt} dt = \lim_{b \to \infty} (-e^{-bx} + 1).$$

Hence, $F(x) = 1$ for $x > 0$
$= 0$ for $x = 0$ does not converge for $x < 0$.

485

Now let $S_R(x) = \int_0^R xe^{-xt}dt = -e^{-Rx} + 1$. Then

$$|F(x) - S_R(x)| = |1 + e^{-Rx} - 1| = e^{-xR} < \epsilon \quad \text{for} \quad 0 < x \le 1$$

$$= |0 + 1 - 1| = 0 < \epsilon \quad \text{for} \quad x = 0.$$

Now choose $\epsilon = \frac{1}{2}$ so that if the number N of the definition existed, we should have for $R > N$,

$$|F(x) - S_R(x)| = e^{-xR} < \frac{1}{2} \quad \text{for} \quad 0 < x \le 1. \tag{2}$$

However, since for every $R > 0$,

$$\lim_{x \to 0^+} e^{-xR} = 1,$$

this shows (2) to be false. Thus, the given integral is not uniformly convergent in the interval $0 \le x \le 1$.

● **PROBLEM 10-20**

Using the Weierstrass M-test, show that

a) $\int_1^\infty \dfrac{\cos xt}{1+t^2} dx$ converges uniformly in the interval $A \le x \le B$.

b) $\int_{0^+}^1 \dfrac{\sin(t/x)}{\sqrt{t}} dt$ converges uniformly in any interval $0 < A \le x \le B$.

Solution: a) $\int_1^\infty \dfrac{\cos xt}{1+t^2} dx$. \hfill (1)

To test the integral (1) for uniform convergence, apply the Weirstrass M-test which states: If for $a \le t < \infty$, $A \le x \le B$, we have:

(i) $f(t,x)$ and $M(t)$ are continuous
(ii) a function $M(t) \ge 0$ such that $|f(t,x)| \le M(t)$
(iii) $\int_a^\infty M(t)dt$ converges

then $\int_a^\infty f(t,x)dt$ is uniformly and absolutely convergent in $A \le x \le B$.

To apply this test to the given integral, let $M(t) = 1/t^2$ since

$$\left|\dfrac{\cos xt}{1+t^2}\right| = \dfrac{|\cos xt|}{1+t^2} \le \dfrac{1}{1+t^2} \le \dfrac{1}{t^2} \quad \text{for} \quad 1 \le t < \infty, A \le x \le B. \text{ Then,}$$

$$\int_1^\infty M(t)dt = \int_1^\infty \dfrac{dt}{t^2},$$

which converges. Hence, since conditions (i)-(iii) are satisfied, for $A \le x \le B$, the integral (1) is uniformly convergent in $A \le x \le B$.

b) $\int_{0^+}^1 \dfrac{\sin(t/x)}{\sqrt{t}} dt$. Although the Weierstrass M-test was given in terms of improper integrals of the first kind in part a) of this problem, there is an analogous test for improper integrals of the second (and third) kind (i.e., substitute $a < t \le b$ for $a \le t < \infty$ and determine whether or not $\int_{a^+}^b M(t)dt$ is convergent for $a < t \le b$, $A \le x \le B$). Hence, since $\left|\dfrac{\sin(t/x)}{\sqrt{t}}\right| = \dfrac{|\sin(t/x)|}{\sqrt{t}} \le \dfrac{1}{\sqrt{t}}$ for

$0 < A \le x \le B$ and $0 < t \le 1$, take $M(t) = 1/\sqrt{t}$. Then $\int_{0^+}^1 \frac{dt}{\sqrt{t}}$ is of the form $\int_a^b \frac{dt}{(t-a)^p}$ which converges for $p < 1$. Here, $p = \frac{1}{2} < 1$, so $\int_{0^+}^1 \frac{dt}{\sqrt{t}}$ converges. Thus, since the conditions of the test are satisfied, the integral $\int_{0^+}^1 \frac{\sin(t/x)}{\sqrt{t}} dt$ converges uniformly in any interval $0 < A \le x \le B$.

• PROBLEM 10-21

Show that the integral
$$\int_0^\infty e^{-xy} \frac{\sin x}{x} dx \qquad (1)$$
converges uniformly on $[0,\infty]$ by:
a) Dirichlet's test for uniform convergence.
b) Abel's test.

Solution: a) Dirichlet's test for uniform convergence states: If $g(x,y)$ is continuous on $\{(x,y) | c \le x, a \le y \le b\}$ and
$$\left| \int_c^R g(x,y) dx \right| < k$$
(k a constant) for all $R \ge c$ and all y on $[a,b]$, if $f(x,y)$ is a decreasing function of x for $x \ge c$ and each fixed y on $[a,b]$, and if $f(x,y)$ approaches zero uniformly in y as $x \to \infty$, then
$$\int_c^\infty f(x,y) g(x,y) dx$$
converges uniformly on $a \le y \le b$. To apply this to the integral (1), let $g(x,y) = e^{-xy} \sin x$ and $f(x,y) = x^{-1}$. Now, $f(x,y) = x^{-1}$ is a decreasing function of x for each fixed y and for $x \ge 0$, and $\lim_{x \to \infty} x^{-1} = 0$. In addition, $g(x,y)$ is continuous on $\{(x,y) | 0 \le x, 0 \le y\}$. Then
$$\left| \int_c^R g(x,y) dx \right| = \left| \int_0^R e^{-xy} \sin x \, dx \right|.$$
However, using the theorem
$$\left| \int_{n\pi}^R g(x) \sin x \, dx \right| \le g(n\pi) \int_{n\pi}^{(n+1)\pi} |\sin x| dx = 2g(n\pi), \qquad (2)$$
let $n = 0$ and $g(x) = e^{-xy}$ in (1) to obtain
$$\left| \int_0^R e^{-xy} \sin x \, dx \right| \le g(0) \int_0^\pi |\sin x| dx = 2g(0) = 2.$$
Hence,
$$\left| \int_0^R e^{-xy} \sin x \, dx \right| \le 2 \quad \text{if} \quad R \ge 0, \; y \ge 0.$$
Thus, by Dirichlet's test, the integral (1) converges uniformly on $y \ge 0$ (i.e., $[0,\infty]$).

b) Abel's test states: If $g(x,y)$ is continuous on $[c,\infty] \times [a,b]$ and $\int_c^\infty g(x,y) dx$ converges uniformly on $[a,b]$, if $f(x,y)$ is a decreasing function of x for $x \ge c$ and each fixed y on $[a,b]$, and if $f(x,y)$ is bounded on $[c,\infty] \times [a,b]$, then

$$\int_c^\infty f(x,y)g(x,y)dx$$

converges uniformly on $[a,b]$.
To apply this test, again let

$$g(x,y) = e^{-xy} \sin x \quad \text{and} \quad f(x,y) = x^{-1}.$$

Then $f(x,y)$ is a decreasing function for $x \geq c$ and since

$$\lim_{x \to \infty} x^{-1} = 0, \quad f(x,y) \text{ is bounded on}$$

$[0,\infty] \times [0,\infty]$. Now, all that is left to show is that $\int_0^\infty e^{-xy}\sin x\, dx$ converges uniformly on $[0,\infty]$. To do this, apply the Weierstrass M-test. Since e^{-xy} is positive for all values of x and y, we have $|e^{-xy} \sin x| = e^{-xy}|\sin x| \leq e^{-xy}$ for $y > 0$, $x > 0$. Hence, let $M(x) = e^{-xy}$ so that

$$\int_0^\infty M(x)dx = \int_0^\infty e^{-xy} dx = \lim_{R \to \infty} \int_0^R e^{-xy} dx$$

$$= \lim_{R \to \infty} \left(\frac{-e^{-Ry}}{y} + \frac{1}{y}\right) = \frac{1}{y}$$

for $y > 0$. Hence, since $\int_0^\infty M(x)dx$ converges for $y > 0$ and $|f(x,y)| \leq M(x)$ for $y > 0$, $x > 0$, the integral $\int_0^\infty e^{-xy}\sin x\, dx$ converges uniformly on $[0,\infty]$. Hence, by Abel's theorem,

$$\int_c^\infty g(x,y)f(x,y)dx = \int_0^\infty e^{-xy}\frac{\sin x}{x} dx \text{ converges uniformly on}$$

$[0,\infty]$.

• **PROBLEM 10-22**

Show that: a) $F(x) = \int_0^\infty e^{-xt^2} dt$ is continuous for $x \geq 1$.

b) $\dfrac{d}{dx}\int_0^\infty e^{-xt^2} dt = -\int_0^\infty t^2 e^{-xt^2} dt$ for $x \geq 1$.

c) $\lim_{x \to 0^+} \int_0^\infty xe^{-xt} dt \neq \int_0^\infty \left(\lim_{x \to 0^+} xe^{-xt}\right)dt$ and explain the result.

<u>Solution</u>: a) To do this problem we need the following theorem:
If $f(t,x)$ is continuous in t and x for $t \geq a$, $c \leq x \leq d$ and $\int_a^\infty f(t,x)dt$ is uniformly convergent for $c \leq x \leq d$, then the function $F(x) = \int_a^\infty f(t,x)dt$ is continuous for $c \leq x \leq d$. Hence, to show that $F(x) = \int_0^\infty e^{-xt^2} dt$ is continuous for $x \geq 1$, all that is required is to prove that $\int_0^\infty e^{-xt^2} dt$ is uniformly convergent for $x \geq 0$ since $f(t,x) = e^{-xt^2}$ is continuous on any interval. To do this, apply the Weierstrass M-test. Since

for $x \geq 1$, let $M(t) = e^{-t^2}$. Then
$$0 \leq e^{-xt^2} \leq e^{-t^2}$$

$$\int_0^\infty e^{-t^2} dt \leq \int_0^\infty e^{-t} dt = \lim_{b \to \infty} \int_0^b e^{-t} dt = 1.$$

Since $e^{-t^2} \leq e^{-t}$ for $t \geq 1$, $\int_0^\infty e^{-t^2} dt = \int_0^1 e^{-t^2} dt + \int_1^\infty e^{-t^2} dt \leq \int_0^1 e^{-t^2} dt + \int_1^\infty e^{-t} dt$

but, $\int_0^1 e^{-t^2} dt$ is a finite integral which can be found to be equal to

$[\pi(1-\frac{1}{e})]^{1/2}$. Now, $\int_1^\infty e^{-t} dt = \lim_{b \to \infty} \int_1^b e^{-t} dt = \lim_{b \to \infty} [-e^{-t}]_1^b = \lim_{b \to \infty} [-e^{-b} + e^{-1}]$

$$= [-0 + e^{-1}] = \frac{1}{e}.$$

Therefore, $\int_0^\infty e^{-t^2} dt$ is uniformly convergent, as is $\int_0^\infty e^{-xt^2} dt$ for $x \geq 1$.

Thus by the theorem, $F(x) = \int_0^\infty e^{-xt^2} dt$ is continuous for $x \geq 1$.

b) To prove the two integrals are equal, first note that $f(t,x) = e^{-xt^2}$ is continuous in any interval. In addition,

$$\frac{\partial f}{\partial x}(t,x) = -t^2 e^{-xt^2}$$

is continuous on the interval $0 \leq t < \infty$ and for all x. Also, note that $\int_0^\infty e^{-xt^2} dt$ converges and that $\int_0^\infty \frac{\partial f}{\partial x}(t,x) dt = \int_0^\infty -t^2 e^{-xt^2} dt$

converges uniformly for $x \geq 1$. This can be shown by using the Weierstrass M-test, letting $M(t) = e^{-t}$ (i.e., $0 \leq t^2 e^{-xt^2} \leq t^2 e^{-t^2} < e^{-t}$).
Now, use the theorem: If $f(t,x)$ is continuous in t and x and has a derivative $\partial f/\partial x$ which is continuous in t and x for $a \leq t < \infty$ and $c \leq x \leq d$, and if $\int_a^\infty f(t,x) dt$ converges and $\int_a^\infty \frac{\partial f}{\partial x} f(t,x) dt$ converges uniformly for $c \leq x \leq d$, then $F(x) = \int_a^\infty f(t,x) dt$ has a continuous derivative for $c \leq x \leq d$ and $F'(x) = \frac{d}{dx} \int_a^\infty f(t,x) dt = \int_a^\infty \frac{\partial f}{\partial x}(t,x) dt$. Hence, applying this theorem, we see that all of the conditions are satisfied, so

$$F'(x) = \frac{d}{dx} \int_0^\infty e^{-xt^2} dt = \int_0^\infty \frac{\partial}{\partial x} e^{-xt^2} dt = -\int_0^\infty t^2 e^{-xt^2} dt.$$

c) Since
$$\lim_{x \to 0^+} \int_0^\infty xe^{-xt} dt = \lim_{x \to 0^+} \left[\lim_{b \to \infty} \int_0^b xe^{-xt} dt\right] = \lim_{x \to 0^+} 1 = 1$$

and
$$\int_0^\infty \left(\lim_{x \to 0^+} xe^{-xt}\right) dt = \int_0^\infty 0 \, dt = 0,$$ the two given integrals are not

equal. The reason for this is that since $F(x) = \int_0^\infty xe^{-xt} dt$ is not

uniformly convergent for $x \geq 0$, we cannot be sure that $F(x)$ is continuous for $x \geq 0$. Thus $\lim_{x \to 0^+} F(x)$ may not equal $F(0)$. Note,

however, that if $f(t,x)$ is continuous for $t \geq a$ and $c \leq x \leq d$, and if $\int_a^\infty f(t,x) dt$ is uniformly convergent for $c \leq x \leq d$, then

$F(x) = \int_a^\infty f(t,x)dt$ is continuous on $c \leq x \leq d$. In addition, if x_0 is any point of $c \leq x \leq d$, then

$$\lim_{x \to x_0} F(x) = \lim_{x \to x_0} \int_a^\infty f(t,x)dt = \int_a^\infty \lim_{x \to x_0} f(t,x)dt .$$

● **PROBLEM 10-23**

a) Given $r = [(x-x_0)^2 + (y-y_0)^2]^{\frac{1}{2}}$, show that the integral $\iint_D \frac{1}{r^m} dA$, where (x_0, y_0) is a point of D and $m > 0$ (so that the integral is improper), is convergent if $m < 2$ (D is a circle with center at (x_0, y_0) and with radius c).

b) Show that $\iint_D \sin(y/x) dA$ is absolutely convergent, where D is the square domain $0 < x < 1$, $0 < y < 1$.

<u>Solution</u>: a) This is a problem concerning an improper multiple integral in a finite region. We have an integral of the form $\iint_D f(x,y)dA$, where D is a closed and bounded region and f is continuous in D except at one point (x_0, y_0) and the function is not integrable over D. To determine if an integral of this type is convergent, let D^* be a region derived from D by removing a small region ΔD having the point (x_0, y_0) in its interior. Now let d be the maximum diameter of ΔD. Then, if $\iint_{D^*} f(x,y)dA$ approaches a limiting value as $d \to 0$, then the integral $\iint_D f(x,y)dA$ equals this value and therefore is convergent. Hence,

$$\iint_D f(x,y)dA = \lim_{d \to 0} \iint_{D^*} f(x,y)dA .$$

For the given problem, $\iint_D \frac{1}{r^m} dA$ and $r = [(x-x_0)^2 + (y-y_0)^2]^{\frac{1}{2}}$, so that the only singularity is at (x_0, y_0) (i.e., at this point, $r = 0$). Hence, delete a small concentric circle of radius d so that D^* is the annulus between these circles. Since the value of the integral is not affected by the location of the axes, we can assume that (x_0, y_0) is the origin.

By the use of polar coordinates we have,

$$\iint_{D^*} \frac{1}{r^m} dA = \int_0^{2\pi} \int_d^c \frac{r}{r^m} dr \, d\theta$$

$$= \int_0^{2\pi} \frac{c^{2-m} - d^{2-m}}{2-m} d\theta = \frac{2\pi}{2-m} [c^{2-m} - d^{2-m}].$$

Consequently,

$$\lim_{d \to 0} \iint_{D^*} \frac{1}{r^m} dA = \lim_{d \to 0} \frac{2\pi}{2-m} [c^{2-m} - d^{2-m}] = \frac{2\pi}{2-m} c^{2-m} \quad \text{if} \quad m < 2 .$$

Thus the integral, $\iint_D \frac{1}{r^m} dA$ converges, where D is the given region, for $m < 2$.

b) $\iint_D \sin(y/x) dA$, where D is the square domain $0 \le x \le 1$, $0 \le y \le 1$.

The function $\sin(y/x)$ is discontinuous on the y-axis. However, note that $|\sin(y/x)| \le 1$ in D. Now, a theorem states: For improper double integrals, if
$$|f(x,y)| \le g(x,y) \text{ and if } \iint_D g(x,y) dA$$
is convergent, then $\iint_D f(x,y) dA$ is absolutely convergent. Therefore, for this problem let $g(x,y) = 1$ so that $\iint_D g(x,y) dA = \iint_D dA$, which is convergent. Hence, the given integral converges absolutely.

EVALUATION OF IMPROPER INTEGRALS

• **PROBLEM 10-24**

Show that

a) if $0 < a < b$, $\int_0^\infty \frac{e^{-at} - e^{-bt}}{t} dt = \ln b/a$.

b) if $a > 0$, $\int_0^\infty \frac{e^{-at} \sin xt}{t} dt = \tan^{-1} x/a$.

Solution: a) The integral is of the first kind, since the integrand approaches a finite limit as $t \to 0$ (that is,
$$\lim_{t \to 0} \frac{e^{-at} - e^{-bt}}{t} = \lim_{t \to 0}(-ae^{-at} + be^{-bt}) = b-a, \text{ by L'Hospital's}$$
rule). To evaluate the given integral start with $F(x) = \int_0^\infty e^{-xt} dt =$
$$\lim_{b \to \infty} \int_0^b e^{-xt} dt = 1/x \text{ for } x > 0. \tag{1}$$

Now (1) is uniformly convergent on the interval $a \le x \le b$ (apply Weierstrass M-test with $M(t) = e^{-at}$ when $t \ge 0$). Then, note the following theorem: Let $f(t,x)$ be continuous for $t \ge a$ and $a \le x \le b$, and let $\int_c^\infty f(t,x) dt$ be uniformly convergent in the interval $a \le x \le b$; then

$$\int_a^b F(x) dx = \int_c^\infty dt \int_a^b f(t,x) dx, \tag{2}$$

where $F(x) = \int_0^\infty f(t,x) dt$. To apply this theorem, note that here

$f(t,x) = \frac{e^{-at} - e^{-bt}}{t}$ is continuous for $t \ge 0$ and $a \le x \le b$. Then

since $\int_0^\infty e^{-xt} dt$ is uniformly convergent for $t \ge 0$ and $a \le x \le b$, we have by (2), $F(x) = \int_0^\infty e^{-xt} dt = -\frac{1}{x} e^{-xt} \Big|_0^\infty = \frac{1}{x}$, and so,

$$\int_a^b \frac{dx}{x} = \int_0^\infty dt \int_a^b e^{-xt} dx \qquad (3)$$

Integration yields

$$\ln b - \ln a = \int_0^\infty \left[-\frac{e^{-xt}}{t} \Big|_a^b \right] dt$$

or

$$\ln(b/a) = \int_0^\infty \frac{e^{-at} - e^{-bt}}{t} dt$$

which is the desired result.

b) The theorem to be used for part (b) is as follows: Let $F(x) = \int_c^\infty f(t,x) dt$ be convergent when $a \le x \le b$. Let the partial derivative $\partial f / \partial x$ be continuous for t and x such that $c \le t$ and $a \le x \le b$. Let $\int_c^\infty \partial F/\partial x \, dt$ converge uniformly on $[a,b]$. Then $F(x)$ has a derivative given by $F'(x) = \int_c^\infty \frac{\partial f(t,x)}{\partial x} dt$. Let

$$F(x) = \int_0^\infty \frac{e^{-at} \sin xt}{t} dt$$

and $f(t,x) = \frac{e^{-at} \sin xt}{t}$ so that $\frac{\partial f}{\partial x}(t,x) = e^{-at} \cos xt$ is continuous for $0 \le t$, and all values of x. The integral $\int_0^\infty e^{-at} \cos xt \, dt$ converges uniformly in all x for $a > 0$ and $F(x) = \int_0^\infty \frac{e^{-at} \sin xt}{t} dt$ is convergent for all x if $a > 0$. Hence, as a result of these conditions (by the given theorem)

$$F'(x) = \int_0^\infty \frac{\partial f}{\partial x}(t,x) dt = \int_0^\infty e^{-at} \cos xt \, dt. \qquad (4)$$

Now, by an integration formula,

$$\int_0^M e^{-at} \cos xt \, dt = \left[\frac{e^{-at}(x \sin xt - a \cos xt)}{a^2 + x^2} \right]_0^M$$

$$= \frac{e^{-aM}(x \sin xM - a \cos xM) + a}{a^2 + x^2}$$

Hence, as $M \to \infty$, we have $\lim_{M \to \infty} \frac{x \sin xM - a \cos xM}{e^{aM}} = 0$, so that

$$\int_0^\infty e^{-at} \cos xt \, dt = \frac{a}{a^2 + x^2} \quad \text{or by (4), } F'(x) = \frac{a}{a^2 + x^2}. \text{ Thus,}$$

$$F(x) = \int \frac{a}{a^2 + x^2} dx = \tan^{-1} x/a + C.$$ To determine the constant of integration C, observe that $F(0) = 0$ by the definition of $F(x)$. In addition, since $\tan^{-1} 0 = 0$, we conclude that $C = 0$. Thus,

$$F(x) = \int_0^\infty \frac{e^{-at} \sin xt}{t} dt = \tan^{-1} x/a.$$

This is the desired result.

• PROBLEM 10-25

Show that

a) $\int_0^\infty x^{-1} \sin x \, dx = (1/2)\pi$ (1)

b) $\int_0^\infty \frac{e^{-ax} - e^{-bx}}{x \sec rx} \, dx = \frac{1}{2} \ln \frac{b^2 + r^2}{a^2 + r^2}$ (2)

where $a, b > 0$.

Solution: a) To show (1) let

$$F(u) = \int_0^\infty e^{-xu} \frac{\sin x}{x} \, dx, \quad u \geq 0 \quad (3)$$

This converges for all $u \geq 0$ and $F(0)$ is the value to be found. Now $F(u)$ is uniformly convergent for all u with $0 \leq u < \infty$. To prove this note that by definition, the integral $F(\alpha) = \int_c^\infty f(x, \alpha) dx$ is uniformly convergent in $[a, b]$ if for each $\epsilon > 0$ there is an N depending on ϵ but not on α such that $\left| F(\alpha) - \int_c^R f(x, \alpha) dx \right| < \epsilon$ for all $R > N$ and all α in $[a, b]$. This can be rewritten by noting that $\left| F(\alpha) - \int_c^R f(x, \alpha) dx \right| = \left| \int_R^\infty f(x, \alpha) dx \right|$. Hence, using integration by parts we have (integrating $\sin x$ and differentiating the remaining factor):

$$\int_R^\infty f(x, u) dx = \int_R^\infty e^{-xu} x^{-1} \sin x \, dx = \left[-e^{-xu} x^{-1} \cos x \right]_R^\infty - \int_R^\infty \frac{(1+xu) e^{-xu}}{x^2} \cos x \, dx.$$

The integrand in the second integral is always less than $1/x^2$ in absolute value, for any $u \geq 0$. Hence,

$$\left| \int_R^\infty e^{-xu} x^{-1} \sin x \, dx \right| \leq \frac{e^{-uR} |\cos R|}{R} + \int_R^\infty x^{-2} dx \leq \frac{e^{-uR}}{R} + \frac{1}{R} \leq \frac{1}{R} + \frac{1}{R}$$

$$\leq \frac{2}{R} < \epsilon \quad \text{for} \quad R > N > \frac{2}{\epsilon}.$$

Therefore

$$\left| F(u) - \int_0^R e^{-xu} \frac{\sin x}{x} dx \right| = \left| \int_R^\infty e^{-xu} \frac{\sin x}{x} dx \right| < \epsilon$$

so that $F(u)$ is uniformly convergent for all u with $0 \leq u < \infty$. Consequently,

$$\lim_{u \to \infty} \int_0^\infty e^{-xu} \frac{\sin x}{x} dx = \int_0^\infty \lim_{u \to \infty} e^{-xu} \frac{\sin x}{x} dx = \int_0^\infty 0 \, dx = 0 \quad (4)$$

and

$$\lim_{u \to 0} \int_0^\infty e^{-xu} \frac{\sin x}{x} dx = \int_0^\infty \lim_{u \to 0} e^{-xu} \frac{\sin x}{x} dx = \int_0^\infty \frac{\sin x}{x} dx. \quad (5)$$

Now, differentiating (3) yields

$$F'(u) = \int_0^\infty \frac{\partial}{\partial u} \left(e^{-xu} \frac{\sin x}{x} \right) dx = -\int_0^\infty e^{-xu} \sin x \, dx \quad (6)$$

which may be integrated by parts twice, to yield $F'(u) = -(1 + u^2)^{-1}$. This is valid for all u with $u > 0$ since the integrand in (6) is dominated by e^{-xu} and the integral of this is uniformly convergent for all u with $d \leq u \leq \infty$ and any $d > 0$. Integrating yields $F(u) = C - \arctan u$, for all $u > 0$. Now, let u increase, so that from (4), $\lim_{u \to \infty} F(u) = 0$, and

$$0 = C - \lim_{u \to \infty} \arctan u = C - \tfrac{1}{2}\pi \quad \text{and} \quad C = \tfrac{1}{2}\pi.$$

Hence, $F(u) = \pi/2 - \arctan u$, so that $F(0) = \int_0^\infty x^{-1} \sin x \, dx = \tfrac{1}{2}\pi$.

b) Start with the integral $\int_0^\infty e^{-\varphi x} \cos rx \, dx$. (7)

Then, $\lim_{M \to \infty} \int_0^M e^{-\varphi x} \cos rx \, dx = \lim_{M \to \infty} e^{-\varphi x} \frac{(r \sin rx - \varphi \cos rx)}{\varphi^2 + r^2} \Big|_0^M = \frac{\varphi}{\varphi^2 + r^2}$

(integration by parts). Furthermore, this integral converges uniformly for $a \leq \varphi \leq b$, where $0 < a < b$ and any r (use the Weierstrass M-test, taking $M(x) = e^{-\varphi x}$ since

$$|e^{-\varphi x} \cos rx| \leq e^{-\varphi x} \quad \text{and} \quad \int_0^\infty e^{-\varphi x} \, dx$$

converges). Now by theorem, if $f(t,x)$ is continuous for $t \geq a$, and $c \leq x \leq d$ and if $\int_a^\infty f(t,x)dt$ is uniformly convergent for $c \leq x \leq d$, then we can integrate $F(x) = \int_a^\infty f(t,x)dt$ with respect to x from c to d to obtain

$$\int_c^d F(x)dx = \int_c^d \left\{ \int_a^\infty f(t,x)dt \right\} dx = \int_a^\infty \left\{ \int_c^d f(t,x)dx \right\} dt.$$

Applying this to (7) yields

$$\int_{x=0}^\infty \left\{ \int_{\varphi=a}^b e^{-\varphi x} \cos rx \, d\varphi \right\} dx = \int_{\varphi=a}^b \left\{ \int_{x=0}^\infty e^{-\varphi x} \cos rx \, dx \right\} d\varphi$$

or

$$\int_{x=0}^\infty \frac{e^{-\varphi x} \cos rx}{-x} \Big|_{\varphi=a}^b dx = \int_{\varphi=a}^b \frac{\varphi}{\varphi^2 + r^2} d\varphi,$$

which is equivalent to

$$\int_0^\infty \frac{e^{-ax} - e^{-bx}}{x \sec rx} dx = \frac{1}{2} \ln \frac{b^2 + r^2}{a^2 + r^2}.$$

This is (2), the desired result.

• **PROBLEM 10-26**

Show that

$$\int_0^\infty \frac{\sin t}{\sqrt{t}} dt = \left(\frac{\pi}{2}\right)^{\frac{1}{2}} \tag{1}$$

given that

$$\int_0^\infty e^{-x^2} dx = \frac{\sqrt{\pi}}{2}, \tag{2}$$

and that

$$\int_0^\infty \frac{ds}{1+s^4} = \frac{\pi}{2\sqrt{2}}. \tag{3}$$

<u>Solution</u>: To show (1), take $x = s\sqrt{t}$ in (2). This yields $dx = \sqrt{t} \, ds$ so that (2) becomes

$$\sqrt{t} \int_0^\infty e^{-s^2 t} ds = \frac{\sqrt{\pi}}{2},$$ which can be rewritten as

$$\frac{2}{\sqrt{\pi}} \int_0^\infty e^{-s^2 t} ds = \frac{1}{\sqrt{t}}. \qquad (4)$$

Therefore, multiplying each side of (4) by sin t and integrating with respect to t yields

$$\int_0^\infty \frac{\sin t}{\sqrt{t}} dt = \frac{2}{\sqrt{\pi}} \int_0^\infty dt \int_0^\infty e^{-s^2 t} \sin t \, ds \qquad (5)$$

$$= \frac{2}{\sqrt{\pi}} \int_0^\infty ds \int_0^\infty e^{-s^2 t} \sin t \, dt. \qquad (6)$$

This because $\int_0^\infty e^{-s^2 t} \sin t \, ds$ is uniformly convergent for $s \geq 0$, $t \geq 0$ and $f(s,t) = e^{-s^2 t} \sin t$ is continuous for $s \geq 0$ and $0 \leq t < \infty$. Hence, by these conditions equation (5) is valid and in addition, it is legitimate to invert the order of integration on the right of (5) to obtain (6).

Now, by an elementary integrating formula,

$$\lim_{M \to \infty} \int_0^M e^{-s^2 t} \sin t \, dt =$$

$$\lim_{M \to \infty} \left[\frac{-e^{-s^2 M}(\cos M + s^2 \sin M) + 1}{s^4 + 1} \right] = \frac{1}{s^4 + 1} \qquad (7)$$

Thus, substituting (7) into (6) yields

$$\int_0^\infty \frac{\sin t}{\sqrt{t}} dt = \frac{2}{\sqrt{\pi}} \int_0^\infty \frac{ds}{s^4 + 1}.$$

Consequently by (3),

$$\int_0^\infty \frac{\sin t}{\sqrt{t}} dt = \frac{2}{\sqrt{\pi}} \cdot \frac{\pi}{2\sqrt{2}} = \left(\frac{\pi}{2}\right)^{\frac{1}{2}}$$

which is the desired result.

● **PROBLEM 10-27**

Evaluate, for any constant $c > 0$,

$$\int_{-\infty}^\infty e^{-cx^2} dx \qquad (1)$$

and using this, evaluate

$$\int_{R^n} e^{-\langle Px, x \rangle} dx_1 dx_2 \ldots dx_n, \qquad (2)$$

where $x \in R^n$, P is a positive definite symmetric matrix, and $\langle \, \rangle$ denotes the Euclidean inner product (i.e., $\langle Tx, x \rangle$ is a positive definite quadratic form).

Solution: To evaluate (1) note that

$$\int_{-\infty}^\infty e^{-cx^2} dx = \int_{-\infty}^\infty e^{-cy^2} dy$$

and hence

$$\left(\int_{-\infty}^{\infty} e^{-cx^2} dx\right)^2 = \int_{-\infty}^{\infty} e^{-cx^2} dx \int_{-\infty}^{\infty} e^{-cy^2} dy$$

$$= \int_{-\infty}^{\infty} \int_{-\infty}^{\infty} e^{-c(x^2+y^2)} dx\, dy$$

$$= \int_{R^2} e^{-c(x^2+y^2)} dx\, dy \, . \qquad (3)$$

Changing to polar coordinates:

$$\int_{R^2} e^{-c(x^2+y^2)} dx\, dy = \int_{R^2} e^{-cr^2} |r|\, dr\, d\theta$$

$$= \int_0^{2\pi} d\theta \int_0^{\infty} re^{-cr^2} dr$$

$$= 2\pi \int_0^{\infty} re^{-cr^2} dr$$

$$= 2\pi \frac{1}{-2c} \lim_{t \to \infty} \int_0^t -2cr\, e^{-cr^2}$$

$$= -\frac{\pi}{c} \lim_{t \to \infty} e^{-cr^2} \Big|_0^t$$

$$= -\frac{\pi}{c} (\lim_{t \to \infty} e^{-ct^2} - e^0)$$

$$= \frac{\pi}{c}$$

Therefore, from (3),

$$\int_R e^{-cx^2} dx = \int_{-\infty}^{\infty} e^{-cx^2} dx = \left(\frac{\pi}{c}\right)^{\frac{1}{2}} \, . \qquad (4)$$

To evaluate (2), the Change of Variables Theorem may be applied. This theorem is as follows: Suppose T is a 1-1 C^1 function of an open set $E \subseteq R^n$ into R^n such that $\det J_T(x) \neq 0$ for all $x \in E$. If f is real-valued and continuous on $T(E)$ with compact support [where support of f, denoted supp f, is the closure of $\{x \in R^n | f(x) \neq 0\}$] contained in $T(E)$, then

$$\int_{T(E)} f = \int_E f \circ T |\det J_T| \qquad (5)$$

In light of this theorem, choose the function T consisting of a rotation p (a composition of $n-1$ simple rotations) and a translation z that will change the quadratic form $<Px,x>$ to one of the form $<Dy,y>$ where D is a diagonal positive definite matrix where $\{\lambda_i | i = 1,\ldots,n\}$ are the eigenvalues of D and hence of P. Since $T = p \circ z$, $J_T = J_{p \circ z} = J_p(z) J_z$ and hence $|\det J_T| = 1$. Clearly, T is C^1 on R^n. Let $A = [-t,t]^n$ and $T(E) = A$. Then $f(x) = e^{-<Px,x>}$ is real-valued and continuous on $A = T(E)$. Also, supp $f = A = T(E)$ is compact. Hence, the conditions of the theorem are satisfied and (5) yields

$$\int_A e^{-\langle Px,x\rangle} = \int_E e^{-\langle P\circ T(x), T(x)\rangle}|J_T|$$

$$= \int_E e^{-\langle Dy,y\rangle}.$$

But since T is volume invariant,

$$\int_{R^n} e^{-\langle Px,x\rangle} = \lim_{t\to\infty} \int_A e^{-\langle Px,x\rangle}$$

$$= \lim_{t\to\infty} \int_E e^{-\langle Dy,y\rangle}$$

$$= \int_{R^n} e^{-\langle Dy,y\rangle}$$

$$= \int_{R^n} e^{-(\lambda_1 y_1^2 + \ldots + \lambda_n y_n^2)}$$

$$= \int_R e^{-\lambda_1 y_1^2} \ldots \int_R e^{-\lambda_n y_n^2}$$

$$= \left(\frac{\pi}{\lambda_1}\right)^{\frac{1}{2}} \ldots \left(\frac{\pi}{\lambda_n}\right)^{\frac{1}{2}}$$

$$= \left(\frac{\pi^n}{\lambda_1 \ldots \lambda_n}\right)^{\frac{1}{2}}$$

$$= \left(\frac{\pi^n}{\det D}\right)^{\frac{1}{2}}$$

$$= \left(\frac{\pi^n}{\det P}\right)^{\frac{1}{2}}.$$

GAMMA AND BETA FUNCTIONS

• **PROBLEM 10-28**

a) Prove that $\Gamma(p+1) = p\Gamma(p)$, $p > 0$.
b) Show that
   $\Gamma(p+k+1) = (p+k)(p+k-1)\ldots(p+2)(p+1)\Gamma(p+1)$.

Solution: a) The gamma function, $\Gamma(p)$, is defined for any positive real number p by

$$\Gamma(p) = \int_0^\infty x^{p-1} e^{-x} dx. \qquad (1)$$

Replacing p by $(p+1)$, (1) becomes

$$\Gamma(p+1) = \int_0^\infty x^p e^{-x} dx. \qquad (2)$$

We must show, by evaluating the improper integral (the upper limit is unbounded) in (2), that

$$\int_0^\infty x^p e^{-x} dx = p \int_0^\infty x^{p-1} e^{-x} dx. \qquad (3)$$

We use integration by parts on the integral on the left. Let $u = x^p$, $dv = e^{-x} dx$. Hence, $du = px^{p-1} dx$ and $v = -e^{-x}$. Then

$$\int_0^\infty x^p e^{-x} dx = \lim_{s\to\infty} \int_0^s x^p e^{-x} dx$$

$$= \lim_{s\to\infty} \left[-x^p e^{-x} \Big|_0^s \right] + p \int_0^\infty x^{p-1} e^{-x} dx$$

$$= \lim_{s\to\infty} \left[-s^p e^{-s} \right] + p\Gamma(p). \tag{4}$$

We must now evaluate the limit as $s \to \infty$ of $-s^p e^{-s}$.
This may be rewritten as

$$\lim_{s\to\infty} \frac{-s^p}{e^s}. \tag{5}$$

Expression (5) has the form ∞/∞; hence we may use L'Hospital's rule,

$$\lim_{x\to\infty} \frac{f(x)}{g(x)} = \lim_{x\to\infty} \frac{f'(x)}{g'(x)} = \ldots = \lim_{x\to\infty} \frac{f^{(n)}(x)}{g^{(n)}(x)}$$

to evaluate (5). Each derivative of e^s will yield e^s; that is, $g(s) = g'(s) = \ldots = g^{(n)}(s)$. If p is finite then there exists an integer $n > p$ and by differentiating s^p n times we get

$$f'(s) = ps^{p-1}, \quad f''(s) = p(p-1)s^{p-2} \quad \text{and}$$

$$f^{(n)}(s) = p(p-1)(p-2) \ldots (p-n+1)s^{p-n}.$$

Now $(p - n) < 0$; therefore

$$f^{(n)}(s) = \frac{\text{const}}{s^\alpha}, \quad \alpha = |p - n|.$$

Then,

$$\lim_{s\to\infty} \frac{s^p}{e^s} = \lim_{s\to\infty} \frac{(\text{const}/s^\alpha)}{e^s} = 0.$$

Therefore (4) reduces to $p\Gamma(p)$. Thus we have shown

$$\Gamma(p + 1) = p\Gamma(p).$$

We obtain an interesting extension of the above results as follows. Note that when p is an integer, repeated evaluation of the gamma function leads to the result

$$\Gamma(p + 1) = p\Gamma(p) = p(p - 1) \Gamma(p-1)$$

$$= \ldots = p(p - 1) \ldots 2 \cdot 1 \Gamma(1) = p! .$$

This result holds since $\Gamma(1) = 1$ as may be seen by letting $p = 1$ in (1) and evaluating the resulting improper integral.
It is for this reason that the gamma function is called the generalized factorial function.

b) Using the property $\Gamma(p+1) = p\Gamma(p)$, applying this property to $\Gamma(p+k+1)$ we obtain

$$\Gamma(p+k+1) = (p+k) \Gamma(p+k) \tag{6}$$

$$\Gamma(p+k) = (p+k-1) \Gamma(p+k-1) . \tag{7}$$

Continuing this procedure $(k-3)$ more times,

$$\Gamma(p+k-1-(k-3)) = \Gamma(p+2) = (p+1) \Gamma(p+1). \tag{8}$$

Substituting (7) into (6) and then substituting the remaining $(k-3)$ evaluations into (1):

$$\Gamma(p+k+1) = (p+k)(p+k-1) \ldots (p+2)(p+1)\Gamma(p+1),$$

which was to be shown.

• **PROBLEM 10-29**

By substituting $y = e^{-x}$ in the expression:
$$\Gamma(n) = \int_0^\infty x^{n-1} e^{-x} \, dx \quad (n > 0).$$
Obtain another form of $\Gamma(n)$.

Solution: Integrating by parts, it can be shown (with $u = x^n$, $dv = e^{-x} dx$ and the formula $\int u \, dv = uv - \int v \, du$) that
$$\int_0^\infty x^n e^{-x} \, dx$$
is equal to
$$-x^n e^{-x} \Big|_0^\infty + n \int_0^\infty x^{n-1} e^{-x} \, dx \quad (n = \text{integer or real number}),$$
where, at $x = 0$, $x^n e^{-x} = 0$. But $\lim_{x \to \infty} x^n e^{-x} = \lim_{x \to \infty} \frac{x^n}{e^x}$, which is of the indeterminate form ∞/∞, for $n > 0$. To use L'Hospital's rule, we let $x^n = f(x)$ and $e^x = g(x)$, and, by definition:
$$\lim_{x \to \infty} \frac{x^n}{e^x} = \lim_{x \to \infty} \frac{f(x)}{g(x)} \equiv \lim_{x \to \infty} \frac{f^{(p)}(x)}{g^{(p)}(x)}$$
where the operation $f^{(p)}(x) \equiv \frac{d^p f(x)}{dx^p}$ and $p \geq n$ is an integer.

Hence we have:
$$\lim_{x \to \infty} \frac{x^n}{e^x} = \lim_{x \to \infty} \frac{n(n-1)(n-2)\ldots(n-p+1)}{x^{p-n} e^x} = 0.$$
Therefore, under these conditions,
$$\int_0^\infty x^n e^{-x} \, dx = n \int_0^\infty x^{n-1} e^{-x} \, dx = n(n-1)(n-2)\ldots(n-k+1) \int_0^\infty x^{n-k} e^{-x} \, dx$$
where $k \leq n$ is another integer. If we let $k = n$, then
$$\int_0^\infty x^n e^{-x} \, dx = n(n-1)(n-2) \ldots 3 \cdot 2 \cdot 1 = n!,$$
which means that integration has been carried out n times, and
$$\int_0^\infty e^{-x} \, dx = -\frac{1}{e^x} \Big|_0^\infty = 1 \quad \text{for} \quad k = n.$$
This process leads to the concept of a generalization of the factorial called the gamma function represented by the capital Greek letter Γ (gamma). Thus, the expression for the gamma function is
$$\int_0^\infty x^n e^{-x} \, dx = \Gamma(n+1) = n\Gamma(n) = n(n-1)!.$$
We now set $y = e^{-x}$. Then $dy = -e^{-x} dx$, or $dx = -e^x dy$. By taking the logarithm, $x = -\ln y = \ln \frac{1}{y}$, and $x^n = \left(\ln \frac{1}{y}\right)^n$, with corresponding limits $1 < y < 0$ for $0 < x < \infty$. Substitution yields:

$$\int_0^\infty x^n e^{-x} dx = -\int_1^0 \left(\ln \frac{1}{y}\right)^n dy = \int_0^1 \ln\left(\frac{1}{y}\right)^n dy = \Gamma(n+1) .$$

By replacing n by n-1, we obtain:

$$\int_0^\infty x^{n-1} e^{-x} dx = \Gamma(n) = \int_0^1 \left(\ln \frac{1}{y}\right)^{n-1} dy .$$

This final integration is convenient for tabulating $\Gamma(t)$, where t is between 1 and 2. The values of $\Gamma(n)$ for $n < 1$ and $n > 2$ can be obtained by use of the following relationships:

$$\Gamma(n) = \frac{\Gamma(n+1)}{n} \quad \text{for} \quad n < 1$$

$$\Gamma(n+1) = n\Gamma(n) \quad \text{for} \quad n > 2 .$$

● **PROBLEM** 10-30

Show that a) $\int_0^\infty e^{-\alpha^2 x^2} dx = \frac{1}{2} \frac{\sqrt{\pi}}{\alpha}$; and hence, b) for n any positive integer,

$$\int_0^\infty x^{2n} e^{-\alpha^2 x^2} dx = \frac{\sqrt{\pi}}{2} \cdot \frac{1 \cdot 3 \cdot 5 \ldots (2n-1)}{2^n \alpha^{2n+1}} .$$

Solution: a) Let $\alpha^2 x^2 = u$, then $dx = \frac{du}{2\alpha^2 x} = \frac{du}{2\alpha^2 (\sqrt{u}/\alpha)} = \frac{du}{2\alpha\sqrt{u}}$.

Substitution in the integrand gives

$$\int_0^\infty e^{-\alpha^2 x^2} dx = \frac{1}{2\alpha} \int_0^\infty u^{-1/2} e^{-u} du .$$

But from the theory of gamma functions we have, for any positive number n,

$$\int_0^\infty x^{n-1} e^{-x} dx = \Gamma(n), \text{ and,}$$

for an integer n we have $\Gamma(n) = (n-1)!$
To apply the integral, we rewrite the integrand as

$$\frac{1}{2\alpha} \int_0^\infty u^{-1/2} e^{-u} du = \frac{1}{2\alpha} \int_0^\infty u^{1/2-1} e^{-u} du = \frac{1}{2\alpha} \Gamma(\tfrac{1}{2})$$

$$= \frac{1}{2\alpha} \sqrt{\pi} ,$$

since $\Gamma(\tfrac{1}{2}) = \sqrt{\pi}$.

b) Using the same procedure, since $x^2 = \frac{u}{\alpha^2}$, $x^{2n} = \frac{u^n}{\alpha^{2n}}$.

Hence,

$$\int_0^\infty x^{2n} e^{-\alpha^2 x^2} dx = \frac{1}{2\alpha} \int_0^\infty \frac{u^n}{\alpha^{2n}} \cdot u^{-1/2} e^{-u} du = \frac{1}{2\alpha^{2n+1}} \int_0^\infty u^{n-\frac{1}{2}} e^{-u} du$$

$$= \frac{1}{2\alpha^{2n+1}} \int_0^\infty u^{(n-\frac{1}{2}+1)-1} e^{-u} du$$

$$= \frac{1}{2\alpha^{2n+1}} \Gamma(n-\tfrac{1}{2}+1)$$

$$= \frac{1}{2\alpha^{2n+1}} \Gamma\left(\frac{2n+1}{2}\right) .$$

By definition, $\Gamma(n+1) = n!$; hence

$$\Gamma\left(\frac{2n+1}{2}\right) = \left(\frac{2n-1}{2}\right)! = \frac{2n-1}{2} \cdot \frac{2n-2}{2} \cdot \frac{2n-3}{2} \cdots \frac{5}{2} \cdot \frac{3}{2} \cdot \tfrac{1}{2}\Gamma(\tfrac{1}{2}).$$

Since the 2's in the denominators exist n times, we can factor them out and obtain

$$\frac{1}{2^n}(2n-1)(2n-2)(2n-3)\cdots 3\cdot 1\cdot \Gamma(\tfrac{1}{2}).$$

In this case,

$$\Gamma\left(\frac{2n+1}{2}\right) = \frac{1}{2^n} 1\cdot 3\cdot 5\cdots(2n-1)\Gamma(\tfrac{1}{2}),$$

which yields the final result. By substitution,

$$\int_0^\infty x^{2n} e^{-\alpha^2 x^2} dx = \frac{1}{2\alpha^{2n+1}}\int_0^\infty u^{n-\tfrac{1}{2}} e^{-u} du = \frac{1}{2\alpha^{2n+1}} \Gamma\left(\frac{2n+1}{2}\right)$$

$$= \frac{1}{2\alpha^{2n+1}} \cdot \frac{1}{2^n} \cdot 1\cdot 3\cdot 5\cdots(2n-1)\sqrt{\pi},$$

where, again, $\Gamma(\tfrac{1}{2}) = \sqrt{\pi}$. Furthermore,

$$\int_0^\infty e^{-\alpha^2 x^2} dx = \lim_{n\to 0}\int_0^\infty x^{2n} e^{-\alpha^2 x^2} dx = \lim_{n\to 0} \frac{(2n-1)!\sqrt{\pi}}{2\alpha\, \alpha^{2n} 2^n} = \frac{\sqrt{\pi}}{2\alpha}$$

as obtained above.

• **PROBLEM 10-31**

Verify the relation

$$B(x,y) = \frac{\Gamma(x)\Gamma(y)}{\Gamma(x+y)}, \quad x,y > 0. \tag{1}$$

where $B(x,y)$ is the beta function and $\Gamma(x)$ is the gamma function of x.

Solution: The beta function is defined by

$$B(x,y) = \int_0^1 t^{x-1}(1-t)^{y-1} dt. \tag{2}$$

If $x \geq 1$ and $y \geq 1$, the integral (1) is proper. However, if $x > 0$ and $y > 0$ and either $x < 1$ or $y < 1$ (or both), the integral is improper, but convergent.
The gamma function is defined by

$$\Gamma(x) = \int_0^\infty t^{x-1} e^{-t} dt \tag{3}$$

This improper integral is convergent for $x > 0$.
To verify (1), begin by letting $z = t^2$ in (3), to yield

$$\Gamma(x) = \int_0^\infty z^{x-1} e^{-z} dz = 2\int_0^\infty t^{2x-1} e^{-t^2} dt.$$

Similarly, letting $v = s^2$ in (3),

$$\Gamma(y) = \int_0^\infty v^{y-1} e^{-v} dv = 2\int_0^\infty s^{2y-1} e^{-s^2} ds.$$

Consequently,

$$\Gamma(x)\Gamma(y) = 4\left[\int_0^\infty t^{2x-1} e^{-t^2} dt\right]\left[\int_0^\infty s^{2y-1} e^{-s^2} ds\right]$$

$$= 4 \int_0^\infty \int_0^\infty t^{2x-1} s^{2y-1} e^{-(t^2+s^2)} dt \, ds . \tag{4}$$

Now transform to polar coordinates, letting $t = r \cos \theta$, $s = r \sin \theta$, so that (4) becomes

$$\Gamma(x)\Gamma(y) = 4 \int_{\theta=0}^{\pi/2} \int_{r=0}^\infty (r \cos \theta)^{2x-1} (r \sin \theta)^{2y-1} e^{-r^2} r \, dr \, d\theta$$

$$= 4 \int_{\theta=0}^{\pi/2} \int_{r=0}^\infty r r^{2(x+y-1)} e^{-r^2} \cos^{2x-1}\theta \sin^{2y-1}\theta \, dr \, d\theta$$

$$= 4 \left[\int_0^\infty r r^{2(x+y-1)} e^{-r^2} dr \right] \left[\int_0^{\pi/2} \cos^{2x-1}\theta \sin^{2y-1}\theta \, d\theta \right] . \tag{5}$$

Let $p = r^2$ in the first integral in (5) to give

$$= \left[2 \int_0^\infty p^{(x+y)-1} e^{-p} dp \right] \left[\int_0^{\pi/2} \cos^{2x-1}\theta \sin^{2y-1}\theta \, d\theta \right]$$

or

$$\Gamma(x)\Gamma(y) = 2\Gamma(x+y) \int_0^{\pi/2} \cos^{2x-1}\theta \sin^{2y-1}\theta \, d\theta . \tag{6}$$

Now, let $t = \sin^2 \theta$ in (2) to give,

$$B(x,y) = \int_0^1 t^{x-1} (1-t)^{y-1} dt = \int_0^{\pi/2} (\sin^2\theta)^{x-1} (\cos^2\theta)^{y-1} 2 \sin \theta \cos \theta \, d\theta .$$

$$B(x,y) = 2 \int_0^{\pi/2} \sin^{2x-1}\theta \cos^{2y-1}\theta \, d\theta .$$

Hence, (6) becomes

$$\Gamma(x)\Gamma(y) = \Gamma(x+y) B(y,x) . \tag{7}$$

However, using the transformation $t = 1-s$ in (2) yields

$$B(x,y) = \int_0^1 t^{x-1} (1-t)^{y-1} dt = - \int_1^0 (1-s)^{x-1} s^{y-1} ds = \int_0^1 (1-s)^{x-1} s^{y-1} ds$$

$$= B(y,x).$$

Thus (7) becomes

$$\Gamma(x)\Gamma(y) = \Gamma(x+y) B(x,y)$$

or

$$B(x,y) = \frac{\Gamma(x)\Gamma(y)}{\Gamma(x+y)} , \quad x,y > 0 .$$

• **PROBLEM 10-32**

Prove

$$\int_0^{\pi/2} \sin^p \theta \, d\theta = \int_0^{\pi/2} \cos^p \theta \, d\theta \tag{1}$$

a) $= \dfrac{1 \cdot 3 \cdot 5 \ldots (p-1)}{2 \cdot 4 \cdot 6 \ldots p} \dfrac{\pi}{2}$ if p is an even positive integer.

b) $= \dfrac{2 \cdot 4 \cdot 6 \ldots (p-1)}{1 \cdot 3 \cdot 5 \ldots p}$ if p is an odd positive integer.

c) Evaluate $\int_0^{\pi/2} \cos^6 \theta \, d\theta$.

Solution: Using the fact that

$$B(x,y) = 2\int_0^{\pi/2} \sin^{2x-1}\theta \cos^{2y-1}\theta \, d\theta = \frac{\Gamma(x)\Gamma(y)}{\Gamma(x+y)} \qquad (2)$$

take $p = 2x-1$, $2y-1 = 0$ so that $x = \frac{1}{2}(p+1)$, $y = \frac{1}{2}$, to yield

$$\int_0^{\pi/2} \sin^p\theta \, d\theta = \frac{\Gamma[\frac{1}{2}(p+1)]\Gamma(\frac{1}{2})}{2\Gamma[\frac{1}{2}(p+2)]} \qquad (3)$$

To prove that $\int_0^{\pi/2}\sin^p\theta \, d\theta = \int_0^{\pi/2}\cos^p\theta \, d\theta$, let $\theta = \pi/2 - \varphi$.

Then $\sin\theta = \sin(\frac{\pi}{2} - \varphi) = \sin\frac{\pi}{2}\cos\varphi - \cos\frac{\pi}{2}\sin\varphi = \cos\varphi$. Hence,

$$\int_0^{\pi/2}\sin^p\theta \, d\theta = -\int_{\pi/2}^0 \cos^p\varphi \, d\varphi = \int_0^{\pi/2}\cos^p\varphi \, d\varphi = \int_0^{\pi/2}\cos^p\theta \, d\theta.$$

a) Case 1: p is an even positive integer. Let $p = 2r$, so that (3) becomes

$$\int_0^{\pi/2}\sin^p\theta \, d\theta = \frac{\Gamma(r+\frac{1}{2})\Gamma(\frac{1}{2})}{2\Gamma(r+1)}$$

$$= \frac{(r-\frac{1}{2})(r-3/2)\ldots\frac{1}{2}\Gamma(\frac{1}{2})\cdot\Gamma(\frac{1}{2})}{2r(r-1)\ldots 1}.$$

Using the fact that $\Gamma(\frac{1}{2}) = \sqrt{\pi}$, we have

$$\int_0^{\pi/2}\sin^p\theta \, d\theta = \frac{(2r-1)(2r-3)\ldots 1}{2r(2r-2)\ldots 2} \cdot \frac{\pi}{2}$$

$$= \frac{1\cdot 3\cdot 5\ldots(2r-1)}{2\cdot 4\cdot 6\cdot 8\ldots 2r} \cdot \frac{\pi}{2} = \frac{1\cdot 3\cdot 5\ldots(p-1)}{2\cdot 4\cdot 6\cdot 8\ldots p} \cdot \frac{\pi}{2}.$$

b) Case 2: p is an odd positive integer. Let $p = 2r+1$; from (3) we have

$$\int_0^{\pi/2}\sin^p\theta \, d\theta = \frac{\Gamma(r+1)\Gamma(\frac{1}{2})}{2\Gamma(r+3/2)}$$

$$= \frac{r(r-1)\ldots 1 \cdot \sqrt{\pi}}{2(r+\frac{1}{2})(r-\frac{1}{2})\ldots\frac{1}{2}\sqrt{\pi}} = \frac{2r(2r-2)\ldots 2}{(2r+1)(2r-1)\ldots 1}$$

$$= \frac{2\cdot 4\cdot 6\ldots 2r}{1\cdot 3\cdot 5\ldots(2r+1)} = \frac{2\cdot 4\cdot 6\ldots(p-1)}{1\cdot 3\cdot 5\ldots p}.$$

c) $\int_0^{\pi/2}\cos^6\theta \, d\theta$. Take $2x-1 = 0$, $2y-1 = 6$ in (2), so that $x = \frac{1}{2}$ and $y = 7/2$, to yield

$$\int_0^{\pi/2}\cos^6\theta \, d\theta = \frac{\Gamma(\frac{1}{2})\Gamma(7/2)}{2\Gamma(4)} = \frac{5/2 \cdot 3/2 \cdot 1/2 \sqrt{\pi}\cdot\sqrt{\pi}}{2\cdot 3!} = \frac{5\pi}{32}$$

or by part (a) of this problem, the integral equals

$$\frac{1\cdot 3\cdot 5}{2\cdot 4\cdot 6} \cdot \frac{\pi}{2} = \frac{5\pi}{32}.$$

• **PROBLEM 10-33**

Find the expression, in terms of n, for

$$\int_0^1 \frac{dx}{\sqrt{1-x^n}}.$$

Evaluate the result for $n = 6$.

Solution: Let $1 - x^n = \cos^2\theta$. Then, $dx = \frac{2\sin\theta\cos\theta \, d\theta}{nx^{n-1}} =$

$\frac{2x^{1-n}\sin\theta\cos\theta}{n}d\theta$. Since $x^n = 1-\cos^2\theta = \sin^2\theta$, $x = \sin^{2/n}\theta$, and $x^{1-n} = \sin^{(2/n)(1-n)}\theta$. Hence $dx = 2/n \sin^{(2/n)(1-n)}\theta \sin\theta \cos\theta \, d\theta = 2/n \sin^{(2-n)/n}\theta \cos\theta \, d\theta$. For $x = 0$, $\cos\theta = 1$, $\theta = 0$ and for $x = 1$, $\theta = \pi/2$. Thus

$$\int_0^1 \frac{dx}{\sqrt{1-x^n}} = 2\int_0^{\pi/2} \frac{\sin^{(2-n)/n}\theta \cos\theta \, d\theta}{n\cos\theta}$$

$$= 2/n \int_0^{\pi/2} \sin^{(2-n)/n}\theta \, d\theta.$$

From the Beta functions, we have:

$$\int_0^{\pi/2} \sin^n\theta \, d\theta = \int_0^{\pi/2} \cos^n\theta \, d\theta = \frac{\Gamma((n+1)/2)}{\Gamma((n/2)+1)} \cdot \frac{\sqrt{\pi}}{2}$$

This follows from the relation

$$\int_0^{\pi/2} \sin^{2x-1}\theta \cos^{2y-1}\theta \, d\theta = \frac{\Gamma(x)\Gamma(y)}{2\Gamma(x+y)},$$

choosing $2x-1 = 0$ and $2y-1 = n$, and the fact that $\Gamma(\tfrac{1}{2}) = \sqrt{\pi}$. Hence,

$$\int_0^{\pi/2} \sin^{(2-n)/n}\theta \, d\theta = \frac{\Gamma(1/n)}{\Gamma\left(\frac{2+n}{2n}\right)} \cdot \frac{\sqrt{\pi}}{2} = \frac{\Gamma(1/n)}{\Gamma\left(\frac{1}{n}+\frac{1}{2}\right)} \frac{\sqrt{\pi}}{2}.$$

Therefore,

$$\int_0^1 \frac{dx}{\sqrt{1-x^n}} = \frac{2}{n} \int_0^{\pi/2} \sin^{(2-n)/n}\theta \, d\theta = \frac{\Gamma(1/n)}{n\Gamma\left(\frac{1}{n}+\frac{1}{2}\right)} \sqrt{\pi}.$$

For $n = 6$, we have

$$\int_0^1 \frac{dx}{\sqrt{1-x^6}} = \frac{1}{3}\int_0^{\pi/2} \sin^{-2/3}\theta \, d\theta = \frac{\Gamma(1/6)}{6\Gamma(2/3)}\sqrt{\pi}$$

$$= \frac{\sqrt{\pi}\,\Gamma(0.166)}{6\,\Gamma(0.666)}.$$

Using the relationship:

$$\frac{\Gamma(n+1)}{n} = \Gamma(n),$$

we have:

$$\Gamma(1/6) = \frac{\Gamma(1/6 + 1)}{1/6} = 6\Gamma(1.166)$$

$$\Gamma(2/3) = \frac{\Gamma(2/3 + 1)}{2/3} = \frac{3}{2}\Gamma(1.666).$$

From tables, $\Gamma(1.1666) = 0.93$ and $\Gamma(1.666) = 0.902$. Hence,

$$\int_0^1 \frac{dx}{\sqrt{1-x^6}} = \frac{\sqrt{\pi}}{6} \cdot \frac{6\Gamma(1.166)}{\frac{3}{2}\Gamma(1.666)} = \frac{2}{3}\sqrt{\pi}\,\frac{0.93}{0.902}$$

$$= 1.216.$$

● **PROBLEM 10-34**

Find the moment of inertia, with respect to the x-axis, of the area bounded by one loop of the curve
$$\rho^3 = \sin^2\theta.$$

Solution: The moment of inertia with respect to the x-axis is given by

$$I_x = \int_R y^2 \, dm,$$

where the elementary mass can be represented by the corresponding area in the plane, $dxdy$, or, in polar coordinates, $\rho d\rho d\theta$. $y = \rho \sin\theta$. Hence,

$$I_x = \int_0^\pi \int_0^{\sin^{\frac{2}{3}}\theta} \rho^3 \sin^2\theta \, d\rho \, d\theta.$$

Due to symmetry about the y-axis, we can rewrite the above expression after performing the integral with respect to ρ. This gives a factor of 2.

$$I_x = 2\int_0^{\pi/2} \frac{(\sin^{\frac{2}{3}}\theta)^4}{4} \sin^2\theta \, d\theta = \tfrac{1}{2} \int_0^{\pi/2} \sin^{8/3}\theta \sin^2\theta \, d\theta$$

$$= \tfrac{1}{2} \int_0^{\pi/2} \sin^{14/3}\theta \, d\theta.$$

To carry out the integration, we use the gamma function.

$$\Gamma(n) = \int_0^\infty x^{n-1} e^{-x} \, dx.$$

Using this definition, we can write:

$$\Gamma(m) \cdot \Gamma(n) = \int_0^\infty s^{m-1} e^{-s} ds \cdot \int_0^\infty t^{n-1} e^{-t} dt.$$

Letting $s = x^2$, $ds = 2x \, dx$; $t = y^2$, $dt = 2y \, dy$, we have:

$$\Gamma(m)\Gamma(n) = \int_0^\infty 2x^{2m-1} e^{-x^2} dx \cdot \int_0^\infty 2y^{2n-1} e^{-y^2} dy$$

$$= 4\int_0^\infty \int_0^\infty x^{2m-1} y^{2n-1} e^{-(x^2+y^2)} dy \, dx.$$

Changing to polar coordinates, we have: $x = \rho\cos\theta$, $dx = \rho\sin\theta \, d\theta$, $y = \rho\sin\theta$, $dy = \rho\cos\theta \, d\theta$, and $x^2 + y^2 = \rho^2$. Furthermore, since $dydx = \rho d\rho d\theta$, the area bounded by the curves over the entire region in the first quadrant gives rise to the variations of ρ, from 0 to ∞, and θ, from 0 to $\pi/2$. Hence,

$$\Gamma(m)\Gamma(n) = 4\int_0^{\pi/2} \int_0^\infty (\rho\cos\theta)^{2m-1} (\rho\sin\theta)^{2n-1} e^{-\rho^2} \rho d\rho d\theta$$

$$= 4\int_0^{\pi/2} \cos^{2m-1}\theta \sin^{2n-1}\theta \, d\theta \int_0^\infty \rho^{2m+2n-2} e^{-\rho^2} \rho \, d\rho$$

The second integral can be evaluated by setting $\rho^2 = z$. Hence,

$$\int_0^\infty \rho^{2(m+n-1)} e^{-\rho^2} d\rho = \tfrac{1}{2} \int_0^\infty z^{(m+n)-1} e^{-z} dz = \tfrac{1}{2} \Gamma(m+n).$$

Then

$$\Gamma(m)\Gamma(n) = 2\Gamma(m+n) \int_0^{\pi/2} \cos^{2m-1}\theta \sin^{2n-1}\theta \, d\theta,$$

which holds if and only if

$$\int_0^{\pi/2} \cos^{2m-1}\theta \sin^{2n-1}\theta \, d\theta = \tfrac{1}{2} \frac{\Gamma(m)\Gamma(n)}{\Gamma(m+n)} \equiv B(m,n), \quad (m > 0, n > 0).$$

By the use of the Beta function, the integral $\tfrac{1}{2}\int_0^{\pi/2} \sin^{14/3}\theta \, d\theta$ can be evaluated by setting $m = \tfrac{1}{2}$ and $n = 17/6$. Hence,

$$\tfrac{1}{2}\int_0^{\pi/2} \sin^{14/3}\theta \, d\theta = \tfrac{1}{2} B(1/2, 17/6) = \tfrac{1}{2} \frac{\Gamma(1/2)\Gamma(17/6)}{\Gamma(10/3)}.$$

Now, using the relationship: $\Gamma(n+1) = n\Gamma(n)$, we have:

$$\Gamma(17/6) = \Gamma(11/6+1) = 11/6 \, \Gamma(11/6) = 11/6 \, \Gamma(1.8333)$$

$$\Gamma(10/3) = \Gamma[(1+4/3)+1] = (1+4/3)\Gamma(4/3+1)$$

$$= \tfrac{7}{3} \cdot \tfrac{4}{3} \Gamma(4/3) = \tfrac{28}{9} \Gamma(1.333).$$

For $\Gamma(\tfrac{1}{2})$, we use $\Gamma(n) = \frac{\Gamma(n+1)}{n}$ $\Gamma(\tfrac{1}{2}) = \frac{\Gamma(1/2+1)}{1/2} = 2\Gamma(1.5).$

Using a table for $\Gamma(t)$, where $1 < t \le 2$, we have:

$$\Gamma(1.83) = 0.94$$
$$\Gamma(1.33) = 0.893$$
$$\Gamma(1.50) = 0.886$$

Finally,

$$I_x = \tfrac{1}{2}\int_0^{\pi/2} \sin^{14/3}\theta \, d\theta = \tfrac{1}{2} \frac{\Gamma(\tfrac{1}{2})\Gamma(17.6)}{\Gamma(10/3)} = \tfrac{1}{2} \frac{2(0.886) \; 11/6 \; (0.94)}{28/9 \; (0.893)}$$

$$\simeq 0.275.$$

CHAPTER 11

INFINITE SEQUENCES

By definition a sequence is an ordered set of numbers in one-to-one correspondence with the positive integers. In other words, a sequence is a set of numbers,

$$S_1, S_2, S_3, \ldots$$

in a definite order of arrangement that is formed according to a definite rule. The above sequence is then denoted symbolically by $\{S_n\}$ (or by the less frequently used functional notation $f(n)$). In addition each number in the sequence is called a term, and S_n is the symbol for the nth term.

Sequences can be finite (i.e., containing a finite number of distinct terms) or infinite (i.e., containing an infinite number of distinct terms). For an infinite sequence $\{S_n\}$, if

$$\lim_{n \to \infty} S_n = A,$$

exists and is finite we say that the sequence $\{S_n\}$ is convergent and has limit A; otherwise it is said to be divergent.

Furthermore, infinite sequences are said to be bounded if there exist two constants m,M such that

$$m \leq S_n \leq M$$

for all n. It should be observed, however, that every convergent sequence is bounded, but that every bounded sequence is not necessarily convergent.

The theory concerning infinite sequences is important to many branches of Mathematics. For example, the notions of the least upper bound, greatest lower bound, and limit superior and limit inferior of a sequence are useful in Mathematical Analysis. In addition, the theory on sequences of functions and uniform convergence are also useful to analysis. Finally, many of the concepts and theorems on infinite sequences carry over to the theory on infinite series which will be covered in the next chapter.

CONVERGENCE OF SEQUENCES

• **PROBLEM 11-1**

Define a convergent sequence in a metric space X. Prove:

a) $$\lim_{n \to \infty} \frac{1}{n^p} = 0 \quad \text{for } p > 0$$

b) $$\lim_{n \to \infty} \sqrt[n]{p} = 1 \quad \text{for } p > 0.$$

c) $$\lim_{n \to \infty} \sqrt[n]{n} = 1.$$

d) $$\lim_{n \to \infty} \frac{n^\alpha}{(1+p)^n} = 0 \quad \text{for } p > 0, \; \alpha \text{ real.}$$

e) If $|x| < 1$, $\lim_{n \to \infty} x^n = 0$.

Solution: a sequence $\{p_n\}$ in a metric space X is said to converge (to $p \in X$) if for all $\varepsilon > 0$, there exists a number $N > 0$ such that $n \geq N$ implies that

$$d(p_n, p) < \varepsilon$$

(d is the distance function in X). Note that p is then said to be the limit of $\{p_n\}$, for which we write,

$$\lim_{n \to \infty} p_n = p.$$

Also note that if $\{p_n\}$ does not converge, it is said to diverge.

a) $\lim_{n \to \infty} \frac{1}{n^p} = 0 \; (p > 0)$: Given $\varepsilon > 0$ we know that

$$\varepsilon^{1/p} > 0.$$

There exists N such that

$$N \varepsilon^{\frac{1}{p}} > 1$$

(by the Archimedian property of R). Then

$$N > \left(\frac{1}{\varepsilon}\right)^{1/p}.$$

So, take

$$n > \left(\frac{1}{\varepsilon}\right)^{1/p},$$

508

which is the same as $\varepsilon > \dfrac{1}{n^p}$. This is what we want, since by definition,

$$\lim_{n \to \infty} \dfrac{1}{n^p} = 0$$

if there exists N such that $n \geq N$ implies that

$$\left| \dfrac{1}{n^p} - 0 \right| < \varepsilon \quad \text{or} \quad \left| \dfrac{1}{n^p} \right| < \varepsilon.$$

b) $\lim_{n \to \infty} \sqrt[n]{p} = 1 \quad (p > 0):$ If $p \geq 1$, then $\sqrt[n]{p} \geq 1$

so that

$$\sqrt[n]{p} = 1 + x_n \text{ for } x_n \geq 0.$$

By the binomial theorem,

$$p = (1 + x_n)^n \geq 1 + nx_n$$

Hence,

$$0 < x_n \leq \left(\dfrac{p-1}{n} \right) \quad \text{and} \quad \lim_{n \to \infty} x_n = 0,$$

since

$$\lim_{n \to \infty} \left(\dfrac{p-1}{n} \right) = 0.$$

Therefore,

$$\lim_{n \to \infty} \sqrt[n]{p} = \lim_{n \to \infty} (1 + x_n) = 1.$$

For $0 < p < 1$ we have $\sqrt[n]{p} < 1$ or $\sqrt[n]{1/p} > 1$, so that $\sqrt[n]{1/p} = 1 + x_n$ for $x_n \geq 0$. Again, by the binomial theorem,

$$\dfrac{1}{p} = (1 + x_n)^n \geq 1 + nx_n.$$

Hence,

$$0 < x_n \leq \dfrac{(1/p) - 1}{n} \quad \text{and} \quad \lim_{n \to \infty} x_n = 0,$$

as before.

Therefore, since

$$p = \dfrac{1}{(1 + x_n)^n},$$

$$\lim_{n\to\infty} \sqrt[n]{p} = \lim_{n\to\infty} \frac{1}{1+x_n} = 1.$$

c) $\lim_{n\to\infty} \sqrt[n]{n} = 1$: Let $x_n = \sqrt[n]{n} - 1$. Then $x_n \geq 0$, for $n \geq 1$, and by the binomial theorem,

$$n = (1+x_n)^n \geq \frac{n(n-1)}{2} x_n^2.$$

Hence,

$$0 \leq x_n \leq \sqrt{\frac{2}{n-1}} \quad (n \geq 2) \quad \text{so that}$$

$$\lim_{n\to\infty} x_n = 0.$$

Therefore,

$$\lim_{n\to\infty} \sqrt[n]{n} = \lim_{n\to\infty} (1+x_n) = 1.$$

d)
$$\lim_{n\to\infty} \frac{n^\alpha}{(1+p)^n} = 0$$

for $p > 0$, α real :

Let c be an integer such that $c > 0$ and $c > \alpha$. Then for $n > 2c$, by the binomial theorem,

$$(1+p)^n > \binom{n}{c} p^c = \frac{n(n-1) \cdots (n-c+1)}{c!} p^c > \frac{n^c p^c}{2^c c!}$$

Hence,

$$0 < \frac{n^\alpha}{(1+p)^n} < \frac{2^c c!}{p^c} n^{(\alpha-c)} \qquad (n > 2c)$$

Since $\alpha - c < 0$, by (a)

$$\lim_{n\to\infty} n^{\alpha-c} = 0$$

Therefore

$$\lim_{n\to\infty} \frac{n^\alpha}{(1+p)^n} = 0.$$

e) $\lim_{n\to\infty} x^n = 0$ if $|x| < 1$:

Take $\alpha = 0$ in (d). Then

$$\lim_{n\to\infty} \frac{1}{(1+p)^n} = 0, \text{ for } p > 0.$$

Let
$$|x| = \frac{1}{1+p} ; \text{ since } |x| < 1 \text{ for } p > 0.$$

Then
$$|x^n| = |x|^n = \frac{1}{(1+p)^n} < \frac{1}{(1+np)}$$

since
$$(1+p)^n > 1 + np \quad \text{for} \quad n = 2, 3, \ldots \text{ if}$$
$$p > -1, p \neq 0.$$

Therefore,
$$|x^n| < \frac{1}{(1+np)} < \varepsilon \qquad \text{for all } n > N.$$

Thus
$$\lim_{n \to \infty} x^n \leq \lim_{n \to \infty} \left(\frac{1}{1+np}\right) = 0.$$

● **PROBLEM 11-2**

Let $\{p_n\}$ be a sequence in a metric space X.

Prove:

a) $\{p_n\}$ converges to $p \in X$ if and only if every neighborhood of p contains all but finitely many of the terms of $\{p_n\}$.

b) If $p \in X$, $q \in X$ and if $\{p_n\}$ converges to p and to q, then $q = p$.

c) If $\{p_n\}$ converges, then $\{p_n\}$ is bounded.

d) If $E \subset X$ and if p is a limit point of E, then there is a sequence $\{p_n\}$ in E such that
$$p = \lim_{n \to \infty} p_n.$$

Solution: a) Suppose $p_n \to p$ and let U be a neighborhood of p. There exists some $\varepsilon > 0$ such that the conditions
$$d(q,p) < \varepsilon, \quad q \in X$$
imply $q \in U$. In addition, by the definition of a convergent sequence in a metric space, there exists N such $n \geq N$ implies
$$d(p_n, p) < \varepsilon.$$

Therefore, $n \geq N$ implies $p_n \in U$.

511

To prove the converse, assume that every neighborhood of p contains all but finitely many of the terms of $\{p_n\}$. For $\varepsilon > 0$, let U be the set of all $q \in X$ such that

$$d(p,q) < \varepsilon.$$

Now, there exists N, corresponding to this U, such that $p_n \in U$ if $n \geq N$. Thus,

$$d(p_n, p) < \varepsilon \quad \text{if} \quad n \geq N.$$

This means that $\{p_n\}$ converges to p.

b) For $\varepsilon > 0$, there exist integers N_1, N_2 such that

$$n \geq N_1 \quad \text{implies} \quad d(p_n, p) < \varepsilon/2$$

$$n \geq N_2 \quad \text{implies} \quad d(p_n, q) < \varepsilon/2.$$

Let $N = \max(N_1, N_2)$. Then for $n \geq N$,

$$d(p,q) \leq d(p, p_n) + d(p_n, q) = \varepsilon.$$

Since ε is arbitrary, it is concluded that $d(p,q) = 0$.

Since X is a metric space, this is only possible if $p = q$.

c) By definition, the sequence $\{p_n\}$ is bounded if its range (the set of all points p_n, $n = 1, 2, 3, \ldots$) is bounded. Given that $\{p_n\}$ converges, we know that for $\varepsilon > 0$, there exists N such that for $n \geq N$, $d(p_n, p) < 1$. Let

$$M = \max\{1, d(p_1, p), d(p_2, p), \ldots, d(p_N, p)\}$$

Then

$$d(p_n, p) \leq M \quad \text{for } n = 1, 2, 3, \ldots.$$

Hence, $\{p_n\}$ is bounded.

d) For each integer $n > 0$, there exists a point $p_n \in E$ such that

$$d(p_n, p) < 1/n.$$

Given $\varepsilon > 0$, choose N so that $N\varepsilon > 1$. If $n \geq N$, $n\varepsilon \geq N\varepsilon > 1$ or $1/n < \varepsilon$.

Therefore,

$$d(p_n, p) < \varepsilon,$$

which means that $\{p_n\}$ converges to p. Hence,

$$\lim_{n \to \infty} p_n = p .$$

● **PROBLEM 11-3**

Show that:

a) $$\lim_{n \to \infty} \frac{n^4 + n^3 - 1}{(n^2 + 2)(n^2 - n - 1)} = 1$$

b) $$\lim_{n \to \infty} \left(1 + \frac{C}{n^2}\right)^n = 1 ,$$

C a constant.

Solution: By definition,

$$\lim_{n \to \infty} p_n = p$$

if for all $\varepsilon > 0$, there exists N such that $n \geq N$ implies that

$$|p_n - p| < \varepsilon .$$

a)
$$\lim_{n \to \infty} \frac{n^4 + n^3 - 1}{(n^2 + 2)(n^2 - n - 1)}$$

$$= \lim_{n \to \infty} \frac{n^4(1 + 1/n - 1/n^4)}{n^2(1 + 2/n^2)n^2(1 - 1/n - 1/n^2)}$$

$$= \lim_{n \to \infty} \frac{(1 + 1/n - 1/n^4)}{(1 + 2/n^2)(1 - 1/n - 1/n^2)}$$

$$= \frac{1 + 0 - 0}{(1 + 0)(1 - 0 - 0)} = 1$$

since

$$\lim_{n \to \infty} \frac{1}{n^p} = 0 \text{ for } p > 0 .$$

b) We want to show that

$$\lim_{n \to \infty} \left(1 + \frac{C}{n^2}\right)^n = 1 .$$

However, from the definition of a convergent sequence, this is equivalent to showing that

$$\left|\left(1 + \frac{c}{n^2}\right)^n - 1\right| < \varepsilon$$

where $\varepsilon > 0$, and $n \geq N$.

To do this we make use of the binomial theorem,

$$(a+b)^m = \sum_{r=0}^{m} \binom{m}{r} a^{m-r} b^r,$$

where

$$\binom{m}{r} = \frac{m!}{r!(m-r)!}.$$

Hence,

$$\left|\left(1 + \frac{c}{n^2}\right)^n - 1\right| = \left|\sum_{r=0}^{n} \binom{n}{r}\left(\frac{c}{n^2}\right)^r - 1\right|$$

$$= \left|\sum_{r=1}^{n} \binom{n}{r}\left(\frac{c}{n^2}\right)^r\right|$$

$$\leq \sum_{r=1}^{n} \frac{n^r}{r!} \left(\frac{|c|}{n^2}\right)^r \quad \text{since } \binom{n}{r} \leq \frac{n^r}{r!}$$

$$\leq \sum_{r=1}^{n} \left(\frac{|c|}{n}\right)^r \quad \text{since } \frac{n^r}{r! n^r} < 1.$$

But

$$\sum_{r=1}^{n} \left(\frac{|c|}{n}\right)^r = \frac{1 - \left(\frac{|c|}{n}\right)^{n+1}}{1 - \frac{|c|}{n}} - 1 = \frac{\frac{|c|}{n} - \left(\frac{|c|}{n}\right)^{n+1}}{1 - \frac{|c|}{n}}$$

$$= \frac{\frac{|c|}{n}\left(1 - \left(\frac{|c|}{n}\right)^n\right)}{1 - \frac{|c|}{n}}$$

$$< \frac{\frac{|c|}{n}}{1 - \frac{|c|}{n}}$$

$\left(\text{if } n > |c| \text{ since } 1 - \left(\frac{|c|}{n}\right)^n < 1\right).$

Therefore,

$$\left|\left(1 + \frac{c}{n^2}\right)^n - 1\right| < \frac{\frac{|c|}{n}}{1 - \frac{|c|}{n}} = \frac{|c|}{n - |c|}$$

which $\to 0$ as $n \to \infty$.

Thus,
$$\lim_{n\to\infty} \left(1 + \frac{c}{n^2}\right)^n = 1.$$

● **PROBLEM 11-4**

Find
$$\lim_{x\to\infty} (x\sqrt{x^2+1} - x^2).$$

Solution: We can find this limit by three different methods. For the first method, let

$$\lim_{x\to\infty} (x\sqrt{x^2+1} - x^2) = \lim_{x\to\infty} (x\sqrt{x^2+1} - x^2)\left(\frac{x\sqrt{x^2+1} + x^2}{x\sqrt{x^2+1} + x^2}\right)$$

$$= \lim_{x\to\infty} \frac{x^2}{x\sqrt{x^2+1} + x^2}$$

$$= \lim_{x\to\infty} \frac{1}{\sqrt{1 + \frac{1}{x^2}} + 1} = \frac{1}{2}$$

since
$$\lim_{n\to\infty} \frac{1}{n^2} = 0.$$

For the second method, we use the following theorem:
Let
$$f(x), g(x) \in C^{n+1} \text{ for } a \leq x \leq b.$$

In addition, let $f^{(k)}(a) = g^{(k)}(a) = 0$ for $k = 0, 1, \ldots, n$ and let $g^{(n+1)}(a) \neq 0$. Then,

$$\lim_{x\to a^+} \frac{f(x)}{g(x)} = \frac{f^{(n+1)}(a)}{g^{(n+1)}(a)}.$$

To be able to apply this theorem to the given problem, we must replace x by $\frac{1}{y}$ and let y approach zero. Then

$$\lim_{x\to\infty} (x\sqrt{x^2+1} - x^2) = \lim_{y\to 0} \left(\frac{1}{y}\sqrt{\frac{1}{y^2} + 1} - \frac{1}{y^2}\right) \quad (1)$$

$$= \lim_{y\to 0} \left(\frac{\sqrt{y^2+1} - 1}{y^2}\right).$$

Now, let $f(y) = (y^2+1)^{1/2} - 1$ and $g(y) = y^2$

515

Then
$$f'(y) = y(y^2 + 1)^{-1/2}, \quad g'(y) = 2y$$
$$f''(y) = (y^2 + 1)^{-1/2} - y^2(y^2 + 1)^{-3/2} \quad g''(y) = 2$$

Here, $n = 1$ since
$$f(0), g(0), f'(0), g'(0)$$
each equal zero. Hence,
$$\lim_{y \to 0} \frac{f(y)}{g(y)} = \frac{f''(a)}{g''(a)} \text{ where a is equal to 0}$$
hence,
$$\lim_{y \to 0} \frac{f(y)}{g(y)} = \frac{f''(0)}{g''(0)} = \frac{1}{2}.$$

Thus, by (1),
$$\lim_{x \to \infty} (x\sqrt{x^2 + 1} - x^2) = \frac{1}{2}.$$

For the third method expand the function in powers of $1/x$. Hence,
$$\lim_{x \to \infty} x\sqrt{x^2 + 1} - x^2 = \lim_{x \to \infty} x^2 \left[\left(1 + \frac{1}{x^2}\right)^{\frac{1}{2}} - 1 \right]$$

(by the binomial theorem)
$$= \lim_{x \to \infty} x^2 \left[\frac{1}{2x^2} - \frac{1}{8x^4} + \cdots \right]$$
$$= \lim_{x \to \infty} \left[\frac{1}{2} - \frac{1}{8x^2} + \cdots \right] = \frac{1}{2}$$

Since $\lim_{n \to \infty} \frac{1}{n^p} = 0$ for $p > 0$.

and
$$\lim_{n \to \infty} (S_{n_1} + S_{n_2} + \cdots) = S_1 + S_2 + \cdots$$

given that
$$\lim_{n \to \infty} S_{n_i} = S_i.$$

• **PROBLEM 11-5**

Show that the following sequences are convergent:

a) $\quad a_n = \dfrac{1 \cdot 3 \cdot 5 \cdots (2n-1)}{2 \cdot 4 \cdot 6 \cdots (2n)} \qquad n = 1, 2, 3, \ldots$

b) $\quad a_n = \dfrac{1}{1!} + \dfrac{1}{2!} + \cdots + \dfrac{1}{n!} \qquad n = 1, 2, 3, \ldots$

Solution: To solve this problem, we first must define a monotonically decreasing sequence, a monotonically increasing sequence, and a bounded sequence.

A sequence $\{a_n\}$ of real numbers is said to be:

(i) monotonically increasing if

$$a_n \leq a_{n+1} \quad (n = 1, 2, 3, \ldots);$$

if $a_n < a_{n+1}$, it is called strictly increasing.

(ii) monotonically decreasing if

$$a_n \geq a_{n+1} \quad (n = 1, 2, 3, \ldots);$$

if $a_n > a_{n+1}$, it is called strictly decreasing.

If $a_n \leq M$ for $n = 1, 2, 3, \ldots$ where M is a constant, then the sequence $\{a_n\}$ is bounded above and M is called an upper bound. If

$$a_n \geq m \text{ for } n = 1, 2, 3, \ldots,$$

where m is a constant, the sequence is bounded below and m is a lower bound. A sequence is called bounded if it is bounded both above and below (i.e., $m \leq a_n \leq M$, for $n = 1, 2, 3, \ldots$).

We note an important theorem which states that a monotonic sequence is convergent (has a limit) provided that it is bounded.

a) $\quad a_n = \dfrac{1 \cdot 3 \cdot 5 \cdots (2n-1)}{2 \cdot 4 \cdot 6 \cdots (2n)}$

By substituting $n = 1, 2, 3, \ldots$ successively, we can write out a few terms of the sequence. Hence,

$$a_1 = \dfrac{1}{2}, \quad a_2 = \dfrac{1 \cdot 3}{2 \cdot 4} = \dfrac{3}{8}, \quad a_3 = \dfrac{1 \cdot 3 \cdot 5}{2 \cdot 4 \cdot 6} = \dfrac{5}{16}, \ldots$$

observing the first few terms, we see that

$$a_2 = \dfrac{3}{4} a_1, \quad a_3 = \dfrac{5}{6} a_2, \quad a_4 = \dfrac{7}{8} a_3.$$

517

Therefore, the sequence is monotonically decreasing. In general,

$$a_{n+1} = \frac{2n+1}{2n+2} a_n < a_n.$$

Now, since all the terms of $\{a_n\}$ are positive, we have $0 < a_n \leq a_1 = \frac{1}{2}$. Therefore, the sequence is bounded, and since it is monotonically decreasing, by the theorem, the sequence is convergent (i.e., has a limit).

b)
$$a_n = \frac{1}{1!} + \frac{1}{2!} + \frac{1}{3!} + \ldots + \frac{1}{n!} :$$

The first few terms of $\{a_n\}$ are

$$a_1 = 1, \quad a_2 = 1 + \frac{1}{1 \cdot 2} = \frac{3}{2}, \quad a_3 = 1 + \frac{1}{1 \cdot 1} + \frac{1}{1 \cdot 2 \cdot 3} = \frac{5}{3},$$

etc. ...
Note that,

$$a_2 = a_1 + \frac{1}{2!}, \quad a_3 = a_2 + \frac{1}{3!}, \quad \ldots, \quad a_{n+1} = a_n + \frac{1}{(n+1)!}, \quad \ldots$$

Hence, $a_n < a_{n+1}$ so that the sequence is monotonically increasing. Therefore to show that the sequence is convergent, all that is needed is to show that it is bounded. Now, if

$$n > 2, \quad \frac{1}{n!} = \frac{1}{1 \cdot 2 \cdot 3 \ldots n} < \frac{1}{2^{n-1}}$$

since

$$2^{n-1} < 1 \cdot 2 \cdot 3 \ldots n.$$

Consequently for $n > 2$,

$$a_n < 1 + \frac{1}{2} + \frac{1}{2^2} + \ldots + \frac{1}{2^{n-1}}. \tag{1}$$

Now, using the formula for the sum of a geometric progression

$$\sum_{k=0}^{n-1} ar^k = a\left(\frac{1-r^n}{1-r}\right), \quad \text{we have}$$

$$1 + \frac{1}{2} + \ldots + \frac{1}{2^{n-1}} = \sum_{k=0}^{n-1} 1 \cdot \left(\frac{1}{2}\right)^k = 1 \cdot \left(\frac{1 - \left(\frac{1}{2}\right)^n}{1 - \frac{1}{2}}\right)$$

$$= 2\left[1 - \left(\frac{1}{2}\right)^n\right] < 2. \tag{2}$$

Therefore, by (1) and (2), $a_n < 2$ for all n. Hence,

$$1 \le a_n < 2,$$

and so, by the theorem stated in this problem, the sequence $\{a_n\}$ is convergent. The limit of the sequence plus one is conventionally denoted by e.

Note that we have shown these sequences to be convergent without determining what their limits are.

● **PROBLEM 11-6**

Find the limit of the sequence defined by

$$x_1 = \frac{2}{3}$$

and

$$x_{n+1} = \frac{(x_n + 1)}{(2x_n + 1)}.$$

Solution: We write the first four terms of the sequence

$$\left\{ \frac{2}{3}, \frac{5}{7}, \frac{12}{17}, \frac{29}{41}, \ldots \right\}.$$

Note that we pass from a/b to

$$\frac{\frac{a}{b} + 1}{\frac{2a}{b} + 1} = (a+b)/(2a+b)$$

from one term of the sequence to the next. To find the limit of this sequence, apply Banach's fixed point theorem, which states for the case of one variable:

Let S be a closed nonempty subset of R. Let f be a contraction mapping on S; f maps S into S such that for some k, $0 < k < 1$, and all x and y in S,

$$|f(x) - f(y)| \le k |x - y|. \tag{1}$$

Then, there is one and only one point x in S for which $f(x) = x$. In addition, if $x_1 \in S$ and $x_{n+1} = f(x_n)$ for all n, then $x_n \to x$ as $n \to \infty$.

To apply this theorem, first note that for $x \ge 0$,

$$\frac{1}{2} \le f(x) = \frac{x + 1}{2x + 1} \le 1$$

because

$$1 - \frac{x + 1}{2x + 1} = \frac{x}{2x + 1} \ge 0.$$

Therefore, with
$$S = \left[\frac{1}{2}\,;\,1\right],$$
f maps S to S and is a contraction. To prove this, note that
$$f(x) - f(y) = f'(z)(x - y)$$
where $x < z < y$, by the Mean Value Theorem. Hence,
$$|f(x) - f(y)| = |f'(z)|\,|x-y| \leq k|x-y|$$
if $|f'(z)| \leq k$. Therefore, to prove that f is a contraction mapping it suffices to show that $|f'(x)| \leq k < 1$ where $x \in S$. Since
$$|f'(x)| = \frac{1}{(2x+1)^2} \leq \frac{1}{4},$$
(1) is satisfied. Consequently, for $x, y \in S$ we have
$$|f(x) - f(y)| \leq \frac{1}{4}|x-y|$$
Therefore, by the theorem, $x_n \to x$, where
$$x = f(x) = \frac{x+1}{2x+1} \quad \text{and} \quad \frac{1}{2} \leq x \leq 1.$$
Hence, since
$$x = \frac{x+1}{2x+1},$$
we have
$$2x^2 + x = x + 1$$
or
$$x^2 = \frac{1}{2} \Longrightarrow x = \left(\frac{1}{2}\right)^{1/2} \quad \left(\frac{1}{2} \leq x \leq 1\right).$$
Thus,
$$\lim_{n \to \infty} x_n = x = \left(\frac{1}{2}\right)^{1/2}.$$

• **PROBLEM 11-7**

Let $\{a_n\}$ be the sequence
$$a_n = \left(1 + \frac{1}{n}\right)^n.$$
Show that a_n is convergent.

Solution: To solve this problem, we make use of the binomial theorem,

$$(1+b)^n = 1+nb + \frac{n(n-1)}{1 \cdot 2} b^2 + \frac{n(n-1)(n-2)}{1 \cdot 2 \cdot 3} b^3 + \ldots + b^n. \tag{1}$$

Note that the coefficient of b^k ($1 \leq k \leq n$) in this expansion is

$$\frac{n(n-1)(n-2) \ldots (n-k+1)}{k!}.$$

Substituting $\frac{1}{n}$ in place of b in (1) yields

$$\left(1+\frac{1}{n}\right)^n = 1+n\left(\frac{1}{n}\right) + \frac{n(n-1)}{1 \cdot 2}\left(\frac{1}{n}\right)^2 + \frac{n(n-1)(n-2)}{1 \cdot 2 \cdot 3}\left(\frac{1}{n}\right)^3$$

$$+ \ldots + \left(\frac{1}{n}\right)^n$$

Now, the expression on the right has $n+1$ terms, where $(k+1)^{\text{th}}$ term is

$$\frac{n(n-1)(n-2) \ldots (n-k+1)}{k! n^k} = \frac{1}{k!} \cdot \left(\frac{n-1}{n}\right)\left(\frac{n-2}{n}\right) \ldots$$

$$\left(\frac{n-k+1}{n}\right)$$

$$= \frac{1}{k!} \cdot (1) \cdot \left(1 - \frac{1}{n}\right)\left(1 - \frac{2}{n}\right) \ldots \left(1 - \frac{k-1}{n}\right).$$

Hence,

$$\left(1+\frac{1}{n}\right)^n = 1 + 1 + \frac{1-\frac{1}{n}}{2!} + \frac{\left(1-\frac{1}{n}\right)\left(1-\frac{2}{n}\right)}{3!} + \ldots$$

$$+ \frac{\left(1-\frac{1}{n}\right) \ldots \left(1-\frac{n-1}{n}\right)}{n!}. \tag{2}$$

Now, assume that we follow the same procedure again; however, this time we replace n with $n+1$ to form a corresponding formula for the expression

$$\left(1+\frac{1}{n+1}\right)^{n+1}.$$

Note that on the right hand side of equation (2), each of the numerators after the first two terms increases if n is replaced by $n+1$. In addition, the total number of terms on the right is increased from $n+1$ to $n+2$. Therefore,

$$\left(1+\frac{1}{n}\right)^n < \left(1+\frac{1}{n+1}\right)^{n+1}.$$

Hence, the sequence is monotonically increasing. Furthermore, from (2) we have

$$\left(1 + \frac{1}{n}\right)^n < 1 + 1 + \frac{1}{2!} + \frac{1}{3!} + \ldots + \frac{1}{n!}.$$

However, from a previous problem we have

$$1 + \frac{1}{2!} + \frac{1}{3!} + \ldots + \frac{1}{n!} < 2.$$

Therefore,

$$\left(1 + \frac{1}{n}\right)^n < 1 + 1 + \frac{1}{2!} + \ldots + \frac{1}{n!} < 1 + 2 < 3.$$

Thus, the sequence is bounded above. Consequently, since $\{a_n\}$ is a monotonically increasing sequence that is bounded above, it has a limit and is therefore convergent. Note that since

$$\lim_{n \to \infty} \left(1 + \frac{1}{n}\right)^n = e,$$

the exponential function, this limit is denoted by e.

● **PROBLEM** 11-8

Define a subsequence and a subsequential limit. Prove:

a) If $\{p_n\}$ is a sequence in a compact metric space X, then some subsequence of $\{p_n\}$ converges to a point of X.

b) Every bounded sequence in R^k contains a convergent subsequence.

<u>Solution</u>: Given a sequence $\{p_n\}$, consider a sequence $\{n_j\}$ of positive integers, such as

$$n_1 < n_2 < n_3 < \ldots .$$

Then, by definition, the sequence $\{p_{n_i}\}$ i = 1, 2, 3, ... is a subsequence of $\{p_n\}$. In addition, if a subsequence converges, its limit is called a subsequential limit.

a) Let B be the range of $\{p_n\}$. If B is finite, then there is a p ∈ B and a sequence $\{n_i\}$ with

$$n_1 < n_2 < n_3 \ldots ,$$

such that

$$p_{n_1} = p_{n_2} = \ldots = p.$$

Therefore, the subsequence $\{p_{n_i}\}$ converges to p.

If B is infinite, since an infinite subset of a compact set X has a limit point in X, we know that B has a limit point $p \in X$. Pick n_1 so that

$$d(p, p_{n_1}) < 1.$$

Having chosen $n_1, \ldots, n_{i-1}$,

since every neighborhood of a limit point of a set contains infinitely many points of that set, there exists an integer

$$n_i > n_{i-1}$$

such that

$$d(p, p_{n_i}) < 1/i.$$

For $\varepsilon > 0$, choose N so that $N\varepsilon > 1$. Thus, for $i > N$, we have $1/i < \varepsilon$. Hence

$$d(p, p_{n_i}) < \varepsilon$$

and $\{p_{n_i}\}$ converges to p.

b) Since it is known that every bounded subset of R^k lies in a compact subset of R^k, this theorem follows from (a).

● **PROBLEM 11-9**

Define a Cauchy sequence.

Prove:

a) In any metric space X, every convergent sequence is a Cauchy sequence.

b) Suppose $\{p_n\}$ is a Cauchy sequence in a compact metric space X, then $\{p_n\}$ converges to some point of X.

Solution: Given a sequence $\{p_n\}$ in a metric space X, if for every $\varepsilon > 0$ there is an integer N such that

$$d(p_n, p_m) < \varepsilon$$

(d is the distance function), if $n \geq N, m \geq N$, then by definition $\{p_n\}$ is a Cauchy sequence.

523

a) Assume $\{p_n\}$ is a convergent sequence. Then for $\varepsilon > 0$, there is an integer N such that
$$d(p,p_n) < \varepsilon/2$$
for all $n \geq N$. Therefore, by the triangle inequality,
$$d(p_n,p_m) \leq d(p_n,p) + d(p,p_m) < \varepsilon,$$
where $n \geq N$, $m \geq N$. Therefore,
$$d(p_n,p_m) < \varepsilon$$
which means that $\{p_n\}$ is a Cauchy sequence.

b) Assume $\{p_n\}$ is a Cauchy sequence in the compact metric space X. Then some subsequence of $\{p_n\}$ converges to a point of X (part (a) of the previous problem). Let the limit of the subsequence $\{p_{n_i}\}$, $i = 1, 2, 3, \ldots$, be P. By the Cauchy condition, for $\varepsilon > 0$, choose N so that
$$d(p_m,p_n) < \varepsilon/2 \text{ if } m \geq N, n \geq N.$$
Since the subsequence converges to p, we can choose one of the indices n_i to be large enough so that $n_i > N$ and $d(p_{n_i},p) < \varepsilon/2$. Then if $n \geq N$ we have
$$d(p_n,p) \leq d(p_n,p_{n_i}) + d(p_{n_i},p) < \varepsilon/2 + \varepsilon/2 = \varepsilon$$
Hence $\{p_n\}$ converges to $p \in X$.

Note that part b) of this problem enables us to determine whether or not a given sequence converges without knowing the limit to which it may converge. We also note that in R^k, every Cauchy sequence converges.

• **PROBLEM 11-10**

a) Find
$$\lim_{n \to \infty} \frac{(n!)^{1/n}}{n}.$$

b) Show that
$$\frac{(n+p)!}{n!} \sim n^p$$
as $n \to \infty$, $p = 1, 2, 3, \ldots$ (The symbol $\sim$ is here read as "is asymptotic to").

Solution: To solve this problem, we make use of Stirling's formula, which states that if n is an integer, for large n, n! is approximately equal to

$$\sqrt{2\pi n}\ n^n\ e^{-n}.$$

This can also be written as

$$\lim_{n\to\infty} \frac{(n/e)^n \sqrt{2\pi n}}{n!} = 1. \qquad (1)$$

Now, for n large,

$$n! \sim \left(\frac{n}{e}\right)^n \sqrt{2\pi n}$$

so that

$$\frac{(n!)^{1/n}}{n} \sim \frac{(2\pi)^{1/2n}\ n^{1/2n}}{e}.$$

However,

$$\lim_{n\to\infty} (2\pi)^{1/2n} = 1 \quad \text{and} \quad \lim_{n\to\infty} n^{1/2n} = 1.$$

Hence,

$$\lim_{n\to\infty} \frac{(n!)^{1/n}}{n} = \frac{1}{e}.$$

b) Given two sequences a_n, b_n, by definition, $a_n \sim b_n$ (a_n is asymptotic to b_n) as $n \to \infty$ if and only if

$$\lim_{n\to\infty} \frac{a_n}{b_n} = 1.$$

Hence, we wish to show that

$$\lim_{n\to\infty} \frac{(n+p)!}{n!\,n^p} = 1 \quad \text{for } p = 1, 2, 3, \ldots$$

This can be done by two different methods. For the first method, use Stirling's Formula. However, first note that

$$\lim_{n\to\infty} \frac{(n+p)!}{n!\,n^p} = \lim_{n\to\infty} \frac{n^n \sqrt{2\pi n}}{n!\,e^n} \cdot \frac{(n+p)!\,e^n}{n^n\,n^p \sqrt{2\pi n}}$$

$$= \lim_{n\to\infty} \frac{n^n \sqrt{2\pi n}}{n!\,e^n} \cdot \frac{(n+p)!\,e^{n+p}}{(n+p)^{n+p}\sqrt{2\pi}} \cdot \frac{e^n\,(n+p)^{n+p}}{\sqrt{n}\,e^{n+p}\,n^{n+p}}$$

$$= \lim_{n\to\infty} \frac{n^n \sqrt{2\pi n}}{n!\,e^n} \cdot \frac{(n+p)!\,e^{n+p}}{(n+p)^{n+p}\sqrt{2\pi}\,\sqrt{n+p}} \cdot \frac{(n+p)^{n+p+\frac{1}{2}}}{e^p\,n^{n+p+\frac{1}{2}}}$$

525

$$= \lim_{n\to\infty} \frac{n^n\sqrt{2\pi n}}{n!e^n} \cdot \frac{(n+p)!\, e^{n+p}}{(n+p)^{n+p}\sqrt{2\pi(n+p)}} \cdot \frac{\left(1+\frac{p}{n}\right)^{n+p+\frac{1}{2}}}{e^p} \quad (2)$$

Now, by (1) we see that

$$\lim_{n\to\infty} \frac{n^n\sqrt{2\pi n}}{e^n\, n!} = 1 \quad \text{and} \quad \lim_{n\to\infty} \frac{(n+p)!\, e^{n+p}}{(n+p)^{n+p}\sqrt{2\pi(n+p)}} = 1. \quad (3)$$

Additionally,

$$\lim_{n\to\infty} \frac{(1+p/n)^{n+p+\frac{1}{2}}}{e^p} = \frac{1}{e^p} \lim_{n\to\infty} (1+p/n)^n (1+p/n)^{p+\frac{1}{2}}$$

However,

$$\lim_{n\to\infty} (1+p/n)^n = e^p \quad \text{and} \quad \lim_{n\to\infty} (1+p/n)^{p+\frac{1}{2}} = 1.$$

Therefore,

$$\lim_{n\to\infty} \frac{(1+p/n)^{n+p+\frac{1}{2}}}{e^p} = \frac{e^p}{e^p} = 1. \quad (4)$$

Thus, by (2), (3), and (4),

$$\lim_{n\to\infty} \frac{(n+p)!}{n!\, n^p} = 1,$$

which is what we wanted to show.

For the second method, observe that

$$\frac{(n+p)!}{n!} = (n+p)(n+p-1)\cdots(n+1)$$

(note that we have p terms on the right).

Therefore,

$$\lim_{n\to\infty} \frac{(n+p)!}{n!\, n^p} = \lim_{n\to\infty} \frac{(n+p)(n+p-1)\cdots(n+1)}{n^p}$$

$$= \lim_{n\to\infty} \left(\frac{n+p}{n}\right)\left(\frac{n+p-1}{n}\right)\cdots\left(\frac{n+1}{n}\right)$$

$$= \lim_{n\to\infty} (1+p/n)\left(1+\frac{p-1}{n}\right)\cdots\left(1+\frac{1}{n}\right) = 1$$

since each of the p factors on the right approaches 1 as $n \to \infty$.

Consequently,

$$\lim_{n\to\infty} \frac{(n+p)!}{n!\, n^p} = 1.$$

Thus,

$$\frac{(n+p)!}{n!} \sim n^p \text{ as } n \to \infty, \quad p = 1, 2, \ldots$$

• **PROBLEM 11-11**

Find:

a) $$\lim_{n \to \infty} \left[\frac{(3n)!}{n^{3n}} \right]^{1/n}.$$

b) for the sequence of vectors

$$\left(\frac{m}{m+1},\ 2^{-m},\ \left(1 + \frac{1}{m}\right)^m \right), \quad m = 1, 2, \ldots \text{ in } R^3$$

the limit as $m \to \infty$.

Solution: a) We want to find

$$\lim_{n \to \infty} \left[\frac{(3n)!}{n^{3n}} \right]^{1/n}.$$

To do this, first note that

$$\frac{(3n)!}{n^{3n}} = \frac{(3n)! \, 3^{3n}}{3^{3n} \, n^{3n}} = \frac{(3n)! \, 3^{3n}}{(3n)^{3n}}.$$

Now, the 3nth root of this expression is

$$\left(\left[\frac{(3n)!}{(3n)^{3n}} \right]^{1/3n} \right)^3 = \frac{[(3n)!]^{1/3n}}{3n} \cdot 3.$$

However, in a previous problem, we have shown that

$$\lim_{k \to \infty} \frac{(k!)^{1/k}}{k} = \lim_{k \to \infty} \left(\frac{k!}{k^k} \right)^{1/k} = \frac{1}{e}.$$

Consequently, we take $k = 3n$ so that

$$\left[\frac{(3n)!}{n^{3n}} \right]^{1/3n} = \left[\frac{3^k \, k!}{k^k} \right]^{1/k} = 3 \left(\frac{k!}{k^k} \right)^{1/k}.$$

Thus,

$$\lim_{n \to \infty} \left[\frac{(3n)!}{n^{3n}} \right]^{1/3n} = \frac{3}{e}.$$

Hence, since

$$\lim_{n \to \infty} \left[\frac{(3n)!}{n^{3n}} \right]^{1/n} = \lim_{n \to \infty} \left[\frac{(3n)!}{n^{3n}} \right]^{1/3n} \left[\frac{(3n)!}{n^{3n}} \right]^{1/3n} \left[\frac{(3n)!}{n^{3n}} \right]^{1/3n},$$

we have

$$\lim_{n \to \infty} \left[\frac{(3n)!}{n^{3n}} \right]^{1/n} = \frac{27}{e^3}.$$

b)
$$\lim_{m \to \infty} \left(\frac{m}{m+1}, \ 2^{-m}, \ \left(1 + \frac{1}{m}\right)^m \right), \qquad m = 1, 2, \ldots$$

By definition, a sequence of vectors in R^n is a sequence $U_1, \ldots, U_m, \ldots$, each element of which is a vector of R^n. To determine if a sequence of vectors in R^n converges we have the following theorem:

Let $\vec{U}_m$ ($m = 1, 2, \ldots$) be a sequence of vectors in R^n. Let

$$\vec{U}_m = (U_{m1}, \ldots, U_{mn}), \quad \vec{U} = (U_1, \ldots, U_n).$$

Then

$$\lim_{m \to \infty} \vec{U}_m = \vec{U} \quad \text{if and only if}$$

$$\lim_{m \to \infty} U_{m1} = U_1, \ \lim_{m \to \infty} U_{m2} = U_2, \ \ldots \ \lim_{m \to \infty} U_{mn} = U_n.$$

For the given sequence,

$$\lim_{m \to \infty} \frac{m}{m+1} = \lim_{m \to \infty} \left(\frac{1}{1 + \frac{1}{m}} \right) = 1, \ \lim_{m \to \infty} 2^{-m} = 0, \ \lim_{m \to \infty} \left(1 + \frac{1}{m}\right)^m = e.$$

Hence,

$$\lim_{m \to \infty} \left(\frac{m}{m+1}, \ 2^{-m}, \ \left(1 + \frac{1}{m}\right)^m \right) = (1, 0, e).$$

● **PROBLEM 11-12**

Evaluate the limit

$$\lim_{x \to \infty} \frac{\sqrt{\log x} \ \log(\log x)}{e^{\sqrt{x}}}.$$

Solution: To evaluate this limit, we make use of the notion of orders of infinity. Let $f(x)$ and $g(x)$ be two functions which become positively infinite as the variable x approaches a finite limit or becomes infinite. Then by definition,

$$f(x) \prec g(x)$$

(read $f(x)$ is a lower order infinity than $g(x)$) if and only if

$$\lim_{x \to \infty} \frac{f(x)}{g(x)} = 0 .$$

Now to apply this definition, we make use of the following arrangement of infinities, written in the order of increasing strength:

$$\ldots \prec \log(\log x) \prec \log x \prec x \prec e^x \prec e^{e^x} . \qquad (1)$$

Note that the order of infinity is increased by raising it to a power $p > 1$, and is decreased if $0 < p < 1$.

For the given limit, from (1) we see that the infinity in the denominator is the strongest of the three infinities,

$$\left(\log(\log x) \prec \sqrt{\log x} \prec e^{\sqrt{x}} \right) .$$

Now, since $x^2 \prec e^{\sqrt{x}}$ and $\log(\log x) \prec \sqrt{\log x} \prec e^{\sqrt{x}}$ \qquad (2)

we have,

$$\lim_{x \to \infty} \frac{\sqrt{\log x} \, \log(\log x)}{e^{\sqrt{x}}} = \lim_{x \to \infty} \frac{(\sqrt{\log x}/x)[\log(\log x)/x]}{e^{\sqrt{x}}/x^2} . \qquad (3)$$

However,

$$\lim_{x \to \infty} \frac{\sqrt{\log x}}{x} = 0 , \quad \lim_{x \to \infty} \frac{\log(\log x)}{x} = 0$$

and

$$\lim_{x \to \infty} \frac{e^{\sqrt{x}}}{x^2} = \infty \quad \text{by (2)} .$$

Thus, by (3)

$$\lim_{x \to \infty} \frac{\sqrt{\log x} \, \log(\log x)}{e^{\sqrt{x}}} = \frac{0 \cdot 0}{\infty} = 0 .$$

LIMIT SUPERIOR AND LIMIT INFERIOR

● **PROBLEM 11-13**

Let $\{S_n\}$ be a sequence of real numbers. Define the limit superior and limit inferior of $\{S_n\}$. Then find the limit superior and limit inferior of the following sequences:

a) $\{S_n\} = 1, 0, -1, 2, 0, -2, 3, 0, -3, \ldots$

b) the sequence containing all rationals.

c) $\{S_n\} = \frac{1}{2}, -\frac{1}{3}, \frac{1}{4}, -\frac{1}{5}, \ldots, (-1)^{n-1}/(n+1), \ldots$

d) $\{S_n\} = -\frac{1}{2}, \frac{2}{3}, -\frac{3}{4}, \frac{4}{5}, \ldots, (-1)^n/[1 + (1/n)], \ldots$

<u>Solution</u>: Let G be the set of numbers x (in the extended real number system) such that $\{S_{n_k}\}$ converges to x for some subsequence $\{S_{n_k}\}$. Then this set G contains all subsequential limits (i.e., limits of all the convergent subsequences) plus possibly the numbers $+\infty$ and $-\infty$. Then, the limit superior and limit inferior of $\{S_n\}$ are respectively,

$$\lim_{n \to \infty} \text{Sup } S_n = \text{Sup } G, \quad \lim_{n \to \infty} \inf S_n = \text{Inf } G$$

(also written as $\overline{\lim} S_n$ or $\underline{\lim} S_n$, respectively).

That is, the limit superior is the least upper bound of the set of subsequential limits, and the limit inferior is the greatest lower bound of the set of subsequential limits.

In other words, x = Sup G if $\beta \leq x$ for every $\beta \in G$ and if $\beta < x$, then β is not an upper bound of G. Similarly, y = inf G if $\beta \geq y$ for every $\beta \in G$ and if $\beta > y$, then β is not a lower bound of G. (Note that the terms least upper bound and greatest bound of a sequence were also defined in Chapter 1).

An alternate definition is: A number $\overline{L}$ is the limit superior of the sequence $\{S_n\}$ if infinitely many terms of the sequence are greater than $\overline{L} - \varepsilon$ (where $\varepsilon > 0$), while only a finite number are greater than $\overline{L} + \varepsilon$. A number $\underline{L}$ is the limit inferior of the sequence $\{S_n\}$ if infinitely many terms of the sequence are less than $\underline{L} + \varepsilon$ (where $\varepsilon > 0$) while only finitely many are less than $\underline{L} - \varepsilon$.

Note: If the sequence $\{S_n\}$ has no upper bound, then

$$\lim_{n \to \infty} \text{Sup } S_n = +\infty ;$$

if there is no lower bound then

$$\lim_{n \to \infty} \inf S_n = -\infty .$$

In addition, if

$$\lim_{n\to\infty} S_n = +\infty \quad \text{then} \quad \lim_{n\to\infty} \text{Sup } S_n = \lim_{n\to\infty} \inf S_n = +\infty$$

and if

$$\lim_{n\to\infty} S_n = -\infty \quad \text{then} \quad \lim_{n\to\infty} \text{Sup } S_n = \lim_{n\to\infty} \inf S_n = -\infty.$$

Furthermore, a sequence $\{S_n\}$ converges if and only if

$$\lim_{n\to\infty} \text{Sup } S_n = \lim_{n\to\infty} \inf S_n \text{ is finite.}$$

a) $\{S_n\} = 1, 0, -1, 2, 0, -2, 3, 0, -3$:

This sequence has only one finite limit point, 0. However, it is not bounded above and it is not bounded below; therefore,

$$\lim_{n\to\infty} \text{Sup } S_n = +\infty \quad \text{and} \quad \lim_{n\to\infty} \inf S_n = -\infty.$$

b) Sequence containing all rationals: Here, every real number is a sequential limit; therefore,

$$\lim_{n\to\infty} \text{Sup } S_n = +\infty \quad \text{and} \quad \lim_{n\to\infty} \inf S_n = -\infty.$$

c) $\{S_n\} = \frac{1}{2}, -\frac{1}{3}, \frac{1}{4}, -\frac{1}{5}, \ldots, (-1)^{n-1}/(n+1), \ldots$:

Here, the greatest lower bound is $-\frac{1}{3}$ and the least upper bound is $\frac{1}{2}$. However,

$$\lim_{n\to\infty} \frac{(-1)^{n-1}}{(n+1)} = 0,$$

therefore

$$\lim_{n\to\infty} \text{Sup } S_n = \lim_{n\to\infty} \inf S_n = 0.$$

d) $\{S_n\} = -\frac{1}{2}, \frac{2}{3}, -\frac{3}{4}, \frac{4}{5}, \ldots, (-1)^n/[1 + (1/n)], \ldots$:

This sequence is bounded above by 1 and bounded below by -1. Here,

$$\lim_{n\to\infty} \text{Sup } S_n = 1 \quad \text{and} \quad \lim_{n\to\infty} \inf S_n = -1$$

● **PROBLEM 11-14**

For any sequence $\{a_n\}$ of positive numbers, prove:

a)
$$\lim_{n\to\infty} \text{Sup } a_n^{1/n} \leq \lim_{n\to\infty} \text{Sup } \frac{a_{n+1}}{a_n}$$

b)
$$\lim_{n\to\infty} \inf \frac{a_{n+1}}{a_n} \leq \lim_{n\to\infty} \inf a_n^{1/n}.$$

Solution: a) Let $\alpha = \lim\limits_{n\to\infty} \text{Sup} \dfrac{a_{n+1}}{a_n}$

If $\alpha = +\infty$, clearly

$$\lim_{n\to\infty} \text{Sup } a_n^{1/n} \leq \alpha.$$

Therefore, for α finite, choose $\lambda > \alpha$. There exists an integer N such that

$$\frac{a_{n+1}}{a_n} \leq \lambda$$

for $n \geq N$. In particular, for any $k > 0$,

$$a_{N+p+1} \leq \lambda a_{N+p} \quad (p = 0, 1, \ldots, k-1).$$

Thus,

$$a_{N+1} \leq \lambda a_N$$

$$a_{N+2} \leq \lambda a_{N+1}$$

$$\vdots$$

$$a_{N+p} \leq \lambda a_{N+p-1}$$

Multiplying these inequalities yields

$$(a_{N+1})(a_{N+2}) \cdots (a_{N+k}) \leq \lambda^k (a_N)(a_{N+1}) \cdots (a_{N+k-1})$$

which is equivalent to

$$a_{N+k} \leq \lambda^k a_N$$

or

$$a_n \leq a_N \lambda^{-N} \cdot \lambda^n \quad (n \geq N)$$

(to see this, let $k = n - N$).

Hence,

$$\sqrt[n]{a_n} \leq \sqrt[n]{a_N \lambda^{-N}} \cdot \lambda.$$

However, since

$$\lim_{n\to\infty} \sqrt[n]{p} = 1 \quad \text{if } p > 0, \text{ we have}$$

$$\lim_{n\to\infty} \text{Sup } \sqrt[n]{a_n} \leq \lambda \tag{1}$$

(because

$$\lim_{n\to\infty} \sqrt[n]{a_N \lambda^{-N}} = 1).$$

Since this is true for all $\lambda > \alpha$, we have

$$\lim_{n \to \infty} \text{Sup } \sqrt[n]{a_n} \leq \alpha = \lim_{n \to \infty} \text{Sup } \frac{a_{n+1}}{a_n}.$$

The inequality

$$\lim_{n \to \infty} \text{Sup } \sqrt[n]{a_n} \leq \alpha$$

can be shown to be valid as follows:

Assume that

$$\lim_{n \to \infty} \text{Sup } \sqrt[n]{a_n} > \alpha.$$

Then

$$\lim_{n \to \infty} \text{Sup } \sqrt[n]{a_n} = \alpha + \varepsilon,$$

where $\varepsilon > 0$. Now, since the inequality (1) must be true for all $\lambda > \alpha$, we can let $\lambda = \alpha + \beta$ where $\beta > 0$ and where $\beta < \varepsilon$. Then

$$\lim_{n \to \infty} \text{Sup } \sqrt[n]{a_n} = \alpha + \varepsilon > \alpha + \beta. \text{ So}$$

$$\lim_{n \to \infty} \text{Sup } \sqrt[n]{a_n} > \lambda.$$

However, this contradicts our previous result that

$$\lim_{n \to \infty} \text{Sup } \sqrt[n]{a_n} \leq \lambda,$$

Therefore,

$$\lim_{n \to \infty} \text{Sup } \sqrt[n]{a_n} \leq \alpha.$$

b) Here, let

$$\alpha = \lim_{n \to \infty} \inf \frac{a_{n+1}}{a_n}$$

For $\alpha = -\infty$, there is nothing to prove. For α finite, choose, $\lambda < \alpha$. There exists an integer N such that

$$\frac{a_{n+1}}{a_n} \geq \lambda$$

for $n \geq N$. Hence, for any $k > 0$,

$$a_{N+p+1} \geq \lambda a_{N+p} \qquad (p = 0, 1, \ldots, k-1).$$

Multiplying these inequalities yields

$$a_{N+k} \geq \lambda^k a_N$$

or

$$a_n \geq a_N \lambda^{-N} \cdot \lambda^n \quad (n \geq N).$$

Consequently,

$$\sqrt[n]{a_n} \geq \sqrt[n]{a_N \lambda^{-N}} \cdot \lambda$$

so that

$$\lim_{n \to \infty} \inf \sqrt[n]{a_n} \geq \lambda$$

$$\lim_{n \to \infty} \sqrt[n]{a_N \lambda^{-N}} = 1.$$

Since this inequality is true for every $\lambda < \alpha$, we have

$$\lim_{n \to \infty} \inf \sqrt[n]{a_n} \geq \alpha = \lim_{n \to \infty} \inf \frac{a_{n+1}}{a_n}$$

● **PROBLEM 11-15**

Show that

$$\lim_{n \to \infty} \frac{n}{(n!)^{1/n}} = e.$$

Solution: Since we know that:

(i)
$$\lim_{n \to \infty} \inf \frac{a_{n+1}}{a_n} \leq \lim_{n \to \infty} \inf a_n^{1/n}$$

(ii)
$$\lim_{n \to \infty} \sup a_n^{1/n} \leq \lim_{n \to \infty} \sup \frac{a_{n+1}}{a_n}$$

(iii)
$$\lim_{n \to \infty} \inf \frac{a_{n+1}}{a_n} = \lim_{n \to \infty} \sup \frac{a_{n+1}}{a_n}$$

is finite if

$$\lim_{n \to \infty} \frac{a_{n+1}}{a_n}$$

exists, we can see that given these conditions,

$$\lim_{n \to \infty} a_n^{1/n}$$

also exists and this limit equals

$$\lim_{n \to \infty} \frac{a_{n+1}}{a_n}.$$

For the given limit, take $a_n = \frac{n^n}{n!}$ so that

$$\lim_{n \to \infty} \frac{n}{(n!)^{1/n}} = \lim_{n \to \infty} a_n^{1/n}. \qquad (1)$$

Now,

$$\lim_{n \to \infty} \frac{a_{n+1}}{a_n} = \lim_{n \to \infty} \frac{(n+1)^{n+1}}{(n+1)!} \cdot \frac{n!}{n^n} = \lim_{n \to \infty} \frac{(n+1)^{n+1}}{(n+1)n^n}$$

$$\lim_{n \to \infty} \left(\frac{n+1}{n}\right)^n = \lim_{n \to \infty} \left(1 + \frac{1}{n}\right)^n = e.$$

Therefore,

$$\lim_{n \to \infty} a_n^{1/n} = e \text{ and by (1)}$$

$$\lim_{n \to \infty} \frac{n}{(n!)^{1/n}} = e.$$

• **PROBLEM 11-16**

Given that

$$e = \sum_{n=0}^{\infty} \frac{1}{n!}$$

prove that

$$\lim_{n \to \infty} \left(1 + \frac{1}{n}\right)^n = e.$$

<u>Solution</u>: Let

$$S_n = \sum_{k=0}^{n} \frac{1}{k!}$$

and

$$t_n = \left(1 + \frac{1}{n}\right)^n.$$

By the binomial theorem $\left((a+b)^m = \sum_{r=0}^{m} \binom{m}{r} a^{m-r} b^r\right.$, m a positive integer$\left.\right)$

$$t_n = 1 + 1 + \frac{1}{2!}\left(1 - \frac{1}{n}\right) + \frac{1}{3!}\left(1 - \frac{1}{n}\right)\left(1 - \frac{2}{n}\right) + \dots$$

$$+ \frac{1}{n!}\left(1 - \frac{1}{n}\right)\left(1 - \frac{2}{n}\right) \cdots \left(1 - \frac{n-1}{n}\right).$$

Since
$$S_n = 1 + 1 + \frac{1}{2!} + \frac{1}{3!} + \cdots + \frac{1}{n!},$$

we have $t_n \leq S_n$. Now by a theorem, if

$$t_n \leq S_n \quad \text{for} \quad n \geq N,$$

(where N is fixed), then

$$\lim_{n \to \infty} \sup t_n \leq \lim_{n \to \infty} \sup S_n .$$

Therefore

$$\lim_{n \to \infty} \sup t_n \leq \lim_{n \to \infty} \sup S_n = e . \qquad (1)$$

Now if $n \geq m$,

$$t_n \geq 1 + 1 + \frac{1}{2!}\left(1 - \frac{1}{n}\right) + \cdots + \frac{1}{m!}\left(1 - \frac{1}{n}\right) \cdots \left(1 - \frac{m-1}{n}\right).$$

Keeping m fixed, let $n \to \infty$ to yield

$$\lim_{n \to \infty} \inf t_n \geq 1 + 1 + \frac{1}{2!} + \cdots + \frac{1}{m!},$$

so that
$$S_m \leq \lim_{n \to \infty} \inf t_n .$$

Letting $m \to \infty$ then gives

$$e \leq \lim_{n \to \infty} \inf t_n . \qquad (2)$$

Hence, from (1) and (2) we have

$$e \leq \lim_{n \to \infty} \inf t_n \leq \lim_{n \to \infty} \sup t_n \leq e .$$

Thus
$$\lim_{n \to \infty} \left(1 + \frac{1}{n}\right)^n = e .$$

● **PROBLEM 11-17**

Find the upper and lower limits of the sequence $\{A_n\}$ defined by

$$A_1 = 0; \quad A_{2m} = \frac{A_{2m-1}}{2}; \quad A_{2m+1} = \frac{1}{2} + A_{2m} .$$

Solution: We write the first few terms of the sequence;

$A_1 = 0$; $A_2 = \dfrac{A_1}{2} = 0$; $A_3 = \dfrac{1}{2} + A_2 = \dfrac{1}{2}$; $A_4 = \dfrac{A_3}{2} = \dfrac{1}{4}$;

$A_5 = \dfrac{1}{2} + \dfrac{1}{4}$; $A_6 = \dfrac{1}{4} + \dfrac{1}{8}$; $A_7 = \dfrac{1}{2} + \dfrac{1}{4} + \dfrac{1}{8}$; etc. ...

which can be rewritten,

$A_m = 0, 0, \dfrac{1}{2}, \dfrac{1}{4}, \dfrac{3}{4}, \dfrac{3}{8}, \dfrac{7}{8}, \dfrac{7}{16}, \dfrac{15}{16}, \dfrac{15}{32}, \dfrac{31}{32}$, etc. ...

for m = 1, 2, 3, ... , respectively.

From which we see that

$$A_{2m} = \dfrac{1}{2} - \dfrac{1}{2^m} \quad \text{and} \quad A_{2m+1} = 1 - \dfrac{1}{2^m}.$$

Now, let E be the set of subsequential limits of $\{A_m\}$. That is, E is the set that contains the limits of convergent subsequences of $\{A_m\}$. Since

$$\lim_{m \to \infty} A_{2m} = \lim_{m \to \infty} \left(\dfrac{1}{2} - \dfrac{1}{2^m} \right) = \dfrac{1}{2}$$

and

$$\lim_{m \to \infty} A_{2m+1} = \lim_{m \to \infty} \left(1 - \dfrac{1}{2^m} \right) = 1$$

we have

$$E = \left\{ \dfrac{1}{2}, 1 \right\}.$$

Another way to find the set E is to notice that

$$\{A_{2m}\} = A_4, A_6, A_8, A_{10}, \ldots$$

and

$$\lim_{m \to \infty} A_{2m} = \dfrac{1}{2^2} + \dfrac{1}{2^3} + \dfrac{1}{2^4} + \ldots = \dfrac{1}{2^2}\left[1 + \dfrac{1}{2} + \dfrac{1}{2^2} + \ldots \right]$$

$$= \dfrac{1}{2^2} \cdot 2 = \dfrac{1}{2}$$

by use of the geometric series

$$\sum_{n=0}^{\infty} x^n = \dfrac{1}{1-x}.$$

Similarly, for $\{A_{2m+1}\}$,

$$\lim_{m \to \infty} A_{2m+1} = \frac{1}{2} + \frac{1}{2^2} + \frac{1}{2^3} + \frac{1}{2^4} + \ldots =$$

$$\frac{1}{2}\left[1 + \frac{1}{2} + \frac{1}{2^2} + \ldots\right] = \frac{1}{2} \cdot 2 = 1.$$

Therefore,

$$\lim \text{Sup } A_n = \text{Sup } E = \text{Sup}\left\{\frac{1}{2}, 1\right\} = 1$$

and

$$\lim \inf A_n = \inf E = \inf\left\{\frac{1}{2}, 1\right\} = \frac{1}{2}$$

Thus the upper limit is 1; the lower limit is $\frac{1}{2}$.

SEQUENCE OF FUNCTIONS

● **PROBLEM 11-18**

Find

$$\lim_{n \to \infty} f_n(x) \quad \text{where}$$

a)
$$f_n(x) = \frac{x}{1 + nx^2} \quad \text{for } -1 \leq x \leq 1.$$

Also find,

$$\lim_{n \to \infty} f_n'(x).$$

b)
$$f_n(x) = x^n \quad \text{for } 0 \leq x \leq 1.$$

<u>Solution</u>: For this problem we start with a definition.

Suppose $\{f_n\}$, $n = 1, 2, 3, \ldots$, is a sequence of functions defined on a set E, and suppose that the sequence of numbers $\{f_n(x)\}$ converges for every $x \in E$. Then a function f can be defined by the following:

$$f(x) = \lim_{n \to \infty} f_n(x) \quad (x \in E).$$

Given this, we say that $\{f_n\}$ converges on E and that f is the limit, or the limit function of $\{f_n\}$.

Though the process of finding the limit of a sequence of functions is similar to that of finding the limit of a sequence, an important problem still arises: If the functions f_n are continuous, differentiable, or integrable, will the same be true of the limit function f? In addition, is it always true that

$$f'(x) = \lim_{n \to \infty} f_n'(x)$$

or that

$$\int_a^b f(x)\,dx = \lim_{n \to \infty} \int_a^b f_n(x)\,dx \text{ ?}$$

Furthermore, f is continuous at x means that

$$\lim_{t \to x} f(t) = f(x).$$

However, to say that the limit of a sequence of continuous functions is continuous means that

$$\lim_{t \to x} \lim_{n \to \infty} f_n(t) = \lim_{n \to \infty} \lim_{t \to x} f_n(t)$$

a)
$$f_n(x) = \frac{x}{1 + nx^2} \quad \text{for } -1 \leq x \leq 1.$$

Here,

$$\lim_{n \to \infty} f_n(x) = \lim_{n \to \infty} \frac{x}{1 + nx^2} = 0$$

for all x in [-1, +1]. That is, for $-1 \leq x \leq 1$, $\lim_{n \to \infty} f_n(x) = f(x) = 0$. Observe that

$$f_n'(x) = \frac{1-nx^2}{(1+nx^2)^2} \quad \text{and for}$$

$x = 0$, $f_n'(x) = 1$ so that

$$\lim_{n \to \infty} f_n'(0) = 1.$$

For $x \neq 0$,

$$\lim_{n \to \infty} f_n'(x) = \lim_{n \to \infty} \frac{1-nx^2}{(1+nx^2)^2} = \lim_{n \to \infty} \frac{1}{(1+nx^2)^2} - \lim_{n \to \infty} \frac{nx^2}{(1+nx^2)^2}$$

Now, immediately we see that

$$\lim_{n \to \infty} \frac{1}{(1+nx^2)^2} = 0.$$

However,

$$\lim_{n\to\infty} \frac{nx^2}{(1+nx^2)^2} = \lim_{n\to\infty} \frac{nx^2}{1+n^2x^4+2nx^2} = \lim_{n\to\infty} \frac{x^2}{1/n + nx^4 + 2x^2} = 0.$$

Therefore,

$$\lim_{n\to\infty} f_n'(x) = \begin{cases} 1 & \text{if } x = 0 \\ 0 & \text{if } x \neq 0 \end{cases}.$$

Consequently, since $f(x) = 0$, we have

$$f'(0) = 0 \neq \lim_{n\to\infty} f_n'(0) = 1.$$

b) $f_n(x) = x^n$ for $0 \leq x \leq 1$. We want to find

$$\lim_{n\to\infty} f_n(x) = \lim_{n\to\infty} x^n \quad \text{for} \quad 0 \leq x \leq 1.$$

Now, for $x = 1$,

$$\lim_{n\to\infty} f_n(1) = \lim_{n\to\infty} 1^n = 1.$$

For $0 \leq x < 1$,

$$\lim_{n\to\infty} x^n = 0. \quad \text{Thus}$$

$$\lim_{n\to\infty} f_n(x) = f(x) = \begin{cases} 1 & x = 1 \\ 0 & 0 \leq x < 1 \end{cases}.$$

● **PROBLEM 11-19**

For $m = 1, 2, 3, \ldots$ let

$$f_m(x) = \lim_{n\to\infty} (\cos m!\, \pi x)^{2n}.$$

Find $\lim_{m\to\infty} f_m(x)$.

<u>Solution</u>: Let

$$f(x) = \lim_{m\to\infty} f_m(x) = \lim_{m\to\infty} \lim_{n\to\infty} (\cos m!\, \pi x)^{2n}$$

so that the problem is to find $f(x)$ (i.e., the limit function of $\{f_m\}$). Now, when $m!x$ is an integer,

$$\cos m!\, \pi x = \begin{cases} 1 & m!x = 0 \\ 1 & m!x \text{ even} \\ -1 & m!x \text{ odd} \end{cases}.$$

540

Hence, $(\cos m!\pi x)^{2n} = 1$ for all $n \in Z$ and $f_m(x) = 1$.

For all other values of x, $|\cos m!\pi x| < 1$ so that

$$f_m(x) = \lim_{n \to \infty} (\cos m!\pi x)^{2n} = 0.$$

This is equivalent to the following:

$$f_m(x) = \begin{cases} 1 & m!x \text{ integer} \\ 0 & m!x \text{ noninteger} \end{cases}.$$

Then for irrational x, $f_m(x) = 0$ for every m, and consequently

$$f(x) = \lim_{m \to \infty} f_m(x) = 0.$$

For x rational, (that is, let $x = s/t$, where s and t are integers), we have

$$m!x = \frac{m!s}{t}.$$

Therefore, $m!x$ is an integer if $m \geq t$. (To see this let t be some integer that is equal to $m-r$, where r is some positive integer. Then

$$\frac{m!s}{t} = \frac{m(m-1)(m-2) \ldots (m-r) \ldots 1 \cdot s}{(m-r)}$$

$$= m(m-1) \ldots (m-(r-1))(m-(r+1)) \ldots 1 \cdot s$$

which is an integer value). Therefore,

$$f(x) = \lim_{m \to \infty} 1 = 1.$$

Hence,

$$f(x) = \lim_{m \to \infty} \lim_{n \to \infty} (\cos m!\pi x)^{2n} = \begin{cases} 0 & (x \text{ irrational}). \\ 1 & (x \text{ rational}) \end{cases}$$

We note that this function $f(x)$ is an everywhere discontinuous function which is not Riemann-integrable.

● **PROBLEM 11-20**

a) Let

$$f_n(x) = n^2 x (1-x^2)^n \quad (0 \leq x \leq 1, \ n = 1, 2, 3, \ldots).$$

Show that

$$\lim_{n \to \infty} \int_0^1 f_n(x) \, dx \neq \int_0^1 \left[\lim_{n \to \infty} f_n(x) \right] dx.$$

b) Repeat for $f_n(x) = nx(1-x^2)^n$.

Solution: a) For $x = 0$, $f_n(0) = 0$ and for $0 < x \leq 1$, we have

$$\lim_{n \to \infty} f_n(x) = \lim_{n \to \infty} n^2 x(1-x^2)^n = 0.$$

This by the theorem that if $p > 1$, α real, then

$$\lim_{n \to \infty} \frac{n^\alpha}{p^n} = 0$$

and by the fact that since $x > 0$, we have $1 - x^2 < 1$. Therefore,

$$\lim_{n \to \infty} f_n(x) = 0 \qquad (0 \leq x \leq 1).$$

Now,

$$\int_0^1 x(1-x^2)^n \, dx = \frac{1}{2} \int_0^1 u^n \, du \qquad \text{(where } u = 1 - x^2\text{)}$$

$$= \frac{1}{2n+2}.$$

Hence,

$$\lim_{n \to \infty} \int_0^1 f_n(x) \, dx = \lim_{n \to \infty} \int_0^1 n^2 x(1-x^2)^n \, dx = \lim_{n \to \infty} \frac{n^2}{2n+2} = +\infty.$$

However,

$$\int_0^1 \left[\lim_{n \to \infty} f_n(x) \right] dx = \int_0^1 0 \, dx = 0.$$

Thus

$$\lim_{n \to \infty} \int_0^1 n^2 x(1-x^2)^n \, dx \neq \int_0^1 \left[\lim_{n \to \infty} n^2 x(1-x^2)^n \right] dx,$$

$$(0 \leq x \leq 1).$$

b)
$$f_n(x) = nx(1-x^2)^n \qquad (0 \leq x \leq 1, \; n = 1, 2, 3, \ldots):$$

Here again $f_n(0) = 0$ and for $0 < x \leq 1$ we have

$$\lim_{n \to \infty} f_n(x) = 0.$$

Hence,

$$\lim_{n \to \infty} f_n(x) = 0 \qquad (0 \leq x \leq 1)$$

as before. Now,

$$\int_0^1 f_n(x)\, dx = \int_0^1 nx(1-x^2)^n\, dx = \frac{n}{2n+2}$$

so that

$$\lim_{n\to\infty} \int_0^1 f_n(x)\, dx = \lim_{n\to\infty} \frac{n}{2n+2} = \frac{1}{2}$$

whereas

$$\int_0^1 \left[\lim_{n\to\infty} f_n(x)\right] dx = 0 .$$

Thus these two problems show that the limit of the integral need not be equal to the integral of the limit, even if both have finite values as in part (b).

● **PROBLEM 11-21**

Prove:

a) The sequence of functions $\{f_n\}$, defined on E, converges uniformly on E if and only if for every $\varepsilon > 0$ there exists an integer N such that $m \geq N$, $n \geq N$, $x \in E$ implies

$$|f_n(x) - f_m(x)| \leq \varepsilon .\qquad (1)$$

b) Suppose $\{f_n\}$ is a sequence of functions defined on E, and suppose

$$M_n = \sup_{x \in E} |f_n(x)|$$

is a finite number for all n.

Then

$$\sum_{n=1}^{\infty} f_n$$

converges uniformly on E if ΣM_n converges.

Solution: By definition a sequence of functions $\{f_n\}$, $n = 1, 2, 3, \ldots$ converges uniformly on E to a function f if for every $\varepsilon > 0$ there is an integer N such that $n \geq N$ implies

$$|f_n(x) - f(x)| \leq \varepsilon$$

for all $x \in E$.

a) Part 1:

Suppose $\{f_n\}$ converges uniformly on E. Prove that

$$\forall \varepsilon > 0, \exists N$$

such that $m \geq N$, $n \geq N$, $x \in E$ implies

$$|f_n(x) - f_m(x)| \leq \varepsilon .$$

Let f be the limit function. Then there is an integer N such that $n \geq N$, $x \in E$ implies

$$|f_n(x) - f(x)| \leq \varepsilon/2 .$$

Hence

$$|f_n(x) - f_m(x)| \leq |f_n(x) - f(x)| + |f(x) - f_m(x)| \leq \varepsilon$$

if $n \geq N$, $m \geq N$, $x \in E$.

Part 2:

Suppose the Cauchy condition holds. Prove that $\{f_n\}$ converges uniformly on E.

Since the sequence $\{f_n(x)\}$ is Cauchy we know it converges for every x to a limit $f(x)$. Therefore, the sequence $\{f_n\}$ converges on E to f. To prove that the convergence is uniform, let $m \to \infty$, while n is fixed in (1). Since, by convergence, $f_m(x) \to f(x)$ as $m \to \infty$, this yields

$$|f_n(x) - f(x)| \leq \varepsilon$$

for every $n \geq N$ and every $x \in E$. Hence, $\{f_n\}$ converges uniformly on E.

Thus, by part 1 and part 2, the proof is complete.

b) First note that f_n converges uniformly to 0 on E if and only if $M_n \to 0$ as $n \to \infty$. If ΣM_n converges, then, for arbitrary $\varepsilon > 0$,

$$\left| \sum_{i=n}^{m} f_i(x) \right| \leq \sum_{i=n}^{m} M_i \leq \varepsilon \quad (x \in E)$$

if m and n are large enough. That is (for $m > n$)

$$\left|\sum_{i=1}^{m} f_i(x) - \sum_{j=1}^{n-1} f_j(x)\right| \leq \varepsilon.$$

Hence, uniform convergence follows from theorem (a) of this problem.

● **PROBLEM 11-22**

Prove the following theorems:

a) Let the functions $\{f_n(x)\}$ be defined on the interval $a \leq x \leq b$ and let $f_n \to f$ uniformly on this interval. Then, if each of the functions f_n is continuous at a point x_0, the limit function f is also continuous at x_0. In addition, if each f_n is continuous on the entire interval, so is f.

b) Suppose that the functions $f_n(x)$ are continuous on the closed interval $a \leq x \leq b$, and suppose that they converge uniformly on this interval to the limit function $f(x)$. Then

$$\int_a^b f(x)\,dx = \lim_{n \to \infty} \int_a^b f_n(x)\,dx.$$

c) Let the functions $f_n(x)$ be defined and have cont. derivatives on the interval $a \leq x \leq b$. If the sequence $\{f_n'(x)\}$ is uniformly convergent on the interval, and the sequence $\{f_n(x)\}$ is convergent, with limit function $f(x)$, then f is differentiable and

$$f'(x) = \lim_{n \to \infty} f_n'(x).$$

Solution: a) Since

$$f(x) - f(x_0) = \left(f(x) - f_n(x)\right) + \left(f_n(x) - f_n(x_0)\right)$$
$$+ \left(f_n(x_0) - f(x_0)\right),$$

we have

$$|f(x) - f(x_0)| \leq |f(x) - f_n(x)| + |f_n(x) - f_n(x_0)|$$
$$+ |f_n(x_0) - f(x_0)|. \qquad (1)$$

Now f is continuous at x_0, if given $\varepsilon > 0$ and x_0 is fixed, then

$$|f(x) - f(x_0)| < \varepsilon$$

provided x is sufficiently close to x_0. However, by uniform convergence, n can be chosen independent of x so large that

$$|f_n(x) - f(x)| < \frac{\varepsilon}{3}$$

for every x in [a,b]. Then by (1) we have

$$|f(x) - f(x_0)| < |f_n(x) - f_n(x_0)| + \frac{2}{3}\varepsilon . \quad (2)$$

Now n has been fixed and since f_n is continuous at x_0, this yields

$$|f_n(x) - f_n(x_0)| < \frac{\varepsilon}{3}$$

if x is sufficiently close to x_0. Then by (2) we have

$$|f(x) - f(x_0)| < \varepsilon$$

which shows that f is continuous at x_0.

b) To prove this theorem, since

$$\int_a^b f_n(x)\,dx - \int_a^b f(x)\,dx = \int_a^b [f_n(x) - f(x)]\,dx ,$$

we have

$$\left| \int_a^b f_n(x)\,dx - \int_a^b f(x)\,dx \right| \leq \int_a^b |f_n(x) - f(x)|\,dx . \quad (3)$$

Suppose $\varepsilon > 0$ is now given. Then because the f_n's converge uniformly on [a,b] we can choose N, independent of x, so large that for $n \geq N$ we have

$$|f_n(x) - f(x)| < \frac{\varepsilon}{b-a} \quad \text{if } a \leq x \leq b . \quad (4)$$

Now from (4), if $n \geq N$

$$\int_a^b |f_n(x) - f(x)|\,dx < \int_a^b \frac{\varepsilon}{b-a}\,dx = \varepsilon .$$

Therefore, (3) yields,

$$\left| \int_a^b f_n(x)\,dx - \int_a^b f(x)\,dx \right| < \varepsilon \quad \text{if } n \geq N . \quad (5)$$

However, (5) is just the condition so that,

$$\int_a^b f(x)\,dx = \lim_{n\to\infty} \int_a^b f_n(x)\,dx$$

is true. This completes the proof.

c) Denote the limit of the sequence $\{f_n'(x)\}$ by $g(x)$. Then we wish to prove that $g(x) = f'(x)$.

Since $f_n'(x)$ converges uniformly to $g(x)$, by theorem (b) of this problem we have

$$\int_a^x g(t)\,dt = \lim_{n\to\infty} \int_a^x f_n'(t)\,dt, \qquad (6)$$

since the convergence is also uniform on any subinterval $a \leq t \leq x$, where $a \leq x \leq b$. Since

$$\lim_{n\to\infty} \int_a^x f_n'(t)\,dt = \lim_{n\to\infty} [f_n(x) - f(a)] = f(x) - f(a),$$

(6) becomes,

$$\int_a^x g(t)\,dt = f(x) - f(a)$$

or

$$f(x) = \int_a^x g(t)\,dt + f(a). \qquad (7)$$

Now g is continuous by theorem (a) of this problem.

Then by the fundamental theorem of elementary calculus (i.e., given $g(t)$ is continuous, $a \leq t \leq b$ and that

$$f(x) = \int_a^x g(t)\,dt,$$

$a \leq x \leq b$, then $f(x)$ is differentiable and

$f'(x) = g(x)$),

we have

$$f'(x) = g(x) = \lim_{n\to\infty} f_n'(x),$$

which completes the proof.

● PROBLEM 11-23

a) Show that $f_n(x) = \frac{x^n}{n}$ converges uniformly on the interval $0 \leq x \leq 1$. Then show that

$$\left(\text{for } f_n(x) = \frac{x^n}{n}\right),$$

$$\lim_{n \to \infty} \int_0^1 f_n(x)\,dx = \int_0^1 \lim_{n \to \infty} f_n(x)\,dx$$

and that for $0 \leq x < 1$

$$\lim_{n \to \infty} f'_n(x) = f'(x).$$

b) Show that

$$f_n(x) = \frac{x}{2(n+1)}$$

converges uniformly on the interval $0 \leq x \leq 1$.

c) Show that $f_n(x) = x^n$ does not converge uniformly on the interval $0 < x < 1$.

Solution: a) For $0 \leq x \leq 1$,

$$\lim_{n \to \infty} f_n(x) = \lim_{n \to \infty} \frac{x^n}{n} = 0,$$

so that $f_n(x)$ converges to 0. We wish to show that this convergence is uniform on this interval. That is, for each $\varepsilon > 0$ there corresponds some integer N such that for every x in the interval,

$$|f_n(x) - f(x)| < \varepsilon$$

if $n \geq N$. Of importance is that N is to be independent of x for convergence to be uniform.

Here,

$$|f_n(x) - f(x)| = \left|\frac{x^n}{n}\right| \leq \left|\frac{1}{n}\right|$$

since $x^n \leq 1$.

Then

$$\left|\frac{x^n}{n}\right| < \varepsilon \quad \text{if} \quad \frac{1}{\varepsilon} < n.$$

Therefore, N can be chosen as the smallest integer

548

greater than $\frac{1}{\varepsilon}$. Note that this choice is independent of x. Thus $f_n(x)$ converges uniformly to 0 in this interval.

Now

$$\lim_{n\to\infty} \int_0^1 f_n(x)\,dx = \lim_{n\to\infty} \int_0^1 \frac{x^n}{n}\,dx = \lim_{n\to\infty} \frac{1}{n(n+1)} = 0.$$

Hence

$$\lim_{n\to\infty} \int_0^1 f_n(x)\,dx = \int_0^1 \lim_{n\to\infty} f_n(x)\,dx = \int_0^1 0\,dx = 0.$$

In addition,

$$\lim_{n\to\infty} f_n'(x) = \lim_{n\to\infty} x^{n-1} = 0 \quad \text{since } 0 \leq x < 1.$$

Therefore,

$$\lim_{n\to\infty} f_n'(x) = f'(x) = 0.$$

b) We have that

$$\lim_{n\to\infty} f_n(x) = \lim_{n\to\infty} \frac{x}{2(n+1)} = 0.$$

Therefore, since $0 \leq x \leq 1$,

$$|f_n(x) - f(x)| = \left|\frac{x}{2n+2}\right| \leq \left|\frac{1}{2n+2}\right| < \left|\frac{1}{n}\right| < \varepsilon$$

if $n > \frac{1}{\varepsilon}$.

Hence, choose N as the smallest integer greater than $\frac{1}{\varepsilon}$. Again this choice is independent of x.

c) Here,

$$\lim_{n\to\infty} f_n(x) = \lim_{n\to\infty} x^n = 0 \quad \text{if } 0 \leq x < 1$$

and

$$\lim_{n\to\infty} x^n = 1 \quad \text{if } x = 1.$$

Suppose that $0 < \varepsilon < 1$ and $0 < x < 1$. Then

$$|f_n(x) - f(x)| < \varepsilon$$

is the same as $x^n < \varepsilon$.

Therefore,

$$n \log x < \log \varepsilon$$

which can be rewritten

$$\log\left(\frac{1}{\varepsilon}\right) < n \log\left(\frac{1}{x}\right)$$

since $\log x < 0$ and $\log\left(\frac{1}{x}\right) > 0$ if $0 < x < 1$.

Then

$$\frac{\log\left(\frac{1}{\varepsilon}\right)}{\log\left(\frac{1}{x}\right)} < n .$$

Now to have this true for all $n \geq N$ it is necessary to choose the integer N large enough so that

$$N > \frac{\log\left(\frac{1}{\varepsilon}\right)}{\log\left(\frac{1}{x}\right)} .$$

Here, we see that N depends both on ε and x. That is, as $\varepsilon \to 0$,

$$\log\left(\frac{1}{\varepsilon}\right) \to +\infty \quad \text{so that } N \to \infty .$$

If ε is remained fixed, as $x \to 1^-$, $\log\left(\frac{1}{x}\right) \to 0^+$ and consequently $N \to \infty$. Thus, there is no value of N such that the given inequality holds simultaneously for all values of x in the interval $0 < x < 1$.

● **PROBLEM 11-24**

Show that

$$f_n(x) = \frac{nx}{1 + n^2 x^2}$$

converges uniformly in any closed interval which does not include $x = 0$, but does not converge uniformly in any interval having $x = 0$ in its interior or at one end.

Solution: For this problem,

$$\lim_{n \to \infty} f_n(x) = \lim_{n \to \infty} \frac{nx}{1 + n^2 x^2} = 0$$

for all values of x.

Note that f_n is an odd function. That is $f_n(-x) = -f_n(x)$, so that the graph of $y = f_n(x)$ is symmetric with respect

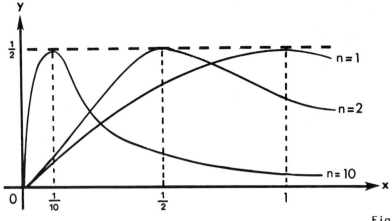

Fig. 1

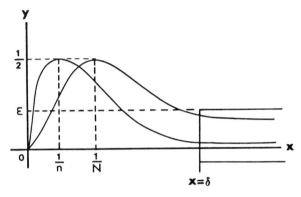

Fig. 2

to the origin. Therefore, we restrict our attention to values of x for which $x \geq 0$. To graph the function use information that is derived from the derivative.

Now,
$$f_n'(x) = \frac{(1 + n^2x^2)n - nx(2n^2x)}{(1 + n^2x^2)^2} = \frac{n(1 - n^2x^2)}{(1 + n^2x^2)^2}.$$

Therefore, for $x > 0$, the graph of $y = f_n(x)$ has a maximum at $x = \frac{1}{n}$ (this from setting $f_n'(x)$ equal to 0 and solving for x), and then diminishes toward zero as $x \to +\infty$. Consequently,

$$f_n\left(\frac{1}{n}\right) = \frac{\left(\frac{n}{n}\right)}{1 + \left(\frac{n^2}{n^2}\right)} = \frac{1}{2}$$

is the maximum value. Fig. 1 shows the graph of $f_n(x)$ for $n = 1, 2, 10$.

From Fig. 1 we now see that the convergence is nonuniform in any interval containing the point $x = 0$, since any such interval will contain the point

$$x = \frac{1}{n}$$

if n is large enough, and so we can never have

$$|f_n(x)| < \varepsilon$$

for all values of x in this interval, no matter how large we take n if $\varepsilon < \frac{1}{2}$.

Now we look at closed intervals that do not include x = 0. That is, we look at all values of x such that $x \geq \delta$, where δ is any fixed positive number. For, $\varepsilon > 0$ being given, if we choose N so large that

$$\frac{1}{n} < \delta \quad \text{and} \quad \frac{n\delta}{1 + n^2\delta^2} < \varepsilon \quad \text{if } N \leq n,$$

then we also have that

$$\frac{nx}{1 + n^2 x^2} < \varepsilon$$

if $N \leq n$ and $x \geq \delta$. Consequently, the convergence is uniform for these latter values of x. (See Fig. 2).

CHAPTER 12

INFINITE SERIES

Since infinite series are widely utilized in many branches of mathematics, an understanding of convergence (or divergence) of these series and some knowledge of the rules governing their use is necessary. By definition, suppose that $a_0, a_1, a_2, \ldots, a_n, \ldots$ is an infinite sequence, then an infinite series is an indicated sum of the form:

$$a_0 + a_1 + a_2 + \ldots + a_n + \ldots \text{ or } \sum_{n=0}^{\infty} a_n.$$

The numbers $a_0, a_1, \ldots$ are called the terms of the series, with a_n defined as the general term.

For an infinite series to be of practical use, it must be convergent. That is, given that S_n is the partial sum

$$S_n = a_0 + a_1 + \ldots + a_n,$$

if

$$\lim_{n \to \infty} S_n = S,$$

then the series

$$\sum_{k=0}^{\infty} a_k$$

converges to S. A series that is not convergent is termed divergent and has no sum assigned to it. Certain tests, such as the ratio test, root test, and integral test are used to determine convergence (or divergence) of some infinite series. If a series has terms which are alternately positive and negative it is called an alternating series, to which the alternating series test is applicable. This test determines convergence (or divergence).

The idea of absolute convergence applies to series that are convergent regardless of the signs assigned to the variable (i.e., the series $\sum a_n$ is absolutely convergent if the series $\sum |a_n|$ is convergent).

Infinite series can also be formed by the sum of an infinite sequence of functions. Here, it is necessary to determine if a series is uniformly convergent. That is, let $\{f_n\}$ be a sequence of functions and let f be the limit function of $\{f_n\}$. Then $\{f_n\}$ converges uniformly to f, if, given $\varepsilon > 0$, there exists an N_0 such that

$$\left| f_n(x) - f(x) \right| < \varepsilon$$

for $n \geq N_0$ and all x in the domain. The most important test to determine uniform convergence is the Weierstrass M-test.

Furthermore, for an infinite series of numbers, operations such as addition are only appropriate for convergent series. The reason for this being, if a series was not convergent but divergent, as extra terms are constantly added a continually different result would be obtained. However, for multiplication and rearrangement of terms to be suitable, it is necessary for the series to be absolutely convergent. In addition, for an infinite series of functions, term by term differentiation and term by term integration can only be applied to uniformly convergent series.

Finally, in computations with series, estimates of error and estimates of sums of convergent series can be determined by various methods.

Note: for divergent series these computations are not valid since the series does not approach a limit. However, a process known as Cesàro summability attaches a sum σ_n (which is the average of the first n partial sums) to certain divergent series.

TESTS FOR CONVERGENCE AND DIVERGENCE

• **PROBLEM 12-1**

Show that the series $\sum_{k=1}^{\infty} \frac{1}{k^p}$ is convergent for $p > 1$ and is divergent for $p \leq 1$.

Solution: To determine if the series is convergent or divergent the following theorem (called the integral test) is used: Let f(x) be a function which is positive, continuous, and nonincreasing as x increases for all values of $x \geq N$, where N is some fixed positive integer. Let the terms of an infinite series be given by $u_n = f(n)$ when $n \geq N$. If

$$\lim_{R \to \infty} \int_N^R f(x)\,dx < \infty \quad (=\infty),$$

then

$$\sum_{n=N}^{\infty} u_n < \infty \quad (=\infty).$$

554

The proof of this theorem starts with the given relationship:

$$f(k+1) \le f(x) \le f(k), \quad k \ge N, \quad k < x \le k+1$$

(since $f(x)$ is nonincreasing).

Integrating $f(x)$ from k to $k+1$ preserves the inequalities:

$$f(k+1) \le \int_{k}^{k+1} f(x)\,dx \le f(k)$$

where

$$k = N, N+1, \ldots, n.$$

Adding these inequalities gives the result

$$\sum_{k=N+1}^{n+1} f(k) \le \int_{N}^{n+1} f(x)\,dx \le \sum_{k=N}^{n} f(k). \tag{1}$$

If the integral

$$\int_{N}^{\infty} f(x)\,dx$$

is convergent, then from (1)

$$\sum_{k=N+1}^{\infty} f(k) \le \int_{N}^{\infty} f(x)\,dx.$$

This shows that the partial sums of the series

$$\sum_{n=N+1}^{\infty} u_n$$

are bounded, and hence that this series is convergent (this from a theorem which states: If $u_n \ge 0$ for every n, then the series

$$\sum_{n=0}^{\infty} u_n$$

is convergent if and only if the sequence $\{S_n\}$ of partial sums is bounded). The series

$$\sum_{n=1}^{\infty} u_n$$

is then convergent also. For the second case, suppose the integral

$$\int_N^\infty f(x)\, dx$$

is divergent. Since $f(x) > 0$ this can happen only if

$$\int_N^{n+1} f(x)\, dx \to +\infty$$

as $n \to \infty$.

It then follows, from the second inequality in (1) that $f(N) + \ldots + f(n) \to +\infty$, and hence that the series is divergent. Thus, the integral test is proven.

For the series

$$\sum_{k=1}^\infty \frac{1}{k^p}$$

consider the integral

$$\int_1^b \frac{dx}{x^p} = \left.\frac{x^{-p+1}}{-p+1}\right|_1^b = \frac{1}{1-p}\left(\frac{1}{b^{p-1}} - 1\right) \quad \text{(for } p \neq 1\text{)}.$$

Then, for $p > 1$

$$\lim_{b \to \infty} \int_1^b \frac{dx}{x^p} = \lim_{b \to \infty} \frac{1}{1-p}\left(\frac{1}{b^{p-1}} - 1\right) = \frac{1}{p-1}.$$

Therefore, the integral converges and by the theorem, so does the series.

For $p < 1$,

$$\lim_{b \to \infty} \int_1^b \frac{dx}{x^p} = \lim_{b \to \infty} \frac{1}{1-p}\left(\frac{1}{b^{p-1}} - 1\right)$$

is unbounded. Therefore, the integral diverges and so does the series.

For $p = 1$,

$$\lim_{b \to \infty} \int_1^b \frac{dx}{x^p} = \lim_{b \to \infty} \int_1^b \frac{dx}{x} = \lim_{b \to \infty} \log b.$$

This grows without bounds. Therefore the integral and thus the series diverges. This shows that the series, called the p-series, converges for p > 1 and diverges for p ≤ 1. Note: If p = 1, the series is called the harmonic series.

• **PROBLEM 12-2**

Test the following series for convergence:

a) $\sum_{n=2}^{\infty} \frac{1}{n(\log n)^2}$

b) $\sum_{n=2}^{\infty} \frac{1}{n(\log n)}$

c) $\sum_{n=4}^{\infty} \frac{1}{n(\log n)[\log(\log n)]^2}$.

Solution: a) To test this series for convergence, apply the integral test. Since the series is

$$\sum_{n=2}^{\infty} \frac{1}{n(\log n)^2} ,$$

set up the integral

$$\int_2^b \frac{dx}{x(\log x)^2} .$$

Then, if the limit of the integral as $b \to \infty$ exists, it follows that the series converges. If the limit of the integral as $b \to \infty$, goes to infinity, then the series diverges. Therefore, to solve

$$\int_2^b \frac{dx}{x(\log x)^2}$$

let $u = \log x$, $du = \frac{dx}{x}$. This yields

$$\int_{\log 2}^{\log b} \frac{du}{u^2} = -u^{-1} \Big|_{\log 2}^{\log b} = \frac{1}{\log 2} - \frac{1}{\log b}$$

Hence

$$\lim_{b\to\infty} \int_2^b \frac{dx}{x(\log x)} = \lim_{b\to\infty} \left(\frac{1}{\log 2} - \frac{1}{\log b}\right) = \frac{1}{\log 2}$$

and thus the series converges.

b) To test the series

$$\sum_{n=2}^{\infty} \frac{1}{n(\log n)},$$

again use the integral test. Therefore set up the integral

$$\int_2^b \frac{dx}{x(\log x)} dx.$$

To solve this integral let $u = \log x$, $du = \frac{dx}{x}$. This yields

$$\int_{\log 2}^{\log b} \frac{du}{u} = \log u \Big|_{\log 2}^{\log b} = \log(\log b) - \log(\log 2).$$

Hence

$$\lim_{b\to\infty} \int_2^b \frac{dx}{x(\log x)} = \lim_{b\to\infty} \big(\log(\log b) - \log(\log 2)\big)$$

which is unbounded and thus the series diverges.

c) To test the series,

$$\sum_{n=4}^{\infty} \frac{1}{n \log n [\log(\log n)]^2}$$

also use the integral test. Therefore to solve

$$\int_4^b \frac{dx}{x(\log x)[\log(\log x)]^2}, \quad \text{let } u = \log(\log x), \quad du = \frac{dx}{x \log x}.$$

Since $\int \frac{du}{u^2} = -u^{-1}$, this yields

$$-\left.\frac{1}{\log(\log x)}\right|_4^b = -\frac{1}{\log(\log b)} + \frac{1}{\log(\log 4)}.$$

Hence

$$\lim_{b\to\infty} \int_4^b \frac{dx}{x \log x [\log(\log x)]^2}$$

$$= \lim_{b\to\infty} \left(\frac{-1}{\log(\log b)} + \frac{1}{\log(\log 4)}\right)$$

$$= \frac{1}{\log(\log 4)},$$

and the series converges.

• **PROBLEM 12-3**

Show that:

a)
$$\sum_{n=2}^{\infty} \frac{(\log n)^2}{n^3} \text{ is a convergent series.}$$

b)
$$\sum_{n=2}^{\infty} \frac{1}{n(\log n)^p} \text{ is convergent if } p > 1, \text{ divergent}$$

for $p \leq 1$.

Solution: a) To show

$$\sum_{n=2}^{\infty} \frac{(\log n)^2}{n^3}$$

is a convergent series apply the integral test. Hence, it must be shown that

$$\lim_{b\to\infty} \int_2^b \frac{(\log x)^2}{x^3} dx < \infty.$$

To do this, use the method of integration by parts. Therefore, take

$$dv = x^{-3}\,dx;\quad v = \frac{-1}{2x^2},\quad u = (\log x)^2,\quad du = \frac{2(\log x)}{x}\,dx.$$

Then,

$$\int_2^b \frac{(\log x)^2}{x^3}\,dx = \left.\frac{-(\log x)^2}{2x^2}\right|_2^b + \int_2^b \frac{2\log x}{2x^3}\,dx. \quad (1)$$

Now, applying integration by parts again, let

$$dv = x^{-3}\,dx;\quad v = \frac{-1}{2x^2},\quad u = \log x,\quad du = \frac{dx}{x}.$$

Then (1) becomes

$$\int_2^b \frac{(\log x)^2}{x^3}\,dx = \left.\frac{-(\log x)^2}{2x^2}\right|_2^b - \left.\frac{\log x}{2x^2}\right|_2^b + \int_2^b \frac{dx}{2x^3}.$$

Hence

$$\lim_{b\to\infty}\int_2^b \frac{(\log x)^2}{x^3}\,dx \quad \text{equals}$$

$$\lim_{b\to\infty}\left(\frac{-(\log b)^2}{2b^2} - \frac{(\log b)}{2b^2} - \frac{1}{4b^2} + \frac{(\log 2)^2}{8} + \frac{(\log 2)}{8} + \frac{1}{16}\right)$$

Then, by L'Hospital's Rule the first three terms tend to 0 as $b\to\infty$.

Therefore

$$\lim_{b\to\infty}\int_2^b \frac{(\log x)^2}{x^3}\,dx = \frac{(\log 2)^2}{8} + \frac{(\log 2)}{8} + \frac{1}{16} < \infty$$

Thus, the series is convergent.

b) To show

$$\sum_{n=2}^{\infty} \frac{1}{n(\log n)^p}$$

is convergent if $p > 1$, divergent for $p \leq 1$, again apply the integral test. Therefore set up the integral

$$\int_2^b \frac{dx}{x(\log x)^p}.$$

To solve this integral let $u = \log x$, $du = \frac{dx}{x}$. This yields

$$\int_{\log 2}^{\log b} \frac{du}{u^p} = \left. \frac{u^{-p+1}}{-p+1} \right|_{\log 2}^{\log b}$$

Therefore,

$$\lim_{b \to \infty} \int_2^b \frac{dx}{x(\log x)^p} = \lim_{b \to \infty} \left(\frac{(\log b)^{-p+1}}{-p+1} - \frac{(\log 2)^{-p+1}}{-p+1} \right) \quad (p \neq 1).$$

If $p > 1$, the first term goes to 0 as $b \to \infty$ and the integral equals

$$\frac{(\log 2)^{-p+1}}{p-1} < \infty.$$

Thus, the series is convergent.

If $p < 1$, the first term grows without bound as $b \to \infty$ and the integral is unbounded. Thus the series is divergent. For $p = 1$,

$$\lim_{b \to \infty} \int_2^b \frac{dx}{x(\log x)^p} = \lim_{b \to \infty} \int_2^b \frac{dx}{x(\log x)} = \lim_{b \to \infty} (\log(\log b) - (\log(\log 2))$$

Thus the series is divergent. This shows that the series is convergent for $p > 1$; and divergent for $p \leq 1$.

● **PROBLEM** 12-4

Given that

$$\zeta(p) = \sum_{k=1}^{\infty} \frac{1}{k^p},$$

Riemann's zeta function (for real p), converges for $p > 1$ and diverges elsewhere, show that

$$\lim_{n \to \infty} \frac{1}{\sqrt{n}} \sum_{k=1}^{n} \frac{1}{k^{3/4}} < \infty.$$

Solution: To show that

$$\lim_{n\to\infty} \frac{1}{\sqrt{n}} \sum_{k=1}^{n} \frac{1}{k^{3/4}} < \infty,$$

use the comparison test by showing this series to be less than $\zeta(p)$, which is convergent for $p > 1$. To do this use Cauchy's Inequality, which states

$$\left(\sum_{k=1}^{n} |a_k b_k|\right)^2 \leq \sum_{k=1}^{n} a_k^2 \sum_{k=1}^{n} b_k^2 \quad . \tag{1}$$

Therefore, taking $a_k = k^{-3/4}$, $b_k = 1$ in equation (1) yields

$$\left(\sum_{k=1}^{n} \left|\frac{1}{k^{3/4}}\right|\right)^2 \leq \sum_{k=1}^{n} \left(\frac{1}{k^{3/4}}\right)^2 \sum_{k=1}^{n} (1)^2,$$

i.e.,

$$\left(\sum_{k=1}^{n} \frac{1}{k^{3/4}}\right)^2 \leq n \sum_{k=1}^{n} \frac{1}{k^{3/2}} \quad .$$

Then

$$\frac{1}{\sqrt{n}} \sum_{k=1}^{n} \frac{1}{k^{3/4}} \leq \left(\sum_{k=1}^{n} \frac{1}{k^{3/2}}\right)^{1/2} \tag{2}$$

However,

$$\sum_{k=1}^{n} \frac{1}{k^{3/2}} = \zeta\left(\frac{3}{2}\right) < \infty \quad .$$

Hence (2) is rewritten

$$\frac{1}{\sqrt{n}} \sum_{k=1}^{n} \frac{1}{k^{3/4}} \leq \left[\zeta\left(\frac{3}{2}\right)\right]^{1/2} < \infty \quad .$$

From which

$$\lim_{n\to\infty} \frac{1}{\sqrt{n}} \sum_{k=1}^{n} \frac{1}{k^{3/4}} < \infty \quad .$$

● **PROBLEM 12-5**

Given that $\sum_{k=1}^{\infty} |a_k|^{3/2} < \infty$

show that

$$\sum_{k=1}^{\infty} |a_k| k^{-2/5} < \infty \qquad (1)$$

Solution: To show that

$$\sum_{k=1}^{\infty} |a_k| k^{-2/5} < \infty$$

(from the given), the following theorem is used. For a given vector $\vec{a}$, the p-norm is defined as

$$|| \vec{a} ||_p = \left(\sum_{k=1}^{n} |a_k|^p \right)^{1/p} .$$

If $p > 1$ and $\frac{1}{p} + \frac{1}{q} = 1$,

then

$$| \vec{a} \cdot \vec{b} | \leq \sum_{k=1}^{n} |a_k b_k| \leq || \vec{a} ||_p \, || \vec{b} ||_q . \qquad (2)$$

Equation (2) is called Hölder's Inequality.

To apply the theorem to this problem, take

$$b_k = k^{-2/5}, \quad p = \frac{3}{2}$$

Then, since

$$\frac{1}{p} + \frac{1}{q} = 1,$$

take $q = 3$. By substitution into equation (2), this yields

$$\sum_{k=1}^{n} |a_k| k^{-2/5} \leq || \vec{a} ||_{3/2} \, || \vec{b} ||_3 . \qquad (3)$$

By definition

$$|| \vec{a} ||_{3/2} = \left(\sum_{k=1}^{n} |a_k|^{3/2} \right)^{2/3} ;$$

$$|| \vec{b} ||_3 = \left(\sum_{k=1}^{n} \left(k^{-2/5} \right)^3 \right)^{1/3} .$$

Therefore (3) becomes

$$\sum_{k=1}^{n} |a_k| k^{-2/5} \le \left(\sum_{k=1}^{n} |a_k|^{3/2} \right)^{2/3} \left(\sum_{k=1}^{n} k^{-6/5} \right)^{1/3} \qquad (4)$$

But, it is given

$$\sum_{k=1}^{\infty} |a_k|^{3/2} < \infty \quad .$$

Hence, since

$$\left(\sum_{k=1}^{n} |a_k|^{3/2} \right)^{2/3} < \left(\sum_{k=1}^{\infty} |a_k|^{3/2} \right)^{2/3} < \infty$$

and

$$\left(\sum_{k=1}^{\infty} k^{-6/5} \right)^{1/3} < \infty$$

(this is because

$$\sum_{k=1}^{\infty} \frac{1}{k^p} < \infty \quad \text{for } p > 1; \text{ here } p = \frac{6}{5})$$

it is seen that the right-hand side of equation (4) remains bounded as $n \to \infty$. Thus

$$\sum_{k=1}^{\infty} |a_k| k^{-2/5} \le \left(\sum_{k=1}^{\infty} |a_k|^{3/2} \right)^{2/3} \left(\sum_{k=1}^{\infty} k^{-6/5} \right)^{1/3} < \infty$$

which shows

$$\sum_{k=1}^{\infty} |a_k| k^{-2/5} < \infty \quad .$$

● **PROBLEM 12-6**

Determine if the series

$$\frac{1}{2} + \frac{1}{3} + \frac{1}{2^2} + \frac{1}{3^2} + \frac{1}{2^3} + \frac{1}{3^3} + \ldots$$

is convergent or divergent by applying the ratio and the root tests.

<u>Solution</u>: The series can be rewritten as the sum of the sequence of numbers given by

$$a_n = \begin{cases} \dfrac{1}{2^{(n+1)/2}} & \text{if } n \text{ is odd } (n > 0) \\ \\ \dfrac{1}{3^{n/2}} & \text{if } n \text{ is even } (n > 0) \end{cases}$$

Now the ratio test states: If $a_k > 0$ and $\lim_{k \to \infty} \dfrac{a_{k+1}}{a_k} = \ell < 1$,

then $\sum_{k=1}^{\infty} a_k$

converges. Similarly, if

$$\lim_{k \to \infty} \dfrac{a_{k+1}}{a_k} = \ell \quad (1 < \ell \le \infty) \qquad \text{then}$$

$\sum_{k=1}^{\infty} a_k$ diverges.

If $\ell = 1$, the test fails. Therefore applying this test gives:

If a_n is odd,

$$\lim_{n \to \infty} \dfrac{a_{n+1}}{a_n}$$

$$= \lim_{n \to \infty} \dfrac{\dfrac{1}{3^{n/2}}}{\dfrac{1}{2^{(n+1)/2}}} = \lim_{n \to \infty} \dfrac{2^{(n+1)/2}}{3^{n/2}} = \lim_{n \to \infty} \left(\dfrac{2}{3}\right)^{n/2} 2^{1/2} = 0 .$$

If a_n is even,

$$\lim_{n \to \infty} \dfrac{a_{n+1}}{a_n} = \lim_{n \to \infty} \dfrac{\dfrac{1}{2^{(n+1)/2}}}{\dfrac{1}{3^{n/2}}} = \lim_{n \to \infty} \left(\dfrac{3}{2}\right)^{n/2} 2^{1/2}$$

and no limit exists.

Hence, the ratio test gives two different values, one < 1 and the other > 1, therefore the test fails to determine if the series is convergent. Thus, another test, known as the root test is now applied. This test states: Let

$\sum_{k=1}^{\infty} a_k$

be a series of nonnegative terms, and let

$$\lim_{n \to \infty} \left(\sqrt[n]{a_n} \right) = S, \text{ where } 0 \leq S \leq \infty \quad . \quad \text{If:}$$

1) $0 \leq S < 1$, the series converges

2) $1 < S \leq \infty$, the series diverges

3) $S = 1$, the series may converge or diverge.

Applying this test yields, if a_n is odd

$$\lim_{n \to \infty} \sqrt[n]{a_n} = \lim_{n \to \infty} \sqrt[n]{\frac{1}{2^{(n+1)/2}}} = \lim_{n \to \infty} \sqrt[n]{\frac{1}{2^{n/2}} \cdot \sqrt[n]{\frac{1}{2^{\frac{1}{2}}}}}$$

$$= \lim_{n \to \infty} \frac{1}{\sqrt{2}} \cdot \frac{1}{2^{1/2n}} = \frac{1}{\sqrt{2}} < 1 \quad .$$

If a_n is even

$$\lim_{n \to \infty} \sqrt[n]{a_n} = \lim_{n \to \infty} \sqrt[n]{\frac{1}{3^{n/2}}} = \frac{1}{\sqrt{3}} < 1 \quad .$$

Thus, since for both cases, the

$$\lim_{n \to \infty} \sqrt[n]{a_n} < 1 \quad ,$$

the series converges.

● **PROBLEM 12-7**

Determine if the series:

a)

$$\sum_{k=1}^{\infty} \frac{(k+1)^{\frac{1}{2}}}{(k^5 + k^3 - 1)^{1/3}}$$

converges or diverges.

b)

$$\sum_{k=1}^{\infty} \frac{k \log k}{7 + 11k - k^2}$$

converges or diverges.

Solution: a) To determine if the given series converges or diverges, the following test called the limit test for convergence is used. This test states:

If
$$\lim_{k \to \infty} k^p U_k = A \text{ for } p > 1,$$

then
$$\sum_{k=1}^{\infty} U_k$$

converges absolutely.

To apply this test let $p = \frac{7}{6} > 1$. Then since

$$U_k = \frac{(k+1)^{1/2}}{(k^5 + k^3 - 1)^{1/3}},$$

$$\lim_{k \to \infty} k^p U_k = \lim_{k \to \infty} \frac{k^{7/6}(k+1)^{1/2}}{(k^5+k^3-1)^{1/3}} = \lim_{k \to \infty} \frac{k^{7/6} k^{1/2}(1+1/k)^{1/2}}{(k^5+k^3-1)^{1/3}}$$

$$= \lim_{k \to \infty} \frac{k^{5/3}(1+1/k)^{1/2}}{(k^5+k^3-1)^{1/3}} = \lim_{k \to \infty} \frac{(1+1/k)^{1/2}}{[k^{-5}(k^5+k^3-1)]^{1/3}}$$

$$= \lim_{k \to \infty} \frac{(1+1/k)^{1/2}}{(1+1/k^2 - 1/k^5)^{1/3}} = \frac{(1)^{1/2}}{(1)^{1/3}} = 1$$

Therefore, the series converges.

b) For the series

$$\sum_{k=1}^{\infty} \frac{k \log k}{7 + 11k - k^2},$$

the following test called the limit test for divergence is used. This test states, If

$$\lim_{k \to \infty} k U_k = A \neq 0 \text{ (or } \pm \infty)$$

then
$$\sum_{k=1}^{\infty} U_k$$

diverges. If $A = 0$, the test fails.

567

Since

$$U_k = \frac{k \log k}{7 + 11k - k^2},$$

$$\lim_{k \to \infty} k U_k = \lim_{k \to \infty} \frac{k^2 \log k}{7 + 11k - k^2} = \lim_{k \to \infty} \frac{k^2 \log k}{k^2\left(\frac{7}{k^2} + \frac{11}{k} - 1\right)}$$

$$= \lim_{k \to \infty} \frac{\log k}{\frac{7}{k^2} + \frac{11}{k} - 1}.$$

As $k \to \infty$, $\log k \to \infty$

while

$$\frac{7}{k^2} + \frac{11}{k} - 1 \to -1.$$

Thus

$$\sum_{k=1}^{\infty} \frac{k \log k}{7 + 11k - k^2} \quad \text{diverges.}$$

● **PROBLEM 12-8**

Determine if the following series are absolutely convergent, conditionally convergent or divergent.

a)
$$\sum_{n=1}^{\infty} \frac{(-1)^{n+1}}{n}$$

b)
$$\sum_{n=2}^{\infty} (-1)^n \left(\frac{n}{1+n^2}\right)^n$$

c)
$$\sum_{n=1}^{\infty} \frac{(-1)^n 2^n}{n!}$$

Solution: a) To determine if the series

$$\sum_{n=1}^{\infty} \frac{(-1)^{n+1}}{n}$$

is convergent or divergent, the following test called the alternating series test is used. This test states:

An alternating series

$$a_1 - a_2 + a_3 - a_4 + \ldots = \sum_{n=1}^{\infty} (-1)^{n+1} a_n, \quad a_n > 0,$$

converges if the following two conditions are satisfied:

i) its terms are decreasing in absolute value:

$$|a_{n+1}| \leq |a_n| \text{ for } n = 1, 2, \ldots$$

ii)
$$\lim_{n \to \infty} a_n = 0$$

For this series, the terms are decreasing in absolute value since

$$1 > \frac{1}{2} > \frac{1}{3} \ldots .$$

Also, the nth term approaches zero so

$$\lim_{n \to \infty} a_n = 0 .$$

Hence, the series converges. Next, if $\Sigma |a_n|$ converges also, then the series Σa_n is absolutely convergent. But the series of absolute values is the harmonic series

$$\sum_{n=1}^{\infty} \frac{1}{n}$$

which is known to diverge. Hence,

$$\sum_{n=1}^{\infty} |a_n| \text{ does not converge and the series } \sum_{n=1}^{\infty} \frac{(-1)^{n+1}}{n}$$

is conditionally convergent.

b) For the series

$$\sum_{n=2}^{\infty} (-1)^n \left[\frac{n}{n^2 + 1} \right]^n$$

use the root test, which states: Let a series

$$\sum_{n=1}^{\infty} a_n$$

be given and let

$$\lim_{n\to\infty} \sqrt[n]{|a_n|} = R$$

Then if $R < 1$, the series is absolutely convergent. If $R > 1$, the series diverges. If $R = 1$, the test fails.

For this series

$$\lim_{n\to\infty} \sqrt[n]{|a_n|} = \lim_{n\to\infty} \sqrt[n]{\left(\frac{n}{1+n^2}\right)^n} = \lim_{n\to\infty} \frac{n}{1+n^2} = \lim_{n\to\infty} \frac{1}{\frac{1}{n}+n} = 0$$

Therefore the series converges absolutely.

c) For the series

$$\sum_{n=1}^{\infty} \frac{(-1)^n 2^n}{n!},$$

use the ratio test. This states that if $a_n \neq 0$ for $n = 1, 2, \ldots$ and

$$\lim_{n\to\infty} \left|\frac{a_{n+1}}{a_n}\right| = L$$

then if $L < 1$,

$$\sum_{n=1}^{\infty} a_n$$

is absolutely convergent, if $L = 1$, the test fails, if $L > 1$,

$$\sum_{n=1}^{\infty} a_n \quad \text{is divergent.}$$

Here

$$\lim_{n\to\infty} \left|\frac{a_{n+1}}{a_n}\right| = \lim_{n\to\infty} \frac{2^{n+1}}{(n+1)!} \cdot \frac{n!}{2^n} = \lim_{n\to\infty} \frac{2^{n+1} \, 2^{-n}(1\cdot 2\cdot 3 \,\ldots\, n)}{1\cdot 2\cdot 3\cdot n\cdot(n+1)}$$

$$= \lim_{n\to\infty} \frac{2}{n+1} = 0.$$

Hence $L = 0$ and the series converges absolutely.

• **PROBLEM 12-9**

Using

$$\sum_{n=2}^{\infty} \frac{1}{(n-1)^p}$$

for ratio comparison, first with $p = 1$ and then with $p > 1$, prove Raabe's Test:

If a series of positive terms a_n has $a_{n+1}/a_n \geq 1 - \frac{1}{n}$ for all large n, then Σa_n diverges; but if

$$\frac{a_{n+1}}{a_n} < 1 - \frac{p}{n}$$

for some constant $p > 1$ for all large n, then the series converges.

Solution: Let $a_{n+1}/a_n \geq 1 - \frac{1}{n}$ for all $n \geq N_0$.

Let $n = N_0 + p$

Then $na_{n+1} \geq (n-1)a_n$ gives

$$(N_0 + p)a_{N_0+p+1} \geq (N_0 + p - 1)a_{N_0+p}$$

$$\geq (N_0 + p - 2)a_{N_0+p-1} \geq$$

$$\cdots \geq (N_0 - 1)a_{N_0}$$

i.e., $na_{n+1} \geq (N_0 - 1)a_{N_0}$ for all $n \geq N_0$

Therefore, $a_{n+1} \geq \frac{(N_0 - 1)a_{N_0}}{n}$

so

$$\sum_{n=N_0}^{\infty} a_{n+1} \geq (N_0 - 1)a_{N_0} \sum_{n=N_0}^{\infty} \frac{1}{n} = \infty \, .$$

Now, let

$$\frac{a_{n+1}}{a_n} < 1 - \frac{p}{n}$$

for all $n \geq N_0$.

Since

$$1 - \frac{p}{n} \leq \left(1 - \frac{1}{n}\right)^p \quad \text{for every } n \geq 1,$$

$$\frac{a_{n+1}}{a_n} < \frac{(n-1)^p}{n^p}$$

or

$$a_{n+1} n^p < a_n (n-1)^p < a_{n-1} (n-2)^p < \ldots < (N_0 - 1)^p a_{N_0}$$

for all $n \geq N_0$.

Therefore

$$\sum_{n=N_0}^{\infty} a_{n+1} < (N_0 - 1)^p a_{N_0} \sum_{n=N_0}^{\infty} \frac{1}{n^p} < \infty$$

i.e.

$$\sum_{n=1}^{\infty} a_n < \infty.$$

Remark: The following inequality was used in the above proof

(*) $1 - p/n \leq (1 - 1/n)^p \quad (p > 1)$.

To prove this inequality consider the function

$f(x) = (1 - x)^p - (1 - px)$.

$f(0) = 0$, $f'(x) = -p(1-x)^{p-1} + p = p[1 - (1-x)^{p-1}] \geq 0$

if $p > 1$ and $0 \leq x \leq 1$. This shows that $f(x)$ increases on $(0,1)$ so

$f(1/n) \geq f(0) = 0$ for every n which is in (*).

● **PROBLEM 12-10**

For $b_n = 1/\{(n-1) \log (n-1)\}$ prove that b_{n+1}/b_n equals $1 - 1/n - (n \log n)^{-1} + 0\left((n^2 \log n)^{-1}\right)$, and then deduce Gauss's test:

If a series of positive terms a_n has

$$a_{n+1}/a_n = 1 - n^{-1} + 0(n^{-q})$$

for some $q > 1$, then Σa_n is divergent.

Solution: Let

$$\sum_{n=1}^{\infty} a_n$$

denote a series of positive terms. It is well known that if

$$\frac{a_{n+1}}{a_n} < k < 1$$

for all large n or if

$$a_n^{1/n} < k < 1$$

for all large n, then the series converges. On the other hand if

$$\frac{a_{n+1}}{a_n} > k > 1$$

or if

$$\sqrt[n]{a_n} > k > 1$$

for all large n, then the series diverges. However these two tests fail if a_{n+1} and a_n are of the same order i.e., if

$$\frac{a_{n+1}}{a_n} \to 1 \quad \text{as } n \to \infty .$$

In this case we must try to estimate the difference

$$a_{n+1}/a_n - 1$$

and see how the next term looks like if we wish to make some statements regarding convergence on the divergence of the series. The problem at hand asks us to carry out this procedure.

$$\frac{b_{n+1}}{b_n} = \frac{(n-1)\log(n-1)}{n \log n}$$

so

$$\frac{b_{n+1}}{b_n} - 1 = \{(n-1)\log(n-1) - n \log n\}/n \log n$$

$$= \frac{1}{n \log n} \{n \log(n-1) - n \log n - \log(n-1)\}$$

$$= \frac{1}{n \log n} \{n(\log(n-1) - \log n) - \log(n-1)\}$$

$$= \frac{1}{n \log n} \{n \log(1 - 1/n) - \log(n(1-1/n))\}$$

$$= \frac{1}{n \log n} \{n \log (1 - 1/n) - \log n - \log (1 - 1/n)\}$$

$$= -\frac{1}{n} + \frac{1}{n \log n} \{n \log (1 - 1/n) - \log (1 - 1/n)\} \quad (1)$$

Recall now the power series expansion for $\log (1 - x)$,

$$\log (1 - x) = -x - \frac{x^2}{2} - \frac{x^3}{3} - \cdots \qquad |x| < 1.$$

If $x = \frac{1}{n}$, then for large n, x is very small and

$$\log (1 - 1/n) \sim -\frac{1}{n},$$

so

$$n \log (1 - 1/n) \sim -1. \quad \text{This implies}$$

$$\frac{b_{n+1}}{b_n} - 1 \sim -\frac{1}{n} - \frac{1}{n \log n} + \frac{1}{n^2 \log n},$$

i.e.,

$$\frac{b_{n+1}}{b_n} \sim 1 - \frac{1}{n} - \frac{1}{n \log n} + \frac{1}{n^2 \log n}.$$

This approximate identity will now be made precise as follows.

Consider

$$\log (1 + x) = \int_0^x \frac{d}{dy} \log (1 + y) \, dy \qquad |x| < 1.$$

Integrate by parts to get

$$\log (1 + x) = x - \int_0^x y(1 + y)^{-1} \, dy \quad (2)$$

The second term on the right side of (2) is less than or equal to

$$\int_0^{|x|} |y| \, |1 + y|^{-1} \, dy$$

let $x = -1/n$ and let $n \geq 2$ so that if $|y| \leq |x| \leq 1/n$

574

$$1 + y \geq 1 - |y| \geq \tfrac{1}{2} \quad \text{and}$$

$$\left| \int_0^x y(1+y)^{-1} \, dy \right| \leq |x| \int_0^{|x|} \tfrac{1}{2} \, dy = \frac{|x|^2}{2} = \frac{1}{2n^2}$$

Thus
$$\log(1 - 1/n) = -\tfrac{1}{n} + R_n$$

where
$$|R_n| \leq \frac{1}{2n^2},$$

so
$$n \log(1 - 1/n) = -1 + R'_n$$

where
$$|R'_n| = n|R_n| \leq \tfrac{1}{2n}.$$

Substituting these into (1) gives
$$b_{n+1}/b_n = 1 - \tfrac{1}{n} + \frac{1}{n \log n} \{-1 + R'_n + \tfrac{1}{n} - R_n\}$$

$$= 1 - \tfrac{1}{n} - \frac{1}{n \log n} + \frac{1}{n^2 \log n} + (R'_n - R_n)/(n \log n)$$

$$= 1 - 1/n - 1/(n \log n) + Q_n$$

where
$$Q_n = \frac{1}{n^2 \log n} + (R'_n - R_n)/(n \log n)$$

and
$$(n^2 \log n)|Q_n| \leq 1 + n(|R'_n| + |R_n|)$$

$$\leq 1 + \tfrac{1}{2} + \tfrac{1}{2n}$$

$$\leq 2.$$

This implies that $Q_n = 0\left((n^2 \log n)^{-1}\right)$
and one can write
$$b_{n+1}/b_n = 1 - 1/n - 1/(n \log n) + 0\left((n^2 \log n)^{-1}\right).$$

This proves the first part of the problem.

In order to prove Gauss's test write

$$b_{n+1} = \beta_n b_n$$

$$a_{n+1} = \alpha_n a_n$$

where

$$\alpha_n = 1 - \frac{1}{n} + \frac{M_n}{n^q} \qquad |M_n| \leq C \text{ for all } n$$

$$\beta_n = 1 - \frac{1}{n} - \frac{1}{n \log n} + \frac{k_n}{n^2 \log n} \qquad |k_n| \leq C \text{ for all } n.$$

Then

$$\alpha_n - \beta_n = 1/(n \log n) + \frac{M_n}{n^q} - k_n/(n^2 \log n)$$

or

$$\alpha_n - \beta_n = \frac{1}{n \log n} \left[1 + \frac{M_n}{n^{q-1}} \log n - \frac{k_n}{n} \right]$$

$$\geq \frac{1}{n \log n} \left[1 - \frac{C}{n^{q-1}} \log n - \frac{C}{n} \right]$$

since

$$q > 1 \qquad \frac{\log n}{n^{q-1}} \to 0 \qquad \text{as } n \to \infty \qquad \text{by}$$

L'Hospital's rule. Hence there exists an integer N_0 so that for every $n \geq N_0$,

$$1 - \frac{C}{n^{q-1}} \log n - \frac{C}{n} > 0 .$$

Then

$$\alpha_n > \beta_n \qquad \text{for every } n \geq N_0 .$$

Now

$$a_{N_0 + k + 1} = \alpha_{N_0 + k} \; \alpha_{N_0+k-1} \cdots \alpha_{N_0} \; a_{N_0}$$

$$\geq \beta_{N_0+k} \; \beta_{N_0+k-1} \cdots \beta_{N_0} \; (a_{N_0}/b_{N_0}) \; b_{N_0}$$

$$= \left(\frac{a_{N_0}}{b_{N_0}} \right) b_{N_0+k+1} \qquad \text{for all } k \geq 0 .$$

Therefore

576

$$\sum_{k=0}^{\infty} a_{N_0+k+1} \geq \left(\frac{a_{N_0}}{b_{N_0}}\right) \sum_{k=0}^{\infty} b_{N_0+k+1} = \infty$$

since

$$\sum \frac{1}{n \log n}$$

diverges by the integral test.

● **PROBLEM 12-11**

Determine if the series

$$1 - \frac{1}{2} + \frac{1 \cdot 3}{2 \cdot 4} - \frac{1 \cdot 3 \cdot 5}{2 \cdot 4 \cdot 6} + \ldots + (-1)^n \frac{1 \cdot 3 \ldots (2n-1)}{2 \cdot 4 \ldots 2n} + \ldots$$

is absolutely convergent, conditionally convergent, or divergent.

Solution: To determine if the given series is absolutely convergent, the following test, known as Raabe's test is used. The test states, let

$$t = \lim_{n \to \infty} n \left(1 - \left|\frac{U_{n+1}}{U_n}\right|\right).$$

Then the series

$$\sum_{n=1}^{\infty} U_n$$

is absolutely convergent if $t > 1$, and is divergent or conditionally convergent if $t < 1$. If $t = 1$ the test fails.

Since

$$U_n = \frac{(-1)^n \, 1 \cdot 3 \ldots (2n-1)}{2 \cdot 4 \ldots 2n}$$

$$\frac{U_{n+1}}{U_n} = \frac{1 \cdot 3 \ldots (2n-1)(2n+1)}{2 \cdot 4 \cdot 6 \ldots 2n(2n+2)} \cdot \frac{2 \cdot 4 \ldots 2n}{1 \cdot 3 \ldots (2n-1)}$$

$$= \frac{2n+1}{2n+2}$$

Then

$$t = \lim_{n \to \infty} n \left(1 - \frac{2n+1}{2n+2}\right) = \lim_{n \to \infty} n \left(\frac{1}{2n+2}\right)$$

i.e.,

$$t = \lim_{n \to \infty} \frac{n}{2n+2} = \frac{1}{2}$$

Therefore, the series is not absolutely convergent. To determine if the series is conditionally convergent or divergent apply the alternating series test. Here, it is seen

$$|U_{n+1}| \le |U_n|$$

since

$$1 > \frac{1}{2} > \frac{1 \cdot 3}{2 \cdot 4} > \cdots .$$

Then, for the series to converge,

$$\lim_{n \to \infty} U_n$$

must be 0; that is, it must be shown that

$$\lim_{n \to \infty} \frac{1 \cdot 3 \cdots (2n-1)}{2 \cdot 4 \cdots 2n} = 0 . \tag{1}$$

To prove (1) let

$$C_n = \frac{1 \cdot 3 \cdots (2n-1)}{2 \cdot 4 \cdots 2n}$$

Then

$$C_n < \frac{2 \cdot 4 \cdots 2n}{3 \cdot 5 \cdots (2n-1)(2n+1)}$$

since

$$\frac{1 \cdot 3 \cdot 5 \cdots (2n-1)}{2 \cdot 4 \cdots 2n} < \frac{2 \cdot 4 \cdots 2n}{3 \cdot 5 \cdots (2n+1)}$$

follows from

$$(1 \cdot 3)(3 \cdot 5)(5 \cdot 7) \cdots (2n-1)(2n+1)$$

$$< (2 \cdot 2)(4 \cdot 4)(6 \cdot 6) \cdots (2n)(2n) .$$

But

$$\frac{2 \cdot 4 \cdots 2n}{3 \cdot 5 \cdots (2n+1)} = \frac{1}{C_n} \frac{1}{2n+1}$$

so that

$$C_n^2 < \frac{1}{2n+1} \quad \text{or} \quad C_n < \frac{1}{\sqrt{2n+1}} .$$

Then

578

$$\lim_{n \to \infty} C_n = 0$$

since

$$\lim_{n \to \infty} \frac{1}{\sqrt{2n+1}} = 0 \, .$$

Therefore, since both conditions of the alternating series test are satisfied, and since Raabe's test proved the series to be not absolutely convergent, the series is conditionally convergent.

● **PROBLEM 12-12**

Determine for what values of m (where m is not zero or a positive integer) that the series

$$1 + \frac{mx}{1!} + \frac{m(m-1)}{2!}x^2 + \ldots + \frac{m(m-1)\ldots(m-n+1)}{n!}x^n + \ldots$$

will converge absolutely when $|x| = 1$.

Solution: To solve this problem, the following test, known as Gauss's test, is applied to the series.

Suppose that

$$\frac{U_{n+1}}{U_n} = 1 - \frac{L}{n} + \frac{C_n}{n^q} \tag{1}$$

where $|C_n| < P$ and $q > 1$. Then the series U_n is absolutely convergent if $L > 1$, and diverges or converges conditionally if $L \leq 1$.

Since

$$U_n = \frac{m(m-1)\ldots(m-n+1)}{n!}x^n ,$$

one has for $|x| = 1$

$$\left|\frac{U_{n+1}}{U_n}\right| = \frac{m(m-1)\ldots(m-n+1)(m-n)}{(n+1)!}$$

$$\times \frac{n!}{m(m-1)\ldots(m-n+1)}$$

or

$$\left|\frac{U_{n+1}}{U_n}\right| = \left|\frac{m-n}{n+1}\right| \, .$$

Then for sufficiently large values of n,

and
$$|m - n| = n - m$$

$$\left|\frac{U_{n+1}}{U_n}\right| = \frac{n-m}{n+1} = \frac{1 - \frac{m}{n}}{1 + \frac{1}{n}}$$

Then by long division

$$\left|\frac{U_{n+1}}{U_n}\right| = 1 - \frac{m+1}{n} + \frac{m+1}{n^2\left(1 + \frac{1}{n}\right)}$$

By substitution into (1) this gives

$$L = m + 1, \quad q = 2, \quad C_n = \frac{m+1}{\left(1 + \frac{1}{n}\right)}.$$

For the series to be absolutely convergent, the test states the condition $L > 1$. Therefore the given series is absolutely convergent when $|x| = 1$ if and only if $m + 1 > 1$, (i.e., if and only if $m > 0$).

SERIES OF FUNCTIONS

● **PROBLEM 12-13**

Show that a convergent series of continuous functions may have a discontinuous sum. What condition is sufficient to ensure that the limit function of a series of continuous functions will be continuous?

Solution: Let $\{f_n\}$ be a sequence of functions defined on $[a,b]$ and let the sequence of partial sums be

$$\{S_n(x)\} = \sum_{i=1}^{n} f_i(x) .$$

Let

$$\sum_{n=1}^{\infty} f_n(x)$$

be the function that forms the limit of the sequence $S_n(x)$ for each $x \in [a,b]$. Then $S_n(x)$ is said to converge (pointwise) to

$$\sum_{n=1}^{\infty} f_n(x)$$

if, given $\varepsilon > 0$, there exists a positive integer N such that

$$\left| \sum_{i=1}^{n} f_i(x) - \sum_{k=1}^{\infty} f_k(x) \right| < \varepsilon$$

for all $n \geq N$, for a given point $x \in [a,b]$.

Suppose

$$f_n(x) = \frac{x^2}{(1 + x^2)^n}$$

$(x \in R; n = 0, 1, 2, \ldots)$

and define

$$f(x) = \sum_{n=0}^{\infty} f_n(x) = \sum_{n=0}^{\infty} \frac{x^2}{(1 + x^2)^n}$$

At $x = 0$, $f_n(0) = \dfrac{0}{(1 + 0)^n} = 0$

and thus

$$f(0) = 0 \, .$$

However, at $x \neq 0$, $f(x) = \sum_{n=0}^{\infty} f_n(x)$

$$= \sum_{n=0}^{\infty} \frac{x^2}{(1 + x^2)^n}$$

$$= x^2 \sum_{n=0}^{\infty} \left(\frac{1}{1 + x^2} \right)^n . \tag{1}$$

The series in (1) is a geometric series of the form

$$\sum_{k=0}^{\infty} y^n$$

with sum

$$\frac{1}{1 - y} \, .$$

Thus (1) equals

$$x^2 \left(\frac{1}{1 - \frac{1}{1+x^2}} \right) = x^2 \left(\frac{1}{\frac{1 + x^2 - 1}{1 + x^2}} \right) = x^2 \frac{1}{x^2} (1 + x^2)$$

$$= 1 + x^2 \, .$$

Thus, the series defined by

$$\sum_{n=0}^{\infty} \frac{x^2}{(1+x^2)^n}$$

yields a convergent series of continuous functions that has a discontinuous sum since

$$f(x) = \begin{cases} 0 & (x = 0) \\ 1 + x^2 & (x \neq 0) \end{cases}.$$

The above series shows that a sequence of continuous partial sums need not converge to a continuous function, although for each value of x the sequence converges. Similarly, it may be shown that the limit of a convergent series of differentiable functions need not be a differentiable function and that the limit function of a convergent series of integrable functions need not be integrable.

A stronger mode of convergence called uniform convergence allows the limit processes outlined above to be interchanged.

• **PROBLEM 12-14**

a) Define uniform convergence for a series of functions.

b) Show

$$\sum_{k=0}^{\infty} x^k$$

converges uniformly in the interval $-a \leq x \leq a$, if $0 < a < 1$.

Solution: a) By definition, the series

$$\sum_{n=1}^{\infty} U_n(x)$$

is uniformly convergent to $S(x)$ in the interval $a \leq x \leq b$ if for each $\varepsilon > 0$, an integer N independent of x, can be found such that

$$|S(x) - S_n(x)| < \varepsilon \text{ for } n \geq N$$

in $a \leq x \leq b$ $\left(\text{where } S_n(x) \text{ represents } \sum_{k=1}^{n} U_k(x)\right)$.

b) For the series

$$\sum_{k=0}^{\infty} x^k,$$

the nth partial sum is $S_n(x) = 1 + x + x^2 + \ldots + x^{n-1}$ \hfill (1)

Multiplying each side by $(1-x)$, equation (1) becomes

$$(1-x)S_n(x) = (1 + x + x^2 + \ldots + x^{n-1})(1 - x)$$

$$(1-x)S_n(x) = (1 + x + x^2 + \ldots + x^{n-1}) + (-x - x^2 - \ldots - x^{n-1} - x^n)$$

$$(1-x)S_n(x) = (1 - x^n)$$

$$S_n(x) = \frac{1 - x^n}{1 - x}.$$

Therefore as $n \to \infty$, in the interval

$$-a \leq x \leq a \quad (0 < a < 1), \quad x^n \to 0$$

and the sum of the series

$$\sum_{k=0}^{\infty} x^k$$

is

$$S(x) = \frac{1}{1 - x} \quad \text{(these are plotted in figure 1).}$$

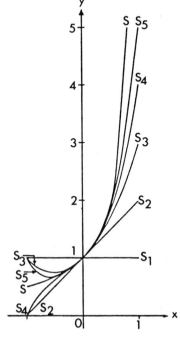

SEQUENCE OF PARTIAL SUMS OF THE GEOMETRIC SERIES.

Fig. 1

To show that the given series converges uniformly to $\frac{1}{1-x}$ in the interval $-a \leq x \leq a$ (if $0 < a < 1$) find an N such that

$$|S_n(x) - S(x)| < \varepsilon$$

for $n \geq N$ for each $\varepsilon > 0$.

Now

$$|S(x) - S_n(x)| = \frac{|x|^n}{|1-x|}.$$

But

$$\frac{|x|^n}{|1-x|} \leq \frac{a^n}{1-a} \quad \text{since} \quad |x| \leq a.$$

Therefore, if $\varepsilon > 0$, it is needed only to choose N so that

$$\frac{a^n}{1-a} < \varepsilon$$

and

$$\frac{|x|^n}{|1-x|} < \varepsilon.$$

We may choose for N any integer greater than

$$-\frac{\log[\varepsilon(1-a)]}{\log(1/a)}. \tag{2}$$

To see this, note that

$$n \log a < \log[\varepsilon(1-a)],$$

but since $0 < a < 1$, $\log a < 0$ and $\log a = -\log\left(\frac{1}{a}\right)$. Hence,

$$-n \log\left(\frac{1}{a}\right) < \log[\varepsilon(1-a)]$$

$$-n < \frac{\log[\varepsilon(1-a)]}{\log(1/a)} \quad \text{(since } \log\left(\frac{1}{a}\right) > 0\text{)}$$

$$n > -\frac{\log[\varepsilon(1-a)]}{\log(1/a)}$$

If N is chosen greater than (2) the condition

$$\frac{a^n}{1-a} < \varepsilon$$

will hold.

Thus, because (2) does not depend on x, the series is uniformly convergent in the given interval.

● **PROBLEM 12-15**

a) Show that

$$\sum_{k=1}^{\infty} \frac{\cos kx}{k^2}$$

converges uniformly in the interval $-\infty < x < \infty$.

b) Repeat for

$$\sum_{k=1}^{\infty} \frac{\cos kx}{2^k}$$

Solution: One important test that can be applied to a large number of series to determine uniform convergence is the Weierstrass M-test. This test states:

Let

$$\sum_{k=1}^{\infty} U_k(x)$$

be a series of functions all defined in some interval $a \leq x \leq b$. If there is a convergent series of constants

$$\sum_{k=1}^{\infty} M_k,$$

such that $|U_k(x)| \leq M_k$ for $a \leq x \leq b$ and $k = 1, 2, \ldots$, then the series

$$\sum_{k=1}^{\infty} U_k(x)$$

converges absolutely for each x in the interval $a \leq x \leq b$ and is uniformly convergent in this same interval. Note that if a series converges uniformly by this test, it also converges absolutely. However, not all uniformly convergent series are absolutely convergent. This shows that for some uniformly convergent series, the sequence M_k required for the Weierstrass test cannot be found and therefore this test cannot be applied.

a) $$\sum_{k=1}^{\infty} \frac{\cos kx}{k^2} \; .$$

Now, to apply the test, it is needed to find a series of constants such that

$$\sum_{k=1}^{\infty} M_k < \infty$$

and

$$|U_k(x)| \le M_k \; .$$

For the given series choose $M_k = k^{-2}$. Then $\sum_{k=1}^{\infty} \frac{1}{k^2} < \infty$ because of the fact that

$$\sum_{k=1}^{\infty} \frac{1}{k^p}$$

is convergent for $p > 1$.

Next,
$$\left| \frac{\cos kx}{k^2} \right| \le \frac{1}{k^2}$$

because $|\cos kx| \le 1$ for all values of x. Thus, since the conditions of the test are satisfied,

$$\sum_{k=1}^{\infty} \frac{\cos kx}{k^2}$$

converges uniformly in the interval $-\infty < x < \infty$.

b) To apply the test to the series

$$\sum_{k=1}^{\infty} \frac{\cos kx}{2^k}$$

choose
$$M_k = \frac{1}{2^k} \; .$$

Then using the root test,

$$\lim_{k \to \infty} \left(\sqrt[k]{\frac{1}{2^k}} \right) = \frac{1}{\sqrt{2}} < 1 \; .$$

586

Therefore, the series

$$\sum_{k=0}^{\infty} \frac{1}{2^k}$$

is convergent. Now

$$\left| \frac{\cos kx}{2^k} \right| \leq \frac{1}{2^k}$$

because $|\cos kx| \leq 1$ for all values of x. Thus, by the Weierstrass M-test, the series is uniformly convergent for $-\infty < x < \infty$.

● **PROBLEM 12-16**

Show that the series

$$\frac{\cos x}{1} + \frac{\cos 3x}{3} + \frac{\cos 5x}{5} + \ldots$$

is convergent if x is not one of the values 0, $\pm\pi$, $\pm 2\pi$, ...

Solution: To determine if the series, rewritten as

$$\sum_{n=1}^{\infty} \frac{\cos(2n-1)x}{2n-1},$$

is convergent, the following theorem (known as Dirichlet's test) is used:

Consider a series of the form

$$a_0 b_0 + a_1 b_1 + a_2 b_2 + \ldots + a_n b_n + \ldots \qquad (1)$$

which satisfies the following conditions:

(i) the terms b_n are positive, decreasing in value (i.e., $b_{n+1} \leq b_n$) and

$$\lim_{n \to \infty} b_n = 0.$$

(ii) there is some constant M independent of n such that

$$|a_0 + a_1 + \ldots + a_n| \leq M$$

for all values of n.

Then the series (1) is convergent.

Hence, to apply this theorem, take

$$a_0 = 0, \ a_1 = \cos x, \ a_2 = \cos 3x, \ a_3 = \cos 5x, \ \ldots,$$

$$a_n = \cos(2n-1)x \qquad n = 1, 2, \ldots$$

and

$$b_0 = 1, \quad b_1 = 1, \quad b_2 = \frac{1}{3}, \quad b_3 = \frac{1}{5} \ldots,$$

$$b_n = (2n-1)^{-1}, \qquad n = 1, 2, \ldots$$

Therefore, since

$$1 > \frac{1}{3} > \frac{1}{5}, \qquad b_{n+1} \leq b_n.$$

Also, as $n \to \infty$, $b_n \to 0$ and condition (i) is satisfied. Now, all that remains is to show that condition (ii) is satisfied. To do this use the trigonometric identity

$$2 \cos A \sin B = \sin(A+B) - \sin(A-B).$$

Taking $B = x$ and A successively equal to $x, 3x, 5x, \ldots$ yields

$2 \cos x \sin x = \sin 2x - 0$

$2 \cos 3x \sin x = \sin 4x - \sin 2x$

$2 \cos 5x \sin x = \sin 6x - \sin 4x$

$\vdots$

$2 \cos(2n-1)x \sin x = \sin[(2n-1)x + x] - \sin[(2n-1)x - x].$

Adding these results gives

$2 \sin x (\cos x + \cos 3x + \ldots + \cos(2n-1)x) =$

$\qquad \sin[(2n-1)x + x]$

which can be rewritten as

$$\cos x + \cos 3x + \ldots + \cos(2n-1)x = \frac{\sin 2nx}{2 \sin x}.$$

Now if x is not one of the values $0, \pm\pi, \pm 2\pi, \ldots$, then $\sin x \neq 0$ and since

$$|\sin 2nx| \leq 1,$$

$$|\cos x + \cos 3x + \ldots + \cos(2n-1)x| \leq \frac{1}{2|\sin x|}.$$

Hence, condition (ii) of the theorem is satisfied with

$$M = \frac{1}{2|\sin x|}.$$

Thus if

$$x \neq 0, \pm\pi, \pm 2\pi, \ldots,$$

the series is convergent.

● PROBLEM 12-17

a) Show that the series

$$\sum_{n=1}^{\infty} n^{-p} \sin(n\theta) \quad (p > 0)$$

is convergent by Dirichlet's test for convergence.

b) Show that

$$\sum_{n=1}^{\infty} (-1)^n \frac{\cos(x/n)}{n}$$

converges by Abel's test.

Solution: a) Dirichlet's test states:

If

$$\sum_{n=0}^{\infty} a_n$$

is a series whose partial sums

$$A_n = \sum_{k=1}^{n} a_k$$

form a bounded sequence, and if b_n decreases to zero as $n \to \infty$, then

$$\sum_{n=0}^{\infty} a_n b_n$$

converges.

To apply this theorem to the series

$$\sum_{n=1}^{\infty} n^{-p} \sin n\theta$$

take $a_n = \sin n\theta$ and $b_n = n^{-p}$. Then using the trigonometric identity

$$-2 \sin \tfrac{1}{2}(\alpha - \beta) \sin \tfrac{1}{2}(\alpha + \beta) = \cos \alpha - \cos \beta \qquad (1)$$

let

$$\theta = \alpha - \beta \; ; \quad 2k\theta = \alpha + \beta \; . \qquad (2)$$

Adding the two values in (2) gives

$$\theta(2k + 1) = 2\alpha \quad \text{or} \quad \alpha = \theta(k + \tfrac{1}{2}) \; .$$

Then
$$2k\theta = \theta(k + \tfrac{1}{2}) + \beta \text{ or } \beta = \theta(k - \tfrac{1}{2}).$$

This yields

$$-2 \sin \tfrac{1}{2}\theta \sum_{k=1}^{n} \sin(k\theta) = \sum_{k=1}^{n} \{\cos[\theta(k+\tfrac{1}{2})] - \cos[\theta(k-\tfrac{1}{2})]\}$$

$$2 \sin\tfrac{1}{2}\theta \sum_{k=1}^{n} \sin(k\theta) = \sum_{k=1}^{n} \{\cos[\theta(k-\tfrac{1}{2})] - \cos[\theta(k+\tfrac{1}{2})]\} \quad (3)$$

$$= \cos \tfrac{\theta}{2} + \cos \tfrac{3\theta}{2} + \cos \tfrac{5\theta}{2} + \ldots + \cos(n-\tfrac{1}{2})\theta - \cos \tfrac{3\theta}{2}$$

$$- \cos \tfrac{5\theta}{2} - \ldots - \cos(n-\tfrac{1}{2})\theta - \cos(n+\tfrac{1}{2})\theta$$

$$= \cos \tfrac{\theta}{2} - \cos(n+\tfrac{1}{2})\theta \quad .$$

Reapplying (1); with $\alpha = \tfrac{\theta}{2}$; $\beta = (n + \tfrac{1}{2})\theta$

yields

$$\cos \tfrac{\theta}{2} - \cos(n + \tfrac{1}{2})\theta = 2 \sin\left((n+1)\tfrac{\theta}{2}\right) \sin\left(n\tfrac{\theta}{2}\right) .$$

Hence

$$A_n = \sum_{1}^{n} \sin k\theta = \frac{\sin\left(\tfrac{n\theta}{2}\right) \sin\left((n+1)\tfrac{\theta}{2}\right)}{\sin\left(\tfrac{\theta}{2}\right)}$$

When θ is not a multiple of 2π (in that case each $a_n = 0$ and the series converges)

$$|A_n| \le \frac{1}{\left|\sin \tfrac{\theta}{2}\right|}$$

since

$$\left|\sin\left(\tfrac{n\theta}{2}\right) \sin\left((n+1)\tfrac{\theta}{2}\right)\right| \le 1 .$$

Then A_n is bounded independently of n for all θ. Also, as $n \to \infty$ $b_n = n^{-p} \to 0$ since $p > 0$.

Therefore

$$\sum_{n=1}^{\infty} n^{-p} \sin n\theta$$

is convergent by Dirichlet's test.

b) Abel's test states:

If

$$\sum_{n=0}^{\infty} a_n$$

is convergent, and if b_n is a bounded monotonic sequence, then

$$\sum_{n=0}^{\infty} a_n b_n$$

converges.

For the series $\sum_{n=1}^{\infty} (-1)^n \frac{\cos\left(\frac{x}{n}\right)}{n}$

take

$$a_n = \frac{(-1)^n}{n} \; ; \quad b_n = \frac{\cos\left(\frac{x}{n}\right)}{n} \; .$$

Now the alternating series test states: Given a series ΣU_n whose terms are alternately positive and negative, if $|U_{n+1}| \leq |U_n|$ for all n and if $U_n \to 0$ as $n \to \infty$, then the series is convergent. Therefore, the series

$$\sum_{n=1}^{\infty} \frac{(-1)^n}{n}$$

is convergent because

$$\left|\frac{1}{n+1}\right| \leq \left|\frac{1}{n}\right| \quad \text{and} \quad \lim_{n \to \infty} \frac{1}{n} = 0 \; .$$

The sequence $b_n = \cos(x/n)$ is monotone and bounded for all x such that

$$\left|\frac{x}{n}\right| < \frac{\pi}{2} \; .$$

Thus, by Abel's test,

$$\sum_{n=1}^{\infty} (-1)^n \frac{\cos\left(\frac{x}{n}\right)}{n}$$

converges.

OPERATIONS ON SERIES

• PROBLEM 12-18

Let a_n be the sequence

$$a_n = \frac{(-1)^{n+1}}{n}.$$

Then

$$\sum_{n=1}^{\infty} \frac{(-1)^{n+1}}{n}$$

is the alternating harmonic series. Given that this series is convergent and that its sum is denoted by S, determine if the sums of the following sequences are convergent, and if so, find their sum in terms of S.

a)
$$b_n = \frac{(-1)^{n+1}}{2n} \qquad n = 1, 2, \ldots$$

b) The sequence formed by

$$c_{2j-1} = 0, \; j = 1, 2, \ldots$$

$$c_{2j} = b_j, \quad j = 1, 2, \ldots$$

c)
$$d_n = a_n + c_n \qquad n = 1, 2, \ldots$$

<u>Solution</u>: a) It is given that the series

$$\sum_{n=1}^{\infty} a_n = \sum_{n=1}^{\infty} \frac{(-1)^{n+1}}{n}$$

is convergent to S. Therefore

$$\sum_{n=1}^{\infty} b_n = \sum_{n=1}^{\infty} \frac{(-1)^{n+1}}{2n} = \frac{1}{2} \sum_{n=1}^{\infty} \frac{(-1)^{n+1}}{n},$$

which converges to $\frac{S}{2}$.

b) The series

$$\sum_{n=1}^{\infty} c_n$$

of the sequence formed by

$$c_{2j-1} = 0, \; j = 1, 2, \ldots, \quad c_{2j} = b_j, \; j = 1, 2, \ldots,$$

is not the same series as

592

$\sum_{n=1}^{\infty} b_n$, but since

$$\sum_{k=1}^{2m} c_k = \sum_{k=1}^{m} b_k \qquad m = 1, 2, \ldots$$

and

$$\sum_{k=1}^{2m+1} c_k = \sum_{k=1}^{m} b_k \qquad m = 1, 2, \ldots \text{ (because } c_{2j-1} = 0\text{)}$$

the series

$$\sum_{n=1}^{\infty} c_n$$

must also be convergent because as $m \to \infty$, the series

$$\sum_{k=1}^{\infty} b_k$$

converges to $\frac{S}{2}$. Therefore, as $m \to \infty$,

$$\sum_{k=1}^{2m} c_k$$

converges to $\frac{S}{2}$ so that the series

$$\sum_{n=1}^{\infty} c_n$$

is convergent to $\frac{S}{2}$.

c) The last sequence is formed by taking

$$d_n = a_n + c_n \qquad n = 1, 2, \ldots$$

Then for $j = 0, 1, 2, \ldots$,

$$d_{4j+1} = a_{4j+1} + c_{4j+1} = \frac{(-1)^{4j+2}}{4j+1} + 0$$

($4j+2$ is always even)

$$= \frac{1}{4j+1} + 0 = \frac{1}{4j+1}.$$

$$d_{4j+2} = a_{4j+2} + c_{4j+2} = a_{4j+2} + b_{2j+1}$$

$$= \frac{(-1)^{4j+3}}{4j+2} + \frac{(-1)^{2j+2}}{2(2j+1)} = \frac{-1}{4j+2} + \frac{1}{4j+2} = 0$$

$$d_{4j+3} = a_{4j+3} + c_{4j+3}$$

$$= \frac{1}{4j+3} + 0 = \frac{1}{4j+3}$$

$$d_{4j+4} = a_{4j+4} + c_{4j+4} = a_{4j+4} + b_{2j+2}$$

$$= \frac{-1}{4j+4} - \frac{1}{2(2j+2)} = \frac{-1}{2j+2}$$

Hence, the numbers in the range of the sequence given by

$$\{d_n \mid n = 1, 2, \ldots \}$$

include the reciprocals of all the odd positive integers, the negatives of the reciprocals of all the even positive integers, and the number zero repeated infinitely often. Therefore, forming a new series by dropping all the zero terms in this sequence, this new series is a rearrangement of the alternating harmonic series. This because, by definition, the series

$$\sum_{n=1}^{\infty} d_n$$

is a rearrangement of the series

$$\sum_{n=1}^{\infty} a_n$$

provided there exists a one-to-one function g of the positive integers onto the positive integers such that for

$$n = 1, 2, \ldots, \qquad d_n = a_{g(n)} ;$$

and for every positive integer n there exists a positive integer m > n such that g(m) ≠ m .

i.e.

$$\sum_{n=1}^{\infty} \frac{(-1)^{n+1}}{n} = 1 - \frac{1}{2} + \frac{1}{3} - \frac{1}{4} + \ldots - \frac{1}{2j+2} + \ldots$$

$$+ \frac{1}{4j+1} - \frac{1}{4j+2} + \frac{1}{4j+3} - \frac{1}{4j+4} + \ldots$$

$$\sum_{n=1}^{\infty} d_n = 1 - \frac{1}{3} - \frac{1}{4} - \frac{1}{2} + \ldots$$

$$+ \frac{1}{4j+1} + \frac{1}{4j+3} - \frac{1}{2j+2} + \frac{1}{4j+5} - \ldots$$

Hence, since

$$\sum_{k=1}^{n} d_k = \sum_{k=1}^{n} a_k + \sum_{k=1}^{n} c_k, \qquad n = 1, 2, 3, \ldots,$$

this rearrangement of the alternating harmonic series must be convergent and its sum is $3S/2$.

● **PROBLEM 12-19**

Given that

$$\log(1+x) = x - \frac{1}{2}x^2 + \frac{1}{3}x^3 - \ldots$$

$$\ldots + (-1)^{n-1} \frac{1}{n} x^n + \ldots \qquad (1)$$

Prove that

$$\frac{3}{2}\log 2 = 1 + \frac{1}{3} - \frac{1}{2} + \frac{1}{5} + \frac{1}{7} - \frac{1}{4} + \frac{1}{9} + \frac{1}{11} - \frac{1}{6} + \ldots$$

Solution: To do this problem let $x = 1$ in equation (1). This yields

$$\log 2 = 1 - \frac{1}{2} + \frac{1}{3} - \frac{1}{4} + \frac{1}{5} - \frac{1}{6} + \frac{1}{7} - \ldots \qquad (2)$$

From which

$$\frac{1}{2}\log 2 = \frac{1}{2} - \frac{1}{4} + \frac{1}{6} - \frac{1}{8} + \frac{1}{10} - \frac{1}{12} + \frac{1}{14} - \ldots \qquad (3)$$

Thus in (3), zero terms may be inserted without affecting the value. Hence (3) becomes

$$\frac{1}{2}\log 2 = 0 + \frac{1}{2} + 0 - \frac{1}{4} + 0 + \frac{1}{6} + 0 - \ldots \qquad (4)$$

Now, it is known that if s and t are the values of two convergent series (i.e.,

$$s = r_1 + r_2 + \ldots + r_n + \ldots ,$$

$$t = t_1 + t_2 + \ldots t_n, \ldots),$$

these two series may be combined by adding corresponding terms, and the result will be a new convergent series whose value is s + t. That is,

$$s + t = U_1 + U_2 + \ldots + U_n + \ldots$$

where

$$U_1 = r_1 + t_1 , \quad U_2 = r_2 + t_2 , \quad \ldots , \quad U_n = r_n + t_n , \quad \ldots .$$

Applying this fact to the problem at hand lets us add (2) and (4) term by term. Therefore, adding gives

$$\tfrac{3}{2} \log 2 = 1 + 0 + \tfrac{1}{3} - \tfrac{1}{2} + \tfrac{1}{5} + 0 + \tfrac{1}{7} - \tfrac{1}{4} + \tfrac{1}{9} - \ldots .$$

Then, the zero terms may be deleted without affecting the value. This yields

$$\tfrac{3}{2} \log 2 = 1 + \tfrac{1}{3} - \tfrac{1}{2} + \tfrac{1}{5} + \tfrac{1}{7} - \tfrac{1}{4} + \tfrac{1}{9} + \ldots$$

which is exactly what was required to be proved.

Note: a different value of the series was reached on rearrangement of the terms because the series is conditionally convergent. Also note that if a series is conditionally convergent, rearrangement of the terms may result in a divergent series. However, if a series is absolutely convergent with sum S, any rearrangement of terms will also be convergent with sum S.

● **PROBLEM 12-20**

Show that if $|x| < 1$,

$$\tfrac{1}{2} \log \tfrac{1 + x}{1 - x} = x + \tfrac{1}{3} x^3 + \tfrac{1}{5} x^5 + \ldots .$$

Solution: Using the fact that log a/b = log a - log b, we have

$$\log \tfrac{1 + x}{1 - x} = \log (1 + x) - \log (1 - x) \tag{1}$$

Now, given that

$$\log (1 + x) = x - \tfrac{1}{2} x^2 + \tfrac{1}{3} x^3 - \tfrac{1}{4} x^4 + \tfrac{1}{5} x^5 - \ldots , \tag{2}$$

if $-1 < x \leq 1$; if x is replaced by -x, then

$$\log(1-x) = -x - \frac{1}{2}x^2 - \frac{1}{3}x^3 - \frac{1}{4}x^4 - \frac{1}{5}x^5 - \ldots,$$

if $-1 \leq x < 1$. (3)

Then, remembering that two convergent series may be combined by adding (or subtracting) corresponding terms (and that the result will be a new convergent series), the two series may be combined by subtraction, which is valid if $-1 < x < 1$. This is because both series are convergent in this interval. Therefore, subtracting (3) from (2) yields

$$\log(1+x) - \log(1-x) = 2x + 0 + \frac{2}{3}x^3 + 0 + \frac{2}{5}x^5 + \ldots$$

$$\log \frac{1+x}{1-x} = 2x + \frac{2}{3}x^3 + \frac{2}{5}x^5 + \ldots$$

$$\log \frac{1+x}{1-x} = 2\left(x + \frac{1}{3}x^3 + \frac{1}{5}x^5 + \ldots\right).$$

Thus

$$\frac{1}{2}\log \frac{1+x}{1-x} = x + \frac{1}{3}x^3 + \frac{1}{5}x^5 + \ldots$$

(if $-1 < x < 1$).

● **PROBLEM 12-21**

Find the series expansion of

$$\frac{1}{1-x}\log \frac{1}{1-x}$$

in powers of x. Is the series absolutely convergent?

Solution: The solution of this problem uses the theorem:

Suppose that each of the series

$$\Sigma u_n, \quad \Sigma v_n$$

is absolutely convergent, with sums U and V respectively:

$$U = u_0 + u_1 + u_2 + \ldots,$$

$$V = v_0 + v_1 + v_2 + \ldots.$$

Let

$$w_0 = u_0 v_0, \quad w_1 = u_0 v_1 + u_1 v_0,$$

and in general

$$w_n = u_0 v_n + u_1 v_{n-1} + \ldots u_n v_0$$

Then the series Σw_n is absolutely convergent, and its sum is UV:

$$UV = w_0 + w_1 + w_2 + \ldots .$$

Also, any infinite series which has as its terms the products $u_i v_j$ (i and j $\geq$ 0) arranged in any order, each product occurring once and only once, is absolutely convergent, with sum UV.

We know that

$$\frac{1}{1-x} = 1 + x + x^2 + \ldots = \sum_{n=0}^{\infty} x^n \text{ if } |x| < 1 \qquad (1)$$

and that

$$\log(1-x) = -x - \frac{x^2}{2} - \frac{x^3}{3} - \ldots = -\sum_{n=1}^{\infty} \frac{x^n}{n}$$

$$\text{if } -1 \leq x < 1 \qquad (2)$$

However, because $\log a = -\log \frac{1}{a}$, (2) can be rewritten as

$$\log\left(\frac{1}{1-x}\right) = x + \frac{x^2}{2} + \frac{x^3}{3} + \ldots = \sum_{n=1}^{\infty} \frac{x^n}{n}$$

$$\text{if } -1 \leq x < 1 . \qquad (3)$$

Since both of these series are absolutely convergent if $|x| < 1$, the aforesaid theorem can be applied. The coefficients in (1) are

$$a_n = 1, n = 0, 1, 2, \ldots$$

(i.e., $a_0 = 1$, $a_1 = 1$, $a_2 = 1$, ...)

Those in (2) are

$$b_0 = 0, b_n = \frac{1}{n}, n = 1, 2, \ldots$$

(i.e., $b_1 = 1$, $b_2 = \frac{1}{2}$, $b = \frac{1}{3}$, ...) .

Consequently,

$$a_0 b_0 = 0 \text{ and if } n \geq 1$$

$$a_0 b_n + a_1 b_{n-1} + \ldots + a_n b_0 = \frac{1}{n} + \frac{1}{n-1} + \ldots + \frac{1}{2} + 1 .$$

Thus, by the theorem,

$$\frac{1}{1-x} \log\left(\frac{1}{1-x}\right) = \sum_{n=1}^{\infty} \left(1 + \frac{1}{2} + \ldots + \frac{1}{n-1} + \frac{1}{n}\right) x^n$$

$$= x + \left(1 + \frac{1}{2}\right) x^2 + \left(1 + \frac{1}{2} + \frac{1}{3}\right) x^3 + \ldots .$$

As the series for

$$\frac{1}{1-x} \quad \text{and} \quad \log \frac{1}{1-x}$$

are absolutely convergent for $|x| < 1$, by the theorem, so is the series for

$$\frac{1}{1-x} \log\left(\frac{1}{1-x}\right)$$

(i.e., it is absolutely convergent for $|x| < 1$).

DIFFERENTIATION AND INTEGRATION OF SERIES

● **PROBLEM 12-22**

Given that the series

$$\sum_{k=0}^{\infty} x^k = \frac{1}{1-x}$$

for $-1 < x < 1$, determine the value of the differentiated series.

<u>Solution</u>: The solution of this problem is facilitated by the use of the following theorem:

A convergent series can be differentiated term by term, provided that the functions of the series have continuous derivatives and that the series of derivatives is uniformly convergent; that is if

(i)
$$U_k'(x) = \frac{dU_k}{dx}$$

is continuous for $a \leq x \leq b$,

(ii) the series

$$\sum_{k=1}^{\infty} U_k(x)$$

converges for $a \leq x \leq b$ to $f(x)$ (i.e.,

$$\sum_{k=1}^{\infty} U_k(x) = f(x)),$$

and

(iii) the series

$$\sum_{k=1}^{\infty} U_k'(x)$$

converges uniformly for $a \leq x \leq b$,

then

$$f'(x) = \sum_{k=1}^{\infty} U_k'(x), \quad a \leq x \leq b.$$

(The derivatives at a and b are understood as right-handed and left-handed derivatives respectively).

In this problem the functions of the series have continuous derivatives in $-1 < x < 1$. Since

$$f(x) = \frac{1}{1-x} = \sum_{k=0}^{\infty} x^k$$

the derived series is

$$\sum_{k=0}^{\infty} k x^{k-1}. \tag{1}$$

To determine if (1) converges uniformly, apply the Weierstrass M-test. Hence, take

$$M_k = k a^{k-1},$$

because

$$|k x^{k-1}| \leq k a^{k-1}$$

for $|x| \leq a$.

Then, using the ratio test for

$$\sum_{k=0}^{\infty} k a^{k-1},$$

$$\lim_{k \to \infty} \left| \frac{a_{k+1}}{a_k} \right| = \lim_{k \to \infty} \left| \frac{(k+1) a^k}{k a^{k-1}} \right| = \lim_{k \to \infty} \left| \frac{(k+1) a}{k} \right| = |a|$$

Therefore, the series

$$\sum_{k=0}^{\infty} ka^{k-1}$$

is convergent for $|a| < 1$.

Hence (1) is uniformly convergent because

$$|U_k(x)| < M_k$$

and

$$\sum_{k=0}^{\infty} M_k < \infty \quad \text{for} \quad -a \leq x \leq a, \quad a < 1.$$

Thus, since the conditions of the theorem (stated at the beginning of this problem) are satisfied, we have

$$\sum_{k=0}^{\infty} U_k'(x) = f'(x),$$

which yields

$$\sum_{k=0}^{\infty} kx^{k-1} = \frac{1}{(1-x)^2}$$

for $|x| \leq a, a < 1$.

● **PROBLEM 12-23**

If $U_k(x) = \dfrac{1}{k^3(1+kx^3)}$ for $0 \leq x \leq 1$

show that

$$\frac{d}{dx}\left[\sum_{k=1}^{\infty} U_k(x)\right] = -3x^2 \sum_{k=1}^{\infty} \frac{1}{k^2(1+kx^3)^2}.$$

Solution: The theorem on differentiability of series states that if:

1) $U_k(x)$ has continuous derivatives for $a \leq x \leq b$,

   $k = 1, 2, \ldots$

2)

$$\sum_{k=1}^{\infty} U_k(x) = S(x) \qquad\qquad a \leq x \leq b$$

3)
$$\sum_{k=1}^{\infty} U_k'(x) \text{ converges uniformly in } a \leq x \leq b$$

Then

$$\sum_{k=1}^{\infty} U_k'(x) = S'(x) \quad \text{or}$$

$$\frac{d}{dx}\left[\sum_{k=1}^{\infty} U_k(x)\right] = \sum_{k=1}^{\infty} \frac{d}{dx} U_k(x)$$

for $\quad a \leq x \leq b$

In this problem the functions of the series have continuous derivatives for $0 \leq x \leq 1$. The series

$$\sum_{k=1}^{\infty} \frac{1}{k^3(1+kx^3)}$$

converges (uniformly) to some value $S(x)$. This is because

$$\left|\frac{1}{k^3(1+kx^3)}\right| \leq \frac{1}{k^3} \quad (0 \leq x \leq 1)$$

and

$$\sum_{k=1}^{\infty} \frac{1}{k^3}$$

converges.

Hence, by the Weierstrass M-test the series converges (uniformly).

Next,

$$U_k'(x) = \frac{-3x^2}{k^2(1+kx^3)^2}$$

and the series

$$\sum_{k=1}^{\infty} U_k'(x) = \sum_{k=1}^{\infty} \frac{-3x^2}{k^2(1+kx^3)^2}$$

converges uniformly in $0 \leq x \leq 1$. This is also shown by the Weierstrass M-test because

$$\left| \frac{-3x^2}{k^2(1+kx^3)^2} \right| \leq \frac{3}{k^2}$$

and

$$\sum_{k=1}^{\infty} \frac{3}{k^2} \quad \text{converges.}$$

Thus, the conditions of the theorem are satisfied and

$$\frac{d}{dx} \sum_{k=1}^{\infty} \frac{1}{k^3(1+kx^3)} = \sum_{k=1}^{\infty} \frac{d}{dx}\left(\frac{1}{k^3(1+kx^3)}\right)$$

$$= -3x^2 \sum_{k=1}^{\infty} \frac{1}{k^3(1+kx^3)^2} .$$

● **PROBLEM 12-24**

a) Show that $\sum_{k=0}^{\infty} x^k \log x$ converges uniformly to $\frac{\log x}{1-x}$ for $0 < x \leq a$ $(0 < a < 1)$.

b) Given that

$$\frac{\pi^2}{6} = 1 + \frac{1}{2^2} + \frac{1}{3^2} + \ldots$$

Show

$$\int_{0+}^{1} \frac{\log x}{1-x} \, dx = \frac{-\pi^2}{6}$$

using the results from (a).

Solution: The partial sum of the series is

$$S_n(x) = \sum_{k=0}^{n} x^k \log x = \log x + x \log x + x^2 \log x + \ldots$$

$$+ x^n \log x$$

which can be rewritten

$$S_n(x) = (\log x)(1 + x + x^2 + \ldots + x^n) . \qquad (1)$$

Multiplying each side of equation (1) by $(1-x)$ yields

$$(1-x) S_n(x) = (\log x)(1-x)(1 + x + x^2 + \ldots + x^n)$$

$$(1-x) S_n(x) = (\log x)(1 + x + x^2 + \ldots + x^n - x - x^2 - x^3 - \ldots - x^{n+1})$$

$$S_n(x) = \frac{(\log x)(1 - x^{n+1})}{(1 - x)}$$

Then

$$S(x) = \lim_{n \to \infty} S_n(x) = \lim_{n \to \infty} \frac{(\log x)(1 - x^{n+1})}{(1 - x)} ,$$

$$S(x) = \frac{\log x}{(1-x)}$$

for $0 < x \leq a$; $0 < a < 1$.

Now to show that the series converges uniformly to $S(x)$, apply the Weierstrass M-test. Take

$$M_k = a^k \log a$$

because

$$|x^k \log x| < a^k |\log a| \quad \text{for } |x| < a .$$

Then using the ratio test for

$$\sum_{k=0}^{\infty} a^k |\log a| , \qquad (2)$$

$$\lim_{k \to \infty} \left| \frac{a_{k+1}}{a_k} \right| = \lim_{k \to \infty} \left| \frac{a^{k+1} \log a}{a^k \log a} \right| = |a|$$

Therefore, the series (2) is convergent for $|a| < 1$.

Hence, the series

$$\sum_{k=0}^{\infty} x^k \log x$$

is uniformly convergent for $0 < x \leq a$; $0 < a < 1$.

Also, at the point $x = 1$ the series has sum 0 and therefore is convergent here also.

b) Since the series is uniformly convergent, term-by-term integration is justified. This is because of a theorem which states: A uniformly convergent series of continuous functions can be integrated term by term. That is, if

$U_k(x) \in C$ for $a \leq x \leq b$, $n = 1, 2, \ldots$

and if

$$f(x) = \sum_{k=1}^{\infty} U_k(x)$$

uniformly in $a \leq x \leq b$ then

$$\int_a^b f(x)\,dx = \sum_{k=1}^{\infty} \int_a^b U_k(x)\,dx$$

$$= \int_a^b U_1(x)\,dx + \int_a^b U_2(x)\,dx + \ldots + \int_a^b U_k(x)\,dx + \ldots$$

Therefore, since $\dfrac{\log x}{1-x} = \sum_{k=0}^{\infty} x^k \log x$,

$$\int_{0+}^{1} \frac{\log x}{(1-x)}\,dx = \sum_{k=0}^{\infty} \int_{0+}^{1} x^k \log x\,dx. \qquad (3)$$

Integrating by parts, letting $u = \log x$; $du = \dfrac{dx}{x}$

and

$$dv = x^k\,dx \text{ so that } v = \frac{x^{k+1}}{k+1}$$

transforms the right side of equation (3) into

$$\sum_{k=0}^{\infty} \left[(\log x) \frac{x^{k+1}}{k+1} \Big|_{0+}^{1} - \int_{0+}^{1} \frac{x^k}{k+1}\,dx \right]. \qquad (4)$$

Then noticing that $x^{k+1} \to 0$ faster than $\log x \to -\infty$ as $x \to 0$, the left hand portion of (4) evaluated at the limits is seen to be 0. The integral is then evaluated and equals

$$\frac{-1}{(k+1)^2}$$

Therefore (3) can be rewritten as

$$\int_{0+}^{1} \frac{\log x}{(1-x)}\,dx = \sum_{k=0}^{\infty} \frac{-1}{(k+1)^2}$$

$$= -\sum_{k=1}^{\infty} \frac{1}{k^2} \quad \text{(by substituting } k = k+1\text{)}$$

$$= -\left(1 + \frac{1}{2^2} + \frac{1}{3^2} + \ldots\right) = \frac{-\pi^2}{6}$$

as was given.

Hence,

$$\int_{0+}^{1} \frac{\log x}{(1-x)} dx = \frac{-\pi^2}{6}.$$

• **PROBLEM 12-25**

a) Show that

$$\sum_{k=0}^{\infty} (-t)^k$$

converges uniformly to $\frac{1}{1+t}$ on any interval $-r \leq t \leq r$, for $(0 < r < 1)$.

Then integrate the series to develop the series expansion of $\log(1 + x)$. ($|x| < 1$).

b) Extend to the end point $x = 1$ by consideration of the series

$$\sum_{k=1}^{\infty} (-1)^{k-1} \frac{x^k}{k},$$

to derive the series expansion for $\log 2$.

Solution: For the series

$$\sum_{k=0}^{\infty} (-t)^k,$$

the nth partial sum is

$$S_n(t) = 1 - t + t^2 - t^3 + t^4 - \ldots + (-1)^n t^n.$$

Multiplying each side by $(1+t)$ gives

$$(1 + t) S_n(t) = (1 + t)(1 - t + t^2 - \ldots + (-1)^n t^n)$$

$$(1 + t) S_n(t) = (1 - t + t^2 - \ldots + (-1)^n t^n)(t - t^2 + t^3 - t^4 + \ldots + (-1)^n t^{n+1})$$

This yields

$$S_n(t) = \frac{1 + t^{n+1}}{1 + t} \quad \text{(if n is even)}$$

$$S_n(t) = \frac{1 - t^{n+1}}{1 + t} \quad \text{(if n is odd)}$$

Therefore, as $n \to \infty$ in the interval $-r \leq t \leq r$ for $0 < r < 1$, $t^{n+1} \to 0$ so that the sum of the series is

$$S(t) = \frac{1}{1 + t}.$$

To show that the series converges uniformly to $S(t)$ in the interval $-r \leq t \leq r$ $(0 < r < 1)$, find an N such that $|S(t) - S_n(t)| < \varepsilon$ for $n \geq N$ for each $\varepsilon > 0$.

Hence

$$|S(t) - S_n(t)| = \frac{|t|^{n+1}}{|1 + t|} \quad \text{and}$$

$$\frac{|t|^{n+1}}{|1 + t|} \leq \frac{r^{n+1}}{1 + r} \quad \text{since} \quad |t| \leq r.$$

Therefore, if $\varepsilon > 0$, it is needed only to choose N so that

$$\frac{r^{n+1}}{1 + r} < \varepsilon \quad \text{and} \quad \frac{|t|^{n+1}}{|1 + t|} < \varepsilon.$$

To do this, notice that

$$r^{n+1} < \varepsilon (1 + r).$$

Thus, $(n + 1) \log r < \log [\varepsilon(1 + r)]$.

However, since

$0 < r < 1$, $\log r < 0$ and $\log r = -\log\left(\frac{1}{r}\right)$

$$-(n + 1) \log\left(\frac{1}{r}\right) < \log [\varepsilon(1 + r)]$$

$$-(n + 1) < \frac{\log [\varepsilon(1 + r)]}{\log\left(\frac{1}{r}\right)} \quad \left(\text{since } \log\left(\frac{1}{r}\right) > 0\right)$$

$$n > \frac{-\log [\varepsilon(1 + r)]}{\log\left(\frac{1}{r}\right)} - 1 \tag{1}$$

Therefore, if N is chosen greater than (1), the condition

$$\frac{r^{n+1}}{(1+r)} < \varepsilon$$

will hold. Hence because (1) depends in no way on t, the series is uniformly convergent in the given interval.

Since the series is uniformly convergent for

$$-r \leq t \leq r \quad (0 < r < 1) ,$$

it can be integrated termwise between 0 and x, $|x| < 1$, to yield

$$\int_0^x \frac{dt}{1+t} = \sum_{k=0}^{\infty} (-1)^k \int_0^x t^k \, dt$$

$$\log(1+x) = \sum_{k=0}^{\infty} (-1)^k \frac{x^{k+1}}{k+1} .$$

Thus, for any x with $|x| < 1$

$$\log(1+x) = x - \frac{x^2}{2} + \frac{x^3}{3} - \ldots + (-1)^{n-1} \frac{x^n}{n} + \ldots .$$

b) If $x > 0$, the series

$$\sum_{k=1}^{\infty} (-1)^{k-1} \frac{x^k}{k}$$

is an alternating series. Since

$$\frac{x^{k+1}}{k+1} < \frac{x^k}{k} \quad (0 \leq x \leq 1)$$

and

$$\lim_{k \to \infty} \frac{x^k}{k} = 0 \quad (\text{for } |x| \leq 1)$$

the series converges for $0 \leq x \leq 1$ by the alternating series test. Now by the fact that the partial sums of an alternating series constantly approach the sum and alternatively lie above and below, so that the error at any stage does not exceed the next term, we may write

$$\left| \log(1+x) - \sum_{1}^{n} (-1)^{k-1} \frac{x^k}{k} \right| \leq \frac{x^{n+1}}{n+1} \leq \frac{1}{n+1}$$

for any x with $0 \leq x \leq 1$. The series is then uniformly convergent on the whole closed interval [0,1] so that term-by-term integration is valid for all x with $0 \leq x \leq 1$. Therefore, setting $x = 1$ yields

$$\log 2 = 1 - \frac{1}{2} + \frac{1}{3} - \frac{1}{4} + \frac{1}{5} - \frac{1}{6} + \frac{1}{7} - \frac{1}{8} + \ldots .$$

• **PROBLEM 12-26**

Justify the equation

$$\sum_{n=1}^{\infty} \frac{\sin n\theta}{n} = \frac{1}{2}(\pi - \theta) \quad \text{for } 0 < \theta < 2\pi$$

given that

$$\frac{r \sin\theta}{1 - 2r \cos\theta + r^2} = \sum_{n=1}^{\infty} r^n \sin(n\theta) \qquad (1)$$

if $|r| < 1$.

Solution: Dividing through by r in equation (1) gives the new equation

$$\frac{(\sin\theta)}{\{1 - 2r \cos\theta + r^2\}} = \sum_{n=1}^{\infty} r^{n-1} \sin n\theta \qquad (2)$$

Now, using the Weierstrass M-test for uniform convergence, take

$$M_n = r^{n-1}$$

so that

$$\sum_{n=1}^{\infty} M_n = \sum_{n=1}^{\infty} r^{n-1} < \infty \quad \text{for } |r| < 1$$

and

$$|r^{n-1} \sin(n\theta)| \le |r^{n-1}| \quad \text{since} \quad |\sin n\theta| \le 1$$

for all values of θ. Therefore, the series

$$\sum_{n=1}^{\infty} r^{n-1} \sin n\theta$$

converges uniformly on the interval $0 \le r \le c$ if $c < 1$.

Now, by a theorem on uniformly convergent series, a uniformly convergent series of continuous functions can be integrated term by term.

Therefore, (2) can be integrated and becomes

$$\int_0^r \frac{\sin\theta}{(1 - 2u \cos\theta + u^2)} du = \sum_{n=1}^{\infty} \int_0^r u^{n-1} \sin n\theta \, du \qquad (3)$$

But $(1 - 2u \cos\theta + u^2) = (u - \cos\theta)^2 + \sin^2\theta$, so that (3) becomes

$$\int_0^r \frac{\sin\theta \, du}{(u - \cos\theta)^2 + \sin^2\theta} = \sum_{n=1}^{\infty} \frac{r^n}{n} \sin n\theta \qquad (4)$$

if $|r| < 1$

However, since

$$\int \frac{a \, dx}{a^2 + x^2} = \text{Arctan} \frac{x}{a},$$

the integral

$$\int_0^r \frac{\sin\theta \, du}{(u - \cos\theta)^2 + \sin^2\theta}$$

$$= \text{Arctan} \left. \frac{u - \cos\theta}{\sin\theta} \right|_0^r$$

$$= \text{Arctan} \left(\frac{r - \cos\theta}{\sin\theta}\right) - \text{Arctan} \left(\frac{-\cos\theta}{\cos\theta}\right)$$

$$= \text{Arctan} \left[\frac{((r-\cos\theta)/\sin\theta) - (-\cos\theta/\sin\theta)}{1 + \left(\frac{r-\cos\theta}{\sin\theta}\right)\left(\frac{-\cos\theta}{\sin\theta}\right)} \right] + k\pi$$

$$\left(\text{since arctan } x - \text{arctan } y = \text{arctan} \left(\frac{x-y}{1+xy}\right)\right)$$

$$= \text{Arctan} \left(\frac{r \sin\theta}{1 - r \cos\theta}\right) + k\pi$$

and $k = 0$ since the integral is 0 when r is 0.

Hence (4) becomes

$$\text{Arctan} \left(\frac{r \sin\theta}{1 - r \cos\theta}\right) = \sum_{n=1}^{\infty} \left(\frac{r^n}{n}\right) \sin n\theta \qquad (5)$$

if $|r| < 1$

Next,

$$\sum_{n=1}^{\infty} \frac{r^n}{n} \sin n\theta$$

$(|r| < 1)$ is uniformly convergent on the closed interval $0 \leq r \leq 1$. This can be proven by Dirichlet's test for uniform convergence, which states: If

(i)
$$\left| \sum_{n=0}^{N} a_n(x) \right| \leq k$$

for all N and all x in U

(ii) $b_n(x)$ decreases as n increases, for all x in U

and

(iii) $b_n(x) \to 0$ uniformly on U as $n \to \infty$, then

$$\sum_{n=0}^{\infty} a_n(x) b_n(x)$$

converges uniformly on U.

Here,

$$\left| \sum_{n=1}^{N} \sin n\theta \right| \leq \left| \sin \tfrac{1}{2} \theta \right|^{-1}$$

if θ is not a multiple of 2π (because

$$\sum_{n=1}^{N} \sin n\theta = \frac{\sin(N\theta/2) \sin((N+1)\theta/2)}{\sin(\theta/2)}$$

where θ is not a multiple of 2π), is otherwise zero, and is therefore bounded uniformly in r. Next,

$$\frac{r^n}{n}$$

is a decreasing function of n tending uniformly to zero as $n \to \infty$ since

$$0 \leq \frac{r^n}{n} \leq \frac{1}{n} \ .$$

Therefore, the conditions for Dirichlet's test are satisfied, thus proving the series to be uniformly convergent. Now we can let $r \to 1$ termwise in equation (5) and since here $\lim \Sigma = \Sigma \lim$, the right side yields the series

$$\sum_{n=1}^{\infty} \frac{\sin n\theta}{n}$$

which is convergent by Dirichlet's test. For the left, θ must be restricted to the range $0 < \theta < 2\pi$ so as to avoid having $1 - r\cos\theta = 0$. Then the limit is

$$\lim_{r \to 1} \text{Arctan} \left(\frac{r \sin\theta}{1 - r\cos\theta} \right) = \text{Arctan} \left(\cot \tfrac{\theta}{2} \right) = \tfrac{1}{2}(\pi - \theta) \ .$$

Note that

$$-\frac{1}{2}\pi < \frac{1}{2}(\pi-\theta) < \frac{\pi}{2}$$

because $0 < \theta < 2\pi$.

Thus

$$\sum_{n=1}^{\infty} \frac{(\sin n\theta)}{n} = \frac{1}{2}(\pi-\theta) \text{ if } 0 < \theta < 2\pi$$

by substitution into equation (5) in the limit case (as $r \to 1$).

ESTIMATES OF ERROR AND SUMS

● **PROBLEM** 12-27

Evaluate the harmonic series of order 6 with an error less than 0.0001.

Solution: For a convergent series with sum S, a function $N(\varepsilon)$ can be determined such that $|S_n - S| < \varepsilon$ for $n \geq N$ (i.e., the first N terms are sufficient to give the desired sum S with an error less than ε). This can also be stated:

Given $S = S_n + R_n$ where R_n is the remainder, an $N(\varepsilon)$ is then needed such that $|R_n| < \varepsilon$ for $n \geq N(\varepsilon)$.

Furthermore, a useful explicit upper estimate for R_n can be found for certain convergent series. More precisely, a sequence K_n converging to 0 can be found for which $|R_n| \leq K_n$ ($n \geq n_1$). If the sequence K_n is monotone decreasing, then $N(\varepsilon)$ can be chosen as the smallest integer n for which $K_n < \varepsilon$. That is, if $N(\varepsilon)$ is so chosen and $n \geq N(\varepsilon)$, then

$$|R_n| \leq K_n \leq K_{N(\varepsilon)} < \varepsilon.$$

The solution to the problem requires the employment of the two following theorems:

(i) For $n \geq n_1$, if $|a_n| \leq b_n$ and

$$\sum_{n=1}^{\infty} b_n$$

converges, then

$$|R_n| \leq \sum_{m=n+1}^{\infty} b_m = K_n$$

for $n \geq n_1$; the sequence K_n is monotonic decreasing and converges to 0.

(ii) Given that the series

$$\sum_{n=1}^{\infty} a_n$$

converges by the integral test, with the function $f(x)$ decreasing for $x \geq C$, then

$$|R_n| < \int_n^{\infty} f(x) \, dx = K_n$$

for $n \geq C$; the sequence K_n is monotone decreasing and converges to zero.

Given the harmonic series, with order $p > 1$, this yields

$$0 < R_n = \sum_{m=n+1}^{\infty} \frac{1}{m^p} < \int_n^{\infty} \frac{1}{x^p} \, dx = \frac{1}{x^{p-1}(-p+1)} \bigg|_n^{\infty} = \frac{1}{(p-1)n^{p-1}}$$

This result can now be used by Theorem (i) for any series whose convergence is established by comparison with a harmonic series of order p.

Here $p = 6$, so that

$$K_n = \frac{1}{(p-1)n^{p-1}} = \frac{1}{5n^5} = 0.2 \, n^{-5}.$$

From this it follows that $N(\varepsilon)$ can be chosen as the smallest integer n such that $K_n < \varepsilon$, i.e.,

$$0.2 n^{-5} < \varepsilon \quad \text{or} \quad n^5 > 0.2\varepsilon^{-1}.$$

Accordingly, to evaluate the harmonic series of order 6 with an error less than 0.0001, 5 terms are sufficient. The reason for this being,

$$|R_n| = K_n < \varepsilon$$

implies, when $\varepsilon = .0001$,

$$0.2\varepsilon^{-1} < n^5 \quad \text{or} \quad .2(10,000) < n^5 \quad \text{or} \quad n^5 > 2,000.$$

The smallest integer satisfying this inequality is $n = 5$. Hence,

$$\sum_{n=1}^{\infty} \frac{1}{n^6} \approx 1 + \frac{1}{2^6} + \frac{1}{3^6} + \frac{1}{4^6} + \frac{1}{5^6} = 1.01735.$$

● **PROBLEM 12-28**

Evaluate the series

$$\sum_{n=1}^{\infty} \frac{n+1}{n \cdot 2^n}$$

with an error less than 0.037.

Solution: To determine the value of the given series with an error less than that specified, the following theorem is needed:

If

$$\left| \frac{a_{n+1}}{a_n} \right| \le r < 1 \quad \text{for} \quad n > n_1$$

(i.e., the series Σa_n converges by the ratio test), then

$$|R_n| \le \frac{|a_{n+1}|}{1-r} = K_n, \quad n \ge n_1 ;$$

the sequence K_n is monotonously decreasing and converges to 0. In addition, if

$$\lim_{n \to \infty} \left| \frac{a_{n+1}}{a_n} \right| = L < 1,$$

then r will be at least equal to L.

For this problem

$$a_n = \frac{n+1}{n \cdot 2^n}.$$

Consequently,

$$\left| \frac{a_{n+1}}{a_n} \right| = \left| \frac{(n+2)}{(n+1) 2^{n+1}} \cdot \frac{n \, 2^n}{(n+1)} \right|$$

$$= \left| \frac{n^2 + 2n}{n^2 + 2n + 1} \cdot \frac{1}{2} \right|$$

Then

$$\lim_{n \to \infty} \left| \frac{a_{n+1}}{a_n} \right| = \lim_{n \to \infty} \left| \frac{n^2 + 2n}{n^2 + 2n + 1} \cdot \frac{1}{2} \right|$$

$$= \lim_{n\to\infty} \left| \frac{1 + 2/n}{1 + 2/n \cdot 1/n^2} \cdot \frac{1}{2} \right| = \frac{1}{2} .$$

(i.e., the series converges by the ratio test since $L = \frac{1}{2} < 1$).

Hence, since it is needed that

$$\left| \frac{a_{n+1}}{a_n} \right| \leq r < 1, \quad r = \frac{1}{2} \text{ can be used.}$$

Accordingly, by the theorem

$$|R_n| \leq \frac{|a_{n+1}|}{1-r} = K_n$$

$$R_n \leq \left| \frac{\frac{(n+2)}{(n+1) 2^{n+1}}}{\frac{1}{2}} \right| = K_n .$$

Thus, for an error less than .037, 5 terms are sufficient. This is because

$$|R_5| \leq \left| \frac{7}{6 \cdot 2^6} \cdot 2 \right| = .03645 < .037 .$$

That is

$$R_5 = \frac{7}{6 \cdot 2^6} + \frac{8}{7 \cdot 2^7} + \ldots < \frac{7}{6 \cdot 2^6} + \frac{7}{6 \cdot 2^7} + \frac{7}{6 \cdot 2^8} + \ldots$$

$$R_5 < \frac{7}{6 \cdot 2^6} \left(1 + \frac{1}{2} + \frac{1}{4} + \ldots \right) = \frac{7}{6 \cdot 2^6} (2) .$$

Since $\sum_{n=0}^{\infty} \frac{1}{2^n} = 2$ (geometric series),

$$R_5 < \frac{7}{192} = 0.037.$$

Finally,

$$\sum_{n=1}^{\infty} \frac{n+1}{n \cdot 2^n} \approx 1 + \frac{3}{2 \cdot 2^2} + \frac{4}{3 \cdot 2^3} + \frac{5}{4 \cdot 2^4} + \frac{6}{5 \cdot 2^5} = 1.6573 .$$

● PROBLEM 12-29

Evaluate the series

$$\sum_{n=2}^{\infty} \frac{1}{(\log n)^n}$$

with an error less than 0.06.

Solution: For the solution to the given problem the following theorem is needed:

If $\sqrt[n]{|a_n|} \leq r < 1$ for $n > n_1$,

(i.e., the series Σa_n converges by the root test), then

$$|R_n| \leq \frac{r^{n+1}}{1-r} = K_n, \text{ for } n \geq n_1.$$

In addition, if

$$\lim_{n \to \infty} \sqrt[n]{|a|} = R < 1,$$

then r will be at least equal to R. If

$$1 > |a_{n+1}|^{1/(n+1)} \geq |a_{n+2}|^{1/(n+2)}$$

for $n \geq n_1$, then

$$|R_n| \leq \frac{|a_{n+1}|}{1 - |a_{n+1}|^{1/(n+1)}} = K_n^*, \text{ for } n \geq n_1$$

Here, the sequences K_n and K_n^* are both monotonously decreasing and converge to zero.

To apply this theorem to the series, first note that

$$\sqrt[n]{|a_n|} = \frac{1}{\log n}$$

which is decreasing and less than 1 for $n = 3, 4, \ldots$
Then

$$\lim_{n \to \infty} \sqrt[n]{|a_n|} = \lim_{n \to \infty} \frac{1}{\log n} = 0 = R < 1.$$

However, for this problem

$$1 > |a_{n+1}| \overset{1/(n+1)}{=} \left|\frac{1}{\log(n+1)}\right| \geq |a_{n+2}| \overset{1/(n+2)}{=} \left|\frac{1}{\log(n+2)}\right|$$

for $n \geq n_1$; then, using $n = 5$ yields

$$|R_n| \leq \frac{|a_{n+1}|}{1 - |a_{n+1}|^{1/(n+1)}} = \frac{\frac{1}{(\log 6)^6}}{1 - \frac{1}{\log 6}} = 0.06 = K_n^*$$

which gives the value of the series with an error less than 0.06 as

$$\sum_{n=2}^{\infty} \frac{1}{(\log n)^n} = \frac{1}{(\log 2)^2} + \frac{1}{(\log 3)^3} + \frac{1}{(\log 4)^4} + \frac{1}{(\log 5)^5}$$

$$= 3.816 .$$

● **PROBLEM 12-30**

For the series

$$\sum_{n=1}^{\infty} \frac{(-1)^{n+1}}{n}$$

show that

a) $-\frac{1}{2} < R_1 < 0$, $0 < R_2 < \frac{1}{3}$, $-\frac{1}{4} < R_3 < 0$.

b) If 3 terms are used, use the theorem to find two values the sum is between.

c) How many terms are needed to compute the sum with an error less than 0.01.

Solution: a) The solution to the problem requires the following theorem:

If the series

$$a_1 - a_2 + a_3 - a_4 + \ldots = \sum_{n=1}^{\infty} (-1)^{n+1} a_n, \quad a_n > a_{n+1} > 0$$

converges by the alternating series test, then

$$0 < |R_n| < a_{n+1} = K_n .$$

Therefore, $N(\varepsilon)$ can be chosen as the smallest integer such that

617

$$a_{n+1} < \varepsilon$$

That is, when a series converges by the alternating series test, the error made in stopping at n terms is in absolute value less than the first term neglected.

To apply this theorem to the given problem, first determine if the series converges by the alternating series test. Here

$$a_{n+1} \leq a_n \quad \text{since} \quad \frac{1}{n+1} \leq \frac{1}{n}$$

(i.e., $1 > \frac{1}{2} > \frac{1}{3} > \frac{1}{4} > \ldots$)

and

$$\lim_{n \to \infty} a_n = \lim_{n \to \infty} \frac{1}{n} = 0.$$

Hence, the series converges by the alternating series test. Then, by the theorem

$$0 < |R_1| < a_2 \quad \text{or} \quad 0 < |R_1| < \frac{1}{2}$$

$$0 < |R_2| < a_3 \quad \text{or} \quad 0 < |R_2| < \frac{1}{3}$$

$$0 < |R_3| < a_4 \quad \text{or} \quad 0 < |R_3| < \frac{1}{4}.$$

The above can be written more precisely as:

$$-\frac{1}{2} < R_1 < 0 \quad \text{since} \quad n = 2 \quad \text{gives} \quad -\frac{1}{2}$$

$$0 < R_2 < \frac{1}{3} \quad \text{since} \quad n = 3 \quad \text{gives} \quad +\frac{1}{3}$$

$$-\frac{1}{4} < R_3 < 0 \quad \text{since} \quad n = 4 \quad \text{gives} \quad -\frac{1}{4}.$$

b) The partial sums are

$$S_1 = 1, \quad S_2 = 1 - \frac{1}{2} = \frac{1}{2}, \quad S_3 = \frac{5}{6}, \quad S_4 = \frac{7}{12}, \ldots$$

If 3 terms are used, the sum is between $\frac{5}{6}$ and $\frac{5}{6} - \frac{1}{4} = \frac{7}{12}$ since $-\frac{1}{4} < R_3 < 0$.

c) To compute the sum with an error less than 0.01, one would need 100 terms, for the 101st term: 1/101 is the first term less than 0.01.

• **PROBLEM 12-31**

Use integral estimates to find upper and lower bounds on n! [Hint:
$$\log N! = \sum_{n=1}^{N} \log n] .$$

Solution: If $f(x)$ is continuous and decreases as x increases on $[n, n+1]$, then $f(n) \geq f(x) \geq f(n+1)$. Consequently,

$$f(n) \geq \int_{n}^{n+1} f(x)\, dx \geq f(n+1) \tag{1}$$

Thus for such f the partial sums of the series

$$\sum_{n=n_0}^{\infty} f(n)$$

can be estimated and if f is positive, the problem of convergence of the series is replaced by that of the integral. If f is increasing with n, then this method can be used with the inequalities reversed for estimating finite sums

$$\sum_{n=1}^{N} f(n) .$$

For this problem, the following theorem is needed:

If $f(x)$ is a real-valued continuous function which decreases as x increases, for all $x \geq n_0 - 1$, then

$$\int_{n_0}^{N+1} f(x)\, dx \leq \sum_{n=n_0}^{N} f(n) \leq \int_{n_0-1}^{N} f(x) \tag{2}$$

for all $N \geq n_0$.

If f is continuous but increasing, then the inequalities are reversed.

To prove this theorem add the inequalities from (1)

$$f(n) \geq \int_{n}^{n+1} f(x)dx \quad \text{for } n = n_0, \ldots, N$$

to yield the left inequality of (2). For the right side add the inequalities

$$f(n) \leq \int_{n-1}^{n} f(x)dx \quad \text{for } n_0 \leq n \leq N.$$

The case of increasing f follows by replacing f by -f in the above.

To apply the theorem to the given problem first notice that log n increases with n. Therefore, the inequalities in (2) must be reversed.

Hence,

$$\int_{n_0}^{N+1} f(x) \, dx \geq \sum_{n=n_0}^{N} f(n) \geq \int_{n_0-1}^{N} f(x) \, dx \qquad (3)$$

for all $N \geq n_0$.

Here, $n_0 = 2$ and $f(n) = \log n$. This yields by (3)

$$\sum_{n=2}^{N} \log n \geq \int_{1}^{N} \log x \, dx.$$

Integration by parts then gives

$$\sum_{n=2}^{N} \log n \geq N \log N - N + 1. \qquad (4)$$

However,

$$\sum_{n=2}^{N} \log n = \log N + \sum_{n=2}^{N-1} \log n \leq \log N + \int_{2}^{N} \log x \, dx \qquad (5)$$

since

$$\int_{2}^{N} \log x \, dx \geq \sum_{n=2}^{N-1} \log n \quad \text{by the left inequality}$$

in (3)

Solving the integral in (5) by parts gives

$$\sum_{n=2}^{N} \log n \leq \log N + N \log N - N - 2 \log 2 + 2. \qquad (6)$$

Then since $\log 1 = 0$, this yields

$$N \log N - N + 1 \leq \sum_{n=1}^{N} \log n \leq N \log N - N + \log N$$

$$- 2 \log 2 + 2.$$

Now

$$\sum_{n=1}^{N} \log n = \log 1 + \log 2 + \ldots + \log N$$

$$= \log N!$$

Thus

$$N \log N - N + 1 \leq \log N! \leq N \log N - N + \log N$$

$$- 2 \log 2 + 2.$$

Writing all terms as powers of e:

$$e^{\log\left(N^N\right)} e^{-N} e \leq e^{\log N!} \leq e^{\log\left(N^N\right)} e^{-N} e^{\log N} e^{-\log 2^2} e^2$$

or

$$N^N e^{-N} e \leq N! \leq N^N e^{-N} (N) \left(\tfrac{1}{4}\right) e^2$$

or

$$(N/e)^N e \leq N! \leq (N/e)^N \tfrac{1}{4} N e^2$$

This is the required interval estimate for $N!$.

CESARO SUMMABILITY

● **PROBLEM 12-32**

> Show that the series $1 - 1 + 1 - 1 + 1 \ldots$ is summable $(C, 1)$ to $\tfrac{1}{2}$.

Solution: There is a process of attaching a sum to a divergent series which is variously known as the method of arithmetic means, Cesàro 1-summability or $(C, 1)$-summability. For this method set

$$S_n = \sum_{k=1}^{n} U_k, \quad \sigma_n = \frac{1}{n} \sum_{k=1}^{n} S_k \qquad n = 1, 2, \ldots.$$

Here, σ_n is the average of the first n partial sums, and is, accordingly, the following linear combination of the first n terms:

$$\sigma_n = \sum_{k=1}^{n} \left(1 - \frac{k-1}{n}\right) U_k$$

By definition the series

$$\sum_{k=1}^{\infty} U_k$$

is summable (C, 1) to A if and only if

$$\lim_{n \to \infty} \sigma_n = A \qquad (1)$$

(1) can also be written as

$$A = \sum_{k=1}^{\infty} U_k \qquad (C, 1).$$

For the given Series:

$$S_1 = 1;\ S_2 = 0;\ S_3 = 1;\ S_4 = 0;\ \ldots\ .$$

Consequently, letting

$$t_n = \sum_{k=1}^{n} S_k \qquad \text{yields}$$

$$t_1 = 1;\ t_2 = 1;\ t_3 = 2;\ t_4 = 2;\ \ldots\ .$$

Therefore

$$\sigma_n = \frac{t_n}{n},$$

which is the sequence

$$1,\ \frac{1}{2},\ \frac{2}{3},\ \frac{1}{2},\ \frac{3}{5},\ \frac{1}{2},\ \ldots,$$

with the nth term being

$$\sigma_n = \begin{cases} \dfrac{1}{2} & \text{if } n \text{ is even} \\[1em] \dfrac{n}{2n-1} & \text{if } n \text{ is odd} \end{cases}$$

Hence

$$\lim_{n \to \infty} \sigma_n \text{ is } \begin{cases} \lim_{n \to \infty} \dfrac{1}{2} = \dfrac{1}{2} & \text{if } n \text{ is even} \\[1em] \lim_{n \to \infty} \dfrac{n}{(2n-1)} = \dfrac{1}{2} & \text{if } n \text{ is odd} \end{cases}$$

Thus by (1) the sequence is summable (C, 1) to A.

● **PROBLEM 12-33**

Show that

a) the series $1 + 0 - 1 + 1 + 0 - 1 + 1 + 0 - \ldots$

is summable (C, 1) to $\frac{2}{3}$.

b) the series

$$\sum_{k=1}^{\infty} (-1)^k k$$

is not summable (C, 1) but is summable (C, 2).

Solution: a) The specified series has the partial sums

$S_1 = 1$; $S_2 = 1$; $S_3 = 0$; $S_4 = 1$, $S_5 = 1$, $S_6 = 0$, $S_7 = 1$;

$S_8 = 1$; $S_9 = 0$; $\ldots$.

Therefore, letting

$$t_n = \sum_{k=1}^{n} S_k \quad \text{yields}$$

$t_1 = 1$, $t_2 = 2$, $t_3 = 2$, $t_4 = 3$, $t_5 = 4$, $t_6 = 4$, $t_7 = 5$,

$t_8 = 6$, $t_9 = 6$, $\ldots$.

From which

$$\sigma_n = \frac{t_n}{n} \quad \text{gives the sequence}$$

$$\sigma_n = 1, 1, \frac{2}{3}, \frac{3}{4}, \frac{4}{5}, \frac{2}{3}, \frac{5}{7}, \frac{3}{4}, \frac{2}{3}, \ldots \quad (1)$$

Accordingly, (1) can be rewritten as

$$\sigma_{3n+1} = \frac{2n+1}{3n+1}, \quad \sigma_{3n+2} = \frac{2n+2}{3n+2},$$

$$\sigma_{3n+3} = \frac{2n+2}{3n+3}$$

By definition, the series is summable (C, 1) to A if and only if

$$\lim_{n \to \infty} \sigma_n = A .$$

For this problem,

$$\lim_{n\to\infty} \sigma_{3n+1} = \lim_{n\to\infty} \sigma_{3n+2} = \lim_{n\to\infty} \sigma_{3n+3} = \lim_{n\to\infty} \sigma_n = \frac{2}{3}.$$

Consequently, the series is summable (C, 1) to $\frac{2}{3}$.

b) The series

$$\sum_{k=1}^{\infty} (-1)^k k = -1 + 2 - 3 + 4 - 5 + 6 \ldots \tag{2}$$

has the partial sums:

$$S_1 = -1,\ S_2 = 1,\ S_3 = -2,\ S_4 = 2,\ S_5 = -3,\ S_6 = 3,\ \ldots.$$

Letting $t_n = \sum_{k=1}^{n} S_k$

yields

$$t_1 = -1,\ t_2 = 0,\ t_3 = -2,\ t_4 = 0,\ t_5 = -3,\ t_6 = 0,$$
$$\ldots \text{ or } t_{2n-1} = -n,\ t_{2n} = 0 \quad n = 1, 2, \ldots.$$

Letting

$$\sigma_n = \frac{t_n}{n}$$

furnishes the following sequence

$$\sigma_n = -1,\ 0,\ \frac{-2}{3},\ 0,\ \frac{-3}{5},\ 0,\ \ldots$$

or similarly written as

$$\sigma_{2n} = 0,\quad \sigma_{2n-1} = \frac{-n}{2n-1} \qquad n = 1, 2, \ldots$$

from which

$$\lim_{n\to\infty} \sigma_{2n} = 0,\quad \lim_{n\to\infty} \sigma_{2n-1} = -\frac{1}{2}.$$

This shows the series is not summable (C, 1) due to the different values in the two limits. Hence a more powerful method of summation is needed. This second method states:

$$\text{If} \quad \lim_{n\to\infty} \frac{2}{n(n+1)} \sum_{k=1}^{n} t_k = A \tag{3}$$

then

$$A = \sum_{k=1}^{\infty} U_k \qquad (C,\ 2)$$

(i.e., the series is summable (C, 2) to A). Therefore letting

$$C_n = \sum_{k=1}^{n} t_k$$

yields the sequence

C_n: $-1, -1, -3, -3, -6, -6, \ldots$

or,

$$C_n = \frac{\frac{-(n+1)}{2}\left(\frac{(n+1)}{2} + 1\right)}{2}$$

for $n = 1, 3, 5, 7, \ldots$

and

$$C_n = \frac{\frac{-n}{2}\left(\frac{n}{2} + 1\right)}{2}$$

for $n = 2, 4, 6, 8, \ldots$.

Hence,

$$\lim_{n \to \infty} \frac{2}{n(n+1)} \sum_{k=1}^{n} U_k = \lim_{n \to \infty} \frac{2}{n(n+1)} \left[\frac{\frac{-(n+1)}{2}\left(\frac{(n+1)}{2} + 1\right)}{2}\right]$$

(for n odd)

$$= \lim_{n \to \infty} \frac{-\left[\frac{(n+1)}{4} + \frac{1}{2}\right]}{n} = \lim_{n \to \infty} -\left(\frac{n + 1 + 2}{4n}\right)$$

$$= \lim_{n \to \infty} -\left(\frac{1 + \frac{1}{n} + \frac{2}{n}}{4}\right) = -\frac{1}{4} \quad \text{(for n odd)}$$

and

$$\lim_{n \to \infty} \frac{2}{n(n+1)} \left[\frac{-\frac{n}{2}\left(\frac{n}{2} + 1\right)}{2}\right] \quad \text{(for n even)}$$

$$= \lim_{n \to \infty} \frac{-\left(\frac{n^2 + 2n}{4}\right)}{n^2 + n} = \lim_{n \to \infty} -\frac{\left(1 + \frac{2}{n}\right)}{4\left(1 + \frac{1}{n}\right)} = -\frac{1}{4} .$$

Thus by (3) the series (2) is summable (C, 2) to $-\frac{1}{4}$.

INFINITE PRODUCTS

● **PROBLEM 12-34**

Prove that the infinite product:

a)
$$\prod_{k=2}^{\infty}\left(1 + \frac{1}{k^2-1}\right)$$

converges and has the value 2,

b)
$$\prod_{k=1}^{\infty}\left(1 - \frac{1}{k+1}\right)$$

diverges.

Solution: An infinite product is a product of the form

$$\prod_{k=1}^{\infty}(1 + U_k) = (1 + U_1)(1 + U_2)(1 + U_3) \ldots$$

where $U_k \neq -1$ for $k = 1, 2, 3, \ldots$). To determine if an infinite product converges or diverges, let

$$P_n = \prod_{k=1}^{n}(1 + U_k),$$

then if there exists a number $P \neq 0$ such that

$$\lim_{n \to \infty} P_n = P,$$

the infinite product, given by

$$\prod_{k=1}^{\infty}(1 + U_k)$$

converges to P. An infinite product that does not converge is said to diverge.

We note that theorems on infinite products often depend upon theorems for infinite series.

a)
$$\prod_{k=2}^{\infty}\left(1 + \frac{1}{k^2-1}\right) = \prod_{k=2}^{\infty} \frac{k^2}{k^2-1}$$

Now

$$P_n = \prod_{k=2}^{n} \frac{k^2}{k^2-1} = \left(\frac{2^2}{2^2-1}\right)\left(\frac{3^2}{3^2-1}\right)\cdots\left(\frac{n^2}{n^2-1}\right)$$

$$= \left(\frac{4}{3}\right)\left(\frac{9}{8}\right)\left(\frac{16}{15}\right)\cdots\left(\frac{n^2}{n^2-1}\right) = \frac{2n}{n+1} \ .$$

Then

$$\lim_{n\to\infty} P_n = \lim_{n\to\infty} \frac{2n}{n+1} = 2 \ .$$

Therefore

$$\prod_{k=2}^{\infty}\left(1 + \frac{1}{k^2-1}\right)$$

converges to 2.

b)

$$\prod_{k=1}^{\infty}\left(1 - \frac{1}{k+1}\right) = \prod_{k=1}^{\infty} \frac{k}{k+1} \ .$$

Then

$$P_n = \prod_{k=1}^{n} \frac{k}{k+1} = \left(\frac{1}{2}\right)\left(\frac{2}{3}\right)\left(\frac{3}{4}\right)\cdots\left(\frac{n-1}{n}\right)\left(\frac{n}{n+1}\right) = \frac{1}{n+1} \ .$$

Hence

$$\lim_{n\to\infty} P_n = \lim_{n\to\infty} \frac{1}{n+1} = 0 \ .$$

Consequently,

$$\prod_{k=1}^{\infty}\left(1 - \frac{1}{k+1}\right)$$

diverges.

• **PROBLEM 12-35**

Prove that the infinite product

$$\prod_{k=1}^{\infty} (1 + U_k)$$

(where $U_k > 0$), converges if

$$\sum_{k=1}^{\infty} U_k \text{ converges.}$$

<u>Solution</u>: Writing the Taylor formula for $f(x) = e^x$

with remainder after n terms gives,

$$e^x = 1 + x + \frac{x^2}{2!} + \ldots + \frac{x^n}{n!} + \frac{x^{n+1} e^\varepsilon}{(n+1)!}$$

where

$$0 < \varepsilon < 1 \text{ and } R_n = \frac{x^{n+1}}{(n+1)!} e^\varepsilon$$

is the error term.

Therefore $1 + x \leq e^x$ for $x > 0$ and consequently,

$$P_n = \prod_{k=1}^{n} (1 + U_k) = (1 + U_1)(1 + U_2)(1 + U_3) \ldots$$

$$\times (1 + U_n) \leq e^{U_1} e^{U_2} e^{U_3} \ldots e^{U_n}$$

which can be rewritten

$$P_n \leq e^{(U_1 + U_2 + U_3 + \ldots + U_n)}.$$

However, it is given that

$$\sum_{k=1}^{\infty} U_k = U_1 + U_2 + U_3 + \ldots$$

converges. Then P_n is a bounded sequence that is also monotonically increasing. Thus since a bounded monotonic sequence has a limit, P_n has a limit and is therefore convergent. This is the required result.

CHAPTER 13

POWER SERIES

Since the use of power series to represent functions is a technique applied in many branches of Mathematics, it is a topic that needs to be covered and understood by all students of Advanced Calculus. Therefore, in this chapter, through the problems solved, some of the more important concepts concerning the study of power series are covered.

Before proceeding, however, one must define what is meant by a power series. By definition, a power series in powers of (x-c) is a series of the form

$$\sum_{n=0}^{\infty} a_n (x-c)^n = a_0 + a_1(x-c) + \ldots + a_n(x-c)^n + \ldots .$$

However, letting c = 0, we obtain a power series in powers of x, which we write as:

$$\sum_{n=0}^{\infty} a_n x^n = a_0 + a_1 x + a_2 x^2 + \ldots a_n x^n + \ldots .$$

Of importance in the study of power series is the notion of radius of convergence (we will denote this value by R). That is, in general, a power series converges for $|x| < R$, diverges for $|x| > R$ and may or may not converge for $|x| = R$. Note that in some cases we may have R = 0 or R = ∞. If R = 0, the series converges only for x = 0 and if R = ∞ the series converges for all values x.

Certain theorems on power series, such as Abel's Limit Theorem are covered as well as operations with power series. For example, under certain conditions a power series can be integrated or differentiated term by term and two power series can be added, subtracted, divided, or multiplied to yield a third power series for each value of x common to their intervals of convergence.

Finally, some frequently used power series and their intervals of convergence are:

$$\sin x = x - \frac{x^3}{3!} + \frac{x^5}{5!} - \ldots + (-1)^{n-1} \frac{x^{2n-1}}{(2n-1)!} + \ldots \quad -\infty < x < \infty$$

$$\cos x = 1 - \frac{x^2}{2!} + \frac{x^4}{4!} - \ldots + (-1)^{n-1} \frac{x^{2n-2}}{(2n-2)!} + \ldots \quad -\infty < x < \infty$$

$$\tan^{-1} x = x - \frac{x^3}{3} + \frac{x^5}{5} - \ldots + (-1)^{n-1} \frac{x^{2n-1}}{2n-1} + \ldots \quad -1 \leq x \leq 1$$

$$e^x = 1 + x + \frac{x^2}{2!} + \ldots + \frac{x^n}{n!} + \ldots \quad -\infty < x < \infty .$$

INTERVAL OF CONVERGENCE

● **PROBLEM 13-1**

a) Define the general form, radius of convergence, and interval of convergence of a power series.

b) Given the power series $\Sigma a_n x^n$, show that the series converges, if given that:

(i) $\quad R = 1/L$ where $L = \lim_{n \to \infty} \left| \frac{a_{n+1}}{a_n} \right|$

exists and $|x| < R$ or

(ii) $\quad R = 1/\alpha$ where $\alpha = \lim_{n \to \infty} \sup \sqrt[n]{|a_n|}$ and $|x| < R$.

Solution: By definition the general form of a power series in powers of $x - x_0$, where x_0 is fixed and x is variable, is

$$\sum_{n=0}^{\infty} a_n (x-x_0)^n = a_0 + a_1(x-x_0) + a_2(x-x_0)^2 + \ldots$$
$$+ a_n (x-x_0)^n + \ldots . \tag{1}$$

However, in the study of power series it is convenient to look at a series of the form,

$$a_0 + a_1 x + a_2 x^2 + \ldots = \sum_{n=0}^{\infty} a_n x^n ;$$

this is a power series in x (i.e., $x_0 = 0$). The reason for this is that equation (1) can be reduced to the above form by the substitution $t = x - a$.

Furthermore, in general a power series converges absolutely for $|x| < R$ and diverges for $|x| > R$, where the constant R is called the radius of convergence of the series. However the series may or may not converge at the end points $x = \pm R$. That is, it may converge at both, at just one, or at neither. In addition, the interval of convergence of the series is the interval $-R < x < R$, with the possible inclusion of the endpoints.

b) For the power series $\Sigma a_n x^n$ if we are given that

(i) $\quad R = \frac{1}{L}$

where
$$L = \lim_{n \to \infty} \left| \frac{a_{n+1}}{a_n} \right|$$

exists, for this case let $U_n = a_n x^n$. Then

$$\lim_{n \to \infty} \left| \frac{U_{n+1}}{U_n} \right| = \lim_{n \to \infty} \left| \frac{a_{n+1}}{a_n} \right| |x| = L|x|.$$

However, by the ratio test, the series ΣU_n converges absolutely if

$$\lim_{n \to \infty} \left| \frac{U_{n+1}}{U_n} \right| < 1,$$

and diverges if

$$\lim_{n \to \infty} \left| \frac{U_{n+1}}{U_n} \right| > 1.$$

Consequently, for convergence

$$L|x| = \frac{|x|}{R} < 1.$$

Therefore, the power series converges absolutely if $|x| < R$, diverges if $|x| > R$. Also if $L = 0$ we have $R = \infty$, and if $L = \infty$ then $R = 0$.

(ii) For this case, again let $U_n = a_n x^n$. Then

$$\lim_{n \to \infty} \sup \sqrt[n]{|U_n|} = \lim_{n \to \infty} \sup \sqrt[n]{|a_n x^n|}$$

$$= |x| \lim_{n \to \infty} \sup \sqrt[n]{|a_n|}$$

$$= |x| \alpha = \frac{|x|}{R}$$

However, by the root test, the series ΣU_n converges absolutely if

$$\lim_{n \to \infty} \sup \sqrt[n]{|U_n|} < 1$$

and diverges if

$$\lim_{n \to \infty} \sup \sqrt[n]{|U_n|} > 1.$$

Therefore, for convergence of the power series

$$\frac{|x|}{R} < 1 \quad \text{or} \quad |x| < R;$$

if $|x| > R$, the power series diverges. In addition, if $\alpha = 0$, $R = +\infty$; if $\alpha = +\infty$, $R = 0$.

We make note that the second method of finding R is more powerful than the first. That is the limit,

$$\lim_{n \to \infty} \left| \frac{a_{n+1}}{a_n} \right|$$

does not exist for certain power series.

• **PROBLEM 13-2**

Find the radius of convergence of the following power series:

a) $\sum_{n=1}^{\infty} \frac{x^n}{n}$

b) $\sum_{n=1}^{\infty} \frac{(2n)!}{(n!)^2} x^n$

c) $\sum_{n=1}^{\infty} \frac{(3n)!}{(n!)^2} x^n$

Solution: a)

$$\sum_{n=1}^{\infty} \frac{x^n}{n} = x + \frac{x^2}{2} + \frac{x^3}{3} + \ldots .$$

Here we let

$$U_n = \frac{x^n}{n}$$

so that

$$\lim_{n \to \infty} \left| \frac{U_{n+1}}{U_n} \right| = \lim_{n \to \infty} \left| \frac{x^{n+1}}{n+1} \cdot \frac{n}{x^n} \right| = \lim_{n \to \infty} |x| \frac{n}{n+1} = |x|.$$

Therefore, $R = 1$ and the series converges for $-1 < x < 1$. Note that the series may or may not converge at the end points $x = \pm 1$. That is, it may converge at both $x = \pm 1$, at just one, or neither. For example, given the endpoints ± 1, we have at $x = 1$ the series,

$$\sum_{n=1}^{\infty} \frac{1}{n},$$

which is the divergent harmonic series. For $x = -1$ we have the alternating series,

$$\sum_{n=1}^{\infty} \frac{(-1)^n}{n}$$

where

$$\left| \frac{(-1)^{n+1}}{n+1} \right| < \left| \frac{(-1)^n}{n} \right|$$

and

$$\lim_{n \to \infty} \frac{(-1)^n}{n} = 0$$

Hence this series is convergent. Thus, $R = 1$ and the series converges absolutely for $-1 \leq x < 1$.

b)
$$\sum_{n=1}^{\infty} \frac{(2n)!}{(n!)^2} x^n = \frac{2!}{(1)^2} x + \frac{4!}{(2!)^2} x^2 + \frac{6!}{(3!)^2} x^3 + \dots .$$

Here we have

$$a_n = \frac{(2n)!}{(n!)^2} \quad \text{and} \quad U_n = \frac{(2n)!}{(n!)^2} x^n$$

$$\lim_{n \to \infty} \left| \frac{U_{n+1}}{U_n} \right| = \lim_{n \to \infty} \left| \frac{(2n+2)!}{[(n+1)!]^2} x^{n+1} \cdot \frac{(n!)^2}{(2n)! x^n} \right| =$$

$$= \lim_{n \to \infty} |x| \frac{(2n+2)(2n+1)}{(n+1)^2} = 4|x| .$$

Therefore $R = \frac{1}{4}$ since we want $4|x| < 1$.

c)
$$\sum_{n=1}^{\infty} \frac{(3n)!}{(n!)^2} x^n = 3! x + \frac{6!}{(2!)^2} x^2 + \frac{9!}{(3!)^2} x^3 + \dots .$$

Here,

$$a_n = \frac{(3n)!}{(n!)^2} \quad \text{and} \quad U_n = \frac{(3n)!}{(n!)^2} x^n$$

$$\lim_{n \to \infty} \left| \frac{U_{n+1}}{U_n} \right| = \lim_{n \to \infty} \left| \frac{(3n+3)! \, x^{n+1}}{[(n+1)!]^2} \cdot \frac{(n!)^2}{(3n)! \, x^n} \right|$$

$$= \lim_{n \to \infty} x \frac{(3n+3)(3n+2)(3n+1)}{(n+1)^2} = \infty$$

Therefore, the series does not converge for any value of

x except x = 0. This means that
$$R = 0 \left(\text{i.e.,} \ R = \frac{1}{\infty} \right).$$

● **PROBLEM 13-3**

Considering all possibilities, find the radius of convergence of the series

$$\sum \frac{(pn)!}{(n!)^q} x^n,$$

where p is a positive integer and q > 0.

Solution: To find R for the given series, three cases must be considered. They are the cases (i) p = q, (ii) p > q, (iii) p < q. However, for each case we will need the ratio test. Hence, in general,

$$\lim_{n \to \infty} \left| \frac{U_{n+1}}{U_n} \right| = \lim_{n \to \infty} \left| \frac{(pn+p)! \ x^{n+1}}{[(n+1)!]^q} \cdot \frac{(n!)^q}{(pn)! \ x^n} \right|$$

$$= |x| \lim_{n \to \infty} \frac{(pn+p)(pn+p-1) \ldots (pn+p-(p-1))}{(n+1)^q}$$

$$= |x| \lim_{n \to \infty} \frac{p(n+1) p\left(n+1 - \frac{1}{p}\right) \ldots p\left(n+1 - \frac{p-1}{p}\right)}{(n+1)^q}.$$

Now since there are p number of terms in the numerator (this because $\frac{(n+x)!}{n!}$ has, after cancellations, x terms in the numerator), we have

$$\lim_{n \to \infty} \left| \frac{U_{n+1}}{U_n} \right| = |x| \ p^p \lim_{n \to \infty} \frac{(n+1)\left(n+1 - \frac{1}{p}\right) \ldots \left(n + \frac{1}{p}\right)}{(n+1)^q} \quad (1)$$

Case (1): Assume p = q, then

$$\lim_{n \to \infty} \left| \frac{U_{n+1}}{U_n} \right| = |x| \ p^p \lim_{n \to \infty} \frac{(n+1)\left(n+1 - \frac{1}{p}\right) \ldots \left(n + \frac{1}{p}\right)}{(n+1)^p}$$

$$= x \ p^p \lim_{n \to \infty} \frac{(n+1)}{(n+1)} \frac{\left(n+1 - \frac{1}{p}\right)}{(n+1)} \ldots \frac{\left(n + \frac{1}{p}\right)}{(n+1)}$$

$$= |x| \ p^p \lim_{n \to \infty} \frac{(n+1)}{(n+1)} \lim_{n \to \infty} \frac{\left(n+1 - \frac{1}{p}\right)}{(n+1)} \ldots \lim_{n \to \infty} \frac{\left(n + \frac{1}{p}\right)}{(n+1)} \quad (2)$$

634

since
$$\lim_{n\to\infty} (a_1 \cdot a_2 \cdot a_3 \cdot \ldots \cdot a_k) = \lim_{n\to\infty} a_1 \lim_{n\to\infty} a_2 \ldots \lim_{n\to\infty} a_k .$$

Therefore,
$$\lim_{n\to\infty} \left| \frac{U_{n+1}}{U_n} \right| = |x| \, p^p$$

since each limit in (2) equals 1. Consequently, the series converges if $|x| < p^{-p}$ therefore $R = p^{-p}$.

Case (ii): Assume $p > q$, then from (1),

$$\lim_{n\to\infty} \left| \frac{U_{n+1}}{U_n} \right| = |x| \, p^p \lim_{n\to\infty} \frac{(n+1)\left(n+1-\frac{1}{p}\right)\ldots\left(n+\frac{1}{p}\right)}{(n+1)^q} \qquad (3)$$

$$= |x| \, p^p \lim_{n\to\infty} \frac{(n+1)}{(n+1)} \frac{\left(n+1-\frac{1}{p}\right)}{(n+1)} \ldots \frac{\left(n+1-\frac{q-1}{p}\right)}{(n+1)} \frac{\left(n+1-\frac{q}{p}\right)}{(n+1)}$$

$$\times \ldots \times \left(n+\frac{1}{p}\right)$$

$$= |x| \, p^p \lim_{n\to\infty} \frac{(n+1)}{(n+1)} \lim_{n\to\infty} \frac{\left(n+1-\frac{1}{p}\right)}{(n+1)} \ldots \lim_{n\to\infty} \frac{\left(n+1-\frac{q-1}{p}\right)}{(n+1)}$$

$$\lim_{n\to\infty} \left(n+1-\frac{q}{p}\right) \ldots \left(n+\frac{1}{p}\right) .$$

Now the first q limits each equal 1, however, the last limit
$$\lim_{n\to\infty} \left(n+1-\frac{q}{p}\right) \ldots \left(n+\frac{1}{p}\right) = \infty ,$$

hence,
$$\lim_{n\to\infty} \left| \frac{U_{n+1}}{U_n} \right| = \infty , \text{ therefore } R = 0 .$$

Case (iii): Assume $p < q$, then from (3),

$$\lim_{n\to\infty} \left| \frac{U_{n+1}}{U_n} \right| = |x| \, p^p \lim_{n\to\infty} \frac{(n+1)}{(n+1)} \frac{\left(n+1-\frac{1}{p}\right)}{(n+1)} \ldots \frac{\left(n+\frac{1}{p}\right)}{(n+1)} \frac{1}{(n+1)^{q-p}}$$

$$\lim_{n\to\infty} \left| \frac{U_{n+1}}{U_n} \right| = |x| \, p^p \lim_{n\to\infty} \frac{1}{(n+1)^{q-p}} = |x| \, p^p \cdot 0 = 0$$

Therefore, for this case $R = \infty$.

Summarizing, we have

$$R = \begin{cases} p^{-p} & \text{for } p = q \\ 0 & \text{for } p > q \\ \infty & \text{for } p < q \end{cases}$$

● **PROBLEM 13-4**

Prove that

$$f(x) = \sum_{n=0}^{\infty} a_n x^n$$

and

$$f'(x) = \sum_{n=0}^{\infty} n a_n x^{n-1}$$

have the same radius of convergence.

Solution:

$$f(x) = \sum_{n=0}^{\infty} a_n x^n = a_0 + a_1 x + a_2 x^2 + a_3 x^3 + \ldots \quad (1)$$

$$f'(x) = \sum_{n=0}^{\infty} n a_n x^{n-1} = a_1 + 2a_2 x + 3a_3 x^2 + 4a_4 x^3 + \ldots \quad (2)$$

Let R and R' denote the radius of convergence of (1) and (2), respectively. Suppose $|x| < R$, and choose x_0 so that

$$|x| < |x_0| < R.$$

Then (1) is convergent with $x = x_0$ and consequently $a_n x_0^n \to 0$ as $n \to \infty$, since for a convergent series

$$\sum_{n=0}^{\infty} U_n$$

we have

$$\lim_{n \to \infty} U_n = 0.$$

Therefore a number $A > 0$ may be chosen such that

$$\left| a_n x_0^n \right| \leq A$$

for all n. Then

$$na_n x^{n-1} = \frac{x_0^n}{x_0^n} \, n \, a_n x^{n-1} = \frac{n}{x_0} \, a_n x_0^n \left(\frac{x}{x_0}\right)^{n-1}$$

Therefore

$$\left| na_n x^{n-1} \right| \leq \frac{A}{|x_0|} \, n \left| \left(\frac{x}{x_0}\right)^{n-1} \right| \tag{3}$$

or

$$\left| na_n x^{n-1} \right| \leq \frac{A}{|x_0|} \, nr^{n-1} \tag{3}$$

where

$$r = \frac{|x|}{|x_0|} < 1$$

The series

$$\sum_{n=0}^{\infty} \frac{A}{|x_0|} \, nr^{n-1}$$

is convergent since

$$\lim_{n \to \infty} \left| \frac{U_{n+1}}{U_n} \right| = \lim_{n \to \infty} \left| \frac{(n+1)r^n}{n r^{n-1}} \right| = \lim_{n \to \infty} \left| \frac{n+1}{n} \right| r = r < 1 \, ;$$

here

$$U_n = \frac{A}{|x_0|} \, nr^{n-1}$$

Thus, by (3), the series

$$\sum_{n=0}^{\infty} na_n x^{n-1} \, ,$$

which is the series (2), is also convergent. Hence, (2) converges if $|x| < R$. It follows that the radius of convergence of (2) is not less than that of (1). That is $R' \geq R$ and if $R = \infty$ this means that $R' = \infty$.

Now we assume that $R' > R$ and we choose x so that $R < |x| < R'$. Then for this x, the series (2) is absolutely convergent, but the series (1) is divergent. Now

$$|a_n x^n| = |n a_n x^{n-1}| \left| \frac{x}{n} \right| < |na_n x^{n-1}|$$

as soon as $n > |x|$. Hence, this shows that the series

(1) must be convergent by the comparison test. However this is a contradiction. Therefore $R' \not> R$, and since we have already shown that $R' \geq R$ (i.e., $R' \not< R$), this means that $R' = R$. This completes the proof.

• **PROBLEM 13-5**

Find the radius of convergence of the power series

$$\sum_{n=1}^{\infty} \frac{x^n}{n^2}.$$

Then determine if the convergence is uniform for $-R \leq x \leq R$.

Solution: Here the ratio gives

$$\lim_{n \to \infty} \left| \frac{U_{n+1}}{U_n} \right| = \lim_{n \to \infty} |x| \frac{n^2}{(n+1)^2} = |x|,$$

so that $R = 1$. That is, the series converges absolutely for $-1 < x < 1$ and diverges for $|x| > 1$. Note we could have also found this result by the relation,

$$R = \frac{1}{\alpha} \quad \text{where} \quad \alpha = \lim_{n \to \infty} \sup \sqrt[n]{|a_n|}$$

where $a_n = \frac{1}{n^2}$ which yields

$$\alpha = \lim_{n \to \infty} \sup \sqrt[n]{|1/n^2|} = \lim_{n \to \infty} \sup \sqrt[n]{\frac{1}{n^2}}$$

$$= \lim_{n \to \infty} \sup \frac{1}{n^{2/n}} = \lim_{n \to \infty} \sup \frac{1}{e^{(2/n) \log n}}$$

$$= \frac{1}{e^{\lim_{n \to \infty} \sup (2/n) \log n}} = \frac{1}{e^0} = 1,$$

therefore,

$$\frac{1}{R} = 1,$$

so that as before $R = 1$. For $x = \pm 1$ the series converges by comparison with the harmonic series of order 2, that is

$$\left| \frac{(\pm 1)^n}{n^2} \right| \leq \frac{1}{n^2}.$$

Hence the series converges for $-1 \leq x \leq 1$. To determine if the convergence is uniform in this interval, we need the Weierstrass M-test for uniform convergence. That is, we must find a convergent series of constants

$$\sum_{n=1}^{\infty} M_n$$

such that

$$\left| \frac{x^n}{n^2} \right| \leq M_n$$

for all x in $-1 \leq x \leq 1$ if we are to determine that

$$\sum_{n=1}^{\infty} \frac{x^n}{n^2}$$

is uniformly convergent in this interval.

Since $-1 \leq x \leq 1$ we have

$$\left| \frac{x^n}{n^2} \right| \leq \frac{1}{n^2}$$

for all x in the range.

Therefore since

$$\sum_{n=1}^{\infty} M_n$$

converges, this shows the given power series converges uniformly on $-1 \leq x \leq 1$.

• **PROBLEM 13-6**

Prove Abel's theorem which states:

If the series

$$\sum_{n=0}^{\infty} a_n x^n$$

converges at R, then it converges uniformly on the closed interval $0 \leq x \leq R$. (A like conclusion holds for $-R \leq x \leq 0$ if the series converges at $x = -R$).

Solution: Suppose ε is any positive number. To show that the series is uniformly convergent for $0 \leq x \leq R$ it is necessary and sufficient to show that to each $\varepsilon > 0$ there corresponds some integer N independent of x such that

$$\left| f_n(x) - f_m(x) \right| < \varepsilon$$

whenever $N \leq n$, $N \leq m$; where $f_n(x) = \sum_{j=0}^{n} a_j x^j$.

Therefore we need to show that there is some integer N such that

$$|a_m x^m + a_{m+1} x^{m+1} + \ldots + a_{m+p} x^{m+p}| < \varepsilon \tag{1}$$

if $N \leq m$, $0 < p$ and $0 \leq x \leq R$. Now set

$$U_k = a_{m+k} R^{m+k} \quad \text{and} \quad V_k = \left(\frac{x}{R}\right)^{m+k}$$

that is

$$U_0 = a_m R^m, \quad U_1 = a_{m+1} R^{m+1}, \quad \ldots, \quad U_p = a_{m+p} R^{m+p}.$$

$$V_0 = \left(\frac{x}{R}\right)^m, \quad V_1 = \left(\frac{x}{R}\right)^{m+1}, \quad \ldots, \quad V_p = \left(\frac{x}{R}\right)^{m+p}$$

Notice that in $0 \leq x \leq R$ we have

$$V_0 \geq V_1 \geq \ldots \geq V_p \geq 0 \quad \text{and} \quad V_0 \leq 1.$$

Then (1) can be written

$$-\varepsilon < U_0 V_0 + U_1 V_1 + \ldots + U_p V_p < \varepsilon \tag{2}$$

To show (1) (since the series is convergent when $x = R$), N can be chosen so that

$$-\varepsilon < a_m R^m + a_{m+1} R^{m+1} + \ldots + a_{m+p} R^{m+p} < \varepsilon$$

if $N \leq m$ and $0 < p$.

However,to proceed further we need the following lemma:

If $m \leq S_k \leq M$, $k = 0, 1, 2, \ldots, p$, \hfill (3)

where

$$S_k = U_0 + U_1 + \ldots + U_k$$

and if

$$V_0 \geq V_1 \geq \ldots \geq V_p \geq 0; \tag{4}$$

then $mV_0 \leq U_0 V_0 + U_1 V_1 + \ldots + U_p V_p \leq MV_0$.

Therefore, for the proof of the theorem, since condition (4) is satisfied by the V's and (3) is satisfied with $m = -\varepsilon$ and $M = \varepsilon$, we have, by the lemma,

$$-\varepsilon V_0 \leq U_0 V_0 + U_1 V_1 + U_2 V_2 + \ldots + U_p V_p \leq \varepsilon V_0 .$$

However, since $V_0 \leq 1$, we have $\varepsilon V_0 \leq \varepsilon$ and $-\varepsilon \leq -\varepsilon V_0$ so that

$$-\varepsilon \leq U_0 V_0 + U_1 V_1 + \ldots + U_p V_p \leq \varepsilon$$

which is precisely inequality (2). This completes the proof for $0 \leq x \leq R$. The case of convergence at $x = -R$ is reduced to the first case by considering $g(x) = f(-x)$ at $x = R$.

• **PROBLEM 13-7**

Prove Abel's limit theorem which states:

If

$$\sum_{n=0}^{\infty} a_n x^n$$

converges at $x = x_0$, where x_0 may be an interior point or an endpoint of the interval of convergence, then

$$\lim_{x \to x_0} \left\{ \sum_{n=0}^{\infty} a_n x^n \right\} = \sum_{n=0}^{\infty} \lim_{x \to x_0} a_n x^n = \sum_{n=0}^{\infty} a_n x_0^n . \qquad (1)$$

If x_0 is a left hand endpoint it is proper to use $x \to x_0^+$ and for a right hand endpoint $x \to x_0^-$.

Solution: To simplify the proof assume the power series to be

$$\sum_{n=0}^{\infty} a_n x^n$$

with the endpoint of its interval of convergence at $x = 1$, so that the series surely converges for $0 \leq x \leq 1$. Now we need to show that

$$\lim_{x \to 1^-} \sum_{n=0}^{\infty} a_n x^n = \sum_{n=0}^{\infty} \lim_{x \to 1^-} a_n x^n = \sum_{n=0}^{\infty} a_n .$$

However, from the previous problem we know that

$$\sum_{n=0}^{\infty} a_n x^n$$

is uniformly convergent for $0 \leq x \leq 1$ since it is convergent at $x = 1$. In addition, by another theorem, if the functions

$$\{U_n(x)\}, \quad n = 0, 1, 2, 3, \ldots$$

are continuous in [a,b] and if

$$\sum_{n=0}^{\infty} U_n(x)$$

converges uniformly to the sum S(x) in [a,b], then S(x) is continuous in [a,b].

To apply this theorem to the proof of the given problem let

$$S(x) = \sum_{n=0}^{\infty} a_n x^n .$$

Then this means that S(x) is continuous at $x = 1$, since S(x) is the uniform limit of continuous functions in [0,1].

Therefore

$$\lim_{x \to 1^-} S(x) = S(1).$$

That is

$$\lim_{x \to 1^-} S(x) = \lim_{x \to 1^-} \sum_{n=0}^{\infty} a_n x^n = S(1) = \sum_{n=0}^{\infty} a_n = \sum_{n=0}^{\infty} \lim_{x \to 1^-} a_n x^n .$$

This completes the proof since extensions to other power series are easily made.

OPERATIONS ON POWER SERIES

• **PROBLEM 13-8**

a) Using power series, show that

$$\frac{d(\sin x)}{dx} = \cos x \quad \text{and} \quad \frac{d(\cos x)}{dx} = -\sin x.$$

b) Then show that $\sin a \cos b + \cos a \sin b = \sin(a+b)$

and

$$\cos a \cos b - \sin a \sin b = \cos(a+b).$$

<u>Solution</u>: The power series expansion for sin x and cos x are for all values x

$$\sin x = x - \frac{x^3}{3!} + \frac{x^5}{5!} - \frac{x^7}{7!} + \cdots \qquad (1)$$

$$\cos x = 1 - \frac{x^2}{2!} + \frac{x^4}{4!} - \frac{x^6}{6!} + \cdots . \qquad (2)$$

Since, by theorem, a power series can be differentiated term-by-term in any interval lying entirely within its radius of convergence, we have by (1) and (2), for all values x

$$\frac{d(\sin x)}{dx} = 1 - \frac{3x^2}{3!} + \frac{5x^4}{5!} - \frac{7x^6}{7!} + \ldots$$

$$= 1 - \frac{x^2}{2!} + \frac{x^4}{4!} - \frac{x^6}{6!} + \ldots = \cos x.$$

and

$$\frac{d(\cos x)}{dx} = \frac{-2x}{2!} + \frac{4x^3}{4!} - \frac{6x^5}{5!} + \ldots$$

$$= -x + \frac{x^3}{3!} - \frac{x^5}{5!} + \ldots = -\sin x$$

b) Using the result from part (a) we have

$$\{ \sin x \cos(h-x) + \cos x \sin(h-x) \}'$$

$$= \cos x \cos(h-x) + \sin x \sin(h-x)$$

$$- \sin x \sin(h-x) - \cos x \cos(h-x)$$

$$= 0$$

Thus sin x cos(h-x) + cos x sin(h-x) is constant and this constant equals the value sin h (this was found by letting x=0). Hence

$$\sin x \cos(h-x) + \cos x \sin(h-x) = \sin h.$$

Now replacing x by a and h by a+b yields

$$\sin a \cos b + \cos a \sin b = \sin(a+b).$$

From which differentiation with respect to a yields

$$\cos a \cos b - \sin a \sin b = \cos(a+b).$$

Note that from this result we obtain

$$\cos^2 x - \sin^2 x = \cos 2x .$$

● **PROBLEM 13-9**

Show that the series representation

$$e^x = 1 + x + \frac{x^2}{2!} + \frac{x^3}{3!} + \ldots + \frac{x^n}{n!} + \ldots$$

is valid for all values of x.

Solution: To do this problem we make use of Taylor's formula. That is if

$$f'(x), f''(x), \ldots, f^{(n)}(x)$$

exist and are continuous in the interval $a \leq x \leq b$ and if $f^{(n+1)}(x)$ exists in the interval $a < x < b$, then

$$f(x) = f(a) + f'(a)(x-a) + \frac{f''(a)(x-a)^2}{2!} + \ldots$$

$$+ \frac{f^{(n)}(a)(x-a)^n}{n!} + R_n \tag{1}$$

where R_n is the remainder and which is written in the Lagrange form as

$$R_n = \frac{f^{(n+1)}(\xi)}{(n+1)!}(x-a)^{n+1}$$

where $a < \xi < x$. Note that as n changes, ξ also changes in general. In addition if for all x and ξ in $[a,b]$ we have

$$\lim_{n \to \infty} R_n = 0,$$

then (1) can be written in the form

$$f(x) = f(a) + f'(a)(x-a) + \frac{f''(a)}{2!}(x-a)^2$$

$$+ \frac{f'''(a)(x-a)^3}{3!} + \ldots \tag{2}$$

Note that (2) is called the Taylor series or expansion of $f(x)$.

For the given problem we have $f(x) = e^x$, so that $f^{(n)}(x) = e^x$ for all orders n. Now taking $a = 0$ in (1) we have

$$f(x) = f(0) + f'(0)x + \frac{f''(0)x^2}{2!} + \ldots + \frac{f^{(n)}(0)x^n}{n!} + R_n$$

which yields upon substitution of $f(x) = e^x$,

$$e^x = 1 + x + \frac{x^2}{2!} + \ldots + \frac{x^n}{n!} + R_n$$

where

$$R_n = \frac{f^{(n+1)}(\xi)}{(n+1)!} x^{n+1} = \frac{e^\xi}{(n+1)!} x^{n+1}$$

where

$0 < \xi < x$.

Now we must prove that

$$\lim_{n \to \infty} R_n = 0,$$

so that we will have the Taylor expansion for e^x. However,

$$-|x| < \xi < |x| \quad \text{and} \quad 0 < e^\xi < e^{|x|};$$

hence

$$|R_n| = \left| \frac{e^\xi}{(n+1)!} x^{n+1} \right| \leq \frac{e^{|x|}|x|^{n+1}}{(n+1)!}. \tag{3}$$

Thus by (3) to prove

$$\lim_{n \to \infty} R_n = 0,$$

it is sufficient to prove that

$$\lim_{n \to \infty} \frac{|x|^n}{n!} = 0.$$

To do this choose an integer N such that $N \geq 2|x|$. Then if $n > N$,

$$\frac{|x|^n}{n!} = \frac{|x|^N}{N!} \frac{|x|^{n-N}}{(N+1)(N+2)\ldots n}$$

$$= \frac{|x|^N}{N!} \left(\frac{|x|}{N+1}\right)\left(\frac{|x|}{N+2}\right) \ldots \left(\frac{|x|}{n}\right)$$

$$\leq \frac{|x|^N}{N!} \left(\frac{N}{N+1}\right)\left(\frac{N}{N+2}\right) \ldots \left(\frac{N}{n}\right) \frac{1}{2^{n-N}} \leq \frac{|x|^N}{N!} \left(\frac{1}{2}\right)^{n-N}$$

Therefore

$$\frac{|x|^n}{n!} \leq \frac{|x|^N}{N!} \left(\frac{1}{2}\right)^{n-N}. \tag{4}$$

Now keeping N fixed we have

$$\lim_{n \to \infty} \left(\frac{1}{2}\right)^{n-N} = 0.$$

Therefore by (4)

$$\lim_{n \to \infty} \frac{|x|^n}{n!} = 0.$$

This means that the series representation

$$e^x = 1 + x + \frac{x^2}{2!} + \ldots + \frac{x^n}{n!} + \ldots$$

is valid for all values of x.

• **PROBLEM 13-10**

Derive the series expansion

$$\sin^{-1} x = x + \frac{1}{2} \frac{x^3}{3} + \frac{1}{2} \cdot \frac{3}{4} \frac{x^5}{5} + \frac{1 \cdot 3 \cdot 5}{2 \cdot 4 \cdot 6} \frac{x^7}{7} + \ldots \quad .$$

Solution: To derive the given series for $\sin^{-1} x$ we need the following theorem:

Let f be a function defined by

$$f(x) = \sum_{n=0}^{\infty} a_n x^n$$

where $R \neq 0$ for the power series. Then f is continuous in the open interval of convergence of the series. Moreover, if a and b are points of this interval,

$$\int_a^b f(x)\,dx = \sum_{n=0}^{\infty} a_n \frac{b^{n+1} - a^{n+1}}{n+1} \quad . \tag{1}$$

That is, the integral of the function is equal to the series obtained by integrating the original power series term by term, i.e.,

$$\int_a^b f(x)\,dx = \sum_{n=0}^{\infty} a_n \int_a^b x^n\,dx \quad .$$

To apply this theorem we start with the fact that

$$\sin^{-1} x = \int_0^x \frac{dt}{\sqrt{1-t^2}} \quad . \tag{2}$$

Now (2) is valid if $|x| \leq 1$; however the integral is improper if $x = \pm 1$ since the integrand becomes infinite at $t = \pm 1$. Now by the binomial theorem,

$$(1+x)^r = 1 + rx + \frac{r(r-1)}{2!} x^2 + \ldots + \frac{r(r-1) \cdots (r-n+1)}{n!} x^n + \ldots$$

when $|x| < 1$, and where r is any real number.

We replace x by $-t^2$ and let $r = -\frac{1}{2}$ to yield

$$(1-t^2)^{-\frac{1}{2}} = 1 - \frac{1}{2}(-t^2) + \frac{\left(-\frac{1}{2}\right)\left(-\frac{3}{2}\right)}{2!} (-t^2)^2 +$$

$$+ \frac{\left(-\frac{1}{2}\right)\left(-\frac{3}{2}\right)\left(-\frac{5}{2}\right)}{3!} (-t^2)^3 + \ldots$$

$$= 1 + \frac{1}{2} t^2 + \frac{1 \cdot 3}{2 \cdot 4} t^4 + \frac{1 \cdot 3 \cdot 5}{2 \cdot 4 \cdot 6} t^6 + \ldots ;$$

where this result is valid if $|t| < 1$. Therefore the series has radius of convergence $R = 1$. Thus, by the theorem we may integrate the series from $t = 0$ to $t = x$ if $|x| < 1$. That is

$$\sin^{-1} x = \int_0^x \frac{dt}{\sqrt{1-t^2}} = \int_0^x dt + \int_0^x \frac{t^2}{2} dt + \int_0^x \frac{1 \cdot 3 t^4}{2 \cdot 4} dt + \ldots$$

$$= x + \frac{1}{2} \frac{x^3}{3} + \frac{1 \cdot 3}{2 \cdot 4} \frac{x^5}{5} + \frac{1 \cdot 3 \cdot 5}{2 \cdot 4 \cdot 6} \frac{x^7}{7} + \ldots$$

if $|x| < 1$. This is the desired expansion.

● **PROBLEM 13-11**

a) Find an expansion in powers of x of the function

$$f(x) = \int_0^1 \frac{1 - e^{-tx}}{t} dt .$$

b) Use the result from part (a) to find $f\left(\frac{1}{2}\right)$ approximately.

Solution: a) Using the fact that for all values x the series representation for e^x is

$$e^x = \sum_{n=0}^{\infty} \frac{x^n}{n!} = 1 + x + \frac{x^2}{2!} + \ldots + \frac{x^n}{n!} + \ldots$$

we have

$$e^{-tx} = 1 - tx + \frac{t^2 x^2}{2!} - \frac{t^3 x^3}{3!} + \ldots .$$

Hence,

$$1 - e^{-tx} = tx - \frac{t^2 x^2}{2!} + \frac{t^3 x^3}{3!} - \ldots$$

so that

$$\frac{1 - e^{-tx}}{t} = x - \frac{tx^2}{2!} + \frac{t^2 x^3}{3!} - \ldots + (-1)^{n-1} \frac{t^{n-1} x^n}{n!} + \ldots .$$

Now this series representation is valid for all values of x and t. In addition, the radius of convergence of the power series in t is $R = \infty$. This is because,

$$\lim_{n \to \infty} \left| \frac{a_{n+1}}{a_n} \right| = \lim_{n \to \infty} |tx| \frac{1}{n+1} = 0$$

so that $R = \infty$.

Therefore, we can integrate the series term by term (this by the theorem in the previous problem) to obtain,

$$f(x) = xt - \frac{t^2 x^2}{2 \cdot 2!} + \frac{t^3 x^3}{3 \cdot 3!} - \ldots + (-1)^{n-1} \frac{t^n x^n}{n \cdot n!} + \ldots \Big|_0^1$$

or

$$f(x) = x - \frac{x^2}{2 \cdot 2!} + \frac{x^3}{3 \cdot 3!} - \ldots + (-1)^{n-1} \frac{x^n}{n \cdot n!} + \ldots \quad . \quad (1)$$

b) From (1) we have,

$$f\left(\frac{1}{2}\right) = \int_0^1 \frac{1 - e^{-t/2}}{t} \, dt = \frac{1}{2} - \frac{1}{2 \cdot 2!}\left(\frac{1}{2}\right)^2 + \frac{1}{3 \cdot 3!}\left(\frac{1}{2}\right)^3 - \ldots$$

which approximately equals 1.13 .

● **PROBLEM 13-12**

a) Prove that

$$\tan^{-1} x = x - \frac{x^3}{3} + \frac{x^5}{5} - \frac{x^7}{7} + \ldots$$

where the series is uniformly convergent in $-1 \le x \le 1$.

b) Prove that

$$\frac{\pi}{4} = 1 - \frac{1}{3} + \frac{1}{5} - \frac{1}{7} + \ldots \quad .$$

Solution: By the geometric series

$$\sum_{n=1}^{\infty} ar^{n-1} = a + ar + ar^2 + \ldots$$

which converges to $\frac{a}{(1-r)}$ if $|r| < 1$ (it diverges for $r \ge 1$), we have, with $r = -x^2$ and $a = 1$, that

$$\frac{1}{1 + x^2} = 1 - x^2 + x^4 - x^6 + \ldots \qquad -1 < x < 1 \quad . \quad (1)$$

Now

$$\tan^{-1} x = \int_0^x \frac{dx}{1 + x^2} \qquad -1 < x < 1 \quad .$$

Therefore by (1),

$$\tan^{-1} x = \int_0^x (1 - x^2 + x^4 - x^6 + \ldots) \, dx$$

or

$$\tan^{-1} x = x - \frac{x^3}{3} + \frac{x^5}{5} - \frac{x^7}{7} + \ldots \qquad -1 < x < 1 \qquad (2)$$

since, if $\Sigma V_n(x)$ converges uniformly to the sum $S(x)$ in $[a,b]$ and if $V_n(x)$, $n = 1, 2, 3, \ldots$, are continuous in $[a,b]$ then

$$\int_a^b S(x)\,dx = \sum_{n=1}^{\infty} \int_a^b V_n(x)\,dx$$

Now, at $x = 1$, we have on the right of (2). the series

$$1 - \frac{1}{3} + \frac{1}{5} - \frac{1}{7} + \ldots \; .$$

However, this series is a convergent alternate series. Similarly, for $x = -1$ we also have a convergent alternate series. Therefore by Abel's Theorem we know that the interval of uniform convergence includes the endpoints $x = \pm 1$. Hence, the series is uniformly convergent for $-1 \leq x \leq 1$ and in this interval

$$\tan^{-1} x = x - \frac{x^3}{3} + \frac{x^5}{5} - \frac{x^7}{7} + \ldots \; . \qquad (3)$$

b) From part (a), using (3) we have,

$$\lim_{x \to 1^-} \tan^{-1} x = \lim_{x \to 1^-} (x - \frac{x^3}{3} + \frac{x^5}{5} - \frac{x^7}{7} + \ldots)$$

which means, since $\tan \frac{\pi}{4} = 1$, that

$$\frac{\pi}{4} = 1 - \frac{1}{3} + \frac{1}{5} - \frac{1}{7} + \ldots \; .$$

This is the desired result.

● **PROBLEM 13-13**

Find a power series in x for:

a) $\tan x$

b) $\frac{\sin x}{\sin 2x}$ $(x \neq 0)$

Solution: To find the power series of the given functions, we need the following theorem:

Given the two power series

$$\sum_{n=0}^{\infty} a_n x^n = a_0 + a_1 x + a_2 x^2 + \ldots + a_n x^n + \ldots$$

and

$$\sum_{n=0}^{\infty} b_n x^n = b_0 + b_1 x + b_2 x^2 + \ldots + b_n x^n + \ldots,$$

where $b_0 \neq 0$, and where both of the series are convergent in some interval $|x| < R$, let f be a function defined by

$$f(x) = \frac{a_0 + a_1 x + a_2 x^2 + \ldots + a_n x^n + \ldots}{b_0 + b_1 x + b_2 x^2 + \ldots + b_n x^n + \ldots}.$$

Then for sufficiently small values of x the function f can be represented by the power series

$$f(x) = c_0 + c_1 x + c_2 x^2 + \ldots + c_n x^n + \ldots,$$

where the coefficients c_0, c_1, c_2, ..., c_n, ... are found by long division or equivalently by solving the following relations successively for each c_i (i = 0 to ∞):

$$b_0 c_0 = a_0$$
$$b_0 c_1 + b_1 c_0 = a_1$$
$$\vdots$$
$$b_0 c_n + b_1 c_{n-1} + \ldots + b_n c_0 = a_n$$
$$\vdots$$

a) To find the power series expansion of tan x we need the Taylor's series for sin x and cos x. That is

$$\sin x = x - \frac{x^3}{3!} + \frac{x^5}{5!} - \ldots$$

and

$$\cos x = 1 - \frac{x^2}{2!} + \frac{x^4}{4!} - \ldots$$

Then,

$$\tan x = \frac{\sin x}{\cos x} = \frac{x - \frac{x^3}{3!} + \frac{x^5}{5!} - \ldots}{1 - \frac{x^2}{2!} + \frac{x^4}{4!} - \ldots} \tag{1}$$

Therefore, by the theorem we can find the power series expansion of tan x by dividing the numerator by the denominator on the right side of (1). Hence, using long division we have

$$1 - \frac{1}{2}x^2 + \frac{1}{24}x^4 - \cdots \overline{\big)\, \begin{array}{l} x + \frac{1}{3}x^3 + \frac{2}{15}x^5 + \cdots \\ x - \frac{1}{6}x^3 + \frac{1}{120}x^5 - \cdots \end{array}}$$

$$\underline{x - \frac{1}{2}x^3 + \frac{1}{24}x^5 - \cdots}$$
$$\frac{1}{3}x^3 - \frac{1}{30}x^5 + \cdots$$
$$\underline{\frac{1}{3}x^3 - \frac{1}{6}x^5 + \cdots}$$
$$\frac{2}{15}x^5 - \cdots$$
$$\underline{\frac{2}{15}x^5 - \cdots}$$

Thus
$$\tan x = x + \frac{1}{3}x^3 + \frac{2}{15}x^5 + \cdots$$

b) Since
$$\sin x = x - \frac{x^3}{3!} + \frac{x^5}{5!} - \cdots$$

we have
$$\sin(2x) = 2x - \frac{(2x)^3}{3!} + \frac{(2x)^5}{5!} - \cdots ,$$

so that
$$\frac{\sin x}{\sin 2x} = \frac{x - \frac{x^3}{3!} + \frac{x^5}{5!} - \cdots}{2x - \frac{(2x)^3}{3!} + \frac{(2x)^5}{5!} - \cdots} . \qquad (2)$$

Now multiplying the numerator and denominator on the right side of (2) by 1/x yields

$$\frac{\sin x}{\sin 2x} = \frac{1 - \frac{x^2}{6} + \frac{x^4}{120} - \cdots}{2 - \frac{4}{3}x^2 + \frac{4}{15}x^4 - \cdots} .$$

Now by long division

$$2 - \frac{4}{3}x^2 + \frac{4}{15}x^4 \, \overline{\big)\, \begin{array}{l} \frac{1}{2} + \frac{1}{4}x^2 + \frac{5}{48}x^4 + \cdots \\ 1 - \frac{1}{6}x^2 + \frac{1}{120}x^4 - \cdots \end{array}}$$

$$\underline{\frac{1}{2} - \frac{4}{6}x^2 + \frac{4}{30}x^4 - \cdots}$$
$$\frac{1}{2}x^2 - \frac{15}{120}x^4 + \cdots$$
$$\underline{\frac{1}{2}x^2 - \frac{1}{3}x^4 + \cdots}$$
$$\frac{25}{120}x^4 - \cdots$$
$$\underline{\frac{25}{120}x^4 - \cdots}$$

Thus, for $x \neq 0$,

$$\frac{\sin x}{\sin(2x)} = \frac{1}{2} + \frac{1}{4}x^2 + \frac{5}{48}x^4 + \ldots \quad .$$

● **PROBLEM 13-14**

Starting from the power series expansion for $\tan^{-1} x$ find the power series expansion for $\tan x$. That is, derive the series expansion by a different method from the one used in the previous problem.

Solution: We are given that

$$\tan^{-1} x = x - \frac{x^3}{3} + \frac{x^5}{5} - \frac{x^7}{7} + \ldots + (-1)^{n-1} \frac{x^{2n-1}}{2n-1} + \ldots$$

for $-1 \leq x \leq 1$ and are asked to derive the power series expansion for $\tan x$. To do this we need the following theorem:

If given the power series

$$y = f(x) = \sum_{n=0}^{\infty} c_n x^n \quad , \quad \text{with } c_0 = 0$$

and if

$$|x| < R_0 \quad \text{and } c_1 \neq 0 ,$$

then there is an inverse function

$$x = g(y) = \sum_{n=1}^{\infty} b_n y^n$$

where $|y| < R_1$, $R_1 > 0$.

In addition, the coefficients b_n are determined from the identity

$$x \equiv \sum_{n=1}^{\infty} b_n \left[\sum_{m=1}^{\infty} c_m x^m \right]^n .$$

For the given problem, we have

$$y = \tan^{-1} x = x - \frac{x^3}{3} + \frac{x^5}{5} - \frac{x^7}{7} + \ldots + (-1)^{n-1} \frac{x^{2n-1}}{2n-1} + \ldots$$

Therefore

$$x = \tan y \equiv \sum_{n=1}^{\infty} b_n y^n = \sum_{n=1}^{\infty} b_n \left(x - \frac{x^3}{3} + \frac{x^5}{5} + \ldots \right)^n$$

or

$$x \equiv b_1 \left(x - \frac{x^3}{3} + \ldots\right) + b_2 \left(x - \frac{x^3}{3} + \ldots\right)^2 +$$
$$+ b_3 \left(x - \frac{x^3}{3} + \ldots\right)^3 + \ldots,$$

or

$$x \equiv b_1 x + b_2 x^2 + x^3 \left(-\frac{1}{3} b_1 + b_3\right) + \ldots$$

Hence $\quad b_1 = 1, \; b_2 = 0, \; b_3 - \frac{1}{3} b_1 = 0, \; \ldots$,

Thus
$$x = \tan y = y + \frac{y^3}{3} + \ldots \; .$$

● **PROBLEM 13-15**

Show that

$$\int_0^1 \frac{\log(1-t)}{t} \, dt = -\left(\frac{1}{1^2} + \frac{1}{2^2} + \frac{1}{3^2} + \ldots + \frac{1}{n^2} + \ldots\right)$$

Solution: We first observe that the integral is improper at $t = 1$, but not at $t = 0$, since, by L'Hospital's rule, we have the integrand approaching the finite limit -1 as $t \to 0$.

Now to show that the integral is equal to the given series, we need to start from Taylor's formula with remainder for

$$\log(1+x) = x - \frac{1}{2} x^2 + \frac{1}{3} x^3 - \ldots + (-1)^{n-1} \frac{1}{n} x^n + R_{n+1}$$

if $1 > x > -1$ and where

$$|R_{n+1}| \leq \begin{cases} \dfrac{|x|^{n+1}}{n+1} & \text{if } 0 \leq x \leq 1 \\[2mm] \dfrac{|x|^{n+1}}{1+x} & \text{if } -1 < x \leq 0 \end{cases} \quad (1)$$

However (1) shows that $R_{n+1} \to 0$ as $n \to \infty$ when x is limited as indicated. Therefore

$$\log(1+x) = x - \frac{1}{2} x^2 + \frac{1}{3} x^3 - \ldots + (-1)^{n-1} \frac{1}{n} x^n + \ldots \; . \quad (2)$$

Now letting $x = -t$ in (2) yields

$$\log(1-t) = -t - \frac{1}{2} t^2 - \frac{1}{3} t^3 - \ldots - \frac{1}{n} t^n - \ldots \; .$$

Dividing by t gives,

$$\frac{\log(1-t)}{t} = -1 - \frac{1}{2}t - \frac{1}{3}t^2 - \ldots - \frac{1}{n}t^{n-1} - \ldots \, .$$

Note that the series diverges at t = 1. However, we can still integrate from 0 to x. This yields,

$$\int_0^x \frac{\log(1-t)}{t}dt = \int_0^x \left(-1 - \frac{1}{2}t - \frac{1}{3}t^2 - \ldots\right)dt$$

$$= -x - \frac{1}{2^2}x^2 - \frac{1}{3^2}x^3 - \ldots - \frac{1}{n^2}x^n - \ldots \, . \quad (3)$$

However, this series converges when x = 1. That is, the series

$$-1 - \frac{1}{2^2} - \frac{1}{3^2} - \ldots - \frac{1}{n^2} - \ldots$$

$$= -\left(1 + \frac{1}{2^2} + \frac{1}{3^2} + \ldots + \frac{1}{n^2} + \ldots\right) = -\sum_{n=1}^{\infty}\frac{1}{n^2},$$

which we know to be convergent. Therefore, as a special case of (3) we have,

$$\int_0^1 \frac{\log(1-t)}{t}dt = -\left(\frac{1}{1^2} + \frac{1}{2^2} + \frac{1}{3^2} + \ldots + \frac{1}{n^2} + \ldots\right).$$

● **PROBLEM 13-16**

Let the functions $J_0(x)$, $J_1(x)$ be defined as follows:

$$J_0(x) = 1 - \frac{x^2}{(1!)^2 2^2} + \frac{x^4}{(2!)^2 2^4} - \ldots + (-1)^n \frac{x^{2n}}{(n!)^2 2^{2n}} + \ldots \quad (1)$$

$$J_1(x) = \frac{x}{2}\left[1 - \frac{x^2}{1!2!2^2} + \ldots + (-1)^n \frac{x^{2n}}{n!(n+1)!2^{2n}} + \ldots\right]. \quad (2)$$

Show that $J_0(x)$, $J_1(x)$ are defined for all values of x and that $J_0{}'(x) = -J_1(x)$.

Solution: Before proceeding we make note that $J_0(x)$ is called the Bessel function of order zero of first kind, that $J_1(x)$ is called the Bessel function of order one of first kind, and that these functions arise in many kinds of physical problems.

Now to show that $J_0(x)$, $J_1(x)$ are defined for all values of x, apply the ratio test to each function. Hence for $J_0(x)$, (setting $U_n = (-1)^n x^{2n}/(n!)^2 2^{2n}$)

$$\lim_{n \to \infty} \left| \frac{U_{n+1}}{U_n} \right| = \lim_{n \to \infty} \left| \frac{(-1)^{n+1} x^{2n+2}}{[(n+1)!]^2 2^{2n+2}} \cdot \frac{(n!)^2 2^{2n}}{x^{2n}(-1)^n} \right|$$

$$= |x^2| \lim_{n \to \infty} \frac{1}{4(n+1)^2} = 0 .$$

Therefore $J_0(x)$ is convergent for all values x. Similarly for $J_1(x)$, we have,

$$\left(\text{setting } U_n = (-1)^n x^{2n}/(n!(n+1)!2^n) \right)$$

$$\lim_{n \to \infty} \left| \frac{U_{n+1}}{U_n} \right| = \lim_{n \to \infty} \left| \frac{(-1)^{n+1} x^{2n+2}}{(n+1)!(n+2)! 2^{n+1}} \cdot \frac{n!(n+1)!2^n}{(-1)^n x^{2n}} \right|$$

$$= \frac{x^2}{2} \lim_{n \to \infty} \frac{1}{(n+1)(n+2)} = 0$$

so that $J_1(x)$ is convergent for all values of x.

Now to show that $J_0'(x) = -J_1(x)$ we need to write the series $J_0(x)$ and $J_1(x)$ in the forms

$$J_0(x) = \sum_{n=0}^{\infty} (-1)^n \frac{x^{2n}}{n!n!2^{2n}}$$

$$J_1(x) = \sum_{n=0}^{\infty} (-1)^n \frac{x^{2n+1}}{n!(n+1)!2^{n+1}}$$

Now by theorem, the power series

$$f(x) = a_0 + a_1 x + a_2 x^2 + a_3 x^3 + \ldots$$

can be differentiated term by term over any interval lying entirely within the interval of convergence; thus

$$f'(x) = a_1 + 2a_2 x + 3a_3 x^2 + \ldots .$$

Therefore,

$$J_0'(x) = \sum_{n=1}^{\infty} (-1)^n \frac{2nx^{2n-1}}{n!n!2^{2n}} = \sum_{n=1}^{\infty} (-1)^n \frac{x^{2n-1}}{(n-1)!n!2^{2n-1}} .$$

Notice that in ther series for $J_0(x)$ the term with n = 0 is a constant, so that the series for $J_0'(x)$ begins with the term for which n = 1. Now writing n + 1 in place of n in the last summation means the new index n will go from 0 to ∞. Consequently we have

$$J_0'(x) = \sum_{n=0}^{\infty} (-1)^{n+1} \frac{x^{2n+1}}{n!(n+1)!2^{2n+1}}$$

$$= - \sum_{n=0}^{\infty} (-1)^n \frac{x^{2n+1}}{n!(n+1)!2^{2n+1}} = - J_1(x)$$

since

$$(-1)^{n+1} = -(-1)^n.$$

This is the desired result.

● **PROBLEM 13-17**

Find the sum of the series

$$\sum_{n=1}^{\infty} n^2 (x+3)^n.$$

Solution: To find the sum of the given power series, we first need to determine its interval of convergence. By the ratio test, if $U_n = n^2(x+3)^n$

$$\lim_{n \to \infty} \left| \frac{U_{n+1}}{U_n} \right| = \lim_{n \to \infty} \left| \frac{(n+1)^2 (x+3)^{n+1}}{n^2 (x+3)^n} \right| = |x+3| \lim_{n \to \infty} \frac{n^2 + 2n + 1}{n^2}$$

$$= |x + 3|.$$

Therefore, the series converges for $|x + 3| < 1$ (i.e., $-4 < x < -2$) and diverges for $|x + 3| > 1$.

Now to proceed further, we consider the geometric series

$$\sum_{n=0}^{\infty} x^n,$$

which converges if $|x| < 1$. That is $R = 1$ and

$$\sum_{n=0}^{\infty} x^n = \frac{1}{(1-x)}$$

where $|x| < 1$. By differentiating this series, which we can do since a power series can be differentiated term by term over any interval lying entirely within the interval of convergence, we obtain

$$\sum_{n=1}^{\infty} nx^{n-1} = \frac{1}{(1-x)^2} \qquad -1 < x < 1 \qquad (1)$$

Differentiating again yields

$$\sum_{n=2}^{\infty} (n^2 - n) x^{n-2} = \frac{2}{(1-x)^3}$$

or

$$\sum_{n=2}^{\infty} n^2 x^{n-2} - \sum_{n=2}^{\infty} n x^{n-2} = \frac{2}{(1-x)^3} \quad . \qquad (2)$$

Multiplying each side of (2) by x^2 gives

$$\sum_{n=2}^{\infty} n^2 x^n - \sum_{n=2}^{\infty} n x^n = \frac{2x^2}{(1-x)^3} \qquad -1 < x < 1$$

or

$$\sum_{n=2}^{\infty} n^2 x^n - \sum_{n=1}^{\infty} n x^n + x = \frac{2x^2}{(1-x)^3}$$

which by (1) means

$$\sum_{n=1}^{\infty} n^2 x^n - \frac{x}{(1-x)^2} = \frac{2x^2}{(1-x)^3} \qquad -1 < x < 1$$

Hence

$$\sum_{n=1}^{\infty} n^2 x^n = \frac{2x^2}{(1-x)^3} + \frac{x}{(1-x)^2} \qquad -1 < x < 1$$

Then substituting x by x+3 we obtain

$$\sum_{n=1}^{\infty} n^2 (x+3)^n = \frac{2(x+3)^2}{(-x-2)^3} + \frac{(x+3)}{(-x-2)^2} \qquad -1 < x+3 < 1$$

or

$$\sum_{n=1}^{\infty} n^2 (x+3)^n = - \frac{x^2 + 7x + 12}{(x+2)^3} \qquad -4 < x < -2 \; .$$

● **PROBLEM 13-18**

Using Power series, show that

$$\log 2 = 1 - \frac{1}{2} + \frac{1}{3} - \frac{1}{4} + \ldots \; .$$

<u>Solution</u>: Starting with the fact that

$$e^x = 1 + x + \frac{x^2}{2!} + \frac{x^3}{3!} + \ldots$$

we have

$$e^x \to \infty \quad \text{as } x \to \infty \quad \text{and}$$

$$e^{-x} = \frac{1}{e^x} \to 0 \quad \text{as } x \to \infty.$$

Therefore e^x increases from 0 to ∞ as x increases from $-\infty$ to ∞. Now by the implicit function theorem we know that $x = e^y$ defines y as a continuous, differentiable function of x, increasing from $-\infty$ to ∞ as x increases from 0 to ∞. This means $y = \log x$ if and only if $x = e^y$. In addition

$$\frac{d(\log x)}{dx} = \frac{1}{x}$$

and so for $x > -1$ we have

$$\log(1+x) = \int_0^x \frac{dt}{1+t}$$

$$= \int_0^x \sum_{n=0}^{\infty} (-1)^n t^n \, dt$$

if $|x| < 1$. This because, for

$$\sum_{n=0}^{\infty} (-1)^n t^n$$

we have

$$S_n = 1 - t + t^2 - t^3 + \ldots + (-1)^n t^n$$

$$tS_n = t - t^2 + t^3 - t^4 + \ldots + (-1)^n t^{n+1}$$

Adding yields

$$S_n + tS_n = 1 + (-1)^n t^{n+1}$$

or

$$S_n = \frac{1 + (-1)^n t^{n+1}}{1+t}$$

and

$$\lim_{n \to \infty} S_n = \frac{1}{1+t}$$

for $|t| < 1$.

Thus,

$$\log(1+x) = x - \frac{x^2}{2} + \frac{x^3}{3} - \ldots$$

for $|x| < 1$.

Abel's limit theorem states that if

$$\sum_{n=0}^{\infty} a_n$$

converges, then

$$\sum_{n=0}^{\infty} a_n x^n$$

converges uniformly on $[-r, 1]$ if $0 \leq r < 1$. In particular

$$\lim_{\substack{x \to 1 \\ x < 1}} \sum_{n=0}^{\infty} a_n x^n = \sum_{n=0}^{\infty} \lim_{\substack{x \to 1 \\ x < 1}} a_n x^n = \sum_{n=0}^{\infty} a_n .$$

Therefore, since

$$\log(1+x) = \sum_{n=1}^{\infty} \frac{(-1)^{n-1}}{n} x^n$$

and since the alternating series

$$\sum_{n=1}^{\infty} \frac{(-1)^{n-1}}{n}$$

converges, we have

$$\lim_{\substack{x \to 1 \\ x < 1}} \sum_{n=1}^{\infty} \frac{(-1)^{n-1}}{n} x^n = \sum_{n=1}^{\infty} \frac{(-1)^{n-1}}{n}$$

Thus,

$$\log 2 = \sum_{n=1}^{\infty} \frac{(-1)^{n-1}}{n} = 1 - \frac{1}{2} + \frac{1}{3} - \frac{1}{4} + \ldots .$$

● **PROBLEM 13-19**

By the use of power series, show that for $x, y \in R$:

a) $e^x e^y = e^{x+y}$

b) $\sin x \cos x = \frac{1}{2} \sin 2x$.

Solution: a) The power series expansion for e^x, where

x is any real number is

$$e^x = 1 + x + \frac{x^2}{2!} + \frac{x^3}{3!} + \ldots = \lim_{n \to \infty} \sum_{k=0}^{n} \frac{x^k}{k!}. \qquad (1)$$

Now by theorem, given the two power series

$$\sum_{n=0}^{\infty} a_n x^n, \quad \sum_{n=0}^{\infty} b_n x^n,$$

the two series can be multiplied to obtain

$$\sum_{n=0}^{\infty} c_n x^n$$

where

$$c_n = a_0 b_n + a_1 b_{n-1} + a_2 b_{n-2} + \ldots + a_n b_0;$$

the result being valid for each x within the common interval of convergence.

Therefore, by (1)

$$e^x e^y = \lim_{n \to \infty} \sum_{k=0}^{n} \frac{x^k}{k!} \sum_{k=0}^{n} \frac{y^k}{k!}$$

$$= \lim_{n \to \infty} \sum_{k=0}^{n} \sum_{j=0}^{k} \frac{x^j}{j!} \frac{y^{k-j}}{(k-j)!}$$

$$= \lim_{n \to \infty} \sum_{k=0}^{n} \frac{1}{k!} \sum_{j=0}^{k} \frac{k!}{j!(k-j)!} x^j y^{k-j} \qquad (2)$$

However, since the binomial theorem states that

$$\sum_{r=0}^{m} \frac{m!}{r!(m-r)!} a^r b^{m-r} = (a+b)^m,$$

(2) becomes

$$e^x e^y = \lim_{n \to \infty} \sum_{k=0}^{n} \frac{1}{k!} (x+y)^k = e^{x+y}$$

That is $e^x e^y = e^{x+y}$, this is the desired result.

b) Since

$$\sin x = x - \frac{x^3}{3!} + \frac{x^5}{5!} - \ldots = \lim_{n \to \infty} \sum_{k=0}^{n} \frac{(-1)^k x^{2k+1}}{(2k+1)!}$$

and
$$\cos x = 1 - \frac{x^2}{2!} + \frac{x^4}{4!} - \ldots = \lim_{n\to\infty} \sum_{k=0}^{n} \frac{(-1)^k x^{2k}}{(2k)!}$$

for all real x, we have

$$\sin x \cos x = \lim_{n\to\infty} \sum_{k=0}^{n} \frac{(-1)^k x^{2k+1}}{(2k+1)!} \sum_{k=0}^{n} \frac{(-1)^k x^{2k}}{(2k)!}$$

$$= \lim_{n\to\infty} \sum_{k=0}^{n} \sum_{j=0}^{k} \frac{(-1)^j x^{2j}}{(2j)!} \frac{(-1)^{k-j} x^{2k+1-2j}}{(2k+1-2j)!}$$

$$= \lim_{n\to\infty} \sum_{k=0}^{n} (-1)^k x^{2k+1} \sum_{j=0}^{k} \frac{1}{(2j)!(2k+1-2j)!}$$

$$= \lim_{n\to\infty} \sum_{k=0}^{n} \frac{(-1)^k x^{2k+1}}{(2k+1)!} \cdot \frac{1}{2} \left[\sum_{i=0}^{2k+1} \frac{(2k+1)!}{i!(2k+1-i)!} + \sum_{i=0}^{2k+1} \frac{(2k+1)!}{i!(2k+1-i)!} (-1)^i \right]$$

$$= \lim_{n\to\infty} \sum_{k=0}^{n} \frac{(-1)^k x^{2k+1}}{(2k+1)!} \cdot \frac{1}{2} \left[(1+1)^{2k+1} + (1-1)^{2k+1} \right]$$

since

$$\sum_{i=0}^{2k+1} \frac{(2k+1)!}{i!(2k+1-i)!} = (1+1)^{2k+1}$$

and

$$\sum_{i=0}^{2k+1} \frac{(2k+1)!}{i!(2k+1-i)!} (-1)^i = (1-1)^{2k+1}$$

by the binomial theorem.

Hence,

$$\sin x \cos x = \frac{1}{2} \lim_{n\to\infty} \sum_{k=0}^{n} \frac{(-1)^k x^{2k+1}}{(2k+1)!} 2^{2k+1}$$

or

$$\sin x \cos x = \frac{1}{2} \lim_{n\to\infty} \sum_{k=0}^{n} \frac{(-1)^k (2x)^{2k+1}}{(2k+1)!} = \frac{1}{2} \sin 2x.$$

This is the desired result.

● **PROBLEM 13-20**

Approximate the value of the integral

$$\int_0^1 \frac{1 - e^{-x^2}}{x^2} \, dx$$

<u>Solution</u>: Starting from the fact that

$$e^a = 1 + a + \frac{a^2}{2!} + \frac{a^3}{3!} + \frac{a^4}{4!} + \ldots$$

for $-\infty < a < \infty$,

and upon setting $a = -x^2$ gives

$$e^{-x^2} = 1 - x^2 + \frac{x^4}{2!} - \frac{x^6}{3!} + \frac{x^8}{4!} - \ldots$$

for $-\infty < x < \infty$.

Therefore

$$1 - e^{-x^2} = x^2 - \frac{x^4}{2!} + \frac{x^6}{3!} - \frac{x^8}{4!} + \ldots$$

which means that

$$\frac{1 - e^{-x^2}}{x^2} = 1 - \frac{x^2}{2!} + \frac{x^4}{3!} - \frac{x^6}{4!} + \ldots$$

Now this series converges for all x and in addition converges uniformly for $0 \le x \le 1$.

This because

$$\frac{1 - e^{-x^2}}{x^2} = \sum_{n=1}^{\infty} \frac{(-1)^{n+1} x^{2n-2}}{n!} \tag{1}$$

and since

$$\left| \frac{x^{2n-2}}{n!} \right| \le \left| \frac{1}{n!} \right|$$

for $0 \le x \le 1$,

from the fact that

$$\sum_{n=1}^{\infty} \frac{1}{n!}$$

converges, it follows that the series (1) converges

uniformly for $0 \leq x \leq 1$ by the Weierstrass M-test.
Integrating both the sides of (1) from 0 to 1 gives

$$\int_0^1 \frac{1 - e^{-x^2}}{x^2} dx = x - \frac{x^3}{3 \cdot 2!} + \frac{x^5}{5 \cdot 3!} - \frac{x^7}{7 \cdot 4!} + \cdots \Big|_0^1$$

$$= 1 - \frac{1}{3 \cdot 2!} + \frac{1}{5 \cdot 3!} - \frac{1}{7 \cdot 4!} + \cdots$$

$$= 1 - 0.166666 + 0.033333 - 0.005952 + \cdots$$

$$\approx 0.8607$$

● **PROBLEM 13-21**

Show that

$$\sum_{n=0}^{\infty} \frac{a_n x^n}{1-x} = \sum_{n=0}^{\infty} (a_0 + a_1 + \cdots + a_n) x^n . \quad (1)$$

Then use the result to find the function represented by the following series:

a)
$$\sum_{n=0}^{\infty} \left(1 + \frac{1}{1!} + \cdots + \frac{1}{n!}\right) x^n$$

b)
$$\sum_{n=1}^{\infty} \left(0 + 1 - \frac{1}{2} + \frac{1}{3} - \cdots + (-1)^{n+1} \frac{1}{n}\right) x^n .$$

Solution: Starting from the right side of equation (1) it follows that

$$\sum_{n=0}^{\infty} (a_0 + a_1 + \cdots + a_n) x^n = \sum_{n=0}^{\infty} (a_0 x^n + a_1 x^n + \cdots + a_n x^n)$$

$$= a_0 + a_0 x + a_1 x + a_0 x^2 + a_1 x^2 + a_2 x^2$$

$$+ a_0 x^3 + a_1 x^3 + a_2 x^3 + a_3 x^3 + \cdots$$

or

$$\sum_{n=0}^{\infty} (a_0 + a_1 + \cdots + a_n) x^n$$

$$= a_0 (1 + x + x^2 + x^3 + \cdots) + a_1 (x + x^2 + x^3 + \cdots)$$

$$+ a_2 (x^2 + x^3 + x^4 + \cdots) + a_3 (x^3 + x^4 + x^5 + \cdots)$$

$$+ \cdots . \quad (2)$$

Now multiplying (2) by (1-x) gives

$$(1-x) \sum_{n=0}^{\infty} (a_0 + \ldots + a_n)x^n =$$

$$(1-x)\left[a_0(1 + x + x^2 + \ldots) + a_1(x + x^2 + x^3 + \ldots) \right.$$
$$+ a_2(x^2 + x^3 + x^4 + \ldots)$$
$$\left. + a_3(x^3 + x^4 + x^5 + \ldots) + \ldots\right] - \left[a_0(x + x^2 + x^3 + \ldots)\right.$$
$$\left. + a_1(x^2 + x^3 + x^4 + \ldots) + a_2(x^3 + x^4 + x^5 + \ldots) + \ldots\right]$$

$$= a_0 + a_1 x + a_2 x^2 + a_3 x^3 + \ldots$$

This means that

$$(1-x) \sum_{n=0}^{\infty} (a_0 + a_1 + \ldots + a_n)x^n = \sum_{n=0}^{\infty} a_n x^n$$

or

$$\sum_{n=0}^{\infty} \frac{a_n x^n}{1-x} = \sum_{n=0}^{\infty} (a_0 + a_1 + \ldots + a_n)x^n$$

a)

$$\sum_{n=0}^{\infty} \left(1 + \frac{1}{1!} + \ldots + \frac{1}{n!}\right)x^n .$$

For this series

$$a_0 = 1, \quad a_1 = \frac{1}{1!}, \ldots, \quad a_n = \frac{1}{n!}$$

Therefore by equation (1),

$$\sum_{n=0}^{\infty} \left(1 + \frac{1}{1!} + \ldots + \frac{1}{n!}\right)x^n = \sum_{n=0}^{\infty} \frac{x^n}{n!} \cdot \frac{1}{1-x} = \frac{1}{1-x} \sum_{n=0}^{\infty} \frac{x^n}{n!}$$

However,

$$\sum_{n=0}^{\infty} \frac{x^n}{n!} = e^x ,$$

so the function represented by the series is

$$\frac{e^x}{1-x}$$

b)
$$\sum_{n=1}^{\infty} \left(0 + 1 - \frac{1}{2} + \frac{1}{3} - \ldots + (-1)^{n+1} \frac{1}{n}\right) x^n .$$

By (1) this series equals

$$\sum_{n=1}^{\infty} (-1)^{n+1} \frac{x^n}{n} \cdot \frac{1}{1-x} .$$

Since

$$\log(1+x) = \sum_{n=1}^{\infty} (-1)^{n+1} \frac{x^n}{n} ,$$

this means that the function represented by the series is

$$\frac{\log(1+x)}{1-x} .$$

CHAPTER 14

FOURIER SERIES

The subject of Fourier series has many and varied applications in mathematics and physics, but finds its most important applications in areas where periodic functions are widely used, especially in electrical circuits and electromagnetic waves. A function f is said to be periodic with period T if $f(x+T) = f(x)$ for all x. An example is the alternating voltage supplied by an electrical wall socket in which case x represents time.

The fundamental question in Fourier series, as in power series, is "when can a function be represented by its Fourier series?" The Fourier series of a function with period 2c (and also a function defined only on an interval of length 2c) is an infinite trigonometric sum of the form

$$\alpha_0 + \sum_{n=1}^{\infty} \alpha_n \cos\left(\frac{n\pi x}{c}\right) + \beta_n \sin\left(\frac{n\pi x}{c}\right).$$

Since each term in the sum has period 2c, so will any function it represents, provided it is convergent. The association between f and this series will be seen to lie in the evaluation of the coefficients and with these coefficients defined in terms of f, the problem will be to find when this series adequately represents f.

Three types of convergence will be discussed: Pointwise Convergence, Uniform Convergence and Convergence in the Mean. Each has its own advantages in the determination of when a function's Fourier series adequately represents it. For instance, the Uniform Convergence Theorem has the advantage that if certain conditions are satisfied, then the Fourier series of f will converge to f at each point where f is defined. However, these conditions on f are very stringent. On the other hand, f's Fourier series will be found to converge in the mean for all but the most pathological of functions. However, this is a statistical type of convergence in that it might not be possible to say that the Fourier series of f at x_o converges to $f(x_o)$ for any x_o! Finally, Pointwise Convergence will be seen to be somewhere in between the other two both in adequacy of representation and stringency of conditions on f. However, Convergence in the Mean is the most widely used for physical applications.

DEFINITIONS AND EXAMPLES

• **PROBLEM** 14-1

Consider the infinite trigonometric series

$$\frac{a_0}{2} + \sum_{n=1}^{\infty} (a_n \cos nx + b_n \sin nx)$$ and assume that it converges uniformly for all $x \in (-\pi, \pi)$. It can then be considered as a function f of x with period 2π, i.e.

$$f(x) = \frac{a_0}{2} + \sum_{n=1}^{\infty} a_n (\cos nx + b_n \sin nx). \quad (1)$$

Determine the values of a_n, b_n in terms of $f(x)$.

Solution: It is this computation which leads to the definition of the Fourier Series of a given function $f(x)$. First multiply both sides of (1) by $\cos mx$ where m is a positive integer which we will vary later. This yields

$$f(x) \cos mx = \frac{a_0}{2} \cos mx + \sum_{n=1}^{\infty} a_n \cos nx \cos mx$$

$$+ \sum_{n=1}^{\infty} b_n \sin nx \cos mx. \quad (2)$$

The next step is to integrate both sides of equation (2) from $-\pi$ to π. In order to integrate the two series on the right term by term these two series would have to be uniformly convergent, but since this exercise is intended only to motivate a definition, we will simply assume that termwise integration is valid. Thus, (2) becomes

$$\int_{-\pi}^{\pi} f(x) \cos mx \, dx = \frac{a_0}{2} \int_{-\pi}^{\pi} \cos mx \, dx + \sum_{n=1}^{\infty} \left(a_n \int_{-\pi}^{\pi} \cos nx \cos mx \, dx \right)$$

$$+ \sum_{n=1}^{\infty} \left(b_n \int_{-\infty}^{\infty} \sin nx \cos mx \, dx \right) \quad (3)$$

This rather formidable expression yields useful information if one recalls the trigonometric identities

$$\sin nx \sin mx = \frac{1}{2} \cos(n-m)x - \frac{1}{2} \cos(n+m)x \quad (4)$$

$$\cos nx \cos mx = \frac{1}{2} \cos(n+m)x + \frac{1}{2} \cos(n-m)x \quad (5)$$

$$\sin nx \cos mx = \frac{1}{2}\sin(n+m)x + \frac{1}{2}\sin(n-m)x. \qquad (6)$$

Using these three identities, the following equations may be verified by carrying out the integrations:

$$\int_{-\pi}^{\pi} \sin nx \cos mx \, dx = 0 \qquad \text{(for all } n,m>0) \qquad (7)$$

$$\int_{-\pi}^{\pi} \cos nx \cos mx \, dx = \begin{cases} 0 & \text{(if } n \neq m) \\ \pi & \text{(if } n = m) \end{cases} \qquad (8)$$

$$\int_{-\pi}^{\pi} \sin nx \sin mx \, dx = \begin{cases} 0 & \text{(if } n \neq m) \\ \pi & \text{(if } n = m) \end{cases} \qquad (9)$$

For instance, using the identity (4) in the integral of equation (9) yields

$$\int_{-\pi}^{\pi} \sin nx \sin mx \, dx = \frac{1}{2}\int_{-\pi}^{\pi} \cos(n-m)x \, dx - \frac{1}{2}\int_{-\pi}^{\pi} \cos(n+m)x \, dx. \qquad (10)$$

If $n \neq m$, then

$$\int_{-\pi}^{\pi} \cos(n-m)x \, dx = \left. \frac{\sin(n-m)x}{n-m} \right|_{-\pi}^{\pi} = 0$$

and if $n = m$, then

$$\int_{-\pi}^{\pi} \cos(n-n)x \, dx = \int_{-\pi}^{\pi} dx = 2\pi.$$

Also,

$$\int_{-\pi}^{\pi} \cos(n+m)x \, dx = \left. \frac{\sin(n+m)x}{n+m} \right|_{-\pi}^{\pi} = 0 \qquad \text{(for all } n,m>0).$$

Using these results in (10) yields the result quoted in (9) and the other formulas are established in a similar fashion. These formulas are called the orthogonality properties of sin and cos.

Returning to the series in (3), it is seen that all terms in the second sum are zero (by equation (7)) and that for any m, only one term in the first sum is nonzero by equation (8). That is, for m>0,

$$\int_{-\pi}^{\pi} \cos mx \, dx = 0$$

so that (3) gives

$$\int_{-\pi}^{\pi} f(x) \cos mx \, dx = a_m \pi \quad (m > 0). \tag{11}$$

The coefficients b_n are treated similarly, that is the expansion (1) is multiplied by sin mx and integrated. Again the orthogonality properties (7)-(9) are employed to yield

$$\int_{-\pi}^{\pi} f(x) \sin mx \, dx = b_m \pi. \tag{12}$$

Finally, to obtain a_0, simply integrate the expansion (1) as it stands from $-\pi$ to π. This results in

$$\int_{-\pi}^{\pi} f(x) \, dx = a_0 \pi. \tag{13}$$

The results of equations (11), (12), (13) may be summarized as

$$a_n = \frac{1}{\pi} \int_{-\pi}^{\pi} f(x) \cos nx \, dx \quad (n \geq 0) \tag{14}$$

$$b_n = \frac{1}{\pi} \int_{-\pi}^{\pi} f(x) \sin nx \, dx \quad (n > 0). \tag{15}$$

Thus, it has been proved that if a function f is representable by a uniformly convergent trigonometric series then that series must have the coefficients of equations (14) and (15).

● **PROBLEM 14-2**

Let $f(x)$ be a real valued function of one variable, $f: R \to R$, with a period 2π (i.e. $f(x+2\pi) = f(x)$ for all $x \in R$). Define the Fourier Series of f.

Solution: In the previous problem it was determined that if a function f was represented by a uniformly convergent trigonometric series then the coefficients in that series must have certain values related to $f(x)$. Now, motivated by that discovery, the desired definition is made as follows: The series

$$\frac{a_0}{2} + \sum_{n=1}^{\infty} (a_n \cos nx + b_n \sin nx)$$

where the coefficients are given by

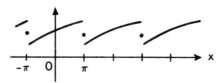

y = FOURIER SERIES OF f

PERIODIC EXTENSION OF A FUNCTION
DEFINED BETWEEN $-\pi$ AND π.   Fig. 1

$$a_n = \frac{1}{\pi} \int_{-\pi}^{\pi} f(x) \cos nx\, dx \qquad (n \geq 0) \qquad (1)$$

$$b_n = \frac{1}{\pi} \int_{-\pi}^{\pi} f(x) \sin nx\, dx \qquad (n > 0) \qquad (2)$$

is called the Fourier Series of f(x) and we write

$$f(x) \sim \frac{a_0}{2} + \sum_{n=1}^{\infty} (a_n \cos nx + b_n \sin nx) \qquad (3)$$

to indicate this association. Note that there is no equality sign in (3) since the series there is not necessarily convergent to f(x) at any point. Questions of convergence will be discussed later and it will be found that a large class of functions are representable by their Fourier series. For now, the Fourier series' of functions will be calculated without regard to their convergence and the symbol ~ will be used.

It is important to note that since f has period 2π, the coefficients in equations (1) and (2) may just as easily be evaluated as

$$a_n = \frac{1}{\pi} \int_a^{a+2\pi} f(x) \cos nx\, dx, \quad b_n = \frac{1}{\pi} \int_a^{a+2\pi} f(x) \sin nx\, dx \qquad (4)$$

for any a such that the interval (a, a+2π) is in the domain of definition of f. It may be that f is defined only in (a, a+2π) for some a in which case (4) and not (1) and (2) must be used. Of course if f is defined only on (-π,π), then (4) may be used with a = -π. In either case, (3) will represent the periodic extension of f on the whole real axis. That is, it will be discussed later that under very "loose" conditions, the series in (3) will converge to

$$L_0 = \frac{1}{2} \left[\lim_{x \to x_0^+} f(x) + \lim_{x \to x_0^-} f(x) \right] \text{ at } x_0 \text{ in its domain of}$$

definition, say [a,a+2π]. (If these limits are equal, the series converges to $f(x_o)$). But then by the periodicity of (3), at all points x_o+2nπ, the series will converge to L_o and will thus have the same graph between a+2nπ and a+2(n+1)π as f does in [a,a+2π]. This is illustrated in the Figure.

• **PROBLEM 14-3**

Let f(x) be a real valued function of one variable, f:R→R, with a period 2c (i.e. f(x+2c) = f(x) for all x∈R). Define the Fourier series of f.

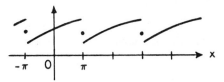

y= FOURIER SERIES OF f

PERIODIC EXTENSION OF A FUNCTION
DEFINED BETWEEN −c AND c.   Fig. 1

Solution: Consider the trigonometric series

$$\frac{a_o}{2} + \sum_{n=1}^{\infty}\left(a_n \cos \frac{\pi n x}{c} + b_n \sin \frac{\pi n x}{c}\right)$$

and assume that it converges uniformly for all x∈R. Then it represents a function of x, i.e.

$$f(x) = \frac{a_o}{2} + \sum_{n=1}^{\infty}\left(a_n \cos \frac{\pi n x}{c} + b_n \sin \frac{\pi n x}{c}\right) \quad (1)$$

and by procedures analogous to those used in a previous problem of this chapter, it can be shown that

$$a_n = \frac{1}{c}\int_{-c}^{c} f(x)\cos\left(\frac{\pi n x}{c}\right)dx, \quad b_n = \frac{1}{c}\int_{-c}^{c} f(x)\sin\left(\frac{\pi n x}{c}\right)dx. \quad (2)$$

These relations will be proven now. To find the b_n, multiply each term of (1) by sin(kπx/c)dx and integrate from −c to c. Thus (1) becomes

$$\int_{-c}^{c} f(x)\sin\frac{k\pi x}{c} dx = \frac{1}{2}a_o \int_{-c}^{c} \sin\frac{k\pi x}{c} dx$$

$$+ \sum_{n=1}^{\infty} a_n\left[\int_{-c}^{c} \cos\frac{n\pi x}{c} \sin\frac{k\pi x}{c} dx\right.$$

$$+ b_n \int_{-c}^{c} \sin \frac{n\pi x}{c} \sin \frac{k\pi x}{c} dx \Bigg] \qquad (3)$$

Note that we interchanged

$$\int_{-a}^{b} \left[\sum_{n=1}^{\infty} g(x) \right] dx$$

to

$$\sum_{n=1}^{\infty} \left(\int_{-a}^{b} g(x) dx \right).$$

This is valid if a series converges uniformly. We have assumed that this is so in the given problem. To evaluate (3) we note that the set of functions

$$\{\sin x, \cos x, \sin 2x, \cos 2x, \ldots, \sin kx, \cos kx \ldots\}$$

is an orthogonal set. This means that for any two members of the set,

$$\int_{a}^{b} \sin mx \cos mx \, dx = 0, \qquad (4)$$

$$\int_{a}^{b} \sin mx \sin nx \, dx = 0 \qquad (m \neq n), \qquad (5)$$

$$\int_{a}^{b} \cos mx \cos nx \, dx = 0 \qquad (m \neq n), \qquad (6)$$

$$\int_{a}^{b} [\sin mx]^2 dx \neq 0, \qquad (7)$$

$$\int_{a}^{b} [\cos mx]^2 dx \neq 0. \qquad (8)$$

Using (4) to (7), (3) is reduced to

$$\int_{-c}^{c} f(x) \sin \frac{k\pi x}{c} dx = b_k \int_{-c}^{c} \sin^2 \frac{k\pi x}{c} dx \qquad (9)$$

From (3) we determine b_k in the following manner. First we evaluate

$$\int_{-c}^{c} \sin^2 \frac{k\pi x}{c} \, dx \tag{10}$$

Now (10) is not zero. Using the trigonometric identities cited in this chapter's first problem, it will equal the number c. Then (9) may be rewritten as

$$b_k = \frac{1}{c} \int_{-c}^{c} f(x) \sin \frac{k\pi x}{c} \, dx, \quad k = 1,2,3,\ldots \tag{11}$$

Thus the coefficients b_n in (1) are

$$b_n = \frac{1}{c} \int_{-c}^{c} f(x) \sin \frac{n\pi x}{c} \, dx, \quad n = 1,2,3,\ldots \tag{12}$$

We now find the coefficient a_n of the cosine terms in (1). Therefore, multiplying (1) by $\cos \frac{n\pi x}{c}$ and integrating from $-c$ to c,

$$\int_{-c}^{c} f(x) \cos \frac{n\pi x}{c} \, dx = \frac{1}{2} a_0 \int_{-c}^{c} \cos \frac{n\pi x}{c} \, dx$$

$$+ \sum_{n=1}^{\infty} \left[\int_{-c}^{c} \cos \frac{n\pi x}{c} \cos \frac{k\pi x}{c} \, dx \right.$$

$$\left. + \int_{-c}^{c} \cos \frac{n\pi x}{c} \sin \frac{k\pi x}{c} \, dx \right].$$

Using facts (4) - (8), and $k \neq 0$,

$$\int_{-c}^{c} f(x) \cos \frac{n\pi x}{c} \, dx = a_n \int_{-c}^{c} \left(\cos \frac{n\pi x}{c} \right)^2 dx.$$

Solving for a_n,

$$a_n = \frac{1}{c} \int_{-c}^{c} f(x) \cos \frac{n\pi x}{c} \, dx \tag{13}$$

where

$$c = \int_{-c}^{c} \left[\cos \frac{n\pi x}{c}\right]^2 dx.$$

Finally, we must determine a_o. From (1), we have

$$\int_{-c}^{c} f(x)\, dx = \frac{1}{2} a_o \int_{-c}^{c} dx + \sum_{n=1}^{\infty} \left[a_n \int_{-c}^{c} \cos \frac{n\pi x}{c} dx + b_n \int_{-c}^{c} \sin \frac{n\pi x}{c} dx \right] \quad (14)$$

Now $\int_{-c}^{c} \cos \frac{n\pi x}{c}\, dx = 0$ for $n \geq 1$ and $\int_{-c}^{c} \sin \frac{n\pi x}{c}\, dx$

$= \frac{2c}{n\pi}\left[-\cos \frac{n\pi x}{c}\right]_{-c}^{c} = 0$, for $n \geq 1$. Thus (14) is reduced to

$$\int_{-c}^{c} f(x)\, dx = \frac{1}{2} a_o \int_{-c}^{c} dx$$

$$\int_{-c}^{c} f(x)\, dx = \frac{1}{2} a_o (2c)$$

and hence $a_o = \frac{1}{c} \int_{-c}^{c} f(x)\, dx.$

Thus $f(x) \sim \frac{1}{2} a_o + \sum_{n=1}^{\infty} \left[a_n \cos \frac{n\pi x}{c} + b_n \sin \frac{n\pi x}{c} \right]$ with a_o, a_n and b_n given by (2). Thus, the coefficients of (1) have been found for the case in which the series there represents some function and converges uniformly to it.

As was done for functions of period 2π, the series (1) will be used to motivate the definition of the Fourier series of any function $f(x)$ of period $2c$. Thus, this series is defined as the series in (1) with coefficients given in (2) but we write

$$f(x) \sim \frac{a_o}{2} + \sum_{n=1}^{\infty} \left[a_n \cos \frac{\pi n x}{c} + b_n \sin \frac{\pi n x}{c} \right] \quad (15)$$

since it is not known in general whether this series is convergent to f(x) at any point. It should be noted that the limits of integration in (2) may be a and a+2c where a is any number such that (a,a+2c) is contained within the domain of definition of f. If f is defined only for some interval (a,a+2c) with some specific (a), then these limits must be used, i.e.

$$a_n = \frac{1}{c} \int_a^{a+2c} f(x) \cos\left(\frac{\pi n x}{c}\right) dx,$$

$$b_n = \frac{1}{c} \int_a^{a+2c} f(x) \sin\left(\frac{\pi n x}{c}\right) dx. \tag{16}$$

Of course if f is defined only on (-c,c) then (16) is used with a = -c.

When the phenomena of a particular problem are periodic in time with period T, then the variable x is usually replaced by t, and the following substitutions are usually made:

$$\frac{\pi}{c} \to \frac{2\pi}{T} = \omega, \quad A_n = \sqrt{a_n^2 + b_n^2}, \quad \phi_n = \arctan \frac{b_n}{a_n} \quad (n>0). \tag{17}$$

The series (15) then becomes

$$f(t) \sim \frac{a_o}{2} + \sum_{n=1}^{\infty} A_n \cos(n\omega t - \phi_n). \tag{18}$$

In any case, always remember that a Fourier series can only be associated with a periodic function. In addition, in a case where the function f is only defined on a certain interval (a,a+2c), (16) rather than (12) and (13) must be used. Then (15) represents the "periodic extension" of f to the whole real axis. That is, it will be discussed later that except for some very pathological functions, the series in (15) will converge to

$$L_o = \frac{1}{2}\left[\lim_{x \to x_o^+} f(x) + \lim_{x \to x_o^-} f(x)\right] \quad \text{in its domain of defini-}$$

tion, say [a,a+2c] (if these limits are equal the series converges to $f(x_o)$). But then by the periodicity of (15), at all points $x_o + 2nc$, the series will converge to L_o and will thus have the same graph between a+2nc and a+2(n+1)c as f does in [a,a+2c]. This is illustrated in the Figure. In most applications in this book the most general equations, (15) and (2) will be used even if $2c = 2\pi$, for then these equations will be reduced to those of the previous problem. Therefore, equations (15) and (2) should be committed to memory.

● **PROBLEM 14-4**

Find the Fourier series of the function $f(x) = e^x$, $-\pi < x < \pi$.

Solution: The given function is defined on the familiar interval $(-\pi, \pi)$. A good habit to acquire is to recall the most general definition of a Fourier series of a function and then use the particular values given in the problem in this definition. Thus, recall that the Fourier series of a function $f(x)$ which is periodic of period $2c$ and defined on some domain D of the real numbers, $f: D \to R$, is given by

$$f(x) \sim \frac{a_0}{2} + \sum_{n=1}^{\infty} \left[a_n \cos\left(\frac{\pi n x}{c}\right) + b_n \sin\left(\frac{\pi n x}{c}\right) \right] \quad (1)$$

where the coefficients are given by

$$a_n = \frac{1}{c} \int_a^{a+2c} f(x) \cos\left(\frac{\pi n x}{c}\right) dx \quad (2)$$

$$b_n = \frac{1}{c} \int_a^{a+2c} f(x) \sin\left(\frac{\pi n x}{c}\right) dx, \quad (3)$$

where a is any number such that the interval $(a, a+2c)$ is contained in D. In the case at hand, $f(x) = e^x$, $x \in (-\pi, \pi)$ is only defined on an interval of length 2π. Hence, (1) represents the periodic extension of f over the whole real axis. (Note that f is not periodic within that interval so 2π is the only period that can be assigned any meaning here.) Therefore, in equations (1)-(3) the substitutions to be made for this particular function are

$$c = \pi; \quad a = -\pi; \quad f(x) = e^x. \quad (4)$$

Using (4), the coefficients become

$$a_0 = \frac{1}{\pi} \int_{-\pi}^{\pi} e^x dx = \frac{2}{\pi} \frac{e^\pi - e^{-\pi}}{2} = \frac{2}{\pi} \sinh \pi.$$

$$a_n = \frac{1}{\pi} \int_{-\pi}^{\pi} e^x \cos nx \, dx = \frac{1}{\pi} \left[e^x \frac{(\cos nx + n \sin nx)}{n^2 + 1} \right]_{-\pi}^{\pi}$$

$$= (-1)^n \frac{e^\pi - e^{-\pi}}{(n^2+1)\pi} = (-1)^n \frac{2 \sinh \pi}{(n^2+1)\pi}.$$

676

$$b_n = \frac{1}{\pi} \int_{-\pi}^{\pi} e^x \sin nx \, dx = \frac{1}{\pi} \left(e^x \frac{(\sin nx - n\cos nx)}{n+1} \right) \Big|_{-\pi}^{\pi}$$

$$= (-1)^{n+1} \frac{e^\pi - e^{-\pi}}{(n^2+1)\pi} n$$

$$= -\frac{(-1)^n \, 2n \sinh \pi}{(n^2+1)\pi}.$$

Substituting these values back into the Fourier expression, (1), with the substitutions of (3) gives

$$f(x) \sim \frac{a_o}{2} + \sum_{n=1}^{\infty} \left(a_n \cos nx + b_n \sin nx \right)$$

$$= \frac{\sinh \pi}{\pi} + \sum_{n=1}^{\infty} \left[\frac{(-1)^n 2 \sinh \pi}{(n^2+1)\pi} \cos nx - \frac{(-1)^n 2n \sinh \pi}{(n^2+1)\pi} \sin nx \right]$$

$$= \frac{\sinh \pi}{\pi} \left\{ 1 + 2 \sum_{n=1}^{\infty} \left[\frac{(-1)^n}{n^2+1} (\cos nx - n \sin nx) \right] \right\}. \quad (5)$$

Recall that the symbol $\sim$ is used in (5) since no determination of the convergence of the series to the function has yet been made. This will be done later and it will be shown that $\sim$ may be replaced by $=$.

● **PROBLEM 14-5**

Determine the Fourier series of the function

$$f(x) = \begin{cases} x, & 0 < x < \pi \\ 0, & \pi < x < 2\pi \end{cases} \quad (1)$$

Solution: Recall the most general definition of the Fourier series of a function f. That is, the Fourier series of a periodic function f of period 2c defined on the interval D is given by

$$f(x) \sim \frac{a_o}{2} + \sum_{n=1}^{\infty} \left[a_n \cos\left(\frac{\pi nx}{c}\right) + b_n \sin\left(\frac{\pi nx}{c}\right) \right] \quad (2)$$

where

$$a_n = \frac{1}{c} \int_a^{a+2c} f(x) \cos\left(\frac{\pi nx}{c}\right) dx \quad (3)$$

$$b_n = \frac{1}{c} \int_a^{a+2c} f(x) \sin\left(\frac{\pi n x}{c}\right) dx, \tag{4}$$

where a is any number such that the interval (a,a+2c) is contained in D. In the case at hand, the function in (1) is defined only on the interval (0,2π) and is not periodic within this interval so 2c = 2π and (2) represents the periodic extension of f. Therefore, in equations (2)-(4) the substitutions

$$2c = 2\pi, \quad a = 0, \quad f(x) = \begin{cases} x, & x \in (0,\pi) \\ 0, & x \in (\pi, 2\pi) \end{cases}$$

are made. For a_n, n>0, this results in

$$a_n = \frac{1}{\pi} \int_0^{2\pi} f(x) \cos nx \, dx = \frac{1}{\pi} \int_0^{\pi} x \cos nx \, dx$$

$$= \frac{1}{\pi} \left[\frac{x}{n} \sin nx + \frac{\cos nx}{n^2} \right]_0^{\pi}$$

$$= \frac{1}{n^2 \pi} (\cos n\pi - 1).$$

For n = 2,4,6,... cos nπ = 1 and hence a_n = 0, and for n = 1,3,5,... cos nπ = -1 and $a_n = \frac{-2}{n^2 \pi}$, n = odd. Similarly, to evaluate b_n, the same substitutions are made yielding

$$b_n = \frac{1}{\pi} \int_0^{2\pi} f(x) \sin nx \, dx = \frac{1}{\pi} \int_0^{\pi} x \sin nx \, dx$$

$$= \frac{1}{\pi} \left[\frac{\sin nx}{n^2} - \frac{x \cos nx}{n} \right]_0^{\pi}$$

$$= \frac{-\pi \cos n\pi}{\pi n} \quad \text{for all } n = 1,2,3,\ldots$$

$$= \frac{-\pi (-1)^n}{\pi n} = \frac{(-1)^{n+1}}{n}.$$

Finally,

$$a_0 = \frac{1}{\pi} \int_0^{2\pi} f(x) \, dx = \frac{1}{\pi} \int_0^{\pi} x \, dx = \left. \frac{x^2}{2\pi} \right|_0^{\pi} = \frac{\pi}{2}.$$

So far, the following results have been obtained:

$$a_o = \frac{\pi}{2}$$

$$a_n = \frac{-2}{n^2 \pi}, \quad n = 1, 3, 5 \ldots$$

$$b_n = \frac{(-1)^{n+1}}{n}, \quad n = 1, 2, 3, \ldots .$$

To make a_n valid for all n, we ensure that n is odd by replacing n by (2n-1). Finally, we obtain

$$a_{2n-1} = \frac{-2}{(2n-1)^2 \pi}, \quad n = 1, 2, 3, \ldots .$$

Now, referring back to the Fourier expression,

$$f(x) \sim \frac{a_o}{2} + \sum_{n=1}^{\infty} (a_n \cos nx + b_n \sin nx)$$

and substituting back their respective values,

$$f(x) \sim \frac{\pi}{4} + \sum_{n=1}^{\infty} \frac{-2}{(2n-1)^2 \pi} \cos(2n-1)x + \frac{(-1)^{n+1}}{n} \sin nx$$

or

$$f(x) \sim \frac{\pi}{4} - \frac{2}{\pi} (\cos x + \frac{\cos 3x}{3^2} + \frac{\cos 5x}{5^2} + \ldots) + \sin x$$

$$- \frac{\sin 2x}{2} + \frac{\sin 3x}{3} \ldots .$$

Again, the symbol "$\sim$" is to be read "has the Fourier series" since convergence questions have not been dealt with yet. It is easy to see that this series is convergent by the Weierstrass M-test and Dirichlet's test, but rather than dealing with the convergence of each Fourier series by such tests, a more powerful general method will be developed later. At that time all convergence questions about previous problems will be answered.

● **PROBLEM 14-6**

Determine the Fourier series of the function given by

$$\begin{cases} f(x) = x^2, \quad x \in (-\pi, \pi) \\ f(x+2\pi) = f(x), \quad \text{all } x \end{cases} . \quad (1)$$

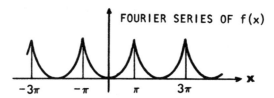

FOURIER SERIES OF f(x)

Solution: The most general definition of the Fourier series of a function f which is periodic with period 2c and defined on the interval D is given by

$$f(x) \sim \frac{a_0}{2} + \sum_{n=1}^{\infty}\left[a_n\cos\left(\frac{\pi nx}{c}\right) + b_n\sin\left(\frac{\pi nx}{c}\right)\right] \qquad (2)$$

where

$$a_n = \frac{1}{c}\int_a^{a+2c} f(x)\cos\left(\frac{\pi nx}{c}\right)dx \qquad (3)$$

$$b_n = \frac{1}{c}\int_a^{a+2c} f(x)\sin\left(\frac{\pi nx}{c}\right)dx . \qquad (4)$$

Here, a is any number such that $(a, a+2c) \subseteq D$. For the function in (1), D is the whole real axis, $2c = 2\pi$ and since any a can be used, the most convenient is $a = -\pi$. Then

$$a_0 = \frac{1}{\pi}\int_{-\pi}^{\pi} x^2 dx = \frac{2\pi^2}{3} \qquad (5)$$

$$a_n = \frac{1}{\pi}\int_{-\pi}^{\pi} x^2 \cos nx\, dx \qquad (6)$$

Integrating (6) by parts twice gives

$$a_n = \frac{1}{\pi}\left[\frac{x^2}{n}\sin nx\right]_0^{\pi} - \frac{2}{n\pi}\int_{-\pi}^{\pi} x \sin nx\, dx$$

$$= 0 + \frac{2}{\pi n^2}[x \cos nx]_{-\pi}^{\pi} - \frac{2}{\pi n^2}\int_{-\pi}^{\pi} \cos nx\, dx$$

$$= \frac{4}{n^2}[\cos n\pi] - \frac{2}{\pi n^3}\left(\sin nx\right)\Big|_{-\pi}^{\pi}$$

$$= \frac{4}{n^2}(-1)^n . \qquad (n > 0) \qquad (7)$$

Finally,

$$b_n = \frac{1}{\pi} \int_{-\pi}^{\pi} x^2 \sin nx \, dx.$$

Note that the integrand, $F(x) = x^2 \sin nx$, is an odd function of x, that is

$F(-x) = (-x)^2 \sin(-nx) = -x^2 \sin nx = -F(x)$. Thus

$$b_n = \frac{1}{\pi} \int_{-\pi}^{0} x^2 \sin nx \, dx + \frac{1}{\pi} \int_{0}^{\pi} x^2 \sin nx \, dx$$

$$= \frac{-1}{\pi} \int_{0}^{\pi} y^2 \sin ny \, dy + \frac{1}{\pi} \int_{0}^{\pi} x^2 \sin nx \, dx = 0 \quad (8)$$

where the change of variables $x = -y$ was made in the first integral of (8). Using these values of a_n and b_n in (2) gives

$$f(x) \sim \frac{\pi^2}{3} + \sum_{n=1}^{\infty} (-1)^n \left(\frac{4}{n^2}\right) \cos nx. \quad (9)$$

The graph of this series is shown in the Figure. Note that even if f had been defined only on $(-\pi, \pi)$, it would have the same Fourier series (9) and this series would have the same graph. Thus, the series would represent a periodic extension of the values of x^2 in the interval $(-\pi, \pi)$ if it converged (it does by the Weierstrass M-test, but convergence questions will be handled more generally later on in this chapter).

● **PROBLEM 14-7**

Determine the Fourier series of the function given by

$$f(x) = \begin{cases} -1 & x \in (-\pi, 0) \\ 1 & x \in [0, \pi] \end{cases}. \quad (1)$$

Solution: This is the familiar square wave used in electronics (see Figure 1). First notice that this function is discontinuous at 0 but it will still have a Fourier series since the Fourier coefficients can be determined for any integrable function. The Fourier series of a function with period $2c$ defined on a domain D is given by

$$f(x) \sim \frac{a_0}{2} + \sum_{n=1}^{\infty} \left[a_n \cos\left(\frac{\pi nx}{c}\right) + b_n \sin\left(\frac{\pi nx}{c}\right) \right] \quad (2)$$

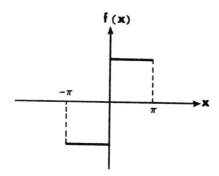

Fig. 1

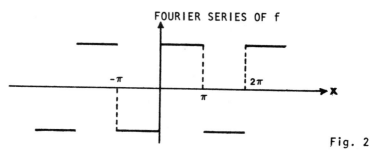

Fig. 2

where

$$a_n = \frac{1}{c} \int_a^{a+2c} f(x) \cos\left(\frac{\pi n x}{c}\right) dx \qquad (3)$$

$$b_n = \frac{1}{c} \int_a^{a+2c} f(x) \sin\left(\frac{\pi n x}{c}\right) dx.$$

Here a is any number such that the interval (a,a+2c) is in D. But since f is only defined on $(-\pi,\pi)$, $c = 2\pi$ and a is $-\pi$ by necessity. Thus, the Fourier coefficients are

$$a_o = \frac{1}{\pi} \int_{-\pi}^{0} (-1) dx + \frac{1}{\pi} \int_{0}^{\pi} (+1) dx = -1 + 1 = 0$$

$$a_n = \frac{1}{\pi} \int_{-\pi}^{0} -\cos nx\, dx + \frac{1}{\pi} \int_{0}^{\pi} \cos nx\, dx = 0 + 0 = 0.$$

$$b_n = \frac{1}{\pi} \int_{-\pi}^{0} (-\sin nx) dx + \frac{1}{\pi} \int_{0}^{\pi} (\sin nx) dx.$$

But

$$\frac{1}{\pi} \int_{-\pi}^{0} (-\sin nx) dx = \frac{1}{n\pi} [1 - \cos n\pi] = \frac{1}{\pi} \int_{0}^{\pi} \sin nx\, dx.$$

Thus,

$$b_n = \frac{2}{\pi} \int_0^\pi \sin nx\, dx = \begin{cases} \frac{4}{n\pi} & (n = \text{odd}), \\ 0 & (n = \text{even}) \end{cases}$$

And so the Fourier series reads

$$f(x) \sim \frac{4}{\pi} \sum_{n=\text{odd}}^\infty \frac{\sin nx}{n}. \qquad (4)$$

It can be proven by Dirichlet's test that this series converges for all x, but convergence questions will be attended to later. It is particularly interesting to note that at x = 0 the series does not converge to f(x), and this will also be examined later. For now, note that the series in (4) represents a periodic extension of the values of f in the interval $(-\pi, \pi)$ as graphed in Figure 2. Thus, (4) is a representation of the square wave used in electronics.

● **PROBLEM 14-8**

Find the Fourier series of the function $f(x) = |x|$, $-\pi < x \leq \pi$.

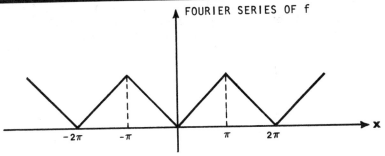

FOURIER SERIES OF f

Solution: The Fourier series of a function f with period 2c (or defined only on the interval 2c) is given by

$$f(x) \sim \frac{a_0}{2} + \sum_{n=1}^\infty \left[a_n \cos\left(\frac{\pi nx}{c}\right) + b_n \sin\left(\frac{\pi nx}{c}\right) \right] \qquad (1)$$

where

$$a_n = \frac{1}{c} \int_a^{a+2c} f(x) \cos\left(\frac{\pi nx}{c}\right) dx \qquad (2)$$

$$b_n = \frac{1}{c} \int_a^{a+2c} f(x) \sin\left(\frac{\pi nx}{c}\right) dx. \qquad (3)$$

Here a is any number such that (a, a+2c) is contained within the domain of definition of f. In the case at hand, $2c = 2\pi$, and $a = -\pi$ by necessity.

In solving the given problem we make use of the notions of odd functions and even functions. A function $f(x)$, is said to be even if $f(-x) = f(x)$. Some examples of even functions are $y = \cos x$, $y = x^2$, and $y = |x|$. One characteristic of such functions is that they are symmetric with respect to the y-axis.

A function $f(x)$ is called odd if $f(-x) = -f(x)$. Examples of odd functions are $y = \sin x$, $y = x$, $y = x^3$. We note further that the product of two even functions is itself an even function while the product of an odd with an even function is an odd function. Finally, if $f(x)$ is even on the interval $-A \leq x \leq A$, then

$$\int_{-A}^{A} f(x)\,dx = 2\int_{0}^{A} f(x)\,dx,$$

and if $f(x)$ is odd over the same interval,

$$\int_{-A}^{A} f(x)\,dx = 0.$$

In the given problem, $f(x) = |x|$ is an even function. The evaluation of the Fourier coefficients is now shown to be considerably simplified. First, consider

$$a_n = \frac{1}{c}\int_{-c}^{c} |x| \cos \frac{n\pi x}{c}\,dx$$

for $-\pi \leq x \leq \pi$ and $a = -c$. Since $|x| = x$ for $0 \leq x \leq \pi$, this may be written as

$$a_n = \frac{2}{c}\int_{0}^{c} x \cos \frac{n\pi x}{c}\,dx.$$

Next, since $\sin x$ is odd, the coefficients

$$b_n = \frac{1}{c}\int_{-c}^{c} f(x)\sin \frac{n\pi x}{c}\,dx, \quad (n = 1,2,3,\ldots)$$

are all equal to zero. Hence we need consider only the a_n in the series (a).

Now,

$$a_n = \frac{2}{\pi}\int_{0}^{\pi} x \cos \frac{n\pi x}{\pi}\,dx = \frac{2}{\pi}\int_{0}^{\pi} x \cos nx\,dx.$$

Integrating by parts,

$$a_n = \frac{2}{\pi} \left[\frac{\cos nx}{n^2} + \frac{x \sin nx}{n} \right]_0^\pi$$

$$= \frac{2}{\pi} \left[\frac{\cos n\pi - 1}{n^2} \right] = \frac{2}{\pi} \left[\frac{(-1)^n - 1}{n^2} \right]$$

(since $\cos n\pi = -1$ for odd n and 1 for even n).

$$\frac{2}{\pi} \left[\frac{(-1)^n - 1}{n^2} \right] = \begin{cases} \frac{-4}{\pi n^2}, & n \text{ odd} \\ 0, & n \text{ even} \end{cases} \quad n = 1, 2, 3, \ldots.$$

The above integration and evaluation is not valid for $n = 0$. To find a_0,

$$a_0 = \frac{2}{\pi} \int_0^\pi f(x) \cos(0) \frac{\pi x}{\pi} dx = \frac{2}{\pi} \int_0^\pi f(x) dx = \frac{2}{\pi} \int_0^\pi x \, dx$$

$$= \frac{x^2}{\pi} \Big|_0^\pi = \pi.$$

Thus, the required series is

$$\frac{\pi}{2} - \frac{4}{\pi} \sum_{\substack{n=1 \\ (n \text{ odd})}}^\infty \frac{\cos nx}{n^2} = \frac{\pi}{2} - \frac{4}{\pi} \sum_{n=1}^\infty \frac{\cos(2n-1)x}{(2n-1)^2}.$$

That is,

$$f \sim \frac{\pi}{2} - \frac{4}{\pi} \sum_{n=1}^\infty \frac{\cos(2n-1)x}{(2n-1)^2}$$

is the Fourier series of f. The graph of this series is shown in the Figure and it can be seen that the triangular waveform there represents a periodic extension of the values of $|x|$ in the interval $[-\pi, \pi]$.

● **PROBLEM 14-9**

Find the Fourier series for the sawtooth waveform shown in Fig. 1.

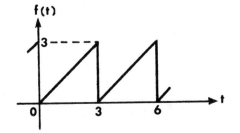

Fig. 1

Solution: We note that this waveform repeats every 3 seconds. Thus the period $\tau = 3s$ and since $\omega_o = \frac{2\pi}{\tau}$, $\omega_o = \frac{2\pi}{3}$.

A periodic function has a Fourier series in the form

$$f(t) \sim a_o + \sum_{n=1}^{\infty} (a_n \cos n\omega_o t + b_n \sin n\omega_o t) \qquad (1)$$

where a_o is the average value of the function $f(t)$ and is defined as

$$a_o = \frac{1}{\tau} \int_0^\tau f(t)\, dt. \qquad (2)$$

$f(t)$ represents one period of the entire periodic function, thus $f(t) = t;\ 0 < t < 3$.

Substituting these values yields:

$$a_o = \frac{1}{3} \int_0^3 t\, dt$$

$$a_o = \frac{1}{3} \left(\frac{t^2}{2}\right)\Big|_0^3$$

$$a_o = \frac{1}{3}\left(\frac{9}{2} - 0\right) = \frac{3}{2}.$$

The coefficients b_n and a_n are defined as

$$b_n = \frac{2}{\tau} \int_0^\tau f(t) \sin n\omega_o t\, dt \qquad (3)$$

$$a_n = \frac{2}{\tau} \int_0^\tau f(t) \cos n\omega_o t\, dt. \qquad (4)$$

Hence,

$$b_n = \frac{2}{3} \int_0^3 t \sin n\frac{2\pi}{3} t\, dt$$

$$b_n = \frac{2}{3} \left[\frac{1}{\left(\frac{n2\pi}{3}\right)^2} \sin n\left(\frac{2\pi}{3}\right)t - \frac{t}{\frac{n2\pi}{3}} \cos n\frac{2\pi}{3} t\right]_0^3$$

$$b_n = \frac{2}{3}\left[\frac{1}{\left(\frac{n2\pi}{3}\right)^2} \sin n2\pi - \frac{9}{n2\pi} \cos n2\pi\right]$$

$$b_n = \frac{3}{n^2 2\pi^2} \sin n2\pi - \frac{3}{n\pi} \cos n2\pi.$$

The sine term is zero for all n since any multiple of 2π in the sine term is zero.

Hence

$$b_n = -\left[\frac{3}{\pi}\frac{1}{n}\right] \qquad n = 1,2,3,\ldots$$

since the cos term is 1 for any multiple of 2π.

$$a_n = \frac{2}{3}\int_0^3 t \cos n\omega_o t\, dt$$

$$a_n = \frac{2}{3}\left[\frac{1}{\left(\frac{n2\pi}{3}\right)^2} \cos \frac{n2\pi}{3} t + \frac{t}{\frac{n2\pi}{3}} \sin \frac{n2\pi}{3} t\right]\Bigg|_0^3$$

$$a_n = \frac{2}{3}\left[\frac{9}{n^2 4\pi^2} \cos n2\pi - \frac{9}{n^2 4\pi^2} \cos 0\right]$$

But $\cos n2\pi = \cos 0$ for all n; therefore, $a_n = 0$ for all n.

The Fourier representation of this waveform is written

$$f(t) \sim \frac{3}{2} - \frac{3}{\pi} \sum_{n=1}^{\infty} \frac{1}{n} \sin n \frac{2\pi}{3} t$$

$$f(t) \sim \frac{3}{2} - \frac{3}{\pi}\left(\sin \frac{2\pi}{3} t + \frac{1}{2} \sin \frac{4\pi}{3} t + \frac{1}{3} \sin \frac{6\pi}{3} t + \ldots\right).$$

CONVERGENCE QUESTIONS

● **PROBLEM 14-10**

Define the following properties of a real valued function f of a real variable:
(a) The "limit from the right" of f at x_o
(b) The "limit from the left" of f at x_o
(c) f is piecewise continuous on (a,b).
(d) The right and left hand derivatives of f at x_o.

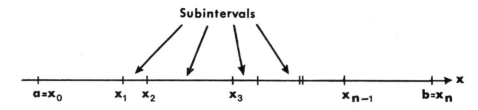

Fig. 1

Solution: The limit from the right of $f(x)$ as x approaches x_o is denoted by

$$f(x_o+) = \lim_{x \to x_o+} f(x) \equiv \lim_{\substack{h \to 0 \\ h > 0}} f(x_o + h) = L. \qquad (1)$$

This definition may be stated in words as follows: the limit from the right of $f(x)$ as x approaches x_o is L if for any positive number ε, there exists a positive number δ such that for all x satisfying $x_o < x < x_o + \delta$ one has

$$|f(x) - L| < \varepsilon.$$

That is, for x sufficiently close to x_o (and greater than x_o), $f(x)$ is as close to L as desired. This definition is especially important when f is only defined for $x > x_o$ or when f is discontinuous at x_o.

(b) The limit from the left of $f(x)$ as x approaches x_o is denoted by

$$f(x_o-) = \lim_{x \to x_o-} f(x) \equiv \lim_{\substack{h \to 0 \\ h > 0}} f(x_o - h) = L. \qquad (2)$$

This equation means that the required limit is L if for any positive ε there exists a positive δ such that for all x satisfying $x_o - \delta < x < x_o$ one has

$$|f(x) - L| < \varepsilon.$$

That is, for x sufficiently close to x_o (and less than x_o), $f(x)$ is as close to L as desired.

(c) A function in a closed interval is called piecewise continuous if the interval can be split up into a finite number of subintervals such that in each subinterval (see Fig. 1) the two following conditions hold:

    i) $f(x)$ is continuous at every interior point

    ii) $f(x)$ possesses (finite) limits at the left and right hand ends.

Condition (ii) means that $f(x_o-)$ and $f(x_o+)$ exist for all

$i = 1, 2, \ldots, n$, and also that $f(a+)$ and $f(b-)$ exist.

(d) If $f(x_o+)$ exists at a point x_o, then f is said to have a right hand derivative at x_o if

$$f'_+(x_o) = \lim_{\substack{t \to 0 \\ t > 0}} \frac{f(x_o + t) - f(x_o+)}{t}$$

exists. Similarly, the left hand derivative of f at x_o is the limit

$$f'_-(x_o) = \lim_{\substack{t \to 0 \\ t < 0}} \frac{f(x_o + t) - f(x_o-)}{t}.$$

● **PROBLEM 14-11**

State the most general Pointwise Convergence Theorem for Fourier series (i.e. the one with the weakest premises). Discuss its meaning.

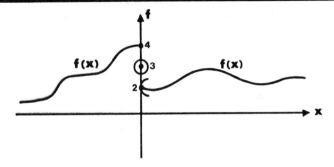

Fig. 1

Solution: Suppose that f is periodic of period $2c$ and is piecewise continuous on the real axis. Then the Pointwise Convergence Theorem states that if the right and left hand derivatives of f at some point x_o exist then the Fourier series of f converges to
$\frac{1}{2}\{f(x_o+) + f(x_o-)\}$ at x_o. In symbols,

$$\frac{1}{2}\{f(x_o+) + f(x_o-)\} = \frac{1}{2} a_o + \sum_{n=1}^{\infty} \left(a_n \cos\left(\frac{\pi n x_o}{c}\right) + b_n \sin\left(\frac{\pi n x_o}{c}\right) \right). \quad (1)$$

Note that if $f(x_o+) = f(x_o-) = f(x_o)$ (that is, if f is continuous at x_o), then

$$f(x_0) = \frac{1}{2} a_0 + \sum_{n=1}^{\infty} \left(a_n \cos\left(\frac{\pi n x_0}{c}\right) + b_n \sin\left(\frac{\pi n x_0}{c}\right) \right),$$

i.e. the Fourier series of f converges to $f(x_o)$ at x_o.
Note that equation (2) is valid on any interval where f is continuous if we replace x_o by x. However, if the interval includes discontinuities then "=" must be replaced by "$\sim$". Hence, equation (2) may be used at points of continuity of f and equation (1) at points of discontinuity. Thus, a function like that graphed in Figure 1 has a convergent Fourier series for all x and this series converges to f except at x = 0 where it converges to the average value of $f(0+)$ and $f(0-)$, namely $\frac{1}{2}(4+2) = 3$.

Note that if the function is defined only on the interval (a,a+2c) for some real a, then the conclusion still holds at all points in the domain of definition of f and the Fourier series is the periodic extension of f on the rest of the real axis. In particular, at points a+2nc, the Fourier series will converge to $\frac{1}{2}\{f(a) + f(a+2c)\}$.

Finally, this theorem is called the "Pointwise" convergence theorem to distinguish it from other theorems dealing with stricter convergence such as uniform convergence and convergence in the mean. The convergence is pointwise in the sense that if the sum of the series is $g(x)$ and the partial sums are denoted by $S_n(x)$, then if given an $\varepsilon > 0$ there exists a K such that for all $n \geq K$,

$$\|S_n(x) - g(x)\| < \varepsilon$$

if x is also specified. That is, the error ε resulting from approximating g by S_n can only be made arbitrarily small for each individual x. The importance of this restriction will be seen later. In any case, the theorem proved here is sufficient for virtually any function with any physical significance and hence almost any application.

● **PROBLEM 14-12**

State the most general Uniform Convergence Theorem for Fourier series (i.e. the one with the weakest premises). Discuss its meaning.

Solution: Let f be continuous on an interval (a,b) with period 2c and suppose that f' is piecewise continuous on (a,b). Then the Uniform Convergence Theorem states that the Fourier series for f converges uniformly to f on (a,b).

Note that if the function is defined only on the interval (a,a+2c) for some real a, then the conclusion still holds at all points in the domain of definition of f and the Fourier series is the periodic extension of f on the rest of the real axis. In this case the series will converge uniformly in any interval (a+2nc, a+2(n+1)c) but not at the endpoints where f is not even defined.

The importance of uniform convergence arises from the

manipulations that can be performed when it exists. It can be proved that if a series of functions Σf_n converges uniformly to f and all of the f_n are Riemann integrable and differentiable then

$$\int_a^b f\,dx = \sum_{n=1}^{\infty} \int_a^b f_n(x)\,dx \tag{1}$$

and

$$f'(x) = \sum_{n=1}^{\infty} f'_n(x). \tag{2}$$

That is, interchange of summation with integration and differentiation is valid on any interval where the series is uniformly convergent. This is useful in determining the Fourier series of $f'(x)$ if the series for $f(x)$ is known.

Finally, in most cases of physical application, especially those in electrical engineering, the functions dealt with are only piecewise continuous so that the theorem stated here is only valid on intervals of continuity and the Pointwise Convergence Theorem is of more practical use.

● **PROBLEM 14-13**

The Fourier series for $f(x) = |x|$ $\quad -\pi < x \leq \pi$ and $f(x+2\pi) = f(x)$ is

$$\frac{\pi}{2} - \frac{4}{\pi} \sum_{n=1}^{\infty} \frac{\cos(2n-1)x}{(2n-1)^2}. \tag{1}$$

Without computing any Fourier coefficients, find the Fourier series for

$$g(x) = \begin{cases} -1 & -\pi < x < 0 \\ +1 & 0 \leq x \leq \pi \end{cases}$$

and $g(x+2\pi) = g(x)$.

Solution: The function f may be written as

$$f(x) = \begin{array}{l} -x \quad \text{if} \quad -\pi < x < 0 \\ x \quad \text{if} \quad 0 \leq x \leq \pi \end{array} \qquad f(x+2\pi) = f(x).$$

In the intervals $(-\pi, 0)$ and $(0, \pi)$, f is continuous and $f'(x) = -1$ and $+1$ respectively, so that f' is piecewise continuous in each interval (in fact it is continuous). Therefore, by the Uniform Convergence Theorem the Fourier series of f converges to f uniformly in these intervals and termwise differentiation is therefore valid, i.e.,

$$f'(x) = \sum_{n=1}^{\infty} f'_n$$

where f_n is the nth term in the Fourier expansion. But note that on $(-\pi, 0) \cup (0, \pi)$, $f'(x) = g(x)$ so that

$$g(x) = \frac{d}{dx}\frac{\pi}{2} - \frac{4}{\pi}\sum_{n=1}^{\infty}\frac{d}{dx}\frac{\cos(2n-1)x}{(2n-1)^2} \quad \begin{array}{l} x \in (-\pi, 0), \\ x \in (0, \pi) \end{array}$$

or

$$g(x) = \frac{4}{\pi}\sum_{n=1}^{\infty}\frac{\sin(2n-1)x}{2n-1}, \quad x \in (-\pi, 0) \cup (0, \pi). \tag{2}$$

At the point 0 (and hence at all points $0 + 2n\pi = 2n\pi$) the Piecewise Convergence Theorem states that the Fourier series must converge to

$$\frac{1}{2}(g(0+) + g(0-)) = \frac{1}{2}(1 - 1) = 0$$

and the series of (2) satisfies this. The same condition must hold at all points $(2n+1)\pi$, that is the Fourier series of g must converge to

$$\frac{1}{2}(g[(2n+1)\pi+] + g[(2n+1)\pi-]) = \frac{1}{2}(1-1) = 0$$

and again (2) satisfies this condition. Thus, the series in (2) is the Fourier series of g and the equality sign is valid for all points such that $x \neq n\pi$.

• **PROBLEM 14-14**

Define convergence in the mean and discuss its connection with ordinary convergence and with the concept of mean squared deviation (or variance) used in statistics.

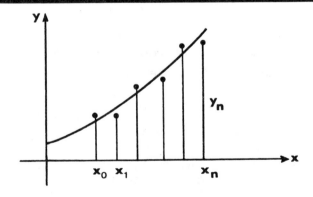

Solution: A sequence of functions $\{f_n(x)\}$ is said to converge in the mean to a function $f(x)$ on an interval (a,b) if

$$\lim_{n\to\infty} \int_a^b [f(x) - f_n(x)]^2 dx = 0. \qquad (1)$$

Recall that ordinary convergence requires that, for

$$f: R \to R, \quad \lim_{n\to\infty} \|f_n(x) - f(x)\| = 0$$

where $\|g(x)\|$ = absolute value of $g(x)$ for real valued g and if g is a vector function,

$$\|g(x)\| = \|g_1(x), g_2(x), \ldots, g_m(x)\| =$$

$$= \sqrt{g_1^2 + g_2^2 + \ldots + g_m^2}. \qquad (2)$$

The function on the number or vector $g(x)$ defined in (2) is an example of a norm on a vector space and the expression in equation (1) is simply another example of a norm on the vector space of these functions, i.e.

$$\int_a^b [g(x)]^2 dx = \|g(x)\|_2 \qquad (3)$$

and $\|\cdot\|_2$ is called the mean square norm. Thus (1) may be written as

$$\lim_{n\to\infty} \|f(x) - f_n(x)\|_2 = 0 \qquad (4)$$

and (4) is now the condition for $\{f_n(x)\}$ to converge in the mean to $f(x)$. In the case of Fourier series, f_n is replaced by the partial sum

$$g_n(x) = \frac{a_o}{2} + \sum_{k=1}^{n} a_k \cos kx + b_k \sin kx \qquad (5)$$

and it is said that the Fourier series for $f(x)$ converges to f in the mean if

$$\lim_{n\to\infty} \int_a^b [f(x) - g_n(x)]^2 dx = 0. \qquad (6)$$

Finally, the concept of variance is approached as follows. Suppose a set of n measurements, $y_1, y_2, \ldots,$

y_n are made to try to determine a value for the fixed quantity y. Then the best estimate of y is taken to be the mean of this set of measurements

$$\langle y \rangle \equiv \frac{1}{n} \sum_{i=1}^{n} y_i \tag{7}$$

and the precision of this estimate is usually described by the mean square deviation,

$$d^2 \equiv \frac{1}{n} \sum_{i=1}^{n} [y_i - \langle y \rangle]^2 \tag{8}$$

which is simply the average value of the quantities $[y_i - \langle y \rangle]^2$. Now suppose that there is a theoretical relationship in which the value of y depends on another measurable quantity, x (e.g. y might be the voltage across a resistor, R, and x might be the current through R in which case y = Rx). Then

$$y = y(x) \tag{9}$$

is the theoretical relationship and each measurement y_i is accompanied by a measurement of x, namely x_i. Then to test how well the data fits (9), one would form the mean square deviation

$$D = \frac{1}{n} \sum_{i=0}^{n} [y_i - y(x_i)]^2 \tag{10}$$

(see Figure). If the measurements y_i were taken at equal intervals Δx, then $\Delta x \cdot n = L$ where $L = x_n - x_o$ is the length of interval. In this case

$$D = \frac{1}{L} \sum_{i=0}^{n} [y_i - y(x_i)]^2 \Delta x$$

As n now increases the sum approaches an integral. It seems logical then to measure the extent of fit of two functions $y(x)$ and $\bar{y}(x)$ by the integral

$$D = \frac{1}{L} \int_a^b [y(x) - \bar{y}(x)]^2 dx \tag{11}$$

or if the functions are $f(x)$ and $g_n(x)$

$$D_n = \frac{1}{L} \int_a^b [f(x) - g_n(x)]^2 dx. \tag{12}$$

Thus from (6) we see that convergence in the mean of a Fourier Series Σg_n to f is equivalent to

$$\lim_{n\to\infty} D_n = 0$$

where D_n is defined in (12).

● **PROBLEM 14-15**

A piecewise continuous function $f(x)$ is to be approximated in the interval $(-\pi,\pi)$ by a trigonometric polynomial of the form

$$g_n(x) = \frac{A_o}{2} + \sum_{k=1}^{n} A_k \cos kx + B_k \sin kx, \tag{1}$$

where A_k, B_k, A_o are undetermined. Prove that the total square deviation

$$D_n = \int_{-\pi}^{\pi} [f(x) - g_n(x)]^2 dx \tag{2}$$

is minimized by choosing A_o, A_k, B_k to be the Fourier coefficients of f, a_k, b_k, a_o.

Solution: Straightforward calculation leads to

$$D_n = \int_{-\pi}^{\pi} [f(x) - g_n(x)]^2 dx \tag{3}$$

$$= \int_{-\pi}^{\pi} [f(x)]^2 dx - 2\int_{-\pi}^{\pi} f(x) g_n(x) dx$$

$$+ \int_{-\pi}^{\pi} [g_n(x)]^2 dx. \tag{4}$$

Now,

$$\int_{-\pi}^{\pi} f(x) g_n(x) dx = \frac{1}{2} A_o \int_{-\pi}^{\pi} f(x) dx + \sum_{k=1}^{n} A_k \int_{-\pi}^{\pi} f(x) \cos kx \, dx$$

$$+ \sum_{k=1}^{n} B_k \int_{-\pi}^{\pi} f(x) \sin kx \, dx. \tag{5}$$

Also, the orthogonality relations for $\sin kx$ and $\cos kx$ give

$$\int_{-\pi}^{\pi} [g_n(x)]^2 dx = \int_{-\pi}^{\pi} \left[\frac{A_o^2}{4} + \sum_{k=1}^{n} A_k^2 \cos^2 kx \right.$$

$$\left. + B_k^2 \sin^2 kx \right] dx + \int_{-\pi}^{\pi} \left[\sum_{\substack{i,j \\ i \neq j}}^{n} A_i B_j \cos ix \sin jx \right] dx. \quad (6)$$

But

$$\int_{-\pi}^{\pi} \left[\sum_{\substack{i,j=1 \\ i \neq j}}^{n} A_i B_j \cos ix \sin jx \right] dx$$

$$= \int_{-\pi}^{\pi} [A_1 B_2 \cos x \sin 2x + A_1 B_3 \cos x \sin 3x + \ldots$$

$$+ A_1 B_n \cos x \sin nx + \ldots + A_n B_{n-1} \cos nx \sin (n-1)x] dx$$

Since $\int_a^b \cos mx \sin nx\, dx = 0$ ($m \neq n$, m, n integers)

the integral of the sum reduces to zero also. Hence,

$$\int_{-\pi}^{\pi} [g_n(x)]^2 dx = \int_{-\pi}^{\pi} \frac{A_o^2}{4} dx + \sum_{k=1}^{n} \left[A_k^2 \int_{-\pi}^{\pi} \cos^2 kx\, dx \right.$$

$$\left. + B_k^2 \int_{-\pi}^{\pi} \sin^2 kx\, dx \right].$$

But

$$\sum_{k=1}^{n} \left[A_k^2 \int_{-\pi}^{\pi} \cos^2 kx\, dx + B_k^2 \int_{-\pi}^{\pi} \sin^2 kx\, dx \right]$$

$$= \pi \sum_{k=1}^{n} \left(A_k^2 + B_k^2 \right). \text{ Thus } \int_{-\pi}^{\pi} [g_n(x)]^2 dx$$

$$= \left\{ \pi \left[\frac{1}{2} A_o^2 + \sum_{k=1}^{n} (A_k^2 + B_k^2) \right] \right\}. \quad (7)$$

Thus, using (5) and (7) in (4):

$$D_n = \int_{-\pi}^{\pi} [f(x)]^2 dx + \left\{ \frac{A_o^2 \pi}{2} - A_o \int_{-\pi}^{\pi} f(x) dx \right\}$$

$$+ \sum_{k=1}^{n} \left\{ \pi A_k^2 - 2A_k \int_{-\pi}^{\pi} f(x) \cos kx\, dx \right\}$$

$$+ \sum_{k=1}^{n} \left\{ \pi B_k^2 - 2B_k \int_{-\pi}^{\pi} f(x) \sin kx\, dx \right\}. \tag{8}$$

This whole expression is minimized if each $\{\cdot\}$ is minimized. For instance the quantity

$$\delta_k = \pi A_k^2 - 2A_k \int_{-\pi}^{\pi} f(x) \cos kx\, dx \qquad (k = 1, 2, \ldots, n)$$

has an extremum if

$$\frac{d\delta_k}{dA_k} = 2\pi A_k - 2 \int_{-\pi}^{\pi} f(x) \cos kx\, dx = 0$$

or

$$A_k = \frac{1}{\pi} \int_{-\pi}^{\pi} f(x) \cos kx\, dx = a_k \quad (k=1,\ldots,n) \tag{9}$$

In this fashion we also obtain

$$B_k = \frac{1}{\pi} \int_{-\pi}^{\pi} f(x) \sin kx\, dx = b_k \quad (k=(1,\ldots,n) \tag{10}$$

$$A_o = \frac{1}{\pi} \int_{-\pi}^{\pi} f(x) dx = a_o. \tag{11}$$

Thus (9), (10) and (11) are statements of the fact that we have proven the Fourier coefficients of f(x) to be those coefficients which minimize the total square deviation of equation (2).

● **PROBLEM 14-16**

a) Use the concept of convergence in the mean to define another way in which the Fourier series of a function f can be said to satisfactorily represent f(x).
b) In this connection, state and prove Parseval's equality.

c) For what general class of functions does the Fourier series converge in the mean to its function?

Solution: (a) The Fourier series of a function f with period 2c may be considered to be an accurate representation of f on the interval (a, a+2c) if the total square deviation approaches zero

$$\lim_{n \to \infty} D_n = 0 \tag{1}$$

where

$$D_n = \int_a^{a+2c} [f(x) - g_n(x)]^2 dx \tag{2}$$

and $g_n(x)$ is the nth partial sum of the Fourier series

$$g_n(x) = \frac{a_0}{2} + \sum_{k=1}^{n} \left[a_k \cos \frac{k\pi x}{c} + b_k \sin \frac{k\pi x}{c} \right]. \tag{3}$$

As discussed earlier, if (1) holds it is said that the Fourier series of f converges in the mean to f.

(b) It was shown in the previous problem that the trigonometric polynomial g_n which minimized the expression (2) (the mean square deviation) was the one with Fourier coefficients and that in this case (for f with period 2π)

$$[D_n]_{min} = \int_{-\pi}^{\pi} [f(x)]^2 dx - \left[\frac{\pi}{2} a_0^2 + \pi \sum_{k=1}^{n} (a_k^2 + b_k^2) \right]. \tag{4}$$

Now if the Fourier series for f converges in the mean to f, then by definition

$$\lim_{n \to \infty} [D_n]_{min} = 0$$

so that from (4)

$$\frac{a_0^2}{2} + \sum_{k=1}^{\infty} (a_k^2 + b_k^2) = \frac{1}{\pi} \int_{-\pi}^{\pi} [f(x)]^2 dx. \tag{5}$$

This is known as Parseval's equality and is useful in the summation of series.

(c) It can be proven that if f is defined on the interval (a, a+2c) and is piecewise continuous there, then the Fourier series of f converges in the mean to f. This requirement is less stringent than that imposed upon f in order that it be pointwise convergent and, of course, much less stringent than the uniform convergence conditions.

This is so because the integral in (2) exists even if the integrand has a countably infinite number of discontinuities and its value is unaltered by any change in the functional values at these points. Thus, mean convergence is a statistical convergence and it must be kept in mind that it does not imply pointwise convergence. On the other hand, in most physical applications the shape of the function over the whole interval (or whole real axis) is the important consideration rather than the value at each point. Therefore, mean convergence is adequate in most physical situations.

• **PROBLEM 14-17**

The Fourier series of the following functions have been found in previous problems but no convergence questions were discussed. Determine now which of these functions has Fourier series which are (i) pointwise convergent (ii) uniformly convergent and (iii) convergent in the mean:

(a) $f(x) = e^x \quad (-\pi < x < \pi)$

(b) $f(x) = \begin{cases} x, & 0 < x < \pi \\ 0, & \pi < x < 2\pi \end{cases}$

(c) $f(x) = x^2, \quad x \in (-\pi, \pi)$
$f(x+2\pi) = f(x), \quad \text{all } x$

(d) $f(x) = \begin{cases} -1 & x \in (-\pi, 0) \\ 1 & x \in (0, \pi) \end{cases}$
$f(x+2\pi) = f(x) \quad \text{all } x$

(e) $f(x) = |x| \quad -\pi < x < \pi$

(f) $f(t) = t, \quad 0 < t < 3$
$f(t+3) = f(t) \quad \text{all } t.$

Solution: (a) (i) The function f is continuous on $(-\pi, \pi)$ (and therefore, of course, piecewise continuous) and has right and left hand derivatives at every point in $(-\pi, \pi)$ (in fact $f'_-(x_o) = f'_+(x_o) = f'(x_o)$) so that for all $x_o \in (-\pi, \pi)$ the Fourier series converges pointwise to $\frac{1}{2}\{f(x_o+) + f(x_o-)\}$, i.e.

$$\frac{1}{2}\{f(x_o+) + f(x_o-)\} = \frac{1}{2}a_o + \sum_{n=1}^{\infty}\left[a_n \cos\left(\frac{\pi n x_o}{c}\right) + b_n \sin\left(\frac{\pi n x_o}{c}\right)\right] \quad (1)$$

where $c = \pi$ in this case.

(ii) Since f is continuous on $(-\pi, \pi)$ and $f'(x) = e^x$ is continuous on $(-\pi, \pi)$ (and hence piecewise continuous), the Fourier series of f converges uniformly to f.

(iii) The function is continuous and hence piecewise continuous. Therefore, its Fourier series converges in the mean to f.

(b)(i) The function is piecewise continuous on $(0,2\pi)$ and has right and left hand derivatives at all $x_o \in (0,2\pi)$ so that its Fourier series converges pointwise to $\frac{1}{2}\{f(x_o+) + f(x_o-)\}$ as in (1). (ii) The function is discontinuous at $x = \pi$ and hence does not have a uniformly convergent Fourier series in $(0,2\pi)$. However, in each of the intervals $(0,\pi)$ and $(\pi,2\pi)$, f is continuous and f' is piecewise continuous (in fact, continuous) so that f's Fourier series converges uniformly to f in each of these intervals. (iii) Since f is piecewise continuous in $(0,2\pi)$, the Fourier series of f converges in the mean to f.

(c)(i) Since f is piecewise continuous on R and has right and left hand derivatives at all $x \in R$ the Fourier series of f at x_o converges to

$\frac{1}{2}\{f(x_o+) + f(x_o-)\}$ for all x.

(ii) f is discontinuous at all points $(2n+1)\pi$ so that its Fourier series is not uniformly convergent on R. However, on any interval $((2n-1)\pi, (2n+1)\pi)$ f is continuous and f' is piecewise continuous (in fact, continuous) so that the Fourier series of f converges uniformly to f on any such interval. (iii) f is piecewise continuous on R so the Fourier series of f converges in the mean to f on any interval in R.

(d)(i) Pointwise convergence for all $x \in R$; (ii) Uniform convergence in any interval $(n\pi, (n+1)\pi)$; (iii) convergence in the mean on any interval in R.

(e)(i) Pointwise convergence on $(-\pi,\pi)$; (ii) Uniform convergence on $(-\pi,\pi)$; (iii) Convergence in the mean on $(-\pi,\pi)$.

(f) (i) Pointwise convergence on R; (ii) Uniform convergence on any interval $(3n, 3(n+1))$; (iii) Convergence in the mean on any interval.

FURTHER REPRESENTATIONS

• **PROBLEM** 14-18

(a) Suppose a function is defined on $(-c,+c)$. Find the Fourier coefficients of f for the case where f is even and also for the case where f is odd.
(b) Motivated by (a), define the Fourier sine and cosine series of a function defined on an interval $(0,c)$ and interpret each graphically.

Solution: (a) The Fourier coefficients of f are given by

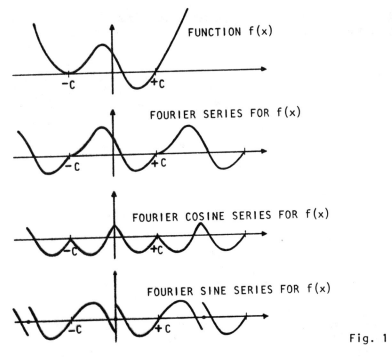

Fig. 1

$$a_n = \frac{1}{c} \int_{-c}^{c} f(x) \cos \frac{n\pi x}{c} \, dx \tag{1}$$

$$b_n = \frac{1}{c} \int_{-c}^{c} f(x) \sin \frac{n\pi x}{c} \, dx. \tag{2}$$

Now if $f(x)$ is even, i.e. $f(-x) = f(x)$, then $f(x) \cos \frac{n\pi x}{c}$ is even since it is the product of even functions, and $f(x) \sin \frac{n\pi x}{c}$ is odd since it is the product of an even function and an odd function. Furthermore,

$$\int_{-c}^{c} g(x) \, dx = \begin{cases} 0, & \text{if } g \text{ is odd} \\ 2 \int_{0}^{c} g(x) \, dx, & \text{if } g \text{ is even} \end{cases} \tag{3}$$

This can be seen in the following way. If g is odd, then

$$\int_{-c}^{c} g(x) \, dx = \int_{-c}^{0} g(x) \, dx + \int_{0}^{c} g(x) \, dx$$

$$= -\int_{0}^{-c} g(x) \, dx + \int_{0}^{c} g(x) \, dx. \tag{4}$$

Now, changing the variable of integration in the first

integral of (4) does not change its value so we make the substitution $x = -y$ to yield

$$\int_{-c}^{c} g(x)\,dx = -\int_{-y=0}^{-y=-c} g(-y)\,d(-y) + \int_{0}^{c} g(x)\,dx$$

$$= \int_{y=0}^{y=c} g(-y)\,dy + \int_{0}^{c} g(x)\,dx$$

$$= -\int_{0}^{c} g(y)\,dy + \int_{0}^{c} g(x)\,dx$$

$$= -\int_{0}^{c} g(x)\,dx + \int_{0}^{c} g(x)\,dx = 0$$

where the fact that $g(-y) = -g(y)$ was used in the third step. Similarly, if g is even then

$$\int_{-c}^{c} g(x)\,dx = \int_{-c}^{0} g(x)\,dx + \int_{0}^{c} g(x)\,dx$$

$$= -\int_{0}^{-c} g(x)\,dx + \int_{0}^{c} g(x)\,dx$$

$$= -\int_{-y=0}^{-y=-c} g(-y)\,d(-y) + \int_{0}^{c} g(x)\,dx$$

$$= \int_{y=0}^{y=c} g(-y)\,dy + \int_{0}^{c} g(x)\,dx$$

$$= \int_{0}^{c} g(y)\,dy + \int_{0}^{c} g(x)\,dx$$

$$= 2\int_{0}^{c} g(x)\,dx.$$

Using the result of (3) in (1) and (2) gives

$$a_n = \frac{2}{c}\int_{0}^{c} f(x)\cos\left(\frac{n\pi x}{c}\right), \quad b_n = 0 \text{ if } f \text{ is even} \tag{5}$$

$$a_n = 0, \quad b_n = \frac{2}{c} \int_0^c f(x) \sin\left(\frac{n\pi x}{c}\right) \text{ if } f \text{ is odd.} \tag{6}$$

(b) Let $f(x)$ be defined on an interval $(0,c)$. Then the Fourier cosine series of $f(x)$ is defined by

$$f(x) \tilde{c} \frac{a_0}{2} + \sum_{n=1}^{\infty} a_n \cos\left(\frac{n\pi x}{c}\right) \tag{7}$$

where

$$a_n = \frac{2}{c} \int_0^c f(x) \cos\left(\frac{n\pi x}{c}\right) dx. \tag{8}$$

From the above discussion it is seen that (7) reproduces the function

$$g(x) = \begin{cases} f(x) & (0 < x < c) \\ f(-x) & (-c < x < 0) \\ f(x+2c) = f(x) & \text{all } x. \end{cases} \tag{9}$$

which may be called the symmetric periodic extension of $f(x)$. Similarly, the Fourier sine series of f is defined by

$$f(x) \tilde{s} \sum_{n=1}^{\infty} b_n \sin\left(\frac{n\pi x}{c}\right) \tag{10}$$

where

$$b_n = \frac{2}{c} \int_0^c f(x) \sin\left(\frac{n\pi x}{c}\right) dx. \tag{11}$$

The series in (10) will reproduce the antisymmetric periodic extension of f

$$g(x) = \begin{cases} f(x) & (0 < x < c) \\ -f(-x) & (-c < x < 0) \\ f(x+2c) = f(x) & \text{all } x \end{cases} . \tag{12}$$

The three different trigonometric series for a given function are represented in the Figure.

Finally, note that both the pointwise convergence theorem and the convergence in the mean theorem apply to these series with the same premises (except that the interval is now restricted to $(0,c)$ rather than $(a,a+2c)$). This is because the functions of (9) and (12) will have the same general characteristics (piecewise continuity, right and left hand derivatives at each point) as their generating function, f. Hence even at $x = 0$, if f has a right hand derivative then g will have both right and left hand derivatives and the pointwise convergence conditions may be applied directly to g.

• **PROBLEM 14-19**

Find the Fourier full-range series for the function

$$f(x) = \frac{x}{2L} + \frac{1}{2} \tag{1}$$

in the interval $(-L,L)$. Also find f's Fourier cosine and sine series in the interval $(0,L)$ and graph all three series.

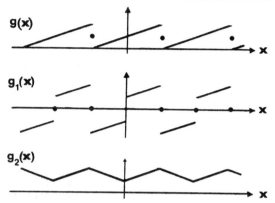

Solution: In each case the problem is just a matter of computing the coefficients. Thus, the full range series is given by

$$f(x) \sim g(x) = \frac{a_o}{2} + \sum_{n=1}^{\infty} a_n \cos \frac{n\pi x}{L} + b_n \sin \frac{n\pi x}{L} \tag{2}$$

where $g(x)$ is used to represent the Fourier series of f and the coefficients are given by

$$a_o = \frac{1}{L} \int_{-L}^{L} f(x)\,dx = \frac{1}{L} \int_{-L}^{L} \left[\frac{x}{2L} + \frac{1}{2}\right] dx$$

$$= \frac{1}{L} \left[\frac{x^2}{4L} + \frac{x}{2}\right]_{-L}^{L} = \frac{1}{L}(0 + L) = 1,$$

$$a_n = \frac{1}{L} \int_{-L}^{L} f(x) \cos\left(\frac{n\pi x}{L}\right) dx$$

$$= \frac{1}{L} \int_{-L}^{L} \frac{x}{2L} \cos\left(\frac{n\pi x}{L}\right) dx + \frac{1}{L} \int_{-L}^{L} \frac{1}{2} \cos\left(\frac{n\pi x}{L}\right) dx$$

$$= 0 + \frac{1}{L} \int_{-L}^{L} \frac{1}{2} \cos\left(\frac{n\pi x}{L}\right) dx$$

since the integrand of the first integral is odd. Thus,

$$a_n = \frac{1}{2L} \int_{-L}^{L} \cos\left(\frac{n\pi x}{L}\right) dx$$

$$= \frac{1}{2n\pi} \int_{-L}^{L} d\left(\sin \frac{n\pi x}{L}\right) dx$$

$$= \frac{1}{2n\pi} \left[\sin \frac{n\pi x}{L}\right]_{-L}^{L} = \frac{1}{2n\pi} [\sin n\pi - \sin(-n\pi)] = 0 - 0$$

$$= 0.$$

Finally,

$$b_n = \frac{1}{L} \int_{-L}^{L} f(x) \sin\left(\frac{n\pi x}{L}\right) dx$$

$$= \frac{1}{L} \int_{-L}^{L} \frac{x}{2L} \sin\left(\frac{n\pi x}{L}\right) dx + \frac{1}{2L} \int_{-L}^{L} \sin\left(\frac{n\pi x}{L}\right) dx$$

$$= \frac{1}{L} \int_{-L}^{L} \frac{x}{2L} \sin\left(\frac{n\pi x}{L}\right) dx + 0$$

since the integrand of the second integral is odd. Then, letting $u = \frac{n\pi x}{L}$, $du = \frac{n\pi}{L} dx$ we find

$$b_n = \frac{1}{L} \int_{-L}^{L} \frac{x}{2L} \sin\left(\frac{n\pi x}{L}\right) dx = \frac{1}{2(n\pi)^2} \int_{-n\pi}^{n\pi} u \sin u \, du.$$

Applying integration by parts

$$b_n = \frac{1}{2(n\pi)^2} \left[-u \cos u \Big|_{-n\pi}^{n\pi} + \int_{-n\pi}^{n\pi} \cos u \, du \right]$$

$$= \frac{1}{2(n\pi)^2} \Big(-\{n\pi \cos n\pi - (-n\pi \cos(-n\pi))\}$$

$$+ \{\sin(n\pi) - \sin(-n\pi)\} \Big)$$

$$= \frac{-2n\pi \cos n\pi}{2(n\pi)^2} = \frac{-\cos n\pi}{n\pi} = \frac{(-1)^{n+1}}{n\pi}.$$

Thus from (2), the Fourier full range series for f is

$$g(x) = \frac{1}{2} + \sum_{n=1}^{\infty} \frac{(-1)^{n+1}}{n\pi} \sin\left(\frac{n\pi x}{L}\right). \tag{3}$$

The Fourier cosine series for f is given by

$$f(x) \underset{c}{\sim} g_2(x) = \frac{a_o}{2} + \sum_{n=1}^{\infty} a_n \cos\left(\frac{n\pi x}{L}\right) \tag{4}$$

where the coefficients are given by

$$a_o = \frac{2}{L} \int_0^L f(x)\,dx = \frac{2}{L} \int_0^L \left[\frac{x}{2L} + \frac{1}{2}\right] dx = \frac{2}{L} \left[\frac{x^2}{4L} + \frac{x}{2}\right]_0^L$$

$$= \frac{3}{2},$$

$$a_n = \frac{2}{L} \int_0^L f(x) \cos\left(\frac{n\pi x}{L}\right) dx$$

$$= \frac{2}{L} \int_0^L \frac{x}{2L} \cos\left(\frac{n\pi x}{L}\right) dx + \frac{1}{L} \int_0^L \cos\left(\frac{n\pi x}{L}\right) dx.$$

Now let $u = \frac{n\pi x}{L}$ and integrate by parts so that

$$a_n = \frac{2}{L} \int_0^{n\pi} \frac{uL}{2(n\pi)^2} \cos u\,du + \frac{1}{L} \int_0^{n\pi} \frac{L}{n\pi} \cos u\,du$$

$$= \frac{1}{(n\pi)^2} \int_0^{n\pi} u \cos u\,du + \frac{1}{n\pi} [\sin u]_0^{n\pi}$$

$$= \frac{1}{(n\pi)^2} \left[[u \sin u]_0^{n\pi} - \int_0^{n\pi} \sin u\,du \right]$$

$$= \frac{1}{(n\pi)^2} (\cos n\pi - \cos 0) = \frac{\cos n\pi - 1}{(n\pi)^2}.$$

Therefore, from (4), the Fourier cosine series for f is

$$g_2(x) = \frac{3}{4} - \frac{2}{\pi^2} \sum_{n=1,3,5}^{\infty} \frac{1}{n^2} \cos \frac{n\pi x}{L}. \tag{5}$$

Finally, the Fourier sine series for f is given by

$$f(x) \tilde{s} g_1(x) = \sum_{n=1}^{\infty} b_n \sin\left(\frac{n\pi x}{L}\right) \qquad (6)$$

where the coefficient b_n is given by

$$b_n = \frac{2}{L} \int_0^L f(x) \sin\left(\frac{n\pi x}{L}\right) dx$$

$$= \frac{2}{L} \int_0^L \frac{x}{2L} \sin\left(\frac{n\pi x}{L}\right) dx + \frac{2}{L} \int_0^L \frac{1}{2} \sin\left(\frac{n\pi x}{L}\right) dx.$$

Now let $u = \frac{n\pi x}{L}$ and integrate by parts to find

$$b_n = \frac{1}{(n\pi)^2} \int_0^{n\pi} u \sin u \, du + \frac{1}{n\pi} \int_0^{n\pi} \sin u \, du$$

$$= \frac{1}{(n\pi)^2} \left\{ (-u \cos u)\Big|_0^{n\pi} - \int_0^{n\pi} (-\cos u) \, du \right\}$$

$$- \frac{1}{n\pi} [\cos n\pi - \cos 0]$$

$$= \frac{1}{(n\pi)^2} \left(-n\pi \cos n\pi + (\sin n\pi - \sin 0) \right)$$

$$+ \frac{1 - \cos n\pi}{n\pi}$$

$$= \frac{1 - 2\cos n\pi}{n\pi}.$$

Therefore, from (6), the Fourier sine series for f is

$$g_1(x) = \frac{3}{\pi} \left[\sum_{n=1,3,5}^{\infty} \frac{1}{n} \sin\left(\frac{n\pi x}{L}\right) \right] - \frac{1}{\pi} \left[\sum_{n=2,4,6,\ldots}^{\infty} \frac{1}{n} \sin\left(\frac{n\pi x}{L}\right) \right]. \qquad (7)$$

All three series are graphed in the figure.

● **PROBLEM 14-20**

Find the Fourier sine series of $f(x) = x^2$ over the interval $(0,1)$.

Solution: The Fourier sine series of a function defined on an interval $(0,c)$ is given by

$$f(x) \sim_s \sum_{n=1}^{\infty} b_n \sin\left(\frac{n\pi x}{c}\right) \tag{1}$$

where

$$b_n = \frac{2}{c} \int_0^c f(x) \sin\left(\frac{n\pi x}{c}\right) dx \tag{2}$$

We now turn to the given problem. Here $f(x) = x^2$, and $c = 1$. Thus we obtain the Fourier series

$$x^2 \sim_s \sum_{n=1}^{\infty} b_n \sin n\pi x \tag{3}$$

where

$$b_n = 2 \int_0^1 x^2 \sin n\pi x \, dx.$$

We must evaluate the b_n. Using integration by parts, we obtain

$$b_n = 2 \int_0^1 x^2 \sin n\pi x \, dx$$

$$= 2 \left\{ \left[\frac{-x^2}{n\pi} \cos n\pi x \right]_0^1 + \frac{2}{n\pi} \int_0^1 x \cos n\pi x \, dx \right\}$$

$$= 2 \left\{ \frac{-(-1)^n}{n\pi} + \frac{2}{n^2 \pi^2} x \sin n\pi x \Big|_0^1 \right.$$

$$\left. - \int_0^1 \frac{2}{n^2 \pi^2} \sin n\pi x \, dx \right\}$$

$$= 2 \left\{ \frac{(-1)^{n+1}}{n\pi} + \frac{2}{n^3 \pi^3} [(-1)^n - 1] \right\}. \tag{4}$$

Substituting (4) for b_n in (3), the required Fourier sine series over $0 < x < 1$ is

$$x^2 \sim_s 2 \sum_{n=1}^{\infty} \left[\frac{(-1)^{n+1}}{n\pi} - \frac{2\{1-(-1)^n\}}{n^3 \pi^3} \right] \sin n\pi x.$$

● **PROBLEM 14-21**

Find the Fourier sine series for the function defined by

$f(x) = 0 \qquad 0 \leq x \leq \pi/2$

$f(x) = 1 \qquad \pi/2 < x \leq \pi.$

Solution: The Fourier sine series of a function defined on $0 \leq x \leq L$

$$f(x) \sim_s \sum_{n=1}^{\infty} b_n \sin \frac{n\pi x}{L} \qquad (n = 1, 2, \ldots) \qquad (1)$$

where

$$b_n = \frac{2}{L} \int_0^L f(n) \sin nx \, dx. \qquad (2)$$

Recall that the Fourier sine series is equivalent to finding the Fourier trigonometric series of an odd function, i.e., a function such that $f(-x) = -f(x)$.

In the given problem, since $f(x) = 0$ for $0 \leq x \leq \pi/2$ we need find its series development only for the interval $\pi/2 < x \leq \pi$. Taking $L = \pi$,

$$b_n = \frac{2}{\pi} \int_{\pi/2}^{\pi} \sin nx \, dx = -\frac{2}{n\pi} \left(\cos nx \Big|_{\pi/2}^{\pi} \right)$$

$$= \frac{2}{n\pi} \left(\cos \frac{n\pi}{2} - \cos n\pi \right) = \frac{2}{n\pi} \left(\cos \frac{n\pi}{2} + (-1)^{n+1} \right). \qquad (3)$$

Substituting (3) into (1),

$$f(x) \sim_s \frac{2}{\pi} \left(\frac{\sin x}{1} - \frac{2\sin 2x}{2} + \frac{\sin 3x}{3} + \frac{\sin 5x}{5} - \frac{2\sin 6x}{6} + \ldots \right).$$

● **PROBLEM 14-22**

Find a cosine series which represents $f(x)$ in $0 \leq x \leq \pi$ if $f(x)$ is defined as

$f(x) = 0 \qquad 0 \leq x \leq \pi/2$

$f(x) = 1 \qquad \pi/2 < x \leq \pi.$

Solution: The Fourier cosine series for a function $f(x)$, defined over $0 \leq x \leq L$, is given by the formula

$$f(x) \sim_c \frac{1}{2} a_0 + \sum_{n=1}^{\infty} a_n \cos \frac{n\pi x}{L} \qquad (1)$$

where

$$a_n = \frac{2}{L} \int_0^L f(x) \cos \frac{n\pi x}{L} \, dx. \qquad (2)$$

In the given problem $0 \leq x \leq \pi$. Thus we may take L as equal to π. Then the coefficients as given by (2) are

$$a_0 = \frac{1}{\pi} \int_{\pi/2}^{\pi} dx = 1/2 , \tag{3}$$

$$a_n = \frac{2}{\pi} \int_{\pi/2}^{\pi} \cos nx\, dx = \frac{2}{n\pi} \sin nx \Big|_{\pi/2}^{\pi}$$

$$= -\frac{2}{n\pi} \sin \frac{n\pi}{2} , \quad n > 0 . \tag{4}$$

Since $\sin \frac{n\pi}{2} = \pm 1$ for $n = 1,3,5,\ldots$, we have, upon substituting (3) and (4) into (1),

$$f(x) \underset{c}{\sim} \frac{1}{2} - \frac{2}{\pi} \left(\frac{\cos x}{1} - \frac{\cos 3x}{3} + \frac{\cos 5x}{5} - \cdots \right)$$

$$= \frac{1}{2} - \frac{2}{\pi} \sum_{n=0}^{\infty} (-1)^n \frac{\cos(2n+1)x}{2n+1} . \tag{5}$$

● **PROBLEM 14-23**

Find the Fourier cosine series over the interval $0 < x < c$ for the function $f(x) = x$.

Solution: If $f(x)$ has a Fourier cosine series over $0 < x < c$, it is of the form

$$f(x) \underset{c}{\sim} \frac{1}{2} a_0 + \sum_{n=1}^{\infty} a_n \cos \frac{n\pi x}{c} . \tag{1}$$

where

$$a_n = \frac{2}{c} \int_0^c f(x) \cos \frac{n\pi x}{c}\, dx. \quad (n = 0,1,\ldots) \tag{2}$$

In the given problem, $f(x) = x$; hence (2) may be rewritten

$$a_n = \frac{2}{c} \int_0^c x \cos \frac{n\pi x}{c}\, dx. \quad (n = 0,1,2,\ldots) \tag{3}$$

The problem at hand is the evaluation of the coefficients a_n. We first find a_n for $n \neq 0$. The integral in (3) may be evaluated using integration by parts. Thus,

$$a_n = \frac{2}{c} \left[\frac{c}{n\pi} x \sin \frac{n\pi x}{c} \Big|_0^c - \frac{c}{n\pi} \int_0^c \sin \frac{n\pi x}{c}\, dx \right], n \neq 0$$

$$= \frac{2}{c} \left[\left(\frac{c}{n\pi}\right)^2 \cos \frac{n\pi x}{c} \Big|_0^c \right] = \frac{2c}{(n\pi)^2} \left[(-1)^n - 1 \right]$$

$$= -\frac{2c}{(n\pi)^2}\left(1-(-1)^n\right).$$

We now find the coefficient a_0 separately.

$$a_0 = \frac{2}{c}\int_0^c x \cos\frac{(0\pi x)}{c}\,dx = \frac{2}{c}\left.\frac{x^2}{2}\right|_0^c = c.$$

Thus the Fourier cosine series over the interval $0 < x < c$ for the function $f(x) = x$ is

$$f(x) \underset{c}{\sim} \frac{1}{2}c - \frac{2c}{\pi^2}\sum_{n=1}^{\infty}\frac{1-(-1)^n}{n^2}\cos\frac{n\pi x}{c}$$

or,

$$f(x) \underset{c}{\sim} \frac{1}{2}c - \frac{4c}{\pi^2}\sum_{k=0}^{\infty}\frac{\cos[(2k+1)\pi x/c]}{(2k+1)^2} \qquad (4)$$

since when n is even, $1 - (-1)^n = 0$.

• **PROBLEM 14-24**

Show that the Fourier series for $f(x)$, where f is periodic with period $2L$, given by

$$f(x) \sim \frac{a_0}{2} + \sum_{n=1}^{\infty}\left[a_n\cos\left(\frac{n\pi x}{L}\right) + b_n\sin\left(\frac{n\pi x}{L}\right)\right] \qquad (1)$$

can be written in complex form as

$$f(x) \sim \sum_{n=-\infty}^{\infty} c_n e^{i\left(\frac{n\pi x}{L}\right)}. \qquad (2)$$

Solution: Recall that Euler's theorem states that

$$e^{iy} = \cos y + i\sin y \qquad (3)$$

so that $e^{-iy} = \cos(-y) + i\sin(-y) = \cos y - i\sin y.$

Therefore,

$$\cos\frac{n\pi x}{L} = \frac{1}{2}\left[e^{i(n\pi x/L)} + e^{-i(n\pi x/L)}\right] \qquad (3a)$$

$$\sin\frac{n\pi x}{L} = \frac{1}{2i}\left[e^{i(n\pi x/L)} - e^{-i(n\pi x/L)}\right]$$

$$= \frac{-i}{2}\left[e^{i(n\pi x/L)} - e^{-i(n\pi x/L)}\right]. \qquad (3b)$$

Using these results in (1) yields

711

$$f(x) \sim \frac{a_0}{2} + \sum_{n=1}^{\infty} \left\{ \frac{a_n}{2} \left[e^{i\left(\frac{n\pi x}{L}\right)} + e^{-i\left(\frac{n\pi x}{L}\right)} \right] \right.$$

$$\left. - \frac{ib_n}{2} \left[e^{i\left(\frac{n\pi x}{L}\right)} - e^{-i\left(\frac{n\pi x}{L}\right)} \right] \right\}$$

$$= \frac{a_0}{2} + \sum_{n=1}^{\infty} \left\{ \left[\frac{a_n - ib_n}{2} \right] e^{i\left(\frac{n\pi x}{L}\right)} + \left[\frac{a_n + ib_n}{2} \right] e^{-i\left(\frac{n\pi x}{L}\right)} \right\}. \quad (4)$$

Now define coefficients c_n as follows

$$c_n = \begin{cases} \frac{1}{2}(a_n - ib_n) & (n > 0) \\ \frac{1}{2}(a_{-n} + ib_{-n}) & (n < 0) \\ \frac{1}{2} a_0 & (n = 0) \end{cases} \quad (5)$$

Then (4) may be written as

$$f(x) \sim \sum_{n=-\infty}^{\infty} c_n e^{i\frac{n\pi x}{L}} \quad (6)$$

and this is the desired complex form of the Fourier series of f. Recall that the Fourier coefficients a_n and b_n are defined by

$$a_n = \frac{1}{L} \int_a^{a+2L} f(x) \cos\left(\frac{n\pi x}{L}\right) dx$$

and

$$b_n = \frac{1}{L} \int_a^{a+2L} f(x) \sin\left(\frac{n\pi x}{L}\right) dx,$$

where a is any number such that (a, a+2L) is in the domain of definition of f. Hence, for $n > 0$

$$c_n = \frac{1}{2}(a_n - ib_n) = \frac{1}{2L} \int_a^{a+2L} f(x) \left[\cos\left(\frac{n\pi x}{L}\right) - i \sin\left(\frac{n\pi x}{L}\right) \right] dx$$

$$= \frac{1}{2L} \int_a^{a+2L} f(x) e^{-i\left(\frac{n\pi x}{L}\right)} dx.$$

For $n < 0$,

$$a_{-n} = \frac{1}{L} \int_a^{a+2L} f(x) \cos\left(\frac{-n\pi x}{L}\right) dx = \frac{1}{L} \int_a^{a+2L} f(x) \cos\left(\frac{n\pi x}{L}\right) dx$$

and

$$b_{-n} = \frac{1}{L} \int_a^{a+2L} f(x) \sin\left(\frac{-n\pi x}{L}\right) dx = \frac{-1}{L} \int_a^{a+2L} f(x) \sin\left(\frac{n\pi x}{L}\right) dx$$

so that

$$c_n = \frac{1}{2}(a_{-n} + ib_{-n}) = \frac{1}{2L} \int_a^{a+2L} f(x) \left[\cos\left(\frac{n\pi x}{L}\right) - i\sin\left(\frac{n\pi x}{L}\right)\right] dx$$

$$= \frac{1}{2L} \int_a^{a+2L} f(x) e^{-i\left(\frac{n\pi x}{L}\right)} dx.$$

Note that this shows $c_n = \frac{1}{2}(a_n - ib_n), n<0$. Finally, for $n=0$

$$c_0 = \frac{a_0}{2} = \frac{1}{2L} \int_a^{a+2L} f(x) dx = \frac{1}{2L} \int_a^{a+2L} f(x) e^{i\left(\frac{0\pi x}{L}\right)} dx.$$

One thus has the concise statement

$$f(x) \sim \sum_{n=-\infty}^{\infty} c_n e^{i\left(\frac{n\pi x}{L}\right)} \quad \text{and} \quad c_n = \frac{1}{2L} \int_a^{a+2L} f(x) e^{-i\left(\frac{n\pi x}{L}\right)} dx \quad (7)$$

which is the required complex form of the Fourier series for f. It is very important to note that although (7) has a complex form, if f(x) is real valued so is the series. How could it be otherwise if we have only used identities relating real numbers to imaginary numbers ((3a),(3b),(5))? If the complex form is used in a problem it may not be obvious how to convert the final answer from complex to real form in order to make numerical approximations. In such a case the relations in (5) must be used to find a_n and b_n in terms of c_n. Even so, in complicated problems the simplicity of manipulations with the complex form make it convenient to carry it along until the end and then convert back to the real form using (5).

• **PROBLEM 14-25**

Represent the function

$$f(x) = \begin{cases} 0 & -\pi < x \leq 0 \\ 1 & 0 < x \leq \pi \end{cases} \quad (1)$$

by a complex Fourier series.

Solution: The complex form of the Fourier series of f is given by

$$f(x) \sim n = \sum_{-\infty}^{\infty} c_n e^{\frac{in\pi x}{L}} \quad (2)$$

where

$$c_n = \frac{1}{2L} \int_a^{a+2L} f(x) e^{-i\frac{n\pi x}{L}} dx \qquad (3)$$

where f is periodic with period 2L and a is any number such that (a,a+2L) is in the domain of definition of f. In the case at hand, L = π and a must be -π because of the definition in (1). Hence

$$c_0 = \frac{1}{2\pi} \int_0^{\pi} dx = \frac{1}{2}$$

and

$$c_n = \frac{1}{2\pi} \int_0^{\pi} e^{-inx} dx = \frac{1-e^{-in\pi}}{2\pi ni}$$

$$= \begin{cases} 0 & (n = \text{even}) \\ \frac{1}{n\pi i} & (n = \text{odd}) \end{cases}.$$

Using these results in (2) gives

$$f(x) \sim \frac{1}{2} + \frac{1}{\pi i} \sum_{\substack{n=-\infty \\ n=\text{odd}}}^{\infty} \frac{1}{n} e^{inx} . \qquad (4)$$

Since it is just a different notation, the complex Fourier series of f given in (4) has the same properties as the ordinary Fourier series, i.e. it represents the periodic extension of f into the whole real axis.

• **PROBLEM 14-26**

Find the exponential Fourier series of the periodic waveform from the graph shown below.

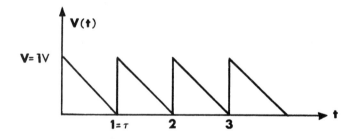

Solution: This is a typical example of the way Fourier series are used to represent functions in electrical engineering. The function to be represented here is v(t) = the voltage at a certain point in a circuit at time t, where v is periodic in time with period $\tau = 2\frac{T}{2} = 1$. The complex form of the Fourier series of such a function is given by

$$v(t) \sim n = \sum_{-\infty}^{\infty} c_n e^{i\left(\frac{n2\pi t}{\tau}\right)}, \quad c_n = \frac{1}{\tau} \int_a^{a+\tau} v(t) e^{-i\left(\frac{n2\pi t}{\tau}\right)} dt. \quad (1)$$

It is convenient to define $\omega = \frac{2\pi}{\tau}$, take the limits of integration from 0 to τ and let j stand for the number whose square is -1, (i.e. j = i). This is convenient in electrical applications since i is usually used for current. In this case (1) becomes

$$v(t) \sim n = \sum_{-\infty}^{\infty} c_n e^{jn\omega t}, \quad c_n = \frac{1}{\tau} \int_0^{\tau} v(t) e^{-jn\omega t} dt. \quad (2)$$

where

$$v(t) = 1 - t, \quad t \in (0, \tau)$$
$$v(t+\tau) = v(t) \quad \text{all } t .$$

First evaluate c_0 by

$$c_0 = \frac{1}{\tau} \int_0^{\tau} v(t) dt$$

(note that this may be interpreted as the average value of the voltage v over the interval $(0, \tau)$). This gives

$$c_0 = \frac{1}{\tau} \int_0^{\tau} (1-t) dt = \left[\frac{1}{\tau} t - \frac{t^2}{2}\right]_0^{\tau}$$

and evaluating yields

$$c_0 = \frac{1}{\tau}\left[\tau - \frac{\tau^2}{2}\right] = 1 - \frac{\tau}{2} = \frac{1}{2} \quad (3)$$

since $\tau = 1$ second.

Similarly, evaluate equation c_n, $n \neq 0$:

$$c_n = \frac{1}{\tau} \int_0^{\tau} (1-t) e^{-jn\omega t} dt,$$

$$c_n = \frac{1}{\tau} \int_0^{\tau} e^{-jn\omega t} dt - \frac{1}{\tau} \int_0^{\tau} t e^{-jn\omega t} dt$$

but

$$\int e^{ax} dx = \frac{e^{ax}}{a}$$

and, using integration by parts,

$$\int x e^{ax} dx = \frac{x e^{ax}}{a} - \frac{e^{ax}}{a^2} .$$

Using these results in the integrals above yields

$$c_n = \frac{1}{\tau}\left[\frac{e^{-jn\omega t}}{(-jn\omega)} - \frac{te^{-jn\omega t}}{(-jn\omega)} + \frac{e^{-in\omega t}}{(-jn\omega)^2}\right]\Bigg|_0^\tau,$$

so that

$$c_n = \frac{1}{\tau}\left[\frac{e^{-jn\omega\tau}-e^0}{(-jn\omega)} - \frac{\tau e^{-jn\omega\tau}-0}{(-jn\omega)} + \frac{e^{-jn\omega\tau}-e^0}{(-jn\omega)^2}\right]. \quad (4)$$

We would seek ways to simplify equation (4) before proceeding further. One way is recalling that

$$\omega = \frac{2\pi}{\tau} \quad \text{rad/s.}$$

Since $\tau = 1$ in this instance, from the graph.

$$\omega = 2\pi \quad \text{rad/s.} \quad (5)$$

Substituting equation (5) into equation (4),

$$c_n = \left[\frac{e^{-j2\pi n}-1}{(-j2\pi n)} - \frac{e^{-j2\pi n}}{(-j2\pi n)} + \frac{e^{-j2\pi n}-1}{(-j2\pi n)^2}\right]. \quad (6)$$

A further simplification may be made by evaluating the exponential quantity with the use of Euler's equation,

$$e^{jx} = \cos x + j \sin x, \quad (7)$$

$$e^{-j2\pi n} = \cos(-2\pi n) + j \sin(-2\pi n) = +1 \quad (8)$$

since the cosine or sine of $\pm 2\pi$ or any whole-number multiple thereof is +1 or zero, respectively. Now, substituting equation (8) into equation (6),

$$c_n = \left[\frac{1-1}{(-j2\pi n)} - \frac{1}{(-j2\pi n)} + \frac{1-1}{(-j2\pi n)^2}\right] \quad \text{and}$$

$$c_n = \frac{1}{j2\pi n}. \quad (9)$$

A further step in simplification can be made by recalling that

$$\frac{1}{j} = -j = 0 - j = e^{-j\frac{\pi}{2}} \quad (10)$$

which can be proven with equation (7). Substituting equation (10) into equation (9),

$$c_n = \frac{1}{2\pi n} e^{-j\frac{\pi}{2}} \quad (11)$$

Finally, substitute equations (3), (5) and (11) into equation (2):

$$v(t) = \frac{1}{2} + \sum_{n=-\infty}^{\infty} (\frac{1}{2\pi n}) e^{-j\frac{\pi}{2}} e^{j2\pi nt} .$$

This can be rewritten as

$$v(t) = \frac{1}{2} + \sum_{n=-\infty}^{\infty} (\frac{1}{2\pi n}) e^{j(2\pi nt - \frac{\pi}{2})} \quad \text{volts}$$

since $e^a e^b = e^{a+b}$.

● **PROBLEM 14-27**

Convert the Fourier series

$$f(t) = \sum_{k=-\infty}^{\infty} \frac{1}{1+jk} e^{jkt} \qquad (1)$$

to trigonometric form. (In (1) we have used $j^2 = -1$ instead of $i^2 = -1$ as is frequently done in electrical engineering.)

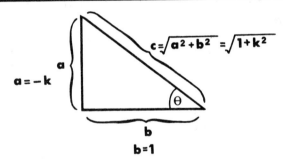

Solution: We know that the Fourier series is

$$f(t) = \frac{a_o}{2} + \sum_{n=1}^{\infty} (a_n \cos n\omega_o t + b_n \sin n\omega_o t)$$

$$= \sum_{n=-\infty}^{+\infty} c_n e^{jn\omega_o t} \qquad (1)$$

where $\qquad c_n = \frac{1}{2}(a_n - jb_n) \qquad$ (2)

$$c_o = \frac{a_o}{2} , \qquad (3)$$

and $\qquad \omega_o = \frac{2\pi}{T} \quad$ where T is the period of the function f (i.e. $f(t+T) = f(t)$). The expression on the left in Eq.(1) is known as the trigonometric form of the Fourier series. We must convert

$$f(t) = \sum_{k=-\infty}^{+\infty} \frac{1}{1+jk} e^{jkt} \quad \text{to trigonometric form.}$$

Making use of Eq.(2) we have

$$\frac{1}{1+jk} = \frac{1}{2} a_n - \frac{1}{2} jb_n ,$$

which can be written in polar form as

$$\frac{1}{\sqrt{1+k^2} \, e^{j \tan^{-1} k}} = \frac{1}{2} a_n - \frac{1}{2} jb_n \quad \text{where } k = \tan\theta,$$

so that

$$\frac{1}{\sqrt{1+k^2}} e^{-j \tan^{-1} k} = \frac{1}{\sqrt{1+k^2}} e^{j \tan^{-1}(-k)} = \frac{1}{2} a_n - \frac{1}{2} jb_n$$

or

$$\frac{1}{2} a_n - \frac{1}{2} jb_n = \frac{1}{\sqrt{1+k^2}} [\cos(\tan^{-1}(-k)) + j \sin(\tan^{-1}(-k))].$$

The figure shows some convenient identities. Thus if $\theta = \tan^{-1}(-k)$, then $\tan\theta = -k = \frac{a}{b}$, thus we may let $b = 1$, $a = -k$ and compute $c = \sqrt{1+k^2}$. Hence

$$\cos(\tan^{-1}(-k)) = \frac{1}{\sqrt{1+k^2}} ; \quad \sin(\tan^{-1}(-k)) = \frac{-k}{\sqrt{1+k^2}}$$

giving

$$\frac{1}{2} a_n - \frac{1}{2} jb_n = \frac{1}{1+k^2} - j \frac{k}{1+k^2}$$

$$a_n = \frac{2}{1+k^2} ; \quad \frac{a_0}{2} = c_0 = 1 ; \quad b_n = \frac{2k}{1+k^2}.$$

Substituting into Eq.(1) yields

$$f(t) = 1 + 2 \sum_{k=1}^{\infty} \frac{\cos kt + k \sin kt}{1+k^2}.$$

APPLICATIONS

• **PROBLEM 14-28**

Find the Fourier series for the waveform shown in the Figure. What is this series if the origin is shifted to 0'? (See figure)

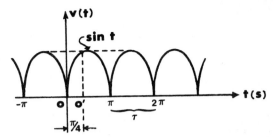

Solution: This is an example of the way Fourier series are used to represent functions in electrical engineering. It is interesting to note that this waveform is called a full-wave rectified wave and is a primitive form of the voltage output of say the rectifier which is attached between a battery which is being recharged and a wall socket. The function to be represented here is $v(t)$ = the voltage at a certain point in a circuit at time t where v is periodic in time with period $\tau = \pi$. The Fourier series of such a function is given by

$$v(t) \sim \frac{a_o}{2} + \sum_{n=1}^{\infty}\left(a_n \cos\frac{n2\pi t}{\tau} + b_n \sin\frac{n2\pi t}{\tau}\right) \quad (1)$$

where

$$a_n = \frac{2}{\tau}\int_a^{a+\tau} v(t)\cos\left(\frac{n2\pi t}{\tau}\right)dt \quad (2)$$

$$b_n = \frac{2}{\tau}\int_a^{a+\tau} v(t)\sin\left(\frac{n2\pi t}{\tau}\right)dt. \quad (3)$$

It is convenient to define $\omega = \frac{2\pi}{\tau} = 2$ and take the limits of integration from 0 to $\tau = \pi$. In this case, (1), (2) and (3) become

$$v(t) \sim \frac{a_o}{2} + \sum_{n=1}^{\infty}(a_n \cos 2nt + b_n \sin 2nt) \quad (4)$$

$$a_n = \frac{2}{\pi}\int_0^\pi v(t)\cos 2nt\, dt \quad (5)$$

$$b_n = \frac{2}{\pi}\int_0^\pi v(t)\sin 2nt\, dt. \quad (6)$$

Now the function $v(t)$ is given by

$$v(t) = \sin(t) \qquad 0 \leq t < \pi$$

$$v(t+\pi) = v(t) \qquad \text{all } t$$

so that from (5) and (6)

$$a_o = \frac{2}{\pi}\int_0^\pi \sin t\, dt = \frac{4}{\pi}$$

719

$$b_n = \frac{2}{\pi} \int_0^\pi \sin t \sin 2nt \, dt$$

$$= \frac{2}{\pi} \int_0^\pi \left[\frac{1}{2} \cos(1-2n)t - \frac{1}{2} \cos(1+2n)t \right] dt$$

$$= \frac{1}{\pi} \frac{\sin(1-2n)t}{(1-2n)} \bigg|_0^\pi - \frac{1}{\pi} \frac{\sin(1-2n)t}{(1+2n)} \bigg|_0^\pi$$

$$= 0 - 0 = 0.$$

And finally

$$a_n = \frac{2}{\pi} \int_0^\pi \sin t \cos 2nt \, dt.$$

Again, using trigonometric identities, this integral is equivalent to

$$a_n = \frac{2}{\pi} \int_0^\pi \left[\frac{1}{2} \sin(1+2n)t + \frac{1}{2} \sin(1-2n)t \right] dt$$

$$= \frac{1}{\pi} \left[-\frac{1}{1+2n} \cos(1+2n)t - \frac{1}{1-2n} \cos(1-2n)t \right] \bigg|_0^\pi$$

$$= \frac{1}{\pi} \left[-\frac{1}{1+2n} \cos(2n+1)\pi - \frac{1}{1-2n} \cos(1-2n)\pi \right]$$

$$- \frac{1}{\pi} \left[-\frac{1}{1+2n} \cos 0 - \frac{1}{1-2n} \cos 0 \right]$$

$$a_n = \frac{1}{\pi} \left(\frac{4}{1-4n^2} \right)$$

The expression for $v(t)$ is, thus,

$$v(t) \sim \frac{2}{\pi} \left[1 + \sum_{n=1}^\infty \frac{2}{1-4n^2} \cos 2nt \right] \quad \text{or since}$$

$$\frac{1}{1-4n^2} = -\frac{1}{4n^2-1}$$

$$v(t) \sim \frac{2}{\pi} [1 - \frac{2}{3} \cos 2t - \frac{2}{15} \cos 4t - \frac{2}{35} \cos 6t \ldots]$$

Now if the origin is shifted to O' as indicated, the time functions are shifted a quarter period. The maximum value of $f(t)$ will now occur at $\frac{\pi}{4}$ which is $\frac{\pi}{4}$ less than the original peak. Hence each term in the series must be shifted a like amount, and

$$v(t) \sim \frac{2}{\pi} [1 - \frac{2}{3} \cos(2t + \frac{\pi}{4}) - \frac{2}{15} \cos(4t + \frac{\pi}{4}) - \ldots].$$

• **PROBLEM 14-29**

The electrical circuit shown in the Figure is driven by a variable electromotive force E(t) which is periodic (but not necessarily sinusoidal) in time. The response of the system is the current I(t) and is known from electromagnetic theory to satisfy

$$L \frac{d^2 I}{dt^2} + R \frac{dI}{dt} + \frac{1}{C} I = \frac{dE}{dt}. \quad (1)$$

Find I(t) in terms of R, L. C and E(t).

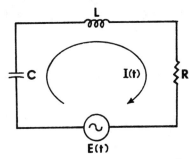

Solution: The solution of a differential equation such as (1) is the sum of a transient term (the general solution to the homogeneous equation $L \frac{d^2 I}{dt^2} + R \frac{dI}{dt} + \frac{1}{C} I = 0$) and a steady state term (a particular solution of (1)). The solution of the homogeneous equation is straightforward, does not require Fourier methods, and dies out rapidly with time. Therefore, assuming that enough time has elapsed so that steady state conditions prevail, (1) will be solved and the solution will give I(t) quite accurately.

Now, under steady state conditions the function I(t) is also periodic with the same period T as for E(t). Let us assume that E(t) and I(t) possess Fourier expansions given in complex form by

$$E(t) = \sum_{n=-\infty}^{\infty} E_n e^{in\omega t}, \quad I(t) = \sum_{n=-\infty}^{\infty} C_n e^{in\omega t} \quad (2)$$

where $\omega = \frac{2\pi}{T}$. Furthermore, let us assume that the series may be differentiated the necessary number of times (these assumptions will be discussed later). Then

$$\frac{dE}{dt} = \sum_{n=-\infty}^{\infty} in\omega E_n e^{in\omega t} \quad (3)$$

721

$$\frac{dI}{dt} = \sum_{n=-\infty}^{\infty} in\omega C_n e^{in\omega t} \qquad (4)$$

$$\frac{d^2I}{dt^2} = \sum_{n=-\infty}^{\infty} (-n^2\omega^2) C_n e^{in\omega t}. \qquad (5)$$

Now substitute (2), (3), (4), and (5) into (1) and note that the coefficients with the same exponential $e^{in\omega t}$ may be equated due to the fact that the set $\{e^{in\omega t}\}_n$ is orthogonal over $(\frac{-T}{2}, \frac{T}{2})$. Recall that orthogonality of complex functions is defined by $\int f_n(x) \overline{f_m(x)} dx = 0$ for $n \neq m$. That is

$$\int_{-\frac{T}{2}}^{\frac{T}{2}} e^{in\omega t} e^{-im\omega t} dt = \begin{cases} 0 & n \neq m \\ T & n = m \end{cases}$$

so to prove that we may equate the coefficients of $e^{in\omega t}$, multiply through the equation

$$L \sum_{n=-\infty}^{\infty} (-n^2\omega^2) C_n e^{in\omega t} + R \sum_{n=-\infty}^{\infty} in\omega C_n e^{in\omega t} + \frac{1}{C} \sum_{n=-\infty}^{\infty} C_n e^{in\omega t}$$

$$= \sum_{n=-\infty}^{\infty} in\omega E_n e^{in\omega t} \qquad (6)$$

by $e^{-in\omega t}$ and integrate from $\frac{-T}{2}$ to $\frac{T}{2}$. Then all terms with $m \neq n$ will vanish and all coefficients of $e^{in\omega t}$ will be multiplied by $2L$. Thus, the equality will be proven. Hence, equating like terms in (6) gives

$$(-n^2\omega^2 L + in\omega R + \frac{1}{C}) C_n = in\omega E_n.$$

Therefore

$$C_n = \frac{i(n\frac{\omega}{L})}{(\omega_o^2 - n^2\omega^2) + 2\alpha n\omega i} E_n \qquad (7)$$

where $\omega_o^2 = \frac{1}{LC}$ is called the natural frequency of the circuit and $2\alpha = \frac{R}{L}$ is called the attenuation factor of the circuit. Thus the problem is solved for a given $E(t)$ since (2) gives the Fourier expansion of $I(t)$ where C_n are given in terms of the Fourier coefficients of $E(t)$ which may be computed. I.e.,

$$C_n = \frac{i(n\frac{\omega}{L})}{(\omega_o^2 - n^2\omega^2) + 2\alpha n\omega i} \cdot \frac{1}{T}\int_{-\frac{T}{2}}^{\frac{T}{2}} E(t)e^{-in\omega t}\,dt. \qquad (8)$$

Once the C_n are computed the series in (2) may be converted into a real Fourier series

$$I(t) = \frac{a_o}{2} + \sum_{n=1}^{\infty}\left(a_n \cos n\omega t + b_n \sin n\omega t\right) \qquad (9)$$

by recalling the relations

$$C_n = \begin{cases} \frac{1}{2}(a_n - ib_n), & n \neq 0 \\ \frac{1}{2}a_o, & n \neq 0 \end{cases}$$

A few terms of (9) may then be used to make good approximations of I at any time t.

Finally, the validity of term by term differentiation was assumed valid. But this procedure has only been shown to be valid for uniformly convergent series so far and this is a fairly stringent requirement. However, as will be discussed in the chapter on transform methods, this procedure can be shown to be valid for a wide range of functions, including almost any physically significant function, by using the theory of distributions. In any case, when solving such problems it is usually best to assume such procedures valid and then justify them later after the solution has been developed.

● **PROBLEM 14-30**

A beam supported at each of its ends is shown in the Figure. It is uniformly loaded by a load q per unit length and the deflection of the beam y(x) is sought. If we choose the direction of the y-axis as downward as indicated in the Figure, the function y(x) is known to satisfy the equation

$$\frac{d^4y}{dx^4} = \frac{1}{EI} q(x) \qquad (1)$$

where q(x) is the load per unit length at point x (q = constant in our case) and $\frac{1}{EI}$ is the rigidity of the beam (also constant). Find y(x).

Solution: Note that since the function y(x) must vanish at x = 0 and x = L, it may be conveniently expanded into a Fourier sine series

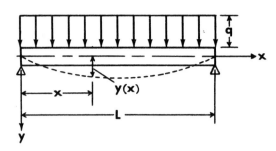

$$y(x) = \sum_{n=1}^{\infty} b_n \sin\left(\frac{n\pi x}{L}\right). \tag{3}$$

Assuming the validity of the fourfold term-by-term differentiation, (3) yields

$$\frac{d^4 y(x)}{dx^4} = \sum_{n=1}^{\infty} \left(\frac{n\pi}{L}\right)^4 b_n \sin\left(\frac{n\pi x}{L}\right). \tag{4}$$

(This is a reasonable assumption since on physical grounds one would expect y(x) to be "infinitely smooth" in which case each of the series for $y^{(n)}(x)$ would converge uniformly, justifying term-by-term differentiations.) Also expand $q(x) = q$ into the Fourier sine series

$$q = \sum_{n=1}^{\infty} q_n \sin\left(\frac{n\pi x}{L}\right). \tag{5}$$

where

$$q_n = \frac{2}{L} \int_0^L q \sin\left(\frac{n\pi x}{L}\right) dx = \begin{cases} \frac{4q}{n\pi} & (n=\text{odd}) \\ 0 & (n=\text{even}) \end{cases}. \tag{6}$$

Substitute both series, (4) and (5), into (1) and note that due to the orthogonality of the set of functions $\{\sin\frac{n\pi x}{L}\}_{n=1}^{\infty}$ we may equate the coefficients of terms with the same n. I.e. since

$$\int_0^L \sin\left(\frac{n\pi x}{L}\right) \sin\left(\frac{m\pi x}{L}\right) dx = \begin{cases} 0 & n \neq m \\ \frac{L}{2} & n = m \end{cases}.$$

we may multiply through the resulting equation

$$\sum_{n=1}^{\infty} \left(\frac{n\pi}{L}\right)^4 b_n \sin\left(\frac{n\pi x}{L}\right) = \frac{1}{EI} \sum_{\substack{n=1 \\ n=\text{odd}}}^{\infty} \frac{4q}{n\pi} \sin\left(\frac{n\pi x}{L}\right) \tag{7}$$

by $\sin\left(\frac{n\pi x}{L}\right)$ and integrate from 0 to L. Then all terms with m≠n will vanish while the coefficients of the nth terms will be multiplied by $\frac{L}{2}$. When this is done, we find

$$\frac{L}{2}\left(\frac{n\pi}{L}\right)^4 b_n = \begin{cases} \frac{L}{2}\frac{1}{EI}\frac{4q}{n\pi}, & n \text{ odd} \\ 0, & n \text{ even} \end{cases}$$

or

$$b_n = \begin{cases} \frac{4qL^4}{EI\pi^5}\frac{1}{n^5}, & n \text{ odd} \\ 0, & n \text{ even} \end{cases} \tag{8}$$

so that

$$y(x) = \frac{4qL^4}{EI\pi^5}\left(\sum_{n=1,3,5,\ldots}^{\infty}\frac{1}{n^5}\sin\frac{n\pi x}{L}\right). \tag{9}$$

A practical advantage of (5) is the rapid convergence of the series due to the fifth power of n in the denominator. for instance, at $x = \frac{L}{2}$ (maximum deflection) the second term in the series represents only $\frac{1}{3^5} \cong .00412$ or .4% of the first term. Thus, the first term may be used for calculations with about 99% accuracy.

Finally, since the series for q(x) (and, therefore, for $\frac{d^4y}{dx^4}$) is not uniformly convergent, the validity of the procedure remains in doubt. However it is easily justified by the theory of distributions which will be discussed in the section on integral transforms.

• **PROBLEM 14-31**

Use Parseval's equality and the results of previous problems concerned with calculation of Fourier coefficients of particular functions to find

(a) $\sum_{n=1}^{\infty} \frac{1}{(2n-1)^2} = 1 + \frac{1}{3^2} + \frac{1}{5^2} + \ldots$

(b) $\sum_{n=1}^{\infty} \frac{1}{(2n-1)^4} = 1 + \frac{1}{3^4} + \frac{1}{5^4} + \ldots$

Solution: Recall that Parseval's equality holds for any periodic function f with period 2c whose Fourier series converges in the mean to f. Since all piecewise continuous functions have convergent Fourier series, Parseval's equality holds for all such functions and is written

$$\frac{1}{c}\int_a^{a+2c} [f(x)]^2 dx = \frac{a_0^2}{2} + \sum_{k=1}^{\infty} (a_k^2 + b_k^2) \qquad (1)$$

where a_0, a_k, b_k are the Fourier coefficients of f and a is any number such that $(a, a+2c)$ is in the domain of definition of f.

(a) Earlier it was found that the Fourier series of the piecewise continuous function

$$f(x) = \begin{cases} -1 & -\pi < x < 0 \\ +1 & 0 < x < \pi \end{cases}$$

$$f(x+2\pi) = f(x) \qquad \text{all } x$$

is

$$f \sim \frac{4}{\pi} \sum_{n=1}^{\infty} \frac{1}{(2n-1)} \sin(2n-1)x \ .$$

Now $\displaystyle \frac{1}{\pi}\int_{-\pi}^{\pi} [f]^2 dx = \frac{1}{\pi}\int_{-\pi}^{0} (-1)^2 dx + \frac{1}{\pi}\int_0^{\pi} (1)^2 dx$

$$= \left.\frac{x}{\pi}\right|_{-\pi}^{0} + \left.\frac{x}{\pi}\right|_0^{\pi} = 2$$

so that Parseval's equality gives

$$2 = \frac{a_0^2}{2} + \sum_{n=1}^{\infty} a_k^2 + b_k^2 = \sum_{n=1}^{\infty} \left(\frac{4}{\pi(2n-1)}\right)^2$$

so that

$$\frac{\pi^2}{8} = \sum_{n=1}^{\infty} \frac{1}{(2n-1)^2} = 1 + \frac{1}{3^2} + \frac{1}{5^2} + \ldots$$

(b) Earlier it was found that the Fourier series of the piecewise continuous function

$$f(x) = |x| \qquad -\pi < x < \pi$$

is

$$f \sim \frac{\pi}{2} - \frac{4}{\pi} \sum_{n=1}^{\infty} \frac{\cos(2n-1)x}{(2n-1)^2}$$

Now

$$\frac{1}{\pi} \int_{-\pi}^{\pi} [f]^2 dx = \frac{1}{\pi} \int_{-\pi}^{\pi} x^2 dx = \frac{1}{\pi} \frac{x^3}{3}\Big|_{-\pi}^{\pi} = \frac{2\pi^2}{3}$$

so that Parseval's equality gives

$$\frac{2\pi^2}{3} = \frac{\pi^2}{2} + \sum_{n=1}^{\infty} \left(\frac{-4}{\pi(2n-1)^2}\right)^2$$

$$= \frac{\pi^2}{2} + \frac{16}{\pi^2} \sum_{n=1}^{\infty} \frac{1}{(2n-1)^4}$$

so that

$$\sum_{n=1}^{\infty} \frac{1}{(2n-1)^4} = \frac{\pi^4}{96} \; .$$

CHAPTER 15

COMPLEX VARIABLES

COMPLEX NUMBERS

● PROBLEM 15-1

Put the following complex numbers into the form $z = x + iy$:

(a) $\dfrac{-1 + 3i}{2 - i}$;

(b) $\dfrac{1 + 2i}{3 - 4i} + \dfrac{2 - i}{5i}$;

(c) $\dfrac{5}{(1-i)(2-i)(3-i)}$.

Solution: In each of these three problems, the key step will be to "real-ize the denominator." This procedure can be generalized as follows:

$$\frac{z_1}{z_2} = \frac{x_1 + iy_1}{x_2 + iy_2} = \frac{x_1 + iy_1}{x_2 + iy_2}\left(\frac{x_2 - iy_2}{x_2 - iy_2}\right)$$

$$= \frac{x_2 x_1 + i(-i)y_1 y_2 + iy_1 x_2 - ix_1 y_2}{x_2^2 - (i \cdot i)y_2^2 + ix_2 y_2 - ix_2 y_2}$$

$$= \frac{x_1 x_2 + y_1 y_2 + i(y_1 x_2 - x_1 y_2)}{x_2^2 + y_2^2}$$

Therefore,

$$\frac{z_1}{z_2} = \frac{x_1 x_2 + y_1 y_2}{x_2^2 + y_2^2} + \frac{i(y_1 x_2 - x_1 y_2)}{x_2^2 + y_2^2} = x_3 + iy_3 \qquad (1)$$

where we have defined the real numbers

$$x_3 = \frac{x_1 x_2 + y_1 y_2}{x_2^2 + y_2^2}, \quad y_3 = \frac{y_1 x_2 - x_1 y_2}{x_2^2 + y_2^2}.$$

The answer is now in the desired form.

Note that if $z = x + iy$, then $\bar{z} = x - iy$ is the complex conjugate of z and that

728

$$z\bar{z} = x^2 + y^2 = (\sqrt{x^2 + y^2})^2 = |z|^2$$

where $|z| = \sqrt{x^2 + y^2}$ is the modulus of z.

Then our method simply says to do the following:

$$\frac{z_1}{z_2} = \frac{z_1}{z_2}\left(\frac{\bar{z}_2}{\bar{z}_2}\right) = \frac{z_1 \bar{z}_2}{|z|^2} \qquad (2)$$

This formula is less cumbersome than (1)
We now employ these results.

(a) $\dfrac{-1 + 3i}{2 - i} = \dfrac{-1 + 3i}{2 - i}\left(\dfrac{2 + i}{2 + i}\right) = \dfrac{-2 + 3 \cdot i \cdot i + 6i - i}{2 \cdot 2 - i \cdot i - 2i + 2i}$

$= \dfrac{-5 + 5i}{5} = -1 + i$

and this is the desired form.

(b) $\dfrac{1 + 2i}{3 - 4i} + \dfrac{2 - i}{5i} = \dfrac{1 + 2i}{3 - 4i}\left(\dfrac{3 + 4i}{3 + 4i}\right) + \dfrac{2 - i}{5i}\left(\dfrac{-5i}{-5i}\right)$

$= \dfrac{3 + 8i^2 + 6i + 4i}{3^2 + 4^2} + \dfrac{-10i + 5i^2}{5^2}$

$= \dfrac{-5 + 10i}{25} + \dfrac{-5 - 10i}{25}$

$= \dfrac{-5 - 5 + 10i - 10i}{25} = \dfrac{-10}{25}$

$= \dfrac{-2}{5}$

(c) $\dfrac{5}{(1-i)(2-i)(3-i)} = \dfrac{5}{(2 + (-i)^2 - i - 2i)(3-i)}$

$= \dfrac{5}{(1-3i)(3-i)} = \dfrac{5}{3 + 3i^2 - 9i - i}$

$= \dfrac{5}{-10i} = \dfrac{5}{-10i}\left(\dfrac{10i}{10i}\right) = \dfrac{50i}{10^2}$

$= \dfrac{1i}{2}$

● **PROBLEM 15-2**

Find a polar representation of $z_0 = 1 + i$ and graph this point in the complex plane.

Solution: To plot a point, $z = x + iy$, in the complex

plane, we simply associate the horizontal Cartesian axis with the real part x, of z, and the vertical axis with the imaginary part y. This is illustrated below in Figure 1. From that figure we can see that a set of polar coordinates

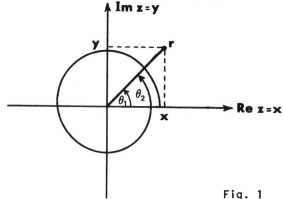

Fig. 1

may also be associated with z, with the identification

$$x = r\cos\theta, \quad y = r\sin\theta, \quad r = \sqrt{x^2 + y^2} = |z|, \tag{1}$$

where $|z|$ is called the modulus of z. Note that $r = |z| \geq 0$ for all complex numbers z. Evidently then,

$$z = x + iy = r\cos\theta + ir\sin\theta = r(\cos\theta + i\sin\theta). \tag{2}$$

Note that θ can take on an infinite number of values,

$$\theta_1 + 2n\pi, \quad n = \ldots -2, -1, 0, 1, 2, \ldots$$

The smallest positive value of θ will be used in this problem. Note also that $\tan\theta = y/x$. Now, for $z_0 = 1 + i$, by (1), $r = |z| = \sqrt{1^2 + 1^2} = \sqrt{2}$ and $\tan\theta = \frac{1}{1} = 1$. This implies that $\theta = \tan^{-1}1 = \frac{\pi}{4} + 2n\pi$.

We will use that θ for which $n = 0$ so that from (2),

$$z_0 = 1 + i = \sqrt{2}\left(\cos\left(\frac{\pi}{4}\right) + i\sin\left(\frac{\pi}{4}\right)\right). \tag{3}$$

This is shown graphically in Figure 2.

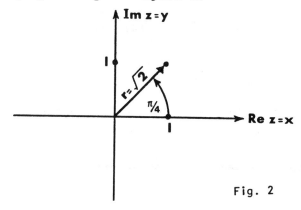

Fig. 2

● **PROBLEM 15-3**

Find all of the nth roots of 1. That is, find the values of z which satisfy the equation

$$z^n = 1, \; z \neq 0.$$

Solution: First recall that if a complex number z is represented in polar form by $z = |z|(\cos\theta + i\sin\theta)$, then we may identify the expression $\cos\theta + i\sin\theta$ with $e^{i\theta}$. (This is Euler's identity.) That is $z = |z|e^{i\theta}$. Note that the relations

$$e^{i\theta_1} e^{i\theta_2} = e^{i(\theta_1+\theta_2)} \tag{1}$$

and

$$\frac{e^{i\theta_1}}{e^{i\theta_2}} = e^{i(\theta_1-\theta_2)} \tag{2}$$

each of which may be proved from the definition,

$$e^{i\theta} = \cos\theta + i\sin\theta,$$

are ample justification of the manipulative features of this notation. It can also be proved by using induction on (1) that

$$(e^{i\theta})^n = e^{in\theta}, \; (n \neq 0, \pm 1, \pm 2, \ldots) \tag{3}$$

so that

$$(\cos\theta + i\sin\theta)^n = \cos n\theta + i\sin n\theta, \; (n = 0, \pm 1, \ldots). \tag{4}$$

This is called De Moivre's Theorem and will be used in this problem. Now, to find z such that

$$z^n = 1,$$

we write $z = re^{i\theta}$ and look for r and θ such that $(re^{i\theta})^n = 1$, or

$$r^n e^{in\theta} = 1 \cdot e^{i2k\pi}, \; (k = 0, \pm 1, \pm 2, \ldots), \; (n = 0, \pm 1, \pm 2, \ldots) \tag{5}$$

since $1 = 1 \cdot (\cos 2k\pi + i\sin 2k\pi)$. Therefore, the nth roots of 1 are those numbers satisfying

$$r^n = 1$$

and $n\theta = 2k\pi$. Hence, $r = 1$ (since $r = |z| \geq 0$) and

$$\theta = \frac{2k\pi}{n}$$

for any integer k, and

$$z = e^{i\left(\frac{2k\pi}{n}\right)} \quad (k = 0, \pm 1, \pm 2, \ldots) \tag{6}$$

are nth roots of 1. However, because of the periodicity of cosine and sine, and therefore of $e^{i\theta}$, we know that

$$e^{i\theta} = e^{i(\theta+2n\pi)} \quad (n = 0, \pm 1, \ldots)$$

so that for $k \geq n$, say $k = n + j$ (where $j \geq 0$). We have

$$e^{i\left(\frac{2k\pi}{n}\right)} = e^{i\left(\frac{2(n+j)\pi}{n}\right)} = e^{i\left(\frac{2j\pi}{n}\right)} e^{i2\pi} = e^{i\left(\frac{2j\pi}{n}\right)}$$

and we may terminate (6) at $k = n-1$ to get n distinct roots of 1, namely

$$z = e^{i\left(\frac{2k\pi}{n}\right)} \quad (k = 0, 1, 2, \ldots, n-1)$$

or

$$z = \cos\left(\frac{2k\pi}{n}\right) + i\sin\left(\frac{2k\pi}{n}\right) \quad (k = 0, 1, 2, \ldots, n-1).$$

COMPLEX FUNCTIONS AND DIFFERENTIATION

● **PROBLEM 15-4**

Show that $f(z) = \bar{z}$ is nowhere differentiable using the definition of differentiability. ($\bar{z}$ is the complex conjugate of z.)

Solution: The definition of the derivative of a complex function, $f(z)$, is

$$\frac{d}{dz} f(z) \equiv \lim_{\Delta z \to 0} \frac{f(z + \Delta z) - f(z)}{\Delta z} \quad (1)$$

It is easily proved that if such a limit exists, it is unique. Therefore, if we can show that the limit in equation (1) has different values for different ways in which Δz approaches zero, we will have shown that this limit is not unique and therefore does not exist.

Write $z = x + iy$. The $\bar{z} = x - iy$ and for $f(z) = \bar{z}$, (1) becomes

$$\frac{d}{dz} \bar{z} = \lim_{\Delta z \to 0} \frac{\overline{z + \Delta z} - \bar{z}}{\Delta z},$$

$$= \lim_{\substack{\Delta x \to 0 \\ \Delta y \to 0}} \frac{\overline{x + iy + \Delta x + i\Delta y} - \overline{x + iy}}{\Delta x + i\Delta y}$$

$$= \lim_{\substack{\Delta x \to 0 \\ \Delta y \to 0}} \frac{x - iy + \Delta x - i\Delta y - (x - iy)}{\Delta x + i\Delta y}$$

$$= \lim_{\substack{\Delta x \to 0 \\ \Delta y \to 0}} \frac{\Delta x - i\Delta y}{\Delta x + i\Delta y} \qquad (2)$$

Now we may choose to approach the point z along the line for which $\Delta y = 0$ (parallel to the x-axis). Then from (2) we have

$$\frac{d}{dz} \bar{z} = \lim_{\Delta x \to 0} \frac{\Delta x}{\Delta x} = \lim_{\Delta x \to 0} 1 = 1 . \qquad (3)$$

But if we choose the line for which $\Delta x = 0$ (parallel to the y-axis) then

$$\frac{d}{dz} \bar{z} = \lim_{\Delta y \to 0} \frac{-i\Delta y}{i\Delta y} = \lim_{\Delta y \to 0} -1 = -1 . \qquad (4)$$

From equations (3) and (4) we see that the limit depends on the way in which we let Δz approach 0, and this is true for any z we choose. We conclude that since a limit must be unique, this limit does not exist for any z. That is, $f(z) = \bar{z}$ is nowhere differentiable.

● PROBLEM 15-5

The function $f(z) = z^2 = x^2 - y^2 + i2xy = u(x,y) + iv(x,y)$ can be shown to be differentiable for all z with

$$\frac{d}{dz}(z^2) = 2z.$$

Prove that this function satisfies the Cauchy-Riemann Theorem for all z.

Solution: The theorem in question states that a function which is differentiable at a point $z = x + iy$ satisfies the Cauchy-Riemann equations at this point and that

$$f'(z) = \frac{\partial u(x,y)}{\partial x} + \frac{i \partial v(x,y)}{\partial x} \qquad (1)$$

The Cauchy-Riemann equations are

$$\frac{\partial u(x,y)}{\partial x} = \frac{\partial v(x,y)}{\partial y} \quad \text{and} \quad \frac{\partial u(x,y)}{\partial y} = \frac{-\partial v(x,y)}{\partial x} . \qquad (2)$$

In the case at hand, $u(x,y) = x^2 - y^2$ and $v(x,y) = 2xy$. Therefore, we have

$$\frac{\partial u}{\partial x} = 2x , \quad \frac{\partial u}{\partial y} = -2y , \quad \frac{\partial v}{\partial x} = 2y , \quad \frac{\partial v}{\partial y} = 2x. \qquad (3)$$

We see immediately from this that

$$\frac{\partial u}{\partial x} = \frac{\partial v}{\partial y} \quad \text{and} \quad \frac{\partial u}{\partial y} = \frac{-\partial v}{\partial x}$$

so that the Cauchy-Riemann conditions hold. Also from equations (3) and (1) we have

$$f'(z) = \frac{\partial u}{\partial x} + \frac{i\partial v}{\partial x} = 2x + i2y = 2z ,$$

which is the result deduced from the definition of a derivative. Therefore, we have verified the theorem for $f(z) = z^2$ for any z.

• **PROBLEM 15-6**

Prove that the function $f(z) = \frac{1}{z}$ is differentiable for all z except z=0 by using the Cauchy-Riemann equations. Also find $f'(z)$.

Solution: We use here the theorem which states that for $f(z) = u(x,y) + iv(x,y)$, if u and v have continuous first partial derivatives at a point $z = x + iy$ and satisfy the Cauchy-Riemann equations at (x,y), then f(z) is differentiable at $z = x + iy$. The Cauchy-Riemann equations are

$$\frac{\partial u}{\partial x} = \frac{\partial v}{\partial y} \quad \text{and} \quad \frac{\partial u}{\partial y} = \frac{-\partial v}{\partial x} . \quad (1)$$

In order to use this theorem for $f(z) = \frac{1}{z}$, we must first write $\frac{1}{z}$ as $u(x,y) + iv(x,y)$. That is, we must solve the equation

$$\frac{1}{x + iy} = u + iv \quad (2)$$

for u and v. This is easily accomplished by writing

$$\frac{1}{x + iy} = \frac{1}{x + iy} \left(\frac{x - iy}{x - iy}\right) = \frac{x - iy}{x^2 + y^2}$$

or

$$\frac{1}{x + iy} = \frac{x}{x^2 + y^2} + \frac{-iy}{x^2 + y^2} . \quad (3)$$

We see from (3) and (2) that

$$u(x,y) = \frac{x}{x^2 + y^2} , \quad v(x,y) = \frac{-y}{x^2 + y^2}$$

from which it follows that

$$\frac{\partial u}{\partial x} = \frac{y^2 - x^2}{(x^2 + y^2)^2} = \frac{\partial v}{\partial y} \qquad (4)$$

and

$$\frac{\partial u}{\partial y} = \frac{-2xy}{(x^2 + y^2)^2} = \frac{-\partial v}{\partial x} \qquad (5)$$

Thus $f(z) = u + iv$ satisfies the Cauchy-Riemann equations by virtue of (4) and (5). Also note that the partial derivatives calculated in (4) and (5) are continuous except for $x^2 + y^2 = |z| = 0$, or $z = 0$. We have therefore satisfied the conditions of the theorem and thereby have proven that

$$\frac{df(z)}{dz} = \frac{d(1/z)}{dz}$$

exists for all $z \neq 0$.

To find $f'(z)$, recall that if the conditions of the above theorem are met,

$$f'(z) = \frac{\partial u}{\partial x} + \frac{i \partial v}{\partial x}$$

In our case, this implies that for $z \neq 0$,

$$\frac{d(1/z)}{dz} = \frac{y^2 - x^2}{(x^2 + y^2)^2} + \frac{i\,2xy}{(x^2 + y^2)^2}. \qquad (6)$$

Now note that

$$\frac{-1}{z^2} = \frac{-1}{x^2 - y^2 + 2ixy} = \frac{-1}{(x^2 - y^2) + i(2xy)} \left(\frac{x^2 - y^2 - 2ixy}{x^2 - y^2 - 2ixy} \right)$$

$$= \frac{y^2 - x^2 + 2ixy}{(x^2 - y^2)^2 + 4x^2y^2} = \frac{y^2 - x^2}{(x^2 + y^2)^2} + \frac{2ixy}{(x^2 + y^2)^2}. \qquad (7)$$

Equating (6) and (7) we see that

$$\frac{d\left(\frac{1}{z}\right)}{dz} = \frac{-1}{z^2}, \quad z \neq 0.$$

• **PROBLEM 15-7**

Let $f(z) = \begin{cases} 0 & , \; z = 0 \\ \dfrac{x^3 - y^3}{x^2 + y^2} + \dfrac{i(x^3 + y^3)}{x^2 + y^2} & , \; z \neq 0 \end{cases}$

Show that the Cauchy-Riemann equations are satisfied at $z = 0$ but that $f'(0)$ does not exist.

Solution: This problem demonstrates the importance of

the continuity conditions in the theorem which states that if a function, $f(z) = u(x,y) + iv(x,y)$ satisfies the Cauchy-Riemann equations at a point z, and if the first partial derivatives, u_x, u_y, v_x, v_y, are continuous then $f'(z)$ exists. Notice that if we prove what the problem asks, we will have shown rather circuitously that by virtue of the theorem, not all of the first partial derivatives are continuous at $z = 0$.

Now, applying the definition for a partial derivative of a real valued function yields

$$u_x(0,0) \equiv \lim_{x \to 0} \frac{u(x,0) - u(0,0)}{x - 0} = \lim_{x \to 0} \frac{(x^3/x^2) - 0}{x} = 1 \quad (1)$$

$$u_y(0,0) \equiv \lim_{y \to 0} \frac{u(0,y) - u(0,0)}{y - 0} = \lim_{y \to 0} \frac{(-y^3/y^2) - 0}{y} = -1 \quad (2)$$

$$v_x(0,0) \equiv \lim_{x \to 0} \frac{v(x,0) - v(0,0)}{x - 0} = \lim_{x \to 0} \frac{(x^3/x^2) - 0}{x - 0} = 1 \quad (3)$$

$$v_y(0,0) \equiv \lim_{y \to 0} \frac{v(0,y) - v(0,0)}{y - 0} = \lim_{y \to 0} \frac{(y^3/y^2) - 0}{y} = 1 \quad (4)$$

The Cauchy-Riemann conditions are $u_x = v_y$ and $u_y = -v_x$ and it is clear from the above equations, (1) - (4), that these conditions are satisfied in this case for $z = x + iy = 0$.

To show that $f'(z)$ does not exist at $z = 0$, first compute the limit of the difference quotient along two different paths of approach to the origin. Then since different results will be obtained along the two different paths and since if a limit exists it must be unique, we will have shown that this limit does not exist. Thus, along the line $y = x$, observe that $z = x + ix$ and

$$f'(0) = \lim_{z \to 0} \frac{f(z) - f(0)}{z - 0} = \lim_{x \to 0} \frac{ix - 0}{x + ix} = \frac{i}{1 + i} = \frac{1}{2} + \frac{1}{2} i \quad (5)$$

Approaching $z = 0$ along the line $y = 0$ (the x axis) note that $z = x$ and

$$f'(0) = \lim_{z \to 0} \frac{f(z) - f(0)}{z - 0} = \lim_{x \to 0} \frac{x + ix}{x} = 1 + i . \quad (6)$$

Equations (5) and (6) show that two different paths of approach to 0 give two distinct results for $f'(0)$. Therefore, $f'(0)$ does not exist.

• **PROBLEM 15-8**

Find the harmonic conjugates of the function

$$u(x,y) = y^3 - 3x^2y .$$

<u>Solution</u>: A function $(h(x,y))$ is harmonic in a domain (an

open connected subset of the complex plane) D of the xy-plane if throughout D it satisfies Laplace's Equation,

$$h_{xx}(x,y) + h_{yy}(x,y) = 0 \qquad (1)$$

where the second partials, $h_{xx}(x,y)$ and $h_{yy}(x,y)$ are continuous on D. If in addition to being harmonic, two functions u and v satisfy the Cauchy-Riemann equations

$$u_x = v_y \quad \text{and} \quad u_y = -v_x \qquad (2)$$

in D, the $v(x,y)$ is said to be a harmonic conjugate of $u(x,y)$. These definitions involving only real valued functions have importance in complex analysis due to a theorem which states that a necessary and sufficient condition for a function $f(z) = u(x,y) + iv(x,y)$ to be analytic in D is that v be a harmonic conjugate of u in D.

To find a harmonic conjugate of $u(x,y) = y^3 - 3x^2y$ first note that it is easily checked that u is itself harmonic. Now in view of the definition, if v is a harmonic conjugate of u it must satisfy equation (2). In particular it is needed that

$$u_x(x,y) = -6xy = v_y(x,y). \qquad (3)$$

Holding x fixed and integrating both sides of the second equality in (3) with respect to y, yields

$$v(x,y) = -3xy^2 + \phi(x) \qquad (4)$$

where $\phi(x)$ is an unknown function of integration. Differentiating (4) with respect to x gives

$$v_x(x,y) = -3y^2 + \phi'(x) \qquad (5)$$

and in view of the second of equations (2) we must have $u_y(x,y) = 3y^2 - \phi'(x)$ or

$$3y^2 - 3x^2 = 3y^2 - \phi'(x) \qquad (6)$$

if v is to be a harmonic conjugate of u. This implies that $\phi'(x) = 3x^2$ or $\phi(x) = x^3 + c$ where c is any arbitrary real number. Using this result in (4), the following form for v is obtained:

$$v(x,y) = x^3 - 3xy^2 + c. \qquad (7)$$

It is easily proven by using the definition that two harmonic conjugates of a function $u(x,y)$ can differ at most by a constant. Therefore, equation (7) gives all possible harmonic conjugates of u. The corresponding analytic function is

$$f(z) = y^3 - 3x^2y + i(x^3 - 3xy^2 + c)$$

or

$$f(x) = i(z^3 + c).$$

• **PROBLEM 15-9**

Find the real and imaginary parts of sin z for complex z. That is, write sin z in the form Re(sin z) + i Im(sin z).

Solution: The sin function for complex numbers is defined as

$$\sin z = \frac{e^{iz} - e^{-iz}}{2i}. \tag{1}$$

From this definition, it follows that if $z = x + iy$, then

$$\sin z = \frac{e^{i(x+iy)} - e^{-i(x+iy)}}{2i}$$

or,

$$\sin z = \frac{e^{ix} e^{i^2 y} - e^{-ix} e^{-i^2 y}}{2i} \tag{2}$$

Recall that $e^{ix} \equiv \cos x + i \sin x$ so that (2) becomes

$$\sin z = (\cos x + i \sin x) \frac{e^{-y}}{2i} - (\cos x - i \sin x) \frac{e^{y}}{2i}.$$

Collecting like terms of sin and cos, yields

$$\sin z = i \sin x \left(\frac{e^{-y} + e^{y}}{2i} \right) + \cos x \left(\frac{e^{-y} - e^{y}}{2i} \right). \tag{3}$$

Recalling that $\frac{1}{i} = -i$ (to prove this, multiply each side by i to obtain an identity) we may write (3) as

$$\sin z = \sin x \frac{e^{y} + e^{-y}}{2} + i \cos x \frac{e^{y} - e^{-y}}{2}$$

so that

$$\sin z = \sin x \cosh y + i \cos x \sinh y, \tag{4}$$

since by definition $\sinh z = \frac{e^{z} - e^{-z}}{2}$ and

$\cosh z = \frac{e^{z} + e^{-z}}{2}$. Equation (4) displays sin z in the desired form so that we have found that

Re(sin z) = sin x cosh y , Im(sin z) = cos x sinh y.

● **PROBLEM 15-10**

Determine whether the following equation is true for all complex z:

$$\log e^z = z.$$

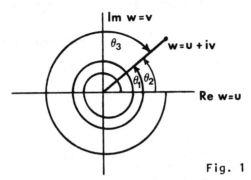

Fig. 1

Solution: The logarithm of a complex number, $w = re^{i\theta}$, is defined by

$$\log w = \text{Log } r + i\theta \quad (1)$$

where $r = |w|$, the modulus or absolute value of w, Log denotes the natural logarithm of a positive real number and $\theta = \arg w$ is an argument of w. Recall that θ is any angle that w makes with the positive real axis when w is interpreted as a directed line segment from the origin as in Figure 1. In that figure, each of θ_1, θ_2 and θ_3 is a possible value of θ. We can see that all values of θ can be written as

$$\theta = \Theta + 2n\pi \quad (n = 0, \pm 1, \pm 2, \ldots) \quad (2)$$

where Θ is called the principle value of arg z and is denoted $\Theta = \text{Arg } z$ (Note that equation (2) would also be true if Θ were any particular value of θ, not necessarily the principle value.) Arg $z = \Theta$ is the unique value of θ such that $-\pi < \Theta \leq \pi$. We can now write (1) as

$$\log w = \text{Log } r + i(\Theta + 2n\pi) \quad (n = 0, \pm 1, \pm 2, \ldots) \quad (3)$$

Now, for $z = x + iy$, $e^z = e^{(x+iy)} = e^x e^{iy}$ so that

$$\log e^z = \text{Log } |e^z| + i \arg e^z$$

$$= \text{Log } |e^x||e^{iy}| + i \arg e^x e^{iy} \quad (4)$$

Since $e^{iy} = \cos y + i \sin y$, $|e^{iy}| = |\cos y + i \sin y| = 1$, (4) becomes

$$= \text{Log } e^x + i(y + 2n\pi)$$

$$= x + iy + i2n\pi.$$

739

Therefore,

$$\log e^z = z + i2n\pi \qquad (n = 0, \pm 1, \pm 2, \ldots)$$

and we conclude that $\log e^z = z$ is not true in general.

• **PROBLEM 15-11**

Determine the values of log (1-i) and specify the principal value.

Solution: Recall the definition of the logarithm of a complex number, $z = re^{i\theta}$, namely

$$\log z = \text{Log } r + i\theta \qquad (1)$$

where $r = |z|$, Log represents the natural logarithm of a real positive number and $\theta = \arg z$ is any angle that z makes with the positive real axis when z is interpreted as a directed line segment from the origin. We may also write

$$\log z = \text{Log } r + i(\theta + 2n\pi) \qquad (n = 0, \pm 1, \pm 2, \ldots) \qquad (2)$$

where θ is the principal value of arg z, that is, $-\pi < \theta \leq \pi$. The value of log z with n = 0 in equation (2) is called the principal value.

Now for $z_0 = x_0 + iy_0 = 1 - i$, we have

$$r = |z_0| = \sqrt{x_0^2 + y_0^2} = \sqrt{1^2 + 1^2} = \sqrt{2} \quad \text{and} \quad \theta = -\pi/4$$

so that

$$z_0 = re^{i(\theta + 2n\pi)} = \sqrt{2} \, e^{i\left(\frac{-\pi}{4} + 2n\pi\right)}$$

so that by (2)

$$\log z_0 = \text{Log } \sqrt{2} + i\left(\frac{-\pi}{4} + 2n\pi\right) \qquad n = 0, \pm 1, \pm 2, \ldots$$

or

$$\log (1-i) = \frac{1}{2} \text{Log } 2 - \frac{\pi i}{4} + 2n\pi i \qquad n = 0, \pm 1, \pm 2, \ldots$$

We see from this that the principal value of log (1-i), written Log (1-i), is

$$\text{Log } (1-i) = \frac{1}{2} \text{Log } 2 - \frac{\pi}{4} i.$$

• **PROBLEM 15-12**

Define branches for the multiple-valued complex function of z, $z^{1/n}$. Also determine the branch point and the principal branch of this function. (n is a positive integer.)

Solution: For any real number n, we have the relation

$$z^{1/n} = \exp\left(\frac{1}{n} \log z\right) \quad z \neq 0 \tag{1}$$

where the complex logarithm is given by

$$\log z = \text{Log } r + i(\Theta + 2k\pi), \quad k = 0, \pm 1, \pm 2, \ldots \tag{2}$$

In this equation, $r = |z|$, Log represents the natural logarithm of a positive real number, and Θ is the principal value of the argument of z. That is, if $z = re^{i\Theta}$, then Θ is that value of θ which satisfies $-\pi < \Theta \leq \pi$. Using the definition (2), this yields

$$z^{1/n} = \sqrt[n]{r} \exp \frac{i(\Theta + 2k\pi)}{n}, \quad k = 0, 1, 2, \ldots, n-1, \quad z \neq 0, \tag{3}$$

where we may terminate the values of k at $n-1$ due to the periodicity of the exponential function. That is

$$\exp \frac{i(\Theta + 2(k+n)\pi)}{n} = \exp\left[\frac{i(\Theta + 2k\pi)}{n} + \frac{2n\pi i}{n}\right]$$

$$= \exp\left(\frac{i(\Theta + 2k\pi)}{n}\right) \exp(i2\pi)$$

$$= \exp\left(\frac{i(\Theta + 2k\pi)}{n}\right).$$

Perusal of equation (3) shows that there are n distinct values of $z^{1/n}$ corresponding to the n values of k. We say that $z^{1/n}$ is a multiple-valued function of z.

Branches of a multiple-valued function $f(z)$ are defined as single-valued functions $F_k(z)$ which are analytic in some domain at each point z of which the value $F_K(z)$ is one of the values $f(z)$. It is clear that the simplest way to define n branches of $f(z) = z^{1/n}$, $z \neq 0$ is to let

$$F_k(z) = \sqrt[n]{r} \exp \frac{i(\Theta + 2k\pi)}{n} \tag{4}$$

$$(r > 0, \ -\pi < \Theta \leq \pi, \quad k = 0, 1, 2, \ldots, n-1).$$

The principal branch of $z^{1/n}$ is the analytic function obtained from equation (1) by using the principal branch of $\log z$ (i.e., the branch for which $k = 0$ in (4)). Thus,

$$F_0(z) = \sqrt[n]{r} \exp\left(\frac{i\Theta}{n}\right), \quad -\pi < \Theta < \pi \tag{5}$$

is the principal branch. Actually, in this case, Θ is not exactly the principal value of arg z since $-\pi < \Theta \leq \pi$. This is because none of the functions $F_K(z)$

are analytic on the ray $\theta = \pi$ since not one is even continuous there. Hence, by the definition of a branch, no branch can contain any point on this ray.

Finally, the branch point of a multiple-valued function is defined as that point common to all possible branch cuts, a branch cut being a line of singular points introduced in defining branches of the function. It is obvious that all possible branch cuts in the case of $f(z) = z^{1/n}$ must pass through the origin, so that $z = 0$ is the only branch point of $f(z)$.

SERIES

● **PROBLEM 15-13**

Justify the definition $e^{ix} = \cos x + i \sin x$ by the formal manipulation of the real power series expansions of these functions.

Solution: The word formal is used in the statement of the problem to indicate that $i = \sqrt{-1}$ may be treated as if it was a real number with the property $(i)^2 = -1$, and obtain the result desired. Thus, we needn't yet introduce the concept of a complex infinite series and in fact will be helping to motivate such a concept with this example.

The real valued Taylor series expansion of e^y is given by

$$e^y = \sum_{k=0}^{\infty} \frac{(y)^k}{k!} \quad \text{so that}$$

if $y = ix$, this will yield

$$e^{ix} = \sum_{k=0}^{\infty} \frac{(ix)^k}{k!}$$

$$= \sum_{j=0}^{\infty} \frac{(ix)^{2j}}{(2j)!} + \sum_{j=0}^{\infty} \frac{(ix)^{2j+1}}{(2j+1)!}$$

$$= \sum_{j=0}^{\infty} \frac{(-1)^j x^{2j}}{(2j)!} + i \sum_{j=0}^{\infty} \frac{(-1)^j x^{2j+1}}{(2j+1)!} \,. \tag{1}$$

Now recall that the two Taylor series expansions on the right in equation (1) are those of $\cos x$ and $\sin x$, so that we deduce

$$e^{ix} = \cos x + i \sin x. \tag{2}$$

Note that since $(e^{ix})^n = e^{inx}$, we instantly obtain from (2) that

$$(\cos x + i \sin x)^n = \cos nx + i \sin nx. \tag{3}$$

Equation (3) is called de Moivre's theorem.

● **PROBLEM 15-14**

> Expand the function $f(z) = \dfrac{1}{z^2}$ in a complex Taylor series about the point $z_0 = 2$. What is the largest circle centered at z_0 in the interior of which this expansion is valid?

Solution: Taylor's theorem for a complex function f states that if f is analytic everywhere inside a circle C_0 centered at z_0 and with radius r_0 (i.e., C_0 consists of all points z such that $|z - z_0| = r_0$), then at each point z in the domain bounded by C_0 we have

$$f(z) = \sum_{n=0}^{\infty} \frac{f^{(n)}(z_0)(z-z_0)^n}{n!} = f(z_0) + f'(z_0)(z-z_0) + \frac{f''(z_0)(z-z_0)^2}{2!}$$

$$+ \ldots + \frac{f^{(n)}(z_0)(z-z_0)^n}{n!} + \ldots . \qquad (1)$$

That is, the power series in (1) converges to f(z) for all z such that $|z - z_0| < r_0$. It is also proven that the largest circle centered at z_0 within which this series, called the Taylor series for f, converges to f is the one whose radius is the distance from z_0 to the nearest singular point of f. (A singular point of f is a point at which f is not analytic.) Therefore, we can immediately answer the second part of the question since the only singular point of

$$f(z) = \frac{1}{z^2}$$

is $z = 0$. Hence, if our circle is centered at $z_0 = 2$, the largest possible radius of convergence is 2.

Consequently, we wish to expand $f(z) = \dfrac{1}{z^2}$ in the domain $D = \{z: |z-2| < 2\}$. Now to calculate the Taylor series we differentiate f n times to obtain

$$f^{(n)}(z) = \frac{(-1)^n (n+1)!}{z^{n+2}} \qquad (n = 0,1,2,\ldots) \qquad (2)$$

Therefore

$$f^{(n)}(2) = \frac{(-1)^n (n+1)!}{2^{n+2}} \qquad (n = 0,1,2,\ldots) . \qquad (3)$$

Using this result in equation (1) with

$z_0 = 2$ we obtain

$$\frac{1}{z^2} = \sum_{n=0}^{\infty} \frac{(-1)^n (n+1)!(z-2)^n}{2^{n+2} n!} \quad . \tag{4}$$

Note that

$$\frac{(z-2)^n}{2^{n+2}} = \frac{1}{4} \frac{(z-2)^n}{2^n} = \frac{1}{4} \left(\frac{z-2}{2}\right)^n$$

and

$$\frac{(n+1)!}{n!} = \frac{(n+1)n(n-1) \cdots 1}{n(n-1) \cdots 1} = n+1$$

so that (4) becomes

$$\frac{1}{z^2} = \frac{1}{4} \sum_{n=0}^{\infty} (-1)^n (n+1) \left(\frac{z-2}{2}\right)^n \quad \text{valid for}$$

$$|z-2| < 2 \quad (\text{i.e., for } z \in D)$$

● **PROBLEM 15-15**

Obtain series expansions of the function

$$f(z) = \frac{-1}{(z-1)(z-2)}$$

about the point $z_0 = 0$ in the regions $|z| < 1$, $1 < |z| < 2$, $|z| > 2$.

Solution: The function f has singular points, or points at which it is not analytic, at $z_1 = 1$, $z_2 = 2$. Taylor's theorem tells us that f has a valid Taylor series expansion about $z_0 = 0$ within the circle $|z| = 1$ since it is analytic there. Hence

$$f(z) = \sum_{n=0}^{\infty} \frac{f^{(n)}(z_0)}{n!} (z-z_0)^n \qquad |z| < 1 \, .$$

It is convenient, however, to use partial fractions to write

$$f(z) = \frac{-1}{(z-1)(z-2)} = \frac{1}{z-1} - \frac{1}{z-2} = \frac{1}{2} \frac{1}{1-z/2} - \frac{1}{1-z} \, . \tag{1}$$

For $|z| < 1$, we have $|z/2| < 1$ and note that these are the conditions under which each of the terms in (1) can be represented by a geometric series. Hence, since

$$\frac{1}{1-w} = \sum_{n=0}^{\infty} w^n, \; (|w| < 1) \, ,$$

$$f(z) = \frac{1}{2} \sum_{n=0}^{\infty} \left(\frac{z}{2}\right)^n - \sum_{n=0}^{\infty} z^n = \sum_{n=0}^{\infty} \left[\frac{1}{2}\left(\frac{z}{2}\right)^n - z^n\right]$$

or
$$f(z) = \sum_{n=0}^{\infty} (2^{-n-1} - 1) z^n \qquad (|z| < 1). \qquad (2)$$

Since the Taylor coefficients are unique this must be the Taylor expansion for $f(z)$ in the region $|z| < 1$ and as a by-product of our calculations we have deduced that

$$\frac{f^{(n)}(0)}{n!}$$

must be equal to the coefficients in (2). Thus

$$f^{(n)}(0) = n! (2^{-n-1} - 1).$$

In the region $1 < |z| < 2$ we must use Laurent's theorem which states that if f is analytic in the region bounded by two concentric circles C_1 and C_2 centered at z_0 then at each point z in that region $f(z)$ is represented by its Laurent series expansion

$$f(z) = \sum_{n=0}^{\infty} a_n (z-z_0)^n + \sum_{n=1}^{\infty} \frac{b_n}{(z-z_0)^n} \qquad (3)$$

where the radius of C_1 is assumed to be less than that of C_2, and

$$a_n = \frac{1}{2\pi i} \int_{C_1} \frac{f(s) ds}{(s-z_0)^{n+1}} \qquad n = 0,1,2,\ldots \qquad (4)$$

$$b_n = \frac{1}{2\pi i} \int_{C_2} \frac{f(s) ds}{(s-z_0)^{-n+1}} \qquad n = 1,2,\ldots \qquad (5)$$

Formulas (4) and (5) are not very useful since the integrals are usually difficult to do. We therefore note that in the region $1 < |z| < 2$, $|1/z| < 1$ and $|z/2| < 1$ so that we may use the geometric series expansion to write

$$f(z) = \frac{1}{z} \frac{1}{1-1/z} + \frac{1}{2} \frac{1}{1-z/2} = \sum_{n=0}^{\infty} \frac{1}{z^{n+1}} + \frac{1}{2} \sum_{n=0}^{\infty} \frac{z^n}{2^{n+1}}, \qquad (6)$$

which is valid for $1 < |z| < 2$. Now the Laurent coefficients in (4) and (5) are unique, i.e., any representation such as that in (6) must be the Laurent expansion regardless of how we arrived at it. Therefore, the coefficients in our expansion must be equal to those of (4) and (5) so that as a by-product of our calculations we have been able to find the values of the integrals there. E.g., for $n = 1$ in (5) we see from (6) that $b_1 = 1$ so

$$\int_{|z|=2} f(z)\,dz = 2\pi i.$$

Finally, for $|z| > 2$ we may write

$$f(z) = \frac{1}{z}\left(\frac{1}{1-1/z} - \frac{1}{1-2/z}\right) \tag{7}$$

and $|1/z| < 1$ as well as $|2/z| < 1$ in the region $|z| > 2$. Therefore, we may use the geometric series in this region to find from (7) that

$$f(z) = \sum_{n=0}^{\infty} \frac{1}{z^{n+1}} - \sum_{n=0}^{\infty} \frac{2^n}{z^{n+1}}$$

or

$$f(z) = \sum_{n=0}^{\infty} \frac{1 - 2^n}{z^{n+1}} \tag{8}$$

Again, since the Laurent coefficients are unique we deduce that the coefficients in (8) are the Laurent coefficients. In particular since $b_1 = 0$ in (8) we find from (4) that

$$\int_C f(s)\,ds = 0,$$

where in this case, C can be any circle centered at $z_0 = 0$ with radius $r > 2$.

● **PROBLEM 15-16**

Find the first three nonzero terms in the Laurent series expansion of csc z about $z_0 = 0$.

Solution: The Laurent series expansion of a function $f(z)$ which is analytic in an annulus bounded by the concentric circles C_1 and C_2 with centers at z_0 is given by

$$f(z) = \sum_{n=0}^{\infty} a_n (z-z_0)^n + \sum_{n=1}^{\infty} \frac{b_n}{(z-z_0)^n} \tag{1}$$

where

$$a_n = \frac{1}{2\pi i} \int_{C_1} \frac{f(s)\,ds}{(s-z_0)^{n+1}} \qquad (n = 0,1,2,\ldots) \tag{2}$$

$$b_n = \frac{1}{2\pi i} \int_{C_2} \frac{f(s)\,ds}{(s-z_0)^{-n+1}} \qquad (n = 1,2,\ldots). \tag{3}$$

Here the radius of C_2 is assumed to be larger than the radius of C_1. The integrals in (2) and (3) are usually difficult and sometimes impossible to do by elementary methods so that a simpler way to proceed is to find the Taylor series expansion for sin z and use long division to obtain a series representing csc z = 1/sin z. Then observing that the Laurent expansion is unique, we can state that the resulting series is indeed the Laurent series.

Now since sin z is analytic $\forall z$ with $|z| < \infty$ we can write its Taylor expansion as

$$\sin z = \sum_{n=0}^{\infty} \frac{f^{(n)}(0) z^n}{n!} = \sum_{n=1}^{\infty} \frac{(-1)^{n-1}}{(2n-1)!} z^{2n-1} \tag{4}$$

Now sin z = 0 only for z = nπ, n = 0, ±1, ±2, ...

Hence, csc z = $\frac{1}{\sin z}$ is analytic in the annulus $0 < |z| < \pi$ can be represented by a series expansion there. Thus

$$\csc z = \frac{1}{\sin z} = \left[\frac{1}{z - \frac{z^3}{3!} + \frac{z^5}{5!} - \frac{z^7}{7!} + \cdots} \right]. \tag{5}$$

Thus, by long division

$$z - \frac{z^3}{3!} + \frac{z^5}{5!} - \frac{z^7}{7!} + \cdots \quad \overline{\left| \begin{array}{l} 1/z + z/3! + z^3\left(\frac{1}{3! \cdot 3!} - \frac{1}{5!}\right) + \cdots \\ 1 + 0 + 0 + 0 + 0 \\ -\left(1 - z^2/3! + z^4/5! - z^6/7! + \cdots\right) \\ \qquad z^2/3! - z^4/5! + z^6/7! - \cdots \\ \qquad -\left(z^2/3! - z^4/3! \cdot 3! + z^6/3! \cdot 5! - \cdots\right) \\ \qquad \qquad z^4\left(\frac{1}{3! \cdot 3!} - \frac{1}{5!}\right) - \cdots \end{array} \right.}$$

Therefore, from equation (5),

$$\csc z = \frac{1}{z} + \frac{1}{3!} z + \left(\frac{1}{3! \cdot 3!} - \frac{1}{5!}\right) z^3 + \cdots$$

or

$$\csc z = \frac{1}{z} + \frac{1}{6} z + \frac{7}{360} z^3 + \cdots \quad (0 < |z| < \pi). \tag{6}$$

As noted before, this must be the Laurent series for csc z since that series is unique. As a by-product of this calculation, we may therefore equate the coefficients in (6) with the corresponding Laurent coefficients given by (2) and (3) with C_2 being the circle $|z| = \pi$, and C_1 the degenerate circle $|z| = 0$. This provides a useful way to calculate the integrals of equations (2) and (3).

• **PROBLEM 15-17**

Find the principal part of the function $f(z) = \dfrac{e^z \cos z}{z^3}$ at its singular point and determine the type of singular point it is.

Solution: The given function has an isolated singular point at $z = 0$. Such a function may be represented by a Laurent series

$$f(z) = \sum_{n=0}^{\infty} a_n (z-z_0)^n + \sum_{n=1}^{\infty} \frac{b_n}{(z-z_0)^n} \tag{1}$$

where a_n and b_n are the Laurent coefficients. Since a Laurent expansion is unique, we needn't calculate these coefficients directly but may proceed as follows, noting that any series of the form (1) that we obtain must be the Laurent series. First, expanding the numerator of f in a Taylor series about $z_0 = 0$ yields

$$e^z \cos z = \left(1 + z + \frac{z^2}{2!} + \ldots\right)\left(1 - \frac{z^2}{2!} + \ldots\right)$$

$$= 1 + z - \frac{z^3}{3} + \ldots \qquad |z| < \infty \tag{2}$$

This is valid by Taylor's theorem since $e^z \cos z$ is analytic for all z. Hence

$$\frac{e^z \cos z}{z^3} = \frac{1}{z^3} + \frac{1}{z^2} - \frac{1}{3} + \ldots \qquad 0 < |z| < \infty \tag{3}$$

As noted above, this must be the Laurent series for $f(z)$.

The principal part of a function f at a point z_0 is defined as the portion of its Laurent series involving negative powers of $z-z_0$. Hence, the principal part of

$$f(z) = \frac{e^z \cos z}{z^3}$$

at 0, its only singular point, can be seen from (3) to be

$$\frac{1}{z^3} + \frac{1}{z^2} .$$

If the principal part of f at z_0 contains at least one nonzero term but the number of such terms is finite, the isolated singular point z_0 is then called a pole of order m, where m is the largest of the powers

$$\frac{b_j}{(z-z_0)^j}$$

in the principal part of f. Hence, in our case, it is said that f has a pole of order 3 at $z_0 = 0$.

● **PROBLEM 15-18**

Find the principal part of the function

$$f(z) = \frac{z}{(z+1)^2(z^3+2)}$$

at $z_0 = -1$. What type of singular point is z_0?

Solution: Since the singular point $z_0 = -1$ is isolated, the function has a Laurent series expansion about -1 which is valid at every point except -1 in the circular domain centered at -1 with radius r equal to the distance between -1 and the next closest singularity, i.e., $r = |(2)^{1/3} - 1|$. To find this expansion we first expand

$$f_1(z) = \frac{z}{z^3+2}$$

in a Taylor series about $z_0 = -1$ to obtain

$$f_1(z) = \sum_{n=0}^{\infty} \frac{f^{(0)}(z_0)}{n!} (z - z_0)^n \quad . \tag{1}$$

Now

$$f_1(z) = \frac{z}{z^3+2}$$

so

$$f_1^{(1)}(z) = \frac{2 - 2z^3}{(z^3+2)^2}$$

and

$$f_1^{(2)} = \frac{6(z^5 - 4z^2)}{(z^3+2)^3}$$

etc. Using these results in (1) we find

$$\frac{z}{z^3+2} = -1 + 4(z+1) - 15(z+1)^2 + \ldots \quad . \tag{2}$$

Now divide equation (2) by $(z + 1)^2$ to obtain

$$f(z) = \frac{z}{(z+1)^2(z^3+2)} = \frac{-1}{(z+1)^2} + \frac{4}{z+1} - 15 + \ldots \quad . \tag{3}$$

The principal part of a function at a point z_0 is defined as that part of its Laurent series involving negative powers of $(z - z_0)$. Since (3) is the Laurent series representing

$$f(z) = \frac{z}{(z+1)^2(z^3+2)}$$

for $0 < |z+1| < 2^{1/3}-1$, we conclude that the principal part of f at $z_0 = -1$ is

$$\frac{-1}{(z+1)^2} + \frac{4}{z+1} .$$

If the principal part of f at z_0 contains at least one nonzero term but the number of such terms is finite, then the isolated singular point z_0 is called a pole of order m where m is the largest of

$$\frac{b_j}{(z-z_0)^j}$$

in the principal part of f. Hence, in our case,

$$f(z) = \frac{z}{(z+1)^2(z^3+2)}$$

is said to have a pole of order 2 at $z_0 = -1$.

INTEGRATION

● **PROBLEM 15-19**

Evaluate the following definite integrals where t is a real number:

a) $\int_1^2 (t + it^2)\,dt$;

b) $\int_0^1 e^{(a+bi)t}\,dt$.

<u>Solution</u>: Any complex valued function, F(t), of a real variable t may be written as

$$F(t) = U(t) + iV(t) . \qquad (1)$$

We say that F(t) is integrable on the interval $a \le t \le b$ if each of U(t) and V(t) is integrable on that interval and we define

$$\int_a^b F(t)\,dt = \int_a^b U(t)\,dt + i\int_a^b V(t)\,dt . \qquad (2)$$

Evaluation of the above integrals then becomes a simple application of real variable integration theory.

a)
$$\int_1^2 (t + it^2)\,dt = \int_1^2 t\,dt + i\int_1^2 t^2\,dt$$

$$= \left.\frac{t^2}{2}\right|_1^2 + \left.\frac{it^3}{3}\right|_1^2$$

$$= \frac{3}{2} + \frac{7}{3}i \;.$$

b)
$$\int_0^1 e^{(a+bi)t}\,dt = \int_0^1 e^{at}e^{ibt}\,dt$$

$$\int_0^1 e^{at}(\cos bt + i\sin bt)\,dt$$

$$\int_0^1 e^{at}\cos bt\,dt + i\int_0^1 e^{at}\sin bt\,dt\;.$$

We may use integration by parts to evaluate each of these real integrals. Thus,

$$\int_0^1 e^{(a+bi)t}\,dt = \left.e^{at}\frac{(a\cos bt + b\sin bt)}{a^2+b^2}\right|_0^1 + \left.ie^{at}\frac{(a\sin bt - b\cos bt)}{a^2+b^2}\right|_0^1$$

$$= e^a\frac{(a\cos b + b\sin b)}{a^2+b^2} - \frac{a}{a^2+b^2}$$

$$+ ie^a\frac{(a\sin b - b\cos b)}{a^2+b^2} - \frac{i(-b)}{a^2+b^2}$$

$$= \frac{ae^a(\cos b + i\sin b) + be^a(\sin b - i\cos b)}{a^2+b^2}$$

$$- \frac{a - ib}{a^2+b^2}$$

$$= \frac{ae^a e^{ib} - ibe^a e^{ib} - (a - ib)}{a^2 + b^2}$$

$$= \frac{(a - ib) e^{(a+ib)} - (a - ib)}{(a + ib)(a - ib)}$$

$$= \frac{e^{a+ib} - 1}{a + ib}, \quad a + ib \neq 0.$$

● **PROBLEM 15-20**

Evaluate the following integrals along the contours indicated where $z = x + iy$.

a) $\int_C z^2 dz$ where C is the path $x = 2t$, $y = 3t$, $1 \leq t \leq 2$.

b) $\int_C \frac{1}{z} dz$ where C is the circular path

$x = \cos t$, $y = \sin t$, $0 \leq t \leq 2\pi$.

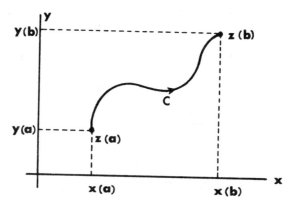

Fig. 1

Solution: The definite integral of a complex valued function f of a complex variable z along a contour C extending from a point $z = \alpha$ to a point $z = \beta$ is defined as follows. Let the contour be represented by the parametric equation

$$z(t) = x(t) + iy(t) \quad (a \leq t \leq b) \tag{1}$$

where C extends from $\alpha = z(a)$ to $\beta = z(b)$ (see Figure 1). Then by definition

$$\int_C f(z)\,dz = \int_\alpha^\beta f(z)\,dz = \int_a^b f[z(t)]z'(t)\,dt \ . \qquad (2)$$

This formula is analogous to that for a line integral in real analysis and we call

$$\int_C f(z)\,dz$$

the line integral of f along C. This formula will be used to evaluate the given integrals.

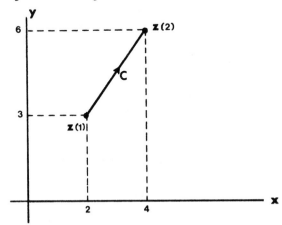

Fig. 2

a) Figure 2 shows the indicated contour and we note

$$z(t) = x(t) + iy(t) = 2t + i3t$$

so that $z'(t) = 2 + 3i$. Also

$$f[z(t)] = [z(t)]^2 = (2t + i3t)^2$$

so that

$$f[z(t)] = (2 + 3i)^2 t^2 \ .$$

Hence, from equation (2) this yields

$$\int_C f(z)\,dz = \int_1^2 (2 + 3i)^2 t^2 \cdot (2 + 3i)\,dt$$

$$= (2 + 3i)^3 \int_1^2 t^2\,dt$$

$$= (2 + 3i)^3 \left. \frac{t^3}{3} \right|_1^2$$

$$= \frac{-322}{3} + 21i.$$

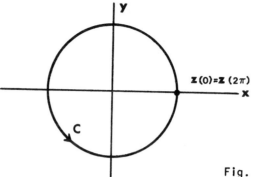

Fig. 3

b) Figure 3 shows the indicated contour and we note that

$$z(t) = x(t) + iy(t) = \cos t + i \sin t = e^{it}$$

so that

$$z'(t) = x'(t) + iy'(t) = -\sin t + i \cos t = ie^{it}.$$

Finally,

$$f(z(t)) = \frac{1}{z(t)} = \frac{1}{e^{it}} = e^{-it}.$$

Therefore, we have from equation (2) that

$$\int_C f(z)dz = \int_0^{2\pi} e^{-it} ie^{it} dt = i \int_0^{2\pi} dt$$

$$= 2\pi i.$$

● **PROBLEM 15-21**

Evaluate

$$\int_{C_1} z^2 dz \quad \text{and} \quad \int_{C_2} z^2 dz$$

where C_1 and C_2 are the contours OB and OAB, respectively, shown in figure 1.

<u>Solution</u>: The contour integral of a function f along a contour C with end points α and β is defined by

$$\int_C f(z)dz \equiv \int_a^b f[z(t)]z'(t)dt \qquad (1)$$

where $z(t) = x(t) + iy(t)$ is a parametric representation of C, $z(a) = \alpha$, $z(b) = \beta$, and $z'(t) = x'(t) + iy'(t)$.

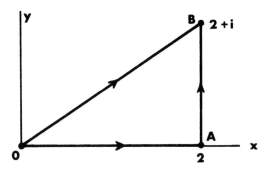

Fig. 1

In this problem, $f(z) = z^2$ and C_1 is the line segment $x = 2y$ $\alpha = 0$ to $\beta = 2 + i$.

Letting y be our parametric variable, we have

$$z(y) = 2y + iy, \quad 0 \le y \le 1 \tag{2}$$

and

$$z'(y) = 2 + i. \tag{3}$$

Also,

$$f[z(y)] = [z(y)]^2 = 3y^2 + i4y^2.$$

Therefore

$$\int_{C_1} z^2 dz = \int_0^1 [z(y)]^2 z'(y) dy$$

$$= \int_0^1 (3y^2 + i4y^2)(2 + i) dy$$

$$= (3 + 4i)(2 + i) \int_0^1 y^2 dy$$

$$= (2 + 11i) \left. \frac{y^3}{3} \right|_0^1 = \frac{2}{3} + \frac{11i}{3}. \tag{4}$$

In order to evaluate the second integral, we break up the contour OAB into two paths, OA and AB. Thus,

$$\int_{C_2} z^2 dz = \int_{OA} z^2 dz + \int_{AB} z^2 dz. \tag{5}$$

We may represent OA parametrically by $z(x) = x$,

$(0 \leq x \leq 2)$, so that $z'(x) = 1$, and we can write the contour AB as $z(y) = 2 + iy$, $(0 \leq y \leq 1)$ with $z'(y) = i$. Hence (5) becomes

$$\int_{C_2} z^2 dz = \int_0^2 x^2 dx + \int_0^1 (2 + iy)^2 i\, dy$$

$$= \frac{x^3}{3} \Big|_0^2 + i \int_0^1 (4 - y^2 + 4iy)\, dy$$

$$= \frac{8}{3} + i \left[\int_0^1 4\, dy - \int_0^1 y^2 dy + \int_0^1 4iy\, dy \right]$$

$$= \frac{8}{3} + i \left[4y - \frac{y^3}{3} + 2iy^2 \right]_0^1$$

$$= \frac{2}{3} + \frac{11}{3} i \qquad (6)$$

Thus we see from (4) and (6) that

$$\int_{C_1} z^2 dz = \int_{C_2} z^2 dz.$$

This fact is predicted by the Cauchy-Goursat Theorem which will be stated and used in later problems.

● **PROBLEM 15-22**

Use the Cauchy-Goursat Theorem to decide whether or not the following integrals have the value zero.

(a)
$$\int_{C_1} (z^2 + z + 2)\, dz$$ where C_1 is a circle of radius R centered at z_0 taken in either sense.

(b)
$$\int_{C_2} \frac{z^2 dz}{z - 3}$$ where C_2 is the circle $|z| = 1$, taken in the positive sense.

(c) $$\int_{C_3} \frac{dz}{z^2 + 2z + 2}$$ where C_3 is the circle $|z| = 2$ taken in the positive sense.

(d) $$\int_B \frac{zdz}{1 - e^z}$$ where B is the boundary of the region given by the circle $|z| = 4$, the square with sides along the lines $x = \pm 1$, $y = \pm 1$, and where B is described so that the region lies to the left of B.

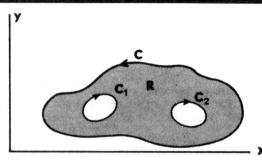

Fig. 1

Solution: The Cauchy-Goursat Theorem for a simple closed contour C states that if a function f is analytic within and on C, then

$$\int_C f(z)\,dz = 0.$$

We will employ this result in (a), (b) and (c).

(a) The function $f(z) = z^2 + z + 2$ is a polynomial and is therefore analytic in the entire complex plane. Hence, in particular, it is analytic within and on any circle of radius R, centered at z_0, thus

$$\int_{C_1} (z^2 + z + 2)\,dz = 0.$$

(b) The function $f(z) = \frac{z^2}{z - 3}$ is analytic everywhere in the complex plane except at the point $z = 3$. However, since this point is neither on nor contained in the circle $|z| = 1$, f is analytic both on and within C_2 so by the Cauchy-Goursat Theorem

$$\int_{C_2} \frac{z^2}{z - 3}\,dz = 0.$$

(c) The function $f(z) = \frac{1}{z^2 + 2z + 2}$ has singularities where $z^2 + 2z + 2 = 0$. That is, where $(z + 1)^2 + 1 = 0$, which implies that $z + 1 = \pm\sqrt{-1}$, so that

$$z_1 = i - 1, \quad z_2 = -i - 1.$$

Now $|z_1| = \sqrt{2}$ and $|z_2| = \sqrt{2}$ so that each of these singularities is contained within the circle $|z| = 2$. Hence it may not be concluded that

$$\int_{C_3} f(z)\,dz = 0.$$

We cannot definitely say that

$$\int_{C_3} f(z)\,dz \neq 0$$

however, since the converse of the Cauchy-Goursat Theorem, namely Morera's Theorem, applies only to functions which are continuous on and within C_3 and

$$f(z) = \frac{1}{z^2 + 2z + 2}$$

is not such a function.

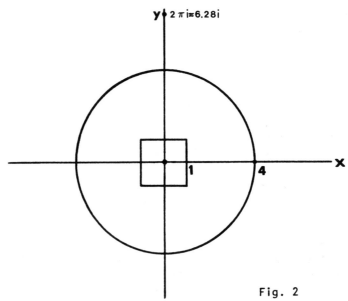

Fig. 2

(d) The Cauchy-Goursat Theorem can be modified to include any contour B which is the boundary of a region R described as follows (see Figure 1). Let C_j (j = 1, 2, ..., n) be a finite number of simple closed contours interior to a simple closed contour C. Then R is the closed region consisting of all points within and on C but not interior to any C_j. However, B is the boundary of R described in a direction such that points of R lie to the left of B.

The region described in the problem is obviously the type described above so that if we can prove that

$$f(z) = \frac{z}{1 - e^z}$$

is analytic within and on B, we may use the theorem. Now $f(z)$ has singular points only when $1 - e^z = 0$ or $e^z = 1$, i.e., when $z + i2n\pi = 0$, $n = 0, \pm 1, \pm 2, \ldots$. But note that none of the points

$$z = 0, \ 2\pi i, \ -2\pi i, \ 4\pi i, \ -4\pi i, \ \ldots$$

is in or on B (see Figure 2). Hence, $f(z)$ is analytic in and on B so that

$$\int_B \frac{z}{1 - e^z} \, dz = 0 \ .$$

● **PROBLEM 15-23**

Evaluate $\int_{C_1} \bar{z} \, dz$ and $\int_{C_2} \bar{z} \, dz$ where $\bar{z}$ is the conjugate of z, C_1 is the upper half of the circle $|z| = 1$ from $z = -1$ to $z = +1$ and C_2 is the lower half of $|z| = 1$ from $z = -1$ to $z = 1$ (see Figure 1).

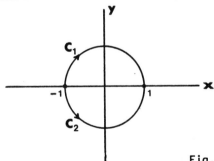

Fig. 1

Solution: A parametric equation for $-C_1$ is

$$z(\theta) = \cos \theta + i \sin \theta = e^{i\theta} \quad (0 \leq \theta \leq \pi),$$

and

$$z'(\theta) = -\sin \theta + i \cos \theta = ie^{i\theta}.$$

Also, if $z = |z|e^{i\psi}$, then $\bar{z} = |z|e^{-i\psi}$ so that on the unit circle, C_1, we may write $\bar{z} = e^{-i\theta}$. Hence

$$\int_{C_1} \bar{z}\,dz = -\int_{-C_1} \bar{z}\,dz = -\int_0^\pi \overline{z(\theta)}\, z'(\theta)\,d\theta$$

$$= -\int_0^\pi e^{-i\theta} ie^{i\theta}\,d\theta = -i\int_0^\pi d\theta$$

$$= -\pi i \qquad (1)$$

C_2 may be represented parametrically by

$$z(\theta) = \cos\theta + i\sin\theta = e^{i\theta} \quad (\pi \le \theta \le 2\pi)$$

so that $z'(\theta) = -\sin\theta + i\cos\theta = ie^{i\theta}$. Also, $\bar{z} = e^{-i\theta}$ on C_2 so that

$$\int_{C_2} \bar{z}\,dz = \int_\pi^{2\pi} \overline{z(\theta)} z'(\theta)\,d\theta = \int_\pi^{2\pi} e^{-i\theta} ie^{i\theta}\,d\theta$$

$$= i\int_\pi^{2\pi} d\theta = i(2\pi - \pi)$$

$$= \pi i \qquad (2)$$

Examination of (1) and (2) shows that

$$I_1 = \int_{C_1} \bar{z}\,dz \ne \int_{C_2} \bar{z}\,dz = I_2 \ .$$

In fact,

$$\int_{-C_1+C_2} \bar{z}\,dz = -I_1 + I_2 = 2\pi i \qquad (3)$$

where $-C_1 + C_2$ is the unit circle described in the positive sense. If we note that on the unit circle we have

$$\bar{z} = \frac{\bar{z}}{|z|^2} = \frac{1}{z},$$

we may conclude that

$$\int_C \frac{dz}{z} = 2\pi i \qquad (4)$$

where C is the unit circle. This is an example of the Cauchy Integral Formula which will be stated and employed in future problems.

● **PROBLEM 15-24**

Evaluate $\int_{-1}^{1} z^{1/2} dz$ where $z^{1/2} = \sqrt{|z|}\, e^{i\theta/2}$

where $0 < \theta < 2\pi$ along any contour lying in the upper half plane and also along any contour lying in the lower half plane. Use indefinite integrals to do this.

Solution: If a function $f(z)$ is analytic in a simply connected domain, D, and z_0 is a fixed but arbitrary point in D, then

$$F(z) = \int_{z_0}^{z} f(s)\, ds \qquad (1)$$

is called an indefinite integral of f, where z is any point in D. Notice from (1) that for each z_0 $F(z)$ is changed by an additive constant and can be written as

$$F(z) = \int f(z)\, dz \qquad (2)$$

to indicate the set of indefinite integrals of f, all differing by a constant. Apparently, then, a definite integral in a region where f is analytic may be written

$$\int_{\alpha}^{\beta} f(z)\, dz = \int_{z_1}^{\beta} f(z)\, dz - \int_{z_1}^{\alpha} f(z)\, dz$$

$$= F(\beta) - F(\alpha) = F(z)\Big|_{\alpha}^{\beta}. \qquad (3)$$

It is also true that for $F(z)$ so defined, $F'(z) = f(z)$ and this fact is usually used to find $F(z)$ for a given $(f(z))$.

Now, the given function is analytic everywhere except on the ray $\theta = 0$ and since $z = 1$ lies on this ray, the conditions to use (2) and (3) are not satisfied. However, we may define

$$f(z) = \sqrt{|z|}\, e^{i\theta/2} \qquad \left(\frac{-\pi}{2} < \theta < \frac{3\pi}{2}\right).$$

This branch of the multiple-valued function $z^{1/2}$ is analytic everywhere except on the ray

$$\Theta = \frac{-\pi}{2}$$

and its values coincide with the values of the given branch in the upper half plane. Hence, we may use it as our integrand and get the same result. An indefinite integral of $f(z)$ is

$$F(z) = \frac{2}{3} z^{3/2} = \frac{2}{3} |z|^{3/2} e^{i3\Theta/2} \qquad \left(\frac{-\pi}{2} < \Theta < \frac{3\pi}{2}\right)$$

as can be seen by differentiating each side of this equation using the Cauchy-Riemann conditions in polar coordinates to find $F'(z) = f(z)$.

Therefore by (3)

$$\int_{-1}^{1} z^{1/2} dz = \frac{2}{3} |z|^{3/2} e^{i3\Theta/2} \Big|_{-1=e^{i\pi}}^{1=e^{0}}$$

$$= \frac{2}{3} (1)^{3/2} e^{i3 \cdot 0/2} - \frac{2}{3} (1)^{3/2} e^{(i3\pi/2)}$$

$$= \frac{2}{3} [1 + i] \qquad \text{(upper half plane)}. \qquad (4)$$

In order to calculate

$$\int_{-1}^{1} z^{1/2} dz$$

for the given branch of $z^{1/2}$ for a contour below the real axis, similar considerations suggest that we use the branch

$$g(z) = \sqrt{|z|} \ e^{i\Theta/2} \qquad \left(\frac{\pi}{2} < \Theta < \frac{5\pi}{2}\right) \qquad (5)$$

which has the same values as the given branch below the real axis and is analytic for all points along any possible path contained therein. The function

$$F(z) = \frac{2z^{3/2}}{3} = \frac{2}{3} |z|^{3/2} e^{i3\Theta/2} \qquad \left(\frac{\pi}{2} < \Theta < \frac{5\pi}{2}\right)$$

is an indefinite integral of $g(z)$ so that from (3) we may write

$$\int_{-1}^{1} z^{1/2} dz = \frac{2}{3} |z|^{3/2} e^{i3\Theta/2} \Big|_{-1=e^{i\pi}}^{1=e^{i2\pi}}$$

$$= \frac{2}{3}(1)^{3/2} e^{\frac{i3\cdot 2\pi}{2}} - \frac{2}{3}(1)^{3/2} e^{i3\pi/2}$$

$$= \frac{2}{3}(-1 + i) \qquad \text{(lower half plane)}. \tag{6}$$

Note that the answers in (4) and (6) are different.

● **PROBLEM 15-25**

Evaluate the integral
$$\int_C \frac{zdz}{(9-z^2)(z+i)}$$
where C is the circle $|z| = 2$ described in the positive sense.

Solution: We will use the Cauchy Integral Formula to solve the integral. This formula states that if a complex valued function, f, is differentiable in a domain, D, and C is a simple closed curve in D, then for any ζ inside C we have

$$f(\zeta) = \frac{1}{2\pi i} \int_C \frac{f(z)dz}{z - \zeta}. \tag{1}$$

If we define
$$f(z) = \frac{z}{9 - z^2},$$

and let C be the circle $|z| = 2$, then f has singularities at $z = \pm 3$ only and is therefore differentiable in a domain around C. Now let $\zeta = -i$ and note that this point is within the circle $|z| = 2$. We have then satisfied the conditions for (1) and write

$$\frac{1}{2\pi i} \int_{|z|=2} \frac{z/(9-z^2)}{z-(-i)} dz = f(-i) = \frac{-i}{9-(-i)^2}$$

so that

$$\int_{|z|=2} \frac{zdz}{(9-z^2)(z+i)} = 2\pi i \frac{(-i)}{9-(i)^2} = \frac{2\pi(-i^2)}{9-(-1)}$$

$$= \frac{\pi}{5}$$

• **PROBLEM 15-26**

Directly verify the Cauchy-Goursat Theorem and Cauchy's integral formula for the appropriate functions by evaluating

$$\int_C (z - z_0)^n \, dz$$

along the circle C with center at z_0 and radius r described in the counterclockwise direction (n is any integer).

Solution: First, let us simply evaluate the integral. A parametric representation of the given circle is

$$z(\theta) = z_0 + re^{i\theta} \quad (0 \leq \theta \leq 2\pi) \ .$$

Hence $z'(\theta) = ire^{i\theta}$ and letting $G(z) = (z - z_0)^n$ we have

$$G[z(\theta)] = (z(\theta) - z_0)^n = (re^{i\theta})^n = r^n e^{in\theta}$$

Using these results, the integral may be written as

$$\int_C (z - z_0)^n \, dz = \int_0^{2\pi} G[z(\theta)] \, z'(\theta) \, d\theta$$

$$= \int_0^{2\pi} r^n e^{in\theta} \, ire^{i\theta} \, d\theta$$

$$= ir^{n+1} \int_0^{2\pi} e^{i(n+1)\theta} \, d\theta$$

$$= ir^{n+1} \int_0^{2\pi} [\cos(n+1)\theta + i \sin(n+1)\theta] \, d\theta. \quad (1)$$

If $n = -1$, equation (1) yields

$$\int_C (z - z_0)^{-1} \, dz = i \int_0^{2\pi} d\theta = 2\pi i \ .$$

If $n \neq -1$, then (1) yields

$$\int_C (z - z_0)^n \, dz = \frac{ir^{n+1}}{n+1} [\sin(n+1)\theta - i\cos(n+1)\theta] \Big|_0^{2\pi}$$

$$= 0$$

Combining the results of (1) and (2) we may write

$$\int_C (z - z_0)^n \, dz = \begin{cases} 2\pi i & \text{if } n = -1 \\ 0 & \text{if } n \neq -1 \end{cases}. \tag{3}$$

Now the Cauchy-Goursat Theorem states that if a function f is analytic at all points within and on a simple closed contour C, then

$$\int_C f(z) \, dz = 0 \tag{4}$$

The circle C given by $z(\theta) = z_0 + re^{i\theta}$, $(0 \leq \theta \leq 2\pi)$, is certainly a simple closed contour and each of the functions $f(z) = (z - z_0)^n$ for $n \geq 0$ is analytic within and on C so that by the Cauchy-Goursat Theorem we would expect

$$\int_C (z - z_0)^n \, dz = 0, \quad n \geq 0.$$

Inspection of (3) shows that this is true, thus we have directly verified this theorem for

$$f(z) = (z - z_0)^n, \quad n \geq 0.$$

Cauchy's Integral Formula says that under the same premises as those for the Cauchy-Goursat Theorem, if z_0 is interior to C, then

$$\int_C \frac{f(z) \, dz}{(z-z_0)^{n+1}} = \frac{2\pi i f^{(n)}(z_0)}{n!} \quad (n = 0, 1, 2, \ldots). \tag{5}$$

In particular, if $f(z) = 1$, then

$$\int_C \frac{dz}{(z-z_0)^{n+1}} = \begin{cases} 2\pi i, & n = 0 \\ 0, & n = 1, 2, \ldots \end{cases}. \tag{6}$$

Thus we see from (6) and (3) that we have indeed verified the Cauchy Integral formula for $f(z) = 1$ on the given circle.

• **PROBLEM 15-27**

Compute the value of the following integral:

$$I = \int_0^{2\pi} \cos^6 \theta \, d\theta.$$

Solution: We will first change the variable of integration to enable us to use an important integral formula. This formula states that if a complex valued function f is differentiable in a domain D and C is a simple closed curve in D, then for any ζ inside C we have

$$f^{(n)}(\zeta) = \frac{n!}{2\pi i} \int_C \frac{f(z) \, dz}{(z - \zeta)^{n+1}}. \tag{1}$$

Recall that $e^{i\theta} = \cos \theta + i \sin \theta$ and

$$e^{-i\theta} = e^{i(-\theta)} = \cos(-\theta) + i \sin(-\theta) = \cos \theta - i \sin \theta.$$

Hence,

$$\cos \theta = \frac{e^{i\theta} + e^{-i\theta}}{2} \tag{2}$$

Now, change the variable of integration by letting $z = e^{i\theta}$. That is, z is a complex number on the unit circle. Since

$$e^{-i\theta} = \bar{z} = \frac{\bar{z}}{z \bar{z}} = \frac{1}{z}$$

on the unit circle, we have from (2)

$$\cos \theta = \frac{1}{2}\left(z + \frac{1}{z}\right) \tag{3}$$

so that

$$\cos^6 \theta = \left(\frac{z + 1/z}{2}\right)^6 \tag{4}$$

Also,

$$\frac{dz}{d\theta} = \frac{d(e^{i\theta})}{d\theta} = ie^{i\theta}$$

so that

$$dz = ie^{i\theta} \, d\theta. \tag{5}$$

As θ varies from 0 to 2π, $z = e^{i\theta}$ traverses the unit circle so that the integral may now be written as one on the unit circle. From (4) and (5) we have

$$I = \int_0^{2\pi} \cos^6\theta \, d\theta = \int_{|z|=1} \left(\frac{z + 1/z}{2}\right)^6 \frac{dz}{iz}$$

$$= \frac{1}{2^6 i} \int_{|z|=1} \left((z + 1/z) \cdot \frac{z}{z}\right)^6 \frac{dz}{z}$$

$$= \frac{1}{2^6 i} \int_{|z|=1} \left(\frac{z^2 + 1}{z}\right)^6 \frac{dz}{z} = \frac{1}{2^6 i} \int_{|z|=1} \frac{(z^2 + 1)^6}{z^7} \, dz \, .$$

Hence

$$\frac{2^6 6!}{2\pi} I = \frac{6!}{2\pi i} \int_{|z|=1} \frac{(z^2 + 1)^6}{(z - 0)^7} \, dz \, . \tag{6}$$

This integral has the same form as (1) with

$$\zeta = 0, \quad f(z) = (z^2 + 1)^6 \quad , \quad n = 6.$$

Since $(z^2 + 1)^6$ is everywhere differentiable (since it is a polynomial) and $\zeta = 0$ is contained in the unit circle, $|z| = 1$, we may use (1) and (6) to write

$$\frac{2^6 6!}{2\pi} I = f^{(n)}(\zeta) = \left.\frac{d^6 (z^2 + 1)^6}{dz^6}\right|_{z=\zeta=0} . \tag{7}$$

We may use the binomial theorem to expand $(z^2 + 1)^6$. This gives

$$(z^2 + 1)^6 = 1 + 6z^2 + \frac{6 \cdot 5}{2!}(z^2)^2 + \frac{6 \cdot 5 \cdot 4}{3!}(z^2)^3$$

$$+ \frac{6 \cdot 5 \cdot 4 \cdot 3}{4!}(z^2)^4 + \frac{6 \cdot 5 \cdot 4 \cdot 3 \cdot 2}{5!}(z^2)^5$$

$$+ \frac{6 \cdot 5 \cdot 4 \cdot 3 \cdot 2 \cdot 1}{6!}(z^2)^6 \, .$$

Hence

$$(z^2 + 1)^6 = 1 + 6z^2 + 15z^4 + 20z^6 + 15z^8 + 6z^{10} + z^{12}$$

and

$$\frac{d^6(z^2 + 1)^6}{dz^6} = 0 + 0 + 0 + 20 \cdot (6 \cdot 5 \cdot 4 \cdot 3 \cdot 2 \cdot 1)$$

$$+ 15 \cdot (8 \cdot 7 \cdot 6 \cdot 5 \cdot 4 \cdot 3) z^2 + 6 \cdot (10 \cdot 9 \cdot 8 \cdot 7 \cdot 6 \cdot 5) z^4$$

$$+ (12 \cdot 11 \cdot 10 \cdot 9 \cdot 8 \cdot 7) z^6 \, .$$

Therefore,

$$\left.\frac{d^6(z^2+1)^6}{dz^6}\right|_{z=\zeta=0} = 20 \cdot (6 \cdot 5 \cdot 4 \cdot 3 \cdot 2 \cdot 1) = 20 \cdot 6!$$

and referring back to (7), we find

$$\frac{2^6 \cdot 6!}{2\pi} I = 20 \cdot 6!$$

so,

$$I = \frac{2\pi \cdot 20}{2^6} = \frac{\pi}{2^5} \cdot 20 = \frac{5\pi}{8}$$

That is,

$$\int_0^{2\pi} (\cos\theta)^6 \, d\theta = \frac{5\pi}{8}$$

● **PROBLEM 15-28**

Compute the value of the following integral

$$\int_0^{2\pi} \frac{d\theta}{1 - 2a\cos\theta + a^2} \, ,$$

for $|a| < 1$.

Solution: We will first change the variable of integration to enable us to use Cauchy's Integral Formula. This formula states that if a complex valued function f is differentiable in a domain D and C is a simple closed curve in D, then for any ζ inside C we have

$$f(\zeta) = \frac{1}{2\pi i} \int_C \frac{f(z) \, dz}{z - \zeta} \, .$$

Recall that

$$e^{i\theta} = \cos\theta + i\sin\theta$$

and

$$e^{-i\theta} = e^{i(-\theta)} = \cos(-\theta) + i\sin(-\theta)$$

so that

$$e^{-i\theta} = \cos\theta - i\sin\theta \, .$$

Hence,

$$\cos\theta = \frac{e^{i\theta} + e^{-i\theta}}{2} \tag{1}$$

Now change the variable of integration by letting $z = e^{i\theta}$. That is, z is a complex number on the unit circle. Since

$$e^{-i\theta} = z^{-1} = \frac{1}{z},$$

we have from (1)

$$\cos\theta = \frac{1}{2}\left(z + \frac{1}{z}\right) \tag{2}$$

Also,

$$\frac{dz}{d\theta} = \frac{d(e^{i\theta})}{d\theta} = ie^{i\theta}$$

so that

$$dz = ie^{i\theta} d\theta. \tag{3}$$

As θ varies from 0 to 2π, $z = e^{i\theta}$ traverses the unit circle so that the integral may now be written as one on the unit circle. From (2) and (3) we have

$$I = \int_0^{2\pi} \frac{d\theta}{1 - 2a\cos\theta + a^2} = \int_{|z|=1} \left(1 - 2a\left[\frac{z + \frac{1}{z}}{2}\right] + a^2\right)^{-1} \frac{dz}{iz} \tag{4}$$

$$= \frac{1}{i}\int_{|z|=1} \frac{dz}{(1 - az - \frac{a}{z} + a^2)z}$$

$$= \frac{1}{i}\int_{|z|=1} \frac{dz}{z - az^2 - a + a^2 z}$$

$$= -\frac{1}{ia}\int_{|z|=1} \frac{dz}{z^2 - (a + (1/a))z + 1}$$

$$I = -\frac{1}{ia}\int_{|z|=1} \frac{dz}{(z-a)(z-a^{-1})}.$$

If we let

$$f(z) = \frac{1}{z - a^{-1}}$$

and $\zeta = a$, we see that

$$-\frac{aI}{2\pi} = \frac{1}{2\pi i}\int_{\text{unit circle}} \frac{f(z)\,dz}{z - \zeta}. \tag{5}$$

Since

$$|a| < 1, \quad \frac{1}{|a|} > 1$$

and so f(z) has no singularities on the unit disk. Hence, f is analytic on the unit circle and therefore differentiable there. Since the conditions of the Cauchy Integral Formula are then satisified, we use it to determine from (5) that

$$-\frac{aI}{2\pi} = \frac{1}{2\pi i} \int_{|z|=1} \frac{f(z)\,dz}{z - \zeta} = f(\zeta) = f(a) = \frac{1}{a - a^{-1}}.$$

Therefore, we have

$$\int_0^{2\pi} \frac{d\theta}{1 - 2a\cos\theta + a^2} = I = \frac{2\pi}{-a}\left(\frac{1}{a - a^{-1}}\right) = \frac{2\pi}{1 - a^2}.$$

● **PROBLEM 15-29**

Use the residue theorem to evaluate

$$\int_C \frac{5z - 2}{z(z - 1)}\,dz$$

where C is the circle $|z| = 2$ described counterclockwise.

<u>Solution</u>: The residue theorem states that if C is a simple closed contour within and on which a function f is analytic except for a finite number of isolated singular points $z_1, z_2, \ldots, z_n$ interior to C, then

$$\int_C f(z)\,dz = 2\pi i\,(B_1 + B_2 + \ldots + B_n) \qquad (1)$$

where B_j is the residue of f at z_j. Here, C must be described in the positive (counterclockwise) sense.

The residue of f at an isolated singular point z_0 is defined as the coefficient of the term

$$\frac{1}{z - z_0}$$

in the Laurent expansion of f about z_0. Laurent's theorem states that this expansion

$$f(z) = \sum_{n=0}^{\infty} a_n (z - z_0)^n + \sum_{n=1}^{\infty} \frac{b_n}{(z - z_0)^n} \qquad (2)$$

has coefficients

$$a_n = \frac{1}{2\pi i} \int_{C_1} \frac{f(s)\,ds}{(s - z_0)^{n+1}} \qquad n = 0, 1, 2, \ldots \qquad (3)$$

$$b_n = \frac{1}{2\pi i} \int_{C_2} \frac{f(s)\,ds}{(s - z_0)^{-n+1}} \qquad n = 1, 2, \ldots \qquad (4)$$

where C_1 and C_2 are any two circles centered at z_0 with radii r_1 and r_2 such that $r_1 < r_2$ and f is analytic in the region between C_1 and C_2. It is equation (4) that leads to the residue theorem. The important thing to note for use in problems is that since the Laurent expansion, (2), is unique, any representation of f of the form (2) that we find (by whatever method) must in fact be the Laurent expansion so that equations (3) and (4) must hold for this series. Thus, we use any convenient method to obtain the form (2) and extract from this the value b_1, i.e., the residue of f at z_0.

The given function,

$$f(z) = \frac{5z - 2}{z(z - 1)},$$

has two isolated singularities, $z_1 = 0$ and $z_2 = 1$ each of which lies within the circle C given by $|z| = 2$. To use (1) it is necessary to find the residues of f at these two points.

To find B_1 at z_1, we write

$$f(z) = \frac{5z - 2}{z(z - 1)} = \left(5 - \frac{2}{z}\right)\left(\frac{-1}{1 - z}\right) = \left(-5 + \frac{2}{z}\right)(1 + z + z^2 + \ldots)$$

$$= \frac{2}{z} - 3 - 3z - 3z^2 - \ldots \qquad (0 < |z| < 1) \qquad (5)$$

where we have used the geometric series for $\frac{1}{z-1}$ which is valid for $|z| < 1$. Recall that f has a Laurent expansion (2) about z_0 if f is analytic in any neighborhood of z_0 and the expansion is valid in any neighborhood we choose as long as f is analytic there (except at z_0 of course). Thus, (5) is the Laurent series for f about $z_1 = 0$ and we see that $B_1 = 2$.

To find B_2 at $z_2 = 1$ use the Taylor series

$$\frac{1}{z} = 1 - (z-1) + (z-1)^2 - \ldots \quad (|z-1| < 1)$$

to find

$$f(z) = \frac{5z-2}{z(z-1)} = \left(5 + \frac{3}{z-1}\right)\frac{1}{z} = \left(5 + \frac{3}{(z-1)}\right)[1-(z-1)+(z-1)^2\ldots] \quad (6)$$

for $0 < |z-1| < 1$. This must be the Laurent series for f about $z_2 = 1$ and we see that the coefficient of $(z-1)^{-1}$ in (6) is 3, i.e., $B_2 = 3$. Thus, from (1) we have

$$\int_C \frac{5z-2}{z(z+1)} dz = 2\pi i (B_1 + B_2) = 10\pi i. \quad (7)$$

• **PROBLEM 15-30**

Use the residue theorem to evaluate

$$\int_C \frac{(1+z^5)\sinh z}{z^6} dz$$

where C is the unit circle $|z| = 1$ described in the positive sense (i.e., counterclockwise).

Solution: The residue theorem states that if C is a simple closed contour within and on which a function f is analytic except for a finite number of isolated singular points $z_1, z_2, \ldots z_n$ interior to C, then

$$\int_C f(z) dz = 2\pi i (B_1 + B_2 + \ldots + B_n) \quad (1)$$

where B_j is the residue of f at z_j. Here, C must be described in the positive sense. The residue of f at an isolated singular point z_0 is defined as the coefficient of the term $(z-z_0)^{-1}$ in the Laurent expansion of f about z_0

$$f(z) = \sum_{n=-\infty}^{\infty} C_n (z-z_0)^n. \quad (2)$$

The given function

$$f(z) = \frac{(1+z^5)\sinh z}{z^6}$$

has only one isolated singularity, namely $z_1 = 0$. To find the Laurent expansion of f about $z_1 = 0$ we use the

Taylor series about 0 for sinh z which is valid for all z, i.e.,

$$\sinh z = \sum_{n=0}^{\infty} \frac{\sinh^{(n)}(0)}{n!} z^n = z + \frac{1}{3!} z^3 + \frac{1}{5!} z^5 + \frac{1}{7!} z^7 + \ldots .$$

Hence

$$f(z) = \frac{(1 + z^5) \sinh z}{z^6}$$

$$= \frac{(1 + z^5)}{z^6} \left(z + \frac{1}{3!} z^3 + \frac{1}{5!} z^5 + \frac{1}{7!} z^7 + \ldots \right)$$

$$= (1 + z^5) \left(z^{-5} + \frac{1}{3!} z^{-3} + \frac{1}{5!} z^{-1} + \frac{1}{7!} z + \ldots \right)$$

$$= z^{-5} + \frac{1}{3!} z^{-3} + \frac{1}{5!} z^{-1} + 1 + \frac{1}{7!} z + \frac{1}{3!} z^2 + \ldots$$

$$(0 < |z| < \infty) .$$

This is a valid expansion of the form (2) and since a Laurent expansion is unique, this must be the Laurent expansion of f about 0. We see that the residue of f at $z_1 = 0$ is

$$\frac{1}{5!} , \text{ i.e., } B_1 = \frac{1}{5!} ,$$

so that from equation (1) we conclude

$$\int_C \frac{(1 + z^5) \sinh z}{z^6} = 2\pi i \left(\frac{1}{5!} \right) = \frac{\pi i}{60} .$$

• **PROBLEM** 15-31

Find the residues of the function f defined by

$$f(z) = \frac{\cosh z}{z^2 (z + \pi i)^3} .$$

Solution: The residue of f at an isolated singular point z_0 is defined as the coefficient of the term

$$\frac{1}{z - z_0}$$

in the Laurent expansion of f about z_0. We say that z_0 is a pole of order m if in that same expansion the largest term of the form

$$\frac{1}{(z-z_0)^j}$$

has j = m. In such a case, we may apply a theorem which states that the residue of f at z_0 (call it B_0), is given by

$$B_0 = \frac{1}{(m-1)!} \lim_{z \to z_0} \frac{d^{m-1}}{dz^{m-1}} [(z-z_0)^m f(z)] . \quad (1)$$

We can see that

$$f(z) = \frac{\cosh z}{z^2 (z + \pi i)^3}$$

has isolated singularities at $z_1 = 0$ and $z_2 = -\pi i$. Furthermore, cosh z is analytic everywhere so it has Taylor expansions about both z = 0 and z = $-\pi i$ given by

$$\cosh z = \sum_{n=0}^{\infty} \frac{\cosh^{(n)} 0}{n!} (z - 0)^n = 1 + \sum_{n=1}^{\infty} \frac{z^{2n}}{(2n)!} \quad (2)$$

and

$$\cosh z = \sum_{n=0}^{\infty} \frac{\cosh^{(n)}(-\pi i)}{n!} (z + \pi i)^n =$$

$$-1 + \sum_{n=1}^{\infty} \frac{\cosh^n(-\pi i)}{n!} (z + \pi i)^n \quad (3)$$

each valid for all z. We can therefore see that f(z) has a pole of order 2 at $z_1 = 0$ and a pole of order 3 at z = $-\pi i$. By equation (1) we can compute

$$B_1 = \frac{1}{(2-1)!} \lim_{z \to 0} \frac{d^{(1)}}{dz^{(1)}} \left[z^2 \frac{\cosh z}{z^2 (z+\pi i)^3} \right]$$

$$= \lim_{z \to 0} \frac{d}{dz} \left(\frac{\cosh z}{(z+\pi i)^3} \right)$$

$$= \lim_{z \to 0} \left[\frac{(z+\pi i) \sinh z - 3 \cosh z}{(z+\pi i)^4} \right] = \frac{-3}{\pi^4} .$$

Also from (1) we see that

$$B_2 = \frac{1}{2!} \lim_{z \to -\pi i} \frac{d^2}{dz^2} \left[(z+\pi i)^3 \frac{\cosh z}{z^2 (z+\pi i)^3} \right]$$

$$= \frac{1}{2} \lim_{z \to -\pi i} \frac{d^2}{dz^2} \left(\frac{\cosh z}{z^2} \right)$$

$$= \frac{1}{2} \lim_{z \to -\pi i} \left[\frac{-4z \sinh z + (z^2 + 6) \cosh z}{z^4} \right]$$

$$= \frac{\pi^2 - 6}{2\pi^4} .$$

Thus, we have found that the residue of f at $z_1 = 0$ is $\frac{-3}{\pi^4}$ and the residue of f at $z_2 = -\pi i$ is

$$\frac{\pi^2 - 6}{2\pi^4} .$$

● **PROBLEM 15-32**

Evaluate $\int_C \frac{ze^z}{z^2 - 1} dz$ where C is the circle $|z| = 2$ taken in the counterclockwise direction.

Solution: We will use the residue theorem to evaluate this integral. This theorem states that

$$\int_C f(z) dz = 2\pi i (B_1 + B_2 + \ldots) \qquad (1)$$

where B_j is the residue of the isolated singular point of f, z_j; where each z_j is contained in the interior of C. A useful theorem states that if z_j is a pole of order m,

$$B_j = \frac{1}{(m-1)!} \lim_{z \to z_j} \frac{d^{m-1}}{dz^{m-1}} [(z-z_j)^m f(z)] . \qquad (2)$$

Now the given function may be written

$$f(z) = \frac{ze^z}{(z+1)(z-1)}$$

so that f has poles of order 1 at $z_1 = +1$, $z_2 = -1$ each contained within the circle $|z| = 2$. From (2)

$$B_1 = \frac{1}{0!} \lim_{z \to 1} (z-1) \frac{ze^z}{(z+1)(z-1)} = \lim_{z \to 1} \frac{ze^z}{z+1} = \frac{e}{2} \qquad (3)$$

775

and

$$B_2 = \frac{1}{0!} \lim_{z \to -1} (z+1) \frac{ze^z}{(z+1)(z-1)} = \lim_{z \to -1} \frac{ze^z}{z-1} = \frac{e^{-1}}{2}. \quad (4)$$

Using the results of (3) and (4) in equation (1) we find

$$\int_{|z|=2} \frac{ze^z}{(z^2-1)} = 2\pi i \left(\frac{e}{2} + \frac{e^{-1}}{2} \right) = 2\pi i \cosh 1.$$

● **PROBLEM 15-33**

Find the partial fraction expansion of

$$f(z) = \frac{z^2 + 1}{z^3 + 4z^2 + 3z}$$

using the theory of residues.

Solution: In this problem we use a theorem which states that for any rational function

$$f(z) = \frac{a_0 z^n + \ldots + a_n}{b_0 z^m + \ldots + b_m} \quad (a_0 \neq 0, \ b_0 \neq 0) \quad (1)$$

we may write

$$f(z) = p_1(z) + p_2(z) + \ldots + p_n(z) \quad (2)$$

where p_j is the principal part of f at the pole z_j. Note that z_j will simply be a zero of the denominator of f. Recall now that the principal part of a function $f(z)$ at z_0 is that part of its Laurent series involving negative powers of $(z - z_0)$. Now note that

$$f(z) = \frac{z^2 + 1}{z^3 + 4z^2 + 3z} = \frac{z^2 + 1}{z(z+1)(z+3)} \quad (3)$$

so that f has simple poles or poles of order 1 at $z = 0$, $z = -1$, $z = -3$; a simple pole at z_0 being one which yields only one term in the principal part of f at z_0, namely

$$\frac{a}{(z - z_0)}.$$

Here a is called the residue of f at z_0 and it can be proven that for a rational function such as

$$f(z) = \frac{A(z)}{B(z)}$$

a simple pole at z_0,

$$a = \frac{A(z_0)}{B'(z_0)} \ .$$

We may now use these results along with equation (2) to find that

$$f(z) = p_1(z) + p_2(z) + p_3(z)$$
$$= \frac{A(0)}{B'(0)} \frac{1}{z} + \frac{A(-1)}{B'(-1)} \frac{1}{(z+1)} + \frac{A(-3)}{B'(-3)} \frac{1}{(z+3)} \quad (4)$$

Where from (3) we have $A(z) = z^2 + 1$, $B(z) = z^3 + 4z^2 + 3z$, $B'(z) = 3z^2 + 8z + 3$. Therefore, from (4) we find

$$f(z) = \frac{1}{3}\frac{1}{z} - \frac{1}{z+1} + \frac{5}{3}\frac{1}{z+3} \ .$$

● **PROBLEM 15-34**

Evaluate $I = \int_0^{2\pi} \frac{1}{\cos\theta + 2} d\theta$.

Solution: We will first change the variable of integration to enable us to use the residue theorem. Recall that

$$e^{i\theta} = \cos\theta + i\sin\theta$$

and

$$e^{-i\theta} = \cos\theta - i\sin\theta$$

so that

$$\cos\theta = \frac{e^{i\theta} + e^{-i\theta}}{2} \ . \quad (1)$$

We then have

$$I = \int_0^{2\pi} \frac{1}{\cos\theta + 2} d\theta = \int_0^{2\pi} \frac{1}{\frac{e^{i\theta} + e^{-i\theta}}{2} + 2} d\theta = \int_0^{2\pi} \frac{2d\theta}{e^{i\theta} + e^{-i\theta} + 4} \quad (2)$$

Thus, I is just a parameterized form of the integral

$$I = \int_{|z|=1} \frac{-2i(z^{-1})dz}{z+z^{-1}+4} = \int_{|z|=1} \frac{-2i}{z^2+4z+1} dz \quad (3)$$

obtained by letting $z = e^{i\theta}$, $dz = ie^{i\theta}d\theta$

(which is true on the unit circle). The integration is taken, of course, in the counterclockwise direction. Now according to the quadratic formula for the roots of a quadratic equation,

$$z^2 + 4z + 1$$

has roots at

$$z_1 = \frac{-4 + \sqrt{16-4}}{2} = -2 + \sqrt{3}$$

and

$$z_2 = \frac{-4 - \sqrt{16-4}}{2} = -2 - \sqrt{3}.$$

This means that

$$f(z) = \frac{-2i}{z^2+4z+1} = \frac{-2i}{(z-z_1)(z-z_2)}$$

has poles of order 1 at z_1 and z_2. Note, however, that since $|z_2| > 1$ and $|z_1| < 1$, only z_1 is contained in the unit circle. Thus, if B_1 is the residue of $f(z)$ at z_1 then the residue theorem tells us that

$$I = \int_{|z|=1} f(z)dz = 2\pi i\, B_1. \quad (4)$$

To find B_1 use the formula which states that if z_j is a pole of order m of $f(z)$, then the residue of f at z_j, B_j, is given by

$$B_j = \frac{1}{(m-1)!} \lim_{z \to z_j} \frac{d^{m-1}}{dz^{m-1}}[(z-z_j)^m f(z)]. \quad (5)$$

For B_1, (5) yields

$$B_1 = \frac{1}{0!} \lim_{z \to z_1} (z-z_1)\left[\frac{-2i}{(z-z_1)(z-z_2)}\right] = \lim_{z \to z_1} \frac{-2i}{z-z_2}$$

$$= \frac{-2i}{z_1 - z_2} = \frac{-2i}{2\sqrt{3}} = \frac{-i}{\sqrt{3}}$$

Thus, from (4), we conclude that

$$I = \int_0^{2\pi} \frac{1}{\cos\theta + 2} d\theta = \int_{|z|=1} \frac{-2i}{z^2 + 4z + 1} dz = (2\pi i)\left(\frac{-i}{\sqrt{3}}\right)$$

Hence

$$I = \frac{2\pi}{\sqrt{3}}.$$

● **PROBLEM 15-35**

Evaluate $I = \int_{-\pi}^{\pi} e^{2\cos\theta} d\theta$.

Solution: Our method will be to change the variable of integration to enable us to use complex integration. Since

$$\cos\theta = \frac{e^{i\theta} + e^{-i\theta}}{2}$$

and

$$z = e^{i\theta}$$

on the unit circle, we see that I is just a parameterized form of the complex integral

$$I = \int_{|z|=1} \exp(e^{i\theta} + e^{-i\theta}) d\theta = \int_{|z|=1} \exp\left(z + \frac{1}{z}\right) \frac{dz}{iz}$$

$$= \int_{|z|=1} \frac{e^z e^{1/z}}{iz} dz = \int_{|z|=1} f(z) dz, \quad (1)$$

since if $z = e^{i\theta}$ then $dz = ie^{i\theta} d\theta = iz d\theta$.

The only singularity of $f(z)$ is at $z_1 = 0$ and this is contained within the unit circle. Therefore, if we can find the residue of $f(z)$ at $z_1 = 0$, i.e., B_1, then by the residue theorem we will have

$$I = \int_{|z|=1} f(z) dz = 2\pi i B_1 \quad (2)$$

The residue of f at z_1 is the coefficient of the term $(z - z_1)^{-1}$ in the Laurent expansion of $f(z)$. To obtain this expansion, recall the Taylor series

$$e^z = \sum_{n=0}^{\infty} \frac{z^n}{n!} \quad \text{and} \quad e^{1/z} = \sum_{n=0}^{\infty} \frac{z^{-n}}{n!}$$

so that we may write

$$f(z) = \frac{1}{iz} e^z e^{1/z} = \frac{1}{iz} \sum_{n=-\infty}^{\infty} \left(\sum_{\substack{k-j=n \\ k \geq 0 \\ j \geq 0}} \frac{1}{k!} \frac{1}{j!} \right) z^n$$

$$= \frac{1}{i} \sum_{-\infty}^{\infty} \left(\sum_{\substack{k-j=n \\ k \geq 0 \\ j \geq 0}} \frac{1}{k!} \frac{1}{j!} \right) z^{n-1}$$

Thus, B_1 is the coefficient with $n=0$, i.e.,

$$B_1 = \frac{1}{i} \sum_{\substack{k=j \\ k \geq 0}} \frac{1}{(i!)(j!)} = \frac{1}{i} \sum_{n=0}^{\infty} \frac{1}{(n!)^2}$$

We then have from (2) that

$$I = \int_{-\pi}^{\pi} e^{2\cos\theta} d\theta = 2\pi \sum_{n=0}^{\infty} \frac{1}{(n!)^2} \quad .$$

● **PROBLEM 15-36**

Evaluate $\quad I_0 = \int_0^{\infty} \frac{dx}{x^2 + 1} \quad .$

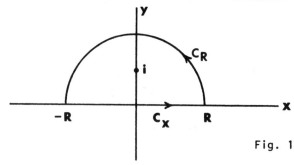

Fig. 1

<u>Solution</u>: First note that since $f(x) = \frac{1}{x^2 + 1}$ is an even function

$$I_0 = \frac{1}{2} \int_{-\infty}^{\infty} \frac{dx}{x^2 + 1} \quad . \tag{1}$$

I_0 may therefore be interpreted as a parameterized form of the complex integral of

$$f(z) = \frac{1}{z^2 + 1} = \frac{1}{(z+i)(z-i)}$$

along the contour described by the entire real axis, C_x, by

$$I_0 = \frac{1}{2} \int_{C_x} \frac{dz}{(z+i)(z-i)} \tag{2}$$

Let C_R denote the upper half of the circle $|z| = R$ where $R > 1$ as shown in figure 1.

Since $f(z)$ has a pole of order 1 at $z_0 = i$, the residue theorem yields

$$\int_C f(z)\,dz = \int_{C_x} f(z)\,dz + \int_{C_R} f(z)\,dz = 2\pi i\, B_1 \tag{3}$$

where B_1 is the residue of $f(z)$ at $z_0 = i$. The residue B_j of a function f at a pole z_0 of order m is given by

$$B_j = \frac{1}{(m-1)!} \lim_{z \to z_0} \frac{d^{m-1}}{dz^{m-1}} \left[(z - z_0)^m f(z) \right]$$

so that in the case at hand

$$B_1 = \frac{1}{0!} \lim_{z \to i} (z - i) \left[\frac{1}{(z-i)(z+i)} \right] = -\frac{i}{2}. \tag{4}$$

Therefore, by (3)

$$\int_{-R}^{R} \frac{dx}{x^2 + 1} = \pi - \int_{C_R} \frac{dz}{z^2 + 1} \tag{5}$$

Now $|z| = R$ when z is on C_R so that

$$|z^2 + 1| \geq |z^2| - 1 = R^2 - 1$$

so that

$$\left| \int_{C_R} \frac{dz}{z^2 + 1} \right| \leq \int_{C_R} \frac{|dz|}{|z^2 + 1|} \leq \int_{C_R} \frac{|dz|}{R^2 - 1}$$

$$= \frac{\pi R}{R^2 - 1} \quad . \tag{6}$$

Taking the limit as $R \to \infty$ on both sides of (6) yields

$$\lim_{R \to \infty} \left| \int_{C_R} \frac{dz}{z^2 + 1} \right| = 0$$

Therefore, if the limit as $R \to \infty$ is taken on both sides of (5), it is found that

$$\int_{-\infty}^{\infty} \frac{dx}{x^2 + 1} = \pi$$

so that

$$I_0 = \frac{1}{2} \int_{-\infty}^{\infty} \frac{dx}{x^2 + 1} = \frac{\pi}{2} \quad .$$

● **PROBLEM 15-37**

Evaluate $I_0 = \displaystyle\int_0^{\infty} \frac{x^2 \, dx}{(x^2 + 9)(x^2 + 4)^2}$.

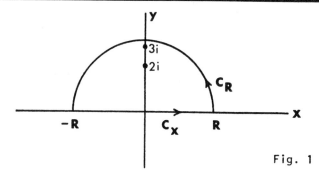

Fig. 1

Solution: First note that since the integrand of I_0 is even,

$$I_0 = \frac{1}{2} \int_{-\infty}^{\infty} \frac{x^2 \, dx}{(x^2 + 9)(x^2 + 4)^2} \quad . \tag{1}$$

I_0 may therefore be interpreted as a parameterized form of the complex integral

$$I_0 = \int_{C_x} \frac{z^2}{(z^2 + 9)(z^2 + 4)^2} \, dz = \int_{C_x} f(z) \, dz \quad , \tag{2}$$

where C_x is the contour described by the entire real axis. Let C_R denote the upper half of the circle $|z| = R$ where $R > 3$ as in Figure 1. Since $f(z)$ has a pole of order 1 at $z_1 = 3i$ and a pole of order 2 at $z_2 = 2i$, each contained in the contour $C = C_x \cup C_R$, the residue theorem states that

$$\int_C f(z)\,dz = \int_{C_x} f(z)\,dz + \int_{C_R} f(z)\,dz = 2\pi i(B_1 + B_2), \tag{3}$$

where B_j is the residue of f at z_j. Since the singular points are poles of order m, the residues may be found from the formula

$$B_j = \frac{1}{(m-1)!} \lim_{z \to z_j} \frac{d^{m-1}}{dz^{m-1}} [(z - z_j)^m f(z)], \tag{4}$$

so that in the case at hand,

$$B_1 = \frac{1}{0!} \lim_{z \to 3i} (z - 3i) \frac{z^2}{(z+3i)(z-3i)(z^2+4)^2} = -\frac{3}{50i} \tag{5}$$

and

$$B_2 = \frac{1}{1!} \lim_{z \to 2i} \frac{d}{dz} (z-2i)^2 f(z) = \lim_{z \to 2i} \frac{d}{dz} \frac{z^2}{(z^2+9)(z+2i)^2}$$

$$= \frac{-13i}{200}. \tag{6}$$

Using (5) and (6) in (3) yields

$$\int_{-R}^{R} f(x)\,dx = \frac{\pi}{100} - \int_{C_R} f(z)\,dz. \tag{7}$$

Now let $R \to \infty$ on each side of equation (7) and note that since

$$|f(z)| = \frac{|z^2|}{|z^2+9||z^2+4|^2} \le \frac{R^2}{(R^2-9)(R^2-4)^2}$$

on the circle $z = Re^{i\theta}$, it is true that

$$\left| \int_{C_R} f(z)\,dz \right| \leq \int_{C_R} |f(z)||dz| \leq \int_{C_R} \frac{R^2 |dz|}{(R^2 - 9)(R^2 - 4)^2}$$

$$= \frac{\pi R^3}{(R^2 - 9)(R^2 - 4)^2} \quad . \tag{8}$$

Since (8) goes to zero as $R \to \infty$, equation (7) yields

$$\int_{-\infty}^{\infty} f(x)\,dx = \frac{\pi}{100}$$

so that from (1)

$$I_0 = \frac{1}{2} \int_{-\infty}^{\infty} \frac{x^2\,dx}{(x^2 + 9)(x^2 + 4)^2} = \frac{\pi}{200} \quad .$$

● **PROBLEM 15-38**

Evaluate

$$I_1 = \int_{-\infty}^{\infty} \frac{x \cos x}{x^2 + 1}\,dx$$

and

$$I_2 = \int_{-\infty}^{\infty} \frac{x \sin x}{x^2 + 1}\,dx \quad .$$

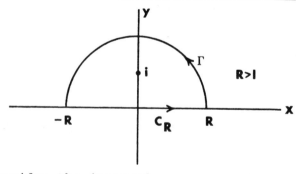

Fig. 1

Solution: Consider the integral

$$I = \int_{-\infty}^{\infty} \frac{xe^{ix}}{x^2 + 1}\,dx = \int_{-\infty}^{\infty} \frac{x \cos x}{x^2 + 1}\,dx + i \int_{-\infty}^{\infty} \frac{x \sin x}{x^2 + 1}\,dx \tag{1}$$

so that
$$I = I_1 + iI_2 \tag{2}$$
where I_1 and I_2 are real.

Now let
$$I_R = \int_C \frac{ze^{iz}\,dz}{z^2+1} = \int_{-R}^{R} \frac{xe^{ix}}{x^2+1}\,dx + \int_\Gamma \frac{ze^{iz}\,dz}{z^2+1} \tag{3}$$

where $C = C_R \cup \Gamma$ is shown in Figure 1. The integrand of I_R, $f(z) = \frac{ze^{iz}}{z^2+1}$, has poles of order 1 at $z_1 = i$ and $z_2 = -i$ but only $z_1 = i$ is contained within C. Therefore, by the residue theorem

$$I_R = \int_C f(z)\,dz = 2\pi i\, B_1, \tag{4}$$

where B_1 is the residue of f at $z_1 = i$.

To compute B_1, recall that if B_j is the residue of f at a pole of order m, i.e., z_j, then

$$B_j = \frac{1}{(m-1)!} \lim_{z \to z_j} \frac{d^{m-1}}{dz^{m-1}} [(z-z_j)^m f(z)]. \tag{5}$$

In the case at hand, $m = 1$ and $z_j = z_1 = i$ so

$$B_1 = \frac{1}{0!} \lim_{z \to i} (z-i) \frac{ze^{iz}}{(z+i)(z-i)} = \frac{ie^{i \cdot i}}{i+i} = \frac{e^{-1}}{2}$$

and from (4),
$$I_R = \frac{\pi i}{e}.$$

Using this result in equation (3) gives
$$\int_{-R}^{R} \frac{xe^{ix}}{x^2+1}\,dx + \int_\Gamma \frac{ze^{iz}\,dz}{z^2+1} = \frac{\pi i}{e}$$

and letting $R \to \infty$ in this equation, it is found that

$$\int_{-\infty}^{\infty} \frac{x\cos x}{x^2+1}\,dx + i \int_{-\infty}^{\infty} \frac{x\sin x}{x^2+1}\,dx + \lim_{R \to \infty} \int_\Gamma \frac{ze^{iz}}{z^2+1}\,dz = \frac{\pi i}{e} \tag{6}$$

Now on the semicircle Γ, $z = Re^{i\theta}$ with $0 \leq \theta \leq \pi$ and $dz = iRe^{i\theta}d\theta$ so that

$$\left| \int_\Gamma \frac{ze^{iz}}{z^2 + 1} dz \right| = \left| \int_0^\pi \frac{Re^{i\theta} e^{iRe^{i\theta}} iRe^{i\theta}}{R^2 e^{i2\theta} + 1} d\theta \right|$$

$$\leq \int_0^\pi \left| \frac{Re^{i\theta} e^{iRe^{i\theta}} iRe^{i\theta}}{R^2 e^{i2\theta} + 1} \right| d\theta$$

$$= \int_0^\pi \left| e^{iR\cos\theta} e^{-R\sin\theta} \right| \left| iRe^{i\theta} \right| \left| \frac{Re^{i\theta}}{R^2 e^{i2\theta} + 1} \right| d\theta$$

$$= \int_0^\pi e^{-R\sin\theta} R \left(\frac{R}{|R^2 e^{i2\theta} + 1|} \right) d\theta$$

$$\leq R \int_0^\pi e^{-R\sin\theta} \left(\frac{R}{R^2 - 1} \right) d\theta .$$

$$= \frac{R^2}{R^2 - 1} \int_0^\pi e^{-R\sin\theta} d\theta .$$

Hence,

$$\left| \int_\Gamma f(z) dz \right| \leq \frac{2R^2}{R^2 - 1} \int_0^{\pi/2} e^{-R\sin\theta} d\theta . \quad (7)$$

It can be seen from Figure 2 and it can be proved analytically that if $0 \leq \theta \leq \pi/2$ then

$$\sin\theta \geq \frac{2\theta}{\pi} .$$

This can be used in (7) to show that

$$\left| \int_\Gamma f(z)\,dz \right| \leq \frac{2R^2}{R^2 - 1} \int_0^{\pi/2} e^{-2R\theta/\pi}\,d\theta = \frac{\pi R^2}{R^3 - R}(1 - e^{-R}). \quad (8)$$

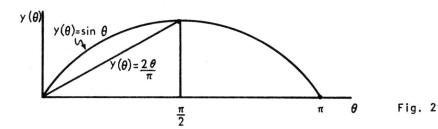

Fig. 2

Letting $R \to \infty$ on both sides of (8), it is seen that

$$\lim_{R \to \infty} \int_\Gamma \frac{z e^{iz}}{z^2 + 1}\,dz = 0$$

so that equation (6) yields

$$\int_{-\infty}^{\infty} \frac{x \cos x\,dx}{x^2 + 1} + i \int_{-\infty}^{\infty} \frac{x \sin x}{x^2 + 1} = \frac{i\pi}{e}. \quad (9)$$

Equating real and imaginary parts in (9), the final result is obtained, namely

$$\int_{-\infty}^{\infty} \frac{x \cos x}{x^2 + 1}\,dx = 0, \quad \int_{-\infty}^{\infty} \frac{x \sin x}{x^2 + 1}\,dx = \frac{\pi}{e}.$$

CHAPTER 16

LAPLACE TRANSFORMS

Because of the analogy between infinite series and improper integrals, one might expect to find an improper integral corresponding to a power series. Indeed, a power series

$$\sum_{n=0}^{\infty} c_n x^n$$

may be written as

$$\sum_{n=0}^{\infty} f(n) x^n , \qquad (1)$$

where f is a function whose value of each n is given by $f(n) = c_n$. The natural analogue of (1) is the improper integral

$$\int_0^{\infty} f(t) x^t \, dt = F(x) , \qquad (2)$$

and except for a minor change in notation, (2) is the Laplace transform of the function f(t). The notational change is accomplished by making the substitution

$$x = e^{-s}$$

in (2) to give

$$F(s) = \int_0^{\infty} f(t) e^{-st} \, dt , \qquad (3)$$

and F(s) is called the Laplace transform of f(t). Letting the expression $L\{f(t)\}$ represent the operation of multiplying f(t) by e^{-st} and integrating from $t = 0$ to $t = \infty$, one could write

$$F(s) = L\{f(t)\} \qquad (4)$$

which makes clear the interpretation of (3) as an operator, L, transforming the function f(t) into the function F(s). This way of interpreting the Laplace transform proves very convenient in applications and allows its inclusion in the class of linear operators which has been investigated thoroughly in mathematical literature. (The linearity property states that

$$L\{af_1(t) + bf_2(t)\} = aL\{f_1(t)\} + bL\{f_2(t)\}$$

and will be proved later in this chapter). Note that just as the series (1) has a radius of convergence, so does the integral (2). That is, (2) converges for

$$0 < x < r$$

for some r so that the Laplace transform, (3), converges for

$$s > \log\left(\frac{1}{r}\right).$$

The most important property of Laplace transforms is the simple relationship between $L\{f(t)\}$ and $L\{f^{(n)}(t)\}$ where $f^{(n)}(t)$ is the nth derivative of f. It will be proved in the chapter that this relationship is

$$L\{f^{(n)}(t)\} = s^n L\{f(t)\} - \sum_{k=1}^{n} s^{k-1} f^{(n-k)}(0)$$

$$= s^n L\{f(t)\} - f^{(n-1)}(0) - s f^{(n-2)}(0) -$$

$$\ldots - s^{n-1} f(0).$$

Thus, to find

$$L\{f^n(t)\}$$

one needs only know $L\{f(t)\}$ and the initial conditions,

$$f^{(k)}(0) \quad (k = 0, 1, \ldots, n-1).$$

This makes the Laplace transform ideal for dealing with differential equations with initial conditions. For instance, one could take the Laplace transform of

$$ay''(t) + by'(t) + cy(t) = f(t),$$
$$y(0) = y_0, \quad y'(0) = y_1,$$

to get an equation involving only $L\{y(t)\}$ and $L\{f(t)\}$, i.e., using the linearity property,

$$aL\{y''(t)\} + bL\{y'(t)\} + cL\{y(t)\} = L\{f(t)\} = F(s),$$

and using the derivative property,

$$a\left[s^2 L\{y(t)\} - sy(0) - y'(0)\right] +$$

$$b\left[sL\{y(t)\} - y(0)\right] + cL\{y(t)\} = F(s).$$

This can be solved for $L\{y(t)\}$ to give

$$L\{y(t)\} = G(s) ,$$

where G(s) involves F(s), s, and the initial values. Thus, if it is known what function g(t) has the Laplace transform G(s), the problem is solved by setting y(t) = g(t). This is the type of problem for which Laplace transforms are useful and the problems connected with their solution are the subject of this chapter.

DEFINITIONS AND SIMPLE EXAMPLES

● **PROBLEM** 16-1

(a) Define the Laplace transform L{f(t)} for a real valued function f(t).

(b) State conditions under which L{f(t)} exists.

(c) Show that L{f_1(t)} exists where

$$f_1(t) = 2te^{t^2} \cos e^{t^2}$$

but that f_1 does not satisfy all conditions in (b).

<u>Solution</u>: (a) Let f(t) be defined on some domain of the real axis containing [0,∞). Then the function G(s) defined by

$$G(s) = \int_0^\infty e^{-st} f(t) dt \tag{1}$$

is called the Laplace transform of f(t) and we write

$$G(s) = L\{f(t)\}$$

to emphasize the point of view that G(s) is the result of a certain operation (defined in (1)) performed on f(t). In general, s can be complex but since most applications use real s, it will be assumed that s is real unless otherwise stated.

(b) The most widely applied existence theorem involves two definitions which will be stated now.

Definition 1: If f is defined on a ≤ t < ∞, then f is piecewise continuous on [a,∞) if for every finite interval a ≤ t ≤ b f has a finite number of discontinuities such that at each discontinuity t = t_0 the limits $f(t_0^+)$ and $f(t_0^-)$ exist.

Definition 2: A function f(t) defined on [a,∞) is said to be exponential order e^α on [a,∞) if

$$|e^{-\alpha t} f(t)| \leq M \qquad (\alpha \text{ is real}) \tag{2}$$

for some real positive constant M and all $t \in [a, \infty)$.

Expressed more qualitatively, this means that $f(t)$ does not increase faster than $e^{\alpha t}$ as $t \to \infty$.

The existence theorem may now be stated.

Theorem: If $f(t)$ is piecewise continuous on $[0, \infty)$ and is of exponential order α, then the Laplace integral

$$G(s) = L\{f(t)\} = \int_0^\infty e^{-st} f(t) dt$$

converges for $s > \alpha$. Moreover, the integral is absolutely and uniformly convergent for $s \geq \alpha'$ where $\alpha' > \alpha$.

Note that if s is allowed to be complex, the theorem still holds but with the conditions

$$\text{Re } s > \alpha \quad \text{and} \quad \text{Re } s \geq \alpha'.$$

(c)
$$L\{f_1(t)\} = \int_0^\infty e^{-st} 2t e^{t^2} \cos e^{t^2} dt . \tag{3}$$

Integrating by parts gives

$$L\{f_1(t)\} = e^{-st} \sin e^{t^2} \Big|_0^\infty + s \int_0^\infty e^{-st} \sin e^{t^2} dt$$

$$= -\sin 1 + s \int_0^\infty e^{-st} \sin e^{t^2} dt . \tag{4}$$

Since

$$\left| \sin e^{t^2} \right| < 1 \quad , \quad \left| e^{-st} \sin e^{t^2} \right|$$

is bounded by e^{-st} for all $s > 0$, and the integral in (4) converges by the Weierstrass M-test for all $s > 0$. Note however that $f_1(t)$ is not of any exponential order since

$$e^{t^2} > Me^{\alpha t}$$

for t suffiently large no matter how large M and α are. Thus, $f_1(t)$ illustrates the fact that the theorem of (b) provides sufficient conditions for the existence of the Laplace integral of (1), but not necessary conditions. However, almost all functions which appear in physical problems do satisfy the theorem.

● **PROBLEM 16-2**

(a) Prove that $f(t) = t^n$, $n > 0$, is of exponential order α on $[0, \infty)$ for all $\alpha > 0$.

(b) Prove that $f(t) = \sin kt$ is of exponential order α on $[0, \infty)$ for all $\alpha > 0$.

Solution: (a) The function $f(t)$ is said to be of exponential order α on $[0, \infty)$ if there exists a real constant α and a positive real constant M such that

$$e^{-\alpha t}|f(t)| < M \text{ for all } t > 0.$$

For positive t

$$|t^n| = t^n.$$

Thus,

$$e^{-\alpha t}|f(t)| = e^{-\alpha t} t^n.$$

Iterative use of L'Hospital's rule shows that, for $\alpha > 0$,

$$\lim_{t \to \infty} e^{-\alpha t} t^n = 0.$$

Arbitrarily setting $M = 1$, it follows from the definition of a limit that there exists a positive constant t_0 such that

$$e^{-\alpha t}|f(t)| < M, \text{ for all } t > t_0.$$

Now let

$$M_1 = \max \{e^{-\alpha t} t^n, t \in [0, t_0]\}$$

(such a max exists since the function $g(t) = e^{-\alpha t} t^n$ is bounded on the closed interval $[0, t_0]$). Then

$$e^{-\alpha t}|f(t)| < M'$$

for all $t > 0$, where

$$M' = \max \{M, M_1\}.$$

Hence, $f(t)$ is of exponential order α for any $\alpha > 0$. Therefore, by the existence theorem, the Laplace transform of f

$$L\{f(t)\} = \int_0^\infty e^{-st} f(t) dt$$

exists for all $s > 0$.

(b) The function $f(t)$ is said to be of exponential order α on $[0,\infty)$ if there exists a real constant α and a positive real constant M such that

$$e^{-\alpha t}|f(t)| < M \quad \text{for all } t > 0.$$

Notice that for $\alpha = 0$,

$$\lim_{t \to \infty} e^{-\alpha t} = 1,$$

and for $\alpha > 0$,

$$\lim_{t \to \infty} e^{-\alpha t} = 0.$$

Thus, for $\alpha \geq 0$,

$$\lim_{t \to \infty} e^{-\alpha t} < 2. \tag{1}$$

Also note that

$$|\sin kt| \leq 1, \tag{2}$$

for all t.

Multiplying inequalities (1) and (2) together, we obtain

$$\left(\lim_{t \to \infty} e^{-\alpha t}\right)|\sin kt| < 2,$$

for $\alpha \geq 0$.

It follows from the definition of a limit that there exists a positive constant t_0 such that

$$e^{-\alpha t}|\sin kt| < 2,$$

for all $t > t_0$. Since the function

$$g(t) = |e^{-\alpha t} \sin kt|$$

is bounded on the interval $[0, t_0]$ it has a maximum M' there, and we may take

$$M = \max\{2, M'\}$$

and conclude that

$$|g(t)| = e^{-\alpha t}|\sin kt| < M$$

for all $t > 0$.

Hence, $f(t)$ is of exponential order α for any $\alpha > 0$. Therefore, by the existence theorem, the Laplace transform of f

$$L\{f(t)\} = \int_0^\infty e^{-st} f(t)\,dt$$

exists for all $s > 0$.

• **PROBLEM 16-3**

(a) Prove that $f(t) = e^{at} \sin kt$ is of exponential order a on $[0,\infty)$.

(b) Prove that e^{t^2} is not of any exponential order on the interval $[c,\infty)$ for an arbitrary c.

Solution: (a) The function $f(t)$ is said to be of exponential order a on $[0,\infty)$ if there exists a real constant a and a positive real constant M such that

$$e^{-at}|f(t)| < M \text{ for all } t > 0.$$

Substituting our expression for $f(t)$,

$$e^{-at}|f(t)| = e^{-at}|e^{at} \sin kt|.$$

We can remove the absolute value sign from e^{at}, since e^{at} is positive for real a. Thus,

$$e^{-at}|f(t)| = e^{-at} e^{at}|\sin kt|$$

$$= |\sin kt|.$$

But

$$|\sin kt| \leq 1$$

for all t; hence,

$$e^{at}|f(t)| \leq 1.$$

From this inequality it follows that

$$e^{-at}|f(t)| < 2$$

for all $t > 0$. (1)

Hence, taking M = 2, (1) shows that $f(t)$ is of exponential order a.

(b) It follows easily from the definition of exponential order already given that a function f(t) is of exponential order if there exists some number b such that

$$\lim_{t\to\infty} \frac{f(t)}{e^{bt}} = 0 .$$

In our case

$$f(t) = e^{t^2} .$$

Thus we examine

$$\lim_{t\to\infty} \frac{e^{t^2}}{e^{bt}} = \lim_{t\to\infty} e^{t(t-b)} .$$

For any value of b, t-b > 0 if t > b. Thus the exponent of e in $e^{t(t-b)}$ becomes and stays positive as $t\to\infty$. Thus

$$\lim_{t\to\infty} e^{t(t-b)} = \infty \neq 0$$

which shows $\exp(t^2)$ is not of any exponential order.

• **PROBLEM 16-4**

Determine those values of s for which the improper integral

$$G(s) = \int_0^\infty e^{-st} dt$$

converges, and find the Laplace transform of f(t) = 1.

Solution: By definition, the improper integral in the problem is a limit of a proper integral

$$G(s) = \lim_{R\to\infty} \int_0^R e^{-st} dt . \qquad (1)$$

For s = 0 this becomes

$$G(s) = \lim_{R\to\infty} \int_0^R e^{-(0)t} dt = \lim_{R\to\infty} \int_0^R dt$$

$$= \lim_{R\to\infty} t \Big|_{t=0}^R = \lim_{R\to\infty} R = \infty .$$

Hence, the integral G(s) diverges for s = 0. For s ≠ 0 equation (1) becomes

$$G(s) = \lim_{R \to \infty} -\left(\frac{1}{s} e^{-st}\right)\Big|_{t=0}^{R}$$

$$= \lim_{R \to \infty} \frac{1}{s}\left(1 - e^{-Rs}\right) = \frac{1}{s} - \lim_{R \to \infty} \frac{e^{-Rs}}{s}. \quad (2)$$

If $s < 0$, then $-Rs > 0$ for positive R; hence, e^{-Rs} approaches infinity as R approaches infinity, and the integral diverges.

If $s > 0$, then $-Rs < 0$ for positive R; hence, e^{-Rs} approaches zero as R approaches infinity, and the integral converges to 1/s.

Extending the domain of the Laplace transform to complex values of s, we evaluate the expression of e^{-Rs} by Euler's formula:

$$e^{-Rs} = e^{-R(\mathrm{Re}\{s\}) - iR(\mathrm{Im}\{s\})}$$

$$= e^{-R(\mathrm{Re}\{s\})}\left[\cos[-R(\mathrm{Im}\{s\})] + i\sin[-R(\mathrm{Im}\{s\})]\right], \quad (3)$$

where Re{s} is the real part of s, Im{s} is the imaginary part of s, and $i \equiv \sqrt{-1}$ is the imaginary constant. The cosine and sine functions are bounded; hence expression (3) diverges, as $R \to \infty$, for Re{s} < 0 and converges to zero for Re{s} > 0. In the case where $s \neq 0$ and Re{s} = 0, we have

$$e^{-R(\mathrm{Re}\{s\})} = 1;$$

hence,

$$e^{-Rs} = \cos[-R(\mathrm{Im}\{s\})] + i\sin[-R(\mathrm{Im}\{s\})],$$

which is a nonconstant periodic function, so e^{-Rs} does not converge to any value as $R \to \infty$.

Since e^{-Rs} is the only expression that varies with R in equality (2), its convergence properties for $R \to \infty$ determine the convergence properties of G(s). Thus, in general, G(s) converges to 1/s for Re{s} > 0 and diverges otherwise.

Using L as the Laplace transform operator, with

$$L\{f(t)\} = F(s) \equiv \int_0^\infty e^{-st} f(t)\, dt,$$

we take the Laplace transform of f(t) for $t \geq 0$:

$$L\{1\} = \int_0^\infty (e^{-st} \cdot 1)\,dt = \int_0^\infty e^{-st}\,dt ,$$

which is the same integral as G(s) in the problem. As we already showed, this integral converges to 1/s when Re{s} > 0 and diverges otherwise. Thus, the required Laplace transform is

$$L\{1\} = \frac{1}{s}, \text{ for } Re\{s\} > 0 .$$

Note that the function f(t) of which we take a Laplace transform needs only be defined for positive real values of its argument t since the integral,

$$\int_0^\infty e^{-st} f(t)\,dt ,$$

is in a region in which $t \geq 0$.

• **PROBLEM 16-5**

Find the Laplace transform of

$$f(t) = t^n ,$$

where n is a positive integer.

Solution: Using L as the Laplace transform operator with

$$L\{f(t)\} = \int_0^\infty e^{-st} f(t)\,dt ,$$

and first considering real s, we find

$$L\{t^n\} = \int_0^\infty (e^{-st})(t^n)\,dt = \lim_{R\to\infty} \int_0^R t^n e^{-st}\,dt .$$

For s = 0,

$$L\{t^n\} = \lim_{R\to\infty} \int_0^R t^n e^{0t}\,dt = \lim_{R\to\infty} \int_0^R t^n\,dt$$

$$= \lim_{R\to\infty} \frac{R^{n+1}}{n+1} = \infty ;$$

hence,

$$L\{t^n\}$$

does not exist for $s = 0$. For $s \neq 0$, integrating by parts,

$$L\{t^n\} = \lim_{R \to \infty} \left\{ \left. \frac{-t^n e^{-st}}{s} \right|_{t=0}^{R} + \frac{n}{s} \int_0^R e^{-st} t^{n-1} \, dt \right\}$$

$$= \lim_{R \to \infty} \left\{ \frac{-R^n e^{-sR}}{s} + \frac{n}{s} \int_0^R e^{-st} t^{n-1} \, dt \right\}$$

$$= \lim_{R \to \infty} \left(\frac{-R^n e^{-sR}}{s} \right) + \frac{n}{s} \lim_{R \to \infty} \left(\int_0^R e^{-st} t^{n-1} \, dt \right)$$

$$= \lim_{R \to \infty} \left(\frac{-R^n e^{-sR}}{s} \right) + \frac{n}{s} L\{t^{n-1}\} . \tag{1}$$

For $s \leq 0$ the argument of the limit in expression (1) diverges as $R \to \infty$; hence,

$$L\{t^n\}$$

does not exist. For $s > 0$ rewrite (1) as

$$L\{t^n\} = \lim_{R \to \infty} \left(\frac{-R^n}{se^{sR}} \right) + \frac{n}{s} L\{t^{n-1}\} . \tag{2}$$

Since both the numerator and denominator in the argument of the limit in equation (2) approach ∞ as $R \to \infty$, we can apply L'Hospital's rule:

$$\lim_{R \to \infty} \left(\frac{-R^n}{se^{sR}} \right) = \lim_{R \to \infty} \left[\frac{d/dR(-R^n)}{d/dR(se^{sR})} \right] = \lim_{R \to \infty} \left(\frac{-nR^{n-1}}{s^2 e^{sR}} \right) \tag{3}$$

As long as the numerator and denominator of the argument of our limit approach ∞ as $R \to \infty$, we can iteratively apply L'Hospital's rule to equality (3):

$$\lim_{R \to \infty} \left(\frac{-R^n}{se^{sR}} \right) = \lim_{R \to \infty} \left(\frac{-nR^{n-1}}{s^2 e^{sR}} \right) = \lim_{R \to \infty} \left[\frac{-n(n-1)R^{n-2}}{s^3 e^{sR}} \right]$$

$$= \ldots = \lim_{R \to \infty} \left[\frac{-n(n-1)(n-2)\ldots(2)(1)R^{n-n}}{s^{n+1} e^{sR}} \right]$$

$$= (-1) \lim_{R \to \infty} \left[\frac{n!}{s^{n+1} e^{sR}} \right] = 0.$$

Substituting this result in equation (2),

$$L\{t^n\} = \frac{n}{s} L\{t^{n-1}\} \quad \text{for } s > 0. \tag{4}$$

Substituting (n-1) for n in equation (4), we find

$$L\{t^{n-1}\} = \frac{n-1}{s} L\{t^{n-2}\}.$$

Substituting this result back into equation (4),

$$L\{t^n\} = \frac{n}{s}\left(\frac{n-1}{s} L\{t^{n-2}\}\right) = \frac{n(n-1)}{s^2} L\{t^{n-2}\}.$$

By iterating this process, we obtain

$$L\{t^n\} = \frac{n(n-1)(n-2)\ldots(2)(1)}{s^n} L\{t^0\} = \frac{n!}{s^n} L\{1\}$$

$$= \frac{n!}{s^n} \lim_{R \to \infty} \int_0^R e^{-st}(1) dt,$$

which converges, since $s > 0$, to

$$L\{t^n\} = \frac{n!}{s^n} \lim_{R \to \infty} \left(\frac{1 - e^{-sR}}{s}\right) = \frac{n!}{s^n} \cdot \frac{1}{s}$$

$$= \frac{n!}{s^{n+1}}, \quad \text{for } s > 0. \tag{5}$$

If s is complex, a similar computation will give the same result, (5), except that the condition on s will now be Re s > 0.

● **PROBLEM 16-6**

Find the Laplace transforms of

(a) $f(t) = e^{kt}$,

where k is a complex constant of the form

$$k = Re\{k\} + i\, Im\{k\}$$

with Re{k} the real part of k, Im{k} the imaginary part of k, and

$i \equiv \sqrt{-1}$.

Use this Laplace transform to find the Laplace transforms of

$$f(t) = e^{-kt} \quad \text{and} \quad f(t) = 1.$$

(b) $f(t) = \sin kt$ where k is a real constant.

Solution: (a) Using L as the Laplace transform operator with

$$L\{f(t)\} = \int_0^\infty e^{-st} f(t) dt,$$

for complex s of the form $s = \text{Re}\{s\} + i\text{Im}\{s\}$, where $\text{Re}\{s\}$ is the real part of s and $\text{Im}\{s\}$ is the imaginary part of s,

$$L\{e^{kt}\} = \int_0^\infty e^{-st}(e^{kt}) dt = \lim_{R\to\infty} \int_0^R e^{(k-s)t} dt.$$

We chose to solve the complex case in this problem because complex numbers are necessary when we determine

$$L\{\cos kt\} \quad \text{and} \quad L\{\sin kt\}$$

from $L\{e^{kt}\}$.

For $s = k$,

$$L\{e^{kt}\} = \lim_{R\to\infty} \int_0^R e^{(k-k)t} dt = \lim_{R\to\infty} \int_0^R dt$$

$$= \lim_{R\to\infty} \left(t \Big|_{t=0}^R \right) = \lim_{R\to\infty} R = \infty;$$

hence

$$L\{e^{kt}\}$$

does not exist for $s = k$. For $s \neq k$,

$$L\{e^{kt}\} = \lim_{R\to\infty} \int_0^R e^{(k-s)t} dt$$

$$= \lim_{R\to\infty} \left[\frac{e^{(k-s)t}}{k-s} \Big|_{t=0}^R \right]$$

$$= \lim_{R\to\infty} \left[\frac{e^{(k-s)R} - 1}{k-s} \right]$$

800

$$= \frac{1}{s-k} + \frac{1}{k-s} \lim_{R\to\infty} e^{(k-s)R},$$

which diverges for

$$\text{Re}\{s\} \leq \text{Re}\{k\}$$

and converges to

$$\frac{1}{s-k}$$

for

$$\text{Re}\{s\} > \text{Re}\{k\}.$$

Thus, the Laplace transform of e^{kt} (for $t > 0$) is

$$L\{e^{kt}\} = \frac{1}{s-k} \quad \text{for Re}\{s\} > \text{Re}\{k\}. \tag{1}$$

Using the constant $(-k)$ in place of k in formula (1), we obtain a new formula:

$$L\{e^{-kt}\} = \frac{1}{s-(-k)} = \frac{1}{s+k}$$

for

$$\text{Re}\{s\} > -\text{Re}\{k\}.$$

If the special case $k = 0$ is used, equation (1) gives

$$L\{1\} = L\{e^{0t}\} = \frac{1}{s-0} = \frac{1}{s}, \quad \text{Re}\{s\} > 0.$$

(b) Using L as the Laplace transform operator with

$$L\{f(t)\} = \int_0^\infty e^{-st} f(t)\, dt,$$

and first considering real s,

$$L\{\sin kt\} = \int_0^\infty (e^{-st})(\sin kt)\, dt = \lim_{R\to\infty} \int_0^R e^{-st} \sin kt\, dt.$$

In the above equality, we substitute the exponential formula for the sine function:

$$\sin kt = \frac{e^{ikt} - e^{-ikt}}{2i},$$

where $i \equiv \sqrt{-1}$, so that

$$L\{\sin kt\} = \lim_{R\to\infty} \int_0^R e^{-st}\left(\frac{e^{ikt} - e^{-ikt}}{2i}\right) dt$$

$$= \frac{1}{2i} \lim_{R\to\infty} \left[\int_0^R e^{(ik-s)t} dt - \int_0^R e^{(-ik-s)t} dt\right].$$

(2)

For $k = 0$ equation (2) gives us

$$L\{\sin kt\} = \frac{1}{2i} \lim_{R\to\infty} \left[\int_0^R e^{(0-s)t} dt - \int_0^R e^{(0-s)t} dt\right]$$

$$= \frac{1}{2i} \lim_{R\to\infty} [0] = 0.$$

When $k \neq 0$, s cannot equal $\pm ik$, since $\pm ik$ is nonzero imaginary; thus, equation (2) gives us

$$L\{\sin kt\} = \frac{1}{2i} \lim_{R\to\infty} \left[\frac{e^{(ik-s)R}}{ik-s} - \frac{1}{ik-s} + \frac{e^{(-ik-s)R}}{ik+s} - \frac{1}{ik+s}\right]$$

$$= \frac{1}{2i} \lim_{R\to\infty} \left[\frac{(ik+s)e^{(ik-s)R} - (ik+s) + (ik-s)e^{(-ik-s)R} - (ik-s)}{(ik-s)(ik+s)}\right]$$

$$= \frac{1}{2i} \lim_{R\to\infty} \left[\frac{(ik+s)e^{(ik-s)R} + (ik-s)e^{(-ik-s)R} - 2ik}{-s^2 - k^2}\right]$$

$$= \frac{k}{s^2+k^2} + \frac{1}{2i} \lim_{R\to\infty} \left[\frac{(ik+s)e^{(ik-s)R} + (ik-s)e^{(-ik-s)R}}{-s^2 - k^2}\right]. \quad (3)$$

When the real part of $(ik-s)$ is negative (i.e., when $-s < 0$) and the real part of $(-ik-s)$ is negative (i.e., when $-s < 0$), the argument of the limit in expression (3) approaches zero as $R\to\infty$; hence

$$L\{\sin kt\} = k/(s^2 + k^2) \text{ for } s > 0.$$

Otherwise, the argument of the limit diverges, and the Laplace transform of sin kt does not exist.

If s is complex, the result

$$L\{\sin kt\} = \frac{k}{s^2 + k^2}$$

still holds, but the condition on s becomes Re s > 0.

BASIC PROPERTIES OF LAPLACE TRANSFORMS

• **PROBLEM 16-7**

Prove the following properties of the Laplace transform denoted by $L\{f(t)\}$

(a) $L\{c_1 f_1(t) + c_2 f_2(t) + \ldots + c_n f_n(t)\} = c_1 L\{f_1(t)\}$

$\qquad + c_2 L\{f_2(t)\} + \ldots + c_n L\{f_n(t)\}$,

where all c_j are constants.

(b) $L\{f^{(n)}(t)\} = s^n L\{f(t)\} - \sum_{k=1}^{n} s^{k-1} f^{(n-k)}(0)$

if $f^{(k)}(t)$ are of some finite exponential orders for $k = 1, 2, \ldots, n-1$ and if

$$L\{f^{(n)}(t)\}$$

exists.

(c) $L\{e^{-at} f(t)\} = G(s+a)$

where $G(s) = L\{f(t)\}$ and a is a real constant.

(d)
$$L\{t^n f(t)\} = (-1)^n \frac{d^n F}{ds^n}$$

where $F(s) = L\{f(t)\}$.

(e)
$$L\{\tfrac{1}{t} f(t)\} = \int_s^\infty F(\sigma) d\sigma$$

where

$\qquad F(s) = L\{f(t)\}$.

Solution: (a) This is called the linearity property and is the defining characteristic of the so-called linear operators of which differentiation, integration, and all integral operators are examples. To prove this property, simply use the definition of

$\qquad L\{f(t)\}$ for $f(t) = c_1 f_1(t) + \ldots + c_n f_n(t)$

to calculate

$$L\{f(t)\} = L\{c_1 f_1(t) + \ldots + c_n f_n(t)\}$$

$$= \int_0^\infty e^{-st}\left[c_1 f_1(t) + \ldots + c_n f_n(t)\right]dt$$

$$= c_1 \int_0^\infty e^{-st} f_1(t)\,dt + \ldots + c_n \int_0^\infty e^{-st} f_n(t)\,dt$$

$$= c_1 L\{f_1(t)\} + c_2 L\{f_2(t)\} + \ldots + c_n L\{f_n(t)\}.$$

Thus we conclude that L is a linear operator.

(b) This is the derivative property of Laplace transforms, and it is this property which makes Laplace transforms useful in solving differential equations and in many other applications. The property will be proven by induction. Thus, suppose $f(t)$ is continuous and integrate the Laplace integral of f by parts:

$$\int_0^\infty e^{-st} f(t)\,dt = \left(\frac{-1}{s}\right) e^{-st} f(t)\Big|_0^\infty + \frac{1}{s}\int_0^\infty e^{-st} f'(t)\,dt$$

or

(if $\lim_{R\to\infty} e^{-sR} f(R) = 0$)

$$s\int_0^\infty e^{-st} f(t)\,dt = f(0) + \int_0^\infty e^{-st} f'(t)\,dt.$$

Assuming that $L\{f'(t)\}$ exists, this yields

$$L\{f'(t)\} = s\,L\{f(t)\} - f(0). \tag{1}$$

Now assume that the derivative property holds for an integer k, i.e., assume that

$$L\{f^{(k)}(t)\} = s^k L\{f(t)\} - \sum_{i=1}^{k} s^{i-1} f^{(k-i)}(0).$$

Then (1) can be used to show that

$$L\{f^{(k+1)}(t)\} = L\{\left(f^{(k)}(t)\right)'\}$$

$$= s\,L\{f^{(k)}(t)\} - f^{(k)}(0)$$

$$= s\left[s^k L\{f(t)\} - \sum_{i=1}^{k} s^{i-1} f^{(k-i)}(0)\right] - f^{(k)}(0)$$

$$= s^{k+1} L\{f(t)\} - \sum_{i=1}^{k+1} s^{i-1} f^{(k+1-i)}(0) \, . \qquad (2)$$

Thus in (1) and (2) it has been shown that the derivative property holds for n=1 and that if it holds for n=k then it holds for n = k + 1. Therefore, by induction we conclude that the derivative property,

$$L\{f^{(n)}(t)\} = s^n L\{f(t)\} - \sum_{k=1}^{n} s^{k-1} f^{n-k}(0)$$

holds for all n provided that, for k = 0, 1, ..., n-1,

$$\lim_{R \to \infty} f^{(k)}(R) e^{-sR} = 0$$

for s large enough (that is, $f^{(k)}(t)$ are of finite exponential orders), and that

$$L\{f^{(n)}(t)\}$$

exists for s large enough.

(c) This is known as the attenuation property or substitution property: If f(t) is "attenuated" by the exponential factor e^{-at}, then the transform is shifted (to the left) with respect to the variable s. To prove this, just recall the definition of the Laplace transform of a function f(t),

$$L\{f(t)\} = G(s) = \int_0^\infty e^{-st} f(t) dt \, ,$$

and compute

$$L\{e^{-at} f(t)\} = \int_0^\infty e^{-st} e^{-at} f(t) dt$$

$$= \int_0^\infty e^{-(s+a)t} f(t) dt \, .$$

Setting r = (s+a) yields

$$\int_0^\infty e^{-(s+a)t} f(t) dt = G(r) = G(s+a)$$

so that

$$L\{e^{-at} f(t)\} = G(s+a).$$

Note that if the region of validity of $L\{f(t)\}$ is $\text{Re}\{s\} > \alpha$ for some α, then the region of validity of $L\{e^{-at} f(t)\}$ is $\text{Re}\{s\} > \alpha - a$. In this case (and almost all cases of "physical" functions) α is the smallest number such that f is of exponential order α.

(d) The expression

$$F(s) = \int_0^\infty e^{-st} f(t)\,dt \tag{3}$$

is a uniformly convergent integral for a piecewise continuous function f and a suitable range of values of s (see problem 1 of this chapter). Therefore, interchange of differentiation and integration is allowed so that

$$\frac{dF(s)}{ds} = -\int_0^\infty t e^{-st} f(t)\,dt$$

or

$$\frac{dF}{ds} = -L\{tf(t)\}.$$

This formula can be generalized to find the nth derivative (provided $L\{t^n f(t)\}$ exists). Thus,

$$\frac{d^n F(s)}{ds^n} = (-1)^n \int_0^\infty t^n e^{-st} f(t)\,dt$$

or

$$L\{t^n f(t)\} = (-1)^n \frac{d^n F(s)}{ds^n},$$

and this will have the same region of validity, $s > \alpha$ for some α, as (3).

(e) Let $F(s) = L\{f(t)\}$ and consider

$$G(s) = \int_s^\infty F(\sigma)\,d\sigma \tag{4}$$

By definition of $F(s)$,

$$G(s) = \int_s^\infty d\sigma \int_0^\infty e^{-\sigma t} f(t) dt .$$

Now, since F is uniformly convergent for all Re{s} ≥ α' for any α' > α (where Re{s} > α is the region of convergence of L{f(t)}), the order of integration may be interchanged to yield

$$G(s) = \int_0^\infty f(t) dt \int_s^\infty e^{-\sigma t} d\sigma = \int_0^\infty f(t) \frac{e^{-st}}{t} dt$$

$$= L\left\{\frac{1}{t} f(t)\right\} .$$

Therefore, recalling the definition of G(s) in (4),

$$\int_s^\infty F(\sigma) d\sigma = L\left\{\frac{1}{t} f(t)\right\}$$

provided that the transform of $\frac{1}{t} f(t)$ exists.

• **PROBLEM 16-8**

Use the Laplace transform of

$$f(t) = e^{kt} , \qquad (1)$$

where k is a complex constant of the form

$$k = \text{Re}\{k\} + i \, \text{Im}\{k\}$$

with Re{k} the real part of k, Im{k} the imaginary part of k, and

$$i \equiv \sqrt{-1} ,$$

to find the Laplace transforms of

$$f(t) = \cosh kt, \sinh kt, \cos kt, \text{ and } \sin kt .$$

<u>Solution</u>: It was found in problem 6 that

$$L\{e^{kt}\} = \frac{1}{s-k} \quad \text{for} \quad \text{Re}\{s\} > \text{Re}\{k\} . \qquad (2)$$

This result could also be looked up in a table of Laplace transforms. In either case, we use the definitions

$$\cosh kt \equiv \frac{e^{kt} + e^{-kt}}{2},$$

and

$$\sinh kt \equiv \frac{e^{kt} - e^{-kt}}{2},$$

and the additional formula

$$L\{c_1 f_1(t) + c_2 f_2(t)\} = c_1 L\{f_1(t)\} + c_2 L\{f_2(t)\}, \qquad (3)$$

to find

$$L\{\cosh kt\} = L\left\{\frac{e^{kt} + e^{-kt}}{2}\right\}$$

$$= \frac{1}{2}\left(L\{e^{kt}\} + L\{e^{-kt}\}\right)$$

$$= \frac{1}{2}\left(\frac{1}{s-k} + \frac{1}{s+k}\right) = \frac{1}{2}\left(\frac{2s}{s^2-k^2}\right)$$

$$= \frac{s}{s^2-k^2}, \quad \text{for} \quad \operatorname{Re}\{s\} > |\operatorname{Re}\{k\}|, \qquad (4)$$

and

$$L\{\sinh kt\} = L\left\{\frac{e^{kt} - e^{-kt}}{2}\right\}$$

$$= \frac{1}{2}\left(L\{e^{kt}\} - L\{e^{-kt}\}\right)$$

$$= \frac{1}{2}\left(\frac{1}{s-k} - \frac{1}{s+k}\right) = \frac{1}{2}\left(\frac{2k}{s^2-k^2}\right)$$

$$= \frac{k}{s^2-k^2}, \quad \text{for} \quad \operatorname{Re}\{s\} > |\operatorname{Re}\{k\}|. \qquad (5)$$

The condition $\operatorname{Re}\{s\} > |\operatorname{Re}\{k\}|$ in formulas (4) and (5) comes from the fact that we derived those formulas for $L\{e^{kt}\}$ and $L\{e^{-kt}\}$, which require $\operatorname{Re}\{s\} > \operatorname{Re}\{k\}$ and $\operatorname{Re}\{s\} > -\operatorname{Re}\{k\}$, respectively. To insure that both $\operatorname{Re}\{s\} > \operatorname{Re}\{k\}$ and $\operatorname{Re}\{s\} > -\operatorname{Re}\{k\}$, it is necessary that $\operatorname{Re}\{s\}$ be greater than the greater of $\operatorname{Re}\{k\}$ and $-\operatorname{Re}\{k\}$. Since one of these ($\operatorname{Re}\{k\}$ or $-\operatorname{Re}\{k\}$) must be positive and the other negative, the greater of the two is the positive one, which is equal to the absolute value of $\operatorname{Re}\{k\}$.

Using the exponential formulas for the cosine and sine functions

$$\cos kt = \frac{e^{ikt} + e^{-ikt}}{2},$$

and

$$\sin kt = \frac{e^{ikt} - e^{-ikt}}{2i},$$

and again the addition formula (3), we find

$$L\{\cos kt\} = L\left\{\frac{e^{(ik)t} + e^{-(ik)t}}{2}\right\}$$

$$= \tfrac{1}{2}(L\{e^{(ik)t}\} + L\{e^{-(ik)t}\}),$$

and a similar expression holds for $L\{\sin kt\}$.

By substituting (ik) for k in formulas (1) and (2),

$$L\{\cos kt\} = \tfrac{1}{2}\left(\frac{1}{s-ik} + \frac{1}{s+ik}\right)$$

$$= \tfrac{1}{2}\left(\frac{2s}{s^2+k^2}\right)$$

$$= \frac{s}{s^2+k^2}, \qquad (6)$$

and

$$L\{\sin kt\} = L\left\{\frac{e^{(ik)t} - e^{-(ik)t}}{2i}\right\}$$

$$= \tfrac{1}{2i}(L\{e^{(ik)t}\} - L\{e^{-(ik)t}\})$$

$$= \tfrac{1}{2i}\left(\frac{1}{s-ik} - \frac{1}{s+ik}\right) = \tfrac{1}{2i}\left(\frac{2ik}{s^2+k^2}\right)$$

$$= \frac{k}{s^2+k^2}. \qquad (7)$$

Laplace transforms (6) and (7) are both subject to the same two existence conditions from the Laplace transforms of e^{kt} and e^{-kt} (which were the base of (6) and (7)). Since we used ik instead of k, the conditions are

$$\text{Re}\{s\} > \text{Re}\{ik\} \quad ,$$

and

$$\text{Re}\{s\} > -\text{Re}\{ik\} \quad .$$

Combining these two conditions as we did for the cosh and sinh Laplace transforms,

$$\text{Re}\{s\} > |\text{Re}\{ik\}| \quad .$$

But

$$|\text{Re}\{ik\}| = |\text{Re}\{i(\text{Re}\{k\} + i\text{Im}\{k\})\}|$$

$$= |\text{Re}\{-\text{Im}\{k\} + i\text{Re}\{k\}\}|$$

$$= |-\text{Im}\{k\}| = |\text{Im}\{k\}| \quad ;$$

hence, the condition for the existence of Laplace transforms (6) and (7) is

$$\text{Re}\{s\} > |\text{Im}\{k\}| \quad ,$$

which, for s and k real, is equivalent to s > 0 .

● **PROBLEM 16-9**

Find the Laplace transform, $L\{f(t)\} = F(s)$, of

(a) $f(t) = 2 \sin t + 3 \cos 2t$

(b) $g(t) = \dfrac{1 - e^{-t}}{t}$.

Solution: (a) We shall use the addition formula

$$L\{c_1 f_1(t) + c_2 f_2(t)\} = c_1 L\{f_1(t)\} + c_2 L\{f_2(t)\} \quad , \tag{1}$$

where c_1 and c_2 are constants, and the formulas which were derived in the previous problems:

$$L\{\sin kt\} = \frac{k}{s^2+k^2} \quad (\text{for } s > 0) \tag{2}$$

$$L\{\cos kt\} = \frac{s}{s^2+k^2} \quad (\text{for } s > 0) \quad , \tag{3}$$

where k is a real constant and s is a real variable.

By formula (1),

$$L\{2 \sin t + 3 \cos 2t\} = 2L\{\sin t\} + 3L\{\cos 2t\}.$$

Applying formulas (2) and (3) to the above equality,

$$L\{2 \sin t + 3 \cos 2t\} = 2\left(\frac{1}{s^2+1^2}\right) + 3\left(\frac{s}{s^2+2^2}\right)$$

$$= \frac{2}{s^2+1} + \frac{3s}{s^2+4}, \text{ for } s > 0.$$

(b) Expanding e^{-t} as an infinite series,

$$e^{-t} = \sum_{n=0}^{\infty} \frac{(-1)^n t^n}{n!};$$

hence,

$$\frac{1 - e^{-t}}{t} = \frac{1 - \left[\sum_{n=0}^{\infty} \frac{(-1)^n t^n}{n!}\right]}{t}$$

$$= \frac{1 - \left[1 + \sum_{n=1}^{\infty} \frac{(-1)^n t^n}{n!}\right]}{t}$$

$$= -\frac{1}{t} \sum_{n=1}^{\infty} \frac{(-1)^n t^n}{n!} = \sum_{n=1}^{\infty} \frac{(-1)^{n-1} t^{n-1}}{n!}.$$

Letting $k = n-1$ in the summation, this becomes

$$\frac{1 - e^{-t}}{t} = \sum_{k=0}^{\infty} \frac{(-1)^k t^k}{(k+1)!};$$

thus

$$L\left\{\frac{1 - e^{-t}}{t}\right\} = L\left\{\sum_{k=0}^{\infty} \frac{(-1)^k t^k}{(k+1)!}\right\} \tag{4}$$

Since L is a linear operator (i.e., since

$$L\{c_1 f_1(t) + c_2 f_2(t) + \ldots\} = c_1 L\{f_1(t)\}$$
$$+ c_2 L\{f_2(t)\} + \ldots, \; c_j = \text{constant}),$$

equality (4) becomes

$$L\left\{\frac{1 - e^{-t}}{t}\right\} = \sum_{k=0}^{\infty} \frac{(-1)^k}{(k+1)!} L\{t^k\}. \tag{5}$$

From the results of a previous problem, or from a table of Laplace transforms, we find that

$$L\{t^k\} = \frac{k!}{s^{k+1}}, \quad s > 0,$$

where k is a nonnegative integer. Substituting this result into equality (5), we obtain

$$L\left\{\frac{1 - e^{-t}}{t}\right\} = \sum_{k=0}^{\infty} \frac{(-1)^k}{(k+1)!}\left(\frac{k!}{s^{k+1}}\right)$$

$$= \sum_{k=0}^{\infty} \frac{(-1)^k}{(k+1)}\left(\frac{1}{s}\right)^{k+1}.$$

This last summation is the infinite series for the natural logarithm of $(1 + 1/s)$, where $|1/s| < 1$, i.e., where $|s| > 1$. Thus,

$$L\left\{\frac{1 - e^{-t}}{t}\right\} = \log\left(1 + \frac{1}{s}\right).$$

An alternative method of solving this problem is to use the formula

$$L\left\{\frac{1}{t} f(t)\right\} = \int_{s}^{\infty} F(\sigma) d\sigma,$$

where $F(s) = L\{f(t)\}$ (we obtained it in problem 7). Here

$$f(t) = 1 - e^{-t},$$

and

$$L\{1 - e^{-t}\} = L\{1\} - L\{e^{-t}\} = \frac{1}{s} - \frac{1}{s+1}$$

since

$$L\{e^{kt}\} = \frac{1}{s-k} \quad \text{(for } k = 0, e^{kt} \equiv 1\text{).}$$

Then

$$L\left\{\frac{1 - e^{-t}}{t}\right\} = \int_{s}^{\infty}\left(\frac{1}{\sigma} - \frac{1}{\sigma+1}\right) d\sigma$$

$$= \left(\log |\sigma| - \log |\sigma + 1|\right)\Big|_{s}^{\infty}.$$

$$= \log \left| \frac{\sigma}{\sigma+1} \right| \Big|_s^\infty = \log 1 - \log \frac{s}{s+1}$$

$$= \log \frac{s+1}{s} \quad \text{(we have } s > 0\text{)}.$$

In fact, the condition $|s| > 1$, or $s > 1$ (since $s > 0$ for $L\{t^k\}$ to exist) is too restrictive, and the second solution shows that for any $s > 0$

$$L\left\{\frac{1 - e^{-t}}{t}\right\}$$

exists.

● **PROBLEM 16-10**

Use the derivative property of Laplace transforms to solve the differential equation

$$y' - y = e^{-x} \tag{1}$$

where $y(0) = 0$ is the initial value of y.

Solution: First multiply equation (1) by e^{-sx} and integrate from 0 to ∞ to get

$$\int_0^\infty e^{-sx} y'(x)\,dx - \int_0^\infty e^{-sx} y(x)\,dx = \int_0^\infty e^{-sx} \cdot e^{-x}\,dx \tag{2}$$

or

$$L\{y'(x)\} - L\{y(x)\} = L\{e^{-x}\}. \tag{3}$$

Now the fact that $L\{e^{-ax}\} = \frac{1}{s+a}$ can be obtained from a previous problem in this chapter or can be looked up on a table of Laplace transforms and is valid for $s > a$. Also, the derivative property states that

$$L\{y'(x)\} = sL\{y(x)\} - y(0), \quad \text{so that}$$

denoting

$$Y(s) = L\{y(x)\}$$

and using (3), one obtains

$$[s Y(s) - y(0)] - Y(s) = \frac{1}{s+1}$$

or since

$$y(0) = 0, \quad (s-1)\,Y(s) = \frac{1}{s+1},$$

and

$$Y(s) = \frac{1}{(s+1)(s-1)} = \frac{1}{2}\frac{1}{s-1} - \frac{1}{2}\frac{1}{s+1}, \qquad (4)$$

where we have used rational fraction decomposition of

$$\frac{1}{(s+1)(s-1)}.$$

The equality in (4) is valid only for $s > 1$, but this will turn out to be insignificant since we wish to invert (4) anyway. Thus, recall that

$$L\{e^x\} = \frac{1}{s-1}, \quad L\{e^{-x}\} = \frac{1}{s+1}$$

so that (4) reads

$$L\{y(x)\} = \frac{1}{2}L\{e^x\} - \frac{1}{2}L\{e^{-x}\}$$

$$= L\left\{\frac{1}{2}e^x - \frac{1}{2}e^{-x}\right\} \qquad (5)$$

since L is a linear operator. Thus $y(x)$ and

$$\frac{1}{2}e^x - \frac{1}{2}e^{-x}$$

have the same Laplace transforms. As will be discussed later, under very general conditions, if two functions have the same Laplace transform, they are identical. This is not true for all functions, but is true for a class of almost all functions of any applicability, and we assume $y(x)$ and

$$\frac{1}{2}e^x - \frac{1}{2}e^{-x}$$

to be such functions. Thus (5) implies that

$$y(x) = \frac{1}{2}e^x - \frac{1}{2}e^{-x} = \sinh x$$

is the solution to the problem. Questions concerning the validity of our "inversion" of the Laplace transform will be discussed more rigorously later in the chapter.

• **PROBLEM 16-11**

Find the Laplace transforms of

(a) $\quad g(t) = e^{-2t} \sin 5t,$

(b) $\quad h(t) = e^{-t}\, t \cos 2t.$

Solution: (a) We shall use the formula

$$L\{\sin kt\} = \frac{k}{s^2+k^2}, \qquad (1)$$

where k is a real constant and s is a real variable. We also use the theorem that states that if f(t) is defined for all nonnegative t, is piecewise continuous on every closed interval [0,b] for b > 0, and is of exponential order e^{at}, then

$$L\{e^{at} f(t)\} = F(s-a), \qquad (2)$$

for $s > \alpha + a$, where a is a real constant and

$$F(s) = L\{f(t)\}.$$

Using formula (1), with f(t) = sin 5t, we find

$$F(s) = L\{\sin 5t\} = \frac{5}{s^2+5^2} = \frac{5}{s^2+25}.$$

In order to use the theorem associated with formula (2), we must first demonstrate that the function f(t) = sin 5t is of exponential order; i.e., there exists a real constant α and positive real constants M and t_0 such that

$$e^{-\alpha t}|f(t)| < M$$

for all $t > t_0$. For $\alpha > 0$,

$$\lim_{t \to \infty} e^{-\alpha t} = 0,$$

and for $\alpha = 0$,

$$\lim_{t \to \infty} e^{-\alpha t} = 1.$$

Thus, for $\alpha \geq 0$,

$$\lim_{t \to \infty} e^{-\alpha t} < 2. \qquad (3)$$

Also note that

$$|\sin 5t| \leq 1, \qquad (4)$$

for all t. Multiplying inequalities (3) and (4) together, we obtain

$$\left(\lim_{t \to \infty} e^{-\alpha t}\right)|\sin 5t| < 2,$$

for $\alpha \geq 0$. It follows from the definition of a limit that there exists a positive constant t_0 such that, for $t > t_0$,

$$e^{-\alpha t}|\sin 5t| < 2,$$

which, taking $M = 2$, shows that $f(t) = \sin 5t$ is of exponential order $e^{\alpha t}$, for $\alpha \geq 0$.

Since $f(t) = \sin 5t$ is defined for all nonnegative t, is piecewise continuous on every closed interval $[0,b]$ for $b > 0$, and is of exponential order $e^{\alpha t}$ for $\alpha \geq 0$, we can use formula (1) in formula (2);

$$L\{g(t)\} = L\{e^{-2t} f(t)\}$$

$$= F(s-a)$$

$$= \frac{5}{(s+2)^2 + 25}, \text{ for } s > -2.$$

(b) Using the property that

$$L\{e^{bt} f(t)\} = F(s-b),$$

where b is a constant and $F(s) = L\{f(t)\}$, provided that $L\{f(t)\}$ exists, we obtain

$$L\{e^{-t} t \cos 2t\} = G(s+1), \qquad (5)$$

where $G(s) = L\{t \cos 2t\}$.

We could find, in a table of Laplace transforms, that

$$L\{t \cos kt\} = \frac{s^2 - k^2}{(s^2 + k^2)^2}, \ s > 0.$$

However, this result may be established easily without consulting such a table. Using the fact that

$$L\{\cos kt\} = \frac{s}{s^2 + k^2}$$

which was established in a previous problem and the theorem which states that

$$L\{t^n f(t)\} = (-1)^n \frac{d^n}{ds^n} L\{f(t)\},$$

it is apparent that

$$L\{t \cos kt\} = (-1)\frac{d}{ds} L\{f(t)\}$$

$$= (-1) \frac{d}{ds} \left(\frac{s}{s^2 + k^2}\right)$$

$$= \frac{s^2 - k^2}{(s^2 + k^2)^2}.$$

The region of validity of this formula is the same as that for $L\{\cos kt\}$, i.e., $s > 0$ for real s.

Hence, taking k = 2, we find

$$L\{t \cos 2t\} = \frac{s^2 - 4}{(s^2 + 4)^2}.$$

Thus,

$$G(s) = \frac{s^2 - 4}{(s^2 + 4)^2},$$

and, from equality (5),

$$L\{e^{-t} t \cos 2t\} = \frac{(s+1)^2 - 4}{[(s+1)^2 + 4]^2}.$$

● **PROBLEM 16-12**

Find the Laplace transform of

(a) $\quad g(t) = te^{4t}$

(b) $\quad f(t) = t^{7/2}$

Solution: (a) We shall use the formula

$$L\{e^{kt}\} = \frac{1}{s-k}, \text{ for } s > k, \tag{1}$$

where k is a real constant, and s is a real variable. We also use the theorem that states that if f(t) is defined for all nonnegative t, is piecewise continuous on every closed interval [0,b] for b > 0, and is of exponential order $e^{\alpha t}$, then

$$L\{t^n f(t)\} = (-1)^n \frac{d^n}{ds^n} L\{f(t)\}, \tag{2}$$

817

for $s > \alpha$, where n is a positive integer.

Using formula (1), we find that

$$L\{e^{4t}\} = \frac{1}{s-4}, \text{ for } s > 4.$$

In order to use the theorem associated with formula (2), we must first demonstrate that the function

$$f(t) = e^{4t}$$

is of exponential order; i.e., there exists a real constant α and positive real constants M and t_0 such that

$$e^{-\alpha t}|f(t)| < M \quad \text{for all } t > t_0.$$

For $\alpha = 4$,

$$\lim_{t \to \infty} e^{-\alpha t}|e^{4t}| = \lim_{t \to \infty} e^{(4-\alpha)t} = 1.$$

For $\alpha > 4$,

$$\lim_{t \to \infty} e^{-\alpha t}|e^{4t}| = \lim_{t \to \infty} e^{(4-\alpha)t} = 0.$$

Thus, for $\alpha \geq 4$,

$$\lim_{t \to \infty} e^{-\alpha t}|e^{4t}| < 2.$$

It follows from the definition of a limit that there exists a positive constant t_0 such that, for $t > t_0$,

$$e^{-\alpha t}|e^{4t}| < 2,$$

which, taking M = 2, shows that $f(t) = e^{4t}$ is of exponential order $e^{\alpha t}$, for $\alpha \geq 4$.

Since $f(t) = e^{4t}$ is defined for all nonnegative t, is piecewise continuous on every closed interval [0,b] for $b > 0$, and is of exponential order $e^{\alpha t}$ for $\alpha \geq 4$, we can substitute formula (1) for $L\{f(t)\}$ in formula (2):

$$L\{g(t)\} = L\{te^{4t}\}$$

$$= (-1)\frac{d}{ds}L\{e^{4t}\} = -\frac{d}{ds}\left(\frac{1}{s-4}\right)$$

$$= \frac{1}{(s-4)^2}, \text{ for } s > 4.$$

(b) Using the property that

$$L\{t^n f(t)\} = (-1)^n \frac{d^n}{ds^n} L\{f(t)\},$$

where n is a positive integer constant, provided that $L\{f(t)\}$ exists, we obtain

$$L\{t^{7/2}\} = L\{t^3 \sqrt{t}\}$$

$$= (-1)^3 \frac{d^3}{ds^3} L\{\sqrt{t}\}$$

$$= -\frac{d^3}{ds^3} L\{\sqrt{t}\}. \qquad (3)$$

Now $L\{\sqrt{t}\}$ may be found using the Γ function whose values are as well tabulated as, say, log x so that an answer expressible in terms of $\Gamma(x)$ is very useful. The definition of the Γ function is

$$\Gamma(k) = \int_0^\infty t^{(k-1)} e^{-t} dt$$

so that

$$\Gamma(k+1) = \int_0^\infty t^k e^{-t} dt. \qquad (4)$$

Now make the substitution $t = sx$ where s is some constant so that $dt = sdx$, $e^{-t} = e^{-sx}$, and $t^k = s^k x^k$. Then (d) becomes

$$\Gamma(k+1) = \int_0^\infty s^k x^k e^{-sx} s\, dx$$

$$= s^{k+1} \int_0^\infty x^k e^{-sx} dx \qquad (5)$$

$$= s^{k+1} L\{x^k\}$$

so that

$$L\{x^k\} = \frac{\Gamma(k+1)}{s^{k+1}}, \quad k > -1, \quad s > 0,\qquad(6)$$

where the restrictions on s and k are imposed to insure the convergence of the integrals in (4) and (5) respectively. Therefore, using

$k = \frac{1}{2}$ in (6) gives

$$L(t^{\frac{1}{2}}) = L(\sqrt{t}) = \Gamma(3/2) s^{-3/2}.$$

Now look up $\Gamma\{3/2\}$ in a table of $\Gamma(x)$ to find $\Gamma(3/2) = \sqrt{\pi}/2$ and

$$L(\sqrt{t}) = \frac{1}{2}\sqrt{\pi}\, s^{-3/2}.$$

Substituting this result in equality (3),

$$L\{t^{7/2}\} = -\frac{d^3}{ds^3}\left(\frac{1}{2}\sqrt{\pi}\, s^{-3/2}\right)$$

$$= -\frac{d^2}{ds^2}\left(-\frac{3}{2}\cdot\frac{1}{2}\sqrt{\pi}\, s^{-5/2}\right)$$

$$= -\frac{d}{ds}\left[\left(-\frac{5}{2}\right)\left(-\frac{3}{2}\right)\cdot\frac{1}{2}\sqrt{\pi}\, s^{-7/2}\right]$$

$$= -\left(-\frac{7}{2}\right)\left(-\frac{5}{2}\right)\left(-\frac{3}{2}\right)\cdot\frac{1}{2}\sqrt{\pi}\, s^{-9/2}$$

$$= \frac{105}{16}\sqrt{\pi}\, s^{-9/2}.$$

● **PROBLEM 16-13**

Find the Laplace transform

$$L\left\{\frac{\sin 3t}{t}\right\}.$$

Solution: Using the property that

$$L\left\{\frac{1}{t}f(t)\right\} = \int_s^\infty F(x)\,dx,$$

where $F(s) = L\{f(t)\}$, provided that $L\{f(t)\}$ and

$$\lim_{\substack{x \to 0 \\ x > 0}} [f(x)/x^r]$$

exist, for some $r > 0$, we obtain

$$L\left\{\frac{\sin 3t}{t}\right\} = \int_s^\infty G(x)\,dx, \qquad (1)$$

where $G(s) = L\{\sin 3t\}$. Now either recall the results of a previous problem or look in a table of Laplace transforms to find that

$$L\{\sin bt\} = \frac{b}{s^2+b^2}, \quad s > 0;$$

hence, taking $b = 3$, we find

$$G(s) = L\{\sin 3t\} = \frac{3}{s^2+3^2}.$$

Thus,

$$G(x) = \frac{3}{x^2+3^2}.$$

Substituting this result into equality (1),

$$\begin{aligned}
L\left\{\frac{\sin 3t}{t}\right\} &= \int_s^\infty \left(\frac{3}{x^2+3^2}\right) dx \\
&= \lim_{R \to \infty} \int_s^R \left(\frac{3}{x^2+3^2}\right) dx \\
&= \lim_{R \to \infty} 3 \int_s^R \frac{dx}{x^2+3^2} \\
&= \lim_{R \to \infty} 3 \left(\frac{1}{3} \operatorname{Arctan} \frac{x}{3}\right)\Bigg|_{x=s}^R \\
&= \lim_{R \to \infty} \left(\operatorname{Arctan} \frac{R}{3} - \operatorname{Arctan} \frac{s}{3}\right) \\
&= \frac{\pi}{2} - \operatorname{Arctan} \frac{s}{3},
\end{aligned}$$

where the capitalized A on the Arctangent function indicates that the range of that function is the open interval $(-\pi/2, \pi/2)$.

STEP FUNCTIONS AND PERIODIC FUNCTIONS

• **PROBLEM 16-14**

Prove that

$$L\{f(t-a)\,\alpha(t-a)\} = e^{-as} L\{f(t)\} \qquad (1)$$

for any function f which has a Laplace transform, and where

$$\alpha(t) = \begin{cases} 0, & t < 0 \\ 1, & t \geq 0 \end{cases}$$

is the unit step function, and $a > 0$. This property of Laplace transforms is called the shifting property.

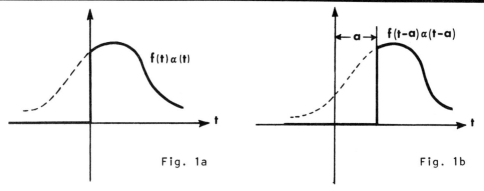

Fig. 1a   Fig. 1b

Solution: Consider the function $f(t)\alpha(t)$ shown in Figure 1a where the dotted curve represents f(t) for t < 0. Note that any function f(t) has the same Laplace transform as $f(t)\alpha(t)$ since the Laplace integral

$$F(s) = L\{f(t)\} = \int_0^\infty e^{-st} f(t) dt = \int_0^\infty e^{-st} f(t) \alpha(t) dt$$

ignores values of f for t < 0. Thus, it is often convenient to treat F(s) as the Laplace transform of $f(t)\alpha(t)$ rather than of f(t) and, in fact, (1) might be written as

$$L\{f(t-a)\alpha(t-a)\} = e^{-as} L\{f(t)\alpha(t)\} . \qquad (2)$$

It can be seen in Figure 1b that $f(t-a)\alpha(t-a)$ where

$$\alpha(t-a) = \begin{cases} 0, & t < a \\ 1, & t \geq a \end{cases}$$

represents a shifting of $f(t)\alpha(t)$ by a length a to the right. Hence (2) will prove to be a useful theorem, and it can be proven by direct calculation:

$$L\{f(t-a)\alpha(t-a)\} = \int_0^\infty e^{-st} f(t-a)\alpha(t-a) dt$$

$$= \int_0^a e^{-st} f(t-a)\alpha(t-a) dt + \int_a^\infty e^{-st} f(t-a)\alpha(t-a) dt. \quad (3)$$

Now,

$$\alpha(t-a) = \begin{cases} 0, & t < a \\ 1, & t \geq a \end{cases}$$

so that (3) becomes

$$L\{f(t-a)\alpha(t-a)\} = \int_a^\infty e^{-st} f(t-a)\alpha(t-a) dt. \quad (4)$$

Now make the change of variable $t = \tau + a$ in (4) to obtain

$$\int_a^\infty e^{-st} f(t-a)\alpha(t-a) dt = \int_0^\infty e^{-s(\tau+a)} f(\tau)\alpha(\tau) d\tau$$

$$= e^{-as} \int_0^\infty e^{-s\tau} f(\tau)\alpha(\tau) d\tau = e^{-as} L\{f(t)\}$$

or

$$L\{f(t-a)\alpha(t-a)\} = e^{-as} L\{f(t)\}.$$

● **PROBLEM 16-15**

Find the Laplace transform $L\{g(t)\}$, where

$$g(t) = \begin{cases} 0, & t < 4 \\ (t-4)^2, & t \geq 4 \end{cases}$$

Solution: The function $g(t)$ can be expressed as $(t-4)^2 \times \alpha(t-4)$, where α is the unit step function, defined as follows:

$$\alpha(x) = \begin{cases} 0, & x < 0 \\ 1, & x \geq 0 \end{cases}$$

so that

$$\alpha(t-4) = \begin{cases} 0, & t < 4 \\ 1, & t \geq 4 \end{cases}.$$

Using the shifting property,

$$L\{f(t-c)\alpha(t-c)\} = e^{-cs} L\{f(t)\},$$

where c is a nonnegative constant, provided that $L\{f(t)\}$ exists, we obtain (taking $c = 4$)

$$L\{g(t)\} = L\{(t-4)^2 \alpha(t-4)\}$$

$$= e^{-4s} L\{t^2\}. \qquad (1)$$

By recalling the result of a previous problem or by looking in a table of Laplace transforms, we find that

$$L\{t^n\} = \frac{n!}{s^{n+1}}, \quad s > 0,$$

where n is a nonnegative integer constant; hence, taking $n = 2$, we find

$$L\{t^2\} = \frac{2}{s^3}.$$

Substituting this result into equality (1),

$$L\{g(t)\} = e^{-4s}\left(\frac{2}{s^3}\right) = \frac{2e^{-4s}}{s^3}$$

• **PROBLEM 16-16**

Find the Laplace transform of the function $f(t)$ shown in the accompanying figure and defined by

$$f(t) = \begin{cases} t, & 0 < t < 4 \\ 5, & t > 4 \end{cases}.$$

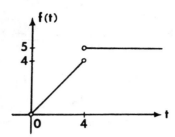

Solution: It is possible to handle some step-function problems without using the shifting property of the Laplace integral in the problem is fairly easy to solve. Thus, note that f(t) need not be defined at t = 0 or at t = 4 since such information is not essential in performing the integration to determine the Laplace transform.

Using L as the Laplace transform operator, with

$$L\{f(t)\} = \int_0^\infty e^{-st} f(t)\,dt,$$

and considering real s, we find

$$L\{f(t)\} = \lim_{R\to\infty} \int_0^R f(t)\,e^{-st}\,dt$$

$$= \int_0^4 te^{-st}\,dt + \lim_{R\to\infty} \int_4^R 5e^{-st}\,dt.$$

For s = 0,

$$L\{f(t)\} = \int_0^4 te^{0t}\,dt + \lim_{R\to\infty} \int_4^R 5e^{0t}\,dt$$

$$= \int_0^4 t\,dt + \lim_{R\to\infty} \int_4^R 5\,dt$$

$$= \frac{4^2}{2} + \lim_{R\to\infty} 5(R-4) = \infty \;;$$

hence L{f(t)} does not exist for s = 0. For s ≠ 0,

$$L\{f(t)\} = \int_0^4 te^{-st}\,dt + \lim_{R\to\infty} \int_4^R 5e^{-st}\,dt$$

$$= \int_0^4 te^{-st} dt + \lim_{R\to\infty}\left(\frac{-5e^{-sR}+ 5e^{-4s}}{s}\right)$$

$$= \int_0^4 te^{-st} dt + \frac{5e^{-4s}}{s} - \frac{5}{s}\lim_{R\to\infty} e^{-sR} .$$

Integrating by parts,

$$L\{f(t)\} = \left[-\frac{t}{s}e^{-st} - \frac{1}{s^2}e^{-st}\right]\Big|_{t=0}^{4} + \frac{5e^{-4s}}{s} - \frac{5}{s}\lim_{R\to\infty} e^{-sR}$$

$$= -\frac{4}{s}e^{-4s} - \frac{1}{s^2}e^{-4s} + \frac{1}{s^2} + \frac{5}{s}e^{-4s} - \frac{5}{s}\lim_{R\to\infty} e^{-sR} .$$

For $s > 0$, the above expression converges to

$$L\{f(t)\} = \frac{e^{-4s}}{s} - \frac{e^{-4s}}{s^2} + \frac{1}{s^2} ;$$

otherwise, $L\{f(t)\}$ does not exist.

Thus, the required Laplace transform is

$$L\{f(t)\} = \frac{e^{-4s}}{s} - \frac{e^{-4s}}{s^2} + \frac{1}{s^2}, \text{ for } s > 0 .$$

● **PROBLEM 16-17**

Find the Laplace transform $L\{f(t)\}$ of the function shown in Figure 1 and defined by

$$f(t) = \begin{cases} t^2, & 0 < t < 2 \\ 6, & t > 2 \end{cases} .$$

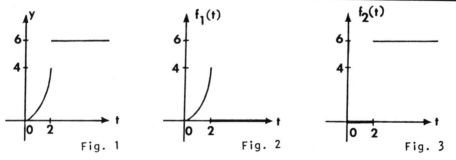

Fig. 1   Fig. 2   Fig. 3

<u>Solution</u>:   We wish to express $f(t)$ in terms of the unit step

function $\alpha(t)$, defined as

$$\alpha(t) = \begin{cases} 0, & t < 0 \\ 1, & t \geq 0 \end{cases}.$$

We can then use the shifting property which states that $L\{g(t-c)\alpha(t-c)\} = e^{-cs}L\{g(t)\}$, where c is a nonnegative constant, provided that $L\{g(t)\}$ exists.

Let us build $f(t)$ out of continuous functions and unit step functions. First, we need a function $f_1(t)$ that is equal to t^2 when $0<t<2$ and zero when $t>2$. The function $t^2\alpha(t-2)$ is equal to t^2 when $t>2$ and zero elsewhere, so that if we subtract it from t^2, we obtain the desired function (see Fig. 2):

$$f_1(t) = t^2 - t^2\alpha(t-2).$$

Second, we need a function $f_2(t)$ that is equal to 6 when $t>2$ and zero when $0<t<2$. We obtain this by multiplying the constant 6 by $\alpha(t-2)$, since $\alpha(t-2) = 0$ when $t<2$ and $=1$ when $t \geq 2$. Thus (see Fig. 3),

$$f_2(t) = 6\alpha(t-2).$$

The function $f(t)$ is obtained by adding together $f_1(t)$ and $f_2(t)$:

$$\begin{aligned} f(t) &= f_1(t) + f_2(t) \\ &= t^2 - t^2\alpha(t-2) + 6\alpha(t-2) \\ &= t^2 + (6-t^2)\alpha(t-2). \end{aligned} \quad (1)$$

This is still not the form we require in order to use the property mentioned at the beginning of this solution. We need to express $(6-t^2)$ as a function of $(t-2)$. We know that

$$(t-2)^2 = t^2 - 4t + 4;$$

hence,

$$\begin{aligned} 6 - t^2 &= -(t^2-4t+4) - 4t + 10 \\ &= -(t-2)^2 - 4t + 10 \\ &= -(t-2)^2 - 4t + 8 + 2 \\ &= -(t-2)^2 - 4(t-2) + 2. \end{aligned}$$

Substituting this last expression into equality (1),

$$f(t) = t^2 + [-(t-2)^2 - 4(t-2) + 2]\alpha(t-2)$$

$$= t^2 - (t-2)^2 \alpha(t-2) - 4(t-2)\alpha(t-2) + 2\alpha(t-2).$$

Taking the Laplace transform of this,
$$L\{f(t)\} = L\{t^2 - (t-2)^2 \alpha(t-2) - 4(t-2)\alpha(t-2) + 2\alpha(t-2)\}.$$

Since L is a linear operator,
$$L\{f(t)\} = L\{t^2\} - L\{(t-2)^2 \alpha(t-2)\} - 4L\{(t-2)\alpha(t-2)\} + 2L\{(1)\alpha(t-2)\}.$$

To the above equality we apply the property mentioned at the beginning of this solution:

$$L\{f(t)\} = L\{t^2\} - e^{-2s} L\{t^2\} - 4e^{-2s} L\{t\} + 2e^{-2s} L\{1\}. \qquad (2)$$

Now recall the result of a previous problem or use a table of Laplace transforms to find that

$$L\{t^n\} = \frac{n!}{s^{n+1}}, \quad s>0,$$

where n is a nonnegative integer constant. Taking n = 0, 1, and 2, respectively, we obtain

$$L\{1\} = \frac{1}{s}, \quad L\{t\} = \frac{1}{s^2},$$

and
$$L\{t^2\} = \frac{2}{s^3}.$$

Substituting these results into equality (2), we obtain

$$L\{f(t)\} = \frac{2}{s^3} - \frac{2e^{-2s}}{s^3} - \frac{4e^{-2s}}{s^2} + \frac{2e^{-2s}}{s}.$$

● **PROBLEM 16-18**

Let f be a periodic function with period T, i.e.,
$$f(t+T) = f(t)$$
for all t (see Figure), and let

$$g_1(t) = \begin{cases} f(t) & (0 \leq t \leq T) \\ 0 & (t > T) \end{cases}.$$

Show that

$$F(s) = \frac{G(s)}{1 - e^{-sT}}$$

where F and G are the Laplace transforms of f and g_1 respectively.

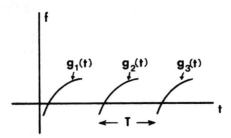

Solution: The Laplace transform of $g_1(t)$ is

$$G(s) = L\{g_1(t)\}.$$

Suppose that $g_1(t)$ is shifted to the right along the t-axis by an amount T. The resulting function, $g_2(t)$, may be written in the form

$$g_2(t) = g_1(t-T)\alpha(t-T)$$

where $\alpha(t-T)$ is the unit step function

$$\alpha(t-T) = \begin{cases} 1, & t \geq T \\ 0, & t < T \end{cases}.$$

If g_1 is shifted to the right by 2T, then the resulting function, g_3, may be written as

$$g_3(t) = g_1(t-2T)\alpha(t-2T)$$

where

$$\alpha(t-2T) = \begin{cases} 1, & t \geq 2T \\ 0, & t \leq 2T \end{cases}.$$

Thus, f(t) may be written as

$$f(t) = g_1(t) + g_2(t) + \ldots$$

$$= g_1(t) + g_1(t-T)\alpha(t-T) + g_1(t-2T)\alpha(t-2T) + \ldots . \quad (1)$$

Therefore, since L is a linear operator, taking the Laplace transform of each side of (1) gives

$$F(s) = L\{f(t)\} = L\{g_1(t)\}$$

$$+ L\{g_1(t-T)\alpha(t-T)\} + L\{g_1(t-2T)\alpha(t-2T)\} + \ldots \quad (2)$$

Now recall the shifting property of L which states that any f which posses a Laplace transform satisfies

$$L\{f(t-a)\alpha(t-a)\} = e^{-as} L\{f(t)\} .$$

Using this in (2) yields

$$F(s) = L\{g_1(t)\} + e^{-Ts} L\{g_1(t)\} + e^{-2Ts} L\{g_1(t)\} + \ldots .$$

Using $G(s) = L\{g_1(t)\}$,

$$F(s) = G(s) \sum_{n=0}^{\infty} e^{-nsT} . \quad (3)$$

The series in (3) is just a geometric series whose sum is known to be

$$\sum_{n=0}^{\infty} e^{-nsT} = \frac{1}{1 - e^{-sT}} , \quad s > 0 ,$$

so that (3) yields

$$F(s) = \frac{G(s)}{1 - e^{-sT}} , \quad s > 0 , \quad (4)$$

and the theorem is proved. An alternative way of stating (4) is that if $f(t)$ is periodic with period T, then

$$F(s) = L\{f(t)\} = \frac{\int_0^T e^{-sT} f(t) dt}{1 - e^{-sT}}$$

since $f(t) = g(t)$ on $(0,T)$.

• **PROBLEM** 16-19

Find the Laplace transform $L\{f(t)\}$, where $f(t)$ is the function shown in figures (a) and (b).

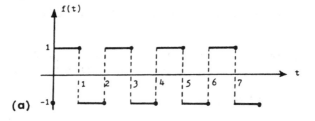

(a)

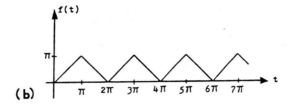

(b)

Solution: (a) Note that f(t) is a periodic function with period $P = 2$; i.e., $f(t+2) = f(t)$ for all t. On the interval $0 < t \leq 2$, f(t) can be expressed in analytic form:

$$f(t) = \begin{cases} 1, & 0 < t \leq 1 \\ -1, & 1 < t \leq 2 \end{cases}$$

Using the property that

$$L\{g(t)\} = \frac{\int_0^P e^{-st} g(t)\, dt}{1 - e^{-Ps}},$$

where g(t) is periodic with period P, we obtain

$$L\{f(t)\} = \frac{\int_0^2 e^{-st} f(t)\, dt}{1 - e^{-2s}}$$

$$= \frac{\int_0^1 e^{-st}(1)\, dt + \int_1^2 e^{-st}(-1)\, dt}{1 - e^{-2s}}$$

$$= \frac{\left.\frac{-e^{-st}}{s}\right|_{t=0}^{1} + \left.\frac{e^{-st}}{s}\right|_{t=1}^{2}}{1 - e^{-2s}}$$

$$= \frac{\frac{1}{s}(1 - 2e^{-s} + e^{-2s})}{1 - e^{-2s}}.$$

Factoring the numerator and denominator of this last fraction,

$$L\{f(t)\} = \frac{\frac{1}{s}(1 - e^{-s})^2}{(1+e^{-s})(1-e^{-s})}$$

$$= \frac{1}{s}\left(\frac{1 - e^{-s}}{1 + e^{-s}}\right).$$

Multiplying the numerator and denominator by $e^{s/2}$,

$$L\{f(t)\} = \frac{1}{s}\left(\frac{e^{s/2} - e^{-s/2}}{e^{s/2} + e^{-s/2}}\right)$$

$$= \frac{1}{s} \tanh \frac{s}{2}.$$

(b) Note that $f(t)$ is a periodic function with period $P = 2\pi$; i.e., $f(t+2\pi) = f(t)$ for all t. On the interval $0 \leq t \leq 2\pi$, $f(t)$ can be expressed in analytic form:

$$f(t) = \begin{cases} t, & 0 \leq t \leq \pi \\ 2\pi - t, & \pi < t \leq 2\pi \end{cases}.$$

Using the property that

$$L\{g(t)\} = \frac{\int_0^P e^{-st} g(t) dt}{1 - e^{-Ps}},$$

where $g(t)$ is periodic with period P, we obtain

$$L\{f(t)\} = \frac{\int_0^{2\pi} e^{-st} f(t) dt}{1 - e^{-2\pi s}}$$

$$= \frac{\int_0^{\pi} e^{-st}(t) dt + \int_\pi^{2\pi} e^{-st}(2\pi - t) dt}{1 - e^{-2\pi s}}$$

$$= \frac{\int_0^\pi t e^{-st} dt - \int_\pi^{2\pi} t e^{-st} dt + 2\pi \int_\pi^{2\pi} e^{-st} dt}{1 - e^{-2\pi s}}.$$

Integrating by parts

$$L\{f(t)\} = \frac{\left(\frac{-te^{-st}}{s} - \frac{e^{-st}}{s^2}\right)\Big|_{t=0}^{\pi} - \left(\frac{-te^{-st}}{s} - \frac{e^{-st}}{s^2}\right)\Big|_{t=\pi}^{2\pi} - \left(\frac{2\pi e^{-st}}{s}\right)\Big|_{t=\pi}^{2\pi}}{1 - e^{-2\pi s}}$$

$$= \frac{\left(\dfrac{-\pi e^{-\pi s}}{s} + \dfrac{1-e^{-\pi s}}{s^2}\right) + \left(\dfrac{2\pi e^{-2\pi s} - \pi e^{-\pi s}}{s} + \dfrac{e^{-2\pi s} - e^{-\pi s}}{s^2}\right)}{1 - e^{-2\pi s}}$$

$$+ \frac{\left(\dfrac{2\pi e^{-\pi s} - 2\pi e^{-2\pi s}}{s}\right)}{1 - e^{-2\pi s}}$$

$$= \frac{1 - 2e^{-\pi s} + e^{-2\pi s}}{s^2(1 - e^{-2\pi s})}$$

$$= \frac{\left(1 - e^{-\pi s}\right)^2}{s^2(1+e^{-\pi s})(1-e^{-\pi s})}$$

$$= \frac{1}{s^2}\left(\frac{1 - e^{-\pi s}}{1 + e^{-\pi s}}\right).$$

Multiplying both the numerator and denominator in the last fraction by $e^{\pi s/2}$,

$$L\{f(t)\} = \frac{1}{s^2}\left(\frac{e^{\pi s/2} - e^{-\pi s/2}}{e^{\pi s/2} + e^{-\pi s/2}}\right)$$

$$= \frac{1}{s^2} \tanh \frac{\pi s}{2}.$$

● **PROBLEM** 16-20

Prove that if

$$f(x+b) = -f(x) \qquad (1)$$

for all x, where b is a constant, then

$$L\{f(t)\} = \frac{\displaystyle\int_0^b e^{-st} f(t)\,dt}{1 + e^{-bs}}$$

where L is the Laplace transform operator. Functions satisfying (1) are often called antiperiodic and are very important in electrical engineering.

Solution: By an application of relation (1), setting $x = t + b$,

$$f(t+2b) = f([t+b] + b) = -f(t+b) . \qquad (2)$$

Now setting $x = t$ in relation (1),

$$-f(t+b) = -[-f(t)] = f(t) . \qquad (3)$$

Combining equalities (2) and (3),

$$f(t+2b) = f(t) ;$$

hence $f(t)$ is periodic with a period of $2b$. Using the property that

$$L\{g(t)\} = \frac{\int_0^P e^{-st} g(t) dt}{1 - e^{-Ps}} ,$$

where $g(t)$ is periodic with period P, we obtain

$$L\{f(t)\} = \frac{\int_0^{2b} e^{-st} f(t) dt}{1 - e^{-2bs}}$$

$$= \frac{\int_0^b e^{-st} f(t) dt + \int_b^{2b} e^{-st} f(t) dt}{1 - e^{-2bs}} . \qquad (4)$$

Making the substitution $y + b = t$ in the second integral, we find that

$$\int_b^{2b} e^{-st} f(t) dt = \int_0^b e^{-s(y+b)} f(y+b) dy$$

$$= e^{-bs} \int_0^b e^{-sy} [-f(y)] dy$$

$$= -e^{-bs} \int_0^b e^{-sy} f(y) dy .$$

When we change the dummy variable from y to t in this last integral, the integral with its coefficient becomes

$$-e^{-bs} \int_0^b e^{-st} f(t) dt .$$

Substituting this result back into equality (4), we obtain

$$L\{f(t)\} = \frac{\int_0^b e^{-st}f(t)\,dt - e^{-bs}\int_0^b e^{-st}f(t)\,dt}{1 - e^{-2bs}}$$

$$= \frac{(1-e^{-bs})\int_0^b e^{-st}f(t)\,dt}{1 - e^{-2bs}}.$$

Factoring the denominator,

$$L\{f(t)\} = \frac{(1-e^{-bs})\int_0^b e^{-st}f(t)\,dt}{(1 - e^{-bs})(1 + e^{-bs})}$$

$$= \frac{\int_0^b e^{-st}f(t)\,dt}{1 + e^{-bs}}$$

● **PROBLEM 16-21**

Find the Laplace transform $L\{h(t)\}$, where

$$h(t) = \begin{cases} 1, & 0 < t < c \\ -1, & c < t < 2c \end{cases}$$

and $h(t+2c) = h(t)$ for all t, with c a constant in the following two ways. (See Figure 1.)

(a) Use the fact that

$$L\{g(t)\} = \frac{1}{s(1 + e^{-cs})}, \qquad (1)$$

where

$$g(t) = \begin{cases} 1, & 0 < t < c \\ 0, & c < t < 2c \end{cases}$$

and

$g(t+2c) = g(t)$, for all t.

835

(see Figure 2.)

(b) Use the result developed in the previous problem, i.e., that for antiperiodic functions of the form

$$f(t+c) = -f(t),$$

the Laplace transform of f is given by

$$L\{f(t)\} = \frac{\int_0^c f(t)e^{-st}\,dt}{1 + e^{-cs}}. \qquad (2)$$

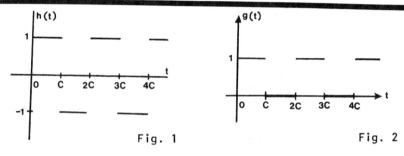

Fig. 1   Fig. 2

Solution: (a) The crucial observation here is that

$$h(t) = 2g(t) - 1$$

for all t. Since L is a linear operator,

$$L\{h(t)\} = L\{2g(t) - 1\}$$
$$= 2L\{g(t)\} - L\{1\} . \qquad (3)$$

From a table of Laplace transforms, we find that

$$L\{t^n\} = \frac{n!}{s^{n+1}},$$

where n is a nonnegative integer constant; hence, taking n = 0,

$$L\{1\} = 1/s .$$

Substituting this result and equality (1) into equality (3) we find that

$$L\{h(t)\} = 2\left[\frac{1}{s(1+e^{-cs})}\right] - \frac{1}{s}$$

$$= \frac{1}{s}\left(\frac{2}{1 + e^{-cs}} - 1\right)$$

$$= \frac{1}{s} \left[\frac{2 - (1 + e^{-cs})}{1 + e^{-cs}} \right]$$

$$= \frac{1}{s} \left(\frac{1 - e^{-cs}}{1 + e^{-cs}} \right).$$

Multiplying the numerator and denominator by $e^{cs/2}$,

$$L\{h(t)\} = \frac{1}{s} \left(\frac{e^{cs/2} - e^{-cs/2}}{e^{cs/2} + e^{-cs/2}} \right)$$

$$= \frac{1}{s} \tanh\left(\frac{cs}{2}\right).$$

(b) In this case the relation (2) takes the form

$$L\{h(t)\} = \frac{\int_0^c h(t) e^{-st} dt}{1 + e^{-cs}} = \frac{\int_0^c 1 \cdot e^{-st} dt}{1 + e^{-cs}}$$

since $h(t) = 1$ in $(0,c)$. Hence

$$L\{h(t)\} = \frac{\left. \frac{-1}{s} e^{-st} \right|_0^c}{1 + e^{-cs}} = \frac{\frac{1}{s}(1 - e^{-cs})}{1 + e^{-cs}}.$$

Again, multiply numerator and denominator by $e^{cs/2}$ to get

$$L\{h(t)\} = \frac{1}{s} \left(\frac{e^{cs/2} - e^{-cs/2}}{e^{cs/2} + e^{-cs/2}} \right) = \frac{1}{s} \tanh\left(\frac{cs}{2}\right).$$

THE INVERSION PROBLEM

• PROBLEM 16-22

Find the inverse Laplace transforms

(a) $$L^{-1}\left\{\frac{2s}{(s^2 + 1)^2}\right\},$$

(b) $$L^{-1}\left\{\frac{1}{\sqrt{s}}\right\}$$

Solution: (a) In a table of Laplace transforms we find

$$L\{t \sin bt\} = \frac{2bs}{(s^2 + b^2)^2},$$

where b is a constant; hence, taking b = 1, we find

$$L\{t \sin t\} = \frac{2s}{(s^2 + 1)^2}.$$

Therefore, by definition of the inverse Laplace transform,

$$L^{-1}\left\{\frac{2s}{(s^2 + 1)^2}\right\} = t \sin t.$$

(b) In a table of Laplace transforms, we find

$$L\left\{\frac{1}{\sqrt{t}}\right\} = \frac{\sqrt{\pi}}{\sqrt{s}}.$$

Thus, by definition of the inverse Laplace transform,

$$L^{-1}\left\{\frac{\sqrt{\pi}}{\sqrt{s}}\right\} = \frac{1}{\sqrt{t}}$$

Since L^{-1} is a linear operator,

$$L^{-1}\left\{\frac{1}{\sqrt{s}}\right\} = \frac{1}{\sqrt{\pi}} L^{-1}\left\{\frac{\sqrt{\pi}}{\sqrt{s}}\right\} = \frac{1}{\sqrt{\pi}} \cdot \frac{1}{\sqrt{t}}$$

● **PROBLEM 16-23**

State a theorem which gives conditions under which two functions must be identical if they have the same Laplace transform, i.e., within what class of functions is the inverse Laplace transform unique? What does this imply about the functions

$$f(t) = L^{-1}\{aF_1(s) + bF_2(s)\}$$

and

$$\bar{f}(t) = aL^{-1}\{F_1(s)\} + bL^{-1}\{F_2(s)\}, \qquad (1)$$

where $F_1(s)$ and $F_2(s)$ are some Laplace transforms?

Solution: Consider all real valued functions which satisfy the following three conditions:

1) each function is piecewise smooth (i.e., has a piecewise continuous derivative) on $(0, \infty)$,

2) each function is of some exponential order (see problem 1 in this chapter),

3) each function is defined by

$$\frac{1}{2}[f(x_0+) + f(x_0-)]$$

at each jump discontinuity x_0.

Then two such functions possessing the same Laplace transform, $F(s)$, must be identical. In other words, the inverse Laplace transform is unique within the class of such functions.

Now the two functions in (1), $f(t)$ and $\bar{f}(t)$ have the same Laplace transform since L is a linear operator. That is

$$L\{f(t)\} = L\{L^{-1}\{aF_1(s) + bF_2(s)\}\} = aF_1(s) + bF_2(s)$$

and

$$L\{\bar{f}(t)\} = L\{aL^{-1}\{F_1(s)\} + bL^{-1}\{F_2(s)\}\}$$

$$= aL\{L^{-1}\{F_1(s)\}\} + bL\{L^{-1}\{F_2(s)\}\}$$

$$= aF_1(s) + bF_2(s)$$

so that

$$L\{f(t)\} = L\{\bar{f}(t)\} \quad . \tag{2}$$

According to the theorem, if these functions satisfy the three conditions above, then (2) implies that they are equal, i.e.,

$$f(t) = \bar{f}(t)$$

or

$$L^{-1}\{aF_1(s) + bF_2(s)\} = aL^{-1}\{F_1(s)\} + bL^{-1}\{F_2(s)\} \quad . \tag{3}$$

Thus, under these conditions L^{-1} is a linear operator. The practical significance of this arises in the following way. Suppose that the Laplace transform has been applied to some functions in a particular problem, used for manipulative purposes, and is now to be inverted. In particular, suppose that we are faced with the problem of inverting a transform which can be split into two parts

$$F(s) = aF_1(s) + bF_2(s)$$

so that

$$f(t) = L^{-1}\{F(s)\} = L^{-1}\{aF_1(s) + bF_2(s)\} \quad .$$

Then in almost all cases, in order for the problem to be soluble, one would have to assume that L^{-1} is linear so that

$$L^{-1}\{aF_1(s) + bF_2(s)\} = aL^{-1}\{F_1(s)\} + bL^{-1}\{F_2(s)\}$$
$$= af_1(t) + bf_2(t) \qquad (4)$$

and

$$f(t) = af_1(t) + bf_2(t).$$

Thus, assuming the linearity of L^{-1} (as is done almost without exception in Laplace transform problems) amounts to assuming that the original function, $f(t)$, satisfies the three conditions stated earlier. Since the great majority of physical problems deal with such functions, this assumption is usually justifiable.

● **PROBLEM 16-24**

Find and sketch the function $g(t)$ which is the inverse Laplace transform

$$g(t) = L^{-1}\left\{\frac{3}{s} - \frac{4e^{-s}}{s^2} + \frac{4e^{-3s}}{s^2}\right\}.$$

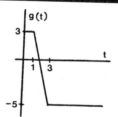

Solution: L^{-1} is a linear operator under the very weak restrictions on $g(t)$ given in an earlier problem. Since these conditions are assumed to hold for all functions dealt with in this book,

$$g(t) = 3L^{-1}\left\{\frac{1}{s}\right\} - 4L^{-1}\left\{\frac{e^{-s}}{s^2}\right\} + 4L^{-1}\left\{\frac{e^{-3s}}{s^2}\right\}. \qquad (1)$$

Recalling the property that

$$L^{-1}\{e^{-ks}F(s)\} = f(t-k)\alpha(t-k),$$

where $f(t) = L^{-1}\{f(s)\}$, k is nonnegative constant, and α is the unit step function

$$\alpha(x) = \begin{cases} 0, & x < 0 \\ 1, & x \geq 0 \end{cases},$$

provided that $L^{-1}\{e^{-ks}F(s)\}$ and $L^{-1}\{F(s)\}$ exist, equality (1) becomes

$$g(t) = \left[3L^{-1}\left\{\frac{1}{s}\right\}\bigg|_{t} - 4L^{-1}\left\{\frac{1}{s^2}\right\}\bigg|_{t-1} \cdot \alpha(t-1) + 4L^{-1}\left\{\frac{1}{s^2}\right\}\bigg|_{t-3} \times \right.$$

$$\left. \times \alpha(t-3)\right]. \tag{2}$$

From a table of Laplace transforms, or by recalling a previous problem, we find that

$$L^{-1}\left\{\frac{1}{s^n}\right\}\bigg|_{x} = \frac{x^{n-1}}{(n-1)!},$$

where n is a positive integer; hence,

$$L^{-1}\left\{\frac{1}{s}\right\}\bigg|_{t} = 1, \quad L^{-1}\left\{\frac{1}{s^2}\right\}\bigg|_{t-1} = t-1,$$

and

$$L^{-1}\left\{\frac{1}{s^2}\right\}\bigg|_{t-3} = t-3.$$

Substituting these results into equality (2),

$$g(t) = 3 - 4(t-1)\alpha(t-1) + 4(t-3)\alpha(t-3). \tag{3}$$

To remove the α function from expression (3), consider first the case $t < 1$. In that case,

$$\alpha(t-1) = \alpha(t-3) = 0,$$

so that

$$g(t) = 3 - 4(t-1)(0) + 4(t-3)(0) = 3. \tag{4}$$

When $1 \leq t < 3$, we have $\alpha(t-1) = 1$ and $\alpha(t-3) = 0$, so that

$$g(t) = 3 - 4(t-1)(1) + 4(t-3)(0) = 7 - 4t. \tag{5}$$

When $t \geq 3$, we have $\alpha(t-1) = \alpha(t-3) = 1$, so that

$$g(t) = 3 - 4(t-1)(1) + 4(t-3)(1) = -5. \tag{6}$$

Grouping together results (4), (5), and (6), we have

$$g(t) = \begin{cases} 3, & t < 1 \\ 7-4t, & 1 \leq t < 3 \\ -5, & t \geq 3 \end{cases}.$$

The graph of g(t) is shown in the figure.

● **PROBLEM 16-25**

Find the inverse Laplace transform

$$f(t) = L^{-1}\left\{\log \frac{s+1}{s-1}\right\}, \quad s > 1.$$

Solution: From a table of infinite series, we can find

$$\log \frac{1+x}{1-x} = 2 \sum_{n=0}^{\infty} \frac{x^{2n+1}}{2n+1}, \qquad (1)$$

for $|x| < 1$. We wish to put $\log[(s+1)/(s-1)]$ into some form for which the series (1) will be useful. Dividing the numerator and denominator by s does not change the value:

$$\log \frac{s+1}{s-1} = \log \frac{1+1/s}{1-1/s}.$$

Now, $|1/s| < 1$, since $s > 1$. Thus,

$$\log \frac{1+1/s}{1-1/s} = 2 \sum_{n=0}^{\infty} \frac{(1/s)^{2n+1}}{2n+1},$$

and

$$L^{-1}\left\{\log \frac{s+1}{s-1}\right\} = L^{-1}\left\{2 \sum_{n=0}^{\infty} \frac{(1/s)^{2n+1}}{2n+1}\right\}.$$

Since L^{-1} is a linear operator, under the very weak restrictions on f(t) given in an earlier problem and which are assumed here,

$$L^{-1}\left\{\log \frac{s+1}{s-1}\right\} = 2 \sum_{n=0}^{\infty} \left(\frac{1}{2n+1}\right) L^{-1}\left\{\left(\frac{1}{s}\right)^{2n+1}\right\}. \qquad (2)$$

From a table of Laplace transforms or from the result of a previous problem, it is found that

$$L^{-1}\left\{\left(\frac{1}{s}\right)^k\right\} = \frac{t^{k-1}}{(k-1)!},$$

where k is a positive integer. Substitution (2n+1) for k in the above formula,

$$L^{-1}\left\{\left(\frac{1}{s}\right)^{2n+1}\right\} = \frac{t^{2n}}{(2n)!},$$

where (2n+1) is a positive integer; i.e., where n is a

nonnegative integer. Substituting this result into equality (2),

$$f(t) = L^{-1}\left\{\log \frac{s+1}{s-1}\right\} = 2 \sum_{n=0}^{\infty} \left(\frac{1}{2n+1}\right) \frac{t^{2n}}{(2n)!}$$

$$= 2 \sum_{n=0}^{\infty} \frac{t^{2n}}{(2n+1)!}$$

$$= \frac{2}{t} \sum_{n=0}^{\infty} \frac{t^{2n+1}}{(2n+1)!} = \frac{2}{t} \sinh t \ .$$

An alternative way of solving this problem is to use the equality

$$L\{tf(t)\} = -\frac{d}{ds} L\{f(t)\}$$

proved in problem 7. Differentiating our function,

$$\frac{d}{ds}\left(\log \frac{s+1}{s-1}\right) = -\frac{2}{s^2-1} \ .$$

If we find $h(t)$ such that

$$L\{h(t)\} = -\frac{2}{s^2-1} \ ,$$

then

$$L\{h(t)\} = L\{t \cdot \frac{h(t)}{t}\} = -\frac{d}{ds} L\left\{\frac{h(t)}{t}\right\}$$

$$= -\frac{2}{s^2-1} = \frac{d}{ds}\left(\log \frac{s+1}{s-1}\right) ,$$

and

$$L\left\{-\frac{h(t)}{t}\right\} = \log \frac{s+1}{s-1} + \text{const.}$$

Constant we will show to be equal zero. We know

$$L\{\sinh t\} = \frac{1}{s^2-1} \ ,$$

so

$$L^{-1}\left\{-\frac{2}{s^2-1}\right\} = -2 \sinh t \ ,$$

843

and we have

$$L\left\{-\left(-\frac{2\sinh t}{t}\right)\right\} = L\left\{\frac{2\sinh t}{t}\right\} = \log\frac{s+1}{s-1} + \text{const.}$$

Now, as $s \to +\infty$ both

$$L\left\{\frac{2\sinh t}{t}\right\} \quad \text{and} \quad \log\frac{s+1}{s-1} \to 0,$$

so const = 0 and

$$L^{-1}\left\{\log\frac{s+1}{s-1}\right\} = \frac{2\sinh t}{t}.$$

● **PROBLEM 16-26**

Find the inverse Laplace transforms

(a)
$$L^{-1}\left\{\frac{1}{s^2 - 2s + 9}\right\},$$

(b)
$$L^{-1}\left\{\frac{s+1}{s^2 + 6s + 25}\right\}.$$

Solution: (a) Completing the square in the denominator $(s^2 - 2s + 9)$,

$$\frac{1}{s^2 - 2s + 9} = \frac{1}{(s-1)^2 + (\sqrt{8})^2} = \frac{1}{\sqrt{8}}\left[\frac{\sqrt{8}}{(s-1)^2 + (\sqrt{8})^2}\right].$$

Thus,

$$L^{-1}\left\{\frac{1}{s^2 - 2s + 9}\right\} = L^{-1}\left\{\frac{1}{\sqrt{8}} \cdot \frac{\sqrt{8}}{(s-1)^2 + (\sqrt{8})^2}\right\}. \qquad (1)$$

Noting the property that

$$L^{-1}\{F(s-k)\} = e^{kt} L^{-1}\{F(s)\}$$

for any function F and constant k, provided that $L^{-1}\{F(s-k)\}$ and $L^{-1}\{F(s)\}$ exist, equality (1) becomes

$$L^{-1}\left\{\frac{1}{s^2 - 2s + 9}\right\} = e^t L^{-1}\left\{\frac{1}{\sqrt{8}} \cdot \frac{\sqrt{8}}{s^2 + (\sqrt{8})^2}\right\}.$$

Since L^{-1} is a linear operator under the very weak restrictions on f(t) given in an earlier problem and which are assumed here,

$$L^{-1}\left\{\frac{1}{s^2 - 2s + 9}\right\} = \frac{1}{\sqrt{8}} e^t L^{-1}\left\{\frac{\sqrt{8}}{s^2 + (\sqrt{8})^2}\right\}. \quad (2)$$

From a table of Laplace transforms or from the results of a previous problem we find that

$$L^{-1}\left\{\frac{b}{s^2 + b^2}\right\} = \sin bt,$$

where b is a constant; hence, b = $\sqrt{8}$, we find

$$L^{-1}\left\{\frac{\sqrt{8}}{s^2 + (\sqrt{8})^2}\right\} = \sin(\sqrt{8}t).$$

Substituting this result into equality (2),

$$L^{-1}\left\{\frac{1}{s^2 - 2s + 9}\right\} = \frac{1}{\sqrt{8}} e^t \sin(\sqrt{8}t).$$

(b) Completing the square in the denominator ($s^2 + 6s + 25$),

$$\frac{s+1}{s^2 + 6s + 25} = \frac{s+1}{(s+3)^2 + 4^2} = \frac{(s+3)-2}{(s+3)^2 + 4^2}$$

$$= \frac{(s+3)}{(s+3)^2 + 4^2} - \frac{1}{2}\left[\frac{4}{(s+3)^2 + 4^2}\right].$$

Thus, since L^{-1} is a linear operator under the restrictions mentioned earlier

$$L^{-1}\left\{\frac{s+1}{s^2 + 6s + 25}\right\} = L^{-1}\left\{\frac{(s+3)}{(s+3)^2 + 4^2}\right\} - \frac{1}{2} L^{-1}\left\{\frac{4}{(s+3)^2 + 4^2}\right\}.$$

(3)

Noting the property that

$$L^{-1}\{F(s-k)\} = e^{kt} L^{-1}\{F(s)\}$$

for any function F and constant k, provided that $L^{-1}\{F(s-k)\}$ and $L^{-1}\{F(s)\}$ exist, equality (3) becomes

$$L^{-1}\left\{\frac{s+1}{s^2 + 6s + 25}\right\} = e^{-3t} L^{-1}\left\{\frac{s}{s^2 + 4^2}\right\} - \frac{1}{2} e^{-3t} L^{-1}\left\{\frac{4}{s^2 + 4^2}\right\}.$$

(4)

From a table of Laplace transforms or from the results of a previous problem we find that

$$L^{-1}\left\{\frac{s}{s^2+b^2}\right\} = \cos bt, \text{ and } L^{-1}\left\{\frac{b}{s^2+b^2}\right\} = \sin bt,$$

where b is a constant; hence, taking b = 4, we find

$$L^{-1}\left\{\frac{s}{s^2+4^2}\right\} = \cos 4t, \text{ and } L^{-1}\left\{\frac{4}{s^2+4^2}\right\} = \sin 4t.$$

Substituting these results into equality (4),

$$L^{-1}\left\{\frac{s+1}{s^2+6s+25}\right\} = e^{-3t}\cos 4t - \frac{1}{2}e^{-3t}\sin 4t.$$

• **PROBLEM 16-27**

Outline the general method of rational fraction decomposition used to find the inverse Laplace transform of a rational function

$$F(s) = P(s)/Q(s),$$

where Q(s) is of higher degree than P(s).

Solution: The decomposition of a rational function is based on the knowledge of roots of the denominator Q(s). If the coefficients of Q(s) are real, then it is known from algebra that Q(s) can be factorized (apart from a constant factor) into factors of the type

$$(s-c)^m$$

for each root c (c is complex in general) of multiplicity m. It is also known that Q(s) can be always factorized into factors of two types

1. $(s-r)^m$ for each real root r of multiplicity m;

2. $(s^2 + as + b)^n$ for each pair of complex conjugate roots (satisfying

$$s^2 + as + b = 0 \text{ with a, b real})$$

of multiplicity n.

For the first decomposition, corresponding to each complex root c of multiplicity m there is a set of m terms in the decomposition of P(s)/Q(s) of the type

$$\frac{A_1}{s-c} + \frac{A_2}{(s-c)^2} + \ldots + \frac{A_m}{(s-c)^m} \quad (A_i = \text{constant}) . \tag{1}$$

Thus, if $Q(s)$ is factorized as follows,

$$\frac{P(s)}{Q(s)} = \frac{P(s)}{(s-c_1)^m (s-c_2)^n (s-c_3)^p} ,$$

the complex decomposition is

$$\frac{P(s)}{(s-c_1)^m (s-c_2)^n (s-c_3)^p} = \frac{A_1}{(s-c_1)} + \ldots + \frac{A_m}{(s-c_1)^m} + \frac{B_1}{(s-c_2)}$$
$$+ \ldots + \frac{B_n}{(s-c_2)^n} + \frac{C_1}{(s-c_3)} + \ldots + \frac{C_p}{(s-c_3)^p} \tag{2}$$

and the constants A_i, B_i, C_i are determined by putting the right hand side of (2) into the form of the left hand side

$$\frac{P(s)}{(s-c_1)^m (s-c_2)^n (s-c_3)^p} = \frac{A_1 (s-c_1)^{m-1} (s-c_2)^n (s-c_3)^p}{(s-c_1)^m (s-c_2)^n (s-c_3)^p}$$
$$+ \ldots + \frac{C_p (s-c_1)^m (s-c_2)^n}{(s-c_1)^m (s-c_2)^n (s-c_3)^p} .$$

Then the coefficients can be solved for by using the resulting equation

$$P(s) = A_1 (s-c_1)^{m-1} (s-c_2)^n (s-c_3)^p + \ldots + C_p (s-c_1)^m (s-c_2)^n$$

which must be true for all s.

For the second decomposition, corresponding to each real root r of multiplicity m there is a set of m terms in the decomposition of $P(s)/Q(s)$ of the form

$$\frac{A_1}{(s-r)} + \frac{A_2}{(s-r)^2} + \ldots + \frac{A_m}{(s-r)^m} , \quad (a_i = \text{const}) , \tag{3}$$

and corresponding to each pair of complex conjugate roots of multiplicity n there is a set of terms of the type

$$\frac{B_1 s + C_1}{s^2 + as + b} + \frac{B_2 s + C_2}{(s^2 + as + b)^2} + \ldots + \frac{B_n s + C_n}{(s^2 + as + b)^n}$$

$$(B_i, C_i = \text{const}) . \tag{4}$$

As in the case of complex roots, the coefficients in (3) and (4) are obtained by equating

$$\frac{P(s)}{Q(s)}$$

with its decomposition, getting the decomposition over a common denominator $Q(s)$ and setting $P(s)$ equal to the resulting numerator.

Once all the constants in (1) or in (3) and (4) are determined, it is a simple matter to invert each individual term in the decomposition. Thus, for terms of the type used in (1) and (3) it is known from previous problems (or from looking in a table of Laplace transforms) that

$$L^{-1}\left\{\frac{A_i}{(s-c)^k}\right\} = A_i \frac{t^{k-1}}{(k-1)!} e^{ct} \tag{5}$$

since by the attenuation property with

$$F(s) = L\{t^{k-1}\} = \frac{(k-1)!}{s^k}$$

we know that

$$L\left\{\frac{A_i t^{k-1}}{(k-1)!} e^{ct}\right\} = \frac{A_i}{(k-1)!} L\{t^{k-1} e^{ct}\}$$

$$= \frac{A_i}{(k-1)!} F(s-c)$$

$$= \frac{A_i}{(k-1)!} \cdot \frac{(k-1)!}{(s-c)^k} = \frac{A_i}{(s-c)^k} \quad .$$

For quatratic terms of the type in (4), it is convenient to "complete the square":

$$s^2 + as + b = (s + a/2)^2 + (b - a^2/4) = (s + a/2)^2 + d^2 \quad ,$$

and then write

$$\frac{Bs + C}{s^2 + as + b} = \frac{B(s + a/2) + \left(C - \left(\frac{a}{2}\right)B\right)}{(s + a/2)^2 + d^2}$$

$$= B \frac{s + a/2}{(s + a/2)^2 + c^2} + \frac{[C - (a/2) B]}{(s + a/2)^2 + d^2} \quad .$$

It is then easily checked that

$$L^{-1}\left\{B \frac{(s + a/2)}{(s + a/2)^2 + d^2}\right\} = Be^{-(a/2)t} \cos dt \tag{6}$$

and

$$L^{-1}\left\{\frac{C - \frac{a}{2}B}{(s + a/2)^2 + d^2}\right\} = \frac{C - \left(\frac{a}{2}\right)B}{d} e^{-(a/2)t} \sin dt \qquad (7)$$

by using the attenuation property (i.e., if

$$L\{g(t)\} = F(s)$$

then

$$L\{g(t)e^{at}\} = F(s-a)$$

and the results

$$L\{\sin kt\} = \frac{k}{s^2 + k^2}, \quad L\{\cos kt\} = \frac{s}{s^2 + k^2}.$$

Using this procedure and equations (5), (6), (7), the inverse Laplace transform of any rational function $P(s)/Q(s)$ can be found.

• **PROBLEM 16-28**

Find the inverse Laplace transform

$$f(t) = L^{-1}\{F(s)\} = L^{-1}\left\{\frac{3s^2 + 17s + 47}{(s + 2)(s^2 + 4s + 29)}\right\}.$$

Solution: Since the degree of the numerator polynomial is less than the degree of the denominator polynomial, factor the denominator and expand the rational fraction by partial fractions.

Thus,

$$F(s) = \frac{3s^2 + 17s + 47}{(s + 2)(s + r_1)(s + r_2)}$$

where

$$r_1 = \frac{-4 + \sqrt{16-116}}{2} \quad ; \quad r_2 = \frac{-4 - \sqrt{16-116}}{2}$$

$$r_1 = -2 + j5 \quad ; \quad r_2 = -2 - j5$$

and

$$j = \sqrt{-1}.$$

Expanding by partial fractions gives

$$F(s) = \frac{K_1}{s+2} + \frac{K_2}{s+2+j5} + \frac{K_2^*}{s+2-j5} \quad (1)$$

or, finding a common denominator for the right hand side of (1)

$$F(s) = \frac{K_1(s+2+j5)(s+2-j5) + K_2(s+2)(s+2-j5) + K_2^*(s+2)(s+2+j5)}{(s+2)(s+2+j5)(s+2-j5)}$$

(2)

But

$$F(s) = \frac{3s^2 + 17s + 47}{(s+2)(s+2+j5)(s+2-j5)}$$

and using this in (2) gives

$$3s^2 + 17s + 47 = K_1(s+2+j5)(s+2-j5) + K_2(s+2)(s+2-j5)$$

$$+ K_2^*(s+2)(s+2+j5) . \quad (3)$$

Now (3) must hold for all s. In particular, when $s = -2$, (3) yields

$$(3s^2 + 17s + 47)\Big|_{s=-2} = K_1(s+2+j5)(s+2-j5)\Big|_{s=-2}$$

so that

$$K_1 = \frac{3s^2 + 17s + 47}{s^2 + 4s + 29}\Big|_{s=-2} = \frac{25}{25} = 1 .$$

Similarly at $s = -2 -j5$, (3) must hold so that

$$K_2 = \frac{3s^2 + 17s + 47}{(s+2)(s+2-j5)}\Big|_{s=-2-j5} = \frac{3(-21 + j20) - 34 - j85 + 47}{(-j5)(-j10)}$$

$$= \frac{-50 - j25}{-50} = 1 + j(0.5)$$

and

$$K_2^* = 1 - j(0.5) .$$

Substituting K_1, K_2, and K_2^* into the partial fraction expansion gives

$$F(s) = \frac{1}{s+2} + \frac{1+j0.5}{s+2+j5} + \frac{1-j0.5}{s+2-j5} .$$

Taking the inverse Laplace transform of F(s) gives

$$f(t) = e^{-2t} + (1+j0.5)e^{-2t-j5t} + (1-j0.5)e^{-2t+j5t}.$$

Multiplying out:

$$f(t) = e^{-2t} + e^{-2t}e^{-j5t} + j0.5e^{-2t}e^{-j5t} + e^{-2t}e^{j5t}$$

$$-j0.5e^{-2t}e^{j5t},$$

and factoring to obtain the form:

$$\cos \omega t = \frac{e^{j\omega t} + e^{-j\omega t}}{2}, \quad \sin \omega t = \frac{e^{j\omega t} - e^{-j\omega t}}{2j},$$

$$f(t) = e^{-2t} + 2e^{-2t}\left| \frac{e^{j5t} + e^{-j5t}}{2} \right| + (2j)(-j0.5e^{-2t})$$

$$\times \left| \frac{e^{j5t} - e^{-j5t}}{2j} \right|$$

$$f(t) = [e^{-2t} + 2e^{-2t}\cos 5t + e^{-2t}\sin 5t].$$

● **PROBLEM 16-29**

Use partial fractions to decompose

(a) $\dfrac{1}{(s+1)(s^2+1)}$;

(b) $\dfrac{1}{(s^2+1)(s^2+4s+8)}$.

Solution: (a) From the linear factor (s+1) in the denominator, we obtain A/(s+1), where A is a constant, undetermined as yet. From the quadratic factor (s^2+1), we obtain $(Bs+C)/(s^2+1)$, where B and C are constants, undetermined as yet. Now set

$$\frac{1}{(s+1)(s^2+1)} = \frac{A}{s+1} + \frac{Bs+C}{s^2+1}. \tag{1}$$

Multiplying both sides of equation (1) by

$$(s+1)(s^2+1),$$

we obtain

$$1 = A(s^2+1) + (Bs+C)(s+1).$$

Collecting like terms,

$$(0)s^2 + (0)s + 1 = (A+B)s^2 + (B+C)s + (A+C) .$$

Equating coefficients of like powers of s on both sides of the above equation,

$$0 = A + B , \qquad (2)$$

$$0 = B + C , \qquad (3)$$

and

$$1 = A + C . \qquad (4)$$

From equation (2), $A = -B$. From equation (3), $C = -B$. Substituting these two results into equation (4),

$$1 = (-B) + (-B)$$
$$= -2B .$$

Thus, $B = -\frac{1}{2}$. From this, it follows that $A = -(-\frac{1}{2}) = \frac{1}{2}$, and

$$C = -(-\frac{1}{2}) = \frac{1}{2} .$$

Substituting these results into equation (1),

$$\frac{1}{(s+1)(s^2+1)} = \frac{(\frac{1}{2})}{s+1} + \frac{(-\frac{1}{2})s + (\frac{1}{2})}{s^2+1} .$$

(b) From the quadratic factors (s^2+1) and (s^2+4s+8) in the denominator, we obtain $(As+B)/(s^2+1)$ and $(Cs+D)/(s^2+4s+8)$, respectively, where A, B, C, and D are constants, undetermined as yet. Now set

$$\frac{1}{(s^2+1)(s^2+4s+8)} = \frac{As+B}{s^2+1} + \frac{Cs+D}{s^2+4s+8} . \qquad (5)$$

Multiplying both sides of equation (5) by

$$(s^2+1)(s^2+4s+8) ,$$

we obtain

$$1 = (As+B)(s^2+4s+8) + (Cs+D)(s^2+1) .$$

Collecting like terms,

$$(0)s^3 + (0)s^2 + (0)s + 1 = (A+C)s^3 + (4A+B+D)s^2 + (8A+4B+C)s + (8B+D) .$$

Equating coefficients of like powers of s on both sides of the above equation,

$$0 = A + C, \tag{6}$$

$$0 = 4A + B + D, \tag{7}$$

$$0 = 8A + 4B + C, \tag{8}$$

and

$$1 = 8B + D. \tag{9}$$

The system of equations (6) through (9) can be solved by Cramer's rule, yielding

$$A = -4/65, \ B = 7/65, \ C = 4/65, \text{ and } D = 9/65.$$

Substituting these results into equation (5),

$$\frac{1}{(s^2+1)(s^2+4s+8)} = \frac{(-4/65)s+(7/65)}{s^2+1} + \frac{(4/65)s+(9/65)}{s^2+4s+8}$$

● **PROBLEM 16-30**

Find the inverse Laplace transforms

(a)

$$L^{-1}\left\{\frac{s+3}{(s-2)(s+1)}\right\},$$

(b)

$$L^{-1}\left\{\frac{8}{s^3(s^2-s-2)}\right\}.$$

<u>Solution</u>: (a) We use the method of partial fractions to decompose $(s+3)/[(s-2)(s+1)]$. Setting

$$\frac{s+3}{(s-2)(s+1)} = \frac{A}{s-2} + \frac{B}{s+1}$$

for all s, with A and B constant, we find $A = 5/3$ and $B = -2/3$ by the following method. Multiply through this equation by $(s-2)(s+1)$ giving

$$s+3 = A(s+1) + B(s-2).$$

Now since this equation must be true for all s, it is true for $s = 2$ in particular. Substituting $s = 2$ gives

$$2 + 3 = A(2+1) + B(2-2) = 3A$$

or
$$A = 5/3.$$

Substituting $s = -1$ gives

$$-1 + 3 = A(-1 + 1) + B(-1 - 2) = -3B$$

or
$$B = -2/3.$$

Thus,
$$L^{-1}\left\{\frac{s+3}{(s-2)(s+1)}\right\} = L^{-1}\left\{\frac{5}{3}\left(\frac{1}{s-2}\right) - \frac{2}{3}\left(\frac{1}{s+1}\right)\right\}.$$

Since L^{-1} is a linear operator under the very weak restrictions imposed on

$$f(t) = L^{-1}\left\{\frac{s+3}{(s-2)(s+1)}\right\}$$

in a previous problem and which are assumed throughout this chapter,

$$L^{-1}\left\{\frac{s+3}{(s-2)(s+1)}\right\} = \frac{5}{3}L^{-1}\left\{\frac{1}{s-2}\right\} - \frac{2}{3}L^{-1}\left\{\frac{1}{s+1}\right\}. \quad (1)$$

From a table of Laplace transforms, or from a previous problem, we find that

$$L^{-1}\left\{\frac{1}{s-a}\right\} = e^{at},$$

where a is a constant; hence,

$$L^{-1}\left\{\frac{1}{s-2}\right\} = e^{2t}$$

and
$$L^{-1}\left\{\frac{1}{s+1}\right\} = e^{-t}.$$

Substituting these results into equality (1),

$$L^{-1}\left\{\frac{s+3}{(s-2)(s+1)}\right\} = \frac{5}{3}e^{2t} - \frac{2}{3}e^{-t}.$$

(b) We use the method of partial fractions, employing only real roots, to decompose

$$\frac{8}{s^3(s^2-s-2)}, \quad \text{or} \quad \frac{8}{s^3(s-2)(s+1)}.$$

Setting

$$\frac{8}{s^3(s^2-s-2)} = \frac{A}{s} + \frac{B}{s^2} + \frac{C}{s^3} + \frac{D}{s-2} + \frac{E}{s+1}$$

for all s, with A,B,C,D, and E constant, we find, using the method of the previous problem,

$A = -3$, $B = 2$, $C = -4$, $D = 1/3$, and $E = 8/3$.

Thus,

$$L^{-1}\left\{\frac{8}{s^3(s^2-s-2)}\right\} = L^{-1}\left\{\frac{-3}{s} + \frac{2}{s^2} - \frac{4}{s^3} + \frac{1/3}{s-2} + \frac{8/3}{s+1}\right\}.$$

Since L^{-1} is a linear operator under the conditions stated earlier,

$$L^{-1}\left\{\frac{8}{s^3(s^2-s-2)}\right\} = \left[-3L^{-1}\left\{\frac{1}{s}\right\} + 2L^{-1}\left\{\frac{1}{s^2}\right\} - 4L^{-1}\left\{\frac{1}{s^3}\right\}\right.$$

$$\left. + \frac{1}{3}L^{-1}\left\{\frac{1}{s-2}\right\} + \frac{8}{3}L^{-1}\left\{\frac{1}{s+1}\right\}\right]. \quad (2)$$

From a table of Laplace transforms, we find that

$$L^{-1}\left\{\frac{1}{s^n}\right\} = \frac{t^{n-1}}{(n-1)!},$$

where n is a positive integer constant; hence,

$$L^{-1}\left\{\frac{1}{s}\right\} = 1, \quad L^{-1}\left\{\frac{1}{s^2}\right\} = t,$$

and

$$L^{-1}\left\{\frac{1}{s^3}\right\} = \frac{t^2}{2}.$$

Substituting these results into equality (2),

$$L^{-1}\left\{\frac{8}{s^3(s^2-s-2)}\right\} = \left[-3+2t-2t^2 + \frac{1}{3}L^{-1}\left\{\frac{1}{s-2}\right\}\right.$$

$$\left. + \frac{8}{3}L^{-1}\left\{\frac{1}{s+1}\right\}\right]. \quad (3)$$

From a table of Laplace transforms, we find that

$$L^{-1}\left\{\frac{1}{s-a}\right\} = e^{at},$$

where a is a constant; hence,

$$L^{-1}\left\{\frac{1}{s-2}\right\} = e^{2t}$$

and

$$L^{-1}\left\{\frac{1}{s+1}\right\} = e^{-t}.$$

Substituting these results into equality (3),

$$L^{-1}\left\{\frac{8}{s^3(s^2-s-2)}\right\} = -3+2-2t^2 + \frac{1}{3}e^{2t} + \frac{8}{3}e^{-t}.$$

● **PROBLEM 16-31**

(a) Define the convolution of two functions $f(t)$ and $g(t)$.

(b) State the convolution theorem for Laplace transforms.

(c) Find the inverse Laplace transform

$$f(t) = L^{-1}\{F(s)\} = L^{-1}\left\{\frac{1}{(s^2+c^2)^2}\right\},$$

(c = constant).

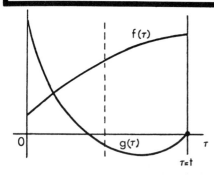

Fig. 1

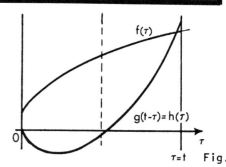

Fig. 2

Solution: (a) A convolution of the two functions $f(t)$ and $g(t)$ is denoted by $(f*g)$ and is defined by

$$(f*g) \equiv \int_0^t f(\tau)g(t-\tau)d\tau. \qquad (1)$$

Thus $(f*g)$ is the integral of the product $f(\tau)h(\tau)$ where $h(\tau)$ is the reflection of $g(\tau)$ with respect to the line $\tau = t/2$ (see Figures 1 and 2). It can be verified that

$(f*g) = (g*f)$ by making the change of variable $u = t - \tau$.

(b) The convolution theorem states that if $f(t)$ and $g(t)$ are piecewise continuous and of exponential orders α and β (see the first problem of this chapter) then

$$L\{f*g\} = F(s)G(s) \quad [\text{Re } s > \max(\alpha,\beta)] \qquad (2)$$

where $F(s) = L\{f(t)\}$ and $G = L\{g(t)\}$. This theorem is very important in the application of Laplace transform technique to differential equations since at the end of a problem one might be faced with inverting $F(s)G(s)$, i.e., finding

$$L^{-1}\{F(s)G(s)\} \quad \text{where} \quad L^{-1}\{F(s)\} = f(t)$$

and

$$L^{-1}\{G(s)\} = g(t)$$

are known functions. But by the convolution theorem,

$$L^{-1}\{F(s)G(s)\} = (f*g) = \int_0^t f(\tau)g(t-\tau)d\tau . \qquad (3)$$

(c) Since the inverse transform of

$$\frac{1}{s^2+c^2}$$

is known to be

$$L^{-1}\left\{\frac{1}{s^2+c^2}\right\} = \frac{1}{c} \sin ct$$

from the previous problem or a table of Laplace transforms, the inverse of

$$F(s) = \left(\frac{1}{s^2+c^2}\right)^2$$

can be found by the convolution theorem. Thus letting

$$F(s) = \frac{1}{s^2+c^2} = G(s)$$

and

$$f(t) = \frac{1}{c} \sin ct = g(t)$$

in (3) yields

$$L^{-1}\left\{\frac{1}{s^2+c^2} \cdot \frac{1}{s^2+c^2}\right\} = \int_0^t \frac{1}{c}\sin c\tau \frac{1}{c}\sin c(t-\tau)d\tau \quad . \quad (4)$$

Using the trigonometric identity

$$\sin(ct-c\tau) = \sin ct \cos c\tau - \cos ct \sin c\tau$$

to evaluate the integral in (4) gives

$$\int_0^t \sin c\tau \sin c(t-\tau)d\tau = \sin ct \int_0^t \sin c\tau \cos c\tau \, d\tau$$

$$- \cos ct \int_0^t \sin^2 c\tau \, d\tau$$

$$= \sin ct \left(\frac{1-\cos 2ct}{4c}\right) - \cos ct \left(\frac{2ct - \sin 2ct}{4c}\right)$$

Using this result in (4) yields

$$L^{-1}\left\{\frac{1}{(s^2+c^2)^2}\right\} = \frac{1}{2c^3}(\sin ct - ct \cos ct) \quad .$$

• **PROBLEM 16-32**

Solve the following problems by employing the convolution theorem:

(a) Find the inverse Laplace transform

$$x_2(t) = L^{-1}\{X_2(s)\} = L^{-1}\left\{\frac{F(s)}{s^2 + 2\lambda s + \omega_0^2}\right\}$$

which arises in the solution of the motion of a forced damped harmonic oscillator.

(b) Find the Laplace transform

$$L\left\{\int_0^t \sinh 2x \, dx\right\} \quad .$$

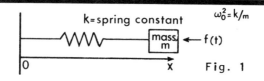

Fig. 1

Solution: (a) A brief sketch of the way in which this problem arises is as follows. The equation of motion of a damped harmonic oscillator which is subject to a time dependent force f(t) (see Fig. 1) is given by

$$\ddot{x} + 2\lambda \dot{x} + \omega_0^2 x = f(t) \qquad \frac{dx}{dt} = \dot{x}, \quad \frac{d^2x}{dt^2} = \ddot{x} \qquad (1)$$

with initial conditions $x(0) = x_0$, $\dot{x}(0) = v_0$. The Laplace transform method calls for taking the transform of (1) giving (using the derivative property of L)

$$s^2 X(s) - sx_0 - v_0 + 2\lambda s X(s) - 2\lambda x_0 + \omega_0^2 X(s) = F(s),$$

where

$$F(s) = L\{f(t)\} \text{ is known and } X(s) = L\{x(t)\}$$

is sought. When solved for $X(s)$, this equation becomes

$$X(s) = \frac{2\lambda x_0 + v_0 + sx_0}{s^2 + 2\lambda s + \omega_0^2} + \frac{F(s)}{s^2 + 2\lambda s + \omega_0^2}$$

$$= X_1(s) + X_2(s). \qquad (2)$$

The first term is inverted by completing the square in the denominator so that

$$\frac{2\lambda x_0 + v_0 + sx_0}{s^2 + 2\lambda s + \omega_0^2} = \frac{x_0(s+\lambda)}{(s+\lambda)^2 + (\omega_0^2 - \lambda^2)} + \frac{v_0 + \lambda x_0}{(s+\lambda)^2 + (\omega_0^2 - \lambda^2)},$$

which, upon term by term inversion and making use of a table of Laplace transforms, yields

$$x_1(t) = L^{-1}\{X_1(s)\} = L^{-1}\left\{\frac{x_0(s+\lambda)}{(s+\lambda)^2 + (\omega_0^2 - \lambda^2)}\right\}$$

$$+ L^{-1}\left\{\frac{v_0 + \lambda x_0}{(s+\lambda)^2 + (\omega_0^2 - \lambda^2)}\right\}$$

$$= x_0 e^{-\lambda t} \cos \omega t + \frac{v_0 + \lambda x_0}{\omega} e^{-\lambda t} \sin \omega t \text{ where } \omega = \sqrt{\omega_0^2 - \lambda^2}.$$

This brings us to the problem at hand, i.e., that of inverting

$$\frac{F(s)}{s^2 + 2\lambda s + \omega_0^2}$$

where
$$F(s) = L\{f(t)\} \ .$$

The relevant pieces of information to be used here are

$$L^{-1}\{F(s)\} = f(t),$$

$$L^{-1}\left\{\frac{1}{s^2 + 2\lambda s + \omega_0^2}\right\} = \frac{1}{\omega} e^{-\lambda t} \sin \omega t \ ,$$

where

$$\omega = \sqrt{\omega_0^2 - \lambda^2} \ ,$$

(this can be deduced by writing

$$\frac{1}{s^2 + 2\lambda s + \omega_0^2} = \frac{1}{(s+\lambda)^2 + (\omega_0^2 - \lambda^2)} = \frac{1}{(s+\lambda)^2 + \omega^2} \ ,$$

recalling the fact that

$$L\{\sin \omega t\} = \frac{\omega}{s^2 + \omega^2}$$

and using the attenuation rule) and the convolution theorem

$$L^{-1}\{F(s)G(s)\} = \int_0^t f(\tau) g(t-\tau) d\tau \ . \qquad (3)$$

Thus, using

$$G(s) = \frac{1}{s^2 + 2\lambda s + \omega_0^2}$$

so that

$$g(t) = L^{-1}\{G(s)\} = \frac{1}{\omega} e^{-\lambda t} \sin \omega t \ ,$$

(3) becomes

$$L^{-1}\left\{\frac{F(s)}{s^2 + 2\lambda s + \omega_0^2}\right\} = \int_0^t \frac{1}{\omega} e^{-\lambda(t-\tau)}$$

$$\times \sin \omega(t-\tau) f(\tau) d\tau \ . \qquad (4)$$

Thus, given $f(t)$, if the integral in (4) can be evaluated

the problem can be solved and the complete solution reads

$$x(t) = x_0 e^{-\lambda t} \cos \omega t + \frac{v_0 + \lambda x_0}{\omega} e^{-\lambda t} \sin \omega t$$

$$+ \frac{1}{\omega} \int_0^t e^{-\lambda(t-\tau)} \sin \omega(t-\tau) f(\tau) d\tau \quad .$$

(b) Recall that the convolution theorem states that if f and g are piecewise continuous and of exponential orders α and β respectively, then

$$L\{(f*g)\} = L\left\{\int_0^t f(\tau) g(t-\tau) d\tau\right\}$$

$$= L\{f(t)\} L\{g(t)\} \, , \quad s > \max(\alpha, \beta) \, . \tag{5}$$

In particular, if

$$L\{g(t)\} = \frac{1}{s}$$

then g = 1 and (5) yields

$$L\left\{\int_0^t f(\tau) d\tau\right\} = \frac{1}{s} L\{f(t)\} = \frac{1}{s} F(s) \, , \tag{6}$$

where $F(s) = L\{f(t)\}$. Using this property in the given problem yields

$$L\left\{\int_0^t \sinh 2x \, dx\right\} = \frac{1}{s} L\{\sinh 2t\} \, . \tag{7}$$

Now either look in a table of Laplace transforms or a previous problem in this chapter to find that

$$L\{\sinh bt\} = \frac{b}{s^2 - b^2} \, , \quad s > |b| \quad ;$$

Hence, taking b = 2 yields

$$L\{\sinh 2t\} = \frac{2}{s^2 - 4} \, , \quad s > 2 \, .$$

Substituting this result into (7),

$$L\left\{\int_0^t \sinh 2x \, dx\right\} = \frac{1}{s} \cdot \frac{2}{s^2-4} = \frac{2}{s(s^2-4)}.$$

APPLICATIONS

● **PROBLEM 16-33**

Find the solution to the initial value problem

$$y''(t) + 4y'(t) + 8y(t) = \sin t, \qquad (1)$$

where

$$y(0) = 1 \text{ and } y'(0) = 0.$$

Solution: This type of differential equation arises quite often in physical problems. One case is that of a damped harmonic oscillator subject to a sinusoidal force sin t (Fig. 1).

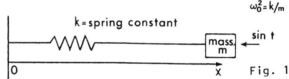

Fig. 1

The differential equation governing the motion of the mass is known to be (with initial conditions)

$$x''(t) + 2\lambda x'(t) + \omega_0^2 x = \sin t, \quad x(0) = x_0, \quad x'(0) = v_0,$$

and where λ is the damping constant. It is seen that (1) is a special case of this. Equations such as (1) also arise in electric circuit theory. In fact, the equation governing the charge q on the capacitor in Figure 2 is known to be

$$L\frac{d^2q}{dt} + R\frac{dq}{dt} + \frac{q}{C} = \sin t$$

where L, R, C are constants and $E(t) = \sin t$ is a sinusoidally

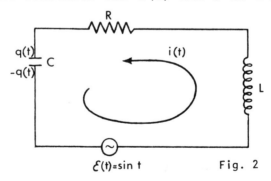

Fig. 2

varying voltage. Thus, equation (1) is also a particular case of this situation. Now to the solution.

Taking the Laplace transform of both sides of equation (1),

$$L\{y''(t) + 4y'(t) + 8y(t)\} = L\{\sin t\}. \qquad (2)$$

From a table of Laplace transforms, we find that

$$L\{\sin bt\} = \frac{b}{s^2 + b^2}, \qquad s > 0,$$

where b is a constant; hence, taking b = 1, we obtain

$$L\{\sin t\} = \frac{1}{s^2 + 1}, \qquad s > 0.$$

Substituting this result into equation (b),

$$L\{y''(t) + 4y'(t) + 8y(t)\} = \frac{1}{s^2 + 1}.$$

Since L is a linear operator,

$$L\{y''(t)\} + 4L\{y'(t)\} + 8L\{y(t)\} = \frac{1}{s^2 + 1}. \qquad (3)$$

Using the properties that

$$L\{y'(t)\} = sL\{y(t)\} - y(0),$$

and

$$L\{y''(t)\} = s^2 L\{y(t)\} - sy(0) - y'(0),$$

provided that $y'(t)$, $y''(t)$, and $L\{y(t)\}$ exist, equation (3) becomes

$$[s^2 L\{y(t)\} - sy(0) - y'(0)] + 4[sL\{y(t)\} - y(0)]$$

$$+ 8L\{y(t)\} = \frac{1}{s^2 + 1}.$$

Solving the above equation for $L\{y(t)\}$,

$$L\{y(t)\} = \frac{(s^2 + 1)(s + 4)y(0) + (s^2 + 1)y'(0) + 1}{(s^2 + 1)(s^2 + 4s + 8)}, \qquad (4)$$

and since $y(0) = 1$ and $y'(0) = 0$, equation (4) becomes

$$L\{y(t)\} = \frac{(s^2 + 1)(s + 4) + 1}{(s^2 + 1)(s + 4s + 8)}$$

$$= \frac{s^3 + 4s^2 + s + 5}{(s^2 + 1)(s^2 + 4s + 8)} \tag{5}$$

Inverting equation (5),

$$y(t) = L^{-1}\left\{\frac{s^3 + 4s^2 + s + 5}{(s^2 + 1)(s^2 + 4s + 8)}\right\}. \tag{6}$$

We use the method of partial fractions to decompose

$$\frac{(s^3 + 4s^2 + s + 5)}{[(s^2 + 1)(s^2 + 4s + 8)]}.$$

Setting

$$\frac{s^3 + 4s^2 + s + 5}{(s^2 + 1)(s^2 + 4s + 8)} = \frac{As + B}{s^2 + 1} + \frac{Cs + D}{s^2 + 4s + 8},$$

for all s, with A, B, C, and D constant, we find upon multiplying through by

$$(s^2 + 1)(s^2 + 4s + 8)$$

that

$$s^3 + 4s^2 + s + 5 = (As + B)(s^2 + 4s + 8) + (Cs + D)(s^2 + 1).$$

Now choose any 4 values for s and substitute them in this equation to obtain four equations in four unknowns (A,B,C,D) Solving these equations will yield

$$A = -\frac{4}{65}, \quad B = \frac{7}{65}, \quad C = \frac{69}{65}, \quad \text{and } D = \frac{269}{65}.$$

Thus equation (6) becomes

$$y(t) = L^{-1}\left\{\frac{-\frac{4}{65}s + \frac{7}{65}}{s^2 + 1} + \frac{\frac{69}{65}s + \frac{269}{65}}{s^2 + 4s + 8}\right\}. \tag{7}$$

Completing the square in the right-hand denominator in equation (4),

$$y(t) = L^{-1}\left\{\frac{-\frac{4}{65}s + \frac{7}{65}}{s^2 + 1} + \frac{\frac{69}{65}s + \frac{269}{65}}{(s + 2)^2 + 2^2}\right\}$$

$$= L^{-1}\left\{\frac{-\frac{4}{65}s + \frac{7}{65}}{s^2 + 1} + \frac{\frac{69}{65}(s+2) + \frac{131}{130} \cdot (2)}{(s+2)^2 + 2^2}\right\}.$$

Since L^{-1} is a linear operator under the conditions assumed in this chapter,

$$y(t) = \left[-\frac{4}{65} L^{-1}\left\{\frac{s}{s^2+1}\right\} + \frac{7}{65} L^{-1}\left\{\frac{1}{s^2+1}\right\}\right.$$

$$+ \frac{69}{65} L^{-1}\left\{\frac{(s+2)}{(s+2)^2 + 2^2}\right\}$$

$$\left. + \frac{131}{130} L^{-1}\left\{\frac{2}{(s+2)^2 + 2^2}\right\}\right]. \qquad (8)$$

Using the property that

$$L^{-1}\{f(s-k)\} = e^{kt} L^{-1}\{f(s)\},$$

where k is a constant, provided that $L^{-1}\{f(s)\}$ exists, equation (8) becomes

$$y(t) = \left[-\frac{4}{65} L^{-1}\left\{\frac{s}{s^2+1}\right\} + \frac{7}{65} L^{-1}\left\{\frac{1}{s^2+1}\right\}\right.$$

$$+ \frac{69}{65} e^{-2t} L^{-1}\left\{\frac{s}{s^2 + 2^2}\right\}$$

$$\left. + \frac{131}{130} e^{-2t} L^{-1}\left\{\frac{2}{s^2 + 2^2}\right\}\right]. \qquad (9)$$

From a table of Laplace transforms, we find that

$$L^{-1}\left\{\frac{s}{s^2 + c_1^2}\right\} = \cos c_1 t,$$

and

$$L^{-1}\left\{\frac{c_2}{s^2 + c_1^2}\right\} = \sin c_2 t,$$

where c_1 and c_2 are constants; hence, taking $c_1 = 1$ and 2 and $c_2 = 1$ and 2, respectively, we obtain

$$L^{-1}\left\{\frac{s}{s^2+1}\right\} = \cos t, \quad L^{-1}\left\{\frac{s}{s^2+2^2}\right\} = \cos 2t,$$

$$L^{-1}\left\{\frac{1}{s^2+1}\right\} = \sin t, \quad L^{-1}\left\{\frac{2}{s^2+2^2}\right\} = \sin 2t.$$

Substituting these results into equation (9),

$$y(t) = -\frac{4}{65}\cos t + \frac{7}{65}\sin t + \frac{69}{65}e^{-2t}\cos 2t$$

$$+ \frac{131}{130}e^{-2t}\sin 2t. \tag{10}$$

Since we made some assumptions about existence of $y(t)$, $y'(t)$, $y''(t)$, and $L\{y(t)\}$, we must check our result (10) against the original problem. Differentiating twice, we obtain

$$y'(t) = \frac{7}{65}\cos t + \frac{4}{65}\sin t + \frac{e^{-2t}}{65}(-7\cos 2t - 269\sin 2t),$$

$$y''(t) = \frac{4}{65}\cos t - \frac{7}{65}\sin t + \frac{e^{-2t}}{65}$$

$$\times (-524\cos 2t + 552\sin 2t).$$

Checking in equation (1),

$$y''(t) + 4y'(t) + 8y(t)$$

$$= \left[\frac{4}{65}\cos t - \frac{7}{65}\sin t\right.$$

$$\left. + \frac{e^{-2t}}{65}(-524\cos 2t + 552\sin 2t)\right]$$

$$+ 4\left[\frac{7}{65}\cos t + \frac{4}{65}\sin t\right.$$

$$\left. + \frac{e^{-2t}}{65}(-7\cos 2t - 269\sin 2t)\right]$$

$$+ 8\left[-\frac{4}{65}\cos t + \frac{7}{65}\sin t + \frac{69}{65}e^{-2t}\cos 2t\right.$$

$$\left. + \frac{131}{130}e^{-2t}\sin 2t\right]$$

$$= \sin t,$$

$$y(0) = -\frac{4}{65} \cos 0 + \frac{7}{65} \sin 0 + \frac{69}{65} e^0 \cos 0 + \frac{131}{130} e^0 \sin 0$$

$$= 1,$$

$$y'(0) = \frac{7}{65} \cos 0 + \frac{4}{65} \sin 0 + \frac{e^0}{65} (-7 \cos 0 - 269 \sin 0)$$

$$= 0;$$

hence, equation (10) is the solution.

• **PROBLEM 16-34**

A beam of rigidity EI is clamped at one end and is loaded as shown in the figure; the weight of the beam is neglected. Find the deflection of the beam, y(x), where, for the coordinate system shown, the following differential relations are known to hold.

(a)

$$\frac{d^2 y(x)}{dx^2} = -\frac{1}{EI} m(x),$$

where y(x) is the deflection of the beam at point x, and m(x) is the bending moment [counterclockwise torque of all (external) forces to the right of point x],

(b)

$$\frac{dm(x)}{dx} = t(x),$$

where t(x) is the shearing force (resultant of all vertical forces to the right of point x),

(c)

$$\frac{dt(x)}{dx} = -q(x),$$

where q(x) is the load per unit length at point x.

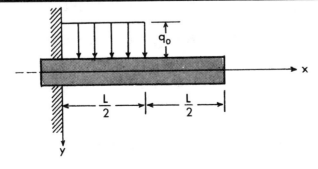

Solution: From these relations it follows that

$$EI \frac{d^4 y(x)}{dx^4} = q(x)$$

Since the beam is clamped, we have $y(0) = 0$, $y'(0)$. Also from the definition of t, since the weight, q_0, per unit length from 0 to L/2 is the only vertical force, it is seen that

t(0) = resultant or "sum" of all vertical forces to the right of

$$0 = q_0 \frac{\text{units of forces}}{\text{unit length}} \times \frac{L}{2} \text{ units of length.}$$

That is,

$$t(0) = q_0 (L/2).$$

Finally, to find m(0) note the following relations for any small interval dx

$q_0 \cdot dx$ = force at points in (x, x+dx), where

$$x \in (0, L/2),$$

$(q_0 \cdot dx) \cdot x$ = torque about 0 due to load at point x.

Therefore,

$$m(0) = \text{total counterclockwise torque} = - \int_0^{L/2} q_0 x \, dx$$

$$= - q_0 \frac{x^2}{2} \Big|_0^{L/2} = - \frac{q_0 L^2}{8}.$$

Using these results for m(0) and t(0) in the conditions (a) and (b) gives

$$y''(0) = q_0 L^2 / 8EI, \quad y'''(0) = - q_0 L / 2EI.$$

Now, take the Laplace transform of the equation (1) and use the derivative property

$$L\{y^{(n)}(t)\} = s^n L\{y(t)\} - \sum_{k=1}^{n} s^{k-1} y^{(n-k)}(0)$$

to find

$$EI[s^4 L\{y(x)\} - y^{(3)}(0) - sy''(0) - s^2 y'(0) - s^3 y(0)] =$$

or, letting $Y(s) = L\{y(t)\}$,

$$EI\left[s^4 Y(s) + \frac{q_0 L}{2EI} - \frac{s q_0 L^2}{8EI}\right] = L\{q_0[1 - \alpha(x - L/2)]\}$$

$$= L\{q_0\} - L\{q_0 \alpha(x - L/2)\} \qquad (2)$$

where

$$\alpha(x) = \begin{cases} 1, & x \geq 0 \\ 0, & x < 0 \end{cases}$$

is the unit step function so that

$$1 - \alpha\{x - L/2\} = \begin{cases} 0, & x \geq L/2 \\ 1, & x < L/2 \end{cases}$$

From a table of Laplace transforms or the results of previous problems it can be found that

$$L\{1\} = \frac{1}{s} \quad , \quad s > 0$$

$$L\{\alpha(t-k)\} = \frac{e^{-ks}}{s} \quad , \quad s > 0 \ .$$

Using these results with $k = L/2$ and rearranging (2) gives

$$s^4 Y(s) - s \frac{q_0 L^2}{8EI} + \frac{q_0 L}{2EI} = \frac{q_0}{EI} \frac{1 - e^{-sL/2}}{s} \ ,$$

so that

$$Y(s) = \frac{q_0}{EI} \frac{1}{s^5} - \frac{q_0}{EI} \frac{e^{-sL/2}}{s^5} + \frac{q_0 L^2}{8EI} \frac{1}{s^3} - \frac{q_0 L^2}{2EI} \frac{1}{s^4} \ .$$

Inversion yields

$$y(x) = L^{-1}\{Y(s)\} = L^{-1}\left\{\frac{q_0}{EI} \frac{1}{s^5}\right\} - L^{-1}\left\{\frac{q_0}{EI} \frac{e^{-sL/2}}{s^5}\right\}$$

$$+ L^{-1}\left\{\frac{q_0 L^2}{8EI} \frac{1}{s^3}\right\} - L^{-1}\left\{\frac{q_0 L}{2EI} \frac{1}{s^4}\right\}$$

or

$$y(x) = \frac{q_0}{24EI} x^4 - \frac{q_0}{24EI}\left(x - \frac{L}{2}\right)^4 \alpha\left(x - \frac{L}{2}\right) + \frac{q_0 L^2}{16EI} x^2 - \frac{q_0 L}{12EI} x^3.$$

It is more convenient to rewrite this solution in the form

$$y(x) = \begin{cases} \dfrac{q_0}{EI}\left(\dfrac{x^4}{24} - \dfrac{Lx^3}{12} + \dfrac{L^2 x^2}{16}\right) & \left(0 \leq x < \dfrac{L}{2}\right) \\ \dfrac{q_0}{EI}\left(\dfrac{L^3 x}{48} - \dfrac{L^4}{384}\right) & \left(\dfrac{L}{2} < x \leq L\right) \end{cases}$$

from which, for instance, it is clearly seen that the right half of the beam will remain straight, a fact anticipated on physical grounds.

● **PROBLEM** 16-35

Solve the initial value problem

$$y''(t) + 2y'(t) + 5y(t) = H(t) \tag{1}$$

$$y(0) = y'(0) = 0,$$

where

$$H(t) = \begin{cases} 1, & 0 \leq t < \pi \\ 0, & t \geq \pi, \end{cases}$$

as shown in the accompanying graph.

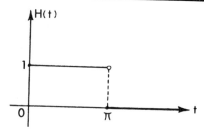

Solution: The function $H(t)$ can be expressed as

$$H(t) = 1 - \alpha(t - \pi),$$

where α is the unit step function

$$\alpha(x) = \begin{cases} 0, & x < 0, \\ 1, & x \geq 0. \end{cases}$$

Thus, equation (1) is equivalent to

$$y''(t) + 2y'(t) + 5y(t) = 1 - \alpha(t - \pi).$$

Taking the Laplace transform of both sides,

$$L\{y''(t) + 2y'(t) + 5y(t)\} = L\{1 - \alpha(t - \pi)\}.$$

Since L is a linear operator,

$$L\{y''(t)\} + 2L\{y'(t)\} + 5L\{y(t)\} = L\{1\} - L\{\alpha(t - \pi)\}. \tag{2}$$

From a table of Laplace transforms, we find that

$$L\{1\} = \frac{1}{s}, \qquad s > 0,$$

and

$$L\{\alpha(t - k)\} = \frac{e^{-ks}}{s}, \qquad s > 0,$$

where k is a nonnegative constant. Taking $k = \pi$, the latter result becomes

$$L\{\alpha(t - \pi)\} = \frac{e^{-\pi s}}{s}, \qquad s > 0.$$

Substituting these results into equation (2),

$$L\{y''(t)\} + 2L\{y'(t)\} + 5L\{y(t)\} = \frac{1}{s} - \frac{e^{-\pi s}}{s}. \tag{3}$$

Using the properties that

$$L\{y'(t)\} = sL\{y(t)\} - y(0)$$

and

$$L\{y''(t)\} = s^2 L\{y(t)\} - sy(0) - y'(0),$$

provided that $y'(t)$, $y''(t)$, and $L\{y(t)\}$ exist, equation (3) becomes

$$[s^2 L\{y(t)\} - sy(0) - y'(0)] + 2[sL\{y(t)\} - y(0)]$$

$$+ 5L\{y(t)\} = \frac{1 - e^{-\pi s}}{s}.$$

Solving for $L\{y(t)\}$,

$$L\{y(t)\} = \frac{1 - e^{-\pi s} + s(s+2)y(0) + sy'(0)}{s(s^2 + 2s + 5)},$$

and since

$$y(0) = y'(0) = 0,$$

$$L\{y(t)\} = \frac{1 - e^{-\pi s}}{s(s^2 + 2s + 5)}.$$

Inverting the above equation,

$$y(t) = L^{-1}\left\{\frac{1 - e^{-\pi s}}{s(s^2 + 2s + 5)}\right\}.$$

Since L^{-1} is a linear operator under the conditions assumed in this chapter

$$y(t) = L^{-1}\left\{\frac{1}{s(s^2 + 2s + 5)}\right\} - L^{-1}\left\{\frac{e^{-\pi s}}{s(s^2 + 2s + 5)}\right\}.$$

Let

$$G(t) = L^{-1}\left\{\frac{1}{s(s^2 + 2s + 5)}\right\}.$$

Thus,

$$y(t) = G(t) - L^{-1}\left\{\frac{e^{-\pi s}}{s(s^2 + 2s + 5)}\right\}. \qquad (4)$$

Using the property that

$$L^{-1}\{e^{-ks} F(s)\} = f(t - k)\alpha(t - k),$$

where k is a nonnegative constant and $f(t) = L^{-1}\{F(s)\}$, provided that $L^{-1}\{F(s)\}$ exists, we find (with $k = \pi$) that

$$L^{-1}\left\{\frac{e^{-\pi s}}{s(s^2 + 2s + 5)}\right\} = G(t - \pi)\alpha(t - \pi);$$

hence, equation (4) is equivalent to

$$y(t) = G(t) - G(t - \pi)\alpha(t - \pi). \qquad (5)$$

We now evaluate the function $G(t)$. We use the method of partial fractions to decompose

$$\frac{1}{[s(s^2 + 2s + 5)]}.$$

Setting

$$\frac{1}{s(s^2 + 2s + 5)} = \frac{A}{s} + \frac{Bs + C}{s^2 + 2s + 5},$$

for all s, with A, B, and C constant, we find

$$A = \frac{1}{5}, \quad B = -\frac{1}{5}, \quad \text{and } C = -\frac{2}{5}.$$

Thus,

$$G(t) = L^{-1}\left\{\frac{1}{5}\left(\frac{1}{s}\right) - \frac{1}{5}\left(\frac{s + 2}{(s^2 + 2s + 5)}\right)\right\}.$$

Completing the square in the denominator of the fraction

$$\frac{(s + 2)}{(s^2 + 2s + 5)}$$

above,

$$G(t) = L^{-1}\left\{\frac{1}{5}\left(\frac{1}{s}\right) - \frac{1}{5}\left[\frac{s + 2}{(s + 1)^2 + 2^2}\right]\right\}$$

$$= L^{-1}\left\{\frac{1}{5}\left(\frac{1}{s}\right) - \frac{1}{5}\left[\frac{s + 1}{(s + 1)^2 + 2^2}\right]\right.$$

$$\left. - \frac{1}{10}\left[\frac{2}{(s + 1)^2 + 2^2}\right]\right\}.$$

Since L^{-1} is a linear operator,

$$G(t) = \frac{1}{5}L^{-1}\left\{\frac{1}{s}\right\} - \frac{1}{5}L^{-1}\left\{\frac{s + 1}{(s + 1)^2 + 2^2}\right\}$$

$$- \frac{1}{10}L^{-1}\left\{\frac{2}{(s + 1)^2 + 2^2}\right\}. \tag{6}$$

Using the property that

$$L^{-1}\{f(s - b)\} = e^{bt} L^{-1}\{f(s)\},$$

873

where b is a constant, provided that $L^{-1}\{f(s)\}$ exists, equation (6) becomes (with b = -1)

$$G(t) = \frac{1}{5} L^{-1}\left\{\frac{1}{s}\right\} - \frac{1}{5} e^{-t} L^{-1}\left\{\frac{s}{s^2 + 2^2}\right\}$$

$$- \frac{1}{10} e^{-t} L^{-1}\left\{\frac{2}{s^2 + 2^2}\right\} . \qquad (7)$$

From a table of Laplace transforms, we find that

$$L^{-1}\left\{\frac{1}{s}\right\} = 1, \quad L^{-1}\left\{\frac{s}{s^2 + c^2}\right\} = \cos ct,$$

and

$$L^{-1}\left\{\frac{c}{s^2 + c^2}\right\} = \sin ct,$$

where c is a constant. Taking c = 2, the last two of the above results become

$$L^{-1}\left\{\frac{s}{s^2 + 2^2}\right\} = \cos 2t,$$

and

$$L^{-1}\left\{\frac{2}{s^2 + 2^2}\right\} = \sin 2t .$$

Substituting these results into equation (7),

$$G(t) = \frac{1}{5} - \frac{1}{5} e^{-t} \cos 2t - \frac{1}{10} e^{-t} \sin 2t, \qquad (8)$$

and thus,

$$G(t - \pi) = \frac{1}{5} - \frac{1}{5} e^{-(t-\pi)} \cos 2(t - \pi)$$

$$- \frac{1}{10} e^{-(t-\pi)} \sin 2(t - \pi) . \qquad (9)$$

Note that

$$\cos 2(t - \pi) = \cos (2t - 2\pi) = \cos 2t,$$

and

$$\sin 2(t - \pi) = \sin (2t - 2\pi) = \sin 2t;$$

874

therefore, equation (9) is equivalent to

$$G(t - \pi) = \frac{1}{5} - \frac{1}{5} e^{-(t-\pi)} \cos 2t - \frac{1}{10} e^{-(t-\pi)} \sin 2t . \quad (10)$$

Substituting equations (8) and (10) into equation (5),

$$y(t) = \frac{1}{5} - \frac{1}{5} e^{-t} \cos 2t - \frac{1}{10} e^{-t} \sin 2t$$

$$- \left[\frac{1}{5} - \frac{1}{5} e^{-(t-\pi)} \cos 2t - \frac{1}{10} e^{-(t-\pi)} \sin 2t \right] \alpha(t - \pi) . \quad (11)$$

Since we made some assumptions about existence of $y(t), y'(t), y''(t)$ and $L(y(t))$, we must check our result (10) against the original problem. Differentiating twice, we obtain

$$y'(t) = \frac{1}{2} \left[1 - e^{\pi} \alpha(t - \pi) \right] e^{-t} \sin 2t$$

$$y''(t) = \frac{1}{2} \left[1 - e^{\pi} \alpha(t - \pi) \right] e^{-t} (- \sin 2t + 2 \cos 2t) .$$

Checking in equation (1),

$$y''(t) + 2y'(t) + 5y(t)$$

$$= \frac{1}{2} [1 - e^{\pi} \alpha(t - \pi)] e^{-t} (- \sin 2t + 2 \cos 2t)$$

$$+ 2 \cdot \frac{1}{2} [1 - e^{\pi} \alpha(t - \pi)] e^{-t} \sin 2t$$

$$+ 5 \left[\frac{1}{5} - \frac{1}{5} e^{-t} \cos 2t - \frac{1}{10} e^{-t} \sin 2t \right.$$

$$\left. - \left(\frac{1}{5} - \frac{1}{5} e^{-(t-\pi)} \cos 2t - \frac{1}{10} e^{-(t-\pi)} \sin 2t \right) \alpha(t-\pi) \right]$$

$$= 1 - \alpha(t - \pi) = H(t),$$

$$y(0) = \frac{1}{5} - \frac{1}{5} e^{0} \cos 0 - \frac{1}{10} e^{0} \sin 0$$

$$- \left[\frac{1}{5} - \frac{1}{5} e \cos 0 - \frac{1}{10} e^{\pi} \sin 0 \right] \alpha(0 - \pi)$$

$$= 0$$

$$y'(0) = \frac{1}{2} [1 - e^{\pi} \alpha(0 - \pi)] e^{0} \sin 0 = 0;$$

hence, equation (11) is the solution.

• **PROBLEM 16-36**

Two circuits are coupled magnetically, as shown in Fig. 1. Find the currents $i_1(t)$ and $i_2(t)$ after the switch S is closed, given that the currents are known to obey the differential equations

$$L_1 \frac{di_1}{dt} + M \frac{di_2}{dt} + R_1 i_1 = e_0 S(t), \tag{1}$$

$$M \frac{di_1}{dt} + L_2 \frac{di_2}{dt} + R_2 i_2 = 0, \tag{2}$$

where

$$S(t) = \begin{cases} 1, & t > 0 \\ 0, & t < 0 \end{cases}$$

is the unit step function and the initial currents are assumed to be zero.

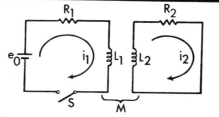

Fig. 1

Solution: The first step is to take the Laplace transform of each of equations (1) and (2) to get

$$L\left\{L_1 \frac{di_1}{dt} + M \frac{di_2}{dt} + R_1 i_1\right\} = L\{e_0 S(t)\} \tag{3}$$

$$L\left\{M \frac{di_1}{dt} + L_2 \frac{di_2}{dt} + R_2 i_2\right\} = L\{0\}. \tag{4}$$

Since L is a linear operator (3) and (4) can be written as

$$L_1 L\left\{\frac{di_1}{dt}\right\} + ML\left\{\frac{di_2}{dt}\right\} + R_1 L\{i_1\} = e_0 L\{S(t)\} \tag{5}$$

$$ML\left\{\frac{di_1}{dt}\right\} + L_2 L\left\{\frac{di_2}{dt}\right\} + R_2 L\{i_2\} = L\{0\}. \tag{6}$$

Note the derivative property states that

876

$$L\{f'(t)\} = sL\{f(t)\} - f(0).$$

Using this in (5) and (6) yields

$$L_1\left[sL\{i_1\} - i_1(0)\right] + M\left[sL\{i_2\} - i_2(0)\right]$$

$$+ R_1 L\{i_1\} = e_0 \cdot \frac{1}{s} \tag{7}$$

$$M\left[sL\{i_1\} - i_1(0)\right] + L_2\left[sL\{i_2\} - i_2(0)\right]$$

$$+ R_2 L\{i_2\} = 0 \tag{8}$$

where a table of Laplace transforms may be consulted to find

$$L\{S(t)\} = 1/s$$

and

$$L\{0\} = 0.$$

Using the fact that

$$i_1(0) = i_2(0) = 0$$

and letting

$$I_1(s) = L\{i_1(t)\}, \quad I_2(s) = L\{i_2(t)\}, \quad (7) \text{ and } (8)$$

become a system of two equations in two unknowns I_1, I_2:

$$(L_1 s + R_1) I_1 + M s I_2 = e_0/s,$$

$$M s I_1 + (L_2 s + R_2) I_2 = 0. \tag{9}$$

The determinant of the system is

$$\Delta = \det \begin{pmatrix} L_1 s + R_1 & Ms \\ Ms & L_2 s + R_2 \end{pmatrix}$$

$$= (L_1 s + R_1)(L_2 s + R_2) - (Ms)(Ms)$$

$$= (L_1 L_2 - M^2) s^2 + (L_1 R_2 + L_2 R_1) s + R_1 R_2.$$

It is known from electromagnetic theory that $L_1 L_2 \geq M^2$. The usual condition in practice is $L_1 L_2 > M^2$. Thus we have, using Cramer's rule on the system (9),

$$I_1(s) = \frac{\begin{vmatrix} e_0/s & Ms \\ 0 & L_2s + R_2 \end{vmatrix}}{\Delta} = \frac{e_0(L_2s + R_2)}{s \cdot \Delta}$$

$$I_2(s) = \frac{\begin{vmatrix} L_1s + R_1 & e_0/s \\ Ms & 0 \end{vmatrix}}{\Delta} = -\frac{Me_0}{\Delta}$$

which may be written in the form

$$I_1(s) = \frac{e_0}{(L_1L_2 - M^2)} \frac{L_2s + R_2}{s(s - r_1)(s - r_2)}, \tag{10}$$

$$I_2(s) = -\frac{Me_0}{(L_1L_2 - M^2)} \frac{1}{(s - r_1)(s - r_2)}, \tag{11}$$

where r_1 and r_2 are the roots of

$$\Delta = (L_1L_2 - M^2)s^2 + (L_1R_2 + L_2R_1)s + R_1R_2 = 0. \tag{12}$$

Since the discriminant of (12) is

$$(L_1R_2 + L_2R_1)^2 - 4R_1R_2(L_1L_2 - M^2) = (L_1R_2 - R_1L_2)^2$$
$$+ 4R_1R_2M^2 > 0,$$

both roots are real.. Moreover, they are both negative since

$$\frac{L_1R_2 + L_2R_1}{L_1L_2 - M^2} > 0$$

and

$$\frac{R_1R_2}{L_1L_2 - M^2} > 0.$$

Rational fraction decomposition yields

$$I_1(s) = \frac{e_0}{L_1L_2 - M^2} \left(\frac{R_2}{r_1r_2s} + \frac{L_2r_1 + R_2}{r_1(r_1 - r_2)(s - r_1)} \right.$$

$$+ \frac{L_2 r_2 + R_2}{r_2(r_2 - r_1)(s - r_2)} \Bigg), \tag{13}$$

$$I_2(s) = -\frac{Me_0}{L_1 L_2 - M^2} \cdot \frac{1}{r_1 - r_2} \left(\frac{1}{s - r_1} - \frac{1}{s - r_2} \right). \tag{14}$$

Inverting [using $r_1 r_2 = R_1 R_2 / (L_1 L_2 - M^2)$], we find that since L^{-1} is a linear operator under the assumptions made in this chapter, (13) and (14) become

$$i_1(t) = L^{-1}\{I_1(s)\} = \frac{e_0}{L_1 L_2 - M^2} \left[\frac{R_2(L_1 L_2 - M^2)}{R_1 R_2} L^{-1}\left\{\frac{1}{s}\right\} \right.$$

$$+ \frac{L_2 r_1 + R_2}{r_1(r_1 - r_2)} L^{-1}\left\{\frac{1}{s - r_1}\right\} + \frac{L r_2 + R_2}{r_2(r_2 - r_1)} L^{-1}\left\{\frac{1}{s - r_2}\right\}$$

$$i_2(t) = \frac{-Me_0}{L_1 L_2 - M^2} \frac{1}{r_1 - r_2} \left[L^{-1}\left\{\frac{1}{s - r_1}\right\} - L^{-1}\left\{\frac{1}{s - r_2}\right\} \right].$$

Recalling that

$$L^{-1}\left\{\frac{1}{s - c}\right\} = e^{ct},$$

$$i_1(t) = \frac{e_0}{R_1} + \frac{e_0}{L_1 L_2 - M^2} \left(\frac{L_2 r_1 + R_2}{r_1(r_1 - r_2)} e^{r_1 t} \right.$$

$$\left. - \frac{L_2 r_2 + R_2}{r_2(r_1 - r_2)} e^{r_2 t} \right),$$

$$i_2(t) = -\frac{Me_0}{(L_1 L_2 - M^2)} \frac{1}{(r_1 - r_2)} (e^{r_1 t} - e^{r_2 t}).$$

It is seen that $i_1 \to e_0/R$ and $i_2 \to 0$ as $t \to \infty$ as expected on physical grounds. Also, $i_1(0) = 0$ and $i_2(0) = 0$.

CHAPTER 17

FOURIER TRANSFORMS

It has been seen in previous chapters that there are often great advantages in representing a function $f(x)$ by its power series

$$f(x) = \sum_{n=0}^{\infty} a_n x^n$$

or by its Fourier series

$$f(x) = \sum_{n=-\infty}^{\infty} c_n e^{i\frac{n\pi x}{L}} \qquad (-L < x < L).$$

These representations are useful (provided f satisfies the necessary conditions for the expansions) for approximating values of f and in solving differential equations. The important thing to note here is that the set of coefficients $\{a_n\}$ "define" f in a sense, since, once they are known, f is known. The same statement is true for the set $\{c_n\}$.

In the previous chapter the concept of the discrete power series representation defined by the a_n's was extended to a continuous Laplace integral defined by the equation

$$F(s) = L\{f(t)\} = \int_0^{\infty} f(t) e^{-st} \, dt \; .$$

The analogy between $a(n)$ (the a_n's considered as a function of n) and $F(s)$ can be brought forth by noting the theorem which states that

$$f(t) = \frac{1}{2\pi i} \int_{\gamma-i\infty}^{\gamma+i\infty} F(s) e^{ts} \, ds$$

where the limits of integration $\gamma-i\infty$ to $\gamma+i\infty$ represent the line in the complex plane defined by $\text{Im}(s) = \gamma$. This equation shows that $f(x)$ is defined by $F(s)$ in the same sense that it is defined by the power series coefficients and the Fourier series coefficients. Thus, one would expect a similar extension of the Fourier series expansion to a continuous "Fourier integral representation" of f, and that is the subject of this chapter. It will be seen that Fourier integrals (or "transforms") have applications similar to those of Laplace transforms in solving differential equations.

DEFINITION OF FOURIER TRANSFORMS

• PROBLEM 17-1

Develop the definition of the Fourier transform of a function $f(x)$ by extending the definition of the Fourier series of f to the case where the discrete spectrum of Fourier coefficients becomes a continuous spectrum.

Solution: Let f be representable by its Fourier series on $(-L, L)$. Then f can be written as

$$f(x) = \sum_{n=-\infty}^{\infty} C_n e^{i\frac{n\pi x}{L}} \qquad (-L < x < L) \qquad (1)$$

where C_n are the complex Fourier coefficients of f given by

$$C_n = \frac{1}{2L} \int_{-L}^{L} f(x) e^{-i\frac{n\pi x}{L}} dx. \qquad (2)$$

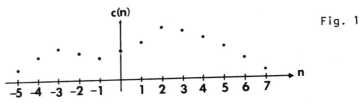

Fig. 1

To emphasize the functional dependence of C_n on n, write $C(n) = C_n$. The function $C(n)$ is called the Fourier spectrum of $f(x)$ and a typical example is plotted in Figure 1 (for convenience, $f(x)$ is assumed real and even so that the $C(n)$ are real and the graph may be made). Now make the substitutions

$$k = \frac{n\pi}{L}, \quad \left(\frac{L}{\pi}\right) C_n = C_L(k)$$

where k is called the wave number of the nth term (or kth term) in the Fourier series expansion of f. Then equations (1) and (2) may be written as

$$C_L(k) = \frac{1}{2\pi} \int_{-L}^{+L} f(x) e^{-ikx} dx,$$

$$f(x) = \sum_{\frac{Lk}{\pi}=-\infty}^{\infty} C_L(k) e^{ikx} \Delta k \qquad (3)$$

881

since

$$\Delta k = \frac{\pi}{L} \Delta n = \frac{\pi}{L} .$$

With the change of scale from n to k the Fourier spectrum may be plotted versus wave number as in Figure 2.

Evidently as L approaches infinity, the wave number spectrum approaches a continuous spectrum, i.e., trigonometric functions of all wave numbers must be summed to represent f. Hence, as $L \to \infty$ (the function is no longer periodic) the sum in the second equation of (3) becomes an integral since $\Delta k \to 0$ and we write

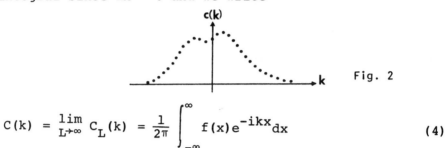

Fig. 2

$$C(k) = \lim_{L \to \infty} C_L(k) = \frac{1}{2\pi} \int_{-\infty}^{\infty} f(x) e^{-ikx} dx \tag{4}$$

and

$$f(x) = \int_{-\infty}^{\infty} C(k) e^{ikx} dk . \tag{5}$$

With a slight change in notation, formulas (4) and (5) become the Fourier transformation. Thus, define a new function F by

$$F(k) = \sqrt{2\pi}\, C(-k)$$

so that the formulas now read

$$F(k) = \frac{1}{\sqrt{2\pi}} \int_{-\infty}^{\infty} f(x) e^{ikx} dx \tag{6}$$

$$f(x) = \frac{1}{\sqrt{2\pi}} \int_{-\infty}^{\infty} F(k) e^{-ikx} dk \tag{7}$$

F(k) is known as the Fourier transform of the function f(x) and, conversely, f(x) is called the inverse Fourier transform of F(k). To emphasize the interpretation of the Fourier transform as an operator operating on f(x), we will write

$$F(k) = \Phi\{f(x)\}$$

where Φ denotes the operation on f described in equation (6). Finally, note that (6) is merely a definition of F(k), just as (2) is a definition of the coefficients C_n, so that there is no question about the validity of this formula. That is, if f(x) is integrable and the integral in (6) converges then F(k) exists. On the other hand there is a question about whether the original function f(x) can be retrieved by the formula (7) since this formula was obtained by a limiting process on the Fourier series of f and there is a question as to whether this series represents f. The theory surrounding the convergence of the integral in (7) to f(x) is very similar to that of the convergence of Fourier series and will be discussed later in this chapter.

● **PROBLEM 17-2**

Find the Fourier transform, $F(k) = \Phi\{f(x)\}$ of the Gaussian probability function

$$f(x) = Ne^{-\alpha x^2} \qquad (N, \alpha = \text{constant}). \qquad (1)$$

Show directly that f(x) is retrievable from the inverse transform. I.e., show that

$$f(x) = \frac{1}{\sqrt{2\pi}} \int_{-\infty}^{\infty} F(k) e^{-ikx} dx = \Phi^{-1}\{\Phi\{f(x)\}\} .$$

Solution: The definition of the Fourier transform of f(x) is

$$F(k) = \Phi\{f(x)\} = \frac{1}{\sqrt{2\pi}} \int_{-\infty}^{\infty} f(x) e^{ikx} dx . \qquad (2)$$

Using the Gaussian function of (1) yields

$$F(k) = \frac{1}{\sqrt{2\pi}} \int_{-\infty}^{\infty} Ne^{-\alpha x^2} e^{ikx} dx = \frac{N}{\sqrt{2\pi}} \int_{-\infty}^{\infty} e^{(-\alpha x^2 + ikx)} dx . \qquad (3)$$

It is convenient to complete the square in the integrand of (3) to get

$$-\alpha x^2 + ikx = -\left(x\sqrt{\alpha} - \frac{ik}{2\sqrt{\alpha}}\right)^2 - \frac{k^2}{4\alpha} . \qquad (4)$$

Now make the change of variables

$$x\sqrt{\alpha} - \frac{ik}{2\sqrt{\alpha}} = u \text{ in (4) to obtain}$$

$$-\alpha x^2 + ikx = -u^2 - \frac{k^2}{4\alpha}, \quad dx = \frac{1}{\sqrt{\alpha}} du \qquad (5)$$

and substitute (5) into (3) to find

$$F(k) = \frac{N}{\sqrt{2\pi\alpha}} e^{-k^2/4\alpha} \int_{-\infty}^{\infty} e^{-u^2} du = N \frac{1}{\sqrt{2\alpha}} e^{-k^2/4\alpha}. \qquad (6)$$

Note that F(k) is also a Gaussian probability function with a peak (the mean) at x = 0. Also note that if f(x) is sharply peaked due to a large α, then F(k) is broadened and vice versa (see Figure 1). This has important applications in Quantum physics where

$$|f(x)|^2$$

represents the probability of finding a (one-dimensional

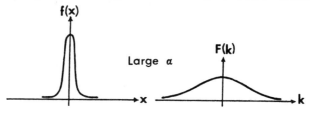

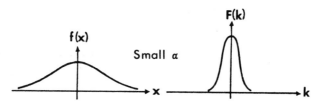

Fig. 1

particle at point x and $|F(k)|^2$ represents the probability of finding this particle with momentum p = ℏk where ℏ is a known constant. Thus, the better one is able to predict the location of the particle (narrow f(x)), the harder it is to predict its momentum (broad F(k)) and vice versa. This is the basic idea behind what is known as the Heisenberg's uncertainty principle and the fact that "narrow" functions have "broad" Fourier transforms and vice versa is a property of all Fourier transforms.

The inverse Fourier integral is given by

$$\Phi^{-1}\{F(k)\} = \frac{1}{\sqrt{2\pi}} \int_{-\infty}^{\infty} F(k) e^{-ikx} dx = \frac{1}{\sqrt{2\pi}} \frac{N}{\sqrt{2\alpha}}$$

$$\times \int_{-\infty}^{\infty} e^{-k^2/4\alpha} e^{-ikx} dk \qquad (7)$$

and it is desired to see whether this equals f(x) as one would expect from the theory. The integral in (7) is calculated in the same way as that in (3). In fact, as a short-cut to that calculation, set

$$\alpha^1 = \frac{1}{4\alpha} \quad \text{and} \quad x^1 = -x$$

to deduce

$$\frac{1}{\sqrt{2\pi}} \int_{-\infty}^{\infty} e^{-\alpha^1 k^2} e^{ix^1 k} \, dk = \frac{1}{\sqrt{2\alpha^1}} e^{-(x^1)^2/4\alpha^1}$$

$$= \sqrt{2\alpha} \, e^{-\alpha x^2}$$

so that from (7)

$$\Phi^{-1}\{F(k)\} = \frac{1}{\sqrt{2\pi}} \int_{-\infty}^{\infty} F(k) e^{-ikx} \, dk = \frac{N}{\sqrt{2\alpha}} \sqrt{2\alpha} \, e^{-\alpha x^2}$$

$$= N e^{-\alpha x^2}$$

or

$$\Phi^{-1}\{F(k)\} = f(x)$$

as expected.

● **PROBLEM 17-3**

Find the Fourier transform, $F(k) = \Phi\{f(x)\}$, of the function

$$f(x) = \frac{a}{x^2 + a^2} \quad (a > 0)$$

by using the residue theorem of complex analysis.

Solution: The Fourier transform of a function is defined by

$$F(k) = \Phi\{f(x)\} = \frac{1}{\sqrt{2\pi}} \int_{-\infty}^{\infty} f(x) \, e^{ikx} \, dx$$

and for the given function this definition becomes

$$F(k) = \int_{-\infty}^{\infty} \frac{a}{\sqrt{2\pi}} \frac{e^{ikx}}{a^2 + a^2} \, dx = \frac{a}{\sqrt{2\pi}} \int_{-\infty}^{\infty} \frac{e^{ikx}}{(x + ai)(x - ai)} \, dx \quad (1)$$

where the denominator has been factored to make clearer

the applicability of the calculus of residues to the problem. First, for the case where $k \geq 0$, make use of the contour shown in Figure 1 and consider the integral

$$J = \oint g(z)\,dz = \oint_C \frac{a}{\sqrt{2\pi}} \frac{e^{ikx}\,dz}{z^2 + a^2}$$

$$= \int_{-R}^{R} \frac{a}{\sqrt{2\pi}} \frac{e^{ikx}\,dx}{(x+ai)(x-ai)} + \int_{C_R} \frac{a}{\sqrt{2\pi}} \frac{e^{ikz}\,dz}{z^2 + a^2}. \quad (2)$$

The residue theorem states that

$$J = 2\pi i \sum_{j=1}^{n} \text{Res } g(b_j) \quad (3)$$

where there are n isolated singularities b_j within the given contour and Res $g(b_j)$ is the residue of g at b_j. In this case $n = 1$ and $b_1 = ai$. The residue of g at ai

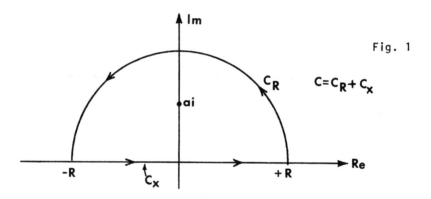

Fig. 1

$C = C_R + C_x$

is easy to calculate in this case since ai is a simple pole of g and we may use the formula

$$\text{Res } g(ai) = \lim_{z \to ai} (z-ai)g(z)$$

so that

$$\text{Res } g(ai) = \lim_{z \to ai} (z-ai) \frac{a}{\sqrt{2\pi}} \frac{e^{ikz}}{(z^2 + a^2)} =$$

$$= \lim_{z \to ai} (z-ai) \frac{a}{\sqrt{2\pi}} \frac{e^{ikz}}{(z-ai)(z+ai)}$$

$$= \frac{a}{\sqrt{2\pi}} \frac{e^{-ka}}{2ai}$$

886

and using (3) gives

$$J = 2\pi i \frac{a}{\sqrt{2\pi}} \frac{e^{-ka}}{2ai} = \sqrt{\frac{\pi}{2}} e^{-ka} \qquad (k \geq 0). \qquad (4)$$

Now, to show that the second integral in (2) goes to zero as $R \to \infty$, make the estimates

$$\left| \int_{C_R} \frac{e^{iz} dz}{z^2 + a^2} \right| \leq \int_{C_R} \frac{|e^{iz} dz|}{|z^2 + a^2|} \leq \max \frac{1}{|z^2 + a^2|}$$

$$\times \int_{C_R} e^{-y} (dz)$$

$$= \max \frac{1}{|z^2 + a^2|} \cdot \int_0^\pi e^{-\cos\theta} R d\theta \leq \max \frac{1}{|z^2 + a^2|} \int_0^\pi eR d\theta$$

$$= eR\pi \max \frac{1}{|z^2 + a^2|} < \frac{eR\pi}{R^2 - a^2} \qquad (5)$$

where

$$\max \frac{1}{|z^2 + a^2|}$$

is the maximum value of $\frac{1}{z^2 + a^2}$ over the contour C_R.

Equation (5) shows that

$$\lim_{R \to \infty} \int_{C_R} \frac{e^{iz} dz}{z^2 + a^2} = \lim_{R \to \infty} \frac{eR\pi}{R^2 - a^2} = 0$$

so that taking the limit as $R \to \infty$ in (2) and recalling (4) gives

$$\lim_{R \to \infty} J = \lim_{R \to \infty} \sqrt{\frac{\pi}{2}} e^{-ka} = \int_{-\infty}^\infty \frac{a}{\sqrt{2a}} \frac{e^{ikx} dx}{x^2 + a^2} + 0$$

or

$$F(k) = \int_{-\infty}^{\infty} \frac{a}{\sqrt{2\pi}} \frac{e^{ikx}}{x^2 + a^2} dx = \sqrt{\frac{\pi}{2}} e^{-ka} \quad (k \geq 0) . \tag{6}$$

For the case where $k < 0$, make use of the contour shown in Figure 2. Again the residue theorem is employed via equation (2) except that C_R is now the curve shown in Figure 2, and equation (3) is now written as

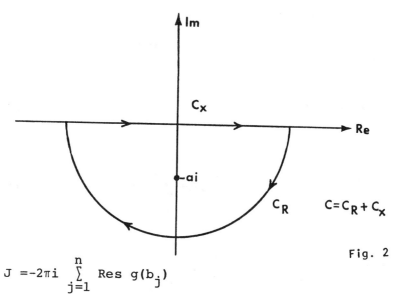

Fig. 2

$$J = -2\pi i \sum_{j=1}^{n} \text{Res } g(b_j)$$

(since C is taken as clockwise) with $n=1$, $b_1 = -ai$. Using the formula

$$\text{Res } g(-ai) = \lim_{z \to -ai} (z + ai) g(z)$$

yields

$$\text{Res } g(ai) = \lim_{z \to -ai} (z + ai) \frac{a}{\sqrt{2\pi}} \frac{e^{ikz}}{(z+ai)(z-ai)}$$

$$= \frac{a}{\sqrt{2\pi}} \frac{e^{ka}}{(-2ai)}$$

so that from (7)

$$J = -2\pi i \frac{a}{\sqrt{2\pi}} \frac{e^{ka}}{2} = \sqrt{\frac{\pi}{2}} e^{ka}$$

for $k < 0$. The proof that the integral over C_R goes to

0 as $R \to \infty$ is completely analogous to that for the C_R of Figure 1 so that (2) yields

$$\lim_{R \to \infty} J = \lim_{R \to \infty} \sqrt{\frac{\pi}{2}} e^{ka} = \int_{-\infty}^{\infty} \frac{a}{\sqrt{2\pi}} \frac{e^{ikx}}{x^2 + a^2} dx + 0$$

so that

$$F(k) = \int_{-\infty}^{\infty} \frac{a}{\sqrt{2\pi}} \frac{e^{ikx}}{x^2 + a^2} dx = \sqrt{\frac{\pi}{2}} e^{ka} \qquad (k < 0) \qquad (8)$$

and equations (6) and (8) may be combined to yield

$$F(k) = \sqrt{\frac{\pi}{2}} e^{-|k|a} \qquad \text{(all } k\text{)}. \qquad (9)$$

As in the previous problem, if $f(x)$ has a sharp peak (small a) then $F(k)$ is broadened and vice versa, a general feature of Fourier transforms.

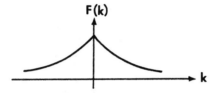

● **PROBLEM 17-4**

Find the Fourier transforms, $F_i(k) = \Phi\{f_i(t)\}$, of the functions whose graphs are shown in the accompanying figure.

Solution: The Fourier transform of a function $f(t)$ is defined by

$$F(k) = \frac{1}{\sqrt{2\pi}} \int_{-\infty}^{\infty} f(t) e^{jkt} dt \qquad (1)$$

where $j = \sqrt{-1}$.

a) The "box" function graphed in Figure 1(a) may be written analytically as

$$f_a(t) = \begin{cases} 1 & |t| \leq d \\ 0 & |t| > d \end{cases} \qquad (d > 0).$$

889

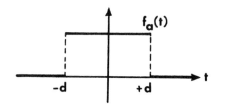

Fig. 1(a)

Using (1), its Fourier transform is found to be

$$F_a(k) = \frac{1}{\sqrt{2\pi}} \int_{-d}^{d} e^{jkt} \, dt = \frac{1}{\sqrt{2\pi}} \frac{e^{jkd} - e^{-jkd}}{jk} \quad (2)$$

Using the formulas

$$e^{j\theta} = \cos\theta + j\sin\theta$$
$$e^{-j\theta} = \cos\theta - j\sin\theta$$

in (2) yields

$$F_a(k) = \sqrt{\frac{2}{\pi}} \frac{\sin dk}{k} \quad (3)$$

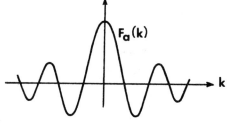

Fig. 2

$F_a(k)$ is graphed in Figure 2. Note the property that as f_a becomes narrower (small d) F_a becomes "broader" and vice versa.

b) The function $f_b(t)$ of Figure 1(b) can be written in analytic form as

$$f_b(t) = \begin{cases} \frac{-V_0}{b}(t-b) & ; \ 0 \le t \le b \\ 0 & ; \ t < 0, \ t > b \end{cases}$$

Hence, from (1)

$$F_b(k) = \frac{-V_0}{b\sqrt{2\pi}} \int_0^b (t-b) e^{-jkt} \, dt$$

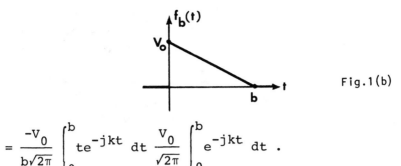

Fig.1(b)

$$= \frac{-V_0}{b\sqrt{2\pi}} \int_0^b t e^{-jkt} dt + \frac{V_0}{\sqrt{2\pi}} \int_0^b e^{-jkt} dt.$$

Using integration by parts on the first integral yields

$$F_b(k) = \left[\frac{-V_0}{b\sqrt{2\pi}} \left(\frac{t}{-jk} e^{-jkt} \right) \Big|_0^b - \frac{V_0}{b\sqrt{2\pi}} \int_0^b \frac{1}{jk} e^{-jkt} dt \right]$$

$$+ \left(\frac{V_0}{\sqrt{2\pi}} \frac{1}{-jk} e^{-jkt} \right) \Big|_0^b$$

$$= \frac{V_0}{b\sqrt{2\pi}} \left(\frac{t}{-jk} e^{-jkt} - \frac{1}{(jk)^2} e^{-jkt} \right) \Big|_0^b$$

$$+ \left[\frac{V_0}{\sqrt{2\pi}} \frac{1}{-jk} e^{-jkb} + \frac{1}{jk} \right]$$

$$\frac{V_0}{\sqrt{2\pi}} \frac{1}{jk} e^{-jkb} - \frac{V_0}{\sqrt{2\pi} \, bk^2} e^{-jkb} +$$

$$\frac{V_0}{\sqrt{2\pi} \, bk^2} - \frac{V_0}{\sqrt{2\pi} \, jk} e^{-jkb} + \frac{V_0}{\sqrt{2\pi} \, jk}$$

$$= \frac{V_0}{\sqrt{2\pi}} \left[\frac{-1}{bk^2} e^{-jkb} + \frac{1}{bk^2} + \frac{1}{jk} \right].$$

c) For the functions in Fig. 1(c) use ω as the transfer parameter instead of k. Then

$$f_c(t) = \sin t; \quad -\pi \le t \le \pi.$$

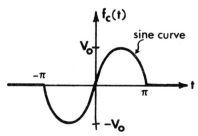

Fig.1(c)

Hence,

$$F_c(\omega) = V_0 \int_{-\pi}^{\pi} \sin t \, e^{-j\omega t} \, dt$$

$$= V_0 \int_{-\pi}^{\pi} \left(\frac{e^{jt} - e^{-jt}}{2j} \right) e^{-j\omega t} \, dt$$

$$= V_0 \int_{-\pi}^{\pi} \frac{e^{jt(1-\omega)} - e^{-jt(1+\omega)}}{2j} \, dt$$

$$= V_0 \left[\frac{e^{jt(1-\omega)}}{2j(j(1-\omega))} + \frac{e^{-jt(1+\omega)}}{2j(j(1+\omega))} \right]\Bigg|_{-\pi}^{\pi}$$

$$= V_0 \left[-\frac{e^{jt(1-\omega)}}{2(1-\omega)} - \frac{e^{-jt(1+\omega)}}{2(1+\omega)} \right]\Bigg|_{-\pi}^{\pi}$$

$$= V_0 \left[-\frac{e^{j\pi(1-\omega)}}{2(1-\omega)} + \frac{e^{-j\pi(1+\omega)}}{2(1+\omega)} - \frac{e^{-j\pi(1-\omega)}}{2(1-\omega)} \right.$$

$$\left. - \frac{e^{j\pi(1+\omega)}}{2(1+\omega)} \right]$$

$$= \frac{V_0 j}{(1-\omega)} \left[\frac{-e^{j\pi(1-\omega)} + e^{-j\pi(1-\omega)}}{2j} \right]$$

$$+ \frac{V_0 j}{(1+\omega)} \left[\frac{e^{j\pi(1+\omega)} - e^{-j\pi(1+\omega)}}{2j} \right]$$

$$= -\frac{jV_0}{1-\omega} \sin \pi(1-\omega) + \frac{jV_0}{1+\omega} \sin \pi(1+\omega)$$

This can be simplified by combining the two terms

$$F_c = \frac{-jV_0(1+\omega)\sin\pi(1-\omega) + jV_0(1-\omega)\sin\pi(1+\omega)}{(1-\omega)(1+\omega)}$$

$$= \frac{-jV_0(1+\omega)\sin\pi\omega - jV_0(1-\omega)\sin\pi\omega}{1-\omega^2}$$

or

$$F_c(\omega) = \frac{-j2V_0 \sin\pi\omega}{1-\omega^2}.$$

PROPERTIES OF FOURIER TRANSFORMS

● **PROBLEM 17-5**

Prove the following properties of Fourier transforms:

a) The Fourier transform of f(x) exists if f is absolutely integrable over $(-\infty, +\infty)$,

b) If f(x) is real valued then

$$F(-k) = F^*(k)$$

where $F^*(k)$ is the complex conjugate of $F(k)$.

Solution: By definition, if f is absolutely integrable over $(-\infty, +\infty)$ then

$$\int_{-\infty}^{\infty} |f(x)| dx \tag{1}$$

exists. The Fourier transform of f is given by

$$F(k) = \frac{1}{\sqrt{2\pi}} \int_{-\infty}^{\infty} f(x) e^{ikx} dx . \tag{2}$$

Now, recalling that $|e^{iy}| = 1$ for all y we have that

$$|f(x) e^{ikx}| = |f(x)|$$

so that

$$\frac{1}{\sqrt{2\pi}} \int_{-\infty}^{\infty} \left| f(x) \, e^{ikx} \right| dx = \frac{1}{\sqrt{2\pi}} \int_{-\infty}^{\infty} |f(x)| \, dx ,$$

and we know that this second integral exists. Thus,

$$f(x) e^{ikx}$$

is absolutely integrable over $(-\infty, \infty)$ and is therefore integrable over $(-\infty, \infty)$. Hence, $F(k)$ exists.

b) This proof is immediate. From (2)

$$F(-k) = \frac{1}{\sqrt{2\pi}} \int_{-\infty}^{\infty} f(x) \, e^{-ikx} \, dx \tag{3}$$

and, recalling the identity $e^{iy} = \cos y + i \sin y$,

$$F^*(k) = \left(\frac{1}{\sqrt{2\pi}} \int_{-\infty}^{\infty} f(x) \, e^{ikx} \, dx \right)^*$$

$$= \left(\frac{1}{\sqrt{2\pi}} \int_{-\infty}^{\infty} f(x) \, \cos kx \, dx + i \, \frac{1}{\sqrt{2\pi}} \int_{-\infty}^{\infty} f(x) \, \sin kx \, dx \right)^*$$

$$= \frac{1}{\sqrt{2\pi}} \int_{-\infty}^{\infty} f(x) \, \cos kx \, dx - i \, \frac{1}{\sqrt{2\pi}} \int_{-\infty}^{\infty} f(x) \, \sin kx \, dx \quad \text{(f real)}$$

$$= \frac{1}{\sqrt{2\pi}} \int_{-\infty}^{\infty} f(x) \, \cos(-kx) \, dx + i \, \frac{1}{\sqrt{2\pi}} \int_{-\infty}^{\infty} f(x) \, \sin(-kx) \, dx$$

$$= \frac{1}{\sqrt{2\pi}} \int_{-\infty}^{\infty} f(x) \, (\cos(-kx) + i \sin(-kx)) \, dx$$

$$= \frac{1}{\sqrt{2\pi}} \int_{-\infty}^{\infty} f(x) \, e^{-ikx} \, dx \ . \tag{4}$$

In the third step the facts that $\cos(-y) = \cos y$ and $\sin(-y) = -\sin y$ were used. Equating (3) and (4) yields

$$F(-k) = F^*(k) \, , \text{for real } f(x) \ .$$

● **PROBLEM 17-6**

a) Prove the attenuation property of Fourier transforms:

$$\Phi\{f(x) \, e^{ax}\} = F(k-ai)$$

where

$$F(k) = \Phi\{f(x)\} \ .$$

b) Prove the shifting property of Fourier transforms:

$$\Phi\{f(x-a)\} = e^{ika} \, F(k) \ .$$

c) Prove the derivative properties of Fourier transforms:

$$\Phi\{f'(x)\} = -ik\Phi\{f(x)\}$$

$$\Phi\{f''(x)\} = -k^2\Phi\{f(x)\} \ .$$

Solution: a) The Fourier transform of $f(x)$ is defined by

$$F(k) = \Phi\{f(x)\} = \frac{1}{\sqrt{2\pi}} \int_{-\infty}^{\infty} f(x) \, e^{ikx} \, dx \ . \tag{1}$$

The Fourier transform of $g(x) = f(x) \, e^{ax}$ is then

$$G(k) = \Phi\{f(x) \, e^{ax}\} = \frac{1}{\sqrt{2\pi}} \int_{-\infty}^{\infty} f(x) \, e^{ax} \, e^{ikx} \, dx$$

$$= \frac{1}{\sqrt{2\pi}} \int_{-\infty}^{\infty} f(x) \, e^{(a+ik)x} \, dx$$

$$= \frac{1}{\sqrt{2\pi}} \int_{-\infty}^{\infty} f(x) \, e^{i(k-ia)x} \, dx \, . \tag{2}$$

Now make the change of variable $r = k - ia$ in (2) to give

$$\Phi\{f(x) \, e^{ax}\} = \frac{1}{\sqrt{2\pi}} \int_{-\infty}^{\infty} f(x) \, e^{irx} \, dx = F(r) = F(k-ia) \, .$$

b) Suppose $f(x)$ is shifted a length a to the right. Then the Fourier transform of this new function $f(x-a)$ is

$$\Phi\{f(x-a)\} = \frac{1}{\sqrt{2\pi}} \int_{-\infty}^{\infty} f(x-a) \, e^{ikx} \, dx \tag{3}$$

Now make the substitution $x' = x - a$. Then (3) becomes

$$\Phi\{f(x-a)\} = \frac{1}{\sqrt{2\pi}} \int_{-\infty}^{\infty} f(x') \, e^{ik(x'+a)} \, dx'$$

$$= e^{ika} \frac{1}{\sqrt{2\pi}} \int_{-\infty}^{\infty} f(x') \, e^{ikx'} \, dx'$$

$$= e^{ika} \, F(k) \, .$$

c) Suppose that $\Phi\{f'(x)\}$ exists. Then by definition (1),

$$\Phi\{f'(x)\} = \frac{1}{\sqrt{2\pi}} \int_{-\infty}^{\infty} f'(x) \, e^{ikx} \, dx \, . \tag{4}$$

Integrating (4) by parts gives

$$\Phi\{f'(x)\} = \frac{1}{\sqrt{2\pi}} \, f(x) \, e^{ikx} \bigg|_{-\infty}^{\infty} - \frac{ik}{\sqrt{2\pi}} \int_{-\infty}^{\infty} f(x) \, e^{ikx} \, dx \, . \tag{5}$$

If the Fourier transform of $f(x)$ exists, this usually implies that $f(x) \to 0$ as $x \to \pm\infty$ (this is sometimes not the case, but then $f(x)$ can be treated as a distribution so that the derivative formula is then still valid). Thus, the first term in (5) is zero and

$$\Phi\{f'(x)\} = -ik\, \Phi\{f(x)\} \, . \tag{6}$$

If

$$\Phi\{f''(x)\}$$

exists, it is given by

$$\Phi\{f''(x)\} = \frac{1}{\sqrt{2\pi}} \int_{-\infty}^{\infty} f''(x)\, e^{ikx}\, dx \, .$$

Again integrating by parts yields

$$\Phi\{f''(x)\} = \frac{1}{\sqrt{2\pi}} f'(x)\, e^{ikx} \Big|_{-\infty}^{\infty} - \frac{ik}{\sqrt{2\pi}} \int_{-\infty}^{\infty} f'(x)\, e^{ikx}\, dx \, . \tag{7}$$

For reasons mentioned above, it is expected that $f'(x) \to 0$ as $x \to \pm\infty$ so that (7) gives

$$\Phi\{f''(x)\} = -ik\, \Phi\{f'(x)\} \, .$$

Using (6) gives

$$\Phi\{f''(x)\} = -k^2\, \Phi\{f(x)\} \, .$$

The obvious extension is made by using $f^{(n-1)}(x) = g(x)$ in the formula of (6) to give

$$\Phi\{g'(x)\} = -ik\, \Phi\{g(x)\}$$

or

$$\Phi\{f^{(n)}(x)\} = -ik\, \Phi\{f^{(n-1)}(x)\} \, ,$$

which yields upon iteration

$$\Phi\{f^{(n)}(x)\} = (-ik)^n\, \Phi\{f(x)\} \, .$$

● **PROBLEM 17-7**

a) Prove that if the functions $g(x)$ and $F(k)$ are absolutely integrable on $(-\infty, +\infty)$ and that the Fourier inversion integral for $f(x)$ is valid for all x except possibly at a countably infinite number of points, then

$$\int_{-\infty}^{\infty} F(k)\, G(-k)\, dk = \int_{-\infty}^{\infty} f(x)\, g(x)\, dx \tag{1}$$

where

$$F(k) = \Phi\{f(x)\}, \quad G(k) = \Phi\{g(x)\}.$$

This is known as the second Parseval theorem of Fourier transform theory.

b) From the above equation (1), prove the first Parseval theorem of Fourier transform theory,

$$\int_{-\infty}^{\infty} |F(k)|^2 \, dk = \int_{-\infty}^{\infty} |f(x)|^2 \, dx. \qquad (2)$$

<u>Solution</u>: a) The Fourier transform of a function $f(x)$ is defined by

$$F(k) = \frac{1}{\sqrt{2\pi}} \int_{-\infty}^{\infty} f(x) e^{ikx} \, dx \qquad (3)$$

so that by definition

$$G(-k) = \frac{1}{\sqrt{2\pi}} \int_{-\infty}^{\infty} g(x) e^{-ikx} \, dx. \qquad (4)$$

Therefore,

$$\int_{-\infty}^{\infty} F(k) G(-k) \, dk = \int_{-\infty}^{\infty} F(k) \, dk \int_{-\infty}^{\infty} \frac{1}{\sqrt{2\pi}} g(x) e^{-ikx} \, dx. \qquad (5)$$

Now $F(k)$ and $g(x)$ are absolutely convergent on $(-\infty, +\infty)$, that is, the integrals

$$\int_{-\infty}^{\infty} |F(k)| \, dk, \quad \int_{-\infty}^{\infty} |g(x)| \, dx$$

are convergent, so that

$$\int_{-\infty}^{\infty} F(k) e^{-ikx} \, dx, \quad \int_{-\infty}^{\infty} g(x) e^{-ikx} \, dx$$

are absolutely convergent (since

$$|F(k)e^{-ikx}| = |F(k)||e^{-ikx}| = |F(k)|$$

and

$$|g(x)e^{-ikx}| = |g(x)|) .$$

Hence, the order of integration in (5) may be interchanged giving

$$\int_{-\infty}^{\infty} F(k)G(-k)\,dk = \int_{-\infty}^{\infty} g(x)\,dx \frac{1}{\sqrt{2\pi}} \int_{-\infty}^{\infty} F(k)e^{-ikx}\,dk . \tag{6}$$

Since the Fournier inversion integral is valid,

$$\frac{1}{\sqrt{2\pi}} \int_{-\infty}^{\infty} F(k)e^{-ikx}\,dk = f(x) \tag{7}$$

and using this result in (6) gives the second Parseval theorem:

$$\int_{-\infty}^{\infty} F(k)G(-k)\,dk = \int_{-\infty}^{\infty} g(x)f(x)\,dx . \tag{8}$$

The validity of (8) is insured even if the Fourier inversion integral for f(x) has a countably infinite number of discrepancies with f(x) since this will not affect the equality of the integrals

$$\int_{-\infty}^{\infty} g(x)(f(x))\,dx \quad \text{and} \quad \int_{-\infty}^{\infty} g(x) \frac{1}{\sqrt{2\pi}} \int_{-\infty}^{\infty} F(k)e^{-ikx}\,dk\,dx .$$

b) The first Parseval theorem is a corollary to the second Parseval theorem stated in equation (8) which follows by letting $f(x) = g(x)$ so that $F(k) = G(k)$ and recalling that (assuming $f = g$ real) $G(-k) = G^*(k)$ where $G^*(k)$ is the complex conjugate of $G(k)$. Noting that

$$G(k)G^*(k) = |G(k)|^2$$

and using these results in (8) gives

$$\int_{-\infty}^{\infty} G(k)G^*(k)\,dk = \int_{-\infty}^{\infty} [g(x)]^2\,dx$$

or

$$\int_{-\infty}^{\infty} |G(k)|^2\,dk = \int_{-\infty}^{\infty} |g(x)|^2\,dx .$$

● **PROBLEM 17-8**

Let $F(k) = \Phi\{f(x)\}$, $G(k) = \Phi\{g(x)\}$ and suppose $F(k)G(k) = \Phi\{h(x)\}$. Prove the convolution theorem for Fourier transforms:

If $g(x)$ and $F(k)$ are absolutely integrable on $(-\infty, \infty)$ and if the Fourier inversion integral for $f(x)$ is valid for all x except possibly a countably infinite number of points, then

$$h(x) = (f * g),$$

where $(f * g)$ is the convolution of f and g defined by

$$(f * g) = \frac{1}{\sqrt{2\pi}} \int_{-\infty}^{\infty} f(\xi)\, g(x-\xi)\, d\xi. \tag{1}$$

Solution: Let Φ^{-1} denote the inverse Fourier transform. Then by definition

$$\Phi^{-1}\{F(k)G(k)\} = \frac{1}{\sqrt{2\pi}} \int_{-\infty}^{\infty} F(k)G(k)\, e^{-ikx}\, dk. \tag{2}$$

Using the definition of the Fourier transform of $g(\xi)$,

$$G(k) = \Phi\{g(\xi)\} = \frac{1}{\sqrt{2\pi}} \int_{-\infty}^{\infty} g(\xi)\, e^{ik\xi}\, d\xi$$

in (2) gives

$$\Phi^{-1}\{F(k)G(k)\} = \frac{1}{2\pi} \int_{-\infty}^{\infty} F(k)\, e^{-ikx}\, dk$$

$$\times \int_{-\infty}^{\infty} g(\xi)\, e^{ik\xi}\, d\xi. \tag{3}$$

The assumption that $g(x)$ and $F(k)$ are absolutely integrable on $(-\infty, \infty)$ means that the integrals

$$\int_{-\infty}^{\infty} |F(k)|\, dk, \quad \int_{-\infty}^{\infty} |g(x)|\, dx$$

are convergent so that

$$\int_{-\infty}^{\infty} F(k)\, e^{-ikx}\, dx, \quad \int_{-\infty}^{\infty} g(x)\, e^{-ikx}\, dx$$

are absolutely convergent (since

$$|F(k)e^{-ikx}| = |F(k)|$$

and

$$|g(x)e^{-ikx}| = |g(x)|).$$

Under these conditions the order of integration in (3) may be interchanged giving

$$\phi^{-1}\{F(k)G(k)\} = \frac{1}{2\pi}\int_{-\infty}^{\infty} g(\xi)d\xi \int_{-\infty}^{\infty} F(k)e^{-ik(x-\xi)}dk. \qquad (4)$$

Since the Fourier inversion integral is valid,

$$\frac{1}{\sqrt{2\pi}}\int_{-\infty}^{\infty} F(k)e^{-ik(x-\xi)}dk = f(x-\xi)$$

and using this result in (4) gives

$$\phi^{-1}\{F(k)G(k)\} = \frac{1}{\sqrt{2\pi}}\int_{-\infty}^{\infty} g(\xi)f(x-\xi)d\xi. \qquad (5)$$

Noting that

$$\phi^{-1}\{F(k)G(k)\} = h(x)$$

and using the definition of convolution in (1), (5) becomes

$$h(x) = (f * g)$$

and the theorem is proved. The validity of (5) is assured even if the Fourier inversion integral for $f(x)$ has a countably infinite number of discrepancies with $f(x)$ since this will not affect the equality of the integrals

$$\frac{1}{\sqrt{2\pi}}\int_{-\infty}^{\infty} g(\xi) f(x-\xi)d\xi$$

and

$$\frac{1}{\sqrt{2\pi}}\int_{-\infty}^{\infty} g(\xi) \left(\frac{1}{\sqrt{2\pi}}\int_{-\infty}^{\infty} F(k)e^{-ik(x-\xi)}dk\right)d\xi.$$

● PROBLEM 17-

a) State conditions under which the Fourier integral formula

$$f(x) = \frac{1}{2\pi} \int_{-\infty}^{\infty} e^{-ikx} dk \int_{-\infty}^{\infty} f(\xi) e^{ik\xi} d\xi \qquad (1)$$

is valid. Discuss its validity for the examples in problems 2 and 3 of this chapter.

b) Recast (1) in real form assuming that $f(x)$ is real.

Solution: The most widely used sufficient conditions for pointwise convergence are as follows:

If $f(x)$ is absolutely integrable and piecewise very smooth on $(-\infty, \infty)$, then the Fourier integral theorem is valid in the sense that

$$\frac{1}{2\pi} \int_{-\infty}^{\infty} e^{-ikx} dk \int_{-\infty}^{\infty} f(\xi) e^{ik\xi} d\xi = \frac{1}{2}[f(x+) + f(x-)]$$

where $f(x+)$ and $f(x-)$ are the right and left hand limits of f at x respectively. Recall that a function f is piecewise very smooth if its second derivative $f''(x)$ is piecewise continuous.

Each of the functions in problems 2 and 3 satisfy these conditions so that the Fourier integral theorem is valid for each. For example,

$$f(x) = Ne^{-\alpha x^2}$$

is absolutely integrable on $(-\infty, \infty)$ since

$$\int_{-\infty}^{\infty} \left| Ne^{-\alpha x^2} \right| dx = \int_{-\infty}^{\infty} Ne^{-\alpha x^2} dx$$

which is a convergent integral. Also, since f has a continuous second derivative it is piecewise very smooth.

b) Write (1) as

$$f(x) = \frac{1}{\sqrt{2\pi}} \int_{-\infty}^{0} F(k) e^{-ikx} dk + \frac{1}{\sqrt{2\pi}} \int_{0}^{\infty} F(k) e^{-ikx} dk .$$

Make the change of variable k' = -k in the first integral to obtain

$$f(x) = \frac{1}{\sqrt{2\pi}} \int_{\infty}^{0} F(-k')e^{ik'x}(-dk') + \frac{1}{\sqrt{2\pi}} \int_{0}^{\infty} F(k)e^{-ikx} dk$$

$$= \frac{1}{\sqrt{2\pi}} \int_{0}^{\infty} F(-k')e^{ik'x} dk' + \frac{1}{\sqrt{2\pi}} \int_{0}^{\infty} F(k)e^{-ikx} dk$$

$$= \frac{1}{\sqrt{2\pi}} \int_{0}^{\infty} [F^*(k)e^{ikx} + F(k)e^{-ikx}] dk \quad (2)$$

where the fact that $F(-k) = F^*(k)$ has been used in the last step. Now since the Fourier integral formula is assumed valid for f,

$$F(k)e^{-ikx} = \frac{1}{\sqrt{2\pi}} \int_{-\infty}^{\infty} f(\xi)e^{ik(\xi-x)} d\xi \quad (3)$$

and taking the complex conjugate of this formula,

$$F^*(k)e^{ikx} = \frac{1}{\sqrt{2\pi}} \int_{-\infty}^{\infty} f(\xi)e^{-ik(\xi-x)} dk. \quad (4)$$

Adding (3) to (4) and recalling that

$$\cos\theta = \frac{e^{i\theta} + e^{-i\theta}}{2}$$

gives

$$F(k)e^{-ikx} + F^*(k)e^{ikx} = \frac{1}{\sqrt{2\pi}} \int_{-\infty}^{\infty} f(\xi) 2\cos k(\xi-x) d\xi$$

which, upon substitution into (2), yields the real form of the Fourier integral formula

$$f(x) = \frac{1}{\pi} \int_{0}^{\infty} dk \int_{-\infty}^{\infty} f(\xi) \cos k(\xi-x) d\xi \quad .$$

● **PROBLEM 17-10**

Find the form of the Fourier integral formula if f(x) is an

a) even function, b) odd function;

thereby motivating definitions for the Fourier cosine and Fourier sine transforms:

$$\Phi_c\{f(x)\} \quad \text{and} \quad \Phi_s\{f(x)\}$$

Solution: The real form of the Fourier integral formula is

$$f(x) = \frac{1}{\pi} \int_0^\infty dk \int_{-\infty}^\infty f(\xi) \cos k(\xi-x) \, d\xi \qquad (1)$$

and using the trigonometric identity

$$\cos(a-b) = \cos a \cos b + \sin a \sin b$$

in (1) yields

$$f(x) = \frac{1}{\pi} \int_0^\infty dk \int_{-\infty}^\infty f(\xi)[\cos kx \cos k\xi + \sin kx \sin k\xi] d\xi$$

$$= \frac{1}{\pi} \int_0^\infty \cos kx \, dk \int_{-\infty}^\infty f(\xi)\cos k\xi \, d\xi + \frac{1}{\pi} \int_0^\infty \sin kx \, dk$$

$$\times \int_{-\infty}^\infty f(\xi)\sin k\xi \, d\xi \,. \qquad (2)$$

a) If f is even then $f(\xi) \sin k\xi$ is odd and the second integral is zero. Hence

$$f(x) = \frac{1}{\pi} \int_0^\infty \cos kx \, dk \int_{-\infty}^\infty f(\xi)\cos k\xi \, d\xi, \qquad (3)$$

f even and since $f(\xi) \cos k\xi$ is even, (3) may be written as

$$f(x) = \frac{2}{\pi} \int_0^\infty \cos kx \, dk \int_0^\infty f(\xi)\cos k\xi \, d\xi, \qquad (4)$$

f even.

This formula suggests the definition of a Fourier cosine transform, analogous to that of the regular Fourier transform, by

$$F_c(k) = \Phi_c\{f(x)\} = \sqrt{\frac{2}{\pi}} \int_0^\infty f(x) \cos kx \, dx \qquad (5)$$

with the inverse

$$\Phi_c^{-1}\{F_c(k)\} = \sqrt{\frac{2}{\pi}} \int_0^\infty F_c(k) \cos kx \, dk = \begin{cases} f(x) & x > 0 \\ f(-x) & x < 0 \end{cases} \qquad (6)$$

Note that this definition applies to all functions defined on $(0,\infty)$ even if undefined elsewhere. In any case, the Fourier cosine integral formula (6) will give back $f(x)$ for $x \in (0,\infty)$ and $f(-x)$ for $x < 0$ so

$$\Phi^{-1}\{\Phi_c\{f(x)\}\}$$

represents the even extension of f into the negative real axis.

b) If f is odd then $f(\xi) \cos k\xi$ is odd and the first integral in (2) is zero leaving

$$f(x) = \frac{2}{\pi} \int_0^\infty \sin kx \, dk \int_0^\infty f(\xi) \sin k\xi \, d\xi, \quad f \text{ odd.} \qquad (7)$$

This formula suggests the definition of a Fourier sine transform by

$$F_s(k) = \Phi_s\{f(x)\} = \sqrt{\frac{2}{\pi}} \int_0^\infty f(x) \sin kx \, dx \qquad (8)$$

with the inverse

$$\Phi_s^{-1}\{F_s(k)\} = \sqrt{\frac{2}{\pi}} \int_0^\infty F_s(k) \sin kx \, dx = \begin{cases} f(x) & x > 0 \\ -f(-x) & x < 0 \end{cases} \qquad (9)$$

As with $\Phi_c\{f(x)\}$ this definition applies to all functions defined on $(0,\infty)$ even if undefined elsewhere. In any case, the Fourier sine integral formula (9) will give back $f(x)$ for $x \in (0,\infty)$ and $-f(-x)$ for $x < 0$. Therefore

$$\Phi^{-1}\{\Phi_s\{f(x)\}\}$$

represents the odd extension of f into the negative real axis.

• **PROBLEM 17-11**

Define the Dirac delta function, $\delta(x)$, and prove the sifting property of $\delta(x)$ for all functions $f(x)$ which are continuous at $x = 0$,

$$\int_{-\infty}^{\infty} \delta(x) f(x) dx = f(0) .$$

Solution: The Dirac delta function may be defined in several ways but its most common definition is as the infinitely sharply peaked function given symbolically by

$$\delta(x) = \begin{cases} 0 & x \neq 0 \\ \infty & x = 0 \end{cases} \tag{1}$$

and with the property that

$$\int_{-\infty}^{\infty} \delta(x) dx = 1 . \tag{2}$$

(i.e., the integral of $\delta(x)$ is normalized to unity).

For this problem start with the integral

$$\int_{-\infty}^{\infty} \delta(x) f(x) \, dx$$

where $f(x)$ is any continuous function. We can evaluate this integral by the following argument:

Since $\delta(x)$ is zero for $x \neq 0$ (by (1)), the limits of integration may be changed to $-\varepsilon$ and $+\varepsilon$, where ε is a small positive number. In addition, since $f(x)$ is continuous at $x = 0$, its values within the interval $(-\varepsilon, +\varepsilon)$ will not differ much from $f(0)$ and we can say, approximately, that

$$\int_{-\infty}^{\infty} \delta(x) f(x) \, dx = \int_{-\varepsilon}^{+\varepsilon} \delta(x) f(x) \, dx \approx f(0) \int_{-\varepsilon}^{+\varepsilon} \delta(x) dx$$

where the approximation improves as ε approaches zero. However, since $\delta(x) = 0$ for $x \neq 0$ and since $\delta(x)$ is normalized we have

$$\int_{-\varepsilon}^{+\varepsilon} \delta(x)\,dx = 1$$

for all values of ε.
It appears then that letting $\varepsilon \to 0$, we have exactly

$$\int_{-\varepsilon}^{-\varepsilon} \delta(x) f(x)\,dx = f(0). \qquad (3)$$

Note that the limits $-\varepsilon$ and $+\varepsilon$ may be replaced by any two numbers a and b provided that $a < 0 < b$. Now the integral (3) is referred to as the sifting property of the delta function, that is $\delta(x)$ acts as a sieve, selecting from all possible values of $f(x)$ its value at the point $x = 0$.

APPLICATIONS OF FOURIER TRANSFORMS

● **PROBLEM** 17-12

Given the current pulse, $i(t) = te^{-bt}$: (a) find the total 1Ω energy associated with this waveform; (b) what fraction of this energy is present in the frequency band from $-b$ to b rad/s?

Solution: Use $j\omega$ as the transfer parameter instead of k. Then the total 1-Ω energy associated with either a current or voltage waveform can be found by use of Parseval's theorem,

$$W_{1\Omega} = \frac{1}{2\pi} \int_{-\infty}^{\infty} |F(j\omega)|^2 d\omega$$

where $F(j\omega)$ is the Fourier transform of the current or voltage waveform.

The Fourier transform of the current is

$$I(j\omega) = \int_0^{\infty} t\, e^{-bt}\, e^{-j\omega t}\, dt$$

$$= \int_0^{\infty} t\, e^{-t(j\omega+b)}\, dt$$

$$= \left(\frac{-te^{-t(j\omega+b)}}{(j\omega+b)} - \int \frac{e^{-t(j\omega+b)}}{-(j\omega+b)} dt \right) \Bigg|_0^\infty$$

$$= \left(-\frac{te^{-t(j\omega+b)}}{j\omega+b} - \frac{e^{-t(j\omega+b)}}{(j\omega+b)^2} \right) \Bigg|_0^\infty$$

$$= \frac{1}{(j\omega+b)^2}$$

$$|I(j\omega)|^2 = \frac{1}{(b^2-\omega^2)^2 + 4\omega^2 b^2} = \frac{1}{(b^2+\omega^2)^2} \quad .$$

The total energy associated with the current is

$$W = \frac{1}{2\pi} \int_{-\infty}^{\infty} |I(j\omega)|^2 d\omega \quad .$$

Since

$$W = \frac{1}{\pi} \int_0^\infty |I(j\omega)|^2 d\omega$$

then

$$W = \frac{1}{\pi} \int_0^\infty \frac{1}{(b^2+\omega^2)^2} d\omega \quad .$$

If we make the trigonometric substitution,

$$\omega = b \tan \theta,$$

then

$$(b^2+\omega^2)^2 = (b^2 \tan^2\theta + b^2)^2 = (b^2 \sec^2\theta)^2 \quad .$$

Also,

$$d\omega = b\sec^2\theta d\theta \quad .$$

Hence

$$W = \frac{1}{\pi} \int \frac{b\sec^2\theta}{b^4 \sec^4\theta} d\theta$$

$$W = \frac{1}{\pi} \int \frac{1}{b^3 \sec^2 \theta} d\theta$$

$$W = \frac{1}{\pi b^3} \int \cos^2 \theta \, d\theta$$

$$W_T = \frac{1}{\pi b^3} (\tfrac{1}{2} \theta + \tfrac{1}{4} \sin 2\theta).$$

Since

$$\omega = b \tan \theta$$

and

$$\theta = \arctan \frac{\omega}{b}$$

then

$$W_T = \left[\frac{1}{\pi b^3} (\tfrac{1}{2} \tan^{-1} \tfrac{\omega}{b} + \tfrac{1}{4} \sin 2(\tan^{-1} \tfrac{\omega}{b})) \right] \Big|_0^\infty .$$

Since

$$\tan^{-1} \infty = \frac{\pi}{2}$$

then

$$W_T = \frac{1}{\pi b^3} (\tfrac{1}{2} \tfrac{\pi}{2})$$

$$W_T = \frac{1}{4b^3} .$$

b) To find the energy present in the frequency band

$$-b < f < b$$

we use Parseval's theorem and integrate:

$$W_b = \frac{1}{2\pi} \int_{-b}^{b} |I(j\omega)|^2 d\omega$$

$$W_b = \frac{1}{\pi} \int_0^b |I(j\omega)|^2 d\omega$$

$$W_b = \left[\frac{1}{\pi b^3} (\tfrac{1}{2} \tan^{-1} \tfrac{\omega}{b} + \tfrac{1}{4} \sin 2(\tan^{-1} \tfrac{\omega}{b})) \right] \Big|_0^b$$

$$W_b = \left[\frac{1}{\pi b^3}\left(\frac{1}{2}\tan^{-1}1 + \frac{1}{4}\sin 2(\tan^{-1}1)\right)\right.$$
$$\left. - \left(\frac{1}{2}\tan^{-1}0 + \frac{1}{4}\sin 2(\tan^{-1}0)\right)\right]$$

$$W_b = \frac{1}{\pi b^3}\left[\frac{\pi}{8} + \frac{1}{4} - 0 - 0\right]$$

$$W_b = \left(\frac{\pi+2}{8}\right)\frac{1}{\pi b^3} .$$

The fraction of energy present is

$$\frac{W_0}{W_T} = \frac{\left[\frac{\pi+2}{8}\right]\frac{1}{\pi b^3}}{\frac{1}{4b^3}} = \frac{\pi+2}{2\pi} = 0.818 .$$

● **PROBLEM** 17-13

a) If $a > 0$ show that the Fourier transform of the function defined by

$$f(t) = e^{-at}\cos\omega dt \quad\quad t \geq 0$$
$$= 0 \quad\quad t < 0$$

is $(a+j\omega)^2/[(a+j\omega)^2 + \omega_d^2]$.

Then find the total 1Ω energy associated with the function

$$f = e^{-t}\cos t \quad\quad t \geq 0$$
$$= 0 \quad\quad t < 0$$

by using:

b) time domain integration. That is, find the total energy by integrating

$$W = \int_0^\infty [f(t)]^2 dt .$$

c) frequency domain integration. That is, find the total energy by integrating

$$W = \frac{1}{2\pi} \int_{-\infty}^{\infty} |F(\omega)|^2 d\omega$$

where $F(\omega)$ is the Fourier transform of the function $f(t)$.

Solution: Using the relation

$$F(\omega) = \int_{-\infty}^{\infty} f(t) e^{-j\omega t} dt$$

we can find the Fourier transform of

$$f(t) = e^{-at} (\cos \omega_d t)$$

Hence,

$$F(\omega) = \int_0^{\infty} e^{-at} (\cos \omega_d t) e^{-j\omega t} dt$$

$$= \int_0^{\infty} e^{-at} \left(\frac{e^{j\omega_d t} + e^{-j\omega_d t}}{2} \right) e^{-j\omega t} dt$$

$$= \int_0^{\infty} \left[\frac{e^{(j\omega_d - j\omega - a)t}}{2} + \frac{e^{(-j\omega_d - j\omega - a)t}}{2} \right] dt$$

$$= \left[\frac{e^{(j\omega_d - j\omega - a)t}}{2(j\omega_d - j\omega - a)} + \frac{e^{-(j\omega_d + j\omega + a)t}}{2(-j\omega_d - j\omega - a)} \right]_0^{\infty}$$

$$= \left[\frac{-1}{2(j\omega_d - j\omega - a)} - \frac{1}{2(-j\omega_d - j\omega - a)} \right]$$

Hence,

$$F(\omega) = \left[\frac{a + j\omega}{(a + j\omega)^2 + \omega_d^2} \right] \cdot$$

b) In the time domain, the total energy is found by integrating

$$[f(t)]^2$$

as follows:

$$W = \int_{-\infty}^{\infty} [f(t)]^2 \, dt$$

$$W = \int_{0}^{\infty} \left[e^{-t} (\cos t) \right]^2 \, dt$$

$$= \int_{0}^{\infty} e^{-2t} \left(\frac{1}{2} + \frac{1}{2} \cos 2t \right) dt$$

$$= \int_{0}^{\infty} \left[\frac{e^{-2t}}{4} + \frac{e^{-2t}}{2} \left(\frac{e^{+j2t}}{2} + \frac{e^{-j2t}}{2} \right) \right] dt$$

$$= \left[-\frac{e^{-2t}}{4} + \frac{e^{-2t} e^{j2t}}{4(-2+j2)} - \frac{e^{-2t} e^{-j2t}}{4(2+j2)} \right]_{0}^{\infty}$$

$$W = \frac{8 - (-2 - j2) + (2 - j2)}{32} = \frac{12}{32} = \frac{3}{8} \, .$$

c) In the frequency domain the total energy is found by integrating

$$W = \frac{1}{2\pi} \int |F(\omega)|^2 d\omega \, .$$

We found in (a)

$$F(\omega) = \frac{a + j\omega}{(a + j\omega)^2 + \omega_d^2} \, .$$

If $a = 1$ and $\omega_d = 1$ then

$$F(\omega) = \frac{1 + j\omega}{(1 + j\omega)^2 + 1}$$

and

$$|F(\omega)|^2 = \frac{1 + \omega^2}{(2-\omega^2)^2 + 4\omega^2} = \frac{1 + \omega^2}{4 + \omega^4}.$$

The energy is

$$W = \frac{1}{2\pi} \int_{-\infty}^{\infty} \frac{1 + \omega^2}{4 + \omega^4} d\omega = \frac{1}{\pi} \int_{0}^{\infty} \frac{1 + \omega^2}{4 + \omega^4} d\omega.$$

Now, by an integral formula,

$$\int_{0}^{\infty} \frac{x^2 + 1}{x^4 + 4} dx = \frac{3\pi}{8}.$$

Hence, we have

$$W = \frac{1}{\pi} \int_{0}^{\infty} \frac{1 + \omega^2}{4 + \omega^2} d\omega = \frac{1}{\pi} \cdot \frac{3\pi}{8}.$$

Thus

$$W = \frac{3}{8}$$

verifying Parseval's identity in this example.

● **PROBLEM 17-14**

Use Fourier transform methods to find the time-domain response of a network having a system function

$j2\omega/(1 + 2j\omega)$,

if the input is

$V(t) = \cos t$

(For a sinusodial input cos t, the Fourier transform is

$\pi[\delta(\omega+1) + \delta(\omega-1)]$).

Solution: The time domain response for a particular input V(t) can be obtained by finding the product of the system function $H(j\omega)$ and the Fourier transform of the input. The inverse Fourier transform of the resulting function is the time-domain response.

For a sinusoidal input cos t, the Fourier transform pair

$\cos t \iff \pi[\delta(\omega+1) + \delta(\omega-1)]$

allows us to find the response

$$f(t) = F^{-1}\left\{\frac{j2\omega}{1+2j}\pi(\delta(\omega+1) + \delta(\omega-1))\right\}$$

$$f(t) = F^{-1}\left\{\frac{j2\pi\omega\,\delta(\omega+1)}{1+2j\omega} + \frac{j2\pi\omega\,\delta(\omega-1)}{1+2j\omega}\right\}$$

$$f(t) = F^{-1}\left\{\frac{j\pi\omega\,\delta(\omega+1)}{\frac{1}{2}+j\omega} + \frac{j\pi\omega\,\delta(\omega-1)}{\frac{1}{2}+j\omega}\right\}$$

Using the sifting property of the unit impulse, we obtain:

$$f(t) = F^{-1}\left\{-\frac{j\pi\,\delta(\omega+1)}{\frac{1}{2}-j} + \frac{j\pi\,\delta(\omega-1)}{\frac{1}{2}+j}\right\}$$

$$f(t) = F^{-1}\left\{-\frac{j\pi\,\delta(\omega+1)(\frac{1}{2}+j)}{\frac{1}{4}+1} + \frac{j\pi\,\delta(\omega-1)(\frac{1}{2}-j)}{\frac{1}{4}+1}\right\}$$

$$f(t) = F^{-1}\left\{\frac{\pi\,\delta(\omega+1)}{\frac{5}{4}} - \frac{j\frac{1}{2}\pi\,\delta(\omega+1)}{\frac{5}{4}} + \frac{\pi\,\delta(\omega-1)}{\frac{5}{4}} + \frac{j\frac{1}{2}\pi\,\delta(\omega-1)}{\frac{5}{4}}\right\}$$

$$f(t) = F^{-1}\left\{\frac{4}{5}\pi(\delta(\omega+1)+\delta(\omega-1)) - \frac{2}{5}\pi(j\,\delta(\omega+1)-j\,\delta(\omega-1))\right\}$$

$$f(t) = \frac{4}{5}\cos t - \frac{2}{5}\sin t.$$

CHAPTER 18

DIFFERENTIAL GEOMETRY

Differential geometry is a consequential branch of mathematics from which important results are obtained. By definition, differential geometry is the study of geometric figures using the methods of calculus. Therefore, when dealing with this area of mathematics it is imperative to remember that a proof to a given hypothesis must not depend only upon our intuition; that is, it is necessary that we also apply the given definitions and previously confirmed theorems (this because our intuition about curves and surfaces, based upon simple examples, may give us an oversimplified visual representation). Specifically, for curves in space, such important concepts as curvature (the rate at which the curve is turning away from its tangent line at a point), torsion (the measure of the rate at which the curve is twisting), and arc length are developed by differential geometry. For example, let a curve be given parametrically by the vector equation $\vec{X} = \vec{X}(t)$. Then using the methods of differential geometry, the arc length between two points $t = a$ and $t = b$ can be found by the equation

$$L = \int_a^b \|\vec{X}'(t)\| \, dt \, .$$

Accordingly, the theory of differential geometry extends to surfaces and curves on a surface. Here we offer a small part of this theory beginning with the concept of a surface patch. Then the first fundamental form is developed and applied to find arc length of a curve on a surface, the angle of intersection of two curves on a surface, and the area of a surface. In addition, the second fundamental form is developed and applied to define such terms as Normal, Gaussian, and Mean, curvature. Furthermore, geodesics, which are arcs of minimal lengths, are also covered.

CURVES

● **PROBLEM 18-1**

a) Define what is meant by a regular parametric representation of a curve.
b) Show that the representation
$$x_1 = (1 + \cos\theta), \, x_2 = \sin\theta, \, x_3 = 3\sin(\theta/3)$$
for $-2\pi \leq \theta \leq 2\pi$ is regular.

Solution: A regular parametric representation of a curve is a vector function
$$\vec{X} = \vec{X}(t) \, , \quad t \in I$$
of t in an interval I with the property that:
 (i) $\vec{X}(t)$ is of class C^1 in I .
 (ii) $\vec{X}'(t) \neq 0$ for all t in I.
The variable t is the parameter of the representation.

Notice that if a basis is chosen in R^3, the equation $\vec{X} = \vec{X}(t)$ can be written as three scalar equations $x_1 = x_1(t)$, $x_2 = x_2(t)$, $x_3 = x_3(t)$, $t \in I$, the components of $\vec{X} = \vec{X}(t)$ with respect to the basis. Accordingly, $\vec{X} = \vec{X}(t)$ is a regular parametric representation if and only if each $x_i(t)$ is in class C^1 and if for each t in I at least one of the $x_i'(t) \neq 0$.

b) For this problem,

$$x_1 = (1 + \cos\theta), \quad x_2 = \sin\theta, \quad x_3 = 3\sin(\theta/3)$$

are all continuous for $-2\pi \leq \theta \leq 2\pi$, and

$$\frac{dx_1}{d\theta} = -\sin\theta, \quad \frac{dx_2}{d\theta} = \cos\theta, \quad \frac{dx_3}{d\theta} = \cos(\theta/3) \tag{1}$$

are continuous.

From (1), $\left[\left(\frac{dx_1}{d\theta}\right)^2 + \left(\frac{dx_2}{d\theta}\right)^2 + \left(\frac{dx_3}{d\theta}\right)^2\right]^{\frac{1}{2}}$

$$= (\sin^2\theta + \cos^2\theta + \cos^2(\theta/3))^{\frac{1}{2}}$$

$$= [1 + \cos^2(\theta/3)]^{\frac{1}{2}} \neq 0.$$

Thus the representation is regular.

• **PROBLEM 18-2**

a) Define a natural representation of a curve.
b) Show that

$$\vec{X} = \left(\tfrac{1}{2}(s + \sqrt{s^2+1}), \tfrac{1}{2}(s + \sqrt{s^2+1})^{-1}, \tfrac{1}{2}\sqrt{2}\log(s + \sqrt{s^2+1})\right). \tag{1}$$

is a natural representation.

Solution: a) Let I be an interval $[a,b]$ in R. Given that $\vec{X} = \vec{X}(t)$ is a regular curve on I, let

$$s = s(t) = \int_{t_0}^{t} \left|\frac{d\vec{X}}{dt}\right| dt, \tag{2}$$

so that if $t \geq t_0$, then $s \geq 0$ and is equal to the length of the arc segment of the curve between $\vec{X}(t_0)$ and $\vec{X}(t)$; if $t < t_0$, then $s < 0$ and is equal to the negative of the length of the arc segment of the curve between $\vec{X}(t_0)$ and $\vec{X}(t)$.

Now (2) has a continuous nonvanishing derivative by the Fundamental Theorem of Calculus, given by

$$\frac{ds}{dt} = \frac{d}{dt}\int_{t_0}^{t}\left|\frac{d\vec{X}}{dt}\right|dt = \left|\frac{d\vec{X}}{dt}\right|. \tag{3}$$

Then, using the fact that a real valued function $s = s(t)$ on an interval I_t is an allowable change of parameter if

(i) $s(t)$ is of class C^1 in I_t

(ii) $ds/dt \neq 0$ for all t in I_t,

by (3), $s = s(t)$ is an allowable change of parameter on I. Furthermore, $s(t)$ is of class C^m on I if $\vec{X}(t)$ is of class C^m. Therefore the arc length s can be introduced along the curve as a parameter. Then, by definition a representation $\vec{X} = \vec{X}(s)$ on I_s is a representation in terms of arc length or a natural representation if $\left|\frac{d\vec{X}}{ds}\right| = 1$. (4)

b) To show (1) is a natural representation we must show that $\left|\frac{d\vec{X}}{ds}\right| = 1$. To do this simplify the calculations, letting $u = s + \sqrt{s^2+1}$ in (1). This gives

$$\vec{X} = (\tfrac{1}{2}u, \tfrac{1}{2}u^{-1}, \tfrac{1}{2}\sqrt{2} \log u) \qquad (5)$$

By the chain rule

$$\frac{d\vec{X}}{ds} = \frac{d\vec{X}}{du}\frac{du}{ds} = \left(\tfrac{1}{2}, -\tfrac{1}{2}u^{-2}, \tfrac{1}{2}\sqrt{2}u^{-1}\right)\left(1 + \frac{s}{\sqrt{s^2+1}}\right).$$

Hence, $\left|\frac{d\vec{X}}{ds}\right| = \left|\frac{d\vec{X}}{du}\right|\left|\frac{du}{ds}\right|$ or

$$\left|\frac{d\vec{X}}{ds}\right| = \left[(\tfrac{1}{2})^2 + \left(-\tfrac{1}{2}u^{-2}\right)^2 + \left(\tfrac{1}{2}\sqrt{2}u^{-1}\right)^2\right]^{\frac{1}{2}} \left(\frac{\sqrt{s^2+1}+s}{\sqrt{s^2+1}}\right)$$

$$= \tfrac{1}{2}\left(1 + u^{-4} + 2u^{-2}\right)^{\frac{1}{2}}\left(\frac{\sqrt{s^2+1}+s}{\sqrt{s^2+1}}\right)$$

Next, factoring and replacing $s + \sqrt{s^2+1}$ with u yields

$$\left|\frac{d\vec{X}}{ds}\right| = \tfrac{1}{2}\left(1 + \frac{1}{u^2}\right)\left(\frac{u}{\sqrt{s^2+1}}\right)$$

$$= \tfrac{1}{2}\left(\frac{u^2+1}{u\sqrt{s^2+1}}\right) = \tfrac{1}{2}\frac{(s+\sqrt{s^2+1})^2 + 1}{(s+\sqrt{s^2+1})\sqrt{s^2+1}}$$

$$= \tfrac{1}{2}\left(\frac{2(s^2+s\sqrt{s^2+1}+1)}{s\sqrt{s^2+1} + s^2 + 1}\right)$$

$$= \tfrac{1}{2}(2) = 1 .$$

Thus (1) is a natural representation.

• **PROBLEM 18-3**

Consider the curve

$$\vec{C} = t\vec{e}_1 + t^2\vec{e}_2 = (t, t^2) , \quad 0 \leq t \leq 1 .$$

Show that it is rectifiable.

Solution: The length of an arc is defined in terms of the lengths of approximating polygonal arcs. Let an arc $\vec{C}$ (not necessarily regular) be given by $\vec{X} = \vec{X}(t)$, $a \leq t \leq b$, and consider a subdivision

$$a = t_0 < t_1 < \ldots < t_n = b$$

of the interval $a \leq t \leq b$. This determines a sequence of points in R^3:

$$x_0 = \vec{X}(t_0), x_1 = \vec{X}(t_1), \ldots, x_n = \vec{X}(t_n)$$

which are joined in sequence to form an approximating polygonal arc P as shown in figure 1.

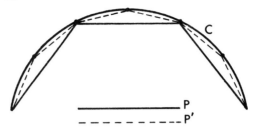

Fig. 1

The length of the line between two adjacent points x_i and x_{i-1} is $\|x_i - x_{i-1}\|$. Therefore the length of P is

$$s(P) = \sum_{i=1}^{n} \|x_i - x_{i-1}\| = \sum_{i=1}^{n} \|\vec{x}(t_i) - \vec{x}(t_{i-1})\|.$$

Now give a better approximating arc P' by introducing additional points as shown. Then the length of P is less than or equal to the length of P' since the length of one side of a polygon is less than or equal to the sum of the lengths of the other two sides. That is $s(P) \leq s(P')$. This leads us to define the length of the arc C as the greatest of lengths of all possible approximating polygonal arcs P. An arc $\vec{X} = \vec{X}(t)$, $a \leq t \leq b$, is said to be rectifiable if the set S of all possible $s(P)$ is bounded from above.
The curve in the given problem is just the piece of the parabola $y = x^2$, for $x \in [0,1]$.

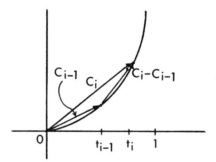

Fig. 2

In fig. 2 we depict a typical polygonal approximation to the curve $\vec{C}$. The length of the ith piece is $|\vec{C}_i - \vec{C}_{i-1}|$ where

$$\vec{C}_i = t_i \vec{e}_1 + t_i^2 \vec{e}_2.$$

The length of the total approximation can be written as

$$s(P) = \sum_{i=1}^{n} \|(t_i \vec{e}_1 + t_i^2 \vec{e}_2) - (t_{i-1} \vec{e}_1 + t_{i-1}^2 \vec{e}_2)\|$$

where the t_i form a subdivision of the interval $[0,1]$, $0 = t_0 < t_1 < \ldots < t_n = 1$. We want to show that the above sum is bounded from above, no matter what subdivision we choose. Separating components we get

$$s(P) = \sum_{i=1}^{n} \|(t_i \vec{e}_1 + t_i^2 \vec{e}_2) - (t_{i-1} \vec{e}_1 + t_{i-1}^2 \vec{e}_2)\|$$

$$= \sum_{i=1}^{n} \|(t_i - t_{i-1}) \vec{e}_1 + (t_i^2 - t_{i-1}^2) \vec{e}_2\|.$$

By the definition of the norm, $\| \cdot \|$, we have $\|\vec{v} + \vec{w}\| \leq \|\vec{v}\| + \|\vec{w}\|$ and $\|c\vec{v}\| = |c|\|\vec{v}\|$. Therefore

$$s(P) \leq \sum_{i=1}^{n} \{|t_i - t_{i-1}|\|\vec{e}_1\| + |t_i^2 - t_{i-1}^2|\|\vec{e}_2\|\}$$

$$= \sum_{i=1}^{n} |t_i - t_{i-1}| + |t_i - t_{i-1}| |t_i + t_{i-1}|$$

(since $\|\vec{e}_1\| = \|\vec{e}_2\| = 1$)

$$= \sum_{i=1}^{n} (t_i - t_{i-1})(1 + t_i + t_{i-1}).$$

$|t_i - t_{i-1}|$ is replaced with $(t_i - t_{i-1})$ since this quantity is always positive. Since $t_i \leq 1$, $1 + t_i + t_{i-1} \leq 3$ so

$$s(P) \leq \sum_{i=1}^{n} 3(t_i - t_{i-1}) = 3 \sum_{i=1}^{n} (t_i - t_{i-1}).$$

But

$$\sum_{i=1}^{n} (t_i - t_{i-1}) = (t_1 - t_0) + (t_2 - t_1) + \ldots + (t_n - t_{n-1})$$

$$= t_n - t_0 = 1 - 0 = 1$$

so $s(P) \leq 3 \cdot 1 = 3$.

We can now conclude that $\vec{C}$ is rectifiable since all polygonal approximations are uniformly bounded.

• **PROBLEM 18-4**

Show that the curve given by

$x_1 = t$

$x_2 = \begin{cases} t \cos(1/t) & \text{for } 0 < t \leq 1 \\ 0 & \text{for } t = 0 \end{cases}$

is not rectifiable.

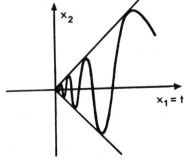

Fig. 1

Solution: For the subdivision $0, \frac{1}{(N-1)\pi}, \ldots, \frac{1}{2\pi}, \frac{1}{\pi}, 1$ consider the polygonal approximation from the point on the curve corresponding to $\frac{1}{(N-1)\pi}$, to the point corresponding to $1/\pi$. This will be less than the length of the curve. The length of the nth piece will be the length of the vector going from the point $(1/n\pi,\ 1/n\pi\ \cos n\pi)$ to $\left(\frac{1}{(n+1)\pi},\ \frac{1}{(n+1)\pi}\cos(n+1)\pi\right)$.

i.e., $\left\|\left(\frac{1}{n\pi}-\frac{1}{(n+1)\pi}\right)\vec{e}_1+\left(\frac{1}{n\pi}\cos n\pi-\frac{1}{(n+1)\pi}\cos(n+1)\pi\right)\vec{e}_2\right\|$.

For $n=1$ we get the point on the curve corresponding to $t=1/\pi$, while for $n=N-2$ we get the point for $t=\frac{1}{(N-1)\pi}$. Therefore, the length of this approximation is, (since the first and last terms have been dropped)

$$S(p) \geq \sum_{n=1}^{N-2}\left\|\left(\frac{1}{n\pi}-\frac{1}{(n+1)\pi}\right)\vec{e}_1+\left(\frac{1}{n\pi}\cos n\pi-\frac{1}{n+1}\cos(n+1)\pi\right)\vec{e}_2\right\|$$

$$\geq \sum_{n=1}^{N-2}\left\|\left(\frac{1}{n\pi}\cos n\pi-\frac{1}{(n+1)\pi}\cos(n+1)\pi\right)\vec{e}_2\right\|$$

since $\|a\vec{e}_1+b\vec{e}_2\|\geq|b|$ because $\vec{e}_1$ and $\vec{e}_2$ are perpendicular. Recognizing that $\cos n\pi=(-1)^n$ and similarly $\cos(n+1)\pi=(-1)^{n+1}$ yields

$$S(p)\geq\sum_{n=1}^{N-2}\left\|(-1)^n\frac{1}{n\pi}-(-1)^{n+1}\frac{1}{(n+1)\pi}\right\|.$$

Replacing $(-1)^{n+1}$ with $(-1)(-1)^n$ and factoring yields

$$S(p)\geq\sum_{n=1}^{N-2}|(-1)^n|\ \left|\frac{1}{n\pi}+\frac{1}{(n+1)\pi}\right|$$

i.e., $S(p)\geq\sum_{n=1}^{N-2}\left|\frac{1}{n\pi}+\frac{1}{(n+1)\pi}\right|\geq\sum_{n=1}^{N-2}\left|\frac{1}{(n+1)\pi}+\frac{1}{(n+1)\pi}\right|$

or

$$S(p)\geq\frac{2}{\pi}\sum_{n=1}^{N-2}\frac{1}{n+1}.$$

However, $\sum_{n=1}^{\infty}\frac{1}{n+1}$ diverges, so $S(p)$ can be made arbitrarily large by allowing N, the number of points in our subdivision, to be large. Therefore the curve is not rectifiable.

● PROBLEM 18-5

If a curve is parametrized by arc length, $\vec{x}=\vec{x}(s)$, the curvature $|k|$ is defined as $\|\ddot{\vec{x}}(s)\|$ (where $\ddot{\vec{x}}(s)=\frac{d^2\vec{x}}{ds^2}$).

Using the chain rule, compute $|k|$ if $\vec{x}$ is parametrized by time

t and show that
$$|k| = \frac{\|\vec{x}' \times \vec{x}''\|}{\|\vec{x}'\|^3}.$$
It is assumed that the curve is regular.

Solution: The curvature of a plane curve is essentially the rate at which the tangent line is turning. For a space curve, this quantity is replaced by two others. The first is again called the curvature and is a measure of the rate at which the curve is turning away from its tangent line at a point, the second is called the torsion and is a measure of the rate at which the curve is twisting out of its osculating plane at a point. Observe that for a plane curve the torsion is zero. (Note: this problem deals only with curvature, the notion of torsion is developed in a latter problem). For the solution to this problem let
$$\dot{\vec{x}} = \frac{d\vec{x}}{ds} \quad \text{and} \quad \vec{x}' = \frac{d\vec{x}}{dt}.$$

By the chain rule
$$\vec{x}' = \frac{d\vec{x}}{dt} = \frac{d\vec{x}}{ds}\frac{ds}{dt} = \dot{\vec{x}}s' \tag{1}$$

and
$$\vec{x}'' = \frac{d(\vec{x}')}{dt} = \frac{d(\dot{\vec{x}}s')}{dt} = \dot{\vec{x}}\frac{ds'}{dt} + \frac{d\dot{\vec{x}}}{dt}s' \tag{2}$$

by the product rule. Reapplying the chain rule (2) can be rewritten
$$\vec{x}'' = \dot{\vec{x}}\frac{ds'}{dt} + \frac{d\dot{\vec{x}}}{ds}\frac{ds}{dt}s' = \dot{\vec{x}}s'' + (s')^2 \ddot{\vec{x}}.$$

Now, using (1), (2),
$$\vec{x}' \times \vec{x}'' = (\dot{\vec{x}}s') \times (\dot{\vec{x}}s'' + \ddot{\vec{x}}(s')^2)$$
$$= (\dot{\vec{x}}s' \times \dot{\vec{x}}s'') + (\dot{\vec{x}}s' \times \ddot{\vec{x}}(s')^2). \tag{3}$$

Since $a\vec{x} \times b\vec{y} = ab\vec{x} \times \vec{y}$ and $\vec{x} \times \vec{x} = 0$, (3) becomes
$$\vec{x}' \times \vec{x}'' = (s')(s')^2 \dot{\vec{x}} \times \ddot{\vec{x}} = (s')^3 \dot{\vec{x}} \times \ddot{\vec{x}}. \tag{4}$$

By the definition of arc-length $s(t) = \int_0^t \|\vec{x}'(\sigma)\|d\sigma$, so that by the fundamental theorem of calculus,
$$\frac{ds}{dt} = s' = \|\vec{x}'(t)\|$$

Substituting this into (4) yields
$$\|\vec{x}' \times \vec{x}''\| = \|\vec{x}'\|^3 \|\dot{\vec{x}} \times \ddot{\vec{x}}\|$$
$$= \|\vec{x}'\|^3 \|\dot{\vec{x}}\|\|\ddot{\vec{x}}\| |\sin \sphericalangle (\dot{\vec{x}},\ddot{\vec{x}})| \tag{5}$$

where $\sin \sphericalangle (\dot{\vec{x}},\ddot{\vec{x}})$ is the sine of the angle between $\dot{\vec{x}}$ and $\ddot{\vec{x}}$.

But $\|\dot{\vec{x}}\| = 1$, i.e., $\dot{\vec{x}} \cdot \dot{\vec{x}} = 1$. Differentiating this relation gives $2\ddot{\vec{x}} \cdot \dot{\vec{x}} = 0$. If $\ddot{\vec{x}} = 0$, $|k| = 0$ and $\|\vec{x}' \times \vec{x}''\| = 0$ so the result is true. If $\ddot{\vec{x}} \neq 0$, then $\dot{\vec{x}}$ and $\ddot{\vec{x}}$ must be perpendicular and hence $|\sin \angle (\dot{\vec{x}}, \ddot{\vec{x}})| = 1$. Furthermore $\|\dot{\vec{x}}\| = 1$ and $\|\ddot{\vec{x}}\| = |k|$. Thus (5) becomes

$$\|\vec{x}' \times \vec{x}''\| = \|\vec{x}'\|^3 |k|$$

or

$$|k| = \frac{\|\vec{x}' \times \vec{x}''\|}{\|\vec{x}'\|^3}$$

which is the desired result.

• **PROBLEM 18-6**

Find the curvature and torsion along the curve
$$\vec{x} = (3t - t^3, 3t^2, 3t + t^3). \qquad (1)$$

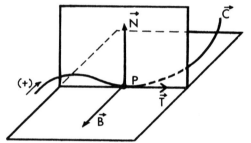

Fig. 1

<u>Solution</u>: By definition the curvature of a curve $\vec{C}$ is a scalar value given by the equation

$$k = \left[\left(\frac{d^2 x}{ds^2}\right)^2 + \left(\frac{d^2 y}{ds^2}\right)^2 + \left(\frac{d^2 z}{ds^2}\right)^2\right]^{1/2}.$$

The torsion along a curve is to some extent a measure of the amount by which the curve is twisted. This quantity is defined using a vector known as the binormal vector (denoted by $\vec{B}$) of a curve $\vec{C}$ at some point P. By definition $\vec{B} = \vec{T} \times \vec{N}$ where $\vec{T}$ is the tangent vector and $\vec{N}$ is the principal normal (see figure 1). Observe that $\vec{T}$, $\vec{N}$, and $\vec{B}$ are mutually perpendicular unit vectors. Since $\frac{d\vec{B}}{ds} = \vec{T} \times \frac{d\vec{N}}{ds} + \frac{d\vec{T}}{ds} \times \vec{N}$ and $\frac{d\vec{T}}{ds} \times \vec{N} = k\vec{N} \times \vec{N} = 0$ (this from the equation $\frac{d\vec{T}}{ds} = k\vec{N}$ and the fact that the cross product of any vector with itself is zero) we have $\frac{d\vec{B}}{ds} = \vec{T} \times \frac{d\vec{N}}{ds}$. Furthermore $\vec{B} \cdot \vec{B} = 1$, since $\vec{B}$ has unit length. Therefore $\vec{B} \cdot \frac{d\vec{B}}{ds} = 0$. Now, if $\frac{d\vec{B}}{ds} \neq 0$, the two preceding equations show that $\frac{d\vec{B}}{ds}$ is perpendicular both to

$\vec{B}$ and to $\vec{T}$; it is therefore a multiple of $\vec{N}$, for $\vec{N}$ is also perpendicular to both $\vec{B}$ and $\vec{T}$. Thus $\frac{d\vec{B}}{ds} = -\tau\vec{N}$ where τ is the proper multiple of $\vec{N}$ to give $\frac{d\vec{B}}{ds}$. The quantity τ thus defined is called the torsion of $\vec{C}$ at the point P.

Curvature and torsion are important values because a curve is completely determined by its curvature and torsion as functions of a natural parameter. That is, if $\vec{C}$ and $\vec{C}'$ are two curves in space such that $k(s) = k'(s)$ and $\tau(s) = \tau'(s)$ for all s, then $\vec{C}$ and $\vec{C}'$ are the same except for their position in space.

To find k and τ apply the following theorems:

$$|k| = \frac{\|\vec{x}' \times \vec{x}''\|}{\|\vec{x}'\|^3} \quad \text{and} \quad \tau = \frac{(\vec{x}' \times \vec{x}'') \cdot \vec{x}'''}{\|\vec{x}' \times \vec{x}''\|^2} \qquad (2)$$

where $|k|$ is the curvature and τ is the torsion. Hence, to apply these theorems, first compute from (1),

$$\vec{x}' = (3 - 3t^2, 6t, 3 + 3t^2) \qquad (3)$$
$$\vec{x}'' = (-6t, 6, 6t) \qquad (4)$$
$$\vec{x}''' = (-6, 0, 6) \qquad (5)$$

Then

$$\vec{x}' \times \vec{x}'' = \begin{vmatrix} \vec{i} & \vec{j} & \vec{k} \\ 3-3t^2 & 6t & 3+3t^2 \\ -6t & 6 & 6t \end{vmatrix}$$

$$= (18t^2 - 18, -36t, 18t^2 + 18). \qquad (6)$$

Now, substitution of (3), (4), (6) into the first equation in (2) yields

$$|k| = \frac{\|(18t^2 - 18, -36t, 18t^2 + 18)\|}{\|(3-3t^2, 6t, 3 + 3t^2)\|^3}$$

$$= \frac{18\|(t^2-1, -2t, t^2 + 1)\|}{27\|(1 - t^2, 2t, 1 + t^2)\|^3}$$

$$= \frac{18}{27} \frac{[(t^2-1)^2 + (-2t)^2 + (t^2+1)^2]^{\frac{1}{2}}}{[(1-t^2)^2 + (2t)^2 + (1+t^2)^2]^{3/2}}$$

$$= \frac{18}{27} \frac{1}{[(1-t^2)^2 + (2t)^2 + (1+t^2)^2]}$$

or

$$|k| = \frac{2}{3} \cdot \frac{1}{2(1 + 2t^2 + t^4)} = \frac{1}{3(1+t^2)^2}.$$

Next substitution of (3), (4), (5), (6) into the second equation in (2) yields

$$\vec{T} = \frac{[18(t^2-1,-2t,1+t^2) \cdot 6(-1,0,1)]}{18^2 \|(t^2-1,-2t,1+t^2)\|^2}$$

$$= \frac{1}{3} \frac{(1-t^2 + 1 + t^2)}{(t^2-1)^2 + (-2t)^2 + (1+t^2)^2}$$

or

$$\tau = \frac{1}{3} \frac{2}{2 + 4t^2 + 2t^4} = \frac{1}{3(1 + t^2)}$$

Note, that in this case $|k| = \tau$.

• **PROBLEM 18-7**

Find the unit principal normal and unit binormal to the curve $\vec{x} = (3t-t^3, 3t^2, 3t + t^3)$.

Solution: The unit tangent vector to a curve $\vec{C}$ can be written as $\vec{T} = \frac{dx}{ds}\vec{i} + \frac{dy}{ds}\vec{j} + \frac{dz}{ds}\vec{k}$ or $\vec{T} = \left(\frac{dx}{ds}, \frac{dy}{ds}, \frac{dz}{ds}\right)$ where s is the arc length measured along $\vec{C}$ in the positive direction from some fixed point. From this we have

$$\frac{d\vec{T}}{ds} = \frac{d^2x}{ds^2}\vec{i} + \frac{d^2y}{ds^2}\vec{j} + \frac{d^2z}{ds^2}\vec{k} \quad . \tag{1}$$

Using the equation $ds^2 = dx^2 + dy^2 + dz^2$ (2)

we have

$$1 = \left(\frac{dx}{ds}\right)^2 + \left(\frac{dy}{ds}\right)^2 + \left(\frac{dz}{ds}\right)^2 \quad .$$

Therefore, differentiating with respect to s,

$$0 = 2\left[\frac{dx}{ds}\frac{d^2x}{ds^2} + \frac{dy}{ds}\frac{d^2y}{ds^2} + \frac{dz}{ds}\frac{d^2z}{ds^2}\right]$$

which is $0 = 2\vec{T} \cdot \frac{d\vec{T}}{ds}$ or $\vec{T} \cdot \frac{d\vec{T}}{ds} = 0$. Hence, when $\frac{d\vec{T}}{ds}$ is not zero, its direction is perpendicular to $\vec{T}$. A unit vector in the direction of $\frac{d\vec{T}}{ds}$ is called the principal normal to $\vec{C}$ at the point in question.

To find the unit normal to this curve, it is necessary to first find the unit tangent. To do this compute

$$\vec{x}' = (3-3t^2, 6t, 3 + 3t^2)$$

and

$$\|\vec{x}'\| = 3\|(1-t^2, 2t, 1 + t^2)\|$$

$$= 3[(1-t^2)^2 + (2t)^2 + (1 + t^2)^2]^{\frac{1}{2}}$$

$$= 3\sqrt{2} (1 + t^2) .$$

The tangent, $\vec{T}$, is in the direction of $\vec{x}'$ but must have unit length. Therefore,

$$\vec{T} = \frac{\vec{x}'}{\|\vec{x}'\|} = \frac{1}{\sqrt{2}(1+t^2)} (1-t^2, 2t, 1+t^2).$$

Now letting $\dot{\vec{T}} = \frac{d\vec{T}}{ds}$ we know that $\dot{\vec{T}}$ is orthogonal to $\vec{T}$ and represents the direction of the normal. The unit normal is then

$$\vec{N} = \frac{\dot{\vec{T}}}{\|\dot{\vec{T}}\|}.$$

Remark Since $\vec{T}' = \frac{d\vec{T}}{dt}$ and $\|\vec{x}'\| = \|\frac{d\vec{x}}{dt}\| = \left[\left(\frac{dx}{dt}\right)^2 + \left(\frac{dy}{dt}\right)^2 + \left(\frac{dz}{dt}\right)^2\right]^{\frac{1}{2}}$

$= \frac{ds}{dt}$ by (2),

we have

$$\frac{d\vec{T}}{ds} = \frac{d\vec{T}}{dt} \cdot \frac{dt}{ds} = \frac{\vec{T}'}{\|\vec{x}'\|}.$$

Hence
$$\dot{\vec{T}} = \frac{\vec{T}'}{\|\vec{x}'\|} = \frac{1}{6(1+t^2)} \frac{d}{dt}\left(\frac{1-t^2}{1+t^2}, \frac{2t}{1+t^2}, 1\right)$$

$$= \frac{1}{6(1+t^2)} \left(\frac{-4t}{(1+t^2)^2}, \frac{2-2t^2}{(1+t^2)^2}, 0\right)$$

or

$$\dot{\vec{T}} = \frac{1}{3(1+t^2)^3} (-2t, 1-t^2, 0)$$

from which
$$\|\dot{\vec{T}}\| = \frac{[(-2t)^2 + (1-t^2)^2]^{\frac{1}{2}}}{3(1+t^2)^3}$$

$$= \frac{(t^2+1)}{3(1+t^2)^3} = \frac{1}{3(1+t^2)^2}.$$

Therefore $\vec{N} = \frac{3(1+t^2)^2}{3(1+t^2)^3} (-2t, 1-t^2, 0) = \frac{1}{(1+t^2)} (-2t, 1-t^2, 0).$

By definition the binormal vector of C at P is given by the equation $\vec{B} = \vec{T} \times \vec{N}$. Hence

$$\vec{B} = \frac{1}{\sqrt{2}(1+t^2)} \cdot \frac{1}{(1+t^2)} (1-t^2, 2t, 1+t^2) \times (-2t, 1-t^2, 0)$$

$$= \frac{1}{\sqrt{2}(1+t^2)^2} \begin{vmatrix} \vec{i} & \vec{j} & \vec{k} \\ 1-t^2 & 2t & 1+t^2 \\ -2t & 1-t^2 & 0 \end{vmatrix}$$

$$= \frac{1}{\sqrt{2}(1+t^2)^2} \left[-(1+t^2)(1-t^2), -2t(1+t^2), (1+t^2)^2\right]$$

or

$$\vec{B} = \frac{1}{\sqrt{2}(1+t^2)} (t^2-1, -2t, 1+t^2).$$

• **PROBLEM 18-8**

Given the circular helix
$$\vec{x}(t) = (a\cos t,\ a\sin t,\ bt),$$
find the tangent vector, the normal vector, the curvature, the binormal vector, and the torsion along this curve.

Solution: Using the fact that the given helix has arc length $s = (a^2 + b^2)^{\frac{1}{2}} t$, let $c = (a^2 + b^2)^{\frac{1}{2}}$ to get $s = ct$; or $t = s/c$. Then the tangent is

$$\vec{T}(s) = \left(-\frac{1}{c} a \sin\left(\frac{s}{c}\right),\ \frac{1}{c} a \cos\left(\frac{s}{c}\right),\ \frac{b}{c}\right)$$

$$= \frac{1}{c}\left(-a \sin\left(\frac{s}{c}\right),\ a \cos\left(\frac{s}{c}\right),\ b\right).$$

Next using $\dfrac{d\vec{T}}{ds} = k\vec{N}$ yields

$$k\vec{N}(s) = \frac{1}{c^2}\left(-a \cos\left(\frac{s}{c}\right),\ -a \sin\left(\frac{s}{c}\right),\ 0\right)$$

$$= \frac{a}{c^2}\left(-\cos\left(\frac{s}{c}\right),\ -\sin\left(\frac{s}{c}\right),\ 0\right).$$

From which $k = \dfrac{a}{c^2}$, $\vec{N} = -\left(\cos\left(\frac{s}{c}\right),\ \sin\left(\frac{s}{c}\right),\ 0\right)$. Now, the binormal is given by

$$\vec{B} = \vec{T} \times \vec{N} = \begin{vmatrix} \vec{i} & \vec{j} & \vec{k} \\ -\frac{a}{c}\sin\left(\frac{s}{c}\right) & \frac{a}{c}\cos\left(\frac{s}{c}\right) & \frac{b}{c} \\ -\cos\left(\frac{s}{c}\right) & -\sin\left(\frac{s}{c}\right) & 0 \end{vmatrix}$$

$$= \frac{1}{c}\left(b \sin\left(\frac{s}{c}\right),\ -b \cos\left(\frac{s}{c}\right),\ a\right).$$

To find the torsion τ, use the formula

$$\frac{d\vec{B}}{ds} = -\tau\vec{N}$$

But

$$\frac{d\vec{B}}{ds} = \frac{1}{c^2}\left(b \cos\left(\frac{s}{c}\right),\ b \sin\left(\frac{s}{c}\right),\ 0\right)$$

or

$$-\tau\vec{N} = \frac{b}{c^2}\left(\cos\left(\frac{s}{c}\right),\ \sin\left(\frac{s}{c}\right),\ 0\right).$$

Recalling

$$-\vec{N} = \left(\cos\left(\frac{s}{c}\right),\ \sin\left(\frac{s}{c}\right),\ 0\right)$$

yields

$$\tau = \frac{b}{c^2}.$$

• **PROBLEM 18-9**

Find the torsion of the helix

$$x_1 = \cos s/\sqrt{2}, \quad x_2 = \sin s/\sqrt{2}, \quad x_3 = s/\sqrt{2}$$

at the point $s = 0$.

Solution: To do this problem the following alternate method of finding the torsion will be used: The torsion of the curve $\vec{x} = \vec{x}(s)$ at the point s_0 is $\tau = \pm \lim_{\Delta s \to 0} \frac{\Delta \varphi}{\Delta s}$ where $\Delta \varphi$ is the angle between the osculating planes at the points s_0 and $s_0 + \Delta s$. (The sign will be determined so that the torsion of a right-handed helix is positive). Before proceeding to apply this method the idea of osculating planes must be defined.

However, to do this, we must first define contact of order n. A curve $\vec{x} = \vec{x}(t)$ and a plane $(\vec{X} - \vec{a}) \cdot \vec{b} = 0$ have contact of order n at a common point $\vec{a} = \vec{x}(t_0)$ if and only if the distance $d(t)$ from a point t of the curve to the plane has a zero of order n+2 at $t = t_0$. Then, by definition a tangent plane to a curve which has contact of order greater than unity with the curve is called an osculating plane.

In the given problem, the vector $\vec{B}$, the normal to the osculating plane, is given by the equation

$$\vec{B} = \vec{T} \times \vec{N} = \frac{\vec{x}'(s) \times \vec{x}''(s)}{\|\vec{x}''(s)\|}.$$

Since $\vec{x}(s) = (\cos s/\sqrt{2}, \sin s/\sqrt{2}, s/\sqrt{2})$,
$\vec{x}'(s) = (-1/\sqrt{2} \sin s/\sqrt{2}, 1/\sqrt{2} \cos s/\sqrt{2}, 1/\sqrt{2})$
$\vec{x}''(s) = (-1/2 \cos s/\sqrt{2}, -1/2 \sin s/\sqrt{2}, 0)$

and

$$\vec{x}'(s) \times \vec{x}''(s) = \begin{vmatrix} \vec{i} & \vec{j} & \vec{k} \\ -\frac{1}{\sqrt{2}}\sin\frac{s}{\sqrt{2}} & \frac{1}{\sqrt{2}}\cos\frac{s}{\sqrt{2}} & \frac{1}{\sqrt{2}} \\ -\frac{1}{2}\cos\frac{s}{\sqrt{2}} & -\frac{1}{2}\sin\frac{s}{\sqrt{2}} & 0 \end{vmatrix}$$

$$= \frac{1}{2\sqrt{2}} (\sin s/\sqrt{2}, -\cos s/\sqrt{2}, 1).$$

Therefore,

$$\vec{B} = \frac{\frac{1}{2\sqrt{2}}(\sin s/\sqrt{2}, -\cos s/\sqrt{2}, 1)}{\frac{1}{2}(\cos^2 s/\sqrt{2} + \sin^2 s/\sqrt{2})^{\frac{1}{2}}}$$

or
$$\vec{B} = (1/\sqrt{2} \sin s/\sqrt{2}, -1/\sqrt{2} \cos s/\sqrt{2}, 1/\sqrt{2}).$$

Now $\vec{B}(s) \cdot \vec{B}(s+\Delta s) = \|\vec{B}(s)\| \cdot \|\vec{B}(s+\Delta s)\| \cos \Delta \varphi$ by definition, where $\cos \Delta \varphi$ is the angle between this vector, $\vec{B}$, at s and the same vector at $s + \Delta s$. Since

$$\|\vec{B}\| = \left(\frac{\sin^2 s/\sqrt{2}}{2} + \frac{\cos^2 s/\sqrt{2}}{2} + \frac{1}{2}\right)^{\frac{1}{2}} = 1 \text{ for all } s,$$

this yields $\cos \Delta \varphi = \vec{B}(s) \cdot \vec{B}(s+\Delta s)$

$$= \frac{1}{\sqrt{2}}(\sin s/\sqrt{2}, -\cos s/\sqrt{2}, 1) \cdot \frac{1}{\sqrt{2}}(\sin(s+\Delta s)/\sqrt{2}, -\cos(s+\Delta s/\sqrt{2}), 1).$$

At the point $s = 0$

$$\cos \Delta \varphi = \frac{1}{2}(\cos \Delta s/\sqrt{2} + 1)$$

or $-2 \cos \Delta \varphi = -\cos \Delta s/\sqrt{2} - 1$, or $2(1 - \cos \Delta \varphi) = 1 - \cos \Delta s/\sqrt{2}$.

Dividing each side by Δs^2 gives

$$\frac{2(1 - \cos \Delta \varphi)}{\Delta s^2} = \frac{1 - \cos \Delta s/\sqrt{2}}{\Delta s^2}$$

so that

$$\tau^2 = \lim_{\Delta s \to 0} \frac{1 - \cos \Delta s/\sqrt{2}}{\Delta s^2} = \frac{1}{4}.$$

Therefore, $\tau = \pm \frac{1}{2}$. We then must choose the positive sign since the present helix is right-handed.

Remark: Note that $\lim_{\Delta s \to 0} \frac{\Delta \varphi}{\Delta s} = \pm \frac{d\vec{B}}{ds}$ and $\frac{d\vec{B}}{ds}$ exists; hence the limit in question exists. One has using Taylor's expansion for $\cos \theta$ near $\theta = 0$,

$$\cos \theta = 1 - \frac{\theta^2}{2} + R(\theta) \text{ where } R(\theta) \leq \text{const} |\theta|^3.$$

Thus $2(1 - \cos \Delta \varphi) = (\Delta \varphi)^2 + 2R(\Delta \varphi)$. Now note that

$$\frac{|R(\Delta \varphi)|}{(\Delta s)^2} \leq \text{const}\left(\frac{|\Delta \varphi|}{|\Delta s|}\right)^2 |\Delta \varphi|$$

$$\therefore \lim_{\Delta s \to 0} \frac{|R(\Delta \varphi)|}{(\Delta s)^2} = 0.$$

Hence

$$\lim_{\Delta s \to 0} \frac{2(1 - \cos \Delta \varphi)}{(\Delta s)^2} = \tau^2.$$

● **PROBLEM 18-10**

Find the order of contact between the helix
$$x_1 = \cos(s/\sqrt{2}), \quad x_2 = \sin(s/\sqrt{2}), \quad x_3 = s/\sqrt{2}$$
and the plane $x_2 = x_3$.

Solution: To solve the given problem, the two following definitions are needed:

i) Let $\varphi(t) \in C^n$, then $\varphi(t)$ has a zero of order n at $t = t_0$ if and only if
$$\varphi^{(k)}(t_0) = 0 \quad k = 0,1,\ldots,n-1$$
$$\varphi^{(n)}(t_0) \neq 0.$$

ii) A curve $\vec{x} = \vec{x}(t)$ and a plane $(\vec{X}-\vec{a})\cdot\vec{Y} = 0$ have contact of order n at a common point $\vec{a} = \vec{x}(t_0)$ if and only if the distance $\varphi(t)$ from a point t of the curve to the plane has a zero of order $n+1$ at $t = t_0$.

To use the two definitions, notice that the plane $x_2 = x_3$ and the given helix intersect at the point $(1,0,0)$, where $s = 0$.
Let $\varphi(s)$ be the distance from a point s of the curve to the plane. Then
$$\varphi(s) = s/2 - 1/\sqrt{2} \sin s/\sqrt{2} \quad \text{(see remark)}$$
so that
$$\varphi'(s) = 1/2 - 1/2 \cos s/\sqrt{2}$$
$$\varphi''(s) = \frac{1}{2\sqrt{2}} \sin s/\sqrt{2}$$
and
$$\varphi'''(s) = 1/4 \cos s/\sqrt{2}.$$

Hence
$$\varphi(0) = \varphi'(0) = \varphi''(0) = 0$$
$$\varphi'''(0) \neq 0.$$

Accordingly, by (i) the zero is of order three and by (ii) the contact is of order two.

Remark: The equation of the plane can be written as $x_2 - x_3 = 0$ and hence the normal has direction $\vec{n} = (0, 1/\sqrt{2}, -1/\sqrt{2})$. Let $P \equiv (x,y,z)$ be any point which is written as $\vec{X}$. Then $\vec{X} = \vec{X} - (\vec{X}\cdot\vec{n})\vec{n} + (\vec{X}\cdot\vec{n})\vec{n}$ and the vector $\vec{X} - (\vec{X}\cdot\vec{n})\vec{n}$ is in the plane (simply because this vector is perpendicular to $\vec{n}$). The distance d from P to this point is $|\vec{X}\cdot\vec{n}|$ and gives the distance from P to the plane. Applying this to the point $P = (\cos s/\sqrt{2}, \sin s/\sqrt{2}, s/\sqrt{2})$ we get
$$\varphi(s) = \text{distance of } P \text{ from the plane} = \left|\frac{1}{\sqrt{2}} \sin \frac{s}{\sqrt{2}} - \frac{s}{2}\right|$$
$$= \frac{s}{2} - \frac{1}{\sqrt{2}} \sin \frac{s}{\sqrt{2}}.$$

This formula can also be obtained using the standard expression of the distance of a point $\vec{X}_0$ from the plane $\vec{a}\cdot\vec{X} = b$, which is given by
$$\frac{|\vec{a}\cdot\vec{X}_0 - b|}{|\vec{a}|}.$$

● PROBLEM 18-11

Show that the intersection of two cylinders with different axes is a curve.

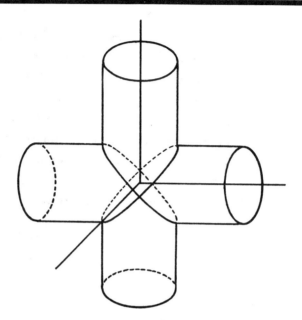

Fig. 1

Solution: Suppose the cylinders are both of radius 1 and one, C_1, has as axis the y-axis, and the other, C_2, has as axis the x-axis. Then C_1 has the equation

$$x^2 + z^2 = 1 \qquad (1)$$

and C_2 has the equation

$$y^2 + z^2 = 1 .$$

The intersection is then the set of points where both equations hold and thus can be written

$$x^2 = 1 - z^2, \; y^2 = 1 - z^2 .$$

Therefore, using z as the parameter, we can parametrize the intersection of the two cylinders by the curve

$$(\pm \sqrt{1 - z^2}, \pm \sqrt{1 - z^2}, z)$$

where $\pm$ alternately gives different parts of the curve.

(Note that the definition of a curve is $f: R \to R^n$).
A simpler parametrization is found by the substitution $x = \cos \theta$. This yields

$$(\pm \cos \theta, \pm \cos \theta, \sin \theta).$$

• **PROBLEM 18-12**

Find the envelope of the family of circles $(x - \alpha)^2 + y^2 = \frac{\alpha^2}{2}$,

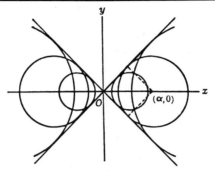

Fig. 1

Solution: If $F(x,y,\alpha) = 0$ is the equation of a one-parameter family of curves, with α the parameter, there may exist a curve C in the plane with the following properties:

i) Each curve of the family is tangent to C,

ii) C is tangent at each of its points to some curve of the family.

Then, by definition (if such a curve exists), C is the envelope of the family.

When the family of curves is given as $F(x,y,\alpha) \equiv 0$, then we can find the envelope analytically by solving the simultaneous equations,

$$F(x,y,\alpha) = 0 \quad \text{and} \quad \frac{\partial F}{\partial \alpha}(x,y,\alpha) = 0. \tag{1}$$

However, note that the preceding exposition does not guarantee that all of the locus thus found will constitute the envelope; it merely establishes that, if there is an envelope, it will be part (perhaps all) of the complete locus defined by (1).

In this case $\quad F = (x-\alpha)^2 + y^2 - \alpha^2/2 \equiv 0 \tag{2}$

and $\quad \frac{\partial F}{\partial \alpha} = 2(x-\alpha)(-1) - \alpha = \alpha - 2x \equiv 0. \tag{3}$

Given two equations in three unknowns we solve for x and y in terms of α. From (3) this yields $x = \alpha/2$. Substitution of this value into (2) gives

$$0 \equiv (\alpha/2 - \alpha)^2 + y^2 - \alpha^2/2$$

$$0 = \alpha^2/4 - \alpha^2 + \alpha^2 + y^2 - \alpha^2/2 = y^2 - \alpha^2/4$$

from which $y = \pm \alpha/2$.

Hence, the envelope is given parametrically by $x = \alpha/2$ and $y = \pm \alpha/2$. Or, by eliminating α, this becomes

$$y^2 - x^2 = 0 \quad \text{as the figure illustrates.}$$

• **PROBLEM 18-13**

Find the envelope of:
  a) the family of curves
$$(y - \alpha)^2 = x^2(1 - x^2). \tag{1}$$
  b) the family of lines, $x/c + y/b = 1$ \hfill (2)
which satisfy the relation $c^2 + b^2 = 1$.

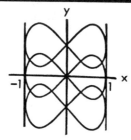

Fig. 1

Solution: For this problem, from (1)
$$F(x,y,\alpha) = (y - \alpha)^2 - x^2(1 - x^2) = 0 \tag{3}$$
and
$$\frac{\partial F}{\partial \alpha}(x,y,\alpha) = 2(y - \alpha)(-1) = 0. \tag{4}$$

Then, since the equation for the envelope is among the simultaneous solutions of (3) and (4), to find the envelope first solve for y in terms of α in (4). This yields $y = \alpha$. Substitution of this value into (3) then gives $x^2(1 - x^2) = 0$. The solutions of this equation are the three lines $x = 0$, $x + 1 = 0$, and $x - 1 = 0$. Note from figure 1 that the y-axis, though obtained as a solution, is not part of the envelope. This is because we are only guaranteed, that if an envelope exists it will appear as a solution to the equations. The converse, all solutions are envelopes, is not true in general.

b) For this problem consider the function
$$F(x,y,c,b) = \frac{x}{c} + \frac{y}{b} - 1 = 0. \tag{5}$$
To find the envelope, simultaneously solve the two equations
$$F(x,y,c,b) = 0 \quad \text{and} \quad \frac{\partial F}{\partial c}(x,y,c,b) = 0 \tag{6}$$
Note that in this problem we consider b as a function of c. From (5)
$$\frac{\partial F}{\partial c} = \frac{-x}{c^2} - \frac{y}{b^2} \frac{db}{dc} = 0 \tag{7}$$
However, since $c^2 + b^2 = 1$, $2c + 2b \frac{db}{dc} = 0$ so that $\frac{db}{dc} = \frac{-c}{b}$, (6) now becomes $\frac{-x}{c^2} - \frac{y}{b^2}\left(\frac{-c}{b}\right) = 0$ or $\frac{yc}{b^3} = \frac{x}{c^2}$. This implies that

$$\frac{x^{\frac{1}{3}}}{c} = \frac{y^{\frac{1}{3}}}{b} \quad \text{or} \quad c = b\frac{x^{\frac{1}{3}}}{y^{\frac{1}{3}}}. \tag{8}$$

Substituting this value into (5) yields

$$\frac{xy^{\frac{1}{3}}}{bx^{\frac{1}{3}}} + \frac{y}{b} = 1.$$

Factoring gives

$$\frac{1}{b} y^{\frac{1}{3}}(x^{\frac{2}{3}} + y^{\frac{2}{3}}) = 1$$

or

$$b = y^{\frac{1}{3}}(x^{\frac{2}{3}} + y^{\frac{2}{3}}). \tag{9}$$

Similarly from (8),

$$b = \frac{cy^{\frac{1}{3}}}{x^{\frac{1}{3}}} \quad \text{so that substitution into}$$

(5) gives

$$\frac{x}{c} + \frac{y^{\frac{2}{3}} x^{\frac{1}{3}}}{c} = \frac{1}{c} x^{\frac{1}{3}}(x^{\frac{2}{3}} + y^{\frac{2}{3}}) = 1$$

or

$$c = x^{\frac{1}{3}}(x^{\frac{2}{3}} + y^{\frac{2}{3}}). \tag{10}$$

Substituting (9),(10) into the relation $b^2 + c^2 = 1$, yields

$$1 = [y^{\frac{1}{3}}(x^{\frac{2}{3}} + y^{\frac{2}{3}})]^2 + [x^{\frac{1}{3}}(x^{\frac{2}{3}} + y^{\frac{2}{3}})]^2$$

$$= (x^{\frac{2}{3}} + y^{\frac{2}{3}})^3.$$

Thus the envelope looks like the curve $x^{\frac{2}{3}} + y^{\frac{2}{3}} = 1$.

SURFACES

● **PROBLEM 18-14**

Show that the mapping
$$\vec{X} = u^2 \vec{e}_1 + uv\vec{e}_2 + v^2 \vec{e}_3$$
is a coordinate patch of class C^∞ on the first quadrant $u > 0$, $v > 0$.

Solution: By definition a coordinate patch of class C^m ($m \geq 1$) in S is a mapping $\vec{X} = \vec{X}(u,v)$ of an open set U into S such that

(i) $\vec{X}$ is of class C^m on U.

(ii) $\vec{X}_u \times \vec{X}_v \neq 0$ for all (u,v) in U.

(iii) $\vec{X}$ is one-to-one and bicontinuous on U.

That is, a coordinate patch is a regular parametric representation of a part of S, which is one-to-one and bicontinuous.

The mapping $\vec{X} = u^2\vec{e}_1 + uv\vec{e}_2 + v^2\vec{e}_3$ is of class C^∞ (condition (i)).

In addition

$$\vec{X}_u = 2u\vec{e}_1 + v\vec{e}_2 + 0\vec{e}_3$$

$$\vec{X}_v = 0\vec{e}_1 + u\vec{e}_2 + 2v\vec{e}_3$$

so that

$$\|\vec{X}_u \times \vec{X}_v\| = \|2v^2\vec{e}_1 - 4uv\vec{e}_2 + 2u^2\vec{e}_3\|$$

$$= (4v^4 + 16u^2v^2 + 4u^4)^{\frac{1}{2}}$$

$$= 2(v^4 + 4u^2v^2 + u^4)^{\frac{1}{2}} \neq 0 \quad \text{(condition (ii))},$$

for $u > 0$ and $v > 0$.

Furthermore, $\vec{X}(u,v) = \vec{X}(u',v')$ implies $u^2 = u'^2$ and $uv = u'v'$ since $x_1 = u^2$ and $x_2 = uv$. However, since $u > 0$ and $u' > 0$, it follows first that $u = u'$ and then $v = v'$. Hence the mapping is one-to-one, and also bicontinuous. That is, the inverse is $u = \sqrt{x_1}$, $v = x_2/\sqrt{x_1}$. Since $u > 0$, this yields $x_1 > 0$ and the inverse is defined and continuous. Thus since the conditions (i)-(iii) are satisfied, the mapping is a coordinate patch of class C^∞.

● **PROBLEM 18-15**

Define a surface patch in R^3. Give some examples of surfaces in R^3.

Solution: A surface patch in R^3 is the image of a domain D in R^2 under a map $\vec{x} = \vec{x}(u,v)$ with the following properties:

(i) $\vec{x}$ is one-to-one,

(ii) $\vec{x}$ is continuously differentiable,

(iii) the vectors $\frac{\partial \vec{x}}{\partial u}$, $\frac{\partial \vec{x}}{\partial v}$ are independent at every point.

(u,v) are called the parameters for the surface patch. The curves $u = $ constant, and $v = $ constant are called the parametric curves. A surface is a set Σ in R^3 which can be covered by surface patches, that is, every point $\vec{p}$ on Σ has a neighborhood N such that $\Sigma \cap N$ is a surface patch.

Notice that if one fixes $u = c$, then the function $\vec{\varphi}(v) = \vec{x}(c,v)$ parametrizes a curve (since $\vec{\varphi}$ is also one-to-one and

$$\frac{d\vec{\varphi}}{dv} = \frac{\partial \vec{x}}{\partial v}(c,v)$$

is everywhere nonzero). The vector $\frac{\partial \vec{x}}{\partial v}$ is thus the tangent vector to the parametric curve $u = $ constant. Condition (iii) asks that the curve $u = c$, $v = c'$ at any point have independent tangents. Another way of phrasing (iii) is that the 2×3 matrix

$$\begin{pmatrix} \frac{\partial \vec{x}}{\partial u} \\ \frac{\partial \vec{x}}{\partial v} \end{pmatrix}$$

has rank 2.

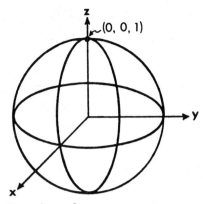

Fig. 1

Example 1: The sphere $x^2 + y^2 + z^2 = 1$ (Fig. 1).
Near the point $(0,0,1)$ one can write z as a function of x and y: $z = (1 - x^2 - y^2)^{1/2}$. Thus one can use x,y to define a surface patch surrounding $(0,0,1)$:

$$\vec{x} = \vec{x}(u,v) = (u, v, (1 - (u^2+v^2))^{1/2})$$

which parametrizes the upper hemisphere as u,v range through the disc $u^2 + v^2 < 1$. In order to see this we must check properties (i), (ii) and (iii). To check (i) suppose

$$\vec{x}(u_1,v_1) = \vec{x}(u_2,v_2) \qquad u_1^2 + v_1^2 < 1, \; u_2^2 + v_2^2 < 1$$

Then $u_1 = v_1$, $u_2 = v_2$, i.e., $\vec{x}$ is one-to-one. Further any point P on the hemisphere satisfies $x^2 + y^2 + z^2 = 1$, $z > 0$, i.e., the point has coordinates (x,y,z) with $x^2 + y^2 = 1 - z^2 < 1$ hence letting $u = x$, $v = y$ we have that P is the image of (u,v) under $\vec{x}$. This shows that $\vec{x}$ maps the open disc $u^2 + v^2 < 1$ onto the upper hemisphere in a one to one manner. Clearly the mapping $\vec{x}$ is continuous and infinitely differentiable except when $u^2 + v^2 = 1$; and these points have been carefully excluded. Finally we compute the vectors $\frac{\partial \vec{x}}{\partial u}$ and $\frac{\partial \vec{x}}{\partial v}$

$$\frac{\partial \vec{x}}{\partial u} = \left(1, \; 0, \; \frac{u}{(1-u^2-v^2)^{1/2}}\right)$$

$$\frac{\partial \vec{x}}{\partial v} = \left(0, \; 1, \; \frac{v}{(1-u^2-v^2)^{1/2}}\right).$$

To see that $\frac{\partial \vec{x}}{\partial u}$ and $\frac{\partial \vec{x}}{\partial v}$ are independent, suppose

$$\alpha_1 \frac{\partial \vec{x}}{\partial u} + \alpha_2 \frac{\partial \vec{x}}{\partial v} = \vec{0}.$$

Then $\alpha_1 \left(1, \; 0, \; \frac{u}{(1-u^2-v^2)^{1/2}}\right) + \alpha_2 \left(0, \; 1, \; \frac{v}{(1-u^2-v^2)^{1/2}}\right) = \vec{0}$. The first component of the above equation gives $\alpha_1 = 0$ and the second component gives $\alpha_2 = 0$. Let us see what the parametric curves are:

$u = c$ is the same as $x = c$ and this is a plane which cuts the hemisphere in a semicircle given by the set of points
$$(c, v, \sqrt{1 - c^2 - v^2}) \quad 0 < v^2 < 1 - c^2 .$$
Similarly $v = c'$ is a semicircle. Finally note that every point on the sphere can be put in a surface patch as above by permuting the roles of (x,y,z). For example, the point $(-1,0,0)$ lies in the surface patch given by
$$\vec{x} = \vec{x}(u,v) = (-(1-u^2-v^2)^{\frac{1}{2}}, u, v) \quad u^2 + v^2 < 1 .$$

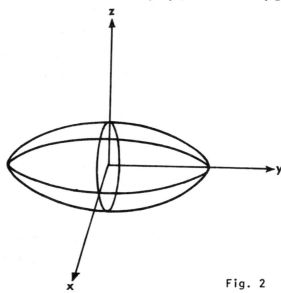

Fig. 2

Example 2: The ellipsoid (Fig. 2)
$$a^2x^2 + b^2y^2 + c^2z^2 = 1 .$$
This can be parametrized by spherical coordinates. Let
$$\vec{x}(u,v) = \left(\frac{\sin u \cos v}{a}, \frac{\sin u \sin v}{b}, \frac{\cos u}{c} \right)$$
$0 < u < \pi$, $0 < v < \pi$. It will be shown that $\vec{x}$ maps one-to-one onto the portion of the ellipsoid $y > 0$. Suppose $\vec{x}(u_1,v_1) = \vec{x}(u_2,v_2)$. Then
$$\cos u_1 = \cos u_2, \quad \sin u_1 \cos v_1 = \sin u_2 \cos v_2 ,$$
since $0 < u_1 < \pi$, $u_1 = u_2$ and further since $\sin u_1 > 0$, $\cos v_1 = \cos v_2$ which gives $v_1 = v_2$. Thus $\vec{x}$ is one-one. Let $P(x,y,z)$ be any point on the ellipsoid with $y > 0$. Now
$$c^2z^2 = 1 - a^2x^2 - b^2y^2 < 1$$
so there exists a unique angle u with $0 < u < \pi$ so that $z = \frac{\cos u}{c}$. Then $a^2x^2 + b^2y^2 = 1 - \cos^2 u = \sin^2 u$. Let $x' = ax/\sin u$, and $y' = by/\sin u$ so that $x'^2 + y'^2 = 1$. There exists a unique angle v,

$0 \leq v < 2\pi$ with $x' = \cos v$ $y' = \sin v$, and since $y' > 0$, $0 < v < \pi$, i.e., $x = \frac{\sin u \cos v}{a}$ and $y = \frac{\sin u \sin v}{b}$. This shows that $\vec{x}$ is onto. Clearly $\vec{x}$ is infinitely differentiable. Finally one has

$$\frac{\partial \vec{x}}{\partial u} = \left(\frac{\cos u \cos v}{a}, \frac{\cos u \sin v}{b}, \frac{-\sin u}{c}\right)$$

$$\frac{\partial \vec{x}}{\partial v} = \left(\frac{-\sin u \sin v}{a}, \frac{\sin u \cos v}{b}, 0\right).$$

Since $\frac{\partial \vec{x}}{\partial u}$ has last component $\frac{-\sin u}{c} < 0$ and $\frac{\partial \vec{x}}{\partial v}$ has the last component 0 it follows that $\partial \vec{x}/\partial u$ and $\partial \vec{x}/\partial v$ are independent. As in example 1, by permuting the roles of (x,y,z) every point on the ellipsoid can be put on a surface patch. Thus the ellipsoid and the sphere are examples of smooth surfaces.

Example 3: The cylinder $x^2 + y^2 = 1$ is a surface. It can be parametrized by using cylindrical coordinates:

$$\vec{x} = \vec{x}(u,v) = (\cos u, \sin u, v) \quad 0 \leq u < 2\pi, \ -\infty < v < \infty,$$

$$\frac{\partial \vec{x}}{\partial u} = (-\sin u, \cos u, 0), \quad \frac{\partial \vec{x}}{\partial v} = (0,0,1).$$

• **PROBLEM 18-16**

a) Define a surface patch surrounding the point $(0,0,1)$ on the sphere $x^2 + y^2 + z^2 = 1$.
b) Show that the paraboloid $z = x^2 + y^2$ is a surface patch.

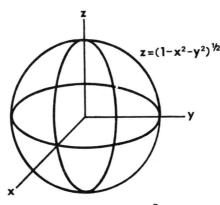

$z = (1-x^2-y^2)^{1/2}$

Fig. 1

Solution: By definition a surface patch in R^3 is the image of a domain D in R^2 under a map $\vec{x} = \vec{x}(u,v)$ with these properties:

(i) $\vec{x}$ is one-to-one,
(ii) $\vec{x}$ is continuously differentiable,
(iii) The vectors $\partial \vec{x}/\partial u$, $\partial \vec{x}/\partial v$ are independent at every point

(i.e., the 2×3 matrix

$$\begin{pmatrix} \frac{\partial \vec{x}}{\partial u} \\ \frac{\partial \vec{x}}{\partial v} \end{pmatrix}$$

has rank 2). Observe that (u,v) are called the parameters for the surface patch and the curves $u = $ constant and $v = $ constant are called the parametric curves.

a) Near the point $(0,0,1)$, z can be written as a function of x and y on the plane $z = (1 - x^2 - y^2)^{\frac{1}{2}}$. Thus we can use x,y to define a surface patch surrounding $(0,0,1)$:

$$\vec{X} = \vec{X}(u,v) = (u, v, (1 - u^2 - v^2)^{\frac{1}{2}}) \tag{1}$$

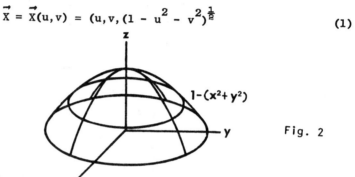

Fig. 2

which parametrizes the upper hemisphere as u,v range through the disk $u^2 + v^2 < 1$. This, because the mapping (1) is continuously differentiable and is also 1-1, for $\vec{X}(u,v) = \vec{X}(u',v')$ implies $(u,v) = (u',v')$, since $x = u$ and $y = v$. Furthermore, the vectors

$$\vec{X}_u = \frac{\partial \vec{x}}{\partial u} = (1, 0, -u(1 - u^2 - v^2)^{-\frac{1}{2}})$$

$$\vec{X}_v = \frac{\partial \vec{x}}{\partial v} = (0, 1, -v(1 - u^2 - v^2)^{-\frac{1}{2}})$$

are independent because the matrix

$$\begin{pmatrix} 1 & 0 & -u(1 - u^2 - v^2)^{-\frac{1}{2}} \\ 0 & 1 & -v(1 - u^2 - v^2)^{-\frac{1}{2}} \end{pmatrix}$$

has rank 2.

Thus, by (i)-(iii), the mapping (1) is a surface patch.

Note: Every point on the sphere can be put in such a surface patch by permuting the roles of (x,y,z) above. For example, the point $(-1,0,0)$ lies in the surface patch given by

$$\vec{X} = \vec{X}(u,v) = \left(-(1 - u^2 - v^2)^{\frac{1}{2}}, u, v\right) \quad u^2 + v^2 < 1.$$

b) The paraboloid $z = x^2 + y^2$ is coordinated by $\vec{X} = \vec{X}(u,v) = (u, v, u^2+v^2)$. This mapping is continuously differentiable and is one-to-one for $\vec{X}(u,v) = \vec{X}(u',v')$ implies $(u,v) = (u',v')$, since $x = u$ and $y = v$.

In addition the vectors $\vec{x}_u = \partial\vec{x}/\partial u = (1,0,2u)$ and $\vec{x}_v = \partial\vec{x}/\partial v = (0,1,2v)$ are independent. That is

$$\|\vec{x}_u \times \vec{x}_v\| = \|(-2u,-2v,1)\|$$
$$= (4u^2 + 4v^2 + 1)^{\frac{1}{2}} \neq 0 \text{ for all } (u,v).$$

Thus by (i)-(iii) the mapping is a surface patch on the paraboloid.

● PROBLEM 18-17

Let Σ be a surface patch parametrized by $\vec{x} = \vec{x}(u,v)$. Let C be a smooth curve on Σ parametrized by $\vec{x} = \vec{x}(u(t),v(t))$, $a < t < b$. Let a_0, b_0 be points in (a,b). Compute the length of the curve C between a_0 and b_0. Define the first fundamental form of Σ.

Solution: Let Σ be a surface, and p a point on the surface. Denote by $T(p)$ the tangent plane to Σ at p. If $\vec{x} = \vec{x}(u,v)$ parametrizes Σ in a neighborhood of p, with $p = \vec{x}(u_0,v_0)$, then the vectors

$$\frac{\partial\vec{x}}{\partial u}(u_0,v_0),\ \frac{\partial\vec{x}}{\partial v}(u_0,v_0)$$

span the plane $T(p)$. The inner product on R^3 induces an inner product on this plane just by restriction. Let

$$\vec{t} = a\vec{x}_u + b\vec{x}_v$$

(where $\vec{x}_u = \frac{\partial\vec{x}}{\partial u}$, $\vec{x}_v = \frac{\partial\vec{x}}{\partial v}$) denote a vector in $T(p)$. By definition the length of $\vec{t}$ is given by

$$\|\vec{t}\|^2 = \langle\vec{t},\vec{t}\rangle = a^2\langle\vec{x}_u,\vec{x}_u\rangle + 2ab\langle\vec{x}_u,\vec{x}_v\rangle + b^2\langle\vec{x}_v,\vec{x}_v\rangle.$$

Now the tangent vector to C is given by

$$\vec{T} = \frac{\partial\vec{x}}{\partial u}\frac{du}{dt} + \frac{\partial\vec{x}}{\partial v}\frac{dv}{dt} = \frac{du}{dt}\vec{x}_u + \frac{dv}{dt}\vec{x}_v$$

so

$$\|\vec{T}\|^2 = \mathcal{E}\left(\frac{du}{dt}\right)^2 + 2\mathcal{F}\frac{du}{dt}\frac{dv}{dt} + \mathcal{G}\left(\frac{dv}{dt}\right)^2$$

where

$$\mathcal{E} = \langle\vec{x}_u,\vec{x}_u\rangle,\ \mathcal{F} = \langle\vec{x}_u,\vec{x}_v\rangle,\ \mathcal{G} = \langle\vec{x}_v,\vec{x}_v\rangle. \qquad (1)$$

By definition the length of C between a_0 and b_0 is

$$\int_{a_0}^{b_0}\|\vec{T}\|dt = \int_{a_0}^{b_0}\sqrt{\mathcal{E}\left(\frac{du}{dt}\right)^2 + 2\mathcal{F}\frac{du}{dt}\frac{dv}{dt} + \mathcal{G}\left(\frac{dv}{dt}\right)^2}\,dt.$$

Let $\vec{t} = \alpha\vec{x}_u + \beta\vec{x}_v$ and $\vec{s} = \gamma\vec{x}_u + \delta\vec{x}_v$ be two tangent vectors at a point p in Σ. Then

$$\langle \vec{t}, \vec{s} \rangle = \mathscr{E}\alpha\gamma + (\alpha\delta + \beta\gamma)\mathscr{F} + \beta\delta\mathscr{G}. \tag{2}$$

The right side of (2) defines a bilinear, nondegenerate, symmetric form on $T(p)$ and is called the first fundamental form of Σ.

Remark 1: By Cauchy-Schwartz inequality

$$\langle \vec{x}_u, \vec{x}_v \rangle^2 \leq \langle \vec{x}_u, \vec{x}_u \rangle \langle \vec{x}_v, \vec{x}_v \rangle$$

i.e., $\mathscr{E}\mathscr{G} - \mathscr{F}^2 \geq 0$.

2. The length of the curve C can be written as $\int_{a_0}^{b_0} ds$ where

$$ds = \sqrt{\mathscr{E}\left(\frac{du}{dt}\right)^2 + 2\mathscr{F}\frac{du}{dt}\frac{dv}{dt} + \mathscr{G}\left(\frac{dv}{dt}\right)^2}\, dt.$$

• **PROBLEM 18-18**

Let Σ be a surface patch parametrized by $\vec{X} = \vec{X}(u,v)$. Let C be a curve on Σ parametrized by $\vec{X} = \vec{X}(u(t), v(t))$. $a \leq t \leq b$. Show that the length of C is

$$\int_a^b \left[\langle \vec{X}_u, \vec{X}_u \rangle \left(\frac{du}{dt}\right)^2 + 2\langle \vec{X}_u, \vec{X}_v \rangle \left(\frac{du}{dt}\frac{dv}{dt}\right) + \langle \vec{X}_v, \vec{X}_v \rangle \left(\frac{dv}{dt}\right)^2\right]^{\frac{1}{2}} dt.$$

Solution: Let p be a point on Σ and let $T(p)$ denote the tangent plane to Σ at p. If $\vec{X} = \vec{X}(u,v)$ parametrizes Σ in a neighborhood of p, with $p = \vec{X}(u_0, v_0)$, then the vectors

$$\frac{\partial \vec{X}}{\partial u}(u_0, v_0), \quad \frac{\partial \vec{X}}{\partial v}(u_0, v_0)$$

span the plane $T(p)$. The inner product on R^3 induces an inner product on this plane just by restriction. Now consider how to express this inner product in terms of the basis $\vec{X}_u, \vec{X}_v$. If $\vec{t} = a\vec{X}_u + b\vec{X}_v$ is a vector in $T(p)$ its length is

$$\|\vec{t}\|^2 = \langle \vec{t}, \vec{t} \rangle = a^2 \langle \vec{X}_u, \vec{X}_u \rangle + 2ab\langle \vec{X}_u, \vec{X}_v \rangle + b^2 \langle \vec{X}_v, \vec{X}_v \rangle \tag{1}$$

Since C is a curve on Σ, choose a parametrization of C:

$$\vec{X} = \vec{F}(t), \quad 0 \leq t \leq L \tag{2}$$

Let $(u(t), v(t))$ be the (u,v) coordinates of $\vec{F}(t)$. Then (2) can be rewritten as $\vec{X} = \vec{X}(u(t), v(t))$ and by the chain rule, the tangent to C is

$$\vec{T} = \vec{X}_u \frac{du}{dt} + \vec{X}_v \frac{dv}{dt}$$

and by (1)

$$\|\vec{T}\|^2 = \langle \vec{T}, \vec{T} \rangle = \langle \vec{X}_u, \vec{X}_u \rangle \left(\frac{du}{dt}\right)^2 + 2\langle \vec{X}_u, \vec{X}_v \rangle \frac{du}{dt}\frac{dv}{dt} + \langle \vec{X}_v, \vec{X}_v \rangle \left(\frac{dv}{dt}\right)^2. \tag{3}$$

The length of C is $\int_a^b \|\vec{T}\| \, dt$, which by (3) is

$$\int_a^b \left[\langle \vec{X}_u, \vec{X}_u \rangle \left(\frac{du}{dt}\right)^2 + 2 \langle \vec{X}_u, \vec{X}_v \rangle \frac{du}{dt}\frac{dv}{dt} + \langle \vec{X}_v, \vec{X}_v \rangle \left(\frac{dv}{dt}\right)^2 \right]^{\frac{1}{2}} dt \quad (4)$$

This is the desired result.

Note: Let $\mathcal{E} = \langle X_u, X_u \rangle$, $\mathcal{F} = \langle X_u, X_v \rangle$, $\mathcal{G} = \langle X_v, X_v \rangle$, and let ds be the integrand which gives arc length along a curve. Then the length of any curve C is $\int_C ds$. According to (4)

$$ds = \left[\mathcal{E}\left(\frac{du}{dt}\right)^2 + 2\mathcal{F}\left(\frac{du}{dt}\frac{dv}{dt}\right) + \mathcal{G}\left(\frac{dv}{dt}\right)^2 \right]^{\frac{1}{2}} dt$$

for any parameter t along C. This can also be written as

$$ds^2 = \mathcal{E}\, du^2 + 2\mathcal{F}\, du\, dv + \mathcal{G}\, dv^2. \quad (5)$$

Then by definition, (5) is called the first fundamental form of Σ.

• **PROBLEM 18-19**

Find the length of the arc $u = e^{k\theta}$, $\theta = \theta$, $0 \le \theta \le \pi$, k = constant on the cone $\vec{X} = (u \cos\theta, u \sin\theta, u)$.

Solution: To solve this problem we make use of the first fundamental form of $X = X(u,v)$,

$$ds^2 = \mathcal{E}\, du^2 + 2\mathcal{F}\, du\, dv + \mathcal{G}\, dv^2$$

where $\mathcal{E} = \vec{X}_u \cdot \vec{X}_u$, $\mathcal{F} = \vec{X}_u \cdot \vec{X}_v$, $\mathcal{G} = \vec{X}_v \cdot \vec{X}_v$.

Then the length of an arc $\vec{X} = \vec{X}(u(t), v(t))$, $a \le t \le b$ on $\vec{X} = \vec{X}(u,v)$ is given by

$$s = \int_a^b \left[\mathcal{E}\left(\frac{du}{dt}\right)^2 + 2\mathcal{F}\frac{du}{dt}\frac{dv}{dt} + \mathcal{G}\left(\frac{dv}{dt}\right)^2 \right]^{\frac{1}{2}} dt. \quad (1)$$

For the given problem, since

$$\vec{X} = (u \cos\theta, u \sin\theta, u)$$

we have

$$\vec{X}_\theta = (-u \sin\theta, u \cos\theta, 0)$$

$$\vec{X}_u = (\cos\theta, \sin\theta, 1).$$

Hence,

$$\mathcal{E} = \vec{X}_\theta \cdot \vec{X}_\theta = u^2 \sin^2\theta + u^2 \cos^2\theta = u^2$$

$$\mathcal{F} = \vec{X}_\theta \cdot \vec{X}_u = -u \sin\theta \cos\theta + u \sin\theta \cos\theta = 0$$

$$\mathcal{G} = \vec{X}_u \cdot \vec{X}_u = \cos^2\theta + \sin^2\theta + 1 = 2.$$

Furthermore, since $u = e^{k\theta}$, $\theta = \theta$, we have

$$\frac{du}{d\theta} = uk, \quad \frac{d\theta}{d\theta} = 1.$$

Then by (1)

$$s = \int_0^\pi \left[u^2\left(\frac{d\theta}{d\theta}\right)^2 + 0 \frac{d\theta}{d\theta}\frac{du}{d\theta} + 2\left(\frac{du}{d\theta}\right)^2 \right]^{\frac{1}{2}} d\theta$$

$$= \int_0^\pi [u^2 + 2u^2 k^2]^{\frac{1}{2}} d\theta = \sqrt{1 + 2k^2} \int_0^\pi u \, d\theta$$

$$= \sqrt{1 + 2k^2} \int_0^\pi e^{k\theta} d\theta = \frac{\sqrt{1 + 2k^2}}{k} e^{k\theta} \Big|_0^\pi$$

$$s = \frac{\sqrt{1 + 2k^2}}{k} (e^{\pi k} - 1).$$

● **PROBLEM 18-20**

Let Σ be a surface. Define the notion of orthogonality of two curves at a point p on Σ. Derive necessary and sufficient conditions for the parametric curves to be orthogonal.

Solution: Let p be parametrized by (u_0, v_0) in some patch surrounding it. Let the patch be given by

$$\vec{x} = \vec{x}(u, v) .$$

Consider two curves C_1 and C_2 passing through p and lying entirely within this patch. Suppose that these curves are given parametrically by

$$C_1: u = u_1(s) \qquad v = v_1(s)$$
$$C_2: u = u_2(s) \qquad v = v_2(s)$$

Then the tangents to these two curves are

$$\vec{T}_1 = \vec{x}_u \frac{du_1}{ds} + \vec{x}_v \frac{dv_1}{ds}, \qquad \vec{T}_2 = \vec{x}_u \frac{du_2}{ds} + \vec{x}_v \frac{dv_2}{ds}.$$

At a point of intersection p the vectors $\vec{T}_1$ and $\vec{T}_2$ lie in the tangent plane at p and their inner product is

$$\langle \vec{T}_1, \vec{T}_2 \rangle = \mathcal{E} \frac{du_1}{ds}\frac{du_2}{ds} + \mathcal{F}\left(\frac{du_1}{ds}\frac{dv_2}{ds} + \frac{du_2}{ds}\frac{dv_1}{ds}\right) + \mathcal{G}\frac{dv_1}{ds}\frac{dv_2}{ds}.$$

The curves are said to be orthogonal at p if $\langle \vec{T}_1, \vec{T}_2 \rangle = 0$.

In particular for the parametric curves $u = c$, $v = c'$

$$\frac{du_1}{ds} = 0 \qquad \frac{dv_1}{ds} = 1 \qquad \frac{du_2}{ds} = 1 \qquad \frac{dv_2}{ds} = 0$$

so $\langle \vec{T}_1, \vec{T}_2 \rangle = \mathcal{F}$ and so the parametric curves are orthogonal at p if and only if $\mathcal{F}$ at p vanishes.

Example 1: Consider the sphere $x^2 + y^2 + z^2 = 1$ parametrized by spherical coordinates

$$\vec{x} = \vec{x}(u,v) = (\sin u \cos v, \sin u \sin v, \cos u).$$

Here $\vec{x}_u = (\cos v \cos u, \cos u \sin v, -\sin u)$

$\vec{x}_v = (-\sin u \sin v, \sin u \cos v, 0)$

and $\mathcal{F} = \langle \vec{x}_u, \vec{x}_v \rangle = 0.$

Example 2: The cylinder parametrized by

$$\vec{x} = \vec{x}(u,v) = (\cos u, \sin u, v)$$

yields

$\vec{x}_u = (-\sin u, \cos u, 0)$

$\vec{x}_v = (0, 0, 1)$

and $\mathcal{F} = \langle \vec{x}_u, \vec{x}_v \rangle = 0.$

• PROBLEM 18-21

Compute the coefficients of the first fundamental form in each of the following cases.

(i) the plane $z = 0$
(ii) the cylinder $x^2 + y^2 = 1$
(iii) the sphere $x^2 + y^2 + z^2 = 1$.

Solution: Let Σ be a surface and let p be a point on Σ. Consider a patch around p given by

$$\vec{x} = \vec{x}(u,v).$$

Then the coefficients of the first fundamental form for Σ are given by

$$\mathcal{E} = \langle \vec{x}_u, \vec{x}_u \rangle, \quad \mathcal{F} = \langle \vec{x}_u, \vec{x}_v \rangle, \quad \mathcal{G} = \langle \vec{x}_v, \vec{x}_v \rangle.$$

(i) The plane can be parametrized by $\vec{x} \equiv \vec{x}(u,v) = (u,v,0)$ so $\vec{x}_u = (1,0,0) \quad \vec{x}_v = (0,1,0)$

$\mathcal{E} = 1, \quad \mathcal{F} = 0, \quad \mathcal{G} = 1.$

(ii) The cylinder can be parametrized by

$\vec{x} = \vec{x}(u,v) = (\cos u, \sin u, v)$

$\vec{x}_u = (-\sin u, \cos u, 0)$

$\vec{x}_v = (0, 0, 1)$

$\mathcal{E} = 1 \quad \mathcal{F} = 0 \quad \mathcal{G} = 1.$

(iii) The sphere can be parametrized by

$\vec{x} = \vec{x}(u,v) = (\sin u \cos v, \sin u \sin v, \cos u)$

$\vec{x}_u = (\cos u \cos v, \cos u \sin v, -\sin u)$

$\vec{x}_v = (-\sin u \sin v, \sin u \cos v, 0)$

$\mathcal{E} = 1 \quad \mathcal{F} = 0 \quad \mathcal{G} = \sin^2 u.$

● **PROBLEM** 18-22

Compute the first fundamental form for the following surfaces:
(i) The paraboloid $z = x^2 + y^2$
(ii) The cone $z^2 = x^2 + y^2$, $z > 0$
(iii) The hyperboloid $z = x^2 - y^2$
(iv) $\Sigma : \vec{x}(u,v) = (u+v^2, v+u^2, uv)$.

Solution: Let Σ be a surface patch given by $\vec{x} \equiv \vec{x}(u,v)$. The first fundamental form is given by

$$ds^2 = \mathcal{E}\, du^2 + 2\,\mathcal{F}\, du\, dv + \mathcal{G}\, dv^2 \tag{1}$$

where $\mathcal{E} = \langle \vec{x}_u, \vec{x}_u \rangle$, $\mathcal{F} = \langle \vec{x}_u, \vec{x}_v \rangle$, $\mathcal{G} = \langle \vec{x}_v, \vec{x}_v \rangle$. If the surface is given by $z = f(x,y)$ where f is a C^1 function then we can take (x,y) as the parameters and the first fundamental form is given by

$$ds^2 = \left[1 + \left(\frac{\partial f}{\partial x}\right)^2\right]dx^2 + 2\,\frac{\partial f}{\partial x}\frac{\partial f}{\partial y}\, dx\, dy + \left[1 + \left(\frac{\partial f}{\partial y}\right)^2\right]dy^2 . \tag{2}$$

(i) Here one can take $f(x,y) = x^2 + y^2$ so

$$ds^2 = (1 + 4x^2)dx^2 + 8xy\, dx\, dy + (1 + 4y^2)dy^2 .$$

(ii) For the cone, $f(x,y) = \sqrt{x^2 + y^2}$ so

$$ds^2 = \left(1 + \frac{x^2}{x^2+y^2}\right)dx^2 + \frac{2xy}{x^2+y^2}\, dx\, dy + \left(1 + \frac{y^2}{x^2+y^2}\right)dy^2 .$$

(iii) In the case of hyperboloid $f(x,y) = x^2 - y^2$ so

$$ds^2 = (1 + 4x^2)dx^2 - 8xy\, dx\, dy + (1 + 4y^2)dy^2 .$$

(iv) In order to compute the first fundamental form in this case use equation (1)

$\mathcal{E} = \langle \vec{x}_u, \vec{x}_u \rangle = (1, 2u, v) \cdot (1, 2u, v) = 1 + 4u^2 + v^2$

$\mathcal{F} = \langle \vec{x}_u, \vec{x}_v \rangle = (1, 2u, v) \cdot (2v, 1, u) = 2v + 2u + uv$

$\mathcal{G} = \langle \vec{x}_v, \vec{x}_v \rangle = (2v, 1, u) \cdot (2v, 1, u) = 1 + 4v^2 + u^2 .$

Hence,

$$ds^2 = (1+4u^2+v^2)du^2 + 2(2u+uv+2v)du\, dv + (1 + u^2 + 4v^2)dv^2 .$$

● **PROBLEM** 18-23

Let $ds^2 = \mathcal{E}\, du^2 + 2\mathcal{F}\, du\, dv + \mathcal{G}\, dv^2$ be the first fundamental form of a surface patch. Show that the family of curves orthogonal to the family defined by $M\, du + N\, dv = 0$ is determined by

$$(\mathcal{E}N - \mathcal{F}M)\, du + (\mathcal{F}N - \mathcal{G}M)\, dv = 0 .$$

Solution: Let the required curves be given by
$$u = u_1(s) \qquad v = v_1(s).$$

The tangent vector at any point to this curve is
$$\vec{T}_1 = \frac{du_1}{ds}\vec{x}_u + \frac{dv_1}{ds}\vec{x}_v.$$

This must be orthogonal to
$$\vec{T}_2 = \frac{du}{ds}\vec{x}_u + \frac{dv}{ds}\vec{x}_v$$

where
$$M\frac{du}{ds} + N\frac{dv}{ds} = 0$$

i.e.,
$$\langle \vec{T}_1, \vec{T}_2 \rangle = \mathscr{E}\frac{du}{ds}\frac{du_1}{ds} + \mathscr{F}\left(\frac{du_1}{ds}\frac{dv}{ds} + \frac{dv_1}{ds}\frac{du}{ds}\right) + \mathscr{G}\frac{dv_1}{ds}\frac{dv}{ds} = 0.$$

Using $\dfrac{du}{ds} = -\dfrac{N}{M}\dfrac{dv}{ds}$ yields
$$\left(-\mathscr{E}\frac{N}{M}\frac{du_1}{ds} + \mathscr{F}\frac{du_1}{ds} - \mathscr{F}\frac{N}{M}\frac{dv_1}{ds} + \mathscr{G}\frac{dv_1}{ds}\right)\frac{dv}{ds} = 0$$

or
$$(\mathscr{E}N - \mathscr{F}M)\frac{du_1}{ds} + (\mathscr{F}N - \mathscr{G}M)\frac{dv_1}{ds} = 0.$$

This can be expressed as
$$(\mathscr{E}N - \mathscr{F}M)du_1 + (\mathscr{F}N - \mathscr{G}M)dv_1 = 0.$$

• PROBLEM 18-24

Let f be a C^1 function defined in a domain D in R^2.
(a) Show that $\Sigma : \{z = f(x,y)\}$ is a surface patch with coordinate x, y.
(b) Compute the first fundamental form.

Solution: (a) Σ is parametrized by $\vec{x} = \vec{x}(u,v)$ with
$$\vec{x}(u,v) = (u, v, f(u,v)).$$

Clearly this mapping is one-to-one, onto Σ. Also $\vec{x}$ is continuously differentiable because by hypothesis f is of class C^1.
$$\vec{x}_u = \left(1, 0, \frac{\partial f}{\partial u}\right) \qquad \vec{x}_v = \left(0, 1, \frac{\partial f}{\partial v}\right)$$

so $\vec{x}_u$ and $\vec{x}_v$ are independent. Thus Σ is a surface patch.

(b) The coefficients of the first fundamental form are given by
$$\mathscr{E} = \langle \vec{x}_u, \vec{x}_u \rangle, \quad \mathscr{F} = \langle \vec{x}_u, \vec{x}_v \rangle, \quad \mathscr{G} = \langle \vec{x}_v, \vec{x}_v \rangle$$
$$\langle \vec{x}_u, \vec{x}_u \rangle = 1 + \left(\frac{\partial f}{\partial u}\right)^2$$

$$\langle \vec{x}_u, \vec{x}_v \rangle = \frac{\partial f}{\partial u} \frac{\partial f}{\partial v}$$

$$\langle \vec{x}_v, \vec{x}_v \rangle = 1 + \left(\frac{\partial f}{\partial v}\right)^2.$$

Therefore the first fundamental form is

$$ds^2 = \mathcal{E} \, du^2 + 2 \mathcal{F} \, dudv + \mathcal{G} \, dv^2$$

$$= \left(1 + \left(\frac{\partial f}{\partial u}\right)^2\right) du^2 + 2 \frac{\partial f}{\partial u} \frac{\partial f}{\partial v} \, dudv + \left(1 + \left(\frac{\partial f}{\partial v}\right)^2\right) dv^2.$$

● **PROBLEM 18-25**

Let u and v be longitude and latitude on a sphere. Find the angle at which the curve $v = u$ cuts the equator. Choose units so that the radius is unity.

Solution: To solve this problem we will make use of the first fundamental form of $\vec{X} = \vec{X}(u,v)$,

$$ds^2 = \mathcal{E} \, du^2 + 2 \mathcal{F} \, dudv + \mathcal{G} \, dv^2$$

where

$$\mathcal{E} = \vec{X}_u \cdot \vec{X}_u, \quad \mathcal{F} = \vec{X}_u \cdot \vec{X}_v, \quad \mathcal{G} = \vec{X}_v \cdot \vec{X}_v.$$

Now a curve on this surface is generally determined by a single relation between u and v,

$$v = h(u). \qquad (1)$$

Then the curve is $\vec{x} = \vec{x}(u, h(u))$ and the direction components of the tangent to this curve will be the components of the vector

$$\vec{X}_u + \vec{X}_v h'. \qquad (2)$$

Similarly, given a second curve with the relation between u and v,

$$v = k(u), \qquad (3)$$

the direction components of the tangent to this curve will be the components of the vector

$$\vec{X}_u + \vec{X}_v k'. \qquad (4)$$

To find the angle θ between the two curves, apply the formula for dot product of two vectors $\vec{A}$ and $\vec{B}$,

$$\vec{A} \cdot \vec{B} = \|\vec{A}\| \, \|\vec{B}\| \cos \theta$$

or

$$\cos \theta = \frac{\vec{A} \cdot \vec{B}}{\|\vec{A}\| \, \|\vec{B}\|}.$$

Hence, for the curves (1), (3), the angle θ can be found by the equation (using (2), (4))

$$\cos \theta = \frac{(\vec{X}_u + \vec{X}_v h') \cdot (\vec{X}_u + \vec{X}_v k')}{\|\vec{X}_u + \vec{X}_v h'\| \, \|\vec{X}_u + \vec{X}_v k'\|}$$

$$= \frac{(\vec{X}_u \cdot \vec{X}_u) + (\vec{X}_u \cdot \vec{X}_v h') + (\vec{X}_u \cdot \vec{X}_v k') + \vec{X}_v h' \cdot \vec{X}_v k'}{\sqrt{(\vec{X}_u \cdot \vec{X}_u) + (2\vec{X}_u \cdot \vec{X}_v h') + (\vec{X}_v h' \cdot \vec{X}_v h')}\sqrt{(\vec{X}_u \cdot \vec{X}_u) + (2\vec{X}_u \cdot \vec{X}_v k') + (\vec{X}_v k' \cdot \vec{X}_v k')}}$$

$$\cos \theta = \frac{\mathcal{E} + \mathcal{F} h' + \mathcal{F} k' + \mathcal{G} h' k'}{\sqrt{\mathcal{E} + 2\mathcal{F} h' + \mathcal{G}(h')^2} \sqrt{\mathcal{E} + 2\mathcal{F} k' + \mathcal{G}(k')^2}} \qquad (5)$$

For the given problem, the equations of the sphere are

$$x_1 = \cos v \cos u, \quad x_2 = \cos v \sin u, \quad x_3 = \sin v,$$

so that $\vec{X}_u = (-\sin u \cos v, \cos v \cos u, 0)$ and

$\vec{X}_v = (-\sin v \cos u, -\sin v \sin u, \cos v)$.

Consequently,

$$\mathcal{E} = \vec{X}_u \cdot \vec{X}_u = \sin^2 u \cos^2 v + \cos^2 v \cos^2 u + 0 = \cos^2 v$$

$$\mathcal{F} = \vec{X}_u \cdot \vec{X}_v = \sin u \sin v \cos u \cos v - \sin v \sin u \cos v \cos u + 0 = 0$$

$$\mathcal{G} = \vec{X}_v \cdot \vec{X}_v = \sin^2 v \cos^2 u + \sin^2 v \sin^2 u + \cos^2 v = 1$$

Then by the first fundamental form

$$ds^2 = \cos^2 v \, du^2 + dv^2.$$

For the angle θ at the point $u = v = 0$, apply formula (5) with $h(u) = u$, $k(u) = 0$ so that $h' = 1$, $k' = 0$.

Proceeding,

$$\cos \theta = \frac{\cos^2 v + 0}{\sqrt{\cos^2 v + 1} \sqrt{\cos^2 v}} = \frac{\cos v}{\sqrt{\cos^2 v + 1}}$$

$$\cos \theta = \frac{\cos 0}{\sqrt{\cos^2 0 + 1}} = \frac{1}{\sqrt{2}}$$

Thus $\theta = \pi/4$.

● **PROBLEM 18-26**

Find the area of the sphere $x^2 + y^2 + z^2 = R^2$, using the coefficients of the first fundamental form.

Solution: Parametrize the sphere as

$$\vec{X} = (R \cos u \cos v, R \cos u \sin v, R \sin u)$$

so that
$$\vec{X}_u = \frac{\partial \vec{X}}{\partial u} = (-R \sin u \cos v, -R \sin u \sin v, R \cos u) \qquad (1)$$
and
$$\vec{X}_v = \frac{\partial \vec{X}}{\partial v} = (-R \cos u \sin v, R \cos u \cos v, 0) . \qquad (2)$$

By definition, given Σ is a surface patch with coordinates u,v ranging through D in R^2, the area of Σ is

$$\iint_D \|\vec{X}_u \times \vec{X}_v\| \, du\, dv . \qquad (3)$$

However,
$$\vec{X}_u \times \vec{X}_v = \begin{vmatrix} \vec{i} & \vec{j} & \vec{k} \\ X_{u_1} & X_{u_2} & X_{u_3} \\ X_{v_1} & X_{v_2} & X_{v_3} \end{vmatrix}$$

$$= (\vec{X}_{u_2}\vec{X}_{v_3} - \vec{X}_{v_2}\vec{X}_{u_3},\ \vec{X}_{u_3}\vec{X}_{v_1} - \vec{X}_{u_1}\vec{X}_{v_3},\ \vec{X}_{u_1}\vec{X}_{v_2} - \vec{X}_{u_2}\vec{X}_{v_1}),$$

so that
$$\|\vec{X}_u \times \vec{X}_v\| = \Big[\vec{X}_{u_2}^2 \vec{X}_{v_3}^2 - 2\vec{X}_{u_2}\vec{X}_{u_3}\vec{X}_{v_2}\vec{X}_{v_3} + \vec{X}_{v_2}^2\vec{X}_{u_3}^2 + \vec{X}_{u_3}^2 \vec{X}_{v_1}^2$$
$$- 2\vec{X}_{u_1}\vec{X}_{u_3}\vec{X}_{v_1}\vec{X}_{v_3} + \vec{X}_{v_3}^2\vec{X}_{u_1}^2 + \vec{X}_{u_1}^2 \vec{X}_{v_2}^2$$
$$- 2\vec{X}_{u_1}\vec{X}_{u_2}\vec{X}_{v_1}\vec{X}_{v_2} + \vec{X}_{v_1}^2 \vec{X}_{u_2}^2 \Big]^{\frac{1}{2}}$$

$$= \Big[(\vec{X}_{u_1}^2 + \vec{X}_{u_2}^2 + \vec{X}_{u_3}^2)(\vec{X}_{v_1}^2 + \vec{X}_{v_2}^2 + \vec{X}_{v_3}^2) - (\vec{X}_{u_1}\vec{X}_{v_1} + \vec{X}_{u_2}\vec{X}_{v_2} + \vec{X}_{u_3}\vec{X}_{v_3})^2 \Big]^{\frac{1}{2}}$$

$$= \Big[(\vec{X}_u \cdot \vec{X}_u)(\vec{X}_v \cdot \vec{X}_v) - (\vec{X}_u \cdot \vec{X}_v)^2 \Big]^{\frac{1}{2}}$$

or
$$\|\vec{X}_u \times \vec{X}_v\| = \sqrt{\mathscr{E}\mathscr{G} - \mathscr{F}^2} . \qquad (4)$$

Therefore, substitution of (4) into (3) yields

$$A = \iint_D \sqrt{\mathscr{E}\mathscr{G} - \mathscr{F}^2} \, du\, dv . \qquad (5)$$

Now, from (1), (2),
$$\mathscr{E} = R^2 \sin^2 u \cos^2 v + R^2 \sin^2 u \sin^2 v + R^2 \cos^2 u$$
$$= R^2$$
$$\mathscr{F} = R^2 \sin u \cos v \cos u \sin v - R^2 \sin u \cos v \cos u \sin v$$
$$= 0$$
$$\mathscr{G} = R^2 \cos^2 u \sin^2 v + R^2 \cos^2 u \cos^2 v = R^2 \cos^2 u .$$

Therefore

$$\sqrt{\mathscr{E}\mathscr{G} - \mathscr{F}^2} = \sqrt{R^2 \cdot R^2 \cos^2 u} = R^2 |\cos u|.$$

Furthermore, since $-\pi/2 \leq u \leq \pi/2$ and $-\pi \leq v \leq \pi$,
($\cos u$ is positive on the interval $-\pi/2 \leq u \leq \pi/2$), (5) becomes,

$$A = \int_{-\pi}^{\pi} \int_{-\pi/2}^{\pi/2} R^2 \cos u \, du \, dv$$

$$= 2\pi R^2 \int_{-\pi/2}^{\pi/2} \cos u \, du = 2\pi R^2 \sin u \Big|_{-\pi/2}^{\pi/2}$$

$$= 2\pi R^2 (1 - (-1))$$

$$A = 4\pi R^2.$$

● **PROBLEM 18-27**

Compute the total area of the torus $x = (a+b \cos \varphi)\cos \theta$, $y = (a+b \cos \varphi)\sin \theta$, $z = b \sin \varphi$, $0 < b < a$ by the first fundamental form.

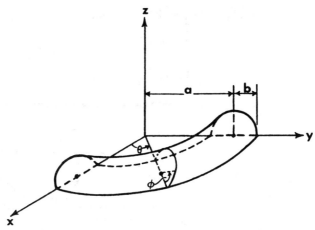

Solution: The part of the surface of the torus in the first octant is shown in fig. 1.
Since $\vec{X} = ((a+b \cos \varphi)\cos \theta, (a+b \cos \varphi)\sin \theta, b \sin \varphi)$

$\vec{X}_\theta = (-(a+b \cos \varphi)\sin \theta, (a+b \cos \varphi)\cos \theta, 0)$

$\vec{X}_\varphi = (-b \cos \theta \sin \varphi, -b \sin \theta \sin \varphi, b \cos \varphi)$

$\mathscr{E} = \vec{X}_\theta \cdot \vec{X}_\theta = (a+b \cos \varphi)^2 \sin^2\theta + (a+b \cos \varphi)^2 \cos^2\theta = (a+b \cos \varphi)^2$

$\mathscr{F} = \vec{X}_\theta \cdot \vec{X}_\varphi = b(a+b \cos \varphi)\sin \theta \cos \theta \sin \varphi - b(a+b \cos \varphi)$
$\qquad \times \sin \theta \cos \theta \sin \varphi = 0$

$\mathscr{G} = \vec{X}_\varphi \cdot \vec{X}_\varphi = b^2 \cos^2\theta \sin^2\varphi + b^2 \sin^2\theta \sin^2\varphi + b^2 \cos^2\varphi = b^2.$

Now to use the definition for area

$$A = \iint_R \sqrt{\mathscr{E}\mathscr{G} - \mathscr{F}^2} \, d\varphi \, d\theta \qquad (1)$$

the integrand $\sqrt{gG - \mathcal{F}^2}$ must be found. Proceeding,

$$\sqrt{gG - \mathcal{F}^2} = ((a+b\cos\varphi)^2 b^2 - 0)^{\frac{1}{2}} = b(a+b\cos\varphi)$$

then

$$A = \iint_R b(a+b\cos\varphi)\,d\varphi\,d\theta .$$

Since we are computing the total area of the torus, the integral must be computed over the range of values, $0 \le \theta \le 2\pi$, $0 \le \varphi \le 2\pi$.

$$A = \int_0^{2\pi} \int_0^{2\pi} b(a+b\cos\varphi)\,d\varphi\,d\theta$$

$$= 2\pi \int_0^{2\pi} (ba+b^2\cos\varphi)\,d\varphi$$

$$= 2\pi[2\pi ba] = 4\pi^2 ab .$$

• **PROBLEM 18-28**

Determine the second fundamental form of the surface represented by

$$\vec{X} = u\vec{e}_1 + v\vec{e}_2 + (u^2 - v^2)\vec{e}_3 \tag{1}$$

where $\vec{e}_1, \vec{e}_2, \vec{e}_3$ are the standard basis:

$$\vec{e}_1 = (1,0,0),\ \vec{e}_2 = (0,1,0),\ \vec{e}_3 = (0,0,1).$$

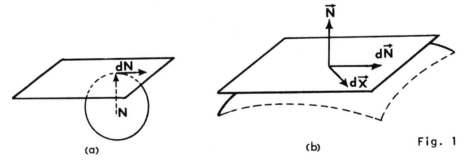

Fig. 1

Solution: Assume $\vec{X} = \vec{X}(u,v)$ is a patch on a surface of class C^n, $n \ge 2$. Then at each point on the patch there is a unit normal

$$\vec{N} = \frac{\vec{X}_u \times \vec{X}_v}{\|\vec{X}_u \times \vec{X}_v\|}$$

which is a function of u and v of class C^{n-1} with differential

$$d\vec{N} = \vec{N}_u\,du + \vec{N}_v\,dv .$$

It can be seen in fig. 1 that $d\vec{N}$ is orthogonal to $\vec{N}$ because it is parallel to the tangent plane of the spherical image of $\vec{N}$. Thus

$d\vec{N}$ is a vector parallel to the tangent plane at $\vec{X}$ as shown in Fig.1 (b). Then the quantity

$$-d\vec{X} \cdot d\vec{N} = -(\vec{X}_u du + \vec{X}_v dv) \cdot (\vec{N}_u du + \vec{N}_v dv)$$

$$= -\vec{X}_u \cdot \vec{N}_u du^2 - \vec{X}_u \cdot \vec{N}_v dudv - \vec{X}_v \vec{N}_u dudv - \vec{X}_v \vec{N}_v dv^2$$

$$= -\vec{X}_u \cdot \vec{N}_u du^2 - (\vec{X}_u \cdot \vec{N}_v + \vec{X}_v \vec{N}_u) dudv - \vec{X}_v \vec{N}_v dv^2 .$$

Let $\mathcal{L} = -\vec{X}_u \cdot \vec{N}_u$, $\mathcal{M} = -\frac{1}{2}(\vec{X}_u \cdot \vec{N}_v + \vec{X}_v \cdot \vec{N}_u)$, $\mathcal{N} = -\vec{X}_v \cdot \vec{N}_v$ to obtain

$$-d\vec{X} \cdot d\vec{N} = \mathcal{L} du^2 + 2\mathcal{M} dudv + \mathcal{N} dv^2 . \tag{2}$$

The function $-d\vec{X} \cdot d\vec{N}$ written in (2) is the second fundamental form of $\vec{X} = \vec{X}(u,v)$. Note that since $\vec{X}_u$ and $\vec{X}_v$ are perpendicular to $\vec{N}$ for all (u,v), this yields

$$0 = (\vec{X}_u \cdot \vec{N})_u = \vec{X}_{uu} \cdot \vec{N} + \vec{X}_u \cdot \vec{N}_u, \quad 0 = (\vec{X}_u \cdot \vec{N})_v = \vec{X}_{uv} \cdot \vec{N} + \vec{X}_u \cdot \vec{N}_v,$$

$$0 = (\vec{X}_v \cdot \vec{N})_u = \vec{X}_{vu} \cdot \vec{N} + \vec{X}_v \cdot \vec{N}_u, \quad 0 = (\vec{X}_v \cdot \vec{N})_v = \vec{X}_{vv} \cdot \vec{N} + \vec{X}_v \cdot \vec{N}_v$$

or

$$\vec{X}_{uu} \cdot \vec{N} = -\vec{X}_u \cdot \vec{N}_u, \quad \vec{X}_{uv} \cdot \vec{N} = -\vec{X}_u \cdot \vec{N}_v = -\vec{X}_v \cdot \vec{N}_u, \text{ and } \vec{X}_{vv} \cdot \vec{N} = -\vec{X}_v \cdot \vec{N}_v .$$

Therefore, $\mathcal{L} = \vec{X}_{uu} \cdot \vec{N}$, $\mathcal{M} = \vec{X}_{uv} \cdot \vec{N}$, $\mathcal{N} = \vec{X}_{vv} \cdot \vec{N}$.

For the given problem, from (1),

$$\vec{X}_u = \vec{e}_1 + 2u\vec{e}_3, \quad \vec{X}_v = \vec{e}_2 - 2v\vec{e}_3, \quad \vec{X}_{uu} = 2\vec{e}_3, \quad \vec{X}_{uv} = 0, \quad \vec{X}_{vv} = -2\vec{e}_3$$

$$\vec{X}_u \times \vec{X}_v = \begin{vmatrix} \vec{e}_1 & \vec{e}_2 & \vec{e}_3 \\ 1 & 0 & 2u \\ 0 & 1 & -2v \end{vmatrix}$$

$$= -2u\vec{e}_1 + 2v\vec{e}_2 + \vec{e}_3$$

and $\|\vec{X}_u \times \vec{X}_v\| = (4u^2 + 4v^2 + 1)^{\frac{1}{2}}$.

Then

$$\vec{N} = \frac{\vec{X}_u \times \vec{X}_v}{\|\vec{X}_u \times \vec{X}_v\|} = \frac{(-2u\vec{e}_1 + 2v\vec{e}_2 + \vec{e}_3)}{(4u^2 + 4v^2 + 1)^{\frac{1}{2}}} .$$

Hence

$$\mathcal{L} = \vec{X}_{uu} \cdot \vec{N} = 2\vec{e}_3 \cdot \vec{N} = 2(4u^2 + 4v^2 + 1)^{-\frac{1}{2}}$$

$$\mathcal{M} = \vec{X}_{uv} \cdot \vec{N} = 0$$

and

$$\mathcal{N} = \vec{X}_{vv} \cdot \vec{N} = -2(4u^2 + 4v^2 + 1)^{-\frac{1}{2}} .$$

Thus the second fundamental form is

$$-d\vec{X} \cdot d\vec{N} = 2(4u^2+4v^2+1)^{-\frac{1}{2}} du^2 + 0 \cdot dudv - 2(4u^2+4v^2+1)^{-\frac{1}{2}} dv^2$$

$$= 2(4u^2 + 4v^2 + 1)^{-\frac{1}{2}}(du^2 - dv^2) .$$

● **PROBLEM 18-29**

Show that the surface

$$\vec{X} = u\vec{e}_1 + v\vec{e}_2 + (u^2 + v^3)\vec{e}_3 \qquad (1)$$

is elliptic when $v > 0$, hyperbolic when $v < 0$, and parabolic for $v = 0$.

Solution: From (1),

$$\vec{X}_u = \vec{e}_1 + 2u\vec{e}_3, \quad \vec{X}_v = \vec{e}_2 + 3v^2 \vec{e}_3 ; \qquad (2)$$

hence

$$\vec{X}_u \times \vec{X}_v = -2u\vec{e}_1 - 3v^2\vec{e}_2 + \vec{e}_3$$

and

$$\|\vec{X}_u \times \vec{X}_v\| = (4u^2 + 9v^4 + 1)^{\frac{1}{2}},$$

so that

$$\vec{N} = \frac{\vec{X}_u \times \vec{X}_v}{\|\vec{X}_u \times \vec{X}_v\|} = \frac{(-2u\vec{e}_1 - 3v^2\vec{e}_2 + \vec{e}_3)}{(4u^2 + 9v^4 + 1)^{\frac{1}{2}}} . \qquad (3)$$

Furthermore, from (2),

$$\vec{X}_{uu} = 2\vec{e}_3, \quad \vec{X}_{uv} = 0, \quad \vec{X}_{vv} = 6v\vec{e}_3 . \qquad (4)$$

Now, by definition, given the coefficients of the second fundamental form ℓ, m, n, the surface is:

(i) elliptic at a point if $\ell n - m^2 > 0$
(ii) hyperbolic at a point if $\ell n - m^2 < 0$
(iii) parabolic at a point if $\ell n - m^2 = 0$

and the coefficients ℓ, m, and n are not all zero.

(iv) planar at a point if $\ell = m = n = 0$.

To apply this definition first find the coefficients of the second fundamental form.

Proceeding, using (3) and (4),

$$\ell = \vec{X}_{uu} \cdot \vec{N} = 2\vec{e}_3 \cdot \frac{(-2u\vec{e}_1 - 3v^2\vec{e}_2 + \vec{e}_3)}{(4u^2 + 9v^4 + 1)^{\frac{1}{2}}}$$

$$= 0 + 0 + 2(4u^2 + 9v^4 + 1)^{-\frac{1}{2}} = 2(4u^2 + 9v^4 + 1)^{-\frac{1}{2}}$$

$$m = \vec{X}_{uv} \cdot \vec{N} = 0$$

$$n = \vec{X}_{vv} \cdot \vec{N} = 6v(4u^2 + 9v^4 + 1)^{-\frac{1}{2}} .$$

Consequently

$$\ell n - m^2 = \frac{12v}{(4u^2 + 9v^4 + 1)^{\frac{1}{2}}} - 0 .$$

Now $(4u^2 + 9v^4 + 1) > 0$ for all (u,v). Therefore,

$$\mathscr{L}\eta - \mathscr{m}^2 > 0 \text{ for } v > 0,$$
$$\mathscr{L}\eta - \mathscr{m}^2 < 0 \text{ for } v < 0,$$
and
$$\mathscr{L}\eta - \mathscr{m}^2 = 0 \text{ for } v = 0.$$

Thus by (i) - (iv), the surface is elliptic for $v > 0$, hyperbolic for $v < 0$, and parabolic for $v = 0$ since $\mathscr{L} \neq 0$ for all (u,v).

● **PROBLEM 18-30**

For the sphere of radius a,
$$\vec{X} = (a\cos\theta\sin\varphi)\vec{e}_1 + (a\sin\theta\sin\varphi)\vec{e}_2 + (a\cos\varphi)\vec{e}_3$$
for $0 < \theta < 2\pi$, $0 < \varphi < \pi$, find:
a) the normal curvature
b) the mean curvature and the Gaussian curvature.

Solution: a) Let P be a point of a surface S of class $C^m (m \geq 2)$. Let $\vec{X} = \vec{X}(u,v)$ be a patch containing P_1 and let $\vec{X} = \vec{X}(u(t),v(t))$ be a regular curve C of class C^2 through P. Given that $\vec{k}$ is the curvature vector of C at P, the normal curvature vector to C at P is the component vector of $\vec{k}$ normal to S.

The component k_n in the direction of $\vec{N}$ is called the normal curvature of C at P and is denoted by $\varkappa_n$, i.e. $\varkappa_n = \vec{k} \cdot \vec{N}$.

The unit tangent to C at P is the vector
$$\vec{T} = \frac{d\vec{X}}{ds} = \frac{d\vec{X}}{dt} \bigg/ \left\|\frac{d\vec{X}}{dt}\right\|$$
and the curvature vector is
$$\vec{k} = \frac{d\vec{T}}{ds} = \frac{d\vec{T}}{dt} \bigg/ \left\|\frac{d\vec{X}}{dt}\right\|.$$

Since $\vec{T}$ is perpendicular to $\vec{N}$ along the curve

Hence
$$0 = \frac{d(\vec{T} \cdot \vec{N})}{dt} = \frac{d\vec{T}}{dt} \cdot \vec{N} + \vec{T} \cdot \frac{d\vec{N}}{dt}.$$
$$\varkappa_n = \vec{k} \cdot \vec{N} = \frac{d\vec{T}}{dt} \cdot \vec{N} \bigg/ \left\|\frac{d\vec{X}}{dt}\right\| = -\vec{T} \cdot \frac{d\vec{N}}{dt} \bigg/ \left\|\frac{d\vec{X}}{dt}\right\|$$
$$= -\frac{d\vec{X}}{dt} \cdot \frac{d\vec{N}}{dt} \bigg/ \left\|\frac{d\vec{X}}{dt}\right\|^2 = -\frac{d\vec{X}}{dt} \cdot \frac{d\vec{N}}{dt} \bigg/ \frac{d\vec{X}}{dt} \cdot \frac{d\vec{X}}{dt}$$
$$= -\left(\vec{X}_u \frac{du}{dt} + \vec{X}_v \frac{dv}{dt}\right) \cdot \left(\vec{N}_u \frac{du}{dt} + \vec{N}_v \frac{dv}{dt}\right) \bigg/ \left(\vec{X}_u \frac{du}{dt} + \vec{X}_v \frac{dv}{dt}\right) \cdot \left(\vec{X}_u \frac{du}{dt} + \vec{X}_v \frac{dv}{dt}\right).$$

Thus, using the coefficients for the first and second fundamental forms, the normal curvature is given by

$$\varkappa_n = \frac{\mathscr{L}(du/dt)^2 + 2\mathscr{m}(du/dt)\cdot(dv/dt) + \eta(dv/dt)^2}{\mathscr{E}(dv/dt)^2 + 2\mathscr{F}(du/dt)(dv/dt) + \mathscr{G}(dv/dt)^2}. \quad (1)$$

To apply (1) to the given problem, compute:

$$\vec{X}_\theta = -(a\sin\theta\sin\varphi)\vec{e}_1 + (a\cos\theta\sin\varphi)\vec{e}_2$$

$$\vec{X}_\varphi = (a\cos\theta\cos\varphi)\vec{e}_1 + (a\sin\theta\cos\varphi)\vec{e}_2 - (a\sin\varphi)\vec{e}_3$$

$$\vec{X}_{\theta\theta} = -(a\cos\theta\sin\varphi)\vec{e}_1 - (a\sin\theta\sin\varphi)\vec{e}_2$$

$$\vec{X}_{\theta\varphi} = -(a\sin\theta\cos\varphi)\vec{e}_1 + (a\cos\theta\cos\varphi)\vec{e}_2$$

$$\vec{X}_{\varphi\varphi} = -(a\cos\theta\sin\varphi)\vec{e}_1 - (a\sin\theta\sin\varphi)\vec{e}_2 - (a\cos\varphi)\vec{e}_3$$

$$\vec{N} = \frac{\vec{X}_\theta \times \vec{X}_\varphi}{\|\vec{X}_\theta \times \vec{X}_\varphi\|} = -(\cos\theta\sin\varphi)\vec{e}_1 - (\sin\theta\sin\varphi)\vec{e}_2 - (\cos\varphi)\vec{e}_3$$

Hence, $\mathscr{E} = \vec{X}_\theta \cdot \vec{X}_\theta = a^2\sin^2\theta\sin^2\varphi + a^2\cos^2\theta\sin^2\varphi = a^2\sin^2\varphi$

$\mathscr{F} = \vec{X}_\theta \cdot \vec{X}_\varphi = -(a^2\sin\theta\sin\varphi\cos\theta\cos\varphi) + (a^2\sin\theta\sin\varphi\cos\theta\cos\varphi) = 0$

$\mathscr{G} = \vec{X}_\varphi \cdot \vec{X}_\varphi = a^2\cos^2\theta\cos^2\varphi + a^2\sin^2\theta\cos^2\varphi + a^2\sin^2\varphi = a^2$

$\mathscr{L} = \vec{X}_{\theta\theta} \cdot \vec{N} = a\cos^2\theta\sin^2\varphi + a\sin^2\theta\sin^2\varphi = a\sin^2\varphi$

$\mathscr{M} = \vec{X}_{\theta\varphi} \cdot \vec{N} = a\sin\theta\cos\varphi\cos\theta\sin\varphi - a\cos\theta\cos\varphi\sin\varphi\sin\theta = 0$

$\mathscr{N} = \vec{X}_{\varphi\varphi} \cdot \vec{N} = a\cos^2\theta\sin^2\varphi + a\sin^2\theta\sin^2\varphi + a\cos^2\varphi = a$.

Consequently
$$\varkappa_n = \frac{\mathscr{L}\,d\theta^2 + 2\mathscr{M}\,d\theta\,d\varphi + \mathscr{N}\,d\varphi^2}{\mathscr{E}\,d\theta^2 + 2\mathscr{F}\,d\theta\,d\varphi + \mathscr{G}\,d\varphi^2}$$

$$= \frac{a\sin^2\varphi\,d\theta^2 + a\,d\varphi^2}{a^2\sin^2\varphi\,d\theta^2 + a^2\,d\varphi^2} = 1/a.$$

Thus $\varkappa_n$ = constant = $1/a$ at every point and in every direction.

b) Given $\varkappa_n$ (the normal curvature), the two perpendicular directions for which the values of $\varkappa_n$ take on maximum and minimum values are called the principal directions, and the corresponding normal curvatures, $\varkappa_1$ and $\varkappa_2$ are called the principal curvatures. By theorem, a number $\varkappa$ is a principal curvature if and only if $\varkappa$ is a solution of the equation $(\mathscr{E}\mathscr{G} - \mathscr{F}^2)\varkappa^2 - (\mathscr{E}\mathscr{N} + \mathscr{G}\mathscr{L} - 2\mathscr{F}\mathscr{M})\varkappa + (\mathscr{L}\mathscr{N} - \mathscr{M}^2) = 0$. (2)

Now divide (2) by $(\mathscr{E}\mathscr{G} - \mathscr{F}^2)$ to obtain

$$\varkappa^2 - \frac{(\mathscr{E}\mathscr{N} + \mathscr{G}\mathscr{L} - 2\mathscr{F}\mathscr{M})}{(\mathscr{E}\mathscr{G} - \mathscr{F}^2)}\varkappa + \frac{(\mathscr{L}\mathscr{N} - \mathscr{M}^2)}{(\mathscr{E}\mathscr{G} - \mathscr{F}^2)} = 0$$

or $\varkappa^2 - 2\mathscr{H}\varkappa + \mathscr{K} = 0$ where

$$\mathscr{H} = \tfrac{1}{2}(\varkappa_1 + \varkappa_2) = \frac{\mathscr{E}\mathscr{N} + \mathscr{G}\mathscr{L} - 2\mathscr{F}\mathscr{M}}{2(\mathscr{E}\mathscr{G} - \mathscr{F}^2)}$$

is the average of the roots $\varkappa_1$ and $\varkappa_2$ and is called the mean curvature at P and

$$K = \kappa_1 \kappa_2 = \frac{\ell n - m^2}{\mathscr{E}\mathscr{G} - \mathscr{F}^2}$$

is the product of the roots and is called the Gaussian curvature at P. Applying these formulas for $\mathscr{N}, \mathscr{K}$ to the given problems yields the mean curvature,

$$\mathscr{N} = \frac{a^3 \sin^2\varphi + a^3 \sin^2\varphi}{2(a^4 \sin^2\varphi)} = \pm 1/a$$

(the sign depending on the orientation of the surface), and the Gaussian curvature

$$\mathscr{K} = \frac{a^2 \sin^2\varphi}{a^4 \sin^2\varphi} = 1/a^2 \quad \text{(constant)}.$$

Remark: By (a), the normal curvature at any point is $1/a$ so every direction is principal. Therefore $\kappa_1 = \pm 1/a = \kappa_2$, so mean curvature is $\pm 1/a$ and Gaussian curvature is $1/a^2$.

● **PROBLEM 18-31**

For the paraboloid of revolution given by $(r\cos\theta, r\sin\theta, 1 - r^2)$ find the normal curvature as a function of direction at the point $(\theta, r) = (\pi/4, 1)$. In addition compute the Gaussian curvature at this point.

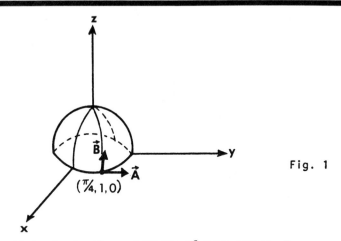

Fig. 1

Solution: Given, that k is the curvature of some curve Γ, resolve k into component vectors $\vec{k}_N$, $\vec{k}_G$ respectively normal and tangential to S. Then $\vec{k}_N$ is called the normal curvature vector and $\vec{k}_G$ is called the geodesic curvature of Γ.

Now, at a point of $\vec{X} = \vec{X}(u,v)$, the normal plane will cut the surface in a normal section whose equation we assume to be $v = \varphi(u)$. The tangent vector of this curve is $d\vec{X}/ds$, and since it is orthogonal to the normal vector $\vec{N}$, $d\vec{X}/ds \cdot \vec{N} = \vec{T} \cdot \vec{N} = 0$. Differentiation with respect to s and use of the Frenet-Serret formulas yields

$$\frac{1}{R}(\vec{B}\cdot\vec{N}) + \left(\frac{d\vec{X}}{ds} \cdot \frac{d\vec{N}}{ds}\right) = 0$$

where $1/R$ is the curvature of the normal section. Since the principal normal $\vec{B}$ lies in the osculating plane (here the plane of section), $\vec{B} = \pm \vec{N}$ and

$$\frac{1}{R} = \pm \frac{(d\vec{X}\cdot d\vec{N})}{(d\vec{X}\cdot d\vec{X})} = \pm \frac{\mathcal{L} + 2\mathcal{M}\varphi' + \mathcal{N}(\varphi')^2}{\mathcal{E} + 2\mathcal{F}\varphi' + \mathcal{G}(\varphi')^2} \quad (1)$$

from the equations of the first and second fundamental forms. The derivative $\varphi'(u)$ might be regarded as a generalized "slope" defining the direction at which the curve leaves the point in question. The curvature of the various normal sections of a given surface at a fixed point depends on this slope. Replace φ' by λ and choose arbitrarily the positive sign in (1). The resulting quantity, termed the normal curvature, $1/r$, of the surface at the point in question as a function of the direction λ is:

$$\frac{1}{r_\lambda} = \frac{\mathcal{L} + 2\mathcal{M}\lambda + \mathcal{N}\lambda^2}{\mathcal{E} + 2\mathcal{F}\lambda + \mathcal{G}\lambda^2} \,. \quad (2)$$

Given the parameterization

$$\vec{X} = (r\cos\theta,\, r\sin\theta,\, 1 - r^2)$$

first compute the values of, $\mathcal{E} = \vec{X}_r \cdot \vec{X}_r$, $\mathcal{F} = \vec{X}_r \cdot \vec{X}_\theta$, $\mathcal{G} = \vec{X}_\theta \cdot \vec{X}_\theta$ (3)

$$\mathcal{L} = \frac{\vec{X}_{rr} \cdot (\vec{X}_r \times \vec{X}_\theta)}{\|\vec{X}_r \times \vec{X}_\theta\|} \qquad \mathcal{M} = \frac{\vec{X}_{r\theta} \cdot (\vec{X}_r \times \vec{X}_\theta)}{\|\vec{X}_r \times \vec{X}_\theta\|}$$

and

$$\mathcal{N} = \frac{\vec{X}_{\theta\theta} \cdot (\vec{X}_r \times \vec{X}_\theta)}{\|\vec{X}_r \times \vec{X}_\theta\|}$$

at the point $(\pi/4,1)$.

The normal curvature, $1/r$, as a function of direction can now be found by (2) with,

$\vec{X}_r = (\cos\theta, \sin\theta, -2r)_{(\pi/4,1)} = (1/\sqrt{2},\, 1/\sqrt{2},\, -2)$

$\vec{X}_\theta = (-r\sin\theta, r\cos\theta, 0)_{(\pi/4,1)} = (-1/\sqrt{2},\, 1/\sqrt{2}, 0)$

$\vec{X}_{rr} = (0,0,-2)_{(\pi/4,1)} = (0,0,-2)$

$\vec{X}_{r\theta} = (-\sin\theta, \cos\theta, 0)_{(\pi/4,1)} = (-1/\sqrt{2},\, 1/\sqrt{2}, 0)$.

$\vec{X}_{\theta\theta} = (-r\cos\theta, -r\sin\theta, 0)_{(\pi/4,1)} = (-1/\sqrt{2},\, -1/\sqrt{2}, 0)$.

Consequently by (3)

$E = 5 \quad F = 0 \quad G = 1$.

To find ℓ evaluate $\dfrac{\vec{X}_{rr} \cdot (\vec{X}_r \times \vec{X}_\theta)}{\|\vec{X}_r \times \vec{X}_\theta\|} = \dfrac{\vec{X}_{rr} \cdot (\vec{X}_r \times \vec{X}_\theta)}{\sqrt{EG - F^2}}$

$= \dfrac{\vec{X}_{rr} \cdot (\vec{X}_r \times \vec{X}_\theta)}{\sqrt{5}}$

$= \dfrac{1}{\sqrt{5}} (0,0,-2) \cdot (\dfrac{1}{\sqrt{2}}, \dfrac{1}{\sqrt{2}}, -2) \times (-\dfrac{1}{\sqrt{2}}, \dfrac{1}{\sqrt{2}}, 0)$

$= \dfrac{1}{\sqrt{5}} (0,0,-2) \cdot (\sqrt{2}, \sqrt{2}, 1) = -\dfrac{2}{\sqrt{5}}$.

Similarly $n = \dfrac{1}{\sqrt{5}}(-\dfrac{1}{\sqrt{2}}, -\dfrac{1}{\sqrt{2}}, 0) \cdot (\sqrt{2},\sqrt{2},1) = -\dfrac{2}{\sqrt{5}}$ and

$m = \dfrac{1}{\sqrt{5}}(-\dfrac{1}{\sqrt{2}}, \dfrac{1}{\sqrt{2}}, 0) \cdot (\sqrt{2},\sqrt{2},1) = 0$.

The expression for the normal curvature now becomes, by (2),

$$\dfrac{1}{r_\lambda} = \dfrac{-2 - 2\lambda^2}{\sqrt{5}(5 + \lambda^2)}.$$

The Gaussian curvature is the product of the maximum and minimum values of $1/r_\lambda$. To find these we first solve

$$\dfrac{d(1/r_\lambda)}{d\lambda} = 0 \quad \text{or} \quad \dfrac{-16\sqrt{5}\,\lambda}{5(5 + \lambda^2)^2} = 0.$$

As solutions we get $\lambda = 0$ and $\lambda = \infty$ (corresponding respectively to $\vec{A}$ and $\vec{B}$ in the figure). The extreme values of $1/r_\lambda$ are found to be

$$\dfrac{1}{r_0} = \dfrac{-2}{5\sqrt{5}} \quad \text{and} \quad \dfrac{1}{r_\infty} = \lim_{\lambda \to \infty} \dfrac{-2 - 2\lambda^2}{\sqrt{5}(5+\lambda^2)} = -\dfrac{2}{\sqrt{5}}.$$

The Gaussian curvature equals $\dfrac{1}{r_0} \cdot \dfrac{1}{r_\infty}$ or $\dfrac{-2}{5\sqrt{5}} \cdot \dfrac{-2}{\sqrt{5}} = \dfrac{4}{25}$.

● **PROBLEM 18-32**

Prove that at each point on a patch

$$\vec{N}_u \times \vec{N}_v = \mathcal{K}(\vec{X}_u \times \vec{X}_v) \qquad (1)$$

where $\mathcal{K}$ is the Gaussian curvature at the point.

Solution: Since $\vec{N}$ is a function of unit length, $\vec{N}_u$ and $\vec{N}_v$ are orthogonal to $\vec{N}$ and therefore are parallel to the tangent plane. From this we can write $\vec{N}_u$ and $\vec{N}_v$ in the following manner, $\vec{N}_u = a\vec{X}_u + b\vec{X}_v$ and $\vec{N}_v = c\vec{X}_u + d\vec{X}_v$ where a,b,c,d are to be determined.

Hence,

$$\vec{N}_u \times \vec{N}_v = (a\vec{X}_u + b\vec{X}_v) \times (c\vec{X}_u + d\vec{X}_v)$$

$$\vec{N}_u \times \vec{N}_v = (ad - bc)(\vec{X}_u \times \vec{X}_v) .\tag{2}$$

Now, to prove (1), all that is needed is to show that in (2)

$$ad - bc = \det\begin{pmatrix} a & b \\ c & d \end{pmatrix} = \mathcal{K} .$$

To do this, first remember that

$$\mathcal{K} = \frac{\mathcal{L}\eta - m^2}{\mathcal{E}\mathcal{G} - \mathcal{F}^2} .\tag{3}$$

Then, proceeding

$$\vec{X}_u \cdot \vec{N}_u = \vec{X}_u \cdot (a\vec{X}_u + b\vec{X}_v) = a\vec{X}_u \cdot \vec{X}_u + b\vec{X}_u \cdot \vec{X}_v = a\mathcal{E} + b\mathcal{F}$$

where $\mathcal{E}, \mathcal{F}$ are coefficients from the first fundamental form. Furthermore, from the second fundamental form

$$\vec{X}_u \cdot \vec{N}_u = -\mathcal{L} .$$

Hence, $a\mathcal{E} + b\mathcal{F} = -\mathcal{L}$. Similarly,

$$\vec{X}_v \cdot \vec{N}_u = a\vec{X}_u \cdot \vec{X}_v + b\vec{X}_v \cdot \vec{X}_v = a\mathcal{F} + b\mathcal{G} = -m$$

$$\vec{X}_u \cdot \vec{N}_v = c\vec{X}_u \cdot \vec{X}_u + d\vec{X}_u \cdot \vec{X}_v = c\mathcal{E} + d\mathcal{F} = -m$$

$$\vec{X}_v \cdot \vec{N}_v = c\vec{X}_u \cdot \vec{X}_v + d\vec{X}_v \cdot \vec{X}_v = c\mathcal{F} + d\mathcal{G} = -\eta .$$

Writing these equations as the matrix product gives

$$\begin{pmatrix} a & b \\ c & d \end{pmatrix}\begin{pmatrix} \mathcal{E} & \mathcal{F} \\ \mathcal{F} & \mathcal{G} \end{pmatrix} = \begin{pmatrix} -\mathcal{L} & -m \\ -m & -\eta \end{pmatrix}$$

From which

$$\det\begin{pmatrix} a & b \\ c & d \end{pmatrix} \det\begin{pmatrix} \mathcal{E} & \mathcal{F} \\ \mathcal{F} & \mathcal{G} \end{pmatrix} = \det\begin{pmatrix} -\mathcal{L} & -m \\ -m & -\eta \end{pmatrix}$$

or $(ad - bc)(\mathcal{E}\mathcal{G} - \mathcal{F}^2) = \mathcal{L}\eta - m^2$

$$(ad - bc) = \frac{\mathcal{L}\eta - m^2}{(\mathcal{E}\mathcal{G} - \mathcal{F}^2)} = \mathcal{K} \quad \text{from (3).}$$

This completes the proof.

• **PROBLEM 18-33**

Find equations for the asymptotic lines for the hyperbolic paraboloid whose equations are

$$bx + ay = 2abu, \quad bx - ay = 2abv, \quad z = 2cuv. \tag{1}$$
$$(a,b,c \neq 0)$$

Solution: The position vector $\vec{r}$ of the point (x,y,z) on the given surface is

$$\vec{r} = a(u+v)\vec{e}_1 + b(u-v)\vec{e}_2 + 2cuv\vec{e}_3 .\tag{2}$$

(Note: (2) is obtained by solving for x,y,z in (1)).

Now by definition a direction at a point on a patch for which
$$-d\vec{X} \cdot d\vec{N} = \mathcal{L}\, du^2 + 2\mathcal{m}\, dudv + \mathcal{n}\, dv^2 = 0 \qquad (3)$$
is called an asymptotic direction. Then a curve on a surface which is tangent to an asymptotic direction at every point is termed an asymptotic line. Therefore, a curve on a surface is an asymptotic line if and only if the direction of the tangent line to the curve at each point satisfies (3) for some patch $\vec{X} = \vec{X}(u,v)$ containing the point.

For this problem from (2),

$\vec{r}_u = (a, b, 2cv)$ $\vec{r}_{uu} = (0,0,0)$

$\vec{r}_v = (a, -b, 2cu)$ $\vec{r}_{vv} = (0,0,0)$

$\vec{r}_{uv} = (0,0,2c)$

and $\vec{r}_u \times \vec{r}_v = (2bc(u+v),\ 2ac(v-u),\ -2ab)$. Consequently,

$\mathcal{L} = \vec{r}_{uu} \cdot \vec{N} = (0,0,0) \cdot (n_1, n_2, n_3) = 0$

$\mathcal{m} = \vec{r}_{uv} \cdot \vec{N} = (0,0,2c) \cdot (n_1, n_2, n_3) = 2cn_3$

$\mathcal{n} = \vec{r}_{vv} \cdot \vec{N} = (0,0,0) \cdot (n_1, n_2, n_3) = 0$.

Since
$$\vec{N} = \frac{\vec{r}_u \times \vec{r}_v}{\|\vec{r}_u \times \vec{r}_v\|} = \frac{(2bc(u+v),\ 2ac(v-u),\ -2ab)}{\|(2bc(u+v), 2ac(v-u), -2ab)\|},$$

we have
$$n_3 = \frac{-2ab}{\|(2bc(u+v), 2ac(v-u), -2ab)\|}$$

Hence,
$$\mathcal{m} = 2cn_3 = \frac{-4abc}{\|(2bc(u+v), 2ac(v-u), -2ab)\|}$$

Then, by (3) the equations for the asymptotic lines are
$$\frac{-8abc}{\|(2bc(u+v), 2ac(v-u), -2ab)\|}\, dudv = 0$$

or $dudv = 0$, since $abc \neq 0$.

● **PROBLEM 18-34**

Find the envelope of:
a) the family of planes
$$x \cos \theta + y \sin \theta + z/\sqrt{3} = 0$$
b) the two-parameter family of spheres

$$(x - a\cos\theta)^2 + (y - a\sin\theta)^2 + z^2 = a^2/4$$

where a, θ are the parameters.

Solution: a) This problem, unlike the previous problems on envelopes, deals with the concept of the envelope of a family of surfaces (the others dealt with a family of curves). However, the analytical procedure for finding envelopes of surfaces is almost identical with that used for finding envelopes of plane curves. Here, α is to be eliminated between the two equations $F(x,y,z,\alpha) = 0$ and

$$\frac{\partial F}{\partial \alpha}(x,y,z,\alpha) = 0 .$$

If this elimination yields an equation in x, y, z, and if there is an envelope, the envelope is part or all of the locus defined by the equation.

In this problem $F(x,y,z,\theta) = x\cos\theta + y\sin\theta + z/\sqrt{3} = 0$ (1)

and $\quad\dfrac{\partial F}{\partial \theta}(x,y,z,\theta) = -x\sin\theta + y\cos\theta = 0$ (2)

Hence, we need to eliminate θ. From (2),

$$x\sin\theta = y\cos\theta \quad \text{so that} \quad \tan\theta = y/x . \quad (3)$$

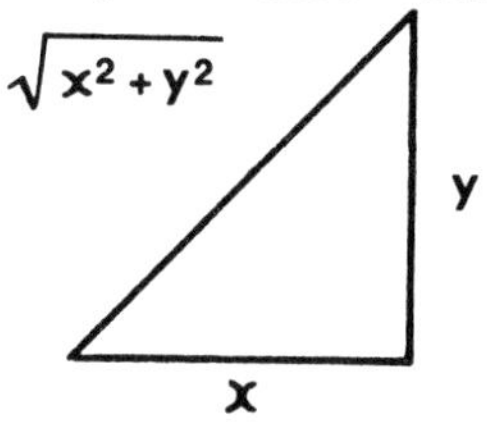

Fig. 1

To find $\sin\theta$ and $\cos\theta$, form a right triangle with one angle θ equal to $\tan^{-1}(y/x)$. Considering the triangle in Fig. 1, we have

$$\sin\theta = \pm\frac{y}{\sqrt{x^2+y^2}} \qquad \cos\theta = \pm\frac{x}{\sqrt{x^2+y^2}} .$$

Placing these values into (1) yields

$$x\left(\pm\frac{x}{\sqrt{x^2+y^2}}\right) + y\left(\pm\frac{y}{\sqrt{x^2+y^2}}\right) + \frac{z}{\sqrt{3}} = 0$$

or

$$\pm\frac{x^2+y^2}{\sqrt{x^2+y^2}} = -\frac{z}{\sqrt{3}} .$$

Squaring both sides to rationalize the denominator, then yields $3(x^2+y^2) = z^2$, which is the equation for a right circular cone with axis along the z-axis, vertex at the origin, and semivertical angle $30°$. Study of the family of planes directly verifies that the cone is the envelope.

b) In the case of two-parameter families, the envelope surface has a single member of the family tangent to it at each point. Given that $F(x,y,z,\alpha,\beta) = 0$ is the equation of the family, if there is an

envelope it is part or all of the locus defined by the simultaneous equations
$$F = 0, \ \partial F/\partial \alpha = 0, \ \partial F/\partial \beta = 0. \tag{4}$$

For this problem
$$F(x,y,z,a,\theta) = (x-a\cos\theta)^2 + (y-a\sin\theta)^2 + z^2 - a^2/4 = 0 \tag{5}$$

and
$$\frac{\partial F}{\partial a} = 2(x-a\cos\theta)(-\cos\theta) + 2(y-a\sin\theta)(-\sin\theta) - a/2 = 0 \tag{6}$$

$$\frac{\partial F}{\partial \theta} = 2(x-a\cos\theta)(a\sin\theta) + 2(y-a\sin\theta)(-a\cos\theta) = 0 \tag{7}$$

The two parameters, a and θ, must now be eliminated from equations (5),(6),(7).

Expanding equation (6) using the identity $2a\cos^2\theta + 2a\sin^2\theta = 2a$ yields
$$x\cos\theta + y\sin\theta = 3a/4. \tag{8}$$

Similarly expanding (7) gives
$$x\sin\theta - y\cos\theta = 0. \tag{9}$$

Hence, from (9) $\tan\theta = y/x$ and
$$\sin\theta = \pm \frac{y}{\sqrt{x^2+y^2}} \quad \cos\theta = \pm \frac{x}{\sqrt{x^2+y^2}}.$$

Substituting these values into (8) yields
$$\pm \frac{x^2+y^2}{\sqrt{x^2+y^2}} = 3a/4$$

or
$$a = \pm 4/3 \sqrt{x^2+y^2}. \tag{10}$$

Now expanding (5) we obtain
$$x^2+y^2+z^2 - 2xa\cos\theta - 2ya\sin\theta + 3/4 \, a^2 = 0$$

or
$$\begin{aligned} x^2+y^2+z^2 &= 2xa\cos\theta + 2ya\sin\theta - 3/4 \, a^2 \\ &= 2a[x\cos\theta + y\sin\theta] - 3/4 \, a^2 \\ &= 2a(3/4 \, a) - 3/4 \, a^2 = 3/4 \, a^2 \ (\text{by (8)}). \end{aligned}$$

Then by (10)
$$x^2+y^2+z^2 = (3/4)(4^2/3^2)(x^2+y^2) = 4/3 \, (x^2+y^2)$$

or
$$3z^2 = (x^2+y^2).$$

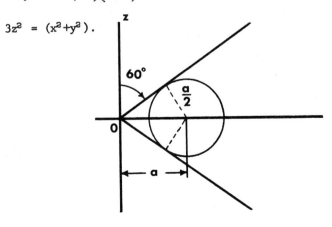

Fig. 2

This is a right circular cone about the z-axis, with semivertical angle $60°$ and vertex at the origin. Then, observing that the spheres (5) have their centers in the xy-plane, at distance a from the origin, and that the radius of such a sphere is a/2 (see figure 2), we may see directly that the cone is the envelope.

● **PROBLEM 18-35**

Prove the following theorem: A one-to-one mapping f of S onto S* is an isometry if and only if on every patch $\vec{X} = \vec{X}(u,v)$ on S the first fundamental coefficients $\mathcal{E}$, $\mathcal{F}$ and $\mathcal{G}$ are equal to the first fundamental coefficients $\mathcal{E}^*$, $\mathcal{F}^*$ and $\mathcal{G}^*$ along its image $\vec{X}^* = \vec{X}^*(u,v) = f(\vec{X}(u,v))$.

Solution: By definition a one-to-one mapping f of a surface S onto a surface S^* is called an isometric mapping or isometry if the length of an arbitrary regular arc $\vec{X} = \vec{X}(t)$ on S is equal to the length of its image $\vec{X}^* = \vec{X}^*(t) = f(\vec{X}(t))$ on S^*. In addition if f is an isometry from S onto S^*, then f^{-1} is an isometry from S^* into S.

To prove the theorem, two things must be shown:

Part 1: Assume f is a one-to-one mapping of S into S* such that on every patch $\vec{X} = \vec{X}(u,v)$ on S the first fundamental coefficients $\mathcal{E}, \mathcal{F}$ and $\mathcal{G}$ are equal to the first fundamental coefficients $\mathcal{E}^*, \mathcal{F}^*$ and $\mathcal{G}^*$ along its image $\vec{X}^* = \vec{X}^*(u,v) = f(\vec{X}(u,v))$. Prove F is an isometry.

To do this, suppose $\vec{X} = \vec{X}(t)$, $a \le t \le b$, is an arbitrary arc C on S. In general C may not lie in any patch on S. However, since C is compact, it will consist of a finite number of arcs C_i, $t_i \le t \le t_{i+1}$, $i = 0,\ldots,n-1$, placed end to end such that each C_i is on some patch $\vec{X}_i = \vec{X}_i(u,v)$. Since the length of an arc on a patch is the integral of the square root of the first fundamental form the length L(C) of C is given by

$$L(C) = \sum_i L(C_i) = \sum_i \left[\int_{t_i}^{t_{i+1}} \sqrt{\mathcal{E}_i \left(\frac{du}{dt}\right)^2 + 2\mathcal{F}_i \left(\frac{du}{dt}\right)\left(\frac{dv}{dt}\right) + \mathcal{G}_i \left(\frac{dv}{dt}\right)^2}\, dt \right].$$

However, by the given, for all i, $\mathcal{E}_i = \mathcal{E}_i^*$, $\mathcal{F}_i = \mathcal{F}_i^*$, and $\mathcal{G}_i = \mathcal{G}_i^*$, where $\mathcal{E}_i^*, \mathcal{F}_i^*, \mathcal{G}_i^*$ are the fundamental coefficients on $\vec{X}_i^* = f(\vec{X}_i(u,v))$. Therefore,

$$L(C) = \sum_i \left[\int_{t_i}^{t_{i+1}} \sqrt{\mathcal{E}_i^* \left(\frac{du}{dt}\right)^2 + 2\mathcal{F}_i^* \left(\frac{du}{dt}\right)\left(\frac{dv}{dt}\right) + \mathcal{G}_i^* \left(\frac{dv}{dt}\right)^2}\, dt \right] = \sum_i L(C_i^*) = L(C^*).$$

Hence the length of any arc C on S is equal to the length of its image C^* on S^*. Thus, by the definition at the beginning of this problem, f is an isometry.

Part 2: Assume f is an isometry from S onto S^* and $\vec{X} = \vec{X}(u,v)$ is a patch on S. Prove that $\mathscr{E} = \mathscr{E}^*$, $\mathscr{F} = \mathscr{F}^*$, $\mathscr{G} = \mathscr{G}^*$ ($\mathscr{E}, \mathscr{E}^*$, etc... defined as in Case 1).

To proceed let (u,v) be an arbitrary point in the domain of $\vec{X} = \vec{X}(u,v)$. In addition, let $u = u(t)$, $v = v(t)$, $a \leq t \leq b$ be an arbitrary arc through (u,v) and let C_T and C_T^* be the images on S and S^* respectively of $u = u(t)$, $v = v(t)$ on the interval $a \leq t \leq T$. Since, by the given, f is an isometry of S onto S^*,

$$L(C_T) = \int_a^T \sqrt{\mathscr{E}\left(\frac{du}{dt}\right)^2 + 2\mathscr{F}\left(\frac{du}{dt}\right)\left(\frac{dv}{dt}\right) + \mathscr{G}\left(\frac{dv}{dt}\right)^2}\, dt = L(C_T^*)$$

$$= \int_a^T \sqrt{\mathscr{E}^*\left(\frac{du}{dt}\right)^2 + 2\mathscr{F}^*\left(\frac{du}{dt}\right)\left(\frac{dv}{dt}\right) + \mathscr{G}^*\left(\frac{dv}{dt}\right)^2}\, dt.$$

Since the above is valid for all T, we have for all t, and in particular at (u,v),

$$\mathscr{E}\left(\frac{du}{dt}\right)^2 + 2\mathscr{F}\frac{du}{dt}\frac{dv}{dt} + \mathscr{G}\left(\frac{dv}{dt}\right)^2 = \mathscr{E}^*\left(\frac{du}{dt}\right)^2 + 2\mathscr{F}^*\frac{du}{dt}\frac{dv}{dt} + \mathscr{G}^*\left(\frac{dv}{dt}\right)^2. \quad (1)$$

However, the curve $u = u(t)$, $v = v(t)$ though (u,v) is arbitrary. Therefore (1) is also valid at (u,v) for all du/dt, dv/dt. Hence $\mathscr{E} = \mathscr{E}^*$, $\mathscr{F} = \mathscr{F}^*$, and $\mathscr{G} = \mathscr{G}^*$ at (u,v). Since (u,v) is arbitrary, the result follows. Thus, by the two cases, the proof of the theorem is complete.

• **PROBLEM 18-36**

Derive the Gauss-Weingarten Equations.

Solution: Assume that $\vec{X} = \vec{X}(u,v)$ is a patch on a surface of class C^m ($m \geq 2$). Then $\vec{X}_u$, $\vec{X}_v$ and $\vec{N}$ are functions of class C^1 and have continuous derivatives $\vec{X}_{uu}$, $\vec{X}_{uv}$, $\vec{X}_{vv}$, $\vec{N}_u$ and $\vec{N}_v$. Using the fact that the vectors $\vec{X}_u$, $\vec{X}_v$, and $\vec{N}$ are linearly independent, we are able to write

$$\vec{X}_{uu} = \Gamma_{11}^1 \vec{X}_u + \Gamma_{11}^2 \vec{X}_v + \alpha_{11} \vec{N}$$

$$\vec{X}_{uv} = \Gamma_{12}^1 \vec{X}_u + \Gamma_{12}^2 \vec{X}_v + \alpha_{12} \vec{N}$$

$$\vec{X}_{vv} = \Gamma_{22}^1 \vec{X}_u + \Gamma_{22}^2 \vec{X}_v + \alpha_{22} \vec{N} \quad (1)$$

$$\vec{N}_u = \beta_1^1 \vec{X}_u + \beta_1^2 \vec{X}_v + \gamma_1 \vec{N}$$

$$\vec{N}_v = \beta_2^1 \vec{X}_u + \beta_2^2 \vec{X}_v + \gamma_2 \vec{N}$$

where the coefficients Γ_{ij}^k, α_{ij}, β_i^j, γ_i are to be determined. The vector $\vec{N}$ is of unit length. Therefore $\vec{N}_u$ and $\vec{N}_v$ are orthogonal to $\vec{N}$. That is,

$$0 = \vec{N}_u \cdot \vec{N} = \beta_1^1 \vec{X}_u \cdot \vec{N} + \beta_1^2 \vec{X}_v \cdot \vec{N} + \gamma_1 \vec{N} \cdot \vec{N}$$

$$0 = \vec{N}_v \cdot \vec{N} = \beta_2^1 \vec{X}_u \cdot \vec{N} + \beta_2^2 \vec{X}_v \cdot \vec{N} + \gamma_2 \vec{N} \cdot \vec{N}.$$

However, since $\vec{X}_u \cdot \vec{N} = \vec{X}_v \cdot \vec{N} = 0$ and $\vec{N} \cdot \vec{N} = 1$, it follows that $\gamma_1 = \gamma_2 = 0$. Furthermore,

$$-\mathcal{L} = \vec{X}_u \cdot \vec{N}_u = \beta_1^1 \vec{X}_u \cdot \vec{X}_u + \beta_1^2 \vec{X}_u \cdot \vec{X}_v = \beta_1^1 \mathcal{E} + \beta_1^2 \mathcal{F} \tag{2}$$

$$-\mathcal{M} = \vec{X}_v \cdot \vec{N}_u = \beta_1^1 \vec{X}_v \cdot \vec{X}_u + \beta_1^2 \vec{X}_v \cdot \vec{X}_u = \beta_1^1 \mathcal{F} + \beta_1^2 \mathcal{G} \tag{3}$$

$$-\mathcal{M} = \vec{X}_u \cdot \vec{N}_v = \beta_2^1 \vec{X}_u \cdot \vec{X}_u + \beta_2^2 \vec{X}_u \cdot \vec{X}_v = \beta_2^1 \mathcal{E} + \beta_2^2 \mathcal{F} \tag{4}$$

$$-\mathcal{N} = \vec{X}_v \cdot \vec{N}_v = \beta_2^1 \vec{X}_v \cdot \vec{X}_u + \beta_2^2 \vec{X}_v \cdot \vec{X}_v = \beta_2^1 \mathcal{F} + \beta_2^2 \mathcal{G}. \tag{5}$$

Then solving (2) and (3) for β_1^1 and β_1^2 yields

$$\beta_1^1 = \frac{\mathcal{M F} - \mathcal{L G}}{\mathcal{E G} - \mathcal{F}^2}, \quad \beta_1^2 = \frac{\mathcal{L F} - \mathcal{M E}}{\mathcal{E G} - \mathcal{F}^2}. \tag{6}$$

Similarly solving (4) and (5) for β_2^1 and β_2^2 gives

$$\beta_2^1 = \frac{\mathcal{N F} - \mathcal{M G}}{\mathcal{E G} - \mathcal{F}^2}, \quad \beta_2^2 = \frac{\mathcal{M F} - \mathcal{N E}}{\mathcal{E G} - \mathcal{F}^2}. \tag{7}$$

In addition

$$\mathcal{L} = \vec{X}_{uu} \cdot \vec{N} = \Gamma_{11}^1 \vec{X}_u \cdot \vec{N} + \Gamma_{11}^2 \vec{X}_v \cdot \vec{N} + \alpha_{11} \vec{N} \cdot \vec{N} = \alpha_{11}$$

$$\mathcal{M} = \vec{X}_{uv} \cdot \vec{N} = \Gamma_{12}^1 \vec{X}_u \cdot \vec{N} + \Gamma_{12}^2 \vec{X}_v \cdot \vec{N} + \alpha_{12} \vec{N} \cdot \vec{N} = \alpha_{12}$$

$$\mathcal{N} = \vec{X}_{vv} \cdot \vec{N} = \Gamma_{22}^1 \vec{X}_u \cdot \vec{N} + \Gamma_{22}^2 \vec{X}_v \cdot \vec{N} + \alpha_{22} \vec{N} \cdot \vec{N} = \alpha_{22}$$

i.e., $\alpha_{11} = \mathcal{L}$, $\alpha_{12} = \mathcal{M}$, $\alpha_{22} = \mathcal{N}$. \tag{8}

Now it is needed to determine Γ_{ij}^k. To do this first note that

$$\vec{X}_u \cdot \vec{X}_{uu} = \tfrac{1}{2} (\vec{X}_u \cdot \vec{X}_u)_u = \tfrac{1}{2} \mathcal{E}_u$$

$$\vec{X}_v \cdot \vec{X}_{vv} = \tfrac{1}{2} (\vec{X}_v \cdot \vec{X}_v)_v = \tfrac{1}{2} \mathcal{G}_v \tag{9}$$

$$\vec{X}_u \cdot \vec{X}_{uv} = \tfrac{1}{2} (\vec{X}_u \cdot \vec{X}_u)_v = \tfrac{1}{2} \mathcal{E}_v$$

$$\vec{X}_v \cdot \vec{X}_{uv} = \tfrac{1}{2} (\vec{X}_v \cdot \vec{X}_v)_u = \tfrac{1}{2} \mathcal{G}_u.$$

In addition

$$\mathcal{F}_u = (\vec{X}_u \cdot \vec{X}_v)_u = \vec{X}_{uu} \cdot \vec{X}_v + \vec{X}_u \cdot \vec{X}_{uv} = \vec{X}_{uu} \cdot \vec{X}_v + \tfrac{1}{2} \mathcal{E}_v$$

$$\mathcal{F}_v = (\vec{X}_u \cdot \vec{X}_v)_v = \vec{X}_{uv} \cdot \vec{X}_v + \vec{X}_u \cdot \vec{X}_{vv} = \tfrac{1}{2} \mathcal{G}_u + \vec{X}_u \cdot \vec{X}_{vv} .$$

Therefore, $\vec{X}_v \cdot \vec{X}_{uu} = \mathcal{F}_u - \tfrac{1}{2} \mathcal{E}_v$ (10)

$$\vec{X}_u \cdot \vec{X}_{vv} = \mathcal{F}_v - \tfrac{1}{2} \mathcal{G}_u .$$

Then from (9)-(10) and the first three equations in (1) we have

$$\tfrac{1}{2} \mathcal{E}_u = \vec{X}_u \cdot \vec{X}_{uu} = \Gamma^1_{11} \vec{X}_u \cdot \vec{X}_u + \Gamma^2_{11} \vec{X}_u \cdot \vec{X}_v = \Gamma^1_{11} \mathcal{E} + \Gamma^2_{11} \mathcal{F} \quad (11)$$

$$\mathcal{F}_u - \tfrac{1}{2} \mathcal{E}_v = \vec{X}_v \cdot \vec{X}_{uu} = \Gamma^1_{11} \vec{X}_v \cdot \vec{X}_u + \Gamma^2_{11} \vec{X}_v \cdot \vec{X}_v = \Gamma^1_{11} \mathcal{F} + \Gamma^2_{11} \mathcal{G} \quad (12)$$

$$\tfrac{1}{2} \mathcal{E}_v = \vec{X}_u \cdot \vec{X}_{uv} = \Gamma^1_{12} \vec{X}_u \cdot \vec{X}_u + \Gamma^2_{12} \vec{X}_u \cdot \vec{X}_v = \Gamma^1_{12} \mathcal{E} + \Gamma^2_{12} \mathcal{F} \quad (13)$$

$$\tfrac{1}{2} \mathcal{G}_u = \vec{X}_v \cdot \vec{X}_{uv} = \Gamma^1_{12} \vec{X}_v \cdot \vec{X}_u + \Gamma^2_{12} \vec{X}_v \cdot \vec{X}_v = \Gamma^1_{12} \mathcal{F} + \Gamma^2_{12} \mathcal{G} \quad (14)$$

$$\mathcal{F}_v - \tfrac{1}{2} \mathcal{G}_u = \vec{X}_u \cdot \vec{X}_{vv} = \Gamma^1_{22} \vec{X}_u \cdot \vec{X}_u + \Gamma^2_{22} \vec{X}_u \cdot \vec{X}_v = \Gamma^1_{22} \mathcal{E} + \Gamma^2_{22} \mathcal{F} \quad (15)$$

$$\tfrac{1}{2} \mathcal{G}_v = \vec{X}_v \cdot \vec{X}_{vv} = \Gamma^1_{22} \vec{X}_v \cdot \vec{X}_u + \Gamma^2_{22} \vec{X}_v \cdot \vec{X}_v = \Gamma^1_{22} \mathcal{F} + \Gamma^2_{22} \mathcal{G} . \quad (16)$$

Then solving (11) and (12) for Γ^1_{11} and Γ^2_{11} yields

$$\Gamma^1_{11} = \frac{\mathcal{G} \mathcal{E}_u - 2\mathcal{F} \mathcal{F}_u + \mathcal{F} \mathcal{E}_v}{2(\mathcal{E}\mathcal{G} - \mathcal{F}^2)}$$
(17)
$$\Gamma^2_{11} = \frac{2\mathcal{E} \mathcal{F}_u - \mathcal{E}\mathcal{E}_v + \mathcal{F} \mathcal{E}_u}{2(\mathcal{E}\mathcal{G} - \mathcal{F}^2)} .$$

Solving (13) and (14) for Γ^1_{12} and Γ^2_{12} gives

$$\Gamma^1_{12} = \frac{\mathcal{G} \mathcal{E}_v - \mathcal{F} \mathcal{G}_u}{2(\mathcal{E}\mathcal{G} - \mathcal{F}^2)} \qquad \Gamma^2_{12} = \frac{\mathcal{E} \mathcal{G}_u - \mathcal{F} \mathcal{E}_v}{2(\mathcal{E}\mathcal{G} - \mathcal{F}^2)} . \quad (18)$$

Solving (15) and (16) for Γ^1_{22} and Γ^2_{22}, we obtain

$$\Gamma^1_{22} = \frac{2\mathcal{G} \mathcal{F}_v - \mathcal{G}\mathcal{G}_u - \mathcal{F} \mathcal{G}_v}{2(\mathcal{E}\mathcal{G} - \mathcal{F}^2)}$$
(19)
$$\Gamma^2_{22} = \frac{\mathcal{E} \mathcal{G}_v - 2\mathcal{F}\mathcal{F}_v + \mathcal{F} \mathcal{G}_u}{2(\mathcal{E}\mathcal{G} - \mathcal{F}^2)} .$$

Thus from (1),(8) and since $\gamma_1 = \gamma_2 = 0$ we have

$$\vec{X}_{uu} = \Gamma^1_{11} \vec{X}_u + \Gamma^2_{11} \vec{X}_v + \ell N$$

$$\vec{X}_{uv} = \Gamma^1_{12} \vec{X}_u + \Gamma^2_{12} \vec{X}_v + m N$$

$$\vec{X}_{vv} = \Gamma^1_{22} \vec{X}_u + \Gamma^2_{22} \vec{X}_v + n N$$

$$\vec{N}_u = \beta_1^1 \vec{X}_u + \beta_1^2 \vec{X}_v$$

$$\vec{N}_v = \beta_2^1 \vec{X}_u + \beta_2^2 \vec{X}_v$$

where the coefficients β_i^j and Γ_{ij}^k are given by equation (6)-(7) and (17)-(19). Thus the first three of the above equations are called Gauss equations and the last two are called the Weingarten equations.

• **PROBLEM 18-37**

Determine the geodesics on the right circular cone
$$\vec{X} = (u \sin \alpha \cos \theta)\vec{e}_1 + (u \sin \alpha \sin \theta)\vec{e}_2 + (u \cos \alpha)\vec{e}_3$$
α = constant, $0 < \alpha < \pi/2$, $u > 0$.

Solution: The geodesic curvature k_g of a curve $\vec{C}$ at P is the vector projection of the curvature vector $\vec{k}$ of $\vec{C}$ at P onto the tangent plane at P. Then a curve $\vec{C}$ along which $k_g = 0$ is called a geodesic. By theorem all "straight lines" on a surface are geodesics. However, a curve not a straight line is a geodesic if and only if the osculating plane of the curve is perpendicular to the tangent plane to the surface at each point. In addition, by theorem, a natural representation of a curve $\vec{X} = \vec{X}(s) = \vec{X}(u(s), v(s))$ of class C^2 on a patch $\vec{X} = \vec{X}(u,v)$ of class C^2 is a geodesic if and only if $u(s)$ and $v(s)$ satisfy

$$\frac{d^2 u}{ds^2} + \Gamma_{11}^1 \left(\frac{du}{ds}\right)^2 + 2\Gamma_{12}^1 \frac{du}{ds}\frac{dv}{ds} + \Gamma_{22}^1 \left(\frac{dv}{ds}\right)^2 = 0$$

$$\frac{d^2 v}{ds^2} + \Gamma_{11}^2 \left(\frac{du}{ds}\right)^2 + 2\Gamma_{12}^2 \frac{du}{ds}\frac{dv}{ds} + \Gamma_{22}^2 \left(\frac{dv}{ds}\right)^2 = 0$$

(1)

where the Christoffel symbols Γ_{ij}^k are given by the equations

$$\Gamma_{11}^1 = \frac{\mathscr{E} \mathscr{E}_u - 2\mathscr{F}\mathscr{F}_u + \mathscr{F}\mathscr{E}_v}{2(\mathscr{E}\mathscr{G} - \mathscr{F}^2)} \qquad \Gamma_{12}^1 = \frac{\mathscr{G}\mathscr{E}_v - \mathscr{F}\mathscr{G}_u}{2(\mathscr{E}\mathscr{G} - \mathscr{F}^2)}$$

$$\Gamma_{11}^2 = \frac{2\mathscr{E}\mathscr{F}_u - \mathscr{E}\mathscr{E}_v + \mathscr{F}\mathscr{E}_u}{2(\mathscr{E}\mathscr{G} - \mathscr{F}^2)} \qquad \Gamma_{12}^2 = \frac{\mathscr{E}\mathscr{G}_u - \mathscr{F}\mathscr{E}_v}{2(\mathscr{E}\mathscr{G} - \mathscr{F}^2)}$$

(2)

$$\Gamma_{22}^1 = \frac{2\mathscr{G}\mathscr{F}_v - \mathscr{G}\mathscr{G}_u - \mathscr{F}\mathscr{G}_v}{2(\mathscr{E}\mathscr{G} - \mathscr{F}^2)} \qquad \Gamma_{22}^2 = \frac{\mathscr{E}\mathscr{G}_v - 2\mathscr{F}\mathscr{F}_v + \mathscr{F}\mathscr{G}_u}{2(\mathscr{E}\mathscr{G} - \mathscr{F}^2)}$$

Therefore, this leads us to regard as a candidate for the geodesic through an arbitrary point $\vec{X}(u_0, v_0)$ in an arbitrary direction

$(du/ds)_0$: $(dv/ds)_0$ the curve $\vec{X}(u(s),v(s))$, where $u(s)$, $v(s)$ are the solutions to (1) satisfying the initial conditions

$$u(0) = u_0, \quad v(0) = v_0, \quad du/ds(0) = (du/ds)_0, \quad dv/ds(0) = (dv/ds)_0.$$

For this problem

$$\vec{X}_u = (\sin\alpha \cos\theta)\vec{e}_1 + (\sin\alpha \sin\theta)\vec{e}_2 + (\cos\alpha)\vec{e}_3$$

$$\vec{X}_\theta = (-u\sin\alpha \sin\theta)\vec{e}_1 + (u\sin\alpha \cos\theta)\vec{e}_2 + 0\,\vec{e}_3.$$

Then

$$\mathscr{E} = \vec{X}_u \cdot \vec{X}_u = \sin^2\alpha \cos^2\theta + \sin^2\alpha \sin^2\theta + \cos^2\alpha = 1$$

$$\mathscr{F} = \vec{X}_u \cdot \vec{X}_\theta = -u\sin^2\alpha \cos\theta \sin\theta + u\sin^2\alpha \sin\theta \cos\theta = 0$$

$$\mathscr{G} = \vec{X}_\theta \cdot \vec{X}_\theta = u^2\sin^2\alpha \sin^2\theta + u^2\sin^2\alpha \cos^2\theta = u^2\sin^2\alpha$$

and by (2)

$$\Gamma^1_{11} = 0; \quad \Gamma^1_{12} = 0; \quad \Gamma^2_{11} = 0; \quad \Gamma^2_{22} = 0$$

$$\Gamma^1_{22} = -\frac{2u^2\sin^2\alpha \, u\sin^2\alpha}{2u^2\sin^2\alpha} = -u\sin^2\alpha$$

$$\Gamma^2_{12} = \frac{2u\sin^2\alpha}{2u^2\sin^2\alpha} = 1/u.$$

Hence the second of equations (1) is

$$\frac{d^2\theta}{ds^2} + \frac{2}{u}\frac{du}{ds}\frac{d\theta}{ds} = 0$$

or

$$\frac{d^2\theta}{ds^2} = -\frac{2}{u}\frac{du}{ds}\frac{d\theta}{ds}. \qquad (3)$$

Let $b = d\theta/ds$, so that (3) becomes

$$\frac{1}{\varphi}\frac{d\varphi}{ds} = -\frac{2}{u}\frac{du}{ds}$$

which is a differential equation that yields as a solution
$\log \varphi = -2\log u + K$ or $\varphi = d\theta/ds = C/u^2\sin^2\alpha$ where $C = e^K \sin^2\alpha$.
Since s is arc length

$$1 = \left\|\frac{d\vec{X}}{ds}\right\|^2 = \left\|\vec{X}_u \frac{du}{ds} + \vec{X}_\theta \frac{d\theta}{ds}\right\|^2$$

$$1 = \mathscr{E}\left(\frac{du}{ds}\right)^2 + 2\mathscr{F}\frac{du}{ds}\frac{d\theta}{ds} + \mathscr{G}\left(\frac{d\theta}{ds}\right)^2. \qquad (4)$$

Remembering that for this problem $\mathscr{E} = 1$, $\mathscr{F} = 0$, $\mathscr{G} = u\sin^2\alpha$,

(4) becomes

$$1 = \left(\frac{du}{ds}\right)^2 + u^2\sin^2\alpha\left(\frac{d\theta}{ds}\right)^2.$$

Substituting $d\theta/ds = C/u^2\sin^2\alpha$ yields

$$1 = \left(\frac{du}{ds}\right)^2 + u^2\sin^2\alpha \frac{C^2}{(u^2\sin^2\alpha)^2}$$

or

967

$$\frac{du}{ds} = \frac{\sqrt{u^2 \sin^2 \alpha - C^2}}{u \sin \alpha}$$

Then by the chain rule

$$\frac{du}{d\theta} = \frac{du}{ds} \cdot \frac{ds}{d\theta} = \frac{1}{C} u \sin \alpha \sqrt{u^2 \sin^2 \alpha - C^2}$$

or

$$\int \frac{du}{u \sin \alpha \sqrt{u^2 \sin^2 \alpha - C^2}} = \int \frac{d\theta}{C} .$$

Then by the formula

$$\int \frac{dx}{x(x^2 - a^2)^{\frac{1}{2}}} = \frac{1}{a} \text{ arc sec}(x/a)$$

we get after the integration is performed

$$u = A \sec [(\sin \alpha)\theta + B]$$

where A = constant, B = constant.

● **PROBLEM 18-38**

Prove the following theorem: Let γ be a geodesic (curve of minimal length) on the surface Σ. Then, at any point $\vec{p}$ on γ, the normal to γ is orthogonal to the tangent plane of Σ.

<u>Solution</u>: Let $\vec{p} \in \gamma$ and let u, v be coordinates for Σ near $\vec{p}$ so that $\vec{p} = (u(0), v(0))$. These coordinates may be chosen so that γ is the curve $v = 0$ and so that the coordinates are everywhere orthogonal. In addition let b be small enough so that the interval from $(-b, 0)$ to $(b, 0)$ in the uv-plane lies on the domain D of the coordinates. If $\Gamma : v = f(u)$ defines a curve lying in D and joining $(-b, 0)$ to $(b, 0)$, then $\vec{X} = \vec{X}(u, f(u))$, $-b \leq u \leq b$ yields another curve on Σ, joining two points of γ (see Figure 1).

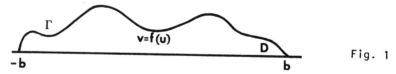

Fig. 1

Since γ is a geodesic, the length of Γ is no less than that of γ. Now to examine the local behavior of γ it is necessary to consider a whole family of curves including γ rather than just one other. To do this let Γ_t be the curve parametrized by $\Gamma_t : \vec{X} = \vec{X}(u, tf(u))$ $-b \leq u \leq b$ for $-1 \leq t \leq 1$. Note that γ is Γ_0 and Γ is Γ_1. Now, let $F(t)$ be the length of Γ_t. Then $F(t)$ has a minimum at $t = 0$, so $F'(0) = 0$ (if F is differentiable). Computing,

$$F(t) = \int_{-b}^{b} \|\vec{X}_u + \vec{X}_v tf'(u)\| du$$

is a differentiable function of t, and

$$F'(t) = \int_{-b}^{b} \frac{\partial}{\partial t}\|\vec{X}_u + \vec{X}_v tf'(u)\| du .$$

At $t = 0$, the integrand is (see remark)

$$\frac{\partial}{\partial t}\langle \vec{X}_u + \vec{X}_v tf'(u), \vec{X}_u + \vec{X}_v tf'(u)\rangle^{\frac{1}{2}}\Big|_{t=0}$$

$$= \tfrac{1}{2} \frac{1}{\|\vec{X}_u\|} 2 \langle \vec{X}_{uv} f(u) + \vec{X}_v f'(u), \vec{X}_u\rangle$$

$$= -\frac{\langle \vec{X}_{uu}, \vec{X}_v\rangle}{\|\vec{X}_u\|} f(u). \tag{1}$$

This last equation because of the assumption that the coordinates are orthogonal $\langle \vec{X}_v, \vec{X}_u\rangle = 0$. First, the second term drops out, second, (1) derives from

$$0 = \frac{\partial}{\partial u}\langle \vec{X}_v, \vec{X}_u\rangle = \langle \vec{X}_{uv}, \vec{X}_u\rangle + \langle \vec{X}_v, \vec{X}_{uu}\rangle .$$

Hence, from $F'(0) = 0$,

$$\int_{-b}^{b} \frac{\langle \vec{X}_v, \vec{X}_{uu}\rangle}{\|\vec{X}_u\|} f(u) du = 0 .$$

This equation must hold for all differentiable functions f such that $f(-b) = f(b) = 0$. This leads to the conclusion $\langle \vec{X}_v, \vec{X}_{uu}\rangle = 0$ along γ. The normal $\vec{N}$ to γ is in the plane spanned by $\vec{X}_u$ and $\vec{X}_{uu}$. Since these are both orthogonal to $\vec{X}_v$, $\vec{N} \perp \vec{X}_v$. Furthermore, $\vec{N}$ is orthogonal to the tangent line of γ which is spanned by $\vec{X}_u$. Thus $\vec{N}$ is orthogonal to both $\vec{X}_u$ and $\vec{X}_v$, so it is orthogonal to the tangent plane of Σ.

Remark: Let $g(t)$ be a function defined on $(0, 1)$ and taking values in R^n. Assume g is differentiable. Then the derivative of the function $\langle \vec{g}(t), \vec{g}(t)\rangle^{\frac{1}{2}}$ at a point t_0 is given by $\langle g(t_0), g(t_0)\rangle^{-\frac{1}{2}} \langle \vec{g}(t_0), \vec{g}'(t_0)\rangle$. In the above example

$$\vec{g}(t) = \vec{X}_u(u, tf(u)) + t\vec{X}_v(u, tf(u))f'(u) .$$

Therefore $\vec{g}'(t) = f(u)\vec{X}_{uv}(u, tf(u)) + \vec{X}_v(u, tf(u))f'(u)$

$$+ tf'(u)f(u)\vec{X}_{vv}(u, tf(u))$$

by the chain rule. Hence $\vec{g}'(0) = f(u)\vec{X}_{uv}(u,0) + \vec{X}_v(u,0)f'(u)$

and

$$\frac{\partial}{\partial t}\langle \vec{g}(t), \vec{g}(t)\rangle^{\frac{1}{2}}\Big|_{t=0} = \frac{1}{\|\vec{X}_u(u,0)\|} \langle f(u)\vec{X}_{uv}(u,0) + \vec{X}_v(u,0)f'(u), \vec{X}_u(u,0)\rangle$$

CHAPTER 19

MISCELLANEOUS PROBLEMS AND APPLICATIONS

MISCELLANEOUS APPLICATIONS

• **PROBLEM** 19-1

Let a, b, c be the lengths of the sides of a triangle and let θ be the angle opposite the side of length c. Find the differential dc and approximate c when $a = 6.20$, $b = 5.90$, and $\theta = 58°$.

Solution: By the law of cosines

$$c^2 = a^2 + b^2 - 2ab\cos\theta. \qquad (1)$$

Recall that the rules analogous to those for derivatives hold for differentials (e.g. $d(fg) = gdf + fdg$ and $df = f'dx$ in the one variable case). Hence,

$2cdc = 2ada + 2bdb - 2b\cos\theta da - 2a\cos\theta db + 2ab\sin\theta d\theta.$

Therefore,

$$dc = \frac{1}{c}(ada + bdb - b\cos\theta da - a\cos\theta db + ab\sin\theta d\theta). \qquad (2)$$

Now let $a = b = c = 6.00$ and $\theta = 60° = \frac{\pi}{3}$. Then $da \approx \Delta a = .20$, $db \approx \Delta b = -.10$, and $d\theta \approx \Delta\theta = -2° = -\frac{\pi}{90}$. Substituting into (2) gives

$$dc = \frac{1}{6}\left[6(.20) + 6(-.10) - 6(\tfrac{1}{2})(.20) - 6(\tfrac{1}{2})(-.10) + 6^2(\tfrac{\sqrt{3}}{2})(-\tfrac{\pi}{90})\right] = .20 - .10 - .10 + .05 - \frac{1}{30}\sqrt{3}\,\pi$$

$\approx -.13.$

Therefore, $\Delta c \approx -.13$ and hence $c = 5.87$ which is exact to two decimal places.

• **PROBLEM** 19-2

Suppose 96 square feet of canvas are available to cover the top, back, and sides of a box-like shelter. What dimensions would maximize the shelter's interior [i.e. volume]?

Solution: Let x, y, and z be the dimensions in feet. Then its interior volume is

$$V(x,y,z) = xyz \tag{1}$$

and the area to be covered by the canvas is

$$S(x,y,z) = 2(xy + xz + yz) = 96 \tag{2}$$

(where z is the length). According to (2), z can be considered to be a function of x and y. Therefore, differentiate both (1) and (2) with respect to x and y to get

$$\frac{\partial V}{\partial x} = yz + xy \frac{\partial z}{\partial x}, \quad \frac{\partial S}{\partial x} = 2\left(y + z + x \frac{\partial z}{\partial x} + y \frac{\partial z}{\partial x}\right) \tag{3}$$

and

$$\frac{\partial V}{\partial y} = xz + xy \frac{\partial z}{\partial y}, \quad \frac{\partial S}{\partial y} = 2\left(x + x \frac{\partial z}{\partial y} + z + y \frac{\partial z}{\partial y}\right). \tag{4}$$

Thus, setting the equations in (3) and (4) to zero, (3) yields

$$\frac{\partial z}{\partial x} = \frac{-yz}{xy} = -\frac{z}{x}, \quad y + z + x \frac{\partial z}{\partial x} + y \frac{\partial z}{\partial x} = 0.$$

Therefore,

$$y + z + x\left(-\frac{z}{x}\right) + y\left(-\frac{z}{x}\right) = 0$$

$$xy + xz - xz - yz = 0$$

$$xy = yz$$

$$x = z.$$

Similarly, (4) yields

$$y = z.$$

Thus x = y = z. Now (2) gives

$$96 = 2(xy + xz + yz) = 6x^2.$$

Hence, x = y = z = 4. That is the shelter should be 4 feet high by 4 feet wide by 4 feet long for a volume of 64 cubic feet.

● **PROBLEM 19-3**

Suppose a student has $90 with which to buy lecture notes at $3 each and packs of beer at $5 each. Let the function

$$f(x,y) = xy \tag{1}$$

express how much satisfaction is derived from buying x lecture notes and y packs of beer. What x and y will maximize the student's pleasure?

Solution: The equation (1), which is called the utility function by economists, is to be maximized under the constraint

$$g(x,y) = 3x + 5y - 90 = 0 \qquad (2)$$

which is derived from the given information (i.e. the student buys x books and y packs for $90). By the method of Lagrange multipliers the maximum occurs when

$$\nabla f(x,y) = \lambda \nabla g(x,y)$$

or

$$(y,x) = \lambda(3,5).$$

Hence $y = 3\lambda$ and $x = 5\lambda$. Substituting these into (2) gives

$$3(5\lambda) + 5(3\lambda) = 30\lambda = 90.$$

So, $\lambda = 3$ and hence $x = 15$, $y = 9$ is a possible maximum. Indeed, this is a maximum by inspection. (E.g. suppose $y = 9.2$ and $x = \frac{1}{3}(90-5(9.2)) = \frac{44}{3} = 14\frac{2}{3}$; $f(14\frac{2}{3}, 9.2) = (14\frac{2}{3})(9.2) = \frac{2024}{15} = 134\frac{14}{15} < 135 = (15)(9) = f(15,9)).$
This answer could also have been found in a more direct manner: Substitute x from (2) into (1) to get

$$f(x,y) = \frac{1}{3}(90-5y)y$$

$$= 30y - \frac{5}{3}y^2$$

$$= F(y),$$

a function of one variable. Proceeding as usual

$$F'(y) = 30 - \frac{10}{3}y = 0$$

and hence $y = 9$ is a critical point. Since

$$F''(9) = -\frac{10}{3},$$

$y = 9$ is a maximum of F. Therefore, $y = 9$ and $x = 15$ must maximize $f(x,y)$. Of course, this agrees with the first answer.

• PROBLEM 19-4

Find the dimensions of the box of largest volume which can be fitted inside the ellipsoid

$$\frac{x^2}{a^2} + \frac{y^2}{b^2} + \frac{z^2}{c^2} = 1,$$

assuming that each edge of the box is parallel to a coordinate axis.

Solution: According to the symmetric nature of the problem, each of the 8 vertices of the box will be on the ellipsoid. Let the vertex in the first octant (x>0, y>0, z>0) be (x,y,z) so that the dimensions of the box must then be 2x, 2y, and 2z. Hence, the volume to be maximized is

$$V(x,y,z) = 8xyz$$

where (x,y,z) is on the ellipsoid. Following Lagrange's method

$$8(yz, xz, xy) = 2\lambda \left(\frac{x}{a^2}, \frac{y}{b^2}, \frac{z}{c^2}\right) \quad (1)$$

or

$$4yz = \lambda \frac{x}{a^2}$$

$$4xz = \lambda \frac{y}{b^2}$$

$$4xy = \lambda \frac{z}{c^2}$$

Thus

$$4xyz = \lambda \frac{x^2}{a^2}$$

$$= \lambda \frac{y^2}{b^2}$$

$$= \lambda \frac{z^2}{c^2}$$

and, hence, $12xyz - \lambda\left(\frac{x^2}{a^2} + \frac{y^2}{b^2} + \frac{z^2}{c^2}\right) = 12xyz - \lambda = 0$,

or $\lambda = 12xyz$. Substituting back into (1) gives

$$(yz, xz, xy) = 3\left(yz \frac{x^2}{a^2}, xz \frac{y^2}{b^2}, xy \frac{z^2}{c^2}\right)$$

or
$$yz(a^2 - 3x^2) = 0$$
$$xz(b^2 - 3y^2) = 0$$
$$xy(c^2 - 3z^2) = 0.$$

Since x,y,z > 0, the only solution is

$$(x,y,z) = \left(\frac{|a|}{\sqrt{3}}, \frac{|b|}{\sqrt{3}}, \frac{|c|}{\sqrt{3}}\right).$$

By inspection, this is a maximum. Hence the box of maximal volume has dimensions

$$\frac{2|a|}{\sqrt{3}} \times \frac{2|b|}{\sqrt{3}} \times \frac{2|c|}{\sqrt{3}}.$$

● **PROBLEM 19-5**

Find the rectangular parallelepiped of surface area a^2 and maximum volume.

Solution: Let the dimensions of the rectangular parallelepiped be x, y, and z. Then its volume and surface area are

$$V(x,y,z) = xyz$$

and

$$S(x,y,z) = 2(xy + xz + yz) = a^2$$

respectively. Applying Lagrange's method yields

$$\nabla V(x,y,z) = \lambda \nabla S(x,y,z)$$

or

$$(yz, xz, xy) = 2\lambda(y+z, x+z, x+y).$$

Since $x, y, z > 0$,

$$\frac{x}{y} = \frac{x+z}{y+z} \quad \text{and} \quad \frac{y}{z} = \frac{x+y}{x+z}.$$

From the first relation one gets $xz = yz$ and from the second $xy = xz$. Thus $x = y = z$ and then the relation $S(x,y,z) = a^2$ yields

$$S(x,y,z) = 2(x^2 + x^2 + x^2) = 6x^2 = a^2$$

or

$$x = \frac{|a|}{\sqrt{6}}.$$

Thus the cube with an edge of length $\frac{|a|}{\sqrt{6}}$ is of maximum volume.

● **PROBLEM 19-6**

Which 3-dimensional rectangular box of a given volume V has the least surface area?

Solution: Let the dimensions of the box be x, y, and z. Hence x, y, and z are positive and $xyz = V$. The surface area is given by

$$S = 2(xy + xz + yz)$$

or, since $xyz = V$,

$$S(x,y) = 2\left(xy + \frac{V}{y} + \frac{V}{x}\right). \tag{1}$$

The problem is, therefore, to minimize (1) on $Q_1 = \{(x,y) \mid x>0, y>0\}$. When $x = y = V^{1/3}$, $S(x,y) = 6V^{2/3}$. Hence, the minimum value of S is no more than $6V^{2/3}$. If $x,y \leq \frac{2}{3}V^{1/3}$, then $S > 2(\frac{V}{x}+\frac{V}{y}) \geq 6V^{2/3}$ and if $x,y,z > \sqrt{3}V^{1/3}$, then $S > 2xy \geq 6V^{2/3}$. Therefore, the minimum of S occurs in the interior of the compact subset $D = \{(x,y) \mid \frac{2}{3}V^{1/3} \leq x,y \leq \sqrt{3}V^{1/3}\}$ of Q_1. Note that $S > 6V^{2/3}$ outside of D and on the boundary of D. The function S must therefore attain its minimum on Q_1 in the interior of D and at this minimum $\frac{\partial S}{\partial x} = 0 = \frac{\partial S}{\partial y}$.

Thus
$$\frac{\partial S}{\partial x}(x,y) = 2y - 2\frac{V}{x^2} = 0 \tag{2}$$

and
$$\frac{\partial S}{\partial y}(x,y) = 2x - 2\frac{V}{y^2} = 0,$$

or
$$x^2y = xy^2 \tag{3}$$

which implies $x = y$. Substituting (3) back into (2) gives

$$2x = 2\frac{V}{x^2}$$

or $x^3 = V$.

Hence, $x = y = V^{1/3}$ minimizes S on Q_1. Furthermore, $z = \frac{V}{xy} = V^{1/3}$ and, therefore, the cube with each edge of length $V^{1/3}$ minimizes the surface area.

• **PROBLEM 19-7**

Minimize the distance from a point $p \in R^n$ to the hyperplane $\langle x,a \rangle + b = 0$ where $a \in R^n$ and $b \in R$. (Assume $a \neq 0$.)

Solution: Let $p = (p_1,\ldots,p_n)$, $a = (a_1,\ldots,a_n)$, and $x = (x_1,\ldots,x_n)$. Then the distance between a point x of the hyperplane and p is defined by

$$d(x) = \|p-x\| = \langle p-x, p-x \rangle^{1/2} = \left(\sum_{i=1}^{n}(p_i - x_i)^2\right)^{1/2}.$$

To minimize d on the hyperplane, the method of Lagrange multipliers will be used. Since $d(x) \geq 0$ for all $x \in R^n$, the method may be applied to $f(x) = d^2(x) = \langle p-x, p-x \rangle$ without any loss of generality. The constraint equation is

$$g(x) = \langle x, a \rangle + b = \left(\sum_{i=1}^{n} x_i a_i \right) + b = 0.$$

According to the method of Lagrange multipliers, set

$$\nabla f(c) = \lambda \nabla g(c)$$

for the critical point $c \in R^n$. Thus

$$-2(p_i - c_i) = \lambda a_i \quad (i=1,\ldots,n)$$

or $\quad p_i - c_i = k a_i \quad (i=1,\ldots,n) \quad (1)$

for $k = -\frac{\lambda}{2}$. Now

$$f(c) = \sum_{i=1}^{n} (p_i - c_i)^2 = \sum_{i=1}^{n} (k a_i)^2 = k^2 \|a\|^2. \quad (2)$$

Furthermore, from (1) $c_i = p_i - k a_i$ $(i=1,\ldots,n)$ and hence

$$g(c) = \left(\sum_{i=1}^{n} (p_i - k a_i) a_i \right) + b$$

$$= \left(\sum_{i=1}^{n} p_i a_i - \sum_{i=1}^{n} k a_i^2 \right) + b$$

$$= g(p) - k \|a\|^2 = 0.$$

Thus $\quad k = \dfrac{g(p)}{\|a\|^2} \quad (3)$

and hence

$$c_i = p_i - g(p) \frac{a_i}{\|a\|^2} \quad (i=1,\ldots,n)$$

or

$$c = p - g(p) \frac{a}{\|a\|^2}. \quad (4)$$

By the nature of this problem it is clear that (4) gives the minimum of f under the constraint g. From (2) and (3), the minimum value of f is

$$d^2(c) = \left(\frac{g(p)}{\|a\|^2} \right)^2 \|a\|^2 = \frac{g^2(p)}{\|a\|^2}.$$

Therefore, the minimum distance from p to the hyperplane $\langle x, a \rangle + b = 0$ is

$$d(c) = \frac{|g(p)|}{\|a\|} = \frac{|\langle p, a \rangle + b|}{\|a\|}.$$

• **PROBLEM 19-8**

Show how to find the semiaxes of the ellipse in which the plane

$$P(x,y,z) = pz + qy + rz = 0 \qquad (pqr \neq 0) \qquad (1)$$

cuts the ellipsoid

$$E(x,y,z) = \frac{x^2}{a^2} + \frac{y^2}{b^2} + \frac{z^2}{c^2} = 1. \qquad (0<a<b<c) \qquad (2)$$

Solution: Since the plane (1) passes through the origin (0,0,0), the endpoints of the semiaxes of the ellipse are the points which extremalize

$$F(x,y,z) = x^2 + y^2 + z^2$$

subject to (1) and (2). By the method of Lagrange

$$\nabla F(x,y,z) = \lambda_1 \nabla P(x,y,z) + \lambda_2 \nabla E(x,y,z)$$

or $\qquad 2(x,y,z) = \lambda_1(p,q,r) + 2\lambda_2\left(\dfrac{x}{a^2}, \dfrac{y}{b^2}, \dfrac{z}{c^2}\right). \qquad (3)$

The system (3) along with equations (1) and (2) will yield one endpoint of each semiaxis, say (x_1, y_1, z_1) and (x_2, y_2, z_2). Then the other endpoints are $(-x_1, -y_1, -z_1)$ and $(-x_2, -y_2, -z_2)$ respectively. Note that this system yields

$$2(x^2, y^2, z^2) = \lambda_1(px, qy, rz) + 2\lambda_2\left(\frac{x^2}{a^2}, \frac{y^2}{b^2}, \frac{z^2}{c^2}\right)$$

and hence

$$(x^2+y^2+z^2) = \tfrac{1}{2}\lambda_1(pz+qy+rz) + \lambda_2\left(\frac{x^2}{a^2} + \frac{y^2}{b^2} + \frac{z^2}{c^2}\right)$$

$$= \lambda_2$$

by (1) and (2). If (x,y,z) were a solution to the system, then $\lambda_2 = \|(x,y,z)\|^2 = (\tfrac{1}{2}d)^2 = \tfrac{1}{4}d^2$ where d is the length of the corresponding semiaxis. Now (3) gives, for $\lambda_3 = \tfrac{1}{2}\lambda_1$

$$\left(x\left(1 - \frac{\lambda_2}{a^2}\right), y\left(1 - \frac{\lambda_2}{b^2}\right), z\left(1 - \frac{\lambda_2}{c^2}\right)\right)$$

$$= (\lambda_3 p, \lambda_3 q, \lambda_3 r) \tag{4}$$

or

$$(x,y,z) = \left(\frac{a^2 p \lambda_3}{a^2 - \lambda_2}, \frac{b^2 q \lambda_3}{b^2 - \lambda_2}, \frac{c^2 r \lambda_3}{c^2 - \lambda_2} \right).$$

So, from (1) assuming $\lambda_3 \neq 0$, (see remark)

$$\frac{a^2 p^2}{a^2 - \lambda_2} + \frac{b^2 q^2}{b^2 - \lambda_2} + \frac{c^2 r^2}{c^2 - \lambda_2} = 0$$

or

$$0 = a^2 p^2 (b^2 - \lambda_2)(c^2 - \lambda_2) + b^2 q^2 (a^2 - \lambda_2)(c^2 - \lambda_2)$$

$$+ c^2 r^2 (a^2 - \lambda_2)(b^2 - \lambda_2).$$

Hence the two solutions of this quadratic equation are the squares of the lengths of the semiaxes.

Remark: The assumption $pqr \neq 0$ shows that λ_3 is not 0. In order to see this observe that λ_2 cannot equal a^2 or b^2 or c^2. For example if $\lambda_2 = a^2$, equation (3) would imply $2x = \lambda_1 p + 2x$ or since $p \neq 0$ λ_1 must be 0. But then (since $a<b<c$, so $\lambda_2 \neq b^2$, $\lambda_2 \neq c^2$) (3) again implies $y=0$, $z=0$. Now (1) would imply $x=0$ which is not compatible with (2). This proves that λ_2 cannot equal a^2 or b^2 or c^2. If $\lambda_3 = 0$, which is the same as $\lambda_1 = 0$, equation (4) would once again imply $x = y = z = 0$ which does not satisfy (2).

● **PROBLEM 19-9**

Construct a Möbius strip and show that its normal is not well-defined (i.e. that it is not orientable).

Fig. 1

Solution: A Möbius strip is a surface where the ends are joined as indicated in Fig. 1 (i.e. with arrows pointed in same direction). To construct a Möbius strip more rigorously start with a circle $x^2 + y^2 = a^2$, $z = 0$, $a > 2$. Take a line segment of length $b < 2a$ and center it at $P = (a \cos\theta, a \sin\theta, 0)$. Now revolve the segment about 0 on the given circle while rotating the segment in the plane determined by O, P, and $(0,0,a)$ according to the following two criteria:

i) at $(a,0,0)$, the segment is vertical (parallel to $(0,0,1)$);
ii) at $P = (a \cos\theta, a \sin\theta, 0)$, its angle with the vertical (PQ, where $Q = (a \cos\theta, a \sin\theta, a)$)

is $\frac{1}{2}\theta$. (See Figure 2.)

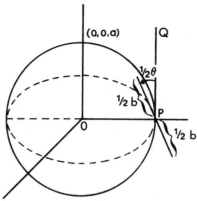

Fig. 2

From the definition and the use of spherical polar coordinates, the Möbius strip is seen to be

$$\{X = a(\cos\theta, \sin\theta, 0) + t(\sin\tfrac{1}{2}\theta\cos\theta, \sin\tfrac{1}{2}\theta\sin\theta, \cos\tfrac{1}{2}\theta) \mid$$

$$0 \leq \theta \leq 2\pi, -\tfrac{1}{2}b \leq t \leq \tfrac{1}{2}b\}.$$

The map $(\theta, t) \to X$ is one-to-one except at $\theta = 0, 2\pi$ since the equations

$$(x,y) = (a + t \sin\tfrac{1}{2}\theta)(\cos\theta, \sin\theta)$$

determine $t \sin\tfrac{1}{2}\theta$ uniquely in terms of x and y. That is, $a + t\sin\tfrac{1}{2}\theta \geq a - \tfrac{1}{2}b > 0$ and hence $t\sin\tfrac{1}{2}\theta = \sqrt{x^2+y^2} - a$; therefore θ is uniquely determined by this and $z = t \cos\tfrac{1}{2}\theta$ for $0 < \theta < 2\pi$. For the normal

$$X_\theta \times X_t = \left[(a + t\sin\tfrac{1}{2}\theta)(-\sin\theta, \cos\theta, 0) + \tfrac{t}{2}(\cos\tfrac{1}{2}\theta\cos\theta,\right.$$

$$\left. \cos\tfrac{1}{2}\theta\sin\theta, -\sin\tfrac{1}{2}\theta)\right] \times (\sin\tfrac{1}{2}\theta\cos\theta, \sin\tfrac{1}{2}\theta\sin\theta,$$

$$\cos\tfrac{1}{2}\theta)$$

$$= (a + t\sin\tfrac{1}{2}\theta)(\cos\theta\cos\tfrac{1}{2}\theta, \sin\theta\cos\tfrac{1}{2}\theta, -\sin\tfrac{1}{2}\theta)$$

$$+ \tfrac{t}{2}(\sin\theta, -\cos\theta, 0).$$

This shows that there are no singularities, for if $X_\theta \times X_t = 0$ then, since $a + t\sin\tfrac{1}{2}\theta > 0$, $\sin\tfrac{1}{2}\theta = 0$. This would mean that $\theta = 0, 2\pi$ for which

$$X_\theta \times X_t = (\pm a, -\tfrac{t}{2}, 0)$$

respectively. Thus, although $\theta = 0, 2\pi$ both determine the segment $X = (a,0,0) + t(0,0,1) = (a,0,t)$ ($|t| \leq \frac{b}{2}$), the normal is not well-defined on this segment. Note that the Möbius strip is one-sided.

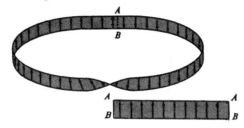

Fig. 3

● **PROBLEM 19-10**

Show that the solid angle is given by

$$\Omega = \iint_S \frac{\cos\phi}{r^2} \, dA = \iint_S \frac{\vec{n} \cdot \vec{OP}}{r^3} \, dA \qquad (1)$$

where r is the distance OP and $\vec{n}$ is the unit normal to S at P in such a direction that the angle ϕ between $\vec{OP}$ and $\vec{n}$ is acute.

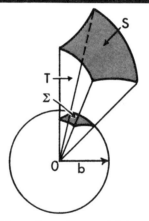

Fig. 1

Solution: First we define Ω, the solid angle. To do this we let S be a surface element and O a fixed point not on S. Assume that S is not intersected more than once by any ray from O, and that no such ray is tangent to S. As a point P varies over S, consider the point Q in which the ray OP (OP can be extended, if necessary) intersects the unit sphere with center at O. The points Q fill out a certain portion of the surface of the unit sphere. By definition, the area Ω of this portion is defined to be the solid angle subtended by S at O.

For the given problem, apply the divergence theorem. To do this first construct the sphere with radius b and center O (choose b small enough so that S lies entirely outside this sphere). Then draw all the rays joining O to S, and let Σ be the portion of the sphere cut by these rays. In addition let T be the solid region formed by the bundle of rays cut off between Σ and S (Fig. 1). For $\vec{F}$ use the vector function $\dfrac{\vec{OP}}{r^3}$ with components $\left(\dfrac{x}{r^3}, \dfrac{y}{r^3}, \dfrac{z}{r^3}\right)$. Now, apply the divergence theorem. Since $r^2 = x^2 + y^2 + z^2$, this yields

$$\frac{\partial}{\partial x}\left(\frac{x}{r^3}\right) = \frac{r^2 - 3x^2}{r^5}, \quad \frac{\partial}{\partial y}\left(\frac{y}{r^3}\right) = \frac{r^2 - 3y^2}{r^5}, \quad \text{and}$$

$$\frac{\partial}{\partial z}\left(\frac{z}{r^3}\right) = \frac{r^2 - 3z^2}{r^5}. \quad \text{Hence}$$

$$\operatorname{div} \vec{F} = \frac{3r^2 - 3(x^2+y^2+z^2)}{r^5} = \frac{3r^2 - 3r^2}{r^5} = 0.$$

Therefore $\iiint_T (\operatorname{div} \vec{F})\, dV = 0.$

Thus, by the divergence theorem, we have

$$\iiint_T \operatorname{div} \vec{F}\, dV = \iint_S \vec{F} \cdot \vec{n}\, dA = 0 \quad \text{where}$$

$\vec{n}$ refers to the outward normal. Now note that the surface of T consists of Σ, S, and the lateral surface formed by the rays joining O to the edge of S. Since $\vec{F}$ has the same direction as $\vec{OP}$, $\vec{F} \cdot \vec{n} = 0$ on this lateral portion and therefore the surface integral over this portion is zero. Hence,

$$\iint_\Sigma \vec{F} \cdot \vec{n}\, dA + \iint_S \vec{F} \cdot \vec{n}\, dA = 0. \tag{2}$$

The integral over s in (2) is the same as the surface integral in (1). Therefore, it remains for us to show that

$$\iint_\Sigma \vec{F} \cdot \vec{n}\, dA = -\Omega. \tag{3}$$

To do this, notice than on Σ the outward normal points away from T, which is toward O. But $\vec{F}$ points away from O, with magnitude $\dfrac{1}{r^2}$. However, since $r = a$ on Σ, $\vec{F} \cdot \vec{n} = -\dfrac{1}{a^2}$ and the integral in (3) becomes

$$-\frac{1}{a^2} \iint_\Sigma dA = -\frac{1}{a^2} (\text{area of } \Sigma).$$

If $a = 1$, the area of Σ is Ω, by the definition of the solid angle. But, in general,

$$\frac{1}{a^2} (\text{area of } \Sigma) = \Omega \qquad \text{since the area}$$

of Σ is proportional to a^2 where a varies. This finishes the proof of (3) and (1).

ELLIPTIC INTEGRALS

● **PROBLEM 19-11**

Express $\int_0^x \sqrt{1-4\sin^2 u}\, du$ in terms of incomplete elliptic integrals where $0 \le x \le \frac{\pi}{6}$.

<u>Solution</u>: The integral of the form

$$y = F(k,\theta) = \int_0^\theta \frac{dt}{\sqrt{1-k^2\sin^2 t}} \qquad 0<k<1, \qquad (1)$$

where θ is the amplitude of y (written θ=am y), and k is its modulus (written k=mod y), is the incomplete elliptic integral of the first kind. If $\theta = \frac{\pi}{2}$ the integral is termed the complete integral of the first kind and is written as $F(k)$.

The incomplete elliptic integral of the second kind is defined by

$$E(k,\theta) = \int_0^\theta \sqrt{1-k^2\sin^2 t}\, dt \qquad 0<k<1. \qquad (2)$$

Note if $\theta = \frac{\pi}{2}$ the integral is denoted by $E(k)$ or E.

Finally the incomplete elliptic integral of the third kind is defined by

$$\Pi(k,n,\theta) = \int_0^\theta \frac{dt}{(1+n\cdot\sin^2 t)\sqrt{1-k'\sin^2 t}} \qquad 0<k<1 \qquad (3)$$

where n is a constant assumed not equal to zero, since if $n = 0$, (3) becomes (1).

For this problem let $\sqrt{4\sin^2 u} = 2\sin u = \sin\theta$ in the given integral. Then $2\cos u\, du = \cos\theta\, d\theta$ or

$$du = \frac{\cos\theta\, d\theta}{2\cos u} = \frac{\cos\theta\, d\theta}{2\sqrt{1-\frac{1}{4}\sin^2 x}}.$$

Hence,

$$\int_0^x \sqrt{1-4\sin^2 u}\, du = \int_0^\theta \cos\theta \cdot \frac{\cos\theta}{2\sqrt{1-\frac{1}{4}\sin^2\theta}}\, d\theta$$

$$= \frac{1}{2}\int_0^\theta \frac{\cos^2\theta\, d\theta}{\sqrt{1-\frac{1}{4}\sin^2\theta}}$$

$$= \frac{1}{2}\int_0^\theta \frac{-3+4(1-\frac{1}{4}\sin^2\theta)}{\sqrt{1-\frac{1}{4}\sin^2\theta}}\, d\theta$$

$$= \frac{-3}{2}\int_0^\theta \frac{d\theta}{\sqrt{1-\frac{1}{4}\sin^2\theta}} + 2\int_0^\theta \sqrt{1-\frac{1}{4}\sin^2\theta}\, d\theta$$

$$= \frac{-3}{2} F(\tfrac{1}{2},\theta) + 2E(\tfrac{1}{2},\theta)$$

where $\theta = \sin^{-1}(2\sin x)$.

● **PROBLEM 19-12**

Evaluate a) $\int_0^2 \frac{dx}{\sqrt{(4-x^2)(9-x^2)}}$

b) $\int_1^\infty \frac{du}{\sqrt{(u^2-1)(u^2+3)}}$

in terms of elliptic integrals.

Solution: a) For this integral use the substitution $x = 2\sin\theta$ in the integrand. This yields

$$\int_0^{\pi/2} \frac{2\cos\theta\, d\theta}{\sqrt{4\cos^2\theta(9-4\sin^2\theta)}} = \int_0^{\pi/2} \frac{d\theta}{\sqrt{9-4\sin^2\theta}}$$

$$= \frac{1}{3}\int_0^{\pi/2} \frac{d\theta}{\sqrt{1-\frac{4}{9}\sin^2\theta}} = \tfrac{1}{3}F(\tfrac{2}{3},\tfrac{\pi}{2}).$$

where $F(k,\emptyset) = \int_0^{\emptyset} \frac{d\theta}{\sqrt{1-k^2\sin^2\theta}}$ $0<k<1$ is the

incomplete integral of the first kind.

b) $\int_1^{\infty} \frac{du}{\sqrt{(u^2-1)(u^2+3)}}$ For this integral let $u = \sec\theta$

in the integrand to yield

$$\int_0^{\pi/2} \frac{\sec\theta \tan\theta \, d\theta}{\sqrt{(\sec^2\theta-1)(\sec^2\theta+3)}} = \int_0^{\pi/2} \frac{\sec\theta \tan\theta \, d\theta}{\sqrt{\tan^2\theta\left(\frac{1+3\cos^2\theta}{\cos^2\theta}\right)}}$$

$$= \int_0^{\pi/2} \frac{d\theta}{\sqrt{1+3\cos^2\theta}} = \int_0^{\pi/2} \frac{d\theta}{\sqrt{1+3(1-\sin^2\theta)}}$$

$$= \int_0^{\pi/2} \frac{d\theta}{\sqrt{4-3\sin^2\theta}} = \frac{1}{2}\int_0^{\pi/2} \frac{d\theta}{\sqrt{1-\frac{3}{4}\sin^2\theta}} = \frac{1}{2}F\left(\frac{\sqrt{3}}{2}, \frac{\pi}{2}\right)$$

$$= \frac{1}{2} F\left(\frac{\sqrt{3}}{2}\right)$$

where $F(k) = F(k, \frac{\pi}{2})$.

● **PROBLEM 19-13**

Evaluate $\int_1^{\infty} \frac{du}{(3u^2+1)\sqrt{(u^2-1)(u^2+3)}}$

in terms of elliptic integrals.

Solution: For the given integral make the substitution $u = \sec\theta$ in the integrand. This yields

$$\int_0^{\pi/2} \frac{\sec\theta \tan\theta \, d\theta}{(3\sec^2\theta+1)\sqrt{(\sec^2\theta-1)(\sec^2\theta+3)}} = \int_0^{\pi/2} \frac{\cos^2\theta \, d\theta}{(3+\cos^2\theta)\sqrt{1+3\cos^2\theta}}$$

$$= \int_0^{\pi/2} \frac{(3+\cos^2\theta) - 3}{(3+\cos^2\theta)\sqrt{1+3\cos^2\theta}} \, d\theta$$

$$= \int_0^{\pi/2} \frac{d\theta}{\sqrt{1+3\cos^2\theta}} - 3\int_0^{\pi/2} \frac{d\theta}{(3+\cos^2\theta)\sqrt{1+3\cos^2\theta}}$$

$$= \int_0^{\pi/2} \frac{d\theta}{\sqrt{4-3\sin^2\theta}} - 3\int_0^{\pi/2} \frac{d\theta}{(4-\sin^2\theta)\sqrt{4-3\sin^2\theta}}$$

$$= \frac{1}{2}\int_0^{\pi/2} \frac{d\theta}{\sqrt{1-\frac{3}{4}\sin^2\theta}} - \frac{3}{8}\int_0^{\pi/2} \frac{d\theta}{(1-\frac{1}{4}\sin^2\theta)\sqrt{1-\frac{3}{4}\sin^2\theta}}$$

$$= \frac{1}{2} F\left(\frac{\sqrt{3}}{2}, \frac{\pi}{2}\right) - \frac{3}{8} \Pi\left(\frac{\sqrt{3}}{2}, -\frac{1}{4}, \frac{\pi}{2}\right)$$

where the first integral is the complete integral of the first kind and the second integral is the complete integral of the third kind with $k = \frac{\sqrt{3}}{2}$, $n = \frac{1}{4}$, and $\emptyset = \frac{\pi}{2}$. That is

$$\Pi(k, n, \emptyset) = \int_0^{\emptyset} \frac{d\theta}{(1+n\sin^2\theta)\sqrt{1-k^2\sin^2\theta}}$$

for $0 < k < 1$, $n \neq 0$.

• **PROBLEM 19-14**

Express the integral:

$$\int_0^{\psi} \frac{\sin^2\psi \, d\psi}{\sqrt{1-k^2\sin^2\psi}} \quad (0 < k < 1)$$

in terms of elliptic integrals. What is the result when the upper limit is $\pi/2$?

Solution: Rewriting the given integral, we attempt to achieve some kind of similarity between the numerator and denominator in the integral. Since, by definition, an integration is nothing more than a summation, multiplication of any individual term by a constant is the same as multiplying the integral by the same constant, Thus,

$$\int_0^{\psi} \frac{\sin^2\psi \, d\psi}{\sqrt{1-k^2\sin^2\psi}} = \frac{1}{k^2}\int_0^{\psi} \frac{k^2\sin^2\psi \, d\psi}{\sqrt{1-k^2\sin^2\psi}} .$$

Also,

$$\frac{1}{k^2} \int_0^\psi \frac{k^2 \sin^2\psi \, d\psi}{\sqrt{1-k^2\sin^2\psi}} = \frac{1}{k^2} \int_0^\psi \frac{1-1+k^2\sin^2\psi}{\sqrt{1-k^2\sin^2\psi}} \, d\psi \ .$$

In this case, we can write:

$$\int_0^\psi \frac{\sin^2\psi \, d\psi}{\sqrt{1-k^2\sin^2\psi}} = \frac{1}{k^2} \int_0^\psi \frac{1-(1-k^2\sin^2\psi)}{\sqrt{1-k^2\sin^2\psi}} \, d\psi$$

$$= \frac{1}{k^2} \int_0^\psi \frac{d\psi}{\sqrt{1-k^2\sin^2\psi}} - \frac{1}{k^2} \int_0^\psi \frac{1-k^2\sin^2\psi}{\sqrt{1-k^2\sin^2\psi}} \, d\psi \ .$$

If we let functions denote the integrals:

$$F(k,\psi) \equiv \int_0^\psi \frac{d\psi}{\sqrt{1-k^2\sin^2\psi}}$$

and

$$E(k,\psi) \equiv \int_0^\psi \sqrt{1-k^2\sin^2\psi} \, d\psi \ ,$$

which is the last integral, we obtain:

$$\int_0^\psi \frac{\sin^2\psi \, d\psi}{\sqrt{1-k^2\sin^2\psi}} = \frac{1}{k^2} [F(k,\psi) - E(k,\psi)] .$$

The new functions introduced are called elliptic integrals: $F(k,\psi)$ is of the first kind and $E(k,\psi)$ is of the second kind for $0 < k < 1$. The number k is called the modulus of the elliptic integral (in conics, k is called the eccentricity); and the upper limit ψ is called the amplitude of the elliptic integral. If the amplitude is $\pi/2$, then, by the known notation, we can write:

$$F(k,\pi/2) \equiv F(k)$$

or just F, and

$$E(k,\pi/2) \equiv E(k) \equiv E \ .$$

Thus, for

$$\psi = \pi/2 \ ,$$

$$\int_0^{\pi/2} \frac{\sin^2\psi \, d\psi}{\sqrt{1-k^2\sin^2\psi}} \equiv \frac{1}{k^2}[F(k) - E(k)].$$

Furthermore, if we are given:

$$\int_0^\psi \frac{\cos^2\psi \, d\psi}{\sqrt{1-k^2\sin^2\psi}}$$

we can use the identity

$$\cos^2\psi = 1-\sin^2\psi,$$

and hence,

$$\int_0^\psi \frac{\cos^2\psi \, d\psi}{\sqrt{1-k^2\sin^2\psi}} = \int_0^\psi \frac{1-\sin^2\psi}{\sqrt{1-k^2\sin^2\psi}} = F(k,\psi)$$

$$- \frac{1}{k^2}[F(k,\psi) - E(k,\psi)]$$

$$= \frac{(k^2-1)F(k,\psi) + E(k,\psi)}{k^2}.$$

• **PROBLEM 19-15**

The integrals:

$$F(k,\psi) = \int_0^\psi \frac{d\psi}{\sqrt{1-k^2\sin^2\psi}},$$

and

$$E(k,\psi) = \int_0^\psi \sqrt{1-k^2\sin^2\psi} \, d\psi,$$

are defined as elliptic integrals of the first and second kind, respectively. If $k > 1$, these integrals may also be expressed in terms of elliptic integrals. Show that

a)

$$\int_0^\psi \frac{d\psi}{\sqrt{1-k^2\sin^2\psi}} = \frac{1}{k} F\left(\frac{1}{k}, x\right),$$

and

b)

$$\int_0^\psi \sqrt{1-k^2\sin^2\psi}\, d\psi$$

$$= \left(\frac{1}{k} - k\right) F\left(\frac{1}{k}, x\right) + kE\left(\frac{1}{k}, x\right),$$

(for $k > 1$), where the amplitude x is expressed in terms of the upper limit ψ by

$$x = \sin^{-1}(k \sin \psi),$$

and the integrals have real values when $k \sin \psi \leq 1$.

<u>Solution</u>: Let $k \sin \psi = \sin x$. Then

$$d\psi = \frac{\cos x}{k\cos\psi} dx,$$

$$1-k^2\sin^2\psi = 1-\sin^2 x = \cos^2 x$$

and

$$x = \sin^{-1}(k \sin\psi).$$

Hence, by substitution,

a)

$$\int \frac{d\psi}{\sqrt{1-k^2\sin^2\psi}} = \int \frac{dx}{k\cos\psi} = \frac{1}{k}\int \frac{dx}{\cos\psi}.$$

By virtue of the fact that $k \sin \psi = \sin x$ and the trigonometric identity
$$\cos^2\psi + \sin^2\psi = 1,$$
we have:
$$\cos \psi = \sqrt{1-1/k^2\sin^2 x}.$$
Hence,

$$\int_0^\psi \frac{d\psi}{\sqrt{1-k^2\sin^2\psi}} = \frac{1}{k}\int_0^x \frac{dx}{\sqrt{1-1/k^2\sin^2 x}}.$$

$F(k,\psi)$ is the elliptic integral:

$$\int_0^\psi \frac{d\psi}{\sqrt{1-k^2\sin^2\psi}} \qquad (0 \le k \le 1).$$

Then we can write:

$$F(1/k,x) \equiv \int_0^x \frac{dx}{\sqrt{1-1/k^2\sin^2 x}} \qquad (k > 1 \text{ or } 0 < 1/k \le 1).$$

Thus, by this definition,

$$\int_0^\psi \frac{d\psi}{\sqrt{1-k^2\sin^2\psi}} = \frac{1}{k}\int_0^x \frac{dx}{\sqrt{1-1/k^2\sin^2 x}} \equiv \frac{1}{k} F(1/k,x)$$

for $k > 1$.

b) By the use of the above argument and another problem, we have

$$\int \sqrt{1-k^2\sin^2\psi}\, d\psi = \int \cos x \left(\frac{\cos x}{k\cos\psi}\, dx\right) = \frac{1}{k}\int \frac{\cos^2 x}{\cos\psi}\, dx,$$

or,

$$\int_0^\psi \sqrt{1-k^2\sin^2\psi}\, d\psi = \frac{1}{k}\int_0^x \frac{\cos^2 x\, dx}{\sqrt{1-1/k^2\sin^2 x}}.$$

If we replace $1/k$ by m in the radical, the relationship proved in another problem shows that:

$$\int_0^x \frac{\cos^2 x\, dx}{\sqrt{1-m^2\sin^2 x}} = \frac{(m^2-1)F(m,x) + E(m,x)}{m^2}$$

for $0 < m < 1$ or $k > 1$.

Substituting the value of m back in the above expression yields:

$$\frac{1}{k}\int \frac{\cos^2 x\, dx}{\sqrt{1-m^2\sin^2 x}} = \frac{1}{k}\cdot\frac{(1/k^2-1)F(1/k,x) + E(1/k,x)}{(1/k)^2}$$

$$= (1/k - k)F(1/k,x) + kE(1/k,x)$$

$$= \int_0^\psi \sqrt{1-k^2\sin^2\psi}\, d\psi.$$

Note that, for k > 1 ,

$$F(1/k, x) \equiv \int_0^x \frac{dx}{\sqrt{1 - 1/k^2 \sin^2 x}} ,$$

the elliptic integral of the first kind.

$$E(1/k, x) \equiv \int_0^x \sqrt{1 - 1/k^2 \sin^2 x} \, dx ,$$

the elliptic integral of the second kind. Both integrals have the modulus 1/k and amplitude x.

• **PROBLEM 19-16**

Given the ellipse: $x^2 + 2y^2 = 8$, find

a) its entire length;

b) the length of the arc from $x = 1$ to $x = 2$.

Solution: The equation of the ellipse can be rewritten as:

$$\frac{x^2}{8} + \frac{y^2}{4} = 1 ,$$

or,

$$\frac{x^2}{a^2} + \frac{y^2}{b^2} = 1 ,$$

from which we infer that the semi-major axis, $a = \sqrt{8} = 2\sqrt{2}$, and the semi-minor axis, $b = \sqrt{2}$. Consequently, the eccentricity,

$$k = \frac{\sqrt{a^2 - b^2}}{a} = \frac{1}{\sqrt{2}} .$$

Using the equation of the ellipse in the polar system, we have

$$x = a \sin \psi, \text{ and } y = \frac{b}{a} \sqrt{a^2 - x^2} = b \cos \psi ,$$

where ψ is the angle which a line from the origin to the point P(x,y) makes with the positive x-axis. Using the relationship:

$$ds = \sqrt{(dx)^2 + (dy)^2},$$

and the above parametric expression, the length of the ellipse is expressed by

$$S = \int \sqrt{dx^2 + dy^2} = \int_0^{2\pi} \sqrt{a^2\cos^2\psi + b^2\sin^2\psi}\, d\psi =$$

$$4\int_0^{\pi/2} \sqrt{a^2 - (a^2-b^2)\sin^2\psi}\, d\psi$$

$$= 4a\int_0^{\pi/2} \sqrt{1 - \frac{a^2-b^2}{a^2}\sin^2\psi}\, d\psi .$$

$$S = 4a\int_0^{\pi/2} \sqrt{1 - k^2\sin^2\psi}\, d\psi ,$$

since, as shown above,

$$k = \frac{\sqrt{a^2-b^2}}{a} = \frac{1}{\sqrt{2}} .$$

For a portion of this, corresponding to a point P(x,y) from the positive x-axis,

$$s = a\int_0^{\psi} \sqrt{1 - k^2\sin^2\psi}\, d\psi .$$

This integral appearing here is the so-called elliptic integral. It cannot be evaluated in finite form in terms of elementary functions. Instead, the values of the integrals for their corresponding k and ψ are given in tabular form.* In most cases, these values are given corresponding to values of $\sin^{-1}k$ in steps of, perhaps 1°, rather than the values of k itself.

From tables of trigonometric functions,
$$\sin^{-1}\frac{1}{\sqrt{2}} = 45°.$$

a) To find the entire length of the ellipse, we have:

$$S = 4a\int_0^{\pi/2} \sqrt{1 - k^2\sin^2\psi}\, d\psi .$$

For k corresponding to sin 45°, the integral value is 1.3506, and

$$s = 4(2\sqrt{2}) \times 1.3506$$
$$\approx 15.28.$$

b) For the length of the arc from $x = 1$ to $x = 2$, we have

$$s\Big|_{x=1}^{x=2} = s\Big|_{x=0}^{x=2} - s\Big|_{x=0}^{x=1},$$

and since

$$x = a \sin \psi = 2\sqrt{2} \sin\psi,$$

we can alternatively write:

$$s\Big|_{\psi=20°45'}^{\psi=45°} = s\Big|_{\psi=0}^{\psi=45°} - s\Big|_{\psi=0}^{\psi=20°45'}.$$

This gives the required length.

$$s = a\int_0^{45°} \sqrt{1 - k^2\sin^2\psi}\, d\psi - a\int_0^{20°45'} \sqrt{1 - k^2\sin^2\psi}\, d\psi.$$

By the use of interpolation we find the elliptic integral corresponding to $\psi' = 20°45'$ under $k = \sin 45°$. From the table, we see that, for $\psi_1 = 20°$, and, the next step, $\psi_2 = 25°$, both under $\alpha = 45°$ for k, we have:

$$E_1 = 0.3456 \text{ and } E_2 = 0.4296,$$

respectively, and hence, by interpolation, for $\psi' = 20°45'$ we have:

$$E' = \frac{E_2 - E_1}{\psi_2 - \psi_1}(\psi' - \psi_1) + E_1 = \frac{0.0420}{(5 \times 60)} \cdot 45' + 0.3456$$

$$= 0.3582.$$

The value of the integral for $\psi = 45°$ under $k = \sin 45°$ is given as 0.7482, and hence the length of the arc for $1 \leq x \leq 2$ is

$$s = a(0.7482) - a(0.3582) = 0.7482(2\sqrt{2}) - 0.3582(2\sqrt{2}),$$

or,

$$s \approx 1.10.$$

---

*For example, see "A Short Table of Integrals" by B. O. Peirce, Ginn and Company, 1929, pages 121, 122 and 123.

PHYSICAL APPLICATIONS

● **PROBLEM 19-17**

Determine the gravitational force field of the earth.

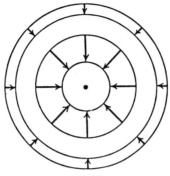

Fig. 1

Solution: Let C be the center of the earth, M the mass of the earth, G the gravitational constant and P any point outside of the earth. Then a particle of unit mass at P is attracted toward C by a force F of magnitude

$$G \frac{M}{\|P-C\|^2} = G \frac{M}{r^2}.$$

Let $\vec{r} = \vec{CP}$. The magnitude of $\vec{r}$ is r. Then a unit vector in the opposite direction of $\vec{r}$ is

$$-\frac{\vec{r}}{r},$$

which is in the same direction as $\vec{F}$. Thus the gravitational force on the unit mass of P is

$$\vec{F} = -G \frac{M}{r^3} \vec{r}.$$

Figure 1 suggests the nature of the vector field F.

● **PROBLEM 19-18**

The electrostatic field produced by a unit positive charge at 0 is

$$E = \frac{1}{r^3} \vec{A}$$

where $\|A\| = r$. Find the divergence of this field (wherever the field is defined, i.e. at all points except 0).

Solution: In a rectangular coordinate system, represented by the orthonormal basis $\{e_1, e_2, e_3\}$, let A have coordinates x, y, and z. Then

$$r^2 = x^2 + y^2 + z^2, \quad A = xe_1 + ye_2 + ze_3.$$

Hence

$$E = (E_1, E_2, E_3) = \left(\frac{x}{r^3}, \frac{y}{r^3}, \frac{z}{r^3}\right).$$

Thus

$$\frac{\partial E_1}{\partial x} = \frac{r^3 - 3xr^2 \frac{\partial r}{\partial x}}{r^6}.$$

Since $r^2 = x^2 + y^2 + z^2$,

$$2r \frac{\partial r}{\partial x} = 2x \quad \text{or} \quad \frac{\partial r}{\partial x} = \frac{x}{r}.$$

So,

$$\frac{\partial E_1}{\partial x} = \frac{r^3 - 3xr^2 \left(\frac{x}{r}\right)}{r^6} = \frac{r^2 - 3x^2}{r^5}.$$

Similarly,

$$\frac{\partial E_2}{\partial y} = \frac{r^2 - 3y^2}{r^5} \quad \text{and} \quad \frac{\partial E_3}{\partial z} = \frac{r^2 - 3z^2}{r^5}.$$

Therefore

$$\begin{aligned}
\text{div } E &= \langle \nabla, E \rangle \\
&= \frac{\partial E_1}{\partial x} + \frac{\partial E_2}{\partial y} + \frac{\partial E_3}{\partial z} \\
&= \frac{3r^2 - 3x^2 - 3y^2 - 3z^2}{r^5} \\
&= \frac{3(r^2 - (x^2+y^2+z^2))}{r^5} = \frac{3(r^2-r^2)}{r^5} \\
&= 0.
\end{aligned}$$

● **PROBLEM 19-19**

Suppose that the electrical potential at the point (x,y,z) is

$$E(x,y,z) = x^2 + y^2 - 2z^2.$$

What is the direction of the acceleration at the point $(1,3,2)$?

Solution: According to the theory of electrodynamics, a

positively charged particle is accelerated in the direction of the maximum decrease of the electrical potential E. The maximum decrease of E is the maximum increase of -E. Since the direction along which -E increases most is the direction of $\nabla(-E) = -\nabla E$, and

$$-\nabla E = -\nabla(x^2+y^2-2z^2) = -(2x, 2y, -4z)$$
$$= (-2x, -2y, 4z).$$

Therefore, the direction of acceleration at $(1,3,2)$ is parallel to $(-2,-6,8)$. Equivalently, the direction of acceleration is the same as the direction of the vector $(-1,-3,4)$.

• **PROBLEM 19-20**

The electrostatic potential at A arising from a dipole of unit strength at 0, oriented in the direction of the unit vector n, is, in polar coordinates,

$$V(r,\theta) = \frac{\cos\theta}{r^2},$$

where $r = \|A\|$ and θ is the angle between n and A (see Figure 1). Find the divergence of the corresponding electrostatic field E.

Fig. 1

Solution: By definition

$$\cos\theta = \frac{n \cdot A}{\|n\|\|A\|} = \frac{n \cdot A}{r};$$

hence

$$V = \frac{n \cdot A}{r^3}.$$

The electrostatic field is $E = -\nabla V$. The problem is to calculate

$$\text{div } E = \text{div}(-\nabla V)$$
$$= -\text{div}\nabla V$$
$$\equiv -\nabla \cdot \nabla V$$
$$\equiv -\nabla^2 V$$

where $\nabla^2 = \dfrac{\partial^2}{\partial x^2} + \dfrac{\partial^2}{\partial y^2} + \dfrac{\partial^2}{\partial z^2}$. Choose a coordinate system with origin O such that $n = (0,0,1)$. Then, if $A = (x,y,z)$,

$$n \cdot A = z \quad \text{and} \quad r^2 = x^2 + y^2 + z^2.$$

Now

$$V = \dfrac{z}{r^3}.$$

Note that

$$\dfrac{\partial r}{\partial x} = \dfrac{x}{r}, \quad \dfrac{\partial r}{\partial y} = \dfrac{y}{r}, \quad \text{and} \quad \dfrac{\partial r}{\partial z} = \dfrac{z}{r}.$$

Calculate the first partial derivatives:

$$\dfrac{\partial V}{\partial x} = -\dfrac{3z}{r^4}\dfrac{\partial r}{\partial x} = -\dfrac{3xz}{r^5},$$

$$\dfrac{\partial V}{\partial y} = -\dfrac{3z}{r^4}\dfrac{\partial r}{\partial y} = -\dfrac{3yz}{r^5},$$

$$\dfrac{\partial V}{\partial z} = \dfrac{1}{r^3} - \dfrac{3z}{r^4}\dfrac{\partial r}{\partial z} = \dfrac{1}{r^3} - \dfrac{3z^2}{r^5}.$$

Next, calculate the necessary second partials:

$$\dfrac{\partial^2 V}{\partial x^2} = -\dfrac{3z}{r^5} + \dfrac{15xz}{r^6}\dfrac{\partial r}{\partial x} = -\dfrac{3z}{r^5} + \dfrac{15x^2 z}{r^7},$$

$$\dfrac{\partial^2 V}{\partial y^2} = -\dfrac{3z}{r^5} + \dfrac{15yz}{r^6}\dfrac{\partial r}{\partial y} = \dfrac{-3z}{r^5} + \dfrac{15y^2 z}{r^7},$$

$$\dfrac{\partial^2 V}{\partial z^2} = -\dfrac{3}{r^4}\dfrac{\partial r}{\partial z} - \dfrac{6z}{r^5} + \dfrac{15z^2}{r^6}\dfrac{\partial r}{\partial z} = -\dfrac{9z}{r^5} + \dfrac{15z^3}{r^7}.$$

Therefore,

$$\text{div } E = -\nabla^2 V = -\left(\dfrac{\partial^2 V}{\partial x^2} + \dfrac{\partial^2 V}{\partial y^2} + \dfrac{\partial^2 V}{\partial z^2}\right)$$

$$= \dfrac{+15z}{r^5} - \dfrac{15(x^2 z + y^2 z + z^3)}{r^7}$$

$$= \dfrac{15zr^2 - 15z(x^2 + y^2 + z^2)}{r^7}$$

$$= 0$$

for all A except $A = 0$.

• PROBLEM 19-21

Let G be the gravitational constant, $r = \|P\|$, and

$$F = -\frac{GM}{r^3} P, \qquad (1)$$

where M is the mass at 0. That is, (1) describes the gravitational field of a mass concentrated at 0. Show that F is irrotational.

Solution: A field is irrotational if its curl is identically 0. The components of F are

$$F_1 = -\frac{GM}{r^3} x, \quad F_2 = -\frac{GM}{r^3} y, \quad F_3 = -\frac{GM}{r^3} z,$$

and $r = \sqrt{x^2+y^2+z^2}$. The objective is to show that

$$\text{curl } F = \nabla \times F$$

$$= \left(\frac{\partial F_3}{\partial y} - \frac{\partial F_2}{\partial z},\ \frac{\partial F_1}{\partial z} - \frac{\partial F_3}{\partial x},\ \frac{\partial F_2}{\partial x} - \frac{\partial F_1}{\partial y} \right)$$

$$= 0.$$

Calculate

$$\frac{\partial F_1}{\partial y} = \frac{3GMx}{r^4} \frac{\partial r}{\partial y}$$

$$= \frac{3GMx}{r^4} \frac{2y}{2\sqrt{x^2+y^2+z^2}}$$

$$= \frac{3GM}{r^5} xy$$

and

$$\frac{\partial F_2}{\partial x} = \frac{3GMy}{r^4} \frac{\partial r}{\partial x}$$

$$= \frac{3GMy}{r^4} \frac{2x}{2\sqrt{x^2+y^2+z^2}}$$

$$= \frac{3GM}{r^5} xy.$$

Thus,

$$\frac{\partial F_2}{\partial x} - \frac{\partial F_1}{\partial y} = 0.$$

By similar calculations

$$\frac{\partial F_3}{\partial y} - \frac{\partial F_2}{\partial z} = 0$$

and

$$\frac{\partial F_1}{\partial z} - \frac{\partial F_3}{\partial x} = 0.$$

Hence, curl $F = 0$, as desired.

• **PROBLEM 19-22**

Find the field due to a uniformly charged circular lamina R of radius a, at some point a distance b from the center of the lamina along the perpendicular. (See Figure 1.)

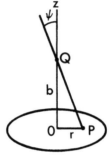

Fig. 1

Solution: In the theory of electrostatics, the field due to a lamina R at Q is defined by

$$\vec{F} = \iint_R \frac{\sigma}{r^3} \vec{PQ}\, dA$$

where $r = PQ = \|P-Q\|$, σ is the charge density, and P varies over R. Hence if $P = (x,y,z)$, $Q = (q_1, q_2, q_3)$, and R is in the xy-plane, then the components of $\vec{F}$ are

$$F_1 = \iint_R \frac{\sigma}{r^3} (q_1 - x)\, dxdy,$$

$$F_2 = \iint_R \frac{\sigma}{r^3} (q_2 - y)\, dxdy,$$

$$F_3 = \iint_R \frac{\sigma}{r^3} q_3 \, dxdy.$$

In the present situation let R be centered at 0 (i.e. R is the region given by $x^2 + y^2 \leq a^2$, $z = 0$). The point in question is then $(0,0,b)$. Let σ be the constant density. Now

$$F_1 = -\iint_R \frac{\sigma x}{\|(x,y,0)-(0,0,b)\|^3} \, dxdy$$

$$= -\iint_R \frac{\sigma x}{(\sqrt{x^2+y^2+b^2})^3} \, dxdy.$$

Switching to polar coordinates gives

$$F_1 = -\sigma \iint_R \frac{r\cos\theta \, r}{(r^2+b^2)^{3/2}} \, drd\theta$$

$$= -\sigma \int_0^{2\pi} \int_0^a \frac{r^2 \cos\theta}{(r^2+b^2)^{3/2}} \, drd\theta$$

$$= \sigma \int_0^a \frac{r^2}{(r^2+b^2)^{3/2}} \, dr \int_0^{2\pi} \cos\theta \, d\theta$$

$$= 0.$$

Similarly, $F_2 = 0$. Also

$$F_3 = \iint_R \frac{\sigma b}{(x^2+y^2+b^2)^{3/2}} \, dxdy$$

$$= b\sigma \iint_R \frac{r}{(r^2+b^2)^{3/2}} \, drd\theta$$

$$= b\sigma \int_0^{2\pi} d\theta \int_0^a \frac{r}{(r^2+b^2)^{3/2}} \, dr$$

$$= 2\pi b\sigma \left[-(r^2+b^2)^{-1/2} \right]\Big|_0^a$$

$$= 2\pi b\sigma \left[-(a^2+b^2)^{-1/2} + (b^2)^{-1/2} \right]$$

$$= 2\pi\sigma \left[1 - \frac{b}{\sqrt{a^2+b^2}} \right].$$

Thus the field at the given point off the lamina is

$$\vec{F} = \left(0,\ 0,\ 2\pi\sigma\left(1 - \frac{b}{\sqrt{a^2+b^2}}\right)\right).$$

● **PROBLEM 19-23**

Suppose a lamina R in the xy-plane bounded by $x = 0$, $x = a$, $y = 0$, $y = b$ has a charge of density $\sigma = xy$. Find the potential at $(0,0,c)$.

Solution: From the theory of electrostatics, the potential at Q produced by a point charge e located at P is defined to be

$$\frac{e}{r},\ \text{where}\ r = \|P-Q\|.$$

The potential produced by several particles is the sum of the individual potentials. Moreover, the potential at Q produced by a lamina in the region R with charge density σ is defined to be

$$\int_R \frac{\sigma}{\|P-Q\|}\, dA$$

where P is the varying point in R. Note that potential is a scalar function. Now in this problem, the potential is

$$V = \iint_R \frac{xy}{\|(x,y,z)-(0,0,c)\|}\, dA$$

$$= \iint_R \frac{xy}{\|(x,y,0)-(0,0,c)\|}\, dx\,dy$$

$$= \iint_R \frac{xy}{\sqrt{x^2+y^2+c^2}}\, dx\,dy$$

$$= \int_0^b y \int_0^a \frac{x}{\sqrt{x^2+y^2+c^2}}\, dx\,dy$$

$$= \int_0^b y\sqrt{x^2+y^2+c^2}\,\Big|_{x=0}^{a}\, dy$$

$$= \int_0^b y\left(\sqrt{a^2+y^2+c^2} - \sqrt{y^2+c^2}\right) dy$$

1000

$$= \int_0^b y\sqrt{a^2+y^2+c^2}\,dy - \int_0^b y\sqrt{y^2+c^2}\,dy$$

$$= \frac{1}{3}(a^2+y^2+c^2)^{3/2}\Big|_0^b - \frac{1}{3}(y^2+c^2)^{3/2}\Big|_0^b$$

i.e. $\quad V = \frac{1}{3}\left[(a^2+b^2+c^2)^{3/2} - (a^2+c^2)^{3/2} - (b^2+c^2)^{3/2} + c^3\right]$

• **PROBLEM 19-24**

A particle moves in the plane according to

$$x = 64\sqrt{3}\,t, \qquad y = 64t - 16t^2$$

and is acted on by a force F which is directly proportional to the velocity but opposite in direction. Find the work done by F from $t = 0$ to $t = 4$.

Solution: The velocity components are

$$v_x = \frac{dx}{dt} = 64\sqrt{3}, \quad v_y = \frac{dy}{dt} = 64 - 32t$$

and hence the components of F are

$$F_x = -64\sqrt{3}\,k, \quad F_y = (32t - 64)k$$

where k is the absolute proportionality constant. The work done is given by

$$W = \int_0^4 F \cdot v\,dt$$

$$= \int_0^4 (-64\sqrt{3}\,k,\ (32t-64k)) \cdot (64\sqrt{3},\ 64-32t)\,dt$$

$$= -k\int_0^4 ((64\sqrt{3})^2 + (64-32t)^2)\,dt$$

$$= -k(32)^2 \int_0^4 ((2\sqrt{3})^2 + (2-t)^2)\,dt$$

1001

$$= -1024k \int_0^4 (16-4t+t^2)dt$$

$$= -1024K \left(16t-2t^2+\frac{1}{3}t^3\right)\Big|_0^4$$

$$= -1024k \left(64-32+\frac{64}{3}\right)$$

$$= -\frac{163,840}{3} k$$

● **PROBLEM 19-25**

Find the center of mass of a hemisphere of radius a>0 assuming the surface is a homogeneous lamina.

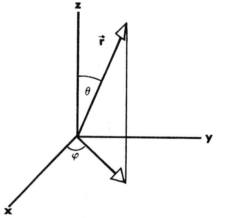

Fig. 1

Solution: Let c be the density of the hemisphere S. The coordinates of the center of mass are given by

$$\bar{x} = \frac{1}{m} \iint_S cx\,dA,$$

$$\bar{y} = \frac{1}{m} \iint_S cy\,dA,$$

$$\bar{z} = \frac{1}{m} \iint_S cz\,dA,$$

where

$$m = \iint_S c\,dA,$$

the mass of S. Calculate m:

$$m = c \iint_S dA$$

i.e. $m = c$ (area of the hemisphere) $= c(2\pi a^2)$.

Due to the symmetric nature of S, it is clear that $\bar{x} = \bar{y} = 0$. Indeed, note that

$$\bar{x} = \frac{c}{m} \iint_S x\, dA.$$

Using spherical polar coordinates $x = a\cos\phi\sin\theta$
$y = a\sin\phi\sin\theta \quad z = a\cos\theta \quad dA = a^2\sin\theta\, d\theta\, d\phi$

$$\bar{x} = \frac{1}{2\pi a^2} \int_0^{2\pi} \int_0^{\pi/2} a\cos\phi\sin\theta \, a^2\sin\theta\, d\theta\, d\phi = 0 \text{ because}$$

$$\int_0^{2\pi} \cos\phi\, d\phi = 0.$$

Similarly, $\bar{y} = 0$. Now calculate $\bar{z}$:

$$\bar{z} = \frac{c}{m} \iint_S z\, dA$$

$$= \frac{1}{2\pi a^2} \int_0^{2\pi} \int_0^{\pi/2} a\cos\theta \, a^2\sin\theta\, d\theta\, d\phi$$

$$= \frac{a}{2\pi} \int_0^{2\pi} d\phi \int_0^{\pi/2} \frac{\sin 2\theta}{2}\, d\theta$$

$$= \frac{a}{2\pi} \, 2\pi \int_0^{\pi/2} \frac{1}{2}\sin 2\theta\, d\theta$$

$$= \frac{a}{2} \left(\frac{-\cos 2\theta}{2}\right)\Big|_0^{\pi/2}$$

$$= \frac{a}{2}$$

So, the center of mass is $(0, 0, \frac{a}{2})$ as might have been expected.

● **PROBLEM 19-26**

Find the center of gravity of a thin uniform sheet of metal which is in the shape of the paraboloid $z = x^2+y^2$, with $x^2+y^2 \leq 1$.

Solution: The center of gravity is the point $(0,0,\bar{z})$ where $M\bar{z} = \iint_S \rho z \, dA$ and $M = \iint_S \rho \, dA$ (ρ is the mass density). Since the sheet of metal is uniform, ρ is constant and $M = \rho A(S)$. To compute this, use the definition

$$\text{Area} = \iint_S dA = \iint_R \left\| \frac{\partial \vec{F}}{\partial x} \times \frac{\partial \vec{F}}{\partial y} \right\| dxdy \quad \text{where}$$

$\vec{F}(x,y) = (x,y,f(x,y))$. Here, $f(x,y) = x^2+y^2$ and $\left\| \frac{\partial \vec{F}}{\partial x} \times \frac{\partial \vec{F}}{\partial y} \right\| = \sqrt{1 + (2x)^2 + (2y)^2}$. Therefore,

$$A(S) = \iint_R \sqrt{1 + 4(x^2+y^2)} \, dxdy \text{ where R is the unit disk,}$$

$x^2+y^2 \leq 1$. To compute this integral, transform to polar coordinates. Accordingly, the integrand becomes $(1+4r^2)^{1/2}$, dxdy is replaced by $rdrd\theta$, and R is replaced by the rectangle $0 \leq \theta \leq 2\pi$, $0 \leq r \leq 1$. Thus,

$$A(S) = \int_0^{2\pi} \int_0^1 \sqrt{1+4r^2} \; rdrd\theta = \frac{\pi}{6}(5\sqrt{5} - 1).$$

In similar fashion,

$$\iint_S z \, dA = \iint_R (x^2+y^2)\sqrt{1+4(x^2+y^2)} \, dxdy$$

$$= \int_0^{2\pi} \int_0^1 r^2\sqrt{1+4r^2} \; rdrd\theta$$

$$= \frac{1}{4} \int_0^{2\pi} \int_0^1 4r^2\sqrt{1+4r^2} \; rdrd\theta$$

$$= \frac{1}{4}\left[\int_0^{2\pi} \int_0^1 (1+4r^2)^{3/2} rdrd\theta\right.$$

$$- \int_0^{2\pi} \int_0^1 \sqrt{1+4r^2} \, r \, dr \, d\theta \Bigg]$$

$$= \left[\frac{1}{4} \cdot \frac{\pi}{10} (25\sqrt{5} - 1) - \frac{\pi}{6} (5\sqrt{5} - 1) \right]$$

$$= \frac{\pi}{12} \left(5\sqrt{5} + \frac{1}{5}\right).$$

Hence since $\bar{z} = \dfrac{\iint_S \rho z \, dA}{\iint_S \rho \, dA} = \dfrac{\iint_S z \, dA}{\iint_S dA}$ (since ρ is constant)

we have

$$\bar{z} = \frac{\frac{\pi}{12}(5\sqrt{5} + \frac{1}{5})}{\frac{\pi}{6}(5\sqrt{5} - 1)} = \frac{1}{2}\left(\frac{5\sqrt{5} + \frac{1}{5}}{5\sqrt{5} - 1}\right) = .56.$$

That is, the center of gravity is the point $(0, 0, .56)$.

● **PROBLEM 19-27**

Find the moment of inertia of the hemispherical surface $z = \sqrt{a^2 - x^2 - y^2}$ ($a > 0$) about the x-axis, assuming the surface to be homogeneous of mass M.

Solution: The moment of inertia is

$$I = \iint_S \sigma(y^2 + z^2) \, dA \qquad (1)$$

where σ is the density and S is the hemispherical surface $z = \sqrt{a^2 - x^2 - y^2}$. This surface integral is equal to

$$I = \iint_R \sigma(y^2 + z^2) \sec\gamma \, dx \, dy \qquad (2)$$

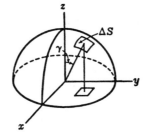

Fig. 1

where R is the circular region $x^2 + y^2 \leq a^2$ and γ is the acute angle between the normal to S at (x,y,z) and the z-axis (see Figure 1). Since

$$\sec\gamma = \frac{a}{z} = \frac{a}{\sqrt{a^2-x^2-y^2}}$$

and

$$y^2 + z^2 = a^2 - x^2,$$

(2) becomes

$$I = \iint_R \sigma a \frac{a^2-x^2}{\sqrt{a^2-x^2-y^2}} \, dxdy.$$

Change (3) to polar coordinates to get

$$I = \sigma a \int_0^{2\pi} \int_0^a \frac{a^2-r^2\cos^2\theta}{\sqrt{a^2-r^2}} \, rdrd\theta.$$

Now make the substitution $r = a\sin\phi$ and evaluate the integral using the trigonometric identities $\sin^2\phi + \cos^2\phi = 1$ and $\cos^2\phi = \frac{1}{2} + \frac{1}{2}\cos 2\phi$:

$$I = \sigma a \int_0^{2\pi} \int_0^{\pi/2} \frac{a^2-a^2\sin^2\phi\cos^2\theta}{\sqrt{a^2-a^2\sin^2\phi}} (a\sin\phi)(a\cos\phi) d\phi d\theta$$

$$= \sigma a \int_0^{2\pi} \int_0^{\pi/2} \frac{a^4(1-\sin^2\phi\cos^2\theta)\sin\phi\cos\phi}{a\cos\phi} d\phi d\theta$$

$$= \sigma a^4 \int_0^{2\pi} \int_0^{\pi/2} \sin\phi(1-(1-\cos^2\phi)\cos^2\theta) d\phi d\theta$$

$$= \sigma a^4 \int_0^{2\pi} \int_0^{\pi/2} (\sin\phi(1-\cos^2\theta)+\sin\phi\cos^2\phi\cos^2\theta) d\phi d\theta$$

$$= \sigma a^4 \int_0^{2\pi} \left(-\cos\phi(1-\cos^2\theta)-\frac{1}{3}\cos^3\phi\cos^2\theta\right)\Big|_{\phi=0}^{\pi/2} d\theta$$

$$= \sigma a^4 \int_0^{2\pi} (1-\cos^2\theta + \frac{1}{3}\cos^2\theta) d\theta$$

$$= \sigma a^4 \int_0^{2\pi} (1 - \frac{2}{3}\cos^2\theta) d\theta$$

$$= \sigma a^4 \int_0^{2\pi} (1 - \frac{2}{3}(\frac{1}{2} + \frac{1}{2}\cos 2\theta))d\theta$$

$$= \sigma a^4 \int_0^{2\pi} (\frac{2}{3} - \frac{1}{3}\cos 2\theta)d\theta$$

$$= \frac{\sigma a^4}{3}(2\theta - \frac{1}{2}\sin 2\theta)\Big|_0^{2\pi}$$

$$= \frac{4\pi\sigma a^4}{3}.$$

Thus the moment of inertia I is $\frac{4\pi\sigma a^4}{3}$. Since $M = 2\pi a^2 \sigma$, we have $I = \frac{2}{3}Ma^2$.

● **PROBLEM** 19-28

A plane lamina is bounded by x=a, x=b, y=0, f(x)>0 where $f \in C[a,b]$. The density is constant along all vertical lines x=c, $c \in [a,b]$, so it can be described as a function of one variable M(x). Find the moment of inertia of the lamina about the y-axis.

Solution: The formula for the moment of inertia about an axis of a set of particles is $I = \sum_{k=1}^{n} m_k r_k^2$ where r_k is the distance of the particle (s_k, m_k) from the axis. This formula has the following facts:

(i) If a total mass is divided into several parts, the moment of inertia of the whole is the sum of the moments of inertia of the parts.

(ii) If additional mass is added, the moment of inertia is increased (if mass is removed, I is decreased).

(iii) If mass is moved farther from the axis, the moment of inertia is increased (if mass is moved nearer the axis, I is decreased).

For the given problem, let the points $\{x_k\}_0^n$ be a subdivision Δ of (a,b) so that erecting ordinates at the points x_k divides the lamina into n vertical strips. Now let I_k and m_k be the moment of inertia and the mass, respectively, of the kth vertical strip. In addition, let $f(x_k')$ and $f(x_k'')$ be the minimum and maximum ordinates of the kth strip. By property (ii), this yields

$$f(x_k')[M(x_k) - M(x_{k-1})] \le m_k \le f(x_k'')[M(x_k) - M(x_{k-1})].$$

Furthermore, by properties (i) and (iii), we have

$$\sum_{k=1}^{n} x_{k-1}^2 f(x_k')[M(x_k) - M(x_{k-1})] \le I$$

$$\le \sum_{k=1}^{n} x_k^2 f(x_k'')[M(x_k) - M(x_{k-1})] \quad (1)$$

where $I = \sum_{k=1}^{n} m_k r_k^2$.

Now, Duhamel's theorem for Stieltjes integrals states: If $f(x), g(x) \in C$ for $a \le x \le b$, if $\alpha(x)$ is a nondecreasing function, if $\{x_k\}_0^n$ is a subdivision Δ of (a,b) and if $x_{k-1} \le \xi_k \le x_k$, $x_{k-1} \le \eta_k \le x_k$ for $k = 1, 2, \ldots, n$ then

$$\lim_{||\Delta|| \to 0} \sum_{k=1}^{n} f(\xi_k) g(\eta_k)[\alpha(x_k) - \alpha(x_{k-1})]$$

$$= \int_a^b f(x) g(x) d\alpha(x) \quad (2)$$

Applying this theorem to (1) yields

$$I = \int_a^b x^2 f(x) dM(x).$$

Note that we have assumed that $a>0$, even though this assumption was not essential to the final result.

● **PROBLEM 19-29**

Find the electrical potential at $(0,0,-a)$ of a uniformly distributed total charge e on the hemisphere S defined by

$$z = \sqrt{a^2 - x^2 - y^2} \quad (a>0).$$

Solution: If c is the constant density of charge, the potential is

$$V = \iint_S \frac{c}{\sqrt{x^2+y^2+(z+a)^2}} dA. \quad (1)$$

Parameterize (x,y,z) as $(a \sin\phi \cos\theta, a \sin\phi \sin\theta, a \cos\phi)$

[i.e. spherical polar coordinates]. Then

$$\sqrt{x^2+y^2+(z+a)^2} = \sqrt{a^2\sin^2\phi\cos^2\theta + a^2\sin^2\phi\sin^2\theta + (a\cos\phi+a)^2}$$

$$= \sqrt{a^2\sin^2\phi(\cos^2\theta+\sin^2\theta) + a^2(\cos^2\phi+2\cos\phi+1)}$$

$$= \sqrt{a^2(\sin^2\phi+\cos^2\phi) + 2a^2\cos\phi + a^2}$$

$$= \sqrt{4a^2 \tfrac{1}{2}(\cos\phi+1)}$$

$$= 2a\cos\tfrac{\phi}{2} \qquad (2)$$

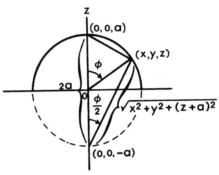

Fig. 1

(See Figure 1.) Let $P(x,y,z)$ represent the change to polar coordinates and $P_\psi = \left(\dfrac{\partial x}{\partial \psi}, \dfrac{\partial y}{\partial \psi}, \dfrac{\partial z}{\partial \psi}\right)$. To compute (1), recall that

$$\iint_S f(u,v)\,dA = \iint_R f(u,v)\sqrt{EG-F^2}\,du\,dv \qquad (3)$$

and

$$ds^2 = E\,du^2 + 2F\,du\,dv + G\,dv^2, \qquad (4)$$

where

$$E = \langle P_\phi, P_\phi \rangle,$$
$$F = \langle P_\phi, P_\theta \rangle,$$
$$G = \langle P_\theta, P_\theta \rangle$$

are the coefficients of the first fundamental form (4). Now

$$P_\phi = (a\cos\phi\cos\theta,\ a\cos\phi\sin\theta,\ -a\sin\phi)$$

and

$$P_\theta = (-a\sin\phi\sin\theta,\ a\sin\phi\cos\theta,\ 0);$$

thus

$$E = a^2\cos^2\phi\cos^2\theta + a^2\cos^2\phi\sin^2\theta + a^2\sin^2\phi$$

$$= a^2[\cos^2\phi(\cos^2\theta + \sin^2\theta) + \sin^2\phi]$$

$$= a^2, \tag{5}$$

$$F = -a^2\sin\phi\cos\phi\sin\theta\cos\theta + a^2\sin\phi\cos\phi\sin\theta\cos\theta + 0$$

$$= 0, \tag{6}$$

and

$$G = a^2\sin^2\phi\sin^2\theta + a^2\sin^2\phi\cos^2\theta$$

$$= a^2\sin^2\phi. \tag{7}$$

Using (2), (3), (5), (6), and (7), (1) becomes

$$V = \iint_R \frac{c}{2a\cos\frac{\phi}{2}} \sqrt{a^2(a^2\sin^2\phi) - 0} \, d\phi d\theta$$

$$= \frac{ac}{2}\int_0^{2\pi} d\theta \int_0^{\frac{\pi}{2}} \frac{\sin\phi}{\cos\frac{\phi}{2}} d\phi, \tag{8}$$

where R is the projection of the hemisphere into the xy-plane (i.e. the circular region given by $x^2+y^2 \le a^2$). Since $\sin\phi = 2\sin\frac{\phi}{2}\cos\frac{\phi}{2}$, (8) becomes

$$V = 2\pi ac \int_0^{\frac{\pi}{2}} \sin\frac{\phi}{2} d\phi$$

$$= 2\pi ac \left. \left(-2\cos\frac{\phi}{2}\right)\right|_0^{\frac{\pi}{2}}$$

$$= 2\pi ac (2-\sqrt{2}).$$

Since the total charge $e = c\int_S dA = 2\pi a^2 c$, this yields

$$V = \frac{(2-\sqrt{2})e}{a}.$$

• **PROBLEM 19-30**

Derive the equation of Continuity for fluid flows:

$$\frac{\partial \rho}{\partial t} = -\text{div } \rho \vec{V}$$

where $\rho(x,y,z,t)$ and $\vec{V}(x,y,z,t)$ are, respectively, the fluid density and velocity at the point (x,y,z) at time t. Conclude $\text{div}\vec{V} = 0$ if the fluid is incompressible.

Solution: A fluid is said to be incompressible if its density is constant, irrespective of time or position. Consider any ball B with boundary ∂B. The total fluid mass in B at time t is

$$\int_B \rho(x,y,z,t)\,dx\,dy\,dz.$$

The rate of increase of the mass is

$$\frac{d}{dt}\int_B \rho(x,y,z,t)\,dx\,dy\,dz = \int_B \frac{\partial \rho}{\partial t}(x,y,z,t)\,dx\,dy\,dz. \quad (1)$$

However, the only way for the mass in B to change is by fluid flow over the boundary of B. Since

$$\int_{\partial B} \vec{V}(x,y,z,t) \cdot d\vec{A}$$

is the volume rate of flow out of B, the rate of flow of mass out of B is

$$\int_{\partial B} \rho(x,y,z,t)\vec{V}(x,y,z,t) \cdot d\vec{A}.$$

The rate of increase of mass in B is then $-\int_{\partial B} \rho\vec{V} \cdot d\vec{A}$. By the divergence theorem, we have

$$-\int_{\partial B} \rho\vec{V} \cdot d\vec{A} = -\int_B \text{div}(\rho\vec{V})\,dx\,dy\,dz.$$

From (1),

$$\int_B \frac{\partial \rho}{\partial t}\,dx\,dy\,dz = -\int_B \text{div}(\rho\vec{V})\,dx\,dy\,dz. \quad (2)$$

Since (2) holds for every ball B, it follows that

$$\frac{\partial \rho}{\partial t} = -\text{div } \rho \vec{v}. \tag{3}$$

If the fluid is incompressible, ρ is constant and (3) yields

$$0 = -\text{div}\rho\vec{v}$$

or

$$0 = -\rho\text{div}\vec{v}.$$

Hence,

$$\text{div}\vec{v} = 0,$$

when the fluid is incompressible.

• **PROBLEM 19-31**

What force field would account for a particle of unit mass moving around in a unit circle in the plane according to the function

$$f(t) = (\cos t, \sin t)? \tag{1}$$

Solution: Differentiating (1) gives

$$f'(t) = (-\sin t, \cos t)$$

and

$$f''(t) = (-\cos t, -\sin t) = -f(t).$$

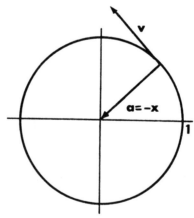

Fig. 1

Thus, the particle is accelerating toward the origin with a constant magnitude of $\|-f(t)\| = \sqrt{\cos^2 t + \sin^2 t} = 1$. (See Figure 1.) Since the force field F satisfies

$$f''(t) = F(f(t), t) = -f(t),$$

this motion can be accounted for by the force field

$$F(x, y, t) = (-x, -y).$$

If a particle has a velocity at time t orthogonal to its position vector then it will continue to move in its circular path about the origin. To see this, solve the initial value problem for $f: R \to R^2$

$$f''(t) = -f(t); \quad f(0) = (x_0, y_0); \quad f'(0) = (-y_0, x_0)$$

The general solution of the homogeneous equation is

$$f(t) = A \sin t + B \cos t$$

where $A, B \in R^2$. Then

$$f(0) = A \sin 0 + B \cos 0$$
$$= B = (x_0, y_0)$$

and

$$f'(0) = A \cos 0 - B \sin 0$$
$$= A = (-y_0, x_0).$$

Hence, the specific solution of the initial value problem is

$$f(t) = (-y_0, x_0) \sin t + (x_0, y_0) \cos t$$
$$= (x_0 \cos t - y_0 \sin t, \; x_0 \sin t + y_0 \cos t),$$

which is just (1) when $(x_0, y_0) = (1, 0)$.

• **PROBLEM 19-32**

Suppose there is a force field defined by

$$F(x, y, z, t) = (-x, -y, 0).$$

If a particle of unit mass is at (1,0,0) with an initial velocity of (0,1,a), what is its path of motion?

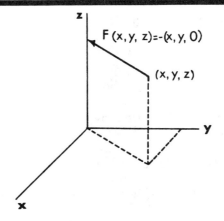

Fig. 1

Solution: This is the same as solving the initial value problem

$$f''(t) = F(f(t), t); \quad f(0) = (x_0, y_0, z_0); \quad f'(0) = (x_0', y_0', z_0'),$$

where $f(t)$ is the function sought, (x_0, y_0, z_0) is the initial position, and (x_0', y_0', z_0') is the initial velocity. Here this becomes

$$f''(t) = (-x, -y, 0); \quad f(0) = (1, 0, 0); \quad f'(0) = (0, 1, a),$$

where $f(t) = (x(t), y(t), z(t))$. This can be separated into the three problems

$$x''(t) = -x(t); \quad x(0) = 1; \quad x'(0) = 0, \tag{1}$$

$$y''(t) = -y(t); \quad y(0) = 0; \quad y'(0) = 1, \tag{2}$$

and $\quad z''(t) = 0; \quad z(0) = 0; \quad z'(0) = a.$ \quad (3)

Problem (1) yields the general solution

$$x(t) = A \sin t + B \cos t$$

where A and B are complex. Moreover

$$x(0) = B = 1$$

and $\quad x'(0) = A = 0.$

Therefore,

$$x(t) = \cos t. \tag{4}$$

Similarly, (2) yields

$$y(t) = \sin t. \tag{5}$$

For (3),

$$z(t) = At + B.$$

Also,

$$z(0) = B = 0$$

and $\quad z'(0) = A = a.$

Hence,

$$z(t) = at \tag{6}$$

Thus, the path followed by the particle is given by (4), (5), and (6) or

$$f(t) = (\cos t, \sin t, at). \tag{7}$$

Depending on the nature of a, (7) will be either a spiral or a circle (see Figure 2).

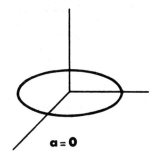

$a = 0$

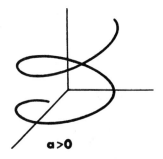

$a > 0$

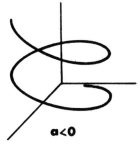

$a < 0$

Fig. 2

● **PROBLEM 19-33**

Suppose

$$A(x,y,z,t) = (-x,0,y)$$

is the acceleration field of a fluid in motion with an initial velocity of $(0,1,0)$. Find the equation of motion and the divergence and curl of the flow.

Solution: Let $f(t) = (x(t), y(t), z(t))$ be the equation of motion and

$$f(0) = (x_0, y_0, z_0). \tag{1}$$

Then

$$f'(0) = (0,1,0) \tag{2}$$

and

$$f''(t) = (-x, 0, y) \tag{3}$$

according to the given information. Hence, in order to find f, the initial value problem defined by (1), (2), and (3) must be solved. The differential equation (3) is the system

$$(x''(t), y''(t), z''(t)) = (-x, 0, y).$$

The first of these has auxilliary equation $m^2 = -1$ and hence it has the general solution

$$x(t) = A \sin t + B \cos t. \tag{4}$$

The second has auxilliary equation $m^2 = 0$ and hence its general solution is

$$y(t) = Ct + D. \tag{5}$$

The third has auxilliary equation $m^2 = 0$ also, but it is not a homogeneous equation. In light of the general solution (5), a general solution for z is

$$z(t) = \tfrac{1}{6}Ct^3 + \tfrac{1}{2}Dt^2 + Et + F. \tag{6}$$

Applying the initial conditions (1) and (2) to solution (4) gives

$$x(0) = B = x_0; \quad x'(0) = A = 0.$$

Thus, $x(t) = x_0 \cos t$. Solution (5) yields

$$y(0) = D = y_0; \quad y'(0) = C = 1$$

under the initial conditions. Thus, $y(t) = t + y_0$. Using this solution and applying the initial conditions to solution (6) gives

$$z(0) = F = z_0; \quad z'(0) = E = 0.$$

Thus, $z(t) = \tfrac{1}{6}t^3 + \tfrac{1}{2}y_0 t^2 + z_0$. Therefore,

$$f(t) = (x_0 \cos t,\ t + y_0,\ \tfrac{1}{6}t^3 + \tfrac{1}{2}y_0 t^2 + z_0).$$

From this, the velocity field is calculated to be

$$V(x,y,z,t) = (-x_0 \sin t,\ 1,\ \tfrac{1}{2}t^2 + y_0 t).$$

Since $x = x_0 \cos t \Rightarrow x_0 = \dfrac{x}{\cos t} \Rightarrow x_0 \sin t = x \tan t.$

Similarly $y_0 = y - t$ so that $y_0 t = yt - t^2$ and hence

$$V(x,y,z,t) = (-x \tan t,\ 1,\ yt - \tfrac{1}{2}t^2)$$

$$\text{div } V(x,y,z,t) = -\tan t + 0 + 0 = -\tan t$$

$$\text{curl } V(x,y,z,t) = (t,0,0).$$

Note that for $t < \tfrac{\pi}{2}$ the fluid is moving in the positive y direction, rotating clockwise around the line parallel to the x axis and spinning away from it ($t<0$), and back again toward it when $t>0$. For $t = \tfrac{\pi}{2}$ $\tan t = \infty$, i.e. div $V = \infty$.

● PROBLEM 19-34

A body falls in a medium offering resistance proportional to the square of the velocity. If the limiting velocity is numerically equal to $g/2 = 16.1$ ft./sec., find a) the velocity at the end of 1 sec.; b) the distance fallen at the end of 1 sec.; c) the distance fallen when the velocity equals 1/2 the limiting velocity; d) the time required to fall 100 ft.

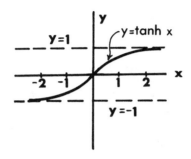

Solution: Taking the frame of reference at a point where the body starts to fall and assuming the downward direction as positive, this particle is acted upon by its weight (positive) and the resisting force (proportional to the square of the velocity and negative with respect to the selected origin). The difference between these forces is characteristic of its motion, and since this quantity equals mass × acceleration, we write:

$$m \frac{d^2 y}{dt^2} = mg - k\left(\frac{dy}{dt}\right)^2$$

where k is a constant of proportionality. Noting that the limiting velocity occurs when the acceleration is zero, i.e. when

$$mg = k\, v_{lim}^2$$

or

$$k = \frac{mg}{v_{lim}^2}$$

where $v_{lim} = dy/dt$ is the maximum velocity and is constant. Hence

$$m \frac{dv}{dt} = mg - \frac{mg}{v_{lim}^2} v^2 .$$

Separating variables and using $v_{lim} = g/2 = 16.1$ ft./sec., we have:

$$\frac{dv}{1 - \frac{v^2}{v_{lim}^2}} = g\, dt.$$

Integrating both sides of the equation with $g/2$ substituted for v_{lim},

$$\int \frac{du}{1-\frac{4v^2}{g^2}} = g \int dt + c_1 .$$

Setting $v = g/2 \tanh x$, $dv = g/2 \operatorname{sech}^2 x\, dx$, and noting that $1-\tanh^2 x = \operatorname{sech}^2 x$ we have:

$$\int \frac{dv}{1-\frac{4v^2}{g^2}} = g/2 \int \frac{\operatorname{sech}^2 x\, dx}{1-\tanh^2 x} = g/2 \int dx = g/2[x] = g/2 \tanh^{-1} \frac{2v}{g}$$

Thus

$$g/2 \tanh^{-1} \frac{2v}{g} = gt + c_1 .$$

Solving for v,

$$v = g/2 \tanh(2t+c_2).$$

Since the body was dropped initially, $v = 0$ for $t = 0$, making $c_2 = 0$. (See the diagram for evaluation.) Furthermore,

$$v = \frac{dy}{dt} = g/2 \tanh 2t.$$

This yields, as a result of integration:

$$y = g/4 \ln(\cosh 2t) + c_3,$$

where, for $t = 0$, $y = 0$, implying that $c_3 = 0$. The equation of motion is therefore:

$$y = g/4 \ln(\cosh 2t).$$

a) To find the velocity at the end of 1 sec.:

$$v_1 = g/2 \tanh(1) = g/2(0.9640) = 15.5 \text{ ft./sec.}$$

b) The corresponding distance covered is:

$$y_1 = g/4 \ln\left[\cosh 2(1)\right] = g/4 \ln(3.762) = g/4(1.32495) = 10.7 \text{ ft.}$$

c) When the speed reaches $1/2\, v_{lim}$, then $v = g/4$ ft./sec., and from the expressions:

$$v = g/4 = g/2 \tanh 2t,$$

and

$$y = g/4 \ln(\cosh 2t),$$

we eliminate t and solve for y in terms of v. Since

$$v = g/2 \tanh 2t = g/2 \frac{\sinh 2t}{\cosh 2t} = g/2 \frac{(\cosh^2 2t-1)^{1/2}}{\cosh 2t},$$

and, solving for $\cosh 2t$, we obtain:

$$\cosh 2t = g/\sqrt{g^2-4v^2}.$$

Hence

$$y = g/4 \ln(\cosh 2t) = g/4 \ln\left[\frac{g}{\sqrt{g^2-4v^2}}\right].$$

Using $v = g/4$

$$y = g/4 \ln \frac{2}{\sqrt{3}} = 1.13 \text{ ft.}$$

d) To calculate the time required for a fall of 100 ft., we use
$$y = g/4 \ln(\cosh 2t).$$
Solving for t,
$$t = 1/2 \cosh^{-1}\left(e^{4y/g}\right) = 1/2 \cosh^{-1}\left(e^{12.4}\right), \text{ for } y = 100.$$
If we neglect the factor $e^{-12.4}$ in comparison to $e^{12.4}$, we can approximate the above equation. Since
$$\cosh 2t = e^{12.4},$$
we let $\cosh 2t = 1/2\, e^{2t}$. Thus,
$$1/2\, e^{2t} = e^{12.4}.$$
Taking the antilogarithm,
$$2t = \ln 2 + 12.4,$$
or,
$$t = 1/2 \ln 2 + 6.2 = 1/2(0.69) + 6.2 = 6.545 \text{ sec.}$$

• **PROBLEM 19-35**

A particle slides freely in a tube which rotates in a vertical plane about its midpoint with constant angular velocity w. If x is the distance of the particle from the midpoint of the tube at time t, and if the tube is horizontal with t = 0, show that the motion of the particle along the tube is given by
$$\frac{d^2 x}{dt^2} - w^2 x = -g \sin wt.$$
Solve this equation if $x = x_0$, $dx/dt = v_0$ when $t = 0$. For what values of x_0 and v_0 is the motion simple harmonic?

Solution: In order to determine the x- and y-components of motion, we choose the frame of reference at the center of mass of the particle. The particle is subjected to horizontal and vertical forces. Restricting our attention to the horizontal components, the two unbalanced forces acting on the particle are a portion of its weight, (proportional to $\sin \theta$), and the centripetal force, the difference between which gives an acceleration along the x-axis. Mathematically,
$$mg \sin \theta + \frac{mv^2}{r} = -m \frac{d^2 x}{dt^2},$$
where $v = wr$, $r = -x$, and $\theta = wt$. Then
$$\frac{d^2 x}{dt^2} = w^2 x - g \sin wt.$$
To find the solution of this non-homogeneous, second order differential equation with constant coefficients, we use the D operator, where $D = d/dt$, hence the differential equation can be rewritten as:

$$(D^2 - w^2)x = -g \sin wt.$$

The auxiliary equation is $m^2 - w^2 = 0$, $m = w$, or $m = -w$, and the complementary solution is:

$$x_c = Ae^{wt} + Be^{-wt}.$$

Furthermore, the particular solution appears as:

$$x_p = C \sin wt + D \cos wt,$$

where A, B, C and D are constants to be determined. Since the general solution is

$$x = Ae^{wt} + Be^{-wt} + C \sin wt + D \cos wt,$$

$$\frac{dx}{dt} = Awe^{wt} - Bwe^{-wt} + Cw \cos wt - Dw \sin wt,$$

and

$$\frac{d^2x}{dt^2} = w^2(Ae^{wt} + Be^{-wt} - C \sin wt - D \cos wt).$$

Substitution in the main differential equation yields:

$$(D^2 - w^2)x = -2Cw^2 \sin wt - 2Dw^2 \cos wt = -g \sin wt,$$

which implies that:

$$-2Cw^2 = -g, \text{ or } C = \frac{g}{2w^2},$$

and $D = 0$. Thus,

$$x = Ae^{wt} + Be^{-wt} + \frac{g}{2w^2} \sin wt.$$

To find the values of A and B, we use the boundary conditions for $t = 0$, $dx/dt = v_0$ and $x = x_0$. And since

$$\frac{dx}{dt} = Awe^{wt} - Bwe^{-wt} + \frac{g}{2w} \cos wt,$$

$$\left.\frac{dx}{dt}\right|_{t=0} = Aw - Bw + \frac{g}{2w} = v_0,$$

and

$$x\Big|_{t=0} = A + B = x_0.$$

These are simultaneous equations with A and B unknown. Substitution of $B = x_0 - A$ in the expression for v_0 yields:

$$A = \frac{v_0 - g/2w + x_0 w}{2w}, \text{ and } B = \frac{x_0 w - v_0 + g/2w}{2w}.$$

Hence,

$$x = \frac{v_0 - g/2w + x_0 w}{2w} e^{wt} + \frac{x_0 w - v_0 + g/2w}{2w} e^{-wt} + \frac{g}{2w^2} \sin wt.$$

After some algebraic manipulation,

$$x = \tfrac{1}{2}x_0(e^{wt} + e^{-wt}) + (v_0/2w - g/4w^2)(e^{wt} - e^{-wt}) + (g/2w^2)\sin wt.$$

The above solution suggests that the motion of the particle is simple harmonic when $x_0 = 0$ and $v_0/2w = g/4w^2$ or $v_0 = \frac{g}{2w}$. In this case $x = \frac{g}{2w^2}\sin wt = \frac{v_0}{w}\sin wt.$

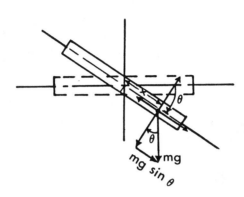

• **PROBLEM 19-36**

(a) A tightly stretched string with fixed end points $x = 0$ and $x = L$ is initially at rest in its equilibrium position. If it is set vibrating by giving each of its points a velocity

$$3(Lx - x^2),$$

find the displacement of any point on the string at any time t. (b) If the string is 2 ft. long, weighs 0.1 lb., and is subjected to a constant tension of 6 lb., find the maximum displacement when t = 0.01 sec.

Solution: The partial differential equation of a stretched string is given by

$$\frac{\partial^2 y}{\partial t^2} = a^2 \frac{\partial^2 y}{\partial x^2}$$

where

$$a^2 = \frac{Fg}{W},$$

F is the tension lb., g is the gravitational acceleration 32.2 ft./sec.^{2}, and w is the weight of the string per unit length in lb./ft.

The given boundary conditions are

$$y(0,t) = y(L,t) = 0 \qquad t \geq 0$$

$$y(x,0) = 0 \qquad x \geq 0$$

and

$$\left.\frac{\partial y}{\partial t}\right|_{t=0} = 3(Lx - x^2).$$

To seek the solution for the partial differential equation, we assume:

$$y(x,t) = X(x) \cdot T(t)$$

i.e., a solution in product form. Hence

$$\frac{\partial^2 y}{\partial t^2} = X(x) T''(t), \text{ and}$$

$$\frac{\partial^2 y}{\partial x^2} = T(t) X''(x).$$

Substitution yields:

$$X(x) T''(t) = a^2 T(t) X''(x).$$

Separating like variables,

$$\frac{X''(x)}{X(x)} = \frac{T''(t)}{a^2 T(t)},$$

where the position of the constant a^2 does not make any difference in the calculation.

Now, we let the left-hand and right-hand sides of the above equation be equal to some constant $(-k^2)$, on the basis that the equality of equations in two different variables holds only if both functions equal the same constant number. Hence,

$$\frac{X''}{X} = -k^2$$

or $X'' + k^2 x = 0$,

the auxiliary equation of which is: $m^2 + k^2 = 0$. Its roots are: $r_1 = ki$, and $r_2 = -ki$; ($i = \sqrt{-1}$), yielding two distinct particular solutions $X(x) = \cos kx$, $X(x) = \sin kx$. Hence, the general solution is given by:

$$X(x) = c_1 \cos kx + c_2 \sin kx.$$

Similarly, for

$$\frac{T''(t)}{a^2 T(t)} = -k^2,$$

the solution is given by

$$T(t) = c_3 \cos akt + c_4 \sin akt.$$

Consequently

$$y(x,t) = X(x) \cdot T(t) = (c_1 \cos kx + c_2 \sin kx)$$

$$(c_3 \cos akt + c_4 \sin akt).$$

The first boundary condition gives:

$$y(0,t) = 0 = c_1 (c_3 \cos akt + c_4 \sin akt)$$

which suggests $c_1 = 0$.

Secondly,

$$y(L,t) = 0 = (c_2 \sin kL)(c_3 \cos akt + c_4 \sin akt),$$

giving $\sin kL = 0$. If we have $kL = n\pi$ with n an integer, we have $\sin kL = \sin n\pi = 0$. Hence, by

$kL = n\pi$, $k = \frac{n\pi}{L}$. Substitution for k in the remaining terms gives:

$$y(x,t) = c_2\left(\sin\frac{n\pi}{L}x\right)\left(c_3 \cos\frac{n\pi a}{L}t\right.$$

$$\left. + c_4 \sin\frac{n\pi a}{L}t\right).$$

Thirdly, the string is initially at rest, which means that:

$$y(x,0) = 0 = c_2\left(\sin\frac{n\pi}{L}x\right)(c_3).$$

Since c_2 cannot be equal to zero - which would show that the string is at rest in its equilibrium position (a trivial result) - c_3 must be equal to zero. Under these conditions, for any x and t:

$$y(x,t) = c_2\left(\sin\frac{n\pi}{L}x\right)c_4\left(\sin\frac{n\pi a}{L}t\right).$$

Since the string is set into motion under a driving force which gives every point of the string the velocity $3(Lx - x^2)$, we have

$$\left.\frac{\partial y(x,t)}{\partial t}\right|_{t=0} = 3(Lx - x^2)$$

$$= c_2\left(\sin\frac{n\pi}{L}x\right)\left(c_4\frac{n\pi a}{L}\right),$$

or, $\quad Lx - x^2 = \dfrac{c_2 c_4}{3} \dfrac{n\pi a}{L} \sin\dfrac{n\pi}{L}x.$

This relationship appears impossible since a sine curve cannot be made to coincide with a parabolic arc over the entire x-range. But this difficulty suggests that the equality can be satisfied if the parabolic arc is approximated by many sine expressions of period 2L, leading to the Fourier sine expansion of

$(Lx - x^2) \quad$ over $\quad 0 < x < L.$

Mathematically,

$$Lx - x^2 = \sum_{n=1}^{\infty} b_n \sin\frac{n\pi x}{L}, \quad 0 < x < L$$

Here, from the theory of Fourier series,

$$b_n = \frac{2}{L} \int_0^L (Lx - x^2) \sin \frac{n\pi x}{L} \, dx$$

($n = 1, 2, 3, \ldots;\ 0 < x < L$).

Expanding the integrand and using integration by parts,

$$b_n = \frac{2}{L} \left\{ \left(-\frac{L^3}{n^2\pi^2} \sin \frac{n\pi x}{L} \right. \right.$$

$$\left. - \frac{xL^2}{n\pi} \cos \frac{n\pi x}{L} \right) \bigg|_0^L$$

$$- \left(-\frac{x^2 L}{n\pi} \cos \frac{n\pi x}{L} + \frac{2xL^2}{n^2\pi^2} \sin \frac{n\pi x}{L} \right.$$

$$\left. \left. + \frac{2L^3}{n^3\pi^3} \cos \frac{n\pi x}{L} \right) \bigg|_0^L \right\}$$

$$= \frac{4L^2}{n^3\pi^3} \left[1 - (-1)^n \right].$$

This shows that, for every even n, $b_n = 0$, and $b_n \neq 0$ for odd n. The odd n can be assured by replacing n by $2n - 1$. Hence we have:

$$b_{2n-1} = \frac{8L^2}{(2n-1)^3 \pi^3}.$$

Thus, the Fourier half-range expansion reads:

$$Lx - x^2 = \frac{8L^2}{\pi^3} \sum_{n=1}^{\infty} \frac{\sin (2n-1)\pi x/L}{(2n-1)^3}.$$

For the initial velocity of the string,

$$Lx - x^2 = \frac{8L^2}{\pi^3} \sum_{n=1}^{\infty} \frac{\sin \frac{(2n-1)\pi x}{L}}{(2n-1)^3}$$

$$= \frac{c_2 c_4}{3} \frac{n\pi a}{L} \sin \frac{n\pi x}{L}.$$

We recall the theorem that says: If $y_1, y_2, \ldots, y_n$ are n solutions of a linear homogeneous partial differential equation, $c_1 y_1 + c_2 y_2 + \ldots + c_n y_n$ is also a solution, where the c's are non-zero constants. Under these conditions, the above equation suggests the solution in a summation form, where the constants $c_2 c_4$ are no longer arbitrary fixed constants but are dependent on n. We can call them $(c_2 c_4)_{2n-1}$ where they are zero for all even n's. Hence we have:

$$\frac{24L^3}{\pi^4 a} \sum_{n=1}^{\infty} \frac{\sin \frac{(2n-1)\pi x}{L}}{(2n-1)^4}$$

$$= \sum_{n=1}^{\infty} (c_2, c_4)_{2n-1} \sin \frac{(2n-1)\pi x}{L},$$

and $(c_2 c_4)_{2n-1} = \frac{24L^3}{\pi^4 a} \cdot \frac{1}{(2n-1)^4}.$

Once the constants are determined,

$$y(x,t) = c_2 c_4 \sin \frac{n\pi x}{L} \sin \frac{n\pi a}{L} t$$

becomes

$$y(x,t) = \frac{24L^3}{\pi^4 a} \sum_{n=1}^{\infty} \frac{1}{(2n-1)^4} \sin \frac{(2n-1)\pi}{L} x \sin \frac{(2n-1)\pi a}{L} t,$$

which is the general solution. We replaced n by $2n - 1$ in the time function because, as we know, the product of zero with a number is zero.

(b) For maximum displacement we have to find the value of x at which

$$\frac{\partial y}{\partial x} = 0:$$

$$\frac{\partial y}{\partial x} = -\frac{24L^2}{\pi^3 a} \sum_{n=1}^{\infty} \frac{1}{(2n-1)^3} \cos \frac{(2n-1)\pi}{L} x$$

$$\sin \frac{(2n-1)\pi a}{L} t = 0,$$

which means that for any n,

$$\cos \frac{(2n-1)\pi}{L} x = 0$$

But we know that

$$\cos \frac{(2n-1)\pi}{2} = 0,$$

that is,

$$\cos \frac{\pi}{2} = \cos \frac{3\pi}{2} = \ldots = 0.$$

Thus, the only possibility for x is the value

$$x = \frac{L}{2}.$$

Hence, for maximum displacement at any time t:

$$y\left(\frac{L}{2}, t\right) = \frac{24L^3}{\pi^4 a} \sum_{n=1}^{\infty} \frac{1}{(2n-1)^4} \sin \frac{(2n-1)\pi}{2} \sin \frac{(2n-1)\pi a}{L} t$$

$$= \frac{24L^3}{\pi^4 a} \sum_{n=1}^{\infty} \frac{(-1)^{n+1}}{(2n-1)^4} \sin \frac{(2n-1)\pi a}{L} t.$$

Recalling the value of a,

$$a = \sqrt{\frac{Fg}{w}},$$

and since the weight of the string wL = 0.1 lb., the tension F = 6 lb., the length of the string L = 2ft. and at time t = 0.01 sec., we have:

$$y\left(\frac{L}{2}, 0.01\right) = \frac{24(2)^3}{\pi^4 \sqrt{\frac{6 \times 32}{0.05}}} \sum_{n=1}^{\infty} \frac{(-1)^{n+1}}{(2n-1)^4} \sin\frac{0.01(2n-1)\pi\sqrt{\frac{6 \times 32}{0.05}}}{2}$$

$$\simeq \frac{1}{26}\left[\sin(0.31\pi) - \frac{\sin(0.93\pi)}{3^4} + \frac{\sin(1.55\pi)}{5^4} - \cdots\right]$$

$$\simeq \frac{1}{26}\left[0.8 - \frac{0.07}{3^4} + \cdots\right]$$

$$\simeq 0.29.$$

INDEX

INDEX

Numbers on this page refer to **PROBLEM NUMBERS**, not page numbers

Abel's:
   test, 10-21, 12-17
   theorem, 10-21, 13-6, 13-7
Absolute:
   continuity, 3-30
   convergence, 10-14, 10-15, 10-17, 10-18, 10-20, 10-23
   maximum, 6-1
   minimum, 6-30
Absolutely convergent, 12-8, 12-11, 12-12, 12-21
Absolutely integrable function, 17-7
Acceleration, 19-19
   field, 19-33
Accumulation point, 1-5, 1-9
Additive inverse, 2-1
Adjoint method, 2-24
Affine transformation, 2-3
Alternating series, 12-25
   harmonic, 12-18
   test, 12-8, 12-30
Analytic function, 15-8, 15-22, 15-24, 15-26
Angle between two vectors, 2-5
Antiperiodic functions, 16-20, 16-21
Anti-symmetric periodic extension, 14-18
Archimedian:
   axiom, 1-4
   property, 11-1
Arc-length, 18-5, 18-8
Area, 9-10 to 9-15
Arithmetic mean, 6-34
Arzela-Ascoli theorem, 3-31
Associative, 2-1, 2-13
Asymptotic:
   direction, 19-33
   lines, 19-33
Attenuation property:
   of Fourier transforms, 17-6
   of Laplace transforms, 16-7, 16-27
Axioms for a vector space over a field, 2-1

Baire category, 3-21
Ball, 3-15
Banach's fixed point theorem, 1-21
Base 2, 1-5
Basis, 1-15, 2-9, 18-1
   definition of, 2-10
Beam, 16-34
Bending moment, 16-34
Bessel:
   function, 13-16
   inequality, 2-29
Beta function, 10-31, 10-33, 10-34
Bijective, 2-19
Binomial:
   series, 11-3, 11-7, 11-16, 15-25
   theorem, 11-1, 11-3, 11-7, 11-16, 15-25
Binormal vector, 18-7, 18-8
Bolzano-Weierstrass Theorem, 1-12
Boundary, 1-10
Boundary surface, 8-36, 8-37
Bounded, 1-7, 1-11, 3-31, 11-2, 11-7
   from above, 1-3, 1-4
   from below, 1-3
   function, 7-4, 7-5, 7-8, 10-1
   monotonic real sequence, 1-7
   sequence, 1-11, 11-5, 11-7
Branch point of function, 15-12

Cantor:
   set, 1-5
   theorem on nested intervals, 1-17, 11-11
Cardinality, 2-8
Cartesian:
   axes, 15-2
   coordinates, 4-8
Cauchy:
   Cauchy-Goursat theorem, 15-21, 15-22, 15-26
   Cauchy-Riemann conditions, 15-5 to 15-18, 15-24
Cauchy-Schwarz inequality, 1-13, 1-15, 3-12, 6-9, 6-32, 6-34, 10-13, 12-4, 18-17

Numbers on this page refer to **PROBLEM NUMBERS**, not page numbers

condition, 1-8, 11-9, 11-21
integral formula, 15-23, 15-25, 15-26, 15-28
mean value theorem, 5-4, 5-5
principal value, 10-7
Center of gravity, 19-25, 19-26
Centripetal force, 19-31
Centroid, 8-40
Cesaro summability, 12-32, 12-33
Chain rule, 4-1, 4-10 to 4-12, 4-15 to 4-17, 4-19, 5-18, 5-19, 7-10, 7-12, 18-2, 18-5
Challenge number, 3-7
Change of basis, 2-25
Change of variables, 4-14
formula, 9-1 to 9-19
theorem, 10-27
Characteristic equation, 2-23, 2-24
Characteristic function, 7-17, 7-18, 7-21
Charge q on the capacitor, 16-33
Christoffel symbols, 18-37
Circuits, 16-36
Circular:
cone, 18-34
helix, 18-8
Classes of differentiable functions, 3-24
Closed set, 1-5, 1-10, 3-31
Closure, 2-1
Coefficients of the first fundamental form, 18-26
Coefficients of the second fundamental form, 18-29
Co-factor, 2-21
Common refinement, 7-19
Commutative, 2-1
operators, 2-4
Compact, 1-17, 1-20, 3-31
metric space, 11-8
Comparison test, 10-3, 10-4, 10-8, 10-14, 10-15, 12-4
Complete metric space, 1-19 to 1-21
Complex:
conjugate, 2-6, 15-1, 15-4, 15-23, 16-27, 17-5

constant, 16-6
decomposition, 16-27
derivative of, 15-4
form, 14-24, 14-26
function, 15-4
infinite series, 15-13
integration, 15-35
logarithm, 15-12
n-tuples, 2-6
numbers, 15-1, 15-9
plane, 15-2,
Taylor series, 15-14
valued function, 15-20, 15-25, 15-27
Composite functions, 4-10, 4-12, 4-15, 4-16, 4-17, 7-10
Concept of partitions, 7-1
Conditionally convergent, 10-14, 10-17, 10-18, 12-8, 12-11, 12-12
divergent, 12-12
Cone, 19-21
Conjugate, 2-17
complex, 2-6, 15-1, 15-4, 15-23, 16-27, 17-5
Connected:
set, 3-22
subset, 1-18
Conservative field, 4-27, 4-30, 8-29
Continuity conditions, 15-7
Continuous function, 5-1
Continuous nonvanishing derivative, 19-2
Contour integral, 15-20
Contraction mapping, 1-21, 1-22
Convergence, 10-3, 10-6, 11-5, 12-1 to 12-3, 12-6, 12-18, 14-14 to 14-17
conditional, 10-14, 10-17, 10-18, 12-8, 12-11, 12-12
sequence, 1-6, 1-22, 11-1, 11-3
subsequence, 11-8
Convergent, (see also convergence), 1-6, 12-7
Convex set, 1-14
Convolution theorem:
for Fourier transforms, 17-8

1032

*...mbers on this page refer to **PROBLEM NUMBERS**, not page numbers*

for Laplace transforms, 16-31, 16-32
...ordinate patch, 18-14
...untable set, 1-2, 3-22
...untably infinite, 1-1
...ramer's rule, 2-21, 16-29
...ritical point, 6-2, 6-4, 6-5, 6-6, 6-7, 6-8, 6-15, 6-30
...url, 9-31 to 9-36, 19-21, 19-33
...urrent pulse, 18-12
...urvature, 19-5, 19-6, 19-8
...ylinder, 18-11, 18-15, 18-20, 18-21
...ylindrical coordinates, 4-8, 18-5

...ecompose, 16-29, 16-30
   complex, 16-27
...ecreasing function, 10-21
...efinite integral, 7-1, 15-19, 15-20
   of a complex valued function, 15-20
...e Moivre's Theorem, 15-3
...e Morgan's Law, 1-11
...ense, 3-21
   subset, 1-5
...erivative:
   directional, 4-24 to 4-26, 5-7, 6-35
   of a real valued function, 5-2
   properties of Fourier transforms, 16-6
   properties of Laplace transforms, 16-7, 16-10
...eterminant, 2-10, 2-15
...iagonalizable, 2-23, 2-24
...ifferentiable, 4-4, 15-4
   series, 12-23
...ifferential, 4-4, 4-5, 4-6, 4-13, 4-18, 21-1
   equation, 4-16, 14-29, 16-36, 19-31, 19-32, 19-34, 19-35
   form, 9-37
   of a vector valued function, 4-5
...ifferentiated series, 12-22
...ifferentiation, 13-8, 13-16, 14-13
   of a series, 12-22 to 12-26
Dimension, 1-9, 2-9
Dipole, 19-20
Dirac delta function, 17-11
Direct elimination, 6-10, 6-11
Directional derivative, 4-24 to 4-26, 5-7, 6-35
Dirichlet's test, 10-21, 12-16, 12-17, 12-26, 14-7
Discontinuities:
   essential, 3-4, 3-18
   of the first kind, 3-4, 3-19
   of the second kind, 3-4, 3-19
   removable, 3-4, 3-18, 3-30
Discontinuous, 14-7, 10-23, 16-1
   function, 3-16
Distance, 2-5
   formula, 6-1
   function, 11-9
Distributive law, 2-1
Divergence, 8-42, 8-43, 19-13, 19-18
   theorem, 9-22 to 9-29, 19-10, 19-30
Divergent:
   conditional, 12-12
   integral, 10-17
   series, 12-1, 12-4, 12-6 to 12-8, 12-11, 12-32
Dot product, 1-13, 2-5, 18-25
Double integral, 9-4
Duhamel's theorem, 19-28
Dummy variable, 16-20

Echelon form, 2-10, 2-11
Eigenvalue, 2-17, 2-23, 6-14, 6-31, 6-37, 10-27
Eigenvector, 2-17, 2-25, 2-24
Electrical engineering and applications, 14-12, 14-26 to 14-29
Electrical potential, 19-19, 19-29
Electric circuit theory, 16-33
Electrodynamics, 19-19
Electromagnetic theory, 16-36
Electromotive force, 14-29

Numbers on this page refer to **PROBLEM NUMBERS**, not page numbers

Electronics, 14-7
Electrostatics, 19-22, 19-23
   field, 19-18
   potential, 19-20
Elementary:
   integrating formula, 10-26
   mass, 10-34
   matrix, 2-20
Ellipsoid, 6-12, 9-7, 18-15, 19-4, 19-8
Elliptic integrals 19-11 to 19-16
Envelope of family of surfaces, 18-12, 18-34, 18-13
Equation of continuity for fluid flows, 19-30
Equicontinuous, 3-31, 3-32
Equivalent metric spaces, 1-16
Estimates of errors & sums, 12-27 to 12-31
Exact differential, 8-10, 8-11, 8-28, 8-29, 9-37
Existence theorem, 16-1
Explicit functions, 4-2
Exponential function, 15-12
Exponential order, 16-1 to 16-3, 16-12, 16-23
External direct sum, 2-29
Extrema, 6-20
Extremal values, 6-29
Euclidean:
   inner product, 6-35, 10-27
   metric, 1-16
   norm, 2-7, 2-16
Euler's:
   formula, 16-4
   identity, 15-3
   theorem, 5-21, 14-24
Even functions, 14-8, 14-18, 15-36, 17-10

Family of curves, 18-12, 18-13, 18-23, 18-38
   circles, 18-12
   lines, 18-13
   planes, 18-34
   spheres, 18-34
Finite:
   dimensional vector space, 2-9
   limit, 11-12

   subcover, 3-31
First partial derivative, 6-1
First Parseval theorem, 17-7
Fixed point, 1-21
Fluid flows, 19-30
Flux in a flow, 8-43
Forced damped harmonic oscillator, 16-32, 16-33
Force field, 19-31
Form, 9-37
Fourier:
   coefficients, 14-1, 14-18, 14-31, 17-1
   cosine, 14-18, 14-19, 14-22, 14-23
   cosine transforms, 17-10
   integral formula, 17-9, 17-10
   inversion integral, 17-7
   series, 14-13, 14-24 to 14-26, 14-28, 17-1, 19-36
   sine expansion, 19-36
   sine series, 14-18 to 14-21, 14-30
   sine transforms, 17-10
   spectrum, 17-1
   transform, 18-1 to 18-24
Free fall in viscous medium, 19-34, 21-34
Frenet-Serret formulas, 18-31
Frequency domain integration, 17-13
Full-wave rectified wave, 14-28
Function, exponential, 3-11
Functionally dependent, 5-28
Fundamental:
   coefficients, 18-35
   criterion of integrability, 7-20
   form, 18-17, 18-19, 18-21 to 18-25, 18-27, 18-32
   theorem of calculus, 7-19, 8-21

Gamma function, 10-28 to 10-31, 10-34
Gaussian curvature, 18-32
   paraboloid, 18-31
   sphere, 18-30
Gaussian probability, 17-2

Numbers on this page refer to **PROBLEM NUMBERS**, not page numbers

Gauss test, 12-10, 12-12
Generalized factorial function, 10-28
Geodesic, 18-37, 18-38
    curvature, 18-31
    on right circular cone, 18-37
Geometric:
    mean, 6-34
    progression, 11-5
    series, 1-7, 12-13, 13-12, 15-15, 15-29, 16-18
Gradient, 4-18, 4-19, 4-21, 4-22, 4-23, 4-25, 4-26, 6-31, 8-16, 9-30
Gram-Schmidt, 2-31, 2-32
Gravitational:
    constant, 19-17
    force field, 19-17, 19-21
Greatest:
    bound, 11-13
    integer function, 3-26
    lower bound, 1-3
Green's theorem, 8-30 to 8-37
    (see also multiple integrals)

Harmonic, 19-35
    conjugates, 15-8
    series, 12-1, 12-8, 12-27, 13-5
Heine-Borel theorem, 1-17
Heisenberg's uncertainty principle 17-2
Helix, 18-9, 18-10
Hermitian, 2-17
    quadratic form, 2-17
Hessian, 6-9, 6-26
Hölder's Inequality, 6-33, 6-34, 10-13, 12-5
Homogeneous:
    function, 2-8
    of degree n, 5-20
Hyperbola, 6-2
Hyperbolic paraboloid, 18-33
Hyperboloid, 18-21
Hyperplane, 19-7

Idempotent matrix, 2-13

Identity transformation, 2-2
Image, 2-11, 18-15
Implicit function:
    method of, 6-11
    theorem, 5-24 to 5-26, 9-11
Implicit partial differentiation, 4-1
Improper integral, 10-1, 10-2, 10-5, 10-10, 10-14, 10-28, 16-4
    of the second kind, 10-7 to 10-9, 10-12, 10-15, 10-18
Improper multiple integral, 10-23
Incomplete elliptic integral:
    of the first kind, 19-11 to 19-13
    of the second kind, 19-11
    of the third kind, 19-11, 19-13
Incompressible, 19-30
Indefinite integrals, 15-24
Independent of the path, 8-22, 8-24, 8-27 to 8-29
Induction, 16-7
Inequality:
    Bessel, 2-29
    Cauchy-Schwarz, 1-13, 1-15, 3-12, 6-9, 6-32, 6-34, 10-13, 12-4, 18-17
    Hölder's, 6-33, 6-34, 10-13, 12-5
    Minkowski, 6-34
    Triangle, 1-13, 1-14, 1-20, 5-9
Infimium, 1-3, 1-11, 3-21
Infinite product, 12-34, 12-35
Infinite series:
    cesaro summability, 12-32 to 12-33
    estimates of error and sums, 12-22 to 12-26
    differentiation and integration of series, 12-18 to 12-21
    operations on series, 12-18 to 12-21
    series of functions, 12-13 to 12-17
    tests for convergence and divergence, 12-1 to 12-12
    (see series)
Initial value problem, 16-33,

Numbers on this page refer to **PROBLEM NUMBERS**, not page numbers

16-35
Injective, 2-19
Inner (or dot) product, 1-13, 2-5, 2-6, 2-17, 2-32, 18-17, 18-18, 18-25
Integral:
   estimates, 12-31
   function over a surface, 9-16 to 9-19
   of the first kind, 10-6
   of the third kind, 10-11
   test, 12-1 to 12-3, 12-10, 12-27
   vector field over a surface, 9-20 to 9-24
Integration:
   area, 9-10 to 9-15
   beta function, 10-27 to 10-34
   change of variables formula, 9-1 to 9-9
   divergence theorem, 9-25 to 9-31
   gamma function, 10-27 to 10-34
   Green's theorem, 8-30 to 8-43
   improper integrals, 10-1 to 10-34
   line integrals, 8-1 to 8-9
   potential function, 8-10 to 8-19
   Riemann integral, 7-1 to 7-13
   Stieltjes integral, 7-14 to 7-21
   Stokes's theorem, 9-31 to 9-36
   surface integral, 9-16 to 9-19
   vector field over a surface, 9-20 to 9-24
Integrator function, 7-21
Intermediate value theorem, 3-25
Interval of convergence, 13-1, 13-16, 13-17
Inverse Fourier transform, 17-1
Inverse function theorem, 5-22 to 5-24, 9-1
Invertible, 2-3, 2-14, 2-16, 2-19, 2-20
   matrix, 2-14
Irrational, 3-20 to 3-22
Irrotational, 19-21
Isolated points, 1-5

Isolated singular points, 15-29 to 15-32
Isometry, 18-35

Jacobian, 4-9, 9-1 to 9-9
   determinant, 4-8
   matrix, 4-7, 4-8, 4-11, 5-18
   Hessian matrices, 5-11, 5-12

Kernel, 2-11

Lagrange:
   method, 6-20 to 6-24, 6-34, 19-4, 19-8
   multipliers, 6-25 to 6-27, 6-31, 6-35, 6-36, 19-7
   theorem, 6-32
Lamina, 8-40, 19-22, 19-23, 19-28
Laplace:
   equation, 15-8
   integral, 16-1, 16-7, 16-14
Laplace transform, 16-1 to 16-36
   convolution theorem for, 16-31
   inverse, 16-22, 16-24 to 16-28, 16-30 to 16-32
   inversion of, 16-10
   operator, 16-20
   properties of, 16-7
Laplacian, 5-21
Latitude, 18-25
Laurent:
   coefficients, 15-15, 15-17
   expansion, 15-29 to 15-31
   series expansion, 15-15 to 15-18
   theorem, 15-15, 15-29
Law of cosines, 19-1
Least square fit, 6-9
Least upper bounds, 1-3, 11-13
Left hand limit, 14-10
Leibniz's rule, 7-11 to 7-13
Length, 2-5

of an arc, 18-3, 18-19
of a curve, 18-4, 18-17
Level curves, 6-2, 6-5, 6-7, 6-13
Level surfaces, 4-23
L'Hospital's Rule, 5-5, 10-5, 10-9, 10-12, 10-29, 12-10, 16-2, 16-5
Limit, 3-1, 14-10
  and continuity, 3-3
  function, 11-18, 11-19, 11-21
  inferior, 11-13
  lower, 11-17
  of integral, 11-20
  of the sequence, 11-6
  superior, 11-13
  test, 12-7
  upper, 11-17
Limit points, 1-10
  neighborhood of, 1-10
Line integral, 15-20
Linear:
  combination, 2-9
  function, 5-2
  mass distribution, 7-19
  operators, 2-18, 16-7, 16-9, 16-10, 16-23
  transformation, 1-15, 2-2, 2-11, 3-12, 5-18, 5-19
Linearity property, 16-7
Linearly:
  dependent set, 1-13, 2-9
  independent set, 2-9
Lipschitz condition, 3-30
Local minimum, 6-37
Logarithm of a complex number, 15-10, 15-11
  natural, 15-12
Long division, 15-16
Longitude, 18-25
Lower bound, 1-3
L_p-norm, 10-13

Magnetic, 16-36
Mapping, 18-14
Matrix, 1-15, 2-10
  arithmetic, 2-13
Maxima, 6-14 to 6-18, 6-22 to 6-25, 6-31, 19-3, 19-4
  value, 3-23
  value theorem, 5-1
  volume, 19-5
Mean:
  curvature, 18-30
  squared deviation, 14-14
  square norm, 14-14
Mean value theorem, 5-2, 5-3, 5-7, 5-8, 5-9, 5-22, 7-19
Mesh of the partition, 7-18
Method of arithmetic means, 12-32
Metric space, 1-16, 3-15, 11-1, 11-2
  compact, 3-32
  complete, 3-33
  equivalence of, 1-16
Minkowski's inequality, 6-34
Minima, 6-14, 6-16, 6-17, 19-6, 19-7
  value, 3-23
Möbius strip, 19-9
Modulus, 15-1, 15-2, 15-10, 19-4
Moment of inertia, 10-34, 19-27, 19-28
Monotonic, 1-7
  function, 3-19, 7-6
  sequence, 1-11, 11-5, 11-7, 12-27, 12-28, 12-35
Monotonically decreasing sequence, 11-5, 12-27, 12-28
Monotonically increasing function, 3-19, 7-6, 7-16, 7-19, 7-20
Monotonically increasing sequence, 11-5, 11-7, 12-35
Morera's Theorem, 15-22
Multiple integrals, 8-30 to 8-37, 8-39 to 8-43
  improper, 10-23
Multiple valued function, 15-12
  complex, 15-12
Multiplication, 2-12
Multiplication of transformation, 2-16
Multiplicity, 16-27
Multiplying series, 12-21

Numbers on this page refer to **PROBLEM NUMBERS**, not page numbers

Natural:
   logarithm, 15-12
   numbers, 1-1
   representation, 18-2
n-dimensional Euclidean space, 2-7
Negative definite, 6-27, 6-29
Neighborhood, 1-9, 1-10
Nested intervals, 1-11, 1-17, 1-19
   theorem, 1-12
Newton's binomial theorem, 3-1
Nilpotent (matrix), 1-13
Non-commutative operators, 2-4
Nonconstant periodic function, 16-4
Non-decreasing sequence, 1-11
Non-increasing:
   function, 10-16, 10-18
   sequence, 1-11
Nonuniform convergence, 11-24
Norm, 1-13
Normal curvature:
   paraboloid, 18-31
   sphere, 18-30
Normal vector, 4-18 to 4-21, 9-21, 9-22, 9-25, 9-27, 9-29, 9-31, 18-8, 19-9
   unit, 9-20
Not rectifiable (curve), 18-4
n-tuples, 2-1
Nullity, 2-11

Odd functions, 14-8, 14-18, 17-10
One to one, 2-18
   correspondence, 1-1
   mapping, 1-2, 18-35
Open:
   connected set, 8-20, 8-21
   cover, 3-31
   interval, 1-5
   set, 1-10
   spherical neighborhood, 1-14
Operations on series, 12-18 to 12-21
Order of contact, 18-10

Orders of infinity, 11-12
Orientable, 9-31, 19-9
Orientation, 8-36, 8-37
Oriented boundary, 9-26
Orthogonal, 4-20, 14-29
   basis of R^n, 2-23, 2-31
   complement, 2-30
   functions, 2-32
   matrix, 2-22, 2-23, 2-27
   set, 14-3
   transformation, 2-28
Orthogonality, 14-1, 14-30, 18-20
Oscillation, 3-21
Osculating planes, 18-9

Pappus theorem, 8-40
Parabola, 18-3
Paraboloid, 9-10, 9-20, 9-32, 18-16, 18-21, 18-31, 19-26
Parallelogram rule (for adding vectors), 4-20
Parallelepiped, 19-5
Parameter, 7-11
Parametric:
   curves, 18-15, 18-20
   equation, 15-20, 15-23
   representation, 15-21
   variable, 15-21
Parametrization, 8-12, 18-11
Parametrized, 8-10
   curve, 18-5
   form, 15-35 to 15-37
   surface, 9-10, 9-12 to 9-18, 9-20, 9-21, 9-24
Parseval's:
   equality, 14-16, 14-31
   theorem, 17-12
Partial derivative, 4-1, 4-3, 4-13, 4-22, 6-3
   definition, 4-3
   implicit, 4-1
Partial differential equation, 19-36
Partial fractions, 16-28 to 16-30, 16-33, 16-35
   expansion, 16-28
   use in decomposition, 16-29, 16-30, 16-33, 16-35

Numbers on this page refer to **PROBLEM NUMBERS**, not page numbers

Partial sums, 1-7, 12-31
Path connected subset, 1-18
Peano's (continuous space filling curve), 3-33
Perfect set, 1-5
Periodic:
 extension, 14-3 to 14-8, 14-11
 function, 16-18, 16-19
Perpendicular, 4-18, 4-19, 4-23 (see orthogonality)
Piecewise:
 continuous, 14-10 to 14-12, 14-16, 16-1
 smooth, 16-23
 very smooth, 17-9
Plane curve:
 curvature of, 18-5
P-norm, 12-5
Pointwise:
 convergence, 17-9
 convergence theorem, 14-11 to 14-13, 14-16, 14-17
Polar:
 coordinates, 4-10, 4-12, 9-3, 9-10, 9-20, 9-25, 9-31, 9-35, 10-23, 10-27, 10-31, 10-34, 15-2, 15-24
 form, 15-3
 representation, 15-2
Pole, 15-17, 15-18
Poles of order m, 15-34, 15-37, 15-38
Polygon, 18-3
Polygonal approximation, 18-4
Position vector, 18-33
Positively homogeneous, 5-20
 functions, 5-21
Positive definite, 6-27, 6-28, 6-37
 quadratic form, 10-27
 symmetric matrix, 10-27
Positive semi-definite, 6-37
Potential functions, 4-27 to 4-30, 8-10 to 8-23, 8-27 to 8-29
Power series:
 complex form, 15-13 to 15-18
 differentiation of, 13-8, 13-17
 division of, 13-13
 integration of, 13-10, 13-11
 interval of convergence, 13-1 to 13-7
 multiplication of, 13-17, 13-19
Principal:
 branch of function, 15-12
 minor, 6-26, 6-27
 part of function, 15-17, 15-18, 15-33
 value, 15-10 to 15-12
Principal axis, 6-12
 theorem, 2-22
Principle of contraction mappings, 1-21, 5-22
Product of complex numbers, 2-6
Projection:
 function, 3-9
 theorem, 2-29
Proper integral, 10-1
p-series, 12-1

Quadratic form, 2-8, 2-26, 5-16, 6-2, 6-14, 6-27, 6-28, 10-27
Quadratic formula, 15-34
Quadric surface, 2-23
Quantum physics, 17-2
Quotient test, 10-4

Raabe's test, 12-9, 12-11
Radius of convergence, 13-1 to 13-5
Rate of change, 4-17, 4-23
Ratio, 13-5
 test, 12-6, 12-24, 12-28, 13-1, 13-3
Rational, 3-20 to 3-22
 fraction decomposition, 16-27
 function, 15-33, 16-27
 numbers, 1-1
Real number, 1-2
"Real-ize the denominator," 15-1
Real power series expansion, 15-13
Real valued functions, 15-8,
 monotonic, 7-15
 partial derivative of, 15-7
Rearrangement, 12-18, 12-19

Numbers on this page refer to **PROBLEM NUMBERS**, not page numbers

Rectangular coordinates, 4-23
Rectifiable curve, 18-3
Region of validity, 16-7
Regular parametric representation, 18-1, 18-14
Relative:
 maximum, 6-1, 6-4 to 6-8, 6-19
 minimum, 6-19, 6-30
Residue theorem, 15-29, 15-30, 15-32, 15-34 to 15-38, 17-3
Residues, 15-33
 of function, 15-31
Riemann:
 integral, 7-1
 integrable function, 7-2, 7-3, 7-5 to 7-9
 zeta function, 12-4
Right circular cone, 19-37
Right hand limit, 14-10
Rolle's theorem, 5-1, 5-2, 5-4, 5-11
Root test, 12-6, 13-1
Roots of equation, 1-22
Rotation, 2-4, 2-28
Row:
 echelon, 2-20
 reduce, 2-11
 reduction (matrix), 2-10

Saddle point, 6-1, 6-2, 6-4 to 6-6, 6-8
Sawtooth waveform, 14-9
Scalar multiplication, 2-1
Schwarz inequality, 1-13, 1-15, 3-12, 6-9, 6-32, 6-34, 10-13, 12-4, 18-17
Second fundamental form, 18-28, 18-32
Second Parseval theorem, 17-7
Second partial derivatives, 4-12, 4-14, 4-15, 4-27
Semi-continuous, 3-26
Separated sets, 3-22
Sequences, 1-6, 1-19, 2-19, 3-1
 convergence of, 11-1 to 11-12
 limit superior and inferior,
  11-13 to 11-17
 of functions, 1-8, 11-18 to 11-21
 of partial sums, 1-7
 of real numbers, 11-13
 of vectors, 11-11
Series (see also Infinite series):
 alternating, 12-8, 12-18, 12-30
 binomial, 11-3, 11-7, 11-16, 15-27
 Fourier series (see Fourier series)
 geometric, 1-7, 12-13, 13-12, 15-15, 15-29, 16-18
 harmonic, 12-1, 12-8, 12-27, 13-5
 Laurent, 15-15 to 15-18
 P-series, 12-1
 power series (see Power series)
 Taylor series, 5-12, 15-13, 15-16, 15-18, 15-30, 15-31, 15-35
Shearing force, 16-34
Shifting property:
 of Fourier transforms, 17-6, 17-11
 of Laplace transforms, 16-15
Simple closed contour, 15-22, 15-26
Simple poles, 15-33
Simply connected domain, 8-24, 15-24
Sine function, 15-9
Singleton set, 7-18
Single-valued function, 15-12
Singularity, 10-5 to 10-8, 15-22, 15-25
Singular points, 15-15, 15-17, 15-18
Sinusoidal force, 16-33
Solid angle, 19-10
Span, 2-9
Sphere, 18-15, 18-16, 18-20, 18-21, 18-25, 18-30
 area of, 18-26
Spherical coordinates, 4-8, 9-3, 9-5, 9-7, 9-18, 9-22, 9-27, 18-15

Numbers on this page refer to **PROBLEM NUMBERS**, not page numbers

Square wave, 14-7
Standard:
    inner product, 2-31
    test integral, 10-15
Step-function, 7-18
Stieltjes integrable function, 7-21
Stieltjes integrals, 7-14 to 7-19, 19-28
Stirling's formula, 11-10
Strict local minimum, 6-37
Strictly decreasing, 11-5
Stokes's theorem, 9-31 to 9-36
Subinterval, 7-1, 7-3
Subsequence, 1-19, 3-1, 11-8
Subsequential limit, 11-8
Substitution property, 16-7
Surface:
    elliptic, 18-29
    hyperbolic, 18-29
    integral, 19-16 to 19-19
    parabolic, 18-29
    patch, 18-15 to 18-18, 18-22 to 18-24, 18-26
Surjective, 2-19, 3-33
    function, 1-5
Symmetric:
    matrix, 2-8, 2-23, 2-27
    periodic extension, 14-18
    positive definite matrix, 2-22
System of equations, 2-21

Tangent plane, 4-20, 4-21, 18-9, 18-18
Tangent vector, 18-6, 18-8
Taylor:
    coefficients, 15-15
    expansion, 5-13 to 5-15
    formula, 12-35, 13-9, 13-15
    series, 5-12
    series expansion, 15-13, 15-16, 15-18, 15-30, 15-31, 15-35
    theorem, 5-11, 5-14 to 5-18, 6-3, 6-30, 6-37, 15-14, 15-15, 15-17
Ternary notation, 1-5
Test for convergence and divergence, 10-9, 12-1 to 12-12
    ence, 10-9, 12-1 to 12-12
Time domain:
    integration, 17-13
    response, 17-14
Topological space, 1-10
Torsion, 18-5, 18-6, 18-8, 18-9
Torus, 9-12, 9-14
    area of, 18-27
Total energy, 17-13
    of current pulse, 17-12
Total mass, 7-19
Total square deviation, 14-15
Trace, 9-23
Transformation, 2-12
Translation, 2-3
Triangle, 19-1
    inequality, 1-13, 1-14, 1-20, 5-9
Trigonometric series, (see Fourier series)
Two complex numbers, 2-6
Two-parameter families, 18-34

Unbounded function, 10-1
Uncountable set, 1-2, 3-22
Uniform:
    continuity, 3-27 to 3-30, 7-11, 7-20
    convergence, 10-19, 10-20, 10-22, 10-25, 11-22 to 11-24, 12-13 to 12-15, 12-22 to 12-26, 13-5, 14-16
    convergence tests, (see Weierstrass M-test)
    convergence theorem for Fourier transforms, 14-12
    convergent sequences, 1-8, 13-12, 13-18, 14-1, 14-5, 14-17
Unit:
    binormal, 18-7
    principal normal, 18-7
    step function, 16-15 to 16-18, 16-35
    tangent vector, 18-7

Numbers on this page refer to **PROBLEM NUMBERS**, not page numbers

Unitary space, 2-7
Upper bound, 1-3, 1-4
Utility function, 19-3

Voltage, 14-26
Volume of the parallelepiped, 2-15

Vector field, 4-27 to 4-30, 8-10 to
  8-14, 8-16, 8-18, 8-23, 8-25,
  8-27, 8-32, 8-34, 8-35, 8-43,
  9-20 to 9-22, 9-26, 9-27, 9-29,
  9-33
Vector function, 4-6, 4-7, 18-1
Vector space, 2-6, 2-7, 2-11,
  7-18
Vector space axioms, 2-1
  closure, 2-1
  commutativity, 2-1
  associativity, 2-1
Vector-valued functions, 4-4, 4-5
Velocity vector, 4-19

Wave equation, 4-14
Wave number, 18-1
Weierstrass:
  m-test, 3-34, 3-35, 10-20 to
  10-22, 10-25, 12-15, 12-22
  to 12-24, 12-26, 13-5, 13-20,
  14-5
  nowhere differentiable function,
  3-24, 3-34
Work, 19-24

Zero element, 2-1
Zero mapping, 2-2

REA's Test Preps
The Best in Test Preparation

- REA "Test Preps" are **far more** comprehensive than any other test preparation series
- Each book contains up to **eight** full-length practice exams based on the most recent exams
- **Every** type of question likely to be given on the exams is included
- Answers are accompanied by **full** and **detailed** explanations

REA has published over 60 Test Preparation volumes in several series. They include:

Advanced Placement Exams (APs)
Biology
Calculus AB & Calculus BC
Chemistry
Computer Science
English Language & Composition
English Literature & Composition
European History
Government & Politics
Physics
Psychology
Statistics
Spanish Language
United States History

College-Level Examination Program (CLEP)
Analyzing and Interpreting Literature
College Algebra
Freshman College Composition
General Examinations
General Examinations Review
History of the United States I
Human Growth and Development
Introductory Sociology
Principles of Marketing
Spanish

SAT II: Subject Tests
American History
Biology
Chemistry
English Language Proficiency Test
French
German

SAT II: Subject Tests (continued)
Literature
Mathematics Level IC, IIC
Physics
Spanish
Writing

Graduate Record Exams (GREs)
Biology
Chemistry
Computer Science
Economics
Engineering
General
History
Literature in English
Mathematics
Physics
Political Science
Psychology
Sociology

ACT - ACT Assessment

ASVAB - Armed Services Vocational Aptitude Battery

CBEST - California Basic Educational Skills Test

CDL - Commercial Driver License Exam

CLAST - College Level Academic Skills Test

ELM - Entry Level Mathematics

ExCET - Exam for the Certification of Educators in Texas

FE (EIT) - Fundamentals of Engineering Exam

FE Review - Fundamentals of Engineering Review

GED - High School Equivalency Diploma Exam (U.S. & Canadian editions)

GMAT - Graduate Management Admission Test

LSAT - Law School Admission Test

MAT - Miller Analogies Test

MCAT - Medical College Admission Test

MSAT - Multiple Subjects Assessment for Teachers

NJ HSPT - New Jersey High School Proficiency Test

PPST - Pre-Professional Skills Tests

PRAXIS II/NTE - Core Battery

PSAT - Preliminary Scholastic Assessment Test

SAT I - Reasoning Test

SAT I - Quick Study & Review

TASP - Texas Academic Skills Program

TOEFL - Test of English as a Foreign Language

TOEIC - Test of English for International Communication

RESEARCH & EDUCATION ASSOCIATION
61 Ethel Road W. • Piscataway, New Jersey 08854
Phone: (732) 819-8880

Please send me more information about your Test Prep books

Name _____

Address _____

City _____ State _____ Zip _____

REA's **Problem Solvers**

The "PROBLEM SOLVERS" are comprehensive supplemental textbooks designed to save time in finding solutions to problems. Each "PROBLEM SOLVER" is the first of its kind ever produced in its field. It is the product of a massive effort to illustrate almost any imaginable problem in exceptional depth, detail, and clarity. Each problem is worked out in detail with a step-by-step solution, and the problems are arranged in order of complexity from elementary to advanced. Each book is fully indexed for locating problems rapidly.

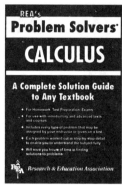

ACCOUNTING
ADVANCED CALCULUS
ALGEBRA & TRIGONOMETRY
AUTOMATIC CONTROL
   SYSTEMS/ROBOTICS
BIOLOGY
BUSINESS, ACCOUNTING, & FINANCE
CALCULUS
CHEMISTRY
COMPLEX VARIABLES
DIFFERENTIAL EQUATIONS
ECONOMICS
ELECTRICAL MACHINES
ELECTRIC CIRCUITS
ELECTROMAGNETICS
ELECTRONIC COMMUNICATIONS
ELECTRONICS
FINITE & DISCRETE MATH
FLUID MECHANICS/DYNAMICS
GENETICS
GEOMETRY
HEAT TRANSFER
LINEAR ALGEBRA
MACHINE DESIGN
MATHEMATICS for ENGINEERS
MECHANICS
NUMERICAL ANALYSIS
OPERATIONS RESEARCH
OPTICS
ORGANIC CHEMISTRY
PHYSICAL CHEMISTRY
PHYSICS
PRE-CALCULUS
PROBABILITY
PSYCHOLOGY
STATISTICS
STRENGTH OF MATERIALS &
   MECHANICS OF SOLIDS
TECHNICAL DESIGN GRAPHICS
THERMODYNAMICS
TOPOLOGY
TRANSPORT PHENOMENA
VECTOR ANALYSIS

*If you would like more information about any of these books,
complete the coupon below and return it to us or visit your local bookstore.*

RESEARCH & EDUCATION ASSOCIATION
61 Ethel Road W. • Piscataway, New Jersey 08854
Phone: (732) 819-8880

Please send me more information about your Problem Solver books

Name _____

Address _____

City _____ State _____ Zip _____